ENCYCLOPEDIA OF STATISTICAL SCIENCES

VOLUME 3

Faà di Bruno's Formula to Hypothesis Testing

ENCYCLOPEDIA OF STATISTICAL SCIENCES

VOLUME 3

FAÀ DI BRUNO'S FORMULA to HYPOTHESIS TESTING

A WILEY-INTERSCIENCE PUBLICATION

John Wiley & Sons

NEW YORK · CHICHESTER · BRISBANE · TORONTO · SINGAPORE

Library of Congress Cataloging in Publication Data:
Main entry under title:
Encyclopedia of statistical sciences.

"A Wiley-Interscience publication."
Contents: v. 1. A to Circular probable error—v. 2. Classification to Eye estimate—v. 3. Faà di Bruno's formula to Hypothesis testing.
1. Mathematical statistics—Dictionaries.
2. Statistics—Dictionaries. I. Kotz, Samuel.
II. Johnson, Norman Lloyd. III. Read, Campbell B.
QA276.14.E5 1982 519.5'03'21 81-10353
ISBN 0-471-05549-2 (v. 3)

Printed in the United States of America

10 9 8 7 6 5 4 3 2 1

CONTRIBUTORS

R. J. Adler, *The Technion, Haifa, Israel.* Hausdorff Dimension

A. G. Agresti, *University of Florida, Gainesville, Florida.* Fallacies, Statistical

F. B. Alt, *University of Maryland, College Park, Maryland.* Hildreth-Lu Scanning Method

L. R. Anderson, *Whitman College, Walla Walla, Washington.* Gain Ratio

C. Anello, *Food and Drug Administration, Rockville, Maryland.* FDA Statistical Programs: An Overview

R. R. Bahadur, *University of Chicago, Chicago, Illinois.* Hodges Superefficiency

L. J. Bain, *University of Missouri, Rolla, Missouri.* Gamma Distribution

R. J. Baker, *Rothamsted Experimental Station, Harpenden, England.* Generalized Linear Models; GLIM

R. E. Barlow, *University of California, Berkeley, California.* Fault Tree Analysis

O. Barndorff-Nielsen, *University of Aarhus, Aarhus, Denmark*, Hyperbolic Distributions

S. Berg, *University of Lund, Lund, Sweden.* Factorial Series Distributions

R. N. Bhattacharya, *University of Arizona, Tucson, Arizona.* Fokker–Planck Equations

P. Blæsild, *University of Aarhus, Aarhus, Denmark.* Hyperbolic Distributions

J. F. Box, *Madison, Wisconsin.* Fisher, Ronald Aylmer

M. C. Bryson, *Los Alamos National Laboratory, Los Alamos, New Mexico.* Heavy-Tailed Distributions

R. J. Buehler, *University of Minnesota, Minneapolis, Minnesota.* Fiducial Inference

D. Burns, *Ferris State College, Big Rapids, Michigan.* Graph Theory

D. G. Chapman, *University of Washington, Seattle, Washington.* Fisheries Research, Statistics in

W. G. Cochran*, *Harvard University, Cambridge, Massachusetts.* Horvitz–Thompson Estimator

D. A. Conway, *University of Chicago, Chicago, Illinois.* Farlie–Gumbel–Morgenstern Distribution

H. C. Copeland, *University of Maryland, College Park, Maryland.* Finance, Statistics in

P. R. Cox, *Mayfield, Sussex, England.* Fertility Measurement

E. L. Crow, *National Telecommunications and Information Administration, Boulder, Colorado.* Freeman–Tukey Test

C. Daniel, *Rhinebeck, New York.* Half-Normal Plots

H. A. David, *Iowa State University, Ames, Iowa.* Gini's Mean Difference

A. W. Davis, *Commonwealth Scientific and Industrial Research Organisation, Glen Osmond, South Australia.* Gram–Charlier Series

*Deceased

D. D. Dorfman, *University of Iowa, Iowa City, Iowa.* Group Testing

S. D. Dubey, *Food and Drug Administration, Rockville, Maryland.* FDA Statistical Programs: Human Drugs

E. J. Dudewicz, *Syracuse University, Syracuse, New York.* Heteroscedasticity

J. W. Duncan, *Department of Commerce, Washington, D.C.* Federal Statistics

J. Edmiston, *Greenhithe, Kent, England.* Human Genetics, Annals of

A. W. F. Edwards, *University of Cambridge, Cambridge, England.* Fermat, Pierre de; Fiducial Distributions

W. J. Ewens, *Monash University, Clayton, Victoria, Australia.* Genetics, Statistics in

W. T. Federer, *Cornell University, Ithaca, New York.* Fractional Factorial Designs

S. E. Fienberg, *Carnegie-Mellon University, Pittsburgh, Pennsylvania.* Hansen Frequencies

T. L. Fine, *Cornell University, Ithaca, New York.* Foundations of Probability

P. R. Fisk, *University of Edinburgh, Edinburgh, Scotland.* Fisher's Ideal Index Number

M. A. Fligner, *University of Arizona, Tucson, Arizona.* Goodman's Y^2

R. A. Fontenot, *Whitman College, Walla Walla, Washington.* Gain Ratio

H. Freudenthal, *Instituut Ontwikkeling Wiskunde Onderqijs, Utrecht, The Netherlands.* Huygens, Christiaan

P. Gaenssler, *University of Munich, Munich, Bayern, German Federal Republic.* Glivenko–Cantelli Theorems

E. A. Gehan, *The University of Texas System Cancer Center, Houston, Texas.* Gehan–Gilbert Test; Historical Controls

J. E. Gentleman, *University of Waterloo, Waterloo, Ontario, Canada.* Graphical Representation, Computer Aided

J. D. Gibbons, *University of Alabama, University, Alabama.* Fisher's Exact Test

W. G. Gilchrist, *Sheffield City Polytechnic, Sheffield, England.* Forecasting

D. C. Gilliland, *Michigan State University, East Lansing, Michigan.* Fair–Jaffee Model

N. C. Giri, *University of Montreal, Montreal, Quebec, Canada.* Hunt–Stein Theorem

R. E. Glaser, *Lawrence Livermore National Laboratory, Livermore, California.* Homogeneity and Tests of Homogeneity

H. L. Gray, *Southern Methodist University, Dallas, Texas.* *G*-Spectral Estimator, The

T. N. E. Greville, *Charlottesville, Virginia.* Graduation

D. Griffiths, *Commonwealth Scientific and Industrial Research Organisation, Lindfield, New South Wales, Australia.* Gambling, Statistics in; Games of Chance (Excluding Blackjack)

A. J. Gross, *Medical University of South Carolina, Charleston, South Carolina.* Hazard Plotting

F. E. Grubbs, *Havre de Grace, Maryland.* Grubbs Estimators

W. C. Guenther, *University of Wyoming, Laramie, Wyoming.* *F* Tests; Hypergeometric Distributions

A. S. Gupta, *University of California, Riverside, California.* Generalized Canonical Variables

M. M. Gupta, *University of Saskatchewan, Saskatoon, Saskatchewan, Canada.* Fuzzy Set Theory

Y. Haitovsky, *Hebrew University, Jerusalem, Israel.* Grouped Data

J. Hannan, *Michigan State University, East Lansing, Michigan.* Fair–Jaffee Model

H. L. Harter, *Wright State University, Dayton, Ohio.* Harter's Adaptive Robust Method

J. A. Hausman, *Massachusetts Institute of Technology, Cambridge, Massachusetts.* Full Information Estimators

D. G. Herr, *University of North Carolina, Greensboro, North Carolina.* Geometry in Statistics

W. G. S. Hines, *University of Guelph, Guelph, Ontario, Canada.* Geometric Mean; Geometric Moving Average

D. C. Hoaglin, *Harvard University, Cambridge, Massachusetts.* Folded Transformations; *g* and *h* Distributions; Generation of Random Variables

J. L. Hodges, *University of California, Berkeley, California.* Hodges–Lehmann Estimators

W. Hoeffding, *University of North Carolina, Chapel Hill, North Carolina.* Hájek's Projection Lemma; Hoeffding's Independence Test

M. Hollander, *Florida State University, Tallahassee, Florida.* Hollander Bivariate Symmetry Test; Hollander Extreme Test; Hollander Parallelism Test; Hollander–Proschan NBU Test

A. Holly, *University of Lausanne, Lausanne, Switzerland.* Hausman Specification Test

L. J. Hubert, *University of California, Santa Barbara, California.* Hierarchical Classification; Hierarchical Cluster Analysis

R. L. Iman, *Sandia Laboratories, Albuquerque, New Mexico.* Harmonic Mean

R. A. Israel, *Bowie, Maryland.* Health Statistics, National Center for

R. B. Israel, *University of British Columbia, Vancouver, British Columbia, Canada.* Gibbs Distributions II

G. C. Jain*, *University of Otago, Dunedin, New Zealand.* Hermite Distributions

N. T. Jazairi, *York University, Downsview, Ontario, Canada.* Gross National Product Deflator; Hedonic Index Numbers

G. M. Jenkins*, *Gwilym Jenkins & Partners, Ltd., Lancaster, England.* Feedforward-Feedback Control Schemes

D. R. Jensen, *Virginia Polytechnic and State University, Blacksburg, Virginia.* Friedman's Chi-Square Test

P. W. M. John, *University of Texas, Austin, Texas.* Group Divisible Designs

A. G. Journel, *Stanford University, Stanford, California.* Geostatistics

C. H. Kapadia, *Southern Methodist University, Dallas, Texas.* Geometric Distribution

A. D. Keedwell, *University of Surrey, Guildford, England.* Graeco-Latin Squares

V. Klemeš, *National Hydrology Research Institute, Ottawa, Ontario, Canada.* Hydrology, Stochastic

G. G. Koch, *University of North Carolina, Chapel Hill, North Carolina.* Hierarchical Kappa Statistics

K. Kocherlakota, *University of Manitoba, Winnipeg, Manitoba, Canada.* Generalized Variance

S. Kocherlakota, *University of Manitoba, Winnipeg, Manitoba, Canada.* Generalized Variance

R. Kolodny, *University of Maryland, College Park, Maryland.* Finance, Statistics in

M. Kuczma, *University of Silesia, Katowice, Poland.* Functional Equations

S. Kullback, *George Washington University, Washington, D.C.* Fisher Information

S. M. Kwerel, *Englewood, New Jersey.* Fréchet Bounds

L. R. LaMotte, *University of Houston, Houston, Texas.* Fixed-, Random-, and Mixed-Effects Models; Fractional Rank Estimators

E. L. Lehmann, *University of California, Berkeley, California.* Hodges–Lehmann Estimators

S. A. Lemeshow, *University of Massachusetts, Amherst, Massachusetts.* Half-Sample Techniques

W. F. Lucas, *Cornell University, Ithaca, New York.* Game Theory

E. Lukacs, *The Catholic University of America, Washington, D.C.* Faà di Bruno's Formula

M. B. McElroy, *Duke University, Durham, North Carolina.* Glahn and Hooper Correlation Coefficients

B. B. Mandelbrot, *IBM Research Center, Yorktown Heights, New York.* Fractals; Fractional Brownian Motions and Gaussian Noises; Fractional Integro-Differentiation; Gibbs Distributions I; Hurst Coefficient (Rescaled Range Analysis)

K. G. Manton, *Duke University, Durham, North Carolina.* Framingham: An Evolving Longitudinal Study

D. W. Matula, *Southern Methodist University, Dallas, Texas.* Graph Theoretical Cluster Analysis

*Deceased

A. E. Maxwell, *University of London, London, England*. Factor Analysis

L. J. Melton III, *Mayo Clinic, Rochester, Minnesota*. Follow-up

P. W. Mielke, *Colorado State University, Fort Collins, Colorado*. Goodman–Kruskal Tau and Gamma

Y. Mittal, *Virginia Polytechnic Institute and State University, Blacksburg, Virginia*. Gaussian Processes

P. A. P. Moran, *Australian National University, Canberra, Australia*. Geometric Probability Theory

G. S. Mudholkar, *University of Rochester, Rochester, New York*. Fisher's z Transformation

J. A. Nelder, *Rothamsted Experimental Station, Harpenden, England*. Generalized Linear Models; GLIM

C. M. Newman, *University of Arizona, Tucson, Arizona*. Fokker–Planck Equations

P. L. Odell, *University of Texas at Dallas, Richardson, Texas*. Gauss–Markov Theorem

J. K. Patel, *University of Missouri, Rolla, Missouri*. Hazard Rate and Other Classifications of Distributions

C. G. Pfeifer, *E. I. Dupont de Nemours and Company, Wilmington, Delaware*. Graphical Representation of Data; Histograms

K. C. S. Pillai, *Purdue University, West Lafayette, Indiana*. Hotelling's T^2; Hotelling's Trace

B. L. Raktoe, *National University of Singapore, Singapore*. Fractional Factorial Designs

P. S. R. S. Rao, *University of Rochester, Rochester, New York*. Hansen–Hurwitz Method for Subsampling Nonrespondents

C. B. Read, *Southern Methodist University, Dallas, Texas*. Fermi–Dirac Statistics; Fieller's Theorem; Fisher–Yates Tests; Five-Number Summaries; Fixed-Width Confidence Intervals; Folded Distributions; Fundamental Identity of Sequential Analysis; Gompertz Distribution; Gosset, William Sealy; Greenwood's Statistic; Helmert, Friedrich Robert

G. W. Reddien, *Southern Methodist University, Dallas, Texas*. Gauss–Seidel Iteration

R. N. Rodriguez, *General Motors Corporation, Warren, Michigan*. Frequency Curves, Systems of; Frequency Surfaces, Systems of

R. M. Royall, *Johns Hopkins University, Baltimore, Maryland*. Finite Populations, Sampling from

S. C. Saunders, *Washington State University, Pullman, Washington*. Fatigue Models

S. R. Searle, *Cornell University, Ithaca, New York*. General Linear Model

E. Seneta, *University of Sydney, Sydney, New South Wales, Australia*. Galton–Watson Process

M. Shaked, *University of Arizona, Tucson, Arizona*. Geometry in Statistics: Convexity

G. A. Shea, *Exxon Research, Linden, New Jersey*. Franklin's Identity; Hoeffding's Lemma

G. R. Shorack, *University of Washington, Seattle, Washington*. Hungarian Constructions of Empirical Processes

M. Sibuya, *Keio University, Yokohama, Japan*. Generalized Hypergeometric Distributions

W. Simpson, *Organization for Economic Cooperation and Development, Paris, France*. Foreign Trade Statistics

C. A. B. Smith, *University College, London, England*. Galton, Francis; Haldane–Smith Test; Human Genetics, Statistics in

W. L. Smith, *University of North Carolina, Chapel Hill, North Carolina*. Generating Functions; Helly–Bray Theorems; Hotelling, Harold

R. D. Snee, *E. I. Dupont de Nemours and Company, Wilmington, Delaware*. Graphical Representation of Data; Histograms

T. P. Speed, *University of Western Australia, Nedlands, Australia*. General Balance

M. D. Springer, *University of Arkansas, Fayetteville, Arkansas*. H-Function Distribution

D. A. Sprott, *University of Waterloo, Waterloo, Ontario, Canada*. Gauss, Carl Friedrich

J. N. Srivastava, *Colorado State University, Boulder, Colorado*. Galois Fields

J. M. Steele, *Stanford University, Stanford, California*. Guessing Models

M. Stone, *University College, London, England*. Fiducial Probability

D. J. Strauss, *University of California, Irvine, California*. Hammersley–Clifford Theorem, The

G. P. H. Styan, *McGill University, Montreal, Quebec, Canada*. Generalized Inverses

E. Sverdrup, *University of Oslo, Oslo, Norway*. Frequency Interpretation in Probability and Statistical Inference

P. Switzer, *Stanford University, Stanford, California*. Geography, Statistics in

G. M. Tallis, *University of Adelaide, Adelaide, South Australia*. Goodness of Fit

P. Tan, *Carleton University, Ottawa, Canada*. Fisher's Problem of the Nile

J. Tiago de Oliveira, *Faculty of Sciences, Lisbon, Portugal*. Gumbel Distribution

D. S. Tracy, *University of Windsor, Windsor, Ontario, Canada*. Fisher's k-Statistics

G. van Belle, *University of Washington, Seattle, Washington*. Fixed Station Networks (Statistical Characteristics of)

K. W. Wachter, *University of California, Berkeley, California*. Haar Distributions

H. Wainer, *Educational Testing Service, Princeton, New Jersey*. Gapping

J. H. Ware, *Harvard University, Cambridge, Massachusetts*. Growth Curves

G. S. Watson, *Princeton University, Princeton, New Jersey*. Geology, Statistics in; Hypothesis Testing

J. T. Webster, *Southern Methodist University, Dallas, Texas*. Factorial Experiments

L. J. Wei, *George Washington University, Washington, D.C.* Friedman's Urn Model; Gehan–Gilbert Test

L. Weiss, *Cornell University, Ithaca, New York*. Generalized Likelihood Ratio Tests; Generalized Maximum Likelihood Estimation; Generalized Sequential Probability Ratio Tests

J. A. Wellner, *University of Rochester, Rochester, New York*. Glivenko–Cantelli Theorems

G. E. Whitehouse, *University of Central Florida, Orlando, Florida*. Flowgraph Analysis

H. O. Wold, *University of Uppsala, Uppsala, Sweden*. Fix-Point Method

M. A. Woodbury, *Duke University, Durham, North Carolina*. Framingham: An Evolving Longitudinal Study

M. G. Yochmowitz, *Air Force School of Aerospace Medicine, Brooks Air Force Base, Texas*. Factor Analysis of Variance Model (FANOVA)

ENCYCLOPEDIA OF STATISTICAL SCIENCES

VOLUME 3

Faà di Bruno's Formula to Hypothesis Testing

F

FAÀ DI BRUNO'S FORMULA

The need for a formula giving an explicit expression for the pth derivative of a function of a function was recognized toward the end of the eighteenth century. Several authors gave partial solutions; references to these are listed in ref. 4.

Let $z = G(y)$ and $y = f(x)$ be two functions such that all derivatives of $G(y)$ and $f(x)$ up to order p exist. The problem is then to find a formula for determining

$$\frac{d^p z}{dx^p} = \frac{d^p}{dx^p} G[f(x)]. \qquad (1)$$

A general solution of this problem was given by Francesco Faà di Bruno [1, 2] and was reproduced in his book [3, p. 3].

Faà di Bruno's formula is given by

$$\frac{d^n}{dt^n} G[f(t)] = \sum \frac{n!}{k_1!\,k_2! \cdots k_n!} \left[\frac{d^p G}{dy^p} \right] \times \left(\frac{f'}{1!} \right)^{k_1} \left(\frac{f''}{2!} \right)^{k_2} \cdots \left(\frac{f^{(n)}}{n!} \right)^{k_n}. \qquad (2)$$

The summation is here extended over all partitions of n such that

$$p = k_1 + k_2 + \cdots + k_n,$$
$$n = k_1 + 2k_2 + \cdots + nk_n. \qquad (3)$$

Faà di Bruno's formula can easily be proved by induction.

As an example, consider the case $G(y) = e^y$, $y = f(t) = \sin t$, and $n = 4$. Then formula (2) becomes

$$\frac{d^4}{dt^4}(e^{\sin t}) = e^{\sin t} \sum \frac{4!}{k_1! \cdots k_4!} \times \left(\frac{f'}{1} \right)^{k_1} \left(\frac{f''}{2!} \right)^{k_2} \cdots \left(\frac{f^{(4)}}{4!} \right)^{k_4}.$$

The summation goes over the following five quadruples (k_1, k_2, k_3, k_4), namely $(0,0,0,2)$, $(1,0,1,0)$, $(0,2,0,0)$, $(2,1,0,0)$, $(4,0,0,0)$. Then one obtains

$$\frac{d^4}{dt^4}(e^{\sin t}) = e^{\sin t}(\cos^4 t - 6\cos^2 t \sin t + 3\sin^2 t - 4\cos^2 t + \sin t).$$

Two cases of (2) are of interest to probabilists and statisticians. These permit establishing relations between the cumulants* and moments of distribution functions, provided that the functions $G(y)$ and $f(t)$ are suitably chosen.

If we select $G(y) = \ln y$, then

$$\frac{d^n}{dt^n} \ln f(t) = \sum \frac{n!}{k_1! \cdots k_n!} \left(\frac{d^p G}{dy^p} \right) \left(\frac{f'}{1} \right)^{k_1} \times \left(\frac{f''}{2!} \right)^{k_2} \cdots \left(\frac{f^{(n)}}{n!} \right)^{k_n}, \qquad (4)$$

and similarly if $G(y) = e^y$, we obtain

$$\frac{d^n}{dt^n} e^{f(t)} = e^{f(t)} \sum \frac{n!}{k_1! \cdots k_n!} \times \left(\frac{f'}{1}\right)^{k_1} \left(\frac{f''}{2!}\right)^{k_2} \cdots \left(\frac{f^{(n)}}{n!}\right)^{k_n}. \tag{5}$$

In (4) and (5) the summation is to be taken over all partitions of n that satisfy (3). We present an example for (4) and also for (5).

In (4) let $f(t)$ be a characteristic function and set $t = 0$. Then we obtain

$$\kappa_n = \sum (-1)^{p-1} \frac{n!(p-1)!}{k_1!, \ldots, k_n!} \alpha_1^{k_1} \cdots \alpha_n^{k_n}. \tag{4a}$$

Next we assume that $f(t)$ is a cumulant generating function (i.e., the logarithm of a characteristic function). In this way we get

$$\alpha_n = \sum \frac{n!}{k_1! \cdots k_n!} \kappa_1^{k_1} \left(\frac{\kappa_2}{2!}\right)^{k_2} \cdots \left(\frac{\kappa_n}{n!}\right)^{k_n} \tag{5a}$$

for $n = 1, 2, \ldots, m$. The summation is extended over all partitions of n that satisfy (3).

Example 1. Put $n = 3$ in (4a) and get $\kappa_3 = \alpha_3 - 3\alpha_1\alpha_2 + 2\alpha_1^3$.

Example 2. In the same way one obtains for $n = 4$ from (5a)

$$\alpha_4 = \kappa_4 + 3\kappa_2^2 + 4\kappa_1\kappa_3 + 6\kappa_1^2\kappa_2 + \kappa_1^4.$$

Another application is the derivation of explicit formulas for the k-statistics (*see* FISHER'S k-STATISTICS). Let $x_1, x_2, \ldots, x_n$ be n independently and identically distributed observations [i.e., a sample of size n from a population with population distribution $F(x)$]. The k-statistic of order p, written as k_p, is a symmetric polynomial statistic whose expectation is equal to the cumulant κ_p of order p.

References

[1] Faà di Bruno, F. (1855). *Ann. Sci. Mat. Fis. Comp. B. Tortolini*, **6**, 479–480.

[2] Faà di Bruno, F. (1855). *Quart. J. Math.*, **1**, 359–360.

[3] Faà di Bruno, F. (1859). *Théorie générale de l'élimination*. De Leiber et Faraquet, Paris (taken over by Gauthier-Villars).

[4] Lukács, E. (1955). *Amer. Math. Monthly*, **62**, 340–348.

(CUMULANTS
FISHER'S k-STATISTICS)

EUGENE LUKÁCS

FACES *See* CHERNOFF FACES

FACTOR *See* FACTORIAL EXPERIMENTS

FACTOR ANALYSIS

Factor analysis is a branch of multivariate analysis* that is concerned with the internal relationships of a set of variables when these relationships can be taken to be linear, or approximately so. Initially, factor analysis was developed by psychometricians and was concerned primarily with hypotheses about the organization of mental ability suggested by the examination of correlation* or covariance* matrices for sets of cognitive test variables (e.g., the set in Table 2). Indeed, Charles Spearman [17], the founder of the subject, postulated that such correlations could be accounted for by a single underlying variable, or factor, which he called g and which later was loosely labeled "general intelligence." But Spearman's hypothesis was soon found to be inadequate and the model was expanded to include several factors. This expansion, which took place gradually over the three or four decades following Spearman's original work, had several features. One was the realization that a battery of cognitive tests, containing subgroups of tests bearing a close family relationship to each other, such as subgroups of verbal or numerical tests, in general required additional group factors (see Table 2) over and above the general Spearman-type factor, to account adequately for their intercorrelations. Another was the somewhat revolutionary step taken by Thurstone [18, 19] of overlooking a possible general factor and of accounting for the intercorrelations of a given set of variables in terms of group (or overlapping group) factors only.

The overall outcome of these developments, largely by nonstatisticians, was a wide variety of different (approximate) methods of carrying out a factor analysis. The methods, too, eventually began to be employed in fields of research other than the cognitive, and today factor analysis is widely used—in subjects as diverse as medicine and meteorology*. A modern account of the work of the psychometricians is given by Harman [5], but for a statistical treatment of factor analysis we must turn mainly to contributions by Lawley and by Jöreskog (see Lawley and Maxwell [12]). An account for the beginner, with an historical introduction, is given by Maxwell [14, Chaps. 1, 2, 5, 6].

Apart from some new methods for exploring the internal relationships of multivariate data, such as cluster analysis, multidimensional scaling*, and nonlinear mapping (for a review, see Everitt [4]), traditionally two models (as distinct from methods within a model) which formally resemble each other but have rather different aims, were available. One is the principal component model and the other the factor analysis model. In the interest of clarity it is advisable at the outset to distinguish between these two approaches. In principal component analysis* a set of p observed correlated variables, denoted by $X_1, \ldots, X_p$, is transformed into an *equal* number of new variables $Y_1, \ldots, Y_p$ that have the property of being uncorrelated. In this model all p components are needed to reproduce accurately the correlation coefficients of the X-variables. In contrast, the aim of factor analysis is to account for these correlations in terms of a much smaller number, k, of hypothetical variables or factors. Put simply, the first question that arises is whether any correlation exists, that is, whether the correlation matrix differs from the unit diagonal matrix*. If there is correlation, the next question is whether a random variable F_1 exists such that all partial correlations* between the X-variables after eliminating the effect of F_1 are zero. If not, two random variables F_1 and F_2 are postulated and the partial correlations after eliminating F_1 and F_2 are examined. The process continues until all partial correlations between the X-variables are zero.

The aims of the two models can also be contrasted by considering the nature of the relationships involved. In component analysis the Y-variables are by definition linear functions of the X-variables and no question of a hypothesis arises. In factor analysis on the other hand, the basic assumption is that

$$X_i = \mu_i + \sum_{r=1}^{k} \lambda_{ir} F_r + E_i \qquad (i = 1, \ldots, p), \tag{1}$$

where μ_i is the mean of X_i, F_r is the rth common factor, the number k $(k < p)$ of such factors being specified, and where E_i is a residual variable representing sources of variation affecting only X_i.

In equations (1) the p random variables E_i are assumed to be independent of one another and of the k factors F_r. The latter may be either correlated (oblique) or uncorrelated (orthogonal). Usually, they are scaled to have unit variances. For convenience, and without loss of generality, we suppose that the means of all variables are zero and henceforth the terms μ_i are dropped from equations (1). The variance of E_i, termed either the residual variance or the unique variance of X_i, is denoted by ψ_i. The coefficient λ_{ir} is known as the *loading* of X_i on F_r or, alternatively, as the loading of F_r in X_i. In practice the λ_{ir} and the ψ_i are usually unknown parameters that require estimation from experimental data. Equations (1) are not capable of direct verification since the p observed variables X_i are expressed in terms of $(p + k)$ other variables that are unobservable, but they imply a hypothesis that can be tested regarding the variances and covariances of the X_i [see expression (3) below].

Having specified the factor model, the next question is to estimate the parameters in it using a sample of observations on the X_i. In the early days approximate methods of estimation only were available, of which the most celebrated was the *centroid* or simple summation method. Other more recent approximate methods are the *principal factor* and *minres* methods; all are described in the book by Harman [5] and by numerous other writers to whom he refers. Efficient statistical procedures were first introduced by Lawley [10] in a paper on the estimation of factor loadings by the method of maximum likelihood*. Difficulties of a computational nature were experienced, however, and it was not until the advent of electronic computers and a new approach to the solution of the basic equations by Jöreskog [7] that the maximum likelihood approach became a feasible proposition.

Currently, factor analysis is considered under two headings. The first is *exploratory* factor analysis, in which a basic problem is that of discovering the number k of common factors necessary and sufficient to account for the intercorrelations of a given set of variables. The second is *confirmatory* factor analysis in which k is assumed to be known and the main problem is to fit a postulated pattern of zero and nonzero loadings to a given correlation matrix (an example is given below).

When a correlation matrix is used in a factor analysis and, in consequence, the variances of all variables are set equal to unity, the quantities $h_i = (1 - \psi_i)$ are known as the *communalities* of the observed variables, that is, the amounts of their respective variances which are accounted for by the common factors.

EXPLORATORY FACTOR ANALYSIS

We return to equations (1) and to the assumptions that follow them. It is convenient to write these equations in matrix form. Let $\mathbf{x}$ and $\mathbf{e}$ denote column vectors with respective elements X_i and E_i $(i = 1, \ldots, p)$, and let $\mathbf{f}$ be the column vector with elements F_r $(r = 1, \ldots, k)$. Then equations (1), omitting the terms μ_i, become

$$\mathbf{x} = \mathbf{\Lambda f} + \mathbf{e}, \tag{2}$$

where $\mathbf{\Lambda} = [\lambda_{ir}]$ is the $p \times k$ matrix of loadings. The covariance (or dispersion) matrix is denoted by $\mathbf{\Sigma} = [\sigma_{ij}]$, where σ_{ii} is the variance of X_i and where σ_{ij} is the covariance of X_i and X_j. In view of the assumptions made about $\mathbf{f}$ and $\mathbf{e}$, we have

$$E(\mathbf{fe}') = \mathbf{0}, \qquad E(\mathbf{ff}') = \mathbf{I}_k, \qquad E(\mathbf{ee}') = \mathbf{\Psi},$$

where $\mathbf{I}_k$ is the unit matrix of order k and where $\mathbf{\Psi}$ is a matrix whose diagonal elements are $\psi_i, \ldots, \psi_p$ and whose nondiagonal elements are zero. Since

$$E(\mathbf{xx}') = E(\mathbf{\Lambda f} + \mathbf{e})(\mathbf{\Lambda f} + \mathbf{e})',$$

we have

$$\mathbf{\Sigma} = \mathbf{\Lambda\Lambda}' + \mathbf{\Psi}. \tag{3}$$

In practice the elements of $\mathbf{\Lambda}$ and $\mathbf{\Psi}$ are unknown parameters that have to be estimated from experimental data. An up-to-date account of how this is done is given by Jöreskog [8], but, as it is very technical, only an introductory outline can be given here.

For a sample of individuals from some population having scores on each of the p variables X_i the sample covariance matrix, $\mathbf{S}$, is estimated in the usual way. The factor estimation problem is to fit a matrix $\mathbf{\Sigma}$ of the form (3) to the matrix $\mathbf{S}$. In many applications the scales of measurement of the observed variables are arbitrary and in such cases one takes $\mathbf{S}$ to be the correlation matrix, $\mathbf{R}$, between the variables. Jöreskog considers three different methods of fitting $\mathbf{\Sigma}$ to $\mathbf{S}$. The first is the unweighted least-squares* method (ULS), which minimizes

$$U = \tfrac{1}{2}\operatorname{tr}(\mathbf{S} - \mathbf{\Sigma})^2,$$

where tr stands for the trace of a matrix. The second is the generalized least-squares method (GLS), which minimizes

$$G = \tfrac{1}{2}\operatorname{tr}\left(\mathbf{I}_p - \mathbf{S}^{-1}\mathbf{\Sigma}\right)^2,$$

while the third is the maximum likelihood method (ML), which minimizes

$$M = \operatorname{tr}(\mathbf{\Sigma}^{-1}\mathbf{S}) - \log|\mathbf{\Sigma}^{-1}\mathbf{S}| - p.$$

Jöreskog points out that the GLS and ML methods are scale free, but both methods require that the matrix $\mathbf{S}$, or $\mathbf{R}$, be positive definite. On the other hand, the ULS method will work even on a matrix that is non-Gramian, and the solution it gives is equivalent to the iterated *principal factor* solution and to the *minres* solution mentioned earlier. When $k > 1$ and there is more than one common factor, it is necessary to remove an element of indeterminacy in the model which arises from the fact that there exist nonsingular linear transformations of the common factors which change $\mathbf{\Lambda}$ but leave $\mathbf{\Sigma}$ unchanged. The usual way to eliminate this indeterminacy is to choose $\mathbf{\Lambda}'\mathbf{\Lambda}$ to be diagonal in ULS and $\mathbf{\Lambda}'\mathbf{\Psi}^{-1}\mathbf{\Lambda}$ to be diagonal in GLS and ML and to estimate the parameters in $\mathbf{\Lambda}$ and $\mathbf{\Psi}$ subject to these conditions.

Factor Rotation and Interpretation

When $k > 1$ the matrix of loadings, $\mathbf{\Lambda}$, is not unique and can be replaced by $\mathbf{\Lambda M}$, where $\mathbf{M}$ is any orthogonal matrix* of order k. This fact is often used by investigators in an effort to simplify the interpretation of their results, the aim in general being to eliminate all or most of the negative signs from $\mathbf{\Lambda}$ and to reduce to zero or near zero as many as possible of its elements. Numerous numerical examples can be found in the textbooks quoted. Originally, $\mathbf{M}$ was obtained by subjective graphical considerations but, more recently, objective analytical procedures (for use on computers) have been developed which achieve much the same ends. The most commonly used techniques are the "varimax method," which retains the orthogonality of the factors, and the "promax method," which allows them to become oblique. For details, see Lawley and Maxwell [12, Chap. 6] and Harman [5, Chaps. 14, 15]. Numerous other procedures for the transformation of factor loadings are also in use and some of these are described by Evans [3], who also provides references.

Standard Errors of Estimates

Although factor analysts, in interpreting their results, frequently assume that several of the elements in $\mathbf{\Lambda}$, either before or after rotation, have true values of zero, it was only recently that methods were found for estimating their standard errors*. These were first given by Lawley [11] as a by-product of maximum likelihood estimation*. Later, Lawley's methods were simplified and extended by Archer and Jennrich [1] and by Jennrich [6].

CONFIRMATORY FACTOR ANALYSIS

On occasion investigators experienced in some field of research may wish to test a specific hypothesis about the factorial composition of a given set of variables in a particular population. In other words, they may feel able to postulate in advance the number of factors they expect and the pattern of zero and nonzero loadings on them. For example, they might postulate a pattern such as

$$\begin{matrix} x & x & x & x & x & x & x & x & x & x \\ x & x & x & x & x & 0 & 0 & 0 & 0 & 0 \\ 0 & 0 & 0 & 0 & x & x & x & x & x & x \end{matrix}$$

for 10 variables having three factors, in which the x's stand for nonzero loadings whose magnitudes are to be estimated. The parameters to which specific values—the zeros in this case—are assigned are referred to as *fixed* parameters, while the others are called *free* parameters. The model used is

$$\mathbf{\Sigma} = \mathbf{\Lambda\Phi\Lambda}' + \mathbf{\Psi},$$

where the matrix $\mathbf{\Phi}$ represents the correlation matrix between the factors. But if the factors are postulated to be orthogonal, this matrix will be unit diagonal. In this model there are restrictions on the number of fixed parameters allowed and on their positioning in $\mathbf{\Lambda}$ (see Lawley and Maxwell [12, Chap. 7]).

Numerical Example

The example below is concerned with the correlational structure of a battery of 10 cognitive tests for measuring the intelligence of young children. The names of the tests

Table 1 Correlation Matrix for 10 Cognitive Tests

Test	1	2	3	4	5	6	7	8	9	10
1	1.00	0.41	0.33	0.27	0.48	−0.05	0.09	−0.10	0.14	0.11
2		1.00	0.29	0.27	0.45	0.14	0.20	−0.06	0.03	0.17
3			1.00	0.32	0.36	0.22	0.18	0.09	0.22	0.41
4				1.00	0.26	0.26	0.16	0.07	0.22	0.38
5					1.00	0.20	0.21	0.17	0.23	0.25
6						1.00	0.39	0.32	0.04	0.27
7							1.00	0.37	0.13	0.47
8								1.00	0.35	0.43
9									1.00	0.33
10										1.00

appear in Table 2. The first five tests are referred to in the literature as "verbal tests" as the children's responses to them are verbal in nature. The remaining five tests are referred to as "performance tests" and responses to them are nonverbal. From prior knowledge of the content of the tests the investigator felt able to postulate that in a confirmatory factor analysis a general factor having loadings on all 10 tests, accompanied by two group factors, one having loadings on the five verbal tests the other on the five performance tests, would be adequate to account for their intercorrelations. The precise pattern of zero and nonzero loadings postulated for three (orthogonal) factors was that given in the previous paragraph. For a sample of 75 children, in a confirmatory study, the intercorrelations of their test scores were found and are given in Table 1.

Table 2 Factor Loadings for 10 Cognitive Tests

	Factor		
Test	I	II	III
1: Information	0.21	0.70	
2: Vocabulary	0.29	0.53	
3: Arithmetic	0.55	0.28	
4: Similarities	0.53	0.19	
5: Comprehension	0.30	0.62	0.24
6: Animal House	0.33		0.33
7: Picture Completion	0.40		0.45
8: Mazes	0.22		0.67
9: Geometric Design	0.30		0.29
10: Block Design	0.68		0.38

The postulated pattern of zero and nonzero loadings was now fitted to the correlation matrix in Table 1 by the method of maximum likelihood (see Lawley and Maxwell [12, Chap. 7]) and the estimates of the nonzero loadings are given in Table 2. A test of significance* showed that these estimated loadings accounted very closely for the observed correlations of the tests and, consequently, their postulated factor content was confirmed.

ESTIMATING FACTOR SCORES

In principal component analysis the components are linear functions of the original variables and there is no difficulty in estimating the scores of an individual on these components. In factor analysis, on the other hand, where the common factors do not fully account for the total variance of the variables, the problem is more difficult and some "minimum variance" or "least-squares"* principle has to be invoked in order that reasonable estimates of factor scores may be obtained. Two estimation procedures in common use are the Thomson or "regression" method, and the Bartlett method (see Lawley and Maxwell [12, Chap. 8]). In the latter the principle used is that of minimizing the sum of squares of the standardized residuals*. In the regression* method the estimate, $\hat{f}$ of f, for orthogonal factors, is given by

$$\hat{\mathbf{f}} = (\mathbf{I} + \boldsymbol{\Gamma})^{-1}\boldsymbol{\Lambda}'\boldsymbol{\Psi}^{-1}\mathbf{x},$$

where $\boldsymbol{\Gamma} = \boldsymbol{\Lambda}'\boldsymbol{\Psi}^{-1}\boldsymbol{\Lambda}$, and for correlated factors by

$$\hat{\mathbf{f}} = \boldsymbol{\Phi}(\mathbf{I} + \boldsymbol{\Gamma}\boldsymbol{\Phi})^{-1}\boldsymbol{\Lambda}'\boldsymbol{\Psi}^{-1}\mathbf{x}.$$

By the Bartlett method the estimate is given by

$$\hat{\mathbf{f}} = \boldsymbol{\Gamma}^{-1}\boldsymbol{\Lambda}'\boldsymbol{\Psi}^{-1}\mathbf{x}.$$

The form of the latter estimate does not depend on whether the factors are correlated or not. Formulae are also available for finding the variances and covariances of the estimated factor scores and the results can be used to assess the degree to which the postulated factors are determined (see Lawley and Maxwell [12, p. 42]).

Factor Analysis and Regression

From the equations above a separate set of weights is derived for estimating the scores on each factor. Each set is applied to the standardized scores on the X-variables and gives an equation similar in form to that obtained in a multiple regression analysis* for estimating scores on a criterion variable. But whereas a criterion variable is one that can be independently observed, a factor is an unobservable variable derived from an *internal* analysis of the X-variables themselves. Yet is worth noting that factor analysis can be employed to advantage when estimating the weights in a multivariate regression analysis where the independent variables are subject to measurement error*, as it enables one to partial out such an error in the estimation procedure (see Lawley and Maxwell [13] and Chan [2]).

COMPARING RESULTS FROM DIFFERENT SOURCES

A question that frequently arises in factor analysis is whether, for a given set of variables, the same factors occur in different populations. This is a difficult question and to make progress with it some simplifying assumptions are generally made. One approach is to assume that, although two covariance matrices $\boldsymbol{\Sigma}_1$ and $\boldsymbol{\Sigma}_2$ involving the same variables may be different, they may still have the same $\boldsymbol{\Lambda}$ matrix. This could occur if the two covariance matrices $\boldsymbol{\Phi}_1$ and $\boldsymbol{\Phi}_2$ between the factors, each of order k, were themselves different. If we assume that the residual variances in the two populations are in each case $\boldsymbol{\Psi}$, the hypothesis to be tested is

$$\boldsymbol{\Sigma}_1 = \boldsymbol{\Lambda}\boldsymbol{\Phi}_1\boldsymbol{\Lambda}' + \boldsymbol{\Psi},$$
$$\boldsymbol{\Sigma}_2 = \boldsymbol{\Lambda}\boldsymbol{\Phi}_2\boldsymbol{\Lambda}' + \boldsymbol{\Psi},$$

and the problem becomes one of estimating the elements of $\boldsymbol{\Lambda}$, $\boldsymbol{\Psi}$, $\boldsymbol{\Phi}_1$, and $\boldsymbol{\Phi}_2$ and of testing for goodness of fit.* More elaborate models have also been proposed (for references and a worked example, see McGaw and Jöreskog [15]).

APPRAISAL

Several attempts have been made to assess the value of factor analysis in research work, typical of which is a series of articles in *The Statistician* in 1962, with rejoinders in succeeding years. On the whole the conclusions are somewhat controversial. Yet, in some measure, factor analysis has proved beneficial: for example, in clarifying our ideas about cognitive abilities, and for giving us useful summary concepts for the description of human personality and of people's attitudes to their social environment. A very informative yet critical article is by Russell et al. [16], concerned with incentives that lead people to smoke tobacco.

Compared with principal component analysis, which is relatively simple to perform, factor analysis is more difficult; but it can be recommended for the analysis of the correlational structure of variables that can only be measured with error, and in situations in which there is reason to believe that this structure can be accounted for in terms of a relatively small number of underlying variables (see Maxwell [14, pp. 11, 17]). Yet Jöreskog, one of the main contributors to the statistical theory of factor analysis, has expressed doubts about the adequacy, in general, of *linear* factor models and has proposed additional models which may prove to be valuable [9].

References

[1] Archer, C. O. and Jennrich, R. I. (1973). *Psychometrika*, **38**, 581–592.

[2] Chan, N. N. (1977). *Biometrika*, **64**, 642–644.

[3] Evans, G. T. (1971). *Brit. J. Math. Statist. Psychol.*, **24**, 22–36.

[4] Everitt, B. S. (1978). *Graphical Techniques for Multivariate Data*. Heinemann, London.

[5] Harman, H. H. (1967). *Modern Factor Analysis*. University of Chicago Press, Chicago. (Has a good historical introduction and deals well with the psychometric development of the subject.)

[6] Jennrich, R. I. (1973). *Psychometrika*, **38**, 593–604.

[7] Jöreskog, K. G. (1967). *Psychometrika*, **32**, 443–482. (A basic theoretical paper with examples of factor estimation by the maximum likelihood method.)

[8] Jöreskog, K. G. (1977). In *Statistical Methods for Digital Computers*, K. Enslein, A. Ralston, and M. S. Wilf, eds., pp. 125–153. Wiley, New York.

[9] Jöreskog, K. G. (1978). *Psychometrika*, **43**, 443–477.

[10] Lawley, D. N. (1940). *Proc. R. Soc. Edinb. A*, **60**, 64–82.

[11] Lawley, D. N. (1967). *Proc. R. Soc. Edinb. A*, **75**, 256–264.

[12] Lawley, D. N. and Maxwell, A. E. (1963; rev. 1971). *Factor Analysis as a Statistical Method*. Butterworth, Kent, England. (The standard textbook for mathematical statisticians.)

[13] Lawley, D. N. and Maxwell, A. E. (1973). *Biometrika*, **60**, 331–338.

[14] Maxwell, A. E. (1977). *Multivariate Analysis in Behavioural Research*. Chapman & Hall, London. (An introductory textbook which relates the development of factor analysis to that of other multivariate techniques.)

[15] McGaw, B. and Jöreskog, K. G. (1971). *Brit. J. Math. Statist. Psychol.*, **24**, 154–168.

[16] Russell, M. A. H., Peto, J., and Patel, U. A. (1974). *J. R. Statist. Soc. A*, **137**, 313–346.

[17] Spearman, C. E. (1904). *Amer. J. Psychol.*, **15**, 201–293.

[18] Thurstone, L. L. (1931). *Psychol. Rev.*, **38**, 406–427.

[19] Thurstone, L. L. (1947). *Multiple Factor Analysis*. University of Chicago Press, Chicago. (One of the classic textbooks but now outdated.)

Bibliography

See the following works, as well as the references just cited, for more information on the topic of factor analysis.

Bartlett, M. S. (1953). In *Essays on Probability and Statistics*. Methuen, London.

Gorsuch, R. L. (1974). *Factor Analysis*. Saunders, Philadelphia. (A comprehensive textbook of intermediate difficulty.)

Hendrickson, A. E. and White, P. O. (1964). *Brit. J. Statist. Psychol.*, **17**, 65–70.

Jöreskog, K. G. and Lawley, D. N. (1968). *Brit. J. Math. Statist. Psychol.* **21**, 85–96.

Kaiser, H. F. (1958). *Psychometrika*, **23**, 187–200.

Lawley, D. N. (1967). *Proc. R. Soc. Edinb. A*, **75**, 171–178.

Maxwell, A. E. (1977). *Psychol. Med.*, **6**, 643–648. (A novel application of factor analysis.)

McDonald, R. P. (1967). *Psychometrika*, **32**, 77–112.

Meredith, W. (1964). *Psychometrika*, **29**, 187–206.

Please, N. W. (1973). *Brit. J. Math. Statist. Psychol.*, **26**, 61–89.

Schönemann, P. M. (1966). *Psychometrika*, **31**, 1–10.

Thomson, G. H. (1951). *The Factorial Analysis of Human Ability*. London University Press, London. (One of the classic textbooks; still useful for beginners.)

(CLASSIFICATION
COMPONENT ANALYSIS
MULTIDIMENSIONAL SCALING
MULTIPLE REGRESSION)

A. E. MAXWELL

FACTOR ANALYSIS-OF-VARIANCE MODEL (FANOVA)

The factor analysis-of-variance model (FANOVA) describes a class of analysis-of-variance* models whose interaction can be specified by a singular value decomposition

Table 1 Special Cases of the FANOVA Model

Model Name	Interaction, γ_{ij}
Additive	0
Concurrent	$K\tau_i\beta_j$
Bundle of lines	
Rows linear	$R_i\beta_j$
Columns linear	$C_j\tau_i$
Combination of concurrent and bundle of lines	$K\tau_i\beta_j + R_i\beta_j$
First sweep of Tukey's vacuum cleaner	$K\tau_i\beta_j + R_i\beta_j + \tau_i C_j$

of residuals. Gollob [6] introduced the FANOVA name to describe models that combine the benefits of a factor analysis* decomposition of residuals with the ease of interpretation permitted in the analysis of variance. Many of these have been identified by Mandel [11–13], who contributed to the development and formulation of the FANOVA model. These special cases are summarized in Table 1.

In a two-way layout, the FANOVA model becomes

$$y_{ij} = \mu + \tau_i + \beta_j + \gamma_{ij} + \epsilon_{ij} \qquad (1)$$

with $i = 1, \ldots, t$; $j = 1, \ldots, b$; $\epsilon_{ij} \sim \text{IN}(0, \sigma^2)$, and interactions γ_{ij} of the form

$$\gamma_{ij} = \sum_{\kappa=1}^{c} \theta_\kappa u_{\kappa i} v_{\kappa j}\,. \qquad (2)$$

This representation for γ_{ij} results from the singular value decomposition of a matrix. The θ_κ's are unknown scalars whose presence indicates interaction. The $u_{\kappa i}$'s and $v_{\kappa j}$'s indicate the structure of this interaction.

In a fixed-effects model*, in addition to the main effects τ_i and β_j summing to zero over their respective subscripts, one has

$$\sum_i \gamma_{ij} = \sum_j \gamma_{ij} = 0.$$

To specify this model uniquely, one adds the constraints

$$\sum_i u_{\kappa i}^2 = \sum_j v_{\kappa j}^2 = 1; \qquad \kappa = 1, \ldots, c.$$

The ability to express γ_{ij} as the sum of multiplicative components $\theta_\kappa u_{\kappa i} v_{\kappa j}$ is the key to deciding if a model belongs to the FANOVA class. Test procedures are given below. Note that the row and column functions, $u_{\kappa i}$ and $v_{\kappa j}$, need not be functions of the main effects τ_i and β_j. In the event that they are functions of the main effects, models such as those listed in Table 1 can result.

The additive model, with no multiplicative components, is the simplest form of the FANOVA model. It has no interaction, so $\theta_1 = \cdots = \theta_c = 0$. The concurrent model has been studied by Ward and Dick [22]. It is a FANOVA model with a single multiplicative component whose interaction can be tested effectively by Tukey's [20] single degree of freedom. The bundle of lines models are also FANOVA models with one multiplicative component. Mandel [11] showed that interaction in the "row linear" or "column linear" bundle of lines model can be tested more effectively with $t - 1$ and $b - 1$ degrees of freedom than by Tukey's [20] single degree of freedom. In the presence of interaction, Mandel showed that the data can be described by a bundle of lines with scatter about the lines measured by the residual mean square. His test measures the equality of slopes of these lines and hence their additivity. Reparameterization of γ_{ij} in Tukey's "vacuum cleaner" model on its first sweep leads to $\gamma_{ij} = (k\tau_i + R_i)\beta_j + \tau_i C_j$. Therefore, it represents a FANOVA model with two multiplicative components. Unlike the FANOVA model, future sweeps of the vacuum cleaner will lead to multiplicative components that are functions of the preceding sweep.

Let $z_{ij} = y_{ij} - y_{i\cdot} - y_{\cdot j} + y_{\cdot\cdot}$ where "$\cdot$" denotes averaging over the appropriate subscripts, and $p = \min((b-1),(t-1))$. The maximum likelihood and least-squares estimates for the parameters in the FANOVA model are:

$$\hat{\mu} = y_{\cdot\cdot}; \quad \tau_i = y_{i\cdot} - y_{\cdot\cdot}; \quad \hat{\beta}_j = y_{\cdot j} - y_{\cdot\cdot}; \quad i = 1, \ldots, t; \; j = 1, \ldots, b.$$

$\hat{\theta}_\kappa^2 = \lambda_\kappa$, the κth largest characteristic root of $\mathbf{Z}'\mathbf{Z}$ (or $\mathbf{Z}\mathbf{Z}'$); $\kappa = 1, \ldots, c$; $\mathbf{Z}$ is the $t \times b$ matrix of z_{ij}'s.

$\hat{u}_\kappa = r_\kappa$, the normalized characteristic vector of $\mathbf{Z}\mathbf{Z}'$ corresponding to root λ_κ; $\kappa = 1, \ldots, c$.

$\hat{v}_\kappa = s_\kappa$, the normalized characteristic vector of $\mathbf{Z}'\mathbf{Z}$ corresponding to root λ_κ; $\kappa = 1, \ldots, c$.

$$\hat{\sigma}^2 = (\lambda_{c+1} + \cdots + \lambda_p)/(bt).$$

Various procedures have been used to determine the number of multiplicative components of interaction in the FANOVA model. To test the presence of a single multiplicative component of interaction, $(H_0 : \theta_1 = 0)$, Johnson and Graybill [9] proposed the likelihood ratio test* statistic

$$U_1 = \lambda_1/(\lambda_1 + \cdots + \lambda_p).$$

They showed that when $b < t$, $\mathbf{Z}'\mathbf{Z}$ has a central Wishart distribution* under H_0 and used a beta approximation with the method of moments* and moment estimates simulated by Mandel [13] to determine the distribution of U_1. Schuurmann et al. [15, 16] found the exact distribution of U_1, and gave percentage points when $(t - b - 1)/2$ is an integer [Schuurmann–Krishnaiah–Chattopadhyah (SKC) tables].

Corsten and van Eijnsbergen [3] considered instead simultaneously testing the hypothesis

$$H_0 : \theta_1 = \theta_2 = \cdots = \theta_c = 0,$$

where $c < \min(b - 1, t - 1)$. Their likelihood ratio statistic is

$$\mathbf{t} = (\lambda_1 + \cdots + \lambda_c)/(\lambda_{c+1} + \cdots + \lambda_p).$$

These λ_i are characteristic roots from the same central Wishart matrix developed by Johnson and Graybill. When $c = 1$, $\mathbf{t} = \lambda_1/(\lambda_2 + \cdots + \lambda_p)$. Therefore, one tests $H_0 : \theta_1 = 0$ when

$$P\left[\lambda_1/(\lambda_2 + \cdots + \lambda_p) \leqslant c_{1\alpha} \mid H_0\right] = 1 - \alpha.$$

Equivalently, inverting the test statistic and critical point,

$$\begin{aligned} 1 - \alpha &= P\left[(\lambda_2 + \cdots + \lambda_p)/\lambda_1 \geqslant c_{2\alpha} \mid H_0\right] \\ &= P\left[(\lambda_1 + \cdots + \lambda_p)/\lambda_1 \geqslant c_{3\alpha} \mid H_0\right] \\ &= P\left[\lambda_1/(\lambda_1 + \cdots + \lambda_p) \leqslant c_{4\alpha} \mid H_0\right] \\ &= P\left[U_1 \leqslant c_{4\alpha} \mid H_0\right]. \end{aligned}$$

Thus critical points for $\mathbf{t}$ can be transformed into critical points for U_1 in testing $H_0 : \theta_1 = 0$. Other simultaneous test procedures have been developed by Krishnaiah and coworkers and are summarized in Krishnaiah and Yochmowitz [10].

Hegemann and Johnson [7] test $H_0 : \theta_2 = 0$ by means of a likelihood ratio statistic, L; L is a special case of Yochmowitz and Cornell's [24] statistic

$$\Lambda_\kappa^* = \lambda_\kappa/(\lambda_\kappa + \cdots + \lambda_p).$$

In particular, Λ_2^* coincides with L, and Λ_1^* with U_1. They use Λ_κ^* to successively test $H_0 : \theta_\kappa = 0$ vs. $H_0 : \theta_\kappa \neq 0$, $\theta_{\kappa+1} = 0$, with $\kappa = 1, \ldots, c$. The motivation for Λ_κ^* is similar to a forward stepwise regression* procedure where the investigator continues to add parameters to the model with $\kappa < p$ until he or she can no longer reject H_0. By assuming that estimates can be treated as parameters once they have been added to the model, they show that under H_0 the roots $\lambda_\kappa, \ldots, \lambda_p$ are characteristic roots of a central Wishart matrix $W_{p-\kappa}^*$ whose order depends on the number of components admitted to the model. This permits successive application of the SKC tables of the distribution of the largest root to the trace of a Wishart matrix. Comparison of these percentiles with Hegemann and Johnson's simulated values in testing $H_0 : \theta_2 = 0$ shows good agreement (42/54 cases were within $\pm 2\%$ and 49/54 cases were within $\pm 3\%$).

Other techniques for examining the form of the FANOVA model include graphical procedures suggested by Yochmowitz [23] and Bradu and Gabriel [1]. Both use the biplot*. Yochmowitz proposed using successive biplots of residuals to examine the clustering about the major principal component axis. If there is clustering, a component is added to the model, residuals are adjusted to eliminate the effect of the component that was added, a biplot of the new residual matrix is constructed, the procedure is repeated. On the other hand, Bradu and Gabriel [1] use the biplot to diagnose models by examining the collinearity* of row and column markers. The additive model results when the two lines are perpendicular. Row linear and column linear models occur when either the row or the column markers are collinear, but not both.

It is important to test for interaction before testing for main effects. In a two-way layout with no replication, standard practice has been to use Tukey's [20] single-degree-of-freedom test for additivity. Ghosh and Sharma [5] have shown that this test has good power in selecting between the concurrent and the additive models. That is, Tukey's test is good for identifying

Table 2 Density of Aqueous Solutions of Ethyl Alcohol: Transformed Data with $p = 3.13897$ Yields

Concentration (% Alcohol by Weight)	Temperature (°C) 10	15	20	25	30	35	40
30.086	0.878730	0.870342	0.861713	0.852771	0.843599	0.834172	0.824545
39.988	0.830132	0.820317	0.813097	0.800329	0.790192	0.779942	0.769575
49.961	0.774201	0.764077	0.753867	0.743600	0.733240	0.722883	0.712377
59.976	0.716709	0.706666	0.696589	0.686450	0.676283	0.666063	0.655797
70.012	0.659941	0.650197	0.640397	0.630575	0.620728	0.610855	0.600942
80.036	0.604593	0.595197	0.585817	0.576421	0.567019	0.557597	0.548148

FANOVA models with a single multiplicative component which is the product of main effects. This need not be true in examining other cases of the FANOVA model. Mandel [11] showed that the row and column linear bundle of line models (single-component FANOVA models that contain one of two main effects in its single multiplicative component) can be more effectively tested by an F-test with $t - 1$ or $b - 1$ degrees of freedom than by Tukey's test. And Hegemann and Johnson [8] have shown that the statistic U_1 (and equivalently, $\mathbf{t}$ and Λ_1^*) has better power than Tukey's test in detecting a single multiplicative component when correlations between interaction effects and main effects are not significant.

Example. To illustrate these methods, we examine a transformation of Osborne et al.'s [14] ethyl alcohol data. This will show the need to test for a particular form of the FANOVA model in transformed as well as untransformed data. Applying Tukey's [20] p-transformation to restore additivity leads to the data given in Table 2. A standard analysis of variance leads to the sum of squares in Table 3, where the concentration × temperature sum of squares is partitioned with 1 degree of freedom for nonadditivity. Tukey's test yields $F = 0.2408$ and is not significant with 1 and 29 degrees of freedom. It appears as though the p-transformation restored additivity. This is not the case, as can be seen in Table 4. Mandel's equality of slopes test leads to an $F = 193.07$ and is highly significant with 5 and 25 degrees of freedom. In this table stepwise tests also support this conclusion, indicating the presence of two multiplicative components of interaction in the transformed data. The remaining three components are pooled together to represent the residual error.

This example illustrates at least two points. First, transformations should not be applied blindly to data. They should be examined to see if they have the desired effect. Second, more studies are needed to determine the power of the procedures used to specify the form of the FANOVA model.

Table 3

Source	d.f.	Sum of Squares	Mean Square	F
		ANOVA		
Concentration	5	0.378463	0.756927×10^{-1}	
Temperature	6	0.161403×10^{-1}	0.269139×10^{-2}	
Concentration × Temperature	30	0.345365×10^{-4}	0.115122×10^{-5}	
		Tukey's [20] Test for Nonadditivity		
Concentration × Temperature	30	0.345365×10^{-4}	0.115122×10^{-5}	
Nonadditivity	1	0.284448×10^{-6}	0.284448×10^{-6}	0.2408
Balance	29	0.342520×10^{-4}	0.118110×10^{-5}	

Table 4

Mandel's [11] Test for Equality of Slopes				
Source	d.f.	Sum of Squares	Mean Square	F
Concentration × temperature	30	0.345365×10^{-4}	0.115122×10^{-5}	
Slopes	5	0.336640×10^{-4}	0.067328×10^{-4}	193.07
Concurrence	1	0.002845×10^{-4}	0.002845×10^{-4}	
Nonconcurrence	4	0.333795×10^{-4}	0.083449×10^{-4}	
Balance	25	0.008724×10^{-4}	0.000349×10^{-4}	

Yochmowitz–Cornell [24] Stepwise Tests for Multiplicative Components		
Source	Sum of Squares	Λ_κ^*
Concentration × temperature	0.345365×10^{-4}	
Component 1	0.339560×10^{-4}	0.98319**
Component 2	0.579325×10^{-6}	0.99810**
Component 3	0.723181×10^{-9}	0.65561
Component 4	0.303165×10^{-9}	0.79806
Component 5	0.767114×10^{-10}	1.00000

**Significant using SKC tables, $\alpha = 0.01$.

The appearance of the FANOVA model is becoming common in many different fields. Besides the psychological data used by Gollob and the chemical data used by Mandel, Snee and co-workers [18, 19] have found it useful in describing both chemical and biological growth curve data. Slater [17] has used it to describe residential sales data and Cornelius et al. [2] have been successful in using FANOVA models to describe agricultural data.

See Gabriel [4] and Tukey [21] for additional information.

References

Letters at the end of reference entries denote one of the following categories:

A: development and formulation of the FANOVA model

B: special cases of the FANOVA model

C: significance tests to determine the number of multiplicative components in the FANOVA model

D: graphical procedures to identify the form of the model

E: applications of FANOVA models

F: power studies

[1] Bradu, D. and Gabriel, K. R. (1978). *Technometrics*, **20**, 47–68. (D)

[2] Cornelius, P. L., Templeton, W. C., and Taylor, T. H. (1979). *Biometrics*, **35**, 849–859. (E)

[3] Corsten, L. C. A. and van Eijnsbergen, C. A. (1972). *Statist. Neerlandica*, **26**, 61–68. (C)

[4] Gabriel, K. R. (1971). *Biometrika*, **58**, 453–467. (D)

[5] Ghosh, M. N. and Sharma, D. (1963). *J. R. Statist. Soc.*, **B**, **25**, 213–219. (F)

[6] Gollob, H. F. (1968). *Psychometrika*, **33**, 73–116. (A)

[7] Hegemann, V. and Johnson, D. E. (1976). *Technometrics*, **18**, 273–281. (C)

[8] Hegemann, V. and Johnson, D. E. (1976). *J. Amer. Statist. Ass.*, **71**, 945–948. (F)

[9] Johnson, D. E. and Graybill, F. A. (1972). *J. Amer. Statist. Ass.*, **67**, 862–868. (C)

[10] Krishnaiah, P. R. and Yochmowitz, M. G. (1980). In *Handbook of Statistics*, Vol. 1: *Analysis of Variance*, North-Holland, Amsterdam, 973–994. (A, B, C, E)

[11] Mandel, J. (1961). *J. Amer. Statist. Ass.*, **56**, 878–888. (B)

[12] Mandel, J. (1969). *J. Res. Nat. Bur. Stand. B*, **73**, 309–328. (A, B)

[13] Mandel, J. (1971). *Technometrics*, **13**, 1–18. (A)

[14] Osborne, N. S., McKelvy, E. C., and Bearce, H. W. (1913). *Bull. Bur. Stand.*, **9**, 327–474.

[15] Schuurmann, F. J., Krishnaiah, P. R., and Chattopadhyah, A. K. (1973). Tables for the distributions of the ratios of the extreme roots to the trace of Wishart matrix. *Aerospace Res. Lab. TR No. 73-0010*, Wright-Patterson AFB, Ohio. (C)

[16] Schuurmann, F. J., Krishnaiah, R. R., and Chattopadhyah, A. K. (1973). *J. Multivariate Anal.*, **3**, 445–453. (C)

[17] Slater, P. B. (1973). *J. Amer. Statist. Ass.*, **68**, 554–561. (E)

[18] Snee, R. D. (1972). *Technometrics*, **14**, 47–62. (E)

[19] Snee, R. D., Acuff, S. K., and Gibson, J. R. (1979). *Biometrics*, **35**, 835–848. (E)

[20] Tukey, J. W. (1949). *Biometrics*, **5**, 232–242. (B)

[21] Tukey, J. W. (1962). *Ann. Math. Statist.*, **33**, 1–67. (B)

[22] Ward, G. C. and Dick, I. D. (1952). *N. Z. J. Sci. Tech.*, **33**, 430–436. (B)

[23] Yochmowitz, M. G. (1974). Testing for Multiplicative Components of Interaction in Some Fixed Models. Ph.D. dissertation, University of Michigan. (C, D)

[24] Yochmowitz, M. G. and Cornell, R. G. (1978). *Technometrics*, **20**, 79–84. (C)

(BIPLOT
EIGENVALUE
LIKELIHOOD RATIO TESTS
SINGULAR VALUE DECOMPOSITION
STEPWISE REGRESSION
WISHART DISTRIBUTION)

MICHAEL G. YOCHMOWITZ

FACTORIAL CUMULANTS

Analogously to the definition of cumulants* (κ_j) in terms of moments* (μ_j') by the formal identity

$$\exp\left[\sum_{j=1}^{\infty} \kappa_j (t^j/j!)\right] = \sum_{j=0}^{\infty} \mu_j' (t^j/j!),$$

(descending) factorial cumulants ($\kappa_{(j)}$) are defined in terms of (descending) factorial moments* ($\mu_{(j)}$) by the formal identity

$$\exp\left[\sum_{j=1}^{\infty} \kappa_{(j)} (t^j/j!)\right] = \sum_{j=0}^{\infty} \mu_{(j)} (t^j/j!).$$

Formulae relating $\kappa_{(j)}$'s and $\mu_{(j)}$'s are exactly parallel to those relating κ_j's and μ_j''s.

Ascending factorial cumulants ($\kappa_{[j]}$) may be defined in the same way, using ascending factorial moments ($\mu_{[j]} = E[X^{[j]}]$).

(CUMULANTS
FACTORIAL MOMENTS)

FACTORIAL EXPERIMENTS

The major purpose of a designed experiment is to investigate the effect on a response of one or more (generally) controllable features germane to the problem. If a condition is specifically changed for portions of an experiment, this condition is designated as a *factor*. The values of the condition that are used in the experiment are referred to as the *levels* of the factor. These terms are quite general but can be best visualized through examples.

1. The purpose of an experiment is to investigate techniques to accelerate the curing of lumber. In process of the experiment each sample is cured at one of four specified temperatures.
2. The purpose of an experiment is to investigate the wear resistance of a rubberized fabric. There are five different methods by which the rubber can be treated; a number of samples are produced under each of the five methods.
3. During the past summer eight varieties of potatoes were raised in an experimental program at a state experiment station.

The factors mentioned in these examples would be Temperature at four levels, Method at five levels, and Variety at eight levels, respectively.

The concept of a *factorial experiment* is that more than one factor is investigated simultaneously in a single experiment. For example, in Example 1 not only is the temperature important but also the amount of time that the sample is submitted to a temperature. Thus Time could well be a second factor. A third factor might be the Type of lumber. Implicit in a factorial experiment is that the same number of samples occur for each combination of levels of the factors. Hence for the example above the same number of samples of each Type of lumber would be submitted to each Temperature–Time combination. Some authors refer to this property as *balance**: a *balanced* experiment.

The varying of several factors simultaneously appeared in agricultural experiments in England during the latter half of the nineteenth century. Its use, however, was quite limited and, in fact, resisted until the excellent presentations by Fisher* [4, Chap. VI] and Yates [6]. In both of these works the authors effectively argued the merits of factorial experimentation over the more familiar method generally referred to as "one-at-a-time" experimentation. (A discussion of this comparison appears later.)

Factorial experimentation has become well accepted in the biological and agricultural sciences. This acceptance in the United States was greatly enhanced by two classic texts: Snedecor [5] and Cochran and Cox [1]. After a lag of perhaps 20 years, the physical sciences began to realize the advantage of investigating several factors simultaneously. Some excellent references are Davies [3, Chaps. 7, 8] and Cox [2, Chap. 6].

Before proceeding, consider the term *factorial design*. *Factorial* implies that more than one factor is being investigated in a single experiment and that each combination of levels of the factors will appear the same number of times. *Design* refers to how the actual experiment is conducted. In particular; physical, budgetary, time, and other constraints may control the nature of the experimental unit (samples) and their assignment to the combinations of levels of the factors (for a detailed discussion, *see* DESIGN OF EXPERIMENTS).

A factorial arrangement can be applied to a number of different experimental designs and the basic properties of the factorial arrangement hold for a wide range of these designs. For the sake of simplicity and with little loss of generality the discussion that follows considers only a completely randomized design.

To illustrate the merits of a factorial experiment, consider the following simple experiment. The purpose is to assess and compare the effectiveness of three types of Styrofoam containers in retaining the temperature of 12-ounce cans of cold drink. Sixteen representative containers of each type were utilized and 48 similar cans of a cold drink. (The temperature of the liquid in a can was reduced to 40°F, the can was then placed in a container, and the increase in temperature was recorded after 20 minutes in a controlled temperature of 70°, 80°, 90° or 100°F.) Table 1 summarizes the responses; the values in the table are the sum of the increase in temperature of four independent cans exposed to the indicated temperature for 20 minutes.

Table 1 Pseudo Data on Increase of Temperature

Type of	Exposure Temperature (°F)				
Material	70	80	90	100	
A	63	66	83	88	
B	73	81	94	99	
C	56	67	85	96	
Total	192	214	262	283	951

Basically, the 48 cans of cold drink were identically treated except for the type of container and the temperature to which they were exposed. These factors were controlled at their respective levels. Consider two modelings of the response, Y: the increase in temperature of the cold drink.

ADDITIVITY OR THE ABSENCE OF INTERACTION*. In this case the relative effectiveness of the three types of containers is the same at all four levels of the temperature of exposure. Symbolically,

$$Y_{ijk} = \left\{\begin{array}{l}\text{overall average}\\ \text{for the experiment}\end{array}\right\} + \left\{\begin{array}{l}\text{effect of the } i\text{th}\\ \text{type of container}\end{array}\right\} + \left\{\begin{array}{l}\text{effect of the } j\text{th}\\ \text{temperature of exposure}\end{array}\right\} + \left\{\begin{array}{l}\text{random experimental}\\ \text{error}\end{array}\right\}$$

or

$$Y_{ijk} = \mu + \alpha_i + \beta_j + \epsilon_{ijk}, \qquad k = 1, 2, 3, 4.$$

with the assumption that the expectation of

the independent random errors, ϵ_{ijk}, is zero with a common variance, σ^2.

$$i = \begin{cases} 1 \rightarrow \textit{type A container} \\ 2 \rightarrow \textit{type B container} \\ 3 \rightarrow \textit{type C container} \end{cases} \qquad j = \begin{cases} 1 \rightarrow 70° \\ 2 \rightarrow 80° \\ 3 \rightarrow 90° \\ 4 \rightarrow 100° \end{cases}$$

The data in Table 1 would tend to support additivity of the effects of these two factors.

The statistic to compare the effectiveness of, say, type A container to type B container would be $\overline{Y}_{1..} - \overline{Y}_{2..}$, the difference in the averages of the 16 samples using container types A and B. This is, in fact, an unbiased estimator of $\alpha_1 - \alpha_2$, with variance $\sigma^2/8$.

The linear effect of the temperature of exposure on the increase in the liquid temperature can be estimated by

$$(-3\overline{Y}_{.1.} - \overline{Y}_{.2.} + \overline{Y}_{.3.} + 3\overline{Y}_{.4.})/20,$$

which has a variance of $\sigma^2/240$, where $\overline{Y}_{.j.}$ is the average of the measurements of all 12 units exposed to the jth temperature (this statistic will be referred to later). In both this and the previous statistic, the factor not being considered has been averaged across its levels. This attribute of a factorial design, in the absence of interaction, is often referred to as *hidden replication*.

To appreciate the value of hidden replication, consider an alternative experimental approach; namely, investigating the characteristics of each factor separately. Continuing the example above, exposure temperature would be held constant at one level and 16 cans could be used for each type of container. A comparison of the effectiveness of type A to type B would be measured by $\overline{Y}_1 - \overline{Y}_2$ (the difference in their sample means), which has a variance of $\sigma^2/8$. Thus utilizing all the experimental units to investigate only one factor results in no more information on that factor than is available from a factorial experiment investigating both factors simultaneously. The investigation of the effect of exposure Temperature using only one of the Type of container leads to a similar conclusion.

NONADDITIVITY* OR THE PRESENCE OF INTERACTION. In this case the relative effectiveness of the types of containers is not the same at all four exposure temperatures. Symbolically, an additional element, $(\alpha\beta)_{ij}$, is introduced into the model to represent this lack of additivity;

$$Y_{ijk} = \mu + \alpha_i + \beta_j + (\alpha\beta)_{ij} + \epsilon_{ijk}.$$

The definition of the main effects, α_i and β_j, are now modified to be the effect averaged over the levels of the other factor *used in this experiment*. The symbol $(\alpha\beta)_{ij}$ is referred to as the *interaction term*.

To illustrate this concept, consider replacing the first row of data in Table 1 with the values

69 72 77 82.

An *analysis of variance** for this modified data is summarized in Table 2. From this, the statistic for testing the null hypothesis of no interaction, $(\alpha\beta)_{ij} = 0$, is

$$F = \text{MS(interaction)}/\text{MS(error)}$$
$$= 9.52/2.76 = 3.45$$

with 6 and 36 degrees of freedom. Assuming normality of the error term, this is significant at the 0.05 level, indicating the presence of interaction between Type of container and Temperature of exposure. The nature of this interaction can be seen from Fig. 1. (A similar type of analysis of variance on the original data yields $F = 1.21$ for the interaction effect, a very reasonable value in the absence of interaction.)

Table 2 Analysis-of-Variance Table

Source	d.f.	Sum of Squares	Mean Square
Type of material	2	84.875	42.44
Exposure temperature	3	314.062	104.69
Interaction	6	57.125	9.52
Error	36	99.480	2.76

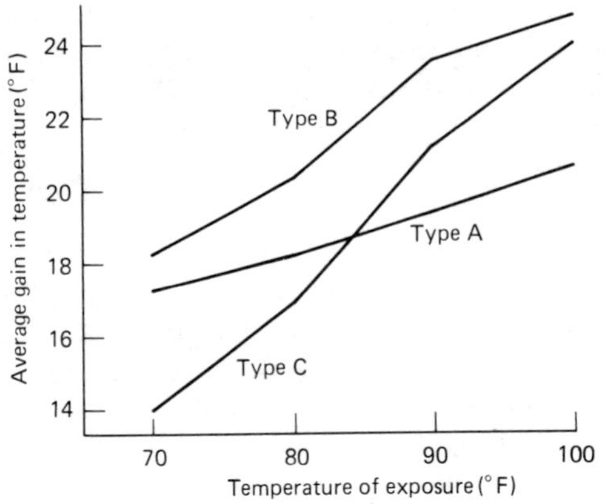

Figure 1 Average gain in temperature for modified data (average of four cans).

The test of hypothesis denotes an interaction, and Fig. 1 indicates that the type C container is more effective (less rise in temperature) than type A at the lower exposure temperature; however, the converse occurs at the higher exposure temperature. Although hidden replication is of little value here, it is also quite apparent that investigating the factors separately could be quite misleading. The conclusion as to the relative effectiveness of the type A and type C containers would be dependent which exposure Temperature was used.

A simple two-factor example has been utilized to illustrate the basic properties of a factorial experiment. These features are retained, however, as the number of factors increase; namely: (1) through hidden replication each factor can be investigated as effectively as if all the experimental units had been employed to investigate a single factor (in the absence of interaction); and (2) interaction, when present, can be detected and investigated.

An additional asset is that the analysis of the data is tractable even in the absence of an electronic computer (*see* ANALYSIS OF VARIANCE). Because of the balance of a factorial experiment the sum of squares for each effect is easily calculated and (under normality) are mutually independent. For example, the sum of squares for Temperature (SST) for the data in Table 1 is

$$\text{SST} = \tfrac{1}{12}(192^2 + 214^2 + 262^2 + 283^2) - 951^2/48$$
$$= 441.0625.$$

Furthermore, these sums of squares can be decomposed into meaningful independent components using orthogonal contrasts. The linear effect mentioned earlier utilized the coefficients from orthogonal polynomials*; $(-3, -1, 1, 3)$. The corresponding coefficients for the quadratic and cubic effects are, respectively, $(1, -1, -1, 1)$ and $(-1, 3, -3, 1)$. The breakdown of SST is then:

Linear:

$$\frac{\left[(-3)(192) + (-1)214 + (1)262 + (3)283\right]^2}{12\left[(-3)^2 + (-1)^2 + (1)^2 + (3)^2\right]}$$
$$= 429.3375.$$

Quadratic:

$$\frac{\left[(1)(192) + (-1)(214) + (-1)262 + (1)283\right]^2}{12\left[(1)^2 + (-1)^2 + (-1)^2 + (1)^2\right]}$$
$$= 11.7042.$$

Cubic:

$$\frac{\left[(-1)192 + (3)214 + (-3)262 + (1)283\right]^2}{12\left[(-1)^2 + (3)^2 + (-3)^2 + (1)^2\right]}$$
$$= 11.7042.$$

The magnitude of "linear" suggests that the Temperature effect is basically linear for the additive example.

In cases where each factor has only two levels the term 2^k *factorial* is often used, where k is the number of factors. In these situations there can be no decomposition of the individual sums of squares; however, simple methods of computation are available.

The major disadvantage of factorial experiments is that at times the number of experimental units necessary to conduct the experiment is exceedingly large. This number will be an integer multiple of the product of the number of levels of each factor; e.g., in the example above, 4(number of Types of containers) · (number of exposure Temperatures) $= 4 \cdot 3 \cdot 4 = 48$. Unfortunately, this required number of experimental units increases rapidly as the number of factors increases.

A basic tenet of a completely randomized factorial experiment is that the experimental units are homogeneous. If the required number of experimental units is available, however, the units are homogeneous only within subgroups; the utilization of a randomized block* may be feasible. When the number of experimental units required for a factorial experiment is not available, a fractional* replication may be possible. Should a number of the factors have quantitative levels, a response surface design* can be employed using fewer experimental units.

Currently, little research is being directed toward factorial experiments per se. Considerable interest is being shown, however, in the related topics: fractional* replications, response surfaces*, and designs for mixtures*.

References

[1] Cochran, W. G. and Cox, G. M. (1950). *Experimental Designs*. Wiley, New York. (A second edition, 1957, is available. Some excellent examples are presented.)

[2] Cox, D. R. (1958). *Planning of Experiments*. Wiley, New York. (Well worth reading by anyone doing experimental work.)

[3] Davies, O. L. (1960). *Design and Analysis of Industrial Experiments*. Oliver & Boyd, Edinburgh. (Excellent on the mechanics of the analyses.)

[4] Fisher, R. A. (1935). *The Design of Experiments*. Collier Macmillan, London. [The reader must be prepared to go slowly and ponder the implications. (Later editions are available.)]

[5] Snedecor, G. W. (1937). *Statistical Methods*. Iowa State University Press, Ames, Iowa. (A seventh edition, 1980, jointly with W. G. Cochran is available. This text teaches by example and is agriculturally oriented.)

[6] Yates, F. (1935). *J. R. Statist. Soc. B*, **2**, 181–223.

(BALANCE IN EXPERIMENTAL DESIGN
DESIGN OF EXPERIMENTS
FACTOR ANALYSIS OF VARIANCE MODEL (FANOVA)
FRACTIONAL FACTORIAL DESIGNS
GENERAL BALANCE
INTERACTION
MIXTURE EXPERIMENTS
RESPONSE SURFACES)

J. T. WEBSTER

FACTORIAL MOMENTS

The rth *factorial moment* of a random variable X (or of its distribution) is usually understood to mean its rth *descending* factorial moment. It is the expected value of the rth descending factorial

$$X^{(r)} = X(X-1)\cdots(X-r+1)$$

and is denoted by $\mu_{(r)}$ or $\mu_{(r)}(X)$.

The rth *ascending* factorial moment of X is the expected value of the rth ascending factorial

$$X^{[r]} = X(X+1)\cdots(X+r-1)$$

and is denoted by $\mu_{[r]}$ or $\mu_{[r]}(X)$.

For the relation

$$X^r = \sum_{j=0}^{r} \frac{\Delta^j 0^r}{j!} X^{(j)},$$

where $\Delta^j 0^r$ is a *difference of zero** and $\Delta^j 0^r / j!$ is a Stirling number of the second kind*), we have the formulae

$$\mu'_r = \sum_{j=0}^{r} \frac{\Delta^j 0^r}{j!} \mu_{(r)} \qquad (r = 1, 2, \ldots),$$

expressing crude moments* in terms of factorial moments. Since

$$X^{(r)} = \sum_{j=0}^{r} S_{r,j} X^j$$

(where $S_{r,j}$ is a Stirling number of the first kind*), we have the formulae

$$\mu_{(r)} = \sum_{j=0}^{r} S_{r,j} \mu'_j,$$

expressing factorial moments in terms of crude moments.

Factorial moments are especially useful for discrete random variables taking only integer values.

(FACTORIAL CUMULANTS
FINITE DIFFERENCES)

FACTORIAL SERIES DISTRIBUTIONS

A large number of mathematical functions can be expanded in series with nonnegative coefficients. By multiplying the terms by a suitable factor, one can make their sum unity and thus produce a discrete probability distribution. We present a class of discrete distributions established in the literature under the name *factorial series distributions*, abbreviated FSD in what follows.

Let f be a real function of the integer-valued variable N. Suppose that f can be expanded in a factorial series in N with nonnegative coefficients; i.e., we assume that $f(N) = \sum a_x N^{(x)}$, where $N^{(x)} = N(N-1)\cdots(N-x+1)$, and $a_x \geqslant 0$. Based on this expansion, the probability function (PF) for an FSD can be given in the form

$$P(x) = N^{(x)}a_x/f(N), \qquad x = 0, 1, \dots, N, \tag{1}$$

where f is the *series function* and N is the integer-valued parameter indexing the family. The set of values for which a_x is >0 is called the *range* of the family (1) of distributions. An equivalent form for (1) is

$$P(x) = \binom{N}{x}\Delta^x f(0)/f(N), \qquad x = 0, 1, \dots, N,$$

where $\Delta^x f(0)$ is the xth forward difference* of the function f, computed at zero.

The factorial moments of a random variable (rv) X having an FSD are given by the formula

$$E(X^{(\nu)}) = \mu^{(\nu)} = N^{(\nu)}[\Delta^\nu f(N-\nu)]/f(N).$$

In particular, the mean of X is

$$E(X) = \mu = N[f(N) - f(N-1)]/f(N).$$

The class of FSD introduced here is the discrete parameter analog of the class of *power series distributions**. When generating the latter class, the starting point is a Taylor expansion of a parametric function. Not surprisingly, the two classes of discrete distributions have many properties in common; see Berg [1] and Johnson and Kotz [5].

GENESIS

Problems associated with a finite collection of events, $E_1, \dots, E_N$, give rise to probability distributions on the set $\{0, 1, \dots, N\}$. For example, we may want to compute the probability that exactly x of the events occur.

Suppose now the events under consideration are symmetric, or *exchangeable*; i.e., the probability of the occurrence of r specific events depends only on r and N, not on the actual set of subscripts chosen. In dealing with problems of the present kind, it is often easier to work with the nonoccurrence of specific events. Denote by $\overline{S}_{r,N}$ the probability that $N-r$ specific events fail to happen, i.e., $\overline{S}_{r,N} = \Pr[\bigcap_{j=1}^{N-r} \overline{E}_{i_j}]$; $r = 0, 1, \dots, N$. The probability that exactly z events do not occur is obtained by applying the *inclusion–exclusion principle** (see Feller [3]):

$$\Pr(Z = z) = \sum_{r=z}^{N} (-1)^{r-z}\binom{r}{z}\binom{N}{r}\overline{S}_{N-r,N} = \binom{N}{z}\Delta^{N-z}\overline{S}_{0,N}.$$

The probability we seek for exactly x events to occur can now be obtained by putting $z = N - x$ in the expression above.

In certain problems the events under consideration are not only symmetric, but have the additional property that the function $\overline{S}_{r,N}$ factorizes into two parts, one depending on r only, the other on N only. In such cases the required probability reduces to that of the standard form for an FSD given in formula (1).

As a simple illustration, let n balls be distributed at random over N cells and without restrictions. Let $\overline{E}_i$; $i = 1, \dots, N$, be the event that cell i is empty. The probability of the event $\overline{E}_{i_1} \cap \cdots \cap \overline{E}_{i_{N-r}}$, i.e., that $N-r$ specified cells are empty, is clearly r^n/N^n. Hence we have both the symmetry and the factorability property required to obtain an FSD. The resulting PF for the number of occupied cells is the well-known classical occupancy distribution*

$$P(x) = \binom{N}{x}\Delta^x 0^n/N^n, \qquad x = 1, 2, \dots, \min(n, N),$$

an FSD with series function $f(N) = N^n$. We discuss this example further later.

From the above we conclude that a random variable having an FSD can be appropriate for describing the number of different units obtained in sampling from a finite population, using a symmetric design which involves replacement of units so that repetitions can occur.

INFERENCE FOR AN FSD

In certain statistical applications the size of a finite population, or the number of classes in a population, conventionally denoted N, is the parameter of interest. This is the case, for instance, in estimating the size of an animal population by means of capture–recapture* sampling. Therefore, it is of interest to develop a theory of estimation for the class of FSDs.

In statistical theory a random sample is usually defined as a sequence of independent, identically distributed rvs. In sampling from a finite population, however, the sample often consists of dependent variables. In our case it is natural to consider a sequence of k dependent variables $Y_1, \ldots Y_i, \ldots Y_k$, such that

$$\Pr(Y_i = y_i/s_{i-1}) = \binom{N - s_{i-1}}{y_i}\Delta^{y_i} f(0 + s_{i-1})/f(N), \quad i = 1, 2, \ldots, k,$$

$y_i = 0, 1, \ldots, N - s_{i-1}$, and where $\{s_i\}$ are the partial sums: $s_i = \sum^i y_\nu$ and $s_0 = 0$. Thus the variables $\{Y_i\}$ are dependent, each having a conditional FSD with the same series function $f(N)$ and the parameter shifted: $N - s_{i-1}$. The sequence of partial sums $\{S_i\}$ forms a *Markov chain* with transition probabilities determined by an FSD (*see* MARKOV PROCESSES).

From the likelihood of the sample $\{y_1, \ldots, y_k\}$, as defined by the joint distribution of the variables $\{Y_i\}$, we see that the sum $s_k = \sum y_i$ is a *sufficient statistic* for the parameter N in the series function $f(N)$ (*see* SUFFICIENCY). Moreover, the marginal distribution of the rv S_k is an FSD in its own right with series function $f(N)^k$, namely

$$P(s_k) = \binom{N}{s_k}\Delta^{s_k} f(0)^k / f(N)^k, \quad s_k = 0, \ldots, N.$$

In certain problems one can obtain a *minimum variance unbiased* (MVU) estimator* of the parameter N. In this case, this estimator must be a function of the complete, sufficient statistic s_k.

In view of the above, it suffices to consider the case when we have a single observation on a rv X distributed according to an FSD. Thus if X has PF

$$P(x) = N^{(x)} a_x / f(N), \qquad x = 0, 1, \ldots, N,$$

it can be shown that

$$\hat{N}_x = (a_{x-1}/a_x) + x \tag{2}$$

is an unbiased estimate of the parameter N, provided that the admissible values of N are included in the range of the distribution, as defined above.

A formula for the variance of $\hat{N}_x$ is not available. However, it is possible to derive an estimate of the variance of $\hat{N}_x$. In fact, by identifying coefficients in factorial series expansions, the result (2) can be extended to more general parametric functions of N. The approach here is that used for power series distributions. It is also possible to use maximum likelihood estimation* and/or the method of moments* in the case of an FSD.

EXAMPLES OF FSD

Matching Distribution

Suppose that a set of N objects, numbered $1, 2, \ldots, N$, respectively, is arranged in random order. A "match" is said to occur if the position of an object coincides with the number assigned to it. The probability of having matches in ν specific positions is clearly $(N - \nu)!/N!$. Thus if X denotes the number of "nonmatches," then X follows an FSD with series function $f(N) = N!$, i.e.,

$$P(x) = \binom{N}{x}\Delta^x 0!/N!. \tag{3}$$

Note that $P(1) = 0$ for all N, because we cannot have $N - 1$ matches without having the Nth match as well.

In this case the factorial moments* are particularly simple:

$$\mu^{(\nu)}(N - X) = 1, \qquad \nu = 1, 2, \ldots, N.$$

This suggests the Poisson distribution* with parameter unity as an approximating distribution, and it does indeed provide an excellent approximation to the FSD (3) even for small values of N.

Binomial Distribution

The binomial distribution*, expressed in terms of the odds ratio* $\theta = p/(1-p)$, $0 < p < 1$, is given by

$$P(x) = \binom{N}{x}\theta^x \Big/ (1+\theta)^N,$$

$$x = 0, 1, \ldots, N; \quad (4)$$

this is actually an FSD with series function $f(N) = (1+\theta)^N$. If we regard θ instead as the parameter, then the above meets the requirements of a power series distribution.

The binomial distribution has important applications in statistics and probability too numerous to mention here. Note, however, that the estimator of the parameter N defined by (2) reduces here to the natural estimate $\hat{N}_x = x/p$. The range of the distribution (4) is the set $\{0, 1, 2, \ldots\}$, so this estimate is always unbiased.

Stevens Craig Distribution

This distribution has PF

$$P(x) = N^{(x)} S_x^n / N^n,$$

$$x = 1, 2, \ldots, \min(n, N), \quad (5)$$

where S_x^n is the *Stirling number of the second kind* with arguments n and x. It can be shown that, if $n \geqslant N$, the estimator

$$\hat{N}_x = S_x^{n+1}/S_x^n,$$

is unbiased for the parameter N.

Apart from being a classical occupancy distribution, the Stevens Craig distribution arises in certain sampling situations. In survey sampling* much interest has focused on estimators based on the distinct units of the sample, disregarding possible repetitions of the same unit; the distribution (5) has played a certain role. Moreover, (5) arises in capture–recapture experiments in which animals are caught, marked, and released one at a time (see Berg [2] and Seber [7]).

Capture–Recapture* Distributions

A population is comprised of $N + A$ elements, where a known number A of elements is marked in a special way. A random sample of size n is taken without replacement. The number of unmarked elements in the sample will be distributed according to

$$P(x) = \binom{N}{x}\binom{A}{n-x} \Big/ \binom{N+A}{n},$$

$$x = \max(0, m - A), \ldots, \min(n, N), \quad (6)$$

a version of the hypergeometric distribution* satisfying the requirements of an FSD.

If it is desired to estimate the parameter N in (6) from a single observation, then by (2),

$$\hat{N}_x = xA/(n - x + 1)$$

is unbiased for N, provided that $n \geqslant N$. This familiar problem is often mentioned in elementary textbooks in probability and statistics, usually in terms of estimating the number of fish in a lake.

The following FSD family is a direct extension of the hypergeometric distribution above:

$$P(x) = \binom{N}{x}\Delta^x \prod_1^k \binom{0}{n_i} \Big/ \prod_1^k \binom{N}{n_i},$$

$$x = \max(n_i), \ldots, \min(N, \textstyle\sum n_i). \quad (7)$$

A rv X distributed according to (7) may be represented as a sum, $X = Y_1 + \cdots + Y_i + \cdots + Y_k$, where Y_i is the number of elements in sample i, not previously observed. Conditionally on the outcome of the preceding samples, each Y_i has a hypergeometric distribution of type (7). In estimation terminology, (7) is the sampling distribution of the sufficient statistic x, the total number of different elements observed in the series of k samples.

Besides being applicable to capture–recapture experiments, (7) arises in *committee size* problems. Given a group of N people and k committees of sizes $n_1, \ldots, n_k$, respectively, to be formed with these people, (7) is the probability that exactly x different people will serve on at least one of the k committees (see Berg [1], Seber [7], and Johnson and Kotz [5]).

CONCLUDING REMARKS

The class of factorial series distributions can be extended in a straightforward way to the multivariate case. The families of discrete distributions mentioned above also have closely related *waiting-time* distributions. The latter arise if, in sampling from a finite population, we fix in advance the number of different elements to be observed and let the number of experiments be a random variable.

RELATED DISTRIBUTIONS

Let the rv X have a probability function of the generic form

$$P(x) = a_x h(x,\theta)/f(\theta), \qquad x = 0, 1, \ldots, \tag{8}$$

where $a_x h(x,\theta) \geqq 0$, θ is a parameter, and $f(\theta) = \sum a_x h(x,\theta)$, it being assumed that the sum converges. If we take $h(x,\theta) = \theta^x$, then (8) defines the class of power series distributions. Similarly, if we set $h(x,N) = N^{(x)}$, we get the class of factorial series distributions.

There are further possibilities, however. If we choose $h(x,\theta) = 1/\theta^{[x]}$, where $\theta^{[x]}$ is the ascending factorial $\theta^{[x]} = \theta(\theta+1)\cdots(\theta+x-1)$, we are led to a class of distributions introduced by Irwin [4], who suggested that for frequency distributions with very long tails an inverse factorial series might be a suitable starting point. The resulting distributions are defined by equations of the form

$$P(x) = a_x\left(\theta^{[x+1]}\right)^{-1} f(\theta), \qquad x = 0, 1, \ldots, \tag{9}$$

where

$$f(\theta) = \int_0^1 t^{\theta-1}\phi(t)\,dt$$

$$\phi(t) = \sum_{j=0}^{\infty} (a_j/j!)(1-t)^j,$$

it being supposed that the series converges.

A particular case is obtained from Waring's expansion:

$$(\theta - a)^{-1} = \sum \left(\theta^{[j+1]}\right)^{-1} a^{[j]},$$

i.e.,

$$P(x) = (\theta - a)a^{[x]}\left(\theta^{[x+1]}\right)^{-1}, \qquad x = 0, 1, \ldots, \tag{10}$$

which is called *Waring's distribution*. The *Yule distribution* is obtained by putting $a = 1$ in (10). Finally, if a, $\theta \to \infty$ in (10) with $a/\theta = p$, constant, the limiting distribution is geometric*:

$$P(x) = (1-p)p^{x-1}, \qquad x = 1, 2, \ldots .$$

The Irwin distributions above should not be confused with *factorial distributions* in the sense of Marlow [6], who starts from the basic expression

$$\sum_{x=0}^{\infty} 1/(x+m)^{(N)} = 1/(N-1)(m-1)^{(N-1)}, \qquad m > N-1,\ N = 2, 3, \ldots .$$

This expansion leads to the following family of type (8):

$$P(x) = (N-1)(m-1)^{(N-1)}/(m+x)^{(N)}, \qquad x = 0, 1, \ldots, \tag{11}$$

where $m > N-1$, $N = 2, 3, \ldots$. The mean is

$$E(X) = \mu = (m - N + 1)/(N - 2).$$

References

[1] Berg, S. (1974). *Scand. J. Statist.*, **1**, 145–152.

[2] Berg, S. (1975). *Scand. J. Statist.*, **2**, 91–94.

[3] Feller, W. (1968). *An Introduction to Probability Theory and Its Applications*, 3rd ed. Wiley, New York.

[4] Irwin, J. O. (1965). *J. R. Statist. Soc. A*, **118**, 394–404.

[5] Johnson, N. L. and Kotz, S. (1977). *Urn Models and Their Applications*. Wiley, New York.

[6] Marlow, (1965). *Ann. Math. Statist.*, **36**, 1066–1068.

[7] Seber, G. A. F. (1973). *The Estimation of Animal Abundance and Related Parameters*. Charles Griffin, London.

Bibliography

See the following works, as well as the references just cited, for more information on the topic of factorial series distributions.

Barndorff-Nielsen, O. (1978). *Information and Exponential Families in Statistical Theory*. Wiley, New York.

Berg, S. (1977). *Scand. J. Statist.*, **4**, 25–30.

David, F. N. and Barton, D. E. (1962). *Combinatorial Chance*. Charles Griffin, London.

Harris, B. (1968). *J. Amer. Statist. Ass.*, **63**, 837–847.

Noack, A. (1950). *Ann. Math. Statist.*, **21**, 127–132.

Patil, G. P. and Joshi, S. W. (1968). *A Dictionary and Bibliography of Discrete Distributions*. Oliver & Boyd, Edinburgh.

Roy, J. and Mitra, S. K. (1957). *Sankhyā*, **18**, 371–379.

(CAPTURE–RECAPTURE METHODS
HYPERGEOMETRIC DISTRIBUTIONS
OCCUPANCY DISTRIBUTIONS
POWER SERIES DISTRIBUTIONS)

SVEN BERG

FACTORIAL SERIES FAMILY *See* EXPONENTIAL FAMILIES

FAILURE RATE CLASSIFICATION OF DISTRIBUTIONS *See* HAZARD RATE AND OTHER CLASSIFICATIONS OF DISTRIBUTIONS

FAIR–JAFFEE MODEL

In 1972, Fair and Jaffee [5] introduced econometric* models that describe the behavior of the quantity Y_t exchanged at time t as the minimum of demand Y_{1t} and supply Y_{2t}, each separately a regression* on certain regressor variables. These are models of a market in disequilibrium and switching regressions with two regimes.

Consider the usual regression model, where at time t the observed random variable is $Y_t = \boldsymbol{\beta}'\mathbf{x}_t + \epsilon_t$ and the ϵ_t are independently and normally distributed with mean zero and variance σ^2. Here $\boldsymbol{\beta} \in R^p$ is the vector of regression coefficients, $\mathbf{x}_t \in R^p$ gives the levels of the p regressor variables at time t, and the model equation is linear in $\boldsymbol{\beta}$. Statistical inference about $(\boldsymbol{\beta}, \sigma^2)$ is based on Y_t, $\mathbf{x}_t$, $t = 1, 2, \ldots, T$. Inference for this model and for variations with less restrictive assumptions on the joint distribution of the ϵ_t is discussed in most standard texts. Inference for nonlinear regression models is attracting attention; see Bunke [4].

The Fair–Jaffee model is a specialization of the following, which we refer to as model $(*)$. For $t = 1, 2, \ldots,$

$$Y_t = \min\{Y_{1t}, Y_{2t}\},$$

$$Y_{it} = \boldsymbol{\beta}_i'\mathbf{x}_t + \epsilon_{it}, \qquad i = 1, 2, \qquad (*)$$

$(\epsilon_{1t}, \epsilon_{2t})$ are independent bivariate normal with mean $(0, 0)$ and covariance matrix

$$\boldsymbol{\Sigma} = \begin{pmatrix} \sigma_1^2 & \sigma_{12} \\ \sigma_{12} & \sigma_2^2 \end{pmatrix}.$$

In $(*)$, we let $\boldsymbol{\theta} = ((\boldsymbol{\beta}_1, \sigma_1^2), (\boldsymbol{\beta}_2, \sigma_2^2), \sigma_{12})$ denote the vector of parameters. Also, we let Θ denote the set of possible $\boldsymbol{\theta}$ values for model $(*)$. Although Fair and Jaffee [5] take $\sigma_{12} = 0$ and $\sigma_1^2, \sigma_2^2 > 0$, we will define the Fair–Jaffee model more generally as model $(*)$ with Θ restricted to full-rank $\boldsymbol{\Sigma}$.

Earlier, Tobin [13] introduced an econometric model that is also a special case of model $(*)$. In the Tobin (Tobit)* model, the observed random variable is $Y_t = \max\{Y_{1t}, 0\}$. This structure results from $(*)$ by restricting Θ to $\sigma_2^2 = \sigma_{12} = 0$, $\boldsymbol{\beta}_2 = \mathbf{0}$, and by replacing $\boldsymbol{\beta}_1$ and ϵ_{1t} by their negatives.

Models such as those of Fair and Jaffee and Tobin, which involve random or fixed truncation* (censoring), arise in many noneconomic situations as well, for example, as models for competing risks* and in life testing*.

Let $\mathbf{X}$ be the design matrix for model $(*)$, that is, the matrix with rows $\mathbf{x}_t'$, $t = 1, 2, \ldots, T$. Let $F_{\theta, \mathbf{X}}$ denote the joint distribution of $Y_1, Y_2, \ldots, Y_T$ given $\boldsymbol{\theta}$ and $\mathbf{X}$. The parameter $\boldsymbol{\theta}$ is identified if $F_{\boldsymbol{\theta}, \mathbf{X}}$ uniquely determines $\boldsymbol{\theta}$. Since Y_t is symmetric with respect to permutation of Y_{1t} and Y_{2t}, parameter values $\boldsymbol{\theta}$ and $\pi\boldsymbol{\theta} = ((\boldsymbol{\beta}_2, \sigma_2^2), (\boldsymbol{\beta}_1, \sigma_1^2), \sigma_{12})$ give rise to the same distribution for $Y_1, Y_2, \ldots, Y_T$; and, therefore, $\boldsymbol{\theta}$ is not identified unless Θ is suitably restricted. A necessary condition which will be assumed throughout all that follows is that Θ satisfies: $\boldsymbol{\theta} \in \Theta$ implies that $\pi\boldsymbol{\theta} \notin \Theta$.

Gilliland and Hannan [6] investigate further the identifiability* of $\boldsymbol{\theta}$ for various subfamilies determined by restrictions on Θ. For the Fair–Jaffee model, if $\mathbf{X}$ has full rank and the sets $\mathscr{U}_i = \{\mathbf{X}\boldsymbol{\beta}_i \mid \boldsymbol{\theta} \in \Theta\}$, $i = 1, 2$, are closed under difference and satisfy $\mathscr{U}_1 \cap \mathscr{U}_2 = \{\mathbf{0}\}$, then $\boldsymbol{\theta}$ is identified. For the Tobin model, $\mathbf{X}$ being full rank is sufficient for identifiability.

The problem of estimation of $\boldsymbol{\theta}$ in model (*) based on $\mathbf{X}$ and $Y_1, Y_2, \ldots, Y_T$ is interesting and not trivial. Here the observations are independently distributed but not identically so, by virtue of the changing levels $\mathbf{x}_t$ of the regressor variables.

Considerable work has been done on estimation in the case of the Tobin model. Amemiya [1] assumes that Θ is a compact subset of $R^p \times (0, \infty)$ and conditions on $\{\mathbf{x}_t\}$ and establishes the local consistency* and asymptotic normality of a local maximizer $\hat{\boldsymbol{\theta}}$ of the likelihood $f_{\boldsymbol{\theta},\mathbf{X}}(y_1, y_2, \ldots, y_T)$. The proof of consistency is an interesting application of an extension of the Jennrich [9] uniform strong law of large numbers*. (Both the extension and original are subsumed in the Ranga Rao [12] version; see Billingsley [3, p. 17, Prob. 8].) Later Olsen [11] shows that the log of the likelihood is concave in the parameter $(\sigma^{-1}\boldsymbol{\beta}, \sigma^{-1})$. In view of this concavity, the two local qualifications above are eliminated, Θ can be taken to be $R^p \times (0, \infty)$ and $\hat{\boldsymbol{\theta}}$ is the global maximizer. For compact Θ the consistency of $\hat{\boldsymbol{\theta}}$ also follows from the Hoadley [8] theorem on maximum likelihood estimation based on independently and nonidentically distributed random variables, although the hypothesis of the theorem is not easily verified in this application. Hoadley's proof is an adaptation of the Wald proof of MLE consistency in the identically distributed case.

Fair and Jaffee [5] for their model discuss estimation of $((\beta_1, \sigma_1^2), (\beta_2, \sigma_2^2))$ based on $Y_1, Y_2, \ldots, Y_T$ and $\mathbf{X}$ and additional information on price changes. Several models for the price change data are presented. Amemiya [2] and Maddala and Nelson [10] discuss the Fair–Jaffee models and, in giving the likelihood $f_{\boldsymbol{\theta},\mathbf{X}}(y_1, y_2, \ldots, y_T)$, allow for $\sigma_{12} \neq 0$ in a full-rank covariance matrix $\boldsymbol{\Sigma}$.

In the formulation of the Fair–Jaffee model presented here, $\boldsymbol{\Sigma}$ is full rank and no auxiliary price information appears in the model. However, if Θ is a compact subset, the maximum likelihood approach does lead to a consistent estimator of $\boldsymbol{\theta}$ under certain conditions. As in the Tobin case, this can be shown to follow from Hoadley [8]. Hartley and Mallela [7] attempt an adaptation of Wald's proof directly to the Fair–Jaffee model.

In the Fair–Jaffee model with certain noncompact Θ, the likelihood $f_{\boldsymbol{\theta},\mathbf{X}}(y_1, y_2, \ldots, y_T)$ is unbounded in $\boldsymbol{\theta}$ and interesting estimation problems are yet to be solved.

References

[1] Amemiya, T. (1973). *Econometrica*, **41**, 997–1016.

[2] Amemiya, T. (1974). *Econometrica*, **42**, 759–762.

[3] Billingsley, P. (1968). *Convergence of Probability Measures*, Wiley, New York.

[4] Bunke, H. (1980). In *Handbook of Statistics*, Vol. 1: *Analysis of Variance*. North-Holland, Amsterdam, pp. 593–615.

[5] Fair, R. C. and Jaffee, D. M. (1972). *Econometrica*, **40**, 497–514.

[6] Gilliland, D. C. and Hannan, J. (1980). *J. Amer. Statist. Ass.*, **75**, 651–654.

[7] Hartley, M. J. and Mallela, P. (1977). *Econometrica*, **45**, 1205–1220.

[8] Hoadley, B. (1971). *Ann. Math. Statist.*, **42**, 1977–1991.

[9] Jennrich, R. I. (1969). *Ann. Math. Statist.*, **40**, 633–643.

[10] Maddala, G. S. and Nelson, F. D. (1974). *Econometrica*, **42**, 1013–1030.

[11] Olsen, R. J. (1978). *Econometrica*, **46**, 1211–1215.

[12] Ranga Rao, R. (1962). *Ann. Math. Statist.*, **33**, 659–680.

[13] Tobin, J. (1958). *Econometrica*, **26**, 24–36.

(COMPETING RISKS
ECONOMETRICS

SWITCHING REGRESSIONS
TOBIT MODEL)

D. C. GILLILAND
J. HANNAN

FALLACIES, STATISTICAL

The rapid growth in the development and application of statistical methodology in this century has been accompanied by a corresponding increase in fallacious statistical reasoning and misuses of statistics. The potential for statistical fallacies has been enhanced by the development of statistical computer packages (creating greater access to complex procedures) and the increasing need for statistical analyses in government agencies, industries, and diverse academic disciplines. The fallacies included in this article were chosen because of their frequent occurrence and because they merit mention in introductory courses in statistical methods. We have not attempted to provide an exhaustive catalog. An indexing of types of fallacies and a discussion of a greater variety of them are given by Good [12, 13]. Other sources with several good examples of statistical fallacies and misuses of statistics include Campbell [8], Cohen [10], Freedman et al. [11], Huff [14, 15], Moran [18], and Wallis and Roberts [23].

We first describe three major types of errors, each of which seems to be responsible for a variety of fallacious arguments, and then we describe briefly several other fallacies which occur in other settings.

FAILURE TO INCLUDE RELEVANT COVARIATES

Many bivariate associations disappear or diminish when appropriate control variables are introduced. Failure to recognize this fact often leads to wrong conclusions about relationships between variables.

For categorical data*, it is misleading to restrict attention to a two-dimensional contingency table when that table is more properly viewed as a marginal distribution of a higher-dimensional array. Bickel et al. [3] illustrate this in their discussion of data concerning admission into graduate school at Berkeley. Investigation of the 2×2 table of admissions decision (admit, do not admit) by sex revealed that relatively fewer women than men were admitted into the graduate school. When admissions were considered separately by academic department, however, the apparent bias against women disappeared. The explanation for the shift in results is that the proportion of women applicants tended to be highest in disciplines that were most competitive for admissions in having the highest rejection rates. Bishop et al. [4, pp. 41–42] present a similar example. An apparent association between amount of prenatal care and infant survival disappears when the data are considered separately for each clinic participating in the study. A pooling of the data from the clinics ignores the dependence of both infant survival and amount of prenatal care on clinic.

The fact that partial associations may be very different in nature from unconditional bivariate associations has been expressed through conditional probabilities as *Simpson's paradox* (see Simpson [20] and Blyth [5, 7]). This paradox states that even if $\Pr(A \mid BC) > \Pr(A \mid B^cC)$ and $\Pr(A \mid BC^c) > \Pr(A \mid B^cC^c)$, it is possible that $\Pr(A \mid B) < \Pr(A \mid B^c)$. In the context of a $2 \times 2 \times 2$ cross-classification of three dichotomous variables X, Y, and Z, Simpson's paradox implies that it is possible to have a positive partial association between X and Y at each level of Z, yet a negative unconditional association between X and Y, due to the nature of the association of Z with both X and Y. For a numerical example, see Blyth [5, p. 264]. Bishop et al. [4, p. 39] give conditions under which partial associations are the same as unconditional associations, so that multidimensional contingency tables* can be meaningfully collapsed.

The same remarks regarding covariates apply to quantitative variables. For example, the mean salary for men may exceed the mean salary for women for the faculty at a particular university. When factors such as department, rank, and number of years in rank are controlled, however, it is possible that the difference in the means may change appreciably. In a related example, Cochran [9] discusses how failure to use age standardization in comparing two populations can lead to fallacious conclusions.

FAILURE TO ADJUST THE ERROR RATE FOR MULTIPLE INFERENCES

The importance of using multiple comparison procedures for making pairwise inferences about several means is emphasized in many statistical methods textbooks (e.g., Snedecor and Cochran [21, p. 272]; *see also* MULTIPLE COMPARISONS). The authors of these books note that the use of a standard error rate (such as 0.05) for each of a large number of inferences may result in an unacceptably large probability of at least one error occuring (e.g., at least one type I error or at least one confidence interval not containing the parameter it is designed to enclose). Fallacious arguments can easily occur in many other contexts from using a single-inference error rate when several inferences have actually been conducted.

This type of error frequently occurs when the need for a multiple-inference approach is not obvious. For example, a researcher analyzing a large data set on several variables may screen it, using computer packages to compute correlations, chi-squares, analyses of variance, regression analyses, etc., on various combinations of the variables. In some cases several competing tests may be conducted to test the same hypothesis. The researcher may select from the computer printout everything achieving significance at the 0.05 level and report those results as if the corresponding hypotheses and analyses were the only ones considered.

A more subtle failure to adjust for multiplicity of inferences results from the tendency of research journals in many fields to publish only those studies that obtain statistical significance at a certain level. If a large number of researchers independently test the same true null hypothesis, there is a good chance that a type I error will be published. Researchers who do not obtain significant results may be discouraged from submitting their findings or feel pressured to find ways of achieving significance (e.g., other tests, more data). Walster and Cleary [24] give a good discussion of this problem. They also emphasize the importance of replication of previously published research so that type I errors are exposed.

Fallacious arguments of a similar nature can occur from treating the maximum or minimum of a set of random variables as if it had the same distribution as an arbitrary one of the variables. This error occurs when a researcher selects the two most distant sample means (out of a collection of several means) as the most interesting finding of a survey and compares those means using standard two-sample tests or confidence intervals. The error also occurs in variable selection for a regression model when at each stage, one tests the significance of the partial effect of X_i on Y *after* having selected X_i because it had the largest such effect out of some set of variables. Many events that seem to be very unusual occurrences when viewed in isolation may seem rather common when considered in proper context. Suppose that a coin shows heads in each of 10 tosses. We would probably suspect that the coin is unbalanced in favor of heads, since the probability of such a rare event if the coin were balanced is only $(\frac{1}{2})^{10} = 0.00098$. However, if we were told that this coin had had the greatest number of heads out of 1000 coins that had been tossed 10 times each, we would be less likely to believe that it was biased (see Wallis and Roberts [23, p. 116]).

Another misuse that occurs when multiple inferences are not recognized as such is the application of fixed sample-size methods after obtaining each new observation in a sequential sampling scheme (see Moran [18] and Armitage et al. [1]). The result we predict may occur if we wait long enough.

CONFUSION OF CORRELATION WITH CAUSATION

Fallacious arguments often result from the belief that correlation* implies causation*. The everyday usage of the word "correlation" in the English language probably contributes to the confusion. The fallacy can often be shown by illustrating the lack of partial association when certain control variables are introduced, as in the sex–graduate admissions study by Bickel et al. [3]. Yule and Kendall [26, Chaps. 4, 15, 16] discuss the problems in detail.

A special case of the correlation–causation fallacy is the "post hoc" fallacy that if A precedes B, it must be a cause of B. Campbell [8, p. 172] illustrates the fallacy by reference to the plane traveler who requests that the captain not turn on the "fasten seat belts" sign, since it always seems to result in a bumpy ride. For other examples of the post hoc fallacy, see Huff [14, Chap. 8].

Other types of misapplications of correlation coefficients abound. Barnard [2] explains why "astonishingly high correlations" need not be especially noteworthy. For example, a correlation arbitrarily close to 1 may be produced by a single outlying observation. Another common error occurs in the generalization of inferences to a different sampling unit. For example, "ecological correlations" computed from rates or totals for units such as counties or states may be very different in magnitude from correlations obtained using data on individuals. Freedman et al. [11, pp. 141–142] state that the correlation between average income and average education computed for nine regions in the United States is about 0.7, whereas it is approximately 0.4 when computed for individuals from census data.

REGRESSION FALLACY

The phenomenon of regression* toward the mean for the bivariate normal regression* model was first noticed by Sir Francis Galton* in his studies of $X =$ father's height and $Y =$ son's height. Based on this phenomenon he made the fallacious conclusion that the variability in heights must decrease with time. A counterexample is given by Good [13], who notes that a reversal of the labeling of Y and X would necessarily force the conclusion that variability in heights is increasing with time. The fact that "highly unusual" observations tend to be followed by more regular observations, when not recognized or not understood, has resulted in various types of fallacies and superstitions (e.g., that a professional athlete having an outstanding first year will have a "sophomore jinx"). For further discussion, see Freedman et al. [11, pp. 158–159].

NEGLECTING ASSUMPTIONS

Misuses of statistical procedures commonly occur from severe violations of basic assumptions concerning method of sampling, required sample size (for use of asymptotically based formulas), measurement scale of variables, and distribution of variables. For example, fallacious conclusions could result from using formulas based on simple random sampling for data collected as a cluster sample, from treating time-series data as an independent identically distributed sequence, from applying the chi-square test* of independence to a contingency table having a small total frequency or having ordered rows or columns, and from blindly applying techniques such as regression, analysis of variance*, and factor analysis* to dichotomous variables. A sociology Master's thesis is rumored to exist which contains about 100 applications of the Kolmogorov–Smirnov* two-sample test, *none* of which attains significance at the 0.05 level. The author apparently applied the standard form of the test designed for continuous variables to highly discrete data, for which that test is highly conservative.

NEGLECTING VARIATION

Fallacies or misleading statements often occur from a failure to consider variation*.

Examples include reporting a percentage without listing its base sample size (and hence its standard error) and attributing importance to a difference between two means which could be explained by sampling error.

MISINTERPRETATION OF STATISTICAL TESTS OF HYPOTHESES

The results of statistical tests are often misinterpreted due to factors such as confusion of statistical significance with practical significance and acceptance of the null hypothesis without consideration of the power function. For good discussions, see Kruskal [17, p. 456] and Kish [16].

FALLACIOUS PROBABILITY REASONING

Many advances in the historical development of probability occurred because of fallacious arguments in gambling situations which led to seemingly contradictory results in the pocketbook (see Freedman et al. [11, pp. 223–225], Huff [15, pp. 63–69], and Todhunter's [22, Chap. XIII] discussion about D'Alembert's fallacies). Among the most common errors are the following: treating sample points as equally likely when they are not; misinterpreting the law of large numbers, as when arguing that in 1,000,000 flips of a fair coin, the number of heads is bound to be within a few units of 500,000; not understanding independent trials or conditional probability, as in the argument that a sequence of 10 consecutive heads in coin flipping is almost sure to be followed by a tail; misuse of the additive law, as in the argument that for n independent trials with probability p of success on each, the probability is np of at least one success. Many fallacious probabilistic arguments result from an unawareness of certain paradoxes. For example, it is tempting to argue that $P(Y > X) > \frac{1}{2}$ and $P(Z > Y) > \frac{1}{2}$ implies that $P(Z > X) > \frac{1}{2}$. The transitivity paradox shows that this need not be the case even if X, Y, and Z are independent (see Blyth [6]).

FALLACIES WITH TIME-SERIES DATA

Failure to recognize the special problems occurring in the analysis of time-series data often leads to fallacious conclusions. For example, strong spurious associations may result from correlating variables which are measured over time and have similar trends (see Huff [14, p. 97]). Errors in statistical analysis often result from applying formulas based on independent observations, such as by using standard regression procedures and disregarding effects of serially correlated error terms (see Wonnacott and Wonnacott [25, pp. 136–147]). Richman and Richman [19] show how fallacious conclusions concerning changes in the level of heroin addiction result from improper analyses of time-series data.

IMPROPER BASE

A common statistical error in the news media is the comparison of frequencies based on different totals. Examples include the statement during the Vietnam war that it was safer to be in the army than driving on the nation's highways due to the lower yearly death total in the war (rates of death should be compared, preferably within age groups).

MAKING AN INFERENCE WITHOUT THE NECESSARY COMPARISON

This error commonly appears through the reporting of only one row of a contingency table. For example, a criminal rehabilitation program might be criticized because participants in it have a recidivism rate of 50%. Without being given the corresponding rate for nonparticipants or for other programs, we would have a difficult time making a judgment.

BIASED DATA THROUGH MEASUREMENT ERROR OR INTERVIEWER EFFECT

Wallis and Roberts [23, p. 96] quote a survey in which the percentage of blacks interviewed who felt the army to be unfair to their race was 35% for those people having a black interviewer and 11% for those having a white interviewer (see also Huff [14, p. 24]).

BIASED DATA DUE TO IMPROPER SAMPLING FRAME

The average number of children in families having students in a particular school would tend to be overestimated by sampling children from that school, since a large family is more likely to be represented in the sample than a small family.

UNCRITICAL RELIANCE ON COMPUTERS

Among the likely consequences of the development of computer-based statistical packages have been a greater relative frequency in the use of statistical procedures that are inappropriate to a problem or which the researcher fails to understand, errors associated with unrecognized multiple inferences due to searching for significant results, and the attributation of greater accuracy to the results than the data warrant (*see* COMPUTERS AND STATISTICS).

References

[1] Armitage, P., McPherson, C. K., and Rowe, B. C. (1969). *J. R. Statist. Soc. A*, **132**, 235–244.

[2] Barnard, G. A. (1977). *J. R. Statist. Soc. A.*, **140**, 200–202.

[3] Bickel, P. J., Hammel, E. A., and O'Connell, J. W. (1975). *Science*, **187**, 398–404.

[4] Bishop, Y. M. M., Fienberg, S. E., and Holland, P. W. (1975). *Discrete Multivariate Analysis*. MIT Press, Cambridge, Mass.

[5] Blyth, C. R. (1972). *J. Amer. Statist. Ass.*, **67**, 364–366.

[6] Blyth, C. R. (1972). *J. Amer. Statist. Ass.*, **67**, 366–373.

[7] Blyth, C. R. (1973). *J. Amer. Statist. Ass.*, **68**, 746.

[8] Campbell, S. K. (1974). *Flaws and Fallacies in Statistical Thinking*. Prentice-Hall, Englewood Cliffs, N.J. (Good source of elementary examples; similar in level to Huff's books.)

[9] Cochran, W. G. (1968). *Biometrics*, **24**, 295–313.

[10] Cohen, J. B. (1938). *J. Amer. Statist. Ass.*, **33**, 657–674.

[11] Freedman, D., Pisani, R., and Purves, R. (1978). *Statistics*. W. W. Norton, New York. (Excellent introductory statistics text which highlights several statistical fallacies.)

[12] Good, I. J. (1962). *Technometrics*, **4**, 125–132.

[13] Good, I. J. (1978). Fallacies, statistical. In *International Encyclopedia of Statistics*, Vol. 1, W. H. Kruskal and J. M. Tanur, eds. Free Press, New York, p. 344. (Contains large bibliography of literature on statistical fallacies.)

[14] Huff, D. (1954). *How to Lie with Statistics*. Norton, New York. (Contains discussions of several basic types of statistical fallacies; written for the layman.)

[15] Huff, D. (1959). *How to Take a Chance*. Norton, New York. (Good examples of fallacious probabilistic arguments.)

[16] Kish, L. (1959). *Amer. Sociol. Rev.*, **24**, 328–338.

[17] Kruskal, W. H. (1978). Significance, tests of. In *International Encyclopedia of Statistics*, Vol. 2, W. H. Kruskal and J. M. Tanur, eds., 944–958. Free Press, New York.

[18] Moran, P. A. P. (1973). *Commun. Statist.*, **2**, 245–257. (Discusses 15 settings in which fallacious statistical analyses commonly occur.)

[19] Richman, A. and Richman, V. (1975). *Amer. Statist. Ass., 1975 Proc. Social. Statist. Sec.*, 611–616.

[20] Simpson, E. H. (1951). *J. R. Statist. Soc. B*, **13**, 238–241.

[21] Snedecor, G. W. and Cochran, W. G. (1967). *Statistical Methods*, 6th ed. Iowa State University Press, Ames, Iowa.

[22] Todhunter, I. (1865). *A History of the Mathematical Theory of Probability*. Macmillan, London.

[23] Wallis, W. A. and Roberts, H. V. (1962). *The Nature of Statistics*. Free Press, New York.

[24] Walster, G. W. and Cleary, T. A. (1970). *Amer. Statist.*, **24**, 16–19.

[25] Wonnacott, R. J. and Wonnacott, T. H. (1970). *Econometrics*. Wiley, New York.

[26] Yule, G. U. and Kendall, M. G. (1937). *An Introduction to the Theory of Statistics*. Charles Griffin, London.

(CAUSATION
LOGIC IN STATISTICAL REASONING)

ALAN AGRESTI

FARLIE–GUMBEL–MORGENSTERN DISTRIBUTIONS

The Farlie–Gumbel–Morgenstern (FGM) system of bivariate distributions includes all cumulative distribution functions of the form

$$F_{X_1,X_2}(x_1, x_2) = F_1(x_1)F_2(x_2)\{1 + \alpha[1 - F_1(x_1)] \times [1 - F_2(x_2)]\} \quad (|\alpha| \leqslant 1),$$

where $F_1(x_1)$ and $F_2(x_2)$ are the marginal distribution functions of X_1 and X_2, respectively, and α is an association parameter. Morgenstern [8] introduced the general class, although a special case with uniform marginals was discussed by Eyraud [2]. Gumbel [4] investigated properties of the bivariate FGM normal distribution and proposed a multivariate generalization. Farlie [3] extended the class to all bivariate distribution functions of the form

$$F_{X_1,X_2}(x_1, x_2) = F_1(x_1)F_2(x_2)[1 + \alpha A(x_1)B(x_2)],$$

where $A(x_1)$ and $B(x_2)$ are bounded functions such that $A(\infty) = B(\infty) = 0$, $d(F_1A)/dF_1$ and $d(F_2B)/dF_2$ are also bounded, and the association parameter α is constrained to lie in an interval about 0.

PROPERTIES OF BIVARIATE FGM DISTRIBUTIONS

Bivariate FGM distributions have considerable analytical appeal since the distribution function is characterized by the univariate marginal distributions. If both $F_1(x_1)$ and $F_2(x_2)$ are absolutely continuous, the joint density function is

$$f_{X_1,X_2}(x_1, x_2) = f_1(x_1)f_2(x_2)\{1 + \alpha[1 - 2F_1(x_1)] \times [1 - 2F_2(x_2)]\}.$$

It follows that the conditional density of X_1 given $X_2 = x_2$ is

$$f_{X_1|X_2}(x_1 | x_2) = f_1(x_1)\{1 + \alpha[1 - 2F_1(x_1)] \times [1 - 2F_2(x_2)]\}.$$

When $\tilde{x}_2 =$ median (X_2), the conditional density of X_1 given $X_2 = \tilde{x}_2$ is the same as the marginal density of X_1.

The regression* curve of X_1 given $X_2 = x_2$ is

$$E(X_1 | X_2 = x_2) = E(X_1) + \alpha[1 - 2F_2(x_2)] \times \int x_1[1 - 2F_1(x_1)]\, dF_1(x_1),$$

which is linear in $F_2(x_2)$. The regression curves for the extended class of bivariate FGM distributions are more general, of the form

$$E(X_1 | X_2 = x_2) = E(X_1) + \alpha \frac{d(F_2B)}{dF_2} \times \int x_1 \frac{d(F_1A)}{dx_1}\, dx_1 .$$

Farlie [3] showed that linear, quadratic, triangular, and discontinuous regression curves are possible by suitable choice of the functions $A(x_1)$ and $B(x_2)$.

Association between random variables (rvs) having a bivariate FGM distribution is characterized by the parameter α. The rvs X_1 and X_2 are independent whenever $\alpha = 0$, positively associated when $\alpha > 0$, and negatively associated when $\alpha < 0$. The maximum likelihood estimate* $\hat{\alpha}$ of α depends on the marginal distributions. For small values of α, the following approximation is useful:

$$\hat{\alpha} \simeq \frac{\sum_{i=1}^{n}[1 - 2F_1(x_{i1})][1 - 2F_2(x_{i2})]}{\sum_{i=1}^{n}[1 - 2F_1(x_{i1})]^2[1 - 2F_2(x_{i2})]^2} .$$

One drawback of bivariate FGM distributions is that they are limited to describing weak dependence between the rvs X_1 and X_2. Measures of dependence vary over a considerably smaller range than for more general classes of bivariate distributions. For example, the Pearson product-moment correlation coefficient depends on the marginal distributions and is given by

$$\rho(X_1, X_2) = \frac{\alpha}{\sigma_1\sigma_2} \int [1 - F_1(x_1)]\, dx_1 \times \int [1 - F_2(x_2)]\, dx_2 ,$$

where σ_i is the standard deviation of X_i $(i = 1, 2)$. When both X_1 and X_2 have a normal distribution, $\rho(X_1, X_2) = \alpha/\pi$ and the range of the correlation coefficient is $[-0.318, 0.318]$ rather than the customary interval $[-1, 1]$. Schucany et al. [9] showed that for continuous bivariate FGM distributions $|\rho(X_1, X_2)| \leqslant \frac{1}{3}$, irrespective of the marginal distributions $F_1(x_1)$ and $F_2(x_2)$.

A number of properties result from the simple analytic form of bivariate FGM distributions. If the marginal distributions of X_1 and X_2 are symmetric, the joint FGM distribution is also symmetric. Random variables having a bivariate FGM distribution are exchangeable whenever the marginal distributions are identical (*see* EXCHANGEABILITY). The FGM system is closed with respect to monotonic increasing functions of the rvs. That is, if X_1 and X_2 have a bivariate FGM distribution and $Y_i = h_i(X_i)$ is a monotonic increasing function of X_i $(i = 1, 2)$, then Y_1 and Y_2 have a bivariate FGM distribution. Also, the system is closed with respect to mixtures of bivariate FGM distributions having the same marginal distributions. Finally, bivariate FGM distributions have an equivalent representation in terms of the survival functions, $S_{X_1,X_2}(x_1, x_2) \equiv \Pr[X_1 > x_1, X_2 > x_2]$ and $S_i(x_i) \equiv 1 - F_i(x_i)$ $(i = 1, 2)$, namely

$$S_{X_1,X_2}(x_1, x_2) = S_1(x_1)S_2(x_2) \times \{1 + \alpha F_1(x_1)F_2(x_2)\} \qquad (|\alpha| \leqslant 1).$$

PROPERTIES OF MULTIVARIATE FGM DISTRIBUTIONS

Johnson and Kotz [6] introduced a general system of multivariate FGM distributions defined by

$$F_{X_1,\ldots,X_s}(x_1, \ldots, x_s) = \prod_{i=1}^{s} F_i(x_i)\left\{1 + \sum_{j_1<j_2} \alpha_{j_1 j_2} \prod_{k=1}^{2} [1 - F_{j_k}(x_{j_k})] + \cdots + \alpha_{12\cdots s} \prod_{k=1}^{s} [1 - 2F_k(x_k)]\right\},$$

where $F_i(x_i)$ is the specified marginal distribution function of X_i $(i = 1, \ldots, s)$. The coefficients α are real numbers, suitably constrained so that $F_{X_1,\ldots,X_s}$ is a multivariate distribution function. Necessary and sufficient conditions that the coefficients α must satisfy are given by Cambanis [1]. Notice that the bivariate FGM distribution includes, as a special case, the generalization due to Gumbel [4],

$$F_{X_1,\ldots,X_s}(x_1, \ldots, x_s) = \prod_{i=1}^{s} F_i(x_i)\left\{1 + \alpha \prod_{k=1}^{s} [1 - F_k(x_k)]\right\}, \qquad (|\alpha| \leqslant 1).$$

A simple derivation shows that the marginal joint distributions of all subsets of $X_1, \ldots, X_s$ also have a multivariate FGM distribution. If the univariate marginal distributions are absolutely continuous, the joint density function is

$$f_{X_1,\ldots,X_s}(x_1, \ldots, x_s) = \prod_{i=1}^{s} f_i(x_i)\left\{1 + \sum_{j_1<j_2} \alpha_{j_1 j_2} \prod_{k=1}^{2} [1 - 2F_{j_k}(x_{j_k})] + \cdots + \alpha_{12\cdots s} \times \prod_{k=1}^{s} [1 - 2F_k(x_k)]\right\}.$$

Johnson and Kotz [7] give expressions for the conditional density functions and regression curves. An interesting result is that the conditional density of X_i given that the other rvs assume their median values is the same as the marginal density of X_i $(i = 1, \ldots, s)$.

Many of the properties of bivariate FGM distributions generalize. In particular, multivariate FGM distributions characterize weak dependence among the random variables $X_1, \ldots, X_s$. The rvs are exchangeable whenever the univariate marginal distributions are identical and the coefficients satisfy $\alpha_{j_i \cdots j_m} \equiv \alpha_m$ $(m = 2, \ldots, s)$. The multivariate FGM system is also closed with respect to monotonic increasing functions of the rvs and mixtures of FGM distributions having the same univariate marginals. The system has the following representation in terms of survival functions:

$$S_{X_1,\ldots,X_s}(x_1,\ldots,x_s) = \prod_{i=1}^{s} S_i(x_i)\left[1 + \sum_{j_i<j_2} \alpha_{j_1j_2} \prod_{k=1}^{2} F_{j_k}(x_{j_k}) - \sum_{j_i<j_2<j_3} \alpha_{j_1j_2j_3} \prod_{k=1}^{3} F_{j_k}(x_{j_k}) + \cdots + (-1)^s \alpha_{12\ldots s} \prod_{k=1}^{s} F_k(x_k)\right].$$

APPLICATIONS OF FGM DISTRIBUTIONS

The FGM distributions are primarily useful as alternatives to the standard multivariate normal distribution. Because they have a simple form, they have been used to assess the efficiency of multivariate nonparametric procedures. Farlie [3] used the extended class of bivariate FGM distributions to study the efficiency of Spearman's rank correlation coefficient, Kendall's τ coefficient, and the probability of concordance relative to the Pearson product-moment correlation coefficient (*see* CORRELATION). His results establish the importance of the underlying bivariate distribution for these coefficients to provide an efficient test for association.

Shaked [10] discussed the analytical appeal of multivariate FGM survival functions in reliability applications. He also gave an example of a multivariate FGM distribution as an appropriate prior in Bayesian survey sampling*. Schucany et al. [9] discussed a bivariate FGM distribution as a model for screening* variables in a quality control* application. Finally, in medical mortality studies, the distributions of baseline characteristics for survivors and nonsurvivors may be completely different. Halperin et al. [5] used a bivariate FGM distribution to demonstrate that inferences about risk factors for the two groups depend strongly on the underlying bivariate distribution of baseline characteristics.

References

[1] Cambanis, S. (1977). *J. Multivariate Anal.*, **7**, 551–559.

[2] Eyraud, H. (1936). *Ann. Univ. Lyon, Sec. A*, **1**, 30–47.

[3] Farlie, D. J. G. (1960). *Biometrika*, **47**, 307–323.

[4] Gumbel, E. J. (1958). *C. R. Acad. Sci. Paris*, **246**, 2717–2719.

[5] Halperin, M., Wu, M., and Gordon, T. (1979). *J. Chronic. Dis.*, **32**, 483–491.

[6] Johnson, N. L. and Kotz, S. (1975). *Commun. Statist.*, **4**, 415–427.

[7] Johnson, N. L. and Kotz, S. (1977). *Commun. Statist. A*, **6**, 485–496.

[8] Morgenstern, D. (1956). *Mitt. Math. Statist.*, **8**, 234–235.

[9] Schucany, W. R., Parr, W. C., and Boyer, J. E. (1978). *Biometrika*, **65**, 650–653.

[10] Shaked, M. (1975). *Commun. Statist.*, **4**, 711–721.

(ASSOCIATION, MEASURES OF
CORRELATION
DEPENDENCE, CONCEPTS OF)

DELORES CONWAY

FATIGUE MODELS

Fatigue is the anthropomorphic name for the failure of materials, principally metals, which occurs after an extended period of service time when it is caused by their sustaining loads, usually of fluctuating magnitude, of less than ultimate stress. As a consequence of the well-accepted Griffith crack theory [5], the fatigue of metals is now understood to mean the propagation of cracks, originating at points of stress concentration, elongating from stress reversal until a critical flaw size is reached by the dominant crack, causing its uncontrolled extension (i.e., brittle fracture of the component). The growth of the dominant crack is a complex phenomenon involving both the geometry and physical constants of the material, as well as the loading spectrum. Fatigue of brittle materials such as glass can take place even under nonfluctuating loads, as it can for composite materials such as wood, where it is called a duration of load effect; see ref. 8 [Vol. 7]. Fatigue is thus one form of fracture (see Argon [2], Freudenthal [3], or Yen [13]). The seven-volume treatise edited by Liebowitz [8] puts this problem in perspective.

Calculation of fatigue life is of great importance in determining the reliability of modern high-performance machines or structures, especially in applications where the strength-to-weight ratio must be high. The interaction between the relevant factors can be seen from the form of a differential equation, well accepted in engineering applications (see ref. 8 [Volume 1] or 1 [STP 415]), between the rate of crack propagation per cycle dc/dN and certain parameters, namely,

$$\frac{dc}{dN} \sim \left(G\gamma\sigma_y^2 H\right)^{-1} K^{\alpha}. \tag{1}$$

Here the elastic modulus G, the strain hardening modulus H, the yield stress σ_y, the surface energy γ, and the exponent $2 < \alpha < 5$ are material parameters that depend on the microstructure, while the stress intensity factor K, which can often be calculated using linear fracture mechanics (see ref. 11), is a design parameter. When fatigue life is treated from a statistical point of view it is often regarded as cumulative damage*. In such cases the engineering aspects, such as total energy, fracture mechanics, or heat (see ref. 1 (STP 459]), are sometimes neglected.

On the other hand, (1) does not reflect the distribution of initial flaw size or variation in the imposed service loads, both of which are stochastic and influential in the determination of the time until the ultimate fracture of the specimen.

The fatigue failure of metallic components is not of practical concern whenever it can be delayed beyond the economic life of the structure, as otherwise determined by obsolescence. This is a desideratum in the design of aluminum structural members, such as those used in aircraft construction; see ref. 1 [STP 404]. Also, the resistance to brittle fracture in high-strength steels, as used for example in high-speed turbines, must be sufficiently high that fatigue cracks can be detected before the critical flaw size is reached; see ref. 8 [Vol. 4]. Moreover, the degree of embrittlement must not be too rapidly advanced by the service environment so as to be deleterious to safety, as, for example, in nuclear reactor containment shells or in ordnance structures; see ref. 8 [Vol. 5].

In both the situations above the utility of probabilistic analysis is apparent in calculating either the residual strength or the optimal times when inspection or tests of fracture toughness are to be made; see Heller [7]. Calculations for any specific material and application would, of course, be proprietary.

One of the first methods proposed for the calculation of fatigue strength was based on the S–N diagram; see Weibull [12, pp. 147ff.]. Initially, this was merely a deterministic functional relationship between fatigue strength S, expressed as the maximum stress during repeated cyclic loads of constant amplitude, and fatigue life N, expressed as the number of cycles (or its logarithm) until failure. When minimum stress was realized to be important, S–S–N diagrams were utilized. Curves were interpolated by various schemes between experimentally determined points to determine this relationship and then extrapolations were made for each application (see Weibull [12, p. 158]).

When scatter in the life observations forced the abandonment of this deterministic hypothesis, probabilistic assumptions were introduced and the S–N diagram was interpreted then as the regression* of fatigue life N on the strength S, and several statistical distributions for life at a given stress were proposed. These included the normal and extreme-smallest-value distributions for both life and log-life. Location, shape, and scale parameters were estimated from sparse data, thus determining a corresponding distribution of strength at a given life. This conditional distribution was, at first, thought to be sufficient to determine the safety of a component after any given service history. Unfortunately, such models entailed concepts such as an absolute minimum fatigue life, an endurance limit, and the probability of permanent survival, (see Freudenthal [4]).

Of course, the ubiquitous difficulties of

statistical estimations in the case of small samples, often truncated, were encountered, but the principal defect of this approach was its failure to include critical engineering factors, including size effect, when testing small specimens. An additional complexity was the realization that the effect of the loading spectrum must include load order and could not be characterized by variables such as maximum and minimum load (see ref. 1 [STP 462]).

There are deterministic models of fatigue crack growth which partially account for the known dependence of incremental crack growth upon the history of its loading spectrum (see ref. 1 [STP 415]). Indeed, fractographic analysis of incremental crack growth demonstrates several phenomena, including crack arrest, deceleration, acceleration or jump, all occurring at the same imposed stress, but depending on the preceding load (see ref. 1 [STP 462]).

Another important practical consideration in fatigue life prediction is size effect; this is necessary to correct the fatigue tests of small material specimens for full-scale application. An exhaustive survey of the literature relating this effect to both fatigue and strength was made by Harter (see ref. 6, and the extensive bibliography given there).

Since the usual stress reversal within a duty cycle will eventually cause a fatigue crack to reach its critical flaw size, at which it will extend uncontrollably, the distribution of the time necessary for this to be accomplished is needed. The *fatigue life distribution* presented in Mann et al. [10, Sec. 4.11], derived from the time necessary to advance a fixed distance by taking a given number of random increments per unit time, was an attempt to do this. If T is the random time until such failure, its distribution for this model is given by

$$F_T(t;\alpha,\beta) = \Phi\{[(t/\beta)^{1/2} - (t/\beta)^{-1/2}]/\alpha\},\quad t > 0,$$

where α and β are shape and scale parameters, respectively, and Φ is the standard normal distribution* function. Thus the density for $t > 0$: $\alpha, \beta > 0$, is

$$f_T(t;\alpha,\beta) = \frac{1}{2\sqrt{2\pi}\,\alpha^2\beta t^2}\,\frac{t^2-\beta^2}{\sqrt{t/\beta}-\sqrt{\beta/t}} \times \exp\left[-\frac{1}{2\alpha^2}\left(\frac{t}{\beta}+\frac{\beta}{t}-2\right)\right].$$

If we define

$$Z = \frac{1}{\alpha}\left(\sqrt{T/\beta} - \sqrt{\beta/T}\right),$$

Z has a standard normal distribution. Solving, we find that

$$T = \beta\left(1 + \frac{\alpha^2 Z^2}{2} + \alpha Z\sqrt{1 + \frac{\alpha^2 Z^2}{4}}\right).$$

Hence

$$ET = \beta\left(1 + \tfrac{1}{2}\alpha^2\right),$$

$$\text{var}(T) = (\alpha\beta)^2\left(1 + (5\alpha^2/4)\right).$$

The maximum likelihood* estimates (MLEs) of α and β, based on a sample $T_1, \ldots, T_n$, are conveniently found by computing the arithmetic and harmonic means*, respectively:

$$S = \frac{1}{m}\sum T_i, \qquad R = \left(\frac{1}{m}\sum T_i^{-1}\right)^{-1},$$

and the translated harmonic mean function defined for $x > 0$ by

$$K(x) = \left[\frac{1}{m}\sum (x + T_i)^{-1}\right]^{-1}.$$

The MLE $\hat{\beta}$ of β is the unique positive solution of the random equation $g(x) = 0$, where

$$g(x) = x^2 - x[2R + K(x)] + R[S + K(x)].$$

Furthermore, $R < \hat{\beta} < S$ always. The MLE $\hat{\alpha}$ of α is then found from

$$\hat{\alpha} = \left(\frac{S}{\hat{\beta}} + \frac{\hat{\beta}}{R} - 2\right)^{1/2}.$$

If α is small (say less than 0.2), a simplified estimate of β may be computed which is virtually the same as the MLE. It is the geometric mean* of the harmonic and arithmetic means, respectively:

$$\tilde{\beta} = \sqrt{SR}\,.$$

This distribution is quite different in its prediction of the low percentiles of life from the extreme-value distributions* utilized earlier.

The stringent reliability requirements of high technology dictate that a design must be utilized which encompasses material properties, stress analysis, material defects, and statistical theory. Such design should avoid discontinuities and notch effects, insofar as possible, and select metallic alloys having the best material properties for the requisite application, as well as calculate the distribution of the fatigue life under the loads anticipated in service. Consideration must also be given to the long-term effects of environmental corrosion as well as to the extreme loads infrequently imposed by rare events which are likely to occur during the design life, such as that necessary in the design of nuclear reactors for seismic events (see ref. 9).

References

[1] American Society for Testing and Materials, Philadelphia, Special Technical Publications: Structural Fatigue in Aircraft, *STP 404* (1955); Fatigue Crack Propagation, *STP 415* (1967); Fatigue at High Temperatures, *STP 459* (1969); Effects of Environment and Complex Load History on Fatigue Life, *STP 462* (1970).

[2] Argon, A. S. (1974). *Composite Materials*, Vol. 5: *Fracture and Fatigue*, L. J. Broutman, ed. Academic Press, New York, pp. 153–190.

[3] Freudenthal, A. M. (1968). *Fracture: An Advanced Treatise*, Vol. 2: *Mathematical Fundamentals*, H. Liebowitz, ed. Academic Press, New York, pp. 592–619.

[4] Freudenthal, A. M., ed. (1956). *Fatigue in Aircraft Structures*. Academic Press, New York.

[5] Griffith, A. A. (1920). *Philos. Trans. R. Soc. Lond. A*, **222**, 168–198.

[6] Harter, H. L. (1977). A Survey of the Literature on the Size Effect on Material Strength, *Air Force Flight Dynamics Tech. Rep. AFFD2-TR-77-11*, Wright-Patterson AFB, Ohio.

[7] Heller, R. A., ed. (1972). Probabilistic Aspects of Fatigue, *American Society for Testing and Materials STP 511*.

[8] Liebowitz, H., ed. (1969). *Fracture: An Advanced Treatise*, Vol. 1: *Microscopic and Macroscopic Fundamentals*; Vol. 2: *Mathematical Fundamentals*; Vol. 3: *Engineering Fundamentals and Environmental Effects*; Vol. 4: *Engineering Fracture Design*; Vol. 5: *Fracture Design of Structures*; Vol. 6: *Fracture of Metals*; Vol. 7: *Fracture of Nonmetals and Composites*. Academic Press, New York.

[9] Liebowitz, H., ed. (1976). *Progress in Fatigue and Fracture*. Pergamon Press, Elmsford, N.Y.

[10] Mann, N. R., Schafer, R. E., and Singpurwalla, N. D. (1974). *Methods for Statistical Analysis of Reliability and Life Data*. Wiley, New York.

[11] Sih, G. C., Wei, R. P., and Erdogan, F., eds. (1975). *Linear Fracture Mechanics*. Env. Publishing Co.

[12] Weibull, W. (1961). *Fatigue Testing and Analysis of Results*. Pergamon Press, Elmsford, N.Y.

[13] Yen, Charles S. (1969). In *Metal Fatigue: Theory and Design*, A. F. Madayag, ed. Wiley, New York, pp. 140–149.

(CUMULATIVE DAMAGE MODELS).

SAM C. SAUNDERS

FAULT TREE ANALYSIS

Fault tree analysis was conceived by H. R. Watson at Bell Telephone Laboratories in the early 1960s as a technique with which to perform a safety evaluation of complex systems. Bell engineers discovered that the method used to describe the flow of logic in data processing equipment could also be used for analyzing the logic that results from component failures. The process for structuring fault trees was further developed by David Haasl [4] at the Boeing Company in Seattle, Washington. Currently it is a widely used analytic tool for engineering safety analysis as well as for the analysis of societal risks, whether nuclear, chemical, or engineering related.

FAULT TREE DEFINITION

Mathematically, a fault tree (or more generally, a logic tree) is an acyclic graph* (U, A), where U is the set of nodes and A is the set of arcs. (That is, a fault tree contains no cycles, i.e., no closed, connected sequence of nodes and arcs.) Any pair of nodes may be joined by at most a single arc, which may be

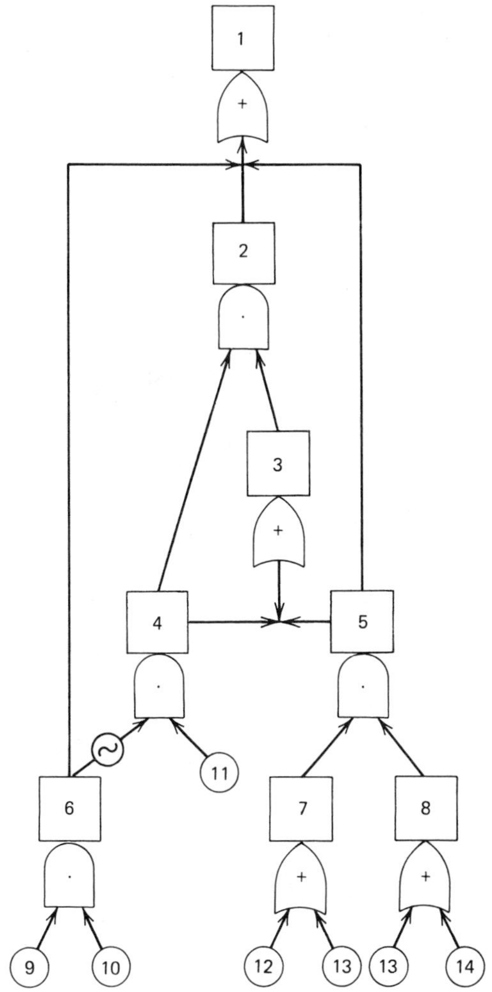

Figure 1

either a *regular arc* or a *complementing arc* (see Fig. 1). Nodes having no entering arcs are *basic nodes*, and those having one or more entering arcs are *gate nodes*. Those which have no leaving arcs are *top nodes*. A fault tree usually has only a single top node. The tree is constructed deductively from the top down by engineers and mathematicians with an intimate knowledge of the system being analyzed. However, in analyzing the tree, the logic flow is directed upward from basic nodes, indicated in Fig. 1 by upward arrows on arcs.

Basic nodes are denoted by circles and gate nodes by rectangles, with node 1 as the top in Fig. 1. A basic node could correspond to a component failure event, for example. It is considered a basic event if there is no intention of analyzing this event in greater depth. Gate nodes correspond to intermediate events, whereas the top node usually corresponds to a very serious system failure event. In Fig. 1, all arcs are regular with the exception of the complementing arc joining nodes 6 and 4, and this arc is distinguished by the symbol "$\sim$."

Associated with each gate is a logic symbol: OR gates have a plus symbol (for set union) while AND gates have a product ($\cdot$) symbol (for set intersection). For example, output event 3 occurs if either input event 4 or 5 (or both) occur. Similarly, output event 5 occurs only if *both* input events 7 and 8 occur. Since the arc connecting gate events 4 and 6 is complemented, gate event 4 occurs only if basic event 11 occurs and gate event 6 does *not* occur.

PURPOSE OF A FAULT TREE ANALYSIS

The fault tree, once constructed, serves as an aid in determining the possible causes of the top event (e.g., a serious accident). When properly used, the fault tree often leads to discovery of failure combinations which otherwise might not have been recognized as causes of the event being analyzed. The fault tree can also be used as a visual tool in communicating and supporting decisions based on the analysis, such as determining the adequacy of a system design. The fault tree provides a convenient and efficient format, helpful for either qualitative or quantitative evaluation of a system, such as determination of the probability of occurrence of the top event. Measures of importance of basic events and combinations of basic events are also used to pinpoint critical failure (accident) scenarios.

Figure 2 is a portion of an industrial-type fault tree. Circles correspond to primary fault events (called PE in Fig. 2). Diamonds correspond to undeveloped fault events (called UE in Fig. 2). Such fault trees are commonly used in nuclear reactor safety analyses.

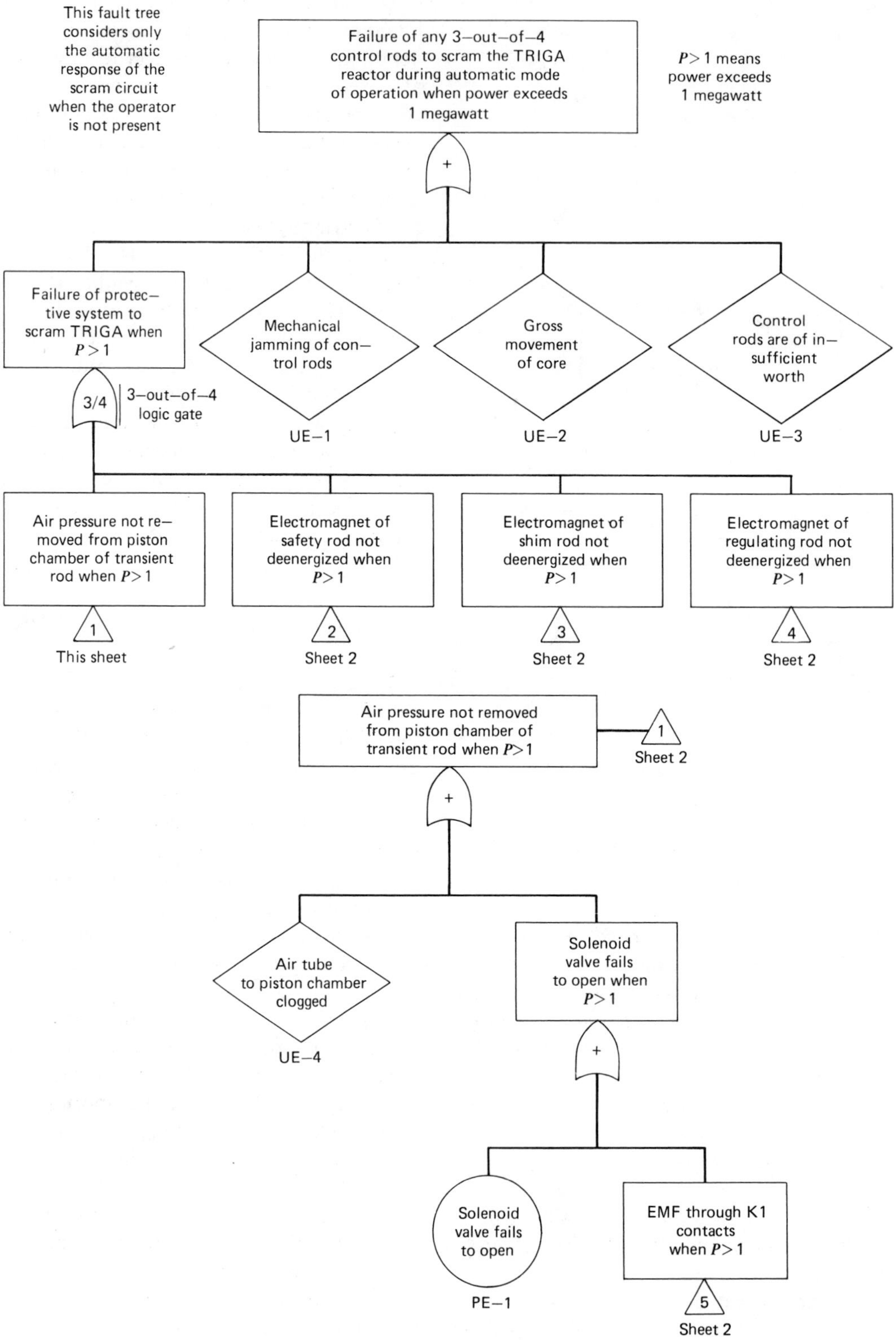

Figure 2 TRIGA nuclear reactor fault tree.

MATHEMATICS OF FAULT TREE ANALYSIS

Boolean switching theory is basic for the mathematics of fault tree analysis. For the fault tree node set $U = [1, 2, \ldots, q]$, let $x_1, x_2, \ldots, x_q$ be Boolean variables assuming values 0 or 1 and let $\mathbf{x} = (x_1, x_2, \ldots, x_q)$. (In Fig. 1, $q = 14$.) For any u in U, let $x_{-u} \equiv 1 - x_u$. The index set for complements is $-U \equiv [-1, -2, \ldots, -q]$ and $(u, -u)$ is a complementary pair of indices.

Expressions may be formed using $x_1, \ldots, x_q, x_{-1}, \ldots, x_{-q}$ and the ordinary Boolean relations of product and sum. An arbitrary nonempty family $\mathscr{I}$ of subsets of $U \cup (-U)$ (not necessarily distinct) is identified with the Boolean sum-of-products expression

$$\sum_{I \in \mathscr{I}} \prod_{i \in I} x_i,$$

where I is a member of the family $\mathscr{I}$. The notation $/\mathscr{I}/\mathbf{x}$ denotes the value of this expression for a given vector $\mathbf{x}$ of 0's and 1's, that is,

$$/\mathscr{I}/\mathbf{x} \equiv \max_{I \in \mathscr{I}} \left(\min_{i \in I} x_i \right) = \sum_{I \in \mathscr{I}} \prod_{i \in I} x_i .$$

Given nonempty families $\mathscr{I}$ and $\mathscr{J}$ of subsets of $U \cup (-U)$, $/\mathscr{I}/ = /\mathscr{J}/$ means that for all $\mathbf{x}$, $/\mathscr{I}/\mathbf{x} = /\mathscr{J}/\mathbf{x}$. It is further assumed that *no set of a family contains a complementary pair*. Whenever a new family is constructed, any set containing complementary pairs is simply eliminated.

A family is said to be *minimal* if all sets are distinct and for any two sets of the family, neither is a subset of the other. For any family $\mathscr{I}$, let $m(\mathscr{I})$ (the "minimization" of $\mathscr{I}$) be the minimal family obtained by eliminating duplicate sets and those which contain another set of $\mathscr{I}$. For instance, $m([\{2,3\}, \{1,2,3\}]) = [\{2,3\}]$. Of course, for any $\mathscr{I}$, $/m(\mathscr{I})/ \equiv /\mathscr{I}/$. The first task of a fault tree analysis is to obtain a certain minimal family of sets of $U \cup (-U)$ called a *prime implicant family*. We are only interested in prime implicant families for fault tree nodes that we wish to analyze, since such families are unique and determine the Boolean expression for the node indicator. For Fig. 1 and node 1,

$$\mathscr{P} = \left[\{9, 10\}, \{12, 14\}, \{13\}, \{11\} \right]$$

is a prime implicant family and

$$x_1 = \sum_{P \in \mathscr{P}} \prod_{i \in P} x_i,$$

where P is a member of the family $\mathscr{P}$ and x_1 is the indicator for the top event in Fig. 1. The first task of a fault tree analysis is to obtain the prime implicant families for fault tree nodes of special interest.

For trees without complemented arcs, the prime implicants are called minimal cut sets. The minimal cut set family for a large fault tree (having, say, more than 100 gate nodes) may consist of millions of sets if the tree has an appreciable number of OR-type gates. Rosenthal [6] has shown that the general problem of finding the complete minimal cut set family associated with a fault tree is a member of the class of NP-complete problems (a class of problems for which it is conjectured that no algorithm exists which will always run on a computer within a polynomial time bound). Hence we cannot expect to devise an algorithm whose running time is bounded for all fault trees by a polynomial in, say, the number of fault tree nodes. The serious analyst should probably not rely on the same method for every fault tree.

ALGORITHMS FOR FINDING MINIMAL CUT SETS

A simple but powerful downward-type algorithm for finding minimal cut sets for fault trees *without* complemented arcs was first announced by Fussell and Vesely [3]. We will illustrate this algorithm for the fault tree of Fig. 1 even though it does contain a complemented arc. We start with node 1 (the top of the tree) and list all input events. Since node 1 corresponds to an OR gate, input events are listed in separate rows, as

6
2
5.

Note that the occurrence of events in any row will cause the top event to occur. Since node 6 corresponds to an AND gate, it is replaced by its input events listed in the same row, so that now we have

9, 10
2
5.

Since 9 and 10 are basic events, they are developed no further. Since 2 is an AND gate, it is replaced by its input events listed in the same row. We now have

9, 10
3, 4
5.

Since gate 3 is an OR gate, the second row is replaced by two rows, so that we have

9, 10
4, 4
5, 4
5.

Eliminating node repetition and supersets, we have

9, 10
4
5.

Supersets may be eliminated either as we go or at the final stage. Continuing in this way, we eventually obtain the following list:

9, 10
12, 14
13
−9, 11
− 10, 11 .

Although this is an implicant family, it is *not* the prime implicant family for this particular tree. If there had been no complemented arcs, the algorithm would have produced the minimal cut sets. The prime implicant family for this tree was given above.

Another algorithm due to Nelson [5] and implemented by Worrell [9] and Willie [8] can be used to find prime implicants for trees with complemented arcs. This algorithm is based on duality theory.

PROBABILITY CALCULATION

The probability of the occurrence of any node fault event can be calculated from the family of prime implicants using the inclusion–exclusion principle*. Another method is to obtain a logically disjoint family $\hat{\mathscr{P}}$ which is equivalent to the family of prime implicants $\mathscr{P}$, meaning that

$$\max_{\hat{P} \in \hat{\mathscr{P}}} \left(\min_{e \in \hat{P}} x_e \right) = \max_{P \in \mathscr{P}} \left(\min_{e \in P} x_e \right).$$

A family $\hat{\mathscr{P}}$ is said to consist of *logically disjoint sets* if for any $\hat{P}_i, \hat{P}_j \in \hat{\mathscr{P}}, \hat{P}_i \cup \hat{P}_j$ contains a complementary pair of elements, that is, a pair $(u, -u)$ for $u \in U$. If $\hat{\mathscr{P}}$ consists of logically disjoint sets, then the events

$$A_i = \left\{ \prod_{e \in \hat{P}_i} x_e = 1 \right\},$$

$1 \leqslant i \leqslant n$, are mutually exclusive and

$$\Pr\left\{ \bigcup_{i=1}^{n} A_i \right\} = \sum_{\hat{P} \in \hat{\mathscr{P}}} \prod_{e \in \hat{P}} p_e ,$$

where p_e is the probability that basic event e occurs. We have assumed that basic events are statistically independent. A method for obtaining $\hat{\mathscr{P}}$ from $\mathscr{P}$ is described in Abraham [1].

The following example illustrates the inclusion–exclusion method* for calculating the probability of the top event. Consider the fault tree in Fig. 1. Then $P_1 = \{9, 10\}$, $P_2 = \{12, 14\}$, $P_3 = \{13\}$, and $P_4 = \{11\}$ are the prime implicants for this tree. Hence

$$\begin{aligned}
&\Pr\{\text{top event occurs}\} \\
&= \Pr\left\{ \bigcup_{i=1}^{4} P_i \right\} \\
&= \sum_{i=1}^{4} \Pr\{P_i\} - \sum_{i<j} \Pr\{P_i \cap P_j\} \\
&\quad + \sum_{i<j<k} \Pr\{P_i \cap P_j \cap P_k\} \\
&\quad - \Pr\{P_1 \cap P_2 \cap P_3 \cap P_4\}.
\end{aligned}$$

If we let each basic event have probability

10^{-3}, we obtain

$$\begin{aligned}\Pr\{\text{top event occurs}\} &= 2\times 10^{-3} + 10^{-1} - 4\times 10^{-9} + 10^{-12}\\ &\quad + 2\times 10^{-15} - 10^{-18} \doteq 2\times 10^{-3}.\end{aligned}$$

If we let p be the number of prime implicants, then the number of terms in our inclusion–exclusion expression is $2^p - 1$. If $p = 20$, we have $2^{20} - 1 \doteq 10^6$ terms!

RELATIONSHIP TO COHERENT STRUCTURES AND NETWORK RELIABILITY

Coherent structures, block diagrams, and two-terminal networks are roughly equivalent system function descriptions, commonly used in reliability theory* (*see* COHERENT STRUCTURES). Each can be translated into an equivalent logic tree description of system events. This is not usually done because of the greater complexity of the equivalent logic tree. The network graph structure can be more effectively used to obtain minimal cuts and system reliability. Recent graph theoretic techniques for calculating network reliability* are described in Satyanarayana and Prabhakar [7].

General logic trees cannot in general be represented as two-terminal networks because of the possibility of complemented arcs. Logic trees are in general more abstract representations of system operation. For both logic trees and networks much research is concerned with developing more efficient computational techniques (*see* NETWORK RELIABILITY).

References

[1] Abraham, J. A. (1979). *IEEE Trans. Rel.*, **R-28**, 58–61. (Presents a method for finding a logically disjoint family of sets from the family of prime implicants. Useful for fault tree probability calculations.)

[2] Barlow, R. E., Fussell, J. B., and Singpurwalla, N. D., eds. (1975). *Reliability and Fault Tree Analysis*. SIAM, Philadelphia. (An edited conference volume containing many basic papers on fault tree analysis.)

[3] Fussell, J. B. and Vesely, W. E. (1972). *Amer. Nucl. Soc. Trans.*, **15**, 262–263.

[4] Haasl, D. (1965). *Advanced Concepts in Fault Tree Analysis*. Systems Safety Symposium, Seattle, Wash., June 8–9. (Engineering oriented.)

[5] Nelson, R. J. (1955). *J. Symb. Logic*, **20**, 105–108.

[6] Rosenthal, A. (1975). In *Reliability and Fault Tree Analysis*, R. E. Barlow, J. B. Fussell, and N. D. Singpurwalla, eds. SIAM, Philadelphia, pp. 133–152. (Discusses the computer complexity of fault tree analysis problems.)

[7] Satyanarayana, A. and Prabhakar A. (1978). *IEEE Trans. Rel.*, **R-27**, 82–100. (Develops a new approach for computing the reliability of two-terminal networks.)

[8] Willie, R. R. (1979). Computer-Oriented Methods for Assessing Reliability of Complex Systems. Ph.D. dissertation, University of California, Berkeley. (A rigorous Boolean approach to fault tree analysis together with algorithms and a computer program.)

[9] Worrell, R. B. (1975). In *Reliability and Fault Tree Analysis*, R. E. Barlow, J. B. Fussell and N. D. Singpurwalla, eds. SIAM, Philadelphia, pp. 165–185. (Presents a Boolean approach to fault tree computations.)

(NETWORK RELIABILITY)

RICHARD E. BARLOW

FDA STATISTICAL PROGRAMS: AN OVERVIEW

At the beginning of the twentieth century [9, 13], the Congress of the United States began enacting legislation aimed at protecting the public health against ineffective and/or harmful products. With each new law, the mission of the U.S. Food and Drug Administration (FDA) became better focused. Today, FDA is an important regulatory agency. Many of the FDA's regulatory decisions depend on the quality of the statistical evidence presented to FDA. Consequently, a variety of statistical programs have emerged which enable FDA to carry out its assigned mission. This article will provide an overview of the FDA's statistical activities in which FDA is engaged. The size of the statistical staff within the various FDA bureaus ranges from one individual in the Bureau of Biologics to over two dozen in the Bureau of Drugs and the Bureau of Foods. Consequently, the level of statistical activity and research varies considerably with the administrative units. For those units where the level of statistical activity is high, examples are presented and a bibliography provided.

In 1906, Congress passed the original Food and Drug Act and a companion bill, the Meat Inspection Act. These laws prohibited the transfer of misbranded and adulterated foods, drinks, and drugs from interstate commerce. This program was administered by Harvey W. Wiley, Chief of the Division of Chemistry, U.S. Department of Agriculture. In 1938, a drug (elixir of sulfanilamide) was associated with the death of 107 persons. This resulted in a new Federal Food, Drug, and Cosmetic Act. The FD&C Act, as it was now called, required the manufacturer of new drugs to show that their products were safe prior to marketing, and required the manufacturer to specify the standards of identity and quality for foods. The FD&C Act allowed FDA to begin the practice of factory inspections and to extend its authority to cover cosmetics and devices. In 1962, Frances O. Kelsey, an FDA medical officer, helped keep the drug Thalidomide off the U.S. market. During the early 1960s, this drug was associated with numerous malformed babies in Western Europe. The Thalidomide tragedy prompted the U.S. Congress to pass the Kefauver–Harris Drug Amendment, which, for the first time, required drug manufacturers to prove to FDA that their products were effective prior to marketing. During the 1970s FDA's role expanded to include regulation of medical devices, diagnostic and laboratory products, radiation from electronic products, and serum and vaccines. FDA also collaborates with other government agencies [notably the Environmental Protection Agency (EPA)] in the area of toxicologic research.

FDA's statistical programs emphasize the protection of human subjects; the design and analysis of studies of controlled experimental studies; epidemiologic* studies, surveys, quality control* techniques; and assessment of animal and laboratory studies. Many areas of statistical theory and applications arise within FDA, and the agency statisticians frequently collaborate with scientists from other disciplines. A heavy emphasis is placed on the effectiveness of experimental design (concepts like statistical power* are important), the quality of statistical evidence (e.g., the potential for nonsampling errors* or bias), and the extent and validity of inference possible. The various FDA statistical programs will be summarized by the organizational unit in which they arise.

The primary responsibility of the statistical staff in the *Bureau of Drugs* is to work with the physicians, chemists, and pharmacologists as a review team member in the evaluation of industry-submitted new drug applications, to design and analyze surveys on the quality of marketed drugs, to participate in postmarketing surveillance programs, and to participate in pharmaceutical research (*see* FDA STATISTICAL PROGRAMS: HUMAN DRUGS).

The *Bureau of Veterinary Medicine* (BVM) is responsible for the safety and efficacy of animal drugs, feed additives, and animal medical devices. Primary among the Bureau's duties is the responsibility to assure that the public is not exposed to harmful residues in edible animal products as a result of drugs given to food-producing animals. This responsibility is carried out in conjunction with the FDA Bureau of Foods and the U.S. Department of Agriculture. The BVM statistical staff are involved in the review and evaluation of protocols and results of studies dealing with the safety and efficacy of marketed veterinary products. Like the Bureau of Drugs, the BVM statisticians are involved in experimental design, and in the appropriate use of statistical methods to reach valid conclusions. Of special interest is the BVM's work on the statistical aspects of low-level exposure to subacute toxicants, such as carcinogens.

The *Bureau of Medical Devices* (BMD) regulates a variety of manufactured products intended for use in medical diagnosis and therapy, including items such as cardiac pacemakers, intraocular lenses, pregnancy test kits, and blood analyzers. The BMD assures their safety and efficacy through the premarket approval, performance standards, and labeling requirements. Each of these requires studies for the development of scientific evidence. The BMD statistics staff have the general responsibility of assuring that evidence presented to or developed by the Bureau is adequate and statistically valid. Premarket approval applications are reviewed and may incorporate in vitro, animal, and clinical studies. Standards for devices incorporate carefully designed test procedures and sampling plans. Diagnostic devices require statistical models to describe their performance characteristics, such as accuracy, precision, linearity, and detection limit. Appropriate statistical techniques are necessary for determining reference values. The statistics staff also design laboratory experiments and protocols and analyze data for testing device performance.

The statistical activities of the *Bureau of Radiological Health* (BRH) are varied and wide ranging. BRH has as its functions the development of criteria and recommendation of standards for safe limits of radiation exposure, the development of methods and techniques for controlling radiation exposure, planning and conduct of research to determine health effects of radiation exposure, and the conduct of studies to ascertain the needs of educational programs to improve the practice of radiation users. BRH uses data generated by surveys, experimental, and epidemiologic* research.

The major statistical work in BRH is in epidemiologic and animal research on the effects of radiation on biologic systems. Mathematical modeling is used in the determinations of dose–response functions and risk assessment of long-term effects. The BRH statisticians participate in the determination of use characteristics of medical radiation procedures by initiating national surveys. The statistical problems associated with engineering research and testing are some of the Bureau's more interesting ones. A high level of expertise is needed to deal with issues of underlying probability distributions of dose, development of extreme value distributions*, the estimation of the potential health implications, and the evaluation of data from experimental and epidemiologic studies.

Two examples of the statistical research undertaken by the BRH are "A Risk/Benefit Analysis by Life Table Modeling of an Annual Breast Cancer Screening Program which Includes X-Ray Mammography" by Chiacchierini et al. [2] and "A Two-Sample Test for Independence in 2 × 2 Contingency Tables with Both Margins Subject to Misclassification" by Chiacchierini and Arnold [1]. The first paper presents a life table model to display and numerically evaluate risk and benefits by comparing breast cancer mortality experience in a hypothetical population of 100,000 women age 35 which differ only by age at first screening. The second paper considers a double sampling scheme for the estimations and testing of 2 × 2 contingency tables* under misclassification in both margins. Maximum likelihood estimates are given for the error-free and misclassification probabilities.

The *National Center for Toxicologic Research* (NCTR) has as its mission the study of the toxic effects of long-term, low-dose exposure to toxic substances, and the study of the biologic mechanisms of toxicity in order to develop more effective protocols to study safety and to improve prediction of human disease from laboratory animal experiments. These objectives are accomplished by experiments aimed at carcinogenesis, teratogenesis, and/or mutagenesis in a variety of laboratory animal species.

The statistics staff of NCTR participates in the design, analysis, and interpretation of these toxicity studies. They utilize a variety of statistical methods and develop new techniques directed at the analysis of long-term toxicologic studies, the adjustment for competing risks, and the estimation of tumor, teratogenic, or mutagenic incidence rates. They often use special statistical procedures required for ratios of discrete random variables (e.g., number of abnormal fetuses divided by the number of viable fetuses per litter) and techniques for estimating risk at low exposure levels from higher observed experimental dose levels.

Research conducted by the staff of NCTR includes the estimation of time-response models such as those presented by Farmer et al. [11] in a paper titled "Dose and Time Response Models for the Incidence of Bladder and Liver Neoplasms in Mice Fed 2-Acetylaminofluorene Continuously" and presented by Kodell et al. [14] in a paper titled "Estimation of Distributions of Time to Appearance of Tumor and Time to Death from Tumor After Appearance in Mice Fed 2-Acetylaminofluorene." Kodell et al. [15] developed a nonparametric joint estimation procedure for estimating disease response and survival functions in experiments where animals are sacrificed [3]. Finally, Gaylor and Kodell [12] developed a method for linear interpolation of low-dose risk assessment of toxic substances.

The *Bureau of Biologics* (BOB) is concerned with toxoids, vaccines, antigens, immune sera, allergenic products, and blood and blood products used to treat, prevent, or diagnose disease. The statisticians in the BOB participate in evaluating studies designed to establish the potency and safety of biologic products; they set standards and develop methods of testing. BOB issues licenses of suitable products, and maintains an updated computerized inventory of each product. The staff of BOB review the quality and integrity of information submitted, particularly clinical, laboratory, and production data. BOB develops methodology, conducts tests of safety, efficacy, and potency of vaccines, and surveys possible long-term effects of immunization against viral and rickettsial disease.

The Division of Mathematics in the *Bureau of Foods* is concerned with mathematical and statistical models in the field of sanitation, health, and economics, and in the design and analysis of studies related to food standards, food, and color additives. The *Office of Health Affairs* in BRH is responsible for the review of methodologies used in the collection of FDA's consumer health statistics (obtained through contracts with private firms, universities, or interagency agreements.) The primary focus of this FDA unit is the methodologic review of consumer surveys and the coordination of survey plans with the Office of Management and Budget (OMB). (OMB has the unique role of limiting the response burden on the public from overlapping surveys.) The surveys conducted by OHA/BRH deal with all agency products (including foods, drugs, biologics, etc.) where a rapid turnaround is required. Several references [3–8, 10] to specific surveys appear in the bibliography. The survey mechanism is a standing contract which provides for alternative sampling designs to meet varying survey needs. The consumer used in this context is broadly defined and includes the general population and speciality groups (e.g., physicians, dentists, and related health professionals). The surveys are used to provide FDA with reliable information concerning consumers' experiences with and attitudes toward FDA-regulated products.

Finally, the *Office of Scientific Liaison Staff* in the Office of the Commissioner of Food and Drugs has scientists with varying expertise, including biostatistics. The statisticians supporting the Commissioner are involved in developing procedures useful in the assessment of human health risks, especially of carcinogenicity and teratogenicity; statistical models and methods appropriate to these investigations are employed.

References

[1] Chiacchierini, R. P., and Arnold, J. C. (1977). *J. Amer. Statist. Ass.*, **72**, 170–174.

[2] Chiaccierini, R. P., Lundin, F. E., and Scheidt, P. C. (1980). In *Prevention and Detection of Cancer*, Part II: *Detection*, Vol. 2: *Cancer Detection in Specific Sites*, H. E. Nieburgs, ed., 1741–1762. Dekker, New York.

[3] DHEW, FDA (1965). *Consumer Attitudes toward Over-the-Counter Drug Labels*, Feb.

[4] DHEW, FDA (1975). *An Investigation of Consumers' Perceptions of Adverse Reaction to Cosmetic Products*, June.

[5] DHEW, FDA (1975). *Survey of Consumers' Perceptions of Patient Package Inserts for Oral Contraceptives*, Sept.

[6] DHEW, FDA (1976). *Influenza Consumer Survey*, Feb.

[7] DHEW, FDA (1978). *An Evaluation of the Awareness and Reactions to Labeling Changes among Users of In Vitro Diagnostic Products*, Sept.

[8] DHEW, FDA (1978). *Comprehension of Selected Labeling Information on Foods, Drugs, and Medical Devices: A Consumer Survey*, Nov.

[9] DHEW, FDA (1979). *Milestones in U.S. Food and Drug Law History*, FDA Consumer Memo No. 79-1063.

[10] DHEW, FDA and CDC (1978). *A National Survey of the Use of Protein Products in Conjunction with Weight Reduction Diets among American Women.*

[11] Farmer, J., Kodell, R. L., and Greenman, D. L. (1979). *J. Environ. Pathol. Toxicol.*, **3**, 55–68.

[12] Gaylor, D. W. and Kodell, R. L. (1980). *J. Environ. Pathol. Toxicol.*, **4**, 305–312.

[13] Janssen, W. F., FDA historian (1978). From the article "The Food and Drug Law of the United States," *Ind. Santé*, Paris; *HEW Publ. (FDA) 79-1054.*

[14] Kodell, R. L., Farmer, J. H., and Greenman, D. L. (1979). *J. Environ. Path. Toxicol.* **3**, 89–192.

[15] Kodell, R. L., Shaw, G. W., and Johnson, A. M. (1982). *Biometrics*, **38**, 43–58.

(FDA STATISTICAL PROGRAMS:
HUMAN DRUGS
FEDERAL STATISTICS)

Acknowledgements

The following individuals contributed information related to their specific bureaus: V. M. Chinchilli, Scientific Liaison Staff; E. W. Gordon, Bureau of Radiologic Health; H. Lee, Bureau of Medical Devices; J. J. Colaianne, Bureau of Veterinary Medicine; S. C. Rastogi, Bureau of Biologics; R. P. Chiacchierini, Bureau of Radiological Health; D. W. Gaylor, National Center for Toxicologic Research.

CHARLES ANELLO

FDA STATISTICAL PROGRAMS: HUMAN DRUGS[1]

OFFICIAL STATISTICAL RESPONSIBILITIES

In this article statistical activities pertaining to human drugs in the Bureau of Drugs of the Food and Drug Administration (FDA) are described with emphasis on official responsibilities, the role of statistics, guidelines, and selected significant problems and research.

The Statistical Program Pertaining to Human Drugs is located in the Division of Biometrics, Office of Biometrics and Epidemiology, Bureau of Drugs, FDA. The Bureau of Drugs develops FDA policy with regard to the safety, effectiveness, and labeling of all drugs for human use. It reviews and evaluates new drug applications (NDAs) and notices of claimed investigational exemption for new drugs (INDs) for human use; develops and implements standards for the safety and effectiveness of all over-the-counter (OTC) drugs; monitors the quality of marketed drugs through product testing, surveillance, and compliance programs; develops and promulgates guidelines on current good manufacturing practices for use by the drug industry; develops and disseminates information and educational material dealing with drugs to the medical community and the public; conducts research and develops scientific standards on the composition, quality, safety and efficacy* of human drugs; collects and evaluates information on the effects and use trend of marketed drugs; monitors prescription drug advertising and promotional labeling to assure their accuracy and integrity; analyzes data on accidental poisonings and disseminates toxicity and treatment information on household products and medicines; evaluates applications for operation of methadone treatment centers and other activities using methadone or other drugs; and directs the FDA antibiotic and insulin certification program.

These activities routinely produce data from which valid inferences are drawn for regulatory decisions. Data are obtained from clinical trials*, laboratory research and experiments, surveys, and epidemiological* studies. Consequently, the Division of Biometrics has responsibility for initiating, planning, developing, and implementing the statistical program in the Bureau of Drugs. It provides comprehensive statistical computational and biomathematical consulting services, which include, on a routine basis:

1. The reviews and evaluations of the statistical adequacy of experimental designs for IND or NDA submissions, medical research reports and evidence submitted in support of drug safety and efficacy claims, critiques of statistical analyses conducted in studies related to clinical trials, medical advertising, OTC drugs, drug surveillance, and quality control*
2. Collaboration on the statistical design and analysis of morbidity and mortality data associated with the use of drugs
3. Evaluation of statistical methods applied in epidemiological studies; application of statistical methods in support of pharmaceutical research and testing
4. Statistical support in the evaluation of experimental design and bioavailability data
5. Development of sampling plans for field inspection programs and statistical methods for evaluating manufacturing quality control procedures for marketed drugs, development of appropriate statistical methods for conducting, monitoring, and evaluating intramural and extramural research contract projects; evaluation of the quality of data and conclusions presented by interested parties
6. Statistical research on methods and applications

Computational activities include development of scientific computer programs, validation or modification of current statistical methodology programs, and preparation of data for executing existing computer programs on a routine basis.

ROLE OF STATISTICS

The 1962 Amendments to the Food, Drug, and Cosmetic Act [6] require that no application to market a new drug shall be approved unless adequate tests show that the drug is safe for use under the conditions presented in the proposed labeling, and substantial evidence from adequate and well-controlled studies demonstrates that the drug is effective for such uses.

Certain essential principles concerning "adequate and well-controlled clinical investigations" appear in the May 8, 1970, Federal Register Statement [1, Sec. 314.111(a)(5)(ii)]. These principles are also summarized in the 1977 HEW publication [7]. They include a clear statement of the objective(s) of the study, patient selection criteria, randomization* procedure, suitable size of a clinical study, comparability of the patient population studied, concurrent comparison groups, nature of the blinding, methods of observations and quantification, suitable length of the study, etc. Adequate documentation of the design, quality, and accuracy of the clinical data; appropriate use of statistical models; analyses (references of publications used in statistical work); and interpretation of results is required. These principles define the role of statisticians in applying sound statistical theory and methodology to ensure the adequacy of well-controlled clinical investigations. Specifically, statisticians are consulted in the planning, design, execution, and analysis of clinical investigations and clinical pharmacology* in order to ensure the validity of estimates of safety and efficacy obtained from these results. They are consulted in drawing valid inferences about drug responses in well-defined target populations.

A statistician in the Bureau of Drugs reviews and evaluates a study protocol to determine whether the proposed study design and statistical procedures:

1. Satisfy the requirements of an adequate and well-controlled clinical investigation
2. Are efficient for the stated objectives without exposing human subjects to unnecessary risks
3. Adequately plan for obtaining valid and accurate data and for monitoring and reporting the study

He or she is a member of the drug review team and is responsible for reviewing and evaluating all submissions that use statistical analyses to make inferences from data. The development of a meaningful review necessitates effective interaction between medical and statistical reviewers and/or other scientists throughout the review process [10, 11]. Outside experts are used selectively to perform statistical reviews of IND/NDA submissions, and of published and unpublished medical reports. Prominent biostatisticians are used as voting members or consultants to the various scientific advisory committees to advise on crucial statistical issues. See refs. 1, 3, 4, and 7–11.

STATISTICAL GUIDELINES

One general statistical guideline has been developed to evaluate the clinical protocols and completed clinical studies done by drug sponsors. Another has been proposed for the use of drug sponsors engaged in the activities of protocol development and NDA submissions. The IND/NDA Statistical Evaluation Guidelines became the official document of the Bureau of Drugs on February 23, 1979 [10], for statistical review and evaluation of protocols, investigational new drugs (INDs), new drug applications (NDAs) including drug efficacy study implementation (DESI), supplemental NDAs, and similar submissions. It is available to outsiders when requested under the Freedom of Information Act. On July 8, 1980, the FDA made available the "General Statistical Documentation Guide for Protocol Development and the NDA Submissions" [8], which sets forth the type of material needed to permit statistical review of protocols and completed clinical studies by the agency.

SELECTED CLINICAL STATISTICAL PROBLEMS

Combination drugs is a fruitful area for innovative statistical research. A combination drug consists of two or more chemical ingredients in a fixed quantity which are combined (or mixed) in one dosage form and administered together in a single tablet, etc. For example, for the treatment of muscle spasm and pain caused by strain or sprain or due to a chronic condition, a muscle-relaxing agent may be combined with a pain-reducing agent.

Two or more drugs are combined to achieve:

1. An increase in efficacy without increasing adverse effects
2. A reduction in adverse effects without any reduction in efficacy
3. An increase in efficacy and decrease in adverse effects simultaneously
4. A decrease in efficacy and decrease in adverse effects
5. An increase in efficacy and an acceptable increase in adverse effects in order to maximize overall benefits to patients (a muscle relaxant plus an analgesic results in increase in efficacy and increase in adverse effects)

Consider a combination drug with two single ingredients, the first intended for fast onset of drug effect and the second for long duration of action. An example would be a topical anesthetic intended for dental use where quick onset and lasting anesthetic effect is desirable.

Sometimes, a third ingredient is developed to provide additionally an increasing percentage of patients experiencing rapid onset of action. Thus combination drug problems offer a challenging area for multivariate* statistical research. *Multiple comparisons** and a multiplicity of statistical hypotheses to be tested is not uncommon in clinical research.

In bioavailability studies, there are four basic variables of interest:

1. Area under the plasma-level curve (AUC)
2. Time to achieve first peak
3. First peak plasma level
4. Maximum peak plasma level

Initially, appropriate univariate analyses of variance are carried out on each of these four variables. Since these variables are generally correlated, a correlation matrix for them is computed; if any of them is not well correlated with the remaining three variables while they themselves are found to be highly correlated, a multivariate analysis of variance is performed utilizing only highly correlated variables. Additionally, in order to determine whether a test agent has any statistically significant effect on certain physiological variables of interest, each such variable is analyzed in the multivariate analysis-of-covariance model. Measurements are taken at the baseline before and after a test agent is administered to selected subjects. Thus the values obtained at n specified time points for each physiological variable are used as the response variables in a **Y** matrix; the baseline value is used as a covariate in a multivariate analysis of covariance model, which is formulated as

$$\mathbf{Y} = \boldsymbol{\mu} + X\mathbf{1} + \mathbf{O}_i + \mathbf{D}_j + \mathbf{S}_{(ij)k} + \mathbf{P}_l + \mathbf{T}_m + \mathbf{DT}_{jm} + \mathbf{PT}_{lm} + \boldsymbol{\epsilon},$$

where

$\mathbf{Y}$ is an $n \times 1$ vector of response variables (measurements taken at n specified time points) for a specified physiological variable of interest (e.g., systolic blood pressure) for a subject assigned to a specific test agent and period,

$\boldsymbol{\mu}$ is an $n \times 1$ vector of n average effects associated with n specified time points,

X is a baseline measurement (scalar), for a specified physiological variable of interest, taken prior to the administration of a specific test agent for a specific period providing $X\mathbf{1}$ as an $n \times 1$ vector with $\mathbf{1}$ being a vector of 1's,

$\mathbf{O}_i$ is the effect due to *order* (sequence) in which the test agents are administered at assigned periods,

$\mathbf{D}_j$ is the effect due to the *day* on which the selected subject starts the study,

$\mathbf{S}_{(ij)k}$ is the effect due to *subjects* nested within order (i) and day (j),

$\mathbf{P}_l$ is the effect due to the *period* of the crossover design,

$\mathbf{T}_m$ is the effect due to the *test agent* selected for the study,

$\mathbf{DT}_{jm}$ is the interaction effect between the day and test agent,

$\mathbf{PT}_{lm}$ is the interaction effect due to period and test agent,

$\boldsymbol{\epsilon}$ is an $n \times 1$ error vector associated with the model, defined above.

The number of selected test agents (m) will generally determine the number of periods (l) and the number of sequences or order (i). The number of days (j) and the number of subjects nested (k) within order (i) and day (j) will be generally determined on practical grounds, keeping in mind that each day a sequence of test agents is repeated at least once. Subjects selected for the study are randomized with respect to day and sequence.

Since healthy human beings are used in *bioavailability* or *bioequivalence studies*, sources and magnitudes of observed variability are generally not as much as for severely sick patients used in clinical studies. Therefore, multiresponse multiperiod crossover design and multivariate covariance analysis often respond well. In many instances, a multiple or sequential range test is applied to make clinically meaningful pairwise comparisons.

In a multiclinic trial different clinics serve different patient subpopulations and use slightly different techniques, despite a common protocol. Analyses based on data combined across investigators should be accompanied by summaries for each investigator and a demonstration of the appropriateness of combining the results.

It is unethical to subject people to risk when a study is too small and cannot be expected to demonstrate the effect desired, or to subject more people to risk than is necessary after the effect has been clearly demonstrated (*see* CLINICAL TRIALS). Clinical investigations are expensive in terms of dollars, time, the commitment of patients and physicians to the study, and physical facilities. During the planning phase the statistician consults with the clinical investigator to determine the size of the difference that is important to detect, the significance level and power of the statistical test, the duration of the study, the acceptable dropout rates, the multiple end points, and the minimum number of patients necessary for the study.

The interpretation of a study's findings require effective consultation with a clinical investigator. A statistically significant finding may have nothing to do with the drug of interest. It could be due to chance, to nonsampling error, to bias, to an indirect effect not specified, etc. Even if the observed drug effect could be attributed to a particular drug, we need to ask: Do the benefits outweigh the risks? Is the observed difference clinically meaningful?

SOME NONCLINICAL STATISTICAL PROBLEMS

FDA has published several regulations (21 CFR 211) which relate to statistical principles and methodology under the title "Current Good Manufacturing Practices for Finished Pharmaceuticals" [1]. The *United States Pharmacopeia* (USP XX) recommends a method for conducting a content uniformity test [15]. The test is conducted in two stages, the first involving 10 and the second 20 tablets (the second stage may not be necessary). Comer, Sampson, and their colleagues have published papers on the statistical aspects of content uniformity test [2]. They discuss the advantages and disadvantages of attribute measures for setting quality standards. They develop mathematical models for determining the operating characteristics of the USP Content Uniformity Test and suggest how other available information can be used to suggest statistically more powerful procedures.

The pyrogen test (see ref. 15), sampling and testing of in-process materials and drug products (Sec. 211.110 of ref. 1), and stability testing (Sec. 211.166 of ref. 1) involve statistical concepts and procedures. Interest in problems in the drug products quality area has steadily increased, while the demand for statistical services in epidemiological projects, such as postmarketing drug surveillance studies, has grown rapidly.

APPLIED STATISTICAL RESEARCH

The Division of Biometrics encounters unusual statistical problems which cannot be readily handled by available statistical approaches. These problems provide an opportunity for staff to conduct applied statistical research and publish it in scientific journals. Selected publications by statisticians of the Division of Biometrics are listed in the references.

O'Neill and Chen [13] developed a statistical model that characterizes the onset and termination of treatment response of a subject over a fixed time period jointly. The model was applied to compare the effect of two bronchodilator drugs in subjects with reversible chronic obstructive lung disease.

Fairweather [5] published a statistical procedure for determining the risk of rejecting a satisfactory product when dissolution tests are substantiated for classic blood level studies. One purpose for this kind of investigation is to determine if currently used in vivo methods could be safely replaced by in vitro methods.

O'Neill and Anello [12] proposed a sequential approach to case-control studies. The case-control method has emerged in recent years as an important epidemiologic tool for studying the potential adverse effects of marketed drugs.

Schuirmann analyzed a collaborative study of the U.S.P. Prednisone Dissolution Calibrator in ref. 14. The study was undertaken to determine the contributions of several factors to the variability of dissolution values and to determine statistically a range of values within which dissolution results may be expected to fall when obtained under certain conditions.

NOTES

1. This work has been produced by Satya D. Dubey, the author, in the capacity of a U.S. federal government employee, as part of his official duty, is in the public domain, and is not subject to copyright.
2. In 1982, the Bureau of Drugs and the Bureau of Biologics of the FDA were merged, and became the *National Center for Drugs and Biologics*.

References

[1] *Code of Federal Regulations, Title 21, Food and Drugs,* Parts 1–499, rev. Apr. 1980. Part 211, pp. 72–89: Sec. 211.110, p. 80; Sec. 211.166, pp. 83–84; and Sec. 314.111(a)(5)(ii), pp. 106–108 (abbrev. as 21 CFR 314.111). U.S. Government Printing Office, Washington, D.C.

[2] Comer, J. P., et al. (1970). *J. Pharmacol. Sci.*, **59**, 210–214.

[3] Daniel, C., Tutle, E. R., and Kadane, J. A., Federal Statistics (1971). *Statistics and Data Analysis in the Food and Drug Administration*. Report of the President's Commission, Vol. 2, pp. 67–95.

[4] Dubey, S. (1981). *Statistics in the Pharmaceutical Industry*, C. Buncher and J. Tsay, eds. Marcel Dekker, New York, Chap. 5.

[5] Fairweather, W. R. (1978). *J. Pharmacokinet. Biopharm.*, **5**(4), 405–418.

[6] Federal Food, Drug, and Cosmetic Act as amended (1962), Sec. 505(d). U.S. Government Printing Office, Washington, D.C.

[7] *General Considerations for the Clinical Evaluation of Drugs* (1977). DHEW Publ. No. (FDA) 77-3040, U.S. Government Printing Office, Washington, D.C., Sept., pp. 2–4.

[8] *General Statistical Documentation Guide for Protocol Development and NDA Submissions* (Draft) (1980). Guideline Leader: S. D. Dubey, Chief, Statistical Evaluation Branch, Division of Biometrics. Statistical Evaluation Branch (HFD-232), Bureau of Drugs, Food and Drug Administration, Department of Health and Human Services, Apr.

[9] *Guidelines for the Clinical Evaluation of Anti-inflammatory Drugs (Adults and Children)* (1977). DHEW Publ. No. (FDA) 78-3054, U.S. Government Printing Office, Washington, D.C., Sept., pp. 24–25.

[10] *IND/NDA Statistical Evaluation Guidelines* (Staff Manual Guide BD 4500.1) (1979). Originator: S. D. Dubey. Food and Drug Administration, Bureau of Drugs, Feb. 23.

[11] O'Fallon, J. R., Dubey, S. D., Salsburg, D. S., Edmonson, J. H., Soffer, A., and Colton, T. (1978). *Biometrics*, **34**, 687–695.

[12] O'Neill, R. T. and Anello, C. (1978). *Amer. Epidemiol.*, **108**(5), 415–424.

[13] O'Neill, R. T. and Chen, C. W. (1978). *Biometrics*, **34**, 411–420.

[14] Schuirmann, D. J. (1980). *Pharmacopeial Forum*, **6**(1), 75–89.

[15] *The United States Pharmacopeia* (20th rev.) (1980). Published by the 1979 United States Pharmacopeial Convention, Inc., 12601 Twinbrook Parkway, Rockville, MD 20852; distributed by Mack Publishing Company, Easton, PA 18042. Pp. 902–903.

(BIOSTATISTICS
CLINICAL TRIALS
FDA STATISTICAL PROGRAMS:
AN OVERVIEW
FEDERAL STATISTICS)

SATYA D. DUBEY

F-DISTRIBUTION

The F-distribution is the distribution of the ratio of two independent "mean χ^2's." A common notation is

$$F_{\nu_1,\nu_2} = \left(\chi^2_{\nu_1}/\nu_1\right)\left(\chi^2_{\nu_2}/\nu_2\right)^{-1}.$$

The probability density function (PDF) of F_{ν_1,ν_2} is

$$\nu_1^{(1/2)\nu_1}\nu_2^{(1/2)\nu_2}\left\{B\left(\tfrac{1}{2}\nu_1,\tfrac{1}{2}\nu_2\right)\right\}^{-1} f^{(1/2)\nu_1-1} \times(\nu_2+\nu_1 f)^{-(1/2)(\nu_1+\nu_2)} \qquad (0<f).$$

The cumulative distribution function (CDF)

is

$$\Pr[F_{\nu_1,\nu_2} \leqslant f_0] = I_{\nu_1 f_0/(\nu_2+\nu_1 f_0)}(\tfrac{1}{2}\nu_1, \tfrac{1}{2}\nu_2),$$

where $I(\cdot, \cdot)$ is the incomplete beta function ratio.

The rth moment about zero of F_{ν_1,ν_2} (for $\nu_2 > 2r$) is

$$\mu_r'(F_{\nu_1,\nu_2}) = \left(\frac{\nu_2}{\nu_1}\right)^r E\left[(\chi^2_{\nu_1})^r\right] E\left[(\chi^2_{\nu_2})^{-r}\right]$$

$$= \left(\frac{\nu_2}{\nu_1}\right)^r \frac{(\frac{1}{2}\nu_1)^{[r]}}{(\frac{1}{2}\nu_2 - 1)^{(r)}}.$$

If $\nu_2 \leqslant 2r$, the rth moment is infinite. The mean and variance are $\nu_2/(\nu_2 - 2)$ for $\nu_2 > 2$ and $2\nu_2^2(\nu_1 + \nu_2 - 2)/[\nu_1(\nu_2 - 2)^2(\nu_2 - 4)]$ for $\nu_2 > 4$, respectively.

The F distribution owes its importance almost entirely to its use in analysis-of-variance* tests when residual variation is normal, and in comparison of variances in two normal populations. In 1922, Fisher* [1] showed F to be the test statistic in goodness-of-fit* tests of regression formulae, obtained its distribution, and in 1925 [2] introduced it with analysis of variance, although in terms of $z = \frac{1}{2}\log F$.

For some appropriate procedures with variance component* models, F_{ν_1,ν_2} distributions are used with fractional values of ν_1 and/or ν_2. The extensive tables of Mardia and Zemroch [5] give values of $F_{\nu_1,\nu_2,1-\alpha}$ (such that $\Pr[F_{\nu_1,\nu_2} > F_{\nu_1,\nu_2,1-\alpha}] = \alpha$) to five significant figures for

$$\nu_1 = 0.1(0.1)1.0(0.2)2.0(0.5)5.0(1)16, 18, 20, 24, 30, 40, 60, 120, \infty$$

$$\nu_2 = 0.1(0.1)3.0(0.2)7.0(0.5)11(1)40, 60, 120, \infty$$

$$\alpha = 0.0001, 0.0005, 0.001, 0.005, 0.01, 0.02, 0.025, 0.03(0.01)0.10, 0.2, 0.25, 0.3, 0.4, 0.5.$$

For analytical development it is usually more convenient to consider the distribution of $G_{\nu_1,\nu_2} = \chi^2_{\nu_1}/\chi^2_{\nu_2}$.

Details of other tables and properties of F_{ν_1,ν_2}, as well as of approximations, are given in Johnson and Kotz [4, Chap. 26]. Guenther [3] describes desk calculators which compute $\Pr(F_{\nu_1,\nu_2} > f)$, and discusses the use of these machines for obtaining percentiles.

References

[1] Fisher, R. A. (1922). *J. R. Statist. Soc.*, **85**, 597–612.

[2] Fisher, R. A. (1925). *Statistical Methods for Research Workers*. Oliver & Boyd, Edinburgh.

[3] Guenther, W. C. (1977). *Amer. Statist.*, **31**, 41–45.

[4] Johnson, N. L. and Kotz, S. (1970). *Distributions in Statistics: Continuous Univariate Distributions—2*. Wiley, New York.

[5] Mardia, K. V. and Zemroch, P. J. (1978). *Tables of the F and Related Distributions with Algorithms*. Academic Press, New York.

(ANALYSIS OF VARIANCE
F-TESTS
NONCENTRAL *F*-DISTRIBUTION)

FECHNER'S THEOREM *See* MEAN–MEDIAN–MODE INEQUALITIES

FEDERAL STATISTICS

DECENTRALIZED STATISTICAL SYSTEM

The statistical system of the U.S. government is highly decentralized. The organizational arrangements for producing federal *statistics* include several statistical collection and analytical agencies, many statistical units, and the statistical activities and outputs of major program agencies. In fact, over 90 federal agencies are authorized to collect, compile, and disseminate statistical data.

The origin of federal data collection activity can be traced back to the Constitution of the United States, which required an enumeration of the population within 3 years after the first meeting of the Congress and every 10 years thereafter. The first *census** of population was conducted in 1790. For a century, the census office was phased out, following its decennial census activity, but subsequent legislation increased responsibilities for data collection activities so that the *Bureau of the Census** was established as a permanent agency in 1902. In 1862, the statutory requirements for *agricultural statistics* were established (Act of May 5, 1862, Chap. 72, Sec. 3, 12 Stat. 387). The *Bureau of Labor Statistics** was given its legislative mandate in 1888 (Act of June 13, 1888, Chap. 89, Sec. 1, 25 Stat. 182). These examples indicate how legislation requiring the collection of statistics is the basis for the formation of new statistical units in government. In addition, as new agencies are created, the statistical activities necessary for informed decision making and for administration of programs are typically developed within new agencies. This process has afforded the United States the best statistical coverage in the world, but it has created a decentralized system, with responsibility and authority for statistical activities divided among many subject matter agencies.

This division of responsibility for statistical activities necessitates that there be a central agency with responsibility and authority for providing general policy guidance on the development of an integrated statistical system to meet the needs of the federal government policy makers and other users of federal *statistics*. The most recent in a series of efforts to coordinate federal statistical activities was the establishment of the *Statistical Policy Division* (SPD) in the *Office of Information and Regulatory Affairs* (OIRA) in the Office of Management and Budget (OMB). For historical background on the central coordinating office, see Appendix 1.

ROLE AND FUNCTIONS OF THE STATISTICAL POLICY DIVISION

As a result of *Public Law 96-511*, the *Paperwork Reduction Act of 1980*, the *Statistical Policy Division* was established effective April 1, 1981, in the *Office of Information and Regulatory Affairs*, Office of Management and Budget. This is the successor of the Office of Federal Statistical Policy and Standards (OFSPS), Department of Commerce. Section 3(a) of that Act requires the President and the Director of OMB to delegate to the Administrator of OIRA all their functions, authority, and responsibility for statistical policy and coordination under Section 103 of the *Budget and Accounting Procedures Act of 1950* (31 U.S.C. 18b). This includes the following responsibilities:

> to develop programs and to issue regulations and orders for the improved gathering, compiling, analyzing, publishing, and disseminating of statistical information for any purpose by the various agencies in the executive branch of the Government. Such regulations and order shall be adhered to by such agencies.

The *statistical policy function* exercised by SPD is government wide and objective in nature, extending to all economic and social statistics throughout the federal government.

Planning and Coordination

As indicated in Section 103, the primary mission of SPD is statistical planning and coordination of statistical programs, agencies, and issues across all departments and all subject areas. SPD takes the lead in formulating recommendations to the Director of OMB on the budgets of the statistical agencies, after considering priorities for improving statistical programs. As part of its coordinating role, the office chairs many interagency committees which consider statistical issues of interest across departments.

Statistical Reports Clearance

One part of *Public Law 96–511* requires review of all new or revised forms for gathering information from 10 or more respondents so as to coordinate federal information requests and to minimize public *reporting burden* and governmental costs associated with federal reports. SPD reviews statistical forms and reporting plans.

Statistical Standards

As the central statistical coordinating agency, SPD issues statistical standards and guidelines as mechanisms for ensuring the quality, comparability, timeliness, and accuracy of federal data. These standards are usually developed through *interagency* technical *committees* under SPD sponsorship. Their use is fostered through the process of forms review under *Public Law 96-511*, the issuance of *Statistical Policy Directives*, and other publications. Some of the well-known standards include the *Standard Industrial Classification*, *Standard Occupational Classification*, and the *Standard Metropolitan Statistical Areas*. Others include the designation of a base year, *race and ethnic categories* for federal reporting, and the *definition of poverty* for statistical purposes. As circumstances require, these standards are revised or new ones are issued. The complete set of directives can be found in the *Statistical Policy Handbook*.

International Statistical Coordination

SPD serves as liaison on statistical matters between U.S. government agencies and *international organizations*. It cooperates with the Department of State by drafting and clearing among interested agencies U.S. position papers and instructions concerning international statistical matters for U.S. delegations attending international conferences; and by formulating recommendations on the nature and extent of U.S. participation at various international conferences. In its liaison capacity, SPD serves as a central point for supplying U.S. data for regular and special publications of the *United Nations**. Under Executive Order 10033 the Division is also responsible for coordinating the handling of requests for statistical data from certain other international organizations to U.S. government agencies, to make sure that such requests are handled in a consistent manner, and that comparable data are supplied by the different agencies. The Division is designated by the *Inter-American Statistical Institute* as the "national focal point" for the United States, to facilitate the exchange of statistical information and publications between the U.S. agencies and other countries in the Western Hemisphere.

Members of the staff of the SPD serve on international bodies concerned with improvement of statistics. The head of this Division has served as the U.S. representative on and chairman of the *United Nations Statistical Commission* and the Commission's Working Group on International Statistical Programs and Coordination since they were established. He is also the U.S. delegate to sessions of the *Committee on Improvement of National Statistics*, the *Conference of European Statisticians*, and the *Economic and Social Commission for Asia and the Pacific*'s *Committee on Statistics*. Staff members have also participated in special projects such as the development of the system of social and demographic statistics and the revision of the *Standard International Trade Classification*.

Publications Program

As part of its coordinating effort, SPD and its predecessor organizations issue publications to inform users of developments affecting the availability of data and statistical programs. A list of these publications can be found in the bibliography.

MAJOR STATISTICAL AGENCIES

Of all the agencies and units that have statistical activities, 38 agencies have a key role in developing and using statistical information. These have a greater impact on the statistical system due to their budget level, number of statistical personnel, and the volume of burden placed on the public in collecting their statistical data. The 38 agencies are listed in Appendix 2. The following highlights the mission and products of six major agencies.

Economics and Statistics Service (ESS), Department of Agriculture

Its mission is to formulate, develop, and administer (1) a program of economic, statistical, and other social science research, analysis, and information related to food, agriculture, and rural resources and communities; and (2) a program to collect and publish statistics related to food, agriculture, and rural resources and communities. The ESS serves as the supplier of current general-purpose agricultural statistics at the national and state levels. Further, it supplies the official national estimates of acreage, yield, and production of crops, stocks and value of farm commodities; and numbers and inventory of livestock. The ESS also develops statistical methodology and standards for agricultural statistics.

Bureau of the Census*, Department of Commerce

This is the largest agency in the statistical system. Its primary mission is that of a general-purpose statistical collection agency meeting a wide range of needs for data. Other functions include demographic analysis; extensive research in statistical methodology, data processing techniques, and equipment; and programs to improve access and utilization of statistical information. Census collects and publishes basic statistics concerning the population and the economy of the nation in order to assist the Congress, the executive branch, and the general public in the development and evaluation of economic and social programs. Periodic censuses include the Census of Population and Housing, which is taken at 10-year intervals, and the economic censuses, including the census of agriculture and the census of governments, which are taken at 5-year intervals. Current surveys and programs are conducted to collect data on various economic activities and demographic changes. Data are collected and published on foreign trade, housing, construction, governments, certain agricultural commodities, industrial output, retail and wholesale trade, and transportation. The Current Population Survey provides data and reports on a variety of demographic characteristics as well as providing employment and unemployment data to the Bureau of Labor Statistics of the Department of Labor.

The Bureau compiles a variety of guides, reference works, and explanatory material on the nature and use of census data. Basic among these is the annual *Statistical Abstract of the United States*, which presents summary statistics on the social, political, and economic activities of the nation.

Bureau of Economic Analysis (BEA), Department of Commerce

Its primary mission is analysis and research related to the preparation and interpretation of the economic accounts of the United States. These accounts provide a quantitative view of the economic process in terms of the production, distribution, and use of the nation's output. The accounts consist of the national income and product accounts, summarized by the gross national product (GNP); wealth accounts, which show the business and other components of national wealth; interindustry accounts, which trace the interrelationships among industrial markets; regional accounts, which provide detail on economic activity by region, state, metropolitan area, and county; and U.S. international transactions accounts, which give detail on U.S. transactions with foreign countries.

The work on the economic accounts is supplemented by the preparation and analysis of other measures of economics activity, including various tools for forecasting economic developments such as surveys of investment outlays and plans of U.S. business, econometric models of the U.S. economy, and a system of economic indicators. The measures and analyses prepared by BEA are disseminated through its monthly publication, the *Survey of Current Business* (including periodic supplements to the *Survey*).

National Center for Education Statistics, Department of Education

Its purpose is to collect, analyze, and disseminate statistics and other education data. Periodically, it publishes a report on the condition of education in the United States.

National Center for Health Statistics*, Department of Health and Human Services

The Center was established to collect and disseminate general-purpose health statistics. It cooperates with other federal agencies, state and local governments, and foreign countries in activities to increase the availability and usefulness of health data. The Center conducts some research on statistical and survey methodology.

The Center conducts several major surveys on a continuing basis to determine such things as health costs, insurance coverage, nutritional status, the supply of health manpower, prevalence of chronic diseases, disability, basic morbidity and mortality data, and utilization of health services. The nation's official statistics on births, deaths, marriages, and divorces are developed from the Center's vital statistics program.

Energy Information Administration (EIA), Department of Energy

This agency's primary responsibility is to collect and analyze data concerning the sources of supply, distribution, and use of energy resources. EIA analytical work includes studies, forecasts, and appraisals of the energy situation, and analyses of policy alternatives. EIA is also concerned with improving the quality in reporting, procedures, and standards as they affect energy statistics.

EIA collects and publishes statistics on petroleum supply, production, and stocks, and statistics on refineries, primary terminals, pipeline companies, and importers. It also collects, compiles, and publishes statistics on all phases of domestic and foreign energy mineral resource developments and on electrical energy production and distribution.

Bureau of Labor Statistics (BLS), Department of Labor

The Bureau of Labor Statistics* (BLS) has the responsibility for collecting and analyzing data on labor and price statistics; that is, employment, unemployment, occupational health and safety, employee compensation, wages, productivity, labor relations, wholesale and retail prices, and export and import prices. Results of analytical studies and information on other BLS data activities are published in the *Monthly Labor Review*.

Additional information on the statistical activities of these six agencies as well as the other federal statistical agencies can be found in the following SPD publications: *A Framework for Planning U.S. Federal Statistics for the 1980's; Revolution in United States Government Statistics 1926–76*; and *Statistical Services of the United States Government, 1975*. See the bibliography for further details.

[*Editorial Note*. The Statistical Policy Branch in the Office of Management and Budget was abolished on April 30, 1982; see *Amstat News*, May 1982. The Committee on Government Statistics of the American Statistical Association* expressed concern that "this action will have serious detrimental effects on Federal statistics" (*Amstat News*, June 1982).]

APPENDIX 1: STATISTICAL POLICY DIVISION—HISTORICAL BACKGROUND

INTERDEPARTMENTAL STATISTICAL COMMITTEE. Established by Executive Order 937, September 10, 1908, it was composed of one member from each of 10 agencies, designated by the President.

CENTRAL BUREAU OF PLANNING AND STATISTICS. Established by the War Industries Board in 1918, it instituted measures to improve the efficiency and quality of the data gathering activities, assembled statistics bearing on the war effort, prepared a catalog of government statistics, advised agencies on statistical methods, promoted adoption of standard definitions, and served as a clearing house of statistical information.

BUREAU OF EFFICIENCY. In existence from 1916 to 1933, it was concerned with statistical coordination for only a brief period (1919–1922).

FEDERAL STATISTICS BOARD. Established in April 1931, it studied the collection, compilation, and use of statistics and recommended economies and means for fuller utilization of statistics and statistical personnel.

CENTRAL STATISTICAL BOARD. Established as an independent agency in 1933, it reviewed plans for the tabulation and classification of statistics needed for purposes of the National Industrial Recovery Act and promoted the coordination and improvement of the statistical services involved. It was given a specific statutory mandate for a 5-year period by Public Law 219, July 1935. Its function and operating methods were basically similar to those now exercised by the Statistical Policy Division of OMB (see last item).

DIVISION OF STATISTICAL STANDARDS, BUREAU OF THE BUDGET. Reorganization Plan I under the Reorganization Act of 1939 transferred the Central Statistical Board to the Bureau of the Budget, where it became the Division of Statistical Standards. The Federal Reports Act of 1942 broadened its scope to cover collection of information generally, made mandatory the review of questionnaires prior to issuance, and made explicit the responsibility of minimizing cost and burden on respondents while maximizing the usefulness of statistics. In a reorganization of the Bureau in April 1952, the name of the Division was changed to Office of Statistical Standards.

STATISTICAL POLICY DIVISION, OFFICE OF MANAGEMENT AND BUDGET. Continuing earlier mandates, but with broader responsibility for statistical policy, the Division was renamed the Office of Statistical Policy in 1969. A further internal reorganization of the Office of Management and Budget resulted in name change to the Statistical Policy Division on November 15, 1971, to parallel other divisions with levels of responsibility.

OFFICE OF FEDERAL STATISTICAL POLICY AND STANDARDS, DEPARTMENT OF COMMERCE. Established in October 1977, by Executive Order 12013, which transferred certain statistical policy functions from the Office of Management and Budget to the Department of Commerce.

STATISTICAL POLICY DIVISION, OFFICE OF MANAGEMENT AND BUDGET. Established effective April 1, 1981, in the Office of Information and Regulatory Affairs following the passage of the Paperwork Reduction Act of 1980, which transferred the statistical policy function to the Office of Management and Budget.

APPENDIX 2: MAJOR AGENCIES IN THE FEDERAL STATISTICAL SYSTEM

General Coordination Agency

Office of Federal Statistical Policy and Standards
  Department of Commerce

Core Multipurpose Collection Agencies

Economics and Statistics Service
  Department of Agriculture

Bureau of the Census
  Department of Commerce

Bureau of Labor Statistics
  Department of Labor

Subject Matter Multipurpose Collection Agencies

National Center for Education Statistics
  Department of Education

National Center for Health Statistics
  Department of Health and Human Services

Office of the Assistant Secretary for

Policy Development and Research
Department of Housing and Urban Development

Bureau of Mines
Department of the Interior

Federal Bureau of Investigation
Department of Justice

Bureau of Justice Statistics
Department of Justice

Law Enforcement Assistance Administration
Department of Justice

Employment and Training Administration
Department of Labor

Internal Revenue Service
Department of the Treasury

Environmental Protection Agency

Energy Information Agency
Department of Energy

Core Multipurpose Analysis Agencies

Economics and Statistics Service
Department of Agriculture

Bureau of Economic Analysis
Department of Commerce

Bureau of Industrial Economics
Department of Commerce

Office of Research and Statistics
Social Security Administration
Department of Health and Human Services

Office of the Assistant Secretary for Planning and Evaluation
Department of Health and Human Services

Research and Special Programs Administration
Department of Transportation

Federal Reserve Board

Program Collection and Analysis Agencies

National Institute of Education
Department of Education

Alcohol, Drug Abuse, and Mental Health Administration
Center for Disease Control
Food and Drug Administration
Health Care Financing Administration
Health Resources Administration
Health Services Administration
National Institutes of Health
Office of the Assistant Secretary for Human Development Services
Department of Health and Human Services

Office of the Assistant Secretary for Community Planning and Development
Office of the Assistant Secretary for Housing
Department of Housing and Urban Development

Federal Highway Administration
National Highway Traffic Safety Administration
Department of Transportation

Office of the Secretary (including Office of Revenue Sharing and office of Tax Analysis)
Department of the Treasury

National Science Foundation

U.S. International Trade Commission

Veterans Administration

BIBLIOGRAPHY: PUBLICATIONS PREPARED BY THE STATISTICAL POLICY DIVISION, OFFICE OF MANAGEMENT AND BUDGET

If GPO is shown, order from the Superintendent of Documents, U.S. Government Printing Office, Washington, DC 20402.

Correlation between the United States and International Standard Industrial Classifications. Tech. Paper 1, Oct. 1979. [Relates the current International Standard Industrial Classification (ISIC) published by the United Nations to the Standard Industrial Classification (SIC). Thus both are concerned with the classification of establishments, i.e., factories, farms, stores, banks, ranches, shops, schools, etc., rather than enterprises, i.e., companies, firms, corporations, etc. The correlation is shown in both ISIC and SIC codes. (GPO Stock No. 003-003-02142-3.)]

Enterprise Standard Industrial Classification Manual, 1974. [Provides a standard for use with statistics about enterprises (i.e., companies, rather than their individual establishments) by kind of economic activity. (Out of print)]

Federal Statistical Directory. 26th ed., 1979. [Lists by organizational units within each agency, the names, office addresses, and telephone numbers of key persons engaged in statistical programs and related activities of agencies of the executive branch of the federal government. (GPO Stock No. 003-005-00184-1. Out of print)]

A Framework for Planning U.S. Federal Statistics for the 1980's. Issued 1978. [Provides a coordinated overview of potential directions for the federal statistical system to take in the coming decade so as to achieve a more integrated set of social and economic statistics. Reviews the organization and operations of the federal statistical agencies, individual statistical programs, and crosscutting issues and makes recommendations for improvement. (GPO Stock No. 003-005-00183-2.)]

Gross National Product Data Improvement Project Report. Issued 1979. [Provides the first comprehensive evaluation of the underlying data used to estimate the national economic accounts and makes recommendations for improving these data. (GPO Stock No. 003-010-00062-7.)]

Revolution in United States Government Statistics, 1926–1976. Issued 1978. [Background document for the *Framework* which traces developments in important areas such as sampling applications, national income accounts, use of computers, and coordinating mechanisms during the 50-year period. Final chapter considers major issues which were unresolved at the end of the 50-year period, leading to the topics which are addressed in the *Framework*. (GPO Stock No. 003-005-00181-6.)]

Standard Industrial Classification Manual, 1972, and 1977 Supplement. [Contains titles and description of industries and alphabetical indexes for both manufacturing and nonmanufacturing industrial establishments (e.g., factories, mills, stores, hotels, mines, farms, banks, depots, warehouses). SIC was developed for use in the classification of establishments by type of activity in which they are engaged; for purposes of facilitating the collection, tabulation, presentation, and analysis of data relating to establishments; and for promoting uniformity and comparability in the presentation of statistical data collected. (GPO Stock No. for the Manual 4101-0066, price $10.25 (hard cover); GPO Stock No. for the 1977 Supplement 003-005-00176-0.)]

Standard Metropolitan Statistical Areas, 1975. [Contains the 1971 criteria used by the Statistical Policy Division in OMB in designating and defining standard metropolitan statistical areas and standard consolidated statistical areas, as well as the titles and definitions of these areas. Also included is a map of the SMSAs. (GPO Stock No. 041-001-00101-8.) The new 1980 standards for metropolitan statistical areas are available from the Statistical Policy Division, OMB, Washington, DC 20503.]

Standard Occupational Classification Manual, 1980. [Contains definitions for over 600 occupational groups to be used in the collection, tabulation, and analysis of data about occupations. This revision of the original 1977 edition includes a direct relation to the *Dictionary of Occupational Titles*, and will be the basis of the Census classification of occupations. (GPO Stock No. 003-005-00187-5 (hard cover).)]

Statistical Policy Handbook. Issued 1978. [Contains the directives for the conduct of federal statistical activities, information about interagency committees which have a significant role in federal statistics, and publications of SPD. Directives provide standards and guidelines for such activities as statistical surveys, publications of statistics, release of statistical information, industrial classification, race and ethnic categories for federal statistics, standard metropolitan statistical areas, and so forth. (GPO Stock No. 003-005-00179-4.)]

Statistical Policy Working Papers. Issued 1978. [Series of technical documents prepared by working groups or task forces under the auspices of SPD. Currently, there are six papers in the series: *Report on Statistics for Allocation of Funds* (GPO Stock No. 003-005-00178-6, price $2.40); *Report on Statistical Disclosure and Disclosure-Avoidance Techniques* (GPO Stock No. 003-005-00177-8, price $2.40); *An Error Profile: Employment as Measured by the Current Population Survey* (GPO Stock No. 003-005-00182-4, price $2.75); *Glossary of Nonsampling Error Terms: An Illustration of a Semantic Problem in Statistics* (available from SPD); *Report on Exact and Statistical Matching Techniques*, issued 1980 (GPO Stock No. 003-005-00186-7, price $3.50); and *Report on Statistical Uses of Administrative Records*, 1980. (GPO Stock No. 003-005-00185-9.)]

Statistical Reporter. [Monthly publication designed for the interchange of information among federal government employees engaged in statistical and research activities. Includes notes on surveys and programs, major organization changes, statistical publications, schedule of release dates for principal federal economic indicators, federal statistical personnel, and a feature article on a current development in federal statistics. (Subscription available through GPO.)]

Statistical Services of the United States Government. 1975 rev. ed. [Basic reference document on the statistical programs of the U.S. government. Describes the federal statistical system and presents brief descriptions of the principal economic and social statistical series. Contains a brief statement of the statistical responsibilities of each agency and a list of its principal statistical publications. (Out of print.)]

1980 Supplement to Economic Indicators. [Contains an explanatory text and historical data for each series which appears in the monthly *Economic Indicators*, which is prepared for the Joint Economic Committee by the Council of Economic Advisers. (GPO Stock No. 052-070-05453-1.)]

(BUREAU OF LABOR STATISTICS
BUREAU OF THE CENSUS, U.S.
FDA STATISTICAL PROGRAMS

HEALTH STATISTICS,
NATIONAL CENTER FOR
LABOR STATISTICS)

JOSEPH W. DUNCAN

FEEDFORWARD–FEEDBACK CONTROL SCHEMES

An important practical problem is one where it is desired to maintain the value of some quality characteristic (e.g., the number of defective components in a batch or the viscosity of a polymer) within acceptable limits. Two devices that have been used for this purpose are the *Shewhart chart** and the *cusum chart* (*see* CUMULATIVE SUM CONTROL CHARTS). Shewhart charts [7, 11] were introduced initially to identify areas where variability is introduced into an industrial process, thus pointing the way to the elimination of this variability. Subsequently, Shewhart charts were used to detect when a shift in the level of a quality characteristic has occurred, leading to an adjustment to some process variable to compensate for the shift in level. It was shown later [1, 10] that, under certain assumptions about the statistical variability of the process, Cusum charts were more efficient for detecting when a change in level had occurred. Such charts imply that there is a "cost" involved in making a change, so that adjustments are made infrequently. However, there are many control problems in industry where the cost of making a change is very small, so that in theory, an adjustment can be made each time a measurement of the quality characteristic is available. Therefore, what needs to be decided in these situations is the magnitude of the adjustment. Such devices are called *process control charts* and require a more precise specification of the characteristics of an industrial process, as explained below. To summarize, Shewhart charts are useful for indicating *which* variables are causing variability in a process. Cusum charts are useful for detecting *when* a "significant" change has occurred and process control charts are useful for deciding by *how much* to make an adjustment.

PROCESS CONTROL PROBLEMS

The development of process control charts was stimulated by the need to improve manual control in the process industries. Figure 1 shows a simplified representation of such a situation in which the fiber content (basis weight) of paper is controlled in a Fourdrinier paper machine. Thick stock (a homogeneous suspension of wood pulp in water plus additives) is fed via a valve (which controls its flow rate) and diluted with backwater from the paper machine to form thin stock, which is then fed to a flow box. The latter distributes the thin stock as it passes through an orifice onto a moving wire mesh, where the paper is formed. After pressing and drying, a beta gauge is used to measure the basis weight (fiber content of the paper). Figure 2 shows process records consisting

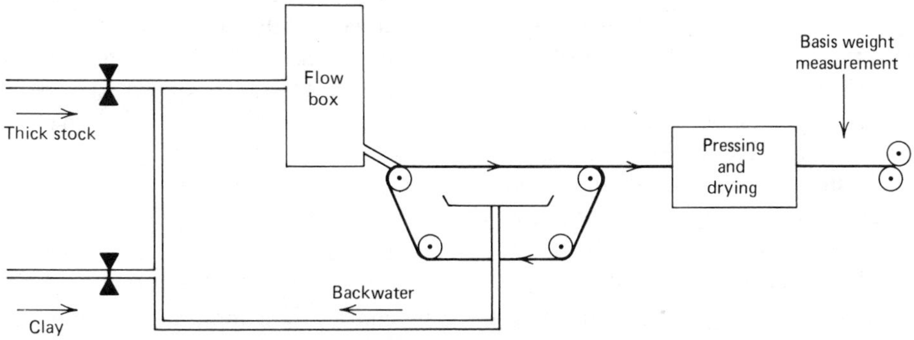

Figure 1 Simplified representation of a Fourdrinier paper machine.

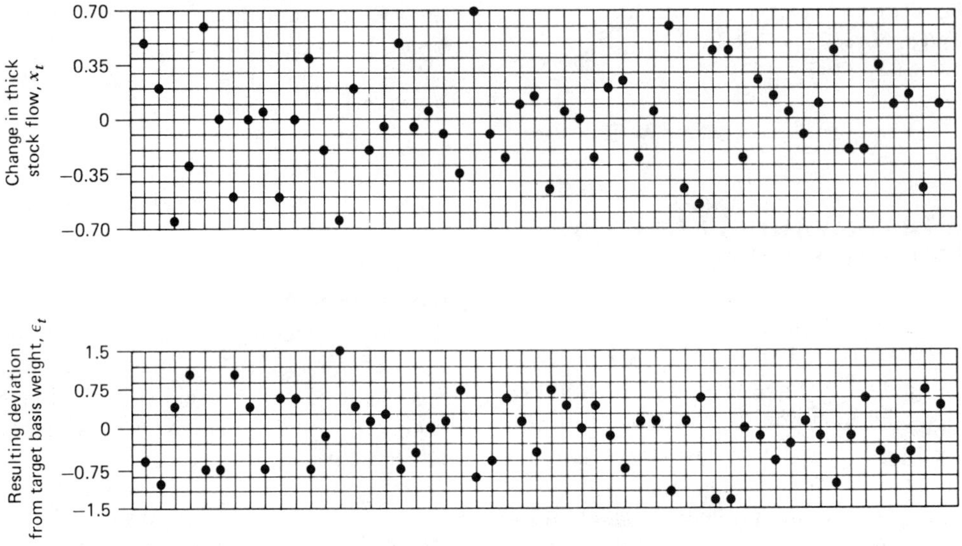

Figure 2 Operating records consisting of changes x_t in thick stock flow and corresponding deviations ϵ_t of basis weight measurements from the target.

of:

ϵ_t: the error, or deviation, at half-minute intervals, of the basis weight from its specified target value, as dictated by the quality of the paper.

x_t: the adjustments made to the valve to control the flow of thick stock—the higher the flow, the higher the basis weight.

The characteristic features of the control problem illustrated in Figs. 1 and 2 are:

1. If no control were applied, the basis weight would wander away from its target value. The resulting time series* is a reconstruction of what would have happened to the basis weight measurement if no control had been applied and represents the *noise* or *disturbance* in the process, a major source of which is the backwater.
2. The dynamic response or transfer function relating the manipulated variable (thick stock flow) and the controlled variable (basis weight) was relatively well understood and could also be estimated from operating records.
3. The cost of making an adjustment was effectively zero because the thick stock valve was to be manipulated by a computer—hence it was possible to make an adjustment after each new deviation ϵ_t comes to hand.

Based on the foregoing three assumptions, a control scheme can be designed that attempts to minimize the mean square of the deviations ϵ_t from the specified target.

DESIGN OF BASIS WEIGHT FEEDBACK CONTROL SCHEME

As indicated above, the structure of the control scheme depends on (a) the noise N_t in the process when no control is exercised, and (b) the dynamic relationship between the manipulated variable and the controlled variable. Whereas the data in Fig. 2 are recorded at intervals of half a minute, some thought needs to be given to the sampling frequency at which control is to be applied (discussed later). The control scheme finally implemented was based on sampling the basis weight measurement every $1\frac{1}{2}$ minutes and taking control action each time such a measurement is made. For intervals of $1\frac{1}{2}$ minutes, it may be shown that the noise N_t if no control were applied is nonstationary and follows an autoregressive–integrated moving average model* of the form

$$\nabla N_t = a_t - \theta a_{t-1} = (1 - \theta B) a_t \qquad (1)$$

with $\theta = 0.31$, where (a) a_t is a random series* or white noise*; (b) ∇ is the backward difference* operator, such that $\nabla N_t = N_t - N_{t-1}$; and (c) B is the backward shift operator, such that $Ba_t = a_{t-1}, B^j a_t = a_{t-j}$. The dynamic response or *transfer function* of the process may be described by the relationship

$$y_{t+1} = g x_t \qquad (2)$$

with $g = 0.17$, where (a) x_t denotes the adjustment made to the manipulated variable at time t and y_{t+1} is the induced change in the level of the controlled variable from time t to $(t+1)$, and (b) the constant $g = 0.17$ is called the *gain* of the system and implies that a unit change in the manipulated variable (thick stock flow rate) at time t has become fully effective by time $t+1$ and increases the controlled variable (basis weight) by 0.17 unit.

The design of the control scheme is based on the principle that the noise could be canceled out exactly if the cumulative effect Y_{t+1} on the controlled variable, as a result of all previous adjustments, is set equal to the negative of the level of the noise at time $(t+1)$, i.e.,

$$Y_{t+1} = -N_{t+1}.$$

However, since the noise N_{t+1} is not known at time $(t+1)$, it has to be replaced by its one-step-ahead forecast $\hat{N}_t(1)$, so that

$$Y_{t+1} = -\hat{N}_t(1) \qquad (3)$$

and the deviation from target ϵ_{t+1} is equal to the one-step-ahead error in forecasting N_{t+1}.

From (3), the change in the level of the controlled variable from time t to $(t+1)$ should be set equal to

$$y_{t+1} = Y_{t+1} - Y_t = -(\hat{N}_t(1) - \hat{N}_{t-1}(1)). \qquad (4)$$

If the noise is of the form (1), it may be shown that

$$\hat{N}_t(1) - \hat{N}_{t-1}(1) = (1 - \theta) a_t. \qquad (5)$$

Substituting (2) and (5) in (4), the optimal control action for this set of assumptions is

$$x_t = \frac{-(1-\theta)}{g} a_t = \frac{-0.69}{0.17} a_t$$
$$\simeq -4a_t = -4\epsilon_t, \qquad (6)$$

where the deviation ϵ_t from target has been set equal to the one-step-ahead forecast error a_t of the noise. The control action (6) can be applied using a manual control chart (Fig. 3) on which the basis weight measurement is plotted. The chart has two scales: an *error scale*, which records the deviations of the basis weight from its target value, which is set as zero; and an *action scale*, calibrated according to (6) (i.e., a unit change in ϵ_t corresponds to a change of -4.0 units in the adjustment x_t). In this example, control action was actually applied by a computer, as described later.

CLOSED-LOOP ESTIMATION

The values of the noise and transfer function parameters can be estimated from operating

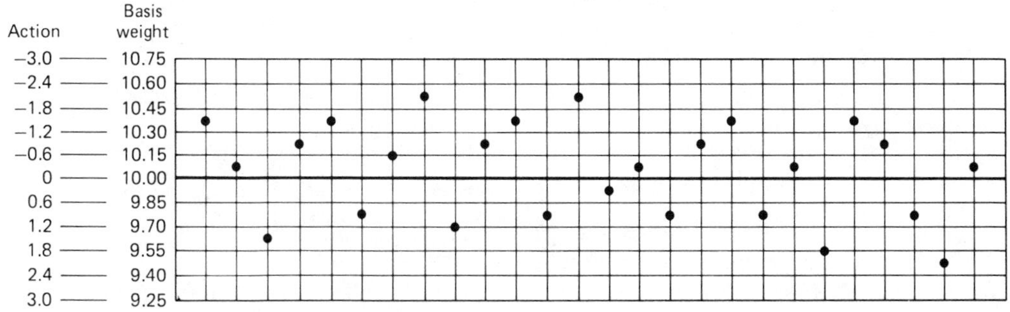

Figure 3 Process control chart showing basis weight measurements and corresponding action scale indicating adjustments to thick stock flow.

records (Fig. 2) corresponding to an imperfect control scheme in which ad hoc adjustments are made to the process. The method is illustrated using the data of Fig. 1, based on a sampling interval of half a minute. A useful preparatory step is to use open-loop records (i.e., measurements of the basis weight when no control action is applied), to identify the noise structure. In this example, for sampling intervals of $\frac{1}{2}$ minute, the noise structure was identified to be of the form (1). Independent tests had shown that a change in the thick stock flow rate took approximately 1.2 minutes before it affected basis weight. Thus the delay lies between two and three sampling intervals of half a minute, leading to a transfer function

$$y_t = gx_{t-3}. \qquad (7)$$

To estimate the parameters from closed-loop operating records obtained from a nonoptimal control scheme, we may argue as follows. The change $\epsilon_t - \epsilon_{t-1}$ in the level of the controlled variable from time $(t-1)$ to time t is made up of (a) the change y_t due to the adjustment x_{t-3} to the manipulated variable at time $(t-3)$, and (b) the change $N_t - N_{t-1}$ in the level of the noise between time $(t-1)$ and time t, that is,

$$\epsilon_t - \epsilon_{t-1} = y_t + N_t - N_{t-1}. \qquad (8)$$

Substituting in (8) from (1) and (7), we obtain

$$\nabla\epsilon_t = gx_{t-3} + a_t - \theta a_{t-1}. \qquad (9)$$

Given operating records consisting of simultaneous values of ϵ_t and x_t, the model (9) can be fitted by nonlinear estimation methods [5]. For one such estimation, based on 183 data points, the following estimates of the parameters and their standard errors were obtained:

$$\hat{\theta} = 0.50 \pm 0.06, \qquad \hat{g} = 0.17 \pm 0.08, \qquad (10)$$

and the estimated residual variance was $\sigma_a^2 = 0.08314$. A general method for estimating transfer function–noise parameters is given in Box and Jenkins [5].

CHOICE OF SAMPLING INTERVAL FOR CONTROL

The control scheme (6) was based on a sampling interval of $1\frac{1}{2}$ minutes and a transfer function (2) which was such that an adjustment x_t made at time t to thick stock flow had become fully effective on basis weight by time $(t+1)$. To illustrate the issues involved when control schemes are designed using different sampling intervals we consider a sampling interval of half a minute. The noise structure is assumed to be of the form (1) and the transfer function of the form (7). The latter implies that the effect of an adjustment x_t made at time t does not become apparent in the basis weight measurement until time $(t+3)$. Therefore, an effective control strategy is one which arranges that the cumulative effect Y_{t+3} up to time $(t+3)$ of all previous adjustments cancels out the forecast $\hat{N}_t(3)$ of the level of the noise made three steps ahead, that is,

$$Y_{t+3} = -\hat{N}_t(3), \text{ or}$$

$$y_{t+3} = Y_{t+3} - Y_{t+2} = -(\hat{N}_t(3) - \hat{N}_{t-1}(3)) = gx_t \qquad (11)$$

using (7). From (4) it follows that

$$N_{t+3} = a_{t+3} + (1-\theta)(a_{t+2} + a_{t+1} + \cdots),$$

$$\hat{N}_t(3) = E_t[N_{t+3}] = (1-\theta)(a_t + a_{t-1} + \cdots),$$

where E_t denotes the conditional expectation at time t. Thus

$$\hat{N}_t(3) - \hat{N}_{t-1}(3) = (1-\theta)a_t \qquad (12)$$

and the deviation ϵ_{t+3} of basis weight from target at time $t+3$ will be equal to the three-steps-ahead forecast error $e_t(3)$ made from origin t, that is,

$$\epsilon_{t+3} = e_t(3) = a_{t+3} + (1-\theta)(a_{t+2} + a_{t+1}) = [1 + (1-\theta)(B + B^2)]a_{t+3}. \qquad (13)$$

Eliminating a_t from (12) and (13) and then substituting in (11), the optimal control action is

$$x_t = \frac{-(1-\theta)}{g}\epsilon_t - (1-\theta)x_{t-1} - (1-\theta)x_{t-2}; \qquad (14)$$

from (13), the variance of the deviations ϵ_t from target is

$$\sigma_\epsilon^2 = \left[1 + 2(1-\theta)^2\right]\sigma_a^2. \qquad (15)$$

Substituting the parameter estimates (10) in (14) and (15), the control action is

$$x_t = -2.9\epsilon_t - 0.5x_{t-1} - 0.5x_{t-2}$$

and $\sigma_\epsilon^2 = 1.5\sigma_a^2$.

To compare the performance of control schemes at different sampling intervals, it is necessary to know the noise structure and transfer function structure at each sampling interval. For the noise structure, use can be made of a result [5, pp. 488, 489] that if the noise (1) is sampled at intervals h, the model of the sampled series is of the same form but with parameters θ_h and residual variance σ_h^2, which can be calculated as follows from the parameters θ and σ_a^2 corresponding to $h = 1$:

$$\frac{(1-\theta_h)^2}{\theta_h} = \frac{h(1-\theta)^2}{\theta}, \qquad \frac{\sigma_h^2}{\sigma_a^2} = \frac{\theta}{\theta_h}.$$

Table 1 shows, for different values of the sampling interval h, (a) the noise parameter θ_h, (b) the transfer function, (c) the optimal control action, and (d) the variance of the deviations from target resulting from these control actions expressed as a multiple of σ_a^2, the variance of the residuals in the noise process corresponding to $h = 1$. The table shows that the variance of the control errors is worse for $h = 2$ than for $h = 1$ but $h = 3$ is better than $h = 2$ and not much worse than $h = 1$. The reason for this is that for $h = 1$, we are forced to forecast three steps ahead compared with one step ahead for $h = 3$. Furthermore, for $h = 2$, because of the transfer function we are forced to wait four sampling intervals before we can assess the effect of a control action on basis weight, whereas for $h = 3$, we can assess the effect of the control after three sampling intervals.

Because of the above and other considerations, the sampling interval chosen for control action in this application was $h = 3$, that is, 90 seconds. Experimental results showed that the noise parameters changed depending on the grade of paper being manufactured. A computer was used to store and implement the control actions for each grade of paper as well as perform other control actions. Its introduction resulted in substantial financial savings [9].

FURTHER PROBLEMS; LITERATURE

The illustration discussed above is an example of *feedback control* where adjustments are made to compensate for the effect of a noise or disturbance which cannot be measured but which nevertheless can be forecast from its past behavior. In some situations it may be possible to measure one or more sources of the noise. Knowing the transfer function between this measured disturbance and the variable being controlled, it is possible to forecast its effect on the variable being controlled and then apply an adjustment to a manipulated variable to compensate for this forecast change. Such control action is referred to as *feedforward control*; in general, it can be applied to compensate

Table 1

Sampling Interval, h	Noise Parameter, θ_h	Transfer Function	Control Action	Variance of Control Error, $\sigma_\epsilon^2/\sigma_a^2$
1 (30 sec)	0.5	$y_t = 0.17x_{t-3}$	$x_t = -2.9\epsilon_t - 0.5x_{t-1} - 0.5x_{t-2}$	1.50
2 (60 sec)	0.38	$y_t = 0.17x_{t-2}$	$x_t = -3.6\epsilon_t - 0.62x_{t-1}$	1.81
3 (90 sec)	0.31	$y_t = 0.17x_{t-1}$	$x_t = -4.0\epsilon_t$	1.59
4 (120 sec)	0.27	$y_t = 0.17x_{t-1}$	$x_t = -4.3\epsilon_t$	1.87

for those disturbances which can be measured and *feedback control* used to "trim" the control action by compensating for those disturbances that cannot be measured. Some examples of feedback control are given in Box and Jenkins [2, 4] and general algorithms for feedback, feedforward, and feedforward-feedback control in Box and Jenkins [5]. A discussion of the modifications needed when the cost of making a control action is nonzero is given in Box and Jenkins [3, 5], where it is shown that charts which are similar in some respects to more conventional Shewhart charts can be justified for specific types of noise, transfer function, and cost assumptions. Problems also occur where the fluctuations in the manipulated variable have to be kept to a minimum in order not to cause too much disruption to the process. Such *constrained control schemes* [5] are important in practice. Box and MacGregor [6] consider statistical problems which arise in building transfer function–noise models from closed-loop operating records. They show that if standard open-loop procedures for iterative building of the models via identification, estimation, and checking [5] are used for closed-loop data, incorrect models may result and lack of fit may not be detected. This problem can be avoided if open-loop methods are used to arrive at the structure of the transfer function and noise models (as in the examples cited above) and parameters are then estimated from closed-loop data. Alternatively, Box and MacGregor suggest that a random or "dither" signal can be added to the manipulated variable, thereby removing some of the ambiguities associated with the identification and checking of models built from closed-loop data. The ideas discussed here may be extended to feedforward–feedback control schemes with several input variables which can be manipulated, other input variables which can be measured but not controlled, and several output variables which are to be controlled. For a discussion of stochastic control problems from the point of view of the control engineer, see Graupe [8].

References

[1] Barnard, G. A. (1959). *J. R. Statist. Soc. B*, **21**, 239–257.

[2] Box, G. E. P. and Jenkins, G. M. (1962). *J. R. Statist. Soc. B*, **24**, 297–331.

[3] Box, G. E. P. and Jenkins, G. M. (1963). *Bull. Int. Statist. Inst.*, **40**, 943–974.

[4] Box, G. E. P. and Jenkins, G. M. (1965). *AIChE Ind. Chem. Eng. Symp. Ser.*, **4**, 61–68.

[5] Box, G. E. P. and Jenkins, G. M. (1976). *Time Series Analysis, Forecasting and Control*, 2nd ed. Holden-Day, San Francisco.

[6] Box, G. E. P. and MacGregor, J. F. (1974). *Technometrics*, **16**, 391–398.

[7] Dudding, B. P. and Jennet, W. J. (1942). Quality Control Charts, *Brit. Stand. 600R*.

[8] Graupe, D. (1976). *Identification of Systems*, 2nd ed. Krieger, Huntington, N.Y.

[9] Oughton, K. D. (1965). *Ind. Electron.*, Aug., 358–362.

[10] Page, E. S. (1961). *Technometrics*, **3**, 1–9.

[11] Shewhart, W. A. (1931). *The Econometric Control of the Quality of Manufactured Product*. Macmillan, New York.

[AUTOREGRESSIVE–INTEGRATED MOVING AVERAGE (ARIMA) MODELS
CONTROL CHARTS
CUMULATIVE SUM CONTROL CHARTS
QUALITY CONTROL, STATISTICAL
SHEWHART CHARTS
TIME SERIES]

G. M. JENKINS

FELLER DIFFUSION PROCESS *See* DIFFUSION PROCESSES

FENCES *See* FIVE-NUMBER SUMMARIES

FERMAT, PIERRE DE

Born: 1601, in Beaumont-de-Lomagne, France.

Died: 1665, in Castres, France.

Contributed to: probability theory.

Pierre de Fermat, Parliamentary Counsellor in Toulouse, France, contributed to the birth of probability theory through his correspondence with Pascal* about the *Problem of Points**, an unsolved gaming problem of his time. Suppose that two players stake equal money on being the first to win n points in a game in which the winner of each point is determined by the toss of a fair coin, heads for one player and tails for the other. If such a game is interrupted when one player still lacks a points and the other b, how should the stakes be divided between them?

On the evidence of the correspondence, which took place in 1654, Fermat and Pascal had independently concluded that the problem could be solved by noting that at most $(a + b - 1)$ more tosses will settle the game, and that if this number of tosses is imagined to have been made, the resulting 2^{a+b-1} possible games (each equally probable) may be classified according to the winner in each case, the stakes then being divided accordingly. Thus the real game, of indeterminate length, is embedded in an imaginary game of fixed length. This is the limit of Fermat's contribution, although Pascal went on to give the general solution, introducing into his argument such novel features as the binomial distribution* for equal probabilities, the notion of expectation, and proof by induction.

When the problem is generalized to three players (each with a probability one-third of winning a point) the enumeration of the games needs more care because the order of the wins is relevant in some cases, and Pascal doubted the method; Fermat, however, explained that if the enumeration were carried out fully there was no problem, and that Pascal's doubts were unjustified.

Fermat is frequently credited with having discovered, in 1636, the relation between adjacent binomial coefficients, but this was in fact well known at the time through the work of Cardano, Faulhaber, and Briggs.

LITERATURE

Mahoney [1] is the standard modern account of Fermat's mathematical work, although it attributes more of the solution to the Problem of Points to him than the evidence seems to justify. For a fuller discussion, and further references, *see* PROBLEM OF POINTS.

Reference

[1] Mahoney, M. S. (1973). *The Mathematical Career of Pierre de Fermat*. Princeton University Press, Princeton, N.J.

A. W. F. EDWARDS

FERMI–DIRAC STATISTICS (WITH MAXWELL–BOLTZMANN AND BOSE–EINSTEIN STATISTICS)

In the development of statistical physics* attempts were made to describe the behavior of particles of gas by supposing the phase space to be divided into a large number of cells, the word "statistics" in this context being used traditionally to mean the underlying model. The three models of interest have applications in other fields, however; we shall describe them in the context of occupancy problems*, and then relate them to statistical physics and other fields. In what follows, the particles may be considered as r balls in n cells.

Feller [3, Chap. II] refers to *Maxwell–Boltzmann* (M-B) *statistics* as the classical occupancy problem. The r balls are distinguishable and distributed at random, so that each of the n^r arrangements is equally likely. The number of balls in any given cell has a binomial distribution* with r "trials" and success probability n^{-1}, and the probability that every cell is occupied is

$$\sum_{i=0}^{n} (-1)^i \binom{n}{i} \left(1 - \frac{i}{n}\right)^r.$$

If the cells are labeled, the probability that they contain $r_1, \ldots, r_n$ balls respectively, where $r_1 + \cdots + r_n = r$, is [3, Sec. II.5]

$$\left\{ r! \Big/ \prod_{i=1}^{n} (r_i!) \right\} n^{-r}.$$

The probability that exactly m cells are empty is

$$\binom{n}{m}\sum_{i=0}^{n-m}(-1)^i\binom{n-m}{i}\left(1-\frac{m+i}{n}\right)^r;$$

if r and n increase so that $\lambda = ne^{-r/n}$ is bounded, this quantity tends to the Poisson* probability $e^{-\lambda}\lambda^m/m!$ [3, Sec. IV.2]. The probability that exactly m cells each contain exactly k particles has a similar limiting value if $\lambda = ne^{-r/n}(r/n)^k/k!$.

In *Bose–Einstein* (B-E) *statistics* the r balls are indistinguishable, and each distinguishable arrangement of balls into cells is equally likely, with probability $1/\binom{n+r-1}{r}$ [3, Secs. II.5, II.11]. The probability that a given cell has exactly k balls is

$$\binom{n+r-k-2}{r-k}\bigg/\binom{n+r-1}{r};$$

if $n\to\infty$ and $r\to\infty$ so that $r/n\to\lambda$, this quantity tends to the geometric* probability $\lambda^k/(1+\lambda)^{k+1}$. The probability that a group of m specified cells contains a total of j balls is

$$\binom{m+j-1}{m-1}\binom{n-m+r-j-1}{r-j}\bigg/\binom{n+r-1}{r},$$

and if $r/n\to\lambda$ as above, this quantity tends to the negative binomial* probability [3, Sec. II.11]

$$\binom{m+j-1}{m-1}\lambda^j\big/(1+\lambda)^{m+j}.$$

The probability that exactly k cells are empty is

$$\binom{n}{m}\binom{r-1}{n-m-1}\bigg/\binom{n+r-1}{r}.$$

Kunte [7] characterizes B-E statistics as a compounding of the multinomial* with the Dirichlet* distribution.

B-E statistics arises if balls are dropped into cells according to Gibrat's law, which says roughly that the probability that a ball is dropped into a specified cell is proportional to the number of balls already in that cell [6]. Specifically, if the "size" of the kth cell is $s_k - 1$, where s_k balls are already in the cell ($k = 1, \ldots, n$) and if $s = \sum_{k=1}^n s_k = n + r$, Gibrat's law here says that

$$\Pr\{(r+1)\text{st ball falls into the }k\text{th cell}\} = s_k/s.$$

Then the probability of the configuration $(s_1, \ldots, s_n)$ is $\binom{n+r-1}{r}^{-1}$.

In *Fermi–Dirac* (F-D) *statistics*, the balls are again indistinguishable, but no cell is permitted to contain more than one ball, so that $r \leqslant n$. Each distinguishable arrangement is equally likely, with probability $1/\binom{n}{r}$ [3, Sec. II.5]. But if the balls are distinguishable, if cells are each divided into m compartments, and no more than one ball can occupy any compartment, with equal probability for each state, then we have the statistics of Brillouin [4, pp. 18–20]. Each configuration occurs with probability $\Gamma(mn - r + 1) \div \{r!\,\Gamma(mn+1)\}$, $r \leqslant mn$. Extensions of this model allow for noninteger and negative values of m.

It is frequently useful to group cells together. Suppose that there are s groups of cells, with n_j cells in the jth group ($j = 1, \ldots, s; n_1 + \cdots + n_s = n$). Then the probability that R_1 balls are arranged in some manner in the first group, R_2 balls in the second group, etc., where $R_1 + \cdots + R_s = r$, is as follows [4, pp. 15–17]:

In M-B statistics:

$$r!\prod_{j=1}^{s}\{(n_j/n)^{R_j}/R_j!\};$$

In B-E statistics:

$$\left\{\prod_{j=1}^{s}(R_j+n_j-1)\right\}\bigg/\binom{n+r-1}{r};$$

In F-D statistics:

$$\left\{\prod_{j=1}^{s}\binom{n_j}{R_j}\right\}^{R_j}\bigg/\binom{n}{r}, \qquad R_j \leqslant n_j.$$

M-B statistics is fundamental in the development of classical statistical mechanics*. A consideration of identical molecules is an insulated container [2, Chaps. 4, 12] leads to the classical *M-B distribution*

$$f(x, y, z, v_x, v_y, v_z) \propto \exp\{-\epsilon/(kT)\},$$

where total energy

$$\epsilon = \tfrac{1}{2}mv^2 + U(x, y, z)$$

$$v^2 = v_x^2 + v_y^2 + v_z^2$$

m = molecular mass

k = Boltzmann's constant

T = temperature of the container

(x, y, z) = coordinates of position

(v_x, v_y, v_z) = coordinates of velocity

$U(x, y, z)$ = potential function

On the left, $f(\cdot)$ is the joint probability density function (PDF) of (x, y, z, v_x, v_y, v_z). For example, $U = mgz$ in a gravitational field.

If U is constant, then $f = f(v_x, v_y, v_z)$ and reduces to a trivariate normal distribution* with independent and identically distributed components of velocity. A source of confusion arises in physics when this is called the Maxwell distribution*; in this work, the latter name refers to the marginal PDF of the speed $|v|$, given by

$$f_v = 4\pi v^2 f(v_x, v_y, v_z) = 4\lambda^{3/2} v^2 e^{-\lambda v^2}, \qquad v > 0;$$

$$\lambda = m/(2kT).$$

The above is a representation in a continuous phase space. This does not apply in quantum physics*, where distributions are developed for groups of quantum states. M-B statistics makes each quantum state equally likely a priori, and leads to the *Boltzmann distribution* [2, Chap. 16]

$$f_i \propto \exp\{-\epsilon_i/(kT)\},$$

where f_i is the probability that a particle is in the ith group of states, ϵ_i is the mean quantized energy in the ith group, and k and T are defined above. This distribution or the M-B distribution describes the phase space adequately for certain gases and solids in low densities or at high temperatures, but may fail otherwise; one must turn then to B-E or F-D statistics.

In quantum physics, particles with integer spin such as photons satisfy B-E statistics; this leads in the notation above to the *Bose–Einstein distribution*, given by [2, Chaps. 5, 19]

$$f_i \propto \{\exp[(\epsilon_i - \mu)/(kT)] - 1\}^{-1},$$

where μ is the chemical potential of the gas. Particles with half-odd-integer spin such as electrons, protons, and neutrons satisfy F-D statistics; this leads in the notation above to the *Fermi–Dirac distribution*, given by [2, Chaps. 5, 19]

$$f_i \propto \{\exp[(\epsilon_i - \mu)/(kT)] + 1\}^{-1}.$$

There are applications of these models outside the field of statistical physics. Thus F-D statistics applies to the distribution of misprints in a book, since each space is occupied by a symbol with at most one misprint [3, Sec. II.5]. Berg and Bjurulf [1] have approached the paradox-of-voting problem using B-E rather than the traditional M-B statistics. Hill [5] shows that with a certain random distribution for the number n of cells, B-E statistics leads to the size of the kth largest cell being approximately proportional to $k^{-(1+\alpha)}$ for some $\alpha > 0$ and for large n; this is a form of Zipf's law*. Hill applies this model to describe the size of cities within regions and the proportion of genera within species.

In B-E statistics the steady-state probability as $n + r \to \infty$ that a cell has j balls follows a Pareto* or geometric law* depending on certain boundary conditions [6]. The distribution of city sizes, for example, is better fitted by the Pareto limiting form.

References

[1] Berg, S. and Bjurulf, B. (1980). *Tech. Rep. No. 3071/1-16*, Dept. of Statistics, University of Lund, Lund, Sweden. (B-E statistics and the paradox of voting.)

[2] Desloge, E. A. (1966). *Statistical Physics*. Holt, Rinehart and Winston, New York. (A clear development of the field at advanced calculus level, with a detailed treatment of statistical models in statistical mechanics and quantum theory and the distributions that arise from these models.)

[3] Feller, W. (1957). *An Introduction to Probability Theory and Its Applications*, Vol. 1, 2nd ed. Wiley,

New York (3rd ed., 1968). (Basic probabilistic properties of the statistical models presented in Feller's lucid style.)

[4] Fortet, R. (1977). *Elements of Probability*. Gordon and Breach, London (English version).

[5] Hill, B. M. (1974). *J. Amer. Statist. Ass.*, **69**, 1017–1026. (Zipf's law* is derived from B-E statistics as a limiting form with applications. The article by Ijiri and Simon [6] should be consulted in parallel.)

[6] Ijiri, Y. and Simon, H. A. (1975). *Proc. Natl. Acad. Sci. USA*, **72**, 1654–1657. (Develops key properties of B-E statistics and limiting distributional forms, with applications.)

[7] Kunte, S. (1977). *Sankhyā A*, **39**, 305–308. (A characterization of B-E statistics.)

(OCCUPANCY PROBLEMS. STATISTICAL PHYSICS)

CAMPBELL B. READ

FERTILITY MEASUREMENT

Animal fertility is studied as an aid to good husbandry. Statistical analyses are made as a branch of ecology; their aim is to find out what factors (such as predators, climate, and food supply) operate to curb natural fertility in the wild; they may lead to the discovery of new methods for controlling pests or preserving rare species.

In human beings the word "fecundity" is used to mean the biological capacity to reproduce; this can hardly be measured directly but it is known to vary with age and, for a woman, before, during, and after pregnancy. Mathematical models involving the use of computers and the technique of simulation* have been constructed, the end products of which, after allowing for fetal loss and contraception where appropriate, are compared with known data of actual births; as a result, it may be possible to test the validity of postulates about, for example, natural wastage or the effectiveness of birth control.

In demography*, the word "fertility" is used to refer to the study of actual births; the demographer's work on the analysis of human fertility is outlined in the article DEMOGRAPHY and is described in more detail here. Fertility can be studied either as a function of "marriage*" or independently; both approaches are necessary for a complete analysis. The stigma associated with illegitimacy varies from place to place and from time to time, so the pressure on a couple to marry in order to have children or to legitimize a conception out of wedlock will vary correspondingly. The relationship between marital fertility and all-persons fertility is thus not constant. The numbers of births depend greatly also on the prevalence of taboos on sexual intercourse and the use and effectiveness of contraception and induced abortion. For these and other reasons, the average number of children born to married couples who have completed their fertile life has changed a lot in some countries, but not in others. In general, countries with highly developed economies today have about enough children for reproduction, whereas most of the developing areas of Africa, Latin America, and Asia have double this or more.

DEVELOPING COUNTRIES

In many developing countries it is not practicable to obtain an accurate record of the number of births occurring year by year. However, nearly all countries have held a census recently, and many censuses have included enquiries about the birth of children. Reliable responses about illegitimate infants cannot be expected, but it has usually been possible to ask married men and women whether they have had a child in the recent past or how many children have been born to them so far. Ideally, this information should include deceased children as well as those alive at the census, but these are not always remembered, especially girls. In typical enquiries, the ratio of girls to boys recorded falls with increasing length of time married, as the following representative figures illustrate:

Married for under 5 years	ratio 0.95
Married for 10–14 years	ratio 0.90
Married for 20–24 years	ratio 0.85

In some African countries, an understatement of fertility by 25% is normal. Clearly, the problems encountered in attempting to measure fertility in developing countries are those associated with putting the right questions in the right way, obtaining the most accurate answers possible, and correctly interpreting the results. Where efforts are made to reduce pressure on resources by encouraging smaller families, an occasional census does not give enough information on progress, and in consequence much use is made of ad hoc sample surveys*. In addition to questions about numbers of children born, couples are asked in these surveys to indicate their attitude to and use of birth control, to say how many children they consider ideal for other couples, and to state how many further births they expect in their own family.

Problems of omission and inaccuracy may be tackled with fuller or more numerous questions: for example, how many children are still living with their mother, how many are living elsewhere, and how many have died. But there can be difficulties over abortions, stillbirths, and adopted children. Some enquiries concentrate on maternity histories rather than number of children, but this alternative approach may not provide a better answer. Too detailed an interview may be counterproductive if resented by the respondent; enumerators are not always diligent enough in putting all the questions, as has been shown where specific enquiries have been repeated precisely with different interviewers.

Comparison of information is one of the chief weapons against error of interpretation of statistics so collected. Data may be tested for internal consistency or against a reasonable external standard—for example, the experience of a similar country for which more reliable information is available. Knowledge of local conditions is essential in this work. Failing a representative set of data as a check, a reasonable mathematical model may be used. Such models may well be required in order to convert the corrected sample or census data into a more convenient form; for instance, the rate at which a population is reproducing itself is not precisely apparent from data of average family sizes for middle-aged couples, but can be derived from a consideration of population mathematics*.

In 1974, a World Fertility Survey was begun by the International Statistical Institute in conjunction with the United Nations. Sample data have been collected in many countries and there have been significant findings on age at marriage, breastfeeding practices, contraceptive knowledge, and use and fertility rates by age, ethnic group, and marriage duration. Emphasis has been placed on good-quality data and on multivariate analysis* of the determinants of fertility.

HISTORICAL STUDIES

In some European countries, parish records of baptisms, marriages, and funerals are available for earlier centuries, notably the seventeenth and eighteenth. As most people then spent all their lives in one locality, demographers have been able from the close study of the names in these records to trace the formation, growth, and dissolution by death of particular families. It has thus been possible to assess completed family size, the ages of the parents at the time of the various births, the time spacing between births, and other measures of fertility. Owing to variations from place to place and from time to time, the addition of data from a number of small areas is essential in order to form a general picture, which even then cannot be on a fully national scale. The earliest censuses did not include questions about fertility; nevertheless it is possible to form a broad idea of the experience of the time indirectly from published statistics: for example, from the ratios of the numbers of children to the numbers of women enumerated. Not long afterward, vital registration began to provide an increasingly accurate count of the number of births, leading to the calculation of the first crude birth rates.

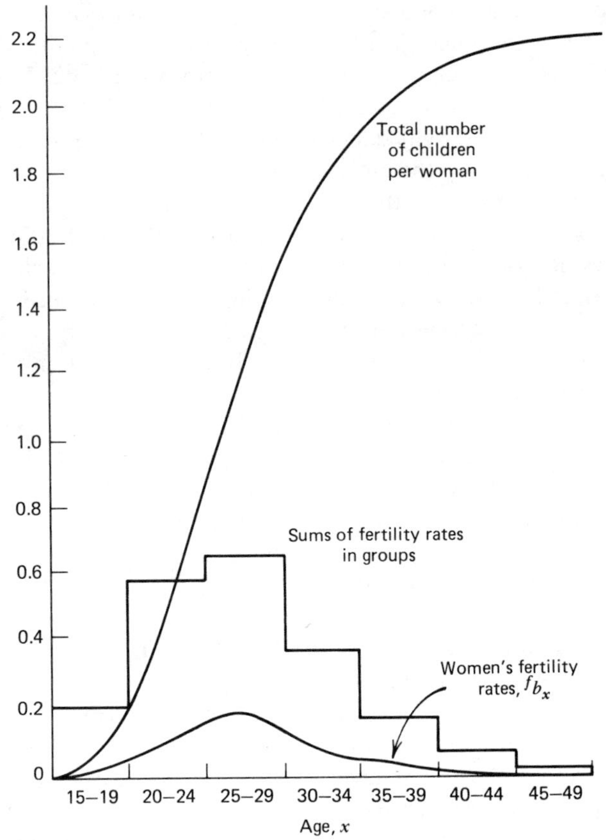

Figure 1

Later, the ages of one or both parents had to be stated on a birth registration, after which fertility rates classified by age could be calculated and the crude rates standardized. Generation analysis also began, independently of marriage, showing how many children a group of women born at the same time have at various ages.

MODERN FERTILITY ANALYSIS IN DEVELOPED COUNTRIES

The ratio of the number of births in a given year to women aged x to the number of those women—$^{f}b_x$ say—can be analyzed by legitimacy, legitimate births being expressed as a proportion of the number of married women and illegitimate births being attributed to spinsters, widows, and divorcees. Similar ratios can sometimes be calculated for men. A curve illustrating the pattern of $^{f}b_x$ as x changes appears in Fig. 1, which shows also the sum of the ratios for all ages up to and including x. This sum gives a measure of the total family size. If the total for all fertile ages is multiplied by the proportion of births of the same sex as the parent, the *gross reproduction rate* for that sex is arrived at; it indicates very neatly the ratio of the size of the next generation to that of the parent generation, ignoring the effect of mortality and assuming that the values of $^{f}b_x$ are constant in time. A correction can be made for mortality, by the use of the formula (in the case of women)

$$\sum_x {}^{f}_{x}p_0 \, {}^{f}b_x ,$$

which is the female *net reproduction rate*. Here are some typical net reproduction rates for a developed country in recent times:

1880	1.50
1920	1.10
1960	1.20
1980	1.00

Such figures demonstrate that any assumption that ${}^{f}b_x$ remains constant in time is unrealistic; it can change even if fertility in marriage is constant, provided that proportions married vary. Another disadvantage of reproduction rates is that the results for men and for women are not normally the same, even when based on the same population at the same time.

In some countries, parents registering a birth have to state how long they have been married or how many children they have had previously. It may then be possible to calculate the ratio of the number of births at marriage duration n to the number of married men or women who have been married for n years. This ratio is the same for either sex and can be summed up to show the total family size, as in Table 1. Such values can be analyzed into those relating to first children, second children, etc., thus showing how the family-building process is varying from time to time; but they relate only to married people. Another form of analysis is to calculate the ratio (independently of age or marriage duration) of the number of couples with m children or more to the number with at least $m - 1$ children; such functions are called *parity progression ratios*.

Information about associations between fertility and social indicators such as education, occupation, and areas of residence is not readily available from birth registration data. Full census data have the volume to permit much subdivision, but it is not always practicable to ask numerous detailed questions about families from the whole population; closer inquiries are sometimes addressed to a sample selected from the census schedules. (The possibilities for analysis are numerous but some difficult matters need resolution; for example, whether attention should be confined to younger or older men or women, to married or single persons, and to all marriages or only unbroken ones.) Large-scale display of the results is often preferable to summarization.

Some typical fertility variations about the national average are illustrated in ratio form in Table 2. Such differences may point to future changes in fashion. Where, however, an average of two children per couple can be born early or late in a marriage that can be fertile for at least 20 years, prediction is difficult because of changes in personal timing, which may represent no real ultimate alteration in family completion. Research workers therefore conduct small sample surveys in order to ascertain people's motivations: for example, their idea of a proper size and composition of family, and how this varies according to economic circumstances. They also enquire into actual intentions for the near future and into the effectiveness of couples' planning methods, such as the use of contraceptives of various kinds or of induced abortion. In this way it may be found out, for instance, how many pregnancies were planned and how many accidental, and by what proportion they have been termi-

Table 1

Number of Years Married	Calendar Year of Marriage 1971	1979
1	0.30	0.25
2	0.70	0.65
3	1.00	0.90
4	1.25	1.10

Table 2

	1950	1970
Type of employment		
Employers and managers	0.90	0.85
Clerical workers	0.95	0.90
Unskilled manual workers	1.25	1.15
Area of residence		
Urban	0.95	0.90
Rural	1.05	1.10
Level of education		
High	0.85	0.80
Minimum	1.20	1.15

nated before birth occurred: this type of information is useful for the determination of social policy as well as for population forecasting.

Bibliography

General demographic textbooks include sections on fertility analysis; *see* DEMOGRAPHY.

Bogue, D. L. (1972). *Demographic Techniques of Fertility Analysis*. University of Chicago, Chicago. (Illustrates the techniques in detail with data for the United States.)

Campbell, A. A. (1963). In *Emerging Techniques in Population Research*. Milbank Memorial Fund, New York.

Holmberg, I. (1970, 1972). *Fecundity, Fertility, and Family Planning*. University of Gothenberg, Gothenburg, Sweden. (Describes the use of electronic computers to study statistically the effects of sterility, fetal wastage, and birth control on fertility.)

Levels and Trends of Fertility throughout the World, 1950–1970. (1977). U.N. Popul. Stud. No. 59. (Discusses methods of analysis and gives a mass of factual information on a global basis.)

Moss, L. and Goldstein, H., eds. (1979). *The Recall Method in Social Surveys*. University of London, London. (Includes a commentary on the accuracy of fertility enquiries in developing countries.)

Peel, J. and Carr, G. (1975). *Contraception and Family Design*. Churchill Livingstone, London. (This describes methods used in a sample fertility survey in Britain and gives the results of the inquiry.)

Srikantan, K. S. (1977). *The Family Program in the Socio-economic Context*. The Population Council, New York. (This book discusses the impact of national family-planning programs on fertility in a number of countries.)

Variables for Comparative Fertility Studies (1967). A Working Paper prepared by a Committee of the International Union for the Scientific Study of Population.

World Fertility Survey: A Summary of Findings, 1977, 1978. International Statistical Institute. *World Fertility Survey: Regional Workshops on Techniques of Analysis*. U.N. Asian Popul. Stud. Ser. 44. (These publications give a useful introduction to the work of the World Fertility Survey.)

(DEMOGRAPHY
MARRIAGE
POPULATION PROJECTIONS
SOCIOLOGY, STATISTICS IN
VITAL STATISTICS)

PETER R. COX

FIDUCIAL DISTRIBUTIONS

The word *fiducial*, like many others, was introduced into statistics by R. A. Fisher*. In 1930 [2], he wrote:

> In many cases the random sampling distribution of a statistic, T, calculable directly from the observations, is expressible solely in terms of a single parameter, of which T is the estimate found by the method of maximum likelihood*. If T is a statistic of continuous variation, and P the probability that T should be less than any specified value, we have then a relation of the form
>
> $$P = F(T, \theta).$$
>
> If now we give to P any particular value such as 0.95, we have a relationship between the statistic T and the parameter θ, such that T is the 95 per cent value corresponding to a given θ, and this relationship implies the perfectly objective fact that in 5 per cent of samples T will exceed the 95 per cent value corresponding to the actual value of θ in the population from which it is drawn. To any value of T there will moreover be usually a particular value of θ to which it bears this relationship; we may call this the "fiducial 5 per cent value of θ" corresponding to a given T. If, as usually if not always happens, T increases with θ for all possible values, we may express the relationship by saying that the true value of θ will be less than the fiducial 5 per cent value corresponding to the observed value of T in exactly 5 trials in 100. By constructing a table of corresponding values, we may know as soon as T is calculated what is the fiducial 5 per cent of θ, and that the true value of θ will be less than this value in just 5 per cent of trials. This is a definite probability statement about the unknown parameter θ, which is true irrespective of any assumption as to its *a priori* distribution.

and:

> Generally, the fiducial distribution of a parameter θ for a given statistic T may be expressed as
>
> $$df = -\frac{\partial}{\partial \theta} F(T, \theta)\, d\theta,$$

while the distribution of the statistic for a given value of the parameter is

$$df = \frac{\partial}{\partial T} F(T, \theta) dT.$$

Fisher was later to observe that the origins of his argument were to be found in Student's famous 1908 paper [14] introducing the *t*-distribution*, for Student noted that "if two observations have been made and we have no other information, it is an even chance that the mean of the (normal) population will lie between them" (*see* GOSSET, WILLIAM SEALY).

The natural interpretation of this statement is that if, on each occasion a sample of 2 is drawn from a normal distribution, the statement "the unknown mean lies between the observations" is made, it will be true with probability $\frac{1}{2}$, whatever the means and variances of the succession of normal distributions involved. It is possible, however, that Student meant no more than that for repeated sampling from any *particular* normal distribution of unknown mean and variance, the observations would straddle the fixed mean with probability $\frac{1}{2}$, and that writing as if the mean itself had a probability distribution was to use an ellipsis. This question of the interpretation of possibly elliptical statements is a recurring difficulty in establishing the origins of fiducial distributions (and hence of confidence intervals). Fisher was later to lay much stress on the careful specification of the reference set with respect to which probability statements were made (a clarification that opened the way for the concepts of conditional inference* and ancillarity*), but without such a specification the ambiguity remains.

In Fisher's 1930 paper [2], however, there is no ambiguity; he saw the distinction that needed to be made, and made it. He conceives of the unknown parameter as having a probability distribution, and is at pains from the outset to distinguish this fiducial distribution from the posterior* (or "inverse") probability distribution that would result from assuming a prior* distribution for the parameter and applying Bayes' theorem*. When in 1932 H. Jeffreys [9] pointed out that the fiducial distribution for the standard deviation σ of a normal distribution was identical to the posterior distribution on the assumption of the prior $df = (1/\sigma)d\sigma$, Fisher [3] replied:

> Probability statements of this [fiducial] type are logically entirely distinct from inverse probability statements, and remain true whatever the distribution *a priori* of σ may actually be. To distinguish them from statements of inverse probability I have called them statements of fiducial probability. This distinction is necessary since the assumption of a given frequency distribution *a priori*, though in practice always precarious, might conceivably be true, in which case we should have two possible probability statements differing numerically, and expressible in a similar verbal form, though necessarily differing in their logical content. The probabilities differ in referring to different populations; that of the fiducial probability is the population of all possible random samples, that of the inverse probability is a group of samples selected to resemble that actually observed.
>
> It is the lack of this distinction that gives a deceptive plausibility to the frequency distribution *a priori*
>
> $$df = d\sigma/\sigma = d(\log \sigma).$$
>
> For this particular distribution *a priori* makes the statements of inverse and of fiducial probability numerically the same, and so allows their logical distinctness to be slurred over.

Many statisticians have failed to appreciate the step that Fisher took in 1930 and have published comments ranging from admitted bewilderment to suggestions that Fisher did not understand conditional probability. The evidence is that the step is indeed difficult to take, although once taken it seems trivially obvious. Thus D. A. S. Fraser, who later constructed the edifice of structural inference* on the foundations of fiducial inference, at first [6] stated of a fiducial distribution: "It is not a probability distribution that represents frequencies in repeated sampling"; by contrast, J. Neyman* initially [11] thought that "the possibility of solving the problems of statistical estimation independently from any knowledge of the *a priori* laws, discovered by R. A. Fisher, makes it superfluous to make any appeals to the Bayes' theorem," but later [12] felt the fiducial argument to be "no more than a misconception born out of an early mistake."

It is likely that much of the adverse comment directed at fiducial probability distributions in general should be reserved for the multiparameter cases, such as Behrens' problem*, in which inconsistencies arise, repeated-sampling interpretations are lost, and a coherent theory seems unachievable; so much so that G. N. Wilkinson [15] has proposed a "noncoherence principle" as salvation for the theory. But the application to single parameters, to which the remainder of this article is confined, not only led to the theory of confidence intervals* but to a rich development of ideas centered on ancillarity, conditional inference, and likelihood*.

THE PIVOTAL QUANTITY

Central to the development of a fiducial probability distribution is the notion of a pivot, or pivotal quantity* [5], which is a function jointly of the parameter and a statistic, whose distribution is independent of the parameter. Thus, if a sample of size n from the normal distribution $N(\mu, 1)$ has mean $\bar{X}$, $(\mu - \bar{X})$ is a pivot, with distribution $N(0, 1/n)$. Evidently, pivots can only be defined if both parameter and sample spaces are continuous (or both discrete, a somewhat artificial situation which has, however, been useful as an aid to discussion). Their essential quality is the possession of a known distribution under repeated sampling.

For a single parameter θ and statistic T,

$$Z = F(T, \theta)$$

itself is evidently a pivotal quantity, since it is a function of T and θ with a known distribution—uniform between 0 and 1. Thus the way seems clear to construct fiducial distributions without difficulty, distributions from which fiducial intervals may be constructed at will in the certain knowledge that they possess the repeated-sampling coverage property (or confidence property).

Another requirement stressed by Fisher was the need for complete absence of information a priori about the parameter in question.

CONFIDENCE INTERVALS

The confidence property has been used as the foundation of a school of statistical inference associated with the name of J. Neyman, but from its inception in 1934 Fisher was at pains to point out that for his purposes of inductive inference the confidence property alone was not enough, because ambiguities would arise unless the statistic T was uniquely chosen. The report of his contribution to the discussion of Neyman's paper introducing confidence intervals [11] says:

> Dr. Fisher's own applications of fiducial probability had been severely and deliberately limited. He had hoped, indeed, that the ingenuity of later writers would find means of extending its application to cases about which he was still in doubt, but some limitations seemed to be essential. Those who had followed the earlier parts of the story would have no difficulty in perceiving these, but there might be pitfalls for those who interested themselves only in the later chapters. In particular, he would apply the fiducial argument, or rather would claim unique validity for its results, only in those cases for which the problem of estimation proper had been completely solved, i.e. either when there existed a statistic of the kind called *sufficient**, which in itself contained the whole of the information supplied by the data, or when, though there was no sufficient statistic, yet the whole of the information could be utilized in the form of *ancillary* information. Both these cases were fortunately of common occurrence, but the limitation seemed to be a

> necessary one, if they were to avoid drawing from the same body of data statements of fiducial probability which were in apparent contradiction.
>
> Dr. Neyman claimed to have generalized the argument of fiducial probability, and he had every reason to be proud of the line of argument he had developed for its perfect clarity. The generalization was a wide and very handsome one, but it had been erected at considerable expense, and it was perhaps as well to count the cost. The first item to which he would call attention was the loss of uniqueness in the result, and the consequent danger of apparently contradictory inferences.
>
> In the second place, Dr. Fisher had limited his application to continuous distributions, hoping with more confidence in this case that the limitation might later be removed. Dr. Neyman removed this limitation, but at the expense of replacing inferences that stated the exact value of the fiducial probability by inequalities, which asserted that it was not less than some assigned value. This also was somewhat a wide departure, for it raised the question whether exact statements of probability were really impossible, and if they were, whether the inequality arrived at was really the closest inequality to be derived by a valid argument from the data.

The differences between the fiducial and confidence viewpoints led to much misunderstanding in the years that followed their introduction (*see* CONFIDENCE INTERVALS AND REGIONS). On the one hand, adherents to fiducial inference sought to make fully informative inductive statements about particular cases, and were not willing to relent over their insistence on the use of fully efficient statistics and exact probability statements (which require the existence of pivots) for this purpose; on the other hand, adherents to confidence intervals, concerned to exploit only the confidence property, and not hesitating to state inequalities where exact statements of probability were not available, were on a much freer rein, and rapidly colonized much of statistics. To the charge that confidence intervals are often absurd as inductive statements in particular cases, they retorted that confidence intervals were not to be given an inductive interpretation, but were to be viewed only as statements asserting long-run coverage probabilities. Usage suggests the contrary. Fiducial distributions faded from the scene, but not before they had given birth to the likelihood principle*.

THE LIKELIHOOD PRINCIPLE

The possibility that the likelihood function should be regarded as carrying all the information* contained in the data (conditional on the particular statistical model) was first suggested in a non-Bayesian context by Fisher in 1934 [4], and is referred to as the *likelihood principle*. Having specified that the estimation procedure to be used to generate a fiducial distribution must be exhaustive, he referred to the two classes of problems admitting exhaustive estimation, the first being where a sufficient statistic* exists, and the second where, although there is no sufficient statistic, conditional inference in effect supplies one.

In the first case the fact that the likelihood function determines the sufficient statistic (and vice versa) supports the likelihood principle, although the relation between the likelihood function and the fiducial distribution is then in want of elaboration; but in the second case, that of the estimation of the parameter in a location-parameter distribution, Fisher, arguing that the fiducial distribution should be conditional on the ancillary statistic which he called the "configuration" or mutual spacing of the sample members, was able to show that this distribution was the normed likelihood function, a circumstance that lent further support to the likelihood principle and paved the way for a fuller understanding of the relationship between a fiducial distribution and a Bayesian posterior distribution.

The phrase "likelihood principle" was not used by Fisher, and his notion was only that the likelihood function "when properly interpreted, must contain the whole of the information." The question of interpretation was thus left open, and repeated-sampling interpretations relying on the likelihood function, such as are provided by fiducial distributions, were not excluded. Later enunciations of the likelihood principle by other authors have sometimes omitted any question of further interpretation, asserting that the likelihood function alone, freed from the model that generated it, contains all the information about the parameter that indexes it.

RELATION TO BAYESIAN POSTERIOR DISTRIBUTIONS

If the fiducial distribution in the location-parameter case is simply the likelihood function normed, it is identical to the Bayesian posterior distribution on the assumption of a uniform prior distribution. Such a posterior distribution, being fiducial, enables intervals to be set up which possess the confidence property, albeit with respect to the reference set of all samples possessing the same configuration as the one actually observed. The notion of a "conditional confidence interval" is thus not recent, and it is no surprise that certain Bayesian posterior distributions have well-defined repeated-sampling interpretations. The question arises, however, as to the extent of the class of problems manifesting similar Bayesian and fiducial solutions. There are two pointers to the answer.

First, for a fiducial distribution to be a valid inference, the making of an observation should not invalidate the statement of the distribution of the pivotal quantity. Fisher did not discuss this requirement, first made explicit by A. P. Dempster [1] and which I. Hacking [8] traced to a remark of H. Jeffreys in 1937 [10]. This *Principle of Irrelevance* requires that the observations be uninformative about the pivot, and in Hacking's version the requirement is satisfied if the likelihood function for the *pivot*, given the observations, is independent of them. Then

$$\frac{\partial Z}{\partial X} = A(X)B(X) \qquad \text{and}$$
$$Z = G(R(X) + S(\theta)),$$

where $G' = A$ and $R' = B$. Thus in the single-parameter case the distribution must either be of the location-parameter form or be transformable to it.

Second, if a fiducial distribution is to possess the property that further data may be incorporated using Bayes' theorem, it must embody the data only through the likelihood function, again implying location-parameter form. This form excludes the general single-parameter case with a sufficient statistic originally included by Fisher.

If the Principle of Irrelevance be admitted, then the application of fiducial distributions for single parameters is reduced to the location-parameter case, or cases transformable thereto, and any wider use of fiducial theory is ruled out. In particular, the use of confidence intervals to reflect confidence in a hypothesis is excluded for all other cases. What then remains of the theory of fiducial distributions is equivalent to the adoption of a uniform prior for the location parameter and the application of Bayes' theorem, the resulting distribution having the conditional repeated-sampling property already alluded to.

RELATION TO STRUCTURAL INFERENCE

One way of looking at the translation-parameter model is to regard the pivotal quantity as an "error variable" e and thus to separate the random variable into two parts,

$$X = \theta + e.$$

D. A. S. Fraser [7] has applied the fiducial argument to such models in his "structural inference"*, employing group-theoretic language to describe the resulting structures. The models are, however, the common statistical models which possess pivotal quantities, with the pivot specified as part of the model. Where, as may happen in multi-parameter cases, a model possesses more than one pivot, Fraser solves the uniqueness problem by asserting that there are as many structural models as there are pivots, and that each has its own interpretation.

CONCLUSION

Fiducial distributions were introduced by R. A. Fisher in 1930 [2] at a time when statisticians held that the only justifiable probability statements about parameters were those based on an agreed Bayesian prior distribution, where such existed. From the outset Fisher insisted that for the avoidance of paradoxes fiducial distributions must utilize all the available information; Neyman dismissed the point, took the fiducial argument's repeated-sampling property as the important principle, and developed the theory of confidence intervals. Fisher, unimpressed, strove to define accurately his notion of "all the available information," and in so doing introduced conditional inference, ancillarity, and the likelihood principle, and added the concept of a pivotal function to the original theory. In multiparameter applications, however, pivots and paradoxes multiplied, leading to unsympathetic comments but no critical analysis. Only after Fisher's death did Hacking, drawing on Jeffreys' work, and Dempster perceive the need for the additional Principle of Irrelevance, which, when applied, reduced the field of application of fiducial distributions to location-and-scale parameter* models. Fraser's structural distributions are essentially fiducial distributions of this kind. The alternative to accepting the limitations imposed by the Principle of Irrelevance is to accept that, in general, fiducial probabilities are not coherent, along the lines suggested by Wilkinson [15]; *see* COHERENCE.

In sum, fiducial distributions have been influential in the development of theories of statistical inference that rely on repeated-sampling considerations, but their early promise has not been fulfilled, nor has any of the theories developed from fiducial theory approached either universal applicability or immunity from criticism. Late in his life Fisher remarked to L. J. Savage [13]: "I don't understand yet what fiducial probability does. We shall have to live with it a long time before we know what it's doing for us. But it should not be ignored just because we don't yet have a clear interpretation."

References

[1] Dempster, A. P. (1964). *J. Amer. Statist. Ass.*, **59**, 56–66.

[2] Fisher, R. A. (1930). *Proc. Camb. Philos. Soc.*, **26**, 528–535.

[3] Fisher, R. A. (1933). *Proc. R. Soc. Lond. A.*, **139**, 343–348.

[4] Fisher, R. A. (1934). *Proc. R. Soc. Lond. A.*, **144**, 285–307.

[5] Fisher, R. A. (1941). *Ann. Eugen. (Lond.)*, **11**, 141–172.

[6] Fraser, D. A. S. (1958). *Statistics: An Introduction*. Wiley, New York.

[7] Fraser, D. A. S. (1968). *The Structure of Inference*. Wiley, New York.

[8] Hacking, I. (1965). *Logic of Statistical Inference*. Cambridge University Press, Cambridge. (Illuminating comments, especially on the Principle of Irrelevance.)

[9] Jeffreys, H. (1932). *Proc. R. Soc. Lond. A.*, **138**, 48–55.

[10] Jeffreys, H. (1937). *Proc. R. Soc. Lond. A.*, **160**, 325–348.

[11] Neyman, J. (1934). *J. R. Statist. Soc.*, **97**, 558–625.

[12] Neyman, J. (1961). *J. Operat. Res. Soc. Japan*, **3**, 145–154.

[13] Savage, L. J. (1964). In C. R. Rao, *Bull. Int. Statist. Inst.*, **40**, 833–939.

[14] "Student" (1908). *Biometrika*, **6**, 1–25.

[15] Wilkinson, G. N. (1977). *J. R. Statist. Soc. B.*, **39**, 119–171.

Bibliography

See the following works, as well as the references just cited, for more information on the topic of fiducial distributions.

Edwards, A. W. F. (1976). *Statistician*, **25**, 15–35. (An elementary account of fiducial probability.)

Fisher, R. A. (1956). *Statistical Methods and Scientific Inference*. Oliver & Boyd, Edinburgh.

Fisher, R. A. (1971–1974). *Collected Papers of R. A. Fisher*, 5 Vols., J. H. Bennett, ed. University of Adelaide, Adelaide, Australia.

Jeffreys, H. (1939). *Theory of Probability*. Clarendon Press, Oxford.

Plackett, R. L. (1966). *J. R. Statist. Soc. A.*, **129**, 249–267. (Excellent review up to 1965, with many references.)

Rao, C. R. (1964). *Bull. Int. Statist. Inst.*, **40**, 833–939. (Proceedings of a symposium on fiducial probability.)

(CONFIDENCE INTERVALS AND REGIONS
FIDUCIAL INFERENCE
FIDUCIAL PROBABILITY
SUFFICIENCY)

A. W. F. EDWARDS

FIDUCIAL INFERENCE

It has been said that fiducial inference as put forward by R. A. Fisher* is not so much a theory as a collection of examples. A gradual evolution of ideas can be seen in Fisher's published work, and he himself may have been less satisfied with his own theories than his writings would lead one to suspect. Joan Box [1, p. 458] writes that he continued to work at these problems but to the end of his days was not satisfied with the further solutions he could arrive at. . . . He did not unravel the puzzle.

In view of this and of the lack of any generally accepted definition of fiducial probability, it is not surprising that the subject has been one of confusion and controversy ever since its introduction in 1930. While interest in fiducial inference has declined since Fisher's death in 1962, there continue to be efforts to clarify and extend his ideas (see, e.g., Fraser [13–18], Hacking [19], Verhagen [34], Bunke [5], Wilkinson [35], Pedersen [26], and Seidenfeld [31]).

Several key ideas can be illustrated by the case of a single observation x from a normal distribution with mean μ and unit variance, for which we use the standard notation $x \sim N(\mu, 1)$. If we put $z = x - \mu$, then $z \sim N(0, 1)$. A quantity like z, which depends on the observation x and the parameter μ and whose distribution is free of the parameter, is called a *pivotal quantity** or *pivot*. The *fiducial argument* consists in writing $\mu = x - z$ and asserting that when we have no knowledge about μ except the value x, our uncertainty about μ is summarized by saying that μ equals x minus an unknown value of a standard normal random variable. In short, $\mu \sim N(x, 1)$. This is called the *fiducial distribution** of μ. The values $x \pm 1.96$ include all but 5% of the distribution and so would be called 95% *fiducial limits* for μ.

If the previously mentioned trial were repeated indefinitely with arbitrarily varying μ values to give (μ_1, x_1), (μ_2, x_2), . . . , and if the μ values were subsequently revealed and plotted on a scale relative to a fixed point x in such a way that $\mu_i - x$ equals the actual ith difference $\mu_i - x_i$, the plotted values would follow a normal distribution centered at x, that is, a $N(x, 1)$ distribution, the fiducial distribution. Thus it can be argued that in the absence of a priori information, our knowledge (or uncertainty) about μ_1 say, given x_1, is summed up by stating that if its value were revealed, it would appear to be a random value from $N(x_1, 1)$.

RELATIONSHIP TO BAYESIAN INFERENCE

If $f(x, \theta)$ is the assumed probability law of data x and $\pi(\theta)$ is a prior density, then Bayes' theorem* yields the posterior density $\pi(\theta \mid x) = \pi(\theta) f(x, \theta) / \int \pi(\theta) f(x, \theta)\, d\theta$. In using the term "inverse probability" Fisher [8] referred to the practice of taking $\pi(\theta)$ to be constant in order to represent prior ignorance (Bayes' postulate). Pointing out that this procedure was inconsistent under transformations of θ, Fisher put forward his own likelihood* theory and fiducial theory to avoid the objectionable postulate.

In more recent times there has been a tendency to regard the prior density $\pi(\theta)$ as a representation of subjective belief. This view was equally distasteful to Fisher, whose constant goal was an objective theory uncontaminated by subjective elements.

RELATIONSHIP TO CONFIDENCE INTERVALS

Like fiducial theory, Neyman's theory of confidence intervals (*see* CONFIDENCE INTERVALS AND REGIONS) leads to probability (or confidence) statements about the value of θ without appealing to any prior density. As we indicate later, in many examples there is a formal correspondence between the two theories in that the fiducial probability of the confidence interval equals the confidence level. Whether or not the two theories give different numerical results, there are differences in their aims and interpretations.

1. In confidence interval theory, θ is considered a fixed constant and the interval is considered random. In fiducial theory, x is considered fixed and θ random, or more accurately, uncertain.
2. Confidence intervals are admittedly nonunique. In Fisher's view it was a fatal defect that different solutions could assign different confidence levels to a single interval. Uniqueness of fiducial distributions was consistently maintained by Fisher but disputed by others. Concepts like sufficiency* and Fisher information* presumably furnish the keys to uniqueness.
3. Fiducial theory yields, through the integral of the density, the fiducial probability of any interval, whereas confidence intervals only assign a prechosen confidence level to a particular interval. This distinction does tend to disappear, however, if one requires confidence intervals for every confidence level γ, $0 < \gamma < 1$, rather than only a single fixed value such as $\gamma = 0.95$.

ESTIMATING A SINGLE PARAMETER

Let x be either a single observation or a sufficient statistic with a CDF $F(x,\theta)$ such that $\partial F/\partial\theta$ is negative. The contours $F = 0.1, 0.2, \ldots, 0.9$ in the (θ, x)-plane slope upward to the right and divide the plane into 10 regions. For any fixed θ, the random value of x has equal probability of falling in each region. If x is fixed, then the set of all θ values typically is divided into 10 intervals. From the fiducial point of view, each of these intervals has fiducial probability (given x) of 0.1. From the confidence interval point of view, the θ values in any k contiguous intervals, say, would constitute a confidence interval with confidence coefficient $k/10$. By refining the subdivision of F values we are led to the expression

$$\varphi(\theta \mid x) = -\partial F(x,\theta)/\partial\theta \qquad (1)$$

for the fiducial density of θ given x, a formula given by Fisher in 1930 [8] when first introducing fiducial probability, and given again in 1956 [12, p. 70]. Fisher's 1930 explanation emphasized frequencies and hardly differs from a description of confidence intervals. Only later when the theories were extended to more complex models did differences become apparent.

Necessary and sufficient conditions for $-\partial F/\partial\theta$ to be a posterior distribution for some prior are given by Lindley [24]: There exist transformations of x to u and θ to τ such that τ is a location parameter for u. The prior on τ must then be uniform, and if regularity conditions require $-\infty < \tau < \infty$, only an improper prior distribution is capable of yielding a posterior density identical with the fiducial density.

To derive (1) by a pivotal argument, let $u = F(x,\theta)$. The transformation from x to u with θ fixed yields the uniform density $g(u) = 1$ $(0 \leqslant u \leqslant 1)$, and thus $u = F(x,\theta)$ is a pivot [10, p. 395]. The transformation from u to θ with x fixed gives $\varphi(\theta \mid x) = g(u) |\partial u/\partial\theta| = -\partial F/\partial\theta$, the fiducial density.

When θ is a location parameter for x, then $F(x,\theta)$ has the form $H(x-\theta)$ and the fiducial density is $\varphi(\theta \mid x) = h(x-\theta)$, where $h(u) = dH(u)/du$. In this case the graphs of $f(x,\theta)$ and $\varphi(\theta \mid x)$ are mirror images.

JOINT PIVOTS

Joint fiducial distributions of two or more parameters can be derived by several methods, and it is not surprising to find uniqueness problems. In this section we describe the method of joint pivots. Other methods, such as the use of conditional and marginal pivots, are mentioned in the next section.

If $u_i = u_i(x_1, x_2, \theta_1, \theta_2)$ $(i = 1, 2)$ where x_1, x_2 are statistics and θ_1, θ_2 are parameters and if the density $g(u_1, u_2) = f(x_1, x_2; \theta_1, \theta_2)$

$|J_{xu}|$ (J_{xu} is the Jacobian* of the transformation with fixed θ_1, θ_2) does not depend on θ_1, θ_2, then u_1, u_2 are joint pivots. Transforming from u_1, u_2 to θ_1, θ_2 with x_1, x_2 fixed (this and the previous transformation must both be one-to-one) gives

$$\varphi(\theta_1, \theta_2 | x_1, x_2) = g(u_1, u_2)|J_{u\theta}|,$$

which is the joint fiducial distribution of θ_1, θ_2, at least if we have chosen legitimate pivots. Fisher [12, p. 172] cautions against an arbitrary choice of pivots but provides no comprehensive rules. An earlier discussion [10, p. 395] ignored consistency problems.

Any fiducial distribution obtained from joint pivots is consistent with a confidence region interpretation by the following argument. Let R be any region in the (u_1, u_2)-plane with $\Pr[R] = \gamma$, and let $S(x_1, x_2)$ be the image of R in the (θ_1, θ_2)-plane depending on fixed observed values x_1, x_2. Then $S(x_1, x_2)$ is a confidence region with confidence level γ, that is,

$$\Pr\left[(\theta_1, \theta_2) \in S(X_1, X_2) | \theta_1, \theta_2\right]$$
$$= \Pr\left[R\right] = \gamma,$$

and $S(x_1, x_2)$ has fiducial probability γ.

STUDENT'S DISTRIBUTION

A well-known fiducial distribution is that of the normal mean μ when the population variance σ^2 is also known. If $\bar{x}$ denotes the mean of a sample of size n, and $s^2 = (n-1)^{-1}\sum(x_i - \bar{x})^2$, then $t = \sqrt{n}(\bar{x} - \mu)/s$ is known to have Student's distribution. In first presenting this example in 1935 [10], Fisher wrote:

> It must now be noticed that t is a continuous function of the unknown parameter, the mean, together with observable values, $\bar{x}$, s and n, only. Consequently the inequality $t > t_1$ is equivalent to the inequality $\mu < \bar{x} - st_1/\sqrt{n}$, so that this last inequality must be satisfied with the same probability as the first. This probability is known for all values of t_1, and decreases continuously as t_1 is increased. Since, therefore, the right-hand side of the inequality takes, by varying t_1, all real values, we may state the probability that μ is less than any assigned value, or the probability that it lies between any assigned values, or, in short, its probability distribution, in the light of the sample observed.

Thus by the pivotal or fiducial argument, the fiducial distribution of μ is the distribution of $\bar{x} - st/\sqrt{n}$ where $\bar{x}$ and s are fixed at their observed values and t has Student's distribution; *see* t-DISTRIBUTION.

The Student example is notable in that many routes converge on the same answer. The fiducial limits are of course identical to the confidence limits found in virtually every statistics textbook. Moreover, Jeffreys [21, pp. 122, 352], using the improper prior $d\mu\, d\sigma/\sigma$ (which he favored for its invariance* properties), noted the correspondence of the fiducial and posterior distributions. In addition, we may mention some consistent variants of the fiducial method:

1. Use joint pivots (t, u) with $u = s/\sigma$ and get the marginal density of μ from the joint density of μ and σ.
2. Use u to get the marginal density of σ from the marginal density of s, then multiply this by the conditional density of μ given σ obtained from the pivot $\bar{x} - \mu$ conditional on σ. (Fisher [12, p. 119] calls this the "rigorous" way.)
3. Use the Fisher–Pitman theory of location and scale parameters* discussed below.
4. Obtain the fiducial distribution of the mean and variance of a future sample of size n' and let n' tend to infinity [10; 12, p. 119].

BEHRENS' DISTRIBUTION

The estimation of the difference of normal means $\delta = \mu_1 - \mu_2$ when the variances are not assumed equal is the *Behrens–Fisher problem**. It is of historical interest as an early example in which fiducial limits are not confidence limits. No entirely satisfactory confidence interval solution is available, and the merits of the Behrens–Fisher solution and its competitors continue to be debated.

In an obvious extension of the notation of the previous section, we can write

$$\delta = \mu_1 - \mu_2 = \bar{x}_1 - \bar{x}_2 - s_1 t_1/\sqrt{n_1} + s_2 t_2/\sqrt{n_2}\ .$$

From this Fisher [10] argued that δ is fiducially distributed like a constant, $\bar{x}_1 - \bar{x}_2$, plus a variable equal to a weighted sum of two independent Student variables (a Behrens distribution). Although the exact coverage probability of the resulting fiducial intervals cannot exactly equal the corresponding fiducial probability (see, e.g., Kendall and Stuart [22, p. 149]), numerical evidence indicates that the procedure is conservative (see Robinson [29] and Savage [30, footnote 28 (by John Pratt)]).

The fiducial distribution of δ is known to equal a posterior distribution corresponding to the improper prior $d\mu_1\, d\mu_2\, d\sigma_1\, d\sigma_2/(\sigma_1\sigma_2)$.

CONDITIONAL PIVOTS

Let **y** denote the vector of $n-1$ spacings* of n ordered observations based on a sample from a location family $f(x-\theta)$. Then the distribution of **y** is free of θ so that **y** is an ancillary statistic*. Moreover, if $\hat{\theta}$ denotes the maximum likelihood* estimator, then ($\hat{\theta}$, **y**) is a sufficient statistic. Fisher favored conditional inference* in problems having this structure, arguing that the fixed value of **y** determines the appropriate reference set for inference about θ. In the location example a fiducial argument based on the distribution of the pivot $\hat{\theta}-\theta$, conditional on **y**, yields a fiducial distribution proportional to the likelihood function: $\varphi(\theta \mid x_1, \ldots, x_n) \propto \prod_{i=1}^{n} f(x_i - \theta)$. The theory remains incomplete because of unanswered problems of existence and uniqueness of ancillaries. The existence question was raised by Fisher [11; 12, p. 118] in stating Fisher's "problem of the Nile"*.

LOCATION AND SCALE MODELS

For a sample of size n from a location-scale model $\sigma^{-1}f((x-\theta)/\sigma)$, the $n-2$ quotients of the $n-1$ spacings of the ordered observations are distributed independently of (θ, σ) and so are jointly ancillary. Conditional joint pivots can be found which yield the Fisher–Pitman fiducial distribution

$$\varphi(\theta, \sigma \mid x_1, \ldots, x_n) \propto \sigma^{-n-1} \prod_{i=1}^{n} f((x_i - \theta)/\sigma).$$

Fisher [9] gave likelihood theory relevant to this model, but the fiducial distribution was first given explicitly by Pitman [27]. The example is discussed again in Fisher [12, pp. 159–163]. The fiducial distribution is evidently equivalent to a posterior distribution corresponding to the improper prior $d\theta\, d\sigma/\sigma$, and marginal distributions of θ and σ can be used to obtain confidence intervals. By transformation the results apply to distributions not initially in location-scale form, such as the Weibull* (see, e.g., Lawless [23]).

DIFFICULTIES

The following examples, paradoxical in varying degrees, show why circumspection is needed in interpretations and manipulations of fiducial probability.

If $x \sim N(\theta, 1)$, the fiducial density of θ^2 derived from $\theta \sim N(x, 1)$ is different from that derived from the density of x^2.

In the estimation of μ_2/μ_1, given a sample from a bivariate normal population, Creasy [6] and Fieller [7] obtained different solutions by using different pivots (the "Fieller–Creasy paradox"; *see* FIELLER'S THEOREM). Mauldon [25] and Tukey [33] give other examples of nonuniqueness involving joint pivots.

Lindley [24] considered two observations from $f(x, \theta) = \theta^2(x+1)e^{-\theta x}/(\theta+1)$ $(x > 0)$ and showed that the fiducial distribution $\varphi(\theta \mid x_1, x_2)$ is not equal to the posterior distribution of θ given x_2 when the prior is taken to equal the fiducial distribution $\varphi(\theta \mid x_1)$. If the fiducial distribution is to be interpreted like a prior distribution, as one might infer for example from Fisher [12, p. 125], then one would have expected equality. A related fact is that $\varphi(\theta \mid x_1)$ is not a posterior density for any prior.

Stein [32] obtained the marginal fiducial distribution of $\sum_1^n \theta_i^2$ from the joint fiducial distribution of $\theta_1, \ldots, \theta_n$ given $x_1, \ldots, x_n$ where $x_i \sim N(\theta_i, 1)$, and showed that there could be arbitrarily large discrepancies between the resulting fiducial probabilities and confidence levels arrived at by using the statistic $\sum_1^n x_i^2$.

Given x_1, x_2 from $N(\mu, \sigma)$ the interval $\min(x_1, x_2) < \mu < \max(x_1, x_2)$ has fiducial probability 0.5, but in the subset of cases where $3|x_1 - x_2| > 2|x_1 + x_2|$ the conditional probability exceeds 0.518 for all μ, σ [4]. Brown [3] gives generalizations; Yates [36] defends Fisher's theory.

INVARIANCE

Many standard parametric models, such as location and scale, have the following invariance property: If x has a distribution in the given family, so does gx, where g is an element of a transformation group G, and there is a one-to-one correspondence between the elements g of G and parameter values θ (*see* INVARIANCE CONCEPTS IN STATISTICS). Fraser [13, 14] set up a rigorous mathematical framework for fiducial theory for such models, and his later theory of structural inference* [16] is a continuation of this work. In these models the *orbit* of any point x is the set of all gx with g ranging over G. The orbit label turns out to be an ancillary statistic, and by using a pivotal argument conditional on the ancillary, one obtains a fiducial distribution that equals the posterior distribution when the prior measure equals the right Haar measure on G. The Haar measure is improper in the most familiar examples, but not for distributions on the circle or sphere [18] (*see* HAAR DISTRIBUTIONS).

If $\psi(\boldsymbol{\theta})$ is a real-valued function of a vector parameter $\boldsymbol{\theta}$, then a sufficient condition for fiducial limits for ψ, obtained from its marginal distribution, to be confidence limits is the following: $\psi(\boldsymbol{\theta}_1) = \psi(\boldsymbol{\theta}_2)$ implies that $\psi(g\boldsymbol{\theta}_1) = \psi(g\boldsymbol{\theta}_2)$ for all g [20]. For further discussion, see Zacks [37, Secs. 7.2, 7.3].

LITERATURE

In view of the unresolved difficulties, it is understandable that textbook authors tend to shy away from fiducial theory. The fiducial advocate Quenouille [28] is one exception. The less partisan writers Kendall and Stuart explain their approach like this:

> There has been so much controversy about the various methods of estimation we have described that, at this point, we shall have to leave our customary objective standpoint and descend into the arena ourselves. [22, p. 152]

Savage [30, p. 467] lists all examples of fiducial distributions in Fisher's published work. For bibliographies, see Tukey [33], Brillinger [2], Savage [30], and Pedersen [26]. The most authoritative source, but not the easiest to read, is Fisher [12—preferably the updated third edition (1973)].

References

[1] Box, J. F. (1978). *R. A. Fisher: The Life of a Scientist*. Wiley, New York.

[2] Brillinger, D. R. (1962). *Ann. Math. Statist.*, **33**, 1349–1355.

[3] Brown, L. (1967). *Ann. Math. Statist.*, **38**, 838–848.

[4] Buehler, R. J. and Feddersen, A. P. (1963). *Ann. Math. Statist.*, **34**, 1098–1100.

[5] Bunke, H. (1975). *Math. Operat. forschung Statist.*, **6**, 667–676.

[6] Creasy, M. A. (1954). *J. R. Statist. Soc. B*, **16**, 186–194.

[7] Fieller, E. C. (1954). *J. R. Statist. Soc. B*, **16**, 175–185.

[8] Fisher, R. A. (1930). *Proc. Camb. Philos. Soc.*, **26**, 528–535.

[9] Fisher, R. A. (1934). *Proc. R. Soc. Lond. A*, **144**, 285–307.

[10] Fisher, R. A. (1935). *Ann. Eugen. (Lond.)*, **6**, 391–398.

[11] Fisher, R. A. (1936). *Proc. Amer. Acad. Arts Sci.*, **71**, 245–258.

[12] Fisher, R. A. (1956). *Statistical Methods and Scientific Inference*. Oliver & Boyd, Edinburgh (3rd ed., Hafner Press, New York, 1973).

[13] Fraser, D. A. S. (1961). *Ann. Math. Statist.*, **32**, 661–676.

[14] Fraser, D. A. S. (1961). *Biometrika*, **48**, 261–280.
[15] Fraser, D. A. S. (1966). *Biometrika*, **53**, 1–9.
[16] Fraser, D. A. S. (1968). *The Structure of Inference*. Wiley, New York.
[17] Fraser, D. A. S. (1976). *J. Amer. Statist. Ass.*, **71**, 99–113.
[18] Fraser, D. A. S. (1979). *Inference and Linear Models*. McGraw-Hill, New York.
[19] Hacking, I. (1965). *Logic of Statistical Inference*. Cambridge University Press, Cambridge.
[20] Hora, R. B. and Buehler, R. J. (1966). *Ann. Math. Statist.*, **37**, 643–656.
[21] Jeffreys, H. (1948). *Theory of Probability*, 2nd ed. Oxford University Press, London.
[22] Kendall, M. G. and Stuart, A. (1961). *The Advanced Theory of Statistics*, Vol. 2. Charles Griffin, London Hafner, New York.
[23] Lawless, J. F. (1978). *Technometrics*, **20**, 353–368.
[24] Lindley, D. V. (1958). *J. R. Statist. Soc. B*, **20**, 102–107.
[25] Mauldon, J. G. (1955). *J. R. Statist. Soc. B*, **17**, 79–85.
[26] Pedersen, J. G. (1978). *Int. Statist. Rev.*, **46**, 147–170.
[27] Pitman, E. J. G. (1939). *Biometrika*, **30**, 391–421.
[28] Quenouille, M. H. (1958). *Fundamentals of Statistical Reasoning*. Charles Griffin, London.
[29] Robinson, G. K. (1976). *Ann. Statist.*, **4**, 963–971.
[30] Savage, L. J. (1976). *Ann. Statist.*, **4**, 441–500.
[31] Seidenfeld, T. (1979). *Philosophical Problems of Statistical Inference*. D. Reidel, Dordrecht, Holland.
[32] Stein, C. (1959). *Ann. Math. Statist.*, **30**, 970–979.
[33] Tukey, J. W. (1957). *Ann. Math. Statist.*, **28**, 687–695.
[34] Verhagen, A. M. W. (1966). The Notion of Induced Probability in Statistical Inference. *Div. Math. Statist. Tech. Paper No. 21*, Commonwealth Scientific Industrial Organization, Melbourne, Australia.
[35] Wilkinson, G. N. (1977). *J. R. Statist. Soc. B*, **39**, 119–171.
[36] Yates, F. (1964). *Biometrics*, **20**, 343–360.
[37] Zacks, S. (1971). *The Theory of Statistical Inference*. Wiley, New York.

(CONFIDENCE INTERVALS AND REGIONS
FIDUCIAL DISTRIBUTIONS
FIDUCIAL PROBABILITY)

ROBERT J. BUEHLER

FIDUCIAL PROBABILITY

(All italics in quotations are ours.)

> *Probability statements derived by arguments of the fiducial type have often been called statements of "fiducial probability." This usage is a convenient one, so long as it is recognized that the concept of probability involved is entirely identical with the classical probability of the early writers, such as Bayes. It is only the mode of derivation which was unknown to them.* [14, p. 51]

Most statistical concepts and theories can be described separately from their historical origins. This is not feasible, without unnecessary mystification, for the case of "fiducial probability." Indeed, it is useful to distinguish two sorts of fiducial probability: an early one, P, and a variety, P^*, that emerged somewhat later. Both were due to Fisher*; the main concern of this entry is simply to narrate their development by Fisher and how they were received and analyzed by other statisticians.

Fisher [9] considered inference about a real parameter θ from a maximum likelihood estimate* T having continuous cumulative distribution function $F(T,\theta)$. If $F(T,\theta) = 1 - P$ has a unique solution $\theta_P(T)$, this is the *fiducial* $100P$ *percent point of* θ. For the case when $\theta_P(T)$ increases with T, $\Pr(\theta < \theta_P(T) \mid \theta) = P$ and P is a *fiducial probability*, equal to the confidence level for the θ-intervals $\{\theta < \theta_P(T)\}$ (*see* FIDUCIAL INFERENCE). If, for fixed T, the definable fiducial probabilities take all values in (0, 1) (a nontrivial condition), then the pairs $\{(P, \theta_P(T)), 0 < P < 1\}$ formally constitute a cumulative distribution function* for what Fisher called the *fiducial distribution*. When $F(T,\theta)$ is differentiable with respect to θ, the fiducial distribution has a formal density $-\partial F(T,\theta)/\partial\theta$. Fisher was keen to promote fiducial probability against *posterior probability*, with the suggested advantage that "the fiducial values are expected to be different in every case [sample], and our probability statements are relative to such variability" [9, p. 535].

The first application of the fiducial argument was the calculation of fiducial 5% points of the correlation coefficient* ρ in a bivariate normal distribution*, for which T is the sample correlation r. The proper interpretation of P was again emphasized: "The value of ρ can then only be less than .765 in the event that r has exceeded its 95 per cent point, an event which is known to occur just once in 20 trials. *In this sense* ρ has a probability of just 1 in 20 of being less than .765."

A further application was provoked by Jeffreys' use [20] of an improper prior for the parameters of a normal distribution to obtain a posterior distribution for the variance. Fisher [10] found that the fiducial distribution for the variance was formally identical, but he drew the distinction: "The probabilities differ in referring to different populations; *that of fiducial probability is the population of all possible random samples*, that of the inverse probability is a group of samples selected to resemble that actually observed" [p. 348].

In the discussion of Neyman* [23], Fisher presented fiducial probability as identical to a confidence coefficient (level) when Neyman's wider approach was restricted to conform to Fisher's theory of *estimation*. In the *rapporteur's* account:

> Dr. Fisher's own applications of fiducial probability had been severely and deliberately limited. He had hoped, indeed, that the ingenuity of later writers would find means of extending its application to cases about which he was still in doubt, but some limitations seemed to be essential. Those who had followed the earlier parts of the story would have no difficulty in perceiving these, but there might be pitfalls for those who interested themselves only in the later chapters. . . . Dr. Neyman claimed to have generalized the argument of fiducial probability. . . . The generalization was a wide and very handsome one, but . . . it was perhaps as well to count the cost. The first item to which he would call attention was the loss of *uniqueness* in the result, and the consequent danger of *apparently* contradictory inferences. . . . Dr. Neyman proposed to extend the fiducial argument from cases where there was only a single unknown parameter, to cases in which there were several. Here, again, there might be serious difficulties in respect to the mutual *consistency* of the different inferences to be drawn; for, with a single parameter, it could be shown that all the inferences might be summarized in a single probability distribution for that parameter, and that, for this reason, all were mutually *consistent*; but it had not yet been shown that when the parameters were more than one any such equivalent frequency distribution could be established.

The twin desiderata, uniqueness and consistency, were thus proposed as additional to the confidence property of fiducial probability. They were repeated in Fisher [11], prior to the author's attempt to generalize the fiducial argument to the multiparameter case. On the basis of a construction for the normal mean and variance, Fisher concluded:

> In general, it *appears* that if statistics $T_1, T_2, T_3, \ldots$ contain jointly the whole of the information available respecting parameters $\theta_1, \theta_2, \theta_3, \ldots$ and if [*pivotal*] functions $t_1, t_2, t_3, \ldots$ of the T's and θ's can be found, the simultaneous distribution of which is independent of $\theta_1, \theta_2, \theta_3, \ldots$, then the fiducial distribution of $\theta_1, \theta_2, \theta_3, \ldots$ *simultaneously* may be found by substitution. [11, p. 395]

For the normal example, Fisher essentially used

$$t_1 = \sqrt{n}\,(\bar{x} - \mu)/\sigma, \qquad t_2 = (n-1)s^2/\sigma^2$$

and, in effect, checked the joint distribution by noting that the marginal distribution of μ and σ would give fiducial probabilities of type P (i.e., with the confidence property). However, Fisher did not extend the check to the marginal distribution of arbitrary functions of μ and σ. Instead, he used the general procedure quoted to derive the now-famous Behrens–Fisher* distribution of the difference of means of two normal populations with unknown, unequal variances.

Bartlett [1] deployed the ingenious device of random pairing of the observations in two samples of size 2 to show that Fisher had developed a new variety of fiducial probability: a P^* type that violated the confidence property of the earlier P type. The remarkable fact that P^* for the Behrens–Fisher problem does have a relevant probabilistic interpretation [12, 30] parallels an alternative interpretation of P for the single normal mean. As derived by Fisher in [11], if $P = \Pr(\tilde{t}_{n-1} < t_P)$, P refers firstly to the *region* $\mu < \bar{x} + st_P/\sqrt{n}$ of the *reference set* consisting of unrestricted values of μ and sample mean and standard deviation, and then, since the probability "does not apply to any special selection of these quantities," P refers to the *interval* picked out by the values $\bar{x}$ and s in the actual sample. The alternative interpretation of P is stated by Yates [30]: *P is the conditional probability of $\mu < \bar{x} + st_P/\sqrt{n}$ given s and the fiducial distribution for σ, i.e., $(n-1)s^2/\sigma^2$ distributed as χ^2_{n-1}*. As Fisher [13] pointed out, the proof is deceptively similar to the classical derivation of the Student distribution. In the parallel result for the Behrens–Fisher case, P^* has s_1/s_2 fixed instead of s and σ_1/σ_2 taking its fiducial distribution based on s_1/s_2. Fisher [15] reiterated this interpretation, while Pedersen [24] gives a fresh proof in an excellent review.

Results such as these for particular examples seem to have provided the impetus for Fisher's continuing advocacy of fiducial probability. Their striking feature—the conditioning on s or s_1/s_2—brought fiducial probability closer to the Bayesian position without actual embroilment in the *a priori*, and further from the Neyman–Pearsonian interpretation that was available for the P-type fiducial probability.

As in the 1935 paper [11], Fisher rarely hesitated to formulate general principles for the proper derivation and use of fiducial probabilities on the basis of particular examples. These principles became an attractive subject of study for the growing band of postwar mathematical statisticians—not always in the spirit of Hacking's [18, p. 152] "The task of future generations is not to confute the genius but to perfect the conjecture." By 1956, when Fisher's *Statistical Methods and Scientific Inference* was published [14], with the philosophy of fiducial probability as a central theme, a number of difficulties had accumulated. These were described by Tukey [28], who refined Fisher's definition of pivotals: A (multivariate) function $t(x,\theta)$ of the data x and the (multivariate) parameter θ is *pivotal** if its distribution for fixed θ is independent of θ; it is *sufficient* if, for every θ', $t(x,\theta')$ is sufficient for θ; it is *smoothly invertible* if, for each x, there is a unique continuous solution $\theta_x(t)$ of the equation $t(x,\theta) = t$.

For example, if x and θ are univariate with $\theta_0 \leqslant \theta \leqslant \theta_1$ and $F(x \mid \theta)$ is continuous in x, then F is pivotal with uniform distribution* on $(0, 1)$. It is smoothly invertible if, for each x, $F(x \mid \theta)$ is strictly decreasing in θ and $F(x \mid \theta_0) = 1$, $F(x \mid \theta_1) = 0$. The case where x is distributed as noncentral chi-squared with given degrees of freedom and noncentrality parameter θ [19] does not satisfy the invertibility condition, with the result that fiducial probability is not defined for the smaller values of P.

The renowned Fieller–Creasy paradox for the ratio of normal means arose when Fieller [8] used a pivotal that was neither sufficient nor smoothly invertible, that nevertheless gave genuine confidence sets, while Creasy [4] employed a sufficient, smoothly invertible pivotal and marginalized the joint fiducial distribution of the means to give conflicting P^*-type fiducial probabilities for the ratio that did *not* have the confidence property! Fieller's confidence level is clearly not interpretable as fiducial probability since, when less than unity, it is inapplicable to an interval consisting of all possible values of the ratio: the data producing such an interval would belong to a *recognizable subset* [14, pp. 57, 109–110] (*see* FIELLER'S THEOREM).

The possibility that sufficiency* and smooth invertibility of pivotals might guarantee uniqueness of fiducial probability was overthrown by the counterexample of Mauldon [22], based on the Wishart distribution*, and by the even simpler counterexamples of Savage [25] and Tukey [28]. Mauldon found large numerical discrepancies in the competing "fiducial probabilities."

Fisher considered such counterexamples artificial [14, p. 120] and thought they could be avoided by a "rigorous step-by-step" construction of the simultaneous fiducial distribution [14, pp. 119, 172] based on factorization properties of the probability density function. Brillinger [2] and Dempster [7] effectively disposed of this loophole and recently Pedersen [24], refining the Dempster example, has analyzed the consequences of the two factorizations for the sufficient statistics s and $r = \bar{x}/s$ from an $N(\mu, \sigma^2)$ sample with $\rho = \mu/\sigma$:

$$\begin{aligned} f(r,s \mid \rho,\sigma) &= f(r \mid \rho) f(s \mid r,\rho,\sigma) \\ &= f(s \mid \sigma) f(r \mid s,\rho,\sigma). \end{aligned}$$

The step-by-step process gives different "fiducial probabilities" for σ conditional on ρ and there appears to be no rationale for claiming that one is correct and the other wrong.

A different critique of fiducial probability stems from the questions: When do fiducial and Bayes posterior probabilities coincide if the latter are generalized to allow unbounded prior measures; and can fiducial probabilities be consistently used as prior probabilities as Fisher [14, pp. 51, 125] indicates? Lindley [21] answered decisively thus: When x and θ are real and $-\partial F(x \mid \theta)/\partial\theta$ is a fiducial density, the fiducial probabilities are Bayes posterior if and only if

$$F(x \mid \theta) = G\left[u(x) - \tau(\theta)\right] \qquad (1)$$

for some G and increasing $u(\cdot)$ and $\tau(\cdot)$, with a uniform prior for $\tau(\theta)$ yielding the identity. Furthermore, when a one-dimensional sufficient statistic exists for random samples of x, the fiducial distribution generated by part of the sample gives the fiducial distribution for the whole, when used as a prior à la Bayes, if and only if (1) obtains. (With the sufficiency, this implies either normality or "gammality"; the latter is the case where fiducial probability is equivalent to Jeffreys' [20] posterior probability for a normal variance.)

Brillinger [2] showed that a similar Bayesian straitjacket does not exist for higher-dimensional fiducial probabilities. However, the group invariance that emerged as necessary for the one-dimensional case provides the general basis for a very wide class of statistical problems in which fiducial probabilities are equivalent to Bayes posterior probabilities. For this class:

1. $x = (a, b)$ where a is ancillary and b is exhaustive.
2. The values of θ constitute a (locally compact topological) group, Θ, of transformations of $\mathscr{B} = \{b\}$ to which the group is isomorphic.
3. Given θ and a, $t = \theta^{-1}b$ is pivotal with probability measure $P(B \mid a)$ for $B \subset \mathscr{B}$.

Then the fiducial probability of bB^{-1} is $P(B \mid a)$, on the grounds that

$$\begin{aligned} P(B \mid a) &= \Pr(\theta^{-1}b \in B \mid a, \theta) \\ &= \Pr(\theta \in bB^{-1} \mid a, \theta) \\ &= \Pr(\theta \in bB^{-1} \mid a) \end{aligned}$$

by independence of θ. The latter probability is attached fiducially to bB^{-1} if there are no recognizable subsets in $\mathscr{B}$, which Hacking's [18] "irrelevance principle" supports since the likelihood function of t, regarded as a transformation of θ, does not depend on b. By the group structure, the fiducial probabilities constitute a fiducial probability measure on Θ that has density $f(\theta^{-1}b \mid a)$ with respect to right-invariant measure on Θ, if t has density $f(t \mid a)$ with respect to the corresponding left-invariant measure on $\mathscr{B}$—thereby establishing an equivalence with Bayes posterior probability for right-invariant prior.

The construction implies that the common fiducial probability attached to the sets $B^{-1}b$, for fixed B, is of P type with the confidence property.

An illustration is provided by fiducial probabilities for $\mu + \alpha\sigma$ from an $N(\mu, \sigma^2)$

sample $x_1, \ldots, x_n$ [14, p. 121]. Here

$$a = ((x_1 - \bar{x})/s, \ldots, (x_n - \bar{x})/s),$$
$$b = (\bar{x}, s), \qquad \theta = (\mu, \sigma),$$
$$\theta^{-1} = (-\mu/\sigma, 1/\sigma),$$
$$t = (t_1, t_2) = ((\bar{x} - \mu)/\sigma, s/\sigma).$$

The choice

$$B = \{t \mid (t_1 - \alpha)/t_2 > -c\},$$

where c is such that $P(B \mid a) = P$, will give the fiducial probability P that $\mu + \alpha\sigma < \bar{x} + cs$.

The Behrens–Fisher problem does not yield to the same treatment. However, it is not necessary, for the confidence property, that group-based fiducial probabilities should refer to bB^{-1} for some fixed B. There is no data-dependent transformation of $\rho = \mu/\sigma$ making it a function of t [27] but, if $\rho(\bar{x}, s)$ is chosen so that each of the intervals $\rho < \rho(\bar{x}, s)$ has the same marginal fiducial probability, it nonetheless happens that the common fiducial probability is of type P [7].

Mathematical intuition suggests that dressing up the fiducial method in group-theoretic clothes is unlikely to dispose of the problem of nonuniqueness; nor does it do so. The simplest case of the Mauldon [22] counterexample corresponds to the possibility that more than one group is available for the same statistical model, giving different fiducial probabilities—in Mauldon's case, groups of upper- and of lower-triangular 2×2 matrices.

The structural probabilities of Fraser [16] are distinguishable from fiducial probabilities mainly because they evade nonuniqueness by supposing that a *transformation model* $x = \theta e$ *is given*, in which Θ is a group that will generate fiducial probabilities in the way described. The rationale of the distinction has been accepted and elaborated by Bunke [3] but questioned by Dawid *et al.* [6].

Group-based fiducial probabilities do not require that x be continuous. Fisher [14, p. 60] maintained that discontinuity was not "suitable for exhibiting the fiducial argument," but Fisher [13] had made use of a discrete pivotal. It is, therefore, possible that he would have accepted the relevance to the fiducial argument of the Flatland example of Stone [26]: the example can be interpreted as demonstrating that, when the group is nonamenable, use of a nonequivariant (no fixed B) region of Θ can produce a P^*-type fiducial probability with a strong inconsistency property.

Fiducial probability has often been written off as a simple aberration (e.g., Good [17]). Less dismissively, it has been labeled "a bold attempt to make the Bayesian omelet without breaking the Bayesian eggs" [25]. However, the generous reception of the boldly incoherent effort by Wilkinson [29] to "reconcile the Fisherian and Neyman–Pearsonian viewpoints" shows that there is a continuing interest in Fisher's "solution of Bayes" problem when knowledge *a priori* is absent" [14, p. 62].

NOTE:

Since the original preparation of this article, Dawid and Stone [5] have considered the role of functional models in an attempt to uncover a general theory of fiducial inference.

References

[1] Bartlett, M. S. (1936). *Proc. Camb. Philos. Soc.*, **32**, 560–566.

[2] Brillinger, D. R. (1962). *Ann. Math. Statist.*, **33**, 1349–1355.

[3] Bunke, H. (1975). *Math. Operat. Forschung Statist.*, **6**, 667–676.

[4] Creasy, M. (1954). *J. R. Statist. Soc. B*, **16**, 186–194.

[5] Dawid, A. P. and Stone, M. (1982). *Ann. Statist.*, **10**, 1054–1067.

[6] Dawid, A. P., Stone, M., and Zidek, J. V. (1973). *J. R. Statist. Soc. B*, **35**, 189–233.

[7] Dempster, A. P. (1963). *Ann. Math. Statist.*, **34**, 884–891.

[8] Fieller, E. C. (1954). *J. R. Statist. Soc. B*, **16**, 175–185.

[9] Fisher, R. A. (1930). *Proc. Camb. Philos. Soc.*, **26**, 528–535.

[10] Fisher, R. A. (1933). *Proc. R. Soc. Lond. A*, **139**, 343–348.
[11] Fisher, R. A. (1935). *Ann. Eugen. (Lond.)*, **6**, 391–398.
[12] Fisher, R. A. (1939). *Ann. Eugen. (Lond.)*, **9**, 174–180.
[13] Fisher, R. A. (1945). *Sankhyā*, **7**, 129–132.
[14] Fisher, R. A. (1956). *Statistical Methods and Scientific Inference*. Oliver & Boyd, Edinburgh.
[15] Fisher, R. A. (1961). *Sankhyā*, **23**, 3–8.
[16] Fraser, D. A. S. (1968). *The Structure of Inference*. Wiley, New York.
[17] Good, I. J. (1978). *Fallacies, Statistical*. In *International Encyclopedia of Statistics*, Vol. 1, W. H. Kruskal and J. M. Tanur, eds. Free Press, New York, p. 344.
[18] Hacking, I. (1965). *Logic of Statistical Inference*. Cambridge University Press, Cambridge.
[19] James, G. S. (1954). *J. R. Statist. Soc. B*, **16**, 175–222.
[20] Jeffreys, H. (1932). *Proc. R. Soc. Lond. A*, **138**, 48–55.
[21] Lindley, D. V. (1958). *J. R. Statist. Soc. B*, **20**, 102–107.
[22] Mauldon, J. G. (1955). *J. R. Statist. Soc. B*, **17**, 79–85.
[23] Neyman, J. (1934). *J. R. Statist. Soc.*, **97**, 558–625.
[24] Pedersen, J. G. (1978). *Int. Statist. Rev.*, **46**, 147–170.
[25] Savage, L. J. (1961). *Proc. 4th Berkeley Symp. Math. Statist. Prob.*, Vol. 1. University of California Press, Berkeley, Calif., p. 578.
[26] Stone, M. (1976). *J. Amer. Statist. Ass.*, **71**, 114–125.
[27] Stone, M. and Dawid, A. P. (1972). *Biometrika*, **59**, 369–375.
[28] Tukey, J. W. (1957). *Ann. Math. Statist.*, **28**, 687–695.
[29] Wilkinson, G. N. (1977). *J. R. Statist. Soc. B*, **39**, 119–171.
[30] Yates, F. (1939). *Proc. Camb. Philos. Soc.*, **35**, 579–591.

(CONFIDENCE INTERVALS AND REGIONS
FIDUCIAL DISTRIBUTIONS
FIDUCIAL INFERENCE
FIELLER'S THEOREM
FISHER, RONALD AYLMER
LIKELIHOOD
LIKELIHOOD PRINCIPLE
POSTERIOR DISTRIBUTIONS)

MERVYN STONE

FIELLER–CREASY PARADOX *See* FIELLER'S THEOREM

FIELLER'S THEOREM

This result provides confidence intervals for the ratio of mean values of two random variables having a bivariate normal distribution*. Originally developed by Fieller [2], it is more clearly explained by Fieller [3].

Let x and y be observed estimates of the means μ_x and μ_y, respectively, and $\gamma = \mu_x/\mu_y$. Let s_{xx}, s_{yy}, and s_{xy} be estimates of the variances and covariance, respectively; in practice, x and y might be observed means of a joint random sample from the underlying bivariate normal distribution. Then the pivotal quantity*

$$\frac{x - \gamma y}{s_{xx} - 2\gamma s_{xy} + \gamma^2 s_{yy}} \tag{1}$$

results from a Student t-distribution* with an appropriate number of degrees of freedom. If the $100(1 - \frac{1}{2}\alpha)$ percentile of this distribution is denoted by t, then Fieller [3] showed that the corresponding $100(1 - \alpha)$ percent confidence limits for γ are given by

$$\frac{xy - t^2 s_{xy} \pm \left[f(x, y, s_{xx}, s_{yy}, s_{xy}) \right]^{1/2}}{y^2 - t^2 s_{yy}},$$

$$f(x, y, s_{xx}, s_{yy}, s_{xy}) = \left(xy - t^2 s_{xy}\right)^2 - \left(x^2 - t^2 s_{xx}\right)\left(y^2 - t^2 s_{yy}\right). \tag{2}$$

If one thinks of x/y as a point estimate of γ, a more natural but lengthier equivalent expression for (2) is

$$\left\{ \frac{x}{y} - \frac{g s_{xy}}{s_{yy}} \pm \frac{t}{y} \left[s_{xx} - 2\frac{x}{y} s_{xy} + \frac{x^2}{y^2} s_{yy} - g\left(s_{xx} - \frac{s_{xy}^2}{s_{yy}} \right) \right]^{1/2} \right\} \Big/ (1 - g), \tag{3}$$

where $g = t^2 s_{yy}/y^2$.

Fieller [3] describes the values in (2) as fiducial limits, as does Finney [4], but Stone points out elsewhere in this encyclopedia (*see* FIDUCIAL PROBABILITY) that "Fieller's confidence level is clearly not interpretable as fiducial probability since, when less than unity, it is inapplicable to an interval consisting of all possible values of the ratio." Fieller demonstrates this difficulty with Cushny and Peebles' data for the effect of two drugs on hours of sleep gained by 10 patients (a set of data used incidentally by Gosset* in the original paper [6] which developed the t-distribution). Figure 1 shows in Fieller's notation the curve leading to the values in (2) as solutions. For this data set, $1 - \alpha > 0.9986$ leads to all values of γ, but this would logically correspond to a fiducial probability of 1.0. Figure 1 also shows that small values of t (or of g) give finite confidence intervals, but for $0.995 \leqslant 1 - \alpha < 0.9986$, the corresponding percentiles t for this set of data lead to confidence regions that *exclude* only a finite interval. These anomalies arise because the function of γ involved is not monotonic. Except when g is reasonably small, the data will not in general be useful for interval estimation of γ.

The Fieller–Creasy paradox arises from a related paper by Creasy [1], who obtains her fiducial distributions of γ from the separate fiducial distributions of μ_x and μ_y. The resulting distribution differs from that of Fieller and so leads to different intervals. Neyman [5] related the paradox to the question of whether or not a meaningful definition of a fiducial distribution exists. "Miss Creasy's solution differs from that of Mr. Fieller and in the past repeated assertions were made that in a given set of conditions a parameter may have only one fiducial distribution."

The principal applications of Fieller's theorem have been in bioassay*, in estimating ratios of regression coefficients and distances between regression lines. Finney [4, pp. 32–35] developed two analogs of the theorem, and gave applications for several types of assay problems [4, p. 663].

References

[1] Creasy, M. (1954). *J. R. Statist. Soc. B*, **16**, 186–194.

[2] Fieller, E. C. (1940). *J. R. Statist. Soc. Suppl.*, **7**, 1–64. (See pp. 45–63.)

[3] Fieller, E. C. (1954). *J. R. Statist. Soc. B*, **16**, 175–185.

[4] Finney, D. J. (1962). *Statistical Method in Biological Assay*. Hafner, New York.

[5] Neyman, J. (1954). *J. R. Statist. Soc. B*, **16**, 216–218.

[6] "Student" (1908). *Biometrika*, **6**, 1–25.

(CONFIDENCE INTERVALS AND REGIONS

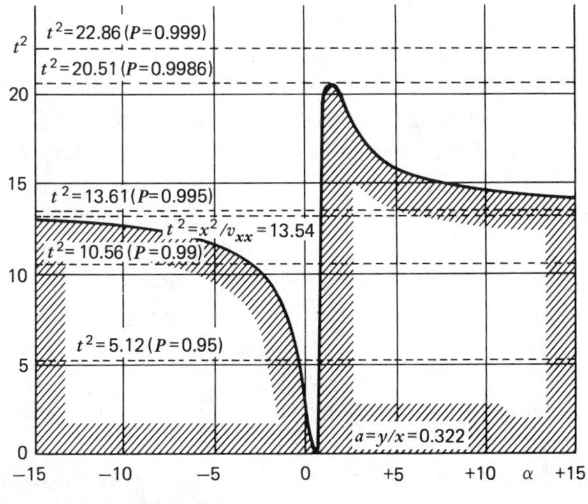

Figure 1 Cushny–Peebles data: fiducial limits for the ratio α = (effect of drug A)/(effect of drug B). (Reproduced by permission of the Royal Statistical Society.)

FIDUCIAL PROBABILITY
PIVOTAL QUANTITIES)

CAMPBELL B. READ

FIGURATE NUMBERS *See* COMBINATORICS

FILTERS *See* STATIONARY PROCESSES

FINAL PREDICTION ERROR *See* AKAIKE CRITERION

FINANCE, STATISTICS IN

Statistical methods have played a critical role in the development of financial theory and its applications to the areas of corporate finance, security and portfolio analysis, and financial institutions and markets. This is particularly so since the origin of modern finance, which began in the late 1950s with the landmark works of Markowitz [35] and Tobin [48] on portfolio selection and Modigliani and Miller [36] on capital structure and firm valuation.

Although finance had emerged as a separate discipline from economics in the first half of this century, in its beginning years it could be best characterized as descriptive; its primary purpose was to describe and to document the rapidly growing number of financial instruments, financial institutions, and financial practices of business firms (see Solomon [44] for a brief history of the evolution of the field). Consequently, only on rare occasions were more than elementary statistics employed. Illustrative of this period is the first scholarly text in financial management [11], which dominated the field for decades.

Since that time the emphasis in financial research has shifted rapidly from descriptive to analytical and the field of finance has undergone a dramatic change. The purpose of this article is to demonstrate how the science of statistics facilitated this change. Selected applications of regression*, factor*, discriminant*, logit*, spectral*, Box–Jenkins*, and Bayesian* analyses are presented as illustrations of the widespread use of statistics in financial research. As of the time of this writing (1981) there is no comprehensive source on this subject.

REGRESSION ANALYSIS AND THE MARKET MODEL

Regression analysis* is one of the most prevalent statistical techniques employed in finance. As discussed below, regression-related methodologies, including serial correlation* and analyses of residuals*, have been applied to virtually all areas of finance, e.g., security analysis, market efficiency studies, financial institutions, mergers, and capital market theory.

The marriage of regression methodology and finance is generally referred to as econometrics*. The objective in econometric modeling is to estimate levels of one or more endogenous variables based on the knowledge of a set of exogenous variables. Relationships among variables are hypothesized from financial theory, whereas estimates are generated via the application of statistical procedures to historic or predicted data. Within the finance literature, Williamson [52] and Lorie and Hamilton [34] present summaries of alternative methodologies and their financial applications. A more sophisticated discourse on the subject which focuses on single- and multifactor market models and on their implications for security market efficiency, portfolio theory, and capital market theory may be found in Fama [16].

In an early study in the investments area, Whitbeck and Kisor [50] suggested that multiple regression analysis could be useful as a tool in security analysis to identify "over-" and "undervalued" securities. Specifically, it was hypothesized that a stock's theoretical ratio of price to earnings (P/E ratio) could be estimated by regressing actual P/E ratios on selected firm financial variables to obtain a function suitable for predictive applications. Subsequently, Bower and Bower [5] applied a similar but more elaborate model to examine common stock valuation.

Cohen et al. [9] and Williamson [52] offer general insights into the usefulness of regression for investment decision making; e.g., Williamson suggests that regression can be useful in evaluating the accuracy of forecasts made by security analysts and for adjusting them for systematic biases.

Regression analysis has been used extensively to test for security market efficiency. One of the most important works on this subject with an excellent summary of previous research is Fama [15]. Fama investigated the daily price behavior of securities from 1958 to 1962. His results, consistent with those of the majority of other studies, indicated insignificant serial correlation and thus supported the random walk* hypothesis. It also may be noted that Fama and others employed runs* and other nonparametric tests to determine whether security price changes conform to a random walk process.

Before considering alternative statistical tests for market efficiency, it is useful to introduce one of the most important applications of regression analysis in finance, the market model, and its theoretical underpinning, capital market theory. In 1959, Markowitz [35] published a theory of portfolio selection based on expected return and risk measures for assets. Initially, the application of this theory required a prohibitively large number of statistical inputs. Sharpe [42] suggested that computational efficiency could be achieved if the correlation of the return of each security with the return of a single market index be used as a proxy for correlations with all possible securities. This resulted in the single-index market model. Based on Markowitz's work, Sharpe [42], Lintner [29], and Mossin [37] developed positive theories of asset valuation in the capital markets under conditions of uncertainty, hereinafter referred to as capital market theory. In the years that have followed, researchers have used this theory and its empirical counterpart, the market model, as vehicles to test statistically and extend existing theories of business finance, and have assessed its normative implications for corporate decision making. Thus capital market theory has provided a theoretical framework for statistical research which cuts across virtually all areas of finance. Introductory descriptions of the market model may be found in Francis and Archer [18] and Tinic and West [47].

One form of the market model can be estimated empirically by regressing an asset's returns on the returns for a market index using historical data. This model provides an advantage over previously developed models in empirical testing because it enables the user to discount information for general market movements and focus on a new set of variables, the statistical residual terms; this set represents changes in security valuation over and above those caused by general market movements. Examining the relationship of a policy change to these terms therefore gives a clearer indication of its effect on security prices. This approach has been taken to test issues ranging from the effect of dividend policy on a firm's cost of capital to the impact of Security and Exchange Commission (SEC) disclosure requirements on the securities market. Important regression studies in which the market model has been used are summarized in Copeland and Weston [10]. These relate to accounting information effects, block trades, insider trading activities, new issues, stock splits, and mutual fund performance.

Although the market model has been applied extensively in several different econometric forms, criticisms have been levied with respect to its restrictive statistical assumptions regarding parameter stationarity, the choice and behavior of alternative market proxies, and violations of homoscedasticity among other potential problems (see Jensen [24], Friend and Bicksler [19], Chen [8], and Bey and Pinches [4]). Larcker et al. [28] have suggested intervention analysis* as a preferable and viable alternative to the use of the market model for residual analysis.

In the aforementioned research, the widespread use of the market model in investments and corporate finance has been emphasized. Regression techniques also are commonly employed in studies of financial institutions (see, e.g., Kane and Buser [26] and Bates and Bradford [3]).

FACTOR, MULTIPLE DISCRIMINANT, AND LOGIT APPLICATIONS

Factor analysis* has been applied most often in conjunction with other statistical or econometric techniques and in "exploratory" studies of large data sets. Pinches and Mingo [39], for example, applied factor analysis in a study of industrial bond ratings, hypothesizing that some combination of financial variables might provide information useful in their prediction. In order to apply a parsimonious discriminant model, factor analysis was used to reduce the dimensionality in the set of explanatory variables. Examples of other studies in which factor analysis has been paired with discriminant analysis* include an examination of the stability of financial patterns in industrial corporations [40] and a model for predicting defaults for small business loans [22]. Factor analysis has also been used together with multiple regression techniques (see, e.g., Herbst [21]).

Most interestingly, Lloyd and Lee [30] employed this method to derive an asset pricing model in terms of block recursive systems of equations. This model, offered as a realistic alternative to Sharpe's single-index model, evaluates large numbers of securities within a simultaneous pricing framework. The advantage of this approach is that significant security interrelationships are considerably smaller.

Multiple discriminant analysis* (MDA) has been applied frequently in various financial contexts; extensive citations of applications can be found in Eisenbeis [14].

Altman's [1] study of corporate bankruptcy, although not the first use of MDA in finance, was perhaps the landmark application of the technique. Using financial ratio data, Altman derived a single linear discriminant function for predicting bankruptcy prior to its occurrence. Because the use of MDA emphasizes related financial implications of sets of ratios, the informational content is much greater when compared to traditional univariate ratio analysis. Altman concluded that MDA could be valuable for business loan evaluation, corporate internal control, and investment analysis.

Subsequent applications have borne this out. Given appropriate discriminant functions, relatively unskilled managers have been able to make reliable business decisions efficiently with the help of MDA predictive classifications.

Other financial activities have been modeled using MDA. As noted earlier, Pinches and Mingo [39] employed the technique to analyze industrial bond ratings. Their primary objective was to determine whether a MDA model with financial statement data could correctly classify subsequent ratings by Moody's. They concluded that bond rating services do not do a significantly better job of evaluating credit worthiness and default risk than can be achieved through the use of MDA with financial data. In another area, Stevens [45] used MDA to analyze characteristics of acquired versus non-acquired firms. After identifying the most likely effective discriminating variables by factor analysis, discriminant coefficients were estimated. Again, MDA was found to be effective in differentiating between acquired and nonacquired firms.

The aforementioned studies are but a few of the applications of MDA. However, its use in finance is not without potential statistical problems. Joy and Tollefson [25] point out several methodological issues, including weaknesses in heretofore employed validation tests, the critical role of properly estimated prior probabilities, and problems with conditional measures of discriminating efficiency. They also propose a Bayesian evaluation approach that directly addresses misclassification costs.

Eisenbeis [14] also evaluates problems in financial MDA applications, including some mentioned above. Additionally, he notes problems in correctly applying linear or quadratic classification* rules and problems caused by unequal group dispersion matrices, variable nonnormality, dimension reduction, and the interpretation of variable significances. He identifies problems that stem from time-series* complications due to the aggregation* of data across time periods and the prediction of classification results into future periods.

Thus, while MDA has been applied in many financial contexts, it appears that difficulties have been ignored in some cases. For this reason, the results of earlier studies are being reassessed, more appropriate guidelines for future applications are being established and alternative statistical procedures are being proposed.

One of these is logit analysis*, a relatively new family of analytic techniques which has been applied in finance and economics in a growing number of studies in lieu of MDA. Press and Wilson [41], examining the choice between logit and discriminant analysis, demonstrated that logit models are more robust than are linear discriminant models since they do not assume that all independent variables are multivariate normal with equal variance–covariance matrices. Because many pertinent variables in financial analysis are dichotomous or otherwise restricted, the use of logit techniques is suggested. Press and Wilson further indicate that, given classification problems involving nonnormal variables, logit outperforms linear MDA, albeit by a small margin.

Wiginton [51] applied both logit and MDA in a recent study of consumer credit applicants. He concluded that the MDA classification results were no better than would be expected by pure chance, but that the logit model correctly classified 62% of the applicants. Dietrich and Sorenson [12] used logit analysis to predict prospective merger candidates. Data were compiled for samples of merged and nonmerged firms for the period 1969–1973. Ten items were used as independent variables, including P/E ratios, operating profit margin, times-interest-earned, current ratio, and asset turnover. After the logit parameters were estimated, a test of the model's accuracy indicated that 91% of a sample of holdout cases were correctly classified. Given these results and the previously mentioned statistical considerations, it is expected that logit analysis will be applied in an increasing number of studies.

TIME-SERIES ANALYSES OF FINANCIAL INFORMATION

Box–Jenkins* (B-J) modeling is regarded as an important technique in analyses of financial time series*. To date, this technique has been applied primarily to study the time-series properties of annual and quarterly corporate earnings, with emphasis on improving earnings forecasts. The prediction of earnings is particularly important in investment analysis and in financial statement evaluation. A distinctive feature of a statistical model approach such as Box–Jenkins is that a detailed examination of the financial series is an integral part of the forecast model choice (*see* FORECASTING). A few examples of its use follow.

Lorek et al. [33] compared B-J earnings estimates with those generated by managers. Their hypothesis was that because managers are privy to more and better information, their forecasts should be superior to those which only use past earnings information. Although it was found that B-J forecasts did not outperform those of a relatively accurate corporate forecast group, for a less accurate group and for the combined sample, B-J forecasts were superior.

Foster [17] applied B-J techniques to study quarterly earnings, sales and expense series from a sample of 69 firms for the period 1946 to 1974. His research focused on the predictive accuracy of these models as well as on the use of B-J procedures in examining the capital market reaction to accounting data. The results obtained with respect to predictive accuracy were similar to those of Lorek et al. Finally, Brown and

Rozeff [6] studied quarterly earnings data for 50 randomly selected firms from 1951 to 1971. Specifically, comparisons were made of prediction errors for seasonal martingale*, seasonal submartingale, and B-J models with those of the Value Line Investment Survey and Standard and Poor's Earnings Forecaster. Results indicated that B-J models are superior to other time-series models but consistently inferior to both of the advisory services.

Additional discussions of quarterly time-series properties of earnings and adaptive expectations models can be found in Lorek [32] and Brown and Rozeff [6], respectively. Stokes and Neuburger [46] provide an economic application of B-J methodology to the study of the effects of monetary changes on interest rates.

Another class of statistical procedure which has been used to examine economic and financial time-series data is spectral analysis*. One major application has been in the study of interest rates and term structure. A bibliography of this literature is provided by Percival [38]. An example of work in this area is Smith and Marcis [43], who applied spectral analysis and cross-spectral techniques to 12 interest rate series on U.S. Treasury issues and selected municipal and corporate bonds for the period March 1951–December 1969. Their objectives were to examine rates on different securities with similar maturities and rates on similar securities with different maturities. Results of spectral analysis led Smith and Marcis to conclude that the individual series studied have dissimilar cyclical and seasonal characteristics, with the former decreasing in significance as term to maturity increases. Further, it was found that municipal bond rates were most affected by these components, as opposed to corporate rates, which appeared to have more modest cyclical and seasonal structures.

Cross-spectral analysis of Treasury issues and municipal bond markets revealed that short- and long-term rates are highly correlated at the cyclical (low-frequency) components, with short-term rates leading long-term rates. Contrarily, seasonal components (higher frequencies) were found to be less correlated and the lead–lag structure reversed.

Spectral analysis also has been enlisted to examine stock price movements and to test for market efficiency. For example, Granger and Morgenstern [20] investigated several series of transactions on the New York Stock Exchange. Their finding that the power at all frequencies is substantially uniform supported the random walk hypothesis. Although some seasonal components were observed, they were so slight as to be of negligible financial consequence for investors.

Finally, in an interesting application, Logue and Sweeney [31] used spectral techniques to study franc/dollar exchange rates. Their tests indicated "white noise"* at the 95% level of confidence, prompting the conclusion that spectral methods are insufficient to demonstrate market efficiency. Their rejection of spectral analysis as appropriate stemmed from the fact that, for the same data, a simple technical filter rule produced trading rules that yielded significant excess profits. Logue and Sweeney attribute this "misperception" of spectral methods to the fact that these methods detect only linear time structures when, in fact, significant nonlinear structures may exist.

STATISTICAL DECISION THEORY AND FINANCE

Although the application of statistical decision theory* in finance is not yet commonplace, Bayesian analysis has been used in the modeling of several notable financial decision problems: e.g., forecasting future security prices, portfolio selection, estimating security betas, and credit analysis (*see* BAYESIAN INFERENCE).

With respect to forecasting security prices, the Bayesian approach was advocated by Winkler [53] in an early application. Prior studies had either examined probability distributions for future price movements at one point in time, or had assumed that distribution parameters are known. In either case the impact of new information is ignored. Given a Bayesian specification, price distributions can be characterized in probabilistic terms, subject to revision as new information becomes available to the decision maker. New information may include market prices, data on transaction volume, dividend and earnings announcements, changes in general economic conditions, etc. Thus the decision theoretic approach views security price forecasting as an adaptive process, a view not incompatible with the rational expectations and efficient market hypotheses.

Although Winkler developed his initial model under limited restrictions, his findings have been generalized and extended to more realistic situations. Winkler and Barry [54] provide a basic model of portfolio choice, extended to the case of nonstationarity in Barry and Winkler [2] and to the consideration of optimal diversification in Klein and Bawa [27].

Other applications of Bayesian statistics include an analysis of market efficiency [23] and the estimation of security betas [49]. With respect to beta estimation, the Bayesian approach is preferable to the standard sampling theory procedure because it produces estimates that minimize the loss due to misestimation.

Finally, Bayesian techniques are useful in commercial and consumer credit analysis. Edmister and Schlarbaum [13] present a methodology for credit evaluation in lending institutions. They demonstrate how available information and subjective probability assessments can be combined to analyze and select a credit analysis system yielding maximum cash flow benefits to the firm. Although not an explicit Bayesian approach, their analysis can readily be extended to allow revisions in policy decisions on the basis of revised probability assessments.

In sum, statistics has been used to test and model security and portfolio return distributions, the role and activities of financial institutions and markets, and the impact of financial decisions on the value of the firm. Many statistical models have become accepted by business and not-for-profit organizations as useful approximations of the behavior of real-world financial data. Statistical science has made and will continue to make important contributions to the theory and practice of finance.

References

[1] Altman, E. I. (1968). *J. Financ.*, **23**, 589–609.

[2] Barry, C. B. and Winkler, R. L. (1976). *J. Financ. Quant. Anal.*, **9**, 217–235.

[3] Bates, T. and Bradford, W. (1980). *J. Financ.*, **35**, 753–768.

[4] Bey, R. P. and Pinches, G. E. (1980). *J. Financ. Quant. Anal.*, **15**, 229–322.

[5] Bower, R. S. and Bower, D. H. (1969). *J. Polit. Econ.*, **77**, 349–362.

[6] Brown, L. D. and Rozeff, M. S. (1978). *J. Financ.*, **33**, 1–16.

[7] Brown, L. D. and Rozeff, M. S. (1979). *J. Account. Res.*, **17**, 341–351.

[8] Chen, S. (1980). *J. Financ. Quant. Anal.*, **15**, 151–174.

[9] Cohen, J. B., Zinbarg, E. D., and Zeikel, A. (1977). *Investment Analysis and Portfolio Management*. Richard Irwin, Homewood, Ill.

[10] Copeland, T. E. and Weston, J. F. (1979). *Financial Theory and Corporate Policy*. Addison-Wesley, Reading, Mass.

[11] Dewing, A. S. (1920). *The Financial Policy of Corporations*. Ronald Press, New York (successive ed.: 1926, 1934, 1941).

[12] Dietrich, J. K. and Sorensen, E. (1980). University of Southern California working paper.

[13] Edmister, R. O. and Schlarbaum, G. G. (1974). *J. Financ. Quant. Anal.*, **9**, 335–356.

[14] Eisenbeis, R. A. (1977). *J. Financ.*, **32**, 875–900.

[15] Fama, E. F. (1965). *J. Bus.*, **38**, 34–105.

[16] Fama, E. F. (1976). *Foundations of Finance*. Basic Books, New York.

[17] Foster, G. (1977). *Account. Rev.*, **52**, 1–21.

[18] Francis, J. C. and Archer, S. H. (1979). *Portfolio Analysis*. Prentice-Hall, Englewood Cliffs, N.J.

[19] Friend, I. and Bicksler, J. L. (1977). *Risk and Return in Finance*. Ballinger, Cambridge, Mass.

[20] Granger, C. and Morgenstern, O. (1970). *Predictability of Stock Market Prices*. Lexington Books, Lexington, Mass.

[21] Herbst, A. F. (1974). *J. Financ.*, **29**, 1087–1103.

[22] Hoeven, J. A. (1979). *J. Commer. Bank Lending*, **61**, 47–60.

[23] Jaffe, J. F. and Winkler, R. L. (1976). *J. Financ.*, **31**, 49–61.

[24] Jensen, M. (1972). *Bell J. Econ. Manag. Sci.*, **3**, 357–398.

[25] Joy, O. M. and Tollefson, J. O. (1975). *J. Financ. Quant. Anal.*, **10**, 723–739.

[26] Kane, E. J. and Buser, S. A. (1979). *J. Financ.*, **34**, 19–34.

[27] Klein, R. W. and Bawa, U. S. (1977). *J. Financ. Econ.*, **5**, 89–111.

[28] Larcker, D. F., Gordon, L. A., and Pinches, G. E. (1980). *J. Financ. Quant. Anal.*, **15**, 267–287.

[29] Lintner, J. (1965). *Rev. Econ. Statist.*, **47**, 13–37.

[30] Lloyd, W. P. and Lee, C. F. (1976). *J. Financ.*, **31**, 1101–1113.

[31] Logue, D. E. and Sweeney, R. J. (1977). *J. Financ.*, **32**, 761–768.

[32] Lorek, K. S. (1979). *J. Account. Res.*, **17**, 190–204.

[33] Lorek, K. S., McDonald, C. L., and Patz, P. H. (1976). *Account. Rev.*, **51**, 321–330.

[34] Lorie, J. H. and Hamilton, J. T. (1973). *The Stock Market: Theories and Evidence*. Richard D. Irwin, Homewood, Ill.

[35] Markowitz, H. M. (1959). *Portfolio Selection: Efficient Diversification of Investments*. Wiley, New York.

[36] Modigliani, F. and Miller, M. H. (1958). *Amer. Econ. Rev.*, **48**, 261–297.

[37] Mossin, J. (1966). *Econometrica*, **34**, 768–783.

[38] Percival, J. (1975). *Rev. Econ. Statist.*, **57**, 107–109.

[39] Pinches, G. E. and Mingo, K. A. (1973). *J. Financ.*, **28**, 1–18.

[40] Pinches, G. E., Mingo, K. A., and Caruthers, J. K. (1973). *J. Financ.*, **28**, 389–396.

[41] Press, S. J. and Wilson, S. (1978). *J. Amer. Statist. Ass.*, **73**, 699–705.

[42] Sharpe, W. F. (1964). *J. Financ.*, **19**, 425–442.

[43] Smith, V. K. and Marcis, R. G. (1972). *J. Financ.*, **27**, 589–605.

[44] Solomon, E. (1967). *The Theory of Financial Management*. Columbia University Press, New York.

[45] Stevens, D. L. (1973). *J. Financ. Quant. Anal.*, **8**, 149–158.

[46] Stokes, H. H. and Neuburger, H. (1979). *Rev. Econ. Statist.*, **61**, 534–548.

[47] Tinic, S. M. and West, R. R. (1979). *Investing in Securities: An Efficient Markets Approach*. Addison-Wesley, Reading, Mass.

[48] Tobin, J. (1958). *Rev. Econ. Stud.*, **25**, 65–86.

[49] Vasicek, O. A. (1973). *J. Financ.*, **28**, 1233–1239.

[50] Whitbeck, V. S. and Kisor, M. (1963). *Financ. Anal. J.*, **19**, 55–62.

[51] Wiginton, J. C. (1980). *J. Financ. Quant. Anal.*, **15**, 757–770.

[52] Williamson, J. P. (1970). *Investments, New Analytic Techniques*. Praeger, New York.

[53] Winkler, R. L. (1973). *J. Financ. Quant. Anal.*, **8**, 387–405.

[54] Winkler, R. L. and Barry, C. B. (1975). *J. Financ.*, **30**, 179–192.

(ECONOMETRICS
MARKETING, STATISTICS IN)

R. KOLODNY
H. C. COPELAND

FINITE DIFFERENCES, CALCULUS OF

The *differences* of a function $y = f(x)$ are the differences between values of y for two different values of the argument x. The calculus of finite differences is concerned with properties of these differences, and their applications. There is a close parallelism between this calculus, which studies the properties of

$$\underset{h}{\Delta} f(x) = [f(x+h) - f(x)]/h,$$

and the differential calculus, which studies the properties of

$$Df(x) = \lim_{h\to 0} \underset{h}{\Delta} f(x).$$

A general class of differences is that of *divided differences*. Using the notation

$$y_i = f(x_i),$$

the *first-order divided differences* are

$$\Delta_{x_i,x_j} f(x) = [f(x_i) - f(x_j)]/(x_i - x_j),$$

the *second-order divided differences* are

$$\Delta_{x_i,x_j,x_h} f(x) = [\Delta_{x_i,x_j} f(x) - \Delta_{x_j,x_h} f(x)]/(x_i - x_h),$$

etc. Calculation of divided differences is exemplified in Table 1, where $y_i = f(x_i)$.

If $f(x)$ is a polynomial of degree n, then for any set of distinct values $x_0, x_1, \ldots, x_n$,

$$f(x) = \sum_{i=0}^{n} a_i f(x_i) \quad \text{with } a_i = \prod_{j=1}^{n} (i) \left\{ \frac{x - x_j}{x_i - x_j} \right\},$$

where $j = i$ is excluded from the product. This is the basis of Lagrange's interpolation* formula.

Table 1

1	x_i	y_i	$\Delta_{x_i,x_{i+1}}$	$\Delta_{x_i,x_{i+1},x_{i+2}}$	$\Delta_{x_i \cdots x_{i+3}}$	
0	5	7.1				
			$0.5/1 = 0.50$			
1	6	7.6		$-0.15/3 = -0.050$		
			$0.7/2 = 0.35$		$0.0125/5 = 0.0025$	
2	8	8.3		$-0.15/4 = -0.0375$		etc.
			$0.4/2 = 0.20$		$0.0175/9 = 0.0019$	
3	10	8.7		$-0.14/7 = -0.020$		
			$0.3/5 = 0.06$			
4	15	9.0				
⋮	⋮	⋮				

In many applications, the arguments are at equal intervals with $x_i = x_0 + ih$. By appropriate change of scale it can be arranged that $h = 1$. The difference

$$\underset{1}{\Delta} f(x) = f(x+1) - f(x) \qquad (1)$$

is conventionally denoted by $\Delta f(x)$ and is called the forward difference*. The symbol Δ is called the forward difference operator. Alternative notations which are also useful in certain circumstances, represent (1) as

$$\delta f(x + \tfrac{1}{2}) \qquad \text{(central difference)}$$

or

$$\nabla f(x+1) \qquad \text{(backward difference*)}.$$

Repeated application of a difference operator is denoted by a "power" of the operator. Thus

$$\begin{aligned}\Delta^2 f(x) &= \Delta(\Delta f(x)) = \Delta(f(x+1) - f(x)) \\ &= \Delta f(x+1) - \Delta f(x) \\ &= f(x+2) - f(x+1) \\ &\quad - (f(x+1) - f(x)) \\ &= f(x+2) - 2f(x+1) + f(x)\end{aligned}$$

and similarly,

$$\begin{aligned}\Delta^3 f(x) = \Delta\big(\Delta^2 f(x)\big) &= f(x+3) - 3f(x+2) \\ &\quad + 3f(x+1) - f(x).\end{aligned}$$

Note that $\Delta^r f(x)$ is *not* the same as the divided difference $\Delta_{x,x+1,\ldots,x+r} f(x)$. The latter is, in fact, equal to $\Delta^r f(x)/r!$.

The *factorials*

$$x^{(a)} = x(x-1)(x-2)\cdots(x-a+1)$$

(descending factorial),

$$x^{\{a\}} = \left(x - \frac{a-1}{2}\right)\left(x - \frac{a-3}{2}\right)\cdots \left(x + \frac{a-3}{2}\right)\left(x + \frac{a-1}{2}\right)$$

(central factorial),

$$x^{[a]} = x(x+1)(x+2)\cdots(x+a-1)$$

(ascending factorial),

play roles in finite difference calculus analogous to those played by the power x^a in differential calculus. Thus, analogously to the formula

$$Dx^a = ax^{a-1},$$

we have

$$\Delta x^{(a)} = ax^{(c-1)}; \qquad \delta x^{\{a\}} = ax^{\{a-1\}};$$
$$\nabla x^{[a]} = ax^{[a-1]}.$$

Analogously to the Taylor expansion

$$f(x+h) = \sum_{j=0}^{\infty} \frac{h^j}{j!} D^j f(x),$$

there are the expansions

$$f(x+h) = \sum_{j=0}^{\infty} \frac{h^{(j)}}{j!} \Delta^j f(x)$$

$$f(x+h) = \sum_{j=0}^{\infty} \frac{h^{\{j\}}}{j!} \delta^j f(x)$$

$$f(x+h) = \sum_{j=0}^{\infty} \frac{h^{[j]}}{j!} \nabla^j f(x).$$

All four formulae are exact if $f(x)$ is a polynomial. The last three formulae (the finite difference ones) are the bases for many interpolation* formulae, used for estimating the values of a function for arguments other than those given in a table.

Finite differences are also used in quadrature formulae (*see* NUMERICAL INTEGRATION) and in summation formulae (*see*, in particular, EULER–MACLAURIN EXPANSION).

Apart from these applications, the methods of finite difference calculus are powerful, and provide elegant solutions for many calculations relating to discrete distributions —notably the evaluation of moments (*see* BINOMIAL DISTRIBUTION; HYPERGEOMETRIC DISTRIBUTIONS; POISSON DISTRIBUTION.)

(BACKWARD DIFFERENCE
BESSEL'S INTERPOLATION FORMULA
DIFFERENCE OF ZERO
EULER–MACLAURIN EXPANSION
EVERETT CENTRAL DIFFERENCE FORMULA
FORWARD DIFFERENCE
INTERPOLATION
LAGRANGE INTERPOLATION
NEWTON'S INTERPOLATION FORMULAE
NUMERICAL INTEGRATION
SUMMATION FORMULAE
WHITTAKER–HENDERSON GRADUATION FORMULA)

FINITE MARKOV CHAINS *See* MARKOV PROCESSES

FINITE POPULATION CORRECTION

The variance of the mean of a random sample of size n, chosen without replacement from a population of size N, is

$$\frac{N-n}{Nn}\text{(population variance)}. \quad (1)$$

For large N this approximates to the commonly used formula

$$n^{-1}\text{(population variance)}. \quad (2)$$

Formula (1) is obtained by multiplying (2) by $(1 - nN^{-1})$. This adjustment is the *finite population correction*.

Although there are other circumstances in which formulae incorporating population size are appropriate (e.g., in expected values of sums of squares in analysis of variance* and in correlations among sample values), the term "finite population correction" is generally understood to have the meaning described above.

(HYPERGEOMETRIC DISTRIBUTIONS
SAMPLING)

FINITE POPULATIONS, SAMPLING FROM

A finite population is a collection of distinct elements, finite in number. Finite population sampling theory is the branch of statistics concerned with choosing samples (subsets) of the elements, observing certain features of the sample elements, and using these observations to make inferences about the whole population. For example, the population elements might be all the short-stay hospitals in a state, and for each hospital in a sample the number of patients discharged during one month might be determined, with the goal of estimating the total number discharged that month in the entire state. Or a sample from a population of accounting records might be audited for the purpose of estimating the proportion of records containing errors.

Finite population sampling, also called *survey sampling*, is distinguished from the rest of statistics by its focus on the *actual population* of which the sample is a part. Outside of sampling theory, the observations are typically represented as realized values of random variables, and the inferences refer, not to any actual population, but to the probability law governing the random variables. For example, a sample of the items coming off an assembly line might be tested to determine how many are defective. One statistical analysis might represent the results as independent random variables, with each item having the same unknown probability θ

of being defective, and seek to estimate θ or to test whether it exceeds a maximum acceptable value. The parameter θ is part of a conceptual probability model describing the variability in the production process. It is not a characteristic of any existing population of produced items.

But the same sample might also be used to estimate the actual proportion of defectives in that day's production, in which case the problem falls within the traditional boundaries of finite population sampling theory.

There are four general stages in a sampling investigation of a finite population:

1. Defining the scope and objectives of the study and choosing tools and techniques for making observations
2. Choosing a sample
3. Gathering data on the sample selected
4. Analyzing the data and making inferences

Sampling* theory has concentrated on stages 2 and 4, whereas success at stages 1 and 3 is relatively more dependent on informal judgment, experience, and administrative skill. (Jessen [16] treats many problems involved in stages 1 and 3, and Stephen [28] traces the early development of sampling procedures.) In the remainder of this article we consider questions of choosing samples and making inferences. It will be assumed that the total number N of population elements is known, and that these elements are identified by the labels $1, 2, \ldots, N$. Associated with element i is the unknown value y_i of a variable of interest. A sample of elements will be selected, the values of y_i will be observed for all sample elements, and inferences must be made concerning a population characteristic such as the total, $T = \sum_1^N y_i$, or the mean, $\bar{y} = T/N$.

A *probability sampling plan* is a scheme that assigns to every subset of elements s a known probability $p(s)$ that s will be the selected sample. These plans, also called *random sampling plans*, serve many useful purposes, especially in protecting against the unconscious biases to which selection procedures based on convenience or informal judgment are vulnerable. But in the approach to finite population sampling theory that has been dominant since Neyman's 1934 paper [23], probability sampling is not only desirable but essential, for the probability distribution used to define basic terms such as the bias and variance of an estimator is the distribution *created* for that estimator by the sampling plan. This approach was adopted by the authors of all the early books on sampling—Yates [31], Deming [7], Cochran [6], Hansen et al. [13]—as well as most of the recent ones, e.g., Jessen [16] and Williams [30]. Its basic ideas are presented most clearly in the short nonmathematical monograph of Stuart [29]. An alternative approach, which bases inferences not on the probability sampling distribution but on prediction models, has made substantial progress during the last decade. Smith [27] traces the historical development of both approaches, and the literature on their foundations is reviewed by Cassel et al. [4].

PROBABILITY SAMPLING THEORY

A probability sampling plan creates a distribution for an estimator as follows: For a given population vector $\mathbf{y}_N = (y_1, y_2, \ldots, y_N)$, every possible sample gives a value to the estimator. The probability that the estimator will assume a particular value is the probability that one of the samples giving that value will be selected. For example, a *simple random sampling* plan of size n assigns to each of the $\binom{N}{n}$ sets of n different elements the same probability $1/\binom{N}{n}$. If y is a zero–one variable, the proportion of 1's in the population, $\bar{y}$, might be estimated by the sample proportion, $\bar{y}_s = \sum_s (y_i/n)$. When a simple random sampling plan is used, the probability that a sample containing k 1's will be selected is

$$\binom{N\bar{y}}{k}\binom{N(1-\bar{y})}{n-k}\bigg/\binom{N}{n},$$

so that this sampling plan generates for $n\bar{y}_s$ a hypergeometric probability distribution*.

In most practical problems estimators have much more complicated sampling distributions than the hypergeometric. Sampling theory has concentrated on examining the estimators' means and variances and on showing that in some important instances the distributions are approximately normal when both n and $(N - n)$ are large.

For example, with a simple random sampling plan the sample mean $\bar{y}_s$ is an unbiased estimator of the population mean $\bar{y}$. The variance of $\bar{y}_s$ is $(1 - n/N)S^2/n$, where $S^2 = \sum_1^N (y_i - \bar{y})^2/(N - 1)$. An unbiased estimator v_s of the variance is obtained when S^2 is estimated by $\sum_s (y_i - \bar{y}_s)^2/(n - 1)$. These statements refer to the distributions created by the simple random sampling plan and are valid whenever such a plan is used. No assumptions about the population vector $\mathbf{y}_N$ are needed. However, if $\mathbf{y}_N$ is considered to belong to a hypothetical sequence of vectors obtained by increasing N, then under some mild assumptions about this sequence, the sampling distribution of $(\bar{y}_s - \bar{y})/v_s^{1/2}$ converges to the standard normal distribution as n and $(N - n)$ grow [11,22]. The last fact allows for the construction of approximate $100(1 - \alpha)$ percent confidence intervals of the form $\bar{y}_s \pm z_\alpha v_s^{1/2}$, where z_α is the $100(1 - \alpha/2)$th percentile of the standard normal distribution*.

The use of simple random sampling and the sample mean for estimating a population mean is intuitively appealing when the population is relatively homogeneous. But in many applications some information about differences among population elements is available. If the elements are villages, for example, each village's location and approximate number of inhabitants might be known. Efforts to use such information to get better estimates at lower cost lead to many different sampling and estimation procedures. Although these procedures can reach bewildering complexity, they often employ only simple combinations of six elementary techniques: stratification, clustering, multistage sampling, probability-proportional-to-size sampling, and ratio or regression adjustment.

Stratification refers simply to dividing the population into disjoint subsets or *strata*, each of which is sampled. The villages, for instance, might be stratified by states; or they might be stratified according to approximate size, with those having fewer than 3000 inhabitants in one stratum, those between 3000 and 5000 in another, etc. If independent simple random sampling plans are used within the strata, the overall scheme is a *stratified random sampling plan*.

Clustering also refers to dividing the population into disjoint subsets. Whether the subsets are strata or clusters depends on how the sample is drawn. When at least one element is sampled from each, the subsets are strata. Otherwise, they are clusters—in effect a new population is defined whose elements are clusters of original elements and a sample of clusters is selected (*see* CLUSTER SAMPLING).

Frequently, the elements within a selected cluster are not all observed, but are themselves sampled. Then the sampling plan is carried out in two stages: a sample of clusters is selected at the first stage, and a sample of elements from within sample clusters is selected at the second stage. In sampling from populations whose elements are arranged in natural clusters, such procedures often have important logistic and statistical advantages.

When the elements vary in size, as in the case of villages, and a measure x of the size of each element is available, stratification is only one of the techniques for using this information. Another way to use the size measures is to make the selection probability of element i proportional to its size x_i. Horvitz and Thompson [15] showed how, for any sampling plan giving every element a nonzero inclusion probability, an unbiased estimator of the population mean can be obtained—if π_i is the probability that element i will appear in the sample, then $\sum_s y_i/N\pi_i$ is an unbiased estimator of $\bar{y}$. For a probability-proportional-to-size (PPS) sampling plan $\pi_i = nx_i/N\bar{x}$, and the Horvitz–Thompson estimator* becomes

$$\bar{x}\sum_s (y_i/x_i)/n.$$

Its sampling variance is small when the ratios y_i/x_i are nearly constant. PPS sampling plans are sometimes used in selecting the first-stage sample in two-stage cluster sampling, where the number of elements in each cluster serves as the measure x of that cluster's size [12].

A size measure can also be used in a *ratio adjustment** $\bar{x}/\bar{x}_s$ applied to the sample mean $\bar{y}_s$ to produce the *ratio estimator** of $\bar{y}$, $\bar{y}_s(\bar{x}/\bar{x}_s)$. The ratio adjustment increases the simple estimate $\bar{y}_s$ when the sample is found to consist of elements which are, on average, smaller than the population average, and decreases $\bar{y}_s$ when the sample elements are larger. Under simple random sampling this estimator has a bias, but if y_i is roughly proportional to x_i, the sampling variance can be much smaller than that of the unadjusted mean $\bar{y}_s$.

Another estimator obtained by using the known values $(x_1, x_2, \ldots, x_N)$ of an auxiliary variable to adjust the sample mean is the *regression estimator** of $\bar{y}$, $\bar{y}_s + b(\bar{x} - \bar{x}_s)$, where

$$b = \sum_s (x_i - \bar{x}_s)(y_i - \bar{y}_s) \Big/ \sum_s (x_i - \bar{x}_s)^2.$$

This estimator also has a bias under simple random sampling, but its sampling variance is small if the N points (x_i, y_i) are concentrated along a straight line [5].

As sampling plans and estimators become more complicated, so do the technical problems. It becomes hard to calculate the sampling bias or to discover when a particular approximation to the bias is useful. It becomes even harder to find and evaluate estimates of the sampling standard error and to determine whether a normal approximation for the sampling distribution is adequate. However, the basic strategy remains simple: Choose a probability sampling plan and estimator which, under that plan, is (approximately) unbiased with small variance. Then express uncertainty by an estimate of the standard error.

This approach has dominated sampling theory since the 1930s, a period that has seen tremendous growth and refinement in the theory and its applications. But this period has also seen a growing awareness that the probability sampling approach has some serious theoretical shortcomings [9]. Moreover, some workers have observed that this approach can produce inferences which are mathematically correct but unreasonable, and have challenged its basic tenet—that inferences should be based on the probability sampling distribution [1, 21]. The weaknesses become apparent after a sample is drawn and its y-values are observed. Now inferences must be made only for the set r of nonsample elements. For example, the population total can be expressed as $T = \sum_s y_i + \sum_r y_i$, and since the first term is known, inference about T is logically equivalent to inference about the unobserved total $\sum_r y_i$. But under the probability sampling approach the only relationship linking the observed elements to the unobserved ones is the fact that the latter might have been chosen for observation but were not. This relationship *alone* appears inadequate as a basis for making useful inferences about the unobserved elements. This judgment is confirmed by the probability sampling distribution's *likelihood** *function*, which sanctions *no* inference concerning the unobserved elements [10].

PREDICTION THEORY

Recent studies have used probability models to express relationships that enable observations on some elements to lead to mathematically precise inferences about others. The numbers $(y_1, y_2, \ldots, y_N)$ are treated as realized values of random variables $(Y_1, Y_2, \ldots, Y_N)$, and it is the joint probability distribution of these random variables that links the observed and unobserved y's. When finite population estimation problems are considered under such models, sometimes called *superpopulation models*, they are recognized as problems of statistical *prediction**. For example, after the sample is observed, estimating $T = \sum_s y_i + \sum_r y_i$ is equivalent to predicting the value $\sum_r y_i$ of the unobserved random variable $\sum_r Y_i$. Solutions to the prediction problem can then be obtained using various general approaches—linear least-squares* techniques, likelihood and Bayesian* methods, etc. Superpopulation models have long been used in probability sampling theory as tools for suggesting estimators and evaluating sampling strategies [5]. But the use of such models to approach finite population estimation as a prediction problem received little attention until the late 1960s.

The prediction approach most easily compared with probability sampling theory is the one based on linear models and *weighted least-squares** prediction techniques. This approach also has been developed in terms of bias, variance, and approximate confidence intervals based on the normal distribution. But here the distribution used in making inferences is the one given by the prediction model, not the distribution created by a probability sampling plan. Thus an estimator is called unbiased for a particular sample when the expected error under the model is zero for that sample, regardless of properties in other samples, and the variance to be estimated is similarly conditioned on the sample actually selected [3, 24].

The Bayesian prediction approach requires a more elaborate model than that used in least-squares prediction. But it too conditions on s and makes its inferences in terms of the model, stating the results in terms of the posterior distribution of the population values, given the sample observations [8].

All the prediction approaches reproduce many key formulas of probability sampling theory. For example, if the Y's are uncorrelated with common mean and variance, standard results from least-squares prediction theory imply that the *best linear unbiased estimator** of $\bar{y}$ is the sample mean, and that an unbiased estimator of the error variance is given by the same formula, v_s, that is used with simple random sampling. When the Y-distribution is normal as well, the sample mean is both the *maximum likelihood** *estimator* [25] and the *fiducial** *estimator* [17], and when the mean of this normal model is itself given a uniform prior distribution, $\bar{y}_s$ is the *Bayes estimator* [8]. Prediction theory also recognizes important reasons for using random sampling plans. Thus in a population where this model for the Y's is reasonable, subscribers to both probability sampling and prediction approaches might find themselves using the same sampling plan and estimator and disagreeing only about how the procedure is justified. The same situation can arise in much more complicated problems, where prediction models lead naturally to stratification and clustering, and to ratio and regression estimators.

Situations where the prediction and probability sampling approaches lead to different results are also important. Examples occur when the inferences are based on systematic samples, or when they refer to subpopulations represented by few if any elements in the sample [14, 20]. They also occur when a simple random sampling plan is used with the ratio or regression estimator—then the approaches lead to substantially different formulas for the standard error. Recent empirical studies of some of these anomalous situations have led to conclusions strongly favoring the prediction approach, but these conclusions have been vigorously challenged by proponents of the probability sampling approach [26, including discussants]. Because the controversy within finite population sampling theory is in fact a facet of the more general continuing debate over the *foundations of statistics** and the role of *randomization** [2, 18, 19], an early resolution is not expected.

References

[1] Basu, D. (1969). *Sankhyā A*, **31**, 441–454.

[2] Basu, D. (1975). *Sankhyā A*, **37**, 1–71.

[3] Brewer, K. R. W. (1963). *Aust. J. Statist.*, **5**, 93–105.

[4] Cassel, C., Sarndal, C., and Wretman, H. H. (1977). *Foundations of Inference in Survey Sampling*. Wiley, New York. (A thorough and dispassionate account of theoretical work before 1976. No exercises.)

[5] Cochran, W. G. (1942). *J. Amer. Statist. Ass.*, **37**, 199–212.

[6] Cochran, W. G. (1953). *Sampling Techniques*. Wiley, New York. [Strong on theory, but with applications clearly in focus. With exercises. The third edition (1977) of this classic includes some predic-

tion theory results, but the probability sampling distribution remains fundamental.]

[7] Deming, W. E. (1950). *Some Theory of Sampling*. Wiley, New York.

[8] Ericson, W. A. (1969). *J. R. Statist. Soc. B*, **31**, 195–233.

[9] Godambe, V. P. (1955). *J. R. Statist. Soc. B*, **17**, 269–278.

[10] Godambe, V. P. (1966). *J. R. Statist. Soc. B*, **28**, 310–319.

[11] Hájek, J. (1960). *Publ. Math. Inst. Hung. Acad. Sci.*, **5**, 361–374.

[12] Hansen, M. H. and Hurwitz, W. N. (1943). *Ann. Math. Statist.*, **14**, 333–362.

[13] Hansen, M. H., Hurwitz, W. N., and Madow, W. G. (1953). *Sample Survey Methods and Theory*, 2 Vols. Wiley, New York. [A comprehensive and authoritative view of the methods (Vol. 1) and theory (Vol. 2) based on probability sampling distributions as of the 1950s. With some exercises.]

[14] Holt, D., Smith, T. M. F., and Tomberlin, T. J. (1979). *J. Amer. Statist. Ass.*, **74**, 405–410.

[15] Horvitz, D. G. and Thompson, D. J. (1952). *J. Amer. Statist. Ass.*, **47**, 663–685.

[16] Jessen, R. J. (1978). *Statistical Survey Techniques*. Wiley, New York. (An up-to-date presentation of the title subject matter, unencumbered by mathematical details or doubts about the probability sampling model. Helpful format ends each chapter with summary, worked examples, references, and exercises.)

[17] Kalbfleisch, J. D. and Sprott, D. A. (1969). In *New Developments in Survey Sampling*, N. L. Johnson and H. Smith, Jr., eds. Wiley-Interscience, New York, pp. 358–389.

[18] Kempthorne, O. (1977). *J. Statist. Plann. Infer.*, **1**, 1–25.

[19] Kiefer, J. (1977). *Synthese*, **36**, 161–176.

[20] Laake, P. (1979). *J. Amer. Statist. Ass.*, **74**, 355–358.

[21] Lahiri, D. B. (1968). Paper presented at a symposium on foundations of survey sampling held at the University of North Carolina, Chapel Hill, N.C. Apr.

[22] Madow, W. G. (1948). *Ann. Math. Statist.*, **19**, 535–545.

[23] Neyman, J. (1934). *J. R. Statist. Soc.*, **97**, 558–606.

[24] Royall, R. M. (1971). In *Foundations of Statistical Inference*, V. P. Godambe and D. A. Sprott, eds. Holt, Rinehart and Winston, Toronto.

[25] Royall, R. M. (1976). *Biometrika*, **63**, 605–614.

[26] Royall, R. M. and Cumberland, W. G. (1978). In *Survey Sampling and Measurement*, N. K. Namboodiri, ed. Academic Press, New York, pp. 293–309.

[27] Smith, T. M. F. (1976). *J. R. Statist. Soc. A*, **139**, 183–195.

[28] Stephan, F. F. (1948). *J. Amer. Statist. Ass.*, **43**, 12–39.

[29] Stuart, A. (1962). *Basic Ideas of Scientific Sampling*. Hafner, New York. [Accessible to readers with no previous statistical training. "The feature of this book is that it leads the reader to grasp the ideas of sampling theory by verifying them upon samples from a small population."—Author's preface to second edition (1976). Probability sampling distributions are eagerly promoted as the only basis for sound inference.]

[30] Williams, W. H. (1978). *A Sampler on Sampling*. Wiley, New York.

[31] Yates, F. (1949). *Sampling Methods for Censuses and Surveys*. Charles Griffin, London. [Results are described, not derived. "On the other hand an attempt has been made to cover all the modern developments of sampling theory which are of importance in census and survey work, and to give an adequate discussion of the complexities that are encountered in their practical application."—Author's preface to third edition (1960).]

(AREA SAMPLING
CLUSTER SAMPLING
DOUBLE SAMPLING
HORVITZ–THOMPSON ESTIMATOR
MULTISTAGE SAMPLING
PREDICTION
PROBABILITY SAMPLING
QUOTA SAMPLING
RANDOMIZED RESPONSE
RATIO ESTIMATORS
SAMPLE SURVEYS
STRATIFIED SAMPLING
SUPERPOPULATION MODELS
SYSTEMATIC SAMPLING)

RICHARD M. ROYALL

FIRST-DIGIT PROBLEM

The problem of identifying the distribution of leading digits in tables or sets of data arises because in practice the first digits 1, 2, . . . , 9 do not occur with equal frequency, but frequently conform to a distribution having a discrete probability density function equal to $\log_{10}\{(d+1)/d\}$ for $d = 1, 2, \ldots, 9$.

This phenomenon is known as Benford's law (see Benford [2]), although it was formulated by Simon Newcomb in 1881 [5], and possibly earlier.

Formally, let A_d denote the set of all non-negative real numbers with leading digit less than or equal to d, so that

$$A_d = \bigcup_{n=-\infty}^{\infty} \left[10^n, (d+1)10^n\right).$$

If "log" denotes logarithms to the base 10, then a logarithmic mapping gives (with obvious notation)

$$\log A_d = \bigcup_{n=-\infty}^{\infty} \left[n, n + \log(d+1)\right).$$

Benford's law then asserts that a random variable X in the underlying population of interest satisfies

$$\Pr(X \in A_d) = \log(d+1) \qquad (d = 1, 2, \ldots, 9). \quad (1)$$

Raimi [7] presents frequencies of first-digit occurrence for a number of data sets, as well as for certain sequences of real numbers, together with those satisfying (1). Some of these sets approximate closely to (1), while others do not.

Explanations of (1) have been produced which bear no relation to actual data sets in the real world, but which derive from the real number system and the decimal system; see Raimi [7] for details.

Pinkham [6] considered a random variable (rv) X with underlying continuous cumulative distribution function (CDF) $F(\cdot)$, so that

$$\begin{aligned}\Pr(X \in A_d) &= \sum_{k=-\infty}^{\infty} \left[F\left((d+1)10^k\right) - F\left(10^k\right)\right] \\ &= H(\log(d+1))\end{aligned}$$

if $H(x) = \sum_{k=-\infty}^{\infty}[F(10^{x+k}) - F(10^x)]$. Thus H is the CDF of $\log X \pmod 1$. Benford's law (1) requires that $H(x) = x$ for $x = \log d (d = 1, 2, \ldots, 9)$; but a sufficient condition for (1) is that

$$H(x) = x, 0 \leqslant x < 1. \qquad (2)$$

Pinkham applied Fourier transform theory to obtain an upper bound to $|H(x) - x|$; this was improved upon by Raimi and then by Kemperman [4] to give

$$\begin{aligned}&|H(x) - x| \\ &\leqslant \min\left[\frac{1}{8} V\left(\frac{d}{dx} F(10^x)\right)\Bigg|_{x=-\infty}^{x=\infty}, \right. \\ &\qquad \left. \frac{1}{64} V\left(\frac{d^2}{dx^2} F(10^x)\right)\Bigg|_{-\infty}^{\infty}\right],\end{aligned}$$

where $V(g(x))|_a^b$ is the total variation in the function g for $a < x < b$.

Example. Let X have a standard Cauchy distribution*, for which [7]

$$V\left(\frac{d^2}{dx^2} F(10^x)\right) = 2.53.$$

Then $|H(x) - x| \leqslant 0.0395$.

A general characterization of distributions yielding small values of $|H(x) - x|$ is unavailable. Certain mixtures of distributions* approach Benford's law, as does the distribution of $\Pi_{i=1}^n X_i$, where the X_i's are uniform independent rvs on $(0, 1)$; also if (2) holds for a rv Y, it holds for cY and for $1/Y$ [1]. The only distribution with $[0.10, 1]$ as support which satisfies (2) exactly is the *reciprocal distribution* with PDF [3]

$$f(x) = (\log e)/x, \qquad 0.10 \leqslant x \leqslant 1. \quad (3)$$

Hamming showed that if a rv X has PDF (3) and Y is a rv with the same support, then XY, X/Y, and Y/X all have PDF (3).

Real-world data sets include many examples which defy Benford's law; telephone directories frequently exclude "1" as a first digit, for example. The aggregate of 20,229 entries from several tables studied by Benford, however, produced very close results (see Raimi [7]). In practice, the scale-invariance property holds very well when it is tried out on data sets which themselves follow (2) closely.

The expository article by Raimi [7], which is the primary source used here, is a very clear survey of the treatment of the problem, and contains a detailed bibliography.

References

[1] Adhikari, A. K. and Sarkar, B. P. (1968). *Sanhkyā B*, **30**, 47–58.

[2] Benford, F. (1938). *Proc. Amer. Philos. Soc.*, **78**, 551–572.

[3] Hamming, R. W. (1970). *Bell Syst. Tech. J.*, **49**, 1609–1625.

[4] Kemperman, J. H. B. (1975). *Bull. Inst. Math. Statist.*, No. 4, 138. Abstr. 75t-47.

[5] Newcomb, S. (1881). *Amer. J. Math.*, **4**, 39–40.

[6] Pinkham, R. S. (1961). *Ann. Math. Statist.*, **32**, 1223–1230.

[7] Raimi, R. (1976). *Amer. Math. Monthly*, **83**, 521–538. (An excellent summary and review.)

FIRST-ORDER MODEL

A regression model which is linear in the parameters and the independent variables (or carriers).

(REGRESSION ANALYSIS
RESPONSE SURFACE ANALYSIS)

FIRST PASSAGE TIME *See* DIFFUSION PROCESSES; RANDOM WALK

FIRST PRODUCT MOMENT

This term is sometimes used for:

1. $n^{-1}\sum_{i=1}^{n} X_i Y_i$, where $(X_1, Y_1), \ldots, (X_n, Y_n)$ are values of two variables X, Y in a random sample of size n (*sample* first product moment), or
2. The population analog of item 1, $E[XY]$.

It is also sometimes applied to the corresponding quantities ("covariances"*) based on deviations from mean values

$$n^{-1}\sum_{i=1}^{n}(X_i - \overline{X})(Y_i - \overline{Y}) \qquad \text{and}$$

$$E[\{X - E[X]\}\{Y - E[Y]\}],$$

respectively.

(CORRELATION)

FISHER, RONALD AYLMER

Born: February 17, 1890, in East Finchley, London.

Died: July 29, 1962, in Adelaide, Australia.

Contributed to: mathematical statistics, probability theory, genetics, design of experiments

Ronald Aylmer Fisher achieved original scientific research of such diversity that the integrity of his approach is masked. Born into the era of Darwin's evolutionary theory and Maxwell's theory of gases, he sought to recognize the logical consequences of a world containing indeterminism, whose certainties were essentially statistical. His interests were those of Karl Pearson*, who dominated the fields of evolution, biometry, and statistics during his youth, but his perspective was very different. His ability to perceive remote logical connections of observation and argument gave his conceptions at once universal scope and coherent unity; consequently, he was little influenced by current scientific vogue at any period of his life.

Fisher was (omitting his stillborn twin) the seventh and youngest child of George Fisher, fine arts auctioneer in the West End, and Katie, daughter of Thomas Heath, solicitor of the City of London. His ancestors showed no strong scientific bent but his uncle, Arthur Fisher, was a Cambridge Wrangler.

In childhood, Fisher met the misfortune first of his poor eyesight, and the eyestrain that was always to limit his private reading, and he learned to listen while others read aloud to him. In 1904, his beloved mother died suddenly of peritonitis. In 1906, his father's business failure required him to become largely self-supporting. His general intelligence and mathematical precocity were apparent early. From Mr. Greville's school in Hampstead, he went on to Stanmore in 1900, and entered Harrow in 1904 with a scholarship in mathematics. In his second year he won the Neeld Medal in mathematical competition with all the school. To avoid eyestrain he received tuition in mathematics under G. H. P. Mayo without pencil, paper, or other visual aids. Choosing spherical trigonometry for the subject of these tutorials, he developed a strong geometrical sense that was greatly to influence his later work. In 1909, he won a scholarship in mathematics to Cambridge University. In 1912, he graduated as a Wrangler and, being awarded a studentship for one year, studied the theory of errors under F. J. M. Stratton and statistical mechanics and quantum theory under J. Jeans.

In April 1912, Fisher's first paper [3] was published, in which the *method of maximum likelihood** was introduced (though not yet by that name). As a result, that summer Fisher wrote to W. S. Gosset* ("Student") questioning his divisor $(n - 1)$ in the formula for the standard deviation. He then reformulated the problem in an entirely different and equally original way, in terms of the configuration of the sample in n-dimensional space, and showed that the use of the sample mean instead of the population mean was equivalent to reducing the dimensionality of the sample space by one; thus he recognized the concept of what he later called *degrees of freedom**. Moreover, the geometrical formulation immediately yielded Student's distribution, which Gosset had derived empirically, and in September Fisher sent Gosset the mathematical proof. This was included in Fisher's paper when, two years later, using the geometrical representation, he derived the general sampling distribution of the correlation* coefficient [4].

Fisher's mathematical abilities were directed into statistical research by his interest in evolutionary theory, especially as it affected man. This interest, developing at Harrow, resulted in the spring of 1911 in the formation of the Cambridge University Eugenics Society at Fisher's instigation. He served on the Council even while he was chairman of the undergraduate committee; he was the main speaker at the second annual meeting of the society. While famous scientists wrangled about the validity either of evolutionary or genetical theory, Fisher accepted both as mutually supportive; for he saw that if natural variation was produced by genetic mechanism, every evolutionary theory except natural selection was logically excluded. While the applicability of genetic principles to the continuous variables in man was disputed on biometrical grounds, Fisher assumed that the observed variations were produced genetically, and in 1916 [5] justified this view by biometrical argument.

In its application to man selection theory raised not only scientific but practical problems. The birthrate showed a steep and regular decline relative to increased social status. This implied the existence throughout society of selection against every quality likely to bring social success. Fisher believed, therefore, that it must result in a constant attrition of the good qualities of the population, such as no civilization could long withstand. It was important to establish the scientific theory on a firm quantitative basis through statistical and genetic research, and, more urgently, to publicize the scientific evidence so that measures should be taken to annul the self-destructive fertility trend.

Fisher accepted at once J. A. Cobb's suggestion [2] in 1913 that the cause of the dysgenic selection lay in the economic advantage enjoyed by the children of small families over those from larger families at every level of society. Later he proposed and urged adoption of various schemes to spread the financial burden of parenthood, so that those who performed similar work should enjoy a similar standard of living, irrespective of the number of their children, but without success; and the family allowance scheme adopted in Great Britain after World War II disappointed his hopes.

To further these aims, on leaving college he began work with the Eugenics Education Society of London, which was to continue for 20 years. From 1914 he was a regular book reviewer for the *Eugenics Review*; in 1920 he became business secretary and in 1930 vice-president of the society; and he pursued related research throughout. Major Leonard Darwin, the president, became a dear and revered friend, a constant encouragement and support while Fisher was struggling for recognition, and a stimulus to him in the quantitative research that resulted in *The Genetical Theory of Natural Selection* [14].

In 1913, Fisher took a statistical job with the Mercantile and General Investment Company in the City of London. He trained with the Territorial Army and, on the outbreak of war in August 1914, volunteered for military service. Deeply disappointed by his rejection due to his poor eyesight, he served his country for the next five years by teaching high school physics and mathematics. While he found teaching unattractive, farming appealed to him both as service to the nation and as the one life in which a numerous family might have advantages. When, in 1917, he married Ruth Eileen, daughter of Dr. Henry Grattan Guinness (head of the Regions Beyond Missionary Union at the time of his death in 1915), Fisher rented a cottage and small holding from which he could bicycle to school, and with Eileen and her sister, began subsistence farming, selling the excess of dairy and pork products to supply needs for which the family could not be self-sufficient. Their evening hours were reserved for reading aloud, principally in the history of earlier civilizations.

In these years Fisher's statistical work brought him to the notice of Karl Pearson*. In 1915, Pearson published Fisher's article on the general sampling distribution of the correlation coefficient in *Biometrika** [4], and went on to have the ordinates of the error of estimated correlations calculated in his department. The cooperative study [21] was published in 1917, together with a criticism of Fisher's paper not previously communicated to its author. Pearson had not understood the method of maximum likelihood Fisher had used, and condemned it as being inverse inference, which Fisher had deliberately avoided. Fisher, then unknown, was hurt by Pearson's highhandedness and lack of understanding, which eventually led to their violent confrontation. Meanwhile, Pearson ignored Fisher's proposal to assess the signficance of correlations by considering not the correlation r itself but

$$z = \tfrac{1}{2}\ln[(1 + r)/(1 - r)],$$

a remarkable transformation that reduces highly skewed distributions with unequal variances to distributions to a close approximation normal with constant variance (*see* FISHER'S z-TRANSFORMATION). Fisher's paper on the correlation* between relatives on the supposition of Mendelian inheritance [5], submitted to the Royal Society in 1916, had to be withdrawn in view of the referee's comments. (Knowing that Pearson disagreed with his conclusions, Fisher had hoped that his new method, using the analysis-of-variance components, might have been persuasive.) In this paper the subject and methodology of biometrical genetics* was created. These facts influenced Fisher's decision in 1919 not to accept Pearson's guarded invitation to apply for a post in his department.

In September 1919, Fisher started work in a new, at first temporary, post as statistician at Rothamsted Experimental Station, where agricultural research had been in progress since 1843. He quickly became established in this work. He began with a study of historical data from one of the long-term experiments, with wheat on Broadbalk, but soon moved on to consider data obtained in current field trials, for which he developed the *analysis of variance**. These studies brought out the inadequacies of the arrangement of the experiments themselves and so led to the evolution of the science of experimental design*. As Fisher worked with experimenters using successively improved designs, there emerged the principles of *randomization**, adequate *replication*, *blocking* and *confounding**, and randomized blocks*, Latin squares*, factorial* arrangements, and other designs of unprecedented efficiency. The statistical methods were incorporated in successive editions of *Statistical Methods for Research Workers* [10]. The 11-page paper on the arrangement of field experiments [11] expanded to the book *The Design of Experiments* [16]. These volumes were supplemented by *Statistical Tables for Biological, Agricultural and Medical Research*, co-authored by Frank Yates [19].

Following up work on the distribution of the correlation coefficient, Fisher derived the sampling distributions of other statistics in common use, including the F-distribution* and the multiple correlation coefficient. Using geometrical representations, he solved for normally distributed errors all the distribution problems for the general linear model, both when the null hypothesis is true and when an alternative hypothesis is true [9, 12].

Concurrently, the theory of estimation* was developed in two fundamental papers in 1922 [7] and in 1925 [8]. Fisher was primarily concerned with the small samples of observations available from scientific experiments, and was careful to draw a sharp distinction between sample statistics (estimates) and population values (parameters to be estimated). In the method of maximum likelihood he had found a general method of estimation that could be arrived at from the mathematical and probabilistic structure of the problem. It not only provided a method to calculate unique numerical estimates for any problem that could be precisely stated, but also indicated what mathematical function of the observations ought to be used to estimate the parameter. It thus provided a criterion for the precise assessment of estimates, a revolutionary idea in 1920.

Using this method to compare two estimates of the spread σ of a normal distribution, Fisher [6] went on to show that the sample standard deviation s was not only better but uniquely best; because the distribution of any other measure of spread conditional on s does not contain the parameter σ of interest, once s is known, no other estimate gives any further information about σ. Fisher called this quality of s *sufficiency**. This finding led to his introduction of the concept of the amount of *information** in the sample, and the criteria of *consistency**, *efficiency**, and sufficiency of estimates, measured against the yardstick of available information. He exploited the asymptotic efficiency of the method of maximum likelihood in 1922, and extending consideration to small samples in 1925, observed that small sample sufficiency, when not directly available, was obtainable via *ancillary statistics** derived from the likelihood function.

Thus, seven years after moving to Rothamsted, Fisher had elucidated the underlying theory and provided the statistical methods that research workers urgently needed to deal with the ubiquitous variation encountered in biological experimentation. Thereafter, he continued to produce a succession of original research on a wide variety of statistical topics. For example, he initiated nonlinear design, invented k-statistics (*see* FISHER'S k-STATISTICS), explored extreme value distributions*, harmonic analysis, multivariate analysis* and the discriminant function, the analysis of covariance*, and new elaborations of experimental design and of sampling survey*.

So diverse and fundamental were Fisher's contributions to mathematical statistics that G. A. Barnard [1] wrote that "to attempt, in a short article, to assess the contributions to the subject by one largely responsible for its creation would be futile." Fisher's central contribution was surely, as Barnard wrote, "his deepening of our understanding of uncertainty" and of "the many types of measurable uncertainty." "He always ascribed to Student the idea that, while there must necessarily be uncertainty involved in statistical procedures, this need not imply any lack of precision—the uncertainty may be capable of precise quantitative assessment. In continuing work begun by Student, of deriving the exact sampling distributions of the quantities involved in the most common tests of significance, Fisher did much to give this idea form and reality." In developing the theory of estimation, he explored the type of uncertainty expressible precisely in terms of the likelihood; and his ideas on the subject never ceased to evolve. From the beginning, however, he distinguished likelihood from mathematical probability, the highest form of scientific inference, which he considered appropriate only for a restricted type of uncertainty. He accepted classical probability theory, of course, and used Bayes' theorem in cases in which there was an observational basis for making probability statements in advance about the population in question; further, he proposed the fiducial argment as leading to true probability statements, at least in one common class of cases.

Fisher introduced the fiducial argument in 1930 [13]. In preparing a paper on the general sampling distribution of the multiple correlation coefficient in 1928, he noticed that in the test of significance the relationship between the estimate and the parameter was of a type he later characterized as pivotal*. He argued that if one quantity was fixed, the distribution of the other was determined; consequently, once the observations fixed the value of the observed statistic, the whole distribution of the unknown parameter was determined. Thus, in cases in which the *pivotal relationship* existed, true probability statements concerning continuous parameters could be inferred from the data. Exhaustive estimates were required [15].

Controversy arose immediately. The fiducial argument was proposed as an alternative to the argument of the title "inverse probability," which Fisher condemned in all cases in which no objective prior probability could be stated. H. Jeffreys led the debate on behalf of less restrictive use of inverse probability, while J. Neyman developed an approach to the theory of estimation through sampling theory, which in some instances led to numerical results different from Fisher's. For many years the debate focused on the case of estimating the difference between two normally distributed populations with unknown variances not assumed to be equal (Behrens' test). This led Fisher to introduce the concept of the *relevant reference set*. He argued that the sampling theory approach ignored information in the sample, and that of all possible samples, only the subset yielding the observed value s_1^2/s_2^2 was relevant. Later problems with the fiducial argument arose in cases of multivariate estimation, because of nonuniqueness of the pivotals. Different pairs of pivotals could be taken that resulted in different and nonequivalent probability statements. Fisher did not achieve clarification of criteria for selection among such alternative pivotals; he was working on the problem at the end of his life (*see* FIDUCIAL DISTRIBUTIONS; FIDUCIAL INFERENCE; FIDUCIAL PROBABILITY).

In proposing the fiducial argument in 1930, Fisher highlighted the issues of scientific inference and compelled a more critical appreciation of the assumptions made, and of their consequences, in various approaches to the problem. In reviewing the subject in *Statistical Methods and Scientific Inference* [18], he distinguished the conditions in which he believed significance tests, likelihood estimates, and probability statements each still had an appropriate and useful role to play in scientific inference.

In his genetical studies, having demonstrated the consonance of continuous variation in man with Mendelian principles, and having thereby achieved the fusion of biometry and genetics [5], in the 1920s Fisher tackled the problems of natural selection, expressed in terms of population genetics*. Of this work, K. Mather [20] wrote:

> Fisher's contribution was basic, characteristic and unique. It is set out in *The Genetical Theory of Natural Selection*. He pointed out that natural selection is not evolution, that in fact evolution is but one manifestation of the operation of natural selection, and that natural selection can and should be studied in its own right. Having delimited his field, he proceeded to cultivate it as only he could, with all his resources of mathematics allied to experimentation carried out in circumstances which most of us would have regarded as prohibitively difficult. Again he went beyond merely harmonizing to fusing the principles of genetics and natural selection. His well-known theory of the evolution of dominance (so sharply criticized, yet so widely accepted) is but one facet of his work: he formulated his fundamental theorem of natural selection, equating the rate of increase in fitness to the genetic variance in fitness; he considered sexual selection and mimicry; and he extended his discus-

sion to man, society and the fate of civilizations. It was a truly great work.

The book was written down by his wife at his dictation during evenings at home; for a while it took the place of the reading and conversation that ranged from all the classics of English literature to the newest archeological research, and centered on human evolution. At home, too, was Fisher's growing family: his oldest son, George, was born in 1919; then a daughter who died in infancy, a second son, and in the end six younger daughters. True to his eugenic ideal, Fisher invested in the future of the race, living simply under conditions of great financial stringency while the children were reared. He was an affectionate father, and especially fond of George, who was soon old enough to join him in such activities as looking after genetic mouse stocks. Wherever possible, he brought the children into his activities, and he answered their questions seriously, with sometimes brilliant simplicity; and he promoted family activities and family traditions. Domestic government was definitely partriarchal, and he punished larger offences against household rules, though with distaste. As long as possible the children were taught at home, for he trusted to their innate curiosity and initiative in exploring their world rather than to any imposed instruction. He had no sympathy with lack of interest, or fear of participation; and if his motive force was itself frightening, learning to deal with it also was a part of the child's education. In fact, he treated his children like his students, as autonomous individuals from the beginning, and encouraged them to act and think on their own responsibility, even when doing so involved danger or adult disapproval.

In 1929, Fisher was elected a Fellow of the Royal Society as a mathematician. The influence of his statistical work was spreading, and he was already concerned that statistics should be taught as a practical art employing mathematical reasoning, not as self-contained mathematical theory. In 1933, Karl Pearson retired and his department at University College London was split; E. S. Pearson* succeeded as head of the statistics department, and Fisher as Galton Professor of Eugenics, housed on a different floor of the same building. For both men, it was an awkward situation. While others gave their interpretation of Fisher's ideas in the statistical department, he offered a course on the philosophy of experimentation in his own. After J. Neyman* joined the statistics department in 1934, relations between the new departments deteriorated, and fierce controversy followed.

Fisher continued both statistical and genetical research. In 1931 and again in 1936, he was visiting professor for the summer sessions at Iowa State University at Ames, Iowa, at the invitation of G. W. Snedecor, director of the Statistical Computing Center. In 1937–1938, he spent six weeks as the guest of P. C. Mahalanobis*, director of the Indian Statistical Institute*, Calcutta. In his department, where Karl Pearson had used only biometrical and genealogical methods, Fisher quickly introduced genetics. Work with mouse stocks, moved from the attic at his home, was expanded, and experimental programmes were initiated on a variety of animal and plant species, for example, to study the problematical tristyly in *Lythrum salicaria*. Fisher was very eager also to initiate research in human genetics.

Sponsored by the Rockefeller Foundation, in 1935 he was able to set up a small unit for human serological research under G. L. Taylor, joined by R. R. Race in 1937. In 1943, Fisher interpreted the bewildering results obtained with the new Rh blood groups in terms of three closely linked genes, each with alleles, and predicted the discovery of two new antibodies and an eighth allele—all of which were identified soon after. Fisher's enthusiasm for blood group polymorphisms continued to the end of his life and he did much to encourage studies of associations of blood groups and disease.

In 1927, Fisher proposed a way of measuring selective intensities on genetic polymorphisms occurring in wild populations, by a combination of laboratory breeding and field observation, and by this method later demonstrated very high rates of selective predation on grouse locusts. E. B. Ford was one of the few biologists who believed in natural selection at the time; in 1928 he planned a long-term investigation of selection in the field, based on Fisher's method. To the end of his life, Fisher was closely associated with Ford in this work, which involved development of capture–recapture* techniques and of sophisticated new methods of statistical analysis. The results were full of interest and wholly justified their faith in the evolutionary efficacy of natural selection alone.

Forcibly evacuated from London on the outbreak of war in 1939, Fisher's department moved to Rothamsted, and finding no work as a unit, gradually dispersed; Fisher himself could find no work of national utility. In 1943, he was elected to the Balfour Chair of Genetics at Cambridge, which carried with it a professorial residence. Lacking other accommodation, he moved his genetic stocks and staff into the residence, leaving his family in Harpenden. Estranged from his wife, separated from home, and deeply grieved by the death in December 1943 of his son George on active service with the Royal Air Force, Fisher found companionship with his fellows at Caius College, and with the serological unit (evacuated to Cambridge for war work with the Blood Transfusion Service), which planned to rejoin his department after the war.

There was little support after the war for earlier plans to build up an adequate genetics department. No bid was made to keep the serological unit. No departmental building was erected. Support for the research in bacterial genetics initiated in 1948 under L. L. Cavalli (Cavalli-Sforza) was withdrawn in 1950 when Cavalli's discovery of the first Hfr strain of *Escherichia coli* heralded the remarkable discoveries soon to follow in bacterial and viral genetics. Fisher cultivated his garden, continued his research, published *The Theory of Inbreeding* [17] following his lectures on this topic, and built a group of good quantitative geneticists. He attempted to increase the usefulness of the university diploma of mathematical statistics by requiring all diploma candidates to gain experience of statistical applications in research, in a scientific department. Speaking as founding president of the Biometric Society*, as president of the Royal Statistical Society*, and as a member or as president of the International Statistical Institute*, he pointed out how mathematical statistics itself owes its origin and continuing growth to the consideration of scientific data rather than of theoretical problems.

His own interests extended to the work of scientists in many fields. He was a fascinating conversationalist at any time, original, thoughtful, erudite, witty, and irreverent; with the younger men, his genuine interest and ability to listen, combined with his quickness to perceive the implications of their research, were irresistible. He encouraged, and contributed to, the new study of geomagnetism under S. K. Runcorn, a fellow of his college. He was president of Gonville and Caius College during the period 1957–1960.

He received many honors and awards: the Weldon Memorial Medal (1928), the Guy Medal of the Royal Statistical Society in gold (1947), three medals of the Royal Society, the Royal Medal (1938), the Darwin Medal (1948), and the Copley Medal (1956); honorary doctorates from Ames, Harvard, Glasgow, London, Calcutta, Chicago, the Indian Statistical Institute*, Adelaide, and Leeds. He was Foreign Associate, United States National Academy of Sciences; Foreign Honorary Member, American Academy of Arts and Sciences; Foreign Member, American Philosophical Society; Honorary Member, American Statistical Association*; Honorary President; International Statistical Institute; Foreign Member, Royal Swedish Academy of Sciences; Member, Royal Danish Academy of Sciences; Member, Pontifical Academy; Member, Imperial German

Academy of Natural Science. He was created Knight Bachelor by Queen Elizabeth in 1952.

After retirement in 1957, Sir Ronald Fisher traveled widely before joining E. A. Cornish in 1959 as honorary research fellow of the C. S. I. R. O. Division of Mathematical Statistics in Adelaide, Australia. He died in Adelaide July 29, 1962.

The collections of Fisher's papers referred to throughout this list are given in full in the subsequent Bibliography.

References

[1] Barnard, G. A. (1963). *J. R. Statist. Soc. A*, **216**, 162–166.

[2] Cobb, J. A. (1913). *Eugen. Rev.*, **4**, 379–382.

[3] Fisher, R. A. (1912). *Messeng. Math.*, **41**, 155–160. [Also in Fisher (1971–1974), paper 1.]

[4] Fisher, R. A. (1915). *Biometrika*, **10**, 507–521. Also in *Bibliography* (4), No. 4 and *Bibliography* (5), No. 1.

[5] Fisher, R. A. (1918). *Trans. R. Soc. Edinb.*, **52**, 399–433. Also in *Bibliography* (4), No. 9.

[6] Fisher, R. A. (1920). *Monthly Not. R. Astron. Soc.*, **80**, 758–770. Also in *Bibliography* (4), No. 12 and *Bibliography* (5), No. 2.

[7] Fisher, R. A. (1922). *Philos. Trans. A*, **222**, 309–368. Also in *Bibliography* (4), No. 18 and *Bibliography* (5), No. 10.

[8] Fisher, R. A. (1925). *Proc. Camb. Philos. Soc.*, **22**, 700–725. Also in *Bibliography* (4), No. 42 and *Bibliography* (5), No. 11.

[9] Fisher, R. A. (1925). *Metron*, **5**, 90–104. Also in *Bibliography* (4), No. 43.

[10] Fisher, R. A. (1925). *Statistical Methods for Research Workers*. Oliver & Boyd, Edinburgh (subsequent editions: 1928, 1930, 1932, 1934, 1936, 1938, 1941, 1944, 1946, 1950, 1954, 1958, 1970).

[11] Fisher, R. A. (1926). *J. Minist. Agric. Gr. Brit.*, **33**, 503–513. Also in *Bibliography* (4), No. 48 and *Bibliography* (5), No. 17.

[12] Fisher, R. A. (1928). *Proc. R. Soc. A*, **121**, 654–673. Also in *Bibliography* (4), No. 61 and *Bibliography* (5), No. 14.

[13] Fisher, R. A. (1930). *Proc. Camb. Philos. Soc.*, **26**, 528–535. Also in *Bibliography* (4), No. 84 and *Bibliography* (5), No. 22.

[14] Fisher, R. A. (1930). *The Genetical Theory of Natural Selection*. Oxford University Press, London (reprinted by Dover, New York, 1958).

[15] Fisher, R. A. (1934). *Proc. R. Soc. A*, **144**, 285–307. Also in *Bibliography* (4), No. 108 and *Bibliography* (5), No. 24.

[16] Fisher, R. A. (1935). *The Design of Experiments*. Oliver & Boyd, Edinburgh.

[17] Fisher, R. A. (1949). *The Theory of Inbreeding*. Oliver & Boyd, Edinburgh.

[18] Fisher, R. A. (1956). *Statistical Methods and Scientific Inference*. Oliver & Boyd, Edinburgh.

[19] Fisher, R. A. and Yates, F. (1938). *Statistical Tables for Biological, Agricultural and Medical Research*. Oliver & Boyd, Edinburgh.

[20] Mather, K. (1963). *J. R. Statist. Soc. A*, **216** 166–168.

[21] Soper, H. E., Young, A. W., Cave, B. M., Lee, A., and Pearson, K. (1916). *Biometrika*, **11**, 328–413.

Bibliography

Box, J. (1978). *R. A. Fisher: The Life of a Scientist*. Wiley, New York. [Complementary to Fisher (1971–1974) in showing the context of his statistical innovations.]

Cochran, W. G. (1967). *Science*, **156**, 1460–1462. (Cochran's encounters with Fisher make an amusing, and telling, character sketch.)

Finney, D. J. (1964). *Biometrics*, **20**, 322–329. [This volume, "In Memoriam R. A. Fisher, 1890–1962," includes articles on his contributions to various branches of statistical science: e.g., Rao (1964) and Yates (1964).]

Fisher, R. A. (1971–1974). *Collected Papers of R. A. Fisher*, 5 vols. J. H. Bennett, ed. University of Australia, Adelaide, Australia. [This compilation includes a complete bibliography of Fisher's published work and the biographical memoir by Yates and Mather (1963); with 291 of Fisher's papers, presented chronologically. A primary source for all students of Fisher or of the historical development of statistical theory through diverse scientific applications, 1912 to 1962.]

Fisher, R. A. (1950). *Contributions to Mathematical Statistics*. Wiley, New York. (Out of print, this volume contains 43 of Fisher's major articles on mathematical statistics, with his introductory comments.)

Rao, C. R. (1964). *Biometrics*, **20**, 186–300.

Savage, L. J. (1976). On rereading R. A. Fisher [J. W. Pratt, ed.], *Ann. Math. Statist.*, **4**, 441–500. (The talk was given in 1970; the article grew ever more comprehensive thereafter, and was published posthumously. A stimulating commentary with references to many others.

Yates, F. (1964). *Biometrics*, **20**, 307–321.

Yates, F. and Mather, K. (1963). *Biogr. Mem. Fellows R. Soc. Lond.*, **9**, 91–120. (An excellent summary.)

(ANALYSIS OF VARIANCE
ANCILLARY STATISTICS
CONDITIONAL INFERENCE

DESIGN OF EXPERIMENTS
DISCRIMINANT ANALYSIS
FIDUCIAL DISTRIBUTIONS
FIDUCIAL INFERENCE
FIDUCIAL PROBABILITY
FISHER INFORMATION
FISHER'S z-TRANSFORMATION
GOSSET, WILLIAM SEALY
LIKELIHOOD
MAXIMUM LIKELIHOOD
PEARSON, KARL
PIVOTAL QUANTITIES
RANDOMIZATION
SUFFICIENCY)

JOAN FISHER BOX

FISHER DISCRIMINANT *See* DISCRIMINANT ANALYSIS

FISHER DISTRIBUTION

This is an alternative name for the logarithmic series distribution*. However, the term refers also to an analog of the normal distribution on the sphere, as used in directional data analysis* (*see* DIRECTIONAL DISTRIBUTIONS).

FISHERIES RESEARCH, STATISTICS IN

Fish comprise about 3% of the world's protein consumption, although its percentage share is much greater in the low-protein less developed countries than in the high-protein developed countries. A major unresolved question is the extent to which production from the oceans (and fresh waters) can be increased to meet the world's increasing food needs. Any success will depend both on rational management of the exploited wild stocks and on the intensive development of aquaculture.

Such rational management is a relatively recent development, although the need for it was expressed during the nineteenth century. This did at least lead to the collection of necessary data on stocks, beginning in particular through the International Council for the Exploration of the Sea formed in 1901, and through a long subsequent series of biological studies. Since World War II a large body of mathematical theory and a variety of computer programs have been developed to analyze the biological and catch data and to form a basis for recommendations for rational management.

The process begins with the fish being caught, which can be brought about by a variety of different devices. Commercial devices have one feature in common—they are selective, sometimes due to the nature of the stock but more usually due to the device. Typically, they capture the larger (and older) fish. Thus a first problem in fisheries statistics is to measure the selection. It is common to define a selection curve $S(x)$ which gives the probability of capture as a function of length x relative to the maximum capture probability. For some catching devices, $S(x)$ is assumed to be a sigmoid function, often the logistic. Special experiments are designed, either comparing the gear to be tested with unselective gear or working with known length subgroups to estimate $S(x)$. If the two-parameter logistic or the integrated normal selection curve is assumed, standard bioassay* methods are available to estimate the parameters. Often in application all that is estimated is the value of x, say τ, such that $S(\tau) = 0.5$; $S(x)$ is then approximated by the *knife-edge selection function*

$$S(x) = 0, \quad x \leqslant \tau, \qquad S(x) = 1, \quad x > \tau.$$

The errors incurred by such an approximation have been investigated in a few instances but are often disregarded.

An alternative type of gear, the gillnet, selects all fish in a central range of lengths (actually girths) with capture probabilities decreasing as length x increases or decreases from this central value. The length at which capture probability is maximized depends on the size of the mesh and is often assumed to be directly proportional to mesh size. If the selection function has the shape of the normal density curve, the size that has maximal capture probability is the mean of this density, say μ. Now with data on numbers caught by length class from nets of different meshes fished in parallel, one can derive the equation

$$\ln(n_{ax}/n_{bx}) = \alpha + \beta x,$$

where

n_{ax} = number of fish of length class x caught by mesh a,

n_{bx} = number of fish of length class x caught by mesh b,

$$\alpha = \frac{\mu_a - \mu_b}{\sigma^2}, \qquad \beta = \frac{\mu_b^2 - \mu_a^2}{2\sigma^2},$$

μ_a = length at which capture probability is maximized with mesh a,

μ_b = length at which capture probability is maximized with mesh b,

σ^2 = variance of selection curve (assumed to be the same for both meshes).

Standard regression procedures are used to obtain $\hat{\alpha}$ and $\hat{\beta}$; then $\mu_a + \mu_b$ is estimated by $-\hat{\beta}/\hat{\alpha}$. If we assume that μ is proportional to mesh size, the parameter of proportionality is estimated by a second regression process.

Other parametric forms can be assumed instead of the normal density and some at least fitted by nonlinear regression procedures. A sophisticated approach is taken by Hamley and Regier [7], who show that the capture process is a result of two processes and then assume the selection curve is a composite of Pearson type I curves.

A second problem at the outset, for stocks in the ocean or large lakes, is identification of the species and/or stock and ascertaining whether the catches involve several separate stocks. A variety of measurements have been used for racial identification-scale counts and morphometric measurements such as body length, head length, and head width. Comparisons between samples and sexes and putative stocks can lead to analysis of covariance* and/or discriminant analysis* [12]. Other methods include protein taxonomy and genetic studies which involve a variety of standard statistical procedures. These and others of a more sophisticated nature are expounded in full by Sokal and Sneath [14].

Having identified the stocks and quantified biases in the fishery catches, it is necessary to set up a basis to sample the catches and to determine what information to obtain on the sampled fish. An introduction to these is given in Gulland [6] and a full discussion of a typical procedure for a particular fishery (North Pacific halibut) is found in Southward [15]. The catch is sampled not only to obtain length and weight information but also for maturity, fecundity, and to make age determinations. Weight–length and fecundity–length are usually expressed as simple power law relationships, determined in the past by linear regression on the log transformed variables. However, depending on the error structure, nonlinear fitting methods should be preferred in some situations. A logistic equation is fitted to maturity–age (or maturity–length) data and often approximated by a knife-edged function similar to that used for selection curves.

Age in fish is often determined by reading layers (annuli) on scales or otoliths (ear bones) or other hard parts. The validation that the layers counted in these structures are actually annuli, has too frequently been neglected, as well as errors that occur in readings within and between readers. Some experiments have been designed and analyzed to estimate or test for such differences as well as variations due to age of the fish, sex, etc. An alternative method of determining age distribution is through an analysis of length, in effect a separation of the length distribution into its components, assumed normal. While graphical methods have often been used, computer procedures are now to be preferred [8].

Since age and length are related and lengths are much easier to determine than ages, age–length keys are created to expand the age determination from a small number of age readings to the whole catch.

Westrheim and Ricker [16] have pointed out that the age–length keys are strongly dependent on the age structure of the population and can give meaningless results when applied to a population with different age composition. For a correct approach to this problem, see Clark [2].

Because there is a relationship not only between body length and age but also between body length and width of the annuli on scales or other structures, it is possible to use these readings also to reconstruct the growth history of individual fish. Such back calculation must be treated with caution because of the effects of selective sampling and selective mortality—a discussion of the problem with interesting statistical questions is given by Ricker [10].

Understanding the impact of a fishery requires estimates of the population "vital rates": i.e., the mortality rates (due to capture by the fishery and to nonfishery causes), recruitment rates, and ultimately population numbers. As noted, most fisheries operate on a part of the total population, typically the older and larger members, and it is the addition to the exploited part in a particular time, in numbers or in weight, that is referred to as recruitment.

Methods used to estimate vital rates or population numbers include capture–recapture methods* and analysis based on changes resulting from the exploitation itself. Most frequently one assumes that the catch is proportional to the population multiplied by the units of effort applied. This is a reasonable assumption when the gear is a trawl hauled along the bottom or through the midcolumn of the water, although in this case adjustments are necessary for the ship's velocity, the size of the mesh, and perhaps other factors. With other types of gear the situation may be more complicated. Where possible, comparative fishing is carried out and the results analyzed by analysis of variance* and covariance* (Robson 1966). In other situations search theory* methods have been applied.

When catch, effort, weight at age, and recruitment data are available, it is straightforward to utilize the *catch equation*, which has the form

$$C = qR\int_{t_r}^{\infty} W(t)f(t)e^{-[Mt+F(t)]}\,dt,$$

where

C = catch from a cohort in units of weight,

$f(t)$ = effort applied at time (age) t,

$W(t)$ = weight of individual in cohort at age t,

M = mortality rate due to all causes other than fishing, assumed to be constant,

$F(t)$ = cumulative effort to time t,

R = recruitment in numbers at age t_r,

t_r = median age of recruitment,

q = catchability coefficient.

While the catch equation applies to a cohort, if the population is stable and stationary, it holds (approximately modified) for the catch in any time period. The integral in the catch equation can be evaluated numerically, but it is usual to assume that $f(t)$ is constant and $W(t)$ has a specified functional form such as

$$W_t = W(1 - ke^{-bt})^3,$$

the Bertalanffly growth model; the integration is carried out in closed form. Tables are available to evaluate the integral if the generalized Bertalanffly growth model is used, i.e.,

$$W_t = W_\infty(1 - e^{-bt})^n.$$

Where data on the age composition of the catch are available, the catch equation by age class provides a method of estimating q and M. The equation is

$$\begin{aligned}\ln U_{t+1} - \ln U_t &= -(M + qf_t)\\ &\quad + \ln\left(\frac{1 - e^{-(M+qf_t)}}{1 - e^{-(M+qf_{t+1})}}\right)\left(\frac{M + qf_{t+1}}{M + qf_t}\right),\end{aligned}$$

where

U_t = catch per unit of effort of an age class in time period t,

f_t = effort applied in time period t,

U_{t+1} = catch per unit of effort of the same age class in time period $t + 1$.

Other symbols are as defined earlier; M and q can be estimated by nonlinear methods but usually the approximation $(1 - e^{-x})/x \doteq e^{-x/2}$ is used to make the equation linear. In practice the regression matrix often turns out to be ill conditioned*, so that the estimates have unsatisfactory properties. One way to meet this problem is to find an independent estimate of M so that only q needs to be estimated from the regression model.

When age data are not available a simpler approach has been taken, assuming that the population biomass satisfies a simple differential equation, e.g.,

$$\frac{d\overline{B}}{dt} = k\overline{B}(M - \overline{B}),$$

where $\overline{B}$ is the mean biomass. The population growth $d\overline{B}/dt$ or, more precisely, $\Delta\overline{B}$ for a unit time period is estimated by the change in biomass (as reflected in catch per unit of effort). If $\Delta\overline{B}_i = \overline{B}_{i+1} - \overline{B}_i$, and recalling that $U_i = q\overline{B}_i$, the equation that results is

$$\Delta\overline{B}_i = \frac{1}{q}\Delta U_i + \frac{1}{2}(C_i + C_{i+1})$$
$$= \frac{k}{q^2}U_i(qM - U_i).$$

This can be rearranged to give a multiple regression* of $C_i + C_{i+1}$ against U_i, U_i^2 and ΔU_i, and standard regression procedures used to estimate the three parameters. Alternative methods of solution have been proposed, starting with Schaefer [13], who developed the procedure, as well as alternative and generalized differential equations for population (or biomass) growth. As soon as estimates of the parameters are available, the population level and the level of catch in weight can be determined that maximizes the latter on a continuing basis. The assumptions of the model have frequently been questioned; it is realized also that environmental and biological fluctuations cause this approach to be rather oversimplified.

The catch from a population depends on recruitment in addition to parameters discussed earlier. Recruitment is difficult to measure and in some cases has been assumed constant over a range of population sizes; hence it can be disregarded by considering not yield but yield per recruitment. This tactic fails, however, if recruitment varies as population varies, which it must over sufficiently wide ranges of population sizes. Furthermore, recruitment is of fundamental ecological importance, so that it is necessary both to measure it and to try to determine what factors affect it, and particularly its relationship to harvest population size. The latter topic has an extensive literature; many aspects were covered in a symposium (see Parrish [9]), and a comprehensive treatment is given by Cushing [3]. The most useful method of estimating recruitment when catch by age information is available is due to Allen [1].

Although a complete understanding of the stock can come only from full biological studies, the need to obtain answers quickly or the inadequacy of the information available has led to widespread use of sonar systems to estimate fish biomass. This approach has led to interdisciplinary investigations utilizing the physics of hydroacoustics, but there are substantial statistical problems of calibration* of different systems as well as converting indices to absolute numbers. This area is changing rapidly under the impact of technology; see Cushing [4].

Most of fisheries science and analysis and most exploitation is on wild stocks. However in some regions, such as southeast Asia, aquaculture, the farming of fish, has been carried on for many centuries. Different types of aquaculture are now practiced in many parts of the world—salmon are raised in pens, mussels on rafts, and many species are cultured in Japan. Clearly, this is the direction required if fisheries are to play a substantial role in the human food problem. So far, little statistical analysis has been carried out, but it is to be expected that the vast body of statistical theory that has been useful in design and analysis of agriculture* experiments will be applicable in aquaculture as a scientific approach is taken to this discipline.

References

[1] Allen, K. R. (1966). *J. Fish. Res. Board Canada*, **23**, 1553–1574.

[2] Clark, W. G. (1981). *J. Fish. Res. Board Canada*, (in press).

[3] Cushing, D. H. (1973). *Recruitment and Parent Stock in Fishes*. Division of Marine Resources, University of Washington, Seattle, Wash.

[4] Cushing, D. H. (1973). *The Detection of Fish*. Pergamon Press, Oxford.

[5] Fox, W. W., Jr. (1975). *Fish Bull.*, **73**, 23–49.

[6] Gulland, J. A. (1962). Manual of Sampling Methods for Fisheries Biology. *FAO Fish. Biol. Tech. Paper 26*.

[7] Hamley, J. M. and Regier, H. A. (1973). *J. Fish. Res. Board Canada*, **30**, 817–830.

[8] Hasselblad, V. (1966). *Technometrics*, **8**, 431–434.

[9] Parrish, B. B., ed. (1973). *Fish Stocks and Recruitment*. Rapports et procès-verbaux des réunions, Vol. 164, Conseil international pour l'exploration de la mer. Charlottenlundslot, Denmark.

[10] Ricker, W. E. (1975). *Computation and Interpretation of Biological Statistics of Fish Populations*. Bull. 191 Environ., Dept. of Fisheries and Marine Service, Ottawa, Canada.

[11] Robson, D. S. (1966). *Res. Bull. Int. Comm. Northeast Atlantic Fish.*, **3**, 5–14.

[12] Royce, W. F. (1954). *Proc. Indo-Pacific Fish Counc.*, **4**, 130–145.

[13] Schaefer, B. (1954). *Bull. Inter-Amer. Trop. Tuna Comm.*, **1**, 27–56.

[14] Sokal, R. R. and Sneath, P. H. A. (1963). *Principles of Numerical Taxonomy*. W. H. Freeman, San Francisco.

[15] Southward, G. M. (1976). *Sampling Landings of Halibut for Age Composition*. Inter-Pacific Halibut Comm. Sci. Rep. No. 58.

[16] Westrheim, S. J. and Ricker, W. E. (1978). *J. Fish. Res. Board Canada*, **35**, 184–189.

(CAPTURE–RECAPTURE METHODS
ECOLOGICAL STATISTICS)

DOUGLAS G. CHAPMAN

FISHER INEQUALITY

This inequality states that for any balanced incomplete block design*, the number of treatments cannot exceed the number of blocks. A simple demonstration is given by Bose [1].

Reference

[1] Bose, R. C. (1949). *Ann. Math. Statist.*, **20**, 619–620.

(BLOCKS, BALANCED INCOMPLETE
INCIDENCE MATRIX)

FISHER INFORMATION

R. A. Fisher's* measure of the amount of information supplied by data about an unknown parameter is the first use of "information" in a technical sense in mathematical statistics. It plays an important role in the theory of statistical estimation and inference, and is closely associated with the concepts of efficiency* and sufficiency*. Wiener [3, p. 76] stated that his definition of information could be used to replace Fisher's definition in the technique of statistics. Savage [2, p. 50] remarked: "The ideas of Shannon and Wiener, though concerned with probability, seem rather far from statistics. It is, therefore something of an accident that the term information coined by them should not be altogether inappropriate in statistics."

Fisher defined the amount of information in a sample about the parameter θ as the reciprocal of the variance of an efficient estimator. For a sample of n independent observations the Fisher information measure is

$$I_F = E_\theta\left\{\left((\partial/\partial\theta)\sum_{i=1}^{n}\ln f(x_i;\theta)\right)^2\right\}, \quad (1)$$

where $f(x;\theta)$ is the common density function of the observed random variables. I_F is clearly nonnegative and measures in some sense the amount of information available about θ in x. For a multinomial distribution* in which the probabilities P_i for $i = 1, \ldots, r$, are functions of a single unknown parameter θ, so that $P_1(\theta) + P_2(\theta) \cdots + P_r(\theta) = 1$, the Fisher measure of information is

$$I_F = \sum_{i=1}^{r} P_i(\theta)((\partial/\partial\theta)\ln P_i(\theta))^2$$

$$= \sum_{i=1}^{r} (1/P_i)(\partial P_i/\partial\theta)^2. \qquad (2)$$

The Fisher information measure is proportional to the precision of an unbiased estimator of θ and in accordance with the classical Cramér–Rao inequality is inversely proportional to the lower bound for the variances of all unbiased estimators of θ. Consider $f(x,\theta)$, a density function corresponding to an absolutely continuous distribution function with parameter θ for a random variable X. Let $T(X)$ be any unbiased estimate of $\phi(\theta)$, a function of θ. Then under suitable regularity conditions the Cramér–Rao* inequality is

$$\operatorname{var} T \geqslant (d\phi/d\theta)^2/I, \qquad (3)$$

where I, the variance of $(d/d\theta)\ln f(x,\theta)$, is the Fisher information measure. Under suitable regularity conditions [5], if X and Y are independent random variables, the Fisher information measure is additive,

$$I_F^{X,Y} = I_F^X + I_F^Y. \qquad (4)$$

If $T(x)$ is a statistic, then

$$I_F^X \geqslant I_F^T \qquad (5)$$

with equality if and only if T is a sufficient statistic. The efficiency of an unbiased estimator for θ is the fraction of information contained in the estimator for estimating θ relative to that contained in an efficient estimator for estimating θ.

Fisher's information matrix is the $k \times k$ matrix whose element in the ith row and jth column is

$$\int f(\omega,\boldsymbol{\theta})\left(\left(\frac{\partial}{\partial\theta_i}\right)\ln f(\omega,\theta)\right) \times \left(\left(\frac{\partial}{\partial\theta_j}\right)\ln f(\omega,\theta)\right) d\omega, \qquad (6)$$

where the probability density $f(\omega,\theta)$ is a function of the vector random variable ω and a k-dimensional parameter $\boldsymbol{\theta}$. Best asymptotically normal (B.A.N.) estimators are those whose asymptotic distribution is normal with covariance matrix the inverse of the Fisher information matrix (*see* ASYMPTOTIC NORMALITY).

Fisher's information measure is related to the information measure discussed in INFORMATION, KULLBACK, which see for concepts and notation, as well as Savage [2], Wilks [4], and Zacks [5]. Consider the parametric case where the probability densities are of the same functional form but differ according to the value of the k-dimensional parameter $\boldsymbol{\theta} = (\theta_1, \theta_2, \ldots, \theta_k)$. Suppose that $\boldsymbol{\theta}$ and $\boldsymbol{\theta} + \boldsymbol{\Delta\theta}$ are neighboring points in the k-dimensional parameter space which is assumed to be an open convex set in a k-dimensional Euclidean space, and set $p(\omega) = f(\omega, \boldsymbol{\theta})$, $\pi(\omega) = f(\omega, \boldsymbol{\theta} + \boldsymbol{\Delta\theta})$. We may write for the Kullback discrimination information

$$I(p:\pi) = \int f(\omega)\ln(f(\omega)/\pi(\omega))\,d\omega,$$

$$I(\theta:\theta+\Delta\theta) = -\int f(\omega,\theta)\Delta\ln f(\omega,\theta)\,d\omega, \qquad (7)$$

where $\Delta f(\omega,\theta) = f(\omega,\theta+\Delta\theta) - f(\omega,\theta)$ and $\Delta \ln f(\omega,\theta) = \ln f(\omega,\theta+\Delta\theta) - \ln f(\omega,\theta)$. Under certain regularity conditions on $f(\omega, \theta)$ [1, p. 26] it may be shown that to within second-order terms

$$I(\boldsymbol{\theta}:\boldsymbol{\theta}+\boldsymbol{\Delta\theta}) = \frac{1}{2}\sum_{\alpha=1}^{k}\sum_{\beta=1}^{k} g_{\alpha\beta}\Delta\theta_\alpha\Delta\theta_\beta$$

$$= I(\boldsymbol{\theta}+\boldsymbol{\Delta\theta}:\boldsymbol{\theta}), \qquad (8)$$

where

$$g_{\alpha\beta} = \int f(\omega,\theta)((\partial/\partial\theta_\alpha)\ln f(\omega,\theta)) \times ((\partial/\partial\theta_\beta)\ln f(\omega,\theta))\,d\omega \qquad (9)$$

and $\mathbf{G}(\boldsymbol{\theta}) = (g_{\alpha\beta})$ is the positive definite Fisher information matrix. Suppose that $x_i(\omega)$, $i = 1, 2, \ldots, k$, are unbiased estimators of the parameters θ_i, $i = 1, 2, \ldots, k$. Let the probability density of the x's be denoted by $g(x)$; then under suitable regularity conditions, to within terms of higher order,

$$2I(\boldsymbol{\theta} : \boldsymbol{\theta} + \boldsymbol{\Delta\theta}; X) = (\boldsymbol{\Delta\theta})'\mathbf{H}(\boldsymbol{\theta})(\boldsymbol{\Delta\theta}) = 2I(\boldsymbol{\theta} + \boldsymbol{\Delta\theta} : \boldsymbol{\theta}; X), \quad (10)$$

where $(\boldsymbol{\Delta\theta})' = (\Delta\theta_1, \Delta\theta_2, \ldots, \Delta\theta_k)$ and $\mathbf{H}(\boldsymbol{\theta})$ is the positive definite matrix $(h_{ij}(\boldsymbol{\theta}))$,

$$h_{ij}(\boldsymbol{\theta}) = \int g(x)\left(\left(\frac{\partial}{\partial\theta_i}\right)\ln g(x)\right) \times \left(\left(\frac{\partial}{\partial\theta_j}\right)\ln g(x)\right) dx \quad (11)$$

$i, j = 1, 2, \ldots, k$. Under suitable regulatory conditions, it may be shown that

$$(\boldsymbol{\Delta\theta})'\mathbf{G}(\boldsymbol{\theta})(\boldsymbol{\Delta\theta}) \geqslant (\boldsymbol{\Delta\theta})'\mathbf{H}(\boldsymbol{\theta})(\boldsymbol{\Delta\theta}) \geqslant (\boldsymbol{\Delta\theta})'\boldsymbol{\Sigma}^{-1}(\boldsymbol{\Delta\theta}), \quad (12)$$

and in particular

$$\det \mathbf{G}(\boldsymbol{\theta}) \geqslant \det \mathbf{H}(\boldsymbol{\theta}) \geqslant \det \boldsymbol{\Sigma}^{-1}, \quad (13)$$

where $(\boldsymbol{\Delta\theta})$ is defined above, $\mathbf{G}(\boldsymbol{\theta})$ is defined in (9), $\mathbf{H}(\boldsymbol{\theta})$ is defined in (11), and $\boldsymbol{\Sigma}$ is the covariance matrix of the unbiased estimators $x_i(\omega)$, $i = 1, 2, \ldots, k$. The first two members in (12) are equal if and only if the unbiased estimators are sufficient, and the last two members are equal if and only if $g(x)$ is of the form of an exponential family*. The results in (12) and (13) are multivariate versions of the Cramér–Rao inequality (3) and the inequality given in (5). Because of the relationship indicated above, properties of Fisher's information and Kullback's discrimination information with respect to additivity, sufficiency*, efficiency*, and grouping of observations are similar and related. We illustrate the foregoing with the following examples [1, pp. 57–59].

Example 1. Consider the binomial with $p(\theta) = \cos^2\theta$, $q(\theta) = \sin^2\theta$, $0 \leqslant \theta \leqslant \pi$ so that $p(\theta) + q(\theta) = 1$. In accordance with (2),

$$I_F = (1/\cos^2\theta)(4\cos^2\theta\sin^2\theta) + (1/\sin^2\theta)(4\sin^2\theta\cos^2\theta) = 4,$$

so that for a sample of n independent observations $I_F = 4n$. In accordance with the inequality (3),

$$\operatorname{var}(x/n) \geqslant (2\cos\theta\sin\theta)^2/4n = \cos^2\theta\sin^2\theta/n = pq/n.$$

Example 2. Consider the normal populations $N(\theta, \sigma^2)$. Let $T(x) = \bar{x}$, the average of n independent observations. It is found that

$$\mathbf{G} = \begin{bmatrix} n/\sigma^2 & 0 \\ 0 & n/(2\sigma^4) \end{bmatrix},$$

$$\mathbf{H} = \begin{bmatrix} n/\sigma^2 & 0 \\ 0 & 1/(2\sigma^4) \end{bmatrix},$$

and the lower bound for the variance of an unbiased estimator of θ, σ^2/n, is attained by the estimator $\hat{\theta} = \bar{x}$.

Example 3. Consider the normal populations $N(\theta, \sigma^2)$. Let $T(x) = (\bar{x}, s^2)$, where $\bar{x}$ is the average and $s^2 = (n-1)^{-1}\sum_{i=1}^n (x_i - \bar{x})^2$ is the unbiased sample variance of the n observations. It is found that

$$\mathbf{G} = \begin{bmatrix} n/\sigma^2 & 0 \\ 0 & n/(2\sigma^4) \end{bmatrix},$$

$$\mathbf{H} = \begin{bmatrix} n/\sigma^2 & 0 \\ 0 & n/(2\sigma^4) \end{bmatrix},$$

$$\boldsymbol{\Sigma} = \begin{pmatrix} \sigma^2/n & 0 \\ 0 & 2\sigma^4/(n-1) \end{pmatrix}.$$

The lower bound for the variance of an unbiased estimator of σ^2, $2\sigma^4/n$, is not attained by the estimator s^2 with a variance $2\sigma^4/(n-1)$. When the population mean is known, say 0, the lower bound for the variance of an unbiased estimator of σ^2 is attained for the estimator $(1/n)\sum_{i=1}^n x_i^2$.

Example 4. Consider bivariate normal populations with zero means, unit variances, and correlation coefficient ρ, and let $T(x) = (y_1, y_2)$, where

$$y_1 = \left(\frac{1}{n}\right)\sum_{i=1}^n (x_{1i} - x_{2i})^2,$$

$$y_2 = \left(\frac{1}{n}\right)\sum_{i=1}^n (x_{1i} + x_{2i})^2.$$

It is found that

$$\mathbf{G}(\rho) = n(1 + \rho^2)/(1 - \rho^2)^2 = \mathbf{H}(\rho);$$

the lower bound for the variance of an unbiased estimator of ρ is $(1 - \rho^2)^2/n(1 + \rho^2)$.

References

[1] Kullback, S. (1959). *Information Theory and Statistics*. Wiley, New York (reprinted by Dover, New York, 1968; Peter Smith Publisher, Magnolia, Mass., 1978). (Begins with a measure-theoretic presentation of theory. Applications are discussed at an intermediate level.)

[2] Savage, L. J. (1954). *The Foundations of Statistics*. Wiley, New York. (A theoretical discussion at the beginning graduate level.)

[3] Wiener, N. (1948). *Cybernetics*. Wiley, New York. (Not primarily statistical.)

[4] Wilks, S. S. (1962). *Mathematical Statistics*. Wiley, New York. (An extensive exposition at the beginning graduate level.)

[5] Zacks, S. (1971). *The Theory of Statistical Inference*. Wiley, New York. (Theoretical presentation at an advanced level.)

Bibliography

See the following works, as well as the references just cited, for more information on the topic of Fisher information.

Cramér, H. (1946). *Mathematical Methods of Statistics*. Princeton University Press, Princeton, N.J. (Primarily theoretical, at the beginning graduate level.)

Cramér, H. (1946). *Skand. Aktuarietidskr.*, **29**, 458–463.

Fisher, R. A. (1922). *Philos. Trans. R. Soc. Lond. A*, **222**, 309–368.

Fisher, R. A. (1925). *Proc. Camb. Philos. Soc.*, **22**, 700–725; *Contributions to Mathematical Statistics*, No. 11. Wiley, New York, 1950.

Rao, C. R. (1945). *Bull. Calcutta Math. Soc.*, **37**, 81–91.

Rao, C. R. (1973). *Linear Statistical Inference and Its Applications*. Wiley, New York. (Intermediate level.)

(CRAMÉR–RAO LOWER BOUND
DISCRIMINANT ANALYSIS
EFFICIENCY
INFORMATION, KULLBACK
INFORMATION MATRIX
MAXIMUM LIKELIHOOD
ESTIMATION
MULTINOMIAL DISTRIBUTION
SUFFICIENCY
UNBIASEDNESS)

S. KULLBACK

FISHER LINEAR DISCRIMINANT *See* DISCRIMINANT ANALYSIS

FISHER–NEYMAN DECOMPOSITION *See* DISCRIMINANT ANALYSIS

FISHER–NEYMAN FACTORIZATION THEOREM *See* SUFFICIENCY

FISHER'S EXACT TEST

Fisher's exact test provides an exact method for testing the null hypothesis of independence for categorical data* in a 2×2 contingency table (*see* CONTINGENCY TABLES) with both sets of marginal frequencies fixed in advance. This test was proposed in Fisher [11, 12], Irwin [17], and Yates [28], and is also known as the *Fisher–Irwin test* for 2×2 tables and as the *Fisher–Yates test**. It is discussed in many books on nonparametric methods, including Siegel [22], Bradley [4], Conover [8], Hays [16], Marascuilo and McSweeney [20], and Daniel [9]. Fisher's [11] example of application is where a human subject's ability to discriminate correctly between two objects is tested. The subject is told in advance exactly how many times each object will be presented and is expected to make that total number of identifications of each object even if guessing.

The sample data for this situation can be presented in a 2×2 table as follows:

	A_1	A_2	Totals
B_1	f_{11}	f_{12}	$f_{1\cdot}$
B_2	f_{21}	f_{22}	$f_{2\cdot}$
Totals	$f_{\cdot 1}$	$f_{\cdot 2}$	$f_{\cdot\cdot}$

The marginal frequencies $f_{1\cdot}$, $f_{2\cdot}$, $f_{\cdot 1}$, and $f_{\cdot 2}$ and the total frequency $f_{\cdot\cdot}$ are all assumed to be fixed.

In the Fisher example, the labels A_1 and A_2 in the 2×2 table might refer to the two objects actually presented, and the labels B_1 and B_2 then would refer to the identification made by the subject (or vice versa). The subject is told the values for $f_{\cdot 1}$ and $f_{\cdot 2}$ and his or her identifications should be allocated so that $f_{1\cdot}$ and $f_{2\cdot}$ agree with these stated values. The null hypothesis to be tested here is that the identification made is in no way influenced by the object actually presented and hence the subject is merely guessing.

In an arbitrary situation where $f_{\cdot\cdot}$ items are classified, A and B refer to some categories or characteristics where each has two subcategories A_1, A_2 and B_1, B_2. The null hypothesis* is that the categories A and B are independent. Under this null hypothesis, the probability of observing any particular sample table with all marginal frequencies fixed is

$$\begin{aligned}
&\Pr[f_{ij} \mid f_{\cdot\cdot}, f_{1\cdot}, f_{\cdot 1}] \\
&\quad = \binom{f_{1\cdot}}{f_{11}}\binom{f_{2\cdot}}{f_{21}} \Big/ \binom{f_{\cdot\cdot}}{f_{\cdot 1}} \\
&\quad = \binom{f_{\cdot 1}}{f_{11}}\binom{f_{\cdot 2}}{f_{12}} \Big/ \binom{f_{\cdot\cdot}}{f_{1\cdot}} \\
&\quad = f_{\cdot 1}!\, f_{1\cdot}!\, f_{2\cdot}!\, f_{\cdot 2}!/(f_{\cdot\cdot}!\, f_{11}!\, f_{12}!\, f_{21}!\, f_{22}!)
\end{aligned} \tag{1}$$

for $i = 1, 2$ and $j = 1, 2$.

Since all the marginal frequencies are fixed, only one of the f_{ij} values is a random variable, say f_{11}, and hence (1) can be regarded as the probability distribution of f_{11}. It should be noted that (1) is the hypergeometric* probability distribution.

In order to test the null hypothesis, a one-tailed P-value can be evaluated as the probability of obtaining a result as extreme as the observed value of f_{11} in a particular direction; the probabilities evaluated from (1) in the appropriate tail ($f_{11}, f_{11} + 1, \ldots$, or $f_{11}, f_{11} - 1, \ldots, 0$) are summed. Alternatively, a critical value of f_{11} in the appropriate tail can be found from (1) such that α is the maximum probability of a type I error. In either case, tables of the hypergeometric distribution such of those of Lieberman and Owen [18] can be used to evaluate (1). Some papers dealing with computer programs for this test are Gregory [14], Robertson [21], and Tritchler and Pedrini [27]. Finney et al. [10] give critical values of f_{11} for $f_{1\cdot} \leqslant 40$.

To illustrate the procedure, consider the following fictitious data:

		Identification by Subject 1	2	Totals
Object Presented	1	4	1	5
	2	1	2	3
Totals		5	3	8

Since $f_{11} = 4$ here, the only possible table result more extreme than that observed in the same direction is $f_{11} = 5$. The one-tailed P-value is then found from (1) as

$$\Pr[f_{11} \geqslant 4] = \left[\binom{5}{4}\binom{3}{1} + \binom{5}{5}\binom{3}{0}\right] \Big/ \binom{8}{5}$$
$$= 16/56 = 0.2857.$$

Exact power calculations for this test appear in Mainland and Sutcliffe [19], Bennett and Hsu [3], Gail and Gart [13], and Haseman [15]. Approximate power based on the arc-sine transformation is reported in Sillitto [23] and Gail and Gart [13]. These tables can be used to determine the sample size required to obtain a specified power for a test of fixed nominal size. Thomas and Gart [25] give a table of confidence interval endpoints for various functions of the true probabilities $\Pr[A_1 \mid B_1]$ and $\Pr[A_1 \mid B_2]$; these confidence interval procedures correspond to Fisher's exact test procedure.

Fisher's exact test can also be used in the case where only one set of marginal frequencies is fixed. This situation would arise, for example, where $f_{1\cdot}$ and $f_{2\cdot}$ are the sizes of samples from two Bernoulli populations and are therefore fixed and A_1 and A_2 refer to the observed numbers of successes and failures, respectively; then $f_{\cdot 1}$ and $f_{\cdot 2}$ are not fixed. The appropriate null hypothesis here is that the probability of a success is the same in the two Bernoulli populations, usually called the null hypothesis of homogeneity*. The test procedure is exactly the same as that outlined above for the test of independence. This test is conditional on the observed values of the unfixed set of marginal frequencies.

This same test can also be used when neither set of marginal frequencies is fixed (called a double dichotomy); the null hypothesis here is that of bivariate independence. The test is again conditional on the observed values of the marginal frequencies.

The two-sample median test, sometimes called the Brown–Mood median test* [5], is a special case of Fisher's exact test where $f_{1.}$ and $f_{2.}$ are the sizes of two independent samples and A_1 refers to the number of observations less than the combined sample median and A_2 is the complement of A_1. The null hypothesis in this case is that the two samples are drawn from populations with the same distribution and therefore the same median.

The normal theory test of equality of proportions and chi-square tests* of independence and homogeneity of proportions are large-sample approximations to Fisher's exact test. The test statistic based on the normal approximation can be written as

$$z = \frac{\sqrt{f_{..}}\,(f_{11}f_{22} - f_{12}f_{21})}{\sqrt{f_{1.}f_{2.}f_{.1}f_{.2}}};$$

the value of z is compared with a value from the standard normal distribution to determine significance. Sweetland [24] compared this approximate test and the exact test and found good agreement when $f_{1.}$ and $f_{2.}$ are close to equal and the test is one-sided.

Fisher's exact test can also be generalized to data presented in an $r \times c$ contingency table. The null probability distribution is the multivariate extension of the hypergeometric distribution. Bedeian and Armenakis [2] deal with a computer program for calculating these probabilities and the coefficient of association λ. The chi-square tests of independence in $r \times c$ tables and homogeneity of k proportions are large-sample approximations to this generalized test. These tests are discussed in Bradley [4], Conover [8], Hays [16], Marascuilo and McSweeney [20], and Daniel [9].

A modification of Fisher's exact test was proposed by Tocher [26]; this is a randomization procedure that can be used irrespective of whether the marginal frequencies are fixed. Some modifications proposed for the case of unfixed marginal frequencies and two-tailed tests are the stepwise procedure of Barnard [1], which he called "progressive conservative," and the binomial test of Burnstein [6], which includes a computer program to find an estimate of the two-tailed P-value or exact significance level. Burnstein [6] compared his procedure with that of Barnard as well as with Fisher's exact test and the normal theory test (or chi-square test). Carr [7] extends Fisher's exact test to an exact test for the equality of k proportions in k samples of equal size.

References

[1] Barnard, G. A. (1947). *Biometrika*, **34**, 123–138.

[2] Bedeian, A. G. and Armenakis, A. A. (1977). *Educ. Psychol. Meas.*, **37**, 253–256.

[3] Bennett, B. M. and Hsu, P. (1960). *Biometrika*, **47**, 393–398; correction, *ibid.*, **48**, 475 (1961).

[4] Bradley, J. V. (1968). *Distribution-Free Statistical Tests*. Prentice-Hall, Englewood Cliffs, N.J.

[5] Brown, G. W. and Mood, A. M. (1951). *Proc. 2nd Berkeley Symp. Math. Statist. Prob.*, University of California Press, Berkeley, Calif., pp. 159–166.

[6] Burnstein, H. (1981). *Commun. Statist. A*, **10**, 11–29.

[7] Carr, W. E. (1980). *Technometrics*, **22**, 269–270.

[8] Conover, W. J. (1971). *Practical Nonparametric Statistics*. Wiley, New York.

[9] Daniel, W. W. (1978). *Applied Nonparametric Statistics*. Houghton Mifflin, Boston.

[10] Finney, D. J., Latscha, R., Bennett, B. M., and Hsu, P. (1963). *Tables for Testing Significance in a 2 × 2 Contingency Table*. Cambridge University Press, Cambridge.

[11] Fisher, R. A. (1935). *The Design of Experiments*. Oliver & Boyd, Edinburgh.

[12] Fisher, R. A. (1935). *J. R. Statist. Soc. A*, **98**, 39–54.

[13] Gail, M. and Gart, J. J. (1973). *Biometrics*, **29**, 441–448.

[14] Gregory, R. J. (1973). *Educ. Psychol. Meas.*, **33**, 697–700.

[15] Haseman, J. K. (1978). *Biometrics*, **34**, 106–109.

[16] Hays, W. L. (1973). *Statistics for the Social Sciences*, 2nd ed. Holt, Rinehart and Winston, New York.

[17] Irwin, J. O. (1935). *Metron*, **12**, 83–94.

[18] Lieberman, G. J. and Owen, D. B. (1961). *Tables*

of the Hypergeometric Probability Distribution. Stanford University Press, Stanford, Calif.

[19] Mainland, D. and Sutcliffe, M. I. (1953). *Canad. J. Med. Sci.*, **31**, 406–416.

[20] Marascuilo, L. A. and McSweeney, M. (1977). *Nonparametric and Distribution-Free Methods for the Social Sciences*. Brooks/Cole, Monterey, Calif.

[21] Robertson, W. H. (1960). *Technometrics*, **2**, 103–107.

[22] Siegel, S. (1956). *Nonparametric Statistics for the Behavioral Sciences*. McGraw-Hill, New York.

[23] Sillitto, S. P. (1949). *Biometrika*, **36**, 347–352.

[24] Sweetland, A. (1972). *A Comparison of the Chi-Square Test for 1 df and the Fisher Exact Test*. Rand Corporation, Santa Monica, Calif.

[25] Thomas, D. G. and Gart, J. J. (1977). *J. Amer. Statist. Ass.*, **72**, 73–76.

[26] Tocher, K. D. (1950). *Biometrika*, **22**, 130–144.

[27] Tritchler, D. L. and Pedrini, D. T. (1975). *Educ. Psychol. Meas.*, **35**, 717–719.

[28] Yates, F. (1934). *J. R. Statist. Soc. Supp.*, **1**, 217–235.

(BROWN–MOOD MEDIAN TEST
CONTINGENCY TABLES)

JEAN DICKINSON GIBBONS

FISHER'S IDEAL INDEX NUMBER

An index number* of prices attempts to measure the average percentage change in prices between two time periods for a collective of heterogeneous items that represent a complete set associated with a recognizable corporate entity or population. Thus a retail price index represents the average percentage price change for all consumer items that enter into the budgets of all households in a city, region, or state. Typically, the quantities to which the prices relate are measured in uncomparable units (cents per pint, cents per dozen, etc.) which must be allowed for in any averaging. A price index number will do this by introducing quantities in weights which represent the relative importance of each item in the collective. A quantity index will also use prices in the relative weights of a weighted average of quantity percentage changes. Thus, if a comparison is to be made between two time periods t_0 and t_1, and if $(\mathbf{p}_i, \mathbf{q}_i)$ $(i = 0, 1)$ represents the vectors of prices and quantities for all items in the index at time t_i, a typical price index will have the general form $100 \times P_{01}(\mathbf{p}_0, \mathbf{p}_1; \mathbf{q}_0, \mathbf{q}_1)$, and a typical quantity index will have the general form $100 \times Q_{01}(\mathbf{q}_0, \mathbf{q}_1; \mathbf{p}_0, \mathbf{p}_1)$. The suffix "01" indicates that the index measures the "level" in period t_1 as a percentage of that in period t_0.

Many different functional forms have been suggested for P_{01} and for Q_{01}. The first systematic study of the relative merits of alternative formulae was undertaken by Irving Fisher [3]. His examination was partly theoretical and partly empirical. The basis of his theoretical comparison was a set of criteria that were supposed appropriate for most index numbers. The four criteria regarded by Fisher as basic may be summarized as follows.

1. Commodity reversal. The value of the index number is unchanged by any permutation of the items in the index.
2. Factor reversal. $P_{01} Q_{01} = \mathbf{p}_1'\mathbf{q}_1/\mathbf{p}_0'\mathbf{q}_0$, where $\mathbf{p}_i'\mathbf{q}_i = \sum_j p_{ij}q_{ij}$ represents the total value of the collective at time t_i.
3. Time reversal. $P_{01}P_{10} = Q_{01}Q_{10} = 1$.

A further criterion, which Fisher argued at length was of no practical importance, is:

4. Circularity. For a comparison between distinct times t_0, t_1, and t_2, not necessarily in temporal order, $P_{01}P_{12} = P_{02}$, $Q_{01}Q_{12} = Q_{02}$.

This criterion implies that information pertaining to t_1 is relevant in any comparison between t_0 and t_2. Fisher rejected this.

Fisher's theoretical study indicated that no index number formula, out of the 134 considered, satisfied all the criteria proposed. The one formula that satisfied all his criteria, save that of circularity, had the form

$$P_{01} = \left[\mathbf{p}_1'\mathbf{q}_1\mathbf{p}_1'\mathbf{q}_0/(\mathbf{p}_0'\mathbf{q}_1\mathbf{p}_0'\mathbf{q}_0)\right]^{1/2}, \quad (1)$$

$$Q_{01} = \left[\mathbf{p}_1'\mathbf{q}_1\mathbf{p}_0'\mathbf{q}_1/(\mathbf{p}_1'\mathbf{q}_0\mathbf{p}_0'\mathbf{q}_0)\right]^{1/2}. \quad (2)$$

This was dubbed by others Fisher's ideal index number, although his empirical study showed that many other formulae gave values which were close to that of the "ideal." Fisher recommended for ordinary practical use Edgeworth's formula:

$$P_{01} = \mathbf{p}_1'(\mathbf{q}_0 + \mathbf{q}_1)/\{\mathbf{p}_0'(\mathbf{q}_0 + \mathbf{q}_1)\}$$

$$Q_{01} = (\mathbf{p}_0 + \mathbf{p}_1)'\mathbf{q}_1/\{(\mathbf{p}_0 + \mathbf{p}_1)'\mathbf{q}_0\}.$$

The circularity criterion has attractive features despite Fisher's strong objections. This criterion can be satisfied by any chain index of the form $I_{0(1)r} = \prod_{j=0}^{r} I_{j,j+1}$, with $t_{j+1} - t_j = h > 0$ for all integer values of j, and $I_{j,j+1}$ an index number defined by a common formula. In particular, when $I_{j,j+1}$ represents Fisher's ideal index number, criteria 1 to 4 plus some of his other criteria will be satisfied. The attractiveness of the chain index has one drawback which Fisher recognized. A comparison of prices between two periods requires that the two corresponding collectives be alike. If the two time periods are sufficiently far apart, the two collectives are likely to be so different that no meaningful comparison can be made. This is evident if a direct comparison is attempted, but will be concealed if the comparison is made via a chain index. This explains Fisher's strong objection to the circularity criterion.

All the index numbers considered by Fisher were explicit functions of prices and quantities. Attention is still confined to explicit functions in more recent publications. An alternative approach to an "ideal" index, adopted by Theil [9], is to seek an implicit function identified by a mapping, which is defined by the criteria to be satisfied, from the space of prices and quantities for all time periods of interest to a lower-dimensional space representing the distinct values of the index numbers at those times. Suppose that the values of a price and a quantity index number are to be calculated for each of n successive points of time t_i, $i = 1(1)n$, with t_1 taken as the comparison base. Write $\Pi_i = P_{1i}$, $\phi_i = Q_{1i}$ and $v_{ij} = \mathbf{p}_i'\mathbf{q}_j/(\mathbf{p}_1'\mathbf{q}_1)$ with $\mathbf{p}_i$ and $\mathbf{q}_j$ defined as before. If we could suppose that $\mathbf{p}_i = \Pi_i\mathbf{p}_1$ and $\mathbf{q}_j = \phi_j\mathbf{q}_1$, then $v_{ij} = \Pi_i\phi_j$. In practice this will not be so. It may be plausible instead to represent $\mathbf{p}_i$ and $\mathbf{q}_j$ by $\mathbf{p}_i = \Pi_i\mathbf{p}_1 + \boldsymbol{\epsilon}_i$ and $\mathbf{q}_j = \phi_j\mathbf{q}_1 + \boldsymbol{\delta}_j$, to give $v_{ij} = \Pi_i\phi_j + e_{ij}$ with $e_{ij} = (\phi_j\boldsymbol{\epsilon}_i'\mathbf{q}_1 + \Pi_i\mathbf{p}_1'\boldsymbol{\delta}_j + \boldsymbol{\epsilon}_i'\boldsymbol{\delta}_j)/(\mathbf{p}_1'\mathbf{q}_1)$. The terms e_{ij} are departures from the "ideal" form of v_{ij} described by $\Pi_i\phi_j$. Theil suggested that the values of the index numbers be calculated as those values of Π_i and ϕ_j that minimize $\sum_{i=1}^{n}\sum_{j=1}^{n}(\Pi_i\phi_j - v_{ij})^2$. This leads to a computing algorithm which defines the implicit function for Π_i and ϕ_j as the elements of eigenvectors* of the matrices $\|\sum_k v_{ik}v_{jk}\|$ and $\|\sum_k v_{ki}v_{kj}\|$. Theil's approach does not lead to values that satisfy the factor reversal criterion. Kloek and de Wit [6] suggested the addition of the constraint $\sum_{r=1}^{n}(\Pi_r\phi_r - v_{rr}) = 0$, which would ensure that this criterion was satisfied on average. Theil [10] reports that Kloek also recommended the use of the n separate constraints $\Pi_r\phi_r - v_{rr} = 0$, $r = 1(1)n$, but no description of the corresponding algorithm was published. This was subsequently described by Fisk [4]. The introduction of these n constraints ensures that criteria 1 to 4, plus most of the others suggested by Fisher, are satisfied for all $2n$ index numbers represented by Π_r and ϕ_r. An advantage of Theil's approach is that, because the values of v_{ij} have to be calculated for all time periods t_1 to t_n, the collectives for all those periods must be alike. Part of Fisher's objection to the circularity condition is avoided.

An alternative and not implausible assumption to that of Theil's is that departures of $\mathbf{p}_i$ and $\mathbf{q}_j$ from $\Pi_i\mathbf{p}_1$ and $\phi_j\mathbf{q}_1$, respectively, are multiplicative; i.e., $\mathbf{p}_i = \Pi_i(\mathbf{p}_1 + \boldsymbol{\epsilon}_i)$ and $\mathbf{q}_j = \phi_j(\mathbf{q}_1 + \boldsymbol{\delta}_j)$. This leads to the relation $v_{ij} = \Pi_i\phi_j(1 + \eta_{ij})$ with

$$\eta_{ij} = (\mathbf{p}_1'\boldsymbol{\delta}_j + \boldsymbol{\epsilon}_i'\mathbf{q}_1 + \boldsymbol{\epsilon}_i'\boldsymbol{\delta}_j)/\mathbf{p}_1'\mathbf{q}_1.$$

Taking logarithms, we may rewrite this as a factorial-type model:

$$y_{ij} = A_i + B_j + \zeta_{ij}$$

with $y_{ij} = \log v_{ij}$, $A_i = \log \Pi_i$, $B_j = \log\phi_j$, and $\zeta_{ij} = \log(1 + \eta_{ij})$. The values of A_i and B_j that minimize $\sum_{i=1}^{n}\sum_{j=1}^{n}\zeta_{ij}^2$, subject to the constraints $A_1 = B_1 = 0$, and $\zeta_{ii} = 0$, $i = 1(1)n$, are

$$\hat{A}_i = \tfrac{1}{2}[(y_{ii} + \bar{y}_{\cdot i} - \bar{y}_{i\cdot}) - (y_{11} + \bar{y}_{\cdot 1} - \bar{y}_{1\cdot})]$$

and

$$\hat{B}_j = \tfrac{1}{2}\left[(y_{jj} + \bar{y}_{j\cdot} - \bar{y}_{\cdot j}) - (y_{11} + \bar{y}_{1\cdot} - \bar{y}_{\cdot 1})\right]$$

with $\bar{y}_{i\cdot} = \sum_{j=1}^{n} y_{ij}/n$ and $\bar{y}_{\cdot j} = \sum_{i=1}^{n} y_{ij}/n$. The required index numbers, which satisfy criteria 1 to 4, are obtained as the antilogarithms of these values. In addition to the ease of calculation, this approach has the interesting property that the expressions for Π_2 and ϕ_2 when $n = 2$ are those of Fisher's ideal index number given by (1) and (2).

Various theoretical studies, based on Fisher's and other criteria have been made (see, e.g., Eichhorn [1, Chap. 11] and Eichhorn et al. [2]). Of more interest is the utility theory* approach to the price index, which seeks an appropriate formula from suitable economic theoretical reasoning (see, e.g., Konüs [7], Wald [12], and Theil [11]). Here the index number is determined by the use to which it is to be put, and is derived from a suitable model of the economic phenomena studied. The connection between method and use was stressed by Mitchell [8, Sec. IV.1] and calls into question the generality of any criteria that may be proposed for an index number. The derivation of index numbers from theoretical models related to the purpose that the index is meant to serve has much to commend it, although it is rarely attempted in practice.

To illustrate the effect of using these different formulae, the U.K. Department of Employment kindly supplied detailed information on expenditure weights and year-on-year indexes of price changes at January in each year for the 96 sections that comprise the U.K. Retail Price Index for the period July 1967–June 1977. These data were used to calculate the matrix of revaluations given in Table 1. The (i, j)th entry in this table corresponds to the revaluation $\mathbf{p}_j'\mathbf{q}_i$ in the notation given above. Similar tables were given by Fowler [5], although the timings for the entries differ. Table 2 gives the values of the index (minus 100) calculated by the different approaches, together with the fixed-base Laspeyres's index (*see* INDEX NUMBERS) and the chained Laspeyres's index. The latter is the index formula used for the calculation of the U.K. Retail Price Index, and the former is a commonly employed formula for price indexes. The Edgeworth index formula gave values that were almost identical with those for the chained Fisher's ideal index number. The largest discrepancy was for January 1977, for which the Edgeworth formula gave the value 168.6.

It is evident from Table 2 that the different formulae give roughly comparable measures of average price changes. The period covered was one of fairly rapid price increases so that the discrepancies are relatively small. The base-weighted Laspeyres's index is the most discrepant, and probably overestimates the rate of average price changes over this period. This comparability of index number values calculated by these different formulas echoes the findings of Fisher's more extended empirical study, and

Table 1 Aggregate of average weekly expenditure, revalued, by "Index" households in the United Kingdom

	Expenditure at Prices Obtaining in January									
Time Period	1968 £	1969 £	1970 £	1971 £	1972 £	1973 £	1974 £	1975 £	1976 £	1977 £
1967–1968	24.222	24.739	24.903	25.210	26.048	27.475	27.377	27.950	26.393	25.850
1968–1969	25.712	26.241	26.411	26.739	27.638	29.164	29.092	29.702	28.038	27.473
1969–1970	26.985	27.526	27.681	28.016	28.935	30.494	30.402	31.026	29.290	28.699
1970–1971	29.279	29.862	30.028	30.362	31.356	33.033	32.893	33.552	31.689	31.039
1971–1972	31.695	32.313	32.467	32.794	33.797	35.531	35.368	36.068	34.067	33.350
1972–1973	34.188	34.845	34.980	35.334	36.382	38.152	37.928	38.671	36.552	35.799
1973–1974	38.376	39.088	39.160	39.563	40.707	42.579	42.115	42.849	40.540	39.702
1974–1975	45.967	46.868	46.956	47.429	48.814	51.081	50.518	51.169	48.387	47.394
1975–1976	56.685	57.743	57.856	58.401	60.056	62.772	62.276	63.024	59.469	58.285
1976–1977	66.635	67.830	67.932	68.512	70.306	73.319	72.692	73.466	69.374	67.844

Table 2 U.K. Retail Price Index: All Items (Percentage Changes from January 1968)

	Unconstrained Indexes		Constrained Indexes		Chained Indexes	
January of:	Laspeyres[a] ($\mathbf{p}_i'\mathbf{q}_1/\mathbf{p}_1'\mathbf{q}_1$)	Theil (Π_i)	Theil (Π_i)	Factorial ($e^{\hat{A}_i}$)	Laspeyres ($I_{1(1)i}$)	Fisher's Ideal ($I_{1(1)i}$)
1969	6.2	6.2	6.3	6.2	6.2	6.1
1970	11.4	11.1	11.7	11.4	11.4	11.3
1971	20.9	20.3	21.2	20.8	20.8	20.6
1972	30.8	29.7	31.0	30.3	30.5	30.2
1973	41.1	39.4	41.1	40.1	40.4	39.9
1974	58.4	55.5	57.2	56.0	56.7	55.8
1975	89.8	86.2	88.4	86.5	88.0	86.4
1976	134.0	129.2	132.1	129.5	131.6	129.4
1977	175.1	168.1	171.4	168.0	170.2	167.3

[a] Weights base 1967–1968.

indicates (as Fisher suggested) that for most practical purposes a simple formula from among this set often suffices.

References

[1] Eichhorn, W. (1978). *Functional Equations in Economics*. Addison-Wesley, Reading, Mass. (Mathematical economics text at postgraduate and research level.)

[2] Eichhorn, W. et al. (1978). *Theory and Applications of Economic Indices*. Physica Verlag, Würzburg. (Proceedings of a conference on economic indices. Papers of varying degrees of mathematical complexity. Mostly postgraduate and research level.)

[3] Fisher, I. (1922). *The Making of Index Numbers: A Study of Their Varieties, Tests, and Reliability*. Houghton Mifflin, Boston. (Pioneering text, the only extensive comparison of alternative formulae ever published. Superseded by later work; mainly of historical interest.)

[4] Fisk, P. R. (1977). *J. R. Statist. Soc. A*, **140**, 217–231. (Technical paper, possibly of some interest to practicing statisticians.)

[5] Fowler, R. F. (1973). *Further Problems of Index Number Construction*. Stud. Off. Statist., Res. Ser. No. 5, H.M. Stationery Office, London. (An empirical examination of the effect of sampling variability in the weights of a Laspeyres index number. Of interest to practicing statisticians.)

[6] Kloek, T. and de Wit, G. M. (1961). *Econometrica*, **29**, 602–616. (Technical paper, possibly of some interest to practicing statisticians.)

[7] Konüs, A. A. (1939). *Econometrica*, **7**, 10–29. (Economic paper, mainly of historical interest.)

[8] Mitchell, W. C. (1921). *The Making and Using of Index Numbers*. Reprinted by Augustus M. Kelley, New York, 1965. (Still worth reading by practicing statisticians, but written well before the computer revolution.)

[9] Theil, H. (1960). *Econometrica*, **28**, 464–480. (Technical paper, possibly of some interest to practicing statisticians.)

[10] Theil, H. (1962). In *Logic, Methodology and Philosophy of Science*, E. Nagel, P. Suppes, and A. Tversky, eds. Stanford University Press, Stanford, Calif.

[11] Theil, H. (1965). *Econometrica*, **33**, 67–87. (Economic paper related to index numbers.)

[12] Wald, A. (1939). *Econometrica*, **7**, 319–331. (Economic paper related to index numbers.)

(INDEX NUMBERS)

P. R. FISK

FISHER'S k-STATISTICS

The problem of "moments of moments"—expressing the sampling moments of sample moments* in terms of population moments—is historically an old one. It is the only way known to attack general sampling distributions* when the parent population is not completely known. Dwyer [9] gives a good account of it.

The straightforward methods soon led to algebraically complex and tedious results (see, e.g., Chuprov [1], Soper [19], and

Church [2]). Craig [5] drew attention to the need for the use of functions other than crude moments if the algebraic formulation was to be made manageable. Fisher [10] provided a simplification by introducing sample symmetric functions whose expected values are the population cumulants*, and called them k-statistics. He also proposed a combinatorial method to obtain their sampling cumulants and product cumulants. Kendall [17] calls this "the most remarkable paper he ever wrote. It forms the basis of most subsequent work on the subject." David [6] calls it "remarkable not only for the brilliance of the statistical technique but also for the condensation of the mathematical argument." According to Tukey [20], the k-statistics are "the most important step so far taken in connection with sampling moments. . . . The sampling behavior of the k's is much simpler than that of the sample moments." Kendall [14] says: "The value of the k-statistics rests chiefly on the relative simplicity which their use imports into certain branches of the theory of sampling."

The *k-statistic* k_p is defined as the symmetric function of the sample observations $x_1, x_2, \dots, x_n$ such that $E(k_p) = \kappa_p$, the population pth cumulant. This definition determines k_p uniquely, for if k_p and k_p' are two sample symmetric functions with expected value κ_p, $k_p - k_p'$ would have expected value zero, which would imply a relationship among the moments. Like the cumulants, the k-statistics enjoy the *semi-invariance property*. They are invariant under a change of origin, except for k_1, which is the sample mean, i.e.,

$$k_p(x_1 + h, x_2 + h, \dots, x_n + h) = k_p, p > 1.$$

Also [14],

$$E[k_p(x_1 + h, x_2, \dots, x_n)] = \kappa_p + h^p/n.$$

The k-statistic k_p is best defined in terms of *partitions* of p. Let a typical partition of p be $P = p_1^{\pi_1} \cdots p_s^{\pi_s}$ with $p_1 > \cdots > p_s$. Here $p = \sum_1^s p_i \pi_i$ and $\pi = \sum_1^s \pi_i$ is the number of parts in P. Also let

$$C(P) = \frac{p!}{(p_1!)^{\pi_1} \cdots (p_s!)^{\pi_s} \pi_1! \cdots \pi_s!}$$

and $[P]$ be the power product sum* (or augmented monomial symmetric function)

$$\sum_{\neq} x_1^{p_1} x_2^{p_1} \cdots x_{\pi_1}^{p_1} x_{\pi_1+1}^{p_2} \cdots x_{\pi_1+\pi_2}^{p_2} \cdots x_{\pi}^{p_s},$$

where the sum is taken over all different values of the subscripts. The symmetric mean or angle bracket* [20] is then

$$\langle P \rangle = [P]/n^{(\pi)}.$$

In terms of these,

$$k_p = \sum_P (-1)^{\pi-1}(\pi - 1)!\, C(P) \langle P \rangle,$$

where the summation extends over all partitions P of p.

Thus when $p = 1$, there is only one partition $P = 1$, and

$$k_1 = \langle 1 \rangle = \frac{[1]}{n} = \frac{1}{n} \sum_1^n x_i = m_1',$$

the sample mean.

When $p = 2$, there are two partitions, 2 and $11 = 1^2$. For $P = 2$, $\pi = 1$, $C(2) = 2!/2! = 1$, and for $P = 11$, $\pi = 2$, $C(11) = 2!/(1!)^2 2! = 1$. Hence $k_2 = \langle 2 \rangle - \langle 11 \rangle$, which can be written as

$$k_2 = \frac{1}{n} \sum_1^n x_i^2 - \frac{1}{n(n-1)} \sum_{i \neq j}^n x_i x_j.$$

In terms of power sums $s_r = \sum_1^n x_i^r$, this is

$$\frac{s_2}{n} - \frac{s_1^2 - s_2}{n(n-1)} = \frac{1}{n-1}\left(s_2 - \frac{s_1^2}{n}\right).$$

Thus $k_2 = [n/(n-1)]m_2$, where m_2 is the sample variance. Since $E(k_2) = \kappa_2 = \mu_2$, we have $E(m_2) = [(n-1)/n]\mu_2$. If sampling is done from a finite population with variance M_2,

$$E_N(k_2) = K_2 = \frac{N}{N-1} M_2,$$

$$E_N(m_2) = \frac{n-1}{n} \frac{N}{N-1} M_2.$$

For N large, this tends to $[(n-1)/n]\mu_2$.

For $P = p_1^{\pi_1} \cdots p_s^{\pi_s}$, $E\langle P \rangle = \mu_{p_1}'^{\pi_1} \cdots \mu_{ps}'^{\pi_s}$, where the μ' are the population raw moments. Thus

$$E(k_p) = \kappa_p$$

$$= \sum_P (-1)^{\pi-1}(\pi - 1)!\, C(P) \mu_{p_1}'^{\pi_1} \cdots \mu_{p_s}'^{\pi_s}.$$

This explains the correspondence between $\kappa_2 = \mu_2' - \mu_1'^2$ and $k_2 = \langle 2\rangle - \langle 11\rangle$.

When $p = 3$, there are three partitions, 3, 21, and $111 = 1^3$.

For $P = 3$, $\pi = 1$, $C(3) = 3!/3! = 1$.

For $P = 21$, $\pi = 2$, $C(21) = 3!/(2!1!) = 3$.

For $P = 1^3$, $\pi = 3$, $C(1^3) = 3!/((1!)^3 3!) = 1$.

Thus $k_3 = \langle 3\rangle - 3\langle 21\rangle + 2\langle 111\rangle$ (cf. $\kappa_3 = \mu_3' - 3\mu_2'\mu_1' + 2\mu_1'^3$). Now $\langle 3\rangle = [3]/n$ and $[3] = s_3$. Also, $\langle 21\rangle = [21]/(n(n-1))$ and $[21] = s_2 s_1 - s_3$; finally, $\langle 111\rangle = [111] \div (n(n-1)(n-2))$ and $[111] = s_1^3 - 3s_2 s_1 + 2s_3$ (see David et al. [8]). Thus

$$k_3 = \frac{1}{n^{(3)}}\left(n^2 s_3 - 3ns_2 s_1 + 2s_1\right).$$

Letting the origin be zero ($s_1 = 0$),

$$k_3 = \frac{n^2}{(n-1)(n-2)} m_3,$$

where m_3 is the sample third moment. Again,

$$E(k_3) = \kappa_3 = \mu_3 \quad \text{and}$$

$$E_N(k_3) = K_3 = \frac{N^2}{(N-1)(N-2)} M_3,$$

so

$$E(m_3) = \frac{(n-1)(n-2)}{n^2}\mu_3 \quad \text{and}$$

$$E_N(m_3) = \frac{(n-1)(n-2)}{n^2}\,\frac{N^2}{(N-1)(N-2)} M_3,$$

which tends to the former for N large.

With $p = 4$, the partitions $P = 4$, 31, 2^2, 21^2, 1^4 have $C(P) = 4!/4!$, $4!/(3!1!)$, $4!/((2!)^2 2!)$, $4!/(2!(1!)^2 2!)$, $4!/((1!)^4 4!)$, i.e., 1, 4, 3, 6, 1, respectively, and $k_4 = \langle 4\rangle - 4\langle 31\rangle - 3\langle 22\rangle + 12\langle 211\rangle - 6\langle 1111\rangle$ (cf. $\kappa_4 = \mu_4' - 4\mu_3'\mu_1' - 3\mu_2'^2 + 12\mu_2'\mu_1'^2 - 6\mu_1'^4$). In terms of power sums, this becomes

$$k_4 = \frac{1}{n^{(4)}}\left\{(n^3 + n^2)s_4 - 4(n^2 + n)s_3 s_1 - 3(n^2 - n)s_2^2 + 12ns_2 s_1^2 - 6s_1^4\right\}.$$

With the origin taken as zero ($s_1 = 0$),

$$k_4 = \frac{n^2}{(n-1)(n-2)(n-3)} \times \left\{(n+1)m_4 - 3(n-1)m_2^2\right\}.$$

Expressions up to k_8 in terms of power sums are given by Kendall and Stuart [18, pp. 280–281]. Expressions for k_9, k_{10}, and k_{11} are given by Ziaud-Din [23, 24] and for k_{12} by Ziaud-Din and Ahmad [25]. Since any rational integral algebraic symmetric function of $x_1, \ldots, x_n$ can be expressed uniquely in terms of the power sums, it can also be expressed uniquely in terms of the k-statistics [14].

A combinatorial method for obtaining the sampling cumulants of k-statistics when sampling from an infinite population is given by Fisher [10]. It involves finding expected values of powers and products of k-statistics. Proofs for the rules employed involve partitions extensively, and are provided by Kendall [15]. A list of formulae for the sampling cumulants and product cumulants of k-statistics is provided by Fisher [10] and by Kendall and Stuart [18, pp. 290–295]. Thus the second cumulant, or variance, of k_2 is

$$\kappa(2^2) = \frac{\kappa_4}{n} + \frac{2\kappa_2^2}{n-1}.$$

When sampling from finite populations*, the formulae become complex. Irwin and Kendall [12] introduce the principle that, if for a function f of the sample values, $E(f) = \sum a_r \kappa_r$, then $E_N(f) = \sum a_r K_r$. They use this principle to derive expressions for the finite population from those for the infinite population and give the first four moments of the mean, the first two moments of the variance, and some product moments. Thus

$$\operatorname{var}(k_1) = \left(\frac{1}{n} - \frac{1}{N}\right)K_2,$$

$$\operatorname{cov}(k_1, k_r) = \left(\frac{1}{n} - \frac{1}{N}\right)K_{r+1},$$

$$\operatorname{var}(k_2) = \frac{(N-n)(Nn - n - N - 1)}{n(n-1)N(N+1)}K_4 + \frac{2(N-n)}{(n-1)(N+1)}K_2^2.$$

If the finite population is symmetrical, k_1 is uncorrelated with any k-statistic of even order and k_2 is uncorrelated with any k-statistic of odd order. If k_1 and k_r are uncorrelated for all $r > 1$, the parent distribution is normal, and if k_1 is independent of any k_r, $r > 1$, the parent is likewise normal. A simpler form for var(k_2) is

$$\left(\frac{1}{n} - \frac{1}{N}\right)K_4 + 2\left(\frac{1}{n-1} - \frac{1}{N-1}\right)K_{22}$$

[20]. This involves K_{22}, a polykay*. Products of k-statistics can be expressed as linear combinations of polykays and their higher moments obtained as in Wishart [22].

The formulae can be used to approximate to a sampling distribution. For example, consider the distribution of $\sqrt{b_1}$ in sampling from a normal population, noting that

$$\sqrt{b_1} = \frac{m_3}{m_2^{3/2}} = \frac{n-2}{\sqrt{n(n-1)}} \frac{k_3}{k_2^{3/2}}.$$

For the normal distribution, the variance of k_3 is

$$\kappa(3^2) = \frac{6n}{(n-1)(n-2)} \kappa_2^3,$$

hence Fisher [10] considers the statistic

$$x = \sqrt{\frac{(n-1)(n-2)}{6n}} k_3 k_2^{-3/2}.$$

By expanding $k_2^{-3/2}$ as

$$\kappa_2^{-3/2}\left(1 + \frac{k_2 - \kappa_2}{\kappa_2}\right)^{-3/2},$$

he obtains the approximate variance of x. Later, Fisher [11] obtains the exact values of the first three even moments of $k_3 k_2^{-3/2}$ by using some recurrence relations; thus

$$\operatorname{var}\left(k_3 k_2^{-3/2}\right) = \frac{6n(n-1)}{(n-2)(n+1)(n+3)}.$$

He also considers $k_4 k_2^{-2}$ similarly. In sampling from normal populations, k_2 is independent of $k_r k_2^{-r/2}$. Such methods allow approximations to the sampling distributions of statistics expressible as symmetric functions, e.g., coefficient of variation* [6], variance-ratio* [7].

The bivariate k-statistic $k_{pp'}$ is defined as the sample symmetric function whose expected value is the bivariate cumulant $\kappa_{pp'}$. That is,

$$k_{pp'} = \sum_P (-1)^{\pi-1}(\pi-1)!\, C(P)\langle P\rangle,$$

where the sum is over all partitions P of the bipartite number pp'. The first few bivariate k-statistics are

$$k_{11} = \frac{1}{n^{(2)}}(ns_{11} - s_{10}s_{01}) = \frac{n}{n-1} m_{11}$$

$$k_{21} = \frac{1}{n^{(3)}}(n^2 s_{21} - 2ns_{10}s_{11} - ns_{20}s_{01} + 2s_{10}^2 s_{01})$$

$$= \frac{n^2}{(n-1)(n-2)} m_{21}.$$

Expressions for k_{31} and k_{22} are given by Fisher [10] and Kendall and Stuart [18]. Their sampling cumulants and product cumulants may be obtained directly by a combinatorial method [10], or by a symbolic operation [16] on the univariate formulae. Cook [3] obtains these, and shows [4] their application in studying the sampling distributions of the correlation and regression coefficients. The covariance of the estimates of variance of two correlated variables is thus

$$\operatorname{cov}(k_{20}, k_{02}) = \kappa\begin{pmatrix}2 & 0\\ 0 & 2\end{pmatrix} = \frac{1}{n}\kappa_{22} + \frac{2}{n-1}\kappa_{11}^2.$$

In samples from a bivariate normal population, $\operatorname{corr}(k_{r0}, k_{0r}) = \rho^r$ and $\operatorname{cov}(k_{tu}, k_{vw}) = 0$ unless $t + u = r + w$ [21].

Multivariate k-statistics may be defined in terms of multivariate power sums $s_{ij\cdots u}$ referring to the products $x_i x_j \cdots x_u$ summed over the sample. Thus

$$k_i = \frac{1}{n} s_i$$

$$k_{ij} = \frac{1}{n^{(2)}}(ns_{ij} - s_i s_j)$$

$$k_{ijl} = \frac{1}{n^{(3)}}\{n^2 s_{ijl} - n(s_i s_{jl} + s_j s_{il} + s_l s_{ij}) + 2s_i s_j s_l\}.$$

One has the four-variate formula

$$\operatorname{cov}(k_{1100}, k_{0011}) = \kappa\begin{pmatrix}10\\10\\01\\01\end{pmatrix} = \frac{1}{n}\kappa_{1111} + \frac{1}{n-1}(\kappa_{1010}\kappa_{0101} + \kappa_{1001}\kappa_{0110})$$

from which trivariate, bivariate, and univariate formulae can be easily generated. Kaplan [13] suggests a compact tensor notation to summarize the multivariate formulae, e.g.,

$$\operatorname{cov}(k_{ab}, k_{ij}) = \frac{1}{n}\kappa_{abij} + \frac{1}{n-1}(\kappa_{ai}\kappa_{bj} + \kappa_{aj}\kappa_{bi}).$$

He gives many such formulae, summarizing those of Cook [3].

References

[1] Chuprov, A. A. (1918). *Biometrika*, **12**, 140–169, 185–210.

[2] Church, A. E. R. (1926). *Biometrika*, **18**, 321–394.

[3] Cook, M. B. (1951). *Biometrika*, **38**, 179–195.

[4] Cook, M. B. (1951). *Biometrika*, **38**, 368–376.

[5] Craig, C. C. (1928). *Metron*, **7**, 3–74.

[6] David, F. N. (1949). *Biometrika*, **36**, 383–393.

[7] David, F. N. (1949). *Biometrika*, **36**, 394–403.

[8] David, F. N., Kendall, M. G., and Barton, D. E. (1966). *Symmetric Function and Allied Tables*. Cambridge University Press, Cambridge. (Contains complete versions of most tables.)

[9] Dwyer, P. S. (1972). In *Symmetric Functions in Statistics*, D. S. Tracy, ed. University of Windsor, Windsor, Ontario, pp. 11–51. (Good historical account with extensive bibliography.)

[10] Fisher, R. A. (1929). *Proc. Lond. Math. Soc.* (2), **30**, 199–238. (Fisher's original paper, introducing k-statistics.)

[11] Fisher, R. A. (1930). *Proc. R. Soc. A*, **130**, 16–28.

[12] Irwin, J. O. and Kendall, M. G. (1944). *Ann. Eugen. (Lond.)*, **12**, 138–142.

[13] Kaplan, E. L. (1952). *Biometrika*, **31**, 319–323.

[14] Kendall, M. G. (1940). *Ann. Eugen.*, **10**, 106–111.

[15] Kendall, M. G. (1940). *Ann. Eugen.*, **10**, 215–222.

[16] Kendall, M. G. (1940). *Ann. Eugen.*, **10**, 392–402.

[17] Kendall, M. G. (1963). *Biometrika*, **50**, 1–15.

[18] Kendall, M. G. and Stuart, A. (1969). *The Advanced Theory of Statistics*, Vol. 1. Charles Griffin, London. (Chapters 12 and 13 give a detailed account of k-statistics and related material.)

[19] Soper, H. E. (1922). *Frequency Arrays*. Cambridge University Press, *Cambridge*.

[20] Tukey, J. W. (1950). *J. Amer. Statist. Ass.*, **45**, 501–519.

[21] Wishart, J. (1929). *Proc. R. Soc. Edin.*, **49**, 78–90.

[22] Wishart, J. (1952). *Biometrika*, **39**, 1–13.

[23] Ziaud-Din, M. (1954). *Ann. Math. Statist.*, **25**, 800–803.

[24] Ziaud-Din, M. (1959). *Ann. Math. Statist.*, **30**, 825–828.

[25] Ziaud-Din, M. and Ahmad, M. (1960). *Bull. Int. Statist. Inst.*, **38**, 635–640.

(ANGLE BRACKETS
CUMULANTS
POLYKAYS
POWER PRODUCT SUM
SAMPLING DISTRIBUTIONS)

D. S. TRACY

FISHER'S PROBLEM OF THE NILE

The *problem of the Nile*, first put forward by R. A. Fisher in 1936 as a challenging, unsolved mathematical problem, reads as follows:

> The agricultural land of a pre-dynastic Egyptian village is of unequal fertility. Given the height to which the Nile will rise, the fertility of every portion of it is known with exactitude, but the height of the flood affects different parts of the territory unequally. It is required to divide the area between the several households of the village, so that the yield of the lots assigned to each shall be in pre-determined proportion, whatever may be the height to which the river rises. [1, p. 257]

This problem arose in his continuing attempts to find, by inductive reasoning, a method for making uncertain inference expressible in mathematical probability, from the observation of a sample about the population from which the sample has been drawn. He felt that the conditions of solvability of this problem would supply the key to the nature of the inductive inferences possible (*see* INFERENCE, STATISTICAL).

Consider the following inference* problem. Let θ be the only unknown parameter in a population with the probability element

$$f(x;\theta)\,dx = \phi(x-\theta)\,dx,$$
$$-\infty < x < \infty, \quad -\infty < \theta < \infty, \qquad (1)$$

where ϕ is a continuous function. It is desired to obtain exact probability statements about θ, an unknown real number.

Fisher would not agree with the frequentist's viewpoint that the probability that θ, an unknown constant, lies within any finite interval must be either 0 or 1. He also rejected the Bayesian approach to obtaining a posterior probability* of θ by using a prior distribution* without objective evidence. Further, he considered the limitations of using purely deductive logic in drawing mathematical conclusions intolerable in statistical inference.

In 1930 he had proposed the fiducial* argument by which the status of θ could be changed from an unknown constant to that of a random variable with a rigorous mathematical probability distribution dependent on the observed data. This argument required, among other conditions, the use of a sufficient estimate of θ which contains all the information (*see* FISHER INFORMATION) supplied by the data. However, he soon realized that for most parametric statistical models sufficient estimates of the parameters do not exist. For example, no single sufficient estimator of θ exists for the model (1) unless the random variable X has a normal distribution or $Y = \exp(X)$ has a gamma distribution*. For the cases when no sufficient estimator of θ exists, the inference problem seemed to be without a solution.

By 1935 he had discovered that the fiducial argument could be extended to the general model (1), using the conditional sampling distribution of the maximum likelihood estimate* of θ, given the values of the set of $n - 1$ differences between successive observations in a sample of size n. These differences form a set of ancillary statistics* whose probability distributions are independent of the unknown parameter θ. This led him to conclude that the condition for further development of the use of fiducial inference* would depend on the solubility of the problem of the Nile.

The following example of the Nile problem was analyzed in detail by Fisher [2] in a paper published in French. (It was later included in his book *Statistical Methods and Scientific Inference* [3], first published in 1956.)

Consider N pairs of observations (x_i, y_i) from a population whose probability element is of the form

$$df = \exp(-\theta x - y/\theta)\, dx\, dy, \qquad x > 0, \quad y > 0, \quad (2)$$

where θ is an unknown positive constant. No single sufficient estimate of θ exists. However, it can be recognized that the statistic U defined by $U^2 = (\sum x_i)(\sum y_i)$ is an ancillary statistic. If the probability element df in (2) represents the fertility of the village at (x, y) as a function of the unknown parameter θ, then the problem of the Nile is solved in the sense that the family of hyperbolas, $U^2 = XY =$ constant, divide the total frequency in fixed proportions independently of the value of θ representing the unknown height to which the Nile will rise. Inference on θ expressible in exact (fiducial) probability statements about θ a posteriori can be made from the conditional distribution of the maximum likelihood estimator of θ, given the observed value of the ancillary statistic U. (See Fisher [3, p. 170].)

The problem of the Nile is then the problem of finding an ancillary statistic $U(X)$ from the original statistical model, $\{f(x;\theta): \theta \in \Omega\}$, such that inference on θ, expressible in mathematical probability, can be made from the reduced model $\{f(x;\theta \mid u): \theta \in \Omega\}$ conditional on the observed value of $U(X) = u$, without loss of information supplied by the data X. The problem is twofold:

1. Given a model $f(x;\theta)$, how can the ancillary statistics be recognized?
2. What types of models admit solutions of the problem of the Nile?

In Fisher's own words: "The problem has not, I believe, in the meanwhile yielded up the conditions of its solubility, upon which, it would appear, the possibility of fiducial inference with two or more parameters, must depend" [3, p. 119].

This problem appears to remain unsolved today.

References

[1] Fisher, R. A. (1936). *Proc. Amer. Acad. Arts Sci.*, **71**, 245–258.

[2] Fisher, R. A. (1948). *Ann. Inst. Henri Poincaré*, **10**, 191–213 (in French).

[3] Fisher R. A. (1959). *Statistical Methods and Scientific Inference*, 2nd ed. Oliver & Boyd, Edinburgh.

Bibliography

See the following works as well as the references just cited, for more information on the topic of Fisher's problem of the Nile.

Darmois, G. (1946). *C. R. Acad. Sci. Paris*, **222**, 266–268 (in French). (This note attempts to characterize all bivariate distributions with a real parameter that admits an ancillary statistic.)

Neyman, J. (1946). *C. R. Acad. Sci. Paris*, **222**, 843–845 (in French). (Neyman remarks that the problem of similar regions is equivalent to Fisher's problem of the Nile.)

Tan, P. (1973). *Commun. Statist.*, **2**, 45–58. (This paper points out that if a model $\{f(x;\theta)\theta \in \Omega\}$ can also be described as a structural model* in the sense of Fraser, the problem of the Nile has a solution.)

(ANCILLARY STATISTICS
FIDUCIAL INFERENCE
INFERENCE, STATISTICAL)

PETER TAN

FISHER'S z-DISTRIBUTION

If F is a random variable having an F-distribution*, this variant is given by $z = \frac{1}{2}\ln F$. Its distribution was studied by R. A. Fisher* [1] in the preparation of tables; it has the advantage of being more closely normal than F.

Reference

[1] Fisher, R. A. (1924). *Proc. Int. Math. Congr. Toronto*, pp. 805–813.

FISHER'S z-TRANSFORMATION

Although the term *correlation* and discussions of related concepts appear in the earlier literature (e.g., F. Galton* [12, 13], W. F. R. Weldon [32, 33], and F. Y. Edgeworth [8]), the product moment correlation* coefficient as presently known was defined by K. Pearson* and W. F. Sheppard in 1897 (see K. Pearson [25]). K. Pearson and L. N. G. Filon [26] suggested that for a sample of size n, n large, r may be treated as normal with standard deviation $(1-\rho^2)/\sqrt{n}$, unless ρ is very close to unity. Student [31], by a laborious process of random sampling with $n = 4, 8$ and $\rho = 0$, empirically but correctly, established the form of the null distribution of r. Soper [30], using asymptotic expansions, obtained higher-order terms in the mean and the standard deviation of r and sought to fit a Pearson curve* for the distribution of r. An intimate account of various activities in the early development of correlation* is given by a participant, F. N. David, in her reminiscence [7] titled "Karl Pearson and Correlation."

The work of Soper [30] brought the problem of exact distribution of the r of normal samples to the attention of R. A. Fisher*, which he is reported to have solved within a week. In the remarkable paper, which stands as a major breakthrough in many ways, Fisher [9] derives the exact distribution of r using geometric reasoning, examines it, and recognizes the severe nonnormality of the distribution. He then argues the consequential inadequacy of using the standard deviation of r in statistical analysis, addresses the estimation of correlation by the method of maximum likelihood, and considers the transformation $t = r/\sqrt{1-r^2}$ and, as an afterthought, $z = \tanh^{-1}(r)$ with an "aim at reducing asymmetry of the curves, or at approximate constancy of the standard deviation." He suspects z to be "possibly superior to t" in this connection. For a recent review of this paper, see Das Gupta [5]. The z-transformation is explored further in Fisher [10] and is highlighted in Chapter 6, on the correlation coefficient, of his epoch-making monograph *Statistical Methods for Research Workers* [11]. Harold Hotelling's observation [15] that "the best present day usage in dealing with the correlation coefficient is based on this chapter that has stood with little or no change since the first edition in 1925" remains valid to date, and the z-transformation is central to this usage.

The genesis and anatomy of the z-transformation for bivariate normal populations is described in the next section. The third section contains a sketch of its behavior in the nonnormal case, and the final section is given to some miscellaneous remarks.

THE z FROM A NORMAL SAMPLE

Consider the product moment correlation coefficient r of a random sample of size n from a bivariate normal population with correlation ρ. The probability density function of r is given by Fisher [9] as

$$f(r) = \frac{(1-\rho^2)^{(n-1)/2}}{\pi(n-3)!}(1-r^2)^{(n-4)/2} \times \left(\frac{\partial}{\sin\theta\,\partial\theta}\right)^{n-2}\frac{\theta}{\sin\theta}, \quad (1)$$

where $\cos\theta = -\rho r$. It has since been expressed in several alternative ways. The following rapidly converging form

$$f(r) = \frac{n-2}{\sqrt{2\pi}}\frac{\Gamma(n-1)}{\Gamma(n-1/2)}(1-\rho^2)^{(n-1)/2} \times (1-r^2)^{(n-4)/2}(1-\rho r)^{-(n-3/2)} \times F\left(\frac{1}{2}, \frac{1}{2}, n-\frac{1}{2}, \frac{1+\rho r}{2}\right), \quad (2)$$

involving the hypergeometric series F, is obtained. The higher-order terms in the following expressions for the expectation and the central moments of r may be found in Hotelling [15]:

$$E(r) = \rho\left\{1 - \frac{1-\rho^2}{2n}\right\} + O\left(\frac{1}{n^2}\right), \quad (3)$$

$$\mu_2(r) = \frac{(1-\rho^2)^2}{n-1}\left\{1 + \frac{11\rho^2}{2n}\right\} + O\left(\frac{1}{n^3}\right), \quad (4)$$

$$\mu_3(r) = \frac{\rho(1-\rho^2)^3}{(n-1)^2}\left\{-6 + \frac{15-88\rho^2}{n}\right\} + O(n^{-4}), \quad (5)$$

$$\mu_4(r) = \frac{(1-\rho^2)^4}{(n-1)^2}\left\{3 + \frac{-6+105\rho^2}{n}\right\} + O(n^{-4}). \quad (6)$$

Consequently, the coefficient of skewness* and the excess of kurtosis* of the distribution are

$$\gamma_1 = \frac{-6\rho}{\sqrt{n}} + o(n^{-1/2}), \quad (7)$$

$$\gamma_2 = \frac{6(12\rho^2-1)}{n} + o(n^{-1}). \quad (8)$$

From these expressions and Fig. 1 it is obvi-

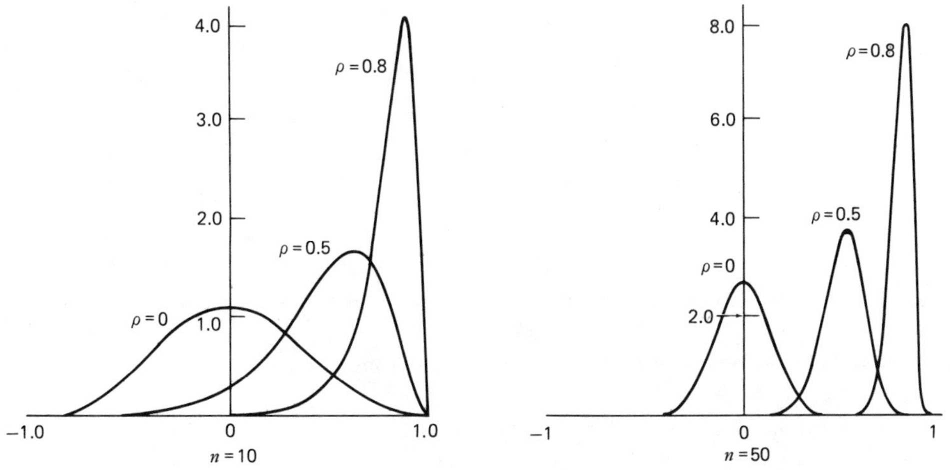

Figure 1 Frequency curves for the correlation coefficient r of samples with $n = 10, 50$.

ous that the distrubution is very asymmetric and its variance is heavily dependent on ρ. This hampers straightforward use of r for drawing conclusions regarding ρ, and one is led to a transformation such as

$$z = \frac{1}{2}\log\frac{1+r}{1-r} = \tanh^{-1}(r) \qquad (9)$$

of r.

Joan Fisher Box [2] in her biography of Fisher remarks that "like many good ideas the z-transformation only gradually revealed its quality even to its originator." Indeed, Fisher [9] notes that r underestimates ρ because the use of r "cramps high values" of ρ "into a small space" near 1, "producing a frequency curve trailing in the negative direction and so tending to reduce its mean." As a remedy he tentatively suggests z as "its superior" for the purpose. In 1915, it seemed to him "not a little attractive, but so far as I have examined it, it does not tend to simplify the analysis." After further exploration [10], in his celebrated monograph [11] Fisher argues more forcefully for z by describing three advantages: (a) Unlike r the standard error of z is practically independent of ρ; (b) the distribution of r is nonnormal even in large samples if ρ is large, whereas z tends rapidly to normality for any ρ; and (c) the distribution of r changes its form rapidly as ρ is changed, making it hopeless to adjust for skewness, whereas because of the nearly constant form of the distribution of z, the accuracy of the tests using z can be improved using small adjustments.

Moments of z

Fisher [10] gave an asymptotic expansion for the probability density function (PDF) of r from normal samples in powers of n^{-1} and of $z - \zeta$, where $\zeta = \tanh^{-1}(\rho)$, and used it to obtain the moments of z. But most of his expressions included minor errors and affected later studies by E. S. Pearson* [24] and F. N. David [6] on the normality of z. However, A. K. Gayen [14], after deriving the correct formulas for the first four moments and the coefficients of skewness and kurtosis of z, finds that these show the normal approximation for z to be "even more accurate" than suggested by the earlier studies. From Gayen [14] we have the following:

$$E(z) = \zeta + \frac{\rho}{2(n-1)}\left\{1 + \frac{5+\rho^2}{4(n-1)}\right\} + o(n^{-2}), \qquad (10)$$

$$\mu_2(z) = \frac{1}{n-1}\left\{1 + \frac{4-\rho^2}{2(n-1)} + \frac{22-6\rho^2-3\rho^4}{16(n-1)^2}\right\} + o(n^{-3}), \qquad (11)$$

$$\mu_3(z) = \frac{\rho^3}{(n-1)^3} + o\left(\frac{1}{n^3}\right), \qquad (12)$$

$$\mu^4(z) = \frac{1}{(n-1)^2}\left\{3 + \frac{14-3\rho^2}{n-1} + \frac{184-48\rho^2-21\rho^4}{4(n-1)^2}\right\} + o(n^{-4}), \qquad (13)$$

$$\beta_1(z) = \frac{\rho^6}{(n-1)^3} + o\left(\frac{1}{n^3}\right), \qquad (14)$$

$$\beta_2(z) = 3 + \frac{2}{n-1} + \frac{4+2\rho^2-3\rho^4}{(n-1)^2} + o(n^{-2}). \qquad (15)$$

These formulae are later confirmed by H. Hotelling* [15], who asserts that Fisher had communicated to him the errors in the formulas in ref. 10 several years prior to 1953.

Distributions of z

The asymptotic normality of r in a sample of size n, as $n \to \infty$, is an immediate consequence of the central limit theorem* and Slutsky's theorem*. Hence by the Mann–Wald theorem z is also asymptotically normal. However, because β_1 and $(\beta_2 - 3)$ in (14) and (15) are small even for moderate n, it can be regarded as normal with mean $\zeta - \tanh^{-1}(\rho)$. In ref. 11 Fisher recommends adjusting for the bias and for the higher-order terms in the variance by treating $z - \zeta$ as normally distributed with mean $\rho/\{2(n-1)\}$ and variance $1/(n-3)$. For details, see Hotelling [15].

The normal approximation for z above is generally recognized as being remarkably accurate when $|\rho|$ is moderate but not as accurate when $|\rho|$ is large, even when n is not small [6]. Mudholkar and Chaubey [23] note that, as the sample size increases, the coefficient of skewness of z decreases much more rapidly than $(\beta_2 - 3)$, the excess of its kurtosis. Hence they suggest approximating the distribution of the standardized variable $\{z - E(z)\}/(\text{s.d.}\, z)$ by a mixture $\lambda N(0,1) + (1-\lambda)L(0,1)$, where $N(0,1)$ and $L(0,1)$ denote the standardized normal and logistic distributions and $\lambda = 1 - 5/\{3(n-1)\}$. Later Chaubey and Mudholkar [3] propose regarding the standardized z as a Student's t variable with $\nu = (3n^2 - 2n + 7)/(n+1)$ degrees of freedom. These two approximations are almost equivalent and compare favorably with the alternatives to z due to Ruben [29] and Kraemer [21] (*see* CORRELATION). Chaubey and Mudholkar [3] also examine the effects of using normal approximations, as, for example, in Prescott [27], for the t-distribution on the approximation for z.

The Stable Variance and Almost Symmetry of z

From (11), (14), and Fig. 2 it is clear that Fisher succeeded remarkably in finding a function of r with these properties. Actually, z can be obtained analytically as a variance stabilizing and a skewness reducing transformation. Bartlett's differential equation in ref. 1 for stabilizing the variance was adapted to this end by Hotelling [15], and by Kendall [19] in the discussion of Hotelling's paper. Now, it is well known (see, e.g., Rao [28, p. 426]) that for any statistic T_n and a differentiable function $g(\cdot)$, $\operatorname{var}(T_n) = \sigma^2(\theta)$ implies that $\operatorname{var}\{g(T_n)\} \doteqdot \sigma^2(\theta)(dg(\theta)/d\theta)^2$. Hence $g(\theta) = \int d\theta/\sigma(\theta)$ is an approximate variance stabilizing transformation.

Using $(1-\rho^2)^2/n$, the leading term in $\operatorname{var}(r)$ in (4) for $\sigma^2(\theta)$ in the paragraph above, we get $\zeta = \tanh^{-1}(\rho)$ and z as the first approximation to the solution for stabilizing the variance. Using the approximate variance $\{1 + [(4-\rho^2)/2n]\}/n$ of z in (11) in the same manner, Kendall et al. [19] obtain $z^* = z - (3z - r)/4n$, the improvement over z obtained earlier by Hotelling [15], using an alternative method. Chaubey and Mudholkar [4] solve a differential equation to demonstrate that z is a skewness reducing transformation of r.

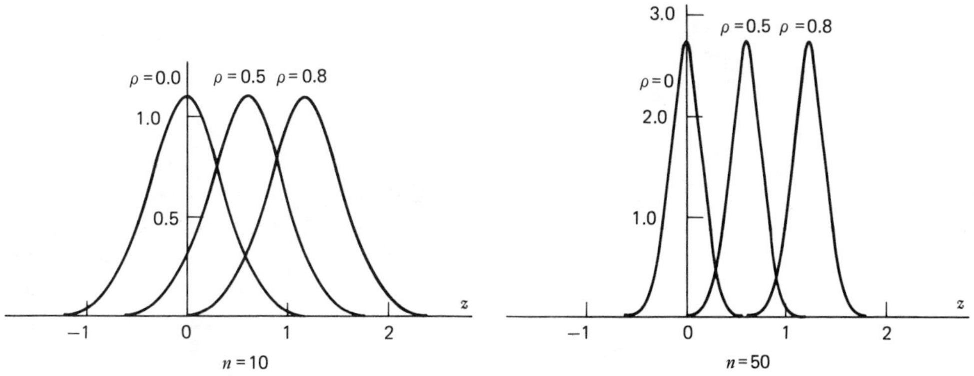

Figure 2 Frequency curves for z of samples with $n = 10, 50$.

THE NONNORMAL CASE: THE ROBUSTNESS OF z

Gayen [14] obtained the nonnull (i.e., $\rho \neq 0$) PDF of the correlation coefficient r, and of z, of a sample from a bivariate type A Edgeworth population, with a view to understanding the effects of nonnormality on the distributions. These PDFs are valid for samples of any size if the population cumulants of type λ_{40}, λ_{31}, λ_{22}, and of higher order, are negligible, but hold good asymptotically for any population. From Gayen [14] it can be seen that, in the general nonnormal case, $\beta_1(z)$ vanishes if $\rho = 0$, but is of order $O(n^{-2})$ compared with the normal theory order $O(n^{-3})$; the leading term in $\beta_2(z) - 3$, which is $O(n^{-1})$, depends on the population skewness and, unlike the normal case, also on ρ. From the analysis and computations with $n = 11$ and $n = 21$, Gayen [14] concludes that for samples of moderate size the normal distribution, with the following mean and variance, is "remarkably good" for approximating z.

$$E(z) = \zeta + \frac{\rho}{2(n-1)} + C_1(z), \qquad (16)$$

$$\operatorname{var}(z) = \frac{1}{n-1}\,\frac{4-\rho^2}{2(n-1)^2} + C_2(z), \qquad (17)$$

where

$$C_1(z) = \{\rho(3-\rho^2)(\lambda_{40}+\lambda_{04}) - 4(1+\rho^2)(\lambda_{31}+\lambda_{13}) + 2\rho(5+\rho^2)\lambda_{22}\} \div \{8(n-1)(1-\rho^2)^2\},$$

$$C_2(z) = \left\{\frac{1}{(n-1)} - \frac{2+\rho^2}{(n-1)^2}\right\}\frac{1}{4\rho(1-\rho^2)^2} \times \{\rho^2(\lambda_{40}+\lambda_{04}) - 4\rho(\lambda_{31}+\lambda_{13}) + 2(2+\rho^2)\lambda_{22}\}$$
$$+ \frac{1}{2(n-1)^2(1-\rho^2)^3}\{-\rho^2(3-\rho^2)(\lambda_{30}^2+\lambda_{03}^2) + 4\rho(\lambda_{30}\lambda_{03}) - (4+13\rho^2+\rho^4)(\lambda_{21}^2+\lambda_{12}^2)$$
$$+ 12\rho(2+\rho^2)(\lambda_{21}\lambda_{12}) - 4(1+2\rho^2)(\lambda_{30}\lambda_{12} + \lambda_{03}\lambda_{21}) + 2\rho(5+\rho^2)(\lambda_{30}\lambda_{21}+\lambda_{03}\lambda_{12})\}.$$

Recent studies of the z-transformation without the normality assumptions are generally in the framework of robust* inference. For a synopsis and references, see Huber [16]. A noteworthy exception is a comparative investigation, by Kocherlakota and Gajjar [20], of the effects of nonnormality on the distributions of the z and of other transformations of the correlation coefficient.

REMARKS

The single most informative reference on the z-transformation in the normal case is Hotelling's [15] report on his decade-long investigation presented to a meeting of the Royal Statistical Society and the follow-up discussion on the paper. It is broad in scope and rich in detail. The discussion is erudite. This study together with Gayen's [14] careful examination of the nonnormal case almost cover the topic.

Irwin [17] in the discussion of Hotelling's report [15] notes that the distribution of the z-transform of r when $\rho = 0$ is that of the variance ratio with $\nu_1 = \nu_2 = n - 2$, and indicates this as the possible reason why Fisher used the symbol z also in the variance ratio distribution.

It may be noted that $z = \tanh^{-1}(R)$ is also the approximate variance stabilizing transformation of a normal multiple correlation coefficient R. However, it is not very useful for approximating the nonnull distribution of R. For a discussion of this and other aspects of the distributions of various correlation coefficients, see Johnson and Kotz [18].

The z-transformation is used by Lin and Mudholkar [22] to construct a test for normality tied to the independence of the mean and variance, a characteristic of a normal sample. The statistic z of this test, termed the z-test, is shown to be practically normal when sample size is as small as 5. The fact that the pairs on which this z is based lack independence is additional evidence of the robust normality of z.

References

[1] Bartlett, M. S. (1946). *Biometrics* **3**, 39–52.

[2] Box, J. F. (1978). *R. A. Fisher: The Life of a Scientist*. Wiley, New York.

[3] Chaubey, Y. P. and Mudholkar, G. S. (1978). *Aust. J. Statist.*, **20**, 250–256.

[4] Chaubey, Y. P. and Mudholkar, G. S. (1981). *Stat. Dept. Tech. Rep.*, University of Rochester, Rochester, N.Y.

[5] Das Gupta, S. (1980). In *R. A. Fisher: An Appreciation*, S. E. Fienberg and D. V. Hinkley, eds. Springer-Verlag, New York.

[6] David, F. N. (1938). *Tables of the Ordinates and Probability Integral of the Distribution of the Correlation Coefficient in Small Samples*. Cambridge University Press, Cambridge.

[7] David, F. N. (1976). *Stat. Dept. Tech. Rep. No. 30*, University of California at Riverside, Riverside, Calif.

[8] Edgeworth, F. Y. (1892). *Philos. Mag.*, **5**, 190–204.

[9] Fisher, R. A. (1915). *Biometrika*, **10**, 507–521.

[10] Fisher, R. A. (1921). *Metron*, **1**, 1–32.

[11] Fisher, R. A. (1925). *Statistical Methods for Research Workers*. Oliver & Boyd, London.

[12] Galton, F. (1885). *J. Anthropol. Inst.*, **15**, 246–263.

[13] Galton, F. (1888).

[14] Gayen, A. K. (1951). *Biometrika*, **38**, 219–247.

[15] Hotelling, H. (1953). *J. R. Statist. Soc.*, **15**, 193–232 (with discussion).

[16] Huber, P. J. (1981). *Robust Statistics*. Wiley, New York.

[17] Irwin, J. O. (1953). *J. R. Statist. Soc. B*, **15**, 225–232.

[18] Johnson, N. L. and Kotz, S. (1970). *Continuous Univariate Distributions*, Vol. 2. Houghton Mifflin, Boston.

[19] Kendall, M. G. et al. (1953). *J. R. Statist. Soc. B*, **15**, 225–232.

[20] Kocherlakota, S. and Gajjar, A. V. (1979). *Stat. Dept. Tech. Rep. No. 92*, University of Manitoba, Canada.

[21] Kraemer, H. C. (1973). *J. Amer. Statist. Ass.*, **68**, 1004–1008.

[22] Lin, C. C. and Mudholkar, G. S. (1980). *Biometrika*, **67**, 455–461.

[23] Mudholkar, G. S. and Chaubey, Y. P. (1976). *Commun. Statist. B*, **5**(4), 163–172.

[24] Pearson, E. S. (1929). *Biometrika*, **21**, 337–360.

[25] Pearson, K. (1920). *Biometrika*, **13**, 25–45.

[26] Pearson, K. and Filon, L. N. G. (1898). *Philos. Trans. R. Soc. Lond.*, **191**, 229–311.

[27] Prescott, P. (1974). *Biometrika*, **61**, 177–180.

[28] Rao, C. R. (1973). *Linear Statistical Inference and Its Applications*, 2nd ed. Wiley, New York.

[29] Ruben, H. (1966). *J. R. Statist. Soc. B*, **28**, 513–525.

[30] Soper, H. E. (1913). *Biometrika*, **9**, 91–115.

[31] "Student" (1908). *Biometrika*, **6**, 1–25.

[32] Weldon, W. F. R. (1890). *R. Soc. Proc.*, **47**, 455.

[33] Weldon, W. F. R. (1892). *R. Soc. Proc.*, **41**, 2

Acknowledgment

Research sponsered by the Air Force Office of Scientific Research, Air Force Systems Command, USAF under Grant AFOSR-77-3360. The U.S. government is authorized to reproduce and distribute reprints for governmental purposes notwithstanding any copyright notation hereon.

The author is grateful to Y. P. Chaubey for some enlightening discussions.

(CORRELATION
FISHER, RONALD AYLMER
PRODUCT MOMENT CORRELATION)

GOVIND S. MUDHOLKAR

FISHER–YATES TESTS

There are two procedures that go by this name. One is a test for independence in 2×2 contingency tables*, described elsewhere (*see* FISHER'S EXACT TEST). The other, summarized here, and also known as the Fisher–Yates–Terry–Hoeffding test, is a nonparametric rank test* of the hypothesis H_0 of equality of k treatment effects in a one-way analysis of variance* (ANOVA) based on normal scores*. The test makes use of normal scores as expected values* of order statistics* from a standard normal distribution, and should not be confused with the van der Waerden test*, which is identical except that "normal scores" are defined as quantiles* of a standard normal distribution.

Let the observations on experimental units be $\{x_{ij}\}$; $j = 1, \ldots, n_i$; $i = 1, \ldots, k$; $N = \sum_{i=1}^{k} n_i$, and let r_{ij} be the rank of x_{ij} in the pooled data. If Φ and ϕ are the cumulative distribution function and probability density function of the standard normal distribution, respectively, and if

$$e_N(r) = \frac{N!}{(r-1)!(N-r)!} \times \int_{-\infty}^{\infty} x[\Phi(x)]^{r-1}[1-\Phi(x)]^{N-r}\phi(x)\,dx,$$

then the test statistic is T, where

$$T = \frac{(N-1)\sum_{i=1}^{k}\left\{n_i^{-1}\left[\sum_{j=1}^{n_i} e_N(r_{ij})\right]^2\right\}}{\sum_{r=1}^{N} e_N^2(r)}.$$

The only available tables of exact percentage points of the null distribution of T are by Lu and Smith [4] for $k = 3$, $n_1 + n_2 + n_3 \leqslant 15$, and $\alpha = 0.10$, 0.05, 0.01, and 0.005; they also give an expression for the variance of T, noting that $E(T \mid H_0) = k - 1$. Fisher and Yates [1] first proposed T as a test statistic, but treat

$$(N-c)T/\{(c-1)(N-1-T)\}$$

as an F-statistic with $c - 1$ and $N - c$ degrees of freedom under H_0, so that effectively the usual ANOVA is carried out on the normal scores rather than on the $\{x_{ij}\}$. Lu and Smith [4] indicate that this is a sufficiently good approximation to use when all sample sizes are 10 or greater, as well as being convenient. Puri [5] proved that the null distribution of T tends asymptotically to χ^2_{k-1} as the group sizes n_i tend to infinity, but the use of this approximation is not recommended. Lu and Smith [4] also discuss two improved but less simple approximations to the null distribution of T.

The asymptotic relative efficiency* (ARE) of the Fisher–Yates test to the classical F-test based on the raw data is

$$\sigma^2\left[\int_{-\infty}^{\infty} \frac{\{f(x)\}^2}{\phi\{\phi^{-1}[F(x)]\}}\,dx\right]^2,$$

where the x_{ij} have common cumulative distribution function (CDF) and probability density function (PDF) F and f, respectively, and common variance σ^2 [5]. This is the ARE relative to the t-test when $k = 2$, always greater than or equal to 1, and equals 1 if F is normal.

When the underlying data are (nearly) normally distributed, the (nearly) optimal rank test is based on normal scores [2]. The Kruskal–Wallis test* is more efficient when the data come from heavy-tailed distributions*; see Hodges and Lehmann [3].

References

[1] Fisher, R. A. and Yates, F. (1938). *Statistical Tables for Biological, Agricultural and Medical Research*. Hafner, New York (6th ed., 1963).

[2] Hajek, J. and Sidak, Z. (1967). *Theory of Rank Tests*. Academic Press, New York.

[3] Hodges, J. L., Jr., and Lehmann, E. L. (1961). *Proc. 4th Berkeley Symp. Math. Statist. Prob.*, Vol. 1. University of California Press, Berkeley Calif., pp. 307–317.

[4] Lu, H. T. and Smith, P. J. (1979). *J. Amer. Statist. Ass.*, **74**, 715–722.

[5] Puri, M. L. (1964). *Ann. Math. Statist.*, **35**, 102–121.

Bibliography

Hoeffding, W. (1950). *Proc. 2nd Berkeley Symp. Math. Statist. Prob.* University of California Press, Berkeley, Calif., pp. 83–92.

Terry, M. E. (1952). *Ann. Math. Statist.*, **23**, 346–366.

(ANALYSIS OF VARIANCE
NORMAL SCORES TESTS
RANKING PROCEDURES
VAN DER WAERDEN TEST)

CAMPBELL B. READ

FITTING EQUATIONS TO DATA *See* CURVE FITTING; FREQUENCY CURVES, SYSTEMS OF

FITTING FREQUENCY CURVES *See* CURVE FITTING; FREQUENCY CURVES, SYSTEMS OF

FIVE-NUMBER SUMMARIES

These are five statistics in exploratory data analysis* which summarize information in a data set of size n. They comprise the two extremes, the median* M and the two *hinges*, which are defined as follows:

1. The *depth d* of a data point is its posi-

tion if we enumerate toward it from the nearer extreme, so that

$$d(M) = (n+1)/2.$$

2. Each *hinge* H is at depth $d(H)$, where

$$d(H) = ([d(M)] + 1)/2,$$

$$[d(M)] = \text{greatest integer less than or equal to } d(M).$$

3. The *H-spread* is the difference between the hinge values, and is close to (if not equal to) the interquartile range*; this is a measure of dispersion of the data. The scatter of data points in the tails is measured by *fences*; if a *step* is $(1.5) \times (H\text{-spread})$, the *inner* and *outer fences* are located one and two steps, respectively, outside the hinges.

Example. For radiation readings in television viewing areas in ten department stores (see ref. 1), given by

0.15, 0.16, 0.36, 0.40, 0.48,
0.50, 0.50, 0.60, 0.80, 0.89,

the five-number summary would be represented as follows:

	d	$n = 13$	
M	5.5	0.49	
H	3	0.36	0.60
Ext	1	0.15	0.89

The H-spread is $0.60 - 0.36 = 0.24$, so that the inner fences lie at 0 and 0.96.

Five-number summaries are represented graphically by box-and-whisker plots, illustrated in Fig. 4 of GRAPHICAL REPRESENTATION OF DATA and in Fig. 2 of EXPLORATORY DATA ANALYSIS.

Reference

[1] *J. Environ. Health* (1969), pp. 359–369.

Bibliography

Tukey, J. W. (1977). *Exploratory Data Analysis*. Addison-Wesley, Reading, Mass. (See Chap. 2.)

Velleman, P. F. and Hoaglin, D. C. (1981). *Applications, Basics and Computing of Exploratory Data Analysis*. Duxbury Press, North Scituate, Mass.

(EXPLORATORY DATA ANALYSIS
GRAPHICAL REPRESENTATION
OF DATA
ICING OF TAILS)

CAMPBELL B. READ

FIXED-, RANDOM-, AND MIXED-EFFECTS MODELS

In a classification by Eisenhart [1], a statistical model is a fixed-, random-, or mixed-effects model if all treatment effects are regarded as fixed effects, all are regarded as random effects, or some are regarded as fixed effects and some as random effects; these terms will be defined presently. The distinction is important because these different types of models correspond to different inferential objectives and lead to different methods of analysis.

Although it is straightforward to classify a given model as fixed, random, or mixed, there may be disagreement as to whether a set of treatment effects should be regarded as fixed or random when the model is applied to a real experimental situation. In many practical situations it is possible to argue with equal persuasiveness that a given treatment effect is fixed or that it is random. In such cases, the decision to designate the effect as fixed or random is based in part on inferential objectives and in part on the credibility of the designation.

MODELS

In an idealized experiment, each experimental *subject* (or *unit*) is exposed to one set of experimental *conditions* and a characteristic of the subject, called the *response*, is recorded. Different subjects may be exposed to different conditions. The objective is to determine how the experimental conditions affect values of the response variable. The experimental conditions are described by combinations of values of one or more *treatment variables* or *factors*; a value of a treatment variable is called a *level*. The *experimental design* specifies the way subjects are to be chosen, the response variable, the treatment variables, the treatment levels to be used or the way in which treatment levels are to be chosen, and the way subjects are to be assigned to conditions, among other things. An *effect* is a change in the response due to a change in level of a treatment or treatments.

Linear statistical models for responses in terms of conditions take the following form:

$$Y = \mu + x_1\beta_1 + x_2\beta_2 + \cdots + x_k\beta_k + e \tag{1}$$

(*see* GENERAL LINEAR MODEL). Here y is the response variable; the variables $x_1, x_2, \ldots, x_k$ are functions of the treatment variables. The term e is included in anticipation that different subjects exposed to the same conditions may yield different responses: it is called the *error* term. The intercept, μ, is regarded as fixed. The coefficients β_1, $\beta_2, \ldots, \beta_k$ are treatment effects. With this model, to determine how the response differs with the treatment variables is to determine the values of, or relations among, the effects $\beta_1, \beta_2, \ldots, \beta_k$.

To facilitate statistical inference, the model includes assumptions about the error term. Typically, e is assumed to be a random variable with expected value 0 and variance uninfluenced by experimental conditions: thus $E(e) = 0$ and $\text{var}(e) = \sigma_e^2$ for all experimental conditions. It is assumed that error terms for different subjects are independent. Of course, other, more complicated, assumptions may be warranted.

Finally, assumptions are made about the nature of the effects. If an effect (a β_i) is assumed to be fixed, it is called a *fixed effect*. If an effect is assumed to be a realized value of a random variable, it is called a *random effect*. Commonly, random effects are assumed to have mean zero and a variance that is the same for all the experimental conditions, and to occur independently. A model in which all treatment effects are fixed is a *fixed-effects model*; in which all treatment effects are random, a *random-effects model*; in which some treatment effects are fixed and some are random, a *mixed-effects model*, or simply a *mixed model*.

EFFECTS: FIXED OR RANDOM?

The problems involved in designating a set of effects as fixed or random may be seen in an example cited by Searle [16]. Litter weights of six 10-day-old litters of mice from each of four dams were measured in order to assess maternal abilities. Here each litter is an experimental unit, litter weight is the response variable, and "dam" is the treatment variable with levels first dam, . . ., fourth dam. The equation for a model relating litter weight to the dam bearing the litter is

$$Y_{ij} = \mu + \delta_i + e_{ij}, \tag{2}$$

where $i = 1, \ldots, 4$ identifies dams and $j = 1, \ldots, 6$ identifies litters. Assume that the error terms are uncorrelated, and that $E(e_{ij}) = 0$ and $\text{var}(e_{ij}) = \sigma_e^2$ for all i and j. The dam (treatment) effects are $\delta_1, \delta_2, \delta_3, \delta_4$.

Common practice is to regard the treatment effects as fixed if those treatment levels used are the only ones about which inferences are sought. If the purpose of the experiment is to compare litter weights among only the four dams used, without regard to any broader collection of dams, then by this practice dam effects would be regarded as fixed.

If inferences are sought about a broader collection of treatment effects than those used in the experiment, or if the treatment levels are not selected purposefully (replication effects often fall into this category), it is common practice to regard the treatment effects as random. If the purpose is to examine maternal abilities of mice of this strain, dam effects would be regarded as random. In this case, the four dams used are of interest only inasmuch as they represent all dams of the strain. Analysis then proceeds as if the four dams resulted as a random sample from this population of dams.

When effects are regarded as random, it is clear that the way in which the treatment levels arise, or are chosen, affects the degree to which the treatment effects present resemble a random sample of treatment effects from the population for which inferences are intended. Carefully designed experiments include randomization* procedures to enhance this resemblance. Even in these situations, though, the target population often is conceptual or inaccessible, so that it is not possible to guarantee that the treatment effects *are* a random sample from the target population. In most cases in which effects are regarded as random, all that can be said is that the treatment levels were not deliberately chosen in such a way as to exclude any portion of the target population. See Searle [16, pp. 376–383] for a discussion of the appropriateness of assuming effects fixed or random.

When treatment effects are assumed to be fixed, individual effects and relations among the effects are of primary interest. When effects are assumed to be random, the portion of the variance of the response variable due to variation of treatment effects often is of primary interest. This leads to problems and procedures for inferences about variance components*. Even in random or mixed models, though, inferences may be sought about the realized values of the random effects. In this case, procedures for estimating the effects may differ from those used for estimating fixed effects.

APPROACHES TO INFERENCE

Different conceptualizations and objectives for an experiment lead to different analytical procedures for a set of treatment effects, depending on whether the effects are fixed or random. To see these differences, consider the general formulation [6] of a model for the vector $\mathbf{Y}$ of values of the response variable:

$$\mathbf{Y} = \mathbf{X}\boldsymbol{\beta} + \mathbf{Z}\mathbf{u} + \mathbf{e}. \tag{3}$$

$\mathbf{X}$ and $\mathbf{Z}$ are matrices of known constants; $\boldsymbol{\beta}$ is a vector of fixed effects; $\mathbf{u}$ is a vector of random effects; and $\mathbf{e}$ is a vector of error terms. Assume that $\mathbf{u}$ and $\mathbf{e}$ are uncorrelated, that $\text{var}(\mathbf{e}) = \sigma_e^2\mathbf{I}$ and that $\text{var}(\mathbf{u}) = \sigma_e^2\mathbf{D}$, where $\mathbf{D}$ is a symmetric nonnegative definite matrix.

Generally, $\mathbf{u}' = (\mathbf{u}_1', \dots, \mathbf{u}_n')$, where $\mathbf{u}_i$ contains the random effects of the ith treatment variable. Typically, it is assumed that $\mathbf{u}_1, \dots, \mathbf{u}_n$ are uncorrelated and that $\text{var}(\mathbf{u}_i) = \sigma_i^2\mathbf{I}$, so that $\text{var}(\mathbf{u})$ is block diagonal with blocks $\sigma_i^2\mathbf{I}$. (See Harville [6] for an analysis of alternative formulations.) The parameters $\sigma_1^2, \dots, \sigma_n^2$ are called *variance components*.

Inferential objectives may involve fixed effects, the variance components, or the random effects. Two approaches that often lead to similar results are likelihood-based and linear models (or least-squares*) procedures. For likelihood-based procedures (e.g., maximum likelihood estimation* and likelihood ratio tests*), distributional assumptions are required, and it is commonly assumed that u and e follow multivariate normal probability distributions. When random effects are present, likelihood-based procedures may be computationally difficult (see Hartley and Rao [3] and Hemmerle and Hartley [7]), and properties of the procedures may be difficult or impossible to evaluate (see also Harville [5]).

Linear models* procedures are more widely used and conveniently available than likelihood-based procedures. These procedures are built around linear (in $\mathbf{Y}$) estimators of effects and analysis of variance (ANOVA*) tables for testing hypotheses. The presence of random effects causes difficulties here, too.

ESTIMATING EFFECTS

Consider estimating a linear combination $\mathbf{c}_1'\boldsymbol{\beta} + \mathbf{c}_2'\mathbf{u}$ of the fixed and random effects, where $\mathbf{c}_1$ and $\mathbf{c}_2$ are vectors of fixed, specified constants. Henderson [8] and Harville [4] have established that if $\mathbf{D}$ is known, $\mathbf{c}_1'\hat{\boldsymbol{\beta}} + \mathbf{c}_2'\hat{\mathbf{u}}$ has minimum mean squared error* $E(\mathbf{a}'\mathbf{y} - \mathbf{c}_1'\boldsymbol{\beta} - \mathbf{c}_2'\mathbf{u})^2$ among linear estimators $\mathbf{a}'\mathbf{y}$ which satisfy $E(\mathbf{a}'\mathbf{y} - \mathbf{c}_1'\boldsymbol{\beta} - \mathbf{c}_2'\mathbf{u}) \equiv 0$. $\mathbf{X}'\mathbf{V}^{-1}\mathbf{X}\hat{\boldsymbol{\beta}} = \mathbf{X}'\mathbf{V}^{-1}\mathbf{Y}$ and $\hat{\mathbf{u}} = \mathbf{D}\mathbf{Z}'\mathbf{V}^{-1}(\mathbf{Y} - \mathbf{X}\hat{\boldsymbol{\beta}})$ [6]. In this formulation $\boldsymbol{\beta}$ and $\mathbf{u}$ are known to be fixed and random effects, respectively. The presence of random effects and the identity of the effects designated as random affect the estimators $\hat{\boldsymbol{\beta}}$ and $\hat{\mathbf{u}}$. If interest were restricted to the fixed effects (i.e., $\mathbf{c}_2 = \mathbf{0}$) and ordinary least-squares used to estimate $\boldsymbol{\beta}$, the estimators apparently would be inefficient. Similar inefficiency would result if the random effects $\mathbf{u}$ are estimated as if they were fixed. Inefficiency may also result if truly fixed effects are regarded as random. This formulation of efficient linear estimators makes it clear that the designation of effects as fixed or random is consequential.

In practice, $\mathbf{D}$ is not known, but the elements of $\text{var}(\mathbf{Y}) = \sigma_e^2(\mathbf{I} + \mathbf{Z}\mathbf{D}\mathbf{Z}')$ are known linear combinations of the variance components $\sigma_1^2, \ldots, \sigma_n^2$. To compute $\hat{\boldsymbol{\beta}}$ and $\hat{\mathbf{u}}$, estimates of the variance components are substituted in $\mathbf{D}$. The resulting estimators of fixed and random effects are then not linear in $\mathbf{Y}$, and general properties of these estimators are not known. See Kakwani [9].

From a different point of view regarding the nature of effects $\boldsymbol{\beta}$ and $\mathbf{u}$, the same estimators arise as Bayes linear estimators* as follows. Let $\boldsymbol{\beta}$ and $\mathbf{u}$ have a joint prior probability distribution such that $\text{cov}(\boldsymbol{\beta}, \mathbf{u}) = \mathbf{0}$, $\text{var}(\mathbf{u}) = \sigma_e^2\mathbf{D}$, and $E(\mathbf{u}) = \mathbf{0}$. Among linear estimators $\mathbf{a}'\mathbf{Y}$ satisfying $\mathbf{X}'\mathbf{a} = \mathbf{c}_1$, $\mathbf{c}_1'\hat{\boldsymbol{\beta}} + \mathbf{c}_2'\hat{\mathbf{u}}$ minimizes $E(\mathbf{a}'\mathbf{Y} - \mathbf{c}_1'\boldsymbol{\beta} - \mathbf{c}_2'\mathbf{u})^2$, with the expectation extending over the prior distribution on $\mathbf{u}$ and $\boldsymbol{\beta}$ as well as the distribution of $\mathbf{e}$. In this approach, $\boldsymbol{\beta}$ and $\mathbf{u}$ are treated similarly, except that the constraint $\mathbf{X}'\mathbf{a} = \mathbf{c}_1$ forces $E(\mathbf{a}'\mathbf{Y}) = \mathbf{c}_1'\boldsymbol{\beta} + \mathbf{a}'\mathbf{Z}\mathbf{u}$ (here the expectation is with respect to $\mathbf{e}$).

VARIANCE COMPONENTS*

Even when only individual effects are of interest, information about the variance components is needed. In other cases, the variance components may be of primary interest. In addition to these reasons, interest in inferential procedures for variance components has been stimulated by some problems not encountered with fixed effects.

Widely used procedures for estimating and testing hypotheses about variance components are based on ANOVA tables. When random effects are present, the expected value (with respect to $\mathbf{u}$ and $\mathbf{e}$) of a sum of squares in an ANOVA table is a quadratic in the fixed effects plus a linear combination of the variance components; see Table 1.

To estimate σ_j^2, a linear combination of the sums of squares is found whose expected value is σ_j^2. To test $H_0: \sigma_j^2 = 0$, an F-statistic is formed as the ratio of two sums of squares (or linear combinations of sums of squares) which have identical expected values, except that the expected value of the numerator includes σ_j^2 and the expected value of the denominator does not. (See Searle [16] and Graybill [2] for a discussion of these methods; *see also* VARIANCE COMPONENTS.)

When Y is normally distributed, these ANOVA-based procedures have good (if not optimal) characteristics for balanced models. In unbalanced models the "standard" ANOVA table may yield poor results and there may not exist an optimal such table (see LaMotte [11]). Other approaches to

Table 1 ANOVA, Random Effects

Source	d.f.	Sum of Squares	E(SS)
1	ν_1	$\mathbf{Y}'\mathbf{A}_1\mathbf{Y}$	$\boldsymbol{\beta}'\mathbf{X}'\mathbf{A}_1\mathbf{X}\boldsymbol{\beta} + a_{11}\sigma_1^2 + \cdots + a_{1n}\sigma_n^2$
2	ν_2	$\mathbf{Y}'\mathbf{A}_2\mathbf{Y}$	$\boldsymbol{\beta}'\mathbf{X}'\mathbf{A}_2\mathbf{X}\boldsymbol{\beta} + a_{21}\sigma_1^2 + \cdots + a_{2n}\sigma_n^2$
$\vdots$	$\vdots$	$\vdots$	$\vdots$
k	ν_k	$\mathbf{Y}'\mathbf{A}_k\mathbf{Y}$	$\boldsymbol{\beta}'\mathbf{X}'\mathbf{A}_k\mathbf{X}\boldsymbol{\beta} + a_{k1}\sigma_1^2 + \cdots + a_{kn}\sigma_n^2$

variance component estimation, not based directly on standard ANOVA tables, are discussed in Searle [16], Rao [14, 15], LaMotte [10, 11], Olsen et al. [12], and Pukelsheim [13].

References

[1] Eisenhart, C. (1947). *Biometrics*, **3**, 1–21. (Distinguishes fixed, random, and mixed models.)

[2] Graybill, F. A. (1976). *Theory and Application of the Linear Model*. Duxbury Press, North Scituate, Mass. (A standard text in linear models. See Sec. 5.5 and Chap. 15.)

[3] Hartley, H. O. and Rao, J. N. K. (1967). *Biometrika*, **54**, 93–108. (Describes an algorithm for maximum likelihood estimation and establishes some properties of the estimators.)

[4] Harville, D. A. (1976). *Ann. Statist.*, **4**, 384–395. (A rigorous and thorough treatment of estimation of random effects.)

[5] Harville, D. A. (1977). *J. Amer. Statist. Ass.*, **72**, 320–338. (A comprehensive survey of maximum likelihood approaches to variance component estimation.)

[6] Harville, D. A. (1978). *Biometrics*, **34**, 441–454. (A general formulation of the two-way mixed model which includes others as special cases. Application of the mixed-model equations.)

[7] Hemmerle, W. J. and Hartley, H. O. (1973). *Technometrics*, **15**, 819–831. (Describes a computational device to reduce the burden of maximum likelihood estimation in mixed models.)

[8] Henderson, C. R. (1975). *Biometrics*, **31**, 423–447. (Describes the mixed-model equations and develops best linear unbiased estimators and predictors.)

[9] Kakwani, N. C. (1967). *J. Amer. Statist. Ass.*, **62**, 141–142. (Shows that, under some conditions, best linear unbiased estimators that use estimated variance components are unbiased.)

[10] LaMotte, L. R. (1973). *Biometrics*, **29**, 311–330. (Uses a general approach to derive "locally best" quadratic estimators of variance components in mixed models.)

[11] LaMotte, L. R. (1976). *Biometrics*, **32**, 793–804. (Examines several alternative estimators of variance components in the unbalanced, random, one-way ANOVA model.)

[12] Olsen, A., Seely, J. and Birkes, D. (1976). *Ann. Statist.*, **4**, 878–890. (Describes a minimal complete class among unbiased quadratic estimators of variance components in models having two variance components. Some fundamental results in linear estimation are also derived.)

[13] Pukelsheim, F. (1976). *J. Multivariate Anal.*, **6**, 626–629. (Observes that invariant quadratics follow linear models in variance components.)

[14] Rao, C. R. (1970). *J. Amer. Statist. Ass.*, **65**, 161–172. [First proposes minimum norm (analogous to least squares) as a criterion for variance component estimation. Develops special Minimum Norm Quadratic Unbiased Estimators (MINQUE*) for a model with unequal error variances.]

[15] Rao, C. R. (1972). *J. Amer. Statist. Ass.*, **67**, 112–115. (Describes minimum norm and minimum variance quadratic unbiased estimators in a general setting.)

[16] Searle, S. R. (1971). *Linear Models*. Wiley, New York. (A standard text in linear models. The discussion beginning Chap. 9 is extensive concerning "fixed or random." Chapters 9–11 are the most complete coverage of mixed models available in one place.)

(ANALYSIS OF VARIANCE
ANOVA TABLE
DESIGN OF EXPERIMENTS
GENERAL LINEAR MODEL
LEAST SQUARES
VARIANCE COMPONENTS)

LYNN ROY LAMOTTE

FIXED-STATION NETWORKS, STATISTICAL CHARACTERISTICS OF

A network of fixed stations is frequently used to determine national characteristics of interest over an extended period of time. Some common examples of fixed station networks are rain gauges at strategic locations across the nation, stations measuring amount of water flow, or stations measuring water quality. A somewhat unusual type of network is the Distant Early Warning (DEW) line across northern Canada. In each case the stations are fixed in location, and measurements of one or more variables are made either continuously, periodically, or at random intervals.

This discussion will be in the context of monitoring the environment, and in particular, water quality. A national or international assessment of water quality is of considerable interest, and frequently, legislation will direct efforts to assess water quality. In the United States, for example, the Water Pollution Control Act of 1972 requires that the U.S. Environmental Protection Agency report annually an assessment of the quality of that nation's navigable waters, and to determine when all navigable waters will be able to support "balanced populations" of fish and wildlife.

Statistical aspects of fixed-station networks include what variables are to be measured, measurement error*, frequency in timing of measurements, and location and number of stations. The measurement of water quality on a national scale presupposes a technology capable of standardized processing of samples, and measurements that can validly be compared across space as well as time. This includes estimation of measurement error and interlaboratory agreement. The location of a station is frequently influenced by extrastatistical considerations such as convenience and demographic or political boundaries. Although fixed stations probabilistically selected are optimal, stations placed according to best professional judgment will provide valid information about changes in water quality. As an analogy, a rain gauge station at an airport will provide some information about rainfall for a region served by the airport, and, more important, will provide reasonably accurate information about the change in rainfall from year to year. The statistical principle involved is that high-order interactions* are not commonly observed. A theoretical-practical problem involves the frequency and timing of measurements. It is impractical to sample randomly over time so that short-term variability cannot be observed at fixed stations. However, a primary purpose of a fixed-station network is to detect long-term trends. The frequency of measurement should be determined accordingly. It will usually be possible to adjust variables with a diurnal component to a common time point. For example, dissolved oxygen levels that display a diurnal sinusoidal pattern can be adjusted to the same time of day.

Fixed-station networks have the advantage of reduction in variability attributable to location. A second advantage is unbiasedness in measuring changes from year to year. Third, there is the practical consideration that a network of fixed stations will be cheaper to maintain than one where equipment and personnel have to be moved from place to place, often in an unpredictable way.

Some drawbacks of networks of fixed stations include the inability to determine short-term cycles at least in space, if not in time. It might be argued that this is not its purpose. Also, such a network will not permit the assessment of cause and effect. Again, it may be argued that fixed-station networks are not designed to do this. Finally, if the stations in a network can be placed according to statistical criteria, there is the problem that optimal allocation for one variable is not optimal for another.

In the United States the General Accounting Office (GAO) [1] has vigorously criticized the U.S. Environmental Protection Agency's network of fixed stations on methodological and design grounds. In its place the GAO has suggested "intensive surveys." An intensive survey studies intensively a small section of a river or stream over a short period of time, say 1 or 2 months. This suggestion has many problems, the major one being lack of generalizability. A better alternative, which complements rather than replaces fixed-station networks, consists of valid sample surveys* of navigable waters. This requires a sampling frame; given such a frame, answers can be obtained to questions not easily answered by fixed-station networks or intensive surveys.

Additional statistical problems associated with networks of fixed stations include the modeling and analysis of autocorrelated data and the combination of data* from various stations [4]. One statistical interpolation technique currently being studied is kriging* (for references, see ref. 2). Additional statistical considerations in network design can be found in ref. 3. For some nonparametric analyses of data from fixed-station networks, see ref. 5.

References

[1] General Accounting Office (1981). *Better Monitoring Techniques Are Needed to Assess the Quality of Rivers and Streams*, Vols. I, II. U.S. General Accounting Office, Gaithersburg, MD 20760. (Fixed-station networks are found wanting, the arguments telling but not convincing.)

[2] Hughes, J. and Lettenmaier, D. P. (1981). *Water Resour. Res.*, **17**, 1641–1650.

[3] Lettenmaier, D. P. (1977). Detections of Trends in Stream Quality: Monitoring Network Design and Data Analysis. *Tech. Rep. No. 51*, C. W. Harris Hydraulics Lab., University of Washington, Seattle, WA 98195.

[4] Switzer, P. (1979). *Water Resour. Res.*, **15**, 1712–1716.

[5] van Belle, G. and Fisher, L. (1977). *J. Water Pollut. Control Fed.*, **49**, 1671–1678. (Nonparametric techniques applicable to data from fixed-station networks.)

(ENVIRONMENTAL STATISTICS
HYDROLOGY, STOCHASTIC
KRIGING)

GERALD VAN BELLE

FIXED-WIDTH AND BOUNDED-LENGTH CONFIDENCE INTERVALS

Let T_n be a statistic based on a random sample of size n from a population with an unknown parameter θ, to be estimated. If for some quantity d, free of θ and of n,

$$\Pr(|T_n - \theta| \leqslant d) \geqslant 1 - \alpha, \qquad (1)$$

then (1) defines a *fixed-width confidence interval* (CI) $I_n = (T_n - d, T_n + d)$ for θ (*see* CONFIDENCE INTERVALS AND REGIONS). If d depends on n and $d = d_n \leqslant D$, say, then I_n is a *bounded-length CI* for θ. The estimation may be achieved with equality in (1), or in an asymptotic sense, as in (3) below.

Although it may be desirable to constrain the width of confidence intervals in such a way in practice, properties such as (1) cannot usually be obtained with predetermined sample sizes; the width of the usual CI for estimating the mean of a normal distribution with unknown variance, for example, depends on the sample variance. However, there may be sequential procedures, incorporating a stopping variable N (*see* SEQUENTIAL ANALYSIS; STOPPING RULES). The problem then is to determine N and T_N so that for given choices of d and of α, one of the following properties holds:

$$\Pr(|T_N - \theta| \leqslant d) \geqslant 1 - \alpha, \qquad (2a)$$

$$\Pr(|T_n - \theta| \leqslant d_N \leqslant d) \geqslant 1 - \alpha, \qquad (2b)$$

$$\lim_{d \to 0} \Pr(|T_n - \theta| \leqslant d) = 1 - \alpha, \qquad (3)$$

where α is fixed. Another asymptotic approach fixes d, while $\alpha \to 0$. See the section "Another Asymptotic Approach."

The asymptotic efficiency* of one procedure relative to another is the limiting ratio of the average sample sizes (say, as $d \to 0$).

In 1945 Stein [19] developed a two-stage procedure for estimating the mean of a normal population (*see* DOUBLE SAMPLING). Most of the asymptotic theory leading to (3) was developed between 1965 and 1971; see refs. 2, 16, and 22. The approach outlined in the section "Another Asymptotic Approach" was developed in the late 1970s.

Fixed-width CIs for population means are described in the four sections following "Notation"; of these, only the second is asymptotic, applicable to distributions other than normal. The two sections following it describe two-stage procedures and the cost of ignorance of the variance; then CIs for the difference of two population means are presented. The final two sections of this article treat distribution-free bounded-length CIs, mainly asymptotic, for the center of location of a continuous symmetric population.

Most of the results that follow are not difficult to apply, and are potentially very useful. But they have not been widely employed, perhaps because of a lack of awareness of them. An application in biostatistics [12] of the asymptotic rule of the section "Asymptotic Fixed-Width CIs for the Mean" is discussed later. The same rule has been used to achieve accuracy in a simulation problem [3]; Stein's rule, to be discussed in the section on two-stage rules, has been applied to estimating the amount of eye movement in watching instrument panels [13].

NOTATION

Let $\Phi(x)$ be the unit normal cumulative distribution function (CDF); let a and a_ν be quantiles such that

$$1 - \Phi(a) = \tfrac{1}{2}\alpha, \qquad a_\nu = t_{\nu,1-(1/2)\alpha},$$

where $t_{\nu,\epsilon}$ is the 100ϵ percent point of a Student t-distribution with ν degrees of freedom. In the random sample $X_1, X_2, \ldots, X_n$,

$$\overline{X}_n = n^{-1}\sum_{i=1}^{n} X_i,$$

$$S_n^2 = (n-1)^{-1}\sum_{i=1}^{n}\left(X_i - \overline{X}_n\right)^2,$$

the sample mean and variance, respectively, and $X_{1,n} \leqslant X_{2,n} \leqslant \cdots \leqslant X_{n,n}$ are the order statistics. When $F(x)$ is the CDF of the parent distribution, $F_n(x)$ is the sample CDF. A normal distribution with mean θ and variance σ^2 is denoted $N(\theta, \sigma^2)$, and $[x]$ is the largest integer less than or equal to x. Let θ be the unknown mean and σ^2 the variance in a normally distributed population. Then

$$\Pr\left(\overline{X}_n - d \leqslant \theta \leqslant \overline{X}_n + d\right) \geqslant 1 - \alpha$$

if $n \geqslant a^2\sigma^2/d^2$, or $n = [a^2\sigma^2/d^2] + 1$. Thus a fixed-width CI for θ is obtained nonsequentially if σ^2 is known.

ASYMPTOTIC FIXED-WIDTH CIs FOR THE MEAN

Let $X_1, X_2, \ldots$ be a random sequence of random variables (rvs) from some unknown population with CDF F, unknown mean θ, and unknown finite variance σ^2, and consider stopping variables N depending at stage n on $(S_2^2, \ldots, S_n^2)$ only. A "good" procedure might have three properties for a fixed-width CI $(\overline{X}_N - d, \overline{X}_N + d)$:

$$\lim_{d\to 0}\left\{d^2N/(a^2\sigma^2)\right\} = 1 \quad \text{almost surely}, \qquad (4)$$

$$\lim_{d\to 0}\Pr\left(\overline{X}_N - d \leqslant \theta \leqslant \overline{X}_N + d\right) = 1 - \alpha \quad \text{(asymptotic consistency)}, \qquad (5)$$

$$\lim_{d\to 0}\left\{d^2E(N)/(a^2\sigma^2)\right\} = 1 \quad \text{(asymptotic efficiency)}. \qquad (6)$$

The rationale behind (4) and (6) is to compare N and $E(N)$ with the ideal procedure when the variance is known, using an appeal to central limit properties.

Theorem 1 (Chow–Robbins [2]). With the assumptions just discussed, let

$$V_n = \left\{(n-1)S_n^2 + c\right\}/n,$$

where $0 \leqslant c \leqslant 1$ and c is constant; and let $b_1, b_2, \ldots$ be a sequence of constants such that $b_n > 0$ and $b_n \to a$ as $n \to \infty$. Let N be the first n for which $V_n \leqslant nd^2/b_n^2$. Then (4), (5), and (6) hold, and $(\overline{X}_N - d, \overline{X}_N + d)$ gives a $100(1-\alpha)$ percent asymptotic fixed-width CI for θ.

Example. Suppose that it is desired to estimate enzyme concentrations in the human pancreas with a fixed-width confidence interval for each enzyme. Schmidt et al. [12] followed the procedure of the theorem above, with $\alpha = 0.95$, $b_n = a = 1.96$, $V_n = S_n^2 + n^{-1}$, and a further requirement that $N \geqslant 3$. Four enzyme concentrations were estimated, each with a prespecified CI length $2d$, measured in suitable units; the four data sets appear in the source paper in full. For instance, in estimating amylase concentration in Lagerlof units, d was chosen to be 0.6; application of the stopping rule led to 58 observations before the inequality $V_n \leqslant nd^2/a^2$ was satisfied, and sampling was then terminated. Since $\bar{x}_{58}$ was observed to be 5.9, the resulting asymptotic 95% fixed-width CI is (5.3, 6.5).

Remark. If F is continuous, we may choose $c = 0$; otherwise, $c > 0$. The original proof required the assumption of a finite fourth moment, but the use of truncated variables makes this unnecessary. The theorem does not establish that (2a) or (2b) holds. See also Khan [8]. Ghosh [6] and Srivastava [17] have extended the theory to interval estimation of regression parameters.

TWO-STAGE RULES (*see* DOUBLE SAMPLING)

Stein's Rule [19] is as follows: Let X_1, $X_2, \ldots$ be a random sequence from a $N(\theta, \sigma^2)$ population, θ and σ^2 unknown, and let m be a fixed integer, $m \geqslant 2$. A sample of size m is drawn and observed. Let

$$N = \max\left\{m, \left[a_{m-1}^2 S_m^2/d^2\right] + 1\right\}.$$

Then

$$\Pr\left(\bar{X}_N - d \leqslant \theta \leqslant \bar{X}_N + d\right) \geqslant 1 - \alpha, \quad (7)$$

so that $(\bar{X}_N - d, \bar{X}_N + d)$ yields a fixed-width CI for θ with coverage probability at least $1 - \alpha$. It can be shown that $E(\bar{X}_N) = \theta$, that the rv $\sqrt{N}(\bar{X}_N - \theta)/S_m$ has a Student t-distribution with $m - 1$ degrees of freedom, and that $\Pr(N < \infty) = 1$. The choice of m is arbitrary (see Moshman [9]), the second stage may not have any observations, and information about σ^2 from the second stage is not used. This is inefficient, because although (5) holds,

$$\lim_{d \to 0}\left\{d^2 E(N)/(a^2\sigma^2)\right\} = a_{m-1}^2/a^2, < 1.$$

Weiss [21] developed a two-stage fixed-width CI for the qth quantile of a continuous distribution, in which (2b) is satisfied for an exact coverage probability, but where the number of observations in the second stage is always positive.

Ghosh [5] derives bounded-length CIs for the difference $\theta_1 - \theta_2$ of the means of two normal populations. A modification of Stein's rule is applied to the random sequence $T_1, T_2, \ldots,$ where $T_i = X_i - Y_i$, and the sequences $X_1, X_2, \ldots$ and Y_1, $Y_2, \ldots$ are drawn from the $N(\theta_1, \sigma_1^2)$ and $N(\theta_2, \sigma_2^2)$ populations, respectively. Ghosh evaluates his CI procedures against others on the basis of two other criteria, Neyman accuracy* and Wolfowitz accuracy*, in addition to the expected sample size.

COST OF IGNORANCE

Let N be a stopping variable such that (7) holds, in sampling the random sequence X_1, $X_2, \ldots$ from a $N(\theta, \sigma^2)$ population with σ^2 unknown. The cost of ignorance of the variance is then $E(N) - a^2\sigma^2/d^2$.

Theorem 2 (Simons [16]). Given n_0, let N be the smallest integer $n \geqslant n_0 \geqslant 3$ such that $n \geqslant a^2 S_n^2/d^2$. Then there is a finite positive integer k such that, for all θ, σ^2, and d,

$$\Pr\left\{|\bar{X}_{N+k} - \theta| < d\right\} \geqslant 1 - \alpha,$$

$$E(N + k) - a^2\sigma^2/d^2 \leqslant n_0 + k.$$

The cost of ignorance is thus bounded by $n_0 + k$ if we take at least n_0 observations initially, and k more observations after the stopping variable N has been observed.

Ray [10] and Starr [18] showed through numerical computations that the procedure above is reasonably consistent and efficient in the sense of (5) and (6) for all σ, even when $n_0 \geqslant 2$ and $k = 0$. Starr has plotted the efficiency and exact coverage probability for $\alpha = 0.05, 0.01$; $n_0 = 3, 5$; $k = 0, 2, 4$.

Without the coverage probability condition in Theorem 2, the cost of ignorance cannot be properly assessed. It is possible to improve on (6), however (see Simons [16]), when $X_1, X_2, \ldots$ comes from any unknown continuous distribution. Let N be the smallest integer $n \geqslant n_0 \geqslant 2$ such that $n \geqslant a^2 S_n^2/d^2$. Then (5) holds, and for all d and α,

$$E(N) - a^2\sigma^2/d^2 \leqslant n_0 + 1.$$

DIFFERENCE OF TWO MEANS (*see* BEHRENS–FISHER PROBLEM)

Let $X_1, X_2, \ldots$ and $Y_1, Y_2, \ldots$ be independent random sequences from continuous distributions having means and variances (θ_1, σ_1^2) and (θ_2, σ_2^2), respectively, where all parameters are unknown. Let

$$S_{X;m}^2 = (m-1)^{-1} \sum_{i=1}^{m} (X_i - \overline{X}_m)^2,$$

$$S_{Y;m}^2 = (m-1)^{-1} \sum_{i=1}^{m} (Y_i - \overline{Y}_m)^2.$$

Robbins et al. [11] derived an asymptotic sequential fixed-width CI for $\theta_1 - \theta_2$ which, in addition to a stopping variable, requires the following sampling rule; take n_0 observations initially in each population ($n_0 \geqslant 2$); if at any stage we have taken l (respectively, m) observations in the X (respectively, Y) population, the next observation is taken in the X (respectively, Y) population according as

$$l/m \leqslant (\text{respectively}, >) S_{X;l}/S_{Y;m}.$$

Let $c_1, c_2, \ldots$ be a sequence of positive constants such that $c_n \to a$ as $n \to \infty$, and if $n = l + m$, let N be the first value of n such that $n \geqslant c_n^2(S_{X;l} + S_{Y;m})^2/d^2$.

Use of the sampling rule leads to almost sure convergence of l/m to σ_1/σ_2 as $l + m \to \infty$. Analogous to (4), (5), and (6),

$$\lim_{d\to 0} \left(d^2N/\{a^2(\sigma_1+\sigma_2)^2\} \right) = 1 \qquad \text{almost surely,}$$

$$\lim_{d\to 0} \left(d^2E(N)/\{a^2(\sigma_1+\sigma_2)^2\} \right) = 1,$$

and if $I_n = (\overline{X}_l - \overline{Y}_m - d, \overline{X}_l - \overline{Y}_m + d)$ when $N = n$,

$$\lim_{d\to 0} \Pr(\theta_1 - \theta_2 \text{ lies in } I_N) = 1 - \alpha.$$

BOUNDED-LENGTH CIs FOR THE MEDIAN OF A SYMMETRIC POPULATION

Let θ be the center of location of a continuous symmetric population with CDF $F(x)$ and probability density function (PDF) $f(x)$. Asymptotic bounded-length CIs for θ have been derived:

1. By Geertsema [4], whose procedures are based on the sign test* and on the Wilcoxon signed-rank test*.
2. By Sen and Ghosh [14], who made use of properties of a general class of rank statistics, including the one-sample normal scores statistic.
3. By Steyn and Geertsema [20], who also attain a prescribed coverage probability in the sense of (2b).

The Sen–Ghosh class of rules is asymptotically as efficient (as $d \to 0$) as the Chow–Robbins rule of Theorem 1 for a broad class of CDFs.

Geertsema's asymptotic bounded-length CI for θ, with the order statistic notation presented earlier, is as follows [4]. Let

$$b(n) = \max\left\{1, \left[\tfrac{1}{2}n - \tfrac{1}{2}a\sqrt{n}\right]\right\},$$

$$c(n) = n - b(n) + 1;$$

$n = 2, 3, \ldots$; and let N be the first integer $n \geqslant n_0$ (n_0 fixed) such that

$$X_{c(n):n} - X_{b(n):n} \leqslant 2d.$$

Choose $(X_{b(n):n}, X_{c(n):n})$ to determine the resulting CI for θ, when $N = n$. Then

$$\lim_{d\to 0} \Pr(X_{b(N):N} \leqslant \theta \leqslant X_{c(N):N}) = 1 - \alpha,$$

$$\lim_{d\to 0} \{d^2E(N)\} = a^2/\left(4\{f(\theta)\}^2\right).$$

ANOTHER ASYMPTOTIC APPROACH

Serfling and Wackerly [15] have developed methods to derive fixed-width CIs which are efficient as the coverage probability $1 - \alpha \to 1$ (rather than as $d \to 0$). While the Chow–Robbins approach of Theorem 1 makes use of central limit theory, the theory of large deviations* is required here.

Using the notation of the preceding section, let I_n be a CI for a parameter θ, based on $X_1, \ldots, X_n (n \geqslant n')$. Then $I_{n'}, I_{n'+1}, \ldots$ is a $2d$-width sequence of CIs if the length $2d_n$ of I_n converges almost surely to $2d$ as $n \to \infty$, and d does not depend on θ. Such sequences are of interest if the coverage probability $1 - \alpha_n$ of I_n converges to 1 as $n \to \infty$; typically, convergence is exponential, with index of exponential convergence $e(d)$, given by

$$e(d) = -\lim_{n\to\infty} \{n^{-1}\log(\alpha_n/2)\}.$$

Given α, d, and all parameters other than θ, then for a given $2d$-width sequence, one could select each interval I_n with the smallest value of n such that $\Pr(I_n \text{ contains } \theta) \geqslant 1 - \alpha$. Call this $n(\alpha, d)$, so that for the sequence $(\overline{X}_n - d, \overline{X}_n + d)$ in the normal case with known variance, $n(\alpha, d) = [a^2\sigma^2/d^2] + 1$. Again, a sequential procedure will be necessary in practice. A "good" stopping variable $N(\alpha, d)$ would require one or both of the properties that

$$N(\alpha,d)/n(\alpha,d) \to 1 \text{ in probability as } \alpha \to 0, \tag{8}$$

$$\lim_{d\to 0} \{EN(\alpha,d)/n(\alpha,d)\} = 1. \tag{9}$$

The following result gives a rule for deriving fixed-width CIs for the median θ of F. Let $\tilde{X}_n$ denote the sample median at stage n, equal to $X_{(n+1)/2,n}$ if n is odd, and to $\frac{1}{2}(X_{n/2,n} + X_{(n+1)/2,n})$ if n is even; and let $I_n = (\tilde{X}_n - d, \tilde{X}_n + d)$. Then the index of exponential convergence $e(d)$ is given by

$$e(d) = -\log\left(2\{F(d) - F^2(d)\}^{1/2}\right).$$

Theorem 3 (Serfling–Wackerly [15]). If F_n is the empirical (sample) CDF at stage n, let $N(\alpha, d)$ be the smallest integer n such that $F_n(\tilde{X}_n + d) < 1$ and

$$n \geqslant \log\alpha/\log\left(2\{F_n(\tilde{X}_n + d) - F_n^2(\tilde{X}_n + d)\}^{1/2}\right).$$

Then, (8) and (9) hold, and for all θ,

$$\lim_{\alpha\to 0} \{N(\alpha,d)/n(\alpha,d)\} = 1 \quad \text{almost surely.}$$

Remark. Geertsema's sign-test rule [4], discussed earlier, does not satisfy (8) or (9). It is not known if the rule just given satisfies (4), (5), or (6); the underlying criteria are essentially different. In ref. 15, a rule based on the sample mean $\overline{X}_n$ is developed, which satisfies (8) and (9). Carroll [1] has studied an approach in which α and d converge to zero together, and has extended the theory given above to intervals based on M-estimators*, which give results superior to those based on sample means or sample medians.

DEVELOPMENTS

It is to be hoped that the methods above will become more well known, both by professional statisticians and analysts using statistical methods in real-world problems. This might lead to more widespread use of fixed-width and bounded-length CIs in practice.

Further work in estimation of other than location parameters may be done. A start has been made on simultaneous estimation of several parameters with *confidence regions of fixed size*; Jones [7] has derived asymptotic confidence regions for a multivariate population mean vector, when the covariance matrix is known except for a constant of multiplication, and an application to multiple comparisons* of several treatments with a control is included.

References

[1] Carroll, R. J. (1977). *J. Amer. Statist. Ass.*, **72**, 901–907.

[2] Chow, Y. S. and Robbins, H. (1965). *Ann. Math. Statist.*, **36**, 457–462. (A mathematical proof of the result in Theorem 1.)

[3] Fishman, G. S. (1977). *Commun. ACM*, **20**, 310–315.

[4] Geertsema, J. C. (1970). *Ann. Math. Statist.*, **41**, 1016–1026.

[5] Ghosh, B. K. (1975). *J. Amer. Statist. Ass.*, **70**, 457–462.

[6] Ghosh, M. (1975). *Ann. Inst. Statist. Math.*, **27**, 57–68.

[7] Jones, E. R. (1977). *Commun. Statist. A*, **6**, 251–264.

[8] Khan, R. A. (1969). *Ann. Math. Statist.*, **40**, 704–709.

[9] Moshman, J. (1958). *Ann. Math. Statist.*, **29**, 1271–1275.

[10] Ray, W. D. (1957). *J. R. Statist. Soc. B*, **19**, 133–143.

[11] Robbins, H., Simons, G., and Starr, N. (1967). *Ann. Math. Statist.*, **38**, 1384–1391.

[12] Schmidt, B., Cornée, J., and Delachaume-Salem, E. (1970). *C. R. Séances Soc. Biol. Filiales*, **164**, 1813–1818. (See the section above, "Asymptotic Fixed-Width CIs for the Mean.")

[13] Seeberger, J. J. and Wierwille, W. W. (1976). *Hum. Factors*, **18**, 281–292. (An example of Stein's Rule [19].)

[14] Sen, P. K. and Ghosh, M. (1971). *Ann. Math. Statist.*, **42**, 189–203.

[15] Serfling, R. J. and Wackerly, D. D. (1976). *J. Amer. Statist. Ass.*, **71**, 949–955. (See the discussion above in the section "Another Asymptotic Approach.")

[16] Simons, G. (1968). *Ann. Math. Statist.*, **39**, 1946–1952.

[17] Srivastava, M. S. (1971). *Ann. Math. Statist.*, **42**, 1403–1411.

[18] Starr, N. (1966). *Ann. Math. Statist.*, **37**, 36–50.

[19] Stein, C. (1945). *Ann. Math. Statist.*, **16**, 243–258. (This is the original presentation of Stein's rule.)

[20] Steyn, H. S. and Geertsema, J. C. (1974). *S. Afr. Statist. J.*, **8**, 25–34.

[21] Weiss, L. (1960). *Naval Res. Logist. Quart.*, **7**, 251–256.

[22] Zacks, S. (1971). *The Theory of Statistical Inference*. Wiley, New York. (An advanced mathematical treatment in Chap. 10 of the procedures described in the sections on asymptotic fixed-width CIs, two-stage rules, and the cost of ignorance.)

(CONFIDENCE INTERVALS AND REGIONS
SEQUENTIAL ANALYSIS
STOPPING RULES)

CAMPBELL B. READ

FIX-POINT METHOD

In simultaneous equations systems* the structural form (SF) and the reduced form (RF) cannot be jointly specified as predictive in the sense of conditional expectations. The fix-point (FP) approach removes this inadequacy by a stochastic reformulation of the SF, without change of the parameters, replacing the explanatory endogenous variables by their conditional expectations. In the reformulated SF both the reformulated variables and their coefficients are unknown; the FP method solves this estimation problem by an iterative sequence of ordinary least-squares* (OLS) regressions*.

Jan Tinbergen's macroeconomic models (1935–1939; see ref. 14) and Trygve Haavelmo's simultaneous equations [8] mark a parting of the ways in the passage from uni- to multirelational models. Dismissing OLS regression as inconsistent, Haavelmo recommended that the parameters of simultaneous equations should be estimated by maximum likelihood* (ML) methods. In the late 1950s the growing size of the macroeconomic models made the ML approach intractable, and reinforced the need for new estimation techniques. The two-stage least-squares* (TSLS) method of Theil [12, 13] and Basmann [2] sacrificed ML optimality aspirations in favor of LS consistency. While simultaneous equations applications prospered under the ever-increasing demand for macroeconomic models, the statistical rationale was subject to serious questioning; see Christ et al. [6]. In the classical ML approach, as emphasized by Wold [17], the transformation from structural form (SF) to reduced form (RF) is made possible by the stringent assumption that the SF relations are deterministic and subject to superimposed "errors." Instead, I argued for a stochastic definition of SF and RF equations in terms of conditional expectations. This last argument was pursued [16, 18, 19] and led to the fix-point (FP) method. The subsequent exposition draws from Wold [22], and in the section "Applications: Comparative Studies" also from Romański and Welfe [11] and Bergström and Wold [4].

SIMULTANEOUS EQUATIONS

A simultaneous equations model is conceptually defined by its SF, while its RF predicts the endogenous variables in terms of the predetermined variables.

Structural Form

The structural form (SF) is a system of n relations,

$$\mathbf{y}_t = \boldsymbol{\beta}\mathbf{y}_t + \boldsymbol{\Gamma}\mathbf{z}_t + \boldsymbol{\delta}_t, \qquad (1)$$

with:

1. variables $\mathbf{y}$, $\mathbf{z}$ observed over time $t = 1, \ldots, T$;
2. n endogenous variables $\mathbf{y}$, $\mathbf{y} = (y_1, \ldots, y_n)$;
3. predetermined variables $\mathbf{z} = (z_1, \ldots, z_m)$, which are either exogenous variables $x_1, \ldots, x_m$ or lagged endogenous variables $y_{i,t-k}$;
4. parameters $\boldsymbol{\beta} = [\beta_{ij}]$, $\boldsymbol{\Gamma} = [\gamma_{ik}]$; $i, j = 1, \ldots, n$; $k = 1, \ldots, m$, where the matrix $\boldsymbol{\beta}$ has zero diagonal, $\beta_{ii} = 0$; $i = 1, \ldots, n$;
5. n residuals $\boldsymbol{\delta} = (\delta_1, \ldots, \delta_n)$.

For the ith relation of the SF we use two equivalent notations:

$$y_{it} = \boldsymbol{\beta}_i \mathbf{y}_t + \gamma_i \mathbf{z}_t + \delta_{it}, \quad (2)$$

$$y_i = \boldsymbol{\beta}_{(i)} \mathbf{y}_{(i)} + \gamma_{(i)} \mathbf{z}_{(i)} + \delta_i, \quad (3)$$

where $\mathbf{y}_{(i)}$ are the endogenous variables with nonzero coefficients $\boldsymbol{\beta}_{(i)}$ in the ith relation, and similarly for $\mathbf{z}_{(i)}$. The SF may involve identities (i.e., equations without residuals) with prespecified parameters $\boldsymbol{\beta}, \boldsymbol{\Gamma}$.

Reduced Form

The reduced form (RF) is obtained by solving the SF for the endogenous variables:

$$\mathbf{y}_t = \boldsymbol{\Omega} z_t + \boldsymbol{\epsilon}_t \quad (4)$$

with parameters and residuals given by

$$\boldsymbol{\Omega} = [\omega_{ij}] = [\mathbf{I} - \boldsymbol{\beta}]^{-1} \boldsymbol{\Gamma}; \quad (5a)$$

$$\boldsymbol{\epsilon}_t = [\mathbf{I} - \boldsymbol{\beta}]^{-1} \boldsymbol{\delta}. \quad (5b)$$

Let $\boldsymbol{\eta}^*$ be the conditional expectation of $\mathbf{y}$ defined by

$$\boldsymbol{\eta}^* = E(\mathbf{y} \mid \mathbf{z}); \qquad \eta^*_{it} = E(y_{it} \mid z_{1t}, \ldots, z_{mt});$$
$$i = 1, \ldots, n; t = 1, \ldots, T. \quad (6)$$

The RF allows the equivalent representation

$$\mathbf{y}_t = E(\mathbf{y}_t \mid \mathbf{z}_t) + \boldsymbol{\epsilon}_t = \boldsymbol{\eta}^*_t + \boldsymbol{\epsilon}_t$$
$$= \sum_{j=1}^{m} (\omega_{ij} z_{jt}) + \epsilon_{it}; \qquad i = 1, \ldots, n. \quad (7)$$

Predictor Specification [18]

To impose this on SF (2) is to assume, in notation (3),

$$E(y_i \mid \mathbf{y}_{(i)}, \mathbf{z}_{(i)}) = \boldsymbol{\beta}_{(i)} \mathbf{y}_{(i)} + \gamma_{(i)} \mathbf{z}_{(i)};$$
$$i = 1, \ldots, n. \quad (8)$$

The predictor specification implies [22, p. 13]:

(a) $$E(\delta_i) = 0. \quad (9a)$$

(b) $$r(\delta_i, \mathbf{y}_{(i)}) = r(\delta_i, \mathbf{z}_{(i)}) = 0. \quad (9b)$$

(c) Under mild supplementary conditions the OLS estimates of $\boldsymbol{\beta}_{(i)}, \gamma_{(i)}$ are consistent in the large-sample sense.

CAUSAL CHAIN (CCh) AND INTERDEPENDENT (ID) SYSTEMS

The simultaneous equations (1) are a CCh system if, after a suitable reordering of SF, the matrix $\boldsymbol{\beta}$ is subdiagonal:

$$\beta_{ij} = 0 \quad \text{when} \quad j > i; \quad i = 1, \ldots, n. \quad (10)$$

If not a CCh system, (1) is an ID system.

CCh Systems

In (1) with (10), both SF and RF allow predictor specification. Hence (8) holds, and

$$E(\mathbf{y}_t \mid \mathbf{z}_t) = E(\mathbf{y}_t \mid z_{1t}, \ldots, z_{mt}) = \boldsymbol{\eta}^*_t. \quad (11)$$

By (8) and (11), the OLS regression estimates of $\boldsymbol{\beta}$, $\boldsymbol{\Gamma}$, and $\boldsymbol{\Omega}$, say $\mathbf{B}$, $\mathbf{G}$, and $\mathbf{W}$, are consistent. In practice, OLS regressions of SF give $\mathbf{B}$ and $\mathbf{G}$, whereas $\mathbf{W}$ is obtained from $\mathbf{B}$ and $\mathbf{G}$, using (5a):

$$\mathbf{W} = [\mathbf{I} - \mathbf{B}]^{-1} \mathbf{G} = [w_{ij}]. \quad (12)$$

Classical ID Systems

It is assumed that the residuals δ_{it} of SF (1) are (a) independent of the predetermined variables z, and (b) independent multivariate normal observations. The SF (1) and RF (4) of a classical ID system do not permit joint predictor specification. For RF we adopt predictor specification, in accordance with its intended operative use. Then OLS is inconsistent if applied to the SF with the classical ID assumptions (see Haavelmo [8]).

Since the SF does not allow predictor specification, the behavioral relations cannot be interpreted as conditional expectations. What then is their operative interpretation and use, if any? A positive answer will be given by the following SF reformulation.

REID (Reformulated ID) Systems

The SF of the simultaneous equations, given by (3), can be rewritten as follows, with $\boldsymbol{\eta}^*$ defined as in (6):

$$y_{it} = \boldsymbol{\beta}_{(i)}\boldsymbol{\eta}^*_{(i)t} + \boldsymbol{\gamma}_{(i)}\mathbf{z}_{(i)t} + \epsilon_{it}\,. \tag{13}$$

The reformulation of $\boldsymbol{\epsilon}_t$ does not change the parameters $\boldsymbol{\beta}_{(i)}, \boldsymbol{\gamma}_{(i)}$ in (13), and is the same as in (4) and (5b):

$$\boldsymbol{\beta}[\mathbf{y}_t - \boldsymbol{\eta}_t^*] + \boldsymbol{\delta}_t = \boldsymbol{\beta}\boldsymbol{\epsilon}_t + [\mathbf{I} - \boldsymbol{\beta}]\boldsymbol{\epsilon}_t = \boldsymbol{\epsilon}_t\,. \tag{14}$$

The RF is the same before and after the reformulation:

$$y_i = \eta_i^* + \epsilon_i = E(y_i \mid z_1, \ldots, z_m) + \epsilon_i \tag{15}$$

$$= \sum_{j=1}^{m} (w_{ij} z_j) + \epsilon_i\,. \tag{16}$$

With assumption (6), both the RF and the reformulated SF allow predictor specification,

$$\text{SF:}\quad E(y_i \mid \boldsymbol{\eta}^*_{(i)}, \mathbf{z}_{(i)}) = \boldsymbol{\beta}_{(i)}\boldsymbol{\eta}^*_{(i)} + \boldsymbol{\gamma}_{(i)}\mathbf{z}_{(i)}\,; \tag{17}$$

$$\text{RF:}\quad E(y_t \mid z_1, \ldots, z_m) = \sum_{j=1}^{m} (\omega_{ij} z_i). \tag{18}$$

If the SF relations contain identities, the reformulated identities will have residuals. On predictor specification (17) the REID system (13) remains formally the same, with or without identities [22, p. 18]. Hence the residuals of the reformulated identities will not influence the FP algorithm.

The reformulation (13) provides causal-predictive interpretation of the SF behavioral relations: each endogenous variable y_i is explained by predetermined variables $\mathbf{z}_{(i)}$ and *expected values* $\boldsymbol{\eta}^*_{(i)}$ of other endogenous variables. At the same time, thanks to (17), (13) implies that consistent estimates $\mathbf{B}$, $\mathbf{G}$ of the SF parameters are provided by the OLS regressions of y_i on $\boldsymbol{\eta}^*_{(i)}$ and $\mathbf{z}_{(i)}$:

$$y_i = \mathbf{b}_{(i)}\boldsymbol{\eta}^*_{(i)} + \mathbf{g}_{(i)}\mathbf{z}_{(i)} + e_i\,; \qquad i = 1, \ldots, n. \tag{19}$$

There is the snag that the $\boldsymbol{\eta}^*_{(i)}$'s are not observed. The fix-point method meets this difficulty by an iterative procedure, say with steps $s = 1, 2, \ldots$. If $\hat{\boldsymbol{\eta}}^*_{(i)}$ is the FP estimate of $\boldsymbol{\eta}^*_{(i)}$, step s gives a proxy (provisional estimate) $y^{(s)}_{(i)}$ for $\hat{\boldsymbol{\eta}}^*_{(i)}$, and in step $s+1$ the proxy parameter estimates $b_i^{(s+1)}$, $g_i^{(s+1)}$ are computed by the multiple OLS regressions:

$$y_i = \mathbf{b}^{(s+1)}_{(i)}\mathbf{y}^{(s)}_{(i)} + \mathbf{g}^{(s+1)}_{(i)}\mathbf{z}_{(i)} + \mathbf{e}_i^{(s+1)}. \tag{20}$$

The classical ID assumptions are more stringent than necessary for the FP iteration (20). What emerges here is the following generalization of REID systems.

GEID (GENERAL ID) SYSTEMS

Given the simultaneous equations (1), let (13) define the reformulated SF. Instead of the classical ID assumptions, we impose only the predictor specification (17). As to the RF, we suspend the predictor specification (18), and use (5a) to obtain $\boldsymbol{\Omega}$ from $\boldsymbol{\beta}$ and $\boldsymbol{\Gamma}$; accordingly,

$$E(\mathbf{y}_t \mid \boldsymbol{\Omega}\mathbf{z}_t) = \boldsymbol{\Omega}\mathbf{z}_t\,. \tag{21}$$

The predictor specification (17) implies as many zero residual correlations (9b) as there are explanatory variables. This is the *parity principle* of GEID systems. For corresponding classical ID systems the residual correlations (9b) are more numerous, as we now illustrate.

One-Loop Model

In equations (1) the *position matrices* $\boldsymbol{\beta}^*$, $\boldsymbol{\Gamma}^*$ serve to illustrate the location of nonzero coefficients β_{ij}, γ_{ik}. We consider the one-loop model with position matrices

$$\boldsymbol{\beta}^* = \begin{bmatrix} 0 & 0 & 0 & \cdots & 0 & 0 & * \\ * & 0 & 0 & \cdots & 0 & 0 & 0 \\ 0 & * & 0 & \cdots & 0 & 0 & 0 \\ \cdot & \cdot & \cdot & \cdots & \cdot & \cdot & \cdot \\ 0 & 0 & 0 & \cdots & * & 0 & 0 \\ 0 & 0 & 0 & \cdots & 0 & * & 0 \end{bmatrix};$$

$$\boldsymbol{\Gamma}^* = \begin{bmatrix} * & 0 & 0 & \cdots & 0 & 0 \\ 0 & * & 0 & \cdots & 0 & 0 \\ 0 & 0 & * & \cdots & 0 & 0 \\ \cdot & \cdot & \cdot & \cdots & \cdot & \cdot \\ 0 & 0 & 0 & \cdots & * & 0 \\ 0 & 0 & 0 & \cdots & 0 & * \end{bmatrix}. \tag{22}$$

The SF of the one-loop model reads, with $i = 2, \ldots, n$:

$$\begin{aligned} y_1 &= \beta_{1n} y_n + \gamma_1 z_1 + \delta_1, \\ y_i &= \beta_{i,i-1} y_{i-1} + \gamma_i z_i + \delta_i. \end{aligned} \tag{23}$$

If (23) is a classical ID system, the corresponding REID system replaces the right-hand variables $\mathbf{y}$ by $\boldsymbol{\eta}^*$. The REID-GEID one-loop model for $i = 2, \ldots, n$ becomes:

$$y_i = \beta_{i,i-1} \eta^*_{i-1} + \gamma_i z_i + \epsilon_i. \tag{24}$$

The one-loop model with n relations has $2n$ parameters $\boldsymbol{\beta}$, $\boldsymbol{\Gamma}$. The classical ID specification implies n^2 zero correlations (9b), against only $2n$ in the GEID specification.

GEID Correlations

The classical ID and REID models impose more zero correlations (9b), often many more than the GEID model. The redundant correlations are *GEID correlations*; clearly, nonzero GEID correlations reduce the SF residual variances, thereby improving the predictive performance of the GEID model. GEID correlations are a source of inconsistency in LIML*, TSLS*, and other estimation methods for classical ID systems. The fix-point algorithm is the only known operative method for consistent estimation of simultaneous equations in the presence of GEID correlations. The FP algorithm is full-information oriented in that estimates of the GEID correlations are obtained, without need to specify their presence in advance.

BASIC FIX-POINT (FP) ALGORITHM

Linking up with (20) for the parameter proxies in step $s + 1$, it remains to indicate (a) the start, $s = 1$, and (b) the proxy for $\hat{\eta}^*_{it}$ in step $s + 1$.

Start, $s = 1$

The FP algorithm starts with a proxy $y^{(1)}_{il}$ for $\hat{\eta}^*_{it}$. The following three alternatives are in frequent use:

$$y \text{ start: } y^{(1)}_{it} = y_{it}, \qquad (i = 1, \ldots, n; \; t = 1, \ldots, T);$$

$$\text{zero start: } y^{(1)}_{it} = 0;$$

$$\text{TSLS start: } y^{(1)}_{it} = \text{the systematic part of } y_{it} \text{ in the OLS regression (4).}$$

In not too large models the choice of start usually does not matter; see May [10]. To explore whether the data are sensitive to the start, it is advisable to run the algorithm twice, say with zero start and y start.

Proxy for $\hat{\eta}^*_{it}$ in Step $s + 1 : y^{(s+1)}_{it}$

Having computed the parameter estimates $b^{(s+1)}_{ij}$, $g^{(s+1)}_{ik}$ by (20), the proxy $y^{(s+1)}_{it}$ can be computed from the SF (13) or from the RF (15).

SF computation: Dropping the residual in (20) gives

$$y_{it} = \sum_{j=1}^{n} \left(b^{(s+1)}_{ij} y^{(s)}_{jt}\right) + \sum_{k=1}^{m} \left(g^{(s+1)}_{ik} z_{kt}\right). \tag{25}$$

RF computation: The RF (15) gives, using (5a):

$$\begin{aligned} \mathbf{y}^{(s+1)}_t &= \mathbf{W}^{(s+1)} \mathbf{z}_t; \\ \mathbf{W}^{(s+1)} &= \left[\mathbf{I} - \mathbf{B}^{(s+1)}\right]^{-1} \mathbf{G}^{(s+1)}. \end{aligned} \tag{26}$$

Passage to the Limit, $s \to \infty$

The superscripts are dropped to denote the limiting PLS estimates:

$$\mathbf{B} = \hat{\boldsymbol{\beta}} = \lim_{s\to\infty} \mathbf{B}^{(s)}; \qquad y_{it}^* = \hat{\eta}_{it}^* = \lim_{s\to\infty} y_{it}^{(s)}; \tag{27}$$

similarly for **G**, **W**, b_{ij}, g_{ik}, w_{ik}, and e_{it}. In regular situations the SF and RF versions (25 and 26) give the same limiting estimates, apart from rounding errors.

The basic FP algorithm has been improved by a family of modified versions, which in regular situations give the same estimates. Some applications will be reported before turning to these and other technical aspects of FP estimation.

APPLICATIONS: COMPARATIVE STUDIES

When comparing the performance of different estimation methods we must distinguish between prediction accuracy and parameter accuracy, as is well known; between predictions with targets inside the period of observation vs. beyond the period (prediction ex post vs. ex ante observing the target); and between comparisons based on real-world models and data vs. simulated data generated by real-world or fictional models. Prediction performance can be directly compared and ranked, since the target of each prediction is directly observable. In joint prediction of n variables y_i over T time points, say with prediction errors e_{it}, Ball's Q^2 is a standard criterion of prediction performance:

$$Q_i^2 = 1 - \sum_{t=1}^{T} (e_{it}^2) / \sum_t (y_{it} - \bar{y}_i)^2; \qquad Q^2 = \frac{1}{n} \sum_{i=1}^{n} Q_i^2. \tag{28}$$

Generally, as we shall illustrate, ML estimation is parameter oriented; FP and other LS methods are prediction oriented. With the classical assumptions of controlled experiments the ML and LS estimates of linear relations are numerically the same. In nonexperimental data there is an either–or: optimal prediction is not parameter optimal, and conversely.

Prediction Performance

Bergström [3] reports applications of FP to seven models that have often been used for illustration and comparison of estimation methods. In prediction ex post, evaluated by Ball's Q^2, FP shows off as distinctly superior. The models of Girshick-Haavelmo (1947), Christ (1966), and Klein (1950) have from 5 to 8 endogenous variables and from 20 to 27 observations; the smallest possible rank sum is 3 and the largest is 24; small rank sums are obtained for FP (3) and IIV (6), and high ones for OLS (20), 3SLS (18), LIML (18), FIML (17), and TSLS (16); an intermediate rank sum is obtained for FIMD (10). For the models of Pavlopoulos (1966), Dutta-Su (1969), Yu (1970), and Suján (1972) the endogenous variables are more numerous, from 17 to 40, than the observations, from 11 to 17; hence the estimation can be performed only for four of the methods; now FP and FIMD give the smallest rank sum (7), and OLS and IIV the highest (13). Table 1 gives similar results.

In prediction ex ante FP loses much of its lead; see Bergström [3]. The comparisons are blurred here, since the estimation performance merges with the quality of the model.

Parameter Accuracy

Bergström [3] gives comparisons of parameter performance. In real-world models parameter accuracy is more diffuse to evaluate than prediction performance—the targets of prediction are observed to be ex ante or ex post, whereas the parameters subject to estimation are unobserved, hypothetical, and at best approximately the same as in applications of the same model to statistical observations elsewhere in time or space. Moreover, for prediction ex ante, the performance of parameter estimation is influenced by the quality of the model. Thus Bergström's comparisons of parameter estimates suggest that,

Table 1 Predictive Performance of Four Estimation Methods on a Polish Model (Observation Periods 1961–1978)[a,b]

	Mean Value of $100Q^2$				Average Absolute Error of Prediction			
	$UW = 0$		$UW = 1$		$UW = 0$		$UW = 1$	
Method	'61–'75	'76–'78	'61–'75	'76–'78	'61–'75	'76–'78	'61–'75	'76–'78
IIV	90.8	88.8	90.8	89.6	7.6	13.8	7.6	13.8
OLS	95.2	90.9	95.2	91.7	6.9	12.4	6.9	12.4
FIMD	95.3	91.0	95.3	91.7	6.9	12.8	6.9	12.8
FP	95.6	91.9	95.6	92.3	6.7	12.8	6.7	11.8

[a]The model has 18 observations, 16 endogenous and 18 predetermined variables, 16 behavioral equations, and 9 identities.
[b]UW, dummy variable for economic policy.
Source. Romański and Welfe [11].

if different methods in a model show large differences in the parameter estimates, this is an indication that the quality of the model is not good.

Simulation Experiments

Simulation experiments on estimation methods are suitable for comparing performances in parameter estimation, since the true parameters are known; Tables 2 and 3 report on the performance of FP under classical ID vs. GEID assumptions [4].

Table 2 reports simulation experiments on the one-loop model (24) for different loop lengths n, with just one exogenous variable z in each structural relation. The data are generated on the REID classical assumptions, with variables standardized to zero mean and unit variance, and with the same parameters $\beta = 0.6$, $\gamma = 1$ throughout. For each model size n and sample size N the experiment comprises RP = 100 replications.

The general counterbalance of prediction vs. parameter accuracy shows off clearly: the performance of FP estimation is superior in prediction, but inferior in parameter accuracy. Relative to TSLS and FIMD the differences are small, only one or two points in the third decimal of Ball's Q^2, and up to 15 points in the third decimal of the standard deviations of the parameter estimates. Results are given for loop models (24) with $n = 3$, 5, or 10 structural relations. We see that model size has little or no influence, either on prediction accuracy or on parameter accuracy.

The simulations were performed with $N = 20$, 40, or 100 observations; Table 2 gives results for $N = 100$. To summarize, FP and TSLS give much the same results, as do OLS and FIMD, whereas the differences are more pronounced between the two pairs of

Table 2 Performance of Four Estimation Methods on Simulated Data for the One-Loop Model (24) with n Structural Equations, One Exogenous Variable per Equation[a]

	$n = 3$			$n = 5$			$n = 10$		
Method	$100Q^2$	$s(b)$	$s(g)$	$100Q^2$	$s(b)$	$s(g)$	$100Q^2$	$s(b)$	$s(g)$
OLS	76.2	0.033	0.059	76.9	0.036	0.054	76.9	0.039	0.054
FIMD	77.1	0.034	0.059	77.0	0.036	0.054	76.9	0.039	0.054
TSLS	77.2	0.038	0.059	77.0	0.041	0.055	76.9	0.043	0.054
FP	77.3	0.039	0.066	77.2	0.041	0.068	77.1	0.044	0.068

[a]$T = 100$ observations, $RP = 100$ replications. Performance criteria: Ball's Q^2; $s(p)$, standard deviation of parameter estimate p.

Table 3 Simulation Experiments As in Table 2, with $n = 3$ Structural Relations and Two GEID Correlations

Method	N	$100Q^2$	N	Equation	$b_i - \beta_i$	$g_i - \gamma_i$	$s(b)$	$s(g)$
OLS	20	73.2	100	1	− 0.151	0.080	0.061	0.103
FIMD	20	71.8	100	1	− 0.207	0.107	0.066	0.105
TSLS	20	73.2	100	1	− 0.009	0.010	0.072	0.105
FP	20	76.6	100	1	− 0.007	0.010	0.067	0.092
OLS	40	73.4	100	2	− 0.144	0.261	0.050	0.096
FIMD	40	72.0	100	2	− 0.190	0.263	0.053	0.096
TSLS	40	73.4	100	2	− 0.064	0.259	0.056	0.097
FP	40	76.4	100	2	− 0.008	0.012	0.064	0.098
OLS	100	74.4	100	3	− 0.142	0.076	0.055	0.103
FIMD	100	72.8	100	3	− 0.196	0.102	0.061	0.103
TSLS	100	74.4	100	3	+ 0.008	0.007	0.067	0.111
FP	100	77.1	100	3	+ 0.007	0.011	0.066	0.088

methods. The prediction performance of FP and TSLS is nearly the same for $N = 100$, and much the same Q^2-values have been obtained for $N = 20$ and $N = 40$. As to parameter accuracy, the standard deviations decrease substantially in the passage from $N = 20$ to $N = 100$, by about 50% for the loop model with $n = 3$, and often more than 65% for $n = 10$. For $n = 3$ and $N = 20$, FIMD and OLS give nearly the same standard deviations, whereas the parameter accuracy of FP here is distinctly inferior.

Table 3 reports simulation experiments on the same model (24) as in Table 2, modified by introducing two GEID correlations:

$$r(\epsilon_1, z_2) = -0.5; \qquad r(\epsilon_1, z_3) = 0.3. \quad (29)$$

These moderate GEID correlations bring moderate inconsistencies into the classical estimation methods. FP is the only known method for consistent estimation of ID systems with GEID correlations.

As to prediction accuracy, Table 3 shows that the performance of FP has lower Q^2 scores than in Table 2, but its lead relative to the classical methods has increased, in accordance with the rationale of FP.

The parameter accuracy is more interesting. The FP estimates show off as unbiased, as expected; the small bias is just noise, about one point in the second decimal.

The GEID correlations are seen to make the classical estimation methods more or less inconsistent. For TSLS the correlations (29) bring no bias in g_1, substantial bias in g_2, and little or no bias in g_3. This accords with FP theory, the bias in g_3 being blurred by noise. The TSLS bias in g_2 brings bias in b_2, again in accordance with theory. As to OLS and FIMD, the GEID correlations bring considerable bias in all the coefficients b_i and g_i.

The standard deviations in Table 3 are computed relative to the average estimates, not the true parameters. For each parameter the standard deviations are nearly of the same order of magnitude for the four estimation methods. The standard deviations relative to the true parameters can readily be deduced from Table 3.

FP FAMILY OF METHODS

The basic FP algorithm has been generalized and improved. The FFP (fractional FP) method [1] and the RFP (recursive FP) method [5] extend and accelerate the FP convergence without changing the limiting FP parameter estimates.

FFP estimation uses the standard relaxation device, the increment of $y_{it}^{(s+1)}$ in step $(s, s+1)$ being cut in proportion α:

$$y_{it}^{(s+1)} = (1-\alpha) y_{it}^{(s)} + \alpha \left[\sum_{j=1}^{n} \left(b_{ij}^{(s+1)} y_{jt}^{(s)} \right) + \sum_{j=1}^{m} \left(y_{ij}^{(s+1)} z_{jt} \right) \right]; \quad i = 1, \ldots, n.$$

RFP estimation improves the convergence in two ways: (a) the SF relations are reordered by an ad hoc algorithm [5], so as to give the matrix $\boldsymbol{\beta}$ a minimum number of nonzero entries above the main diagonal; (b) the step $(s, s+1)$ having n substeps $i = 1, \ldots, n$, the proxies $y_t^{(s+1)}$, $h \leqslant i$, are utilized in the formula for $y_{i+1,t}^{(s+1)}$ [22, pp. 40–44].

Computer programs are available at nominal cost for the FP, FFP, and RFP methods, including Bodin's reordering algorithm [3]. Data input are either the raw data y_{it}, z_{it}, or the cross-products over t of the raw data.

AFP (Algebraic FP) Estimation [9]

The predictor specification (6) is interpreted as n linear constraints, and taken into account by Lagrange multipliers. The ensuing estimation has been performed for some simple models, but is not practicable for large systems.

Autocorrelated Errors

To reduce the standard errors of parameter estimates autoregressive transformation of the data has been implemented for FP estimation [3]. The ensuing prediction performance must be evaluated on the untransformed data.

Nonlinearities in the Endogenous Variables

Edgerton [7] linearizes the model by Taylor expansions around the conditional expectations (8).

Convergence of the FP Algorithm

Exact convergence conditions, local or global, are rarely available for nonlinear algorithms, and FP is no exception. Useful approximate conditions based on the matrix **B** are available [1]. In practice convergence has rarely been a problem. The fastest convergence has as a rule been obtained by RFP estimation.

Robustness

Simulation experiments on robustness* in prediction and parameter performance show that FP on the whole is well on par with the classical estimation methods [15], supporting FP further as a prediction-oriented approach with consistent parameter performance.

References

[1] Ågren, A. (1980). In *The Fix-Point Approach to Interdependent Systems*, H. Wold, ed. North-Holland, Amsterdam, pp. 65–76.

[2] Basmann, R. L. (1957). *Econometrica*, **25**, 77–83.

[3] Bergström, R. (1980). In *The Fix-Point Approach to Interdependent Systems*, H. Wold, ed. North-Holland, Amsterdam, pp. 109–158, 197–224.

[4] Bergström, R. and Wold, H. (1982). *The Fix-Point Method in Theory and Practice*, Vol. 22, *Applied Statistics and Econometrics*, eds. G. Tintner, H. Strecker, and E. Féron, Göttingen: Vandenhoeck and Ruprecht. (About 100 pages.)

[5] Bodin, L. (1980). In *The Fix-Point Approach to Interdependent Systems*, H. Wold, ed. North-Holland, Amsterdam, pp. 37–64, 225–242.

[6] Christ, C., Hildreth, C., and Liu, T.-Ch. (1960). *Econometrica*, **28**, 835–865.

[7] Edgerton, D. (1980). In *The Fix-Point Approach to Interdependent Systems*, H. Wold, ed. North-Holland, Amsterdam, pp. 243–282.

[8] Haavelmo, T. (1943). *Econometrica*, **11**, 1–12.

[9] Lyttkens, E. (1973). *J. R. Statist. Soc. A*, **136**, 353–394.

[10] May, S. (1980). In *The Fix-Point Approach to Interdependent Systems*, H. Wold, ed. North-Holland, Amsterdam, pp. 303–318.

[11] Romański, J. and Welfe, W. (1980). On forecasting efficiency of different estimation methods for interdependent systems. *4th World Congr. Econometric Soc.*, Aix-en-Provence, France, Aug. 28–Sept. 2, 1980.

[12] Theil, H. (1953). Estimation and simultaneous correlation in complete equation systems. Central Planning Bureau, The Hague (mimeo).

[13] Theil, H. (1958). *Economic Forecasts and Policy*. North-Holland, Amsterdam.

[14] Tinbergen, J. (1939). *Statistical Testing of Business Cycle Theories*, Vol. 2: *Business Cycles in the United States of America*: *1919–1932*. United Nations, Geneva.

[15] Westlund, A. (1980). In *The Fix-Point Approach to Interdependent Systems*, H. Wold, ed. North-Holland, Amsterdam, pp. 283–302.

[16] Wold, H. (1959/60). In *Probability and Statistics: The Herald Cramér Volume*, U. Grenander, ed. (Almqvist & Wiksell, Uppsala, 1959; Wiley, New York, 1960), pp. 355–434.

[17] Wold, H. (1960). *Econometrica*, **28**, 443–463.

[18] Wold, H. (1963). *Sankhyā A*, **25**(2), 211–215.

[19] Wold, H. (1963). *Scr. Varia*, **28**, 115–166 (Pontifical Academy of Science, Vatican City, 1965).

[20] Wold, H. (1965). *Arkiv för Matematik*, **6**, 209–240.

[21] Wold, H. (1966). In *Research Papers in Statistics: Festschrift for J. Neyman*, F. N. David, ed. Wiley, New York, pp. 411–444.

[22] Wold, H., ed. (1980). *The Fix-Point Approach to Interdependent Systems*. North-Holland, Amsterdam. (Detailed account of the rationale, technique, and evolution of FP estimation.)

Bibliography

See the following works, as well as the references just cited, for more information on the topic of the fix-point method.

Mosbaek, E. J. and Wold, H., with contributions by E. Lyttkens, A. Ågren, and L. Bodin (1970). *Interdependent Systems: Structure and Estimation*. North-Holland, Amsterdam.

Wold, H. (1969). *Synthèse*, **20**, 427–482. (Discourse on the definition and meaning of causal concepts; see also Chap. 4 in Mosbaek and Wold (1970).)

(LEAST SQUARES
MAXIMUM LIKELIHOOD
PREDICTION
SIMULTANEOUS EQUATIONS SYSTEMS
TWO-STAGE LEAST SQUARES)

HERMAN WOLD

FLOWGRAPH ANALYSIS

A *flowgraph* is a graphical representation which simultaneously displays all the relationships (equations) among the variables of a given system. Basically, a flowgraph is a way of writing linear algebraic equations. The flowgraph may be solved for the variables in the system (represented by nodes) in terms of the relationships that exist between the variables (represented by branches and called *transmittances*). A flowgraph then

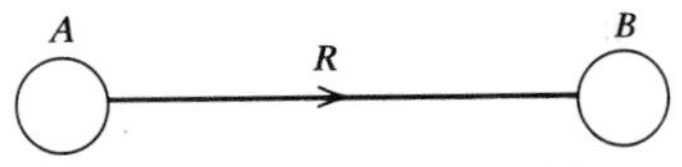

Figure 1 Simple flowgraph element.

consists of a set of nodes and branches, where the nodes represent variables and the branches indicate that a relationship exists between the nodes the branches connect.

Consider the simple flowgraph in Fig. 1, which expresses the relationship $B = RA$. In this flowgraph A and B are nodes; R is a transmittance. The arrow's direction is important because it shows that A is an independent variable while node B is a dependent variable. The transmittance R expresses the relationship between A and B.

GERT (Graphical Evaluation Review Technique) is an extension of flowgraph analysis to solve stochastic problems. For a complete development of the GERT technique, see Whitehouse [2]. A node in a GERT network consists of an input side and output side. The three logical relationships on the input side are given in Table 1a. On the output side, the two relationships are defined in Table 1b.

Associated with a GERT transmittance are two parameters. Each transmittance has an assigned probability conditional upon the realization of the node from which it emanates, and an assigned time conditioned on the selection of the transmittance. To include both parameters in a transformation function which could be treated uniformly throughout a network requires the creation of a new function. The originators of GERT choose to define a w-function, $w(s)$, as the product of the conditional probability p of selecting the branch, and the conditional moment generating function* (MGF) of the time to traverse the branch, $M(s)$. The nature of flowgraph arithmetic requires that the function $w(s)$ be a multiplicative variable. Since probabilities may be multiplied, they have been incorporated into the function $w(s)$ without any modification. Time, however, is an additive quantity, so that some conversion was necessary to incorporate it into the multiplicative transmittance.

Table 1. Logical Relationships
a. Input side

Name	Symbol	Characteristic
Exclusive-or	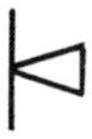	The realization of any branch leading into the node causes the node to be realized; however, one and only one of the branches leading into this node can be realized.
Inclusive-or		The realization of any branch leading into the node causes the node to be realized.
And		The node will be realized only if all the branches leading into the node are realized.

b. Output side

Name	Symbol	Characteristic
Deterministic		All branches emanating from the node are taken if the node is realized, i.e., all branches emanating from this node have a probability parameter equal to 1.
Probabilistic		At most, one branch emanating from the node is taken if the node is realized.

The originators chose the MGF since the MGF of the sum of two independent random variables is equal to the product of the individual MGFs. Thus

$$w(s) = pM(s).$$

It has been shown that GERT networks consisting of only exclusive-or/probabilistic nodes can be solved by flowgraph analysis. Many probabilistic systems can be modeled by these networks [2].

To solve a flowgraph of any significant size requires that the transmittance through the network be determined by Mason's rule. Before stating this rule and illustrating its application, some basic terminology must be defined.

Path: A path through a network is any sequence of nodes that will cause the terminal node to be realized without realizing any given node more than once.

First-Order Loop: A first-order loop is any closed path that will cause the return to the initial node of the loop without realizing any given node in the closed path more than once.

mth-Order Loop: An mth-order loop consists of m nontouching first-order loops.

With the foregoing definitions in mind, Mason's rule may be stated as follows:

$$w_E(s) = \frac{\sum_j P_j(s)\left[1 + \sum_m (-1)^m L(m,s)\right]}{1 + \sum_m \sum_i (-1)^m L_i(m,s)},$$

where

$w_E(s)$ = equivalent transmittance between initial and terminal node,

$L_i(m,s)$ = ith loop of order m,

$P_j(s)$ = w-function of the jth path,

$L(m,s)$ = sum of the w-functions of loops of order m not touching the jth path.

The example shown in Fig. 2 illustrates the application of Mason's rule to a flowgraph.

Application of the preceding formula will provide $w_E(s)$ as the solution to the flowgraph, but some means of separating the information contained in the equivalent transmittance must be provided. It can be

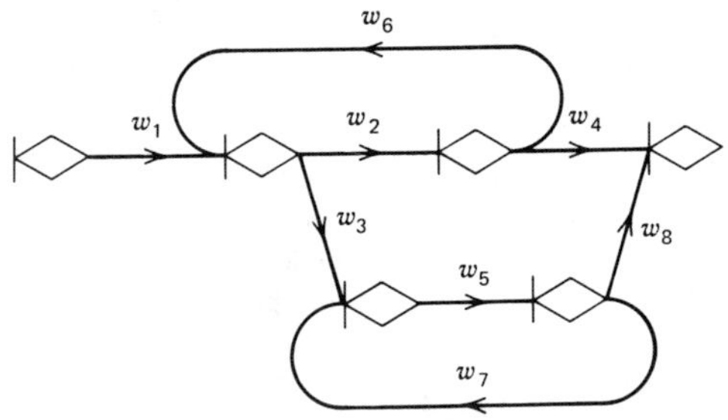

Paths	Loops 1st-order	Loops 2nd-order
$w_1w_2w_4$	w_2w_6	$w_2w_6w_5w_7$
$w_1w_3w_5w_8$	w_5w_7	

$$w_E(s) = \frac{(w_1w_3w_5w_8)(1) + (w_1w_2w_4)(1 - w_5w_7)}{1 - w_2w_6 - w_5w_7 + w_2w_6w_5w_7}$$

Figure 2 An application of Mason's Rule.

observed that:

$$w_E(s) = P_E M_E(s)$$

and

$$M_E(s)|_{s=0} = 1.$$

Therefore,

$$P_E = w_E(s)|_{s=0}$$

and

$$M_E(s) = w_E(s)/P_E.$$

Now that the MGF of the time to the completion of the network has been obtained, any number of moments of the distribution about the origin are found by differentiating the MGF to the degree equivalent to the order of the moment and evaluating the derivative at $s = 0$.

To solve complex systems with GERT a computer is mandatory, since the number of loops and paths expand rapidly.

Consider the following reliability repair problem [1] to which GERT is applicable. Suppose that a device is being developed for a given application. The application is such that the device, when put into operation, either succeeds or fails to accomplish that which it is designed to do. Suppose further that there is only one thing that can go wrong with the device, and the device will eventually fail due to this fault. The purpose of the development effort on the device is to discover what the cause of failure is and then attempt to redesign or fix the device so that it will not fail at all. Assume that repair either fixes the device or not; i.e., the probability "$1 - p$" of a defective operation is constant until the device is completely fixed and always works. The development effort then consists of repeated attempts of the device. If the device operates successfully on any given trial, the designer or development engineer decides to make no redesign action and proceeds to the next trial on the chance that he or she has already fixed the device and that its probability of failure is zero. If it fails on any given trial, the engineer goes to work on it and has probability a of fixing the device permanently prior to the next trial. We can define

$M_r(s)$ = MGF of the repair time, including the trial immediately following the repair;

$M_t(s)$ = MGF of trial time.

There are three outcomes that are possible from a given trial: (1) the trial is successful given the device is faulty; (2) the trial is a failure given the device is faulty; and (3) the trial is successful given that the device is

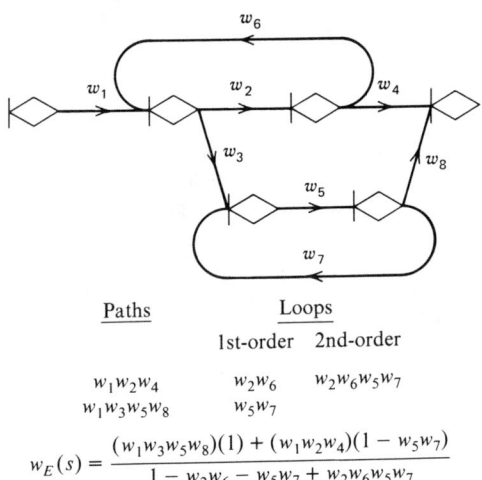

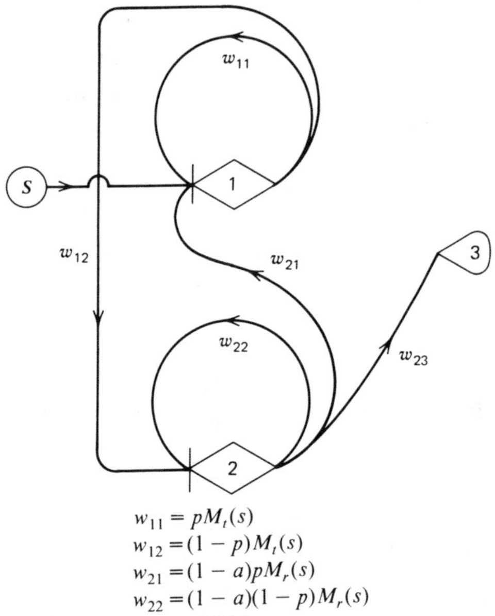

$w_{11} = pM_t(s)$
$w_{12} = (1 - p)M_t(s)$
$w_{21} = (1 - a)pM_r(s)$
$w_{22} = (1 - a)(1 - p)M_r(s)$
$w_{23} = aM_r(s)$

Figure 3 GERT representation of a reliability repair model.

fixed. These outcomes will represent the events of the GERT network. The model is now easily represented by GERT as shown in Fig. 3. The MGF of the time until the device is completely dependable, i.e., will never fail again, is equal to $M_{S,3}(s)$, where

$$M_{s,3}(s) = \frac{a(1-p)M_t(s)M_r(s)}{1 - pM_t(s) - (1-a)(1-p)M_r(s)}.$$

References

[1] Whitehouse, G. E. (1970). *Technometrics*, **12**, 33–48.

[2] Whitehouse, G. E. (1973). *Systems Analysis and Design Using Network Techniques*. Prentice-Hall, Englewood Cliffs, N.J.

(GRAPH THEORY
NETWORK ANALYSIS)

GARY E. WHITEHOUSE

FOKKER–PLANCK EQUATIONS

The Fokker–Planck equation is the differential equation $\partial\rho/\partial t = A^*\rho$, which describes the evolution of $\rho(t)$, the probability distribution at time t of a Markov process* $X(t)$. It was first derived in the context of physical diffusion by Einstein in his classic papers of 1905–1907 (see ref. 4) on the theory of Brownian* movement and its name comes from some of the subsequent physics literature (see ref. 5). Brownian movement refers to the motion of a small particle suspended in a fluid which is caused by random collisions with the individual molecules of the fluid. Under suitable assumptions, including homogeneity, isotropy, and absence of external forces, it can be shown as a consequence of the central limit theorem* that the distribution $\rho(t)$ of the position after time t of a particle which was initially at $\mathbf{y}$ is normal with mean $\mathbf{y}$ and covariance matrix $tD\mathbf{I}$, where $\mathbf{I}$ is the identity matrix and $D > 0$ is called the diffusion constant. It follows that $\rho(t)$ has a density $\rho(t,\mathbf{x})$ which satisfies Einstein's equation,

$$(\partial/\partial t)\rho(t,\mathbf{x}) = (D/2)\sum_{l=1}^{3}(\partial^2/\partial x_l^2)\rho(t,\mathbf{x}). \tag{1}$$

Following Wiener's rigorous construction of a Markov process whose $\rho(t)$ satisfies this equation, the notion of a Fokker–Planck equation for more general Markov processes was developed by Kolmogorov [3] and, in that context, it is usually called Kolmogorov's forward equation in the probability literature. Current mathematical research is based on the semigroup* approach pioneered by Feller and others (see ref. 1), which will also be the basic point of view used here.

In a Markov process with the state space $S = \{x_1, \ldots, x_n\}$, the transition probabilities define a matrix $\mathbf{P}^t$ for $t \geqslant 0$,

$$P_{ij}^t = \Pr[X(t+s) = x_j \mid X(s) = x_i]; \tag{2}$$

we assume here (and analogously throughout) that $X(t)$ is time homogeneous so that the right side of (2) is independent of $s \geqslant 0$. From (2) follows the *Chapman–Kolmogorov equation**, $\mathbf{P}^t\mathbf{P}^s = \mathbf{P}^{t+s}$. Hence $\mathbf{P}^t$ is a semigroup and can be expressed as a matrix exponential, $\mathbf{P}^t = \exp(t\mathbf{A})$, where the *generator* $\mathbf{A}$ is given by $A_{ij} = [(d/dt)P_{ij}^t]_{t=0}$. It follows that

$$P_{ij}^t = \begin{cases} A_{ij}t + o(t), & i \neq j \\ 1 + A_{ii}t + o(t), & i = j \end{cases} \quad \text{as } t\downarrow 0, \tag{3}$$

so that for $i \neq j$, A_{ij} is the rate at which the process makes a transition from x_i to x_j, and $-A_{ii} = \sum_{j\neq i}A_{ij}$ is the total rate at which the process leaves the state x_i. The probability vector $\boldsymbol{\rho}(t)$, defined by $\rho(t)_i = \Pr[X(t) = x_i]$, satisfies $\boldsymbol{\rho}(t) = \mathbf{P}^{t*}\boldsymbol{\rho}(0)$, where $P_{ij}^{t*} = P_{ji}^t$. Consequently, one has the Fokker–Planck equation: $d\boldsymbol{\rho}/dt = \mathbf{A}^*\boldsymbol{\rho}$, or, more explicitly,

$$(d/dt)\rho(t)_i = \sum_{j=1}^{n} A_{ij}^*\rho(t)_j = \sum_{j=1}^{n} A_{ji}\rho(t)_j. \tag{4}$$

For a general Markov process with state space S, the semigroup P^t is defined to operate on an appropriately chosen space C of functions f on S by

$$(P^t f)(x) = E\left[f(X(t+s)) \mid X(s) = x \right],$$

which is a natural generalization of (2). Under appropriate conditions, there is a generator A operating on C such that $P^t = \exp(tA)$ with $A = (d/dt)P^t|_{t=0}$. The adjoint semigroup P^{t*} acting on a dual space C^* of measures ρ on S is defined by

$$\int f(x)(P^{t*}\rho)(dx) = \int (P^t f)(x)\rho(dx)$$

and then the *backward equation*,

$$(d/dt)(P^t f) = A(P^t f),$$

implies the forward, or Fokker–Planck, equation,

$$d\rho(t)/dt = (d/dt)\left(P^{t*}\rho(0)\right) = A^*\rho(t).$$

In the Wiener process* description of k-dimensional Brownian motion with positive definite *diffusion matrix* $\mathbf{D}$ and constant drift *velocity* $\mathbf{b}$, the state space is $S = \mathbb{R}^k$ and the Fokker–Planck equation is

$$(\partial/\partial t)\rho(t, \mathbf{x}) = \frac{1}{2} \sum_{l,m=1}^{k} (\partial^2/\partial x_l \partial x_m)(D_{lm}\rho(t, \mathbf{x})) - \sum_{l=1}^{k} (\partial/\partial x_l)(b_l \rho(t, \mathbf{x})), \qquad (5)$$

where $\rho(t, x)$ is the density of $\rho(t)$, which is necessarily smooth for $t > 0$. Einstein's equation (1) is of course a special case of (5). In a general *diffusion process**, $\mathbf{D}$ and $\mathbf{b}$ depend on $\mathbf{x}$ with various technical assumptions needed to ensure that the formally defined A is really the generator of a Markov semigroup (see ref. 2).

In a Markov *pure jump process** on a general state space S, the rate of transition from x to a set $\Gamma \subset S\backslash\{x\}$ is given by $A(x, \Gamma) = \int_\Gamma A(x, dy)$, where, in the simplest case, $A(x, \cdot)$ is a finite positive measure on $S\backslash\{x\}$. Here $S\backslash\{x\}$ denotes the space S excluding the single point x. If we define $A(x, \{x\}) = -A(x, S\backslash\{x\})$, then the formal Fokker–Planck equation is

$$(\partial/\partial t)(\rho(t, dx)) = \int_S A(y, dx)\rho(t, dy), \qquad (6)$$

which is a natural generalization of (4). By focusing on the backward generator A, more general jump processes can be defined in which $A(x, dy)$ is infinite near $y = x$. More general Fokker–Planck equations arise from processes containing both diffusion and jump components; the resulting A^* is essentially a sum of the two types of operators described above.

References

[1] Feller, W. (1971). In *An Introduction to Probability Theory and Its Applications*, Vol. 2, 2nd ed. Wiley, New York, Chaps. 10, 13, and 14.

[2] Gihman, I. I. and Skorohod, A. V. (1972). *Stochastic Differential Equations*. Springer-Verlag, New York.

[3] Kolmogorov, A. N. (1931). *Math. Ann.*, **104**, 415–458.

[4] Nelson, E. (1967). *Dynamical Theories of Brownian Motion*. Princeton University Press, Princeton, N.J. (Contains a delightful account of Einstein's work.)

[5] Wax, N., ed. (1954). *Selected Papers on Noise and Stochastic Processes*. Dover, New York. (Of several articles of interest, the article by S. Chandrasekhar provides a nice survey of and extensive references to work subsequent to Einstein's, including those of Fokker, Planck, Ornstein and Uhlenbeck, and Smolouchowski.)

(BROWNIAN MOTION
CHAPMAN–KOLMOGOROV EQUATION
DIFFUSION PROCESSES
JUMP PROCESSES
MARKOV PROCESSSES
WIENER PROCESS)

R. N. BHATTACHARYA
C. M. NEWMAN

FOLDED DISTRIBUTIONS

These occur when a random variable X can take positive or negative values, but the absolute value $|X|$ alone is of interest. An example would be measurement of some machine parts which are dropped into place, but if any part is turned around, the sign of the measurement would change. The commonest such folded distribution is the folded normal distribution; note that while the underlying variable X may be normal, the false assumption that $|X|$ is approximately normal can lead to serious errors (see Nelson [5]).

If $X \sim N(\mu, \sigma^2)$, the probability density function $g(\cdot)$, mean μ_f, and standard deviation σ_f of $Y = |X|$ are given by

$$g(y) = \sqrt{2/\pi}\,\sigma^{-1}\cosh(\mu y/\sigma^2) \times \exp\{-(y^2+\mu^2)/(2\sigma^2)\}, \quad y > 0,$$

$$\mu_f = \sigma\sqrt{2/\pi}\exp(-\theta^2/2) - \mu\{1 - 2\Phi(\theta)\},$$

$$\sigma_f^2 = \mu^2 + \sigma^2 - \mu_f^2,$$

where $\theta = \mu/\sigma$ and Φ is the standard normal distribution function. The density $g(\cdot)$ is also that of $\sigma\sqrt{U}$, where U is noncentral chi-square* with 1 degree of freedom and noncentrality parameter μ^2/σ^2. The *half-normal distribution* is the special case for which $\mu = 0$; the normal distribution is folded about its point of symmetry.

Nelson [5] gives tables of the folded normal distribution function to four decimal places for the minimum value 1.3236 of the ratio μ_f/σ_f, and for $\mu_f/\sigma_f = 1.35(0.05)3.00$. If $\mu_f/\sigma_f > 3$, the distribution is approximately normal. See Johnson and Kotz [3, Chap. 28] for a summary of estimation procedures; also Patel and Read [6, Sec. 2.6.1], Leone et al. [4], Elandt-Johnson [1], and Johnson [2].

References

[1] Elandt-Johnson, R. C. (1961). *Technometrics*, **3**, 551–554. (Includes a discussion and tables of higher moments and moment ratios.)

[2] Johnson, N. L. (1962). *Technometrics*, **4**, 249–256. (The accuracy of maximum likelihood estimators is discussed.)

[3] Johnson, N. L. and Kotz, S. (1970). *Distributions in Statistics. Continuous Univariate Distributions*, Vol. 2. Wiley, New York.

[4] Leone, F. C., Nelson, L. S., and Nottingham, R. B. (1961). *Technometrics*, **3**, 543–550. (Includes tables of the distribution function.)

[5] Nelson, L. S. (1980). *J. Quality Tech.*, **12**, 236–238.

[6] Patel, J. K. and Read, C. B. (1982). *Handbook of the Normal Distribution*. Marcel Dekker, New York.

(NORMAL DISTRIBUTION)

CAMPBELL B. READ

FOLDED TRANSFORMATIONS

Analyses of proportions or counted fractions often use transformations that treat a fraction and its complement symmetrically, in addition to the transformation's usual role of reshaping the scale. That is, if u is the fraction and T is the transformation,

$$T(1-u) = -T(u).$$

For example, if u is the proportion of survey respondents who answer "yes" to a question and the remaining $1-u$ answer "no," then, in the transformed scale, restating the analysis in terms of "no" instead of "yes" changes the sign of the result, but not the magnitude.

Exploratory data analysis* [2] has given the name *folded transformation* to any T that treats u and $1-u$ in this way. The most direct method of constructing such transformations is to use a simple monotonic function, g:

$$T(u) = g(u) - g(1-u).$$

Thus the folding is about $u = \frac{1}{2}$, so that $T(\frac{1}{2} - \Delta) = -T(\frac{1}{2} + \Delta)$ and $T(\frac{1}{2}) = 0$. The simplest folded transformation is the *plurality*,

$$\text{plur}(u) = 2u - 1,$$

derived from $g(u) = u$. Another commonly used transformation is the *logit**,

$$\text{logit}(u) = \log_e(u) - \log_e(1-u),$$

which provides a tail-stretching effect that simplifies a number of analyses. Intermediate between the plurality and the logit in its effect on the ends of the proportion scale is the *folded square root* [2],

$$\text{froot}(u) = \sqrt{u} - \sqrt{1-u}\,.$$

As a way of unifying these and other folded transformations, Tukey [1, Chap. 28] introduced the family of transformations, indexed by λ,

$$T_\lambda(u) = \frac{2^\lambda}{2\lambda}\left[u^\lambda - (1-u)^\lambda\right], \qquad 0 < u < 1.$$

The index, λ, can take any real value, and the scaling constant matches the values of $T_\lambda(u)$ in the neighborhood of $u = \frac{1}{2}$, by giving the curves of the T_λ the same slope at $u = \frac{1}{2}$. Ignoring the constant factor, the logit fits into this family as the limiting case $\lambda \to 0$. If U is a random variable uniformly distributed on the interval (0, 1), the symmetric random variables $T_\lambda(U)$ are the members of the family of lambda distributions*.

Two other common transformations for proportions are reasonably closely approximated (after straightforward rescaling) by members of the lambda family of transformations. The first of these is the *anglit*,

$$\text{anglit}(u) = \text{arc}\sin\sqrt{u}\,,$$

expressed in radians, which corresponds approximately to $\lambda = 0.41$. The second is the *probit** or normit,

$$\text{probit}(u) = \Phi^{-1}(u),$$

where Φ is the standard normal cumulative distribution function. Taking $\lambda = 0.14$ yields a good approximation to the probit from $u = 0.001$ to $u = 0.999$. (For u close enough to 0 and 1, it must become unsatisfactory because T_λ has finite range for $\lambda > 0$.)

References

[1] Tukey, J. W. (1971). *Exploratory Data Analysis*, limited prelim. ed., Vol. 3. Addison-Wesley, Reading, Mass. (Available from University Microfilms.)

[2] Tukey, J. W. (1977). *Exploratory Data Analysis*. Addison-Wesley, Reading, Mass.

(EXPLORATORY DATA ANALYSIS)

DAVID C. HOAGLIN

FOLLOW-UP

Follow-up is the process of locating research subjects and determining whether or not some outcome of interest has occurred. Follow-up can be directed at a variety of end points, can be carried out concurrently or through historical records, can be conducted once or sequentially, and can be of short or long duration. The objective always remains the same, however: to maintain observation of the study subjects and to do so in a manner that avoids introducing bias.

Follow-up is an integral part of the investigation when it is required and should not be undertaken casually. The quality of the follow-up effort may determine the success of trials of therapeutic or prophylactic regimens, epidemiologic cohort studies, investigations into the natural history of disease, etc., where subjects must be followed for months, years, or even decades to observe the outcome. Because of its importance, follow-up must be considered when the study is originally designed. The decisions made about the specific follow-up method to be used, the choice of end points to measure, and the duration of follow-up desired may have a major impact on the way the rest of the study is conducted (*see* EPIDEMIOLOGICAL STATISTICS).

CONCURRENT FOLLOW-UP

Follow-up is most often concurrent; i.e., the subjects are identified in the present and then followed into the future. This method permits the nature of the follow-up to be tailored to the needs of the study, as arrangements are made to contact the subjects at intervals to assess their status. The primary disadvantage is that one may have to wait years for the outcome to be known. Further, concurrent follow-up implies a requirement for continuity of the research team (and funding) so that the necessary contact can be made as scheduled.

The actual follow-up involves two main activities: (1) locating the subject and then (2) collecting the data required for the study. Ideally, these activities are carried out using a rigorously established protocol which is thoroughly and systematically applied to everyone in the study. When this is not the case, systematic errors are easily introduced. These can cause both very severe and extremely subtle problems; see Feinstein [2].

The first step in avoiding such bias is to minimize the loss of subjects by locating each one. This is important because the subjects who are successfully followed until the termination of the study are frequently quite unlike those who are lost to follow-up early on or who cannot be traced at all. Individuals may be lost differentially on the basis of important study variables, including exposure to the factor of interest or risk of developing a particular outcome. For example, patients doing poorly on a specific therapeutic regimen may be overrepresented among study dropouts. The subjects who are not followed may also be more or less likely to have the outcome of interest diagnosed or detected, especially if there are major disparities in medical care available geographically or socioeconomically to the lost subjects. Subjects who are not followed have, in addition, a reduced opportunity to report the occurrence of any particular outcome should it occur. Because the characteristics of the subjects who are easily followed may differ from those lost to follow-up, a low percentage of complete follow-up (less than 80 or 90%) may cast doubt on the validity of conclusions drawn from the investigation.

In terms of locating the subjects, concurrent follow-up studies have two important advantages over historical follow-up. First, all the subjects will have been located at least once for the initial evaluation. Detailed data, such as the identity of next of kin who will always know the subject's whereabouts, can be collected at that time specifically to simplify the task of subsequent follow-up. Second, regular contact with each subject can ensure that changes in name (especially important for women) or address can be identified as they occur. After many years, it becomes extremely difficult to relocate some individuals.

After each subject has been located, one must collect the desired information. The objective of good follow-up is to collect comparable data on each individual. Problems arise when follow-up is not conducted in a uniform manner. Following one subgroup of study patients with direct examination and another indirectly with mailed questionnaires may produce distorted results since the reported frequency of even serious and distinct conditions, such as cancer, is likely to differ between the two groups. Many statistical techniques exist which allow adjustment for different durations of follow-up, but these cannot correct for biases introduced by qualitatively unequal ascertainment of outcome. The opportunity for detection of the outcome should be the same for all subjects.

The nature of the specific end point to be measured in determining outcome has a profound effect on the follow-up methods required. The design and implementation of instruments to collect follow-up data, either directly through reexamination of subjects or indirectly through mailed questionnaires, etc., is outside the scope of this entry but some general comments may be useful.

The end points to be measured customarily involve mortality (death), morbidity (illness), or a change in some subclinical physiologic or psychological parameter. Mortality is the easiest endpoint to detect through follow-up and is of great interest when survival is the outcome of interest. Mortality can be a very insensitive indicator of other outcomes, however, and should not be used for that purpose. Many cases of a particular disease can occur, for example, for each death recorded. Further, the use of mortality introduces factors associated with survival in addition to others associated with development of the disease in the first place. Because mortality is a relatively infrequent event, its choice as an end point generally requires a larger study; because determination of alive or dead status is relatively inexpensive, such a large study may be quite feasible. It may also be possible to take advantage of local, state, and national death registration systems to document the occurrence of death, as discussed below in connection with historical follow-up methods.

Morbidity is the end point most often desired in follow-up studies. If the problem of defining and diagnosing the disease can be solved, the difficulty in follow-up is to detect its occurrence. This may require examination of the subject. Reliance on the follow-up letter or questionnaire introduces major questions about the quality of the data collected except in the most specific of disease entities and, even then, steps should be taken to confirm the diagnosis. It is especially hazardous to use surrogates, such as nonspecific physiologic changes or cause of death statements on death certificates, for morbidity determinations.

Change in a physiologic or other parameter may be the most sensitive indicator of the outcome of interest; as such changes may be more frequent than frank clinical disease or death, it may be possible to design a smaller study. However, it is important to realize that an increased sensitivity may involve a concomitantly reduced specificity (*see* MEDICAL DIAGNOSIS, STATISTICS IN) because not all of the measured changes will reflect the disease process of interest.

The nature of the end point to be measured will also dictate the duration of follow-up. If particular outcomes are infrequent or long delayed after the initial event, for example, extended follow-up will be required. The frequency of follow-up is also affected by the choice of end points. More frequent follow-up is required if the outcome of interest is transient, and it can improve the completeness of ascertainment and the quality of information concerning dates of events, the details of recent illness, etc. Frequent contact may also help maintain subject interest in the investigation. Regular follow-up is sometimes sacrificed to reduce costs, but this may be false economy if the overall loss to follow-up is greatly increased or the quality of the data collected is compromised.

Problems associated with prospective follow-up of many years duration have been reported in connection with the Framingham* heart study. It was found [4] that losses to follow-up due to death or refusal to be reexamined were greatest among the elderly and, in the case of death, elderly men. Losses due to emigration, on the other hand, were greater among the young. As might be anticipated, loss by death was greatest among those with certain risk factors such as hypertension. The subjects who emigrated seemed somewhat more healthy than those who remained behind. The rate of loss to follow-up was inversely related to the ease with which patients were recruited into the study in the first place. The impact of these losses varied with the end point being considered: mortality rates were less affected than observations on the changing natural history of hypertension. It has further been shown [3] that the apparent frequency of various cardiovascular disease endpoints depends on the source of the follow-up data. Compared to clinic data, for example, hospital information overemphasized sudden death and underrepresented angina pectoris.

HISTORICAL FOLLOW-UP

In historical follow-up one identifies the study cohort as of some past date and follows it to the present. This method holds the promise of obtaining follow-up of long duration without having to wait. The main limitation, of course, is finding a "natural experiment" or suitable study cohort selected, described, and identified sufficiently well to support the desired investigation.

Although the activities of historical follow-up, i.e., locating the subjects and collecting data, are the same as for concurrent follow-up, the methods involved may be very different. Ascertainment of outcome is limited to measurements on members of the cohort who have survived to the present or to information available in existing records. Thus, mortality is often the most feasible end point for this method of follow-up.

However, the primary problem in historical follow-up is locating subjects. This generally entails finding a current mailing address or telephone number. Although details depend on the specific study environment, certain procedures are commonly employed. The original study records or medical records, etc., are usually reviewed first to obtain the most recent address, as well as additional leads such as the identity of parents or children or next of kin, referring physician, employer, insurance company, addresses of local relatives, names of witnesses on documents, etc., who should be contacted if the subject cannot be found directly. Correspondence in medical records may also reveal the fact that the subject has died.

A follow-up letter is usually then sent to the last known address. It may request the follow-up data or may simply introduce the research project and alert the subject to a subsequent contact. If no reply is received after a suitable length of time (often 30 days), a second letter may be sent. If the letters are unanswered or returned unclaimed, an attempt should be made to contact the patient directly. This effort can be very expensive in both personnel time and telephone tolls but is often extremely helpful in ensuring a high response rate.

If the foregoing measures have failed to locate the patient (i.e., letters returned by the post office), more imaginative steps are indicated. A search should first be made through local telephone books and city directories. Other possible sources of information are the post office, town and county clerks, unions and professional groups, alumni societies, the motor vehicle bureau, churches, the present resident at the subject's former address, former neighbors identified through a street address directory, etc. If the subject is from a small town, it may be possible to contact the bank, school, police department, or even to call every one in the telephone directory with the same last name. The services of the local credit bureau or a commercial "skip-tracing" firm may also be useful. It may be possible, depending on the nature of the investigation, to enlist the aid of a governmental agency such as the Social Security Administration, Veteran's Administration, Internal Revenue Service, etc. Although these organizations are very concerned with protecting individual privacy, follow-up letters will sometimes be forwarded to the subject without the agency providing an address directly to the research team.

If death is the outcome of interest, obituary columns can be reviewed or local funeral home directors contacted. For subjects who may have died prior to 1979, alphabetical listing of lost individuals can be sent to key state vital statistics* bureaus who will usually identify any deaths and provide a death certificate for a fee. For potential deaths in 1979 and after, a similar search can be conducted nationally through the new National Death Index, recently initiated by the Division of Vital Statistics, National Center for Health Statistics.*

Many of the methods above were used in a retrospective cohort study of breast cancer risk after fluoroscopy and have been evaluated in detail by Boice [1]. Vital status was determined most often from city directories or telephone books and somewhat less often from clinical records and searches of state vital statistics data. After addresses were obtained, follow-up information was collected through questionnaires, with an increase in the cumulative response rate with each of three successive mailings and a subsequent telephone call. The characteristics of patients lost to follow-up were substantially different from those of the patients located, and crude mortality was greater for those found early than late. Breast cancer incidence was somewhat greater among those who required a third mailing or telephone call to induce a response than among the subjects who answered the first questionnaire. As Boice points out, however, the direction and magnitude of these biases can be quite different depending on the nature of the cohort being followed as well as on the end point being measured.

References

[1] Boice, J. D. (1978). *Amer. J. Epidemiol.*, **107**, 127–139. (This is a very detailed discussion of the utility of various follow-up methods to locate subjects for a historical follow-up study.)

[2] Feinstein, A. R. (1977). *Clinical Biostatistics*. C. V. Mosby, St. Louis. (This text features extensive coverage of the many potential biases in biomedical research and is highly recommended for additional reading. Chapter 6 is perhaps most relevant to the problems associated with follow-up.)

[3] Friedman, G. D., Kannel, W. B., Dawber, T. R., and McNamara, P. M. (1967). *Amer. J. Public Health*, **57**, 1015–1024.

[4] Gordon, T., Moore, F. E., Shurtleff, D., and Dawber, T. R. (1959). *J. Chronic Dis.*, **10**, 186–206. (This and the previous reference cover many of the important features of concurrent follow-up studies.)

Bibliography

See the following works, as well as the references just cited, for more information on the topic of follow-up.

Fox, J. P., Hall, C. E., and Elveback, L. R. (1970). *Epidemiology, Man, and Disease*. Macmillan, New York. (This and the following reference are basic epidemiology texts and contain discussions of follow-up methods in both observational and experimental studies.)

MacMahon, B. and Pugh, T. F. (1970). *Epidemiology: Principles and Methods*. Little, Brown, Boston.

(CLINICAL TRIALS
EPIDEMIOLOGICAL STATISTICS)

LEE J. MELTON III

FORCE OF MORTALITY

This is a term used in actuarial* science to represent the proportional rate of decrease of a survivor function over time. Let $S(t)$ be the probability that an individual survives beyond time t, where $S(0) = 1$. Then the lifetime of the individual has cumulative distribution function* $F(t)$, where

$$F(t) = 1 - S(t).$$

If F has probability density function (PDF) or mortality function $f(t)$, the *force of mortality* $\gamma(t)$ is given by

$$\gamma(t) = f(t)/S(t).$$

Then $\gamma(t)$ is the PDF of the conditional survival time beyond t, given that the individual has survived until time t. In reliability theory* and survival analysis*, $\gamma(t)$ is known as the *failure rate* or *hazard rate**, and has the property that

$$S(t) = \exp\left[-\int_0^t \gamma(s)\,ds\right].$$

(ACTUARIAL STATISTICS—LIFE
MORTALITY
RISK ANALYSIS)

FORECASTING

Human beings have been trying to forecast the future since earliest times. They have, for example, based their forecasts on the stars, the entrails of chickens, belief in certain moral structures in society, economic theories, and the intuition of panels of experts. The use of statistical ideas as a basis for *forecasting* goes back at least to the early use of *least squares** to fit planetary orbits and forecast the future positions of the planets (*see* STATISTICS IN ASTRONOMY). There has been an interest in economic forecasting throughout this century, particularly in relation to business cycles. The dramatic increase in interest in forecasting came with the advent of the computer, for then routine large-scale forecasting became a possibility. Many methods were developed, such as those of Holt–Winters, that made effective use of the relatively limited storage capacity of the early computers.

The prime motive for forecasting is the awareness that some assessment of the future possibilities is essential to the decision making of the present. The contribution of statistical methods to the art of forecasting is twofold. First, there is a growing range of methods that focus on the use of structures that can be identified in past data as a rational basis for generating forecasts. Second, there have been introduced statistical measures for the "postevent" assessment of the quality and properties of forecasts, however obtained.

THE MODELING APPROACH

The essence of the statistical approach to forecasting is to seek to model the situation using past data. The model may be a simple straight-line trend fitted to a single set of sales figures or it may be a large *econometric** model with hundreds of variables. Considering the simplest case, suppose that an examination of sales figures led to a model for sales x_t in week t, of the form

$$x_t = \alpha + \beta t + \epsilon_t,$$

where $\alpha + \beta t$ represents a linear trend in sales, with a mean change of β per week, and ϵ_t is a random variable with zero mean representing the essential randomness in the data. The forecast, $\hat{x}_{t+h}$, of some future sales, x_{t+h}, will be

$$\hat{x}_{t+h} = \alpha + \beta(t + h).$$

This forecast is based on extrapolating the straight line into the future and setting the unpredictable ϵ_{t+h} at its expected value of zero. In practice α and β will be unknown and estimates $\hat{\alpha}$ and $\hat{\beta}$ will have to be obtained for them using the existing data.

This example raises some aspects of statistical forecasting that are quite general.

I. The forecast is conditional on the two assumptions, that the model being used adequately describes the main structures of the situation, and that these structures will continue to exist for the time covered by the forecast.

II. The forecast will consequently be in error (1) because of the unpredictability of certain components, e.g., the ϵ_t above, (2) because of errors in estimation of the parameters, e.g., in $\hat{\alpha}$ and $\hat{\beta}$ above, (3) because of lack of total correspondence between model and reality, (4) because of unpredicted changes in that reality in the future.

III. The model may arise and be used in a number of different fashions:

1. It may arise on the basis of certain theoretical laws or considerations.
2. It may arise in a purely empirical fashion, appearing from the analysis of the data to provide the best model to describe that data.
3. The best model may be quite complex but for forecasting a short time in the future a simpler model (e.g., $\alpha + \beta t$), is felt to be adequate.
4. In large commercial forecasting exercises, where many thousands of quantities are forecast, an empirical and robust formula for forecasting may be used (e.g., a weighted average of the last few observations). Such a formula will imply a model but will not make explicit use of it.

There are many different forecasting methods (and corresponding computer packages) available. The differences between them lie usually in the types of model used and the methods of estimating the model parameters. Rather than describe the various methods in detail it is perhaps more informative to indicate the components that are used in the models and the ways they may be fitted and forecast. For though the methods differ, they all select from a common but developing pool of components and techniques.

MODEL COMPONENTS

Deterministic Time Components

These include a constant level μ, a linear trend βt, a quadratic trend γt^2, exponential growth $e^{\alpha t}$ (sometimes introduced by a log transformation of other variables), other forms of growth (*see* GROWTH CURVES), regular and seasonal, oscillations, and less regular long-term oscillations (cycles*). These are usually combined additively. A forecast of a future value of these components is obtained by direct substitution of an appropriate value for t; thus βt at time t becomes a future $\beta(t + h)$ at time $t + h$.

Stochastic Components

Almost all models include at least one component, ϵ_t, which represents the unpredictable elements in a situation. The ϵ_t are commonly assumed to be representable by a sequence of independent identically distributed random variables with zero mean and, usually, constant variance. In some situations the current value of a variable may not only contain the contribution from ϵ_t, but it may also be influenced by previous values

$\epsilon_{t-1}, \epsilon_{t-2}, \ldots$. Thus the stochastic element may take the more general "moving average*" form:

$$\epsilon_t - \theta_1\epsilon_{t-1} - \theta_2\epsilon_{t-2} - \cdots - \theta_k\epsilon_{t-k}.$$

Forecasting using these components is based on (1) "forecasting" future unpredictable ϵ's by their mean value of zero, (2) using past data to estimate previous values $\epsilon_t, \epsilon_{t-1}$, etc., and substituting these in the model.

Regressive Components

Many models depend on identifying relationships between the variable being forecast, x_t, and other variables, say y_t and z_t. If the relation is linear, the model would be of linear regression* form

$$x_t = \alpha + \beta y_t + \gamma z_t + \epsilon_t.$$

If good forecasts of future values of y and z are available, the relation may be used to forecast x. Often relations involve time delays and the linear relation might be

$$x_t = \alpha + \beta y_{t-1} + \gamma z_{t-1} + \epsilon_t.$$

Such a relation will enable y_t and z_t to be used to forecast x_{t+1}. Such a model is said to involve *lagged variables* x_{t-1}, z_{t-1} (*see* LAG MODELS). Sometimes an important lagged variable is x itself, the current value x_t depending on previous values $x_{t-1}, x_{t-2}, \ldots$. Thus

$$x_t = \phi_1 x_{t-1} + \phi_2 x_{t-2} + \cdots + \phi_h x_{t-h} + \epsilon_t.$$

Such a relation is referred to as *autoregression*. To forecast with autoregressive components, x_t, x_{t-1}, etc., are used in the model to forecast x_{t+1}, by $\hat{x}_{t+1}$, i.e., $\hat{x}_{t+1} = \phi_1 x_t + \phi_2 x_{t-1} + \cdots + \phi_h x_{t+1-h}$. We then repeat the process to get $\hat{x}_{t+2}$ from $\hat{x}_{t+1}, x_t, x_{t-1}$, etc.

State Components

In many situations the structure is not static but changes in a partially predictable or at least understood fashion. The change may be sudden or continuous. There is thus a need to model the state of the structure. At the simplest level the state changes due to factors, such as advertising campaigns, holidays, price changes, etc., that can be identified and modeled by simple components which only apply during certain time periods. Such factors can be allowed for by multiplying the components that model their effects by an "indicator variable," d_t, where $d_t = 1$ when the factor is operating and $d_t = 0$ when it is not. In a more complex situation a further level of model building may be introduced to describe the changes in the state of the situation. For example, additional equations may allow for the model parameters of the basic model to change either suddenly or in some relatively slow random fashion. Forecasts are simply obtained from models, including indicator variables, provided that one knows which state $d_{t+h} = 0$ or $d_{t+h} = 1$ will hold at the future date. For the more general state models the data are used to follow the state of the situation and assess the probabilities of the possible future states or predict the future parameter values of the model.

Model Composition

The ways in which model components are combined are numerous. *Linear* models are by far the most common. For example, a model might take the form

$$x_t = \alpha + \beta t + \gamma z_{t-1} + \phi x_{t-1} + \delta d_t + \epsilon_t.$$

Terms are thus added together with parameters, the α, β, γ, etc., indicating the magnitude of the influence of each variable. The possibilities are not limited to additive models. In particular, for models involving growth, or for seasonality, model components need to be introduced multiplicatively to give an adequate description of what is observed. For example, a classical seasonal model takes the form

$$x_t = \text{trend} \times \text{seasonal} \times \text{cycle} \times \text{random component},$$

where in this case the random component (termed the "irregular component" in this context) will have unit mean.

A further consideration in model composition relates to the decision as to whether to make certain aspects of the situation explicit or not. Suppose, for example, that we think that the underlying mean, μ, of x varies slowly in time. We may ignore the fact in the model and put

$$x_t = \mu + \cdots,$$

but seek to estimate the current value of μ by using only recent data. We may try to model the wandering of μ, now μ_t, in time, for example by an autoregressive model

$$\mu_t = \psi\mu_{t-1} + \eta_t,$$

where η_t is a random component and ψ a new model parameter. Alternatively, we might eliminate μ_t by differencing the data, i.e., $\nabla x_t = x_t - x_{t-1}$, for ∇x_t will be uninfluenced by a slow change in μ. The model will then be constructed for the differenced data rather than the original data.

MODEL FITTING FOR FORECASTING

In classical statistics the problem of fitting the model is dealt with by *estimation** theory. Although many forecasting methods make use of standard estimation theory, a considerable number of variations have been used in forecasting applications. The underlying reasons for these variations in approach are as follows:

1. Data always become available sequentially as $x_1, x_2, x_3, \ldots$ in the forecasting context and thus to get the best estimates, in the practical sense, we constantly need to update the estimates.
2. Classical estimation seeks estimates that give a close fit between model and existing data. In forecasting, the objective is a close fit of the model to the future data.

We now briefly consider the main methods of estimation used in forecasting.

Classical Estimation

Least squares* is the most common estimation method used in forecasting. It provides direct solutions for linear models and approximate and iterative methods are readily available for nonlinear models. For the linear model the estimates can be expressed in fairly simple recursive forms to update estimates $\hat{\theta}_{t-1}$, obtained at time $t-1$, by using the new data, x_t, to calculate the new estimate $\hat{\theta}_t$. Often such recurrence relations take the form

$$\hat{\theta}_t = \hat{\theta}_{t-1} + \text{correction depending on } x_t.$$

Discounted Estimation

Classical estimation treats all data as equally relevant to parameter estimation. If parameters are slowly changing quantities, it is appropriate to give more weight to recent data than to earlier data. For example, the least-squares criterion

$$\sum(\text{observation} - \text{fitted model})^2$$

may be replaced by a weighted or "discounted" least-squares criterion

$$\sum w_r(x_{t-r} - \text{fitted model at } t-r)^2,$$

where w_r is a weight that decreases as the age, r, of the data increases. A very commonly used weight is the exponential or geometric weight $w_r = a^r$, $0 < a \leqslant 1$, a (or $1-a$ in some books) being referred to as the "discounting factor."

Recursive Estimation

Useful recursive formulae may be derived heuristically without reference to classical estimation. For example, if the model were a "mean" plus "random variation," that is,

$$x_t = \mu + \epsilon_t,$$

and μ was thought to vary slowly, an intuitive recursive estimator $\hat{\mu}_t$ is given by

$$\hat{\mu}_t = (1-a)x_t + a\hat{\mu}_{t-1}, \qquad 0 \leqslant a \leqslant 1.$$

This formula takes a simple weighted average of the latest observed value, x_t, and the previous estimate $\hat{\mu}_{t-1}$. As this formula is applied repeatedly, we are in effect smoothing the data to give at each time an estimate of the local mean. This process is often called "exponential smoothing"; *see* GRADUATION.

Bayes Estimation

The recursive concept of using new data to modify the prior estimators and models, together with the fact that in some forecasting situations the costs associated with forecast errors can be financially measured, both imply a forecasting context for *Bayesian* methods of *estimation* (*see* BAYESIAN INFERENCE). A number of forecasting methods apply this approach to estimation.

Adaptive Estimation

The potential lack of stability of parameter values in many situations has led to a number of methods based on monitoring the success of forecasting and feeding the results back into the estimation process. The revised estimates are then used in producing the next forecasts. For example, a measure of the magnitude of the forecast errors might be used to modify the weight a used in the recursive formula just given. The weight a would be increased when the errors were small and the situation stable, but decreased, to take greater notice of new data, if there were indications of rapid change in mean. The process is thus "adaptive" to change within the limits of the model structure.

FORECASTING METHODS

In an ideal world each problem can be treated as unique, a suitable model is constructed and fitted, and one obtains the best possible forecast from the best possible model. In reality there are rarely sufficient resources available to do this. Thus there has grown a number of standard methods that are well understood and are readily available as computer packages. Some of the more common of these methods are listed below.

Holt–Winters: This uses a trend-seasonal model fitted by recursive or adaptive methods. Although limited, it is one of the earliest and most basic methods.

Brown: This uses polynomials and Fourier-based seasonals in the model, which is fitted by discounted least squares, sometimes with adaptive variations.

Census: This uses trends, seasonals, and cycles in a multiplicative fashion and fits by a range of iterative methods involving recursion and classical ideas.

*Box–Jenkins**: This uses an autoregressive–moving average* stochastic model with trends and seasonal components removed by differencing. The model is fitted by classical estimation and the method also provides ways of identifying the number of autoregressive and moving average terms to be included in the model.

Kalman/Bayesian: This range of methods can include most model components but is built around the use of state variables. The model may be fitted by least squares or discounted least squares and it lends itself to Bayesian estimation.

*Econometric**: These methods refer essentially to models involving regressor variables as their main components, although trends and seasonals may well be included. The models are usually fitted using classical estimation criteria.

FORECAST QUALITY

We now turn briefly to the contribution of statistics to the assessment of the quality of forecasts. Clearly, these methods will apply irrespective of whether the forecast was obtained using the approaches discussed above.

In the real world forecast errors lead to poor decisions which can lead to excessive costs, i.e., "losses." If a forecast $\hat{x}_{t+h}$ is made of a quantity x_{t+h}, the cost of the error will depend on $\hat{x}_{t+h}$ and x_{t+h} and can be represented as $C(x_{t+h}, \hat{x}_{t+h})$. If this quantity can be evaluated for a considerable number of our past forecasts and observations, we will have a fairly clear numerical measure of our forecast quality. By going back over past data we can also find those forecasts that would have been produced by several competing methods (or by using different parameter values with the same method). This information would be of substantial value when deciding on the best method for a specific situation.

Unfortunately, the complexity of situations prevents a single definite cost measure c being formulated. Thus a range of simple statistical measures have been developed to measure various aspects of forecast quality. Most of these are based on an examination of the simple forecast errors, $e_1, e_2, \ldots, e_n$, where $e_t = x_t - \hat{x}_t$, obtained over a series of n trials.

The major measures (see also corresponding entries) are:

Mean Error:

(a) $$\bar{e} = \frac{1}{n} \sum_{i=1}^{n} e_i .$$

This indicates whether the forecasts are under- or overestimating. The forecast "bias" is given by $\bar{e}$, which is positive for underestimating.

*Mean Square Error**:

(b) $$\text{MSE} = \frac{1}{n} \sum_{i=1}^{n} e_i^2 .$$

This is the most common measure of the precision of forecasts.

Mean Absolute Error:

(c) $$\text{MAE} = \frac{1}{n} \sum_{i=1}^{n} |e_i| .$$

As, in forecasting, we are always more interested in the current (indeed future) state rather than the far past these measures are often replaced by weighted or discounted versions using exponential weights c^r, $0 < c \leqslant 1$. Thus if t is the current time and r the age of the forecast errors, the measures above become

(a′) $$\sum_{r=0}^{t-1} c^r e_{t-r} ,$$

(b′) $$\sum_{r=0}^{t-1} c^r e_{t-r}^2 ,$$

(c′) $$\sum_{r=0}^{t-1} c^r |e_{t-r}| .$$

The magnitude of (a′) depends on the magnitudes of the data, so it is sometimes useful to standardize (a′) by dividing by a measure of variability such as the root mean square error*, $\sqrt{(\text{b}')}$ or (c′). We thus define a standardized bias,

$$T_t = \sum_{r=0}^{t-1} c^r e_{t-r} \Big/ \sum_{r=0}^{t-1} c^r |e_{t-r}| .$$

T_t is called a *tracking signal*, as it is used to track the bias in the forecasts as new observations and forecasts are obtained. Since $-1 \leqslant T \leqslant 1$, its numerical value is fairly easy to interpret. If $|T|$ gets too high, it indicates a significant bias (positive or negative) in the forecasts and thus a possible change in the situation. This property has led to T_t, and quantities like it, being used as the means of providing the feedback required for adaptive estimation.

As the forecast error is in many ways analogous to a "residual," methods of the kinds devised for the analysis of residuals (*see* RESIDUAL ANALYSIS) can be applied to give a more detailed study of forecast errors and quality.

Operational procedures for monitoring the quality of forecasts in a working forecasting system have been developed. Some use traditional quality control* methods, including cusum charts*; others are based on tracking signals.

A major lesson learned from the many comparisons between methods that have been carried out is that there is no a priori best method. Each case needs careful consideration and trials with a variety of approaches. Even then, the fact that one method has been best in the past is no guarantee that it will produce the best forecasts in the future. Judgment based on detailed understanding and insights will always be a major factor in the art of forecasting, with statistical techniques providing the best support that they can.

Bibliography

General Methods of Forecasting

Chisholme, R. K. and Whitaker, G. R. (1971). *Forecasting Methods*. Richard D. Irwin, Homewood, Ill.

Fildes, R. (1979). *J. Operat. Res. Soc.*, **30**, 691–710.

Firth, M. (1977). *Forecasting Methods in Business and Management*. Edward Arnold, London.

Gilchrist, W. G. (1974). *Omega*, **2**, 733–750.

Gilchrist, W. G. (1976). *Statistical Forecasting*. Wiley, London.

Makridakis, S. (1976). *Int. Statist. Rev.*, **44**, 29–70.

Makridakis, S. (1978). *Int. Statist. Rev.*, **46**, 255–278.

Montgomery, D. C. and Johnson, L. A. (1976). *Forecasting and Time Series Analysis*. McGraw-Hill, New York.

Robinson, C. (1971). *Business Forecasting: An Economic Approach*. Thomas Nelson, London.

Silk, L. S. and Curley, M. L. (1970). *A Primer on Business Forecasting*. Random House, New York.

Wheelwright, S. C. and Makridakis, S. (1973). *Forecasting Methods for Management*. Wiley, New York.

Special Methods of Forecasting

HOLT–WINTERS

Chatfield, C. (1978). *Appl. Statist.*, **27**, 264–279.

Holt, C. C., Muth, J. F., Modigliani, F., and Simon, H. A. (1960). *Planning Production, Inventories, and Work Force*. Prentice-Hall, Englewood Cliffs, N.J.

Winters, P. R. (1960). *Manag. Sci.*, **6**, 324–342.

BROWN

Brown, R. G. (1959). *Statistical Forecasting for Inventory Control*. McGraw-Hill, New York.

Brown, R. G. (1963). *Smoothing, Forecasting and Prediction of Discrete Time Series*. Prentice-Hall, Englewood Cliffs, N.J.

CENSUS

Durbin, J. and Murphy, M. J. (1975). *J. R. Statist. Soc. A*, **138**, 385–410.

Shishkin, J. et al. (1967). *The X-11 Variant of the Census II Method Seasonal Adjustments Program*, Bur. Census, Tech. Paper No. 15.

BOX–JENKINS

Anderson, O. D. (1975). *Time Series Analysis and Forecasting: The Box–Jenkins Approach*. Butterworth, London.

Box, G. E. P. and Jenkins, G. M. (1970). *Time Series Analysis: Forecasting and Control*. Holden-Day, San Francisco.

Jenkins, G. M. (1979). *Practical Experiences with Modelling and Forecasting Time Series*. Gwilym Jenkins and Partners, St. Helier, Jersey, Channel Islands.

Nelson, C. R. (1973). *Applied Time Series Analysis for Managerial Forecasting*. Holden-Day, San Francisco.

KALMAN–BAYESIAN

Harrison, P. J. and Stevens, C. F. (1976). *J. R. Statist. Soc. B*, **38**, 205–247.

Kalman, R. E. and Bucy, R. S. (1961). *J. Basic Eng.*, **83D**, 95–108.

Mehra, R. K. (1978). Kalman Filters and Their Application in Forecasting, *Manag. Sci.* (Special Issue on Forecasting), S. Makridakis and S. C. Wheelwright, eds. North-Holland, Amsterdam.

Morrison, N. (1969). *Introduction to Sequential Smoothing and Prediction*. McGraw-Hill, New York.

ECONOMETRIC

Granger, C. W. J. and Newbold, P. (1977). *Forecasting Economic Time Series*. Academic Press, New York.

Pindyck, R. S. and Rubinfield, D. L. (1976). *Econometric Models and Economic Forecasts*. McGraw-Hill, New York.

Spencer, M. H., Clark, C., and Hoguet, P. W. (1961). *Business and Economic Forecasting: An Econometric Approach*. Richard D. Irwin, Homewood, Ill.

Theil, H. (1958). *Economic Forecasts and Policy*. North-Holland, Amsterdam.

Theil, H. (1966). *Applied Economic Forecasting*. North-Holland, Amsterdam.

Forecast Quality

Ash, J. C. K. and Smith, D. J. (1973). *Forecasting the U.K. Economy*. Saxon House, Farnborough, Hants, England.

Granger, C. W. J. and Newbold, P. (1973). *Appl. Econ.*, **5**, 35–47.

Groff, G. K. (1973). *Manag. Sci.*, **20**, 22–30.

Makridakis, S. and Horn, M. (1979). *J. R. Statist. Soc. A*, **142**.

McKenzie, E. (1974). *Statistician*, **23**, 107–116.

Reid, D. J. (1972). In *Forecasting in Action*, M. J. Bramson et al., eds. Operational Research Society and Society for Long Range Planning, London.

Trigg, J. W. and Leach, A. G. (1967). *Operat. Res. Quart.*, **18**, 53–59.

Wood, D. and Fildes, R. In *Forecasting and Planning*, R. Fildes and D. Wood, eds. Teakfield, Farnborough, Hants, England.

(BOX–JENKINS MODEL
ECONOMETRICS
LAG MODELS
PROBABILITY FORECASTING
SEASONALITY
TIME SERIES)

WARREN GILCHRIST

FOREIGN TRADE STATISTICS, INTERNATIONAL

International foreign trade statistics are compiled to make trade data of individual countries comparable for purposes of market research and economic analysis. The underlying concepts and definitions accepted by most of the world's countries for collecting and reporting foreign data have been developed by the Statistical Commission of the United Nations* [6].

There are two definitions of a country's statistical territory in common use. Countries reporting *general trade* include all merchandise crossing their national boundary, whether the goods cross the customs frontier or only enter customs-bonded and free areas for storage or for further processing and reexport. Since processing in customs-bonded areas is an important activity in many countries, reporting under the definition of "general trade" is considered to be a more accurate representation of a country's foreign trade. Under *special trade* reporting goods are only recorded when they cross the "customs frontier." Imports for processing in customs-bonded areas should therefore be excluded, but their significance has led to a recommendation to include them where available. Goods stored in customs-bonded warehouses are recorded as imported on the date on which they are withdrawn and clear customs. The difference in definition produces time lags in comparisons of data between countries reporting "general" and "special" trade.

According to the United Nations definition of merchandise, all goods that "add to or subtract from the stock of material resources in a country as a result of their movement into or out of the country" [6, p. 5] should be included in the statistics whether or not a commercial transaction has taken place. This definition excludes such items as goods consigned by a government to its armed forces and diplomatic representatives abroad, transit trade and all temporary trade such as travelers' effects, goods for exhibition only, returnable samples, etc. Transactions involving gold, securities, bank notes, and coins should be differentiated between nonmonetary and monetary operations, the latter being considered to be capital transfers and excluded from foreign trade statistics. There are a number of commercially important transactions which may not fit this definition by not crossing the statistical border, yet whose exclusion would be misleading. Examples are trade in new and used ships and aircraft, trade in gas, electricity, and water, and purchase and sale of bunkers, fish, and salvage.

There are two principal commodity classifications in use today: the Standard International Trade Classification (SITC) developed by the United Nations [7] and the Brussels Tariff Nomenclature (BTN) [2, p. 5]. The SITC (now in its second revision) is a five-digit classification of products by classes of goods and is especially useful for economic analysis. The BTN, which arranges goods according to their content, is considered more suitable for customs purposes. In the 1950s a one-to-one correspondence between these two classifications was established to facilitate comparisons between countries and to allow regrouping of data for economic studies. Current work in this area is concentrated on developing cross classifications with countries having unrelated nomenclatures (especially the Council for Mutual Economic Assistance) and at regrouping data for more specific analytical needs such as the uses to which goods are put [1] (Broad Economic Categories, BEC) and the industries in which goods are normally produced [3, 4] (the International Standard Industrial

Classification of all Economic Activities, ISIC).

Differences in national practices in valuation of goods, units of quantity, conversion factors, and definitions of partner country all pose potential difficulties in making comparisons. It is generally accepted that the most useful valuation for studying trade flows is the transaction value. In the case of imports this covers purchase price of the goods plus the cost of insurance, freight, and unloading (c.i.f.). It excludes any charges imposed in the country of import. For exports the transaction value covers all costs free on board or at the frontier of the exporting country (f.o.b.). Export duties and charges are included.

Establishing a definition of partner country creates perhaps the greatest difficulties in analyzing world trade statistics. Countries from which goods are purchased or to which they are sold may be totally unrelated to the movement of goods. Country of production for imports is useful in establishing the relationship between the producing and importing country but may lead to differences in reporting (e.g., where final destination is not known at the time of export) or can be difficult to determine when goods have undergone successive transformations in several countries. Export data based on country of final consumption are likely to be equally inaccurate as final destination is often unknown at the time of shipment and goods may be resold en route. Imports and exports based on country of last consignment give a clear pattern of the movement of goods.

AVAILABILITY OF DATA

Several international organizations publish International Foreign Trade data with varying coverage. An annotated directory of sources, *A Guide to the World's Foreign Trade Statistics* [5], has been published by the International Trade Centre UNCTAD/GATT, Geneva. The availability of foreign trade statistics continues to improve with developments in data processing. Most countries store full detail on computer files and both international organizations and commercial service bureaus have built up comprehensive historical data banks. Data are also available on microfiches and magnetic tape. It is now also possible to envisage direct computer links between users and these sources of data.

References

[1] *Classification by Broad Economic Categories, Defined in Terms of SITC Rev. 2* (1976). Statist. Papers, Ser. M, No. 53, Rev. 1, United Nations, New York.

[2] Customs Co-operation Council (1972). *Nomenclature for the Classification of Goods in Customs Tariffs* (BTN), Brussels.

[3] *Indexes to the International Standard Industrial Classification of all Economic Activities* (1971). Statist. Papers, Ser. M, No. 4, Rev. 2, Add. 1, United Nations, New York.

[4] *International Standard Industrial Classification of all Economic Activities* (1968). Statist. Papers, Ser. M, No. 4, Rev. 2, United Nations, New York.

[5] International Trade Centre, UNCTAD/GATT, *A Guide to the World's Foreign Trade Statistics*, 1977, Palais des Nations, Geneva.

[6] *International Trade Statistics, Concepts and Definitions* (1970). Statist. Papers, Ser. M, No. 52, United Nations, New York.

[7] *Standard International Trade Classification, Revision 2* (1975). Statist. Papers, Ser. M, No. 34/Rev. 2, United Nations, New York.

(UNITED NATIONS STATISTICAL COMMISSION)

W. SIMPSON

FORESTRY *See* STATISTICS IN FORESTRY

FORMS, BILINEAR

Bilinear forms are functions of two sets of variables $x_1, \ldots, x_m; y_1, \ldots, y_n$ which are linear in each set if the other is regarded as fixed. Such a form can be written

$$a + \sum_{i=1}^{m} b_i x_i + \sum_{i=1}^{n} c_i y_i + \sum_{i=1}^{m} \sum_{j=1}^{n} d_{ij} x_i y_j$$

or, equivalently,

$$a_{00} + \sum_{i=1}^{m} a_{i0}x_i + \sum_{i=1}^{n} a_{0i}y_i + \sum_{i=1}^{m} \sum_{j=1}^{n} a_{ij}x_iy_j .$$

If $a = b_i = c_i = 0$ ($a_{00} = a_{i0} = a_{0i} = 0$) for all i, it is a *homogeneous* bilinear form.

Multilinear forms are defined in a similar way.

FORMS, QUADRATIC

A quadratic form in variables $x_1, \ldots, x_m$ can be written

$$a + \sum_{i=1}^{m} b_i x_i + \sum\sum_{i<j} c_{ij}x_ix_j$$

or, equivalently,

$$a_0 + \sum_{i=1}^{m} a_i x_i + \sum\sum_{i<j} a_{ij}x_ix_j .$$

If $a_0 = b_i = 0$ ($a_0 = a_i = 0$) for all i, it is a *homogeneous* quadratic form. $\mathbf{C} = (c_{ij})$, or $\mathbf{A} = (a_{ij})$ is the *matrix* of the form. If this matrix is positive (definite), the form is a *positive (definite)* form.

(CHI-SQUARE
IDEMPOTENT MATRIX
NONCENTRAL CHI-SQUARE)

FORWARD DIFFERENCE

A term used in finite differences*. The forward difference operator Δ, applied to a function $f(x)$, produces the difference between $f(x + 1)$ and $f(x)$. Symbolically,

$$\Delta f(x) = f(x + 1) - f(x).$$

This quantity is called the first (forward) difference of $f(\cdot)$ at x. Repetition of this operation produces the second (forward) difference

$$\begin{aligned}\Delta^2 f(x) &= \Delta(\Delta f(x)) = \Delta f(x + 1) - \Delta f(x)\\ &= f(x + 2) - f(x + 1)\\ &\quad - (f(x + 1) - f(x))\\ &= f(x + 2) - 2f(x + 1) + f(x).\end{aligned}$$

Generally, for any positive integer n,

$$\begin{aligned}\Delta^n f(x) &= \Delta\left(\Delta^{n-1} f(x)\right)\\ &= \sum_{j=0}^{n} (-1)^j \binom{n}{j} f(x + n - j). \quad (1)\end{aligned}$$

Introducing the *displacement operator** E [with the property $Ef(x) = f(x + 1)$], we have symbolically

$$\Delta \equiv E - 1.$$

Formula (1) can be written

$$\begin{aligned}\Delta^n f(x) &= (E - 1)^n f(x)\\ &= \left\{ \sum_{j=0}^{n} (-1)^j \binom{n}{j} E^{n-j} \right\} f(x).\end{aligned}$$

The *central* difference operator δ has the property

$$\delta f(x) = f(x + \tfrac{1}{2}) - f(x - \tfrac{1}{2}).$$

Symbolically,

$$\Delta \equiv \delta E^{1/2}.$$

(BACKWARD DIFFERENCE
FINITE DIFFERENCES, CALCULUS OF
GRADUATION
INTERPOLATION)

FORWARD SELECTION PROCEDURE *See* ELIMINATION OF VARIABLES; STEPWISE REGRESSION

FOUNDATIONS OF PROBABILITY

Foundations of Probability as a Discipline

Any account of the foundations of probability (hereinafter designated F. of P.) must be read skeptically. The objectives are to understand and develop probabilistic reasoning. The scale of the effort in the F. of P. is broad, usually conceptual, and generally not concerned with the more narrowly focused technical advances in the familiar numerical probability concept. Efforts in the F. of P. are likely to be critical, expository, and/or argumentative, exhibiting a tradition inherited from the philosophy of science. The F.

of P. lacks widely accepted guidelines as to important problems and acceptable solutions; this lack has been hospitable to the development of several schools of thought, each housing a variety of positions.

Contributions to the F. of P. come from many disciplines, including philosophy of science, economics, mathematical probability, physics, and engineering, with philosophers of science in the modal but not the majority position. There is no single journal that is a reliable source for current literature. Some journals that occasionally publish articles on the F. of P. are: *Annals of Probability**, *Annals of Statistics**, *British Journal of the Philosophy of Science*, *Journal of the Royal Statistical Society** (*Series B*), *Synthèse*, *Theory and Decision*, and *Theoria*. *Mathematical Reviews* can be profitably consulted under 60A05.

Brief Historical Background

While we are likely to learn little to improve the mathematics of numerical probability by consulting the past, history is of importance to the F. of P., for it provides a perspective on the conceptual assumptions that lie at the root of our present-day notions of probability. This perspective can enlighten us as to alternative assumptions, as well as enable us to better perceive the consequences of commitments to particular probability concepts and statistical methodologies. Standard historical works include David [8], Maistrov [34], Todhunter [45], and more recently the work of Hacking [21] and Shafer [43]; these latter two works are of interest for their greater concern with key conceptual issues, although their historical accuracy has been challenged.

Probability ideas, as expressed by scholars and not just as implied by gambling* or religious practices, can be traced back to the ancient Egyptians and Greeks and then forward through the Middle Ages (e.g., the writings of Aquinas) to the appearance of recognizably modern probability in the last half of the seventeenth century. Scholarly views up to the onset of the Renaissance held that it did not admit refined degrees and that the basis for a probability assertion was the opinions of qualified individuals. Numerical probability on a scale from 0 to 1 first appears in the seventeenth century (if we omit scattered precursors who were ignored in their time), at a time when a belief in the desirability of mathematical precision is replacing the belief in the possibility of certain knowledge about the world, the latter possibility having been called into question by Cartesian skepticism. At the close of the seventeenth century we find not only numerical probability and expectation but also Bernoulli's law of large numbers* (published posthumously in 1713). The latter supplied the basis for a frequentist view of probability that largely came to supplant the subjective/epistemic view predominant hitherto (*see* FREQUENCY INTERPRETATION OF PROBABILITY).

The eighteenth century saw the introduction of conditional probability* and Bayes'* formula, great advances in the calculus and limit laws, the elimination of the belief held by Bernoulli* that the sum of the probability of an event and the probability of its complement need not be 1, and the beginnings of the scientific use of probability. The use of probability in the physical and social sciences grew greatly in the nineteenth century under the influence of the works of Boltzmann*, Laplace*, Quetelet*, and others, and increasing attention was paid to a frequentist view [48].

Major contributions to the F. of P. in the first part of the twentieth century include the following:

1. A frequentist interpretation by R. von Mises* based on a notion of randomness and repeated experiments modeled by the sample space and collective [49] (a posthumous work with references to earlier work)
2. A. N. Kolmogorov's axiomatization of probability as a normed measure [25]
3. B. de Finetti's and L. J. Savage's development of the subjective* view of probability [9–11, 38]

4. R. Carnap's development of logical probability, as an objective assessment of the degree to which an evidence statement (inductively) supports a hypothesis statement [4]
5. J. M. Keynes' [24], B. O. Koopman's [27], and B. de Finetti's studies of comparative probability.

Role of the F. of P. in Statistics

The gap between the F. of P. and applied probability* and statistics is narrower than might be expected. Knowledge of the different mathematical and interpretive concepts of probability treated in the F. of P. can guide the selection of families of probability models (not necessarily numerical ones) so as to better reflect the indeterminate, uncertain, or chance phenomena being treated, and can clarify a choice among the divergent, conflicting statistical methodologies now current. The relevance of the F. of P. to statistical methodology is established when we realize that the positions of the Neyman–Pearsonians, Bayesians, structuralists, fiducialists, and maximum entropists are better understood when it is seen that these schools rely on different concepts of probability, albeit this difference is obscured by common agreement on the mathematical structure (excepting countable additivity) of probability. Neyman–Pearsonians postulate that a class of uncertain/indeterminate phenomena (i.e., the "unknown" parameter) cannot be given a probability model. Bayesians/subjectivists in turn insist upon giving this class an overly precise numerical probability model, but allow great latitude in the subjectively based choice of model. Structuralists/fiducialists and maximum entropists carry the modeling process one step further by claiming to provide objective, rational grounds for the selection of a unique numerical probability model to describe, say, the "unknown" parameter. (Barnett [1] provides some of the details of the different statistical methodologies we have mentioned.) The F. of P. enables us to understand the actual claims and presuppositions of these methodologies (not just the proffered written and verbal explanations of proponents and opponents), and it provides a basis for the improvement and selective utilization of these methodologies.

Lest the reader misunderstand, the F. of P. is far from a panacea for our statistical and modeling ills. It is an undeveloped subject that, like philosophy, will always promise more than it delivers. Yet what it does deliver is essential to the development of statistical methodology.

Outline of Contents

Our presentation is calculated to broaden the reader's view by first detaching him or her from the notion that probability is only numerical (*see* AXIOMS OF PROBABILITY) and then by supplying some examples of nonnumerical probability concepts. It is also our intent to expose the reader to a variety of interpretations that are available for probability.

In the following section, we comment on probabilistic reasoning, the subject of the F. of P., locate it with respect to science, mathematics, and rationality, and observe that a probability concept lies at the center of probabilistic reasoning. The third section introduces a five-element framework for analyzing and presenting a probability concept. Elements of the framework are illustrated by reference to several current concepts of probability, with details reserved to the References. Most entries in the reference list pertain to specific theories of probability. A broader coverage is available in Black [3], Fine [13], and Kyburg [29]. The References should also assist the reader to establish his or her own perspective on the F. of P.

PROBABILISTIC REASONING

The F. of P. has probabilistic reasoning as its subject. This is a mode of rational thought, engaged in by diverse entities having various aims. Furthermore, rationality itself evolves in time, varies with the reasoner and with the goals of the reasoner [35].

The diversity of conditions under which probabilistic reasoning is engaged in makes a prima facie case for a parallel diversity in its forms.

Probabilistic reasoning, like rationality, has the status of a methodology and as such is neither an empirical discipline nor a branch of deductive or mathematical logic. Probability is neither contained within science nor within mathematics; it is in fact part of the methodology of science, although the appropriateness of specific probability models for categories of empirical phenomena is the subject of scientific investigations. Probability is no more a branch of mathematics than is physics, although it owes a great debt to mathematics for its formulation and development. Hence neither science nor mathematics can be relied on to confirm or refute the broad probabilistic ideas that are the concern of the F. of P. This status makes it difficult to determine generally agreed-on canons of acceptability for proposed or familiar modes of probabilistic reasoning, and it partially explains the controversial nature of the F. of P.

A concept of probability is the armature around which one constructs a form of probabilistic reasoning. We thus turn to present a framework for the classification and analysis of probability concepts and aspects of the associated system of probabilistic reasoning.

FRAMEWORK FOR THE ANALYSIS OF PROBABILITY CONCEPTS

We identify five elements (reasoner, domain of application, semantics/interpretation, syntax/mathematical structure, and working basis) that form a framework for the analysis and selection of a probability concept.

The Reasoner

The first element is the entity having aims or goals that lead it to engage in probabilistic reasoning. Examples of such entities, hereafter referred to as *reasoners*, are:

1. Individual
2. Organization (e.g., a business planning ahead, a family group)
3. Society (e.g., the inductive reasoning and rhetoric of the medical profession differs from that of the legal profession, which in turn differs from that of engineering).

Goals can include:

1. Summarization of data
2. Understanding (e.g., inference in science, generation/selection of a model)
3. Choice of action (e.g., decision making by an individual or formal organization)
4. Communication (e.g., between individuals or between an individual scientist and his or her colleagues, say).

One also needs to take into account the knowledge and information available to the reasoner and the limits to the "intelligence" of the reasoner (bounded rationality).

Domain of Application

The domain of application over which the reasoner seeks some mastery is a milieu isolated from its surroundings by the reasoner's interest in the occurrence of *events* (we shall use "events" to refer both to events as commonly conceived and to propositions describing events) in some collection and by the reasoner's, perhaps tacit, identification of an available setting or background that has a bearing on the collection of events. Analysis of the domain guides the choice of an appropriate form of probabilistic reasoning; such analyses can point the way to the need for new theories. Examples of backgrounds or settings drawn from the logical, mental, and physical realms include:

1. **Logical.** Formally presented information corpus
2. **Mental.** Individual's or group's state of knowledge or belief

3. **Physical.** Category of empirical phenomena having nearly identical characteristics

Descriptors associated with the foregoing settings are, respectively:

1. Indeterminate
2. Uncertain
3. Chance or aleatory.

The term *epistemic* is used to jointly refer to indeterminate and uncertain settings; it also often suggests that the issue is one of inductive reasoning from the setting to the events, with probability inhering in the link between the two. *Aleatory*, on the propensity account, suggests that the reasoner is concerned with a physical property of some empirical phenomenon and not just with his or her knowledge about this property; probability inheres in the phenomenon itself. It is also common for "chance" or "aleatory" to refer to a summarization of empirical data through relative frequencies of occurrence. Further classification of domains can be made along the "subjective"/"objective" axis. Objective domains are those that are in principle available to more than one individual and that can be described independently of the particular observer. Subjective domains are to a significant extent private and cannot be described without reference to a particular individual. In practice, one rarely encounters pure cases of chance, indeterminate, uncertain, objective, or subjective domains (e.g., consider the bases for the choice of a reference class for a frequentist probability model).

Examples of domains of types 1 and 3 often differ more in the view of probability than in the phenomena themselves. For example, the events of interest may concern the outcome of the nth toss of a particular coin when we have recorded the results of the first $(n-1)$ tosses. The explicit corpus would then be the record of $(n-1)$ tosses and perhaps some generalities about coins. One can then construe probability either as a physical property of the coin, with the corpus providing relevant measurement data, or one can construe probability as representing the linkage or relation between the given record of $(n-1)$ tosses and background on coins and the nth toss; the first case is what we mean by type 3 and the second case is of type 1.

Examples of type 2 domains are common in the subjectivist/personalist/Bayesian decision-making area [11, 18, 23, 33, 38, 39]. An individual confronted with a choice of actions introspectively and seriously weights his or her beliefs or strength of conviction as to the probabilities of the consequences attendant on each choice of action. He or she then assesses the expected utility* for each action and decides on an action with maximal expected utility. It is an important feature of the subjective account that there need not be formal knowledge of the contents or even of the extent of the domain.

Semantics/Interpretations

A domain thus contains both a collection of events whose occurrences are of interest to a reasoner and a setting identified by the reasoner as informative about the occurrence of events and as relevant to achieving its goals. In some fashion, the reasoner decides that it can perhaps identify which of the events are, say, probable, or which events are more probable than other events, or even assign a numerical probability to each event. Implicit in this process is an initial determination as to what provides the evaluative basis for the probability concept being invoked (e.g., what weather data and theory can we use to calculate the probability of rain tomorrow). The evaluative basis largely, but not completely, fixes the meaning of the probability concept, which must have meaning extending beyond its evaluative basis if it is to serve a role other than that of data summarization. Nevertheless, we discuss semantics in terms of this basis, for it provides the least controversial access to elements of meaning. Possible evaluative bases for probability are:

1. Past occurrences of other events of the same type

2. The phenomenon/experiment generating the events
3. The "state of mind" of the reasoner as exhibited by his or her behavior
4. The inductive linkage or relation between the setting and the events.

The interpretation of probability most closely associated with Basis 1 is that of converging relative frequencies of occurrence for each event A in the collection; typically, the domain type is 3 (*see* FREQUENCY INTERPRETATION OF PROBABILITY). Reflection on Basis 1, though, reveals that it is less understandable than it seems at first glance. One possibility is that probability is just descriptive of actual occurrences without regard to the random experiment or setup generating the events. Typically, the descriptive approach is elaborated in a frequentist context in terms of the proportion of occurrences of an event in repeated performances of a random experiment. However, probability as a description of actual occurrences is an easily discredited view; we have no grounds for determining such a probability in advance of the actual occurrences. Hence, the common frequentist view of probability is an ensemble one that carefully refrains from committing itself to what will actually occur.

Generally, one would ask of a property that it have known relations to other properties and terms. Yet, if probability is just a property of occurrences of events, and not say, of the underlying random experiment, then what is our basis for determining the value of probability, and how is it related to other probabilities and properties?

Basis 2 is a somewhat clearer basis for, say, the common converging relative-frequency-of-occurrence interpretation (see Giere [15], Gillies [17], and Mellor [36]). A random phenomenon or experiment is now said to have a "physical" characteristic: namely, its disposition, propensity, or tendency to produce, say, A with probability $\Pr(A)$, or its greater tendency to produce A than to produce B, and this characteristic manifests itself through occurrence rates for the events that are the outcomes of the performance of the experiment; i.e., the propensity is large that the observed relative frequency of event A will be close to the propensity of the experiment to produce A. Probability on this account is a property of the apparatus and not of its output, with implications for the outputs that make it easy to travel the road to Basis 1.

Basis 3 is the one held by proponents of the subjectivist/personalist/Bayesian interpretation (*see* DEGREES OF BELIEF) as well as by some holders of a subjective epistemic theory [16, 32, 42]. Subjective probability describes the strength of belief of the reasoner (an individual) concerning the truth of propositions or occurrence of events. Strength of belief is determined through a process of introspection and manifests itself through overt choice/betting behavior.

Basis 3 might also include Toulmin's conception of probability as guarding or qualifying an assertion [46]. Although Toulmin's interpretation has few adherents, it could have a role in communications; an individual can signal reservations about a statement in a graded manner that can, perhaps, be made interpersonally meaningful.

Basis 4 is maintained by proponents of objective epistemic theories [4, 5, 7, 30], who hold that the probability of an event is relative to a given background corpus (domain type 1), and it represents the rational degree of inductive support lent by the corpus to the truth/occurrence of the event. It is perhaps also held by some subjective epistemic theorists (e.g., Shafer [42]). Semantic position 4 and domain type 1 also seem to be appropriate as locations for the following: computational complexity-based induction methods and universal probability constructions proposed by Kolmogorov [26], Solomonoff [44], and Chaitin [6] and developed also by Schnorr [40] and others; Fisher's fiducial inference* (see Seidenfeld [41]); Fraser's structural inference* [14]; and some treatments of likelihood-based statistical support [2, 12, 20]. At any rate, fiducial and structural inference do not seem to directly admit frequentist or subjectivist interpretations.

Other issues in the consideration of the

semantics of probability include those aspects of the domains referred to in the section "Domain of Application" by the descriptions "indeterminate," "uncertain," and "chance," as well as by the objective/subjective classifications. Furthermore, there is the generally overlooked issue of the extent to which distinctions or degrees of precision are meaningful. The prevalent commitment to numerical probability has been supported by, and in turn supports, a view that at least in principle, degrees of indeterminacy, uncertainty, and chance* can be arbitrarily finely graded. This is a remarkable hypothesis having little justification except that of its widespread acceptance. We can address this issue more clearly when we turn to the mathematical structures we adopt for probability.

Syntax/Mathematical Structure

VIEWPOINT OF MEASUREMENT THEORY. Once the reasoner has adopted a concept of probability supported by the domain of application, he or she then wishes to move this system into a formal mathematical domain so as better to determine the implications of the position. From the viewpoint of the theory of measurement [28] the domain and the selected probability concept form an "empirical relational system." The reasoner homomorphically maps this system into a "mathematical relational system." The collection of events becomes a collection $\mathscr{A}$ of subsets $A, B, C, \ldots$, of a sample space Ω. Relations are then identified between elements of the mathematical domain $\mathscr{A}$ that homomorphically represent the relational properties holding between events, and these relations characterize the adopted probability concept. For example, if the reasoner has only classified events as to whether or not they are probable, this is represented by partitioning $\mathscr{A}$ so as to specify those sets that are probable. If the reasoner has instead considered binary comparisons of the likelihood of events, this is represented by a binary relation $\gtrsim$ such that $A \gtrsim B$ represents the claim "the event represented by A is at least as probable as the event represented by B." One might expect that $\gtrsim$ is an order relation for a reasonable probability concept. If, finally, the reasoner holds that his or her goals and setting justify a numerical assignment, he or she may adopt the familiar probability measure representation

$$\mu : \mathscr{A} \to [0, 1],$$

with $\mu(A) \geqslant \mu(B)$ if and only if A represents an event at least as probable as the event represented by B. A numerical probability assignment usually also carries the implications that such statements as "A is x times as probable as B" and "A is more probable than B by x" are meaningful in the empirical domain.

It is common in discussions of the meaning of probability to reverse the process we have described and start with a well-defined mathematical structure for a probability concept. One then attempts to coordinate the given mathematical structure with an application domain through an interpretation so as to give meaning to what is otherwise merely a collection of symbols. The possible interpretations are constrained by the axiomatic structure of the probability concept but are not determined by it. (*See* AXIOMS OF PROBABILITY.) This reverse process has seemed reasonable only because of the prevalent belief that the only mathematical structure for a probability concept is that of Kolmogorov for numerical probability [25]. We hope that our discussion of syntactical variety in the section "Hierarchy of Probability Concepts" will encourage the study and use of new mathematical structures for probability.

EVENT COLLECTIONS. The events of interest in the domain are represented either by sets or by propositions. The collection of events is typically assumed to generate a Boolean algebra or σ-algebra of sets or propositions, although there is some interest, largely motivated by quantum logic, in lattices*. Too little attention has been paid to the mathematical structure of event collection representations [37]. It is generally not possible to enumerate all possible events (complex systems occasionally surprise us by behaving in unforeseen ways), and therefore the sample space is at best a list of practical possibilities. (The device of logically completing the space by introducing a new event, "everything else," has problems of its own.) Furthermore, it has been argued in discussions of quantum logic [22] that representations of event collections need not be closed under the usual Boolean operations. Nevertheless, almost all work in the F. of P. presumes at least a Boolean algebra of sets or propositions.

HIERARCHY OF PROBABILITY CONCEPTS. The recognition that probabilistic reasoning must confront a wide range of domains and levels of information, knowledge, belief, and empirical regularity can lead us to an acceptance of an hierarchy of increasingly precise mathematical concepts of probability. This hierarchy has been little explored, as almost all of the effort, even in the F. of P., has been devoted to the acknowledgedly vital case of familiar numerical probability. That numerical probability may be inadequate to the full range of uses of probabilistic reasoning is suggested by the following observations:

1. There are categories of empirical phenomena (e.g., speech and picture information sources, earthquakes, weather) for which there is no obvious stability of relative frequency for all events of interest.
2. An information corpus may lack information; the resulting indeterminacy should be respected and not papered over by dubious hypotheses (e.g., "If you know nothing about the parameter, then adopt a uniform or maximum entropy* prior for it.").
3. Self-knowledge of individuals is intrinsically limited, and attempts to force belief or conviction to fit the mold of a particular "rational" theory can only yield results of unknown value.

An attempt to accommodate to the preceding observations leads to the following hierarchy of concepts:

1. "Possibly A."
2. "Probably A" or "not improbably A."
3. "A is at least as probable as B."
4. "A has interval-valued probability $(\underline{P}(A), \overline{P}(A)$" where "$\underline{P}(A)$" is called the lower probability of A and "$\overline{P}(A)$" is the upper probability of A.
5. "A has probability x."

Conditional versions of each of the foregoing concepts are also available. The use of a statistical hypothesis (family of probability measures) to specify our knowledge of the domain phenomena can lead to upper and lower probabilities generated by the upper and lower envelopes of the family of measures; a special case of such hypotheses are the so-called "indeterminate probabilities" [31, 32]. It is shown in Walley and Fine [50] that the subclass of concept 4 called belief functions* can serve to represent concept 3, which in turn can represent concept 2. Furthermore, the belief functions are a subclass of lower probabilities and are therefore representable by them. These representations (e.g., "probably A" iff "A is at least as probable as not A") are generally nonunique but establish the hierarchy of concepts.

The probability concepts just introduced must then be given structure through a set of axioms and definitions of significant terms (e.g., independence or unlinkedness, expectation). The most familiar axiom set is that given by Kolmogorov for concept 5. Some discussion of axioms for the other concepts is given in Walley and Fine [50]. Here we will only present the basic axioms for lower and upper probabilities [19]. Let $\mathscr{A}$ be an algebra of subsets of a set Ω and let A, B denote elements of $\mathscr{A}$.

A1. $\underline{P}(A) \geqslant 0$.

A2. $\underline{P}(\Omega) = 1$.

A3. If $A \cap B = \varnothing$ (empty set), then

$$\underline{P}(A) + \underline{P}(B) \leqslant \underline{P}(A \cup B) \qquad \text{(superadditivity)}$$

$$\overline{P}(A) + \overline{P}(B) \geqslant \overline{P}(A \cup B) \qquad \text{(subadditivity)}.$$

A4. $\overline{P}(A) + \underline{P}(A^c) = 1$ (A^c is the complement of A).

Elementary consequences of these axioms include:

1. $\underline{P}(\emptyset) = \overline{P}(\emptyset) = 0$.
2. $\overline{P}(A) \geqslant \underline{P}(A)$.
3. $\overline{P}(\Omega) = 1$.
4. $A \supset B$ implies that $\underline{P}(A) \geqslant \underline{P}(B)$, $\overline{P}(A) \geqslant \overline{P}(B)$.
5. If $A \cap B = \emptyset$, then

$$\underline{P}(A \cup B) \leqslant \underline{P}(A) + \overline{P}(B) \leqslant \overline{P}(A \cup B).$$

If $\overline{P} \equiv \underline{P}$ or if in A3 either superadditivity or subadditivity is replaced by additivity, A1 to A4 reduce to the Kolmogorov axioms for the finite case.

In one is interested in algebras containing infinitely many events, the basis axioms would need to be supplemented. In fact, there is almost no work in the F. of P. that proceeds solely in terms of the basic axiom set we have given (*see* BELIEF FUNCTIONS).

Working Basis

While it is the role of interpretation to coordinate the mathematical, axiomatically constrained concept with the domain of events of interest to the reasoner, this coordination is typically idealized and not itself a working basis for probabilistic reasoning. Statistics is the discipline that supplies the working basis for numerical probability with a frequency-of-occurrence interpretation. Statistics is also of value in supplying the basis for numerical probability in the personalist/subjectivist setting. In this case there is also a literature concerned with the actual capabilities of humans to supply probabilities and consistently/rationally/coherently use probabilities to choose actions [47]. Any truly rational theory of partial belief or conviction must take account of the capabilities of the agent in question. Psychology and business school students of these capabilities have shown that human performance is somewhat at variance with the idealized requirements of current personal probability theories.

Little is known at present about the practical issues connected either with formal concepts of probability other than the numerical one or with interpretations of the objective epistemic variety. Walley and Fine [51] outline the elements of a frequentist basis for interval-valued probabilities.

References

[1] Barnett, V. (1973). *Comparative Statistical Inference*. Wiley, New York.

[2] Birnbaum, A. (1977). *Synthese*, **36**, 19–49.

[3] Black, M. (1967). Probability. In *The Encyclopedia of Philosophy*, Vol. 6, P. Edwards, ed. Macmillan, New York, pp. 464–479.

[4] Carnap, R. (1962). *The Logical Foundations of Probability*, 2nd ed. University of Chicago Press, Chicago.

[5] Carnap, R. and Jeffrey, R. C. (1971). *Studies in Logic and Probability*, Vol. 1. University of California Press, Berkeley, Calif.

[6] Chaitin, G. J. (1969). *J. ACM*, **16**, 145–149.

[7] Cohen, L. J. (1977). *The Probable and the Provable*. Clarendon Press, Oxford.

[8] David, F. N. (1962). *Games, Gods and Gambling*. Charles Griffin, London.

[9] de Finetti, B. (1931). *Fundam. Math.*, **17**, 298–329.

[10] de Finetti, B. (1937). Foresight: its logical laws, its subjective sources, transl. in H. Kyburg and H. Smokler, eds., *Studies in Subjective Probability*. Wiley, New York, 1964, pp. 93–158.

[11] de Finetti, B. (1974). *Theory of Probability*, Vols. 1, 2. Wiley, New York.

[12] Edwards, A. W. F. (1973). *Likelihood*. Cambridge University Press, Cambridge.

[13] Fine, T. L. (1973). *Theories of Probability: An Examination of Foundations*. Academic Press, New York.

[14] Fraser, D. A. S. (1968). *The Structure of Inference*. Wiley, New York.

[15] Giere, R. N. (1976). In *Foundations of Probability Theory, Statistical Inference and Statistical Theories of Science*, Vol. 2, W. L. Harper and C. A. Hooker, eds. D. Reidel, Dordrecht, Holland, pp. 63–101.

[16] Giles, R. (1976). In *Foundations of Probability Theory, Statistical Inference and Statistical Theories of Science*, Vol. 1. W. L. Harper and C. A. Hooker, eds. D. Reidel, Dordrecht, Holland, pp. 41–72.

[17] Gillies, D. A. (1973). *An Objective Theory of Probability*. Methuen, London.

[18] Good, I. J. (1950). *Probability and the Weighing of Evidence*. Charles Griffin, New York.

[19] Good, I. J. (1962). In *Logic, Methodology and Philosophy of Science*, E. Nagel, P. Suppes, and A. Tversky, eds. Stanford University Press, Stanford, Calif., pp. 319–329.

[20] Hacking, I. (1965). *The Logic of Statistical Inference*. Cambridge University Press, Cambridge.

[21] Hacking, I. (1975). *The Emergence of Probability*. Cambridge University Press, Cambridge.

[22] Jauch, J. M. (1968). *Foundations of Quantum Mechanics*. Addison-Wesley, Reading, Mass.

[23] Jeffrey, R. C. (1965). *The Logic of Decision*. McGraw-Hill, New York.

[24] Keynes, J. M. (1921). *A Treatise on Probability*, reprinted 1962. Harper and Row, New York.

[25] Kolmogorov, A. N. (1933). *Grundbegriffe der Wahrscheinlichkeitrechnung*. Transl. by N. Morrison, *Foundations of the Theory of Probability*, 2nd ed. Chelsea, New York, 1956.

[26] Kolmogorov, A. N. (1965). *Problemy peredachii informatsii*, **1**, 4–7 (Soviet journal translated as *Problems of Information Transmission*).

[27] Koopman, B. O. (1940). *Bull. Amer. Math. Soc.*, **46**, 763–774.

[28] Krantz, D. H., Luce, R. D., Suppes, P., and Tversky, A. (1971). *Foundations of Measurement*, Vol. 1. Academic Press, New York.

[29] Kyburg, H. E., Jr. (1970). *Probability and Inductive Logic*. Macmillan, Toronto.

[30] Kyburg, H. E., Jr. (1974). *The Logical Foundations of Statistical Inference*. D. Reidel, Dordrecht, Holland.

[31] Levi, I. (1974). *J. Philos.*, **71**, 391–418.

[32] Levi, I. (1980). *The Enterprise of Knowledge*. MIT Press, Cambridge, Mass.

[33] Lindley, D. (1965). *Introduction to Probability and Statistics from a Bayesian Viewpoint*, Vols. 1, 2. Cambridge University Press, Cambridge.

[34] Maistrov, L. E. (1974). *Probability Theory: A Historical Sketch*, 2nd ed., S. Kotz, trans. Academic Press, New York.

[35] March, J. (1978). *Bell J. Econ.*, Autumn, 587–608.

[36] Mellor, D. H. (1971). *The Matter of Chance*. Cambridge University Press, Cambridge.

[37] Randall, C. H. and Foulis, D. J. (1976). In *Foundations of Probability Theory, Statistical Inference and Statistical Theories of Science*, Vol. 3, W. L. Harper and C. A. Hooker, eds. D. Reidel, Dordrecht, Holland.

[38] Savage, L. J. (1954). *The Foundations of Statistics*. Wiley, New York.

[39] Savage, L. J. (1962). *The Foundations of Statistical Inference: A Discussion*. Wiley, New York.

[40] Schnorr, C. P. (1971). *Zufalligskeit und Wahrscheinlichkeit*. Springer-Verlag, Berlin.

[41] Seidenfeld, T. (1979). *Philosophical Problems of Statistical Inference*. D. Reidel, Dordrecht, Holland.

[42] Shafer, G. (1976). *A Mathematical Theory of Evidence*. Princeton University Press, Princeton, N.J.

[43] Shafer, G. (1978). *Arch. History Exact Sci.*, **19**, 309–370.

[44] Solomonoff, R. J. (1978). *IEEE Trans. Information Th.* **IT-24**, 422–443.

[45] Todhunter, I. (1865). *A History of the Mathematical Theory of Probability: From the Time of Pascal to That of Laplace*. Reprinted by Macmillan, London, 1949.

[46] Toulmin, S. E. (1958). *The Uses of Argument*. Cambridge University Press, Cambridge.

[47] Tversky, A. and Kahneman, D. (1974). *Science*, **185**, 1124–1131.

[48] Venn, J. (1888). *The Logic of Chance*, 3rd ed. Macmillan, London.

[49] von Mises, R. (1964). *Mathematical Theory of Probability and Statistics*, ed. and complemented by H. Geiringer. Academic Press, New York.

[50] Walley, P. and Fine, T. L. (1979). *Synthese*, **41**, 321–374.

[51] Walley, P. and Fine, T. L. (1982). *Ann. Statist.*, **10**, 741–761.

Acknowledgment

Preparation of this article partially supported by National Science Foundation Grant SOC-7812278.

(AXIOMS OF PROBABILITY
BAYESIAN INFERENCE
BELIEF FUNCTIONS
CAUSATION
CHANCE
DEGREES OF BELIEF
FREQUENCY INTERPRETATION OF PROBABILITY AND STATISTICAL INFERENCE
LOGIC IN STATISTICAL REASONING)

TERRENCE L. FINE

FOURIER COEFFICIENTS

Let $f(x)$ be a Lebesgue-integrable function on $(-\pi, \pi)$. The coefficients

$$a_k = \pi^{-1}\int_{-\pi}^{\pi} f(t)\cos(kt)\,dt,$$

$$b_k = \pi^{-1}\int_{-\pi}^{\pi} f(t)\sin(kt)\,dt; \quad k = 0, 1, 2, \ldots,$$

are the *Fourier coefficients* of f. The formal series

$$\tfrac{1}{2}a_0 + \sum_{k=1}^{\infty}\{a_k\cos(kx) + b_k\sin(kx)\} \quad (1)$$

is the *Fourier series* of f [3, Chap. 7]. Since $\sin x$ and $\cos x$ are periodic of period 2π,

$$f(x + 2\pi) = f(x).$$

Let $c_k = \frac{1}{2}(a_k - ib_k)$, $c_{-k} = \frac{1}{2}(a_k + ib_k)$; $i = \sqrt{-1}$; $k = 0, 1, 2, \ldots$. Then

$$f(x) = \sum_{k=-\infty}^{\infty} c_k e^{ikx},$$

$$c_k = \tfrac{1}{2}\pi^{-1} \int_{-\pi}^{\pi} f(t) e^{-ikt}\, dt; \qquad (2)$$

$$k = 0, \pm 1, \pm 2, \ldots .$$

If f is of bounded variation, the formal series (1) converges at every point x to the value $\frac{1}{2}\{f(x+0) + f(x-0)\}$. If, in addition, f is continuous everywhere in a closed interval I, then (1) is uniformly convergent on I. Other conditions for uniform convergence are given in Zygmund [6].

Distributional properties of Fourier coefficients and some applications are discussed in Freedman and Lane [2]. Woods and Posten [5] combined the foregoing concepts with Chebyshev polynomials (see Snyder [4] and CHEBYSHEV–HERMITE POLYNOMIALS) to provide Fourier representations of cumulative distribution functions of random variables with supports over finite ranges; they also provided a numerical computing procedure.

Fourier coefficients are important in the theory of characteristic functions*, indirectly through the part played in that theory by Fourier transforms (*see* INTEGRAL TRANSFORMS). Feller [1, Sec. XIX.4] proves, however, that a continuous function f of period 2π is a characteristic function if and only if $c_k \geqslant 0$ and $f(0) = 1$. In this case (2) holds.

References

[1] Feller, W. (1971). *An Introduction to Probability Theory and Its Applications*, 2nd ed., Vol. 2. Wiley, New York.

[2] Freedman, D. and Lane, D. (1980). *Ann. Statist.*, **8**, 1244–1251.

[3] Kaplan, W. (1973). *Advanced Calculus*. Addison-Wesley, Reading, Mass. (A lucid discussion appears in Chap. 7.)

[4] Snyder, M. A. (1966). *Chebyshev Methods in Numerical Approximations*. Prentice-Hall, Englewood Cliffs, N.J.

[5] Woods, J. D. and Posten, H. O. (1977). *Commun. Statist. B*, **6**, 201–219.

[6] Zygmund, A. (1959). *Trigonometric Series*. 2 vols. Cambridge University Press, Cambridge.

(CHARACTERISTIC FUNCTIONS
INTEGRAL TRANSFORMS)

FOURIER TRANSFORM *See* INTEGRAL TRANSFORMS

FRACTALS

INTUITIVE CHARACTERIZATION

A fractal [2–4] is a geometric shape in which coexist distinctive features of every conceivable linear size, ranging between zero and a maximum that allows for two cases: If the fractal is bounded, the maximal feature size is of the order of magnitude of the fractal's overall size. If the fractal is unbounded, its nonempty intersection with a box of side L has a maximal feature size of the order of L. A "folk theorem" (i.e., a statement phrased sufficiently sharply to be of practical heuristic use, and sufficiently loosely to make it impossible to draw counterexamples) asserts that, when a random set is scaling, meaning it is invariant under either a similarity or affinity transformation, it should be expected to be a fractal. The set may be the graph of a random function in continuous space and time.

MATHEMATICAL DEFINITIONS

A set F is a *fractal set* [3, 4] if its Hausdorff–Besicovitch (= fractal) dimension D and its topological dimension D_T, are related by $D > D_T$. D is defined and discussed in the entry HAUSDORFF DIMENSION. A set F is a *uniformly fractal set* if D_T and $D > D_T$ take on the same values for the intersection of F by every ball centered on F. A random set F is *uniformly fractal* if D_T and $D > D_T$ almost surely take up the same value for the intersection of F with every ball centered on F. Fractals can be *dusts*, (connected) *curves*, and (connected) *surfaces*, these terms being defined by $D_T = 0$, $D_T = 1$, and $D_T > 1$, respectively. A theorem of Szpilrajn [1] establishes that every set in Ω satisfies $D \geqslant D_T$.

COMMENTS

STANDARD SETS OF GEOMETRY. The standard sets of geometry, those of Euclid, satisfy $D = D_T$; they are among the sets excluded by the foregoing definition. A sufficient condition for both D and D_T to be defined is that F be a subset of a separable metric space Ω.

PROTOTYPICAL RANDOM FRACTALS RELATED TO WIENER'S BROWNIAN MOTION. The graph of the scalar Brownian motion* $B(t)$ in $\mathbb{R}^2$ (= time × one space coordinate) is a fractal curve, because $D_T = 1$ while $D = 1.5$. The set of zeros of $B(t)$ is a fractal dust because $D_T = 0$ while $D_T = .5$. The set of values of a Brownian mapping of $\mathbb{R}$ on $\mathbb{R}^E$, with $E \geqslant 2$, is a fractal curve, because $D_T = 1$ while $D = 2$.

OTHER RANDOM FRACTALS. The graph of Lévy's Brownian function of N-dimensional "time" is a fractal surface, because $D_T = N$ while $D = N + 0.5$. A technique useful in modifying D is fractional integrodifferentiation*.

EXAMPLES OF NONRANDOM FRACTALS. Many of the classic counterexamples of analysis designed between 1875 and 1925 are known or suspected fractals. In each instance, the construction involves recursive steps or iteration. The Weierstrass continuous but nondifferentiable function (orally presented in 1872, first mentioned in print in 1875) is a suspected fractal. The Cantor triadic set (1884) is a fractal dust, $D_T = 0$, with $D = \log 2/\log 3$. It is a nonrandom prototype of the zero set of the Brownian $B(t)$. The Koch "snowflake" curve (1904) is a fractal curve, $D_T = 1$, with $D = \log 4/\log 3$. The observation that these sets are not "pathological" but indispensable tools in model making originates in refs. 2–4.

References

[1] Hurewicz, W. and Wallman, H. (1941). *Dimension Theory*. Princeton University Press, Princeton, N.J.

[2] Mandelbrot, B. B. (1975). *Les Objets fractals: forme, hasard et dimension*. Flammarion, Paris. (This work introduced the term *fractal*. Superseded by ref. 4.)

[3] Mandelbrot, B. B. (1977). *Fractals: Form, Chance, and Dimension*. W. H. Freeman, San Francisco. (This work defined the term, *fractal*. Superseded by ref. 4. For a capsule review, see HAUSDORFF DIMENSIONS.)

[4] Mandelbrot, B. B. (1982). *The Fractal Geometry of Nature*. W. H. Freeman, San Francisco. (This is the only current book devoted exclusively to fractals.)

(FRACTIONAL BROWNIAN MOTIONS
AND GAUSSIAN NOISES
FRACTIONAL INTEGRODIFFERENTIATION
HAUSDORFF DIMENSION)

BENOIT B. MANDELBROT

FRACTILES *See* QUANTILES

FRACTIONAL BROWNIAN MOTIONS AND FRACTIONAL GAUSSIAN NOISES

Fractional Brownian motion (fBm) is the term attached by Mandelbrot and Van Ness [5] to a family of Gaussian processes* related to Brownian motion*. These processes are special but very important in hydrology* [*see* HURST COEFFICIENT (RESCALED RANGE ANALYSIS)] and in diverse other sciences where fractal* models are effective [4].

AXIOMATIC DEFINITION

$B_H(t)$ is the Gaussian random function satisfying the scaling conditions $Z(0) = 0$, $EZ(t) = 0$, and $EZ^2(t) = |t|^{2H}$, and the stationary increments condition that the distribution of $Z(t+s)-Z(t)$ is independent of t. It is shown in ref. 5 [Sec. 3.1] that one must have $0<H<1$.

Almost all sample paths $B_H(t)$ are almost everywhere continuous, but not differentiable. In fact, with probability 1

$$\limsup_{t\to t_0}\left|\frac{B_H(t)-B_H(t_0)}{t-t_0}\right|=\infty.$$

DISCRETE FRACTIONAL GAUSSIAN NOISES

The continuous fractional Gaussian noise is the derivative $B_H'(t)$. However, $B_H(t)$ being nondifferentiable, $B_H'(t)$ is not an ordinary but a generalized function (Schwartz distribution). Its covariance is given by

$$r_c(k)=H(2H-1)k^{2H-2}.$$

The simplest approximate random process that avoids Schwartz distributions is the sequence $W_H(t)=B_H(t+1)-B_H(t)$, $-\infty<t<\infty$, called *discrete fractional Gaussian noise* in ref. 5. It is stationary, Gaussian, with mean zero, unit variance, and correlation* given by

$$r(k)=0.5\{|k+1|^{2H}-2|k|^{2H}+|k-1|^{2H}\}.$$

1. In the case $0.5<H<1$, called *persistent*, $r(k)\sim r_c(k)$ as $k\to\infty$. It follows that $\sum_{k=0}^{\infty}r(k)=\infty$. The long-range dependence and resulting low frequencies are so strong that the spectral density $f(\lambda)=\sum_{k=-\infty}^{+\infty}e^{ik\lambda}r(k)$ is infinite at the origin. Figures 1 to 3 show simulated

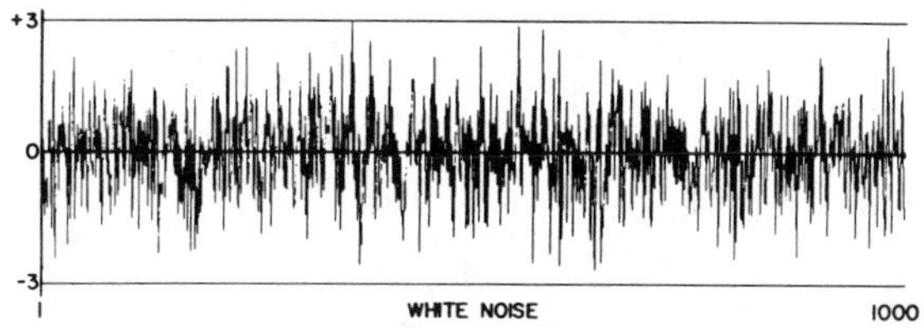

Figure 1 Sample of 1000 values of *white noise*, also called *sequence of independent Gaussian random variables of zero mean and unit variance*, or *increments of Brownian motion*. This is the special case of fractional Gaussian noise for $H=0.5$.

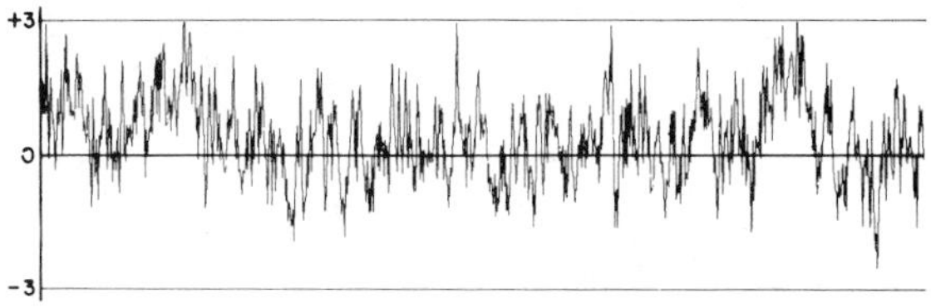

Figure 2 Sample of 1000 values of an approximation to *fractional Gaussian noise*, also called *increments of fractional Brownian motion*. In this instance, $H=J=0.7$, meaning that the strength of long-run persistence is moderate.

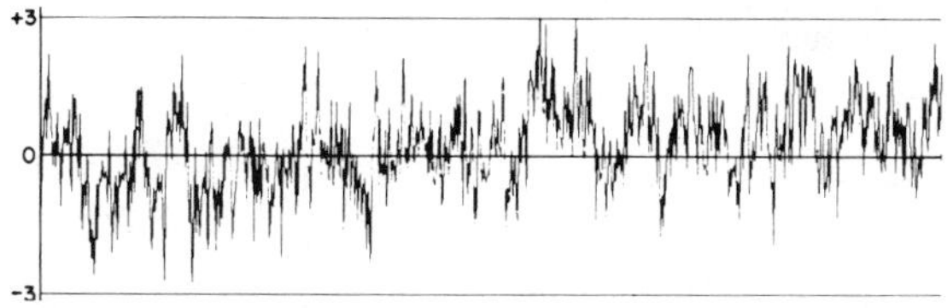

Figure 3 Sample of 1000 values of an approximation to *fractional Gaussian noise*. In this instance, $H=J=0.7$, meaning that the strength of long-run persistence is high.

fractional Gaussian noises for various values of H. Even on a casual glance the effect of the low frequency is obvious [6].

2. When $H = 0.5$, the $Z(t+1) - Z(t)$ are uncorrelated for $k > 1$; hence $B_{0.5}(t)$ is ordinary Brownian motion.
3. In the case $0 < H < 0.5$, called *antipersistent*, $\sum_{k=-\infty}^{+\infty} |r(k)| < \infty$. On the other hand, $\sum_{k=-\infty}^{+\infty} r(k) = 0$. The spectral density is zero at the origin.

It is interesting in this light to recall a measure of the span of memory of a process due to G. I. Taylor: it is the integral of its covariance. If $0.5 < H < 1$, this measure correctly asserts that the memory of $X(t)$ is infinite; if $0 < H < 0.5$, however, Taylor asserts that the memory vanishes, whereas it is in fact infinite.

This feature, and the correlation properties listed above, used to be viewed as "pathological," and to be excluded without comment by almost all of time-series* analysis. In fact (largely through the work of Mandelbrot [4]) they have become indispensable in the modeling of natural phenomena (*see* SCALING PHENOMENA). One class of examples arises in economics. Economic time series "typically" exhibit cycles of all orders of magnitude, the slowest cycles having periods of duration comparable to the total sample size. The sample spectra of such series show no sharp "pure period" but a spectral density with a sharp peak near frequencies close to the inverse of the sample size. Another class of examples arises in the study of fluctuations in solids. Many such fluctuations are called "$1:f$ noises" because their sample spectral density takes the form $\lambda^{-\beta}$ with λ the frequency and an exponent β frequently close to 1. A third class of phenomena with extremely long interdependence is encountered in hydrology*, where the fact that the Hurst coefficient* J satisfies $J > 0.5$ is related to $H > 0.5$ and persistence.

CONSTRUCTIVE DEFINITIONS

USING FRACTIONAL INTEGRODIFFERENTIATION. fBm can be represented in terms of the ordinary Brownian process (Wiener process) $B(t)$ as a fractional integrodifferential in the sense of Leibniz, Abel, Riemann, Liouville, and Weyl. The increment $B_H(t_1) - B_H(t_2)$ is the convergent difference of the values taken for $t = t_2$ and $t = t_1$ by the divergent integral

$$B_H(t) = \frac{1}{\Gamma(H+0.5)} \int_{-\infty}^{t} (t-s)^{H-0.5}\, dB(s).$$

This formula's denominator ensures that, when $H - 0.5$ is an integer, a fractional integral becomes an ordinary repeated integral. The motions with $0.5 < H < 1$ involve fractional *integration*, which is a smoothing operation, which is why they are more persistent than $B(t)$. The motions with $0 < H < 0.5$ involve fractional *differentiation*, which is unsmoothing, which is why they are antipersistent.

Bilateral integrodifferentiation is often more convenient:

$$2\Gamma(H+0.5)B_H(t) = \left\{ \int_{-\infty}^{t} (t-s)^{H-0.5}\, dB(s) - \int_{t}^{\infty} (t-s)^{H-0.5}\, dB(s) \right\}.$$

USING A SPECTRAL FORMULA. As shown by Hunt [1, p. 67], a constant V_H^* exists such that

$$B_H(t_2) - B_H(t_1) = V_H^* \int_0^{\infty} \left(e^{2\pi i\lambda t_2} - e^{2\pi i\lambda t_1}\right)\lambda^{-H-0.5}\, dB(\lambda).$$

EXTRAPOLATION AND INTERPOLATION OF $B_H(t)$

G_1 and G_2 being two dependent Gaussian random variables with zero mean, it is known that

$$E[G_1 \mid G_2]/G_2 = E[G_1 G_2]/E[G_2^2].$$

Thus, knowing $B_H(0) = 0$ and $B_H(T)$, we find for $t = sT$ that

$$\frac{E[B_H(t) \mid B_H(T)]}{B_H(T)} = 0.5\left[|s|^{2H} + 1 - |s-1|^{2H}\right].$$

For Brownian motion, $H = 0.5$, $Q_H(s)$ is represented by a kinked curve made up of sections of straight lines. When $0.5 < H < 1$, $Q_H(s)$ has a continuous derivative $Q'_H(s)$ which satisfies $0 < Q'_H(0) = Q'_H(1) < 1$ and $Q'_H(0.5) > 1$. Finally, for $0 < H < 0.5$, $Q_H(s)$ is differentiable except at $s = 0$ and $s = 1$, where it has cusps. When $0.5 < H < 1$, $Q_H(s) \sim H|s|^{2H-1}$ for s large, and the extrapolation involves a *nonlinear* "*pseudotrend*" that diverges to infinity. When $0 < H < 0.5$, $Q_H(s) \sim 0.5$ for s large, and the extrapolation has a *nonlinear* "*pseudotrend*" that converges to $0.5[B_H(0) + B_H(T)]$.

VARIABLE TRENDS. Time series $X(t)$ without "seasonal effects" are reputed to be decomposable into an "oscillatory component" and a "linear trend." The trend is interpreted as an estimate of $E[X(t+\tau) - X(t)]$, and attributed to major "causal" changes in the mechanism generating $X(t)$. The oscillation, on the contrary, is taken to be an "uncontrollable" stationary process, hopefully free of low-frequency components.

It is obvious that, in the case of fBm with $H \neq 0.5$, the task of distinguishing the linear trend Δt from the nonlinear "trends" just described is practically impossible; fBm falls outside the usual dichotomy between causal trends and random perturbations.

HIERARCHICAL APPROXIMATIONS

The asymptotic correlation of $B_H(t+1) - B_H(t)$ is the weighted sum of exponential correlations. The equation

$$|s|^{2H-2} = \int_0^\infty e^{-|s|u} \frac{u^{1-2H}}{\Gamma(2-2H)}\, du$$

represents $X(t)$ as a continuous sum of Gauss–Markov processes. The *fast fractional Gaussian noise* (FFGN) approximation [2] replaces the integral by a discrete finite sum of independent Gauss–Markov processes, whose number $N(T)$ must depend on the desired sample length T. This dependence extends to all other approximations, and is noteworthy. In the standard processes, which have short-run dependence, samples can be generated using an algorithm *independent* of sample length. When a process is long-run dependent, on the contrary, the approximation consists in replacing an infinite span of dependence by a finite one of the order of the sample size. Hence, the algorithm must depend on T.

A second approximation is the broken-line process advanced by Mejia (see ref. 3).

Either of these mathematical decompositions seems to imply the existence of a multilevel *hierarchy* of underlying mechanisms whose total effect is equivalent to that of self similarity. The idea of hierarchy first entered through turbulence theory by the hand of Richardson, and is useful in many domains of physics; but it is neither easy to demonstrate the physical reality of hierarchical levels nor to explain why their contributions should add up to an asymptotic power law for the correlation; see Mandelbrot [4].

References

[1] Hunt, G. A. (1951). *Trans. Amer. Math. Soc.*, **71**, 38–69.

[2] Mandelbrot, B. B. (1971). *Water Resour. Res.*, **7**, 543–553.

[3] Mandelbrot, B. B. (1972). *Water Resour. Res.*, **8**, 1354–1356.

[4] Mandelbrot, B. B. (1982). *The Fractal Geometry of Nature*. W. H. Freeman, San Francisco. (Applications and references.)

[5] Mandelbrot, B. B. and Van Ness, J. (1968). *SIAM Rev.*, **10**, 422–437. (Basic mathematical properties and past literature.)

[6] Mandelbrot, B. B. and Wallis, J. R. (1969). *Water Resour. Res.*, **5**, 228–267. (Simulations.)

(BROWNIAN MOTION
FRACTALS
FRACTIONAL INTEGRODIFFERENTIATION
GAUSSIAN PROCESSES
HURST COEFFICIENT (RESCALED RANGE ANALYSIS))

BENOIT B. MANDELBROT

FRACTIONAL FACTORIAL DESIGNS

Ever since Fisher* [5] introduced the notion of *factorial experimentation*, a tremendous development of ideas in this area has taken place. In factorial experimentation (originally called "complex experimentation" by Fisher) several factors may be studied simultaneously instead of experimenting with them one at a time (*see* FACTORIAL EXPERIMENTS). For example, in an agricultural experiment we may assess the effects of nitrogen and phosphate fertilizers on the yield of wheat by carrying out an experiment with various combinations of levels of the two fertilizers. If the experimenter specifies k_1 levels of the nitrogen fertilizer and k_2 levels of the phosphate fertilizer and all the $k_1 \cdot k_2$ combinations are used in the experiment, such an experiment is called a *complete factorial*. If fewer than the $k_1 \cdot k_2$ combinations are used, the term *fractional factorial* has been used in the literature for such an experiment.

Yates [19] provided the first comprehensive approach to complete factorials and presented some ideas in fractional factorials. It was Fisher [6], however, who systematically constructed classes of fractional factorials, where each factor had the same prime number of levels. These designs came about as a by-product of the construction of *confounded designs* (*see* CONFOUNDING). The formal approach to fractional factorial designs is due to Finney [4].

Many problems in factorial experimentation turn out to have a geometric, algebraic or combinatorial flavor. As a consequence, mathematical structures, such as finite groups, finite rings, finite fields, and finite geometries, can be successfully used in elucidating and resolving many issues. Fisher [6, 7] used finite Abelian groups and Bose [1] relied heavily on finite Euclidean and finite projective geometries in the construction and enumeration of *regular fractions* of symmetrical prime powered factorials. More recently, a general algebraic-combinatorial theory of fractional factorials has been developed by Pesotan et al. [11]. This theory relied on some invariance results of Srivastava et al. [18] and several unsolved problems associated with it have been reported by Raktoe and Pesotan [13].

In the following sections a systematic discussion is presented on the most important aspects of fractional factorials.

FACTORIAL ARRANGEMENTS AND FRACTIONAL FACTORIAL DESIGNS

In this section use is made of notation and definitions developed in ref. 14 and used in Hedayat et al. [9] and Federer et al. [3].

A distinction will be made between sets and collections; a set lists distinct elements, whereas in a collection repetitions are allowed. In many scientific investigations experimenters are interested in studying the effects of t controllable variables called *factors*, to be denoted by $F_1, F_2, \ldots, F_t$. For each factor there will be a specified range of values of interest to the experimenter. These will be called *levels* of the factors, to be indicated by $G_1, G_2, \ldots, G_t$. A factor will be called *quantitative* if the underlying levels of interest are real numbers and *qualitative* if the levels are specified qualities rather than real numbers. Denote the cardinality of G_i by k_i, and throughout assume that the G_i's are finite sets. The sets $G_1, G_2, \ldots, G_t$ are potential levels and it is not necessarily true that all of them will be used in a particular experiment.

Let G be the Cartesian product of the G_i's, i.e., $G = \times_{i=1}^{t} G_i$, where the symbol $\times$ denotes the Cartesian product. The set G together with the F_i's is often referred to as the $k_1 \times k_2 \times \cdots \times k_t$ *factorial* or $k_1 \times k_2 \times \cdots \times k_t$ *crossed classification*. An element of G is called a *treatment* and G itself is called the *factor space* or *space of treatments*. In the literature, the terms "treatment combinations," "assemblies," "runs," and "subclasses" are also frequently used for treatments.

Let N be the cardinality of G (i.e., $N = \prod_{i=1}^{t} k_i$) and let G be indexed by the index set $\{1, 2, \ldots, N\}$. Then a *factorial arrangement* or *factorial design* with parameters $k_1, k_2, \ldots, k_t, m, n, r_1, r_2, \ldots, r_N$, is defined to be a *collection* of n treatments such that the jth treatment of G has multiplicity $r_j \geqslant 0$, m is the number of nonzero r_j's, and $\sum_{j=1}^{N} r_j = n > 0$. A factorial arrangement is denoted by the symbol FA(k_1, $k_2, \ldots, k_t; m; n; r_1, r_2, \ldots, r_N$) or simply by FA if everything is clear from the context.

In the discipline of statistics the multiplicity r_j is referred to as the *replication number* of the jth treatment, i.e., how many times the jth treatment is repeated in the factorial arrangement. The definition of a factorial arrangement adopted here is in complete agreement with that of a general t-way crossed classification with r_j observations on the jth treatment.

A factorial arrangement is said to be a *complete factorial arrangement* or a *complete replicate* if $r_j > 0$ for all $j = 1, 2, \ldots, N$. It is said to be a *minimal complete factorial arrangement* if $r_j = 1$ for all j. Note that a minimal complete factorial arrangement is a single copy of the factor space G. A complete factorial arrangement such that $r_j = r$ for all j is said to consist of r *complete replicates*.

A factorial arrangement is symmetrical if $k_i = s$ for all $i = 1, 2, \ldots, t$, and otherwise it is *asymmetrical* or *mixed*. An FA is *prime powered* if $k_i = p_i^{u_i}$ such that for each i, p_i is a prime and u_i is a natural number greater than or equal to 1. It follows that a factorial arrangement can be *symmetrical prime powered* or *mixed prime powered*.

A factorial arrangement is said to be an *incomplete factorial arrangement* or a *fractional factorial design*, or more simply, a *fractional replicate*, if some but not all r_j's are equal to zero. A fractional replicate is denoted by FFA($k_1, k_2, \ldots, k_t; m; n; r_1, r_2, \ldots, r_N$) or by FFA if it is clear from the context.

If the levels of the ith factor are made to correspond to the residue classes modulo k_i, i.e., $G_i = \{0, 1, 2, \ldots, k_i - 1\}$, then under componentwise addition modulo the k_i's it can be shown that G is an *Abelian group*. For the symmetrical prime-powered s^t factorial, each of the G_i's can be identified with the *Galois field** GF(s), where $s = p^u$ and p a prime. It can then be established that G is a t-dimensional vector space over GF(s). From a geometric viewpoint such a vector space is known as a *finite Euclidean geometry* $EG(t, s)$ of dimension t over GF(s).

Before proceeding we present an example to illustrate the concepts defined so far.

Example. An industrial experiment was planned to study the effect of both curing time and composition on the tensile strength of plastic compounds. Three times—1 hour, 2 hours, and 4 hours—were selected and four mixes—A, B, C, and D—were prepared. Observations were to be made on combinations of curing times and compositions. This is a 3×4 factorial with the quantitative factor F_1 = curing time, the qualitative factor F_2 = composition, $G_1 = \{1$ hour, 2 hours, 4 hours$\} = \{1, 2, 4\}$, and $G_2 = \{A, B, C, D\}$. The set of treatment combinations is

$$G = \{(1,A),(1,B),(1,C),(1,D),\\ (2,A),(2,B),(2,C),(2,D),\\ (4,A),(4,B),(4,C),(4,D)\}.$$

Note that in using $G_1 = \{1, 2, 4\}$ we have deleted units. Indeed, frequently we use *labels* to indicate the levels of factors. For our example it is common to use the following sets of labels: $G_1^* = \{0, 1, 2\}$ and $G_2^* = \{0, 1, 2, 3\}$. The set of treatment combinations is then depicted as

$$G^* = \{(0,0),(0,1),(0,2),(0,3),(1,0),(1,1),\\ (1,2),(1,3),(2,0),(2,1),(2,2),(2,3)\}$$

with each element having the obvious real meaning, e.g., $(2, 0) = (4$ hours, $A)$. Since $k_1 = 3$ and $k_2 = 4 = 2^2$, all factorial arrangements in this example are mixed prime powered. The factor space G, or its equivalent representation G^*, is a minimal complete factorial arrangement. The following factorial arrangement in terms of G^* is complete but not minimal:

$$\text{FA}(3,4;12;15;2,3,1,1,1,1,1,1,1,1,1,1)\\ = \{(0,0),(0,0),(0,1),(0,1),(0,1),(0,2),\\ (0,3),(1,0),(1,1),(1,2),(1,3),\\ (2,0),(2,1),(2,2),(2,3)\}.$$

An example of a fractional replicate in terms

of G^* is

$$\mathrm{FA}(3,4;5;6;1,1,0,0,0,0,0,1,1,2,0,0) = \{(0,0),(0,1),(1,3),(2,0),(2,1),(2,1)\}.$$

Finally, note that G^* under componentwise addition modulo 3 and modulo 4 is an Abelian group of order 12.

THE LINEAR MODEL AND ESTIMATION OF EFFECTS FOR A FRACTIONAL FACTORIAL EXPERIMENT

In this section the linear model for analyzing data from an experiment using a fractional factorial design is introduced. The approach adopted here can be found in several places in the literature and conforms to the usual linear model notation found in Graybill [8] and Searle [16] (*see* GENERAL LINEAR MODEL).

Let D be a factorial arrangement. With each treatment g in D we associate a random variable Y_g, which is called an *observation* or response. We will assume the *univariate* case; i.e., Y_g will be one-dimensional and assume values in a one-dimensional Euclidean set. Let $\mathbf{Y}_D$ be the $n \times 1$ vector of observations for the factorial arrangement D, where the components of $\mathbf{Y}_D$ are the Y_g's.

In most settings a linear model is associated with a minimal complete factorial arrangement D^* in the following way:

$$\begin{aligned} &\text{(a)} \quad E[\mathbf{Y}_{D^*}] = \mathbf{X}_{D^*}\boldsymbol{\beta}, \quad \text{and} \\ &\text{(b)} \quad \operatorname{cov}[\mathbf{Y}_{D^*}] = \sigma^2 \mathbf{I}_N, \end{aligned} \qquad (1)$$

where $\mathbf{X}_{D^*}$ is a known $N \times N$ *design matrix**, $\mathbf{I}_N$ is the identity matrix of order N, and $\boldsymbol{\beta}$ is the vector of N parameters consisting of $N-1$ *factorial effects* and the *mean*. If $\mathbf{X}_{D^*} = \mathbf{X}_1 \otimes \mathbf{X}_2 \otimes \cdots \otimes \mathbf{X}_t$, where each $\mathbf{X}_i$ is a $k_i \times k_i$ orthogonal matrix with each entry in the first column equal to $1/\sqrt{k_i}$, and $\otimes$ denotes Kronecker product*, then $\boldsymbol{\beta}$ is a vector of factorial effects under the *product definition*. This approach is especially applicable under the *orthogonal polynomial** and *Helmert polynomial* settings. For the symmetrical prime-powered factorial, the entries of $\mathbf{X}_{D^*}$ can be obtained from a fixed basic orthogonal matrix by using the *geometric definition* of effects. The model for any fractional factorial design D is induced by (1) in the sense that the design matrix $\mathbf{X}_D$ is read off from $\mathbf{X}_{D^*}$ taking repetitions of treatment combinations into account.

The most practical partitioning of the parametric vector is $\boldsymbol{\beta}' = (\boldsymbol{\beta}_1' \,\vdots\, \boldsymbol{\beta}_2' \,\vdots\, \boldsymbol{\beta}_3')$, where $\boldsymbol{\beta}_1$ is an $N_1 \times 1$ vector to be estimated, $\boldsymbol{\beta}_2$ is an $N_2 \times 1$ vector not of interest and not assumed to be known, and $\boldsymbol{\beta}_3$ is an $N_3 \times 1$ vector of parameters assumed to be known (which without loss of generality can be taken to be zero) such that $1 \leqslant N_1 \leqslant N$, $0 \leqslant N_2 \leqslant N-1$, and $0 \leqslant N_3 = N - N_1 - N_2 \leqslant N-1$. This partitioning explicitly leads to the following four cases:

$$\begin{aligned} &\text{(a)} \quad N_1 = N, \quad N_2 = N_3 = 0, \\ &\text{(b)} \quad N_2 = 0, \quad N_3 \neq 0, \\ &\text{(c)} \quad N_2 \neq 0, \quad N_3 \neq 0, \text{ and} \\ &\text{(d)} \quad N_2 \neq 0, \quad N_3 = 0. \end{aligned} \qquad (2)$$

Case (a) may be viewed as a special case of (b) by letting $\boldsymbol{\beta}_1$ exhaust $\boldsymbol{\beta}$ so that $N_3 = 0$. Similarly, case (d) can be considered a special case of (c) by letting $\boldsymbol{\beta}_1$ and $\boldsymbol{\beta}_2$ exhaust $\boldsymbol{\beta}$ so that $N_3 = 0$. It thus suffices to analyze cases (b) and (c) in (2).

Denote a parameter in $\boldsymbol{\beta}$ by the symbol $\boldsymbol{\beta}_1^{x_1}\boldsymbol{\beta}_2^{x_2} \cdots \boldsymbol{\beta}_t^{x_t}$, where $(x_1, x_2, \ldots, x_t)$ is an element of $G = \times_{i=1}^{t} G_i$, $G_1 = \{0, 1, 2, \ldots, k_i - 1\}$. Then $\boldsymbol{\beta}_1^0\boldsymbol{\beta}_2^0 \cdots \boldsymbol{\beta}_t^0$ is called the mean, and a factorial effect $\boldsymbol{\beta}_1^{u_1}\boldsymbol{\beta}_2^{u_2} \cdots \boldsymbol{\beta}_t^{u_t}$ is said to be of *order* k if exactly k of the exponents are nonzero. A fractional design D is said to be of *resolution* R if all factorial effects up to order k are *estimable*, where k is the greatest integer less than $R/2$, under the assumption that all factorial effects of order $R-k$ and higher are zero. When $R = 2r$ the design is known as a design of *even resolution*, and for $R = 2r+1$ the design is said to be of odd *resolution*. Thus resolution III designs allow estimation of all main effects (i.e., effects of order 1) under the assumption that all interactions* (i.e., effects of order 2 or higher) are zero. Designs of resolution IV allow estimation of all main effects under the assumption that interactions of order 3 or higher (i.e., effects of order greater than or equal to 3) are zero, without assuming that two-factor interactions (i.e., effects of order 2) are equal to zero. Designs of odd resolution belong to case (b) and designs of even resolution belong to case (c) of (2).

The model for any fractional factorial design D under case (b) is given by

$$E[\mathbf{Y}_D] = \mathbf{X}_{D1}\boldsymbol{\beta}_1 \qquad \operatorname{cov}[\mathbf{Y}_D] = \sigma^2\mathbf{I}_n, \quad (3)$$

where the design matrix $\mathbf{X}_{D1}$ is simply read off from $\mathbf{X}_{D*}$ of (1) taking repetitions of treatment combinations into account. Similarly, the design matrix $[\mathbf{X}_{D1} \vdots \mathbf{X}_{D2}]$ for a design D under case (c) is obtained from $\mathbf{X}_{D*}$ as

$$\begin{aligned} E[\mathbf{Y}_D] &= [\mathbf{X}_{D1} \vdots \mathbf{X}_{D2}][\boldsymbol{\beta}_1' \vdots \boldsymbol{\beta}_2']' \\ \operatorname{cov}[\mathbf{Y}_D] &= \sigma^2\mathbf{I}_n. \end{aligned} \quad (4)$$

The best linear unbiased estimator (BLUE) of $\boldsymbol{\beta}_1$ and the covariance for the two cases are given by (5) and (6), respectively:

$$\left.\begin{aligned} \hat{\boldsymbol{\beta}}_1 &= [\mathbf{X}_{D1}'\mathbf{X}_{D1}]^{-1}\mathbf{X}_{D1}'\mathbf{Y}_D \\ \operatorname{cov}[\hat{\boldsymbol{\beta}}_1] &= [\mathbf{X}_{D1}'\mathbf{X}_{D1}]^{-1}\sigma^2, \end{aligned}\right\} \quad (5)$$

$$\left.\begin{aligned} \hat{\boldsymbol{\beta}}_1 &= [\mathbf{X}_{D1}'\mathbf{X}_{D1} - \mathbf{X}_{D1}'\mathbf{X}_{D2}(\mathbf{X}_{D2}'\mathbf{X}_{D2})^{-}\mathbf{X}_{D2}'\mathbf{X}_{D1}]^{-1} \\ &\quad \times [\mathbf{X}_{D1}' - \mathbf{X}_{D1}'\mathbf{X}_{D2}(\mathbf{X}_{D2}'\mathbf{X}_{D2})^{-}\mathbf{X}_{D2}]\mathbf{Y}_D \\ \operatorname{cov}[\hat{\boldsymbol{\beta}}_1] &= [\mathbf{X}_{D1}'\mathbf{X}_{D1} - \mathbf{X}_{D1}'\mathbf{X}_{D2} \\ &\quad \times (\mathbf{X}_{D2}'\mathbf{X}_{D2})^{-}\mathbf{X}_{D2}'\mathbf{X}_{D1}]^{-1}\sigma^2. \end{aligned}\right\} \quad (6)$$

In expression (6) A^- denotes a *generalized inverse** of A. For brevity the covariances in either case will be written as $\mathbf{M}_D^{-1}\sigma^2$, where $\mathbf{M}_D^{-1}$ is known as the *covariance matrix* and $\mathbf{M}_D$ itself is called the *information matrix*. An unbiased *estimator of* σ^2 is obtained by utilizing the BLUEs in (5) and (6), i.e.,

$$\hat{\sigma}^2 = \frac{1}{n - N_1}[(\mathbf{Y}_D - \mathbf{X}_{D1}\hat{\boldsymbol{\beta}}_1)'(\mathbf{Y}_D - \mathbf{X}_{D1}\hat{\boldsymbol{\beta}}_1)]$$

for case (b), and for case (c),

$$\hat{\sigma}^2 = \frac{1}{n - N_1}\left[(\mathbf{Y}_D - \mathbf{X}_{D1}\hat{\boldsymbol{\beta}}_1 - \mathbf{X}_{D2}\tilde{\boldsymbol{\beta}}_2)' \times (\mathbf{Y}_D - \mathbf{X}_{D1}\hat{\boldsymbol{\beta}}_1 - \mathbf{X}_{D2}\tilde{\boldsymbol{\beta}}_2)\right],$$

where

$$\tilde{\boldsymbol{\beta}}_2 = (\mathbf{X}_{D2}'\mathbf{X}_{D2})^{-}[\mathbf{X}_{D2}'\mathbf{Y}_D - \mathbf{X}_{D2}'\mathbf{X}_{D1}\hat{\boldsymbol{\beta}}_1].$$

Under the assumption of normality, *tests of hypotheses** and *confidence interval** *estimators* for the vector $\boldsymbol{\beta}_1$ can be obtained as indicated, for example, in Graybill [8].

OPTIMAL FRACTIONAL FACTORIAL EXPERIMENTS

The problem of choosing an optimal fractional factorial design is discussed in this section using optimality criteria developed by Kiefer [10] and Hedayat et al. [9].

Let $\mathscr{D}$ be a class of competing fractional factorial designs in either setting (b) or (c) of the partitioning in (2). Assume that each design $D \in \mathscr{D}$ is capable of providing an unbiased estimator for $\boldsymbol{\beta}_1$. There are several optimality criteria based on the covariance matrix $\mathbf{M}_D^{-1}$ of the BLUE of $\boldsymbol{\beta}_1$. The most popular ones are based on the spectrum or set of characteristic roots of $\mathbf{M}_D^{-1}$. Denoting the roots of $\mathbf{M}_D^{-1}$ in increasing order of magnitude by $\lambda_1, \lambda_2, \ldots, \lambda_{N_1}$, we have the following functionals:

$$\begin{aligned} &\text{(a)} \quad \prod \lambda_i = |\mathbf{M}_D^{-1}|, \\ &\text{(b)} \quad \sum \lambda_i = \operatorname{trace} \mathbf{M}_D^{-1}, \\ &\text{(c)} \quad \lambda_{N_1} = \max(\lambda_1, \lambda_2, \ldots, \lambda_{N_1}). \end{aligned} \quad (7)$$

A design that minimizes over $\mathscr{D}$ the criteria in (a), (b), and (c) of (7) is known as a *d-optimal design*, an *a-optimal design*, and an *e-optimal design*, respectively. Statistical interpretations of these criteria are available and they may be found in Kiefer [10]. It should be noted that one may base these optimality criteria on the spectrum of the information matrix $\mathbf{M}_D$ rather than on that of $\mathbf{M}_D^{-1}$.

Another criterion, which does not rely on $\mathbf{M}_D^{-1}$, was developed by Hedayat et al. [9]. If in the settings (b) and (c) of (2) the assumption that $\boldsymbol{\beta}_3 = \mathbf{0}$ is in doubt, then

$$E[\hat{\boldsymbol{\beta}}_1] = \boldsymbol{\beta}_1 + \mathbf{A}_D\boldsymbol{\beta}_3,$$

where $\mathbf{A}_D$ is the *alias matrix* of the design D relative to $\boldsymbol{\beta}_1$ and $\boldsymbol{\beta}_3$ [e.g., for case (c) of (2) $\mathbf{A}_D = (\mathbf{X}_{D1}'\mathbf{X}_{D1})^{-1}\mathbf{X}_{D1}'\mathbf{X}_{D2}$]. The *norm* $\|\mathbf{A}_D\| = [\operatorname{trace}(\mathbf{A}_D'\mathbf{A}_D)]^{1/2}$ was proposed by Hedayat et al. [9] for the selection of an optimal design and a design is said to be *alias optimal* if it minimizes $\|\mathbf{A}_D\|$ over $\mathscr{D}$.

Apart from these criteria one may impose other desirable properties on $\mathscr{D}$ for the selection of a design, such as *orthogonality* (i.e., $\mathbf{M}_D^{-1}$ is diagonal) or *balancedness* (i.e., $\mathbf{M}_D^{-1} = a\mathbf{I} + b\mathbf{J}$, where $\mathbf{J}$ is a square matrix of order N_1 all whose elements are 1's). Orthogonality implies *uncorrelatedness* of the estimators of the elements of β_1 and balancedness implies equal variances and equal covariances of the estimators. These concepts have been generalized to *partial orthogonality* and *partial balancedness* (e.g., see Srivastava and Anderson [17]).

If the mean is the first element in $\boldsymbol{\beta}_1$, then the first element of the vector $\boldsymbol{\beta}_1 + \mathbf{A}_D\boldsymbol{\beta}_3$ is known as the *generalized defining relationship* of D relative to $\boldsymbol{\beta}_1$ and $\boldsymbol{\beta}_3$ and the whole vector itself is called the *aliasing structure* of D relative to $\boldsymbol{\beta}_1$ and $\boldsymbol{\beta}_3$. The aliasing structure for a symmetrical prime-powered fractional factorial design becomes tractable via group-theoretic techniques if the design is *regular*; i.e., it is a *subspace* (or coset of a subspace) when the complete set of treatment combinations is viewed as the t-dimensional vector space over the Galois field GF(s) (see, e.g., Raktoe et al. [15]). For the 2^t factorial the generalized defining relationship is known as the *defining contrast* (see, e.g., Cochran and Cox [2]).

CONSTRUCTION OF FRACTIONAL FACTORIAL DESIGNS

The construction of an optimal design in either setting (b) or (c) of (2) is by no means a simple matter since it involves the design parameters $k_1, k_2, \ldots, k_t, m, n, r_1, r_2, \ldots, r_N, N_1, N_2, N_3$ and selection of the optimality criterion. Indeed, there is a formidable combinatorial problem associated with finding for the $k_1 \times k_2 \times \cdots \times k_t$ factorial all designs which simply lead to unbiased estimation of β_1 in (b) or (c) of (2), let alone obtaining the optimal ones. There is no unique method available for all factorials, and depending on the nature of the factorial one may utilize various techniques to obtain useful designs.

Raktoe et al. list 21 methods in ref. 14. With an additional 3, these are: (1) trial and error and/or computer methods; (2) Hadamard matrix* methods; (3) confounding techniques; (4) group theory methods; (5) finite geometrical methods; (6) algebraic decomposition techniques; (7) combinatorial-topological methods; (8) foldover techniques; (9) collapsing of levels methods; (10) composition (direct product and direct sum) methods; (11) permutation of levels and/or factors methods; (12) coding theory methods; (13) orthogonal array techniques; (14) partially balanced array techniques; (15) orthogonal Latin square* methods; (16) block design techniques; (17) weighing design* techniques; (18) F-square techniques; (19) lattice design* methods; (20) graph-theoretical* methods; (21) one-a-time methods; (22) inspection methods; (23) patterned matrix methods; and (24) cutting and adjoining matrix methods.

In order to demonstrate the complexity of the problems we illustrate the general combinatorial problem for case (b) of (2) for resolution III designs in its simplest possible setting. Assume that the mean is of interest in estimation so that a *minimal* (or *saturated*) resolution III design calls for $N_1 = m = n = \sum k_i - t + 1$ distinct treatment combinations for estimation of $\boldsymbol{\beta}_1$ since there are $N_1 = \sum(k_i - 1) + 1$ parameters in $\boldsymbol{\beta}_1$. We have seen that a necessary and sufficient condition for estimating $\boldsymbol{\beta}_1$ is that the rank of $\mathbf{X}_{D1}$ (or of $\mathbf{X}'_{D1}\mathbf{X}_{D1}$) is equal to m. Denote the class of possible designs by $\mathscr{D}_m$; then clearly its cardinality is equal to $|\mathscr{D}_m| = C_m^N = (N!)/\{(m!)(N-m)!\}$. Denote by $\mathscr{D}_{m,m}$ the class of designs in $\mathscr{D}_m$ which allow estimation of $\boldsymbol{\beta}_1$ and by $\mathscr{D}_{m,0}$ the singular class of designs (i.e., $|\mathbf{X}_{D1}| = 0$). The cardinality of $\mathscr{D}_{m,0}$ (and hence of $\mathscr{D}_{m,m}$) is not known in general at present. For the *2^t factorial* it has been enumerated for $t \leqslant 7$ (see Raktoe [12] and Table 1). It may be shown that the proportion of singular designs goes to zero as t goes to infinity for the 2^t factorial design.

Using Helmert polynomials and corresponding nonnormalized columnwise or-

Table 1 Cardinality of Classes of 2^t Factorial Designs

	t					
	2	3	4	5	6	7
$\lvert\mathscr{D}_m\rvert = C^{2^t}_{t+1}$	4	70	4368	906,192	621,216,192	1,429,702,652,400
$\lvert\mathscr{D}_{m,0}\rvert$	0	12	1360	350,000	255,036,992	571,462,430,224
$\lvert\mathscr{D}_{m,m}\rvert$	4	58	3008	556,192	366,179,200	858,240,222,176
$\lvert\mathscr{D}_{m,0}\rvert / C^{2^t}_{t+1}$	0	0.1714	0.3114	0.3862	0.4105	0.3997
$\lvert\mathscr{D}_{m,m}\rvert / C^{2^t}_{t+1}$	1	0.8286	0.6886	0.6138	0.5895	0.6003

thogonal matrices in (1) for the $k_1 \times k_2 \times \cdots \times k_t$ factorial, one may deduce under the determinant criterion that for a saturated resolution III design $D \in \mathscr{D}_m$:

$$\prod_{i=1}^{t} (k_i!)^2 \leqslant |\det \mathbf{X}'_{D1}\mathbf{X}_{D1}| \leqslant \prod_{i=1}^{t} (k_i!)^2 m^m \prod_{i=1}^{t} k_i^{-k_i}. \quad (8)$$

The lower bound in (7) is attained by the *one-at-a-time* design $D^* = \{(000\ldots0), (100\ldots0), (200\ldots0), \ldots, (k_1 - 100\ldots0), \ldots, (010\ldots0), (020\ldots0), \ldots, (0k_2 - 10\ldots0), \ldots, (000\ldots01), (000\ldots02), \ldots, (000\ldots0k_t - 1)\}$, which is a *least d-optimal* resolution III design. The upper bound is attained if and only if an orthogonal design can be constructed. For the 2^t factorial the bound is achieved whenever a *Hadamard matrix** of order $t+1$ exists. A necessary condition for this is that $t + 1 = 0 \pmod 4$. Hadamard matrices have been constructed for all $t + 1 \leqslant 200$ so that minimal d-optimal resolution III designs are available for all 2^t factorials, where $t = 4s - 1$ and $s = 1, 2, \ldots, 50$. For $t + 1 \neq 0 \pmod 4$ other techniques have been used apart from computer methods. Examples of minimal d-optimal resolution III designs for the 2^t factorial are given in Table 2.) These designs are such that the spectrum of the information matrix is invariant under the group of level permutations.

Some of the references listed below deal with constructions of other types of fractional factorial designs, such as resolution IV and V designs which are not necessarily minimal.

Table 2 Minimal d-optimal Resolution III Designs

t	d-Optimal Design
2	{(00), (10), (01)}
3	{(000), (110), (101), (011)}
4	{(0000), (1110), (1101), (1011), (0111)}
5	{(00000), (11100), (11010), (11001), (10111), (01111)}
6	{(000000), (111000), (110110), (110101), (101100), (101011), (011111)}
7	{(0000000), (1111000), (1100110), (1010101), (1001011), (0110011), (0101101), (0011110)}

References

[1] Bose, R. C. (1947). *Sankhyā*, **8**, 107–166.

[2] Cochran, W. G. and Cox, G. M. (1957). *Experimental Designs*, 2nd ed. Wiley, New York.

[3] Federer, W. T., Hedayat, A., and Raktoe, B. L. (1975). In *A Survey of Statistical Design and Linear Models*, J. N. Srivastava, ed. North-Holland, Amsterdam, pp. 145–153.

[4] Finney, D. J. (1945). *Ann. Eugen.*, **12**, 291–301.

[5] Fisher, R. A. (1926). *J. Minist. Agric.*, **33**, 503–513.

[6] Fisher, R. A. (1942). *Ann. Eugen.*, **11**, 341–353.

[7] Fisher, R. A. (1945). *Ann. Eugen.*, **12**, 283–290.

[8] Graybill, F. A. (1976). *Theory and Application of the Linear Model*. Duxbury Press, North Scituate, Mass.

[9] Hedayat, A., Raktoe, B. L., and Federer, W. T. (1974). *Ann. Statist.*, **2**, 650–660.

[10] Kiefer, J. C. (1959). *J. R. Statist. Soc. B*, **21**, 272–304.

[11] Pesotan, H., Raktoe, B. L., and Federer, W. T. (1975). *Ann. Inst. Statist. Math.*, **27**, 55–80.

[12] Raktoe, B. L. (1979). *Amer. Math. Monthly*, **86**, 49.

[13] Raktoe, B. L. and Pesotan, H. (1974). *Sankhyā B*, **46**, 457–461.

[14] Raktoe, B. L., Hedayat, A., and Federer, W. T. (1981). *Factorial Designs*. Wiley, New York.

[15] Raktoe, B. L., Pesotan, H., and Federer, W. T. (1980). *Canad. J. Statist.*, **8**, 65–77.

[16] Searle, S. R. (1971). *Linear Models*. Wiley, New York.

[17] Srivastava, J. N. and Anderson, D. (1970). *J. Amer. Statist. Ass.*, **65**, 828–843.

[18] Srivastava, J. N., Raktoe, B. L., and Pesotan, H. (1976). *Ann. Statist.*, **4**, 423–430.

[19] Yates, F. (1935). *J. R. Statist. Soc. B*, **2**, 181–247.

Bibliography

See the following works, as well as the references just cited, for more information on the topic of fractional factorial designs.

Addelman, S. (1963). *J. Amer. Statist. Ass.*, **58**, 45–71.

Addelman, S. (1972). *J. Amer. Statist. Ass.*, **67**, 103–111.

Anderson, D. A. and Federer, W. T. (1975). *Utilitas Math.*, **7**, 135–150.

Anderson, D. A. and Thomas, A. M. (1979). *Commun. Statist. A*, **8**, 931–943.

Bose, R. C. and Srivastava, J. N. (1964). *Bull. Int. Inst. Statist.*, **40**, 780–794.

Box, G. E. P. and Hunter, J. S. (1961). *Technometrics*, **3**, 311–351.

Draper, N. R. and Mitchell, T. J. (1967). *Ann. Math. Statist.*, **38**, 1110–1126.

Federer, W. T. and Balaam, L. N. (1972). *Bibliography on Experiment and Treatment Design (Pre-1968)*. Oliver & Boyd, Edinburgh.

Gulati, B. R. (1972). *Ann. Math. Statist.*, **43**, 1652–1663.

Hedayat, A. and Wallis, W. D. (1978). *Ann. Statist.*, **6**, 1184–1238.

John, P. W. M. (1971). *Statistical Design and Analysis of Experiments*. Macmillan, New York.

Margolin, B. H. (1969). *J. R. Statist. Soc. B*, **31**, 514–523.

Margolin, B. H. (1972). *J. R. Statist. Soc. B*, **34**, 431–440.

Pesotan, H., Raktoe, B. L., and Worthley, R. G. (1978). *J. Statist. Plann. Infer.*, **2**, 277–291.

Plackett, R. L. and Burman, J. P. (1946). *Biometrika*, **33**, 305–325.

Srivastava, J. N. and Chopra, D. V. (1975). *J. Indian Soc. Agric. Statist.*, **23**, 124–131.

Webb, S. R. (1968). *Technometrics*, **10**, 291–299.

(ALIAS
CONFOUNDING
DESIGN OF EXPERIMENTS
ESTIMABILITY
FACTORIAL EXPERIMENTS
GENERAL LINEAR MODEL
GEOMETRY IN STATISTICS
OPTIMUM DESIGN OF EXPERIMENTS)

W. T. FEDERER
B. L. RAKTOE

FRACTIONAL INTEGRODIFFERENTIATION

This little known transformation has become of interest in probability theory through fractional Brownian motions*, through fractional Gaussian noises*, and through fractals. It affects statistical modeling through the discussion of the Hurst coefficient*.

The earliest precursor was Leibniz, who, after defining the notions of $(d/dx)f$ and $(d/dx)^2f$, sought to interpolate between them with fractional derivatives. The idea is to transform x^m into $[\Gamma(m+1)/\Gamma(n+1)]\, x^n$, where $m-n$ is not necessarily an integer.

SPECTRAL DOMAIN CHARACTERIZATION

When $f(t)$ has the Fourier transform (FT) $\hat{f}(\lambda)$, it is true under wide conditions that $(d/dt)^k f(t)$ has the FT $\lambda^k\hat{f}(\lambda)$, and that the kth repeated integral of $f(t)$ has the FT $\lambda^{-k}\hat{f}(\lambda)$; *see* INTEGRAL TRANSFORMS. Hence, whenever the inverse FT of $\lambda^k\hat{f}(\lambda)$ is defined when $|k|$ is not an integer, it can be called a fractional differential when $k > 0$, and a fractional integral when $k < 0$.

REAL DOMAIN FORMULATIONS

The kth repeated integral of a fraction $f(t)$ can be written as

$$\int_0^t \ldots dc \int_0^c db \int_0^b f(a)\,da = \frac{1}{\Gamma(k)} \int_0^t (t-s)^{k-1} f(s)\,ds.$$

When the extension of this transformation to noninteger k is meaningful, it defines the *Riemann–Liouville fractional integrodifferential* of $f(t)$. Positive k's involve fractional *integration*, which is a smoothing operation, whereas negative k's involve fractional *differentiation*, which is unsmoothing.

A variant that is more useful in probability theory is the *Weyl fractional integrodifferential*

$$\frac{1}{\Gamma(k)}\int_{-\infty}^{t}(t-s)^{k-1}f(s)\,ds.$$

Another variant, due to B. Mandelbrot, eliminates the asymmetry between past and future by considering

$$\frac{1}{\Gamma(k)}\left\{\int_{-\infty}^{t}(t-s)^{k-1}f(s)\,ds - \int_{t}^{\infty}|t-s|^{k-1}f(s)\,ds\right\}.$$

Bibliography

Ross, B., ed. (1975). *Fractional Calculus and Its Application*. Springer-Verlag, New York.

(FRACTIONAL BROWNIAN MOTIONS AND GAUSSIAN NOISES)

BENOIT B. MANDELBROT

FRACTIONAL RANK ESTIMATORS

In estimating the vector $\boldsymbol{\beta}$ of coefficients in the standard regression* model with $E(\mathbf{Y}) = \mathbf{X}\boldsymbol{\beta}$ (with $\mathbf{X}$ an $n \times k$ matrix of known constants) and $\text{var}(\mathbf{Y}) = \sigma^2\mathbf{I}$, Marquardt [3] described fractional rank estimators and showed that they have properties similar to those of ridge* estimators. In particular, a fractional rank estimator is biased and may have less total mean squared error than the least-squares estimator of β.

Fractional rank estimators are closely related to principal components estimators. Let columns of the $k \times k$ matrix P be the eigenvectors of $X'X$ corresponding to its eigenvalues* $\lambda_1 \geqslant \lambda_2 \geqslant \cdots \geqslant \lambda_k$. Let $\Lambda = \text{diag}(\lambda_i)$. Regressing $\mathbf{Y}$ on the first r principal components* of $\mathbf{XX}'$ ($r < k$) gives the principal components estimator of $\boldsymbol{\beta}$,

$$\tilde{\boldsymbol{\beta}}_r = \left(\sum_{i=1}^{r}\lambda_i^{-1}\mathbf{P}_i\mathbf{P}_i'\right)\mathbf{X}'\mathbf{Y}.$$

Such an estimator is sometimes used when $\lambda_{r+1}, \ldots, \lambda_k$ are so small as to cause large variances of some components of the least-squares* estimator. The fractional rank estimator is

$$\tilde{\boldsymbol{\beta}}_{r,c} = \left(\sum_{i=1}^{r}\lambda_i^{-1}\mathbf{P}_i\mathbf{P}_i' + c\lambda_{r+1}^{-1}\mathbf{P}_{r+1}\mathbf{P}_{r+1}'\right)\mathbf{X}'\mathbf{Y}$$

with $0 \leqslant c < 1$, which appears to be a compromise between $\tilde{\boldsymbol{\beta}}_r$ and $\tilde{\boldsymbol{\beta}}_{r+1}$.

The generalized ridge estimators* [1] are of the form

$$\tilde{\boldsymbol{\beta}}_R = \left[\sum_{i=1}^{k}(\lambda_i + k_i)^{-1}P_iP_i'\right]X'Y,$$

with all $k_i \geqslant 0$, so that the fractional rank estimator is a limiting case of the generalized ridge estimators with $k_i = 0$, $i = 1, \ldots, r$, $k_{r+1} = \lambda_{r+1}(c^{-1} - 1)$ and $k_i = \infty$, $i = r+2, \ldots, k$.

Marquardt showed that if the parameters β and σ^2 are restricted so that $\beta'\beta \leqslant M\sigma^2$ for some positive number M, there exists a fractional rank estimator which is uniformly better than the least-squares estimator. In this situation, LaMotte [2] showed that there exists another biased linear estimator of β which is uniformly better than the fractional rank estimator.

References

[1] Hoerl, A. E. and Kennard, R. W. (1970). *Technometrics*, **12**, 55–67.

[2] LaMotte, L. R. (1978). *Technometrics*, **20**, 281–290.

[3] Marquardt, D. W. (1970). *Technometrics*, **12**, 591–612.

(BAYESIAN INFERENCE
COMPONENT ANALYSIS
RIDGE REGRESSION)

LYNN ROY LAMOTTE

FRAGMENTARY DATA *See* SPARSE DATA

FRAMINGHAM: AN EVOLVING LONGITUDINAL STUDY

The *Framingham Study* is a long-term study designed both to identify the relation of putative risk factors to circulatory and other disease risks, and to characterize the natural history of chronic circulatory disease processes. It is an interesting example of longitudinal research, wherein limited initial study goals were greatly expanded during the course of data collection. According to Gordon and Kannel,

> the Framingham, Massachusetts, study was designed . . . to measure . . . factors in a large number of "normal" persons . . . and to record the time during which these selected factors act and interact before cardiovascular disease results. . . . The purpose of the Framingham program . . . was the development of case-finding procedures in heart disease. The potential of the Framingham program for epidemiological studies soon became apparent, however, and the program turned increasingly in that direction. [3, p. 124]

The potential of the Framingham Study for epidemiological* studies of a wide variety has been amply demonstrated. The fact that such studies have been extended in the range of etiological factors considered and to include studies of other disease processes, such as malignant neoplasia, attests to the value of longitudinal research (given a high level of imagination and energy of the study directors). Our purpose in this article, however, is not to review the range of studies conducted with the Framingham data which have been basic to the identification of serum cholesterol, smoking and elevated blood pressure as important coronary heart disease (CHD) risk factors, but rather briefly to review its initial design and select statistical issues that arise with continuous adaptation of a longitudinal study.

Specifically, we will examine the Framingham Study both in terms of the basic data collected, as well as in terms of specific analytic and statistical issues that arise in (1) the evaluation of continuous disease processes with periodic and limited data collection, and (2) the development of procedures for the analysis of new types of data whose collection was begun after study initiation. We first briefly describe the basic components of the Framingham Study, then describe issues that arise in modeling disease processes with these data, and finally, consider problems inherent in changing and expanding data collection procedures during the course of the study.

DESCRIPTION OF THE FRAMINGHAM POPULATION AND SAMPLE

In 1948, when the study began, Framingham was an industrial and trading center of 28,000 persons located 21 miles west of Boston. The town was selected for study because, besides being the location of the first community study of tuberculosis in 1917, there was community interest in the project [3, p. 126]. There was available for Framingham a published town list containing the names of 10,000 persons between the ages of 30 and 59 inclusive. The Bureau of the Census* matched the January 1, 1950, town list in the age range 30–59 with the census schedules for April 1, 1950. "Some 89 percent of those on the census schedules were found on the town list" [3, p. 127]. This matching led to the projection of a 10% rate of loss to follow-up*. Since it was decided that it was feasible to handle a sample of 6000 persons in the study, this projected low rate dictated that a sample of 6600 persons, or about two-thirds of the eligible population be drawn. Of the 6000 expected to be examined, it was estimated that 5000 of them would be free of cardiovascular disease. It was further expected that 400 would develop cardiovascular disease or die of it by the end of the fifth year of the study, 900 by the end of the tenth year, and 1500 by the end of the twentieth year. It was decided to sample by households and to utilize every person in the chosen age range in each household.

The first major adaption of the study design was forced by a high nonresponse* rate.

The actual nonresponse rate of 31.2% far exceeded the expected 10%, so the study population was augmented by using 740 of a group of volunteers. "It was at first planned to re-examine only those volunteers who were 'normal' on their initial examination. This plan, however, was not rigorously followed. While 13 people were eliminated for hypertensive or coronary heart disease, these omissions modified the clinical character of the group only trivially" [3, p. 129]. Thus, at the very beginning of the study, a second basic change in the sample design was necessitated, i.e., that, instead of collecting data only on disease-free persons, a sample was drawn with persons manifesting limited cardiovascular conditions. As the study progressed and before any attempt was made to screen out diseased persons, it was decided that persons with cardiovascular disease should be retained in the study since manifest disease could be viewed as a stage in the total disease process. To have removed such persons from the study population, it was concluded in retrospect, would have served to seriously compromise the goal of the study of CHD in a human population.

In addition to the difficulties with sample design, a number of basic measurement issues also had to be resolved during the course of the study. For example, the initial characterization of the sample with respect to the precursors of disease suffered somewhat, since serum cholesterol was not in adequate control until nearly the end of the first measurements examination. As a result, the initial serum cholesterol values were taken from the second exam in the majority of cases. Cigarette-smoking history was not ascertained on the first exam and the then unknown pressor effect at the first measurement was found to affect the early exam blood pressures. The measurements made on each exam were improved over the life of the study. New measurements were introduced at subsequent examination. For example, diet history and a physical activities study were initiated at exam 4, protein-bound iodine was measured at exams 4 and 5, and a psychological questionnaire was administered on exams 8 and 9. It was concluded that the Framingham study could have been considered successful even if its only yield was the information derived about measurement phenomena.

The second examination began in earnest in May 1951, more than two and a half years after the first exam, which began September 29, 1948. Repeated examinations continued at an interval of 2 years. As might be expected, there was a considerable difference in the rate of return to subsequent examinations both within and between the sample and the volunteer groups. According to Gordon and Kannel, it was "surprising to find that where the person remains alive the likelihood of re-examination is about the same in one age-sex group as another" and "that essentially permanent loss to follow-up for reasons other than death has been relatively constant from one examination to another" [3, p. 134].

METHODOLOGICAL ISSUES IN THE ANALYSIS OF DISEASE PROCESSES IN LONGITUDINAL POPULATION STUDIES

The prime value of the Framingham study is that a large study population was followed for a period of now over 32 years, yielding important time-series* information on the change in physiology and risk factors as well as follow-up on disease events. Interestingly, much of the temporal information within the Framingham data on physiological changes has not been fully exploited, despite the explicit desire to assess the natural history of chronic disease processes. One of the reasons for this seems to be the lack of effort to develop appropriate time-series models for human epidemiological data. The primary analytic strategy that has been applied to longitudinal measures of risk has been the logistic multiple regression* strategy due to Cornfield [1]. In applying discriminant analysis* to CHD risks, Cornfield recognized that the posterior probability was a logistic function of the discriminant score, leading him to accept the logistic response function as the appropriate model of CHD response to risk factors.

One limitation of Cornfield's procedure was the necessity of making the standard discriminant assumption that the risk factors of two groups were normally distributed. Halperin et al. [4] developed conditional maximum likelihood* procedures which did not depend on a multivariate normality assumption. These logistic regression strategies have been a primary analytic tool for analyzing the relation of CHD to putative risk factors in the Framingham and other longitudinally followed study populations. Recent attention has been focused on the fact that the logistic function is a mathematically inappropriate functional form to model continuous time processes such as the development of a chronic disease [8]. The implication of these logical inconsistencies is that the logistic regression coefficients from studies of different length and risk levels cannot be directly compared. One solution that has been proposed is to apply the logistic regression to subintervals of comparable length in the studies to be compared [2, 9]. Such strategies are still subject to the difficulty that the logistic coefficients can be confounded with differences in the mean risk of the studies and do not offer a solution for the case where either (1) the follow-ups were conducted for different length intervals, or (2) follow-ups were conducted at irregular intervals. More recent efforts have been directed to the development of methods that are consistent with continuous time changes in risk [5] and one that models physiological changes directly [7].

ADAPTIVE NATURE OF LONGITUDINAL STUDIES

One characteristic of the Framingham study is that it provided a learning opportunity for the study directors. That is, knowledge gained from the study has been used to enhance the efficiency and utility of the data collection* process. Although application of insight acquired during the study is extremely important, it raises a number of difficult statistical issues. First, there are issues that arise because of the modification of data collection procedures. For example, laboratory procedures for assessing serum cholesterol changed over the first few waves of the study. Second, various types of information collection may be either initiated or discontinued. For example, measurements of uric acid ceased after the fourth set of measurements. Third, special studies and measurements are conducted on a one-time basis during the course of the study. Finally, although the biennial measurements of physiological risk variables ceased after 20 years, special survival and other follow-up studies have continued. All these deviations from the initial data collection protocol have introduced interesting but difficult statistical problems.

Three basic statistical models might be employed in the development of analytic means for dealing with these problems. The first is the missing information principle (EM algorithm); which would permit the utilization of incomplete data* [6]. The second principle involves measurement error* models and concepts of reliability*—concepts well known to psychometricians, but exploited infrequently in biostatistics*. A third might be the use of empirical Bayes* procedures to integrate data at various levels into a complete model of disease process and to deal with the problem of systematic loss to follow-up and mortality selection.

Applications of such statistical principles and corresponding statistical models are required to make full and efficient use of the rich but sometimes irregular data collected in longitudinal studies of human populations. The central statistical issue seems to be that, in collecting epidemiological data on human health, it is difficult to construct and maintain, in the face of practical exigencies, a study design that permits the application of "simple" statistical procedures. In dealing with human population data, as opposed to data where more stringent experimental controls can be imposed, it appears necessary to apply more general statistical strategies in order to derive inferences from the available data.

References

[1] Cornfield, J. (1962). *Proc. Fed. Amer. Soc. Exper. Biol.*, **21**, 58–61. (Classical reference for logistic risk.)

[2] Cornfield, J. (1978). Personal communication to Max A. Woodbury.

[3] Gordon, T. and Kannel, W. B. (1970). In *The Community as an Epidemiologic Laboratory*, I. I. Kessler and M. L. Levine, eds. Johns Hopkins University Press, Baltimore, Md. (Good general description of the Framingham Study as originally conceived and as it changed over time.)

[4] Halperin, M., Blackwelder, W. C., and Verter, J. I. (1971). *J. Chronic Dis.*, **24**, 125–158. (Alternative method for carrying out logistic regression.)

[5] Lellouch, J. and Rakavato, R. (1976). *Int. J. Epidemiol.*, **5**, 349–352. (Log-linear model of mortality risk depending on covariates.)

[6] Orchard, T. and Woodbury, M. A. (1971). *Proc. 6th Berkeley Symp. Math. Statist. Prob.*, Vol. 1. University of California Press, Berkeley, Calif., pp. 697–715. (Methods for maximum likelihood estimation when data are missing.)

[7] Woodbury, M. A., Manton, K. G., and Stallard, E. (1979). *Biometrics*, **35**, 575–585. (Investigation of physiological changes in CHD risks in Framingham.)

[8] Woodbury, M. A., Manton, K. G., and Stallard, E. (1981). *Int. J. Epidemiol.*, **10**, 187–197. (Examines issues in risk modeling in longitudinal studies.)

[9] Wu, M. and Ware, J. H. (1979). *Biometrics*, **35**, 513–521. (Proposes a strategy for applying logistic regression to longitudinal studies.)

(BIOSTATISTICS
EPIDEMIOLOGICAL STATISTICS
FOLLOW-UP
LOGISTIC REGRESSION)

M. A. Woodbury
K. G. Manton

FRANKLIN'S IDENTITY

Fabian Franklin received a Ph.D. in mathematics from Johns Hopkins University in 1880 and taught there from 1879 to 1895. Afterward he edited the *Baltimore News* and *New York Evening Post*, espousing conservative economic and political causes, and vigorously attacking Prohibition's limitation of personal liberty [1, pp. 206–207]. In 1885 [2] he introduced to the American mathematical community an inequality of Chebyshev's that had appeared in Hermite's 1882 lecture notes. The reader should note that this is not the Chebyshev inequality commonly used in the proof of the weak law of large numbers for finite variance, (*see* CHEBYSHEV INEQUALITY; LAWS OF LARGE NUMBERS). Franklin generalized Chebyshev's result and provided an identity that made the proof almost immediate. Indeed, the identity itself can be easily generalized to allow a more general inequality still.

Franklin presented the identity in two forms. In the first version, let u, v be continuous functions on the interval (a, b). Then

$$\int_a^b \int_a^b [u(x) - u(y)][v(x) - v(y)]\, dx\, dy = 2\left\{(b-a)\int_a^b uv\, dx - \int_a^b u\, dx \int_a^b v\, dx\right\}.$$

In the second version, let $u_1, u_2, \dots, u_n$ and $v_1, v_2, \dots, v_n$ be finite sequences. Then

$$\sum_{j=1}^n \sum_{i=1}^n [(u_i - u_j)(v_i - v_j)] = 2\left[n\sum_{i=1}^n u_i v_i - \sum_{i=1}^n u_i \sum_{j=1}^n v_j\right].$$

Both forms are special cases of a general version provided by Shea [5]: Let u, v be real-valued functions defined on a set R, and suppose that u, v, and uv are integrable with respect to some measure μ with $\mu(R) < \infty$. Then

$$\int_R \left[\int_R \{[u(x) - u(y)] \times [v(x) - v(y)]\}\mu(dx)\right]\mu(dy) = 2\left[\mu(R)\int_R uv\, d\mu - \int_R u\, d\mu \int_R v\, d\mu\right].$$

If μ is a positive measure with $\mu(R) = 1$, we have the identity appearing in Hoeffding's lemma*: If (X_1, Y_1) and (X_2, Y_2) are independent and identically distributed (i.i.d) random vectors, and all expectations exist, then

$$E[(X_1 - X_2)(Y_1 - Y_2)] = 2[E(X_1 Y_1) - EX_1 EY_1].$$

This last version appears in Lehmann [4] and has a multivariate generalization to the i.i.d. triples (X_1, Y_1, Z_1) and $(-X_2, Y_2, Z_2)$, given by Jogdeo [3].

The Chebyshev inequality* proved by Franklin's identity makes the further assumption that u and v are monotonic. If both u and v are nondecreasing or both are nonincreasing, the integrand on the left-hand side of the identity is nonnegative. Therefore,

$$(b-a)\int_a^b uv\,dx \geqslant \int_a^b u\,dx \int_a^b v\,dx,$$

$$n\sum_{i=1}^n u_i v_i \geqslant \sum_{i=1}^n u_i \sum_{j=1}^n v_j,$$

and for μ a positive measure on a totally ordered set R,

$$\mu(R)\int_R uv\,d\mu \geqslant \int_R u\,d\mu \int_R v\,d\mu.$$

If one of u and v is nondecreasing and the other is nonincreasing, the inequalities are reversed. Equality holds if and only if u or v is constant almost everywhere.

A generalization of the inequality to several variables relies on Franklin's identity and an inductive argument provided by Lehmann [4]. Two real-valued functions u and v of n arguments are said to be *concordant* for the ith coordinate if, with all other coordinates fixed, the functions are both nondecreasing or nonincreasing functions of the ith coordinate. If they are monotone in opposite directions for the ith coordinate, they are *discordant* in the coordinate. Let $\mu_1, \ldots, \mu_n$ be positive finite measures for the totally ordered sets $R_1, \ldots, R_n$, respectively. Let $\Omega = R_1 \times \cdots \times R_n$ and let u, v, and uv be integrable over Ω. If for each coordinate, u and v are concordant, then for $\mathbf{x} = (x_1, \ldots, x_n)$

$$\begin{aligned}&[\mu_1(R_1)\cdots\mu_n(R_n)]\\&\quad\times\int_\Omega u(\mathbf{x})v(\mathbf{x})\,d\mu_1(x_1)\cdots d\mu_n(x_n)\\&\geqslant \int_\Omega u(\mathbf{x})\,d\mu_1(x_1)\cdots d\mu_n(x_n)\\&\quad\times\int_\Omega v(\mathbf{x})\,d\mu_1(x_1)\cdots d\mu_n(x_n).\end{aligned}$$

The inequality is reversed if u and v are discordant for each coordinate. Lehmann's proof can be used to extend this result from product measures to measures sharing a certain "quadrant dependency." For further applications of Franklin's identity, *see* HOEFFDING'S LEMMA.

References

[1] American Council of Learned Societies, ed. (1954). *Dictionary of American Biography*, Vol. 11, Suppl. 2. Scribner's, New York.

[2] Franklin, F. (1885). *Amer. J. Math.*, **7**, 377–379. (A readable and straightforward presentation of this Chebyshev inequality and a number of applications to calculus results.)

[3] Jogdeo, K. (1968). *Ann. Math. Statist.*, **39**, 433–441. (A generalization of Lehmann's [4] study of bivariate dependence to several variables.)

[4] Lehmann, E. L. (1966). *Ann. Math. Statist.*, **37**, 1137–1153. (A study of bivariate quadrant dependence and the resultant unbiasedness of tests based on Pearson's r, Spearman's rho, Kendall's tau, etc.)

[5] Shea, G. (1979). *Ann. Statist.*, **7**, 1121–1126. (A study of the use of monotone regression and quadrant dependence in multivariate independence.)

(CHEBYSHEV INEQUALITY
HOEFFDING'S LEMMA)

GERALD A. SHEA

FRASER STRUCTURAL MODEL *See* EQUIVARIANT ESTIMATORS

FRÉCHET BOUNDS

Maurice Fréchet [5] developed bounds on the probabilities of the union and intersection of dependent probability systems of m events. These bounds were designed to improve corresponding bounds on these probabilities obtained by Boole [3] and Bonferroni [2]; in addition, he extended Bonferroni's inequalities*. He expressed his results in sequences of inequalities which comprise a partial set of necessary conditions for consistency of partially macro-specified systems (partially specified by selected sums of probabilities associated with the system). Later Fréchet [6] applied the theory of partially micro-specified systems (partially specified by selected individual probabilities associated with the system) to develop upper and lower bounds on the joint distribution function of two random variables, when only the two individual marginal distribution functions of the two variables are specified. Fréchet's seminal work has motivated extensions which are also presented here.

A probability system of m dependent events $E_1, E_2, \ldots, E_m$ consists of 2^m distinct events with associated probabilities which can be specified in two alternative but equivalent ways. Let

$$p_{[i_1 i_2 \cdots i_r]} = \begin{cases} \text{probability that only} \\ \text{the events } E_{i_1}, E_{i_2}, \\ \ldots, E_{i_r} \text{ simultaneously occur (and} \\ \text{the remaining events} \\ \text{do not occur);} \end{cases} \quad (1)$$

$$p_{i_1 i_2 \cdots i_r} = \begin{cases} \text{probability that the} \\ \text{events } E_{i_1}, E_{i_2}, \ldots, E_{i_r} \\ \text{simultaneously occur} \\ \text{(without regard to the} \\ \text{remaining events).} \end{cases} \quad (2)$$

With $P_{[0]} = p_{[0]} = p_0$, all possible values of (1) or, alternatively (2), each comprise 2^m probabilities and specify the system. We now define

$$\begin{aligned} P_{[r]} &= \sum p_{[i_1 i_2 \cdots i_r]}, \text{ summed over all } \binom{m}{r} \text{ selections} \\ &\quad \text{of } r \text{ indices } i_1, i_2, \ldots, i_r \text{ from } (1, 2, \ldots, m). \\ &= \text{probability that exactly } r \text{ of the } m \text{ events occur;} \end{aligned} \quad (3)$$

$$\begin{aligned} S_r &= \sum p_{i_1 i_2 \cdots i_r}, \text{ summed over all selections of } i_1, \\ &\quad i_2, \ldots, i_r \text{ from } (1, 2, \ldots, m), i_1 < i_1 < \cdots < i_r \\ &= \text{sum of the probabilities of occurrence of all} \\ &\quad \text{possible combinations of } r \text{ of the } m \text{ events;} \end{aligned} \quad (4)$$

$$\begin{aligned} P_r &= P_{[r]} + P_{[r+1]} + \cdots + P_{[m]}, \quad r = 1, 2, \ldots, m \\ &= \text{probability of occurrence of } r \text{ or more} \\ &\quad \text{of the } m \text{ events.} \end{aligned} \quad (5)$$

These definitions imply the following well-known relationship:

$$\begin{aligned} S_r &= \sum_{j=0}^{m-r} \binom{r+j}{r} P_{[r+j]} \\ &= P_{[r]} + \binom{r+1}{r} P_{[r+1]} + \cdots + \binom{m}{r} P_{[m]}, \\ &\qquad r = 1, 2, \ldots, m. \end{aligned} \quad (6)$$

FRÉCHET BOUNDS AND EXTENSIONS: PARTIALLY MACRO-SPECIFIED SYSTEMS

The following set of upper bounds on $p_{1,2,\ldots,m}$, the probability of simultaneous occurrence of all m events, is the most useful and powerful of Fréchet's results for partially macro-specified systems, deriving its power from use of a relatively fine and unaggregated consistency property of partially micro-specified systems.

$$\begin{aligned} 0 &\leqslant S_m \leqslant S_{m-1} \Big/ \binom{m}{m-1} \leqslant \cdots \leqslant S_r \Big/ \binom{m}{r} \\ &\leqslant \cdots \leqslant S_2 \Big/ \binom{m}{2} \leqslant S_1/m \leqslant 1. \end{aligned} \quad (7)$$

Since $S_m = p_{12 \cdots m} = p_{[12 \cdots m]} = P_{[m]}$, (7) can be restated as Fréchet's upper bound on $p_{12 \cdots m}$:

$$p_{12 \cdots m} \leqslant S_r \Big/ \binom{m}{r}, \qquad r = 1, 2, \ldots, m. \quad (8)$$

In (20) following, for example, Fréchet's upper bound on $p_{1,2,\ldots,m}$, when only S_1 is specified, follows a fortiori when the bound $\min\{p_1, \ldots, p_m\}$ is replaced by $S_1/m = \sum_1^m p_i/m \geqslant \min\{p_1, \ldots, p_m\}$.

Fréchet was apparently unaware that bounds (7) and (8) also imply a useful lower bound on P_1, the probability of one or more (i.e., the union) of the m events. This bound follows from application of (7) and (8) to the corresponding system of m complementary events. The extended Fréchet lower bound on P_1, (extended by Kwerel) is stated as

$$P_1 \geqslant \sum_{j=1}^{r} (-1)^{j-1} \binom{r}{j} \left\{ S_j \Big/ \binom{m}{j} \right\}, \qquad r = 1, 2, \ldots, m; \quad (9)$$

for example, from (8) and (9) for systems of m events partially specified only by

$$S_1 : p_{12\cdots m} \leqslant S_1/m; \qquad P_1 \geqslant S_1/m. \qquad (10)$$

$$S_1, S_2 : p_{12\cdots m} \leqslant 2S_2/[m(m-1)];$$

$$P_1 \geqslant 2S_1/m - 2S_2/[m(m-1)]. \qquad (11)$$

$$S_1, S_2, S_3 : p_{12\cdots m} \leqslant 6S_3/[m(m-1)(m-2)];$$

$$P_1 \geqslant \{3S_1/m - 6S_2/[m(m-1)] + 6S_3/[m(m-1)(m-2)]\}. \qquad (12)$$

For certain restricted classes of systems, bounds (8) and (9) are most stringent (i.e., the best attained bounds). However, they do not in general give the most stringent bounds (see Kwerel [7]). For example, for bounds (11), where only S_1 and S_2 are macro-specified (see Kwerel [8]), the upper bound on $p_{12\cdots m}$ given above is most stringent for systems for which $1 \geqslant S_1 \geqslant 0$; while for systems where $m \geqslant S_1 \geqslant 1$, it is most stringent only when

$$\{(m-1)/2\} \geqslant \{S_2/S_1\} \geqslant [\{(m-1)/2\} - \{(m-S_1)/(2S_1)\}].$$

Also (see ref. 8), the lower bound on P_1 is most stringent only for the systems where $(m-1) \geqslant S_1 \geqslant 0$ and

$$\{(m-1)/2\} \geqslant \{S_2/S_1\} \geqslant \{(m-2)/2\};$$

or only for the systems where $m \geqslant S_1 \geqslant (m-1)$ and

$$\{(m-1)/2\} \geqslant \{S_2/S_1\} \geqslant [\{(m-2)/2\} + \{m(S_1-m+1)/(2S_1)\}].$$

Fréchet [5] presented a sequence of lower bounds on $p_{12\cdots m}$ which generalize Boole's inequality* [3]:

$$p_{12\cdots m} \geqslant \left\{ S_r \Big/ \binom{m-1}{r-1} - (m/r) + 1 \right\}, \qquad r = 1, 2, \ldots, m. \qquad (13)$$

(These bounds are not, in general, monotonically increasing with r.)

When $r = 1$, (13) becomes Boole's inequality [see also (20) following]:

$$p_{12\cdots m} \geqslant \{S_1 - m + 1\}. \qquad (14)$$

For the restricted class of systems of m events macro-specified only by S_1 and for which $S_1 \geqslant (m-1)$, Boole's inequality gives the most stringent lower bound. For systems partially macro-specified by S_1, $S_2, \ldots, S_r$ $(1 < r < m)$, bounds (13) and (14) do not in general give the most stringent lower bound on $p_{12\cdots m}$, and tend to be weak because of their highly aggregated basis.

Fréchet [5] also generalized Bonferroni's inequalities, which give bounds on $P_{[r]}$. These *Fréchet bounds on* $P_{[r]}$, the probability of exactly r events, are stated as

$$S_r - (r+1)S_{r+1} \leqslant P_{[r]} \leqslant S_r$$

$$S_r - \binom{r+1}{1}S_{r+1} + \binom{r+2}{2}S_{r+2} - \binom{r+3}{3}S_{r+3} \leqslant P_{[r]} \leqslant S_r - \binom{r+1}{1}S_{r+1} + \binom{r+2}{2}S_{r+2}.$$

$$\vdots \qquad (15)$$

These bounds on $P_{[r]}$ reduce to Bonferroni's inequalities when $r = 0$ (with $S_0 \equiv 1$); and give Bonferroni's inequalities on P_1 when $(1 - P_1)$ is substituted for $P_{[0]}$ in inequalities (15). They tend in general to be weak because of their highly aggregated basis.

The most stringent upper and lower bounds on $p_{12\cdots m}$ and P_1 for systems of m events partially macro-specified by $S_1, S_2, \ldots, S_r$ $(r < m)$, as well as necessary and sufficient conditions for consistency of the $S_1, S_2, \ldots, S_r$, are developed by Kwerel in ref. 7. The most stringent bounds on $P_{[r]}$ also follow directly from ref. 7 and are given in ref. 8, where only S_1 and S_2 are specified.

FRÉCHET BOUNDS AND EXTENSIONS: PARTIALLY MICRO-SPECIFIED SYSTEMS

Fréchet made important application of the theory of dependent systems of two events partially micro-specified only by the probabilities p_1 and p_2 of the individual events. These probabilities imply well-known upper and lower bounds on p_{12}, the probability of simultaneous occurrence of the two events.

For any given system specified only by p_1 and p_2,

$$0 \leqslant p_{12} \leqslant \min\{p_1, p_2\}. \qquad (16)$$

Also, from the bounds on $\overline{P}_1$, the probability of the union of the corresponding system of complementary events, where $\bar{p}_1 = 1 - p_1$, $\bar{p}_2 = 1 - p_2$,

$$\max\{\bar{p}_1, \bar{p}_2\} \leqslant \overline{P}_1 \leqslant \{\bar{p}_1 + \bar{p}_2\}. \qquad (17)$$

Since $p_{12} = 1 - \overline{P}_1$, it follows from (16) and (17) that

$$\max\{p_1 + p_2 - 1, 0\} \leqslant p_{12} \leqslant \min\{p_1, p_2\}. \qquad (18)$$

Fréchet [6] applied these bounds to the random variables X_1, X_2 where only their marginal distribution functions $F_1(x_1) = \Pr\{X_1 \leqslant x_1\}$ and $F_2(x_2) = \Pr\{X_2 \leqslant x_2\}$ are specified. In terms of the preceding notation, for any given fixed values of x_1 and x_2, let $p_1 = F_1(x_1)$ and $p_2 = F_2(x_2)$, while $p_{12} = F_{1,2}(x_1, x_2) = \Pr\{X_1 \leqslant x_1, X_2 \leqslant x_2\}$ is the corresponding joint distribution function (df) of X_1 and X_2. From (18) we obtain Fréchet's bounds on $F_{1,2}(x_1, x_2)$:

1. For any given value of x_1, x_2 and given marginals $F_1(x_1)$, $F_2(x_2)$,

$$\max\{F_1(x_1) + F_2(x_2) - 1, 0\} \leqslant F_{1,2}(x_1, x_2) \leqslant \min\{F_1(x_1), F_2(x_2)\}. \qquad (19)$$

2. The upper bounding function on $F_{1,2}(x_1, x_2)$ characterized by $\min\{F_1(x_1), F_2(x_2)\}$, and the lower bounding function on $F_{1,2}(x_1, x_2)$ characterized by $\max\{F_1(x_1) + F_2(x_2) - 1; 0\}$ are in general proper distribution functions with marginals $F_1(x_1)$ and $F_2(x_2)$ as x_1 and x_2 vary over the entire event space R^2.

Fréchet's bounds on joint distribution functions have been extended by Dall'Aglio [4], Sklar [10], and others [1] to the case of n random variables $X_1, X_2, \ldots, X_n$ when only the n marginal df's $F_1(x_1), F_2(x_2), \ldots, F_n(x_n)$ are specified. This extension is incomplete, being confined to development of upper and lower bounds on the value of the joint df $F_{1,2,\ldots,n}(x_1, x_2, \ldots, x_n) = \Pr\{X_1 \leqslant x_1, X_2 \leqslant x_2, \ldots, X_n \leqslant x_n\}$ in isolation only, without regard to corresponding bounds on (or values of) joint df's of lower order. For given $x_1, \ldots, x_n$, the n events $X_1 \leqslant x_1, \ldots, X_n \leqslant x_n$, comprise a dependent system partially micro-specified only by the probabilities $F_1(x_1), \ldots, F_n(x_n)$. Complete characterization of the function $F_{1,2,\ldots,n}(x_1, x_2, \ldots, x_n)$ for given $\mathbf{x} = (x_1, x_2, \ldots, x_n)$ requires systemically consistent characterization of the probabilities of the associated system of 2^n elementary events (1); or, equivalently, consistent characterization of the values of the associated system (2) of all the $(2^n - n - 1)$ joint df's $F_{i_1, i_2, \ldots, i_s}(x_{i_1}, x_{i_2}, \ldots, x_{i_s})$, where $i_1 < i_2 < \cdots < i_s$ is any selection of s integers from $(1, 2, \ldots, n)$, $s = 2, \ldots, n$. The more general (and complete) extension given by Kwerel [9], which follows, characterizes the most stringent (i.e., best attainable) upper and lower bounding functions on $F_{1,2,\ldots,n}(x_1, x_2, \ldots, x_n)$ as well as the most stringent upper and lower bounds on the specific value of $F_{1,2,\ldots,n}(x_1, x_2, \ldots, x_n)$.

Before proceeding to the more general extension, we give bounds on the individual values, in isolation, of the joint df's of all orders for given $\mathbf{x} = (x_1, x_2, \ldots, x_n)$ and $F_1(x_1), F_2(x_2), \ldots, F_n(x_n)$. By straightforward extension of (16), (17), and (18) we obtain

$$\max\left\{\sum_{i_1}^{i_s} p_i - s + 1, 0\right\} \leqslant p_{i_1, i_2, \ldots, i_s} \leqslant \min\{p_{i_1}, p_{i_2}, \ldots, p_{i_s}\}; \qquad (20)$$

and thus

$$\max\left\{\sum_{i_1}^{i_s} F_1(x_i) - s + 1, 0\right\}$$
$$\leqslant F_{i_1,i_2,\ldots,i_s}(x_{i_1}, x_{i_2}, \ldots, x_{i_s})$$
$$\leqslant \min\{F_{i_1}(x_{i_1}), \ldots, F_{i_s}(x_{i_s})\},$$
$$i_1 < \cdots < i_s \text{ from } (1,2,\ldots,n),$$
$$s = 2, \ldots, n. \quad (21)$$

where $F_{i_1,i_2,\ldots,i_s}(x_{i_1}, x_{i_2}, \ldots, x_{i_s}) = \Pr\{X_{i_1} \leqslant x_{i_1}, X_{i_2} \leqslant x_{i_2}, \ldots, X_{i_s} \leqslant x_{i_s}\}$.

Systemically consistent (and complete) characterization of the function (as against the individual value of) $F_{1,2,\ldots,n}(x_1, x_2, \ldots, x_n)$ at $\mathbf{x} = (x_1, x_2, \ldots, x_n)$ with given marginals $F_1(x_1), F_2(x_2), \ldots, F_n(x_n)$ requires systemically consistent characterization of the following 2^n probabilities of the elementary events (1) associated with the corresponding system of n dependent events $X_1 \leqslant x_1$, $X_2 \leqslant x_2, \ldots, X_n \leqslant x_n$, where x_i and $\bar{x}_i$ represent the events $X_i \leqslant x_i$ and $X_i > x_i$, $i = 1, 2, \ldots, n$, respectively:

$$p_{[0]} = \Pr\{\bar{x}_1, \bar{x}_2, \bar{x}_3, \ldots, \bar{x}_n\}$$
$$p_{[x_1]} = \Pr\{x_1, \bar{x}_2, \bar{x}_3, \ldots, \bar{x}_n\}$$
$$\vdots \qquad \vdots$$
$$p_{[x_{i_1}, x_{i_2}, \ldots, x_{i_s}]} = \Pr\{\bar{x}_1, \bar{x}_2, \ldots, \bar{x}_{i_1-1}, x_{i_1}, \bar{x}_{i_1+1}, \ldots, \bar{x}_{i_s-1}, x_{i_s}, \bar{x}_{i_s+1}, \ldots, \bar{x}_n\}, \quad (22)$$
$$i_1 < i_2 < \cdots < i_s \text{ from } (1, 2, \ldots, n),$$
$$s = 1, 2, \ldots, n$$
$$\vdots \qquad \vdots$$
$$p_{[x_1, x_2, \ldots, x_n]} = \Pr\{x_1, x_2, x_3, \ldots, x_n\},$$

such that the sum of the probabilities of events where x_i appears equals $F_i(x_i)$, $i = 1, 2, \ldots, n$ (and that all the 2^n probabilities are nonnegative and sum to 1). Note that for any given characterization the (individual) value of $F_{1,2,\ldots,n}(x_1, x_2, \ldots, x_n) = P_r\{x_1, x_2, x_3, \ldots, x_n\}$.

An equivalent systemically consistent characterization (2) of the function $F_{1,2,\ldots,n}(x_1, x_2, \ldots, x_n)$ can be obtained from (22) by summing the probabilities of the 2^{n-s} events where $x_{i_1}, x_{i_2}, \ldots, x_{i_s}$ appears, along with all possible combinations of the remaining $n - s$ events and their complements, to obtain the following set of the n given $F_i(x_i)$ values and the corresponding $(2^n - n - 1)$ values of the joint df's of all orders:

$$F_{i_1,i_2,\ldots,i_s}(x_{i_1}, x_{i_2}, \ldots, x_{i_s})$$
$$= \sum_{\substack{\nu=1 \\ j_1 < \cdots < j_\nu \\ j \neq i}}^{n-s} p_{[x_{i_1}, \ldots, x_{i_s}, x_{j_1}, \ldots, x_{j_\nu}]},$$
$$i_1 < \cdots < i_s \text{ from } (1, 2, \ldots, n),$$
$$s = 1, \ldots, n,$$
$$j_1 < \cdots < j_\nu, j \neq i \text{ from } (1, 2, \ldots, n). \quad (23)$$

Since system (22) or, equivalently, (23) is only partially micro-specified by the n marginal probabilities $F_i(x_i)$ at $\mathbf{x}$, there is in general a large family of possible systemically consistent characterizations of the function $F_{1,2,\ldots,n}(x_1, x_2, \ldots, x_n)$ with the given marginal values $F_i(x_i)$, $i = 1, 2, \ldots, n$ at $\mathbf{x}$.

We define two classes of characterizations to which we shall refer. The *most stringent upper* (respectively, *lower*) *bounding function* on $F_{1,2,\ldots,n}(x_1, x_2, \ldots, x_n)$ at $\mathbf{x}$ is a systemically consistent characterization of the function $F_{1,2,\ldots,n}(x_1, x_2, \ldots, x_n)$ in which the joint df's of all orders simultaneously take on their most stringent (i.e., best attainable) upper (respectively, lower) bound values.

Most Stringent Upper Bounding Function on $F_{1,2,\ldots,n}(x_1, x_2, \ldots, x_n)$

For any given value of $\mathbf{x} = (x_1, x_2, \ldots, x_n)$ and given marginals $F_1(x_1), \ldots, F_n(x_n)$, the most stringent upper bounding function on $F_{1,2,\ldots,n}(x_1, x_2, \ldots, x_n)$ always exists and is characterized by the following 2^n elementary event probabilities, where without loss of generality, the x_i and $F_i(x_i)$ are labeled so that $F_1(x_1) \geqslant F_2(x_2) \geqslant \cdots \geqslant F_n(x_n)$:

$$\Pr\{\bar{x}_1, \bar{x}_2, \bar{x}_3, \ldots, \bar{x}_n\} = 1 - F_1(x_1),$$
$$\vdots$$
$$\Pr\{x_1, x_2, x_3, \ldots, x_j, \bar{x}_{j+1}, \ldots, \bar{x}_n\} = F_j(x_j) - F_{j+1}(x_{j+1}), \quad j = 1, \ldots, n-1, \tag{24}$$
$$\vdots$$
$$\Pr\{x_1, x_2, x_3, \ldots, x_n\} = F_n(x_n).$$

The probabilities of the remaining $(2^n - n - 1)$ events are equal to 0. Equivalently, this function is characterized by the following set of simultaneously attained most stringent upper bound values on the joint df's of all orders in addition to the given marginal df values $F_1(x_1), \ldots, F_n(x_n)$:

$$F^+_{i_1, i_2, \ldots, i_s}(x_{i_1}, x_{i_2}, \ldots, x_{i_s}) = \min\{F_{i_1}(x_{i_1}), F_{i_2}(x_{i_2}), \ldots, F_{i_s}(x_{i_s})\},$$
$$i_1 < \cdots < i_s \text{ from } (1, 2, \ldots, n), \quad s = 2, \ldots, n-1, \tag{25}$$
$$F^+_{1,2,\ldots,n}(x_1, x_2, \ldots, x_n) = \min\{F_1(x_1), F_2(x_2), \ldots, F_n(x_n)\}.$$

As x varies over R^n without restriction, the most stringent upper bounding function on $F_{1,2,\ldots,n}(x_1, x_2, \ldots, x_n)$, (25) is a proper joint df. This reflects the fact that in addition to the usual properties, the sequence of joint df values of all orders (25) characterizing it over any interval is always nondecreasing and systemically consistent.

Most Stringent Lower Bounding Function on $F_{1,2,\ldots,n}(x_1, x_2, \ldots, x_n)$

Case 1.

$$\sum_1^n F(x_i) \leqslant 1. \tag{26}$$

For given $\mathbf{x} = (x_1, x_2, \ldots, x_n)$ and given marginals $F_1(x_1), \ldots, F_n(x)$ satisfying (26), the most stringent lower bounding function on $F_{1,2,\ldots,n}(x_1, x_2, \ldots, x_n)$ similarly always exists and is characterized by the following 2^n elementary event probabilities:

$$\Pr\{\bar{x}_1, \bar{x}_2, \bar{x}_3, \ldots, \bar{x}_n\} = 1 - \sum_1^n F_i(x_i),$$
$$\vdots \tag{27}$$
$$\Pr\{\bar{x}_1, \bar{x}_2, \bar{x}_3, \ldots, \bar{x}_{j-1}, x_j, \bar{x}_{j+1}, \ldots, \bar{x}_n\} = F_j(x_j), \quad j = 1, \ldots, n.$$

The probabilities of the remaining $(2^n - n - 1)$ events are equal to 0.

Equivalently, this function is characterized by a set of simultaneously attained most stringent lower bound values on the joint df's of all orders in addition to the marginal df values $F_1(x_1), \ldots, F_n(x_n)$:

$$F^-_{i_1, i_2, \ldots, i_s}(x_{i_1}, x_{i_2}, \ldots, x_{i_s}) = 0,$$
$$i_1 < \cdots < i_s \text{ from } (1, 2, \ldots, n), \quad s = 2, \ldots, n-1, \tag{28}$$
$$F^-_{1,2,\ldots,n}(x_1, x_2, \ldots, x_n) = 0.$$

Given any $\mathbf{x}' = (x_1', x_2', \ldots, x_n')$ such that $\sum_1^n F_i(x_i') < 1$, the most stringent lower bounding function on $F_{1,2,\ldots,n}(x_1, x_2, \ldots, x_n)$, (28), has the properties of a proper joint df on every interval $(\mathbf{x}, \mathbf{x}']$ for all $\mathbf{x} < \mathbf{x}'$ as $\mathbf{x}$ varies down to $-\infty$. Again, the sequence of joint df values of all orders, (28), characterizing it over any interval, is always nondecreasing and systemically consistent in this restricted region.

Case 2.

$$\sum_1^n F_i(x_i) \geqslant (n-1). \tag{29}$$

For given $\mathbf{x} = (x_1, x_2, \ldots, x_n)$ and given marginals $F_1(x_1), \ldots, F_n(x)$ satisfying (29), the most stringent lower bounding function on $F_{1,2,\ldots,n}(x_1, x_2, \ldots, x_n)$ similarly always exists and is characterized by the following 2^n elementary event probabilities:

$$\Pr\{x_1, x_2, \ldots, x_{j-1}, \bar{x}_j, x_{j+1}, \ldots, x_n\} = 1 - F_j(x_j), \quad j = 1, 2, \ldots, n,$$
$$\vdots \tag{30}$$
$$\Pr\{x_1, x_2, \ldots, x_n\} = \sum_1^n F_i(x_i) - n + 1.$$

The probabilities of the remaining $(2^n - n - 1)$ events are equal to 0. [Condition (29) is necessary and sufficient for the existence of characterization (30).]

Equivalently, this function is characterized by a set of simultaneously attained most stringent lower bound values on the joint df's of all orders in addition to the marginal df values $F_1(x_1), \ldots, F_n(x_n)$:

$$F^-_{i_1,i_2}(x_{i_1}, x_{i_2}) = F_{i_1}(x_{i_1}) + F_{i_2}(x_{i_2}) - 1,$$
$$\vdots \qquad\qquad \vdots$$
$$F_{i_1,i_2,\ldots,i_s}(x_{i_1}, x_{i_2}, \ldots, x_{i_s}) = F_{i_1}(x_{i_1}) + \cdots + F_{i_s}(x_{i_s}) - s + 1,$$
$$i_1 < \cdots < i_s \text{ from } (1,2,\ldots,n), \quad s = 2, \ldots, n-1, \qquad (31)$$
$$F^-_{1,2,\ldots,n}(x_1, x_2, \ldots, x_n) = F_1(x_1) + \cdots + F_n(x_n) - n + 1.$$

Given any $\mathbf{x}' = (x'_1, x'_2, \ldots, x'_n)$ such that $\sum_1^n F_i(x'_i) \geqslant (n-1)$, the most stringent lower bounding function, (31), has the properties of a proper joint df on every interval $[\mathbf{x}', \mathbf{x})$ for all $\mathbf{x} > \mathbf{x}'$ as $\mathbf{x}$ varies to $+\infty$. Again, the sequence of joint df values of all orders, (31), characterizing it over any interval, is always nondecreasing and systemically consistent in this restricted region.

CHARACTERIZATION OF THE FUNCTION $F_{1,2,\ldots,n}(x_1, x_2, \ldots, x_n)$ WHEN

$$F_{1,2,\ldots,n}(x_1, x_2, \ldots, x_n) = 0. \qquad (32)$$

Case.

$$1 < \sum_1^n F_i(x_i) < (n-1). \qquad (33)$$

1. For any given $\mathbf{x} = (x_1, x_2, \ldots, x_n)$ and given marginals $F_1(x_1), \ldots, F_n(x_n)$ satisfying (33), at least one systemically consistent characterization of $F_{1,2,\ldots,n}(x_1, x_2, \ldots, x_n)$ always exists for which (32) is satisfied. This characterization is given by the marginals and the following set of joint df values:

$$F_{i_1,i_2,\ldots,i_s}(x_{i_1}, x_{i_2}, \ldots, x_{i_s}) = \alpha\left\{\sum_{i_1}^{i_s} F_i(x_i) - s + 1\right\} + (1-\alpha)\min\{F_{i_1}(x_{i_1}), \ldots, F_{i_s}(x_{i_s})\},$$
$$i_1 < \cdots < i_s \text{ from } (1,2,\ldots,n), \quad s = 2, \ldots, n-1, \qquad (34)$$

$$F_{1,2,\ldots,n}(x_1, x_2, \ldots, x_n) = \alpha\left\{\sum_1^n F_i(x_i) - n + 1\right\} + (1-\alpha)\min\{F_1(x_1), \ldots, F_n(x_n)\} = 0.$$

$$\alpha = \min\{F_1(x_1), \ldots, F_n(x_n)\} \Big/ \left\{(n-1) - \sum_1^n F_i(x_i) + \min\{F_1(x_1), \ldots, F_n(x_n)\}\right\}. \qquad (35)$$

From (33), α always exists such that $1 > \alpha > 0$.

2. The characterization (34) does not comprise a most stringent lower bounding function; in general, no such function exists when (32) and (33) hold.

In general, no sequence of systemically consistent characterizations of $F_{1,2,\ldots,n}(x_1, x_2, \ldots, x_n)$ satisfying (32) and (33) exists for which the constituent joint df's are nondecreasing as $\mathbf{x}$ increases over the restricted region defined by (33).

We illustrate the most stringent upper and lower bounding functions on $F_{1,2,\ldots,n}(x_1, x_2, \ldots, x_n)$, given in the preceding sections, for the case $n = 2$. These lead to and illuminate the Fréchet bounds (19).

From (24) and (25) the most stringent upper bounding function on $F_{1,2}(x_1, x_2)$ for given $\mathbf{x} = (x_1, x_2)$ and given $F_1(x_1) > F_2(x_2)$

is:

$$\Pr\{\bar{x}_1, \bar{x}_2\} = 1 - F_1(x_1)$$
$$\Pr\{x_1, \bar{x}_2\} = F_1(x_1) - F_2(x_2)$$
$$\Pr\{\bar{x}_1, x_2\} = 0$$
$$\Pr\{x_1, x_2\} = F_2(x_2)$$

or

$$\begin{aligned} F_1(x_1) &= \Pr\{x_1, \bar{x}_2\} + \Pr\{x_1, x_2\} \\ F_2(x_2) &= \Pr\{\bar{x}_1, x_2\} + \Pr\{x_1, x_2\} \qquad (36) \\ F_{1,2}^{+}(x_1, x_2) &= \Pr\{x_1, x_2\} \\ &= \min\{F_1(x_1), F_2(x_2)\} \\ &= F_2(x_2). \end{aligned}$$

Equations (36) comprise a proper distribution function as x varies over R^2.

From (27) and (28), the most stringent lower bounding function on $F_{1,2}(x_1, x_2)$ for given $\mathbf{x} = (x_1, x_2)$ and $F_1(x_1), F_2(x_2)$, where $\sum_1^2 F_i(x_i) \leqslant 1$, is

$$\Pr\{\bar{x}_1, \bar{x}_2\} = 1 - \sum_1^2 F_i(x_i)$$
$$\Pr\{x_1, \bar{x}_2\} = F_1(x_1)$$
$$\Pr\{\bar{x}_1, x_2\} = F_2(x_2)$$
$$\Pr\{x_1, x_2\} = 0$$

or

$$\begin{aligned} F_1(x_1) &= \Pr\{x_1, \bar{x}_2\} + \Pr\{x_1, x_2\} \\ F_2(x_2) &= \Pr\{\bar{x}_1, x_2\} + \Pr\{x_1, x_2\} \qquad (37) \\ F_{1,2}^{-}(x_1, x_2) &= \Pr\{x_1, x_2\} = 0. \end{aligned}$$

Equation (37) has the properties of a proper df on every interval $(\mathbf{x}, \mathbf{x}']$, $\mathbf{x} < \mathbf{x}'$, with $\sum_1^2 F_i(x_i') < 1$.

From (30) and (31), the most stringent lower bounding function on $F_{1,2}(x_1, x_2)$ for given $\mathbf{x} = (x_1, x_2)$ and $F_1(x_1), F_2(x_2)$ where $\sum_1^2 F_i(x_i) \geqslant 1$ is:

$$\Pr\{\bar{x}_1, \bar{x}_2\} = 0$$
$$\Pr\{x_1, \bar{x}_2\} = 1 - F_2(x_2)$$
$$\Pr\{\bar{x}_1, x_2\} = 1 - F_1(x_1)$$
$$\Pr\{x_1, x_2\} = \sum_1^2 F_i(x_i) - 2 + 1$$

or

$$\begin{aligned} F_1(x_1) &= \Pr\{x_1, \bar{x}_2\} + \Pr\{x_1, x_2\} \\ F_2(x_2) &= \Pr\{\bar{x}_1, x_2\} + \Pr\{x_1, x_2\} \qquad (38) \\ F_{1,2}^{-}(x_1, x_2) &= \Pr\{x_1, x_2\} \\ &= F_1(x_1) + F_2(x_2) - 1. \end{aligned}$$

Equations (38) have the properties of a proper df on every interval $[\mathbf{x}', \mathbf{x})$, $\mathbf{x} > \mathbf{x}'$ with $\sum_1^2 F_i(x_i') \geqslant 1$.

For $n = 2$, (33) takes the form $F_1(x_1) + F_2(x_2) = 1$, characterizations (37) become identical with characterizations (38), and the ensemble of most stringent lower bounding functions (37) and (38) comprise a proper df as $\mathbf{x}$ varies over R^2 without restriction.

References

[1] American Mathematical Society (1976). *Amer. Math. Monthly*, **85**, 393.

[2] Bonferroni, C. E. (1936). *Ist. Sup. Sci. Econ. Commer. Firenze*, **8**, 1–62.

[3] Boole, G. (1854). *Investigation of Laws of Thought on Which Are Founded the Mathematical Theories of Logic and Probability*. London.

[4] Dall'Aglio, G. (1972). *Symp. Math.*, **9**, 131–150.

[5] Fréchet, M. (1940). *Les Probabilités, associées à un système d'événements compatibles et dépendants*, Première Partie. Hermann & Cie, Paris.

[6] Fréchet, M. (1951). *Ann. Univ. Lyon*, Sec. A, Ser. 3, **14**, 53–77.

[7] Kwerel, S. M. (1975). *J. Appl. Prob.*, **12**, 612–619.

[8] Kwerel, S. M. (1975). *J. Amer. Statist. Ass.*, **70**, 472–479.

[9] Kwerel, S. M. (1980). Most stringent bounding functions on joint distribution functions partially micro-specified by their marginal distribution functions. (In preparation.)

[10] Sklar, A. (1973). *Kybernetika* (*Prague*), **9**, 449–460.

(BONFERRONI INEQUALITIES
BOOLE INEQUALITY
CUMULATIVE DISTRIBUTION
FUNCTION)

S. M. KWEREL

FRÉCHET DERIVATIVE *See* STATISTICAL FUNCTIONALS

FREEMAN–HALTON TEST

This test was designed in part to give exact probabilities in contingency tables, particularly when some expected frequencies are so small that the usual chi-square test* is inappropriate. In this test, roughly speaking, all configurations are considered which give rise to the same marginals; the probabilities of those configurations, which give a joint multinomial probability no greater than that observed, are summed to give the exact P-value of the test. See Freeman and Halton [1]; Gart [2] relates it to a modified likelihood ratio* chi-square test, and Heller [3] gives a FORTRAN computer program which computes exact P-values.

References

[1] Freeman, G. H. and Halton, J. H. (1951). *Biometrika*, **38**, 141–149.

[2] Gart, J. J. (1966). *J. R. Statist. Soc. B*, **28**, 164–179.

[3] Heller, R. (1979). *EDV Med. Biol.*, **10**, 62–63.

(CHI-SQUARE TESTS
CONTINGENCY TABLES
GART'S INDEPENDENCE TEST STATISTIC)

FREEMAN–TUKEY TEST

The Freeman–Tukey test is a test of the goodness of fit* of a hypothesized model for counted data. It is assumed that each individual of a population under study can be classified in one and only one of k categories (*see* CATEGORICAL DATA). A random sample of size N is drawn from the population and results in *observed* counts or frequencies of individuals in the k categories,

$$x_1, x_2, \ldots, x_k,$$

such that

$$\sum_{i=1}^{k} x_i = N.$$

Under some hypothetical probability model the corresponding *expected* counts, denoted by

$$\hat{m}_1, \hat{m}_2, \ldots, \hat{m}_k,$$

such that

$$\sum_{i=1}^{k} \hat{m}_i = N,$$

may be calculated. The accent marks (ˆ) indicate that the true model expected frequencies m_i depend on parameters that may be estimated from the data. It is desired to test whether the model is consistent with the data. One measure of goodness of fit of the model is the Freeman–Tukey statistic [2, pp. 130, 137]

$$T^2 = \sum_{i=1}^{k} \left(\sqrt{x_i} + \sqrt{x_i + 1} - \sqrt{4\hat{m}_i + 1}\right)^2.$$

The Freeman–Tukey statistic is an alternative to the best known statistic for goodness of fit, the Pearson chi-square statistic

$$X^2 = \sum_{i=1}^{k} (x_i - \hat{m}_i)^2 / \hat{m}_i.$$

Another alternative is the (log) likelihood ratio* statistic [2, p. 125]

$$G^2 = 2 \sum_{i=1}^{k} x_i \log_e(x_i / \hat{m}_i)$$

(*see* CHI-SQUARE TESTS). If there are s ($s < k - 1$) parameters in the model, if the model is correct, and if the parameters are estimated by the maximum likelihood* method, it can be shown under certain regularity conditions that T^2, X^2, and G^2 have the same asymptotic distribution as N becomes infinite, and that distribution is the chi-square distribution with $k - s - 1$ degrees of freedom [2, pp. 130, 513–516]. If the model is not correct, the asyptotic behaviors of the three statistics may be quite different from one another. Crow and Miles [3] experienced this contrasting asymptotic behavior in using all three statistics to test whether errors in digital telephone transmissions obeyed a Markov chain model of first, second, or third order. The statistics had very difficult values for first order, but similar (and small) values for second order.

The Freeman–Tukey statistic is so named because it is based on the transformation

$\sqrt{Y} + \sqrt{Y+1}$ of a Poisson* variable Y recommended by Freeman and Tukey [5] for the purpose of stabilizing the variance. An empirical study showed that the variance of the transformed variable is within 6% of unity for $E[Y] = m$ greater than or equal to unity. The transformed variable is asymptotically normally distributed with variance unity and mean $\sqrt{4m+1}$. Thus the residuals from the fitted model that are squared and summed to form T^2 are approximately standard normal deviates under the hypothetical model and may be examined for consistency with the model in the same way that residuals from an analysis of variance or regression analysis may be used; see Anscombe and Tukey [1] or Draper and Smith [4] as well as Bishop et al. [2, pp. 136–141] and Haberman [6, pp. 138–144]. The variance stabilization property would seem to make the Freeman–Tukey deviates more attractive for residual analysis than the signed square roots of the terms in X^2, but Haberman [6] states that "No clear evidence exists that other choices [than those from X^2] are better for use with general log-linear models."

A simple numerical comparison of T^2, X^2, and G^2 may be made using a 2×2 table given by Haberman [6] (following Toivanen and Hirvonen [8]); see Table 1. For these data $T^2 = 7.990$, $X^2 = 7.970$, X^2 (modified by Yates' continuity correction*) = 7.787, and $G^2 = 7.988$. They differ so little because of the large sample size. Since the 1% point of chi-squared for 1 degree of freedom is 6.635, Toivanen and Hirvonen concluded that "The difference between the sex ratio of babies born to toxemic mothers and that of controls is significant."

Larntz [7] invistigated the small-sample behavior of T^2, X^2, and G^2 under the null hypothesis for five different models with a wide range of parameter values. For a nominal significance level of 0.05 (and of 0.10 and 0.01 also in some cases) he obtained by exact calculation or Monte Carlo simulation the actual rejection rates (true significance levels) of each test. Minimum category expected frequencies ranged down to well below 1. He concluded as follows:

> Based on a criterion of the closeness of small sample distribution to the asymptotic chi-squared approximation, the Pearson chi-squared statistic is by far the most desirable. Both the likelihood rato and Freeman–Tukey statistics yield too many rejections under the null distribution. . . . The high Type I error rates for the likelihood ratio and Freeman–Tukey statistics result from the large contributions to the chi-squared value for very small counts in cells with moderate expected values.

Table 1 Newborn Babies

	Males	Females	Ratio
Toxemic Mothers	588	473	1.24
Controls	4196	4061	1.03

References

[1] Anscombe, F. J. and Tukey, J. W. (1963). *Technometrics*, **5**, 141–160.

[2] Bishop, Y. M. M., Fienberg, S. E., and Holland, P. W. (1975). *Discrete Multivariate Analysis: Theory and Practice*. MIT Press, Cambridge, Mass.

[3] Crow, E. L. and Miles, M. J. (1977). *Confidence Limits for Digital Error Rates from Dependent Transmissions*. Office of Telecommunications Rep. 77-118, U.S. Dept. of Commerce, Boulder, Colo.

[4] Draper, N. R. and Smith, H. (1966). *Applied Regression Analysis*. Wiley, New York, Chap. 3.

[5] Freeman, M. F. and Tukey, J. W. (1950). *Ann. Math. Statist.*, **21**, 607–611.

[6] Haberman, S. J. (1974). *The Analysis of Frequency Data*. Univ. of Chicago Press, Chicago, Ill.

[7] Larntz, K. (1978). *J. Amer. Statist. Ass.* **73**, 253–263.

[8] Toivanen, P. and Hirvonen, T. (1970). *Science*, **170**, 187–188.

(CHI-SQUARE TESTS
EQUALIZATION OF VARIANCE)

EDWIN L. CROW

FREEMAN–TUKEY TRANSFORMATIONS

Freeman and Tukey [1] introduced transformations to stabilize the variances of binomial* and Poisson* frequencies (*see* EQUAL-

IZATION OF VARIANCE). If x occurrences of some event are observed in n trials, the Freeman–Tukey (FT) transformation of the binomial variable is

$$\theta = \tfrac{1}{2}\left\{\arcsin\sqrt{x/(n+1)} + \arcsin\sqrt{(x+1)/(n+1)}\right\}.$$

In most cases for which $np \geqslant 1$, p being the probability of the event of interest occurring in any given trial, the variance of θ is within 6% of $(n+\frac{1}{2})^{-1}$ in radians, and within $821(n+\frac{1}{2})^{-1}$ in degrees. If x is a Poisson frequency count, the FT transformation is

$$g = \sqrt{x} + \sqrt{(x+1)}\ .$$

Then the variance of g is approximately 1 if the Poisson parameters exceeds 1.

Mosteller and Youtz [2] found that, as n increases, the set of values of p about $p = \frac{1}{2}$ for which the variance of θ is stable is increasingly wide; it is more stable and generally smaller than that of $\theta' = \arcsin\sqrt{x/n}$. They give a table of values of θ to two decimal places, corresponding to x and n, $1 \leqslant n \leqslant 50$; also tables of g and g^2 for $0 \leqslant x \leqslant 50$.

The FT transformation in testing goodness of fit of models in categorical data* analysis is given by

$$T^2 = \sum_{\text{all cells}}\left\{\sqrt{\text{observed}} + \sqrt{\text{observed} + 1} - \sqrt{4(\text{expected}) + 1}\right\}^2;$$

(*see* FREEMAN–TUKEY TEST).

References

[1] Freeman, M. F. and Tukey, J. W. (1950). *Ann. Math. Statist.*, **21**, 607–611.

[2] Mosteller, F. and Youtz, C. (1961). *Biometrika*, **48**, 433–440. (Tables of transformed variates, and properties of transformations.)

(EQUALIZATION OF VARIANCE
FREEMAN–TUKEY TEST)

FREQUENCY CURVES, SYSTEMS OF

The distribution of observations in a set of univariate data can often be analyzed by fitting a frequency curve to a histogram* of the data and utilizing the probability density or distribution function* corresponding to the curve for subsequent computation or derivation. This article surveys the frequency curve systems which are most commonly applied in current statistical work: the Pearson, Gram–Charlier*, Burr*, Johnson*, and Tukey lambda systems*. The discussion highlights various aspects of these systems, including their theoretical genesis, shape flexibility, techniques for parameter estimation,* tabulation of percentiles (*see* QUANTILES), special applications, and analytical advantages.

An annotated reference list and bibliography are provided as a guide to the literature on systems of frequency curves. The reader should also consult articles in this encyclopedia on specific families and systems of distributions, as well as related statistical methods, such as goodness-of-fit tests* and estimation techniques.

HISTORICAL BACKGROUND

The oldest and most extensively studied system of frequency curves is the *Pearson system*, developed between 1890 and 1900 by Karl Pearson*, one of the founders of modern statistical theory. During the late nineteenth century there had been a tendency to regard all distributions as normal*; data histograms that displayed multimodality were often fitted with mixtures* of normal density curves, and histograms that exhibited skewness* were analyzed by transforming the data so that the resulting histogram could be fitted or "graduated" with a normal curve. As an alternative, Karl Pearson constructed a system of theoretical curves with widely varying shapes for obtaining improved approximations to frequency data. The first reference to this system appeared in Pearson [69]; see also Pearson [70, 71] and the collection of Pearson's early papers edited by E. S. Pearson [60]. A historical account of this period in the early development of statistics has been given by E. S. Pearson [61, 62].

The introduction of the Pearson system was a significant development for two reasons: the system yielded simple mathematical representations, involving a small number of parameters, for histogram data in many applications, and it provided a theoretical framework for various families of sampling distributions discovered subsequently by Pearson and others. An early criticism of the Pearson system was that it had not been derived from a "theory of errors" such as Gauss* and Laplace* had applied in their derivation of the normal distribution; see Särndal [80] and LAWS OF ERRORS. Consequently, a second approach, based on series expansions involving the normal density and its derivatives, was developed for expressing nonnormality in frequency data.

Series expansions of distributions that could deviate from normality had been published by Gram [34] and Thiele [91]. C.V.L. Charlier [16–18], a Swedish astronomer, showed how these curves could be justified through a *hypothesis of elementary errors* and applied them by fitting numerous sets of data. The expansion discussed by Charlier [16] is a *Gram–Charlier type A series*; the first term in the series is the normal density function. A *type B series* was obtained by Charlier [17] for fitting highly skewed data; its dominant term (in a limiting sense) is the Poisson distribution*.

A drawback of the Gram–Charlier series is that the terms are not listed in a monotonically decreasing order of importance. Edgeworth [27] introduced a similar series in which the terms are ordered according to magnitude, and claimed that his results, rather than Charlier's, represented the true generalization of Laplace's central limit theorem*. The two series are occasionally distinguished by referring to Charlier's type A series as type Aa and *Edgeworth's series* as type Ab. In applications, the Gram–Charlier and Edgeworth expressions fitted to data are truncated versions of the corresponding infinite series.

After the early 1900s, the introduction of additional families and systems of frequency curves progressed sporadically. There was a continuing interest in methods of curve fitting (see Pearson [73]), and various attempts were made to extend the available systems to systems of frequency surfaces* for fitting bivariate frequency data. However, newer developments, such as hypothesis testing*, provided alternatives to curve fitting which made more efficient use of sample data in making inferences. Nevertheless, three of the curve systems most commonly applied for fitting were introduced during the 1940s.

The *Burr system* was constructed in 1941 by Irving W. Burr, an American pioneer in the field of quality control* (see Burr [12]). Burr's objective was to fit cumulative distributions rather than density functions to frequency data, to avoid the problems of numerical integration which are encountered when probabilities are evaluated from Pearson curves; *see* BURR DISTRIBUTIONS.

The *symmetric Tukey lambda distributions* were introduced by John W. Tukey in 1947 to facilitate a study of order statistics* (see Hastings et al. [35]). This family and several of its more recent extensions have been applied in simulation* work and in approximating percentiles of theoretical distributions, as well as fitting.

The *Johnson system* (also referred to as the *translation system*) was developed by Norman L. Johnson [38] in his University of London Ph.D. dissertation. The construction utilized four distinct monotone transformations* which, when applied to skewed frequency data, yield approximate normality. Earlier writers had studied various transformations to normality, including the log transformation [33, 51], the power transformation [48], and a monotone polynomial transformation [28, 29]. The advantage gained with Johnson's transformations is that, when inverted and applied to a normally distributed variable, they yield four families of density curves with a high degree of shape flexibility.

Since the 1940s, interest in systems of distributions for curve fitting has diminished due, in part, to the development of nonparametric density estimation* methods which provide alternatives, such as spline* techniques, for smoothing histograms (*see* CURVE FITTING). Numerous new classes of distributions have arisen in theoretical work, but relatively few are potentially applicable for fitting a variety of frequency data. Among the exceptions are the Bessel function* distributions (see McKay [52], Bhattacharyya [4], and McNolty [53]) and the hyperbolic distributions* (see Barndorff-Nielsen [2] and Blæsild [5]). See the Proceedings of the NATO Advanced Study Institute edited by Patil et al. [57] for recent developments of specialized families of distributions and their applications to model building.

Systems of frequency curves continue to play a role in the analysis of distribution data, because they provide functionally simple approximations to observed distributions in situations where it is difficult to derive a model. Moreover, they are finding new applications in simulation* work, as well as providing approximations for theoretical distributions.

THEORETICAL GENESIS OF THE COMMONLY USED SYSTEMS

An understanding of its theoretical development can be helpful in selecting the system that is most appropriate for a given application.

Pearson [69] took as his starting point the skewed binomial* and hypergeometric* distributions, which he smoothed in an attempt to construct skewed continuous density functions. The derivation led to a differential equation

$$\frac{dp}{dx} = \frac{-(a+x)p}{c_0 + c_1 x + c_2 x^2}, \qquad (1)$$

whose solutions $p(x)$ are the density functions of the *Pearson system*. Note that Pearson himself did not begin with the differential equation, although that impression is occasionally conveyed by modern writers. Nonetheless, it is clear that some of the solutions to (1) have a single mode ($dp/dx = 0$ at $x = -a$) and smooth contact with the horizontal axis ($dp/dx = 0$ when $p(x) = 0$), both of which are natural properties for a frequency curve.

The various types or families of curves within the Pearson system correspond to distinct forms of solutions to (1) (*see* Elderton and Johnson [30], Johnson and Kotz [43], Kendall and Stuart [49], and PEARSON SYSTEM OF CURVES). Although the solutions $p(x)$ can be expressed in closed form, some of the types are analytically awkward. Moreover, because the solutions are not obtained by probabilistic modeling, their parameters may be difficult to interpret in practice. On the other hand, the Pearson differential equation (1) is sufficiently general to yield solutions with a wide variety of curve shapes, and consequently the Pearson system has been used successfully for applications in which a close, smooth approximation to a histogram, rather than a simple fitted equation, is the primary objective.

The original derivation of the *Burr system*, unlike that of the Pearson system, did begin with a differential equation. Burr [12] constructed cumulative distribution functions $F(x)$ satisfying the differential equation

$$\frac{dF}{dx} = F(1-F)g(x,F), \qquad (2)$$

an analog of (1); different choices of the function $g(-,-)$ generate various families of solutions $F(x)$. Because many of these solutions are functionally simple, they obviate the analytical difficulties often encountered when a fitted density is integrated to obtain probabilities or percentiles. Moreover, the simplicity of the Burr forms increases the potential for meaningful interpretation of their parameters in modeling applications (*see* BURR DISTRIBUTIONS).

Gram–Charlier and *Edgeworth curves* were derived in an attempt to express nonnormal densities as infinite series involving the normal density and its derivatives. In practice, a finite number of terms from one of these series is taken for the fitted expression, although the truncated series may be negative over certain intervals or may exhibit multimodality. Conditions under which fitted Gram–Charlier and Edgeworth curves are positive and unimodal were obtained by Barton and Dennis [3] and Draper and Tierney [26]. Their work reveals that Gram–Charlier and Edgeworth curves fit histograms which are at most mildly nonnormal (in terms of skewness and kurtosis). On the other hand, the theoretical properties of Edgeworth expansions make them a natural tool for many mathematical studies of nonnormality (*see* CORNISH–FISHER/EDGEWORTH EXPANSIONS; GRAM–CHARLIER SERIES).

As we have noted, transformation to normality is the basis for the *Johnson system*. Johnson [38] discussed four transformations, each of which induces a family of distributions within the system:

$$\begin{aligned} f_N(Y) &= \gamma + \delta Y \\ f_L(Y) &= \gamma + \delta \log Y \\ f_B(Y) &= \gamma + \delta \log(Y/(1-Y)) \\ f_U(Y) &= \gamma + \delta \sinh^{-1} Y. \end{aligned} \tag{3}$$

The Johnson families were derived by obtaining the distributions of Y, assuming that $f_I(Y)$ has a normal distribution, $I = N, L, B, U$. For $I = N$ and $I = L$, the resulting distributions constitute the normal and lognormal* families, respectively. Transformations f_B and f_U were chosen to produce two additional families which extend the curve shape variety of the log-normal family. Although the analytical forms of Johnson distributions are not particularly simple, their extensive variety makes them appropriate for fitting histograms in many situations, particularly when *both* a transformation to normality *and* a fitted curve are needed.

Monotone transformations for achieving distributions other than the normal can be used to build a variety of curve systems. For instance, by inverting the four transformations in (3) and applying them to a double exponentially distributed variable, one obtains an analog of the Johnson system (see Johnson [39]). Systems of distributions can also be constructed by applying a simple transformation directly to a variable with a specified distribution. One example is the class of distributions generated by taking power transformations of a gamma-distributed variable (see Johnson and Kotz [44]). Another example is the Burr system, several of whose families are related by log and power transformations.

A more complicated transformation is used to define the *symmetric Tukey lambda distributions*. These are the distributions of Z, where

$$Z = \begin{cases} \left[U^{\lambda} - (1-U)^{\lambda}\right]/\lambda, & \lambda \neq 0 \\ \log(U/(1-U)), & \lambda = 0, \end{cases} \tag{4}$$

and U is uniformly distributed on $[0, 1]$. Note that the $(100p)$th percentile of Z can be obtained explicitly by substituting p for U in (4). Extensions of this family have been achieved by replacing U with a beta*-distributed variable [45] and by injecting additional parameters [74, 75].

SUMMARY OF THE MAIN FUNCTIONAL FORMS

This section summarizes the main functional forms of the most commonly applied systems of frequency curves. The parameters indicated for each density or distribution function are shape parameters*; a location parameter* (ξ) and a scale parameter* (λ) can be added by replacing x with $(x - \xi)/\lambda$, where $-\infty < \xi < \infty$ and $\lambda > 0$.

Pearson distributions are classified into types that correspond to forms of solutions to the Pearson differential equation (1). The equations of the three main Pearson types are:

Type I:

$$p(x) = \frac{1}{B(m_1, m_2)} x^{m_1}(1-x)^{m_2},$$

$$0 \leqslant x \leqslant 1; m_1 > -1, m_2 > -1.$$

Type IV:

$$p(x) = \frac{1}{K}(1 + x^2)^{-m}\exp(-\nu\tan^{-1}x),$$
$$-\infty < x < \infty;\, m > \tfrac{1}{2},\, \nu \neq 0. \quad (5)$$

Type VI:

$$p(x) = \frac{1}{B(p,q)}\, x^{p-1}/(1 + x)^{p+q},$$
$$0 \leqslant x < \infty;\, p > 0, q > 0.$$

Type I are beta distributions; in a slightly different form, type VI are better known as *F*-distributions*. For additional details concerning the main as well as the transitional Pearson families (among which occur the normal, Student's *t**, and gamma distributions*), see the general works by Johnson and Kotz [43], Elderton and Johnson [30], Ord [56], and Kendall and Stuart [49]. Examples of the main types of Pearson curves are illustrated in Fig. 1.

Although Burr identified over 12 families in his system of distributions, only two, the type XII and the type III families, have been discussed in the literature; their cumulative distribution functions are:

Type III:

$$F(x) = (1 + x^{-c})^{-k}, \qquad x > 0;\, c > 0, k > 0. \quad (6)$$

Type XII:

$$F(x) = 1 - (1 + x^{c})^{-k},$$
$$x > 0;\, c > 0, k > 0.$$

Examples of Burr III curves are shown in Fig. 2. For further details, the reader should consult the original paper by Burr [12], or subsequent papers by Burr [13], Burr and Cislak [15], Rodriguez [76], and Rodriguez [78].

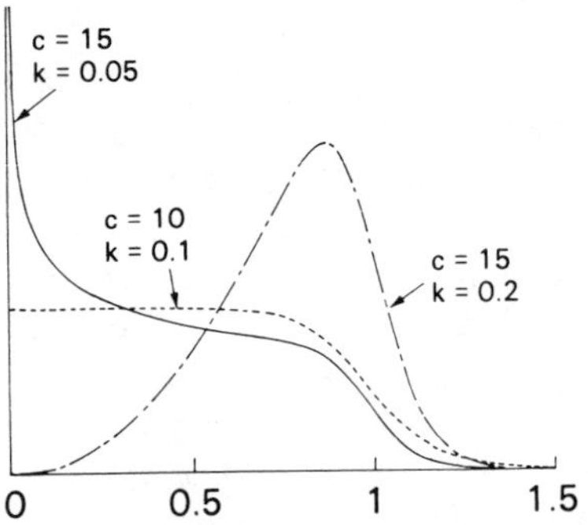

Figure 2 Examples of frequency curves belonging to the type III family of the Burr system.

For the Gram–Charlier and Edgeworth expansions, the truncated expressions generally fitted to histograms are:

Gram–Charlier:

$$f(x) = \left\{1 + \tfrac{1}{6}\sqrt{\beta_1}\,(x^3 - 3x) + \tfrac{1}{24}(\beta_2 - 3)(x^4 - 6x^2 + 3)\right\}z(x) \quad (7)$$

Edgeworth:

$$f(x) = \left\{1 + \tfrac{1}{6}\sqrt{\beta_1}\,(x^3 - 3x) + \tfrac{1}{24}(\beta_2 - 3)(x^4 - 6x^2 + 3) + \tfrac{1}{72}\,\beta_1(x^5 - 10x^3 + 15x)\right\}z(x), \quad (8)$$

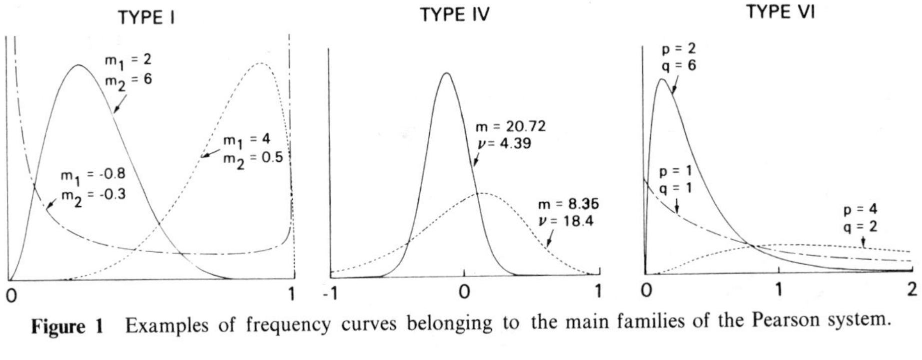

Figure 1 Examples of frequency curves belonging to the main families of the Pearson system.

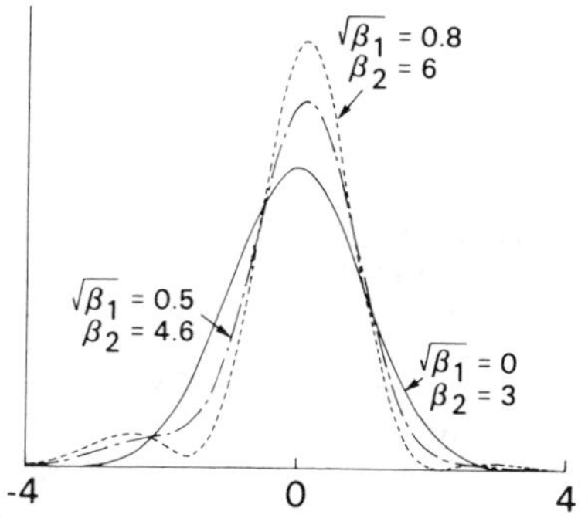

Figure 3 Examples of frequency curves belonging to the Gram–Charlier family. The curve for which $\sqrt{\beta_1} = 0$ and $\beta_2 = 3$ is the normal density curve.

where $z(x)$ is the standard normal density function, and $\sqrt{\beta_1}$ and β_2 are the (estimated) skewness and kurtosis of the distribution to be fitted. See Fig. 3 for examples of Gram–Charlier curves. Additional details concerning series expansion fits are given in Elderton and Johnson [30], Johnson and Kotz [43], Kendall and Stuart [49], and Ord [56].

The four families of the Johnson system are designated by the symbols S_N, S_L, S_B, and S_U, corresponding to the transformations listed in the preceding section.

The equations for the density functions are:

S_L family:

$$f(x) = \frac{\delta}{\sqrt{2\pi}}\,\frac{1}{x}\exp\left[-(\gamma + \delta\log x)^2/2\right],$$
$$0 \leqslant x < \infty.$$

S_B family:

$$f(x) = \frac{\delta}{\sqrt{2\pi}}\,\frac{1}{x(1-x)} \times \exp\left[-\left(\gamma + \delta\log\frac{x}{1-x}\right)^2/2\right],$$
$$0 \leqslant x \leqslant 1. \quad (9)$$

S_U family:

$$f(x) = \frac{\delta}{\sqrt{2\pi}}\,\frac{1}{\sqrt{x^2+1}} \times \exp\left[-\left(\gamma + \delta\log\left\{x + \sqrt{(x^2+1)}\right\}\right)^2\Big/2\right],$$
$$-\infty < x < \infty.$$

In all three forms, $\delta > 0$ and $-\infty < \gamma < \infty$. See Fig. 4 for examples of S_U and S_B curves. Special properties are given by Elderton and Johnson [30], Johnson [38], Johnson and Kotz [43], Kendall and Stuart [49], and Ord [56].

The density functions of the symmetric Tukey lambda family are defined implicitly by the equations

$$f(x(u)) = \left[u^{\lambda-1} + (1-u)^{\lambda-1}\right]^{-1}$$
$$x(u) = \begin{cases}\left[u^{\lambda} - (1-u)^{\lambda}\right]/\lambda, & \lambda \neq 0\\ \log(u/(1-u)), & \lambda = 0,\end{cases} \quad (10)$$

where $0 \leqslant u \leqslant 1$, and the range of variation of $x(u)$ is $[-1/\lambda, 1/\lambda]$ if $\lambda > 0$, and

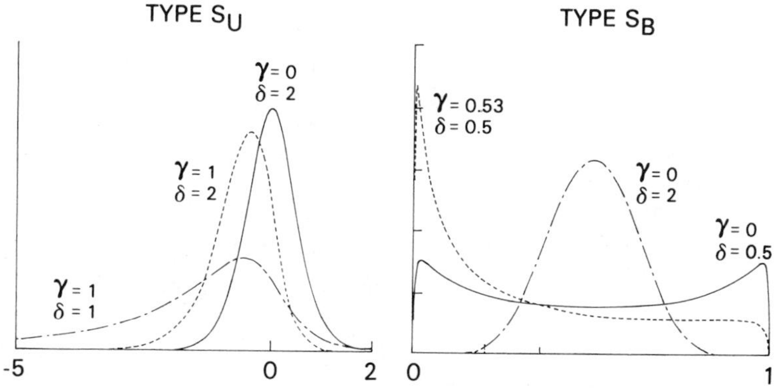

Figure 4 Examples of frequency curves belonging to the main families of the Johnson system.

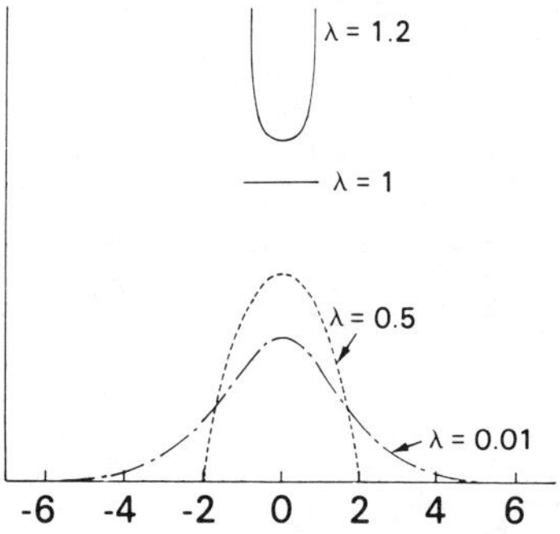

Figure 5 Examples of frequency curves belonging to the symmetric Tukey lambda family. The line labeled $\lambda = 1$ corresponds to the uniform density.

$(-\infty, \infty)$ if $\lambda \leqslant 0$. At the extremes of this range,

$$f(x(0)) = f(x(1)) = \begin{cases} 1, & \lambda > 1, \\ \frac{1}{2}, & \lambda = 1, \\ 0, & \lambda < 1. \end{cases}$$

If $\lambda = 1$ or $\lambda = 2$, the density is uniform. Additional properties of the family are described by Joiner and Rosenblatt [47]. Several curves in this family are illustrated in Fig. 5.

Table 1 summarizes some of the properties of the curve systems commonly used for fitting frequency data. No one system is suitable for all situations, but the variety of features available does provide a selection of useful alternatives in applications.

TECHNIQUES FOR FITTING FREQUENCY CURVES

In any statistical curve-fitting problem, the method of fitting is as crucial as the choice of system in providing a fit with a meaningful interpretation. The oldest technique for fitting frequency curves is the *method of moments**, proposed for use with the Pearson system. By equating sample moments* (up to third or fourth order) with corresponding theoretical moments, one obtains a nonlinear system of equations involving the parameters of the fitted curve. The equations have explicit solutions in the case of the Pearson system, although this is not true for curve systems in general.

When fitting with the Pearson or Johnson systems, the method of moments uniquely identifies both the fitted curve and its family, because the third and fourth moment combinations of the families within these systems do not overlap. This does not occur with the Burr system or with series expansion fits; the user must select the family within the system before applying the method of moments.

Table 1 Summary of Curve Systems Commonly Used for Fitting Frequency Data

System	Theoretical Genesis	Densities in Explicit Form?	Distributions in Explicit Form?	Percentiles in Explicit Form?	Includes Normality?	Range of Skewness and Kurtosis
Pearson	Differential equation	Yes	No	No	Yes	Maximum
Gram–Charlier/ Edgeworth	Series expansion	Yes	No	No	Yes	Moderate
Burr	Differential equation	Yes	Yes	Yes	No	Extensive
Johnson	Transformation to normality	Yes	No	No	Yes	Maximum
Tukey lambda	Transformation	No	No	Yes	No	Restricted (symmetric densities)

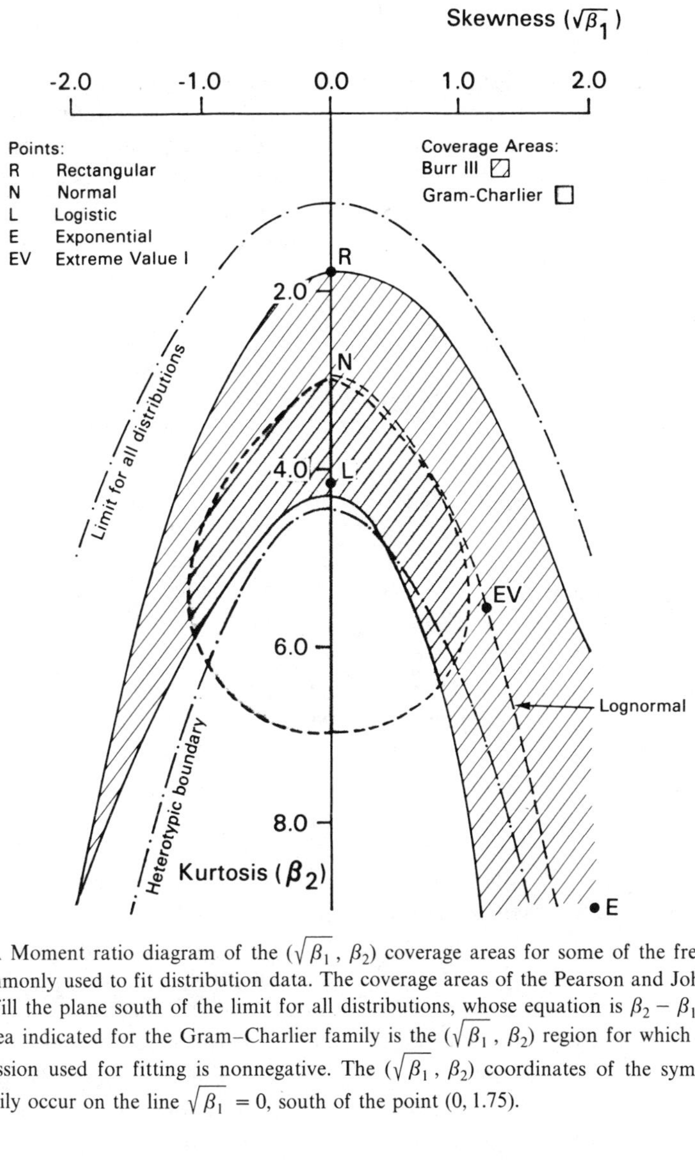

Figure 6 A Moment ratio diagram of the ($\sqrt{\beta_1}$, β_2) coverage areas for some of the frequency curve systems commonly used to fit distribution data. The coverage areas of the Pearson and Johnson systems completely fill the plane south of the limit for all distributions, whose equation is $\beta_2 - \beta_1 - 1 = 0$. The coverage area indicated for the Gram–Charlier family is the ($\sqrt{\beta_1}$, β_2) region for which the truncated series expression used for fitting is nonnegative. The ($\sqrt{\beta_1}$, β_2) coordinates of the symmetric Tukey lambda family occur on the line $\sqrt{\beta_1} = 0$, south of the point (0, 1.75).

An important advantage of the method of moments is that both the observed distribution and the range of moment combinations that can be fitted with a particular system can be represented graphically on a moment ratio* diagram of skewness ($\sqrt{\beta_1} = \mu_3/\mu_2^{3/2}$) and kurtosis* ($\beta_2 = \mu_4/\mu_2^2$). Figure 6 illustrates the coverage areas for several of the more familiar systems of frequency curves. In principle, it is possible to fit any histogram with a Pearson or Johnson curve by the method of moments, whereas the positivity region of the Gram–Charlier curves is fairly restricted. Craig [21] and Boetti [6] proposed alternative moment ratio diagrams.

Tables are available which facilitate the application of the method of moments. Johnson [40], Johnson and Kitchen [41, 42], and Pearson and Hartley [65] present tables for use with the Johnson S_U and S_B families. Burr [14] published a table for fitting Burr XII curves. Tables are generally not needed to fit Pearson curves; see Elderton and Johnson [30] for computational examples, as well as Hoadley [36] and Müller and Vahl [55] for a discussion of the case where the left boundary and first three moments are known. In some systems there is more than one curve with the same combination of $\sqrt{\beta_1}$ and β_2, a phenomenon that complicates the construction and use of tables; an example is the extended Tukey lambda family discussed by Ramberg et al. [75].

Increasingly, systems of frequency curves are fitted with computer* programs that eliminate the interpolation* that is inevitable when using tables (see, e.g., Bouver and Bargmann [7b]). Rodriguez [77] describes an interactive computer graphics approach for fitting Burr III curves which involves a "live" moment ratio diagram.

A major theoretical disadavantage of the method of moments is that it does not generally provide the most efficient estimators of the unknown parameters (see Fisher [31] and Shenton [81]). Karl Pearson questioned the reliability of this method for fitting type IV curves whose moments of order eight and higher are infinite, since the variance of the fourth moment estimate is infinite in such situations. In Fig. 6 the $(\sqrt{\beta_1}, \beta_2)$ coordinates of these curves are represented by the area below (south of) the "heterotypic boundary" whose equation is $8\beta_2 - 15\beta_1 - 36 = 0$. (A transformation that carries type IV distributions into distributions with $(\sqrt{\beta_1}, \beta_2)$ points in the region covered by the type I family is proposed by Bowman and Dusenberry [8].) Other disadvantages of the method of moments are that it does not apply to censored data (*see* CENSORING), and that, as in the case of Johnson S_B curves, it can be difficult to express the theoretical moments explicitly.

In principle, these difficulties can be overcome by the method of maximum likelihood*. In practice, however, likelihood estimates of the parameters are generally impossible to derive explicitly, and the likelihood equations may be difficult to solve numerically, particularly if a threshold parameter is to be estimated.

Additional approaches for fitting frequency curves include the method of percentiles, the minimum chi-square method*, the method of cumulative moments [12], the method of frequency moments [83], and the method of maximum likeness [2]. (See Ord [56] for a general discussion of fitting techniques.) The method of percentiles has been recommended for use with the Johnson S_B family (see Bukač [11], Johnson [38], Mage [50], Slifker and Shapiro [85], and Wheeler [93]), and it is potentially applicable to systems, such as the Burr, for which percentiles are easy to express. In some cases two methods can be combined; for instance, the method of moments can provide a set of parameter estimates that serve as an initial approximation for an iterative maximum likelihood solution procedure.

TECHNICAL PROBLEMS SUBSEQUENT TO FITTING

When selecting a system of frequency curves for a particular application, the analyst should consider the uses to be made of the fitted form and the extent to which computational simplicity is desirable. Often a fitting procedure culminates with the evaluation of probabilities or percentiles. Since most Pearson densities cannot be integrated in closed form, tables and methods for finding percentages and percentile points (generally as functions of $\sqrt{\beta_1}$ and β_2) have been developed by various authors: Amos and Daniel [1], Bouver and Bargmann [7a], Bowman and Shenton [9, 10], Davenport and Herring [24], Dershowitz [25], Johnson et al. [46], Pearson and Hartley [64, 65], and Pearson and Merrington [66]. A particularly accessible table of standardized percentile points of Pearson curves is given by Pearson and Hartley [65]; a special table (see Pearson [72]) or numerical integration* is needed to determine the normalization constant K in the Pearson type IV form given in (5); Woodward [94] discusses the approximation of type IV tail probabilities.

For the Johnson curve families, percentages and percentile points can be found by using the associated transformations together with a table of the normal distribution. The percentiles (but not the cumulative distribution functions) of the symmetric Tukey lambda family and its extensions can be expressed in closed form. Even more attractive in this sense are the Burr distributions, for which both the cumulative distributions and their inverses can be written in closed form.

When working with Gram–Charlier and Edgeworth fits, expansions for percentiles, referred to as Cornish–Fisher expansions*, can be derived as functions of corresponding percentiles of the standard normal distribution. See Cornish and Fisher [20], Fisher and Cornish [32], and Johnson and Kotz [43].

Fitted frequency curves are often accompanied by a test or a measure of goodness of fit. Application of the chi-square test* in this context requires care (see Elderton and Johnson [30]). A fit obtained by the method of moments or maximum likelihood does not generally result in a minimum chi-square value.

The percentile points of frequency curves having the same first four moments provide a convenient basis for comparisons of two or more systems. Studies of this type have been carried out by Johnson [38], Merrington and Pearson [54], Pearson et al. [68], and Pearson and Tukey [67]. The results can be useful for approximating mathematically unknown distributions of statistics, as well as selecting curve systems for fitting frequency data.

SPECIAL APPLICATIONS

As early as 1908, W. S. Gosset* determined the distribution of the sample variance s^2 of a small, normally distributed set of observations by deriving the skewness and kurtosis of s^2, and realizing that these values are those of a Pearson type III distribution (see Student [89]). Although this result can be obtained mathematically, systems of frequency curves continue to be used to approximate the distributions of statistics which are intractable, but whose first four moments (or three moments and an endpoint) can be derived.

A few examples are worth noting. Student [90] used Pearson approximations to calculate upper percentile points of the range in a normal sample; the approximations were later found to be very close to the values obtained directly by Pearson and Hartley [63]. This approach has also been used to approximate the percentile points of the sample skewness and kurtosis of a normal sample; see D'Agostino and Pearson [22], Pearson [58, 59], and Shenton and Bowman [82]. More recently, Solomon and Stephens [87] studied Pearson curve approximations to the distribution of a sum of weighted chi-square variables. Problems of this type are surveyed in a subsequent article by Solomon and Stephens [88]; see also Solomon and Stephens [86].

A second special application of frequency curve systems is in the representation of nonnormal distributions in robustness* studies of testing and estimation procedures. In addition to reawakening interest in older curve systems, this application has stimulated the development of various new families of curves, including those discussed by Ramberg [74] and Johnson et al. [37]. Cooper et al. [19] describe a computer program for generating random numbers with Pearson distributions.

Finally, the theoretical form of a fitted frequency curve can be exploited to enhance subsequent analysis in certain specialized applications, even when the fitted form cannot be derived from assumptions via a modeling argument. Two examples are worth citing:

Example 1. In the analysis of mass-size distribution data, the observed mass density is of the form $x^3f(x)$, where $f(x)$ is the probability density for size (diameter). Fitting the observed distribution with a frequency curve enables one to develop expressions for $f(x)$ and its moments. Although the log-normal family has traditionally been used for this purpose on the basis of modeling arguments, a more flexible system of curves often provides better fits. Among various types that have been substituted for the log-normal are hyperbolic distributions* [2, 5], so called because the logarithm of the density function is a hyperbola.

Example 2. The Gini index has been studied extensively by econometricians as a measure of the extent of inequality in an income distribution (*see* INCOME INEQUALITY MEASURES). Roughly, an index of zero indicates a population in which all have the same income, and an index of 1 indicates a population in which half receive no income and the other half receive an identical income. Kendall and Stuart [49] describe the mathematical relationship between an income distribution and its Gini index; among the families of curves considered for fitting income distribution data are the log-normal and gamma [79], the Burr XII [84], and the Burr III [23]. For a Burr III distribution with threshold parameter ξ, scale parameter λ, and shape parameters c and k, the Gini index is

$$-1+\frac{2kB(2k+1/c,1-1/c)+\xi/\lambda}{kB(k+1/c,1-1/c)+\xi/\lambda}.$$

The histogram in Fig. 7 illustrates an income distribution based on data in ref. 92 for persons 14 and older, with a fitted Burr III curve for which the value of the derived Gini index is 0.45. In this example, neither a log-normal nor a Burr XII curve would provide a fit that matches the twisted reverse-J shape of the histogram. [It should be noted that, at any reasonable significance level, a χ^2 goodness-of-fit test would reject the Burr III fit, as well as log-normal and Burr XII fits, since the sample size (167,262,000 persons) is very large.] Dagum [23] discusses examples of this approach to estimating Gini indices.

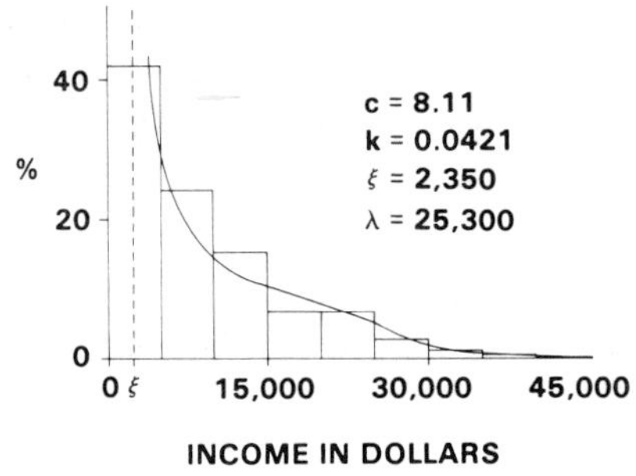

Figure 7 Observed income distribution fitted with a Burr III curve by the method of moments. (Based on 1979 U.S. Census Bureau data for persons 14 and older [92].)

References

The letter following each reference entry denotes its primary content, according to the following scheme:

- G: general reference on systems of frequency curves
- D: detailed reference on a specific family or system
- M: methods of fitting
- A: applications
- T: tabulation or computation
- H: historical reference

[1] Amos, D. E. and Daniel, S. L. (1971). Tables of Percentage Points of Standardized Pearson Distributions. *Res. Rep. No. SC-RR-71 0348*, Sandia Laboratories, Albuquerque, N.M. (T)

[2] Barndorff-Nielsen, O. (1977). *Proc. R. Soc. Lond. A*, **353**, 401–419. (D)

[3] Barton, D. E. and Dennis, K. E. R. (1952). *Biometrika*, **39**, 425–427. (D)

[4] Bhattacharyya, B. C. (1942). *Sankhyā*, **6**, 175–182. (D)

[5] Blæsild, P. (1978). The Shape of the Generalized Inverse Gaussian and Hyperbolic Distributions. *Res. Rep. No. 37*, Dept. of Theoretical Statistics, University of Aarhus, Aarhus, Denmark. (D)

[6] Boetti, G. (1964). *Gi. Isti. Ital. Att.*, **27**, 99–121. (D)

[7a] Bouver, H. and Bargmann, R. E. (1974). Tables of the Standardized Percentage Points of the Pearson System of Curves in Terms of β_1 and β_2.

Themis Tech. Rep. No. 32, University of Georgia, Athens, Ga.

[7b] Bouver, H. and Bargmann, R. E. (1977). *Amer. Statist. Ass., 1977 Proc. Statist. Computing Sec.*, pp. 127–132. (T)

[8] Bowman, K. O. and Dusenberry, W. E. (1974). In *A Modern Course on Statistical Distributions in Scientific Work*, Vol. 1: *Models and Structures*, G. P. Patil, S. Kotz, and J. K. Ord, eds. D. Reidel, Dordrecht, Holland. pp. 381–390. (M)

[9] Bowman, K. O. and Shenton, L. R. (1979). *Biometrika*, **66**, 147–151. (T)

[10] Bowman, K. O. and Shenton, L. R. (1979). *Commun. Statist. B*, **8**, 231–244. (T)

[11] Bukač, J. (1972). *Biometrika*, **59**, 688–690. (M)

[12] Burr, I. W. (1942). *Ann. Math. Statist.*, **13**, 215–232. (D)

[13] Burr, I. W. (1968). *J. Amer. Statist. Ass.*, **63**, 636–643. (D)

[14] Burr, I. W. (1973). *Commun. Statist.*, **2**, 1–21. (T)

[15] Burr, I. W. and Cislak, P. J. (1968). *J. Amer. Statist. Ass.*, **63**, 627–635. (D)

[16] Charlier, C. V. L. (1905). *Ark. Mat. Astron. Fys.*, **2**(8). (H)

[17] Charlier, C. V. L. (1905). *Ark. Mat. Astron. Fys.*, **2**(15). (H)

[18] Charlier, C. V. L. (1906). Researches into the theory of probability, *Medd. Lunds Astron. Obs.* (H)

[19] Cooper, J. D., Davis, S. A., and Dono, N. R. (1965). Pearson Universal Distribution Generator (PURGE), *Proc. 19th Annu. Conf. Amer. Soc. Quality Control*, pp. 402–411. (T)

[20] Cornish, E. A. and Fisher, R. A. (1937). *Rev. Int. Statist. Inst.*, **5**, 307–320. (D)

[21] Craig, C. C. (1936). *Ann. Math. Statist.*, **7**, 16–28. (D)

[22] D'Agostino, R. and Pearson, E. S. (1973). *Biometrika*, **60**, 613–622. (A)

[23] Dagum, C. (1979). The Generation and Distribution of Income, the Lorenz Curve, and the Gini Ratio. *Res. Paper No. 7907*, Faculty of Social Sciences, Dept. of Economics, University of Ottawa, Ottawa. (A)

[24] Davenport, J. M. and Herring, T. A. (1976). *Amer. Statist. Ass. 1976 Proc. Statist. Computing Sec.*, 139–141. (T)

[25] Dershowitz, A. F. (1966). Polynomial Fit to Percentiles of Pearson Curves. *TIS Rep. 66-Ch-SD-511*, General Electric Company. (T)

[26] Draper, N. R. and Tierney, D. E. (1972). *Biometrika*, **59**, 463–464. (D)

[27] Edgeworth, F. Y. (1907). *J. R. Statist. Soc. A*, **70**, 102–106. (H)

[28] Edgeworth, F. Y. (1916). *J. R. Statist. Soc. A*, **79**, 455–500. (H)

[29] Edgeworth, F. Y. (1917). *J. R. Statist. Soc. A*, **80**, 65–83; 266–268; 411–437. (H)

[30] Elderton, W. P. and Johnson, N. L. (1969). *Systems of Frequency Curves*. Cambridge University Press, Cambridge. (G)

[31] Fisher, R. A. (1921). *Philos. Trans. R. Soc. Lond. A*, **222**, 309–368. (H)

[32] Fisher, R. A. and Cornish, E. A. (1960). *Technometrics*, **2**, 209–226. (D)

[33] Galton, F. (1879). *Proc. R. Soc. Lond.*, **29**, 365–367. (H)

[34] Gram, J. P (1879). *On Raekkeudviklinger bestemte ved Hjaelp av de mindste kvadraters Methode*. Gad, Copenhagen. (H)

[35] Hastings, C., Mosteller, F., Tukey, J. W., and Winsor, C. P. (1947). *Ann. Math. Statist.*, **18**, 413–426. (D)

[36] Hoadley, B. (1968). *Biometrika*, **55**, 559–563. (M)

[37] Johnson, M. E., Tietjen, G. L., and Beckman, R. J. (1980). *J. Amer. Statist. Ass.*, **75**, 276–279. (D)

[38] Johnson, N. L. (1949). *Biometrika*, **36**, 149–176. (D)

[39] Johnson, N. L. (1954). *Trabajos Estadist.*, **5**, 283–191. (D)

[40] Johnson, N. L. (1965). *Biometrika*, **52**, 547–558. (T)

[41] Johnson, N. L. and Kitchen, J. O. (1971). *Biometrika*, **58**, 223–226. (T)

[42] Johnson, N. L. and Kitchen, J. O. (1971). *Biometrika*, **58**, 657–668. (T)

[43] Johnson, N. L. and Kotz, S. (1970). Continuous Univariate Distributions, Vol. 1. Houghton Mifflin, Boston. (G)

[44] Johnson, N. L. and Kotz, S. (1972). *Biometrika*, **59**, 226–229. (D)

[45] Johnson, N. L. and Kotz, S. (1973). *Biometrika*, **60**, 655–661. (D)

[46] Johnson, N. L. Nixon, E., Amos, D. E., and Pearson, E. S. (1963). *Biometrika*, **50**, 459–498. (T)

[47] Joiner, B. L. and Rosenblatt, J. R. (1971). *J. Amer. Statist. Ass.*, **66**, 394–399. (D)

[48] Kapteyn, J. C. (1903). *Skew Frequency Curves in Biology and Statistics*. Astronomical Laboratory, Noordhoff, Groningen. (H)

[49] Kendall, M. G. and Stuart, A. (1977). *The Advanced Theory of Statistics*, Vol. 1: *Distribution Theory*, 4th ed. Macmillan, New York. (G)

[50] Mage, D. T. (1980). *Technometrics*, **22**, 247–252. (D)

[51] McAlister, D. (1879). *Proc. R. Soc. Lond.*, **29**, 367–375. (H)

[52] McKay, A. T. (1932). *Biometrika*, **24**, 39–44. (D)

[53] McNolty, F. (1967). *Sankhyā B*, **29**, 235–248. (A)

[54] Merrington, M. and Pearson, E. S. (1958). *Biometrika*, **45**, 484–491. (A)

[55] Müller, P.-H. and Vahl, H. (1976). *Biometrika*, **63**, 191–194. (D)

[56] Ord, J. (1972). *Families of Frequency Distributions*. Charles Griffin, London. (G)

[57] Patil, G. P., Kotz, S., and Ord, J. K., eds. (1974). *A Modern Course on Statistical Distributions in Scientific Work*, Proc. NATO Adv. Study Inst. held at the University of Calgary, Calgary, Alberta, Canada, July 29–Aug. 10, 1974, 3 vols. D. Reidel, Dordrecht, Holland. (G)

[58] Pearson, E. S. (1930). *Biometrika*, **22**, 239–249. (A)

[59] Pearson, E. S. (1931). *Biometrika*, **22**, 423–424. (A)

[60] Pearson, E. S. (1948). *Karl Pearson's Early Statistical Papers*. Cambridge University Press, Cambridge.

[61] Pearson, E. S. (1965). *Biometrika*, **52**, 282–285. (A)

[62] Pearson, E. S. (1967). *Biometrika*, **54**, 341–355. (H)

[63] Pearson, E. S. and Hartley, H. O. (1942). *Biometrika*, **32**, 301–310. (A)

[64] Pearson, E. S. and Hartley, H. O. (1966). *Biometrika Tables for Statisticians*, Vol. 1. Cambridge University Press, Cambridge. (T)

[65] Pearson, E. S. and Hartley, H. O. (1972). *Biometrika Tables for Statisticians*, Vol. 2. Cambridge University Press, Cambridge. (T)

[66] Pearson, E. S. and Merrington, M. (1951). *Biometrika*, **38**, 4–10. (T)

[67] Pearson, E. S. and Tukey, J. W. (1965). *Biometrika*, **52**, 533–546. (G)

[68] Pearson, E. S. Johnson, N. L., and Burr, I. W. (1979). *Commun. Statist. B*, **8**, 191–229. (G)

[69] Pearson, K. (1895). *Philos. Trans. R. Soc. Lond. A*, **186**, 343–414. (H)

[70] Pearson, K. (1901). *Philos. Trans. R. Soc. Lond. A*, **197**, 443–459. (H)

[71] Pearson, K. (1916). *Philos. Trans. R. Soc. Lond. A*, **216**, 429–457. (H)

[72] Pearson, K. ed. (1930). *Tables for Statisticians and Biometricians*, 3rd ed., Part 1. Biometric Laboratory, University College, London. (T)

[73] Pearson, K. (1936). *Biometrika*, **28**, 34–59. (M)

[74] Ramberg, J. S. (1975). In *A Modern Course on Statistical Distributions in Scientific Work*, Vol. 2: *Model Building and Model Selection*, G. P. Patil, S. Kotz, and J. K. Ord, eds. D. Reidel, Dordrecht, Holland, pp. 51–64. (D)

[75] Ramberg, J. S. Tadikamalla, P. R., Dudewicz, E. J., and Mykytka, E. F. (1979). *Technometrics*, **21**, 201–214. (D)

[76] Rodriguez, R. N. (1977). *Biometrika*, **64**, 129–134. (D)

[77] Rodriguez, R. N. (1980). SHAPE: An interactive graphics SAS procedure for fitting frequency data, *Proc. 5th Ann. SAS Users Group Int. Conf.*, SAS Institute, Cary, N.C., pp. 174–179. (D)

[78] Rodriguez, R. N. (1983). *The Moment Ratio Geography of Burr III Distributions*, forthcoming research publication, General Motors Research Laboratories, Warren, Mich. (D)

[79] Salem, A. Z. B. and Mount, T. D. (1974). *Econometrica*, **42**, 1115–1127. (A)

[80] Särndal, C.-E. (1971). *Biometrika*, **58**, 375–391. (H)

[81] Shenton, L. R. (1951). *Biometrika*, **37**, 111–116. (M)

[82] Shenton, L. R. and Bowman, K. O. (1975). *J. Amer. Statist. Ass.*, **70**, 220–228. (A)

[83] Sichel, H. S. (1949). *Biometrika*, **36**, 404–425. (M)

[84] Singh, S. K. and Maddala, G. S. (1976). *Econometrica*, **44**, 963–970. (A)

[85] Slifker, J. F. and Shapiro, S. S. (1980). *Technometrics*, **22**, 239–246. (D)

[86] Solomon, H. and Stephens, M. A. (1975). Pearson Curves Revisited. *Tech. Rep. No. 226*, Dept. of Statistics, Stanford University, Stanford Calif. (A)

[87] Solomon, H. and Stephens, M. A. (1977). *J. Amer. Statist. Ass.* **72**, 881–885. (A)

[88] Solomon, H. and Stephens, M. A. (1978). *J. Amer. Statist. Ass.* **73**, 153–160. (A)

[89] Student [Gosset, W. S.] (1908). *Biometrika*, **6**, 1–25. (H)

[90] Student [Gosset, W. S.] (1927). *Biometrika*, **19**, 151–164. (H)

[91] Thiele, T. N. (1889). *Forelaesninger over almindeling iaktlagelseslaere*. Gad, Copenhagen. (H)

[92] U.S. Census Bureau (1979). *Curr. Popul. Rep., Ser. P-60, No. 116*.

[93] Wheeler, R. E. (1980). *Biometrika*, **67**, 725–728. (D)

[94] Woodward, W. A. (1976). *J. Amer. Statist. Ass.*, **72**, 881–885. (D)

BIBLIOGRAPHY

See the following works, as well as the references just cited, for more information on the topic of systems of frequency curves. Letter symbols are the same as those used with references.

Bhattacharyya, B. C. (1943). *Sankhyā*, **6**, 415–418. (D)

Bouver, H. and Bargmann, R. E. (1978). *Amer. Statist. Ass., 1978 Proc. Statist. Computing Sec.*, pp. 314–319. (T)

Bowman, K. O. and Shenton, L. R. (1973). *Biometrika*, **60**, 155–167. (D)

Bowman, K. O. and Shenton, L. R. (1980). *Commun. Statist. B*, **9**, 127–132. (T)

Chambers, E. and Fowlkes, E. B. (19666). *A Dictionary of Distributions: Comparisons with the Standard Normal.* Bell Telephone Laboratories, Murray Hill, N.J. (G)

Charlier, C. V. L. (1905). *Ark. Mat. Astron. Fys.*, **2**(20). (H)

Charlier, C. V. L. (1914). *Ark. Mat. Astron. Fys.*, **9**, 1–18. (H)

de Fériet, J. K. (1966). The Gram–Charlier Approximation of the Normal Law, etc. *Rep. No. 2013*, Applied Mathematics Laboratory David Taylor Model Basin, Dept. of the Navy, Washington, D.C. (D)

Dunning, K. A. and Hanson, J. N. (1977). *J. Statist. Comp. Simul.*, **6**, 115–128. (D)

Edgeworth, F. Y. (1896). *Phil. Mag. 5th Ser.*, **41**, 90–99. (H)

Elderton, W. P. (1953). *Frequency Curves and Correlation*, 4th ed. Harren, Washington, D.C. (G)

Hahn, G. J. and Shapiro, S. (1967). *Statistical Models in Engineering*. Wiley, New York. (G)

Hill, I. D. (1976). *Appl. Statist.*, **25**, 190–192. (D)

Hill, I. D. Hill, R., and Holder, R. L. (1976). *Appl. Statist.*, **25**, 180–189. (D)

Khamis, S. H. (1958). *Statist. Math.*, **37**, 385–396. (D)

Leslie, D. C. M. (1959). *Biometrika*, **46**, 229–231. (D)

Moore, P. G. (1957). *J. Amer. Statist. Ass.*, **52**, 237–246. (D)

Pearson, E. S. (1963). *Biometrika*, **50**, 95–112. (A)

Pearson, K. (1924). *Biometrika*, **16**, 198–200. (H)

Shenton, L. R. and Carpenter, J. A. (1965). *Biometrika*, **52**, 119–126. (D)

Tukey, J. W. (1960). The Practical Relationship between the Common Transformations of Percentages of Counts and of Amounts. *Tech. Rep. No. 36*, Statistical Techniques Research Group, Princeton University, Princeton, N.J. (D)

Tukey, J. W. (1962). *Ann. Math. Statist.*, **33**, 1–67. (D)

Van der Stok, J. P. (1908). *Proc. R. Acad. Sci. Amst.*, **10**, 799–817. (H)

(BURR DISTRIBUTIONS
CORNISH–FISHER/EDGEWORTH EXPANSIONS
FREQUENCY SURFACES, SYSTEMS OF
GRAM–CHARLIER SERIES
JOHNSON SYSTEM OF DISTRIBUTIONS
LAMBDA DISTRIBUTIONS
METHOD OF MOMENTS
PEARSON SYSTEM OF DISTRIBUTIONS)

ROBERT N. RODRIGUEZ

FREQUENCY INTERPRETATION IN PROBABILITY AND STATISTICAL INFERENCE

AXIOMATIC DEFINITION OF PROBABILITY

An introduction to probability theory could start with a formal mathematical description of basic rules for operating with probabilities. These rules could be declared as the defining properties of the concept of probability.

Let us briefly recall these basic rules. We consider a space Ω of points ω, and a class $\mathscr{A}$ of subsets A, B, etc., of Ω. The class contains Ω itself. The sets in $\mathscr{A}$ may be identified with possible statements or possible events occurring in a statistical experiment. For each set A in $\mathscr{A}$ is defined a *probability* of A [i.e., $\Pr(A)$] obeying the following basic rules. $0 \leqslant \Pr(A) \leqslant 1$, $\Pr(\Omega) = 1$. $\Pr(\bigcup_{i=1}^{\infty} A_i) = \sum_{i=1}^{\infty} \Pr(A_i)$, if the A_i are exclusive, and $\bigcup A_i, A_1, A_2, \ldots, \epsilon\, \mathscr{A}$. For obvious reasons it is convenient to let $\mathscr{A}$ be a σ-algebra, and this is therefore assumed. Now, the *conditional probability*, $\Pr(A \mid B)$, of A given B is defined as $\Pr(A \mid B) = \Pr(A \cap B)/\Pr(B)$ provided that $\Pr(B) > 0$. A and B are said to be *independent* if $\Pr(A \cap B) = \Pr(A) \times \Pr(B)$. More generally, $A_1, \ldots, A_n$ are said to be independent if for any m, $1 \leqslant i_1 < i_2 < \cdots < i_m \leqslant n$ and $j \neq i_1, \ldots, i_m$, the events A_j and $A_{i_1} \cap A_{i_2} \cap \cdots \cap A_{i_m}$ are independent.

These are the very basic elementary axioms and definitions (*see* FOUNDATIONS OF PROBABILITY). This approach to probability by means of axioms is usually attributed to Kolmogoroff [11]. They hardly need any justification; any statistician or physicist can testify to their usefulness through many years of experience. So we can safely go on and in a purely deductive manner develop the apparatus called calculus of probability. In the course of this development it is natural to dwell on a Bernoulli trial sequence where the probability of an event A in each of n independent trials is p. The probability that A shall occur exactly x times in the sequence is $\binom{n}{x} p^x (1-p)^{n-x}$ (*see* BINOMIAL

DISTRIBUTION). From this result, a nice probabilistic relationship between the probability p of A and the relative frequency x/n of A can be studied. It is seen that if n is not too small then x/n is likely to be close to p. By considering an infinite trial sequence, we can derive the famous result of Bernoulli* in 1713 [2] that x/n converges in probability to p, and the famous result of É. Borel in 1909 that $\lim_{n\to\infty} x/n = p$ with probability 1. Of course, such results as they stand, are purely mathematical results. However, it is natural to *use the Bernoulli trial sequence as a model for certain situations from the physical world*, which are among the most elementary from a statistical point of view (e.g., the observations of incidents of deaths within a year among n persons of the same age). We then reach the conclusion that the probability of A in one trial can be estimated by the relative frequency of A. This is a remarkable result, since the axioms and definition of probability do not involve the relative frequency at all. The only thing needed is to take the daring step to *identify the Bernoulli trial sequence with phenomena from real life*. In doing so the statistician ceases being a mathematician and becomes a real statistician. Sooner or later, he or she will have to take that step.

The concept of probability is very abstract, like the notion of gravity. (Isaac Newton had great agonies about accepting such an abstract force.) So it is a great relief to be able to talk about frequencies and probabilities of dying and about falling bodies. The falling body is only one consequence of gravity; the pendulums of Galileo and Huygens, and Kepler's laws for the motions of the planets, are other consequences. Similarly, probabilities have many uses, and are estimated in many ways depending on the statistical experiment, of which the Bernoulli experiment is only one, resulting in the estimate x/n.

HYPOTHETICAL REPETITIONS

Among ordinary people it is common to talk about an event with probability $1/3$ as the event which will occur once out of three times on the average in a hypothetical sequence of repeated trials; many professional people, both statisticians and physicists, have adopted the same attitude. They will confine themselves to probabilities only in connection with hypothetically repeated trials. Thus Jerzy Neyman* [15] states: "I shall be concerned solely with the frequentist theory of probability, intended to provide a mathematical model of relative frequencies. Within this theory the assertion that the probability of A is equal to, say, one-half, is interpreted to mean that, in a specified hypothetical series of instances, the relative frequency of A is equal to one-half." (It goes almost without saying that the famous theory of testing statistical hypotheses due to J. Neyman and E. S. Pearson* [16] does not depend on such an interpretation of probability.) The difficulties arising with such a concept of probability is seen from the following examples.

A convict with a death sentence hanging over his head may have a chance of being pardoned. He is to make a choice between white and black and then draw a ball randomly from an urn containing 999 white balls and 1 black ball. If the color of the ball agrees with his choice, he will be pardoned. His lawyer visiting the convict before the experiment takes place, wants to know which color he will choose. The convict replies that he will choose white, because in that case the probability of being pardoned is 0.999, whereas if he chooses black it will only be 0.001. The lawyer is sceptical and wants to know what these probabilities mean. The convict explains that out of many hypothetical drawings he will in 99.9% of the trials be pardoned and in 0.1% of the trials be executed. The lawyer finds this reasoning highly theoretical (representing the practical life, as lawyers always think they do). He argues that the first trial is binding on the part of the authorities, and that the convict is going to hang at most once. Undoubtedly, the lawyer is not going to convince the convict of the error in his judgment. But that is because the convict, after all, attaches 99.9% *probability to the single trial about to be performed.* That probability is a very real thing to the convict and it is reliably estimated from past experiences concerning urn drawings.

Passengers crossing the Atlantic on a ship from Halifax to Liverpool at the beginning of February 1942 knew that one out of four ships were sunk by German submarines. Of course, it was an important characteristic of that single voyage and the passengers would take precautions with respect to lifesaving equipment, etc., even if they expected to undertake no more crossings. Here again the high probability connected with the single voyage was very real.

The idea of hypothetical repetitions is often used when the "experiment" is a method used in a statistical investigation, and it may be very useful for an intuitive understanding of the performance functions in terms of probabilities (e.g., the level of significance*). However, it is not difficult in the spirit of the two examples given above to construct examples of situations that a statistician will never be faced with again. The statistician would still apply a method that has (say) a nice power function and level of significance.

VIEWS ON THE AXIOMATIC APPROACH

Quite apart from the idea of hypothetical repetition, it is clear that many authors of texts in probability and statistics have used relative frequency as a justification for the axioms of probability. This was done by Kolmogoroff [11] himself, the inventor of the axiomatic method. The same approach is used by Cramér [5], Bartlett [1], and Neyman [14].

In earlier statistics textbooks probability is defined either as the limit of relative frequency of the event in an infinite sequence of trials or as the relative number of outcomes favorable to the event among a number of possible and equally likely outcomes. Whatever definition is used, the author arrives at the operational rules given in the first section of this article and uses these rules as a basis for further treatment of the subject, *forgetting about the definitions and their limitations*.

Hence the axiomatic approach, sometimes thought of as unworldly, *is really the pragmatic approach to the subject. It starts by stating the bare essentials*. The reason for the resistance to the axiomatic approach as unworldly is perhaps founded in the general belief that axioms should represent "self-evident truth," which need not and cannot be proved (as is often explained in dictionaries even today), instead of rules that serve a practical purpose. Everybody knows that the search for the self-evident truth soon becomes very complicated, as the discussion through centuries of the Euclidean parallel axiom shows.

The operational and pragmatic attitude in Kolmogoroff's axiomatic approach [11] is also apparent in his definition of conditional probability* and expectations*, two notions that were hardly used before 1933 (and some years after) except in very elementary contexts. Today these concepts are of invaluable significance in probability theory and the theory of statistics.

The idea in Kolmogoroff's definition of conditional expectation (and hence probability) of X given Y, denoted by $E(X \mid Y)$ is that for any function Z of Y, $E(Z(Y)X)$ should be equal to $EZ(Y)E(X \mid Y)$, for obvious reasons. This determines $E(X \mid Y)$ almost uniquely by the Radon–Nikodym theorem.

"PERSONAL" PROBABILITY IN SINGLE TRIALS

Returning to the argument of the section "Hypothetical Repetitions," it is perhaps natural to take the view of Borel [4] that *a probability is attached to a single "isolated" trial*. Thus in the case of Borel's law of large numbers* the probability that the frequency converges concerns *one* isolated infinite sequence of Bernoulli trials. It is not necessary to embed the sequence in an infinite sequence of Bernoulli sequences in order to give Borel's law a meaning. Such an interpretation is hardly found in any modern textbook of probability theory (although one should not object to using it as an intuitive interpretation). The view today seems to be that probability is justified in its own right; "it is for experience to be interpreted in terms of probability and not for the probability to be interpreted in terms of experience" [12].

It is also reasonable to agree with Keynes [10] that in making statistical decisions we should really not talk about probabilities of events, *but only of the probabilities of a statement made by a person concerning the event.* This accords with views taken by Neyman [15] that statistical theory deals with "inductive behavior" and by Wald* [18, 19] that statistical theory deals with "decisions." It is for the statisticians to behave and decide.

However, even if all probabilities are "personal," some of them may be measured objectively, perhaps by a Bernoulli trial sequence. Thus the ratio of live born boys to live born girls has shown a remarkable stability over time, hence the probability in the statement that the child about to be born should be a male may "objectively" be taken to the 0.516 (according to some statistics).

Often, such a statistical design is not available. A person M's evaluation of the probability p of the statement that X is going to beat Y in a forthcoming game of tennis may be found by a method of betting [4]. M is offered an opportunity of getting a certain sum if an event occurs. He may himself choose if he will bet on the event that "X beats Y" or the event "5 or 6 eyes" is shown by a throw with a die. If he prefers the first event then $p > \frac{1}{3}$. By repeated comparisons with known probabilities, p could be determined accurately.

Another method to determine p is to ask M how much he is willing to pay for the promise of a gain S if X beats Y. If he is willing to pay any amount $\leqslant b$ but not an amount $> b$, he judges the odds to be $b:(S-b)$ and $p = b/S$. By this "personal" (subjective) estimation method the axioms of probability may be justified. If M is willing to pay up to b_1 (respectively, b_2) for the promise of a gain S if the event B_1 (respectively, B_2) occurs, then $\Pr(B_1) = b_1/S$, $\Pr(B_2) = b_2/S$. But obviously M would then be willing to pay $b_1 + b_2$ for the promise to gain S if either B_1 or B_2 occurs, provided that B_1 and B_2 are exclusive. Hence $\Pr(B_1 \cup B_2) = \Pr(B_1) + \Pr(B_2)$, justifying the additive rule.

Regardless of one's opinion concerning the usefulness of the probabilities just mentioned, it simplifies matters to hold the view that there is just one concept of probability (not a "personal" probability, a "degree of rational belief" probability, an "objective" probability, etc.). However, according to what kind of statements the probability is applied, there may be more or less objective and subjective methods of estimating them. There is no doubt that the probabilities applied to the examples just mentioned are acceptable from a mathematical point of view. Presumably, it remains to be seen if such applications will be valuable in the empirical sciences. De Finetti [6, p. 197] points out that "in the field of economics, the importance of probability is, in certain respects, greater than in any other field." He quotes Trygve Haavelmo [9], who stated in a presidential address to the Econometric Society that "subjective probabilities . . . are realities in the minds of people" and hoped that "ways and means can be found to obtain actual measurement of such data."

FREQUENCY DEFINITION OF PROBABILITY

The opinion that *probability has a meaning only in connection with mass phenomena and frequencies* was held by Hans Reichenbach in 1939 and by Richard von Mises* [17]. They drew the conclusion that probability should be defined by means of frequency. Hence they considered the connection between probability and frequency to be more *fundamental than what follows from estimation (decision) theory* and the identification of Bernoulli trials with real phenomena (see the first section).

In order to define the probability of A, they have to start from an infinite sequence of A's and not-A's. The proportion of A's among the n first terms in the sequence is postulated to converge to the probability of A as $n \to \infty$. To make such a definition of probability unambiguous the sequence must have the property of being an "irregular arrangement" ("regellose Anordnung"). In order not to be circular this concept has to be defined without using probability (otherwise, it could just be identified with an infinite Bernoulli sequence). This seems to have created unsurmountable difficulties. Borel [4] comments on the difficulties by stating that the human mind is not capable of imitating randomness; that is why, in order to apply probability to real phenomena, it is necessary to introduce the notion of *probability* of *isolated events* (see the section "'Personal' Probability in Single Trials"). Von Mises' theory once attracted great attention and many well-known statisticians and probabilists (Cramér, Khintchine) have commented on it; see Maistrov [13, p. 254]. They seem to agree that von Mises has not been able to state mathematically meaningful axioms for trial sequences by means of which to define probability. Lately, the problem of the irregular arrangement (collective) has again attracted attention (*see* AXIOMS OF PROBABILITY).

STATISTICAL DECISION FUNCTIONS AND PROBABILITY

An important application is to statistical analysis of observations and estimation of parameters, the general theory of which was treated in important papers by R. A. Fisher* [7, 8]. Fisher's work leads naturally to the decision theoretical view advocated by Neyman and Pearson and by Wald. J. Neyman and E. S. Pearson [16] talked about hypotheses and "rules to govern our behavior with regard to them"; Neyman [15] commented upon rejection of an hypothesis and insisted that this has "nothing in common with reasoning and that this . . . amounts to taking a 'calculated risk,' to an act of will to behave in the future . . . in a particular manner." This accords well with the decision function point of view taken by Wald in his monograph [19]. He describes there the general features of statistical inference problems.

We review a somewhat simplified version of the general theory (*see* DECISION THEORY).

We start with a vector of observations $\mathbf{X} = (X_1, \ldots, X_n)$. A priori discussions of the model for the situation lead us to think that the probability measure of $\mathbf{X}$ can be written $\Pr(\mathbf{X} \in A) = P_\theta(A)$, where $\theta = (\theta_1, \ldots, \theta_p)$ is known to belong to a set Θ. Thus the distribution of $\mathbf{X}$ is one within the class $\{P_\theta(\cdot)\}_{\theta \in \Theta}$.

The second important feature is the decision space Δ consisting of decisions $d \in \Delta$, each of which is a possible comprehensive conclusion of the investigation. We can then define a measurable function $d = \delta(\mathbf{X})$ of the observations $\mathbf{X}$ with values in Δ (having introduced a σ-field in Δ. It may also be convenient to operate with randomized decisions). This function δ is the *decision function* and it defines the *statistical method* to be used. The function tells you that if $\mathbf{X}$ is observed, then take decision $d = \delta(\mathbf{X})$. The statistician's main problem is the functional problem of finding the right δ to be used. To solve this problem, he will study the performance function $\beta(D, \theta; \delta)$, which for each measurable set $D \subset \Delta$ and each θ gives the probability of taking a decision in D if θ is the true parameter value. Hence

$$\beta(D, \theta; \delta) = P_\theta(\{\mathbf{X} \mid \delta(\mathbf{X}) \in D\}).$$

If for a given δ, we study β as a function of D and θ, we know which calculated risk we take by using δ. A certain θ may make it highly desirable (undesirable) to make a decision in a certain D. Then we want β to be large (small). This clarifies the mission of the probability concept in statistical inference. We start with a stochastic model P_θ for the observations **X**. *Then with any tentative choice of statistical method* δ *we can find the probability distribution* $\Pr(d \in D) = \beta(D, \theta; \delta)$ *of* d. Thus, in the spirit of Keynes, we have found the probability of statements about nature; but we have not found the probability of θ itself.

ALTERNATIVE USE OF PROBABILITY IN STATISTICAL INFERENCE

The probability concept has often been assigned to play a role in statistical inference which differs markedly from the one outlined above. Discussing the probabilities of peculiar (queer) events, Bertrand [3] asked how one could define peculiarities which could not be due to randomness. His view is described in the following manner by Neyman and Pearson [16]:

> Bertrand put into statistical form a variety of hypotheses, as for example the hypothesis that a given group of stars with relatively small angular distances between them as seen from the earth, form a "system" or group in space. His method of attack, which is that in common use, consisted essentially in calculating the probability, P, that a certain character, x, of the observed facts would arise if the hypothesis tested were true. If P were very small, this would generally be considered as an indication that the hypothesis, H, was probably false, and vice versa. Bertrand expressed the pessimistic view that no test of this kind could give reliable results.

Lately, similar ideas have been advanced. The question has been raised *whether we could reject a statistical hypothesis after having observed the most probable outcome under the hypothesis*. One might perhaps be captivated by this leading question and answer it in the negative. However, on second thoughts one would make a turnabout.

The question would of course be answered in the affirmative by any statistician. Rejecting an hypothesis after having observed the most probable outcome under the hypothesis is done by statisticians every day in run-of-the-mill statistical work. If the hypothesis is to the effect that in n Bernoulli trials the probability of success is 0.6, then the probability of a given sequence of events is $(0.6)^x(0.4)^{n-x}$, where x is the number of successes. This has its largest value when $x = n$. But certainly with $x = n = 1{,}000{,}000$ successes you would reject 0.6, even if we have observed the most probable outcome.

It might perhaps be objected to this example that we should only consider methods based on a minimal sufficient statistic a priori T [hence considering $\binom{n}{x}(0.6)^x(0.4)^{n-x}$ instead of $(0.6)^x(0.4)^{n-x}$]. Furthermore, in order to get rid of the nuisance parameters under the hypothesis one should consider the conditional density of T given the minimal sufficient statistic U under the hypothesis. The idea now is to reject the hypothesis when this density is small, adjusting it to a level of significance in the usual manner. The following example shows that the principle still does not work.

It is suspected that there are far more nonpaying (cheating) passengers on a tramcar line A than on a tramcar line B. Hence an inspection is made and the inspector finds the first nonpaying passenger on line A after X inspections. On line B the first nonpaying passenger is found after Y_1 inspections and the second after additional Y_2 inspections. Let the probability that a passenger is nonpaying be p_A and p_B, respectively, on the two lines. Then we have

$$\Pr((X = x) \cap (Y_1 = y_1) \cap (Y_2 = y_2)) = p_A(1 - p_A)^{x-1} p_B^2 (1 - p_B)^{y_1 + y_2 - 2}.$$

The sufficient statistic a priori is then $T = (X, Y_1 + Y_2)$. Set $Y = Y_1 + Y_2$. We find that

$$\Pr((X = x) \cap (Y = y)) = (y - 1) p_B^2 (1 - p_B)^{y-2} p_A (1 - p_A)^{x-1}.$$

Under the null hypothesis $p_A = p_B = p$ and $U = X + Y_1 + Y_2 = X + Y$ is a sufficient statistic:

$$\Pr(U = u) = \binom{u-1}{2} p^3 (1 - p)^{u-3}.$$

Hence under the hypothesis

$$\Pr(X = x \mid U = u) = 2(u - x - 1)/\{(u - 1)(u - 2)\}; \quad x = 1, 2, \ldots, u - 2,$$

which decreases from $2/(u - 1)$ to $2/[(u - 1)(u - 2)]$.

We should of course reject when X is small, but get rejection for large X by the principle. This example can be made two-sided. Then we ought of course to reject when X is close to 1 or $u - 2$, but by the principle we get rejection only when X is close to $u - 2$.

It is obvious that the test cannot be constructed from the distribution of the observations under the hypothesis alone. The following example is perhaps illuminating, explaining at the same time the "surprising" fact that 13 spades in bridge is not less likely than any other ordinary "hand."

The hypothesis is that all $N = 52!/(13!)^4$ combinations of hands are equally likely. If the dealer now gives himself or herself (say) 13 spades, one might become suspicious. This is not due to this hand being less likely than others, which it is not, but that according to the rules of the game this hand is favorable to a player. There are, in other words, other circumstances than those which follow from the density under the hypothesis that are taken into account.

To be more precise, we might arrange the hands in equivalence classes, such that all hands in the same equivalence class are equally favorable to the dealer. The classes may then be ordered according to increasing degree of being favorable. (This work of classification and ordering would certainly be formidable, perhaps prohibitive, but in principle it could be done.) Let $T(x)$ be the ordinal number of the group to which hand x belongs. Clearly, the hypothesis (of no cheating) should be rejected if $T(x)$ is large. On the other hand, it seems to be irrelevant whether group number $T(x)$ contains few or many combinations. *Thus it is the rules of the game, and a thorough knowledge of the game of bridge, which are required to determine the test. The test cannot be constructed from the hypothesis alone.*

References

[1] Bartlett, M. S. (1955). *Introduction to Stochastic Processes*. Cambridge University Press, Cambridge (reprinted 1962).

[2] Bernoulli, J. (1713). *Ars conjectandi*. (German version: *Wahrscheinlichkeitsrechnung*, Engelmann, Leipzig, 1899.)

[3] Bertrand, J. (1907). *Calcul de probabilité*, 2nd ed. Gauthier-Villars, Paris.

[4] Borel, E. (1939). *Philosophie des probabilités*. Gauthier-Villars, Paris.

[5] Cramĕr, H. (1945). *Mathematical Methods of Statistics*. Almqvist & Wicksell, Uppsala (reprinted by Princeton University Press, Princeton, N.J., 1946).

[6] De Finetti, B. (1974). *Theory of Probability: A Critical Introductory Treatment*. Wiley, New York (English translation).

[7] Fisher, R. A. (1921). *Philos. Trans. R. Soc.*, **271**, 309.

[8] Fisher, R. A. (1925). *Proc. Camb. Philos. Soc.*, **22**, 700.

[9] Haavelmo, T. (1958). *Econometrica*, **26**, 351–357.

[10] Keynes, J. M. (1921). *A Treatise on Probability*. Macmillan, London (reprinted 1943).

[11] Kolmogoroff, A. (1933). *Grundbegriffe der Wahrscheinlichkeitsrechnung*, Ergebnisse der Mathematik und ihrer Grenzgebiet, Vol. II. Springer-Verlag, Berlin. (English transl: *Foundation of the Theory of Probability*, 2nd ed. Chelsea, New York, 1956.)

[12] Koopman, B. O. (1940). *Ann. Math.*, **41**, 269–292.

[13] Maistrov, L. E. (1974). *Probability Theory: A Historical Sketch*. Academic Press, London.

[14] Neyman, J. (1950). *First Course in Probability and Statistics*. Holt, Rinehart and Winston, New York.

[15] Neyman, J. (1957). *Rev. Int. Statist. Inst.*, **25**, 7.

[16] Neyman, J. and Pearson, E. S. (1933). *Philos. Trans. R. Soc. A*, **231**, 289–337.

[17] von Mises, R. (1928). *Wahrscheinlichkeit, Statistik und Wahrheit*. Springer-Verlag, Berlin. (English transl.: *Probability, Statistics and Truth*. Macmillan, New York, 1939.)

[18] Wald, A. (1939). *Ann. Math. Statist.*, **10**, 299–326.

[19] Wald, A. (1950). *Statistical Decision Functions*. Wiley, New York.

(AXIOMS OF PROBABILITY
DECISION THEORY

FOUNDATIONS OF PROBABILITY
LOGIC IN STATISTICAL REASONING)

ERLING SVERDRUP

FREQUENCY POLYGON

This is a graphical representation of a frequency distribution, sharing many of the properties of a histogram*. It is constructed from grouped data* in intervals, preferably of equal length, the frequencies being plotted against the midpoints of the class intervals. The resulting polygon is usually "tied down" to read zero at the extremes of the class intervals; this may be done by introducing an extra class interval at each end of the range of observed values and labeling these with observed frequencies of zero. The data in Table 1, however, are nonnegative of necessity, so the frequency polygon in Fig. 1 is tied down to zero on the left at the origin (zero age), rather than at the class mark −0.5.

(GRAPHICAL REPRESENTATION OF
DATA
HISTOGRAMS)

FREQUENCY SURFACES, SYSTEMS OF

Most of the classical statistical techniques used to analyze continuous multivariate data are based on the assumption that the data possess a multinormal distribution. However, as early as 1893, Karl Pearson* demonstrated cases in which the bivariate normal distribution* was inadequate for describing the correlation* structure of bivariate data.

Table 1 Reported Suicide Attempts by Telephone Callers to Suicide Prevention of Dallas by Age, 1971

Age	6–15	16–25	26–35	36–45	46–55	56–64
Class Mark	9.5	19.5	29.5	39.5	49.5	59.5
Frequency	4	28	16	8	4	1

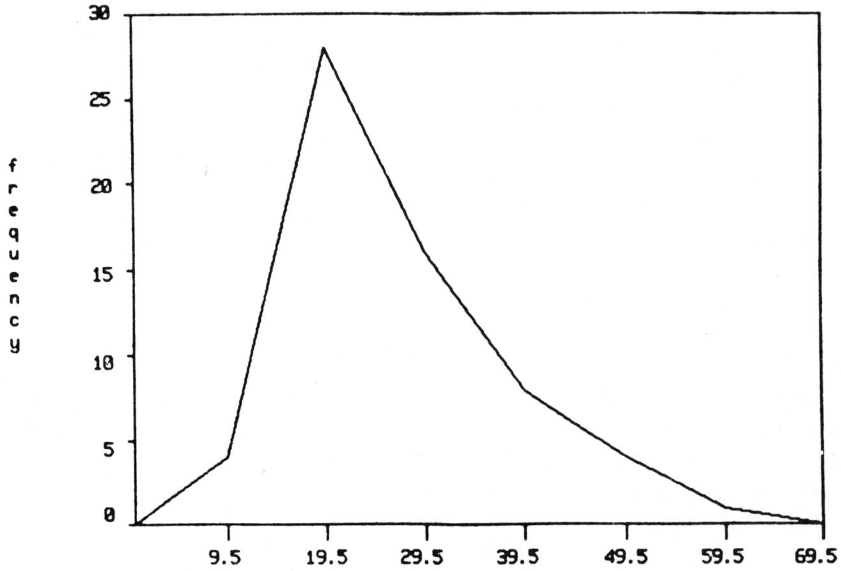

Figure 1.

During the early 1900s, Pearson and others attempted to construct systems of bivariate distributions, or "skewed frequency surfaces," to provide alternatives to the bivariate normal distribution for fitting bivariate frequency data, just as the Pearson curve system* was being used to fit nonnormal univariate frequency data.

Although the development of frequency surfaces has tended to parallel that of frequency curve* systems, the mathematical complications are formidable. Consequently, systems of frequency surfaces are seldom used for fitting observed distributions and are of more theoretical interest. This article summarizes the forms and properties of the better known surface systems, together with construction techniques which can, in some instances, be extended to produce systems of multivariate distributions.

In addition to articles on specific bivariate distributions, the reader should consult articles on related topics, such as correlation*, regression*, and systems of frequency curves*.

HISTORICAL BACKGROUND

The first attempts to construct systems of frequency surfaces were a consequence of the work of Francis Galton*, who in the late 1800s developed the concepts of regression* and correlation and related them to the bivariate normal distribution (see, e.g., Galton [9]). By 1893 Karl Pearson had observed nonnormality in bivariate data and felt the need for "skew frequency surfaces" which would extend Galton's theory of correlation.

Pearson [32] began by attempting to derive bivariate distributions of correlated variables for which a rotation transformation yields independence*, as it does for bivariate normally distributed variables. This attempt was unsuccessful, since, in general, uncorrelated variables obtained by rotation are not necessarily independent.

Pearson then extended the approach he had used to generate his system of frequency curves, namely, smoothing a hypergeometric distribution* to obtain a differential equation for the density functions of the curves. The extension, which involved a double hypergeometric series, led to two differential equations of the form

$$\frac{1}{f}\frac{\partial f}{\partial x} = \frac{C_1(x,y)}{Q(x,y)}, \qquad \frac{1}{f}\frac{\partial f}{\partial y} = \frac{C_2(x,y)}{Q(x,y)}, \tag{1}$$

where C_1 and C_2 are distinct cubic functions in x and y, Q is a quartic function in x and y, and $f = f(x, y)$ is the desired bivariate density function. Pearson did not succeed in finding general solutions, as he had done for the univariate case, and special solutions obtained by other authors did not prove to be practical for fitting data (see Pearson [33]).

Pearson [33, 34] also studied the construction of joint distributions based on broad assumptions concerning their regression and scedastic functions rather than their densities or marginal distributions*. However, the probabilities computed from this "method of nonlinear regression" were not sufficiently accurate in practice. By making specific assumptions concerning the regression and scedastic functions, and requiring that the conditional skewness and kurtosis functions be constant, Narumi [29] derived the functional forms of a number of bivariate distributions. These include the bivariate beta and Student's t families, which are of theoretical interest, but their applicability for fitting frequency data is limited, since their conditional distributions have the same shape (see Rhodes [38], Pretorius [37], and Mardia [26]).

Beginning around 1910, various authors attempted to derive frequency surfaces as bivariate series expansions whose dominant term is the bivariate normal density. This approach paralleled the successful method of constructing frequency curves based on type A series, better known as Gram–Charlier* and Edgeworth expansions (*see*

CORNISH–FISHER/EDGEWORTH EXPANSIONS). The general form of the bivariate density considered is

$$f(x, y) =$$

$$\phi(x, y) + \sum_{i+j \geqslant 3} (-1)^{i+j} \frac{a_{ij}}{i!\, j!} \frac{\partial^{i+j}}{\partial x^i \partial y^j} \phi(x, y),$$

where $\phi(x, y)$ is the standard bivariate normal density function with arbitrary correlation ρ. Since the marginal densities of this expression are Gram–Charlier (or Edgeworth) series, the corresponding surface is designated as type AA. As early as 1896 Edgeworth [4] had studied truncated versions of the series for $f(x, y)$ with $i + j = 3$; properties of this expression were later discussed by Rhodes [39]. Wicksell [51] constructed versions for which $i + j = 4$ and $i + j = 6$. The finite series expressions for which $3 \leqslant i + j \leqslant 4$ are known as type AaAa surfaces, whereas the expressions including terms arranged by order of magnitude (up to $i + j = 6$) are called type AbAb. The "15-constant surface" (also referred to as a type AaAa surface) studied by Pearson [35] involved terms up to $i + j = 4$. This work was summarized by Pretorius [37], who used type AA surfaces to fit several sets of bivariate frequency data.

Another technique that had been employed in the univariate case was the method of translation*. The idea was to examine the joint distribution of variables which, when transformed by fairly simply functions, have a bivariate normal distribution. Early transformations included cubic polynomial transformations [4, 5], log transformations* applied to both variables [53, 55], and the log transformation applied to one variable [20]. A translation system of frequency surfaces was later obtained by Johnson [14], who applied four distinct monotone transformations taken two at a time to achieve bivariate normality.

Although Pearson's differential equation approach did not produce a usable system of frequency surfaces, the method continued to be investigated. In 1947–1948, van Uven [50] derived a system of bivariate Pearson distributions whose density functions $f(x, y)$ satisfy a pair of partial differential equations somewhat simpler in structure than (1). Several of van Uven's solutions were surface families that had been obtained by earlier writers using different methods, as described by Pearson [33]. Bivariate extensions of the Pearson curve system were also considered by Risser [40–42], Sagrista [46], and Risser and Traynard [43].

Other systems of frequency curves have been extended to systems of frequency surfaces. Takahasi [48], Durling [3], and Rodriguez [44] derived families of bivariate Burr distributions*, while Johnson and Kotz [17] developed multivariate Tukey lambda distributions*.

Methods of constructing more general systems of multivariate distributions with arbitrary marginals and correlation structure have been introduced during the last 30 years. These include the Farlie–Gumbel–Morgenstern* approach [8, 10, 28], the canonical variables approach [22, 49], and the method due to Plackett [36].

For further details concerning historical development, see the survey articles by Pretorius [37], Lancaster [23], and Kotz [21], as well as books by Mardia [26], Johnson and Kotz [16], and Ord [31].

BIVARIATE EXTENSIONS OF FREQUENCY CURVE SYSTEMS

Since there is no unique bivariate extension of a given frequency curve system, it is important to specify the method of construction, as well as the original curve system, when referring to frequency surface systems obtained by extension. The discussion that follows describes some of the surface systems which can be derived from the Pearson, Gram–Charlier, Johnson, and Burr curve systems. For additional details, see the general references by Mardia [26], Ord [31], and Johnson and Kotz [16], as well as specialized references cited below.

The Pearson Surface System Constructed by van Uven

The best known bivariate extension of the Pearson curve system is the surface system derived by van Uven [50]. The density functions $f(x, y)$ for this system are obtained as solutions to the differential equations

$$\begin{aligned} \frac{1}{f}\frac{\partial f}{\partial x} &= L_1(x, y)/Q_1(x, y), \\ \frac{1}{f}\frac{\partial f}{\partial y} &= L_2(x, y)/Q_2(x, y), \end{aligned} \tag{2}$$

where L_1 and L_2 are distinct linear functions of x and y, Q_1 and Q_2 are distinct quadratic functions of x and y, and the right-hand sides of the equations are assumed to be irreducible. When x or y is fixed, $f(x, y)$ satisfies the differential equation that generates the Pearson curves, and consequently the conditional distributions corresponding to the bivariate density $f(x, y)$ belong to the univariate Pearson system (*see* FREQUENCY CURVES, SYSTEMS OF). The marginals are also Pearson distributions.

If X and Y have joint density $f(x, y)$, the regression of Y on X is linear, provided that $Q_2 f$ is equal to zero at each end of the range of variation for y. This property restricts the applicability of the van Uven system for fitting bivariate frequency data to situations in which a linear regression structure is observed.

In order to solve (2), van Uven considered all possible forms for Q_1 and Q_2. In the simplest of these Q_1 and Q_2 have no common factor, which leads to the so-called type I solution,

$$f(x, y) = g(x)h(y),$$

where $g(x)$ and $h(y)$ are univariate Pearson densities. Van Uven's derivations of $f(x, y)$ for the remaining cases are lengthy and notationally complicated; see also Mardia [26]. The Roman numeral classification which van Uven assigned to solution types refers to the corresponding structures of Q_1 and Q_2, and should not be confused with the Roman numeral designation for the Pearson marginals of $f(x, y)$.

The van Uven system includes the bivariate normal family, designated as type VI. Other families in the system had been discovered previously; the type IIa distributions were derived by Filon and Isserlis (see Pearson [33]), the type IIIa distributions were discussed by Pearson [34], and the Type IVa distributions were derived by McKay [27].

Table 1 (adapted from Elderton and Johnson [6]) summarizes the forms of the more important families and subfamilies of the van Uven system. Some of these forms are better known by different names; further details are given by Mardia [26] and Johnson and Kotz [16].

Published examples of fitted van Uven surfaces are rare. Elderton and Johnson [6] suggest fitting by the method of moments*, starting with the marginal distributions of the data and then matching product moments to obtain the remaining parameters; formulas for product moments were given by van Uven [50]. However, because the correlation structures of van Uven surfaces are generally determined once their marginals have been fixed, they lack the shape flexibility generally required in fitting bivariate frequency data.

The 15-Constant Pearson Surfaces

When fitting bivariate frequency data with a type AA surface, the fitted equation is a truncated version of the infinite series expression for the bivariate density function. We will discuss only the 15-constant form, sometimes referred to as type AA or type AaAa, which was derived by Pearson [35] and applied by Pretorius [37] (see also Mardia [26]).

The equation of the 15-constant surface is

$$\begin{aligned} f(x, y) = \phi(x, y)(1 &- a_0 + a_1 x + a_2 y + b_1 x^2 \\ &+ 2b_2 xy + b_3 y^2 + c_1 x^3 + c_2 x^2 y \\ &+ c_3 xy^2 + c_4 y^3 + d_1 x^4 + d_2 x^3 y \\ &+ 3d_3 x^2 y^2 + d_4 xy^3 + d_5 y^4), \end{aligned} \tag{3}$$

where $\phi(x, y)$ is the standard bivariate normal density with correlation ρ. The con-

Table 1 Pearson Surfaces Constructed by van Uven

van Uven's Designation	Bivariate Density $f(x, y)$	Restrictions on Parameters and Range	Pearson Types of Marginals		Aliases for Bivariate Form
IIα	$\frac{\Gamma(m_1+m_2+m_3)}{\Gamma(m_1)\Gamma(m_2)\Gamma(m_3)} x^{m_1-1} y^{m_2-1}(1-x-y)^{m_3-1}$	$m_1, m_2, m_3 > 0$ $x, y > 0; x+y \leqslant 1$	I or II	I or II	Dirichlet Bivariate beta
II$a\beta$	$\frac{\Gamma(-m_3+1)x^{m_1-1}y^{m_2-1}(1+x+y)^{m_3-1}}{\Gamma(m_1)\Gamma(m_2)\Gamma(-m_1-m_2-m_3+1)}$	$m_1, m_2 > 0; m_1+m_2+m_3 < 0$ $x, y > 0$	VI	VI	Bivariate F
II$a\gamma$	$\frac{\Gamma(-m_2+1)x^{m_1-1}y^{m_2-1}(-1-x+y)^{m_3-1}}{\Gamma(m_1)\Gamma(m_3)\Gamma(-m_1-m_2-m_3+1)}$	$m_1, m_3 > 0; m_1+m_2+m_3 < 0$ $y-1 > x > 0$	VI	VI	
IIb	$\frac{x^{m_1-1}y^{m_2-1}\exp[-(x+1)/y]}{\Gamma(m_1)\Gamma(-m_1-m_2)}$	$m_1 > 0; m_1+m_2 < 0$ $x, y > 0$	VI	V	
III$a\alpha$	$\frac{-m\sqrt{1-\rho^2}}{\pi k^m}(k+x^2+2\rho xy+y^2)^{m-1}$	$m < 0; \lvert\rho\rvert < 1; k > 0$	VII	VII	Bivariate t (Bivariate Cauchy if $\rho = 0$ and $m = -1/2$)
III$a\beta$	$\frac{m\sqrt{1-\rho^2}}{\pi k^m}(k-x^2+2\rho xy-y^2)^{m-1}$	$m > 0; \lvert\rho\rvert < 1; k > 0$ $x^2-2\rho xy+y^2 < k$	II	II	Bivariate type II
IVa	$\frac{x^{m_1-1}(x-y)^{m_2-1}e^{-y}}{\Gamma(m_1)\Gamma(m_2)}$	$m_1, m_2 > 0$ $0 < x < y$	III	III	McKay's bivariate gamma
VI	$\frac{1}{2\pi\sqrt{1-\rho^2}}\exp[-(x^2-2\rho xy+y^2)/(2(1-\rho^2))]$	$\lvert\rho\rvert < 1$	Normal	Normal	Bivariate normal

stants in (3) can be related to the moments of the bivariate density $f(x, y)$ as follows: Denoting the central product moments of order (i, j) by μ_{ij}, write

$$\sqrt{\beta_{10}} = \mu_{30}/\mu_{20}^{3/2} \quad \text{and} \quad \beta_{20} = \mu_{40}/\mu_{20}^{2}$$

for the skewness and kurtosis of the first marginal distribution, and

$$\sqrt{\beta_{01}} = \mu_{03}/\mu_{02}^{3/2} \quad \text{and} \quad \beta_{02} = \mu_{04}/\mu_{02}^{2}$$

for the skewness and kurtosis of the second marginal distribution. Then, as in Mardia [26], let

$$\begin{aligned} a &= 1/(1-\rho^2) \\ B_{20} &= (\beta_{20}-3)/24 \\ B_{02} &= (\beta_{02}-3)/24 \\ Q_{31} &= (\mu_{31}-3\rho)/6 \\ Q_{13} &= (\mu_{13}-3\rho)/6 \\ Q_{22} &= (\mu_{22}-1-2\rho^2)/12. \end{aligned}$$

It can be shown that

$$a_0 = -3a^2\{B_{20}+B_{02}-\rho(Q_{31}+Q_{13}) + (1+2\rho^2)Q_{22}\},$$

$$a_1 = \tfrac{1}{2}a^2\left\{3\rho\mu_{21}-\sqrt{\beta_{10}}-(1+2\rho^2)\mu_{12} + \rho\sqrt{\beta_{01}}\right\},$$

$$b_1 = -3a^3\{2B_{20}+2\rho^2B_{02}-2\rho Q_{31} - \rho(1+\rho^2)Q_{13} + (1+5\rho^2)Q_{22}\},$$

$$b_2 = \tfrac{3}{2}a^3\{4\rho(B_{20}+B_{02}) - (1+3\rho^2)(Q_{31}+Q_{13}) + 4\rho(2+\rho^2)Q_{22}\},$$

$$c_1 = \tfrac{1}{6}a^3\left\{\rho^2\left(3\mu_{12}-\rho\sqrt{\beta_{01}}\right) - \left(3\rho\mu_{21}-\sqrt{\beta_{10}}\right)\right\},$$

$$c_2 = \tfrac{1}{2}a^3\left\{(1+2\rho^2)\mu_{21}-\rho\sqrt{\beta_{10}} - \rho(2+\rho^2)\mu_{12}+\rho^2\sqrt{\beta_{01}}\right\},$$

$$d_1 = a^4(B_{20}+\rho^4B_{02}-\rho Q_{31}-\rho^3Q_{13} + 3\rho^2Q_{22}),$$

$$d_2 = -a^4\{4\rho B_{20}+4\rho^3B_{02}-(1+3\rho^2)Q_{31} - \rho^2(3+\rho^2)Q_{13} + 6\rho(1+\rho^2)Q_{22}\},$$

$$d_3 = a^4\{2\rho^2(B_{20}+B_{02}) - \rho(1+\rho^2)(Q_{31}+Q_{13}) + (1+4\rho^2+\rho^4)Q_{22}\}.$$

The other constants, a_2, b_3, c_3, c_4, d_4, and d_5, can be found by interchanging subscripts; location and scale parameters can be added to the surface equation by substituting $(x-\mu_1)/\sigma_1$ and $(y-\mu_2)/\sigma_2$ for x and y. Thus the equation for $f(x, y)$ is completely specified by 15 constants: the first four moments of the marginal distributions; the product moments μ_{11}, μ_{12}, μ_{21}, μ_{22}, μ_{13}, μ_{31}; and a normalizing constant. Note that fitting the surface by the method of moments is straightforward, in spite of the complicated form of the equation.

The marginal densities of the 15-constant surface are truncated Gram–Charlier series*. If X and Y have the joint density $f(x, y)$, the density function of X can be expressed as

$$\phi(x)\left\{1+\sqrt{\beta_{10}}\,(x^3-3x)/6 + (\beta_{20}-3)(x^4-6x^2+3)/24\right\},$$

where $\phi(x)$ is the standard normal density function. (The marginal density of Y is obtained by replacing β_{10} and β_{20} with β_{01} and β_{02}, respectively.) The regression function of Y on X is

$$E(Y\,|\,X) = \rho X + B(X)/A(X),$$

where

$$A(X) = 1 + \sqrt{\beta_{10}}\,(X^3 - X)/6 + (\beta_{20} - 3)(X^4 - 6X^2 + 3)/24,$$

$$B(X) = \left(\mu_{21} - \rho\sqrt{\beta_{10}}\right)(X^2 - 1)/2 + (\mu_{31} - \rho\beta_{20})(X^3 - X)/6.$$

The scedastic function of Y on X is

$$\text{var}(Y\,|\,X) = 1 - \rho^2 - C(X)/A(X) - (B(X)/A(X))^2,$$

where

$$C(X) = \left\{\rho\left(\mu_{21} - \rho\sqrt{\beta_{10}}\right) - (\mu_{12} - \rho\mu_{21})\right\}X + \left\{2\rho(\mu_{31} - \rho\beta_{20}) - \mu_{22} + 1 + \rho^2(\beta_{20} - 1)\right\}(X^2 - 1)/2$$

(see also Wicksell [54], Pearson [35], and Pretorius [37]).

The evaluation of probabilities from the 15-constant surface requires numerical integration*. Wicksell [51] found approximations for contours of equal probability centered at the mode of the type AA surface.

Pearson's 1925 paper [35] on the 15-constant surface is particularly valuable for its comments concerning the flexibility of the surface for fitting bivariate frequency data. He recognized that neither "crateroid and J-sectional" bivariate histograms, nor those with supports "theoretically limited by lines or triangles" could be fitted. These limitations were more disturbing to Pearson than the analytical complications posed by 15 constants, which he regarded as absolutely necessary.

Pretorius [37] examined the type AA surface in greater detail by fitting five observed bivariate distributions ranging in degree of skewness. In each case, he applied the method of moments to obtain the fit, and compared regression and scedastic curves estimated from the data with those derived from the fitted surface. Several difficulties included poor agreement between the estimated and derived scedastic curves, multimodality of the fitted surface, and the appearance of negative surface values in outlying regions. The latter problem, illustrated in Fig. 1c, is a consequence of truncating the infinite series expansion for $f(x, y)$. Nonetheless, Pretorius concluded that the 15-constant Pearson surface "remains the most general of the surfaces that have been propounded."

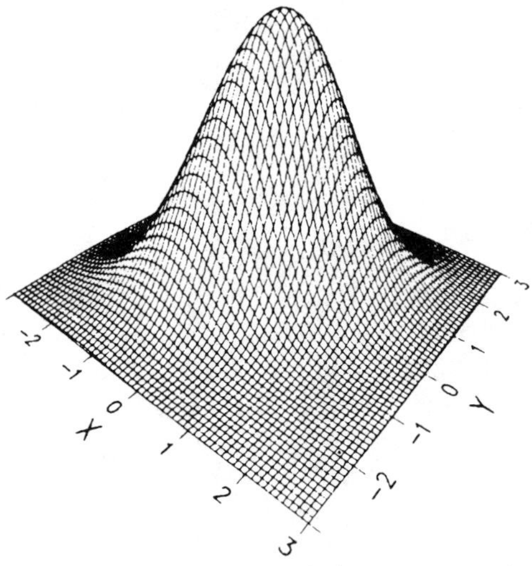

Figure 1a Fifteen-constant Pearson surface for which $\sqrt{\beta_{10}} = 0$, $\beta_{20} = 3$, $\sqrt{\beta_{01}} = 0$, $\beta_{02} = 3$, $\rho = 0.5$, $\mu_{12} = 0$, $\mu_{21} = 0$, $\mu_{13} = 1.5$, $\mu_{31} = 1.5$, and $\mu_{22} = 1.5$. This is simply the standard bivariate normal density surface with $\rho = 0.5$.

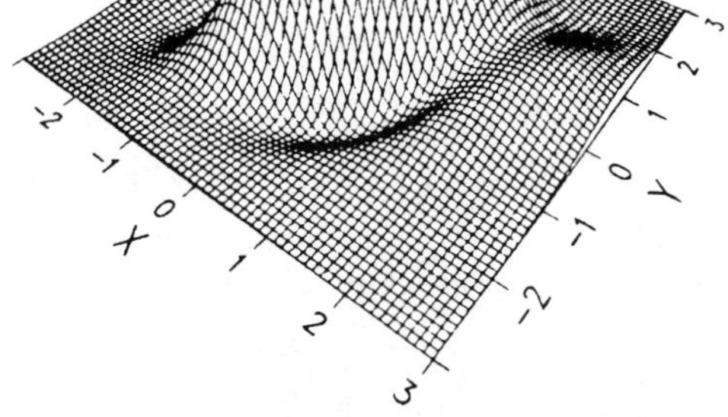

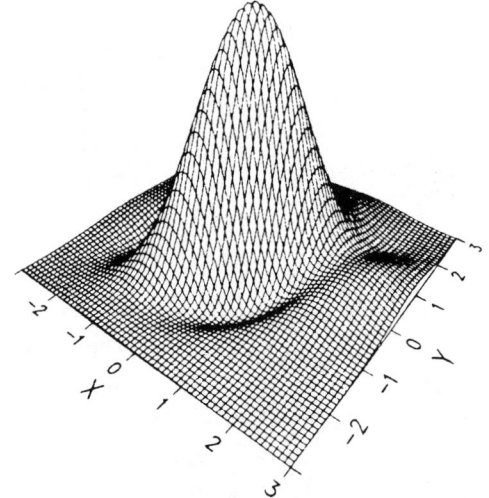

Figure 1b. Fifteen-constant Pearson surface for which $\sqrt{\beta_{10}} = 0.5$, $\beta_{20} = 4.6$, $\sqrt{\beta_{01}} = 0.8$, $\beta_{02} = 6.0$, $\rho = 0.5$, $\mu_{12} = 0$, $\mu_{21} = 0$, $\mu_{13} = 1.5$, $\mu_{31} = 1.5$, and $\mu_{22} = 1.5$.

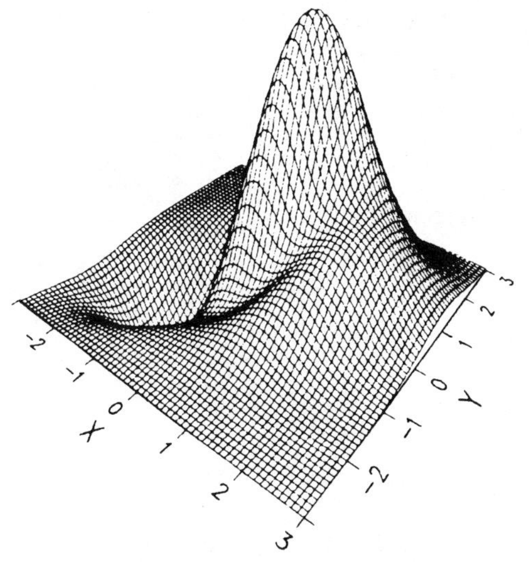

Figure 1c. Fifteen-constant Pearson surface for which $\sqrt{\beta_{10}} = -1$, $\beta_{20} = 2$, $\sqrt{\beta_{01}} = 2$, $\beta_{02} = 5$, $\rho = 0.5$, $\mu_{12} = 0$, $\mu_{21} = 0$, $\mu_{13} = 1.5$, $\mu_{31} = 1.5$, and $\mu_{22} = 1.5$. This example is not a probability density surface, since it is negative for some values of x and y. Both negativity and multimodality can occur when 15-constant surfaces are arbitrarily fitted to bivariate frequency data.

Surfaces Constructed by Translation

The translation system of surfaces was developed by Johnson [13, 14], who extended the approach that he had previously used to obtain a system of frequency curves (*see* FREQUENCY CURVES, SYSTEMS OF). The univariate construction involved four monotone transformations:

$$f_L(y) = \log y$$
$$f_N(y) = y$$
$$f_B(y) = \log(y/(1-y))$$
$$f_U(y) = \sinh^{-1} y.$$

In order to construct a system of frequency surfaces, Johnson [14] defined the S_{IJ} family as the set of bivariate distributions of Y_1 and Y_2 for which

$$\begin{aligned} Z_1 &= \gamma_1 + \delta_1 f_I((Y_1 - \xi_1)/\lambda_1), \\ &\qquad (\delta_1 > 0, \lambda_1 > 0) \\ Z_2 &= \gamma_2 + \delta_2 f_J((Y_2 - \xi_2)/\lambda_2), \\ &\qquad (\delta_2 > 0, \lambda_2 > 0) \end{aligned} \tag{4}$$

have a standard bivariate normal distribution with correlation ρ. Here I and J are chosen from the set L, N, B, and U, resulting in a system of 10 distinct surface families. The S_{LL} family consists of bivariate log-normal distributions which had been discussed by Wicksell [52].

Each S_{IJ} family involves nine parameters, of which five (γ_1, γ_2, δ_1, δ_2, and ρ) are shape parameters and four (ξ_1, ξ_2, λ_1, and λ_2) are location and scale parameters. (It is assumed in what follows that $\xi_1 = \xi_2 = 0$ and $\lambda_1 = \lambda_2 = 1$.) The marginal distributions of Y_1 and Y_2 are univariate S_I and S_J distributions, respectively (described in FREQUENCY CURVES, SYSTEMS OF).

The conditional distribution of Y_2, given Y_1, belongs to the univariate S_J family, and its shape parameters are

$$\gamma = (1-\rho^2)^{-1/2}\left[\gamma_2 - \rho\{\gamma_1 + \delta_1 f_I(Y_1)\}\right],$$

$$\delta = (1-\rho^2)^{-1/2}\delta_2.$$

Thus if $\rho \neq 0$, the shape of the conditional distribution varies with the value of Y_1; this flexibility can be useful when fitting observed distributions. If $I = B$ or $I = U$, the skewness of the conditional distribution changes sign at exactly one point in the range of Y_1. If $I = L$, this skewness is constant for all values of Y_1, but its variance depends on Y_1.

The regression function of Y_2 on Y_1 is analytically complicated, whereas the median regression function* can be expressed in a convenient form. Since the median of a variable with a univariate S_J distribution is given by $f_J^{-1}(-\gamma/\delta)$, the median of Y_2, given Y_1, is

$$f_J^{-1}\left[(\rho\gamma_1 - \gamma_2)\delta_2^{-1} + (\rho\delta_1/\delta_2)f_I(Y_1)\right].$$

Table 2 (adapted from Johnson [14]) lists the 16 possible median regression functions for the translation surface system; all are monotone and have at most one point of inflection. The median regression of Y_2 on Y_1 is linear only in the bivariate normal case, or if $I = J$ and

$$\rho\delta_1 = \delta_2, \qquad \rho\gamma_1 = \gamma_2.$$

Table 2 Median Regressions for S_{IJ} Families

Distribution of: Y_2	Y_1	Median of Y_2 When $Y_1 = y_1$
S_N	S_N	$\log\theta + \phi y_1$
S_N	S_L	$\log\theta + \phi\log y_1$
S_L	S_N	$\theta e^{\phi y_1}$
S_N	S_B	$\log\theta + \phi\log\{y_1/(1-y_1)\}$
S_B	S_N	$[1+\theta^{-1}e^{-\phi y_1}]^{-1}$
S_N	S_U	$\log\theta + \phi\log(y_1+\sqrt{y_1^2+1})$
S_U	S_N	$[\theta e^{\phi y_1} - \theta^{-1}e^{-\phi y_1}]/2$
S_L	S_L	θy_1^ϕ
S_L	S_B	$\theta[y_1/(1-y_1)]^\phi$
S_B	S_L	$[1+\theta^{-1}y_1^{-\phi}]^{-1}$
S_L	S_U	$\theta[y_1+\sqrt{y_1^2+1}]^\phi$
S_U	S_L	$[\theta y_1^\phi - \theta^{-1}y_1^{-\phi}]/2$
S_B	S_B	$\theta y_1^\phi[(1-y_1)^\phi + \theta y_1^\phi]^{-1}$
S_B	S_U	$[1+\theta^{-1}\{\sqrt{y_1^2+1}-y_1\}^\phi]^{-1}$
S_U	S_B	$[\theta y_1^{2\phi} - \theta^{-1}(1-y_1)^{2\phi}]y_1^{-\phi}(1-y_1)^{-\phi}/2$
S_U	S_U	$[\theta\{y_1+\sqrt{y_1^2+1}\}^\phi - \theta^{-1}\{\sqrt{y_1^2+1}-y_1\}^\phi]/2$

Note: $\theta = \exp[(\rho\gamma_1 - \gamma_2)/\delta_2]$; $\phi = \rho\delta_1/\delta_2$.

The $(100p)$th percentile of Y_2, given Y_1, is

$$f_J^{-1}\Big[Z_p(1-\rho^2)^{1/2} - \gamma_2 + \rho\{\gamma_1 + \delta_1 f_I(Y_1)\}\Big]/\delta_2,$$

where Z_p is the $(100p)$th percentile of the standard normal distribution.

Johnson [14] discussed the fitting of translation surfaces to bivariate frequency data. This can be done by first fitting the marginal distributions with appropriate S_I and S_J curves, transforming the data with the corresponding f_I and f_J functions, and setting ρ equal to the correlation of the transformed values. Crofts [2] obtained formulas for maximum likelihood estimators of the parameters of the S_{NL} family.

Additional details regarding translation surfaces are given by Johnson [14], Ord [31], Elderton and Johnson [6], Mardia [26], and Johnson and Kotz [16].

Burr Surfaces Derived from Mixture Representations

Starting with a differential equation, Burr [1] derived a system of univariate distribution functions $F(x)$, among which the type III and type XII families are particularly useful (*see* FREQUENCY CURVES, SYSTEMS OF). In standard form, the distribution function of the Burr III family* is

$$F_{\text{III}}(x) = (1 + x^{-c})^{-k}, \qquad (x > 0;\ c, k > 0)$$

and the distribution function of the Burr XII family* is

$$F_{\text{XII}}(x) = 1 - (1 + x^{c})^{-k}, \quad (x > 0;\ c, k > 0).$$

A multivariate extension of the Burr XII family was constructed by Takahasi [48], who exploited a mixture representation for the univariate Burr XII distributions instead of generalizing the Burr differential equation. The Burr XII distributions can be obtained as a scale mixture of Weibull distributions*

$$1 - \exp(-\theta y^{c}) \qquad (y > 0;\ c > 0)$$

compounded with respect to θ, where θ has a gamma distribution* with parameter k. In the notation of Johnson and Kotz [15], this representation can be expressed as

$$\text{Weibull}\,(\theta, c) \bigwedge_{\theta} \text{gamma}(k) = \text{Burr XII}(c, k).$$

For θ fixed, if X and Y are conditionally independent Weibull variables with

$$P[X \leqslant x \mid \theta] = 1 - \exp(-\theta x^{c_1}),$$
$$P[Y \leqslant y \mid \theta] = 1 - \exp(-\theta y^{c_2}),$$

then the unconditional joint density of X and Y is

$$k(k+1)c_1c_2x^{c_1-1}y^{c_2-1}(1 + x^{c_1} + y^{c_2})^{-k+2},$$
$$(x, y > 0;\ c_1, c_2, k > 0),$$

the bivariate version of Takahasi's extension. Both the marginal and the conditional distributions of X and Y belong to the univariate Burr XII family, and the joint cumulative distribution of X and Y has a closed form,

$$P[X \leqslant x, Y \leqslant y] = 1 - (1 + x^{c_1})^{-k} - (1 + y^{c_2})^{-k} + (1 + x^{c_1} + y^{c_2})^{-k}.$$

Although functionally attractive, Takahasi's bivariate Burr XII form is not generally suitable for fitting bivariate frequency data, because the correlation of X and Y is completely determined by the marginal distributions of X and Y. Durling [3] ameliorated this deficiency with a slight generalization of the joint distribution of X and Y:

$$P[X \leqslant x, Y \leqslant y] = 1 - (1 + x^{c_1})^{-k} - (1 + y^{c_2})^{-k} + (1 + x^{c_1} + y^{c_2} + \gamma x^{c_1}y^{c_2})^{-k},$$

where $0 \leqslant \gamma \leqslant k + 1$. The marginal distributions are unchanged, but for fixed c_1, c_2, and k, the additional parameter γ allows for some variation in the correlation of X and Y, and in the shapes of their conditional distributions.

The univariate Burr XII distributions were the only family studied in detail by Burr, who was apparently unaware that they are dominated in shape flexibility by the Burr III distributions (*see* BURR DISTRIBUTIONS). By applying the methods of Takahasi [48] and Durling [3], Rodriguez [44] derived a family of Burr III surfaces which inherits the property of greater shape flexibility. The starting point is a mixture representation involving the extreme value distributions* of type II:

$$\text{extreme value II}(\theta, c) \bigwedge_{\theta} \text{gamma}(k) = \text{Burr III}(c, k).$$

The bivariate Burr III distribution function obtained is

$$P[X \leqslant x, Y \leqslant y] = (1 + x^{-c_1} + y^{-c_2} + \gamma x^{-c_1}y^{-c_2})^{-k} \qquad (x, y > 0),$$

where c_1, c_2, and k are positive marginal

shape parameters and $0 \leqslant \gamma \leqslant k+1$; X and Y are independent if $\gamma = 1$.

If X and Y have a bivariate Burr III distribution, the regression function of Y on X is nonlinear:

$$E[Y|X] = (k+1)B(k+1+c_1^{-1}, 1-c_1^{-1}) \times [(1+\gamma X^{-c_2})/(1+X^{-c_2})]^{1/c_1} + \gamma(k+1)B(k+c_1^{-1}, 2-c_1^{-1}) \times [(1+\gamma X^{-c_2})/(1+X^{-c_2})]^{(1-c_1)/c_1} - \gamma B(k+1+c_1^{-1}, 1-c_1^{-1}) \times [(1+\gamma X^{-c_2})/(1+X^{-c_2})]^{(1-c_1)/c_1}.$$

[Here $B(-,-)$ is the beta function.] The correlation of X and Y exists if $0 \leqslant \gamma \leqslant 1$, $c_1 > 1$, and $c_2 > 1$. For fixed c_1, c_2, and k, the correlation is monotone decreasing in γ:

$$\mathrm{corr}(X,Y) = B(k+c_1^{-1}, 1-c_1^{-1})B(1-c_2^{-1}, k+c_2^{-1}) \times [F(-c_1^{-1}, -c_2^{-1}; k; 1-\gamma) - 1] \times \{[B(1-2c_1^{-1}, k+2c_1^{-1})/k - B^2(1-c_1^{-1}, k+c_1^{-1})] \times [B(1-2c_2^{-1}, k+2c_2^{-1})/k - B^2(1-c_2^{-1}, k+c_2^{-1})]\}^{-1/2}.$$

[Here $F(-,-;-;-)$ is the Gauss hypergeometric function* defined by the series

$$F(a,b;c;z) = \sum_{n=0}^{\infty} \frac{(a)_n (b)_n}{(c)_n} \frac{z^n}{n!}, \qquad |z| \leqslant 1,$$

where $(a)_n = a(a-1)\cdots(a-n+1)$.] For combinations of c_1, c_2, and k encountered in typical applications, the maximum value of $\mathrm{corr}(X,Y)$ can be as high as 0.8.

Rodriguez [44] utilized computer graphics to illustrate the variety of Burr III surfaces; some of these are bell-shaped, others cross-sectionally J-shaped (see Fig. 2). The surface variety is sufficiently broad to provide adequate fits for many of the bivariate histogram shapes commonly encountered.

In addition, a major advantage is the functional simplicity of the fitted expression. Both the marginal and joint cumulative distributions have closed forms; moreover, X^{c_2/c_1} and Y have the same marginal Burr III distributions. For the bivariate Burr III family, the conditional distribution of Y, given X, is also simple:

$$P[Y \leqslant y | X = x] = (1+\beta y^{-c_2})^{-k-1}(1+\gamma y^{-c_2}),$$

where $\beta = (1+\gamma x^{-c_1})/(1+x^{-c_1})$. On the other hand, the disadvantages of these surfaces are that their moments and correlations are complicated functions of their parameters, and that their marginal distri-

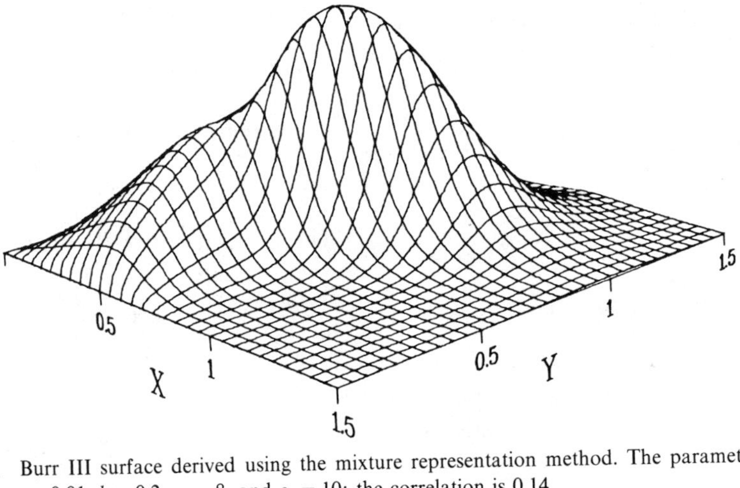

Figure 2a. Burr III surface derived using the mixture representation method. The parameters of this surface are $\gamma = 0.01$, $k = 0.2$, $c_1 = 8$, and $c_2 = 10$; the correlation is 0.14.

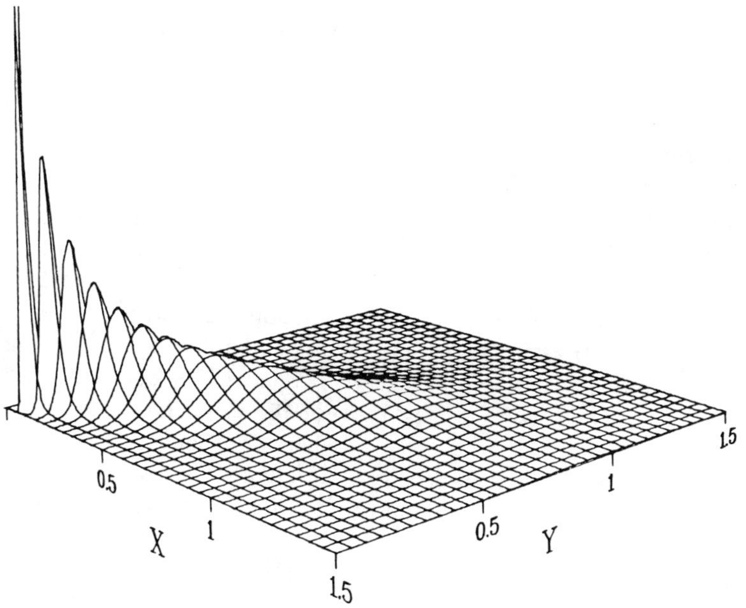

Figure 2b. Burr III surface derived using the mixture representation method. The parameters of this surface are $\gamma = 0$, $k = 0.2$, $c_1 = 5$, and $c_2 = 5$; the correlation is 0.74.

butions must share the same shape parameter k.

Rodriguez and Taniguchi [45] used Burr III surfaces to fit bivariate distribution data by the methods of moments and maximum likelihood (see Fig. 3). In this application, the data consist of gasoline octane requirements for 229 vehicles as determined by their owners (customers) and by expert raters. The fitted surfaces yield expressions for the marginal distributions of customer and rater requirements, as well as the conditional distributions of customer requirements. Although a chi-square* statistic is not a precise indicator of goodness-of-fit in this situation, the value of χ^2 (= 26.6) for the surface in

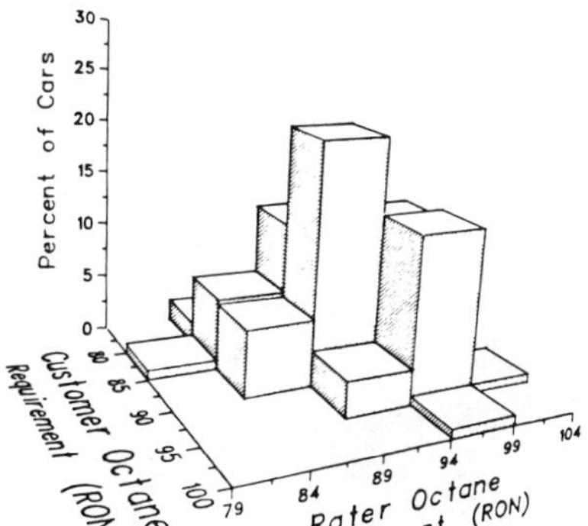

Figure 3a. Bivariate histogram for the octane requirements of 229 vehicles, as determined by their owners (customers) and by expert raters. (Reproduced from Rodriguez and Taniguchi [45] by permission of the Society of Automotive Engineers.)

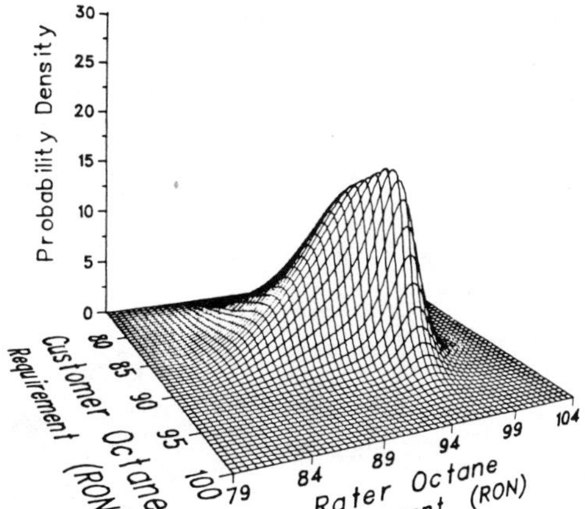

Figure 3b. Burr III surface fit (derived using the mixture representation method) for the distribution of customer and rater octane requirements. (Reproduced from Rodriguez and Taniguchi [45] by permission of the Society of Automotive Engineers.)

Fig. 3b is less than the value of χ^2 (= 30.7) for a bivariate normal fit, indicating an improvement with the bivariate Burr III distribution.

Further properties of Burr surfaces derived from mixture representations are given by Takahasi [48], Durling [3], Johnson and Kotz [16], and Rodriguez [44].

SURFACE SYSTEMS WITH ARBITRARY MARGINAL AND CORRELATION STRUCTURES

A number of methods have been proposed for constructing systems of bivariate and multivariate distributions with general marginal distributions and correlations. These techniques may yield frequency surface systems free of some of the shape limitations in the bivariate extensions of curve systems considered previously. However, although the methods have received considerable attention, relatively few of the surface systems that can be developed have been examined in detail.

Among these methods of construction, the Farlie–Gumbel–Morgenstern (FGM) approach is the most straightforward (*see* FARLIE–GUMBEL–MORGENSTERN DISTRIBUTIONS). The method was introduced by Eyraud [7], and, independently, by Morgenstern [28] and Gumbel [10]. Given two variables, X and Y, with distributions $F(x)$ and $G(y)$, respectively, the family of FGM bivariate distributions (see Morgenstern [28]) is

$$H(x, y) = F(x)G(y)\{1 + \alpha[1 - F(x)][1 - G(y)]\},$$

where $|\alpha| \leqslant 1$. Farlie [8] considered an extended version in which the factors $1 - F(x)$ and $1 - G(y)$ are replaced by more general functions (see also Mardia [25, 26], Nataf [30], and Johnson and Kotz [18, 19]).

Gumbel [10] pointed out that $H(x, y)$ is not the usual bivariate normal distribution when $F(x)$ and $G(y)$ are replaced by univariate normal distributions. Gumbel [11, 12] discussed the bivariate exponential and bivariate logistic families obtained by the FGM construction. Rodriguez [44] explored the FGM bivariate Burr III distributions that result when $F(x)$ and $G(y)$ are univariate Burr III distributions with distinct sets of shape parameters; this procedure avoids the parametric restriction suffered by the Burr III surfaces described in the previous section (see Fig. 4).

A limitation of surfaces constructed by the FGM approach is that their correlations ρ (which are proportional to α) have bounds that are fairly close to zero. Schucany et al. [47] discuss the ratio ρ/α for various FGM families and show that $|\rho| \leqslant \frac{1}{3}$ for FGM families in general. Consequently, systems of FGM surfaces can only be used to fit bivariate frequency data which exhibit mild dependence.

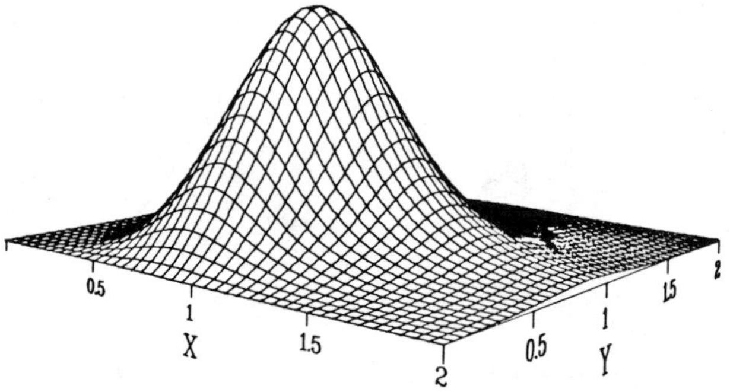

Figure 4a. Burr III surface derived using the Farlie–Gumbel–Morgenstern method. The parameters of this surface are $\alpha = 0.5$, $c_1 = 6$, $k_1 = 0.4$, $c_2 = 5.5$, and $k_2 = 0.8$; the correlation is 0.14.

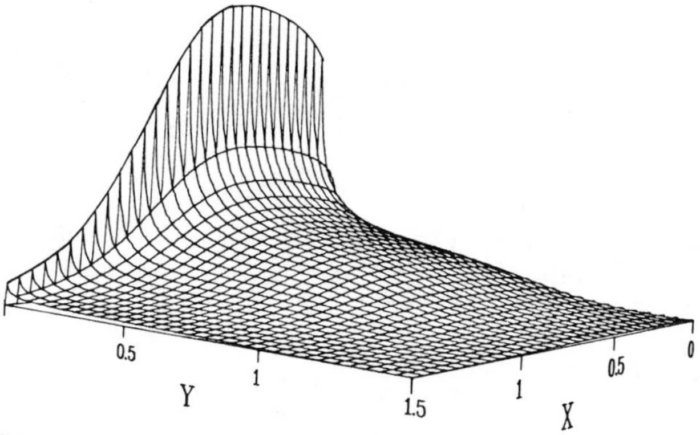

Figure 4b. Burr III surface derived using the Farlie–Gumbel–Morgenstern method. The parameters of this surface are $\alpha = 0$, $c_1 = 6$, $k_1 = 0.1$, $c_2 = 5.5$, and $k_2 = 0.2$; the correlation is zero.

Other methods of construction with arbitrary marginal and correlation structures have been developed by van Uven [49], Lancaster [22], Plackett [36], and Mardia [24]. These approaches, primarily of theoretical interest, are discussed by Mardia [26], Johnson and Kotz [16], and Ord [31].

SPECIAL PROBLEMS IN FITTING SURFACES

When fitting bivariate frequency data with systems of surfaces, a number of problems can occur which do not arise in the univariate case. These difficulties, described by Ord [31], are worth mentioning, since the method of fitting used in an application is as important to proper interpretation of the fit as the choice of surface system.

Because of the large number of parameters involved, fitting by the methods of moments and maximum likelihood is likely to be difficult computationally, particularly if threshold parameters are to be estimated.

Procedures in which the marginals are first fitted with curves may not yield adequate fits for the correlation structure of the data, especially if the number of parameters left over after fitting the marginals is relatively small. Fitting the regression and scedastic functions of the surface has been suggested as an alternative; least-squares estimation can be tried when the regression function is linear, but the estimators may be biased when both variables are random. Moreover, the regression functions for the more flexible surface systems are nonlinear, and their scedastic functions may not be constant.

The maximum likelihood approach is sometimes used with systems involving transformations to bivariate normality. Maximum likelihood estimators obtained from transformed data may present difficulties if the transformations have been determined by the data, and they should be interpreted with care when transforming back to the data.

References

The letter following each reference entry denotes its primary content, according to the following scheme:

- G: general reference on systems of frequency surfaces
- D: detailed reference on a specific family or system
- M: methods of constructing bivariate distributions with arbitrary marginal and correlation structures
- A: applications
- H: historical reference

[1] Burr, I. W. (1942). *Ann. Math. Statist.*, **13**, 215–232. (D)

[2] Crofts, A. E. (1969). An Investigation of Normal Lognormal Distributions. *Tech. Rep. No. 32*, Themis Contract, Dept. of Statistics, Southern Methodist University, Dallas, Tex. (D)

[3] Durling, F. C. (1974). In *A Modern Course on Statistical Distributions in Scientific Work*, Vol. 1: *Models and Structures*, G. P. Patil, S. Kotz, and J. K. Ord, eds. D. Reidel, Dordrecht, Holland, pp. 329–335. (D)

[4] Edgeworth, F. Y. (1896). *Philos. Mag., 5th Ser.*, **41**, 207–215. (H)

[5] Edgeworth, F. Y. (1917). *J. R. Statist. Soc. A*, **80**, 266–288. (H)

[6] Elderton, W. P. and Johnson, N. L. (1969). *Systems of Frequency Curves*. Cambridge University Press, Cambridge, England. (G)

[7] Eyraud, H. (1936). *Ann. Univ. Lyon Sec. A*, **1**, 30–47. (M)

[8] Farlie, D. J. G. (1960). *Biometrika*, **47**, 307–323. (M)

[9] Galton, F. (1877). *R. Inst. Gr. Brit. Proc.*, **8**, 282–301. (H)

[10] Gumbel, E. J. (1958). *C. R. Acad. Sci. Paris*, **246**, 2717–2719. (M)

[11] Gumbel, E. J. (1960). *J. Amer. Statist. Ass.*, **55**, 89–117. (D)

[12] Gumbel, E. J. (1961). *J. Amer. Statist. Ass.*, **56**, 335–349. (D)

[13] Johnson, N. L. (1949). *Biometrika*, **36**, 149–176. (D)

[14] Johnson, N. L. (1949). *Biometrika*, **36**, 297–304. (D)

[15] Johnson, N. L. and Kotz, S. (1970). *Continuous Univariate Distributions*, Vol. 1. Houghton Mifflin, Boston. (G)

[16] Johnson, N. L. and Kotz, S. (1972). *Continuous Multivariate Distributions*. Wiley, New York. (G)

[17] Johnson, N. L. and Kotz, S. (1973). *Biometrika*, **60**, 655–661. (D)

[18] Johnson, N. L. and Kotz, S. (1975). *Commun. Statist.*, **4**, 415–427. (M)

[19] Johnson, N. L. and Kotz, S. (1977). *Commun. Statist. A*, **6**, 485–496. (M)

[20] Jørgensen, N. R. (1916). *Undersøgelser over Frequensflader of Correlation*. Busch, Copenhagen. (H)

[21] Kotz, S. (1974). In *A Modern Course on Statistical Distributions in Scientific Work*, Vol. 1: *Models and Structures*, G. P. Patil, S. Kotz, and J. K. Ord, eds. D. Reidel, Dordrecht, Holland, pp. 247–270. (G)

[22] Lancaster, H. O. (1958). *Ann. Math. Statist.*, **29**, 719–736. (M)

[23] Lancaster, H. O. (1972). *Math. Chronicle (N.Z.)*, **2**, 1–16. Reproduced in *Studies in the History of Statistics and Probability*, Vol. 2, M. Kendall and R. L. Plackett, eds. Macmillan, New York, pp. 293–309. (H)

[24] Mardia, K. V. (1967). *Biometrika*, **54**, 235–249. (M)

[25] Mardia, K. V. (1970). *Sankhyā A*, **32**, 119–121. (M)

[26] Mardia, K. V. (1970). *Families of Bivariate Distributions*. Charles Griffin, London (G)

[27] McKay, A. T. (1934). *J. R. Statist. Soc. B*, **1**, 207–216. (D)

[28] Morgenstern, D. (1956). *Mitteilingsbl. Math. Statist.*, **8**, 234–235. (M)

[29] Narumi, S. (1923). *Biometrika*, **15**, 77–88, 209–221. (H)

[30] Nataf, A. (1962). *C. R. Acad. Sci. Paris*, **225**, 42–43. (M)

[31] Ord, J. K. (1972). *Families of Frequency Distributions*. Charles Griffin, London. (G)

[32] Pearson, K. (1905). *Drapers' Co. Res. Mem. Biom. Ser.*, **2**. (H)

[33] Pearson, K. (1923). *Biometrika*, **15**, 222–230. (H)

[34] Pearson, K. (1923). *Biometrika*, **15**, 231–244. (H)

[35] Pearson, K. (1925). *Biometrika*, **17**, 268–313. (D)

[36] Plackett, R. L. (1965). *J. Amer. Statist. Ass.*, **60**, 516–522. (M)

[37] Pretorius, S. J. (1930). *Biometrika*, **22**, 109–223. (H)

[38] Rhodes, E. C. (1923). *Biometrika*, **14**, 355–377. (H)

[39] Rhodes, E. C. (1925). *Biometrika*, **17**, 314–326. (H)

[40] Risser, R. (1945). *C. R. Acad. Sci. Paris*, **220**, 31–32. (D)

[41] Risser, R. (1947). *C. R. Acad. Sci. Paris*, **225**, 1266–1268. (D)

[42] Risser, R. (1950). *Bull. Actuaires Français*, **191**, 141–232. (D)

[43] Risser, R. and Traynard, C. E. (1957). *Les Principes de la statistique mathématique*, Vol. 2, Part 2. Gauthier-Villars, Paris. (D)

[44] Rodriguez, R. N. (1980). Multivariate Burr III Distributions, Part I: Theoretical Properties. *Res. Publ. GMR-3232*, General Motors Research Laboratories, Warren, Mich. (D)

[45] Rodriguez, R. N. and Taniguchi, B. Y. (1980). *Trans. Soc. Automo. Eng., SAE Tech. Paper* No. 801356, 4213–4240. (A)

[46] Sagrista, S. N. (1952). *Trabajos Estadíst.*, **3**, 273–314 (in Spanish). (D)

[47] Schucany, W. R., Parr, W. C., and Boyer, J. E. (1978). *Biometrika*, **65**, 650–653. (M)

[48] Takahasi, K. (1965). *Ann. Inst. Statist. Math.*, **41**, 1999–2020. (D)

[49] van Uven, M. J. (1925–1926). *Proc. R. Acad. Sci. Amst.*, **28**, 797–811, 919–935; **29**, 580–590. (H)

[50] van Uven, M. J. (1947–1948). *Proc. R. Acad. Sci. Amst.*, **50**, 1063–1070, 1252–1263; **51**, 41–52, 191–196. (D)

[51] Wicksell, S. D. (1917). *Sven. Aktuarieførning. Tidskr.*, **4**, 122–140. (H)

[52] Wicksell, S. D. (1917). *Sven. Aktuarieførning. Tidskr.*, **4**, 141–161. (H)

[53] Wicksell, S. D. (1917). *Ark. Mat. Astron. Fys.*, **12**, (20). (H)

[54] Wicksell, S. D. (1917). *Medd. Lunds Astron. Obs. Ser. 2*(**17**). (H)

[55] Wicksell, S. D. (1923). *Ark. Mat. Astron. Fys.*, **17**(19). (H)

Bibliography

See the following works, as well as the references just cited, for more information on the topic of systems of frequency surfaces. Letter symbols are the same as those used with references.

Fréchet, M. (1951). *Ann. Univ. Lyon, Sec. A, Ser. 3*, **14**, 53–77. (M)

Isserlis, L. (1914). *Philos. Mag.*, **28**, 379–403. (H)

Lancaster, H. O. (1975). *Biometrika*, **44**, 289–292. (M)

Neyman, J. (1926). *Biometrika*, **18**, 257–262. (H)

Pearson, E. S. (1962). Frequency Surfaces. *Tech. Rep. No. 49*, Statistical Techniques Research Group, Princeton University, Princeton, N.J. (G)

Steyn, H. S. (1960). *Proc. R. Acad. Sci., Amst.*, **63**, 302–311. (D)

van Uven, M. J. (1929). *Proc. R. Acad. Sci., Amst.*, **32**, 793–807, 995–1007, 1085–1103. (H)

Wicksell, S. D. (1917c). *Philos. Mag., 6th Ser.*, **24**, 389. (H)

Wicksell, S. D. (1935). Expansion of frequency functions for integer variates in series, *8th Skand. Mat. Kongr., Stockholm, 1934, Comptes Rendus*. H. Ohlssons Boktryckeri, Lund. 306–325. (H)

Yule, G. U. (1897). *Proc. R. Soc. Lond.*, **60**, 447–489. (H)

(BURR DISTRIBUTIONS
CORRELATION
FARLIE–GUMBEL–MORGENSTERN DISTRIBUTIONS
FREQUENCY CURVES, SYSTEMS OF
GRAM–CHARLIER SERIES
JOHNSON SYSTEM OF DISTRIBUTIONS
PEARSON SYSTEM OF DISTRIBUTIONS
REGRESSION
TRANSFORMATIONS)

ROBERT N. RODRIGUEZ

FREQUENCY TABLE

Generally, this term can apply to any table setting out the frequencies of occurrence of various events. It is very often understood to refer to a table showing grouped univariate data (*see* GROUPED DATA). The first column shows the group (or class) limits; the second, the number of individuals ("frequency") with values of the variables within the stated limits. Sometimes the first column gives only central values for the relevant group (especially if the group width is constant). The data in such a frequency table can be exhibited graphically as a histogram*.

(CATEGORICAL DATA
GROUPED DATA)

FRIEDMAN'S CHI-SQUARE TEST

Friedman's test is used in lieu of the analysis of variance* in complete two-way comparative experiments to avoid the assumption of normality and to facilitate the analysis of ordinal data*. Let $\{X_{ij};\ 1 \leqslant j \leqslant k,\ 1 \leqslant i \leqslant n\}$ consist of random observations associated with the jth of k treatments on the ith of n blocks of size k, such that X_{ij} takes values in an ordered set $\mathscr{X}$ and observations in different blocks are independent. The object is to test against general alternatives the hypothesis H that treatments are equally effective. Replace $\{X_{i1}, X_{i2}, \ldots, X_{ik}\}$ by their respective ranks $\{R_{i1}, R_{i2}, \ldots, R_{ik}\}$ in each block, 1 replacing the smallest, 2 the next smallest, . . . , and k the largest and let $R_j = R_{1j} + R_{2j} + \cdots + R_{nj}$ for $j = 1,$

$2, \ldots, k$. Friedman's test rejects H at the significance level α when the statistic (also called Friedman's S-statistic)

$$X_r^2 = \frac{12}{nk(k+1)} \sum_{j=1}^{k} \left[R_j - n(k+1)/2 \right]^2$$

exceeds the $100(1-\alpha)$ percentile, $C_{1-\alpha}(k, n)$, of its distribution under H. The null distribution of X_r^2 may be assembled from tables in Owen [18], Hollander and Wolfe [7], and Odeh [17] for the cases $\{k = 3, 2 \leqslant n \leqslant 15;\ k = 4,\ 2 \leqslant n \leqslant 8;\ k = 5,\ 3 \leqslant n \leqslant 8;\ k = 6,\ 2 \leqslant n \leqslant 6\}$. Extensions to the latter are given by Likeš and Laga [12], Michaelis [13], and by Sacks and Selvin [22]. An approximate test uses the limiting chi-square (χ^2) distribution of X_r^2 (as $n \to \infty$) with $k-1$ degrees of freedom. Equivalent tests were developed independently by Wallis [30] and by Kendall and Babington Smith [10], the latter using an approximating Snedecor–Fisher* distribution which was compared by Friedman [4] with his χ^2 approximation.

As an example, suppose that $n = 5$ surveillance devices are observed under $k = 3$ atmospheric conditions, giving the information on signal strengths by condition within devices: [25, 10, 15], [120, 100, 60], [4, 1, 3], [80, 65, 75], and [strong signal, no signal, weak signal], larger values reflecting stronger signals. Corresponding ranks are [3, 1, 2], [3, 2, 1], [3, 1, 2], [3, 1, 2], and [3, 1, 2]; the needed sums are $R_1 = 15$, $R_2 = 6$, and $R_3 = 9$; and the value of X_r^2 is 8.4. For a test at level $\alpha = 0.024$ the exact critical value is $C_{0.976}(3, 5) = 7.6$; at $\alpha = 0.025$ the χ^2 approximation yields the percentile $\chi^2_{2,0.975} = 7.38$. Significance is declared by the exact and approximate tests, supporting the conclusion that atmospheric conditions do affect signal strengths of the devices.

Friedman's test applies in the analysis of repeated measurements and in certain mixed and random models of the analysis of variance. Here the joint cumulative distribution function (CDF) of $\{X_{i1}, \ldots, X_{ik}\}$ is of the translation-parameter type $F_i(x_1 - \theta_1, \ldots, x_k - \theta_k)$ such that $F_i(\cdot, \ldots, \cdot)$ is symmetric under permutations. The further case that $\{X_{ij}\}$ are mutually independent, having the CDFs $\{F_{ij}(x) = F_i(x - \theta_j)\}$, corresponds to the two-way analysis of variance with k treatments assigned randomly within blocks to kn unrelated subjects. Friedman's test is valid in each case for testing $H\colon \theta_1 = \theta_2 = \cdots = \theta_k$ against general alternatives. Apart from comparative experiments, Friedman's test applies in any two-way data for which (1) the observations in different rows are independent, and (2) the joint distributions within rows are symmetric under permutations when H is true.

Further guidelines in the use of Friedman's test are supplied by its performance relative to the analysis of variance. Compared to the analysis of variance, the asymptotic relative efficiency (ARE) of Friedman's test is the limit of the ratio of sample size of the former to the latter, both tests operating at the same level and achieving the same power at alternatives near H. The following summary applies to CDFs having continuous densities. In heteroscedastic experiments for which block effects induce the scale changes $F_i(x - \theta_j) = F_0((x - \theta_j)/\sigma_i)$, F_0 being some CDF, the ARE is greater for heterogeneous than for homogeneous variances as required in the analysis of variance; even under normality the ARE may exceed unity if variances are sufficiently heterogeneous (see Sen [23]). Similar conclusions apply when the joint CDF of $\{X_{i1}, \ldots, X_{ik}\}$ is $F_0((x_1 - \theta_1)/\sigma_i, \ldots, (x_k - \theta_k)/\sigma_i)$ (see Sen [25]). If both distributions and scale parameters differ among blocks [i.e., $\{F_i((x - \theta_j)/\sigma_i)\}$ obtains], the lower bound $0.864k/(k+1)$ for the ARE holds uniformly in $\{F_1, F_2, \ldots\}$ (see Sen [23]). Under normality and equal variances, where the analysis of variance is fully efficient, the ARE is $0.955k/(k+1)$; compare van Elteren and Noether [28] under independence and Sen [25] when $\{X_{i1}, \ldots, X_{ik}\}$ are equicorrelated Gaussian variables. In summary, Friedman's test may be recommended for wide usage, giving protection against heterogeneous variances even under normality, yet retaining reasonable efficiency when the analysis of variance applies.

The utility of Friedman's test depends in part on the accuracy of the χ^2 approximation to the distribution of X_r^2. Comparisons with exact percentiles have shown the approximation to be adequate in certain cases. Friedman [3] showed that tail probabilities beyond the 80th percentile differ at most by 0.02 unit for $k = 3$ and $n = 9$, and by 0.03 unit for $k = 4$ and $n = 4$, the error diminishing as n increases. More generally, uniform bounds on the error of approximation are available for both central and noncentral distributions (see Jensen [9]); for the central case these yield

$$\sup_{0 \leqslant z < \infty} |G_{n,k}(z) - \Psi_{k-1}(z)| \leqslant B(k)/n^{1/2},$$

where $G_{n,k}(z)$ is the CDF of X_r^2, $\Psi_{k-1}(z)$ is the CDF of the central χ^2 distribution having $k - 1$ degrees of freedom, and $B(k)$ is a finite positive constant depending only on k. Friedman [4] cautions that for $k > 7$ and $n < 6$, the approximation of Kendall and Babington Smith [10] is preferred.

Various extensions of Friedman's test are available. Companion procedures due to Steel [26], Nemenyi [15], and others for comparing treatments with a control and for making all pairwise comparisons among treatments are discussed in Miller [14] and Hollander and Wolfe [7]. An extension due to Durbin [2] applies in balanced incomplete block* experiments, for which efficiency comparisons are given in van Elteren and Noether [28]. These topics are developed in Noether [16], along with procedures for analyzing $m > 1$ observations per cell. Patel [19] generalized Friedman's test to accommodate censorship on the right, where the censoring* process may vary from block to block. A procedure for testing the homogeneity of M collections of Friedman's rank sums was developed by Lehmacher and Wall [11] as a means to comparing M response curves. Extensions due to Gerig [5,6] apply in multivariate comparative experiments with or without covariates; these are given in part in Puri and Sen [20]. Some allied multiple testing procedures are given in Jensen [8] for use when comparisons among treatments are to be made separately for disjoint subsets of the responses.

Several nonparametric competitors to Friedman's [3] test are available. Brown and Mood [1] proposed the statistic

$$B_r = \frac{k(k-1)}{na(k-a)} \sum_{j=1}^{k} (M_j - na/k)^2,$$

where $M_j = \sum_{i=1}^{n} m_{ij}$ and m_{ij} is 1 or 0 according as R_{ij} is not greater than or greater than a for some a in the set $1 \leqslant a < k$, usually taken as the largest integer contained in $(k+1)/2$. Youden [31] and Thompson and Willke [27] developed a procedure, based on the extreme rank sums $R_{\min} = \min\{R_1, R_2, \ldots, R_k\}$ and $R_{\max} = \max\{R_1, R_2, \ldots, R_k\}$, which not only identifies inferior or superior treatments, but supports distribution-free* confidence sets for the location parameters $\{\theta_1, \theta_2, \ldots, \theta_k\}$. Sen [24] gave tests based on general scores functions yielding the Friedman and Brown–Mood tests as special cases, and studied the ARE of such tests, relative to the analysis of variance, with the following conclusions. If $k = 3$ and $\{F_1, F_2, \ldots, F_n\}$ are symmetric about zero, Friedman's test is optimal. If $k \geqslant 4$ and $\{F_1, F_2, \ldots, F_n\}$ are identical, Friedman's test is optimal for the logistic and the Brown–Mood test for the exponential distribution.

Some recent applications of Friedman's test include those of Reinmuth and Geurts [21] and of Wagner [29].

References

[1] Brown, G. W. and Mood, A. M. (1951). *Proc. 2nd Berkeley Symp. Math. Statist. Prob.*, University of California Press, Berkeley, Calif., pp. 159–166. [Gives an alternative to Friedman's (1937) test using contingency tables.]

[2] Durbin, J. (1951). *Brit. J. Psychol.*, **4**, 85–90. (Source paper; extends Friedman's test to balanced incomplete block designs.)

[3] Friedman, M. (1937). *J. Amer. Statist. Ass.*, **32**, 675–701. (Source paper; develops Friedman's test; gives some exact distributions; compares the χ^2 approximation; establishes limiting distribution.)

[4] Friedman, M. (1940). *Ann. Math. Statist.*, **11**, 86–92. [Source paper; investigates accuracy of Kendall and Babington Smith's (1939) approximation; shows it to be quite good; compares it with the χ^2 approximation.]

[5] Gerig, T. M. (1969). *J. Amer. Statist. Ass.*, **64**, 1595–1608. (Source paper; uses Lawley–Hotelling statistics based on ranks; studies large-sample distributions; gives ARE relative to normal-theory test.)

[6] Gerig, T. M. (1975). *J. Amer. Statist. Ass.*, **70**, 443–447. [Source paper; extends Gerig's (1969) test to include covariates; gives asymptotic distributions; studies ARE relative to normal-theory test.]

[7] Hollander, M. and Wolfe, D. A. (1973). *Nonparametric Statistical Methods*. Wiley, New York. (Excellent general reference.)

[8] Jensen, D. R. (1974). *Ann. Statist.*, **2**, 311–322. [Source paper; gives multiple tests using Gerig's (1969) statistics; studies the limiting joint distribution of several statistics; gives bounds on the rate of convergence.]

[9] Jensen, D. R. (1977). *Metrika*, **24**, 75–85. [Source paper; gives bounds on convergence rates for distributions of statistics due to Friedman (1937), Nemenyi (1963), and Steel (1959); treats null and nonnull cases.]

[10] Kendall, M. G. and Babington Smith, B. (1939). *Ann. Math. Statist.*, **10**, 275–287. [Source paper; develops test equivalent to Friedman's (1937); suggests using approximating Snedecor–Fisher distribution.]

[11] Lehmacher, W. and Wall, K. D. (1978). *Biom. J.*, **3**, 261–273. (Source paper; replaces observations from response curves by ranks; compares homogeneity of rank sums over response curves.)

[12] Likeš, J. and Laga, L. (1980). *Biom. J.*, **5**, 443–440.

[13] Michaelis, J. (1971). *Biom. Zeit.*, **13**, 118–129.

[14] Miller, R. G. (1966). *Simultaneous Statistical Inference*. McGraw-Hill, New York. (Excellent reference on parametric and nonparametric approaches to multiple inferences.)

[15] Nemenyi, P. (1963). *Distribution-Free Multiple Comparisons*. Ph.D. dissertation, Princeton University. [Source paper; treats nonparametric multiple inference; summarized in Miller (1966).]

[16] Noether, G. E. (1967). *Elements of Nonparametric Statistics*. Wiley, New York. (Excellent general reference.)

[17] Odeh, R. E. (1977). *Commun. Statist. Simul. Comp. B*, **6**, 29–48. (Gives exact tables of the null distribution of X_r^2 for cases $\{k = 5,\ 6 \leqslant n \leqslant 8;\ k = 6,\ 2 \leqslant n \leqslant 6\}$; describes combinatorial algorithms.)

[18] Owen, D. B. (1962). *Handbook of Statistical Tables*. Addison-Wesley, Reading, Mass. (Gives extensive tables of the exact null distribution of X_r^2.)

[19] Patel, K. M. (1975). *Commun. Statist.*, **4**, 389–394. [Source paper; extends Friedman's (1937) test; gives asymptotic distribution of the statistic.]

[20] Puri, M. L. and Sen, P. K. (1971). *Nonparametric Methods in Multivariate Analysis*. Wiley, New York. (Excellent general reference for multivariate nonparametric statistics.)

[21] Reinmuth, J. E. and Geurts, M. D. (1977). *Decis. Sci.*, **8**, 134–150. (Novel application in model selection and validation using spectral methods in time series.)

[22] Sacks, S. T. and Selvin, S. (1979). *Statist. Neerlandica*, **33**, 51–54.

[23] Sen, P. K. (1967). *Biometrika*, **54**, 677–679. [Source paper; studies efficiency of Friedman's (1937) test under heterogeneous error variances; gives bounds on ARE.]

[24] Sen, P. K. (1968). *J. R. Statist. Soc. B*, **30**, 312–317. [Gives alternatives to Friedman's (1937) test.]

[25] Sen, P. K. (1972). *Metrika*, **18**, 234–237. [Source paper; studies ARE of Friedman's (1937) test relative to the analysis of variance; treats correlated errors arising in some random-effects and mixed models.]

[26] Steel, R. G. D. (1959). *J. Amer. Statist. Ass.*, **54**, 767–775. (Source paper; develops multiple sign tests; studies the limiting distribution of the statistic.)

[27] Thompson, W. A. and Willke, T. A. (1963). *Biometrika*, **50**, 375–383. [Studies an alternative to Friedman's (1937) test.]

[28] van Elteren, Ph. and Noether, G. E. (1959). *Biometrika*, **46**, 475–477. [Source paper; studies ARE of Friedman's (1937) and Durbin's (1951) tests relative to the analysis of variance.]

[29] Wagner, E. E. (1976). *Educ. Psychol. Meas.*, **36**, 615–617. (Application in detecting ranking biases of judges owing to initial position in items to be ranked.)

[30] Wallis, W. A. (1939). *J. Amer. Statist. Ass.*, **34**, 533–538. [Source paper; develops test equivalent to Friedman's (1937).]

[31] Youden, W. J. (1963). *Mater. Res. Stand.*, **3**, 9–13. [Proposes a competitor to Friedman's (1937) test; studied in detail by Thompson and Willke (1963).]

(ANALYSIS OF VARIANCE RANK TESTS)

D. R. Jensen

FRIEDMAN'S URN MODEL

In 1949, Friedman proposed the following urn model, which is a generalization of the classical Pólya's urn* scheme (see Johnson and Kotz [6]). Start with W_0 white and B_0 black balls ($W_0 + B_0 > 0$). A draw is effected as follows:

1. Choose a ball at random from the urn.
2. Observe its color, return the ball to the urn.
3. Add α balls of the same color and β balls of the opposite color.

Let (W_n, B_n) denote the composition of the urn after n successive draws. The stochastic process $\{(W_n, B_n); n = 0, 1, \dots\}$ is called *Friedman's urn process* (FP). When $\beta = 0$ the process reduces to the Pólya urn scheme.

Friedman [5] obtained elegant and almost explicit expressions for the generating functions* of the W_n. The properties of $\{(W_n, B_n); n = 0, 1, \dots\}$ with $\beta = 0$ are well known (see Feller [2, Chap. IV] and Fréchet [3]). For $\beta > 0$, Freedman [4] has investigated the asymptotic properties of the Friedman urn process.

The foregoing urn model has been generalized by Athreya and Karlin [1]. An urn contains balls of p different colors. We start with Y_{0i} balls of color i ($i = 1, 2, \dots, p$). A draw consists of the following operations:

1. Select a ball at random from the urn.
2. Record its color C and return the ball to the urn.
3. If $C = i$, add a random number R_{ij} of balls of color j ($j = 1, 2, \dots, p$) where the vector $R_i = (R_{i1}, \dots, R_{ip})$ has the probability generating function $h_i(s)$.

Let $Y_n = (Y_{n1}, \dots, Y_{np})$ denote the composition of the urn after n successive draws. The stochastic process $\{Y_n; n = 0, 1, 2, \dots\}$ on the p-dimensional integer lattice is called a *generalized Friedman's urn process* (GFP).

An interesting application of the FP is in the design of sequential experiments*. Suppose that two treatments A and B are to be compared. Eligible subjects arrive at an experimental site sequentially and must be assigned immediately to one of these two treatment groups. Now consider the Friedman's urn model with $W_0 = B_0 = \gamma$, $\alpha = 0$, and $\beta > 0$. Each time a white ball is drawn, the next subject will be assigned to treatment A. Similarly, when a black ball is drawn, the subject will be assigned to treatment B. This urn design UD(γ, β) forces a small experiment to be balanced while tending toward a complete randomization scheme as the size of the experiment increases [7, 8].

Wei and Durham [9] also used the GFP to introduce a class of treatment allocation rules which are simple, nondeterministic, and tend to put more subjects on better treatments.

References

[1] Athreya, K. B. and Karlin, S. (1968). *Ann. Math. Statist.*, **39**, 1801–1817.

[2] Feller, W. (1965). *Introduction to Probability Theory and Applications*, Vol. 1, 3rd ed., Wiley, New York.

[3] Fréchet, M. (1943). *Les Probabilités associées à un système d'événements compatibles et dépendants.* Hermann & Cie, Paris.

[4] Freedman, D. (1965). *Ann. Math. Statist.*, **36**, 956–970.

[5] Friedman, B. (1949). *Commun. Pure Appl. Math.*, **2**, 59–70.

[6] Johnson, N. L. and Kotz, S. (1977). *Urn Models and Their Application.* Wiley, New York.

[7] Wei, L. J. (1977). *J. Amer. Statist. Ass.*, **72**, 382–386.

[8] Wei, L. J. (1979). *Ann. Statist.*, **7**, 291–296.

[9] Wei, L. J. and Durham, S. (1978). *J. Amer. Statist. Ass.*, **73**, 840–843.

(PÓLYA'S URN
URN MODELS)

L. J. WEI

FROOT *See* FOLDED TRANSFORMATIONS

F-TESTS

DEFINITION

Let U and V be independent random variables having chi-square* distributions with r_1 and r_2 degrees of freedom, respectively. Then the ratio

$$F = \frac{U/r_1}{V/r_2} \tag{1}$$

is called an F random variable with r_1 and r_2 degrees of freedom. The density of F is

$$g(f) = \frac{\Gamma[(r_1+r_2)/2](r_1/r_2)^{r_1/2}}{\Gamma(r_1/2)\Gamma(r_2/2)} \times \frac{f^{r_1/2-1}}{(1+r_1 f/r_2)^{(r_1+r_2)/2}}, \qquad f > 0. \tag{2}$$

Numerous tables of the quantile* of order p, say $f_{r_1,r_2;p}$, exist but tabulation is sparse for higher degrees of freedom. A number of modern desk and hand-held calculators compute the distribution function of F (some with very large degrees of freedom) from which quantiles can be obtained by trial.

Tests of statistical hypotheses* which are based on an observed value of a statistic of form (1) are called F-tests.

LINEAR HYPOTHESIS ABOUT NORMAL RANDOM VARIABLES

This topic probably leads to the most important use of F-tests. Entire books have been written on this subject, so we cannot hope to give a thorough coverage in a few paragraphs. These include texts by Graybill [2], Scheffé [10], Searle [11], and Seber [12] (*see* GENERAL LINEAR MODEL).

One major result leading to F-tests is the following:

Let $X_1, X_2, \ldots, X_n$ be independent normal* random variables with means $\mu_1, \mu_2, \ldots, \mu_n$ and common variance σ^2. Assume that in the parameter space Ω the μ_i are functions of p mathematically independent parameters, $\beta_1, \beta_2, \ldots, \beta_p$. That is, in Ω,

$$\mu_i = \sum_{j=1}^{p} g_{ij}\beta_j, \qquad i = 1, 2, \ldots, n,$$

where the g_{ij} are known constants (frequently, 0 or 1). Further, assume that under the hypothesis the β_j satisfy q additional equations (or are in the subspace ω)

$$\sum_{j=1}^{p} k_{ij}\beta_j = 0, \qquad i = 1, 2, \ldots, q, \tag{3}$$

where the k_{ij} are known constants. Let

$$Q_\Omega = \min \sum_{i=1}^{n} (x_i - \mu_i)^2 = \sum_{i=1}^{n} (x_i - \hat{\mu}_i)^2,$$

$$Q_\omega = \min \sum_{i=1}^{n} (x_i - \mu_i)^2 = \sum_{i=1}^{n} (x_i - \hat{\hat{\mu}}_i)^2.$$

Then, to test (3) with significance level α, the likelihood ratio* procedure rejects when

$$f_{q,n-p,\theta} = \frac{n-p}{q}\,\frac{Q_\omega - Q_\Omega}{Q_\Omega} > f_{q,n-p;1-\alpha}. \tag{4}$$

Here $F_{q,n-p,\theta}$ is a noncentral* F random variable with noncentrality parameter θ. If (3) is true, then $\theta = 0$ and $F_{q,n-p,0} = F_{q,n-p}$ is distributed as central F. If (3) is false, $\sigma^2\theta$ is obtained by replacing X_i by μ_i in $Q_\omega - Q_\Omega$.

Two important special cases of the foregoing theorem are the following:

One-Way Classification* (or Completely Randomized Design)

Assume that the X_{ij} have independent normal distributions with means μ_j, $i = 1, 2, \ldots, n_j$; $j = 1, 2, \ldots, r$. In the notation of the theorem, n is now $N = \sum_{j=1}^{r} n_j$ and μ_i is $\mu_{ij} = \mu_j$, $i = 1, 2, \ldots, n_j$; $j = 1, 2, \ldots, r$, so that $p = r$, $\beta_j = \mu_j$. The hypothesis is H_0: $\mu_1 = \mu_2 = \cdots = \mu_r$ (or $\mu_2 - \mu_1 = 0, \ldots,$

$\mu_r - \mu_1 = 0$) with $q = r - 1$. The statistic is

$$F_{r-1,N-r,\theta} = \frac{\sum_{j=1}^{r} \sum_{i=1}^{n_j} \left(\bar{X}_{\cdot j} - \bar{X}\right)^2/(r-1)}{\sum_{j=1}^{r} \sum_{i=1}^{n_j} \left(X_{ij} - \bar{X}_{\cdot j}\right)^2/(N-r)}, \quad (5)$$

$$\bar{X}_{\cdot j} = \sum_{i=1}^{n_j} X_{ij}/n_j,$$

$$\bar{X} = \sum_{j=1}^{r} \sum_{i=1}^{n_j} X_{ij}/N,$$

$$\sigma^2\theta = \sum_{j=1}^{r} n_j(\mu_j - \mu)^2,$$

$$\mu = \sum_{j=1}^{r} \mu_j/r.$$

Randomized Blocks

Assume that the X_{ij} have independent normal distributions with means $\mu_{ij} = \mu + \alpha_i + \beta_j$, $i = 1, 2, \dots, n$; $j = 1, 2, \dots, r$; $\sum_{i=1}^{n} \alpha_i = \sum_{j=1}^{r} \beta_j = 0$. The n of the theorem is now nr and $p = 1 + (n-1) + (r-1) = n + r - 1$, which makes the $n - p$ of the theorem $nr - n - r + 1 = (n-1)(r-1)$. The hypothesis is H_0: $\beta_j = 0$, $j = 1, 2, \dots, r$ (or $\mu_{\cdot 1} = \mu_{\cdot 2} = \cdots = \mu_{\cdot r}$ where $\mu_{\cdot j} = \sum_{i=1}^{n} \mu_{ij}/r$) so that $q = r - 1$. The statistic is

$$F_{r-1,(n-1)(r-1),\theta} = \frac{\sum_{j=1}^{r} \sum_{i=1}^{n} \left(\bar{X}_{\cdot j} - \bar{X}\right)^2/(r-1)}{\sum_{j=1}^{r} \sum_{i=1}^{n} \left(X_{ij} - \bar{X}_{i\cdot} - \bar{X}_{\cdot j} + \bar{X}\right)^2/(n-1)(r-1)}, \quad (6)$$

$$\bar{X}_{i\cdot} = \sum_{j=1}^{r} X_{ij}/r,$$

$$\sigma^2\theta = n \sum_{j=1}^{r} \beta_j^2 = n \sum_{j=1}^{r} (\mu_{\cdot j} - \mu)^2,$$

$$\mu = \sum_{j=1}^{r} \sum_{i=1}^{n} \mu_{ij}/(rn).$$

For either of the foregoing models the power of the test for a specific alternative can be calculated with available tables. For the one-way classification with $n_j = n$, $j = 1, 2, \dots, r$ or for randomized blocks, the minimum n which satisfies

$$\text{power} = \alpha \quad \text{if } H_0 \text{ is true},$$
$$\geqslant 1 - \beta, \quad \text{if a specific alternative is true} \quad (7)$$

can be easily found. Some helpful references are a book by Odeh and Fox [9] and several short papers by Guenther [3, 5, 6].

RANDOM-EFFECTS MODELS

Another way to write the assumptions for the completely randomized design of the preceding section is

$$X_{ij} = \mu + \beta_j + E_{ij},$$
$$i = 1, 2, \dots, n_j; \quad j = 1, 2, \dots, r,$$

where the E_{ij} are independently normally distributed with mean 0 and variance σ^2 and for convenience we require $\sum_{j=1}^{r} n_j\beta_j = 0$. With the β_j regarded as a set of constants, the model is said to have *fixed effects*. Suppose it is further assumed that the β_j constitute a random sample from a normal distribution with mean 0 and variance σ_β^2. Then the model is said to have *random effects*. Such an assumption would be appropriate if the r sampled normal distributions are a random sample from a population of normal distributions. A test of H_0: $\sigma_\beta^2 \leqslant R_0\sigma^2$ against H_1: $\sigma_\beta^2 > R_0\sigma^2$, where R_0 is a specific value of $R = \sigma_\beta^2/\sigma^2$, may be of interest. (A frequent choice is $R_0 = 0$, which says that there is no variability between the sampled normal distributions.) If the sample sizes are equal ($n_j = n$, $j = 1, 2, \dots, r$), the appropriate statistic to use is

$$F_{r-1,N-r} = F_{r-1,N-r,\theta} \frac{1}{1 + nR_0}, \quad (8)$$

where $F_{r-1,N-r,\theta}$ is given by (5), H_0 being rejected if the observed value of the statistic is large. The power of the test is a central F

probability

$$\Pr\left[F_{r-1,r(n-1)} > f_{r-1,r(n-1);1-\alpha}\left(\frac{1+nR_0}{1+nR}\right)\right], \tag{9}$$

easily evaluated on calculators. The kind of sample size problem described at the end of the preceding section can be solved by trial, finding the minimum n that satisfies

$$f_{r-1,r(n-1);1-\alpha}\left(\frac{1+nR_0}{1+nR_1}\right) \leqslant f_{r-1,r(n-1);\beta} \tag{10}$$

for a specific $R = R_1$. (Here β is a probability.)

For further information on random effects and mixed models (part fixed, part random), see the texts referenced earlier and FIXED-, RANDOM-, AND MIXED-EFFECTS MODELS.

OTHER *F*-TESTS ARISING FROM NORMAL RANDOM VARIABLES

Tests based on observed values of F random variables arise in other contexts with normal distributions. Three will be briefly considered.

Tests for the Ratio of Two Normal Variances

Suppose that two normal distributions have variances σ_1^2 and σ_2^2 and that it is desired to test hypotheses about $R^2 = \sigma_1^2/\sigma_2^2$. Take random samples of size n_1, n_2 from the distributions with variances σ_1^2, σ_2^2 and denote the sample variances by S_1^2, S_2^2. Then, in (1), let $U = (n_1 - 1)S_1^2/\sigma_1^2$, $V = (n_2 - 1)S_2^2/\sigma_2^2$, so that

$$F_{r_1,r_2} = S_1^2/(S_2^2R^2), \tag{11}$$

where $r_1 = n_1 - 1$, $r_2 = n_2 - 1$, has an F-distribution. The observed value of this statistic when $R = R_0$ can be used to test

1. $H_0: R \leqslant R_0$ against $H_1: R > R_0$
2. $H_0: R \geqslant R_0$ against $H_1: R < R_0$
3. $H_0: R = R_0$ against $H_1: R \neq R_0$

and with $R = R_1$ can be used to test

4. $H_0: R \leqslant R_1, R \geqslant R_2$ against $H_1: R_1 < R < R_2$
5. $H_0: R_1 \leqslant R \leqslant R_2$ against $H_1: R < R_1, R > R_2$

For all five cases uniformly most powerful unbiased tests* are available. For case 1 the critical region is $s_1^2/(s_2^2R_0^2) > f_{r_1,r_2;1-\alpha}$, and for any R the power of the test is

$$\Pr\left(F_{r_1,r_2} > \frac{R_0^2}{R} f_{r_1,r_2;1-\alpha}\right). \tag{12}$$

With $n_1 = n_2 = n$ and a specific alternative value $R = R_1$, the minimum n satisfying

$$\frac{R_0^2}{R_1^2} f_{n-1,n-1;1-\alpha} \leqslant f_{n-1,n-1;\beta} \tag{13}$$

satisfies (7). Obvious changes give the corresponding results for case 2. Some iteration is required to obtain the critical regions for the last three cases (for details, see a paper by Guenther [4]), but power is again expressible as central F integrals.

Test for the Mean Vector of a Multivariate Normal Distribution

Let $\mathbf{X}_1, \mathbf{X}_2, \ldots, \mathbf{X}_n$ be a random sample from a p-variate normal* distribution with mean $\boldsymbol{\mu}$ and covariance matrix $\boldsymbol{\Sigma}$. Then the likelihood ratio test* of $H_0: \boldsymbol{\mu} = \boldsymbol{\mu}_0$ against $H_1: \boldsymbol{\mu} \neq \boldsymbol{\mu}_0$ rejects when $f_{p,n-p} > f_{p,n-p;1-\alpha}$, where

$$F_{p,n-p} = \frac{(n-p)T^2}{p(n-1)} \tag{14}$$

and $T^2 = n(\overline{\mathbf{X}} - \boldsymbol{\mu}_0)'S^{-1}(\overline{\mathbf{X}} - \boldsymbol{\mu}_0)$, $\overline{\mathbf{X}}$ is the vector of sample means, $\mathbf{S}$ is the sample covariance matrix. If the mean is $\boldsymbol{\mu}$ instead of $\boldsymbol{\mu}_0$, then (14) has a noncentral* F distribution with p and $n-p$ degrees of freedom and noncentrality parameter $n(\boldsymbol{\mu} - \boldsymbol{\mu}_0)'\boldsymbol{\Sigma}^{-1}(\boldsymbol{\mu} - \boldsymbol{\mu}_0)$ (where $\boldsymbol{\Sigma}$ is unknown). For further details, see the texts by Anderson [1], Kshirsagar [7], and Morrison [8].

Test That the Multiple Correlation Coefficient is Zero

Let $\mathbf{X} = (X_1, X_2, \ldots, X_p)$ have a p-variate normal distribution and let R be the multiple correlation coefficient between X_1 and the other $p-1$ variables. Denote the maximum likelihood* estimator of R by $\hat{R}$. Then the likelihood ratio test of $H_0: R = 0$ against $H_1: R \neq 0$ rejects when $f_{p-1,n-p} > f_{p-1,n-p;1-\alpha}$, where

$$F_{p-1,n-p} = \frac{\hat{R}^2}{1-\hat{R}^2} \cdot \frac{n-p}{p-1}. \qquad (15)$$

For further details, see the texts cited in the preceding paragraph.

OTHER *F*-TESTS

Tests based on (1) arise in a number of other distributional situations. We will list a few.

Exponential Location Parameter

Let $X_1, X_2, \ldots, X_n$ be a random sample from a two-parameter exponential* distribution with density

$$f(x;\, \mu, \sigma) = \frac{1}{\sigma} \exp\left[-(x-\mu)/\sigma\right],$$

$$x \geqslant \mu, \quad \sigma > 0$$

and let $Y = \min X_i$. Then UMP unbiased tests* of hypotheses 1 to 5 (with R replaced by μ) can be based on

$$F_{2,2n-2} = \frac{n(n-1)(Y-\mu)}{\sum_{i=1}^n (X_i - Y)}.$$

Here

$$U = 2n(Y-\mu)/\sigma,$$

$$V = 2\sum_{i=1}^n (X_i - Y)/\sigma,$$

$$r_1 = 2, \qquad r_2 = 2n-2.$$

Pareto Location Parameter

Let $X_1, X_2, \ldots, X_n$ be a random sample from a two-parameter Pareto* distribution with density

$$f(x;a,b) = \frac{ab^a}{x^{a+1}}, \qquad x \geqslant b > 0, \quad a > 0$$

and let $Y = \min X_i$. Then UMP unbiased tests* of cases 1 to 5 of the preceding section (with R replaced by b) can be based on

$$F_{2,2n-2} = \frac{-n(n-1)\ln(b/Y)}{\ln\left[\prod_{i=1}^n (X_i/Y)\right]}.$$

Here

$$U = -2an\ln(b/Y),$$

$$V = 2a\ln\left[\prod_{i=1}^n (X_i/Y)\right],$$

$$r_1 = 2, \qquad r_2 = 2n-2.$$

Inverse Gaussian Location Parameter

Let $X_1, X_2, \ldots, X_n$ be a random sample from an inverse Gaussian* distribution with density

$$f(x;\, \mu, \lambda) = \left[\lambda/(2\pi x^3)\right]^{1/2} \times \exp\left[-\lambda(x-\mu)^2/(2\mu^2 x)\right],$$

$$x, \mu, \lambda > 0.$$

Then tests for μ can be based on

$$F_{1,n-1} = \frac{n(n-1)(\bar{X}-\mu)^2}{\mu^2 \bar{X} \sum_{i=1}^n (1/X_i - 1/\bar{X})}.$$

Here

$$U = n\lambda(\bar{X}-\mu)^2/(\mu^2 \bar{X}),$$

$$V = \lambda \sum_{i=1}^n (1/X_i - 1/\bar{X}),$$

$$r_1 = 1, \qquad r_2 = n-1.$$

One-Way Analysis of Variance for the Inverse Gaussian

Let $X_{1j}, X_{2j}, \ldots, X_{n_j j}$, $j = 1, 2, \ldots, r$, be r random samples of size $n_1, n_2, \ldots, n_r$ from inverse Gaussian distributions having the same λ and means $\mu_1, \mu_2, \ldots, \mu_r$, respectively. Then the likelihood ratio test of

$H_0: \mu_1 = \mu_2 = \cdots = \mu_r$ rejects when $f_{r-1,N-r} > f_{r-1,N-r;1-\alpha}$, where

$$F_{r-1,N-r} = \frac{\sum_{j=1}^{r}\sum_{i=1}^{n_j}\left(1/\bar{X}_{.j} - 1/\bar{X}\right)/(r-1)}{\sum_{j=1}^{r}\sum_{i=1}^{n_j}\left(1/X_{ij} - 1/\bar{X}_{.j}\right)/(N-r)}$$

and the notation is the same as in the section "One-Way Classification" (see also Tweedie [13]).

References

[1] Anderson, T. W. (1958). *An Introduction to Multivariate Statistical Analysis*. Wiley, New York.

[2] Graybill, F. A. (1976). *Theory and Application of the Linear Model*. Duxbury Press, North Scituate, Mass.

[3] Guenther, W. C. (1977). *Amer. Statist.*, **31**, 117–118.

[4] Guenther, W. C. (1977). *Amer. Statist.*, **31**, 175–177.

[5] Guenther, W. C. (1979). *Commun. Statist. B*, **8**, 169–171.

[6] Guenther, W. C. (1979). *Amer. Statist.*, **33**, 209–210.

[7] Kshirsagar, A. M. (1972). *Multivariate Analysis*. Marcel Dekker, New York.

[8] Morrison, D. F. (1976). *Multivariate Statistical Methods*. McGraw-Hill, New York.

[9] Odeh, R. E. and Fox, M. (1975). *Sample Size Choice*, Marcel Dekker, New York.

[10] Scheffé, H. (1959). *The Analysis of Variance*. Wiley, New York.

[11] Searle, S. R. (1971). *Linear Models*. Wiley, New York.

[12] Seber, G. A. F. (1977). *Linear Regression Analysis*. Wiley, New York.

[13] Tweedie, M. C. K. (1957). *Ann. Math. Statist.*, **28**, 362–377.

(ANALYSIS OF COVARIANCE
ANALYSIS OF VARIANCE
BLOCKS, RANDOMIZED COMPLETE
DESIGN OF EXPERIMENTS
FIXED-, RANDOM-, AND MIXED-EFFECTS MODELS
GENERAL LINEAR MODEL[a]
HOTELLING'S T^2
ONE-WAY CLASSIFICATION
TWO-WAY AND HIGHER CLASSIFICATIONS)

[a] This entry includes discussion of the general linear hypothesis.

WILLIAM C. GUENTHER

FULL-INFORMATION ESTIMATORS

Full information estimation has arisen in econometrics* to denote estimation of a model in which all a priori information is used in obtaining the model estimates. The origin of the term is associated with the seminal work done on estimation of the simultaneous equation model in econometrics in the 1940s and 1950s. Much of this original research was associated with the Cowles Foundation; to my knowledge the first use of the term full-information maximum likelihood (FIML) estimation occurs in Koopmans and Hood [7]. The term "full-information estimation" is now used in a broader sense than FIML, which is associated with a particular likelihood function. Various maximum likelihood*, minimum distance*, or instrumental variable estimation* techniques which all use a priori structural information are referred to as full information, even though they may not be asymptotically equivalent to maximum likelihood procedures.

Possibly the best example of full-information estimators arises with the linear simultaneous equation model in econometrics. Assume that the specification of a system of linear stochastic equations which determine the outcome of M jointly endogenous variables (*see* ECONOMETRICS) over a period of T observations

$$\mathbf{YB} + \mathbf{Z\Gamma} = \mathbf{U}, \tag{1}$$

where $\mathbf{Y}$ is a $T \times M$ matrix of jointly endogenous variables, $\mathbf{Z}$ is a $T \times K$ matrix of predetermined variables, and $\mathbf{U}$ is a $T \times M$ matrix of structural disturbances of the system. We use predetermined variables in the sense that $\mathbf{Z'U}/T$ converges in probability to zero as $T \to \infty$ (*see* LAWS OF LARGE NUMBERS). The matrices of unknown coefficients $\mathbf{B}$ and $\mathbf{\Gamma}$ are subject to a priori linear restrictions that arise from economic theory. The conditional distribution of the residuals (given $\mathbf{Z}$) is assumed to have moments $E_{\mathbf{Z}}\mathbf{U} = \mathbf{0}$, $V_{\mathbf{Z}}(\mathbf{U}) = \mathbf{\Sigma} \otimes \mathbf{I}_T$ where $\otimes$ denotes the direct or Kronecker product* of two matrices. Both $\mathbf{B}$ and $\boldsymbol{\Sigma}$ are assumed to be nonsingular, as are the probability limits of the second-order moment matrices of the predetermined variables and the endogenous variables.

Associated with the linear simultaneous equation structural specification is the reduced form, derived from the structural specification by a nonsingular transformation (multiplication by $\mathbf{B}^{-1}$)

$$\mathbf{Y} = -\mathbf{Z}\boldsymbol{\Gamma}\mathbf{B}^{-1} + \mathbf{U}\mathbf{B}^{-1} = \mathbf{Z}\boldsymbol{\Pi} + \mathbf{V}. \quad (2)$$

The reduced form is used for forecasting* and testing purposes in relation to the structural specification. If the unknown coefficients $\boldsymbol{\Pi}$ in (2) are estimated in unrestricted form, the specification is an example of multivariate least squares* where equation-by-equation estimation with ordinary least squares is numerically identical to maximum likelihood estimation, given an M-variate normality assumption on the distribution of $\mathbf{U}$. However, if the a priori restrictions on the structural model are sufficient in number to create an overidentified model (*see* IDENTIFIABILITY), then the reduced-form coefficients $\boldsymbol{\Pi}$ will be subject to nonlinear equality restrictions. Use of these overidentifying restrictions will lead to full information estimation of $\boldsymbol{\Pi}$ either by maximum likelihood (FIML) or by minimum distance estimation. If the restrictions are correct, asymptotically more efficient estimates of $\boldsymbol{\Pi}$ will be obtained than from the least-squares estimates. A test of the overidentifying restrictions can be conducted with either a Wald-type test or a likelihood ratio test* first proposed by Koopmans and Hood [7].

We return now to estimation of the unknown coefficients of the structural model and consider a single equation, say the first. After choice of a normalization, $\beta_{11} = 1$, we write the first equation as

$$\mathbf{y}_1 = \mathbf{Y}_1\boldsymbol{\beta}_1 + \mathbf{Z}_1\boldsymbol{\gamma}_1 + \mathbf{u}_1, \quad (3)$$

where all a priori known coefficients have been eliminated. $\mathbf{Y}_1$ is a $T \times r_1$ matrix, $\mathbf{Z}_1$ is a $T \times s_1$ matrix, $\mathbf{y}_1$ and $\mathbf{u}_1$ are $T \times 1$ vectors, and $\boldsymbol{\beta}_1$ is a $r_1 \times 1$ vector while $\boldsymbol{\gamma}_1$ is a $s_1 \times 1$ vector. Least-squares estimation of (3) is inconsistent due to the presence of jointly endogenous variables on the right-hand side. Consistent estimation is achieved by the method of instrumental variables* where the matrix of predetermined variables $\mathbf{Z}$ is used to form instruments. Consistent estimation is possible as long as (1) satisfies the identification condition, see Koopmans and Hood [7]. The most widely used instrumental variable estimation technique in this situation is the two-stage least-squares* (2SLS) estimator first proposed by Basmann [1] and by Theil [8].

Again, we may achieve more efficient estimation by the use of full-information estimation. To do so, we make use of overidentifying restrictions on any of the other equations, if these exist. Full-information estimation can be done by FIML again, given an M-variate normality assumption on the distribution of $\mathbf{U}$. Alternatively, full-information minimum distance or instrumental variable estimators exist which take account of all overidentifying restrictions. The most widely used full-information instrumental variable estimator is three-stage least squares (3SLS), first proposed by Zellner and Theil [10]. If a normality assumption on $\mathbf{U}$ is made, these full-information estimators all have identical (first-order) asymptotic distributions as long as $\boldsymbol{\Sigma}$ is specified to be unrestricted. Relationships among the various estimators are investigated by Hausman [4] and Hendry [6].

As with estimation of the reduced form, use of full-information estimation will tend to more efficient estimators asymptotically, but it increases the possibility of inconsistent estimation via misspecification. Assume that the first equation is specified correctly. Then estimation by instrumental variables or 2SLS yields consistent estimates as long as the instruments are valid. Since full-information estimation uses information from all the equations, FIML or 3SLS estimation leads to inconsistent estimates if *any* equation is misspecified. It often occurs in econometrics that there is considerably more knowledge about the proper specification of one equation as opposed to another equation in the system. A likelihood ratio test may be used to test the validity of the overidentifying restrictions or a specification test proposed by Hausman [5] may also be used. These same considerations arise in estimation of the structural form of a nonlinear simultaneous-equation system; for an example, see Fair and Parke [2].

The best known simple example of these issues occurs in the famous Klein model I, an early three-stochastic-equation macroeconometric model described in many textbooks (e.g., Theil [9, pp. 432ff]). The reader should note that in the 30 years since this model's formulation, macroeconometric models have grown to specifications of 100 or more equations.

The model consists of six equations, of which three are identities. The first equation is the consumption function

$$C_t = \alpha_{11}(W_t + W_t') + \alpha_{12}P_t + \alpha_{13}P_{t-1} + \alpha_{10} + u_{1t}, \quad (4)$$

where C_t is aggregate consumption, P_t and P_{t-1} are current and lagged total profits, W_t is the private industry wage bill, and W_t' is the government wage bill. The next equation is the investment function

$$I_t = \alpha_{21}P_t + \alpha_{22}P_{t-1} + \alpha_{23}K_{t-1} + \alpha_{20} + u_{2t}, \quad (5)$$

where I_t is net investment, P_t and P_{t-1} are again current and lagged profits, and K_t is the capital stock. The last stochastic equation is the wage equation

$$W_t = \alpha_{31}Q_t + \alpha_{32}Q_{t-1} + \alpha_{33}(t - 1931) + \alpha_{30} + u_{3t}, \quad (6)$$

where W_t is the wage bill, Q_t and Q_{t-1} are current and lagged private output, and t is the time trend variable. The model is then closed by the three identities

$$K_t = K_{t-1} + I_t,$$
$$Q_t = C_t + I_t + G_t, \quad (7)$$
$$P_t = Q_t - W_t - T_t,$$

where the additional variables G_t and T_t are nonwage government expenditure and business taxes, respectively.

The model is estimated on 21 annual observations for the period 1920–1941. (One observation is dropped because of the lagged variables.) Full-information estimation accounts for all overidentifying restrictions; Klein model I has 12 such restrictions. On the other hand, limited-information estimators treat the stochastic equations one at a time and use only the overidentifying restrictions in a given equation which is being estimated. Klein model I has four such restrictions in each equation. Estimates for this model are given by Theil [9, pp. 517–519] and by Hausman [3, p. 649]. Our first test compares the full-information estimates (FIML) of the reduced-form equation (2) to the unrestricted multivariate least-squares estimates as a test of the overidentifying restrictions. A likelihood ratio test of all the overidentifying restrictions has a value of 39.11 with 12 degrees of freedom which rejects the null hypothesis at conventional significance levels. Another indication of misspecification occurs when we compare the estimates for the wage equation (6). The 2SLS estimates are (0.438, 0.147, 0.130, 1.50), while the FIML estimates are (0.234, 0.285, 0.285, 5.79). A specification test based on these estimates [5] also rejects the null hypothesis of no misspecification.

References

[1] Basmann, R. L. (1957). *Econometrica*, **25**, 77–83.

[2] Fair, R. E. and Parke, W. R. (1980). *J. Econometrics*, **13**, 269–291.

[3] Hausman, J. A. (1974). *Ann. Econ. Social Meas.*, **3/4**, 641–652.

[4] Hausman, J. A. (1975). *Econometrica*, **43**, 727–738.

[5] Hausman, J. A. (1978). *Econometrica*, **46**, 1251–1271.

[6] Hendry, D. F. (1976). *J. Econometrics*, **4**, 51–88.

[7] Koopmans, T. C. and Hood, W. C. (1953). In *Studies in Econometric Method*, W. C. Hood and T. C. Koopmans, eds. Yale University Press, New Haven, Conn.

[8] Theil, H. (1961). *Economic Forecasts and Policy*. North-Holland, Amsterdam.

[9] Theil, H. (1971). *Principles of Econometrics*. Wiley, New York. (A good intermediate-level textbook for econometrics.)

[10] Zellner, A. and Theil, H. (1962). *Econometrica*, **30**, 54–78.

(IDENTIFIABILITY
INSTRUMENTAL VARIABLE ESTIMATION
SIMULTANEOUS-EQUATION MODELS
TWO-STAGE LEAST SQUARES)

JERRY A. HAUSMAN

FUNCTIONAL EQUATIONS

Strictly speaking, a functional equation is any equation that contains an unknown function or functions. But nowadays this expression usually has a narrower meaning, as an equation containing unknown functions on which no infinitesimal operation is performed. (A more formal definition is to be found in refs. 1 and 13.) In this sense, for example, differential equations or integral equations* are not functional equations. This is the point of view adopted also in the American Mathematical Society Subject Classification Scheme.

The modern theory of functional equations splits into two parts:

1. Functional equations in a single variable
2. Functional equations in several variables

Functional equations in a single variable are referred to also as functional equations of rank 1, or *iterative functional equations*. These equations contain only one independent variable (which, however, can lie in a higher dimensional, or even in an abstract space). The name iterative functional equations is due to the fact that, on one hand, such equations play a fundamental role in iteration theory [13, 17, 18], and, on the other hand, the iteration is a main tool in handling such equations. The main reference for this type of equation is ref. 13, a book with an ample bibliography, which may be a source of further references.

The most important, best known, and most often applied functional equation in a single variable is the *Schröder equation*

$$\varphi[f(x)] = s\varphi(x), \tag{1}$$

where φ is the unknown function. (The function f is regarded as given.) Another important example is the *Abel equation*

$$\varphi[f(x)] = \varphi(x) + 1. \tag{2}$$

The Schröder equation (1) and the Abel equation (2) are particular cases of the more general equation

$$\varphi(x) = h(x, \varphi[f(x)]). \tag{3}$$

Equation (3), and also more general equations in a single variable, have been thoroughly investigated (see refs. 5 and 13 for detailed results and further references).

In general, the solution of an equation of type (3) depends on an arbitrary function (i.e., it can be prescribed arbitrarily on a set with a nonempty interior); fairly strong assumptions are necessary to furnish the uniqueness of solutions.

Functional equations in a single variable occur in almost all branches of mathematics, but they play a particularly important role in the theory of dynamical systems (the problem of linearization; see ref. 10, where also much information about the multidimensional analog of (1) can be found), and in the theory of stochastic processes* (see, e.g., refs. 4 and 16).

The most important and most widely known *functional equation in several variables* is the *Cauchy equation*

$$\varphi(x + y) = \varphi(x) + \varphi(y). \tag{4}$$

In dealing with functional equations like (4) we have more freedom; we can keep one of the variables fixed, and manipulate with the other. It follows from (4) by induction that $\varphi(nx) = n\varphi(x)$ for arbitrary positive integers n, whence $\varphi(2x) = 2\varphi(x)$, and consequently $\varphi(0) = 0$. If $y = -x$, (4) yields $\varphi(-x) = -\varphi(x)$. Thus $\varphi(kx) = k\varphi(x)$ for arbitrary integer k. Now, if $q = k/m$, where k is an integer and m a positive integer, then

$$k\varphi(x) = \varphi(kx) = \varphi(mqx) = m\varphi(qx),$$

and finally

$$\varphi(qx) = q\varphi(x)$$

for arbitrary rational q. Hence, if x is real and φ is continuous, then

$$\varphi(x) = x\varphi(1),$$

and thus φ is linear homogeneous. On the other hand, it can be proved that if φ is Lebesgue measurable, or bounded from one side on a set of a positive Lebesgue measure, then φ is continuous (see refs. 6 and 15). On the other hand, discontinuous solutions of (4) can be constructed with the aid of the Hamel basis (see ref. 9; also ref. 1). The extension to more dimensions presents no difficulties. See also ref. 6 for various contexts in which (4) appears.

The functional equations

$$\varphi(x + y) = \varphi(x)\varphi(y), \qquad (5)$$

$$\varphi(xy) = \varphi(x) + \varphi(y), \qquad (6)$$

$$\varphi(xy) = \varphi(x)\varphi(y) \qquad (7)$$

are also referred to as Cauchy's functional equations. They can easily be reduced to (4). Under mild regularity assumptions solutions of (5) have the form $\varphi(x) = \exp(cx)$, whereas solutions of (6) have the form $\varphi(x) = c\log|x|$, $x \neq 0$. [In both these cases c is an arbitrary constant; (5) admits also the trivial solution $\varphi = 0$.] For (7) the regular nonzero solutions ($\varphi = 0$ also is a solution) are

$$\varphi(x) = 1, \qquad \varphi(x) = |x|^c, \quad \text{and}$$

$$\varphi(x) = |x|^c \operatorname{sgn} x,$$

where c is an arbitrary constant and $0^c = 0$ for all c.

Similar is the situation with the *Pexider equations*

$$\alpha(x + y) = \beta(x) + \gamma(y), \qquad (8)$$

$$\alpha(x + y) = \beta(x)\gamma(y), \qquad (9)$$

$$\alpha(xy) = \beta(x) + \gamma(y), \qquad (10)$$

$$\alpha(xy) = \beta(x)\gamma(y). \qquad (11)$$

Here all the functions α, β, and γ are unknown. The general solutions of (8) and (10) are the functions

$$\alpha(x) = \varphi(x) + a + b, \qquad \beta(x) = \varphi(x) + a,$$

$$\gamma(x) = \varphi(x) + b,$$

where φ satisfies (4) and (6), respectively; a and b are arbitrary constants. The general solutions of (9) and (11) are the functions

$$\alpha(x) = ab\varphi(x), \qquad \beta(x) = a\varphi(x),$$

$$\gamma(x) = b\varphi(x),$$

where φ satisfies (5) and (7), respectively; a and b are arbitrary constants. Concerning (4) to (11), see ref. 1.

Functional equations of the Cauchy and Pexider type are central to characterizations* of statistical distributions—in particular the exponential*. In this connection see refs. 8 and 12, where other functional equations used in statistics are mentioned.

Many functional equations play an important role in information theory* (see ref. 2). The fundamental equation of information is

$$\psi(x) + (1 - x)\psi\left(\frac{y}{1 - x}\right)$$

$$= \psi(y) + (1 - y)\psi\left(\frac{x}{1 - y}\right), \qquad (12)$$

postulated for $x, y \in [0, 1)$, such that $x + y \leqslant 1$, and for real-valued ψ defined on $[0, 1]$. Subject to the additional conditions

$$\psi(1 - x) = \psi(x), \quad \psi(1) = 0, \quad \psi(\tfrac{1}{2}) = 1,$$

a function ψ satisfying (12) is called an information function. The general form of such a function is

$$\psi(x) = -x\varphi(x) - (1 - x)\varphi(1 - x),$$

$$x \in [0, 1], \qquad (13)$$

where φ satisfies (6) for $x, y > 0$ and the condition $\varphi(2) = 1$ (see ref. 11). Moreover, we adopt the convention that $0\varphi(0) = 0$. Under some natural additional conditions solution (13) gives rise to the Shannon entropy* (for $n = 2$)

$$\psi(x) = -x\log_2 x - (1 - x)\log_2(1 - x). \qquad (14)$$

The Shannon entropies for higher n are derived from (14) recursively.

Another important functional equation in several variables is the *translation equation* (see ref. 1)

$$\Phi(\Phi(x,t),s) = \Phi(x,t+s), \qquad (15)$$

occurring, in particular, in the theory of Lie groups. Under mild regularity assumptions, the solution to (15) has the form

$$\Phi(x,t) = \varphi^{-1}(\varphi(x)+t), \qquad (16)$$

where φ is an arbitrary invertible function. If we also add to (15) the condition

$$\Phi(x,1) = f(x), \qquad (17)$$

then it follows from (16) that φ must satisfy the Abel equation (2). Conditions (15) and (17) express the fact that f can be embedded in a continuous group of transformations; see ref. 18.

Functional equations of the Cauchy type, such as equations expressing homomorphisms, constitute a large part of algebra. Other types and classes of functional equations in several variables have been studied; see ref. 1, which is the standard reference for this type of equation. Functional equations in several variables occur in almost all branches of mathematics. Many examples of applications can be found in refs. 1 and 6. Applications in the theory of geometric objects are presented in ref. 3, applications in economics in ref. 7, applications in information theory in ref. 2.

Applications of functional equations in probability theory are found in refs. 4, 8, 12, 14, and 16.

References

[1] Aczél, J. (1966). *Lectures on Functional Equations and Their Applications*. Academic Press, New York.

[2] Aczél, J. and Daróczy, Z. (1975). *On Measures of Information and Their Characterizations*. Academic Press, New York.

[3] Aczél, J. and Gołąb, S. (1960). *Funktionalgleichungen der Theorie der geometrischen Objekte, Monogr. Mat. No.* 39. *Polish Scientific Publishers, Warsaw.*

[4] Athreya, K. B. and Ney, P. (1972). *Branching Processes*. Springer-Verlag, New York.

[5] Baron, K. (1978). *Functional Equations of Infinite Order*, Pr. Nauk. Uniw. Śląski. Katowicach No. 265. Uniwersytet Śląski, Katowice, Poland.

[6] Dhombres, J. (1979). *Some Aspects of Functional Equations*. Chulalongkorn University Press, Bangkok, Thailand.

[7] Eichhorn, W. (1978). *Functional Equations in Economics*. Addison Wesley, Reading, Mass.

[8] Galambos, J. and Kotz, S. (1978). *Characterizations of Probability Distributions*. Lect. Notes Math., **675**.

[9] Hamel, G. (1905). *Math. Ann.*, **60**, 459–462.

[10] Hartman, P. (1964). *Ordinary Differential Equations*. Wiley, New York.

[11] Jessen, B., Karpf, J., and Thomp, A. (1968). *Math. Scand.*, **22**, 257–265.

[12] Kagan, A. M., Linnik, Yu. V., and Rao, C. R. (1973). *Characterization Problems in Mathematical Statistics*. Wiley, New York.

[13] Kuczma, M. (1968). *Functional Equations in a Single Variable*. Monogr. Mat. No. 46. Polish Scientific Publishers, Warsaw.

[14] Lukács, E. and Laha, R. G. (1964). *Applications of Characteristic Functions*. Charles Griffin, London.

[15] Ostrowski, A. (1929). *Dtsch. Math. Ver.*, **38**, 54–62.

[16] Seneta, E. (1969). *Adv. Appl. Prob.*, **1**, 1–42.

[17] Szekeres, G. (1958). *Acta Math.*, **100**, 203–258.

[18] Zdun, M. C. (1979). *Continuous and Differentiable* Iteration Semigroups, *Pr. Nauk. Uniw. Śląski. Katowicach No.* 308. *Uniwersytet Śląski, Katowice, Poland.*

MAREK KUCZMA

FUNCTIONS, APPROXIMATIONS TO

See MATHEMATICAL FUNCTIONS, APPROXIMATIONS TO

FUNDAMENTAL IDENTITY OF SEQUENTIAL ANALYSIS

Abraham Wald* [9, pp. 158–160] obtained two results of key importance in sequential analysis* and the theory of random walks*.

Wald's Lemma. *Let Z be a random variable such that*

(a). $\Pr[Z > 0] > 0$ *and* $\Pr[Z < 0] > 0$.

(b). *The moment generating function**

$M_Z(t)$, i.e., $E[e^{tZ}]$, is finite for all real values of t.
Then

1. $E[|Z|^r] < \infty$; $r = 1, 2, \ldots$.
2. *If* $E[Z] = 0$, $t = 0$ *is the only root of the equation* $M_Z(t) = 1$.
3. *If* $E[Z] \neq 0$, *there exists exactly one nonzero root* h *of the equation* $M_Z(t) = 1$; h *and* $E[Z]$ *are then of opposite sign.*

Wald's original result was confined to establishing the uniqueness of h in part 3 of the lemma; the proof depends on showing $M_Z(t)$ to be a convex function of t [7, pp. 482–484].

Wald's Fundamental Identity. *Let Z_1, $Z_2, \ldots$ be a sequence of mutually independent and identically distributed (iid) random variables, such that* $\Pr[|Z_i| > 0] > 0$ $(i = 1, 2, \ldots)$ *and with common moment generating function $M(t)$, assumed to exist for some interval on the real line. Let $S_n = Z_1 + Z_2 + \cdots + Z_n$, and define N, a stopping variable to be the first value of n such that the inequalities*

$$b < S_n < a \qquad (-\infty < b < 0 < a < \infty)$$

are violated. Then

$$E\left[e^{tS_N}\{M(t)\}^{-N}\right] = 1.$$

The proof [3, pp. 372–374] depends on showing that $\Pr[N < \infty] = 1$. Doob [2, pp. 302–303, 350–352] obtained the result under more general stopping rules*, using the property that the sequence $\{e^{tS_n}\{M(t)\}^{-n} : n = 1, 2, \ldots\}$ is a martingale*.

APPLICATIONS

In sequential analysis,

$$Z_i = log\left[f(X_i; \theta_1)/f(X_i; \theta_0)\right],$$

where $X_1, X_2, \ldots$ is a sequence of independent, identically distributed (i.i.d.) random variables, each having probability density function (PDF) $f(x; \theta)$ for some unknown value θ of a parameter. The sum S_n appears in a sequential probability ratio test* or SPRT of the simple hypothesis

$$H_0 : \theta = \theta_0$$

against the simple alternative

$$H_1 : \theta = \theta_1.$$

Wald [8, pp. 134–135] used the fundamental identity and the accompanying lemma to show that, if H_0 or H_1 is accepted when $S_N \leqslant b$ or $S_N \geqslant a$, respectively, then the operating characteristic or OC function (*see* POWER) of the SPRT is, approximately,

$$\Pr[\text{accept } H_0 \mid \theta] \simeq (1 - e^{ah(\theta)})/(e^{bh(\theta)} - e^{ah(\theta)});$$

here $h(\theta)$ is the unique value of t satisfying Wald's lemma when X_i has PDF $f(x; \theta)$, and the equation $M_Z(h(\theta)) = 1$ reduces to

$$\int_{-\infty}^{\infty} \{f(x; \theta_1)/f(x; \theta_0)\}^{h(\theta)} f(x; \theta)\, dx = 1$$

when X is continuous, and

$$\sum_x \{f(x; \theta_1)/f(x; \theta_0)\}^{h(\theta)} f(x; \theta) = 1$$

when X is discrete. Wald [8, 9] also used the lemma and the identity to approximate $E(N \mid \theta)$ (*see* AVERAGE SAMPLE NUMBER) and the characteristic function* of N. The OC approximation above can be derived [5, p. 602] using the lemma and the property that $\{f(x; \theta_1)/f(x; \theta_0)\}^{h(\theta)} f(x; \theta)$ is a PDF.

Example 1. If X_i has a Bernoulli distribution* with success probability p and $H_0: p = p_0$, $H_1: p = p_1 > p_0$, then $h(p)$ satisfies the equation

$$p(p_1/p_0)^{h(p)} + (1 - p)\{(1 - p_1)/(1 - p_0)\}^{h(p)} = 1.$$

Example 2. If X_i has a normal distribution* with unknown mean μ and known variance σ^2, and $H_0: \mu = \mu_0$, $H_1: \mu = \mu_1 > \mu_0$, then $h(\mu)$ is given explicitly by the equation

$$h(\mu) = (\mu_1 + \mu_0 - 2\mu)/(\mu_1 - \mu_0).$$

Example 3. Suppose that $H_0: \lambda = \lambda_0$, $H_1: \lambda = \lambda_1 > \lambda_0$, where X_i has either a Poisson distribution* with mean λ or an exponential distribution* with mean $1/\lambda$. Then [4, pp. 108, 111] $h(\lambda)$ satisfies the equation

$$\lambda\left\{(\lambda_1/\lambda_0)^{h(\lambda)} - 1\right\} = (\lambda_1 - \lambda_0)h(\lambda).$$

The fundamental identity has been used to determine $\Pr(N < \infty)$ in a random walk* with one absorbing barrier [1], a version of the gambler's ruin problem. Miller [6] derives an extension to the Identity and gives a list of further references. Ghosh [4, pp. 179, 275] gives an extension to time-continuous processes with stationary independent increments (*see* STATIONARY PROCESSES).

References

[1] Bahadur, R. R. (1958). *Ann. Math. Statist.*, **29**, 534–543.

[2] Doob, J. L. (1953). *Stochastic Processes*. Wiley, New York. (An advanced measure-theoretic approach; see Chap. 7.)

[3] Ferguson, T. S. (1967). *Mathematical Statistics: A Decision Theoretic Approach*. Academic Press, New York. (A mathematical approach; see Chap. 7.)

[4] Ghosh, B. K. (1970). *Sequential Tests of Statistical Hypotheses*. Addison-Wesley, Reading, Mass. (A mathematical treatment, containing many examples.)

[5] Kendall, M. G. and Stuart, A. (1967). *The Advanced Theory of Statistics*, Vol. 2, 2nd ed. Hafner, New York. (A mathematical and statistical approach, advanced calculus level; see Chap. 34.)

[6] Miller, H. D. (1961). *Ann. Math. Statist.*, **32**, 549–560.

[7] Rao, C. R. (1973). *Linear Statistical Inference and Its Applications*, 2nd ed. Wiley, New York. (A condensed but illuminating mathematical treatment; see Chap. 7.)

[8] Wald, A. (1945). *Ann. Math. Statist.*, **16**, 117–186. (See remarks following ref. 9.)

[9] Wald, A. (1947). *Sequential Analysis*. Wiley, New York. (The results are derived by their author in their very readable original versions.)

[10] Wetherill, G. B. (1975). *Sequential Methods in Statistics*, 2nd ed. Halsted Press, New York. (This book concentrates on applications of sequential techniques. Chapter 2 contains a brief discussion of the fundamental identity.)

(AVERAGE SAMPLE NUMBER
RANDOM WALK
SEQUENTIAL ANALYSIS
SEQUENTIAL PROBABILITY RATIO TEST
STOPPING RULES)

CAMPBELL B. READ

FUNOP, FUNOR-FUNOM

These acronyms describe procedures devised by Tukey [1] for checking data for apparently aberrant values—or "spottiness." An essential feature is the use of ordered reduced deviations from the median*, i.e.,

$$Z_i = \{X_i - \text{med}(X_1, \ldots, X_n)\}/a_{i:n},$$

where $a_{i:n}$ is (an approximation to) the expected value of the numerator of Z_i if the X's are independent unit normal variables—in place of the original order statistics* $X_1 \leqslant X_2 \leqslant \cdots \leqslant X_n$. (Tukey suggests taking $a_{i:n} = \Phi^{-1}((3n-1)/(3n+1))$.)

FUNOP (full normal plot) is defined by the following steps:

1. Calculate the median $\acute{Z}$ of the Z_i's, excluding the middle third—those that use only Z_i with $i \leqslant \frac{1}{3}n$ or $i \geqslant \frac{2}{3}n$.
2. Regard X_i's for which

$$|Z_i| \geqslant \left\{\max\left(B; A|a_{i:n}|^{-1}\right)\right\}\acute{Z}$$

as possibly erroneous.

Note that exclusion of the middle third pretty well ensures that all Z_i's used in step 1 are positive, so $\acute{Z}$ will be positive. A and B are chosen more or less arbitrarily.

FUNOR-FUNOM ("R" stands for "rejection"; "M" for "modification") is a system of analysis that uses FUNOP. The essential idea is to apply FUNOP (with $A = A_R$; $B = B_R$) to residuals from a fitted general linear model* and to adjust entries corresponding to residuals indicated as possibly erroneous, iterating this (FUNOR) procedure until there are no such residuals among the modified set. Then with $A = 0$, $B = B_M$, increase all observed values with Z's greater than $B_M\acute{Z}$ by the amount of this excess multiplied by the appropriate $a_{i:n}$ (the FUNOM cycle). It is suggested that a table of the accumulated adjustments be constructed, in addition to one of the adjusted observed values.

Examples of applications are given in Tukey [1].

Reference

[1] Tukey, J. W. (1962). *Ann. Math. Statist.*, **33**, 1–67.

(EXPLORATORY DATA ANALYSIS
HALF-NORMAL PLOTS
OUTLIERS
PROBABILITY PLOTTING
RESIDUALS)

FUZZY SETS AND SYSTEMS—AN INTERNATIONAL JOURNAL *See* JOURNAL OF FUZZY SETS AND SYSTEMS

FUZZY SET THEORY

Fuzziness of a phenomenon arises because of a lack of well-defined boundaries. Specifically, let A be a subset of X, a universe of discourse, covering a range of objects, where the transition between membership and nonmembership of A is gradual rather than abrupt. This set A obviously has no well-defined boundaries and hence it is a *fuzzy set*, for example, a set of rich people in community X. Naturally, there are members of X who are definitely rich (with membership $\mu_A = 1$), others who are definitely not rich ($\mu_A = 0$). There are, however, members of X which do not belong to either class, having graded membership μ_A somewhere between 0 and 1. This graded membership $\mu_A \in [0, 1]$ represents the tendency index subjectively assigned by an individual, and is context dependent. Also in medical diagnosis*, the relationship between symptoms and diseases is often described qualitatively rather than quantitatively, due to the absence of sharp boundaries characterizing the attributes that may describe a given disease.

Zadeh has pointed out rightly that one of the most important facets of human thinking is the ability to summarize information "into labels of fuzzy sets which bear an approximate relation to the primary data." Linguistic descriptions, which are usually summary descriptions of complex situations, are fuzzy in essence.

The mathematical tools and computer-based techniques for analyzing and solving problems embodied in deterministic and probabilistic (uncertain) environments are very well developed. In a probabilistic environment, uncertainty (as opposed to fuzziness or vagueness) arises from the probabilistic behavior of certain physical phenomena in mechanistic systems, i.e., systems which, in the main, are governed by the physical laws of mechanics, electromagnetism, and thermodynamics. We knew the important role that vagueness and inexactitude played in human decision making, but we did not know until 1965 how the vagueness arising from subjectivity (which is inherent in human thought process) could be modeled or analyzed [12].

Commenting on the issue of fuzziness and randomness, Zadeh writes:

> *To some, fuzziness is a disguised form of randomness. This is a misconception—a deep-seated misconception that has retarded the development of a conceptual framework for dealing with fuzziness as a basic and distinct facet of reality. Indeed, fuzziness is more than a facet of reality; it is one of its most pervasive characteristics—a characteristic rooted in the bounded capacity of the human mind to process and store information.* [*Zadeh, in ref.* 2].

> *The source of imprecision is the absence of sharply defined criteria of class membership rather than the presence of random variables.* [12].

HISTORICAL AND BIBLIOGRAPHICAL REMARKS

In 1965, Lotfi A. Zadeh laid the foundation of fuzzy set theory. Today, these concepts are gaining acceptability among engineers, scientists, mathematicians, linguists, and philosophers. The field has blossomed into a many-faceted field of inquiry, drawing on and contributing to a wide spectrum of areas ranging from pure mathematics to human perception and judgment. Its influence in science, engineering*, and social sciences has been felt already, and is certain to grow in the decade to come [1–4, 8]. It is not a paradox that a science of fuzziness must be precise! Thus "fuzzy analysis" does considerably overlap diverse areas: interval analysis, probability theory, lattice theory, Boolean algebraic analysis, statistical multivariate analysis*, linguistic analysis, cluster analysis, pattern recognition*, analysis of evidence, etc.

There have been several monographs on fuzzy set theory, a tutorial treatise in several volumes by Kaufmann [4–8], a mathematically oriented concise book by Negoita and Ralescu [11], and an excellent book covering both theory and application by Dubois and Prade [1]. There are also three volumes of papers edited by Zadeh et al. [14] and Gupta et al. [2, 3]. An extensive bibliography on the subject may be found in Dubois and Prade [1], and Gupta et al. [2, 3]. There is a regular publication appearing in the area; *see* JOURNAL OF FUZZY SETS AND SYSTEMS.

FUZZY MATHEMATICS

A subset A of X with a *membership function* $\mu(x)$ which takes any value in the interval $[0, 1]$ is called a *fuzzy set*[1]

$$A = \int_{x \in X} (\mu_A(x) \mid x), \tag{1}$$

where $\mu_A : X \to [0, 1]$.

We define the *union* of two fuzzy sets A, B with membership functions μ_A, μ_B as

$$A \cup B \overset{\Delta}{=} \int_{x \in X} \max(\mu_A(x), \mu_B(x) \mid x). \tag{2}$$

The *intersection* A and B is the set

$$A \cap B \overset{\Delta}{=} \int_{x \in X} \min(\mu_A(x), \mu_B(x) \mid x). \tag{3}$$

The *complement* of a set A is the set $\tilde{A}$:

$$\tilde{A} \overset{\Delta}{=} \int_{x \in X} (1 - \mu_A(x) \mid x). \tag{4}$$

Example 1.

$$X = \{x_1, x_2, x_3, x_4, x_5, x_6\}$$

$$A = \{(0.6 \mid x_1), (0.4 \mid x_2), (0.3 \mid x_3), (0.8 \mid x_4), (0.5 \mid x_5), (1 \mid x_6)\}$$

$$B = \{(0.8 \mid x_1), (0.3 \mid x_2), (1 \mid x_3), (1 \mid x_4), (0.4 \mid x_5), (0.9 \mid x_6)\}.$$

Then

$$A \cup B = \{(0.8 \mid x_1), (0.4 \mid x_2), (1 \mid x_3), (1 \mid x_4), (0.5 \mid x_5), (1 \mid x_6)\}$$

$$A \cap B = \{(0.6 \mid x_1), (0.3 \mid x_2), (0.3 \mid x_3), (0.8 \mid x_4), (0.4 \mid x_5), (0.9 \mid x_6)\}$$

$$\tilde{A} = \{(0.4 \mid x_1), (0.6 \mid x_2), (0.7 \mid x_3), (0.2 \mid x_4), (0.5 \mid x_5), (0 \mid x_6)\}.$$

Note that unlike ordinary set theory,

$$A \cup \tilde{A} \neq X \tag{5}$$

$$A \cap \tilde{A} \neq \emptyset \tag{6}$$

in general. Fuzzy sets obey the commutative, distributive, and associative laws as well as de Morgan's laws of set theory, however. An ordinary set is a fuzzy set where membership is either 0 or 1.

A fuzzy set A with membership function $\mu_A(\cdot)$ is a *subset* of a fuzzy set B with membership function $\mu_B(\cdot)$ if

$$\mu_A(x) \leqslant \mu_B(x) \tag{7}$$

for all $x \in X$. In this case, we write $A \subseteq B$.

Let $m_A = \max_{x \in X} \mu_A(x)$ be the *height* of the fuzzy set. If $m_A = 1$, the fuzzy set is a *normal fuzzy set*; for $m_A < 1$ it is a *subnormal fuzzy set*.

Let $B = \alpha A$ denote a fuzzy set obtained by changing memberships proportionately; $\mu_B(x) = \alpha \mu_A(x)$, where $\alpha > 0$. For normal fuzzy sets $\alpha \leqslant 1$; while for subnormal fuzzy sets $\alpha \leqslant 1/m_A$. The partial presence indicated by $\mu(x)$ is also called the *degree* of membership of x in A or its *fuzziness level*, which may be measured in a number of ways.

1. Suppose that one has a subset $B \subseteq X$; then B may be considered a fuzzy subset

of a cover or as a partition $K = \{A_1, A_2, \ldots, A_k\}$ of X, with $\mu_B(A_i)$ given by

$$\mu_B(A_i) = \frac{|A_i \cap B|}{|A_i \cup B|}, \tag{8}$$

where the notation $|A|$ stands for the number of elements in A.

Example 2. Let

$$X = \{1,2,3,4,5,6,7,8,9\},$$
$$K = \{\{1,3,5\},\{3,6,9\},\{2,4,8\},\{1,3,7\},\{2,3,8\}\}$$
$$= \{A_1, A_2, A_3, A_4, A_5\},$$

and

$$B = \{2,3,5,9,8\}.$$

Then considering B as a fuzzy subset of K it may be written as

$$B = \{(\tfrac{1}{3} \mid A_1), (\tfrac{1}{3} \mid A_2), (\tfrac{1}{3} \mid A_3), (\tfrac{1}{7} \mid A_4), (\tfrac{3}{5} \mid A_5)\}$$

or as the 5-tuple of partial membership

$$\mu_B = [\tfrac{1}{3}, \tfrac{1}{3}, \tfrac{1}{3}, \tfrac{1}{7}, \tfrac{3}{5}].$$

2. Consider an optimization problem with a set C of r objective functions, $C = \{f_1, \ldots, f_r\}$, where

$$f_i : \mathscr{R}^n \to \mathscr{R} \tag{9}$$

and each of the objective functions is to be maximized. Any solution x may be considered a fuzzy subset of the set of objectives in the following way: Let f_i^* denote the maximum value, disregarding other objective functions and assuming that $f_i^* < \infty$. Then for an arbitrary x in the feasible region,

$$f_i(x) \leqslant f_i^* .$$

Then any x may be considered as a fuzzy set of C with membership vector $\mu_x = \{\mu_1, \ldots, \mu_r\}$ where

$$\mu_i = (f_i^* - f_i(x)) \mid f_i^* .$$

We can take advantage of partial membership to consider other operations on fuzzy sets not available in ordinary set theory. For this we consider a fuzzy set as a "representational picture" obtained by a set of evaluations and criteria. This permits concepts of picture processing to be applied to a fuzzy set. Such an approach yields a new meaning to the membership of an element $x \in X$. In this one sets up a "gray scale" of membership between full membership to nonmembership. A (black-and-white) television picture is a fuzzy subset of the white screen (or the black screen). Thus one can think of *focusing* a fuzzy set, of *concentrating*, of *dilating*, of *contrast intensifying*, of *blurring*, etc.

Let A be a fuzzy set of X with membership function $\mu_A(\cdot)$. Then its concentration is the set

$$\text{CON}(A) = \int_{x \in X} (\mu_A^2(x) \mid x), \tag{10}$$

while its dilation is the set

$$\text{DIL}(A) = \int_{x \in X} (\mu_A^{0.5}(x) \mid x). \tag{11}$$

The set A^{α} is defined

$$A^{\alpha} = \int_{x \in X} (\mu_A^{\alpha}(x) \mid x), \tag{12}$$

of which the CON and DIL are special cases for $\alpha = 2$ and $\alpha = 0.5$, respectively.

We may build other operations from the operations above.

The *contrast intensification* of a fuzzy set A is the set defined as

$$\text{INT}(A) = \begin{cases} \text{CON}(A) & \text{for all } x \text{ such that } \mu_A(x) < 0.5 \\ \text{DIL}(A) & \text{for all } x \text{ such that } \mu_A(x) \geqslant 0.5. \end{cases} \tag{13}$$

This has the property of increasing membership if greater than 0.5 and decreasing it if less than 0.5.

The *blurring* of a fuzzy set A is the set defined as

$$\text{BLR}(A) = \begin{cases} \text{DIL}(A) & \text{for all } x \text{ such that } \mu_A(x) < 0.5 \\ \text{CON}(A) & \text{for all } x \text{ such that } \mu_A(x) \geqslant 0.5. \end{cases} \tag{14}$$

This has the property of decreasing membership if greater than 0.5, and increasing it if less than 0.5. The level 0.5 is chosen arbitrarily.

We may define some further compositions of two fuzzy sets A and B of X. Their *convex combination* is the fuzzy set

$$\alpha A + (1-\alpha)B = \int_{x \in X} (\alpha\mu_A(x) + (1-\alpha)\mu_B(x) \mid x), \quad 0 \leqslant \alpha \leqslant 1. \tag{15}$$

Their *product* is the fuzzy set

$$AB = \int_{x \in X} (\mu_A(x)\mu_B(x) \mid x). \tag{16}$$

Their *algebraic sum* is the fuzzy set

$$A \oplus B = \int_{x \in X} (\mu_A(x) + \mu_B(x) - \mu_A(x)\mu_B(x) \mid x). \tag{17}$$

This last definition is related to the additive set concept in measure-theoretic terms. However, we can see it as a special case of a k-sum of two fuzzy sets A and B where $k_* \leqslant k \leqslant k^*$,

$$k^* = \frac{\min_x(\mu_A(x) + \mu_B(x))}{\mu_A(x) \cdot \mu_B(x)}, \tag{18}$$

$$k_* = \frac{\max_x(\mu_A(x) + \mu_B(x) - 1)}{\mu_A(x) \cdot \mu_B(x)}, \tag{19}$$

and provided that $\mu_A(x) \neq 0 \neq \mu_B(x)$.

A k-sum ensures the resulting fuzzy set to have a proper characteristic function. It follows that $k = 1$ always satisfies any arbitrary k-sum of fuzzy sets. If fuzzy sets are restricted to a particular family, then (k_*, k^*) is chosen to characterize the family.

Example 3. Consider the fuzzy sets A and B of Example 1. Their characteristic vectors are

$$\mu_A = [0.6, 0.4, 0.3, 0.8, 0.5, 1],$$
$$\mu_B = [0.8, 0.3, 1, 1, 0.4, 0.9].$$

Thus we have

$$\mu_{\mathrm{CON}(A)} = [0.36, 0.16, 0.09, 0.64, 0.25, 1],$$
$$\mu_{\mathrm{DIL}(A)} = [0.77, 0.63, 0.54, 0.89, 0.71, 1],$$
$$\mu_{\mathrm{INT}(A)} = [0.77, 0.16, 0.09, 0.89, 0.71, 1],$$
$$\mu_{\mathrm{BLR}(A)} = [0.36, 0.63, 0.54, 0.64, 0.25, 1].$$

For a convex combination with $\alpha = 0.2$,

$$\mu_{\alpha A + (1-\alpha)B} = [0.76, 0.32, 0.86, 0.96, 0.42, 0.92]$$
$$\mu_{AB} = [0.48, 0.12, 0.3, 0.8, 0.2, 0.9]$$
$$\mu_{A \oplus B} = [0.92, 0.58, 1, 1, 0.7, 1].$$

DISTANCE AND ENTROPY CONCEPTS IN FUZZY SETS

The profile of a fuzzy set may be considered a point in a unit n-cube. One may define both a norm and a distance concept. The *distance* between two fuzzy sets A and B on a finite universe X with membership functions $\mu_A(\cdot)$ and $\mu_B(\cdot)$ is defined by

$$d_p(A, B) = \left\{ \sum_{x \in X} |\mu_A(x) - \mu_B(x)|^p \right\}^{1/p}, \quad p \geqslant 1. \tag{20}$$

For $p = 1$, we obtain a generalized Hamming distance between A and B. For $p = 2$, we obtain a root-square or Euclidean distance. For $p > 2$, we have an l_p distance between A and B. It is easy to show that $d_p(A, B)$ satisfies the necessary properties of a distance function, i.e., for $A \neq B \neq C$,

$$\begin{aligned} &d_p(A, B) \geqslant 0, && \text{nonnegativity} \\ &d_p(A, B) = d_p(B, A), && \text{symmetry} \\ &d_p(A, B) + d_p(B, C) \geqslant d_p(A, C) && \text{triangle inequality.} \end{aligned} \tag{21}$$

By setting $B = \emptyset$ one has $\mu_B(x) = 0$ for all $x \in B$. This yields all distances of fuzzy sets A, $A \neq B$, with respect to $\emptyset$. The distances are termed the *norm* or *length* of the fuzzy set, which we write

$$\|A\|_p = d_p(A, \emptyset) = \left(\sum_{x \in X} |\mu_A(x)|^p \right)^{1/p}, \quad p \geqslant 1. \tag{22}$$

In the case of two ordinary sets the Hamming distance between them is the number of elements in which they differ while the norm is the number of elements in a set.

Example 4. Consider the fuzzy sets of Example 1, with

$$\mu_A = [0.6, 0.4, 0.3, 0.8, 0.5, 1],$$
$$\mu_B = [0.8, 0.3, 1, 1, 0.4, 0.9].$$

The Hamming distance ($p = 1$), written $d_H(A, B)$, is given by

$$d_H(A,B) = |0.8 - 0.6| + |0.4 - 0.3| + |0.3 - 1| + |0.8 - 1| + |0.5 - 0.4| + |1 - 0.9| = 1.4.$$

Similarly, the Hamming lengths of A and B are

$$\|A\|_1 = 3.6, \qquad \|B\|_1 = 4.4.$$

The *focus* of a fuzzy set A on a fuzzy set B is the set FCS(A) $= A^{\alpha^*}$ where

$$d_p(A^{\alpha^*}, B) = \min_{\alpha} d_p(A^{\alpha}, B). \quad (23)$$

This has the property of altering a set A by taking its powers and bringing it close to B. When X has a countable infinity of elements, then the summation in the definition for the Hamming length can be determined if the series is convergent. Likewise, for an uncountable point set X, a convergent integral must be considered.

Consider a fuzzy set as a profile determined on a set of polar scales. It would be a very unclear evaluation if each scale response is right in the middle. On the other hand, if each evaluation is to one extreme or the other, the evaluation is very clear. For example, a "maybe" response is unclear, whereas a response "yes" or "no" is clear. Similarly, saying that a picture is extremely beautiful makes the evaluation clearer than saying it is "average." The *entropy* of a fuzzy set A on a universe X, with membership function $\mu_A(\cdot)$, is given by

$$H_L(A) = \frac{1}{N} \sum_{x \in X} \text{In}(x), \quad (24)$$

where N is the number of elements in x, X being finite, and In(x) is the *incertitude* of the evaluation along scale x given by

$$\text{In}(x) = -\{\mu_A(x)\log_2 \mu_A(x) + (1 - \mu_A(x))\log_2 \mu_A(x)\}. \quad (25)$$

The incertitude of a scale is 1 if $\mu_A(x) = 0.5$ and is 0 if $\mu_A(x) = 1$ or 0. Likewise, the entropy* of a set A is 1 if for every x, $\mu_A(x) = 0.5$ and is 0 if for every x, $\mu_A(x) = 1$ or 0. This notion of entropy is obtained in a nonprobabilistic setting. An interesting property of a fuzzy set A is that

$$H_L(A) \geqslant H_L(\text{INT}(A)); \quad (26)$$

i.e., intensifying a set reduces its incertitude.

TOPOLOGICAL AND PROBABILISTIC CONNECTIONS

Fuzziness has been given the interpretation of subjective probability. This seems to be an unnecessary restriction, as we might point to the existence of literature dealing with subjective probability and its use in Bayesian* methods of decision* analysis. There are other properties of fuzziness, such as the incertitude caused by vague language or by ambiguity; these may be studied by interpretations other than that of subjective probability. There are many interrelations between fuzziness and probability as may be seen in this section.

Let X have n elements. Let A be a fuzzy set on X with membership function $\mu_A(\cdot)$. Then, one can associate a finite-dimensional probability vector with A. Let this vector be denoted by p_A. Then

$$\sum_{x \in X} p_A(x) = 1. \quad (27)$$

Now each $p_A(x)$ can be found from the memberships $\mu_A(\cdot)$. We proceed thus: Define

$$l = d_1(A) = \sum_{x \in X} \mu_A(x), \quad (28)$$

which gives the Hamming norm of A. Then

$$p_A(x) = \mu_A(x) | l \qquad \text{for } l \neq 0; \quad (29a)$$

set

$$p_A(x) = 0 \qquad \text{for } A = \emptyset. \tag{29b}$$

In fact, this association is not unique. The family

$$A = \left\{ A \middle| \begin{matrix} A & \text{is a fuzzy set} \\ p_A & \text{is a fixed vector} \end{matrix} \right\}, \tag{30}$$

where $p_A = \mu_A(x) | l$ defines a collection of fuzzy sets all having the same probability vectors.

It is possible to confuse a fuzzy set with a finite probability vector. A strange reasoning prevails in that any vector dealing with incertitude and imprecision, with numbers ranging from 0 to 1, must be a probability vector. This is erroneous reasoning; *fuzziness is not probability*.

Suppose that one has n normalized random variables, $x_i(\omega)$, $i = 1, \ldots, n$, where $\omega \in \Omega$ is a random parameter with an associated distribution function $dp_i(\omega)$ and joint distribution function $dp_{ij}(\omega)$. Then one can set up a variance–covariance matrix of size $n \times n$ with the (i, j)th entry

$$\xi_{ij} = \exp(x_i x_j). \tag{31}$$

This matrix has $\xi_{ii} = 1$ for all i and $|\xi_{ij}| \leqslant 1$. Define a transformation

$$\mu_{ij} = \frac{1 + \xi_{ij}}{2}. \tag{32}$$

This yields a fuzzy matrix with entries between 0 and 1. From this we can consider a similarity relation on the set of random variables $1, \ldots, n$ or that the ith random variable is a fuzzy set of the variables $x_1, \ldots, x_n$.

There are other ways of combining and treating probabilistic and fuzzy concepts jointly, keeping their intrinsic characterizations of event uncertainty and event labeling ambiguity separate.

FUZZY EVENTS AND FUZZY RANDOM VARIABLES

In 1968, Zadeh introduced the concept of a fuzzy event and studied its basic properties. This was the first attempt at providing a mathematical theory that accounts for two fundamental modes of uncertainty inherent in most practical problems (i.e., statistical uncertainty and fuzziness or ambiguity; see Khalili [9]).

The notions of an event and its probability are basic concepts of probability theory. By contrast, in everyday experiences one frequently encounters situations in which an "event" is a fuzzy rather than a sharply defined collection of points. For example, the ill-defined events "It is very expensive," "x is approximately equal to 25," "Canada is colder than Japan," are fuzzy because of the imprecision of the meaning of the underlined words.

Let Ω be a Euclidean n-space R^n, $\boldsymbol{\beta}$ be a Borel field in R^n, and P a probability measure on $\boldsymbol{\beta}$. A *fuzzy event* in R^n is a fuzzy set F on R^n whose membership function $\mu_F(x)$ is measurable. The probability of the fuzzy event F is defined by the Lebesgue–Stieltjes integral

$$P(F) = \int_{R^n} \mu_F(x)\, dP = E(\mu_F),$$

where $E(\mu_F)$ is the expectation of μ_F and the fuzzy set F in R^n is defined by the membership function $\mu_F : R^n \to [0, 1]$. Two fuzzy events F_1 and F_2 are *independent* iff

$$P(F_1 \cdot F_2) = P(F_1) \cdot P(F_2).$$

For a detailed discussion of these notions, see Khalili [9] and Zadeh [13].

Fuzzy random variables are random variables whose values are not sharply defined but are fuzzy. To illustrate this notion, consider an opinion poll, during which a number of individuals are questioned on their opinion concerning the interest rates in North America in a particular year. The responses are classified into four categories: "very very high," "very high," "high," and "no opinion."

The uncertainties associated with this example are of two kinds. First, fuzziness arises because of "dimness of perception" of an individual, which introduces a sort of uncertainty about the precise meaning of the underlined words. Second, randomness arises because it is not known which response may be expected from any given individual. Therefore, randomness is caused by some chance mechanism, whereas fuzziness is due to the "perception windowing" effect of an individual. A detailed exposition may be found in Kwakernaak [10].

CONCLUSION

A subject in its early developmental stages has many ramifications. Some of these have been introduced; other topics have been left out. We hope that novice readers having obtained a flavor of the theory of fuzzy sets may on their own discover new interpretations.

In this article an interpretation of fuzzy set theory in terms of the semantic differential concept has been introduced. This interpretation has many practical implications. In many areas of decision and information sciences the semantic differential has been successfully used as an "instrument", of measuring the meaning of a concept. Fuzzy set theory provides an excellent framework within which such measurements can be used in conjunction with theoretical studies.

NOTE

1. $\int_{x \in X} (\mu_A(x) | x)$ is Zadeh's notation of a fuzzy set A on a continuous universe X.

$\sum_{x \in X} \mu_A(x) | x$ is Zadeh's notation of a fuzzy set A on a discrete universe X.

References

[1] Dubois, D. and Prade, H. (1979). *Fuzzy Sets and Systems: Theory and Applications*. Academic Press, New York.

[2] Gupta, M. M., ed.; Saradis, G. N. and Gaines, B. R. assoc. eds. (1977). *Fuzzy Automata and Decision Processes*. North-Holland, Amsterdam. (Contains an annotated bibliography and introductory papers.)

[3] Gupta, M. M., ed.; Ragade, R. K. and Yager, R. R. assoc. eds. (1979). *Advances in Fuzzy Set Theory and Applications*. North-Holland, Amsterdam. (Contains an exhaustive bibliography and introductory papers.)

[4] Kaufmann, A. (1972). *Theory of Fuzzy Sets*. Merson, Paris. (A good introduction.)

[5] Kaufmann, A. (1973). *Introduction à la théorie des sous-ensembles flous*, Vol. 1: *Eléments théoretiques de base*. Masson & Cie, Paris.

[6] Kaufmann, A. (1975). *Introduction a la théorie des sous-ensembles flous*, Vol. 2: *Applications à la linguistique et à la sémantique*. Masson & Cie, Paris.

[7] Kaufmann, A. (1975). *Introduction à la théorie des sous-ensembles flous*, Vol. 3: *Applications à la classification et la reconnaissance des dormes, aux automates et aux systèmes, aux choix des critères*. Masson & Cie, Paris.

[8] Kaufmann, A. (1975). *Theory of Fuzzy Subsets*. Academic Press, New York. (A very good introduction.)

[9] Khalili, S. (1979). *J. Math. Anal. Appl.*, **67**, 412–420. (A discussion of independent fuzzy events.)

[10] Kwakernaak, H. (1978). *Inf. Sci.*, **15**, 1–29. (Develops the concept of fuzzy random variables.)

[11] Negoita, C. V. and Ralescu, D. A. (1975). *Applications of Fuzzy Sets to Systems Analysis*. Birkhäuser, Basel, Switzerland.

[12] Zadeh, L. A. (1965). *Inf. Control*, **8**, 338–353.

[13] Zadeh, L. A. (1968). *J. Math. Anal. Appl.*, **23**, 421–427. (A discussion of fuzzy events.)

[14] Zadeh, L. A., Fu, K. S., Tanaka, K., and Shimura, M. eds. (1975). *Fuzzy Sets and Their Applications to Cognitive and Decision Processes*. Academic Press, New York.

Bibliography

See the following works, as well as the references just cited, for more information on the topic of fuzzy set theory.

Bezdek, J. C. and Dunn, J. C. (1975). *IEEE Trans. Computers*, **24**, 835–837.

Good, I. J. (1978). *J. Statist. Comp. Simul.*, **7**, 296–299.

International Journal of Fuzzy Sets and Systems, North-Holland, Amsterdam.

Kickert, W. J. M. (1978). *Fuzzy Theories in Decision Making*. Nijhoff, Leiden, Holland.

(INFORMATION AND CODING THEORY
SAMPLE SPACE
UNCERTAINTY)

MADAN M. GUPTA

G

GAIN RATIO

The notion of *gain ratio* was introduced by Joseph [6] in connection with a card game called Minoru. The term "gain ratio" was coined and used extensively by Downton and Holder [2]. It is used mainly in connection with casino gambling*. The gain ratio is defined as the absolute value of the expected gain per unit stake divided by the standard deviation. More precisely, for $1 \leqslant i \leqslant n$, let X_i represent the gain at the ith play of a gambler who bets in a sequence of n plays and who, on any play, wins α units with probability p and loses 1 unit with probability $q = 1 - p$. Here p, q, and α do not depend on i. Let μ and σ denote the expectation and standard deviation, respectively, of X_i. The gain ratio is $|\mu|\sigma^{-1}$ and is the absolute value of the reciprocal of the coefficient of variation*. The gain ratio is readily computed to be (see Epstein [4, Chap. 3])

$$|\alpha p - q|\left[\alpha^2 p + q - (\alpha p - q)^2\right]^{-0.5}.$$

Interest in the gain ratio seems to be based on its relationship with the central limit theorem*. Indeed, for a player as above with zero probability of ruin, the probability that the player's fortune shows a net gain after n plays is given, for n large, by the approximation

$$\Pr[S_n > 0] \doteq \int_{-\infty}^{x} \frac{1}{\sqrt{2\pi}} e^{-(1/2)t^2} dt,$$

where $S_n = \sum_{i=1}^{n} X_i$ and $x = \mu\sigma^{-1}\sqrt{n}$. Thus, if n is fixed and large and $\mu < 0$, then for the gambler with large initial capital, $\Pr[S_n > 0]$ decreases as a function of the gain ratio. For the gambler with large initial capital who is primarily interested in ending up ahead at the end of a betting session, the "best" bet among several games with negative expectation is the game with the smallest gain ratio. For the gambler as above who has the choice of several games with equal negative expectation (e.g., different bets at a roulette wheel), the "best" bet in the sense above is the bet with the largest variance. In the case of roulette (see Downton [3]), the single-number wager is "best." For a discussion of the various versions of roulette, see Epstein [4, Chap. 4] and GAMBLING, STATISTICS IN.

The analysis above suggested several problems concerning the gain ratio; the two following were considered in Anderson and Fontenot [1].

1. Consider k distinct casino games with negative expectations and with gain ratios $r_1 < r_2 < \cdots < r_k$. By the central

Table 1 Selected Values of N with $\Pr[S_n > 0]$ Not Monotonic in Gain Ratio

	Type of Wager					
N	Single Number	One of Two	One of Three	One of Four	One of Six	One of Twelve
301	0.427	0.462	0.398	0.420	0.390	0.359
601	0.458	0.417	0.388	0.414	0.364	0.312
911	0.417	0.419	0.415	0.369	0.364	0.284
1207	0.427	0.380	0.385	0.351	0.324	0.248
1501	0.431	0.388	0.355	0.359	0.308	0.225

Source. Anderson and Fontenot [1], reprinted with the permission of the *Journal of the Royal Statistical Society*.

limit theorem, $\Pr[S_n > 0]$ will be monotonic in the gain ratio for n sufficiently large. But how large must n be?

2. If the model above takes into account small initial capital, is $\Pr[S_n > 0]$ still monotonic in the gain ratio?

No general solution to **1** is known. However, for particular values of p, q, α, and k, computer iteration (see below) can be used to find how large n must be. In Anderson and Fontenot [1], the authors found that among 1-, 2-, 3-, 4-, 6-, and 12-number wagers at British roulette, $\Pr[S_n > 0]$ is monotonic in the gain ratio for $n > 1790$ but is not monotonic for $n = 1790$ or for many smaller values of n. Table 1 illustrates lack of monotonicity of $\Pr[S_n > 0]$ for various n and various wagers at British roulette.

For casino games with negative expectation played by a gambler with small initial capital, it can be shown that $\Pr[S_n > 0]$ is in general not monotonically decreasing in the gain ratio. In Anderson and Fontenot [1], examples are given that show that for certain values of small initial capital, $\Pr[S_n > 0]$ behaves erratically as a function of the gain ratio. In some cases, $\Pr[S_n > 0]$ is increasing rather than decreasing as in the large-capital case. These effects appear to be caused by the possibility of the gambler being ruined. Figure 1 illustrates these effects. It should be pointed out that there is no simple closed form expression for $\Pr[S_n > 0]$ in terms of the various parameters of interest.

Gambling problems such as **1** and **2** above may be studied in the setting of an unbalanced random walk* with a particle moving on a coordinate line (see Anderson and Fontenot [1]). The particle moves α units to the right with probability p and one unit to the left with probability $q = 1 - p$. The

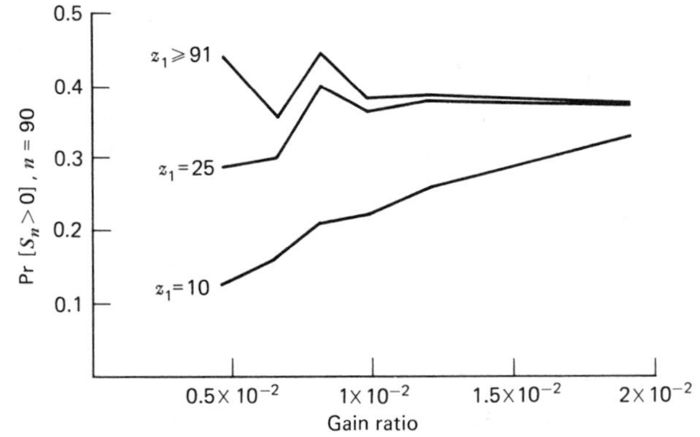

Figure 1 Dependence of $\Pr[S_n > 0]$ on the gain ratio and initial capital z_1 for $n = 90$ plays. (From ref. [1]; reprinted with the permission of the *Journal of the Royal Statistical Society*.)

probability $P_{z,n}$ of the particle being at coordinate z at time n was obtained by solving iteratively on a computer the difference equation* (with appropriate boundary conditions)

$$P_{z,n} = pP_{z-\alpha,n-1} + qP_{z+1,n-1}.$$

Thus, if z is the initial capital, then

$$\Pr[S_n > 0] = \sum_{i=1}^{n\alpha} P_{z+1,n}.$$

The theory of such random walks is well known and understood only in the case $\alpha = 1$ (see Feller [5, Chap. 14]).

References

[1] Anderson, L. R. and Fontenot, R. A. (1980). *J. R. Statist. Soc. A*, **143**, 33–40.

[2] Downton, F. and Holder, R. L. (1972). *J. R. Statist. Soc. A*, **135**, 336–364.

[3] Downton, F. (1980). *J. R. Statist. Soc. A*, **143**, 41–42.

[4] Epstein, R. A. (1977). *The Theory of Gambling and Statistical Logic*. Academic Press, New York.

[5] Feller, W. (1968). *An Introduction to Probability Theory and Its Applications*, 3rd ed. Wiley, New York.

[6] Joseph, A. W. (1933). *J. Inst. Actuaries*, **64**, 172–180.

(COEFFICIENT OF VARIATION
GAMBLING, STATISTICS IN
RANDOM WALK)

LARRY R. ANDERSON
ROBERT A. FONTENOT

GALLUP POLL *See* PUBLIC OPINION POLL

GALOIS FIELDS

A field is a set S of elements among which two operations (i.e., "addition," denoted by $+$, and "multiplication," denoted by $\times$) are defined, such that if a, b, and c are any three (not necessarily distinct) elements of S, then the following conditions are satisfied:

1. Both $(a + b)$ and $(a \times b)$ are elements of S;
2. $(a + b) + c = a + (b + c)$ and $a \times (b \times c) = (a \times b) \times c$;
3. $(a + b) = (b + a)$ and $(a \times b) = (b \times a)$;
4. $a \times (b + c) = (a \times b) + (a \times c)$ and $(b + c) \times a = (b \times a) + (c \times a)$;
5. There exist two elements in S, denoted by 0 and 1 (and called, respectively, the identity with respect to addition and the identity with respect to multiplication) such that $0 + a = a$ and $1 \times a = a \times 1 = a$;
6. There exist elements x and y in S such that $a + x = 0$ and $(b \times y) = 1$ if $b \neq 0$.

[Actually, condition (5) is implied by the remaining ones.]

A *Galois field* is simply a field that has a finite number of elements. These are named after the great mathematician Galois (who, unfortunately, died very young in a duel). The symbol GF(q) is used to denote a Galois field with q elements. The simplest example of a field is GF(2), which consists of the two identity elements 0 and 1 only. Treated under addition and multiplication as in ordinary arithmetic (together with the *additional* rule that $1 + 1 = 0$), these obviously satisfy the six conditions enumerated above for a field. The field implicitly used in the ordinary arithmetic in every day life is the field of rational numbers; this consists of all fractions of the form u/v, where v is nonzero and u and v are integers. This field, obviously, has an infinite number of elements and is thus *not* a Galois field.

It can be proven that GF(q) exists if and only if q is either a prime number, or a power of a prime number. When q is a prime number, the elements of GF(q) are the integers $0, 1, 2, \ldots, (q - 1)$, where addition or multiplication is conducted as usual except that the result is reduced modulo q; i.e., a *suitable* multiple of q is added to the results so as to obtain a number between 0 and $(q - 1)$. Thus, when $q = 11$, we have $(-5) \times (4) = (2)$ since the result of the usual multiplication of (-5) and (4) is (-20), and we must add 2×11 to (-20) to obtain a number between 0 and 10. Finally, $(5) \times (4)$

$= (9)$, since 5×4 is 20, and we must add $(-1) \times 11$ to it.

When $q = p^n$, where p is a prime number and n is an integer, the elements of GF(q) are polynomials of the form $a_0 + a\theta + \cdots + a_{n-1}\theta^{n-1}$, where θ is just a "symbol" (following the usual rules of grade school algebra), and the a_j are elements of GF(q), and where the addition and multiplication of polynomials are done following these same rules. However, whenever we obtain a term with θ^m, where $m \geqslant n$, we replace θ^n by $(b_0 + b_1\theta + \cdots + b_{n-1}\theta^{n-1})$, where the b's are certain elements of GF(q). As an example, it can be proven that when $q = 2$ and $n = 3$, we can take $b_1 = 0$ and $b_0 = b_2 = 1$. The elements of GF(2^3) are 0, 1, θ, $\theta + 1$, θ^2, $\theta^2 + 1$, $\theta^2 + \theta$, and $\theta^2 + \theta + 1$. Also,

$$(\theta^2 + 1)(\theta^2 + \theta) = \theta^4 + \theta^3 + \theta^2 + \theta$$
$$= \theta(1 + \theta^2) + \theta^3 + \theta^2 + \theta = \theta^2.$$

The polynomial in the b's with $(-\theta^n)$ added to it is called a *minimum function*.

Every nonzero element x in GF(q) satisfies the condition $x^{q-1} = 1$. If q is odd, this gives $x^{(q-1)/2} = 1$ or (-1), in which case x is respectively called a (*quadratic*) *residue* or *nonresidue*.

Galois fields arise frequently in the theory of statistical design of scientific experiments and in various branches of combinatorial mathematics.

(FRACTIONAL FACTORIAL DESIGNS)

J. N. SRIVASTAVA

GALTON, FRANCIS

Born: February 16, 1822, near Birmingham, England.

Died: January 17, 1911, in Haslemere, England.

Contributed to: statistical methods, genetics.

Sir Francis Galton came from an intellectual family; his mother was Violetta Darwin (aunt of Charles Darwin), and his grandfather Samuel Galton was a Fellow of the Royal Society (F.R.S.). He was the youngest of a family of nine brought up in a large house near Birmingham, where his father, Samuel Tertius Galton, ran a bank. He showed early intellectual powers; at 5 he could "add and multiply by all numbers up to 8" and "say all the Latin substantives and adjectives and active verbs." But he failed to qualify in medicine, as his father had hoped, or to complete an honors degree in mathematics at Cambridge. This mattered little, as he had independent means. In 1858 he married Louisa, sister of Montague Butler, a future Master of Trinity College, Cambridge.

He was a late developer, with published work beginning only at 28. His first interests were in exploration of South West Africa (later, Egypt and elsewhere), geography*, and meteorology*, subjects that were major interests throughout his life; he was the inventor of the word "anticyclone". He was on the management committee of Kew Observatory, and for much of his life served on the Council of the Royal Geographic Society. He was elected a fellow of the Royal Society for his geographical studies at the age of 38. From the age of 43 genetics* and the use of statistical methods for the study of all kinds of questions became major preoccupations. These statistics were sometimes of measurable quantities, such as the heights of parents and children. But quite often he used subjective evaluation (at one time he carried a device to record whether the women he met were pretty, plain, or ugly, reaching the conclusion that London women were some of the prettiest in Britain). He also used questionnaires, although the wording of the questions might hardly meet modern criteria. In his book *Natural Inheritance* he wrote:

> Some people hate the very name of statistics but I find them full of beauty and interest. Whenever they are not brutalized, but delicately handled by the higher methods, and are warily interpreted, their power of dealing with complicated phenomena is extraordinary. [2, p. 62]

His major contributions to statistics were in connection with genetics and psychology*. He used the normal (Gaussian) distribution* a great deal. Essentially, he supposed that where a character could be ranked but not directly measured, he could usually assign it a value corresponding to a normal deviate with the same expected rank. He first used the word *correlation** in his book *Hereditary Genius* [1] in a general statistical sense, saying that the characteristic of strong morality and moral instability are in no way correlated. In 1877 he introduced a numerical measure of *regression** or *reversion*, effectively the modern regression coefficient, as the average deviation from the mean of children of a given parent as a fraction of that parent's deviation. With help from a Cambridge mathematician, Hamilton Dixon, he explained this in terms of a bivariate (normal) distribution* with elliptical contours.[1] He obtained the axis of the ellipse, thus anticipating principal components. The fact that the regression of y on x differed from that of x on y worried him persistently until in 1877 by a flash of inspiration (at Naworth Castle) he saw that the solution was to normalize x and y in terms of their own variability. The regression coefficient he then called the "co-relation" (or correlation). Since it was a regression he called it r, whence the modern symbol r for correlation. (Galton's coefficient was not quite identical with the modern one, since he measured variability by interquartile distance rather than by standard deviation.)

Galton's mathematical and statistical ideas were often simple, though profound in their implications. In his later years he was a close friend of Karl Pearson*, then Professor of Applied Mathematics at University College, London, while in 1901 he gave a sum of £200 to found the Journal *Biometrika**. His Eugenics Record Office was later combined with Karl Pearson's Biometric Laboratory, to form what came to be the Galton Laboratory. Francis Galton's contributions to the advance of genetics and statistics did not end with his death, for he left £45,000 to found the Galton Chair of National Eugenics. Its first holder was Karl Pearson, who combined the studies of statistics and genetics in a single department at University College (still existing but as two separate departments).

Galton had in later life a very wide range of other interests, including psychology (memory, intelligence, perception, association, etc.), human faculty (visualization, hearing, etc.), education, fingerprints, and others to most of which he tried to adapt statistical methods and mechanical inventions. He unsuccessfully tried to do arithmetic by smell. Many of his ideas were well ahead of their time; he suggested a coin called a "groat" (= a present-day penny), referred to the three dimensions of space and one of time, and speculated on interplanetary communication by coded signals related to simple arithmetic.

He was the author of over 300 publications, including 17 books.

NOTE

1. A diagram used by Galton in discovering bivariate normal elliptical contours of equal density is reproduced as Fig. 2 in the article CORRELATION, together with further discussion.—*Ed.*

References

[1] Galton, F. (1869). *Hereditary Genius*. Macmillan, London. (Uses the word "correlation," but not with precise definition.)

[2] Galton, F. (1889). *Natural Inheritance*. Macmillan, London.

Bibliography

The first four papers by Galton give quantitative values to qualitative characters and the sixth and seventh introduce regression as a statistical concept.

Forrest, D. W. (1974). *Francis Galton, the Life and Work of a Victorian Genius*. Elek, London/Taplinger, New York. (Excellent, semipopular biography.)

Galton, F. (1872). *Fortnightly Rev.*, **12**. (Controversial application of statistical analysis to prayer.)

Galton, F. (1873). *Educational Times*. (First thoughts on the probability of survival of families.)

Galton, F. (1874). *Nature (Lond.)*, **9**, 342–343.

Galton, F. (1875). *Philos. Mag.*, **49**, 33–46.

Galton, F. (1885). *J. Anthropol. Inst.*, **15**, 246–263.

Galton, F. (1886). *Proc. R. Soc.*, **40**, 42–63.

Galton, F. (1889). *Proc. R. Soc.*, **45**, 135–145. (The origin of a precise definition of correlation.)

Galton, F. (1895). *Nature (Lond.)*, **51**, 319.

Galton, F. (1901). *Biometrika*, **1**, 7–10. (Recognition of biometry as a new, useful discipline.)

Galton, F. (1907). *Biometrika*, **5**, 400–404.

Galton, F. (1907). *Probability, the Foundation of Eugenics*. Henry Froude, London.

Pearson, K. (1914, 1924, 1930). *The Life, Letters and Labours of Francis Galton*. 3 vols. Cambridge University Press, Cambridge.

Watson, H. W. and Galton, F. (1874). *J. Anthropol. Inst. Gt. Brit. Ireland*, **4**, 138–144. (Introduces the Galton–Watson* problem of the chance of survival of families when the distribution of the number of children is known.)

(CORRELATION
HUMAN GENETICS, STATISTICS IN
PEARSON, KARL)

CEDRIC A. B. SMITH

GALTON DIFFERENCE PROBLEM

This deals with the determination of the expected value of the difference between the rth and $(r+1)$st order statistics*; it was considered by Karl Pearson* early in the twentieth century.

(ORDER STATISTICS
SPACINGS)

GALTON–WATSON PROCESS

The Galton–Watson process (or, more accurately, the Bienaymé–Galton–Watson process) is the fundamental example of a branching process*. Descendants of a single ancestor at time 0 produce offspring independently, and the probability distribution of the number of offspring of any one individual is identical with that of the initial ancestor. Denoting by p_r $(r = 0, 1, 2, \dots)$ the probability that any one individual has r offspring, then, if Z_n $(n = 0, 1, 2, \dots)$ is the total number of individuals in the nth generation $(Z_0 = 1)$, $p_r = P[Z_1 = r]$, and $G(s) = \sum_{r=0}^{\infty} p_r s^r$, $s \in [0, 1]$, is the probability generating function* of Z_1; the mean number of offspring per individual is, thus, $m = \sum_{r=1}^{\infty} r p_r = G'(1-)$. A more rigorous description of the stochastic process $\{Z_n\}$, $n \geqslant 0$, defines it as a Markov chain on the countable state space $\{0, 1, 2, \dots\}$, where $\{0\}$ is an absorbing state (corresponding to "extinction") and when $Z_n > 0$ the transition mechanism is described by

$$Z_{n+1} = X_1^{(n)} + X_2^{(n)} + \cdots + X_{Z_n}^{(n)},$$

where $X_i^{(n)}$, $i = 1, 2, \dots$; $n = 0, 1, 2, \dots$ are independently and identically distributed random variables, distributed as Z_1 (*see* MARKOV PROCESSES).

If we exclude the trivial case $p_1 = 1$, the criticality theorem* for the process asserts that if $m \leqslant 1$, extinction will occur with probability $q = 1$; but if $m > 1$, the probability q of ultimate extinction is the unique root of the equation $G(s) = s$ in the interval $0 \leqslant s < 1$, and thus $q < 1$. I. J. Bienaymé* gave a completely correct statement of it in 1845 [1], which has passed unnoticed till recently [2]; the usual designation for the process $\{Z_n\}$, $n \geqslant 0$, originates from the partly correct statement of the criticality theorem in 1873–1874 by F. Galton* and H. W. Watson [3].

From a theoretical aspect, the process $\{Z_n\}$, $n \geqslant 0$, is the most extensively studied example of both an absorbing Markov chain on a countable state space, and of a branching process, due to the functional iteration property [4] $G_n(s) = G(G_{n-1}(s))$, where $G_n(s) = \sum_{r=0}^{\infty} P[Z_n = r] s^r$. The theory of the process and its generalizations focuses on elegant limit theorems, which under moderate conditions describe the weak convergence of the distribution functions

$$P\left[\{Z_n / E(Z_n \mid Z_n > 0)\} \leqslant x \mid Z_n > 0\right]$$

as $n \to \infty$, originally developed by R. A. Fisher*, A. N. Kolmogorov, and A. M. Yaglom.

The description underlying the process $\{Z_n\}$, $n \geqslant 0$, has a breadth of applicability (*see* CRITICALITY THEOREM) which makes the

Galton–Watson process one of the most practically useful stochastic processes.

References

[1] Bienaymé, I. J. (1845). *Soc. Philom. Paris Extraits*, **5**, 37–39 (also in *L'Institut, Paris*, **589**, 131–132).

[2] Heyde, C. C. and Seneta, E. (1972). *Biometrika*, **59**, 680–683.

[3] Kendall, D. G. (1966). *J. Lond. Math. Soc.*, **41**, 385–406.

[4] Seneta, E. (1969). *Adv. Appl. Prob.*, **1**, 1–69. (Iteration-theoretic/functional equation standpoint.)

Bibliography

See the following works, as well as the references just cited, for more information on the Galton–Watson process.

Athreya, K. B. and Ney, P. (1972). *Branching Processes*. Springer-Verlag, Berlin. (This work and the book by Harris (1963) are standard references on the theory of branching processes.)

Harris, T. E. (1963). *The Theory of Branching Processes*. Springer-Verlag, Berlin.

Jagers, P. (1975). *Branching Processes with Biological Applications*. Wiley, London.

Mode, C. J. (1971). *Multitype Branching Processes*. American Elsevier, New York.

Sevastyanov, B. A. (1971). *Vetviashchiesia Protsessi. (Branching Processes)*. Nauka, Moscow. (Standard Russian-language reference.)

(BRANCHING PROCESSES
CRITICALITY THEOREM
MARKOV PROCESSES
STOCHASTIC PROCESSES)

E. SENETA

GAMBLER'S RUIN *See* GAMBLING, STATISTICS IN

GAMBLING, STATISTICS IN

THE HISTORY OF PROBABILITY IN GAMBLING

Games of chance* based on tossing astragali (knucklebones) are at least 5000 years old. Gambling on the outcome of these games was common from early times. No records have been found of the calculation of probabilities (nor indeed of the concept of probability) in games of chance until the fifteenth century. There is evidence that in the preceding two hundred years elementary combinatorial calculations as to the number of outcomes in dice-throwing games were attempted. It was Cardano who introduced probabilistic ideas into the mathematical analysis of gambling games. Although his *Liber de Ludo Aleae* was written in about 1526, it was first published with his collected works in 1663, 87 years after his death. For a translation, see Ore [16]. Later, Galileo, Pascal*, Fermat*, Huygens*, the Bernoullis*, De Moivre* [4], and others contributed to the development of modern probability theory through the analysis of gambling problems; to such problems probability theory owes its origins and rapid early growth. For more detailed discussions of the early developments, see Todhunter [24], David [3], and Pearson and Kendall [18]. Much of this early work, while setting the foundations of probability* theory, involved only simple arithmetic calculations. The problems of "duration of play" and "gambler's ruin" arose as natural extensions of the "problem of points*" discussed by Pascal and Fermat in their correspondence over problems raised by the Chevalier de Méré.

In the general form of this problem two players A and B play a series of games with respective probabilities of winning p and $q = 1 - p$. Given, after any game, that A has capital a and B has capital b and that the winner of a game acquires 1 unit of capital from the loser, the question is: What are the probabilities $P_A(a,b,n)$ and $P_B(a,b,n)$ that A or B, respectively, will win (ruin his opponent) in n games? An account of the contributions of Huygens, De Moivre, de Montmort, and the Bernoullis to the solution is given by Thatcher [23]. Most textbooks on probability contain some account of modern approaches (e.g., difference equations*, generating functions*, and martingales*) to the solution of the gambler's ruin problem (see, e.g., Feller [10, Chap. XIV; 11, Chap. VI]).

Further historical information may be found in many books, especially Todhunter [24] and David [3]. For Karl Pearson's fascinating discussion of the history of the subject as he perceived it against the changing background of intellectual scientific and religious thought, see E. S. Pearson [17].

Much of the modern mathematical theory of gambling deals with problems abstracted from those of the real gambler and "in spite of the probabilist's tendency to invoke gambling imagery, works of probability have little influence, for good or bad, on the gambling man" [8]. However, there have been many contributions on gambling problems which are of considerable mathematical interest in their own right (see, e.g., MARTINGALES and ONE- AND TWO-ARMED BANDIT PROBLEMS [22]). Like the "duration of play" problem, the latter has led to significant developments in the study of sequential medical trials*.

Further, it was not until the development of the necessary mathematical framework that some quite basic general results regarding gambling systems were formally proved. Moran [15] points out that Doob [5] was the first to give an explicit proof that a successful gambling system is impossible.

SYSTEMS AND FALLACIES (SEE EPSTEIN [9])

Although no system can lead ultimately to an increase in expected "gain" in a pure gamble, many have been proposed. The fallacy of such systems as the *martingale* (doubling up), the *Labouchère** (cancellation; see Downton [6]) and *D'Alembert* systems has been well exposed. The first of these involves a geometric progression of stakes in a sequence of losing bets; the other two involve arithmetic progression. In a fair game no such system can affect the expected gain of zero. These three and many other systems do, however, lead to a high probability of a small profit as opposed to a low probability of a large loss (the antithesis of a lottery!). Also famous is the *St. Petersburg paradox**, in which the player receives a return of 2^n units should a head first occur at the nth toss of a coin. This leads to an infinite expected return which suggests that a fair entrance fee to play the game should also be infinite. Of course, this ignores the finite wealth of both the player and the bank. Allowance for these resolves the paradox of infinities. Other systems based on the "maturity of chances" or so-called "law of averages" are common; they are fostered by the belief that such an event as a long run of heads in coin tossing should be balanced by a greater chance of a tail on the next toss. The fallacy is due to a popular misconception of the law of large numbers*.

POPULAR CASINO GAMES

Roulette is the oldest and perhaps the most popular of the casino games. A ball is spun around a wheel and eventually lands in one of a set of numbered slots. Bets are placed on numbers or combinations of numbers. There are 36 numbers on which one can bet and the odds offered are based on relative frequencies among these 36 numbers. Thus a bet on a single number attracts odds of 35–1 and a bet on even attracts odds of 1–1. The house advantage is due to a zero (or zeros). Perhaps because of the simplicity and the lack of skill involved in this game of pure chance, roulette more than any other casino game has attracted many systems. As already mentioned, systems cannot affect the expected gain (which is negative). However, they can affect the probability of finishing ahead (or of winning a desired sum) after a session of games (see Dubins and Savage [8], Downton and Holder [7], Anderson and Fontenot [1], and Downton [6]; also, *see* GAIN RATIO).

Craps, especially in the United States, is the most popular of the casino dice games. It is played with two dice. If a player's score (sum of numbers showing on the two dice) is 7 or 11, he or she wins (a natural); if it is 2, 3, or 12, he or she loses. Any other score is designated a point; following such a score a player will continue to throw the dice until he or she either scores 7 (and loses) or repeats the same point (not necessarily with the same combination of numbers (e.g., 4 and 5 represent the same "point," 9, as do 3 and 6), in which case he wins. The basic and most common bet in craps is the pass line bet, which is a bet on winning according to the sequence above. If we designate by p_i the probability of a score of i, then the probability of winning on the pass line is

$$p_7 + p_{11} + \frac{p_4^2}{p_4 + p_7} + \frac{p_{10}^2}{p_{10} + p_7} + \frac{p_5^2}{p_5 + p_7} + \frac{p_9^2}{p_9 + p_7} + \frac{p_6^2}{p_6 + p_7} + \frac{p_8^2}{p_8 + p_7}$$

$$= \frac{1}{36}\left(6 + 2 + 2 \times \frac{3^2}{9} + 2 \times \frac{4^2}{10} + 2 \times \frac{5^2}{11}\right)$$

$$= \frac{244}{495} \approx 0.493.$$

Thus there is a small expected loss from this even-money bet. A variety of other bets is possible in craps. With the exception of the "odds" bet, which may be made only as an additional bet (and has zero expectation), all have negative expected gain.

LOTTERIES

The first public lottery awarding money prizes was the Lotto de Firenze, established in 1530. Most lotteries and games like bingo (or housie) and keno offer no significant decision-making process, no subjective assessment, and are games of pure chance* with substantial negative expectation.

Some lotteries involve the player in choosing a set of, e.g., 6 numbers from 40. The prize numbers are drawn at random. The probabilities of winning a prize are independent of choice of number, but the prizes themselves, being a fixed proportion of the total pool shared among those who chose the winning numbers, are random. Thus astute choice of unpopular numbers may increase the expected return, although the probability of winning remains the same. For a simple but graphic demonstration of this, see Lewis [14].

Keno is the most popular lottery-type casino game, at least in the United States. In Keno there are 80 numbers to choose from and a player may choose from 1 to 15 numbers. The payoff is, of course, a function of how many numbers the player chooses; 10 is the most popular number. The probability of k matches between the n numbers a player selects and the 20 numbers drawn by the casino is given by

$$p(k) = \binom{20}{k}\binom{60}{n-k} \Big/ \binom{80}{n}.$$

House odds increase with k. On a typical bet in which the player selects 10 numbers (a "10-spot ticket") these may vary from 0 (no payoff) for k between 0 and 4 through even money for $k = 5$ to 25,000 to 1 for $k = 10$. The odds vary to some extent among casinos and reflect only qualitatively the relative chances of the various outcomes, as given by the formula above. Keno is very profitable for casinos; the expected loss is usually on the order of 20%.

Football pools bear similarities to such lotteries. Here the "random" mechanism generating prizes is the outcome of a series of football games. There is clearly an element of skill in the prediction of the outcome of a game, although many gamblers generate their predictions through some random or pseudo-random mechanism. For a discussion of the element of skill in this context, see Peterson [20].

HORSE RACING

Like roulette, horse racing has attracted enormous numbers of systems. Unlike roulette, however, there are many unknown elements in horse racing. Accordingly, the usefulness of any system that relies on an assessment of form relative to the odds offered or on subjective assessment that, e.g., short-priced horses will offer better value in the last race of a day (because many gamblers will be attempting to recoup their losses by backing a long-shot) can only be tested empirically. Statistical analysis may well confirm that some of the thousands of systems which have been advocated are successful; of course one would have to be conscious of the type I error* when carrying out the thousands of hypothesis* tests associated with such analyses.

In betting on the parimutuel or totalizator system the payoff odds on a horse are inversely proportional to the amount of money wagered on a horse. The amount of money bet on a horse directly reflects the confidence the betting population as a whole has in that horse. That the consensus of subjective probabilities aggregated in this way closely reflects the true probabilities of a horse winning (or being placed) has been tested and confirmed by Hoerl and Fallin [13], at least in regard to the betting population at two major race tracks in the United States. Any discrepancies between the parimutuel odds and the true odds *could* be turned to advantage in a betting system. An extreme example of this is the manipulation of odds in a betting coup or builder play to artificially lengthen the course odds on a horse and then back it for a large sum off course (see Epstein [9, p. 292]). Of course, this relies on a gambling system in which off-course payoff odds are determined by on-course odds. In some countries bookmakers both on and off course frame their own odds (in such a way that the sum of nominal probabilities exceeds 1; the bookmaker, like the totalizator operator and the government, takes his or her cut). Some bookmakers offer each-way bets (one unit is bet on a win, a second unit is bet on the hope of finishing among the first three). One-fourth of the win odds are offered for this second bet. For a discussion of each-way odds, see Plackett [21] and Barton [2]. Again, any system based on such odds would require empirical testing.

References

[1] Anderson, L. R. and Fontenot, R. A. (1980). *J. R. Statist. Soc. A*, **143**, 33–40.

[2] Barton, N. G. (1980). *Aust. Math. Soc. Gaz.*, **7**, 38–45.

[3] David, F. N. (1962). *Games, Gods and Gambling*. Charles Griffin, London.

[4] De Moivre, A. (1738). *The Doctrine of Chances; or, a Method of Calculating the Probabilities of Events in Play*. (2nd ed., reproduced by Frank Cass, London, 1967.)

[5] Doob, J. L. (1936). *Ann. Math.*, **37**, 363–367.

[6] Downton, F. (1980). *J. R. Statist. Soc. A*, **143**, 41–42.

[7] Downton, F. and Holder, R. L. (1972). *J. R. Statist. Soc. A*, **135**, 336–356.

[8] Dubins, L. E. and Savage, L. J. (1965). *How to Gamble If You Must: Inequalities for Stochastic Processes*. McGraw-Hill, New York.

[9] Epstein, R. A. (1977). *The Theory of Gambling and Statistical Logic*. Academic Press, New York. (An excellent overview of theory and practice and a useful source of references.)

[10] Feller, W. (1950). *An Introduction to Probability Theory and Its Applications*, Vol. 1. Wiley, New York.

[11] Feller, W. (1966). *An Introduction to Probability Theory and Its Applications*, Vol. 2. Wiley, New York.

[12] Graham, V. L. and Tulcea, C. I. (1978). *A Book on Casino Gambling Written by a Mathematician and a Computer Expert*. Van Nostrand Reinhold, New York. (An elementary introduction for the novice at both gambling and mathematics.)

[13] Hoerl, A. E. and Fallin, H. K. (1974). *J. R. Statist. Soc. A*, **137**, 227–230.

[14] Lewis, K. (1979). *Teach. Statist.*, **1**, 91–92.

[15] Moran, P. A. P. (1979). In *Chance in Nature*, Australian Academy of Science, pp. 65–71.

[16] Ore, O. (1953). *Cardano, the Gambling Scholar*. Princeton University Press, Princeton, N.J. (Includes a translation of Cardano's historically significant work *De Ludo Aleae*. Ore also quotes other scholars' views on Cardano's work: de Montmort "here one finds only erudition and moral reflexions"; Todhunter, "so badly written as to be scarcely intelligible.")

[17] Pearson, E. S., ed. (1978). *The History of Statistics in the 17th and 18th Centuries against the Changing Background of Intellectual, Scientific and Religious Thought. Lectures by Karl Pearson Given at University College London during the Academic Sessions 1921–1933*. Charles Griffin, London.

[18] Pearson, E. S. and Kendall, M. G., eds. (1970). *Studies on the History of Statistics and Probability*. Charles Griffin, London.

[19] Pearson, K. (1894). *Fortnightly Rev.*, **55**, 183–193.

[20] Peterson, A. W. (1952). *J. R. Statist. Soc. A*, **115**, 199–218.

[21] Plackett, R. L. (1975). *Appl. Statist.*, **24**, 193–202.

[22] Robbins, H. (1956). *Proc. Natl. Acad. Sci. U.S. A.*, **42**, 920–923.

[23] Thatcher, A. R. (1957). *Biometrika*, **44**, 515–518.

[24] Todhunter, I. (1865). *A History of the Mathematical Theory of Probability from the Time of Pascal to That of Laplace*. Macmillan, Cambridge.

(BLACKJACK
GAMES OF CHANCE,
EXCLUDING BLACKJACK
MARTINGALES)

DAVID GRIFFITHS

GAMES OF CHANCE (EXCLUDING BLACKJACK)

In classical times, the principal instrument of play (or randomizing agent) in games of chance was probably the astragalus or knucklebone. From the astragalus evolved the die. Many of the popular games of chance today involve dice or playing cards; the latter were apparently not introduced into Europe until the fourteenth century, although their place and time of origin is unknown. Certainly it would appear to be much later than that of the die; dice were in use around 3000 B.C. [1, 2, 8].

DICE GAMES

Perhaps the most famous dicing problem (and historically important in terms of the development of probability theory) is that posed by de Méré to Pascal*. Others, including Cardano and Galileo, had used probabilistic arguments earlier to solve simple dicing problems. Correspondence between Pascal and Fermat* led to the answer to de Méré's question concerning the relative odds of throwing at least one six with four throws of one die as compared with throwing at least one pair of sixes with 24 throws of two dice [the probabilities being $1 - (1 - \frac{1}{6})^4$ and $1 - (1 - \frac{1}{36})^{24}$, respectively]. Their correspondence also led to the solution of the "problem of points*" and laid down a basis for the development of probability theory.

The probabilistic analysis of many dice games, e.g., the common casino game *Craps*, is very simple (*see* GAMBLING, STATISTICS IN). The outcome of the game depends entirely on chance*, although players have a choice of bets (all with negative expectations). There do exist dice games with quite complex strategies (e.g., backgammon and the simple children's game PIG). No optimal strategy for either of these games is known, although analysis of end game positions is possible [6, 9]. For a recent note on end-game doubling strategies in backgammon, see Tuck [11].

Backgammon is essentially a "race" game, as are many popular parlor dice games, such as Snakes and Ladders. Unlike backgammon, most such parlor games are games of pure chance. Probabilistic analysis of such games or more complex popular games such as Monopoly seems of little interest.

CARD GAMES

Following the publication of Hoyle's *Short Treatise on Whist* published in 1742 (see ref. 3), there has been much written on the mathematics of card games, particularly blackjack*, poker, and contract bridge (see ref. 10). Many card games involve memory, mixed playing strategies (in which the decision of each play is governed by a randomization* procedure, this being preferable to any single "pure" strategy (*see* GAME THEORY) and assorted other skills; as such, card games are generally more interesting (and more difficult) than most dice games both to the player and the analyst. There are several interesting statistical questions relating to randomness of shuffling, matching problems*, and games of patience (e.g., the probability of "getting out"). Here we shall discuss only games involving two or more players. Some casino games, e.g., *trente et quarante*, are games of pure chance and of little interest. Trente et quarante is played with six packs of cards and involves dealing cards into two rows (noire, rouge) until the cumulative totals exceed 30. Probabilities associated with this game were first calculated by Poisson*. Baccarat Chemin de Fer bears similarities to blackjack* but is a much simpler game [4, 5, 7].

A *poker* hand consists of 5 cards dealt from a pack of 52. Thus the number of distinct poker hands is $\binom{52}{5} = 2{,}598{,}950$. It is a matter of elementary combinatorics* to determine the probabilities of various poker hands. For example, the number of hands with (exactly) one pair is $\binom{13}{1}\binom{4}{2}\binom{12}{3}\binom{4}{1}^3 =$ 1,098,240 and hence the probability of a pair is 0.42257. In draw poker a player may discard some cards and replace these from the pack. The probabilities of various improvements to the hand are also readily calculated. These two sets of probabilities are the basic tools in the analysis of poker. Of course, hands of several players are dependent. This, combined with partial information regarding other players' holdings (either directly, as in stud poker, where some cards are exposed face up or indirectly through drawing and betting—both of which may involve bluffing) renders the determination of an optimum strategy sufficiently complicated that many mathematicians have constructed simpler models of poker. Analyses of these models indicate near-optimal strategies for the real game.

Contract bridge involves dealing 13 cards to each of four players from a 52-card pack. That this may be achieved in some 5×10^{28} ways indicates the impossibility of exhaustive enumeration in analysis of the game. However, knowledge of probability distributions of suit lengths, high card counts, and the like (in a hand and in a partnership) is an invaluable guide to the development of bidding and playing strategies. Also invaluable is a knowledge of residual distributions after a player has taken account of his or her own hand. Evaluation of hands is essential for the development of bidding systems. A good system of evaluation will be characterized by its likelihood of leading to an optimal bidding sequence. The relationship between cards held in a particular suit and trick-winning potential (in offense or defense) is a test of a good scoring system, which should also aim at leading to bids that give maximal information to one's partner but as little as possible to one's opponents. Knowledge of probability distributions of two-hand (or more, as the tricks are played) residues is crucial to the choice of plays, although the sheer weight of combinatorial possibilities together with the impossibility of precisely quantifying the information in bidding and progressive play renders the game immune to complete mathematical exposition.

OTHER GAMES OF CHANCE

For a discussion of roulette, *see* GAMBLING, STATISTICS IN. Many other common games of chance, such as dominoes, mastermind, battleships (and other hide-and-seek games), are amenable to at least partial statistical analysis. Even with a game like Scrabble it is possible to model the distribution of the number of pieces remaining in each opponent's hand when one player goes out and to empirically (if not theoretically) examine the bias due to order of play.

References

[1] David, F. N. (1955). *Biometrika*, **42**, 1–15. (The first in the *Biometrika* series of historical articles.)

[2] David, F. N. (1962). *Games, Gods and Gambling*. Charles Griffin, London. (An excellent review of the history of probability theory and its gambling links.)

[3] Dawson, L. H. (1950). *Hoyle's Games Modernised*, 20th ed. Routledge & Kegan Paul, London.

[4] Downton, F. and Lockwood, C. (1975). *J. R. Statist. Soc. A*, **138**, 228–238.

[5] Downton, F. and Lockwood, C. (1976). *J. R. Statist. Soc. A*, **139**, 356–364.

[6] Epstein, R. A. (1977). *The Theory of Gambling and Statistical Logic*. Academic Press, New York. (A very thorough survey of the field, including interesting discussion of the mathematics of games of pure skill; useful bibliography.)

[7] Kemeny, J. G. and Snell, J. L. (1957). *Amer. Math. Monthly*, **64**, 465–469.

[8] Kendall, M. G. (1957). *Biometrika*, **44**, 260–262.

[9] Ollis, A. and Griffiths, S. (1980). *Math. Gaz.*, **64**, 283–286.

[10] Scarne, J. (1961). *Scarne's Complete Guide to Gambling*. Simon and Schuster, New York. (Comprehensive but full of anecdotes and exaggerated

claims as to Scarne's role in the mathematical development.)

[11] Tuck, E. O. (1981). *Math. Sci.*, **6**, 43–61.

(BLACKJACK
CHANCE
GAMBLING, STATISTICS IN)

DAVID GRIFFITHS

GAME THEORY

The theory of games is concerned with situations involving conflict and/or cooperation. It consists of a variety of mathematical models formulated to study different sorts of competitive interactions. Such events arise naturally with a multiplicity of participants having different preferences or objectives and alternative choices of action available to them. The theory has been used to analyze a number of different types of strategic encounters which arise in fields such as economics, political science*, operations research*, and military science. Although much of the terminology in game theory relates to that used in common "parlor games," the theory is even more useful for modeling other occurrences which frequently take place in social activities and in decision making.

Suggestions of the need for a theory of games as well as a few illustrations of a game theoretical nature were presented in the eighteenth and nineteenth centuries. A few general mathematical theorems and the analysis of some special classes of games appeared between 1912 and 1944. Nevertheless, modern game theory began in 1944 with the publication of the monumental volume *Theory of Games and Economic Behavior* [7] by the mathematician John von Neumann and the economist Oskar Morgenstern.

Three general and abstract formulations for the analysis of game-type problems were presented in their book, and most investigations since then relate rather closely to one or more of these models; the *extensive* (or tree) form, the *normal* (or strategic or matrix) form, and the *characteristic function* (or cooperative or coalitional) form of a game. Games are also classified according to the number of players; the number of strategies or options available to the players; the different states of information involved; whether they are zero-sum or general-sum, cooperative or noncooperative, allow side payments or not; and whether they involve single or repeated play. Game theory is concerned with different kinds of uncertainty* and chance events; it differs from traditional statistics and probability theory in that two or more decision makers with different preferences and goals are involved. It is concerned with determining optimal strategic behavior, equilibrium outcomes, stable results in bargaining, coalitional stability, equitable allocation, and similar resolutions of group differences.

STRATEGIES AND PAYOFFS

An *n-person game in normal form* is merely a list of options, called *pure strategies*, available to each of the *n players*, together with a rule for specifying the payoff to each player once the specific choices are selected. If the n players are denoted by $1, 2, \ldots, n$ and their respective pure strategy spaces by S_1, $S_2, \ldots, S_n$, then a *game* (in normal form) is determined by a real-valued function

$$F : S_1 \times S_2 \times \cdots \times S_n \to \mathbb{R}^n,$$

where $\mathbb{R}^n$ is the set of all n-dimensional vectors with real components. The real number $F_i(s_1, s_2, \ldots, s_n)$ is the *payoff* to player i when each player j selects the particular strategy $s_j \in S_j$, and may very well represent i's "utility*" for some nonnumerical outcome. It is assumed, in the initial theory, that the sets S_i and the payoff function F are known to each player. Chance events can be incorporated by including a *chance player*, denoted by 0, whose strategy is to select from a "strategy space" S_0 according to a known probability distribution. One then works with *expected* payoffs. The goal of each player is to maximize his or her payoff, and the player may select his or her pure

Table 1

				II		
	1	2	...	j	...	n
1	a_{11}	a_{12}	...	a_{1j}	...	a_{1n}
2	a_{21}	a_{22}	...	a_{2j}	...	a_{2n}
I ...			...			
i	a_{i1}	a_{i2}	...	a_{ij}	...	a_{in}
...			...			
m	a_{m1}	a_{m2}	...	a_{mj}	...	a_{mn}

strategy according to some probability distribution in an attempt to optimize the expected payoff in light of the options available to the other players.

There is a well-developed and useful theory for finite *two-person, zero-sum* games (in normal form), also called *matrix* games. The two players will be denoted by I and II, and one player's gain is always the other player's loss, i.e.,

$$F_{\text{I}}(i, j) = -F_{\text{II}}(i, j)$$

for all strategies $i \in S_{\text{I}}$ and $j \in S_{\text{II}}$. The payoff function F can be represented by an $m \times n$ matrix $\mathbf{A} = [a_{ij}]$ (see Table 1). Player I has m (pure) strategies $1, 2, \ldots, i, \ldots, m$, which correspond to the rows of **A**, and II has n strategies $1, 2, \ldots, j, \ldots, n$, which correspond to the columns of **A**. The matrix entry a_{ij} is the payoff from player II (the *column* player) to player I (the *row* player) when I selects strategy i and II selects strategy j.

Example 1. Colonel Blotto has two divisions of troops which he can allocate to either of two strategic mountain passes, P_1 and P_2. The Enemy commander also has two divisions available to deploy. If one sends more divisions to a given pass than his opponent, he wins one point for taking the pass plus one point for each opponent's division he captures at the pass; he loses a point for each pass or division the opponent captures. This game is described by the 3×3 matrix in Table 2, where the three pure strategies (i, j), with $i + j = 2$, indicate that i divisions were sent to the first pass P_1 and j divisions to the second pass P_2. The payoffs in the matrix are gains to Blotto and losses to the Enemy. It is clear that each antagonist should send two divisions to the same pass (i.e., concentrate his forces). The resulting payoff to each player is 0, the *value* of this *symmetric* game, i.e., one with a skew-symmetric payoff matrix **A**. The strategy (2, 0), or (0, 2), is an *optimal strategy* for Colonel Blotto, as well as for the Enemy. A triple consisting of an optimal strategy for the first player, an optimal strategy for the second player, and the value of the game is a *solution* for the game. In this example the value 0 is simultaneously the maximum of the individual row minima (called the *maximin*) and the minimum of the individual column maxima (the *minimax*).

Table 2

			Enemy	
		(2, 0)	(1, 1)	(0, 2)
Blotto	(2, 0)	0	1	0
	(1, 1)	−1	0	−1
	(0, 2)	0	1	0

Example 2. Consider the Colonel Blotto game in which Blotto has four divisions, the Enemy has three divisions, and there are two mountain passes. The strategies and payoffs

Table 3

			Enemy			Row Minima
		(3, 0)	(2, 1)	(1, 2)	(0, 3)	
Blotto	(4, 0)	4	2	1	0	0
	(3, 1)	1	3	0	−1	−1
	(2, 2)	−2	2	2	−2	−2
	(1, 3)	−1	0	3	1	−1
	(0, 4)	0	1	2	4	0
Column Maxima:		4	3	3	4	

are given in Table 3 together with the row minima and the column maxima. Note that the maximin value of 0 for Blotto is less than the minimax value of 3 for the Enemy. Blotto can guarantee a payoff of at least 0 by using pure strategies (4, 0) or (0, 4), whereas the Enemy can keep his *losses* down to 3 by playing his pure strategy (2, 1) or (1, 2). Since the maximin $= 0 < 3 =$ minimax, this game does not have a solution in *pure* strategies. In this case the players can resort to statistical considerations and increase their gains in terms of *expected* values if they *randomize* over their respective strategy sets. The selection of pure strategies according to a given probability distribution is called a *mixed strategy*. If Blotto selects the *optimal* mixed strategy $\mathbf{x}^0 = (\frac{4}{9}, 0, \frac{1}{9}, 0, \frac{4}{9})$, i.e., plays his five listed strategies with these respective probabilities, then his *expected* payoff will be $\frac{14}{9}$. If the Enemy picks his *optimal* mixed strategy $\mathbf{y}^0 = (\frac{1}{18}, \frac{4}{9}, \frac{4}{9}, \frac{1}{18})$, he can hold his expected losses down to at least the game *value* $v = \frac{14}{9}$; each component of the vector $\mathbf{A}(\mathbf{y}^0)^t$, where t denotes the *transpose* (i.e., the column vector obtained from the row vector $\mathbf{y}^0$), is at most $\frac{14}{9}$. The triple $(\mathbf{x}^0, \mathbf{y}^0, v)$ is a *solution* (in mixed strategies) for this game. Note that a pure strategy is a mixed strategy in which all the "weight" is placed on one pure strategy.

The famous *minimax theorem* proved by John von Neumann in 1928 states that for any (two-person, zero-sum) matrix game given by an $m \times n$ matrix $\mathbf{A}$, there are m-dimensional and n-dimensional probability vectors, $\mathbf{x}^0$ and $\mathbf{y}^0$ respectively, and a value $v = \mathbf{x}^0\mathbf{A}(\mathbf{y}^0)^t$ such that

$$\mathbf{x}\mathbf{A}(\mathbf{y}^0)^t \leqslant v \leqslant \mathbf{x}^0\mathbf{A}\mathbf{y}^t$$

for all m- and n-dimensional probability vectors $\mathbf{x}$ and $\mathbf{y}$. That is, there is an optimal mixed strategy for each player so that player I can obtain the payoff of v in expected values, whereas player II can simultaneously hold his expected losses down to v. Solving a given game $\mathbf{A}$ for a solution $(\mathbf{x}^0, \mathbf{y}^0, v)$ is equivalent to solving a pair of dual linear programs, and is usually accomplished by employing one of the known algorithms for this purpose, such as the simplex method of George B. Dantzig (1947) or the method of Leonid G. Khachiyan (1979). In fact, the duality theorem in linear programming*, first observed by von Neumann (1947), is equivalent to the minimax theorem.

In 1950, John F. Nash extended the concept of optimal strategies for matrix games to the notion of *equilibrium strategies* for n-person or non-zero-sum games in normal form. The idea of equilibrium is basic to any notion of solution for such games when they are played *noncooperatively*, i.e., without any communication or without the ability to make binding agreements. A set of n mixed strategies $x_1, x_2, \ldots, x_n$ for each of the players $1, 2, \ldots, n$ is in *equilibrium* if none of the players can unilaterally change his or her strategy to another one which will cause the player to increase his or her expected payoffs while the other players continue to use the same strategies. In other words, each player is playing optimally under the condition that none of the other players will alter their strategy choice. This does not imply that some players may not benefit if two or more of them were to alter their strategies. Nash proved that every n-person general-sum game in normal form has at least one equilibrium "point" in mixed strategies. Such equilibria turn out to be "fixed points" of certain mappings. As such, they can be computed approximately by means of "path-following" algorithms for approximating fixed points. The two-person general-sum games in normal form are often referred to as *bimatrix games*.

Example 3. The best known bimatrix game is the *prisoner's dilemma*, given in Table 4. Each of the two players, I and II, can choose to cooperate (C) or to defect (D). If they

Table 4

		II: C	II: D
I	C	(5, 5)	(0, 10)
	D	(10, 0)	(2, 2)

both cooperate, each obtains a payoff of 5 units; if both defect, each receives 2 units. If one cooperates and the other defects, the former obtains 0 while the latter receives a payoff of 10. Note that in the two-dimensional payoff vectors, the first component goes to player I while the second is for II. For each player, the strategy choice D "dominates" C in the sense that a higher payoff results, independent of what one's opponent chooses. The strategy pair (D, D) is the unique equilibrium outcome for this game (in pure or mixed strategies), and its use results in the payoffs $(2, 2)$. The most likely result of a one-shot play of this frequently occurring noncooperative game is that both players defect and settle for just 2 units each.

On the other hand, if the prisoner's dilemma is played cooperatively in the sense that agreements can be made and enforced, the likely outcome is the strategy pair (C, C) with a payoff of 5 to each player. If the prisoner's dilemma is to be played repeatedly between the same two players, it is likely that they would soon see the benefits of both playing the cooperating strategy C.

Example 4. A man M and a woman W must decide whether to attend a basketball game B or a gymnastics meet G this Saturday night. Each must indicate their choice in advance, and if they do not select the same event, they will instead spend the evening at home, denoted by H. He prefers B to G to H and she prefers G to B to H, as indicated by the bimatrix in Table 5.

The strategy pairs (B, B) and (G, G) are both pure strategy equilibrium points, but M prefers the former and W prefers the latter. There is also a unique symmetric mixed strategy equilibrium in which M plays the probability vector $(\frac{4}{5}, \frac{1}{5})$ and W plays $(\frac{1}{5}, \frac{4}{5})$, with an expected payoff of $\frac{4}{5}$ each. However, there may be little incentive for either to play this mixed strategy, since if either one does so, then any pure (or mixed) strategy for the other player achieves this same expected return to the latter and some mixed strategies actually achieve a higher "maximin" value. On the other hand, if the two players were allowed to cooperate, they would likely decide to play the "correlated" mixed strategy which selects only from the payoffs $(4, 1)$ and $(1, 4)$ with equal probabilities of $\frac{1}{2}$ each.

Table 5

		W: B	W: G
M	B	(4, 1)	(0, 0)
M	G	(0, 0)	(1, 4)

The equilibrium concept of Nash is fundamental in the study of n-person games in normal form with $n > 2$ when such games are played noncooperatively.

The Bargaining Problem

The general two-person cooperative game, or *bargaining problem*, consists of a convex compact set R in Euclidean 2-space together with a *status quo* point (u^0, v^0) in R (see Fig. 1). The points (u, v) in R give (expected) payoffs to the two players, respectively, which they can realize if they cooperate, where (u^0, v^0) is the current payoffs or "fallback position" if they fail to come to an agreement. The players should cooperate to the extent of selecting a point from the "Pareto surface" of R, i.e., a point on the "northeast" boundary A of R. However, their interests are in opposition when they move along this optimal frontier A. Several methods have been proposed for selecting a

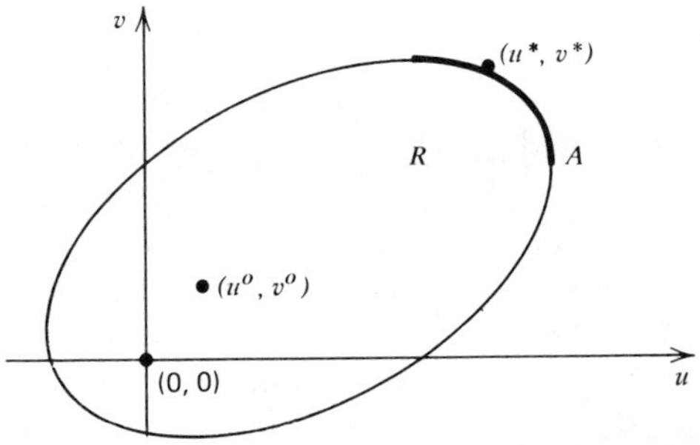

Figure 1

particular point (u^*, v^*) on A, and such theories often have an axiomatic basis. Nash suggested that (u^*, v^*) should be chosen from the set A so as to maximize the product $(u^* - u^0)(v^* - v^0)$, and showed that this is the unique point satisfying a few reasonable conditions, or axioms. Several alternative theories have also been proposed for selecting such a solution point. Some but not all of these theories can be extended to n-person games.

Multiperson Cooperative Games

When three or more players interact in a cooperative manner, the formation of coalitions is a crucial aspect. Economic cartels, international alliances, voting blocs, and other sociological groupings are common phenomena. The worth or power of a potential coalition is of prime importance and can often be represented by a numerical value. The first comprehensive model for analyzing such games was the theory of *solutions* (or *stable sets*) for *n-person games in characteristic function form* presented by von Neumann and Morgenstern in 1944 [7]; several variants and different solution concepts have since been introduced. The von Neumann–Morgenstern model and three other solution concepts will be described. A typical model consists of four parts: a method for assigning numbers or sets of outcomes to coalitions, a set of n-dimensional vectors A which is the set of all realizable distributions of the payoffs among the n players, a "preference relation" or other criteria for selecting the more "desirable" or "stable" outcomes, and some point or subset of A (not necessarily unique) which is the resulting solution concept for the game.

Let $N = \{1, 2, \ldots, n\}$ be a set of *n players* denoted by $1, 2, \ldots, n$; and let $v : 2^N \to \mathbb{R}$ be a real-valued *characteristic function* which assigns a real number $v(S)$ to each nonempty *coalition* in N (i.e., $S \subset N$). 2^N denotes the set of all subsets of N, and one normally assumes that $v(\emptyset) = 0$ for the empty set $\emptyset$. An *n-person game (in characteristic function form)* is merely a pair (N, v). It is not uncommon to assume further that the function v is *superadditive*; i.e., whenever $S \cap T = \emptyset$,

$$v(S \cup T) \geqslant v(S) + v(T),$$

but much of the theory does not require this. Next, one considers the set of all realizable distributions of the value (wealth or power) $v(N)$ among the n players. This set of *imputations* is defined as

$$A = \{\mathbf{x} : x_1 + x_2 + \cdots + x_n = v(N) \text{ and } x_i \geqslant v(\{i\}) \text{ for all } i \in N\},$$

where $\mathbf{x} = (x_1, x_2, \ldots, x_n)$ and each $x_i \in \mathbb{R}$. These conditions state that the amount $v(N)$ is available to be distributed among the n players and no player i will accept less than the amount $v(\{i\})$ which he or she can obtain in the singleton coalition consisting of just himself or herself.

The players in a particular coalition S would prefer an imputation $\mathbf{x}$ over another imputation $\mathbf{y}$ if $x_i > y_i$ for each $i \in S$; they could reasonably argue for x if they were not asking for more than their coalitional value $v(S)$. This leads one to define that $\mathbf{x}$ *dominates* $\mathbf{y}$ if and only if $x_i > y_i$ for each $i \in S$ and $\sum_{i \in S} x_i \leqslant v(S)$, where S is some (nonempty) subset of N. We write $\mathbf{x} \operatorname{dom} \mathbf{y}$, and $\mathbf{x} \operatorname{dom}_S \mathbf{y}$, respectively, when $\mathbf{x}$ dominates $\mathbf{y}$, and when one wishes to mention explicitly that S is one of the acting coalitions in this domination. If $\mathbf{x} \in A$ and $B \subset A$, the following subsets of A are of interest:

$$\operatorname{Dom} \mathbf{x} = \{\mathbf{y} \in A : \mathbf{x} \operatorname{dom} \mathbf{y}\}$$

and

$$\operatorname{Dom} B = \{\mathbf{y} \in A : \mathbf{x} \operatorname{dom} \mathbf{y} \text{ for some } \mathbf{x} \in B\}.$$

The relation "dom" is a binary irreflexive relation on A but is not a symmetric (or asymmetric) or a transitive relation. For a given coalition S, "dom_S" is irreflexive, asymmetric, and transitive.

Example 5. Consider the three-person cooperative game with characteristic function $v(123) = 5$, $v(12) = 4$, $v(13) = 3$, $v(23) = 2$, $v(1) = v(2) = v(3) = 0$. [Note that commas

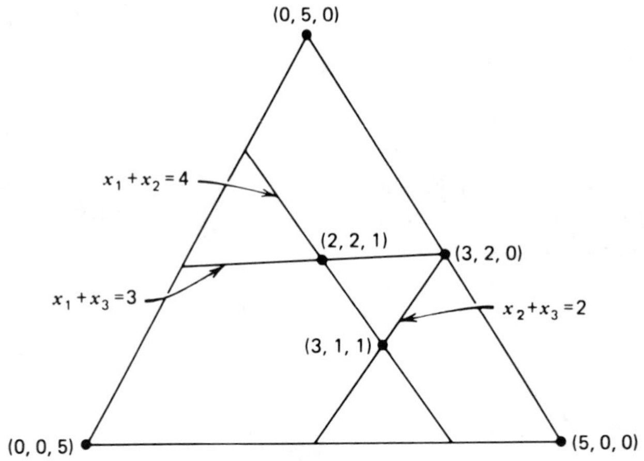

Figure 2

and braces have been deleted; e.g., $v(12)$ stands for the expression $v(\{1,2\})$.] The set of imputations is the simplex

$$A = \{(x_1, x_2, x_3) : x_1 + x_2 + x_3 = 5 \text{ and } x_1 \geqslant 0,\ x_2 \geqslant 0,\ x_3 \geqslant 0\},$$

pictured as the large triangle in Fig. 2. The three lines $x_i + x_j = v(ij)$ for $\{i, j\} \subset N = \{1,2,3\}$ are also indicated in Fig. 2; the resulting inverted triangle with vertices (3, 2, 0), (2, 2, 1), and (3, 1, 1) consists of those points at which every potential coalition in N is satisfied simultaneously; i.e.,

$$\sum_{i \in S} x_i \geqslant v(S)$$

for all nonempty $S \subset N$. This region consists of those imputations which are undominated and is called the *core* C of the game, i.e., $C = A - \text{Dom}\, A$. Note also that the three "corner" regions in A are contained in $\text{Dom}\, C$. The remaining two triangles in the figure are contained in $A - \text{Dom}\, C$ but not in $A - \text{Dom}\, A$, and outcomes in these triangles may occur when either of the coalitions $\{1,3\}$ or $\{2,3\}$ "holds together and squeezes unreasonable amounts from the other player."

The core was first defined for general n-person cooperative games in 1953 by Gillies and Shapley, although it had been discussed for individual examples and for particular classes of games at various times prior to such explicit and general game models. One difficulty with the core as a solution concept is that it is the empty set for many games (e.g., see Example 7). Many games arising in political science* (e.g., voting games) have empty cores, whereas many large classes of games which occur in economics (e.g., markets or cost allocation games) have nonempty cores. The core has proved to be an important solution concept in modern economic theory as well as in game theory.

The first general solution concept proposed for n-person games (in characteristic function form) was the von Neumann–Morgenstern *solution*. These are now often referred to as *stable sets*, since they are only one of the several different "solution concepts" introduced for such games since 1944. A subset V of the imputation set A is a *stable set* if no imputation in V dominates any other one in V, and if any imputation not in V is dominated by at least one imputation in V. This says that

$$V \cap \text{Dom}\, V = \varnothing \quad \text{and} \quad V \cup \text{Dom}\, V = A$$

or alternatively that

$$V = A - \text{Dom}\, V,$$

i.e., V is a "fixed point" under the function $f : 2^A \to 2^A$ given by

$$f(B) = A - \text{Dom}\, B$$

since $f(V) = V$.

Example 6. Consider the three-person game with $v(123) = v(12) = v(13) = 1$ and $v(23) = v(1) = v(2) = v(3) = 0$. The set of imputations for this game is

$$A = \{(x_1, x_2, x_3) : x_1 + x_2 + x_3 = 1 \text{ and } x_1 \geqslant 0,\ x_2 \geqslant 0,\ x_3 \geqslant 0\},$$

and the core is $C = \{(1,0,0)\}$. The unique "symmetric" stable set V^s is the line segment joining $(1,0,0)$ to $(0,\frac{1}{2},\frac{1}{2})$; i.e., the players in $\{2,3\}$ split evenly whatever they can bargain away from player 1. It is unlikely that player 1 can play 2 and 3 off against each other to the extreme of obtaining the full payoff of 1 for himself or herself, i.e., the final outcome may be close to, but not in, the core. *Some* other continuous curves from C to a point on the edge of A with $x_1 = 0$ also form stable sets, and the union of all stable sets for this game gives all of A.

Example 7. Consider the three-person game $v(123) = v(12) = v(13) = v(23) = 1$ and $v(1) = v(2) = v(3) = 0$. This is referred to as the three-person *constant-sum* game since $v(S) + v(N - S) = 1$ for all $S \subset N$. It is also called the three-person *simple-majority* (rule) game since any coalition of size larger than $\frac{3}{2}$ is "winning." The unique symmetric stable set for this game is the finite set $V^s = \{(0,\frac{1}{2},\frac{1}{2}),\ (\frac{1}{2},0,\frac{1}{2}),\ (\frac{1}{2},\frac{1}{2},0)\}$, i.e., some two players form a coalition and divide their gain evenly among just the two of them. Other "discriminatory" stable sets are obtained by picking one of the three players, i, and a number d, $0 \leqslant d < \frac{1}{2}$, and taking the line segment $V_i^d = \{(x_1, x_2, x_3) \in A : x_i = d\}$. Thus player i, the "agent," receives the amount d, and the remaining two players bargain over how to split the remainder $1 - d > \frac{1}{2}$. The union of all such sets V_i^d is all of A, and the intersection of all stable sets for this game is the empty set Ø. Clearly, the core C for this game is Ø.

Many classes of n-person games are known to have stable sets. Von Neumann and Morgenstern [7] proved that every three-person game has at least one stable set, and they described *all* such sets. Bondareva and others in Leningrad proved in 1979 [3] that every four-person game has at least one stable set. The major theoretical problem was to determine whether every n-person game had at least one stable set. In 1968, Lucas published an example of a 10-person game which has no stable set, and in 1980 he and Rabie described a 13-person game which has no stable set and an empty core as well. The examples above show that stable sets are in general not unique. Although some stable sets provide insights into how such games may be resolved, other stable sets may be difficult to interpret in terms of "playing" the actual games. Nevertheless, stable set theory remains challenging, particularly from the purely mathematical point of view.

It is desirable in the n-person cooperative games to have a solution concept which determines a unique imputation in the set A, assuming of course that A is not empty. Several such "one-point" solution concepts have been proposed; the best known and most generally applied is the Shapley value (see Shapley [6]). The one imputation determined by such a value theory can often be interpreted as an a priori evaluation of each player's worth in the game, as an expected value of each player, as a fair distribution of gains, as an equitable allocation of costs among the players, or as a power index in voting games. Another power index, proposed by Banzhaf, has been accepted as a reasonable measure of voting power in several court decisions in the United States. The *Shapley value* ϕ is given by

$$\phi_i = \sum_{\{S \subset N : i \in S\}} \left\{ \frac{(s-1)!\,(n-s)!}{n!} \times \left[v(S) - v(S - \{i\}) \right] \right\}$$

for all $i \in N$, where $s = |S|$ denotes the number of players in coalition S. ϕ will be an imputation if $v(S) - v(S - \{i\}) \geqslant v(\{i\})$ for all $S \subset N$ such that $i \in S$, e.g., if v is superadditive. Such values are normally supported on the grounds that they are some sort of statistical expectation, arise from a type of fair division scheme, or satisfy some set of reasonable criteria or axioms.

Another one-point solution concept for the n-person cooperative games (N, v) is the *nucleolus* introduced by Schmeidler in 1969, and defined in terms of the *excesses*

$$e(x, S) = v(S) - \sum_{i \in S} x_i,$$

where $x \in A$ and $S \subset N$. The excess $e(x, S)$ expresses the "group attitude" of coalition S towards the outcome x: a large excess is undesirable, whereas a negative one "satisfies" this coalition's potential. The *nucleolus* ν is the unique imputation (assuming that $A \neq \emptyset$) which minimizes the largest excess, and then the second largest excess, and in turn the third largest excess, and so on, until ν is uniquely determined. So the nucleolus minimizes the largest "coalitional complaint" and then the second largest, etc.; it is in the core C whenever $C \neq \emptyset$, and can be viewed as a "middle" point in C. It is also located in several other solution concepts, in particular in various "bargaining sets" proposed by Aumann and Maschler in the early 1960s. The nucleolus, like the values, can be viewed as an equitable or fair allocation, e.g., in allocating costs or taxes.

Games in Extensive Form

The third basic formulation for analyzing game-like situations is the extensive or tree form of a game. It is often employed in initial attempts at modeling a game problem. It allows for a succession of moves by different players, including chance moves, and can handle quite intricate information patterns as well. The definition of the extensive form of a game usually make use of some terms from graph theory*. A *graph* consists of a set of points, called *vertices*, together with some (unordered) pair of vertices, called *edges*, which can be viewed as lines connecting the corresponding two vertices. A *tree* is a "connected" graph with no "cycles." An *n-person game in extensive form* is a tree along with the following specifications. One vertex is the starting point of the game, called the *root*. Each "nonterminal" vertex is identified with one of the players, $1, 2, \ldots, n$, or the chance "player" 0, and represents a potential choice point for this player. A *terminal* vertex has only one edge, represents a possible termination point of the game, and thus has associated with it an n-dimensional *payoff* vector. The set of all vertices identified with a particular player is partitioned into subsets called *information sets*; this player knows what information set he or she is in but not the precise vertex he or she is at within this subset. All vertices in the same information set must have the same number of "following" edges, corresponding to the choices that this player has when he or she is at any such vertex. One also assumes that no vertex in an information set "follows" another vertex in the *same* information set as one traces a path from the root to some terminal vertex in the tree. A *pure strategy* for a given player is the selection of a particular edge at each of his or her information sets. A *mixed strategy* is a probability distribution over his or her pure strategies. Each player sets out to maximize the expected payoff which he or she will receive at the end of the game. The extensive form for the game of "matching pennies" is shown in Fig. 3. The optimal mixed strategy is to mix heads H and tails T with probability $\frac{1}{2}$ each.

The first major theorem for games in extensive form was proved by Zermelo in 1912. He showed that every two-person, zero-sum game with perfect information (e.g., chess, checkers, or tic-tac-toe) has an optimal *pure* strategy for each player. A game has *perfect information* when each information set consists of a single vertex; i.e.,

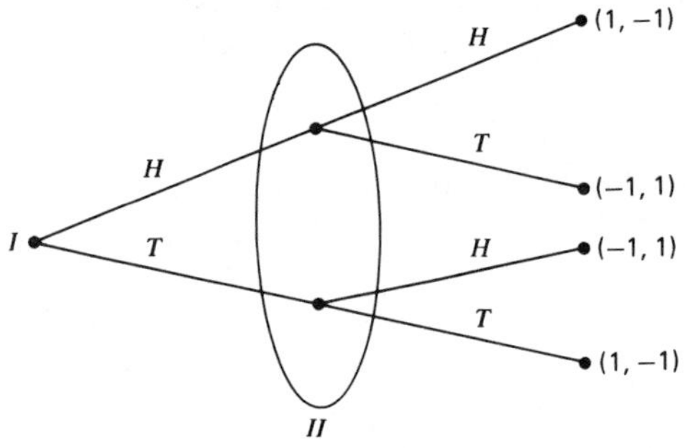

Figure 3

any time a player moves, he or she knows precisely where he or she is in the game tree. Von Neumann's minimax theory (1928) shows that a finite two-person, zero-sum game (in normal or extensive form) has an optimal solution in *mixed* strategies. In the 1950s, Kuhn, Thompson, and others proved important theorems about optimal solutions in "behavioral" or "composite" strategies when games have "perfect recall" or other specific types of information patterns. Nash equilibrium points are again the main solution concept for n-person or general-sum games in extensive, as well as in normal form. Although games in extensive form are most useful for modeling game problems, one often reduces the game to an equivalent normal-form game before solving it. However, this reduction may lead to an enormous number of pure normal-form strategies, even for a relatively small game tree. Nevertheless, there are no really efficient practical algorithms for solving moderate-size games directly from their extensive form. Games in extensive form have been applied to several areas within pure mathematics, e.g., to get "Conway's numbers," in computational complexity, or, in the case of trees of infinite length, to logic and foundations.

APPLICATIONS TO STATISTICS

There is extensive interplay between various game theory models and methods employed in statistics and probability theory. Games in extensive form are concerned mainly with information patterns and the resulting optimal way to randomize over pure strategies. Solution concepts for multiperson cooperative games frequently involve probability theory, particularly when these models are extended to (nonatomic) games with "a continuum (or ocean) of players," as illustrated in Aumann and Shapley [1]. Games in normal form are often played repeatedly and/or with incomplete information. The literature describes various types of recursive, stochastic, ruin, and Markov games; in the "limiting case" one obtains differential games and control theory.

The field of statistical decision theory* can be viewed as two-person games (in normal form) in which one player is "nature," not necessarily playing in an optimal manner. A game-theoretical approach may also be appropriate for problems of hypothesis testing*, e.g., to some situations involving inspections. In auditing bank records, verifying compliance with the law, inspecting for violation of standards for measures, accounting for dangerous or precious materials, etc., the error or loss may be caused by a rational person who should be modeled as a player in a two-person noncooperative game. The inspector can construct payoff functions which take account of the value of recovery or apprehension as well as the cost of additional search or protection. These payoffs may or may not have the zero-sum property, depending on the diverter's objectives and utilities. Nevertheless, an optimal strategy for the resulting game when viewed as a zero-sum one will guarantee the corresponding player his or her maximin value as the expected payoff, independent of the other players' payoff function. Such game models are receiving increased consideration in safeguarding nuclear materials, in preventing prescription drugs from entering the blackmarket, in auditing personal income tax returns, and in other areas. For additional discussions, see Avenhaus [2], Frick [4], and Goldman and Pearl [5].

References

The following books and journal articles are fairly technical.

[1] Aumann, R. J. and Shapley, L. S. (1974). *Values of Non-atomic Games*. Princeton University Press, Princeton, N.J.

[2] Avenhaus, R. (1977). *Materials Accountability: Theory, Verification, Applications*. Wiley, New York.

[3] Bondareva, O. N., Kulakovskaya, T. E., and Naumova, N. I. (1979). *Leningrad Univ. Vestnik*, **6**, 104–105.

[4] Frick, H. (1979). *Int. J. Game Theory*, **8**, 175–192.

[5] Goldman, A. J. and Pearl, M. H. (1976). *J. Res. Natl. Bur. Stand. B. Math. Sci.*, **80B** (Apr.–June), 189–236.

[6] Shapley, L. S. (1953). In *Ann. Math. Studies No. 28*, Princeton, N. J., pp. 307–317.

[7] von Neumann, J. and Morgenstern, O. (1944). *Theory of Games and Economic Behavior*. Princeton University Press, Princeton, N.J. (2nd ed., 1947; 3rd ed., 1953; paperback, Wiley, 1965).

Bibliography

See the following works, as well as the references just cited, for more information on game theory. Some elementary and relatively nontechnical books follow.

Brams, S. J. (1975). *Game Theory and Politics*. Free Press, Glencoe, Ill. (Some applications to political science and voting.)

Davis, M. (1970). *Game Theory: A Nontechnical Introduction*. Basic Books, New York (paperback, 1973).

Dawkins, R. (1976). *The Selfish Gene*. Oxford University Press, London (paperback). (An application to biology and evolution.)

Luce, R. D. and Raiffa, H. (1957). *Games and Decisions: Introduction and Critical Survey*. Wiley, New York. (A thorough survey for its time.)

McDonald, J. (1975). *The Game of Business*. Doubleday, New York (paperback, Anchor Books, 1977). (Games arising in business.)

Rapoport, A. (1960). *Fights, Games, and Debates*. University of Michigan Press, Ann Arbor, Mich.

Williams, J. D. (1954). *The Compleat Strategyst* [sic]. McGraw-Hill, New York (rev. ed., 1966). (Matrix games.)

Some more technical books and journal articles follow.

Dubey, P. and Shapley, L. S. (1979). *Math. Operat. Res.*, **4**, 99–131.

Hamburger, H. (1973). *J. Math. Sociol.* **3**, 27–48.

Isaacs, R. (1965). *Differential Games: A Mathematical Theory with Applications to Warfare and Pursuit, Control and Optimization*. Wiley, New York.

Karlin, S. (1959). *Mathematical Methods in Games, Programming, and Economics*, Vols. 1, 2. Addison-Wesley, Reading, Mass.

Lucas, W. F. (1971). *SIAM Rev.*, **13**, 491–523.

Maschler, M., Peleg, B., and Shapley, L. S. (1979). *Math. Operat. Res.*, **4**, 303–338.

Roth, A. E. (1979). Axiomatic Models of Bargaining. *Lect. Notes Econ. Math. Syst.*, **170**, Springer-Verlag, New York.

Tucker, A. W., et al., eds. (1950, 1953, 1957, 1959, 1964). *Contributions to the Theory of Games*, Vols. 1–4, and *Advances in Game Theory* (Nos. 24, 28, 39, 40, and 52 in *Annals of Mathematics Studies*). Princeton University Press, Princeton, N.J.

(BAYESIAN INFERENCE
DECISION THEORY)

W. F. LUCAS

GAMMA DISTRIBUTION

A random variable Y follows a three-parameter gamma distribution [denoted by $Y \sim \text{GAM}(\theta, \kappa, \eta)$] if its probability density function is given by

$$g(y;\theta,\kappa,\eta) = \frac{(y-\eta)^{\kappa-1}\exp[-(y-\eta)/\theta]}{\theta^{\kappa}\Gamma(\kappa)},$$

$$\eta < y < \infty; \quad \theta, \kappa > 0, \quad -\infty < \eta < \infty,$$

where $\Gamma(\kappa) = \int_0^\infty t^{\kappa-1}e^{-t}\,dt$, the *gamma function*. The parameters η and θ are location-scale parameters and κ is a shape parameter; η is also known as a threshold parameter. In many applications $\eta = 0$ or is assumed known. Setting $\eta = 0$ or letting $X = Y - \eta$ leads to the more common two-parameter gamma distribution,

$$X \sim \text{GAM}(\theta, \kappa, 0) = \text{GAM}(\theta, \kappa);$$

reference to the gamma distribution in the following will mean two-parameter gamma unless otherwise specified.

The gamma distribution is Type III of Pearson's system* of distributions. It is an *Erlang distribution** if κ is a positive integer, and reduces to the *exponential distribution** if $\kappa = 1$. The gamma distribution approaches a normal distribution* as $\kappa \to \infty$. Setting $\theta = 2$ and $\kappa = \nu/2$ gives the very important chi-square distribution* with ν degrees of freedom, $\text{GAM}(2, \nu/2) = \chi^2(\nu)$.

Some important properties of the gamma distribution include the following.

1. If $X \sim \text{GAM}(\theta, \kappa)$, then

$$2X/\theta \sim \chi^2(2\kappa).$$

2. Moment generating function*:

$$M_x(t) = (1 - \theta t)^{-\kappa}, \qquad t < 1/\theta.$$

3. Characteristic function*:

$$C(t) = (1 - i\theta t)^{-\kappa}.$$

4. Moments*:

$$\mu_r' = E(X^r) = \Gamma(\kappa + r)\theta^r/\Gamma(\kappa), \qquad r > -\kappa;$$

$$\mu = E(X) = \kappa\theta,$$

$$\sigma^2 = E(X - \mu)^2 = \kappa\theta^2.$$

5. Cumulants*:

$$k_r = \kappa\theta^r(r-1)!$$

6. Reproductive property: If X_i are independent and $X_i \sim \text{GAM}(\theta, \kappa_i)$, $i = 1, \dots, n$, then

$$\sum_{i=1}^{n} X_i \sim \text{GAM}\left(\theta, \sum_{i=1}^{n} \kappa_i\right).$$

7. Relationship to normal distribution: If U_i are independent standard normal variables, $U_i \sim N(0,1)$, $i = 1, \dots, n$; then

$$U_i^2 \sim \chi^2(1), \qquad \sum_{i=1}^{n} U_i^2 \sim \chi^2(n).$$

8. Distribution of quadratic forms: If $X_1, \dots, X_n$ follow a multivariate normal distribution,

$$f(x_1, \dots, x_n) = C\exp\left[-\tfrac{1}{2}\mathbf{X}^T\mathbf{A}\mathbf{X}\right],$$

where C is a constant and $\mathbf{A}$ is a positive-definite matrix, then

$$\mathbf{X}^T\mathbf{A}\mathbf{X} \sim \chi^2(n).$$

9. Hazard function: The gamma distribution is IFR for $\kappa > 1$ and DFR for $\kappa < 1$ (*see* HAZARD RATE CLASSIFICATION OF DISTRIBUTIONS).

10. Infinitely divisible: It follows from property 6 that if X_i are independent and $X_i \sim \text{GAM}(\theta, \kappa/n)$, $i = 1, \dots, n$, then

$$\sum_{i=1}^{n} X_i \sim \text{GAM}(\theta, \kappa).$$

Bondesson [5] considers infinite divisibility* of powers of gamma variables.

Many of the basic properties and results for the gamma distribution are given by Johnson and Kotz [28], Bury [8], and Patel et al. [44]. Johnson and Kotz [28] discuss approximations for percentage points and probability integrals and they review the numerous tables available. Saunders and Moran [52] consider the monotonicity of differences and ratios of gamma quantiles* with respect to the shape parameter.

Cumulative probabilities for the gamma variable may be expressed in terms of the chi-square distribution as indicated in property 1. The cumulative distribution function (CDF) is sometimes called the *incomplete gamma function* or *incomplete gamma function ratio*, and has been extensively tabulated by Pearson [45] and Harter [24]. Also, suppose that z_α denotes the 100α percentage point of a standard normal* variable and $\chi_\alpha^2(\nu)$ denotes the 100α percentage point of a chi-square distribution with ν degrees of freedom. Wilson and Hilferty [62] derive the following very good approximation:

$$\chi_\alpha^2(\nu) = \nu\left(1 - \frac{2}{9\nu} + z_\alpha\sqrt{\frac{2}{9\nu}}\right)^3.$$

If κ is a positive integer, say k, the gamma CDF may be expressed in terms of the CDF of a Poisson* variable,

$$\begin{aligned} G(x;\theta,k) &= 1 - \sum_{i=0}^{k-1} (x/\theta)^i e^{-x/\theta}/i! \\ &= 1 - P(k-1, x/\theta), \end{aligned} \qquad (1)$$

where $P(n; x/\theta)$ represents the CDF for the number of occurrences, N, in the interval $[0,x]$ from a Poisson process* with intensity $\lambda = 1/\theta$. This is indicative of the relationship between the Poisson process and the gamma distribution. If X denotes the waiting time until the kth occurrence of a Poisson process with intensity $\lambda = 1/\theta$, then the right-hand side of (1) represents

$$\Pr[X \leqslant x] = 1 - \Pr(k-1 \text{ or fewer occurrences in } [0,x]).$$

This result may also be expressed in terms of independent exponential variables. If T_i represents times between occurrences of the Poisson process, then the T_i are independent, exponentially distributed, $T_i \sim \mathrm{GAM}(\theta, 1)$, and

$$X = \sum_{i=1}^{k} T_i \sim \mathrm{GAM}(\theta, k).$$

Partly because of the above, the gamma distribution is an important life-testing* model, providing a rather flexible skewed density over the positive range. For $\kappa = 1$ it becomes the exponential density with constant hazard rate function,

$$\lambda(x) = f(x)/\left[1 - F(x)\right] = 1/\theta.$$

For $\kappa > 1$, $\lambda(x)$ is an increasing function approaching $1/\theta$ asymptotically from below as $x \to \infty$. For $\kappa < 1$, $\lambda(x)$ approaches $1/\theta$ asymptotically from above as $x \to \infty$. For example, the gamma distribution would be a useful failure-time model for a system under continuous maintenance if it experiences some wear-out or degradation initially but reaches a stable state of repair as time goes on, so that the constant hazard function would apply in that stage.

There have been many characterizations* of the gamma distribution, particularly involving the constancy of various regression functions (see, e.g., Johnson and Kotz [28], Patel et al. [44], or Kagan et al. [31]). If X_1 and X_2 are independent positive variables, then the variables $X_1 + X_2$ and $X_1/(X_1 + X_2)$ are independent if and only if X_1 and X_2 have gamma distributions with the same scale parameter [37]. Early applications of the distribution, dating back to Laplace* [33], were involved more with the gamma or chi-square as a derived distribution than as a population model. As suggested by property **8**, the chi-square distribution has received wide usage as the exact or approximate distribution of quadratic forms* in the areas of linear models and analysis of variance*. Through the normal approximation to the multinomial distribution*, these results also apply to goodness-of-fit* and contingency table* problems.

Johnson and Kotz [28] provide a good general review of the gamma distribution, including historical development, properties, characterizations, estimation, and related distributions, such as the truncated gamma, compound gamma, and generalized gamma distribution. Johnson and Kotz [29] review the noncentral chi-square and distributions of quadratic forms.

A different generalization called the Lagrangian gamma is developed by Nelson and Consul [41] as the distribution of the time between occurrences of a generalized Poisson process. In the same volume a multivariate gamma distribution with marginal gammas is discussed by Dussauchoy and Berland [14]. A recent paper on a multivariate gamma distribution of quadratic forms is that by Tan and Wong [59]; see Johnson and Kotz [30] for a more general review.

The gamma distribution has been utilized as a prior density in Bayesian analysis problems and it was derived as a posterior distribution by Laplace [33]. It is the natural conjugate prior for Bayesian analysis of Poisson data (*see* BAYESIAN INFERENCE). The inverted gamma distribution ($T = 1/X$), with density

$$f(t) = \theta^{-\kappa}\left[\Gamma(\kappa)\right]^{-1} t^{-(\kappa+1)} e^{-1/(\theta t)},$$

$$0 < t < \infty, \quad 0 < \theta, \quad \kappa < \infty,$$

is also a useful prior density, for example in estimating variances of normal populations (see also Gleser and Healy [18], Draper and Guttman [13], and Raiffa and Schlaifer [49]).

Gupta and Groll [21] describe the use of the gamma distribution in acceptance sampling* based on life tests. It has been applied in a wide assortment of other areas such as weather analysis (*see* METEOROLOGY, STATISTICS IN), inventory theory*, insurance risk theory, economics (*see* ECONOMETRICS), queuing theory*, and life testing. Explicit references may be found in current indexes.

There has been considerable interest in computer generation of gamma random variables (see, e.g., Tadikamalla [58]), and Ronning [50] discusses random-number generation from multivariate gamma distributions.

The gamma distribution is a member of the regular exponential class (*see* EXPONEN-

TIAL FAMILIES), and it follows that the arithmetic mean*, $\bar{X} = \sum_{i=1}^{n} X_i/n$, and the geometric mean*, $\tilde{X} = (\prod_{i=1}^{n} X_i)^{1/n}$, are a set of complete, sufficient statistics for θ and κ (*see* SUFFICIENCY). Also, $\bar{X}$ is a complete, sufficient statistic for θ for fixed κ; so, for known values of κ, optimum inference procedures for θ are easily obtained based on the pivotal quantity* result,

$$2n\bar{X}/\theta \sim \chi^2(2n\kappa).$$

It also follows that $\bar{X}$ is distributed independently of $\bar{X}/\tilde{X}$ and independently of any other such statistic whose distribution is free of the scale parameter θ (*see* BASU THEOREMS).

The most commonly used method of point estimation is that of maximum likelihood*. The maximum likelihood estimators (MLEs), $\hat{\kappa}$ and $\hat{\theta}$, are the solutions of the equations

$$\hat{\theta} = \bar{X}/\hat{\kappa}$$

$$\ln \hat{\kappa} - \Psi(\hat{\kappa}) = \ln(\bar{X}/\tilde{X}),$$

where $\Psi(z) = \Gamma'(z)/\Gamma(z)$ (*see* DIGAMMA FUNCTION). Since $\hat{\kappa}$ is a function only of $\bar{X}/\tilde{X}$ and does not depend separately on n, $\bar{X}$, and $\tilde{X}$, convenient tables giving $\hat{\kappa}$ as a function of M [9], where $M = \ln(\bar{X}/\tilde{X})$ and $\hat{\kappa}$ as a function of $[1 - (\bar{X}/\tilde{X})^{-1}]^{-1}$ [61] have been computed. Greenwood and Durand [19] give the following very good rational approximation for $\hat{\kappa}$,

$$\hat{\kappa} \doteq (0.5000876 + 0.1648852M - 0.0544276M^2)/M, \qquad 0 \leqslant M \leqslant 0.5772,$$

$$\hat{\kappa} \doteq \frac{8.898919 + 9.059950M + 0.9775373M^2}{M(17.79728 + 11.968477M + M^2)}, \qquad 0.5772 \leqslant M \leqslant 17,$$

$$\hat{\kappa} \doteq 1/M, \qquad M > 17.$$

Note also that $\hat{\kappa}M \to 0.5$ as $M \to 0$ and $\hat{\kappa}M \to 1$ as $M \to \infty$. The means and variances of the MLEs have been extensively tabulated as a function of κ by Bowman and Shenton [6, 7, 53, 54], and they discuss approximate unbiasing methods. Of course, $\hat{\mu} = \bar{X}$ is the uniformly minimum variance unbiased estimator* of $\mu = \kappa\theta$.

As $n \to \infty$ the MLEs are asymptotically efficient, and they are asymptotically normally distributed;

$$\left(\sqrt{n}\,(\hat{\kappa}/\kappa - 1), \sqrt{n}\,(\hat{\theta}/\theta - 1)\right) \xrightarrow{d} MVN(\mathbf{0}, \mathbf{V}),$$

where

$$\mathbf{V} = \frac{1}{D}\begin{pmatrix} 1 & -1 \\ -1 & \kappa\Psi'(\kappa) \end{pmatrix},$$

$$D = \kappa\left[\kappa\Psi'(\kappa) - 1\right].$$

Harter [26] reviews maximum likelihood estimation for the two-parameter and three-parameter gamma distribution and tabulates numerical values of the asymptotic variances and covariances for $\kappa = 1, 2$, and 3, when any one, two, or all three of the parameters are assumed unknown. He also includes the case of singly and doubly censored samples. Cohen and Norgaard [10] study MLEs based on progressively censored samples (*see* CENSORING *and* PROGRESSIVE CENSORING SCHEMES).

Wilk et al. [61] provide tables to aid in computing the MLEs based on type II censored samples. Suppose that the first r-order statistics* $X_{1:n}, \ldots, X_{r:n}$ are available, and let $\tilde{X}_c = (\prod_{i=1}^{r} X_{i:n})^{1/r}$ and $\bar{X}_c = \sum_{i=1}^{r} X_{i:n}/r$; then the MLEs may be tabulated in terms of $P = \tilde{X}_c/\bar{X}_c$, $S = \bar{X}_c/X_{r:n}$, and $f = r/n$. Note that $\bar{X}_c$, $\tilde{X}_c$, and $X_{r:n}$ are jointly sufficient statistics in this case.

Johnson and Kotz [28], Sarhan and Greenberg [51], and David [12] review estimation and other results based on order statistics. Harter [26] also tabulates expected values of gamma order statistics for $\kappa = 0.5(0.5)4.0$ and $n = 1(1)40$ (see also Hassanein [27]).

Tests (or confidence intervals) for κ with θ being a nuisance* scale parameter may be based on the statistic $M = \ln(\bar{X}/\tilde{X})$. These tests will have optimum properties in the sense that M is a sufficient statistic relative to the class of all statistics whose distributions are independent of θ. Unfortunately, the exact distribution of M is very complicated. For large values of κ, say $\kappa \geqslant 2$, approximately,

$$W = 2n\kappa M \sim \chi^2(n-1)$$

(see, e.g., Linhart [35]). More generally, a chi-square approximation valid for all κ is derived by Bain and Engelhardt [2] based on a two-moment fit. Approximately,

$$CW \sim \chi^2(\nu),$$

$$C = \frac{n\phi_1(\kappa) - \phi_1(n\kappa)}{n\phi_2(\kappa) - \phi_2(n\kappa)},$$

$$\nu = \left[n\phi_1(\kappa) - \phi_1(n\kappa) \right] C,$$

$$\phi_1(z) = 2z\left[\ln z - \Psi(z) \right],$$

$$\phi_2(z) = 2z\left[z\Psi'(z) - 1 \right].$$

Values of C and $\nu/(n-1)$ are tabulated for a wide range of κ and n, and approximations are given. Also, $C \to 1$ as $\kappa \to 0$ or $\kappa \to \infty$, $\nu \to n-1$ as $\kappa \to \infty$, and $\nu \to 2(n-1)$ as $\kappa \to 0$. Thus tests of hypotheses* for κ are easily carried out and confidence limits may be obtained from the tables iteratively. The test of $H_0: \kappa = 1$ is a test of exponentiality against a two-parameter gamma distribution as alternative. Shorack [55] shows the foregoing procedure to be uniformly most powerful scale invariant in this case; Moran [40] and Glaser [17] discuss optimum properties of the test, which is generalized to the three-parameter case by Bain [1, p. 342].

Wyckoff and Engelhardt [63] develop tests for κ under type II censored sampling based on $T = \ln(A_r/G_r)$, where A_r and G_r are the Winsorized* means

$$A_r = \left[\sum_{i=1}^{r} X_i + (n-r)X_r \right] / n,$$

$$G_r = \left[\left(\prod_{i=1}^{r} X_i \right) \left(X_r^{n-r} \right) \right]^{1/n}.$$

They obtain percentage points of T by Monte Carlo simulation*; they also provide a chi-square approximation and give asymptotic results.

It is possible to develop tests for θ with κ an unknown nuisance parameter based on the conditional distribution of $\bar{X}$ given $\tilde{X}$. This density is independent of κ since $\tilde{X}$ is sufficient for κ for fixed θ. Also, the conditional density is in the exponential class, so optimum conditional tests exist. The completeness property makes it possible to associate these tests with uniformly most powerful unbiased unconditional tests, due to their uniqueness.

The exact conditional density of $\bar{X}$ given $\tilde{X}$ is extremely complicated. Suppose that $W_n = \bar{X}/\tilde{X}$ and $G_n = \tilde{X}/\theta$; comprehensive tables of percentage points, $U_\gamma(g)$, are included in Engelhardt and Bain [15] or Bain [1] such that

$$P\left[\sqrt{n}\, g(W_n - E(W_n \mid g)) \leqslant U_\gamma(g) \mid G_n = g \right] = \gamma.$$

Values of $E(W_n \mid g)$ are also included. As $g \to \infty$, $E(W_n \mid g) \to 1$ and

$$U_\gamma(g) \to \sqrt{n}\left[\frac{\chi^2_\gamma(n-1)}{2n} - \frac{1-1/n}{2} \right].$$

For large n, $E(W_n \mid g) \doteq m_g$, $ng^2 \mathrm{Var}(W_n \mid g) \doteq c_g$, and $U_\gamma(g) \doteq Z_\gamma \sqrt{c_g}$, where m_g and c_g are tabulated and Z_γ denotes a standard normal percentage point. For example, a UMPU size α test of $H_0: \theta \geqslant \theta_0$ against $H_a: \theta < \theta_0$ is to reject H_0 if

$$\sqrt{n}\, g_0\left[(\bar{x}/\tilde{x}) - E(W_n \mid g_0) \right] < U_\alpha(g_0),$$

$$\text{where } g_0 = \tilde{x}/\theta_0.$$

Miller [39] provides Bayesian inference* procedures for the gamma distribution, and Lingappaiah [34] considers a Bayesian approach to prediction* problems. Barnett and Lewis [4] include tests for outliers* in gamma samples. O'Brien [42] proposes a sequential interval estimation procedure for the shape parameter of the gamma distribution.

Dahiya and Gurland [11] derive a goodness-of-fit statistic for the gamma distribution and Pettitt [46] studies a Cramér–von Mises statistic* for the shape-parameter-known case. Locke [36] develops a gamma model test based on the Lukács characterization. Bain and Engelhardt [3] consider discriminating between the gamma and Weibull* distributions on the basis of probability of correct selection. A different approach is to assume a very general model such as the generalized gamma distribution and then develop procedures or tests to discriminate between special cases within that distribution. The generalized gamma distribution includes the gamma and Weibull distributions as well as others; discrimination procedures within this model have been studied by Parr and Webster [43], Volodin [60], and Farewell and Prentice [16].

The generalized gamma distribution as derived by Stacy [56] is given by

$$f(x; a, d, p) = (p/a^d)x^{d-1}e^{-(x/a)^p}/\Gamma(d/p), \quad x > 0.$$

Setting $p = 1$ gives the gamma and setting $d = p$ gives the Weibull distribution. An interesting reparameterization by Prentice [47, 48] makes it possible to include the normal distribution as a limiting case.

Basic properties of the generalized gamma distribution are given by Stacy and Mihram [57]. Harter [25, 26] considers maximum likelihood* estimation, including the four-parameter case with unknown threshold parameter, and he numerically evaluates asymptotic variances for both complete and censored samples. Some inference procedures are developed by Hager and Bain [22] and Hager et al. [23], including a test that data come from a Weibull distribution against a generalized gamma alternative (see also references in Johnson and Kotz [28]).

References

[1] Bain, L. J. (1978). *Statistical Analysis of Reliability and Life-Testing Models*. Marcel Dekker, New York. (Statistical inference procedures for the gamma distribution.)

[2] Bain, L. J. and Engelhardt, M. (1975). *J. Amer. Statist. Ass.*, **70**, 948–950.

[3] Bain, L. J. and Engelhardt, M. (1980). *Commun. Statist. A*, **9**(4), 375–381.

[4] Barnett, V. and Lewis, T. (1978). *Outliers in Statistical Data*. Wiley, New York.

[5] Bondesson, L. (1978). *Scand. Actuarial J.*, 48–61.

[6] Bowman, K. O. and Shenton, L. R. (1968). *Properties of Estimators for the Gamma Distribution. Rep. CTC-1*, Union Carbide Corp., Oak Ridge, Tenn.

[7] Bowman, K. O. and Shenton, L. R. (1970). *Small Sample Properties of Estimators for the Gamma Distribution. Rep. CTC-28*, Union Carbide Corp., Nuclear Division, Oak Ridge, Tenn.

[8] Bury, K. V. (1975). *Statistical Models in Applied Science*. Wiley, New York.

[9] Choi, S. C. and Wette, R. (1969). *Technometrics*, **11**, 683–690.

[10] Cohen, A. C. and Norgaard, N. J. (1977). *Technometrics*, **19**, 333–340.

[11] Dahiya, R. C. and Gurland, J. (1972). *Technometrics*, **14**, 791–801.

[12] David, H. A. (1970). *Order Statistics*. Wiley, New York.

[13] Draper, N. R. and Guttman, I. (1978). *Commun. Statist. A*, **7**(5), 441–451.

[14] Dussauchoy, A. and Berland, R. (1974). In *Statistical Distributions in Scientific Work*, Vol. 1, G. P. Patil, S. Kotz, and J. K. Ord, eds. D. Reidel, Boston.

[15] Engelhardt, M. and Bain, L. J. (1977). *Technometrics*, **19**, 77–81.

[16] Farewell, V. T. and Prentice, R. L. (1977). *Technometrics*, **19**, 69–75.

[17] Glaser, R. E. (1976). *J. Amer. Statist. Ass.*, **71**, 480–487.

[18] Gleser, L. T. and Healy, J. D. (1976). *J. Amer. Statist. Ass.*, **17**, 977–981.

[19] Greenwood, T. A. and Durand, D. (1960). *Technometrics*, **2**, 55–56.

[20] Gross, A. T. and Clark, V. A. (1975). *Survival Distributions: Reliability Applications in the Biomedical Sciences*. Wiley, New York. (Statistical procedures for the gamma distribution.)

[21] Gupta, S. S. and Groll, P. A. (1961). *J. Amer. Statist. Ass.*, **56**, 942–970.

[22] Hager, H. W. and Bain, L. J. (1970). *J. Amer. Statist. Ass.*, **65**, 1601–1609.

[23] Hager, H. W., Bain, L. J., and Antle, C. E. (1971). *Technometrics*, **13**, 547–557.

[24] Harter, H. L. (1964). *New Tables of the Incomplete Gamma-Function Ratio and of Percentage Points of the Chi-Square and Beta Distributions*. U.S. Government Printing Office, Washington, D.C.

[25] Harter, H. L. (1967). *Technometrics*, **9**, 159–165.

[26] Harter, H. L. (1969). *Order Statistics and Their Use in Testing and Estimation*, Vol. 2: *Estimates Based on Order Statistics from Various Populations*. U.S. Government Printing Office, Washington, D.C.

[27] Hassanein, K. M. (1977). *Scand. Acturial J.*, 88–93.

[28] Johnson, N. L. and Kotz, S. (1970). *Continuous Distributions*, Vol. 1. Houghton Mifflin, Boston (distributed by Wiley, New York). (Excellent general review and bibliography up to 1970.)

[29] Johnson, N. L. and Kotz, S. (1970). *Continuous Distributions*, Vol. 2. Houghton Mifflin, Boston (distributed by Wiley, New York).

[30] Johnson, N. L. and Kotz, S. (1972). *Continuous Multivariate Distributions*. Wiley, New York.

[31] Kagan, A. M., Linnik, Y. V., and Rao, C. R. (1973). *Characterization Problems in Mathematical Statistics*. Wiley, New York.

[32] Lancaster, H. O. (1969). *The Chi-Squared Distribution*. Wiley, New York. (Theoretical development of applications of the chi-squared distribution.)

[33] Laplace, P. S. (1836). *Théorie analytique des probabilités* (supplement to the 3rd ed.).

[34] Lingappaiah, G. S. (1979). *Commun. Statist. A*, **8**, 1403–1424.

[35] Linhart, H. (1965). *Biometrics*, **31**, 733–737.

[36] Locke, C. (1976). *Commun. Statist. A*, **1**, 351–364.

[37] Lukács, E. (1956). *Proc. 3rd Berkeley Symp. Math. Statist. Prob*., Vol. 2. University of California Press, Berkeley, Calif., pp. 195–214.

[38] Mann, N. R., Schafer, R. E., and Singpurwalla, N. D. (1974). *Methods for Statistical Analysis of Reliability and Life Data*. Wiley, New York. (Statistical procedures for the gamma distribution.)

[39] Miller, R. B. (1980). *Technometrics*, **22**, 65–70.

[40] Moran, P. A. P. (1951). *J. R. Statist. Soc. B*, **13**, 147–150.

[41] Nelson, D. L. and Consul, P. C. (1974). In *A Modern Course on Statistical Distributions in Scientific Work*, Vol. 1, G. P. Patel, S. Kotz, and J. K. Ord, eds. D. Reidel, Boston.

[42] O'Brien, P. C. (1973). *Technometrics*, **15**, 563–570.

[43] Parr, V. B. and Webster, J. T. (1965). *Technometrics*, **7**, 1–10.

[44] Patel, J. K., Kapadia, C. H., and Owen, D. B. (1976). *Handbook of Statistical Distributions*. Marcel Dekker, New York. (Basic properties, characterizations.)

[45] Pearson, K. (1951). *Tables of the Incomplete Gamma Function*. Cambridge University Press, Cambridge.

[46] Pettitt, A. N. (1978). *Biometrika*, **65**, 232–235.

[47] Prentice, R. L. (1974). *Biometrika*, **61**, 539–544.

[48] Prentice, R. L. (1975). *Biometrika*, **62**, 607–614.

[49] Raiffa, H. and Schlaifer, R. (1961). *Applied Statistical Decision Theory*. Division of Research, Graduate School of Business Administration, Harvard University, Cambridge, Mass. (Use of gamma in Bayesian problems.)

[50] Ronning, G. (1977). *Technometrics*, **19**, 179–183.

[51] Sarhan, A. E. and Greenberg, B. G., eds. (1962). *Contributions to Order Statistics*. Wiley, New York. (Linear estimation and other results for central and extreme order statistics.)

[52] Saunders, F. W. and Moran, P. A. P. (1978). *J. Appl. Prob*., **15**, 426–432.

[53] Shenton, L. R. and Bowman, K. O. (1969). *Sankhyā B*, **31**, 379–396.

[54] Shenton, L. R. and Bowman, K. O. (1972). *Technometrics*, **14**, 725–733.

[55] Shorack, G. R. (1972). *J. Amer. Statist. Ass*., **67**, 213–214.

[56] Stacy, E. W. (1962). *Ann. Math. Statist*., **28**, 1187–1192.

[57] Stacy, E. W. and Mihram, G. A. (1965). *Technometrics*, **7**, 349–358.

[58] Tadikamalla, P. R. (1978). *Commun. ACM*, **21**, 419–422.

[59] Tan, W. Y. and Wong, S. P. (1978). *Commun. Statist.—Simul. Comput*., **7**, 227–242.

[60] Volodin, I. N. (1974). *Theory Prob. Appl*., **19**, 383–389.

[61] Wilk, M. B., Gnanadesikan, R., and Huyett, M. J. (1962). *Biometrika*, **49**, 524–545.

[62] Wilson, E. B. and Hilferty, M. J. (1931). *Proc. Natl. Acad. Sci*., **17**, 684–688.

[63] Wyckoff, J. and Engelhardt, M. (1979). *J. Amer. Statist. Ass*., **74**, 866–871.

(CHI-SQUARE DISTRIBUTION
EXPONENTIAL DISTRIBUTION)

LEE BAIN

GAMMAIZATION

A term used by some writers (e.g., Mitra and Mahajan [1, p. 480]) to denote a transformation to a gamma distribution* or to an approximation thereof.

Reference

[1] Mitra, S. K. and Mahajan, B. M. (1970). In *Essays in Probability and Statistics*, R. C. Bose et al., eds. University of North Carolina Press, Chapel Hill, N.C.

g-AND-*h*-DISTRIBUTIONS

In fitting a theoretical distribution to data, it is often advantageous to use quantiles* rather than moments* or the method of maximum likelihood*, especially if one is concerned that a few observations may not fit in well with the rest. The family of *g*-and-*h*-distributions, introduced by Tukey [3], is defined in terms of quantiles and encompasses a considerable variety of distribution shapes.

DEFINITION

A distribution shape corresponds to a location-scale family of distributions. That is, two distributions have the same "shape" if the corresponding random variables, X_1 and X_2, are related by $X_2 = a + bX_1$ for some constants a and b with $-\infty < a < +\infty$ and $b > 0$. A shape may be a subfamily of a much richer family of distributions such as the gamma distributions, the symmetric stable laws*, or the Pearson curves*.

In what follows, Y is the "standard" representative of a shape, and other random variables with that shape are related to Y through $X = A + BY$.

The *g*-and-*h*-distributions constitute a two-parameter family of shapes and are defined in terms of their quantile function relative to the standard Gaussian (i.e., normal*) distribution,

$$Q_{g,h}(z) = \frac{e^{gz} - 1}{g} e^{hz^2/2}, \tag{1}$$

so that if z_p is the pth quantile of the standard Gaussian distribution (i.e., $P\{Z \leqslant z_p\} = p$), $0 < p < 1$, then $Q_{g,h}(z_p)$ is the pth quantile of a standard *g*-and-*h*-distribution (with the specified values of g and h). The parameter g controls asymmetry or skewness*, while h controls elongation or the extent to which the tails are stretched (relative to the Gaussian).

When $g = 0$, (1) reduces to

$$Q_{0,h}(z) = ze^{hz^2/2}. \tag{2}$$

The random variable $Y = Q_{0,h}(Z)$ has a symmetric distribution, and when $h = 0$, it is simply Z. Thus in (2), $e^{hz^2/2}$ serves as a tail-stretching operator; for $h > 0$ the farther a quantile is into the tail, the more it is stretched from its standard Gaussian value. (Negative values of h may be used, but $Q_{0,h}$ is no longer monotonic when $z^2 > -1/h$.) Known as the *h*-distributions, the members of this subfamily (i.e., $g = 0$) have Paretian tails (*see* PARETO DISTRIBUTIONS), and the factor of $\frac{1}{2}$ in the exponent of (2) yields a close approximation to the Cauchy distribution* at $h \approx 1$.

The other simple subfamily, the *g*-distributions, comes from setting $h = 0$ in (1):

$$Q_{g,0}(z) = \frac{e^{gz} - 1}{g}. \tag{3}$$

Rewriting this as $Q_{g,0}(z) = [(e^{gz} - 1)/(gz)]z$ makes it possible to regard the expression in brackets as a skewing operator. It produces skewness to the right when $g > 0$, skewness to the left when $g < 0$, and no skewness when $g = 0$. Equation (3) indicates how the *g*-distributions are matched to the standard Gaussian distribution at the median*: $Q_{g,0}(0) = 0$ and $Q_{g,0}(z) \approx z$ for z near 0. Also, from the form of $Q_{g,0}$ it is straightforward to see that the shapes in the lognormal* family of distributions correspond to *g*-distributions with positive values of g. Thus the *g*-distributions are essentially lognormal, but the *g*-and-*h*-distributions provide a much greater range of skewness and elongation.

ESTIMATION

Estimation of g and h requires the median and a set of additional quantiles symmetrically placed about the median. That is, these quantiles form pairs, x_p and x_{1-p}, for suitable values of p, $0 < p < 0.5$. In small samples one can work with the full set of order statistics*, but usually it is necessary to select quantiles. One convenient set is the "letter values" [2], which correspond to taking the integer powers of $\frac{1}{2}$ as the values of p.

Unless $g = 0$, the estimation process begins with g. Because the data quantiles, x_p, are related to the "standard" quantiles, y_p, through $x_p = A + By_p$, it is easy to see that $A = x_{0.5}$. For values of p other than 0.5, straightforward algebra yields the value of g

that exactly fits the spacing of x_p and x_{1-p} about the median. Denoted by g_p, this value is given by

$$g_p = -\frac{1}{z_p}\ln\left(\frac{x_{1-p} - x_{0.5}}{x_{0.5} - x_p}\right), \quad (4)$$

where z_p is the corresponding standard Gaussian quantile. Even though the quantile function in (3) involves constant g, data may give g_p that vary from one value of p to another. Working with several values of g_p provides a basis for resistance (*see* EXPLORATORY DATA ANALYSIS). With as many as 10 values of p, each yielding a value of g_p, one or two unusual values should stand out. A simple (and usually adequate) resistant procedure adopts the median of the available g_p as the estimate of g. A generalization of the g-distributions, discussed later in this article, allows for systematic variation in g.

For a pure g-distribution, only the scale constant, B, remains to be estimated. A reasonable way to approach this is by means of a quantile-quantile plot [4] of the data quantiles, x_p, against the quantiles of the standard g-distribution, $Q_{g,0}(z_p)$, for the fitted value of g (*see* GRAPHICAL REPRESENTATION OF DATA). The slope of this plot is B. This method of estimating B provides a check on the appropriateness of using a constant value of g to describe the skewness of the data. Systematic curvature would mean that such a simple model is not adequate.

Because the elongation factor cancels in calculating g_p, (4) is valid for all values of h. This leaves the task of estimating h, first when $g = 0$ and then when $g \neq 0$. When $g = 0$, the standard quantile function is $Q_{0,h}$ as in (2). The difference between the upper and lower pth quantiles of the data is thus

$$x_{1-p} - x_p = -2B \times Q_{0,h}(z_p), \quad (5)$$

which yields

$$\ln\left(\frac{x_{1-p} - x_p}{-2z_p}\right) = \ln B + h\frac{z_p^2}{2}. \quad (6)$$

Because both h and the scale constant B are unknown, one plots the left-hand side of (6) against z_p^2, one point for each of the values of p. This approach makes reasonably direct use of the data, provides a basis for resistance, and indicates whether a constant value of h is adequate to describe the data.

When $g \neq 0$, it is necessary first to estimate g (as described earlier) and then to adjust the data toward symmetry before making the plot to estimate h. Illustrating the adjustment step in terms of the median and the upper pth quantile, write

$$x_{1-p} - x_{0.5} = \frac{B}{g}(e^{-gz_p} - 1)e^{hz_p^2/2}, \quad (7)$$

divide through by $(e^{-gz_p} - 1)/g$, and take logarithms to return to the situation of (6). Again a plot yields an estimate for h.

NONCONSTANT g AND h

When a constant value of g or h appears inadequate, a generalization of the family of distributions (see ref. 3) regards both g and h as functions of z^2. Then the constant values of g and h are only the constant terms of polynomials such as

$$g(z) = g_0 + g_1 z^2,$$
$$h(z) = h_0 + h_1 z^2.$$

To explore these possibilities, plot g_p from (4) against z_p^2 and look for curvature in the plot based on (6). The greater flexibility that comes from allowing nonconstant g and h seems adequate to describe quite a wide range of distributions, both empirical and theoretical.

MOMENTS

Calculation of the moments of the g-and-h-distributions involves straightforward but tedious algebra. For the standard random variable, Y, obtained from (1), the mean and the variance, respectively, are

$$E(Y) = \frac{1}{g\sqrt{1-h}}\left(e^{(1/2)g^2/(1-h)} - 1\right),$$

$$\operatorname{var}(Y) = \frac{1}{g^2\sqrt{1-2h}}\left(e^{2g^2/(1-2h)} - 2e^{(1/2)g^2/(1-2h)} + 1\right) - \frac{1}{g^2(1-h)}\left(e^{(1/2)g^2/(1-h)} - 1\right)^2.$$

When $h = 0$, these reduce, respectively, to

$$E(Y) = (e^{(1/2)g^2} - 1)/g,$$
$$\mathrm{var}(Y) = e^{g^2}(e^{g^2} - 1)/g^2.$$

In addition,

$$E\left[(Y - E(Y))^3\right] = e^{3g^2/2}(e^{3g^2} - 3e^{g^2} + 2)/g^3,$$
$$E\left[(Y - E(Y))^4\right] = e^{2g^2}(e^{6g^2} - 4e^{3g^2} + 6e^{g^2} - 3)/g^4.$$

When $g = 0$, the first and third moments are zero by symmetry, and

$$\mathrm{var}(Y) = 1/(1 - 2h)^{3/2},$$
$$E(Y^4) = 3/(1 - 4h)^{5/2}.$$

Note that the g-and-h-distributions have finite expectation only when $h < 1$ and finite variance only when $h < \frac{1}{2}$. Thus this family of quantile-based distributions covers a greater range of shapes than can be described in terms of the customary moment-based measures of skewness and kurtosis*.

References

[1] Hoaglin, D. C. and Peters, S. C. (1979). *Proc. Computer Sci. Statist.: 12th Annu. Symp. Interface*, J. F. Gentleman, ed. University of Waterloo, Waterloo, Ontario, pp. 418–423.

[2] Tukey, J. W. (1977). *Exploratory Data Analysis*. Addison-Wesley, Reading, Mass.

[3] Tukey, J. W. (1977). Modern techniques in data analysis. NSF-sponsored regional research conference at Southeastern Massachusetts University, North Dartmouth, Mass.

[4] Wilk, M. B. and Gnanadesikan, R. (1968). *Biometrika*, **55**, 1–17.

(EXPLORATORY DATA ANALYSIS
KURTOSIS
LOCATION AND SCALE PARAMETERS
LOGNORMAL DISTRIBUTION
NORMAL DISTRIBUTION
QUANTILES
SKEWNESS)

DAVID C. HOAGLIN

GAPPING

Gapping is a statistical technique for finding unusual spaces or *gaps* in univariate data strings. It stems from the notation that we expect to find data distributions relatively dense in the middle and relatively sparse in the extremes. Thus a gap in the middle of the distribution that is the same size as one in a tail must be taken more seriously. To allow us to examine gaps and to tell an unusual one from one that is of ordinary size without directly considering where in the distribution it arose, we first weight each empirical gap so that the ones in the middle get larger weights than the ones in the tails. Next we look at the distribution of weighted gaps and determine how large a weighted gap must be in order for us to consider it "unusual." The original work in this was done by Tukey [3], and a study which determined the probability of unusual gaps was completed by Wainer and Schacht [4].

THE DETAILS

Consider an ordered data vector $\mathbf{x}$, where $x_i \geqslant x_{i+1}$, for $i = 1, \ldots, n - 1$. We can then define the gaps g_i as

$$g_i = x_i - x_{i+1}, \qquad i = 1, \quad n - 1$$

and a set of approximately Gaussian (actually logistic*) weights

$$w_i = i(n - i).$$

We then define a weighted gap y_i as

$$y_i = (w_i g_i)^{1/2}.$$

The square-root transformation is used to change what is almost an exponential distribution* to something more nearly symmetric.

It is clear that the values of the y_i's range upward from zero and center themselves around some middle value that depicts the "average" weighted gap. The data analyst has to determine if a particular weighted gap is sufficiently greater than this typical value to warrant more careful attention. A 25% trimmed mean* is used to represent this

Table 1 Percentage of Weighted Gaps above the z^* Value Shown for Samples of Size 50 from a t-Distribution with Inverse Degrees of Freedom as Shown

	D.F.$^{-1}$								
z^*	0	1/8	2/8	3/8	4/8	5/8	6/8	7/8	1
1.50	19.8	20.1	20.0	20.7	21.3	21.9	22.7	23.5	24.4
1.75	11.2	11.3	11.4	12.4	13.2	14.4	15.4	16.4	17.8
2.00	5.9	5.9	6.1	7.1	8.1	9.3	10.6	11.8	13.4
2.25	2.9	2.9	3.2	4.0	5.0	6.2	7.7	8.9	10.5
2.50	1.4	1.3	1.6	2.3	3.3	4.3	5.7	7.0	8.5
2.75	0.6	0.5	0.8	1.4	2.2	3.1	4.4	5.6	7.0
3.00	0.2	0.2	0.4	0.8	1.6	2.3	3.5	4.5	5.8
3.25	0.1	0.1	0.2	0.6	1.2	1.8	2.9	3.8	4.9
3.50	0.0	0.0	0.0	0.4	0.9	1.5	2.4	3.2	4.2
3.75	0.0	0.0	0.0	0.3	0.7	1.2	2.0	2.7	3.7
4.00	0.0	0.0	0.0	0.2	0.6	1.0	1.7	2.4	3.3

Source. Wainer and Schacht [4], reprinted with the permission of *Psychometrika*.

Table 2 Gapping Analysis of Occupational Prestige Scale Values for Italian Males

Occupation	Scale Value	Gap	Weighted Gap	z^*
Physician	1.48			
		0.02	0.77	0.27
Professor	1.46			
		0.26	3.92	1.38
Author	1.20			
		0.05	2.13	0.74
Lawyer	1.15			
		0.19	4.59	1.61
Politician	0.96			
		0.02	1.76	0.62
Banker	0.94			
		0.07	3.28	1.15
Clergyman	0.87			
		0.09	4.03	1.41
Artist	0.78			
		0.09	4.13	1.45
Actor	0.69			
		0.02	1.76	0.62
Missionary	0.67			
		0.13	5.29	1.85
Civil serv. employee	0.54			
		0.06	3.56	1.25
Businessman	0.49			
		0.00	0.00	0.00
School principal	0.49			
		0.02	2.33	0.82
Elem. school teacher	0.47			

Table 2 (*Continued*)

Occupation	Scale Value	Gap	Weighted Gap	z^*
GAP		0.54	11.63	4.08
Bookkeeper	− 0.07			
		0.02	2.37	0.83
Soldier	− 0.09			
		0.21	7.26	2.54
Typographer	− 0.30			
		0.10	5.08	1.78
Electrician	− 0.40			
		0.03	2.61	0.91
Policeman	− 0.48			
		0.06	3.75	1.31
Neutral point	− 0.48			
		0.03	2.59	0.91
Grocer	− 0.51			
		0.06	3.63	1.27
Tailor	− 0.57			
		0.00	0.00	0.00
Carpenter	− 0.57			
		0.01	1.02	0.36
Farmer	− 0.57			
		0.15	5.44	1.90
Butcher	− 0.73			
		0.02	1.92	0.67
Plumber	− 0.75			
		0.33	7.19	2.52
Mailman	− 1.08			
		0.00	0.52	0.18
Truck driver	− 1.08			
		0.00	0.47	0.17
Miner	− 1.08			
		0.03	1.62	0.57
Barber	− 1.11			
		0.08	2.23	0.78
Shoemaker	− 1.20			
		0.20	2.52	0.88
Janitor	− 1.40			

Source. Gulliksen and Gulliksen [2], reprinted with the permission of *Psychometrika*.

middle value. This statistic, denoted y_{T25}, is calculated by trimming off the 25% smallest weighted gaps and the 25% largest, and then taking the mean of the remaining middle 50%. More specifically,

$$y_{T25} = \sum_{i=g+1}^{n-g} y_i/h,$$

where $n = g + h + g$, $(g = n/4)$—shown this way to stress ordering. We then represent each weighted gap as the ratio z_i^* where

$$z_i^* = y_i/y_{T25}\,. \tag{1}$$

This standardizes the distribution of weighted gaps around a middle value of $z^* = 1$. If the original data distribution of the x's was Gaussian we will expect to find about 97% of the values of z^* less than 2.25. Thus we can use this value of z^* as a nominal 97% bound on the significance of any particular gap. This bound holds rather well throughout the entire distribution, indicating that the original weighting scheme, meant to equalize the size of gaps, has worked.

Table 1 shows the nominal type I errors* that one should expect using this scheme, as a function of the original distribution—the distributions used (t's with various degrees of freedom) vary in the fatness of their tails. When the $\text{d.f.}^{-1} = 0$ the underlying distribution is Gaussian; as the degrees of freedom decrease (d.f.^{-1} gets larger), the distribution gets progressively fatter tailed, making it harder and harder to distinguish a gap from a straggly tailed distribution. Nevertheless, the technique is rather robust and for even substantial departures from normality the nominal rule of "$z^* \geqslant 3$ indicates a significant gap" seems reasonable. Although this table shows these values for data strings of length 50, this result generalizes to both shorter and longer strings rather well; however, the power of the test is greater when the sample sizes are smaller.

An important caveat in using this gapping scheme is that the probabilities given in Table 1 are *per gap*, not *whole test* probabilities. Thus one must take into account the total number of gaps in each sample before assigning a probability to a particular one; this is akin to the same problem in multiple pairwise comparisons.

AN EXAMPLE (FROM REF. 4)

In a monograph detailing their study of occupational prestige, Gulliksen and Gulliksen [2] list the scale values for 32 occupations. These scale values were obtained from successive interval judgments and estimated using the *law of categorical judgment* (see Bock and Jones [1]). The data in Table 2 are the scale values for these occupations, from physician as the highest prestige occupation, to janitor as the lowest. A natural question is: Are there any natural breaks in these scale values? Perhaps there is separation of the white-collar from blue-collar jobs, or skilled from unskilled, or professional from nonprofessional. Thus we employed the gapping procedure previously described to these data; the results are shown in Table 2.

The biggest gap is between "elementary school teacher" and "bookkeeper," which seems to separate occupations requiring a college degree from those which do not. A z^* of 4.08 is clearly significant [($p < 0.0005$) for distributions thinner tailed than $t_{(4)}$], even with 31 gaps being considered ($p < 0.02$). There are two other gaps with large z^* levels; between "soldier" and "typographer" ($z^* = 2.54$) and between "plumber" and "mailman" ($z^* = 2.52$). Neither of these achieves nominal significance when the "family of tests" significance is considered, although it does point out that one ought to look out for similar gaps in follow-up studies.

CONCLUSION

Gapping is a scheme that is useful for detecting unusual holes in a data distribution. It is most useful as an exploratory technique with smallish samples since it will tend to point out where a distribution is thinner than one would expect. A piling up of gaps in the region of thinness is the symptom that one should look for. This method will be especially helpful when the data are too limited for something fancier (like Gaussian decomposition); it has been demonstrated that gapping can give hints for hole detection when the data are too sparse for these more distribution-bound techniques to work.

References

[1] Bock, R. D. and Jones, L. V. (1968). *The Measurement and Prediction of Judgment and Choice*. Holden-Day, San Francisco.

[2] Gulliksen, H. and Gulliksen, D. (1972). *Attitudes of Different Groups toward Work, Aims, Goals and Activities*. Multivariate Behavior Research Monograph.

[3] Tukey, J. W. (1971). *Exploratory Data Analysis*. Addison-Wesley, Reading, Mass. (limited prelim. ed.).

[4] Wainer, H. and Schacht, S. (1978). *Psychometrika*, **43**, 203–212.

(EXPLORATORY DATA ANALYSIS
TRIMMED MEAN)

HOWARD WAINER

GART'S INDEPENDENCE TEST STATISTIC

If n_{ij} $(i = 1, \ldots, r; j = 1, \ldots, c)$ are cell frequencies in a $r \times c$ contingency table with the marginals $n_{i\cdot} = \sum_j n_{ij}$, $n_{\cdot j} = \sum_i n_{ij}$, $n_{\cdot\cdot} = \sum_i \sum_j n_{ij}$, the likelihood-ratio* chi-square* statistic G^2 for testing independence can be modified (see Gart [1]) to give M/C, where

$$\begin{aligned} M = {} & \sum_i \sum_j (2n_{ij} + 1)\ln(2n_{ij} + 1) \\ & - \sum_i (2n_{i\cdot} + c)\ln(2n_{i\cdot} + c) \\ & - \sum_j (2n_{\cdot j} + r)\ln(2n_{\cdot j} + r) \\ & + (2n_{\cdot\cdot} + rc)\ln(2n_{\cdot\cdot} + rc); \end{aligned}$$

$$\begin{aligned} C = {} & 1 + \frac{1}{3(r-1)(c-1)} \\ & \times \left[\sum_i \sum_j \frac{1}{2n_{ij} + 1} - \sum_i \frac{1}{2n_{i\cdot} + c} \right. \\ & \left. - \sum_j \frac{1}{2n_{\cdot j} + r} + \frac{1}{2n_{\cdot\cdot} + rc} \right]. \end{aligned}$$

The modified test treats M/C as chi-square with $(r-1)(c-1)$ degrees of freedom. Numerical examples indicate that this test statistic and variants of it give more accurate results than do Pearson's chi-square or G^2. Gart [1] relates his procedure to the Freeman–Halton test* and also describes a logit* approach.

Reference

[1] Gart, J. J. (1966). *J. R. Statist. Soc. B*, **28**, 164–179.

(CHI-SQUARE TESTS
CONTINGENCY TABLES)

GASTWIRTH'S ESTIMATOR *See* QUANTILES

GATEAUX DERIVATIVE *See* STATISTICAL FUNCTIONALS

GAUSS, CARL FRIEDRICH

Born: April 30, 1777, in Brunswick, Germany.

Died: February 23, 1855, in Göttingen, Germany.

Contributed to: statistics, mathematical physics, astronomy.

[*Editors' Note.* Carl Friedrich Gauss was born into a humble family in Brunswick, Germany; he became famous as an astronomer and mathematician before he was 25 years old, being educated at the Universities of Göttingen and Helmstedt, where he received his doctorate in 1799. For a time he was supported by a stipend from the Duke of Brunswick, but in 1807 he moved to Göttingen to become director of the observatory, and remained there for the rest of his life.

Kenneth O. May [11] described Gauss as "one of the greatest scientific virtuosos of all time"; his penetrating research and prolific output bear witness to that. At times his results were produced more rapidly than he could set them down and publish them. One example of this was his accurate prediction in 1801 of the location in the heavens of a supposed new distant planet, Ceres, which for a time had been lost. Gauss was heaped with praise on account of this achievement; but he did not set down until 1809 the processes by which he had made his prediction, namely, the refinement of orbit theory and the method of least squares*.

The range of Gauss's influence in science, particularly in mathematical physics, has been enormous. He made strides in celestial mechanics, geodesy, number theory, optics, electromagnetism, real analysis, theoretical physics, and astronomy, as well as in statistics. It is surprising that he found no collaborators in mathematics, worked alone for most of his life, and never visited Paris (perhaps because he was strongly opposed to political revolutions). May [11] contains a selective but comprehensive bibliography of Gauss, including translations and reprints.]

Gauss's principal contributions to statistics are in the theory of estimation, and are known as *least squares*. The problem is to estimate k unknown parameters $\theta_1, \theta_2, \ldots, \theta_k$ on the basis of $n > k$ observations $y_1, y_2, \ldots, y_n$, where $y_i = \xi_i(\theta_1, \theta_2, \ldots, \theta_k) + e_i$. The quantities e_i are observational errors, assumed to be random and free from systematic error. In matrix notation this is

$$\mathbf{Y} = \boldsymbol{\xi} + \mathbf{e}, \tag{1}$$

where $\mathbf{Y}$, $\boldsymbol{\xi}$, $\mathbf{e}$ are the respective column vectors of the y_i's, ξ_i's, and e_i's. This problem is historically referred to as the *combination of observations*, and was regarded by Gauss as one of the most important in natural philosophy; he attached great importance to his contributions to this field.

For clarity, it is useful to distinguish between the *principle* of least squares and the *theory* of least squares.

PRINCIPLE OF LEAST SQUARES

The principle of least squares chooses θ to minimize $Q = \sum(y_i - \xi_i)^2$. Thus θ is a solution of $\partial Q/\partial\theta_j = 0$, $j = 1, 2, \ldots, k$. These are the least-squares equations

$$\sum(y_i - \xi_i)\partial\xi_i/\partial\theta_j = 0. \tag{2}$$

In the special but widespread case where the ξ_i are linear in the θ's, $\xi_i = \sum_j x_{ij}\theta_j$, or in matrix notation, $\boldsymbol{\xi} = \mathbf{X}\boldsymbol{\theta}$, the x's being known constants, (1) is

$$\mathbf{Y} = \mathbf{X}\boldsymbol{\theta} + \mathbf{e}, \tag{3a}$$

the Gauss linear model* discussed in statistics texts as linear regression*. The least-squares equations are now $\mathbf{X'X} = \mathbf{X'Y}$ and the least-squares estimate is

$$\hat{\boldsymbol{\theta}} = (\mathbf{X'X})^{-1}\mathbf{X'Y}. \tag{3b}$$

There has been much discussion surrounding priority in the development of least squares, since Legendre also considered (1) and (3a) and essentially developed (2) and (3b). Whereas Gauss published in 1809 [5], Legendre published in 1805 [10] and was responsible for the name "least squares" (moindre carrés). Of relevance to this issue of priority is the fact that Legendre confined himself to the *principle* of least squares outlined above, whereas Gauss developed in addition the statistical theory of least squares as outlined below.

STATISTICAL THEORY

Gauss's First Approach [5]

Assuming that $e_i = y_i - \xi_i$ are independent random variables with distribution $f(e_i)$, the joint distribution of the observational errors is $\Omega = \prod f(e_i) = \prod f(y_i - \xi_i)$. Assuming that all values of the θ's are equally probable, the distribution of the θ's given the observed values $\mathbf{y}$ is by Bayes' theorem* proportional to Ω. Gauss chose the most probable value $\hat{\boldsymbol{\theta}}$ (i.e., the mode of Ω) as an estimate of $\boldsymbol{\theta}$. This is obtained as a root of

$$\frac{\sum \partial \log f(y_i - \xi_i)}{\partial \xi_i} \frac{\partial \xi_i}{\partial \theta_j} = 0, \qquad j = 1, 2, \ldots, k. \tag{4}$$

To proceed further, the mathematical form of f must be known. To this end, Gauss assumed that for the special case of (1) in which $y_i = \theta_1 + e_i$ for all i (so that there is only one parameter θ_1), the least-squares estimate should be the arithmetic mean*, $\hat{\theta}_1 = \bar{y}$. It follows that f must be the normal distribution $(\sqrt{2\pi}\,\sigma)^{-1}\exp(-e^2/2\sigma^2)$. The distribution of the θ's is then proportional to

$$\left(\frac{1}{\sqrt{2\pi}\,\sigma}\right)^n \exp(-Q^2/2\sigma^2),$$

where $Q = \sum(y_i - \xi_i)^2$. This probability is maximized by minimizing Q, which is the principle of least squares described above. Gauss also considered the case of unequal variances σ_i^2, leading to weighted least squares* $Q = \sum(y_i - \xi_i)^2/\sigma_i^2$. He then went through arguments, now standard in statistical texts, to show that $\boldsymbol{\theta}$ has the multivariate normal distribution about $\hat{\boldsymbol{\theta}}$ with covariance matrix $(\mathbf{X}'\mathbf{X})^{-1}\sigma^2$ (see Seal [12] for details).

Gauss's Second Approach [7]

The essential feature of this second approach is the assumption that when the true value θ is estimated by $\hat{\theta}$, an error $\theta - \hat{\theta}$ is committed, which entails a loss. The estimate $\hat{\theta}$ is then chosen to minimize the expected loss. He took a *convenient*, although admittedly *arbitrary* loss which is proportional to the squared error $(\theta - \hat{\theta})^2$. Then $\hat{\theta}$ is chosen to minimize the mean square error* (MSE), $E(\theta - \hat{\theta})^2$.

Gauss assumed that the errors e_i were sufficiently small that their squares and higher powers could be ignored, and thus restricted attention to *linear* estimates $\hat{\boldsymbol{\theta}} = \mathbf{CY}$ such that $\mathbf{CX} = \mathbf{I}$, the $(k \times k)$ identity matrix. He then showed that among all such estimates, the least-squares estimate (3b) minimizes the MSE. The resulting MSE is $(\mathbf{X}'\mathbf{X})^{-1}\sigma^2$, thus reproducing the results of the first approach (with the exception, of course, of the normality of $\boldsymbol{\theta}$). In addition, the following results were obtained:

The least-squares estimate of any linear parametric function $\alpha = \sum g_i\theta_i = \mathbf{g}'\boldsymbol{\theta}$ is given by $\hat{\alpha} = \mathbf{g}'\hat{\boldsymbol{\theta}}$, with standard error

$$\sigma_\alpha = \left\{\mathbf{g}'(\mathbf{X}'\mathbf{X})^{-1}\mathbf{g}\right\}^{1/2}\sigma.$$

The minimum of Q is $Q_m = \mathbf{Y}'(\mathbf{Y} - \mathbf{X}\hat{\boldsymbol{\theta}})$.

If $\alpha = \mathbf{g}'\boldsymbol{\theta}$ is held fixed and the θ_i's otherwise allowed to vary, then Q can take on a relative minimum Q_r such that $Q_r - Q_m \leqslant C^2$ implies that $|\hat{\alpha} - \alpha| \leqslant C\sigma_\alpha/\sigma$.

An additional observation y with corresponding x-values

$$\mathbf{x}' = (x_{n+1,1}, x_{n+2,2}, \ldots, x_{n+1,k})$$

can be incorporated into the original least-squares estimate $\hat{\boldsymbol{\theta}}$ to form the updated estimate $\boldsymbol{\theta}^*$ via

$$\boldsymbol{\theta}^* = \hat{\boldsymbol{\theta}} - \mathbf{M}(\mathbf{x}'\hat{\boldsymbol{\theta}} - y)/(1 + w),$$

where $\mathbf{M} = (\mathbf{X}'\mathbf{X})^{-1}\mathbf{x}$, and $w = \mathbf{x}'(\mathbf{X}'\mathbf{X})^{-1}\mathbf{x} = \mathbf{x}'\mathbf{M}$ (which is a scalar). The covariance matrix of $\boldsymbol{\theta}^*$ is $((\mathbf{X}'\mathbf{X})^{-1} - \mathbf{MM}'/(1 + w))\sigma^2$ and the new minimum value of Q is $Q_m^* = Q_m + (\mathbf{x}'\hat{\boldsymbol{\theta}} - y)^2/(1 + w)$. This allows continuous updating of the least-squares estimate in the light of further observations obtained sequentially, without having to invert the new $(\mathbf{X}'\mathbf{X})$ matrix each time, and may be called "recursive least squares."

$E(Q_m) = (n - k)\sigma^2$; thus σ should be estimated by $\hat{\sigma} = \sqrt{Q_m/(n - k)}$, and not by $\sqrt{Q_m/n}$. Gauss also obtained the standard error of the estimate and noted that, when the e_i's are standard normal, this standard error becomes the standard error of the sum of $(n - k)$ independent errors e_i.

If $\mathbf{X} = \mathbf{I}_n$, so that there are n parameters $\xi_i = \theta_i$, n observations $y_i = \theta_i + e_i$, and if there are r linear restrictions $\mathbf{F}\boldsymbol{\theta} = \mathbf{0}$ where $\mathbf{F} = (f_{ij})$, $i = 1, \ldots, r$, then among linear functions the estimate of $\alpha = \mathbf{g}'\boldsymbol{\theta}$ that minimize the mean square error is $\hat{\alpha} = \mathbf{g}'\boldsymbol{\theta}^*$, where $\boldsymbol{\theta}^*$ is the least-squares estimate of $\boldsymbol{\theta}$ subject to $\mathbf{F}\boldsymbol{\theta} = \mathbf{0}$. Its standard error is $\hat{\sigma} = \sqrt{Q_m/r}$.

Discussion

Gauss not only developed the foregoing theory, but also *applied* it to the analysis of observational data, much of which he himself collected. Indeed, the development of the theory was undoubtedly a response to the problems posed by his astronomical and geodetic observations. For example, at the end of *Theoria Motus Corporum Coelestium* [5], Gauss states that he has used this principle since 1795. See also numerical examples in Gauss [4; 7 (1826); 8, 9].

The difference in generality between the first and second approaches should be noted. The first approach allows $\hat{\boldsymbol{\theta}}$ to be any function of the observations, but requires the errors of observation e_i to be normally distributed about a zero mean. The second approach restricts the estimate $\hat{\boldsymbol{\theta}}$ to linear functions of the observations but allows the e_i to have any distribution with a mean of zero and finite variance.

The maximizing of Ω leading to (4) is equivalent to the method of estimation known today as maximum likelihood*, the most useful general method of estimation. The properties of the maximum likelihood estimate were obtained by Fisher (1922 [3] and later).

The only way in which modern textbooks add to the foregoing theory of Gauss is in the explicit setting out of tests of linear hypotheses $\mathbf{A}\boldsymbol{\theta} = \mathbf{0}$ and in the concomitant exact distributional theory associated with normally distributed observational errors, e.g., the t, F, and χ^2 distributions.

An unfortunate feature of the treatment in modern textbooks is their reference to the foregoing as the Gauss–Markov theory, and in their insistence that Gauss was seeking unbiased estimates, $E(\hat{\boldsymbol{\theta}}) = \boldsymbol{\theta}$. Regarding the former, there seems no justification for associating the name of Markov with the foregoing theory (but *see* GAUSS–MARKOV THEOREM). Regarding the latter, it is true that $\mathbf{CX} = \mathbf{I}_k$ implies that $\hat{\boldsymbol{\theta}}$ of (3b) is unbiased. But the *requirement* of unbiasedness is unreasonable and can lead to absurd results. That Gauss did not insist on unbiased estimates is evident from his estimate of σ as $\sqrt{Q/(n-k)}$, which is biased. The condition $\mathbf{CX} = \mathbf{I}_k$ is in fact a "consistency" criterion, the purpose of which is to specify what is being estimated. In the present case, the requirement is that if the observational errors are all $e_i \equiv 0$, then the equations $\mathbf{Y} = \mathbf{X}\boldsymbol{\theta}$ must be consistent and the estimate must be identical to the true value, $\hat{\boldsymbol{\theta}} \equiv \boldsymbol{\theta}$. As stated by Bertrand [1, p. 255], the official translator of Gauss's statistical work into French, "car, sans cela, toutes les mesures étant supposées exact, la valeur qu'on en déduit ne le serait pas."

Also, modern texts do not restrict the domain of application of linear estimates, as did Gauss, by assuming the observational errors to be sufficiently small that their squares and higher powers can be ignored. Similarly, many modern texts overemphasize the criterion of mean square error, regarding it as the foundation on which to base a theory of estimation. In this way they ignore the qualifications imposed by Gauss.

There are other statistical contributions of Gauss. A notable one [6] is the demonstration that for a normal distribution, the most precise estimate of σ^2 among estimates depending on $S_k = \sum |e_i|^k$ is obtained when $k = 2$. His calculations were independently obtained by Fisher [2], who went on to isolate the property of sufficiency*. In another noteworthy paper Gauss [9] exemplifies the use of weighted least squares in determining longitude by the use of a chronometer. He presents there a discussion of the structure of the observation errors e_i obtained in the use of a chronometer, and takes the standard deviation σ_i to be proportional to the square root of the time elapsed between observations. All of the above demonstrate a mixture of theory and of *application* that seems all too rare today.

A more detailed account and discussion of the matters raised in this article appear in Sprott [13].

References

[1] Bertrand, J. (1888). *Calcul des probabilités*. (2nd ed., 1972, Chelsea, New York.)

[2] Fisher, R. A. (1920). *Monthly Notices R. Astron. Soc.*, **80**, 758–770.

[3] Fisher, R. A. (1922). *Philos. Trans. R. Soc. A*, **222**, 309–368.

[4] Gauss, C. F. (1803–1809). Disquisitiones de Elementis Ellipticis Pallidis. *Werke*, **6**, 1–24.

[5] Gauss, C. F. (1809). Theoria Motus Corporum Coelestium. *Werke*, **7**. (English transl: C. H. Davis. Dover, New York, 1963).

[6] Gauss, C. F. (1816). Bestimmung der Genauikgeit der Beobachtungen. *Werke*, **4**, 109–117.

[7] Gauss, C. F. (1821, 1823a, 1826). Theoria Combinationis Erroribus Minimis Obnoxiae, Parts 1, 2 and Suppl. *Werke*, **4**, 1–108.

[8] Gauss, C. F. (1823b). Anwendungen der Wahrscheinlichkeitsrechnung auf eine Aufgabe der praktischen Geometrie. *Werke*, **9**, 231–237.

[9] Gauss, C. F. (1824). Chronometrische Langenbestimmungen. *Astron. Nachr.*, **5**, 227.

[10] Legendre, A. M. (1805). *Nouvelles méthodes pour la determination des orbits des comètes*. (Appendix: Sur la méthode des moindre carrés.)

[11] May, K. O. (1972). In *Dictionary of Scientific Biography*, Vol. 5. Scribners, New York, pp. 298–315.

[12] Seal, H. L. (1967). *Biometrika*, **54**, 1–24.
[13] Sprott, D. A. (1978). *Historia Math.*, **5**, 183–203.

Bibliography

Gauss, C. F. (1957). Gauss's Work (1803–1826). On The Theory of Least Squares. Mimeograph (Dept. of Science and Technology, Firestone Library, Princeton University). English translations of 1821, 1823a, 1826, 1816, 1823b, 1824 and the statistical parts of 1809 and 1803–1809, by H. F. Trotter, based on the French translation by J. Bertrand published in 1855 and authorized by Gauss.

(GAUSS–MARKOV THEOREM
GENERAL LINEAR MODEL
LEAST SQUARES)

D. A. SPROTT

GAUSS–HELMERT MODEL

This model is described by the equation

$$\mathbf{z} = \mathbf{A}\mathbf{x} + \mathbf{B}\mathbf{y}, \quad E(\mathbf{y}) = \mathbf{0}, \quad \mathrm{Var}(\mathbf{y}) = \sigma^2\mathbf{\Sigma},$$

where $\mathbf{A}$ and $\mathbf{B}$ are matrices of known coefficients, $\mathbf{x}$ is a vector of fixed unknown parameters, $\mathbf{y}$ is a vector of random unknown parameters, $\mathbf{z}$ is a vector of random observations, the matrix $\mathbf{\Sigma}$ is assumed to be positive definite, and σ is an unknown scalar parameter.

Let $\mathbf{C}_{zz}$ be the variance–covariance matrix of $\mathbf{z}$, and $\mathbf{C}_{yz}$ the joint covariance matrix of the vectors $\mathbf{y}$ and $\mathbf{z}$; let

$$\hat{\mathbf{x}} = \left(\mathbf{A}'\mathbf{C}_{zz}^{-1}\mathbf{A}\right)\mathbf{A}'\mathbf{C}_{zz}^{-1}\mathbf{z},$$

$$\hat{\mathbf{y}} = \mathbf{C}_{yz}\mathbf{C}_{zz}^{-1}(\mathbf{z} - \mathbf{A}\hat{\mathbf{x}}).$$

Then [1] the best linear unbiased estimator of $\mathbf{a}'\mathbf{x} + \mathbf{b}'\mathbf{y}$ in the mean square error sense is $\mathbf{a}'\hat{\mathbf{x}} + \mathbf{b}'\hat{\mathbf{y}}$, for any arbitrary vectors $\mathbf{a}$ and $\mathbf{b}$. Koch [1] also gives additional properties and some historical remarks.

Reference

[1] Koch, K. R. (1979). *Boll. Geod. Sci. Affini*, **38**(4), 553–563.

(GENERAL LINEAR MODEL)

GAUSSIAN DISTRIBUTION *See* LAWS OF ERROR; NORMAL DISTRIBUTION

GAUSSIAN PROCESSES

Gaussian processes arise in practice when Gaussian modeling is used for a phenomenon evolving in time. Typical examples of these are concentrations of oxidants in the air, flood levels of a river, wind thrust on high-rise buildings, etc. The practical advantage of Gaussian processes is the wide-ranging and varied situations in which they can be used. The theoretical appeal is the relative simplicity of their mathematical structure. Many of the commonly made assumptions on stochastic processes achieve an added strength and become simplified under the Gaussian hypothesis. The most famous Gaussian process studied is *Brownian motion** or the *Wiener process*, which models the erratic behavior of particles suspended in a fluid as observed by the botanist Brown in 1827.

Gaussian processes other than these (especially the stationary ones) have come into prominence only since the 1960s. Following is a review of the trends and results in the field of Gaussian processes (other than Brownian motion and related processes), classified according to the commonly used categories of stochastic processes. To avoid unnecessary complexity, we assume the processes to be real valued.

A stochastic process $\{X(t);\ t \in T\}$ is called a *Gaussian process* if all its finite-dimensional distributions are multivariate Gaussian, i.e.,

$$E\left\{\exp\left(i\sum_{j=1}^{k} s_j X_{t_j}\right)\right\} = \exp\left\{i\sum_{j=1}^{k} m_j s_j - \tfrac{1}{2}\sum_{j=1}^{k}\sum_{l=1}^{k} s_j \lambda_{jl} s_l\right\}$$

for all $t_1, \ldots, t_k \in T$, where $-\infty < m_j < \infty, j = 1, 2, \ldots, k$, and λ_{jl} are elements of some nonnegative definite matrix Λ. Since linear combinations and limits of Gaussian variables are Gaussian, all linear operations on the process (e.g., stochastic integration or differentiation) would lead to Gaussian processes.

A stochastic process $\{X(t);\ t \in T\}$ is *strictly stationary** if all the finite-dimensional distributions are invariant under a time shift. It is *weak* or *covariance stationary* if $E(X(t)) \equiv m$ and the covariance $E((X(t)-m)(X(s)-m)) = r(t-s)$; $t, s \in T$. In the case of a (real-valued) Gaussian process, both these notions are equivalent. Because Gaussian distributions are determined by the first two moments, for every covariance function there exists a stationary Gaussian process (SGP) with this covariance function.

Let $Z_1, \ldots, Z_n$ be any set of complex numbers and $r(t)$ the covariance function of a (weakly) stationary process $\{X(t);\ t \in T\}$. Then

$$\sum_{j=1}^{n}\sum_{k=1}^{n} r(t_j - t_k) Z_j \bar{Z}_k = E\left|\sum_{j=1}^{n} X(t_j) Z_j\right|^2 \geqslant 0$$

for all n and $t_1, \ldots, t_n \in T$. Any continuous function $r(t)$ satisfying the foregoing property is said to be *nonnegative definite*. Bochner's theorem states that $r(t)$ is nonnegative definite if and only if it has the *spectral representation*

$$r(t) = \int_{-\infty}^{\infty} \exp(it\lambda)\, dF(\lambda),$$

where $F(\lambda)$ is real, nondecreasing, and bounded, and when F is properly normalized, it is called the *spectral distribution* (see Cramér and Leadbetter [2, p. 126]).

Pólya's criterion (see Lukács [9, p. 83]) says that the nonnegative, convex functions on $[0, \infty)$ are covariance functions, showing that the class of SGP is very rich indeed. A stationary process itself has a spectral representation given by

$$X(t) = \int_{-\infty}^{\infty} \exp(2\pi it\lambda)\, dy(\lambda);$$

$$E|dy(\lambda)|^2 = dF(\lambda)$$

(see Doob [3, p. 532]). A stationary process is a *moving-average process* if and only if its spectral distribution is absolutely continuous. In the case of a SGP, the absolutely continuous spectrum leads to a representation of the process as a moving average of the white noise* on the line. The general *least-squares approximation* reduces to the linear one (see Doob [3, pp. 533, 563]). The white noise process is thought of as a Gaussian process with a constant spectral density. Since a constant is nonintegrable, the correct mathematical definition of the white noise process can be given only in terms of a stochastic differential equation. For practical purposes, it can be taken to be an *Ornstein–Uhlenbeck process**, i.e., a SGP with covariance $r(t) = (\sigma^2/2\alpha)\exp\{-\alpha|t|\}$ for very large values of constants $\alpha > 0$ and σ such that $C = \sigma^2/2\alpha^2$ remains constant.

Historically, the Ornstein–Uhlenbeck process arose as the solution X_t of the stochastic differential equation $dX_t = -\alpha X_t\, dt + \sigma dW_t$, with the initial condition that X_0 is a Gaussian variable with mean zero, variance $\sigma^2/2\alpha$, and W_t as the Brownian motion. It represents one of the velocity components of a particle described in the first paragraph under the influence of friction but of no other force field. It is the only SGP to possess the Markov property. Many results (e.g., the distributions of the first passage time and the maximum) have been studied in this case, using the Markov property.

As mentioned above, for suitable choices of the constants σ and α it is taken as an approximation to Gaussian white noise. The integral of this process, i.e., $Y_t = Y_0 + \int_0^t X_s\, ds$ (sometimes called an Ornstein–Uhlenbeck position process), gives an approximation to Brownian motion itself. Again, if $\alpha \to \infty$ as $C = \sigma^2/(2\alpha^2)$ remains constant, the finite-dimensional distributions of the Y_t process approach those of $Y_0 + \sqrt{2C}\, W_t$, where W_t is Brownian motion and Y_0 is normally distributed or a constant.

In terms of the Hilbert space H generated by a zero-mean *second-order process* $\{X(t);\ -\infty < t < \infty\}$, $H(t)$ = the Hilbert space generated by $\{X(u);\ u \leqslant t\}$ is called the "past and the present" of $X(t)$. The limiting subspace $H(-\infty)$ ($\subset H(t)$ for all t) is called the "remote past." The process is called *deterministic* (or *nonregular*) otherwise. In case $H(-\infty) = 0$, the process is called *purely nondeterministic*. In the deterministic process, perfect (prediction error zero) least-squares prediction of X_t is possible based on the past. Every zero-mean second-order process can be written as the sum of two mutually orthogonal parts, one deterministic and the other purely nondeterministic. In the case of a stationary process, the discrete and the singular components of the spectral distribution contribute only to the deterministic part. However, the absolutely continuous component can contribute to either of the two parts (see Doob [3, p. 579]). A purely nondeterministic stationary process can be represented as

$$X(t) = \int_{-\infty}^{t} g(t-s)\, d\eta(s)$$

for some *impulse response* function $g(s)$ as a special case of the spectral representation (see Cramér and Leadbetter [2, p. 146]). The least-squares linear predictor of $X(t_1)$ in terms of $\{X(t);\ -\infty < t \leqslant t_0 < t_1\}$ is given explicitly by

$$\int_{-\infty}^{t_0} g(t_1 - s)\, d\eta(s).$$

In the case of a SGP,

$$X(t) = \int_{-\infty}^{t} g(t-s)\xi'(s)\, ds,$$

where ξ' is the white noise *input*, and the best linear predictor $\int_{-\infty}^{t_0} g(t_1 - s)\xi'(s)\, ds$ now becomes the best predictor.

Another large subclass of SGPs with absolutely continuous spectrum can be written as a suitable moving average* of the "noise" in the plane [1, p. 329]. This representation can be used more readily for simulation of SGPs. Multidimensional isotropic Gaussian random fields as moving averages of the appropriate "white noise" are considered by Mittal [10, p. 519] and could lead to simulations of such processes.

Ergodicity of a process refers to its ability to make increasingly accurate estimates as $T \to \infty$, of ensemble averages like $EX(t)$ or $E(X(t)X(s))$ based on the observed segment of a single sample path. In a (covariance) stationary process with zero mean, $(1/T)\int_0^T X(t)\, dt$ converges in quadratic mean to zero if $(1/T)\int_0^T r(t)\, dt \to 0$ as $T \to \infty$. This condition is satisfied if and only if the spectral distribution is continuous at the origin. In the case of a SGP, quadratic mean convergence can be replaced by almost sure convergence (*see* CONVERGENCE OF SEQUENCES OF RANDOM VARIABLES).

Analytic properties like continuity or differentiability of a stochastic process are proved under *local conditions* on the process which control the interdependence of neighboring random variables. In the case of a zero-mean second-order process, these can be stated in terms of the covariance function $r(t,s) = E(X(t)X(s))$. (Note that the processes are assumed real-valued.) For example, if

$$r(t+h, t+h) + r(t,t) - 2r(t, t+h) < \frac{K|h|}{|\ln|h||^q}$$

for some constants $K > 0$ and $q > 3$, then $X(t)$ is equivalent to a process which with probability 1 has continuous sample paths. For a Gaussian process the following much weaker sufficient condition can be given [2, pp. 89, 183]:

$$r(t+h, t+h) + r(t,t) - 2r(t, t+h) < \frac{K}{|\ln|h||^q}$$

for some constants $K > 0$ and $q > 3$. Some alternative sufficient conditions can be stated which are not possible without the Gaussian assumption. For a SGP, X. Fernique has given a necessary and sufficient condition for the continuity of the sample paths. Extensive results about the analytic properties of Gaussian processes are available under considerably weaker conditions than those on the corresponding second-order processes. These are too numerous and technical to state here; see the survey paper by Dudley [4], which also contains an extensive bibliography.

The *mixing conditions* of a process relate to the asymptotic independence of distantly situated random variables. The strength of the Gaussian assumption enables one to formulate these in terms of the covariance function alone. The pioneering work of Berman [1] in SGPs covered two alternative mixing conditions, $\lim_{t\to\infty}\{r(t)\ln t\} = 0$ (rate condition) and $\int_0^\infty r(t)\,dt < \infty$ (integral condition). The rate-type condition is in commoner use than the integral type one, even though neither implies the other.

The following is a sample of asymptotic results available in SGPs. More information and appropriate references can be found in a book by Galambos [5] and in notes by Leadbetter et al. [8]. The results are stated here for the discrete-time case, although all of them extend to continuous time under suitable local conditions. The mean and the variance of the stationary Gaussian sequence $\{X_n;\ n \geqslant 0\}$ are assumed to be zero and 1, respectively, and $r_n = EX_kX_{n+k}$ with $\lim_{n\to\infty}(r_n \ln n) = 0$. If $M_n = \max_{1\leqslant i\leqslant n} X_i$, then, properly normalized, M_n converges in distribution to the double exponential or extreme-value distribution*, i.e.,

$$\lim_{n\to\infty} \Pr\{M_n \leqslant U_n(x)\} = \exp(-\exp(-x)),$$
$$-\infty < x < \infty,$$

where $U_n(x) = b_n + x/c_n$; $c_n = (2\ln n)^{1/2}$ and $b_n = c_n - \ln(4\pi \ln n)/(2c_n)$. If $M_n^{(k)}$ denotes the kth largest maximum in $X_1, \ldots, X_n$ (thus $M_n = M_n^{(1)}$) for a fixed integer k, then

$$\lim_{n\to\infty} \Pr(M_n^{(k)} \leqslant U_n(x))$$
$$= \lim_{n\to\infty} \Pr(\text{exceedances of } U_n(x) \text{ are} \leqslant k)$$
$$= \exp(-\exp(-x)) \sum_{j=0}^{k-1} (\exp(-x))^j/j!.$$

If $W_n = \min_{1\leqslant i\leqslant n} X_i$, then

$$\lim_{n\to\infty} \Pr\{M_n \leqslant U_n(x);\ W_n \leqslant -U_n(y)\}$$
$$= \exp(-\exp(-x))\{1 - \exp(-\exp(-y))\}.$$

That is, the maximum and the minimum are asymptotically independent. If L_n = location of the maximum, it is asymptotically uniformly distributed, i.e., $\lim_{n\to\infty} \Pr(L_n/n \leqslant u) = u$ for $0 \leqslant u \leqslant 1$. Among many other quantities studied are the zero crossings, ranks, range, record values and times, and the joint distributions of various quantities such as the maxima on disjoint intervals or the height and the location of the maximum, etc.

For a SGP with absolutely continuous spectrum, the traditional *strong mixing condition* is equivalent to the continuity of the spectral density (see Kolmogorov and Rozanov [7]). Additional conditions on the spectrum can ensure both the rate and the integral types of conditions. [For example, if the derivative of the spectral density satisfies a Lipschitz condition of order $\alpha > 0$, then $r(t) = O(t^{-\alpha})$.] But in general, these conditions are not comparable with the strong mixing. [One point of interest can be noted; the *Ornstein–Uhlenbeck* process with $r(t) = O(t^{-\alpha})$ has a representation as an *autoregressive process* of order 1.] Leadbetter et al. [8, pp. 29, 35] proved many of the asymptotic results for stationary processes under the conditions $D(U_n)$ and $D'(U_n)$. The rate condition in SGPs can be weakened somewhat by controlling the excursions of $r(t)$ above $f(t)$ where $f(t)$ is $O(1/\ln t)$. However, the condition itself is almost necessary, as would be apparent in the following. One noteworthy feature of a SGP is the availability of results in the strongly dependent case. For example, if $\lim_{n\to\infty} r_n \ln n = C$, then the limit distribution of the normalized maximum is the standard normal when $C = \infty$ and a convolution of the double exponential and the standard normal when $0 < C < \infty$ (see Mittal and Ylvisaker [11]). Extensions of many results are now available for the dependent case. A unified theory of dependence such as this seems impossible without the Gaussian assumption.

Various zero–one* laws could be proved by making strong use of the Gaussian assumption. For example, in a SGP, the sample paths of the process are continuous with probability either zero or 1; when it is zero, the sample paths, with probability 1, are unbounded in each interval. Dudley [4] and Kallianpur [6] have more discussion of zero–

one laws for Gaussian processes. A more traditional zero–one law occurs in connection with Feller's ideas of "upper" and "lower functions." If $\{S_n;\ n \geqslant 1\}$ is a sequence of partial sums of independent, identically distributed standard normal variables and $\{\phi_n;\ n \geqslant 1\}$ a sequence of positive numbers with $\lim_{n\to\infty}\phi_n = \infty$, then $\Pr(S_n > n\phi_n$ i.o. (infinitely often)) is 0 or 1 according as $\sum_n(\phi_n/n)\exp(-\phi_n^2/2)$ is finite or infinite. The classical *law of the iterated logarithm* follows easily from this result. The ideas of upper and lower functions are extended with S_n/n replaced by either the discrete- or continuous-time SGP. The laws of the iterated logarithm for maxima are related to these results and are found in both the weakly and strongly dependent case.

More information and references can be found in the following.

References

[1] Berman, S. M. (1975). *Pacific J. Math.*, **58**(2), 323–329. (An interesting necessary and sufficient condition for absolutely continuous covariance functions. Research level.)

[2] Cramér, H. and Leadbetter, M. R. (1967). *Stationary and Related Stochastic Processes*. Wiley, New York. (Very readable but somewhat dated. Still serves as excellent source for stationary processes. Graduate level.)

[3] Doob, J. L. (1953). *Stochastic Processes*. Wiley, New York. (Classical source. Compact with information. Difficult to read. Graduate level.)

[4] Dudley, R. M. (1973). *Ann. Prob.*, **1**(1), 66–103. (Survey paper of Gaussian processes. Research level.)

[5] Galambos, J. (1978). *The Asymptotic Theory of Extreme Order Statistics*. Wiley, New York. (The only recent book on extreme value theory.)

[6] Kallianpur, G. (1970). *Trans. Amer. Math. Soc.*, **149**, 199–211. (Highly analytical results on zero–one laws of Gaussian processes. Research level.)

[7] Kolmogorov, A. N. and Rozanov, Yu. A. (1960). *Theory Prob.*, **5**, 204–208. (Russian-style compact statements of results on mixing conditions. Research level.)

[8] Leadbetter, M. R., Lindgren, G., and Rootzén, H. (1979). *Statist. Res. Rep. No. 1979-2*, University of Umea, Umea, Sweden. (Useful notes on recent advances in extreme value theory. Graduate level.)

[9] Lukács, E. (1970). *Characteristic Functions*. Hafner, New York. (Excellent book filled with information. Graduate level.)

[10] Mittal, Y. D. (1975). *Pacific J. Math.*, **64**(2), 517–538. (Analytic characterization of covariance functions in n-dimensional space. Research level.)

[11] Mittal, Y. D. and Ylvisaker, N. D. (1975). *Stoch. Proc. Appl.*, **3**, 1–18. (New limit laws arising in very dependent Gaussian processes. Research level.)

(AUTOREGRESSIVE MOVING-AVERAGE MODELS
BROWNIAN PROCESSES
LAWS OF THE ITERATED LOGARITHM
MARKOV PROCESSES
ORNSTEIN–UHLENBECK PROCESS
WHITE NOISE
WIENER PROCESS)

YASHASWINI MITTAL

GAUSS–JORDAN ELIMINATION

This is a procedure for solving the n linear equations

$$a_{i1}x_1 + a_{i2}x_2 + \cdots + a_{in}x_n = b_i, \qquad i = 1, 2, \ldots, n, \qquad (1)$$

which eliminates variables one by one. Let $a_{ij}^{(1)} = a_{ij} - a_{i1}a_{1j}/a_{11}$, where it is assumed that $a_{11} \neq 0$, and let $b_i^{(1)} = b_i - a_{i1}b_1/a_{11}$; $i, j = 2, \ldots, n$. Equations (1) then become

$$\begin{aligned} a_{11}x_1 + a_{12}x_2 + \cdots + a_{1n}x_n &= b_1, \\ a_{22}^{(1)}x_2 + \cdots + a_{n}^{(1)}x_n &= b_2^{(1)}, \\ \vdots \qquad\qquad & \quad \vdots \\ a_{n2}^{(1)}x_2 + \cdots + a_{nn}^{(1)}x_n &= b_n^{(1)}. \end{aligned} \qquad (2)$$

This is the first *forward step*. At the $(k+1)$th forward step, let

$$\begin{aligned} a_{ij}^{(k+1)} &= a_{ij}^{(k)} - a_{ik}^{(k)}a_{kj}^{(k)}/a_{kk}^{(k)}, \\ b_i^{(k+1)} &= b_i^{(k)} - a_{ik}^{(k)}b_k^{(k)}/a_{kk}^{(k)}, \\ & \qquad k = 2, 3, \ldots, n-1; \end{aligned}$$

$i, j = k+1, k+2, \ldots, n$. If the system (1) has a unique solution set, and if $a_{11} \neq 0$, then the *pivots* $a_{kk}^{(k)}$ are nonzero. The *method of pivot selection* allows users in computer programming to choose the largest of the remaining $|a_{ij}^{(k)}|$ as the next pivot, thus avoiding as far as possible problems arising from pivots that may be close to zero [or equal to zero if the matrix of coefficients in (1) is singular].

After the forward steps, we obtain an upper triangular system:

$$\begin{aligned} a_{11}x_1 + a_{12}x_2 + a_{13}x_3 + \cdots + a_{1n}x_n &= b_1, \\ a_{22}^{(1)}x_2 + a_{23}^{(1)}x_3 + \cdots + a_{2n}^{(1)}x_n &= b_2^{(1)}, \\ a_{33}^{(2)}x_3 + \cdots + a_{3n}^{(2)}x_n &= b_3^{(2)}, \\ \vdots \qquad & \quad \vdots \\ a_{nn}^{(n-1)}x_n &= b_n^{(n-1)} \end{aligned} \tag{3}$$

The variables can then be obtained in the order $x_n, x_{n-1}, \ldots, x_1$ by obvious *backward steps*.

For an alternative elimination procedure, *see* SWEEP OPERATIONS. Gauss–Jordan elimination is particularly useful in solving the normal equations in the general linear model*.

Bibliography

Gantmacher, F. R. (1959). *The Theory of Matrices*, Vol. 1. Chelsea, New York. (Contains theoretical generalizations.)

Hanna, S. C. and Saber, J. C. (1978). *Linear Programming and Matrix Algebra*, 2nd ed. Babson College Press, Babson Park, Mass. (Includes a compact flowchart.)

Seber, G. A. F. (1977). In *Linear Regression Analysis*. Wiley, New York, Chap. 11.

(LINEAR ALGEBRA, COMPUTATIONAL
LINEAR PROGRAMMING
SWEEP OPERATIONS)

GAUSS–MARKOV THEOREM

The Gauss–Markov theorem summarizes a basic statistical concept of minimum variance* linear unbiased estimation of parameters in a linear model when the functional form of the probability density function of the errors is not known. The errors are assumed to be independently distributed and their expected values are all zero and they possess the same unknown variance.

A linear model is defined by an observation

$$Y_i = \sum_{j=1}^{p} X_{ij}\beta_j + e_i, \qquad i = 1, 2, \ldots, n$$

written as a linear combination of p unknown parameters β_j and an unobservable random variable e_i, which in turn defines the matrix form of the model, $\mathbf{Y} = \mathbf{X}\boldsymbol{\beta} + \mathbf{e}$; (*see* GENERAL LINEAR MODEL).

Carl Friedrich Gauss (1777–1855) in the first part of his *Theoria Combinationis* [2 (1821)] proved that, among all the unbiased estimates of the parameters in a linear model, the one produced by least-squares procedures has minimum variance. In the second part of this monograph [2 (1823)] Gauss extended the theorem to estimating a linear combination of the parameters [7] (*see* GAUSS, CARL FRIEDRICH).

A modern version of the Gauss–Markov theorem is given in matrix notation as follows [3]:

Theorem. If the linear model of full rank $\mathbf{Y} = \mathbf{X}\boldsymbol{\beta} + \mathbf{e}$ is such that the following two conditions on the random vector $\mathbf{e}$ are

$$E(\mathbf{e}) = \boldsymbol{\phi} \quad \text{and} \quad E(\mathbf{e}\mathbf{e}^T) = \sigma^2\mathbf{I},$$

then the minimum variance linear unbiased estimate of $\boldsymbol{\beta}$ is given by least squares*; i.e.,

$$\hat{\boldsymbol{\beta}} = (\mathbf{X}^T\mathbf{X})^{-1}\mathbf{X}^T\mathbf{Y}$$

is the minimum variance linear unbiased estimate of $\boldsymbol{\beta}$.

Proof. First, the least-squares estimate is formulated. The sums of the errors squared is given by

$$Q = \mathbf{e}^T\mathbf{e} = (\mathbf{Y} - \mathbf{X}\boldsymbol{\beta})^T(\mathbf{Y} - \mathbf{X}\boldsymbol{\beta}),$$

whose derivative with respect to the vector $\boldsymbol{\beta}$ gives the p normal equations

$$\frac{\partial Q}{\partial \boldsymbol{\beta}} = -2\mathbf{X}^T\mathbf{Y} + 2(\mathbf{X}^T\mathbf{X}).$$

On equating the normal equation to the

$p \times 1$ zero vector and solving for $\boldsymbol{\beta}$, the solution is $\hat{\boldsymbol{\beta}} = (\mathbf{X}^T\mathbf{X})^{-1}\mathbf{X}^T\mathbf{Y}$, the least-squares estimator for the parameter vector $\boldsymbol{\beta}$.

Let $\mathbf{A}$ be any $p \times n$ matrix and $\tilde{\boldsymbol{\beta}} = \mathbf{AY}$, a general linear function of $\mathbf{Y}$ and an estimate of $\boldsymbol{\beta}$. Let $\mathbf{A} = (\mathbf{X}^T\mathbf{X})^{-1}\mathbf{X}^T + \mathbf{B}$; then select $\mathbf{B}$ so that $\tilde{\boldsymbol{\beta}}$ is the minimum variance unbiased estimate of $\boldsymbol{\beta}$. The requirement of unbiasedness* implies that

$$\begin{aligned} E[\tilde{\boldsymbol{\beta}}] = E[\mathbf{AY}] &= E\left[\left((\mathbf{X}^T\mathbf{X})^{-1}\mathbf{X}^T + \mathbf{B}\right)\mathbf{Y}\right] \\ &= E\left[(\mathbf{X}^T\mathbf{X})^{-1}\mathbf{X}^T\mathbf{Y} + \mathbf{BY}\right] \\ &= \boldsymbol{\beta} + \mathbf{BX}\boldsymbol{\beta} \\ &= \boldsymbol{\beta}; \end{aligned}$$

i.e., $\mathbf{BX}\boldsymbol{\beta} = \mathbf{0}$ for all $\boldsymbol{\beta}$, or $\mathbf{BX} = \mathbf{0}$.

The property that $\boldsymbol{\beta}$ must be minimum variance implies that the matrix $\mathbf{B}$ must be selected so that the variance of β_j is minimized for each $i = 1, 2, \ldots, p$ subject to the constraint $\mathbf{BX} = \mathbf{0}$. Hence consider the covariance matrix of $\tilde{\boldsymbol{\beta}}$,

$$\begin{aligned} V(\tilde{\boldsymbol{\beta}}) &= E\left[(\tilde{\boldsymbol{\beta}} - \boldsymbol{\beta})(\tilde{\boldsymbol{\beta}} - \boldsymbol{\beta})^T\right] \\ &= E\left\{\left(\left[(\mathbf{X}^T\mathbf{X})^{-1}\mathbf{X}^T + \mathbf{B}\right]\mathbf{Y} - \boldsymbol{\beta}\right)\right. \\ &\qquad \times \left.\left(\left[(\mathbf{X}^T\mathbf{X})^{-1}\mathbf{X}^T + \mathbf{B}\right]\mathbf{Y} - \boldsymbol{\beta}\right)^T\right\} \end{aligned}$$

Substituting $\mathbf{X}\boldsymbol{\beta} + \mathbf{e}$ for $\mathbf{Y}$ and the constraint $\mathbf{BX} = \mathbf{0}$, it follows that

$$V(\tilde{\boldsymbol{\beta}}) = \sigma^2\left[(\mathbf{X}^T\mathbf{X})^{-1} + \mathbf{BB}^T\right].$$

If the variance of $\tilde{\beta}_i$ is to be minimal for each i, the diagonal elements of $V(\tilde{\boldsymbol{\beta}})$ must be minimized. Since the diagonal elements of both $(\mathbf{X}^T\mathbf{X})^{-1}$ and $\mathbf{BB}^T$ are positive and nonnegative, respectively, the diagonal elements of $V(\tilde{\boldsymbol{\beta}})$ will be minimized by selecting the diagonal elements of $\mathbf{BB}^T$ to be zero. The ith diagonal elements of $\mathbf{BB}^T$, if $\mathbf{B} = \{b_{ij}\}$, are given by

$$C_{ii} = \sum_{j=1}^{n} b_{ij}^2, \qquad i = 1, 2, \ldots, p.$$

That is, $C_{ii} = 0$ if and only if $b_{ij} = 0$ for all $i = 1, 2, \ldots, p$ and $j = 1, 2, \ldots, n$, or, equivalently, the matrix $\mathbf{B}$ is composed of elements whose values are all zero (i.e., $\mathbf{B} = \mathbf{0}$).

Therefore, $\mathbf{A} = (\mathbf{X}^T\mathbf{X})^{-1}\mathbf{X}^T$ and the best estimate is the least-squares estimate, that is, $\tilde{\boldsymbol{\beta}} = \hat{\boldsymbol{\beta}}$, where $\hat{\boldsymbol{\beta}} = (\mathbf{X}^T\mathbf{X})^{-1}\mathbf{X}^T\mathbf{Y}$.

Corollary. *Under the linear model hypothesis given in the Gauss–Markov theorem, the minimum variance unbiased estimate of any linear combination of the β_i is the same linear combination of the minimum variance unbiased estimates of the β_i.*

It is important to note that the functional form of the density function of the errors is not known; however, if one assumes normality [i.e., $\mathbf{e} \sim N(\mathbf{0}, \sigma^2\mathbf{I})$], then the estimate $\hat{\boldsymbol{\beta}}$ possesses the following properties: it is

1. Consistent
2. Efficient
3. Unbiased
4. Sufficient
5. Complete
6. Minimum variance unbiased
7. $\hat{\boldsymbol{\beta}} \sim N[\boldsymbol{\beta}, \sigma^2(\mathbf{X}^T\mathbf{X})^{-1}]$

Note that property (6) implies that the minimum variance linear unbiased estimate when the functional form of the density function is unknown is also the minimum variance unbiased estimator when the density function is assumed to be a normal probability density.

If the assumptions concerning the $n \times 1$ error vector $\mathbf{e}$ are modified slightly so that the $n \times n$ covariance matrix is no longer $\sigma^2\mathbf{I}$ ($\mathbf{I}$ is the identity matrix), but the matrix $\boldsymbol{\Sigma} = \{\sigma_{ij}\}$, where $E[e_i e_j] = \sigma_{ij}$, then the Gauss–Markov estimate of $\boldsymbol{\beta}$ is

$$\tilde{\boldsymbol{\beta}} = (\mathbf{X}^T\boldsymbol{\Sigma}^{-1}\mathbf{X})^{-1}\mathbf{X}^T\boldsymbol{\Sigma}^{-1}\mathbf{Y}.$$

Also, if one again assumes that the errors are distributed normally, then $\boldsymbol{\beta}$ is again the minimum variance unbiased estimator for $\boldsymbol{\beta}$ when $\boldsymbol{\Sigma}$ is known. A necessary and sufficient condition for $\hat{\boldsymbol{\beta}}$ to be equal to $\tilde{\boldsymbol{\beta}}$ is that the covariance matrix of $\hat{\boldsymbol{\beta}}$ be of the form [5]

$$(1 - p)\mathbf{I} + p\mathbf{JJ}^T,$$

where p is a scalar such that $0 \leqslant p < 1$ and $\mathbf{J}^T = (1, 1, \ldots, 1)^T$.

For those cases in which $\boldsymbol{\Sigma}$ is unknown, one usually estimates the matrix $\boldsymbol{\Sigma}$ and substitutes the estimate when sufficient data are available. However, this estimator is not optimal and the least-squares estimator is often used and can for some applications prove a better estimator.

Also, since $(\mathbf{X}^T\mathbf{X})^{-1}\mathbf{X}^T = \mathbf{X}^+$, a generalized inverse* for $\mathbf{X}$, a theory of Gauss–Markov estimation has been developed using the theory of generalized inverses. This theory includes those cases in which $\mathbf{X}^T\mathbf{X}$ may or may not be singular. Also, the theory has been extended to the case where $\boldsymbol{\Sigma}$ is a singular covariance matrix, and to the case involving assumption of singular normal probability density functions [6].

There are extensions of the theorem to cases in which there exists a set of known linear constraints on the parameter vector $\boldsymbol{\beta}$, in which there exists a set of linear inequality constraints on the parameter vector $\boldsymbol{\beta}$, and when the vector $\boldsymbol{\beta}$ is random with known moments. Also, there exist coordinate-free versions of the theorem [4].

Others have given proofs for the theorem first proved and reported by Gauss. One of these was Andrei Andreyevich Markov* [1856–1922], who rediscovered the theorem in 1900. Others had rediscovered the same result during the period of years between Gauss and Markov; however, David and Neyman [1] first used the name *Gauss–Markov theorem* and Markov's name has stayed attached to it. Later, A. C. Aitken developed the matrix formulation that is preferred today.

References

[1] David, F. N. and Neyman, J. (1938). *Statist. Res. Mem.*, **2**, 105–116.

[2] Gauss, C. F. (1821, 1823, 1826). Theoria Combinationis Erroribus Minimis Obnoxiae, Parts 1, 2, and Suppl. *Werke*, **4**, 1–108.

[3] Graybill, F. A. (1976). *Theory and Application of the Linear Model.* Duxbury Press, North Scituate, Mass.

[4] Kruskal, W. (1961). *Proc. 4th Berkeley Symp. Math. Stat. Prob.* Vol. 1. University of California Press, Berkeley, Calif., pp. 435–451.

[5] McElroy, F. W. (1967). *J. Amer. Statist. Ass.*, **62**, 1302–1304.

[6] Rao, C. R. and Mitra, S. K. (1971). *Generalized Inverse of Matrices and Its Applications.* Wiley, New York.

[7] Seal, H. L. (1970). In *Studies in the History of Statistics and Probability*, E. S. Pearson and M. G. Kendall, eds. Hafner, New York.

(GAUSS, CARL FRIEDRICH
GENERAL LINEAR MODEL
HAT MATRIX
LEAST SQUARES
MINIMUM VARIANCE UNBIASED ESTIMATOR)

PATRICK L. ODELL

GAUSS–NEWTON ITERATION *See* NEWTON ITERATION EXTENSIONS

GAUSS QUADRATURE *See* NUMERICAL INTEGRATION

GAUSS–SEIDEL ITERATION

Given a linear system of the form

$$\mathbf{Ax} = \mathbf{b}, \tag{1}$$

where $\mathbf{A} = [a_{ij}]$ is an $n \times n$ matrix and $\mathbf{x} = \{x_i\}$ and $\mathbf{b} = \{b_i\}$ are n-vectors, Gaussian elimination is likely to be the most efficient technique to produce a solution when n is not too large (say, $n \leqslant 200$). Even when n is large, Gaussian elimination may still be efficient if the matrix $\mathbf{A}$ has special structure and a large number of zero entries. See, for example, the research monograph by Duff and Stewart [1]. However, for many large problems, it may be desirable to use an iterative method. Such problems quite often arise as a result of discretizing a differential equation [3] or as the normal equations* in least squares*.

The *Jacobi iterative method* (simultaneous displacements) for solving (1) is based on rewriting it (if possible) in the equivalent form

$$x_i = \left(-\sum_{j \neq i} a_{ij} x_j + b_i\right) \Big/ a_{ii}, \quad i = 1, \ldots, n. \quad (2)$$

One chooses an initial approximation $\mathbf{x}^{(0)} = \{x_i^{(0)}\}$, sets $k = 0$, and computes

$$x_i^{(k+1)} = \left(-\sum_{j \neq i} a_{ij} x_j^{(k)} + b_i\right) \Big/ a_{ii}, \quad i = 1, \ldots, n. \quad (3)$$

If $\mathbf{x}^{(k+1)}$ is sufficiently accurate, the process is complete. Otherwise, k is incremented by one and (3) is repeated.

The *Gauss–Seidel method* (successive displacements) uses each new coordinate $x_i^{(k+1)}$ immediately in the computation of the next one. Thus (3) is modified to become

$$x_i^{(k+1)} = \left(-\sum_{j=1}^{i-1} a_{ij} x_j^{(k+1)} - \sum_{j=i+1}^{n} a_{ij} x_j^{(k)} + b_i\right) \Big/ a_{ii}, \quad i = 1, \ldots, n. \quad (4)$$

The Jacobi and Gauss–Seidel iterative methods can be written in matrix form for theoretical purposes. Computations are based on the formulas (3) and (4). Let $-\mathbf{L}$ be the strictly lower triangular part of $\mathbf{A}$, $-\mathbf{U}$ the strictly upper part, and $\mathbf{D}$ the diagonal. Then (3) can be expressed as

$$\mathbf{x}^{(k+1)} = \mathbf{D}^{-1}(\mathbf{L} + \mathbf{U})\mathbf{x}^{(k)} + \mathbf{D}^{-1}\mathbf{b}, \quad k = 0, 1, \ldots, \quad (5)$$

and (4) can be expressed as

$$\mathbf{x}^{(k+1)} = (\mathbf{D} - \mathbf{L})^{-1}\mathbf{U}\mathbf{x}^{(k)} + (\mathbf{D} - \mathbf{L})^{-1}\mathbf{b}, \quad k = 0, 1, \ldots. \quad (6)$$

The Gauss–Seidel method is often successfully accelerated by a modification known as *successive overrelaxation* (SOR). The name is based on a physical interpretation of the method, first used in deriving it. The Gauss–Seidel iterate is computed as before by

$$\bar{x}_i^{(k+1)} = \left(-\sum_{j=1}^{i-1} a_{ij} x_j^{(k+1)} - \sum_{j=i+1}^{n} a_{ij} x_j^{(k)} + b_i\right) \Big/ a_{ii},$$

but the new value of x_i is taken to be

$$x_i^{(k+1)} = x_i^{(k)} + \omega\left(\bar{x}_i^{(k+1)} - x_i^{(k)}\right),$$

where ω is an acceleration parameter. It can be shown that only $0 < \omega < 2$ need be considered, and, of course, $\omega = 1$ gives back the Gauss–Seidel method.

All of these basic iterative schemes have the form $\mathbf{x}^{(k+1)} = \mathbf{M}\mathbf{x}^{(k)} + \mathbf{d}$, as we have shown in (5) and (6) for Jacobi and Gauss–Seidel. It is possible to show that

$$\|\mathbf{x}^{(k)} - \mathbf{x}\| \sim \rho(\mathbf{M})^k \|\mathbf{x}^{(0)} - \mathbf{x}\|,$$

where $\|\cdot\|$ is any vector norm, $\mathbf{x} = \mathbf{A}^{-1}\mathbf{b}$, and $\rho(\mathbf{M})$, the *spectral radius* of $\mathbf{M}$, is given by $\rho(\mathbf{M}) = \max\{|\lambda|: \lambda \text{ an eigenvalue of } \mathbf{A}\}$. Thus the relative error after k iterations is approximately $\rho(\mathbf{M})^k$; an accuracy of 10^{-t} is expected after k iterations if $\rho(\mathbf{M})^k \leqslant 10^{-t}$ or $k \geqslant t/\{-\log_{10}\rho(\mathbf{M})\}$. Thus it is desirable to have $\rho(\mathbf{M})$ as small as possible. Let $\mathbf{M}_J$, $\mathbf{M}_G$, and $\mathbf{M}_\omega$ denote the iteration matrices for the Jacobi, Gauss–Seidel and SOR methods, respectively. In many important cases $\rho(\mathbf{M}_G) = \rho(\mathbf{M}_J)^2$, the optimal $\omega^* = 2/(1 + \sqrt{1 - \rho(\mathbf{M}_J)^2})$, and $\rho(\mathbf{M}_{\omega^*}) = \omega^* - 1$. These formulas are valid, for example, if $\mathbf{A}$ is positive definite and tridiagonal. Then the Gauss–Seidel method converges about twice as fast as the Jacobi method, and the optimal SOR can be an order of magnitude faster still.

This statement is not true in general. In fact, there are linear systems for which the Jacobi method converges and the Gauss–Seidel method does not, and conversely (see Varga [3]). If $\mathbf{A}$ is strictly diagonally dominant, (i.e., $|a_{ii}| > \sum_{j \neq i} |a_{ij}|$, $i = 1, \ldots, n$), then $\mathbf{x}^{(k)}$ converges to $\mathbf{A}^{-1}\mathbf{b}$ for any $\mathbf{x}^{(0)}$ for both (3) and (4). If $\mathbf{A}$ is symmetric and positive definite, the SOR iterates converge to $\mathbf{A}^{-1}\mathbf{b}$ for any $\mathbf{x}^{(0)}$ and $0 < \omega < 2$. This is the classical Ostrowski–Reich theorem. Comprehensive results on convergence can be found in ref. 4.

The calculation of ω^* is difficult and is usually obtained only approximately. An excellent package of computer programs, ITPACK, is available [2], containing SOR routines that provide automatic estimates for the optimal ω. Other acceleration techniques are also available in ITPACK.

Under certain conditions, the iterative methods discussed here can be extended to the case where the matrix of coefficients is singular but consistent. Details can be found in ref. 1. Iterative methods for solving the large sparse linear least-squares problem is still a research topic. A brief discussion and references may be found in ref. 1.

References

[1] Duff, I. S. and Stewart, G. W. (1979). *Sparse Matrix Proceedings 1978*. SIAM, Philadelphia.

[2] ITPACK (1980). International Mathematical and Statistical Libraries, Inc., Houston, Tex.

[3] Varga, R. S. (1962). *Matrix Iterative Analysis*. Prentice-Hall, Englewood-Cliffs, N.J.

[4] Young, D. M. (1971). *Iterative Solution of Large Linear Systems*. Academic Press, New York.

(COMPUTERS AND STATISTICS)

George W. Reddien

GAUSS'S INEQUALITY *See* CAMP–MEIDELL INEQUALITY

GAUSS THEOREM *See* GAUSS–MARKOV THEOREM

GAUSS-WINCKLER INEQUALITY

Let X be a continuous random variable with a single mode* m and a differentiable probability density function; let $\nu_1', \nu_2', \ldots$ be the absolute moments* $E(|X - m|^k)$, $k = 1, 2, \ldots$. Then

$$\{(r+1)\nu_r'\}^{1/r} < \{(s+1)\nu_s'\}^{1/s}, \quad 0 < r < s.$$

(LIAPUNOV'S INEQUALITY)

GEARY'S RATIO *See* DEPARTURES FROM NORMALITY, TESTS FOR

GEARY TEST *See* DEPARTURES FROM NORMALITY, TESTS FOR

GEHAN–GILBERT TEST

A frequent problem in medical follow-up* and industrial life testing* studies is the comparison of two different treatments based on censored survival data (*see* CENSORING). For example, in a clinical trial* conducted by Freireich et al. [4], 6-mercaptopurine (6-MP) was compared to placebo in the maintenance of remissions for patients with acute leukemia. Patients entered the study serially in time and were randomly allocated to one of the two treatments (*see* RANDOMIZATION). When the study was closed (one year after the start of the trial), the following lengths of remission were recorded:

	Length of Remission (weeks)
6-MP	$6, 6, 6, 7, 10, 13, 16, 22, 23, 6^+, 9^+, 10^+, 11^+, 17^+, 19^+, 20^+, 25^+, 32^+, 32^+, 34^+, 35^+$
Placebo	1, 1, 2, 2, 3, 4, 4, 5, 5, 8, 8, 8, 8, 11, 11, 12, 12, 15, 17, 22, 23

where a + sign indicates that the patient was still in remission at one year. It is desired to compare lengths of remission between these two treatment groups.

More generally, let response variables X_i^0, $i = 1, \ldots, m$ ($Y_j^0, j = 1, \ldots, n$) be independent and identically distributed with a continuous distribution function $F(G)$. The null hypothesis* that we wish to test is $H_0 : F \equiv G$ against the alternative $H_1 : X^0$ is stochastically larger than Y^0. In studies involving a comparison of lifetimes, X_i^0 and Y_j^0

may be censored from the right by random variables U_i and V_j (*see* CENSORING, RIGHT), respectively, and hence cannot always be observed. In the example above, U_i, V_j would be the length of observation time for the given patient from time of entry into study until the closing date of one year. Rather, our observations consist of the minima

$$X_1 = \min(X_1^0, U_1), \dots, X_m = (X_m^0, U_m),$$

$$Y_1 = \min(Y_1^0, V_1), \dots, Y_n = (Y_n^0, V_n)$$

and two sequences $\Delta_1, \dots, \Delta_m$ and $E_1, \dots, E_n$, where

$$\Delta_i = \begin{cases} 1 & \text{if } X_i = X_i^0 \\ 0 & \text{otherwise,} \end{cases}$$

$$E_j = \begin{cases} 1 & \text{if } Y_j = Y_j^0 \\ 0 & \text{otherwise.} \end{cases}$$

Motivated by the two sample Wilcoxon–Mann–Whitney test (*see* WILCOXON TEST), Gilbert [6] and Gehan [5] independently proposed an asymptotically distribution-free test (*see* DISTRIBUTION-FREE METHODS) which rejects H_0 for large values of the test statistic defined by

$$W_{m,n} = \sum_{i=1}^{m} \sum_{j=1}^{n} \psi_{ij}/mn,$$

where

$$\psi_{ij} = \begin{cases} 1 & X_i > Y_j, \quad E_j = 1, \\ -1 & X_i < Y_j, \quad \Delta_i = 1, \\ 0 & \text{otherwise.} \end{cases}$$

Under H_0 and as both m and n go to ∞ in such a way that $m/(m+n) \to \lambda$, $0 < \lambda < 1$, the standardized value of $W_{m,n}$ converges in distribution to the standard normal random variable [3, 5, 6] (*see* CONVERGENCE OF SEQUENCES OF RANDOM VARIABLES).

If U and V have the same distributions (i.e., X^0 and Y^0 are subject to the same censoring mechanism, a distribution-free test based on $W_{m,n}$ which is conditional on the censorship actually encountered in the trial, can be obtained for testing H_0 [5]. In this case it is difficult to make a general statement on the size of samples necessary before the asymptotic theory described above holds. But Gehan [5] has found that the normal approximation of the null distribution of $W_{m,n}$ with continuity correction* can be applied when sample sizes are as small as $m = n = 5$, as long as the total number of censored observations is six or less.

Competitors of the Gehan–Gilbert test include the Efron test [3] (*see* KAPLAN–MEIER ESTIMATOR) and the log rank test [2, 10, 11] (*see* MANTEL–HAENSZEL STATISTIC) of the proportional hazards model*. The Efron test procedure is more powerful than the Gehan–Gilbert test [8], but computational difficulties inhibit its use in practice. The choice between the Gehan–Gilbert test and the log rank test has been extensively discussed (see Lee et al. [9] and Tarone and Ware [12]). The former test gives more weight to differences in survival distribution arising early, whereas the latter test gives more weight to later differences. The log rank test is more powerful than the Gehan–Gilbert test when the hazard functions of survival time in the two groups are proportional, (*see* PROPORTIONAL HAZARDS MODEL) but not otherwise [9].

Breslow [1] used the same score function as Gehan and Gilbert and generalized the Kruskal–Wallis test* to the k-sample problem with censored data. When $k = 2$, the test reduces to the Gehan–Gilbert test. Koziol and Reid [7] developed multiple comparisons* procedures for k samples when the observations are subject to censorship. Wei [13] and Woolson and Lachenbruch [14] have extended the Gehan–Gilbert test to paired observations subject to arbitrary right censorship.

References

[1] Breslow, N. (1970). *Biometrika*, **57**, 579–594.

[2] Crowley, J. and Thomas, D. R. (1975). *Large Sample Theory for the Log Rank Test. Tech. Rep. No. 415*, Dept. of Statistics, University of Wisconsin, Madison, Wis.

[3] Efron, B. (1967). *Proc. 5th Berkeley Symp. Math. Statist. Prob.*, Vol. 4, University of California Press, Berkeley, Calif., pp. 831–854.

[4] Freireich, E. J., Gehan, E. A., Frei, E., III, Schroeder, L. R., Wolman, I. J., Anbari, R., Burgert, E. O., Mills, S. D., Pinkel, D., Selawry, O. S.,

Moon, J. H., Gendel, B. R., Spurr, C. L., Storrs, R., Haurani, F., Hoogstraten, B., and Lee, S. (1963). *Blood*, **21**, 699–716.

[5] Gehan, E. A. (1965). *Biometrika*, **52**, 203–223.

[6] Gilbert, J. P. (1962). *Random Censorship*. Ph.D. dissertation, University of Chicago.

[7] Koziol, J. A. and Reid, N. (1977). *Commun. Statist. A*, **6**, 1149–1164.

[8] Latta, R. B. (1977). *Biometrika*, **64**, 633–635.

[9] Lee, E. T., Desu, M. M., and Gehan, E. A. (1975). *Biometrika*, **62**, 425–432.

[10] Mantel, N. (1966). *Cancer Chemother. Rep.*, **50**, 163–170.

[11] Peto, R. and Peto, J. (1972). *J. R. Statist. Soc. A*, **135**, 185–207.

[12] Tarone, R. and Ware, J. (1977). *Biometrika*, **64**, 156–160.

[13] Wei, L. J. (1980). *J. Amer. Statist. Ass.*, **75**, 634–637.

[14] Woolson, R. F. and Lachenbruch, P. A. (1980). *Biometrika*, **67**, 597–606.

(CENSORING, RIGHT
DISTRIBUTION-FREE METHODS
KAPLAN–MEIER ESTIMATOR
LIFE TESTING
LOG RANK SCORES, STATISTICS, AND TESTS
MANTEL–HAENSZEL STATISTIC
PROPORTIONAL HAZARDS MODEL
WILCOXON TESTS)

L. J. Wei
E. A. Gehan

GENERAL BALANCE

The concept of *general balance* arises in the context of the *weighted least-squares** analysis of data under what is usually termed a *mixed model*. If the data $\mathbf{y} = (y_h)$ comprise an $n \times 1$ column vector and $\boldsymbol{\theta} = (\theta_i)$ and $\boldsymbol{\alpha} = (\alpha_j)$ are $s \times 1$ and $v \times 1$ column vectors of parameters, then such models of the *expectation* **E**y and *dispersion* **D**y of y are of the familiar kind:

$$\mathsf{E}\mathbf{y} = \mathbf{X}\boldsymbol{\alpha}, \qquad \mathsf{D}\mathbf{y} = \mathbf{V}(\boldsymbol{\theta}). \tag{1}$$

Here $\mathbf{X}$ is a *known* $n \times v$ *design matrix* and $\mathbf{V}(\boldsymbol{\theta}) = \sum_i \theta_i \mathbf{C}_i$ is a positive definite covariance matrix given in terms of *known symmetric matrices* $\mathbf{C}_1, \ldots, \mathbf{C}_s$.

The main problems which arise here involve the calculation of best linear unbiased estimates (BLUEs) of linear contrasts $\mathbf{t}'\boldsymbol{\alpha}$ in the mean value parameter $\boldsymbol{\alpha}$, the estimation of the dispersion parameters θ_i or linear combinations of them, and the calculation of best linear unbiased predictors (BLUPs) of any random effects which might underlie the dispersion structure of $\mathbf{y}$. This last point is relevant if the data $\mathbf{y}$ are viewed as

$$\mathbf{y} = \mathbf{X}\boldsymbol{\alpha} + \mathbf{Z}_1\boldsymbol{\beta}_1 + \cdots + \mathbf{Z}_s\boldsymbol{\beta}_s, \tag{2}$$

where $\boldsymbol{\beta}_1, \ldots, \boldsymbol{\beta}_s$ are uncorrelated vectors of uncorrelated random effects having zero mean and variances $\boldsymbol{\theta}_1, \ldots, \boldsymbol{\theta}_s$, respectively, and $\mathbf{Z}_1, \ldots, \mathbf{Z}_s$ are further design matrices allocating these random effects to units. In this case $\mathbf{C}_i = \mathbf{Z}_i\mathbf{Z}_i'$ and it is usually supposed that $\mathbf{Z}_s = \mathbf{I}$, the $n \times n$ identity matrix. Given (2), the calculation of BLUPs of the $\boldsymbol{\beta}$ terms frequently arises.

Under a simplifying assumption on the model for the dispersion of $\mathbf{y}$, *general balance* is a condition relating the model for **E**y to that for **D**y; more fully, it is a relation between the column space $\mathscr{R}(\mathbf{X})$ of the basic design matrix and the symmetric matrices $\mathbf{C}_1, \ldots, \mathbf{C}_s$. It was introduced (in a slightly less general form) by Nelder [5] and further developed in Nelder [6].

When the condition of general balance is satisfied, the estimation of $\boldsymbol{\alpha}$ by weighted least squares is direct and readily interpreted, as is the estimation of $\boldsymbol{\theta}$, and in a sense the condition is necessary for these conclusions to hold. General balance is thus a condition under which the usual mixed model analysis of data is particularly straightforward.

Almost all experimental designs ever used (which involve random effects) satisfy the condition of general balance. This includes *all* block designs, Latin* and Graeco-Latin square* designs, and all designs with balanced confounding* or proportional replication. All the row-and-column designs* of Pearce [7] and others (e.g., Preece [8]) satisfy the condition, as do most other multistrata designs in the literature. Exceptions include some multiphase and changeover designs*.

DEFINITIONS

General balance evolved out of efforts to understand the structure and analysis of general designed experiments, and although the discussion that follows is more widely applicable, we will illustrate it only with examples from this field. In this context $\mathbf{X}$ is a binary matrix describing the allocation of v treatments to n experimental units, $\mathbf{X}'$ forms total over units having the same treatment, and $\mathbf{X}'\mathbf{X}$ is the diagonal matrix whose entries are the treatment replications. We will also write $\mathbf{P} = \mathbf{X}(\mathbf{X}'\mathbf{X})^{-1}\mathbf{X}'$ for the (unweighted) orthogonal projection onto $\mathscr{R}(\mathbf{X})$; all this does is *average* over units with the same treatment.

There are two mathematical conditions which seem to be necessary in order that the problems noted above have uniquely defined, direct, and readily interpreted solutions. They are:

(C) The symmetric matrices $\mathbf{C}_1, \ldots, \mathbf{C}_s$ all commute.

(GB) The symmetric matrices $\mathbf{PC}_1\mathbf{P}, \ldots, \mathbf{PC}_s\mathbf{P}$ all commute.

Condition (C) is equivalent to the matrices $\mathbf{C}_1, \ldots, \mathbf{C}_s$ being simultaneously diagonalizable and under it we can regard dispersion matrices of the form $\mathbf{V}(\boldsymbol{\theta})$ in (1) as linear combinations of known *orthogonal* symmetric *idempotent** matrices. *From now on* we suppose that condition (C) holds, in which case we may as well assume that

$$\mathbf{C}_i = \mathbf{C}_i^2 = \mathbf{C}_i', \quad \mathbf{C}_i\mathbf{C}_j = 0, \qquad 1 \leqslant i < j \leqslant s, \tag{C*}$$

$$\mathbf{C}_1 + \cdots + \mathbf{C}_s = \mathbf{I}.$$

Similarly, we will suppose that the parameter space for $\boldsymbol{\theta}$ is $\theta_1 > 0, \theta_2 > 0, \ldots, \theta_s > 0$, this whole process possibly requiring a reparameterization. The range $\mathscr{R}(\mathbf{C}_i)$ of $\mathbf{C}_i$ is termed the ith *stratum*, with $\mathbf{C}_i$ being the orthogonal projection onto this stratum, and we note that $\mathbf{EC}_i\mathbf{y} = \mathbf{C}_i\mathbf{X}\boldsymbol{\alpha}$ and $\mathbf{DC}_i\mathbf{y} = \theta_i\mathbf{C}_i$. As the model for $\mathbf{C}_i\mathbf{y}$ has its dispersion *known* up to a scalar, linear inference concerning $\boldsymbol{\alpha}$ is possible *within* stratum i, and in this context $\mathbf{X}'\mathbf{C}_i\mathbf{X}$ is the *stratum information matrix*.

It is also clear that the condition (GB) of *general balance* is equivalent to $\mathbf{PC}_1\mathbf{P}, \ldots, \mathbf{PC}_s\mathbf{P}$ being simultaneously diagonalizable, and so implies the existence of orthogonal symmetric idempotent $n \times n$ matrices $\mathbf{P}_1, \mathbf{P}_2, \ldots$ and coefficients (λ_{ij}) such that

$$\text{(GB)*} \quad \mathbf{PC}_i\mathbf{P} = \sum_j \lambda_{ij}\mathbf{P}_j, \qquad i = 1, 2, \ldots, s.$$

One readily concludes that $\lambda_{ij} \geqslant 0$, $\sum_i \lambda_{ij} = 1$ for each j, and that $\sum_j \mathbf{P}_j = \mathbf{P}$. The matrices $\{\mathbf{P}_j\}$ are projections onto the components of an orthogonal decomposition of the treatment or regression subspace, and we obtain a corresponding decomposition of the mean-value parameter $\boldsymbol{\alpha}$ by writing $\mathbf{T}_j = (\mathbf{X}'\mathbf{X})^{-1}\mathbf{X}'\mathbf{P}_j\mathbf{X}$. The following facts are easily checked:

1. $\mathbf{T}_j^2 = \mathbf{T}_j$;
2. $\mathbf{T}_j\mathbf{T}_k = 0 = \mathbf{T}_k\mathbf{T}_j \quad$ if $j \neq k$;
3. $\mathbf{T}_j' = (\mathbf{X}'\mathbf{X})\mathbf{T}_j(\mathbf{X}'\mathbf{X})^{-1}$;
4. $\sum_j \mathbf{T}_j = \mathbf{I}$, the identity matrix of size v;
5. $\mathbf{XT}_j = \mathbf{P}_j\mathbf{X}$.

From condition (GB)* and the definition of the $\mathbf{T}_j$, we obtain

$$\text{(GB)**} \qquad (\mathbf{X}'\mathbf{X})^{-1}\mathbf{X}'\mathbf{C}_i\mathbf{X} = \sum_j \lambda_{ij}\mathbf{T}_j, \qquad i = 1, \ldots, s,$$

and, conversely, if condition (GB)** holds for $v \times v$ matrices $\mathbf{T}_1, \mathbf{T}_2, \ldots$ satisfying all of the above, then (GB)* can be recovered by putting $\mathbf{P}_j = \mathbf{XT}_j(\mathbf{X}'\mathbf{X})^{-1}\mathbf{X}'$, $j = 1, 2, \ldots$. In other words, (GB)** is equivalent to (GB), and we find it the most convenient form of the property. Notice that if the treatments are all equally replicated (i.e., if $\mathbf{X}'\mathbf{X} = r\mathbf{I}$ for some positive integer r), then the $\{\mathbf{T}_j\}$ are all symmetric and in this case (GB)** is simply asserting that the strata information matrices can be diagonalized simultaneously. This is the form introduced by Nelder [5] and our modification to cover unequally replicated designs is just a minor extension.

The subspaces $\mathscr{R}(\mathbf{T}_1), \mathscr{R}(\mathbf{T}_2), \ldots$ are *orthogonal* with respect to the inner product $\langle \boldsymbol{\alpha}, \boldsymbol{\beta} \rangle = \boldsymbol{\alpha}'\mathbf{X}'\mathbf{X}\boldsymbol{\beta}$ and define what we will call the *treatment decomposition* of the mean-value parameter $\boldsymbol{\alpha}$, namely $\boldsymbol{\alpha} = \sum_j \mathbf{T}_j \boldsymbol{\alpha}$.

Before describing any further aspects of (GB) we pause to introduce three examples which will provide illustrations of the discussion that follows.

Example 1: Block Designs. Let us suppose that we have b blocks of p plots each, and that there are v treatments. If we denote the incidence matrix* for blocks by $\mathbf{Z}$, then a convenient general model for the dispersion of our responses has $\mathbf{C}_0 = (bp)^{-1}\mathbf{J}$, $\mathbf{C}_1 = p^{-1}\mathbf{Z}\mathbf{Z}' - (bp)^{-1}\mathbf{J}$, and $\mathbf{C}_2 = \mathbf{I} - p^{-1}\mathbf{Z}\mathbf{Z}'$, with $\theta_0 > 0$, $\theta_1 > 0$, and $\theta_2 > 0$ to ensure positive definiteness of $V(\boldsymbol{\theta})$. Here $\mathbf{J}$ is the $bp \times bp$ matrix of 1's. If our model is of the form (2) with $\mathbf{Z}_1 = \mathbf{Z}$, $\mathbf{Z}_2 = \mathbf{I}$, then we would have the constraints $\theta_0 = \theta_1 > \theta_2 > 0$ on our dispersion parameters, something that it is not always necessary or convenient to do.

It is clear that (C) is satisfied for all (equal block size) block designs, and it is also not hard to show that they all satisfy (GB) as well, no matter how the treatments are allocated to plots. Notice in this case that $\mathbf{C}_0$ is the grand mean averaging operator and $\mathbf{C}_1$ the operator that passes to block means adjusted for the grand mean, while $\mathbf{C}_2$ simply adjusts each observation by its block mean. From this it follows that $\mathbf{C}_0\mathbf{P} = \mathbf{C}_0$, whence $\mathbf{P}\mathbf{C}_0\mathbf{P} = \mathbf{C}_0$, $\mathbf{P}\mathbf{C}_1\mathbf{P} = p^{-1}\mathbf{P}\mathbf{Z}\mathbf{Z}'\mathbf{P} - \mathbf{C}_0$, and $\mathbf{P}\mathbf{C}_2\mathbf{P} = \mathbf{P} - p^{-1}\mathbf{P}\mathbf{Z}\mathbf{Z}'\mathbf{P}$, and then (GB) is immediate.

Example 2: Row-and-Column Designs. In this case we suppose that there are r rows and c columns with incidence matrices $\mathbf{Z}_1$ and $\mathbf{Z}_2$, respectively, each row meeting each column in exactly one plot, and that we have v treatments with incidence matrix $\mathbf{X}$. A general model for the dispersion of the responses has $\mathbf{C}_0 = (rc)^{-1}\mathbf{J}$, $\mathbf{C}_1 = c^{-1}\mathbf{Z}_1\mathbf{Z}_1' - (rc)^{-1}\mathbf{J}$, $\mathbf{C}_2 = r^{-1}\mathbf{Z}_2\mathbf{Z}_2' - (rc)^{-1}\mathbf{J}$, and $\mathbf{C}_3 = \mathbf{I} - c^{-1}\mathbf{Z}_1\mathbf{Z}_1' - r^{-1}\mathbf{Z}_2\mathbf{Z}_2' + (rc)^{-1}\mathbf{J}$, with $\theta_0 > 0$, $\theta_1 > 0$, $\theta_2 > 0$, and $\theta_3 > 0$. Once more we note that if our model was of the form (2) with $\mathbf{Z}_1, \mathbf{Z}_2$ as above and $\mathbf{Z}_3 = \mathbf{I}$, then constraints on the θ_i would arise; in this case these are $\theta_2 \geqslant \theta_3$, $\theta_1 \geqslant \theta_3$, and $\theta_0 = \theta_1 + \theta_2 - \theta_3$. Unless it is necessary to use these constraints, we prefer to avoid them.

Table 1

	1	2	3	4
1	B	A	A	A
2	A	C	C	B
3	B	B	D	C
4	D	D	D	C

It is easy to check that (C) holds, but in general (GB) fails. A simple example is Table 1, involving four treatments $A, \ldots, D$ constructed by A. Houtman. To see that (GB)* fails, we note that the contrast $(3, -1, -1, -1)'$ (comparing A with the average of B, C, and D) is an eigenvector of $\mathbf{X}'\mathbf{C}_2\mathbf{X}$ but not of $\mathbf{X}'\mathbf{C}_1\mathbf{X}$. However, all row-and-column designs that have ever been used in practice satisfy (GB)! Once more we note that the matrices $\mathbf{C}_0$, $\mathbf{C}_1$, $\mathbf{C}_2$, and $\mathbf{C}_3$ have simple interpretations in terms of row, column, and overall averages.

Example 3: The Classical Split-Plot Design*. For this design we suppose that there are b blocks of p plots each and that each plot is subdivided into q subplots. If blocks, plots, and subplots have incidence matrices $\mathbf{Z}_1, \mathbf{Z}_2$, and $\mathbf{Z}_3 = \mathbf{I}$, respectively, then our strata projections are $\mathbf{C}_0 = (bpq)^{-1}\mathbf{J}$, $\mathbf{C}_1 = (pq)^{-1}\mathbf{Z}_1\mathbf{Z}_1' - (bpq)^{-1}\mathbf{J}$, $\mathbf{C}_2 = q^{-1}\mathbf{Z}_2\mathbf{Z}_2' - (pq)^{-1}\mathbf{Z}_1\mathbf{Z}_1'$, and $\mathbf{C}_3 = \mathbf{I} - q^{-1}\mathbf{Z}_2\mathbf{Z}_2'$. As before we simply require the parameters θ_0, θ_1, θ_2, and θ_3 to be positive rather than satisfying $\theta_0 = \theta_1 \geqslant \theta_2 \geqslant \theta_3 \geqslant 0$ which result from the model (2), and observe that this block structure certainly satisfies (C).

The traditional treatment allocation associated with a split-plot design involves two crossed factors A and B having p and q levels, respectively, with A being applied to main plots within each block and B to subplots within each plot. If the matrices effecting the usual decomposition of $\boldsymbol{\alpha}$ into its average effect, A and B main effects*, and AB interaction* are denoted by $\mathbf{T}_0$, $\mathbf{T}_A$, $\mathbf{T}_B$, and $\mathbf{T}_{A.B}$, respectively, we find that split-plot designs satisfy (GB) with $\mathbf{X}'\mathbf{C}_0\mathbf{X} = b\mathbf{T}_0$, $\mathbf{X}'\mathbf{C}_1\mathbf{X} = 0$, $\mathbf{X}'\mathbf{C}_2\mathbf{X} = b\mathbf{T}_A$, and with $\mathbf{X}'\mathbf{C}_2\mathbf{X} = b(\mathbf{T}_B + \mathbf{T}_{A.B})$.

EIGENANALYSIS

Suppose now that (GB) [as well as (C)] holds for our model (1). We will show that in this case *contrasts** $\mathbf{t}'\boldsymbol{\alpha}$ of the mean-value parameter $\boldsymbol{\alpha}$ having coefficients $\mathbf{t}$ satisfying $(\mathbf{X}'\mathbf{X})^{-1}\mathbf{t} \in \mathscr{R}(\mathbf{T}_j)$ for some j possess very special properties. This will lead us to a simple description and interpretation of the $\mathbf{V}(\boldsymbol{\theta})$-weighted least-squares estimate $\hat{\boldsymbol{\alpha}}$ of $\boldsymbol{\alpha}$ when $\boldsymbol{\theta}$ is known.

If $(\mathbf{X}'\mathbf{X})^{-1}\mathbf{t} \in \mathscr{R}(\mathbf{T}_j)$ and $\lambda_{ij} > 0$, then $\mathbf{t}'\boldsymbol{\alpha}$ is *estimable* in stratum i and its BLUE is $\mathbf{t}'\hat{\boldsymbol{\alpha}}_i = \mathbf{t}'(\mathbf{X}'\mathbf{C}_i\mathbf{X})^{-}\mathbf{X}'\mathbf{C}_i\mathbf{y}$, simplifying to $\mathbf{t}'[\lambda_{ij}(\mathbf{X}'\mathbf{X})]^{-1}\mathbf{X}'\mathbf{C}_i\mathbf{y}$. This fact follows from the identity (3), which is a consequence of relations (1) to (5) above:

$$(\mathbf{X}'\mathbf{C}_i\mathbf{X})^{-}\mathbf{X}'\mathbf{X} = \sum_j \lambda_{ij}^{-1}\mathbf{T}_j \qquad (3)$$

(sum over j with $\lambda_{ij} > 0$),

where $^{-}$ denotes the Moore–Penrose generalized inverse*. Further, $\mathbf{t}'\hat{\boldsymbol{\alpha}}_i$ has variance $\mathbf{t}'[\lambda_{ij}(\mathbf{X}'\mathbf{X})]^{-1}\mathbf{t}\theta_i$. If $\lambda_{ij} = 0$, then $\mathbf{P}_j$ is orthogonal to $\mathbf{C}_i$ and $\mathbf{t}'\boldsymbol{\alpha}$ is therefore *not* estimable in stratum i. The expression $\lambda_{ij}(\mathbf{X}'\mathbf{X})$ defines the *effective replication* in stratum i associated with the estimation of contrasts having coefficients in $\mathbf{X}'\mathbf{X}\mathscr{R}(\mathbf{T}_j) = \mathscr{R}(\mathbf{X}'\mathbf{P}_j\mathbf{X})$, a notion which is easier to comprehend when $\mathbf{X}'\mathbf{X} = r\mathbf{I}$, for then we have effective replication $r\lambda_{ij}$ for contrasts with coefficients in $\mathscr{R}(\mathbf{T}_j)$.

Every contrast $\mathbf{t}'\boldsymbol{\alpha}$ estimable in stratum i has coefficients $\mathbf{t} \in \mathscr{R}(\mathbf{X}'\mathbf{C}_i\mathbf{X})$, and so by (GB)**, $(\mathbf{X}'\mathbf{X})^{-1}\mathbf{t}$ may be written uniquely as a sum of elements from those subspaces $\mathscr{R}(\mathbf{T}_j)$ for which $\lambda_{ij} > 0$. This implies that we also have a simple expression for its BLUE, namely $\mathbf{t}'\sum_j \mathbf{T}_j[\lambda_{ij}(\mathbf{X}'\mathbf{X})]^{-1}\mathbf{X}'\mathbf{C}_i\mathbf{y}$ with variance $\theta_i\sum_j \mathbf{t}'\mathbf{T}_j[\lambda_{ij}(\mathbf{X}'\mathbf{X})]^{-1}\mathbf{t}$. Note that we do not need to know the θ_i in order to calculate BLUEs *within* strata.

Turning now to the $\mathbf{V}(\boldsymbol{\theta})$-weighted least-squares estimate $\hat{\boldsymbol{\alpha}}$ of $\boldsymbol{\alpha}$, we note that it is given by the formula

$$\hat{\boldsymbol{\alpha}} = \left(\mathbf{X}'\mathbf{V}(\boldsymbol{\theta})^{-1}\mathbf{X}\right)^{-}\mathbf{X}'\mathbf{V}(\boldsymbol{\theta})^{-1}\mathbf{y}$$

$$= \sum_i \sum_j w_{ij}\mathbf{T}_j\left[\lambda_{ij}(\mathbf{X}'\mathbf{X})\right]^{-1}\mathbf{X}'\mathbf{C}_i\mathbf{y}, \qquad (4)$$

where $w_{ij} = \theta_i^{-1}\lambda_{ij}/(\sum_k \theta_k^{-1}\lambda_{kj})$ are the *weights*, here assumed known. If $(\mathbf{X}'\mathbf{X})^{-1}\mathbf{t} \in \mathscr{R}(\mathbf{T}_j)$, then the *overall* BLUE $\mathbf{t}'\hat{\boldsymbol{\alpha}}$ of $\mathbf{t}'\boldsymbol{\alpha}$ is given by the expression

$$\mathbf{t}'\sum_i \left\{w_{ij}\left[\lambda_{ij}(\mathbf{X}'\mathbf{X})\right]^{-1}\mathbf{X}'\mathbf{C}_i\mathbf{y}\right\} = \sum_i w_{ij}\mathbf{t}'\hat{\boldsymbol{\alpha}}_i,$$

the *correctly weighted* (i.e., inversely according to the strata variances) *linear combination* of its estimates from within the strata for which $\lambda_{ij} > 0$. This property provides one of the reasons why we have described (GB) as a condition under which we obtain *directly interpretable* solutions of the problems associated with (1).

In general the weights will not be known and must be estimated. A way of doing so is outlined in the following section and the estimates obtained are then substituted in (4). There does not appear to be any work improving this which applies generally; Brown and Cohen [2] study a particular case.

Our claim that these solutions are *simple* needs more clarification. It depends on the ease with which the eigenvalues λ_{ij} and the associated projections $\mathbf{C}_1, \ldots, \mathbf{C}_s$ and $\mathbf{T}_1, \mathbf{T}_2, \ldots$ can be calculated. In most of the common examples these are combinations of simple averaging operators, and are usually associated with standard decompositions of data and parameters. We have already seen this with the strata projections associated with block and row-and-column and split-plot designs, and will see it later for some typical treatment decompositions.

Example 1 (continued): Balanced Incomplete Block Designs*. We now suppose that each of our v treatments is replicated r times (giving $rv = bp$) and that each pair of distinct treatments appears in exactly λ blocks [whence $\lambda(v-1) = r(p-1)$]. If we write

$\mathbf{T}_0 = v^{-1}\mathbf{J}$ and $\mathbf{T}_1 = \mathbf{I} - v^{-1}\mathbf{J}$ where $\mathbf{I}$ and $\mathbf{J}$ are the identity and "all 1's" matrices of size v, respectively, it is easy to check that $\mathbf{X}'\mathbf{C}_0\mathbf{X} = r\mathbf{T}_0$, $\mathbf{X}'\mathbf{C}_1\mathbf{X} = r(1-e)\mathbf{T}_1$, and $\mathbf{X}'\mathbf{C}_2\mathbf{X} = re\mathbf{T}_1$, where $e = \lambda v/(rp)$ is the *efficiency factor* of the design [10].

As an illustration of the foregoing discussion we note that the *intrablock estimate* of $\boldsymbol{\alpha}$ is $\hat{\boldsymbol{\alpha}}_2 = (re)^{-1}\mathbf{X}'\mathbf{C}_2\mathbf{y}$ and that a contrast $\mathbf{t}'\boldsymbol{\alpha}$ would be estimated using this with variance $(re)^{-1}(\mathbf{t}'\mathbf{t})\theta_2$. On the other hand, if $\theta_2 = \theta_1 = \theta$ (in which case the blocking has no effect) we could estimate $\mathbf{t}'\boldsymbol{\alpha}$ with variance $r^{-1}(\mathbf{t}'\mathbf{t})\theta$. The ratio of these two variances is e, which explains our use of the term "efficiency factor."

Example 2 (expanded): Factorial* and Pseudo-factorial Treatment Structures. Consider the arrangement in two replicates of a row–column design in Table 2.

Table 2

a_0b_0	a_0b_1	a_0b_2		a_0b_0	a_2b_1	a_1b_2
a_1b_0	a_1b_1	a_1b_2		a_2b_2	a_1b_0	a_0b_1
a_2b_0	a_2b_1	a_2b_2		a_1b_1	a_0b_2	a_2b_0
	1				2	

We recognize five strata here: the grand mean stratum, between replications, between rows within replications, between columns within replications, and plots (within rows and columns and replications). The strata projections are just the obvious (adjusted) averaging operators and will not be given here.

Now there is a standard decomposition (see Kempthorne [4]) of the set of nine treatment contrasts into the mean, A, B, AB, and AB^2 contrasts on 1, 2, 2, and 2 degrees of freedom, and these provide the key to our perception of general balance with this design. For we easily perceive the confounding of A and B with rows and columns, respectively, in replicate 1, and likewise AB and AB^2 in replicate 2. In the light of these remarks Table 3 (values of λ_{ij}) may be checked.

Table 3

	Treatment Term				
Stratum	Mean	A	B	AB	AB^2
Grand mean	1	0	0	0	0
Replicates	0	0	0	0	0
Rows	0	$\frac{1}{2}$	0	$\frac{1}{2}$	0
Columns	0	0	$\frac{1}{2}$	0	$\frac{1}{2}$
Plots	0	$\frac{1}{2}$	$\frac{1}{2}$	$\frac{1}{2}$	$\frac{1}{2}$

Finally, we observe that a two-replicate lattice square design for nine unstructured treatments can be completely analyzed by associating pseudo-factors with the treatments as above; *see* LATTICE DESIGNS.

Example 4: Supplemented Balance. Let us suppose that we take a balanced incomplete block design as described above and enlarge each block by one plot to which we apply a single *control* "treatment." Then we would obtain a block design consisting of b blocks of $p+1$ plots each, having $v+1$ "treatments" comprising v original treatments each replicated $r = bp/v$ times and a control replicated b times. What is the "treatment" decomposition in this case?

Not surprisingly, we divide the $(v+1)$-dimensional parameter space into three parts, corresponding to average "treatment" effect, a contrast between average of the original treatments and the control, and $v-1$ contrasts between the original treatments. The matrices that effect this decomposition are

$$\mathbf{T}_0 = (v+1)^{-1}\mathbf{J},$$

$$\mathbf{T}_1 = \left[bp(p+1)\right]^{-1}\boldsymbol{\sigma}\boldsymbol{\sigma}'\mathbf{X}'\mathbf{X},$$

$$\mathbf{T}_2 = \mathbf{I} - \mathbf{T}_0 - \mathbf{T}_1,$$

where $\boldsymbol{\sigma}$ is the $(v+1)\times 1$ vector having the value -1 on original treatments and $+p$ on the control. Note that $\boldsymbol{\sigma}'\mathbf{X}'\mathbf{X}\boldsymbol{\sigma} = (-1)^2 \times bp + p^2 \times b = bp(p+1)$.

The contrast $\mathbf{X}'\mathbf{X}\boldsymbol{\sigma}$ is estimated intrablock (i.e., with efficiency 1), while all the $v-1$ contrasts in $\mathscr{R}(\mathbf{T}_2)$ are estimated with efficiency e^*, where $(1-e^*)(p+1) = p(1-e)$, e being the efficiency factor of the original balanced incomplete block design. This design phenomenon has been called *supplemented balance* by Pearce [7] (see also Alvey et al. [1, Chap. 6, Sec. 2.4.1]). It applies to row-and-column designs and more generally.

OTHER ASPECTS OF (GB)

The term *balance* was originally introduced to describe block designs in which all simple contrasts such as $\alpha_1 - \alpha_2$ are estimated with the same variance in the intrablock stratum. It is clear from the foregoing that this is indeed true for balanced incomplete block designs. *Partially balanced incomplete block designs** have a more subtle form of this property: the set of all treatment contrasts can be decomposed into a number of orthogonal subspaces within each of which (normalized) contrasts are estimated with the same variance. This is the form which justifies the use of the term *general balance*, for within each of the subspaces $\mathscr{R}(\mathbf{T}_j)$, which provide an orthogonal decomposition of the set of all treatment contrasts, we find that under the simplifying assumption of equal replication a normalized contrast is estimated with variance $\lambda_{ij}^{-1}\theta_i$ in stratum i when $\lambda_{ij} > 0$, and overall with variance $\sum_i \lambda_{ij}^{-1}\theta_i$. The general case is similar.

We have already drawn attention to the *combinability* property of estimates in the common eigenspaces of the scaled strata information matrices. This property is in fact equivalent to (GB) in the following sense: If the space of all treatment contrasts can be decomposed into a direct sum of subspaces of contrasts within each of which this combinability property (the correctly weighted BLUEs from within strata coincides with the overall weighted least-squares BLUE) holds, then the model satisfies (GB).

It seems that designs satisfying (GB) are essentially the most general ones for which an analysis-of-variance* table may be calculated. Under normality this enables the significance of treatment terms estimated *within strata* to be tested via an F-test. The details are as shown in Table 4.

The hypothesis that $\mathbf{T}_j\boldsymbol{\alpha} = \mathbf{0}$ may be tested by the obvious F-ratio in any stratum for which $\lambda_{ij} > 0$ and there are sufficiently many degrees of freedom (d.f.) in the residual line. We note that not only is there no known way of *combining the tests* relating to a treatment term estimated in more than one stratum, but the natural *overall test* of $\mathbf{T}_j\boldsymbol{\alpha} = \mathbf{0}$ based on weighted least-squares theory does not seem to have been studied when $\boldsymbol{\theta}$ is unknown.

A natural way in which the *dispersion parameters* $\{\theta_i\}$ can be estimated is the following: Equate the sums of squares of residuals within each stratum to their expected values. This leads to the equations

$$(\mathbf{y} - \mathbf{X}\hat{\boldsymbol{\alpha}})'\mathbf{C}_i(\mathbf{y} - \mathbf{X}\hat{\boldsymbol{\alpha}}) = \theta_i\left[r(\mathbf{C}_i) - \sum_j w_{ij} r(\mathbf{T}_j)\right], \quad i = 1, \ldots, s, \quad (5)$$

Table 4 ANOVA Table for Stratum i

Source	d.f.	Sum of Squares	EMS
⋮	⋮	⋮	
Treatment term j (when $\lambda_{ij} > 0$)	$r(\mathbf{T}_j)$	$\mathbf{y}'\lambda_{ij}^{-1}\mathbf{C}_i\mathbf{X}\mathbf{T}_j(\mathbf{X}'\mathbf{X})^{-1}\mathbf{X}'\mathbf{C}_i\mathbf{y}$	$\theta_i + \dfrac{\lambda_{ij}}{r(T_j)}\boldsymbol{\alpha}'\mathbf{X}'\mathbf{X}\mathbf{T}_j\boldsymbol{\alpha}$
⋮	⋮		
Residual	$r(\mathbf{C}_i) - \sum_{j:\lambda_{ij}>0} r(\mathbf{T}_j)$ (may be zero)	By difference (may be zero)	θ_i (if d.f. > 0)

where $r(\cdot)$ denotes the rank of a matrix. We note that the weights (w_{ij}) involve the $\{\theta_i\}$ and so an iterative method of solution is required (see Nelder [6] for one approach). However, (5) is in fact the set of likelihood equations for the $\{\theta_i\}$ based on error contrasts and an assumption of normality for $\mathbf{y}$, and their solution leads to *restricted maximum likelihood estimates*. In this case we may calculate the *information matrix* $I(\boldsymbol{\theta})$ of the $\{\hat{\theta}_i\}$, and $-2I(\boldsymbol{\theta})$ turns out to have diagonal elements

$$\theta_i^{-2}\left[r(\mathbf{C}_i) - \sum_{j:\lambda_{ij}>0} r(\mathbf{T}_j) + \sum_j (1 - w_{ij})^2 r(\mathbf{T}_j)\right] \tag{6a}$$

and off-diagonal elements

$$\theta_i^{-1}\theta_{i'}^{-1}\sum_j w_{ij}w_{i'j}r(\mathbf{T}_j). \tag{6b}$$

A particular advantage of using this procedure with (GB) designs is the fact that a simple expression can be found for the right-hand side of (5) and Fisher's scoring method can be adopted using (6a) and (6b) without recalculating $\hat{\boldsymbol{\alpha}}$ explicitly (see Nelder [6]). In general this cannot be avoided.

We close with a remark about the calculation of BLUPs of random effects under models such as (2). This is really an aspect of the block structure defined by $\mathbf{C}_1, \ldots, \mathbf{C}_s$, so we do not give any details. It will suffice to observe that if $\mathbf{b}'\mathbf{U}\mathbf{y}$ is the best linear predictor of a contrast of random effects when $\boldsymbol{\alpha} = \mathbf{0}$, then $\mathbf{b}'\mathbf{U}(\mathbf{y} - \mathbf{X}\hat{\boldsymbol{\alpha}})$ is the BLUP of this same contrast for arbitrary $\boldsymbol{\alpha}$. The great advantage of having (GB) hold is the simplicity of the expression for $\hat{\boldsymbol{\alpha}}$ and the ease of estimation of the necessary dispersion parameters.

REMARKS

General balance was introduced by Nelder [5] with the intention of using the notions to permit the writing of very general computer programs for analyzing designed experiments. Many of the ideas described above have been implemented in a modified form in GENSTAT (see Alvey et al. [1, Chap. 6]). The example of a row-and-column design not satisfying (GB) comes from the thesis of Houtman [3], as does the combinability result given in the preceding section. For a recent review of the literature on variance components* and BLUPs which includes a discussion of (GB), see Thompson [9].

References

[1] Alvey, N. G. et al. (1977). *GENSTAT. A General Statistical Program*. Rothamsted Experimental Station, Harpenden, Herts., England.

[2] Brown, L. D. and Cohen, A. (1974). *Ann. Statist.*, **2**, 963–976.

[3] Houtman, A. (1980). *The Analysis of Designed Experiments*. Ph.D. dissertation, Princeton University Press, Princeton, NJ.

[4] Kempthorne, O. (1952). *The Design and Analysis of Experiments*. Wiley, New York.

[5] Nelder, J. A. (1965). *Proc. R. Soc. A*, **283**, 147–178.

[6] Nelder, J. A. (1968). *J. R. Statist. Soc. B*, **30**, 303–311.

[7] Pearce, S. C. (1963). *J. R. Statist. Soc. A*, **126**, 353–377.

[8] Preece, D. A. (1966). *Biometrics*, **22**, 1–25.

[9] Thompson, R. (1980). *Math. Operationsforsch. Statist. Ser. Statist.*, **11**.

[10] Yates, F. (1936). *Ann. Eugen.*, **7**, 121–140.

(BALANCE IN EXPERIMENTAL DESIGN
BLOCKS, BALANCED INCOMPLETE
DESIGN AND ANALYSIS OF EXPERIMENTS
ESTIMABILITY
GENERAL LINEAR MODEL
PARTIALLY BALANCED DESIGNS
SPLIT-PLOT DESIGNS
WEIGHTED LEAST SQUARES)

T. P. SPEED

GENERALIZED CANONICAL VARIABLES

A random vector $\mathbf{X}$ of interest poses both analytical and economical problems in many situations if it has too many components. The large number of correlations associated with the components of $\mathbf{X}$ makes it difficult to comprehend overall or general relationships. A reduction in dimensionality together with some representative measure of relationships among the variables can thus be of practical importance. Generalized canonical variables (GCVs), with their associated correlations, termed the generalized canonical correlations (GCCs), attempt to serve this purpose.

It is assumed that a meaningful subgrouping of $\mathbf{X}: p \times 1$ into several disjoint subvectors $\mathbf{X}_1, \ldots, \mathbf{X}_k$, $\mathbf{X}_i : p_i \times 1$, $\sum_1^k p_i = p$ is already given. For $k = 2$, Hotelling [6] introduced the concept of canonical variables (CVs) and associated canonical correlations (CCs) (*see* CANONICAL ANALYSIS).

DEFINITION

The first GCV, $\mathbf{Y}^{(1)}$ with $\mathbf{Y}^{(1)\prime} = [Y_1^{(1)}, \ldots, Y_k^{(1)}] = \mathbf{f}^{(1)}(\mathbf{X})' = [f_1^{(1)}(\mathbf{X}_1), \ldots, f_k^{(1)}(\mathbf{X}_k)]$, where the $f_j^{(1)}$s are real-valued functions, is a k-dimensional random variable, the components of which are chosen so as to optimize a criterion based on some function of their correlation matrix. For each such function there will be a corresponding GCV as a generalization of CV. The higher-stage GCVs, $\mathbf{Y}^{(i)} = \mathbf{f}^{(i)}(\mathbf{X})$, $i = 1, 2, \ldots$, are also k-dimensional random variables, the components of which are chosen so as to optimize the same criterion with some additional constraints imposed at each stage regarding the relationships among the variables at a given stage with those in the preceding stages. [These constraints may be different for different methods. Also, the $f_j^{(i)}(\mathbf{X}_j)$s are usually linear functions of the $\mathbf{X}_j$s, $j = 1, 2, \ldots, k$, and have unit variances.] For a given GCV, the associated correlation can be loosely termed as GCC, although depending on the method of construction of the GCVs such GCCs may in fact be correlation matrices instead of a one-dimensional summary statistic. It may be reasonable to terminate the procedure at the sth stage if, depending on the method, at the $(s + 1)$th stage the scalar GCC is near zero or the off-diagonal elements of the GCC matrix are close to zeros.

There are two basic problems related to GCV analysis. One is concerned with the construction of GCVs when the population dispersion matrix is known. The other deals with statistical inference when only a sample from the population is available. Although the first problem had been attacked since 1951, the research on the second started only recently with Sen Gupta [12]. Also, a realistic interpretation of GCVs is of great practical importance.

CONSTRUCTION

For construction of GCVs, one faces at least three problems of optimization: (a) selection of the number k of subgroups and their corresponding elements, (b) determination of the compounding functions $f_j^{(i)}$s, and (c) deciding on the optimal stage of stopping for higher-order GCVs. As stated above, it is usually assumed for (a) that the number and the elements of the subgroups are completely specified. Otherwise, one may attempt to use cluster analysis techniques. For (b) several available methods will be discussed. With $k > 2$, for (c), the situation is "somewhat arbitrary" for some of these methods.

The algebraic derivations become simpler and essentially the same as in CC analysis if we can reduce the several groups effectively to two groups. Further, for multivariate normal populations, testing that the CCs are all zeros is equivalent to testing that the two sets are independent. These considerations motivated the construction of various conditional GCCs. A set or sets of variables are held fixed and effectively the two sets of residuals are analyzed. The GCCs can then be obtained through parallel test criteria of (conditional) independence as in the case of two-group CCs. Most recently this concept has led to g_1 and g_2 bipartial GCC analysis (see Lee [8]). For each of the cases above, the GCVs are the normalized eigenvectors associated with the eigenvalues and GCCs corresponding to the determinantal equation

$$\begin{gathered} |{}_k\Sigma^*_{od} - (k-1)\rho_k {}_k\Sigma^*_d| = 0, \\ {}_k\Sigma^* = {}_k\Sigma^*_{od} + {}_k\Sigma^*_d . \end{gathered} \tag{1}$$

${}_k\Sigma^*_d$ is the diagonal supermatrix with elements Σ^*_{ii}, $i = 1, \ldots, k$, and ${}_k\Sigma^*$ is the modified covariance matrix, modified by the particular generalization under consideration. Although computationally quite convenient, these GCVs seem to be of limited practical utility because of their conditional nature.

McKeon [10] has suggested a (first) GCC obtained as a generalization of a modified intraclass correlation coefficient*. For the CVs ($k = 2$) (see, e.g., Anderson [1, Chap. 12, eq. (14)]) and the first GCV by the method of McKeon ($k = k$), ${}_k\Sigma^*$ of (1) is simply Σ, the dispersion matrix of $\mathbf{X}$. Carroll [2], Horst [5], and several other authors have also arrived at a similar solution, although from different viewpoints. Sen Gupta [12] derived new GCVs obtained by modifying with the equicorrelation constraint the criterion of minimizing the generalized variance* of $\mathbf{Y}$ (i.e., $|\Sigma_{\mathbf{Y}}|$; see Anderson [1, p. 305, Prob. 5]). These new GCVs are quite convenient for purposes of statistical inference and will be referred to later in that context.

Let the correlation (also covariance) matrix of $\mathbf{Y}^{(1)}$ be $\Sigma^{(1)} = (\rho_{ij}^{(1)})$. An important property of CC analysis is that the CVs are invariant under nonsingular transformations of either set. Exploiting this property, Horst [5] proposed maximizing $\sum\sum_{i<j=1}^{k} \rho_{ij}^{(1)}$ and also maximizing the largest eigenvalue of $\Sigma^{(1)}$. These are termed the SUMCOR and MAXVAR methods, respectively. Exploiting the same property, Steel [14] suggested the GENVAR method, where GCVs are obtained by minimizing the determinant $|\Sigma^{(1)}|$, the generalized variance of $\mathbf{Y}^{(1)}$. The methods of SSQCOR and MINVAR were proposed by Kettenring [7]. The former attempts to maximize $\sum\sum_{i<j=1}^{k} \rho_{ij}^{(1)2}$ or equivalently trace of $\Sigma^{(1)2}$, while the latter minimizes the smallest eigenvalue of $\Sigma^{(1)}$. Kettenring has given some interesting factor-analytic interpretations for the foregoing five procedures. However, for GENVAR, SSQCOR, and SUMCOR, the corresponding defining equations cannot be presented in simple form as in (1) and iterative procedures are required even for first GCCs and GCVs.

All of the methods above reduce to CC analysis for $k = 2$. Each method emphasizes some aspect of the correlation matrices of the GCVs and hence attempts to detect certain forms of linear relationships among the sets of variables. Thus, depending on the problem, it may be necessary to use several of these methods to better understand the underlying relationships among the sets of variables in $\mathbf{X}$.

For singular ${}_k\Sigma^*$, see Sen Gupta [11].

STATISTICAL INFERENCE

When only a sample is available from the population, the GCVs are usually estimated by the maximum likelihood* method, which effectively replaces the population parameters in (1) by their sample counterparts.

Exact distributions for GCVs are quite complicated and seem to be nearly intractable for those obtainable only by iterative procedures. A large-sample approximation by multivariate normality may be suggested.

Several problems of tests of hypotheses* were formulated in Sen Gupta [12]. Since GCVs obtained by different methods optimize different criteria, the test statistics and associated distributions need to be worked out separately for each method. We illustrate this below using the new GCVs obtained in Sen Gupta [12]. For simplicity, only tests based on the first new GCV, $\mathbf{Y}^{(1)} \equiv \mathbf{Y}$, will be considered. Let the corresponding correlation matrix of $\mathbf{Y}$ be $\Sigma_{j^{\mathbf{Y}}}$. Large samples are assumed.

1. It would be natural to explore to what extent the GCVs optimize the criteria, i.e., what value of the criterion is achieved by a particular GCV. For new GCVs, this leads to a test of $H_0 : |\Sigma_{\mathbf{Y}}| = \sigma_0^{2k}$ (specified) against $H_1 : |\Sigma_{\mathbf{Y}}| < \sigma_0^{2k}$. A test for $H_0 : \rho^{(1)} = \rho_0^{(1)}$

(given) against $H_1 : \rho^{(1)} \neq \rho_0^{(1)}$ can also be proposed, where $\rho^{(1)}$ is the new (first) GCC. These involve curved exponential families*. Likelihood ratio tests* and a locally most powerful test for $\rho^{(1)}$ are available. The latter test is globally one-sided unbiased and the exact distribution of the test statistic is available in terms of the (tabulated) Kummer's function (see Sen Gupta [13]).

2. Next, it would be of interest to determine whether with the same dimension k, reasonable alternative regroupings of $\mathbf{X}$ produce better results (see, e.g., Gnanadesikan [3, p. 77, para. 4]). If there are m reasonable regroupings, each of dimension k, then for the new GCVs, one would test $H_0 : |\mathbf{\Sigma}_{{}_j\mathbf{Y}}|$s all equal against $H_1 : |\mathbf{\Sigma}_{{}_j\mathbf{Y}}|$s are in a given order, where ${}_j\mathbf{Y}$ is the (first) GCV for the jth regrouping, $j = 1, \ldots, m$. If the same set of data is used to estimate the m GCVs, the tests will be quite involved, owing to the mutual dependence of the GCVs. Some judicious transformations are helpful here to yield simpler tests.

3. It may be possible to reasonably regroup $\mathbf{X}$ into various numbers k_i, $i = 1, \ldots, l$ of subsets, yielding GCVs of different dimensions. It is then worthwhile to explore whether a GCV of smaller dimension performs as good as, if not better than, one of a higher dimension. For such a case, the former GCV will naturally be preferred. For l subsets, with new GCVs, one would test $H_0 : |\mathbf{\Sigma}_{{}_j\mathbf{Y}}|^{1/k_j}$ all equal against $H_1 : |\mathbf{\Sigma}_{{}_i\mathbf{Y}}|^{1/k_i} < |\mathbf{\Sigma}_{{}_j\mathbf{Y}}|^{1/k_j}$, $k_i < k_j$, $i \neq j$, $i, j = 1, \ldots, l$, where ${}_i\mathbf{Y}$ is the (first) new GCV of dimension k_i obtained by the ith mode of regrouping, $i = 1, \ldots, l$. Some solutions can be presented here through isotonic regression*. For more details, see Sen Gupta [12].

Example. From Thurstone and Thurstone [15] $k = 3$ sets of scores on three batteries of three tests each (i.e., $p_i = 3$, $i = 1, 2, 3$, $p = 9$) for several individuals are available. The vector variables $\mathbf{Z}_i : 3 \times 1$, $i = 1, 2, 3$, represent different measures of verbal, numerical, and spatial abilities of the subjects tested. A transformation was employed on $\mathbf{Z}' = (\mathbf{Z}_1', \mathbf{Z}_2', \mathbf{Z}_3')$ to give internally "sphericized" standardized variables $\mathbf{X}$. Horst [4], Kettenring, Gnanadesikan, and Sen Gupta have derived GCVs for $\mathbf{X}$ from its covariance (same as correlation) matrix $\mathbf{R}$ (see also McDonald [9]). Let $\mathbf{R} = (\mathbf{R}_{ij})$, the $\mathbf{R}_{ij}$ being 3×3 matrices. Then $\mathbf{R}_{ii} = \mathbf{I}$, $i = 1, 2, 3$, and

$$\mathbf{R}_{12} = \begin{bmatrix} 0.636 & 0.126 & 0.059 \\ -0.021 & 0.633 & 0.049 \\ 0.016 & 0.157 & 0.521 \end{bmatrix};$$

$$\mathbf{R}_{13} = \begin{bmatrix} 0.626 & 0.195 & 0.059 \\ 0.035 & 0.459 & 0.129 \\ 0.048 & 0.238 & 0.426 \end{bmatrix};$$

$$\mathbf{R}_{23} = \begin{bmatrix} 0.709 & 0.050 & -0.002 \\ 0.039 & 0.532 & 0.190 \\ 0.067 & 0.258 & 0.299 \end{bmatrix}.$$

Consider, including MAXVAR, the different approaches discussed above which yield the same result in the first stage as that of McKeon's method. For McKeon's method [e.g., in (1)] we have $\mathbf{\Sigma}^* = \mathbf{R}$, $\mathbf{\Sigma}_d^* = \mathbf{R}_d = \mathbf{I}$, and the first GCV is obtained from the eigenvector $\mathbf{v}^{(1)'} = (\mathbf{v}_1^{(1)'}, \ldots, \mathbf{v}_k^{(1)'})$ corresponding to the largest eigenvalue of the equation $|\mathbf{R} - \lambda\mathbf{I}| = 0$. The first GCC for McKeon's method is $\rho^{(1)} = 0.745$, which is larger than any pairwise correlation in $\mathbf{R}$. In the MAXVAR method, the usual requirement that the components $Y_i^{(1)} = \boldsymbol{\alpha}_i^{(1)'} \mathbf{X}_i$ of the first GCV $\mathbf{Y}^{(1)}$ have unit variance is achieved by letting $\boldsymbol{\alpha}_i^{(1)} = \mathbf{v}_i^{(1)} / \mathbf{v}_i^{(1)'}\mathbf{v}_i^{(1)}$, $i = 1, \ldots, k$. This yields

$$\begin{aligned} \mathbf{Y}^{(1)} = (&0.732X_{11} + 0.514X_{12} + 0.447X_{13}, \\ &0.659X_{21} + 0.625X_{22} + 0.420X_{23}, \\ &0.678X_{31} + 0.640X_{32} + 0.362X_{33}), \end{aligned}$$

where X_{ij} is the jth element of $\mathbf{X}_i$, $i, j = 1, 2, 3$.

It turns out that $\mathbf{Y}^{(1)}$ is "virtually identical" to the first GCVs obtained by iterative procedures for the SUMCOR method by Horst [4] and the SSQCOR and GENVAR methods by Kettenring. The same is true for the first GCC matrices for these methods, which are all almost identical to

$$\mathbf{\Sigma}^{(1)} = \begin{bmatrix} 1.000 & 0.735 & 0.756 \\ & 1.000 & 0.743 \\ & & 1.000 \end{bmatrix}.$$

For higher-stage GCVs further iterative computations are required (see Horst [4] and Kettenring [7] for details).

INTERPRETATIONS

Although all the original variables are needed to obtain the GCVs, the final result would indicate that only a few GCVs need to be retained. All future analysis can be limited to these retained GCVs. For example, these GCVs, instead of **X**, may constitute the variables in future regression analysis. A meaningful reduction of dimensionality is achieved.

The GCVs obtained above as linear combinations of the original variables have in most cases very little practical meaning by themselves. However, one can attempt to interpret the GCVs (see Timm [16] for the case of CCs) via the correlations of the linear compounds with the corresponding elements involved, i.e., correlations of $Y_i^{(t)}$ with X_{ij}s, $i = 1, \ldots, k$, $j = 1, \ldots, p_i$, $t = 1, \ldots, s$.

References

[1] Anderson, T. W. (1958). *An Introduction to Multivariate Statistical Analysis*. Wiley, New York. (An excellent presentation of the theory of CVs.)

[2] Carroll, J. D. (1968). *Proc. Amer. Psychol. Ass.*, 227–228.

[3] Gnanadesikan, R. (1977). *Methods for Statistical Data Analysis of Multivariate Observations*. Wiley, New York.

[4] Horst, P. (1961). *Psychometrika*, **26**, 129–149.

[5] Horst, P. (1965). *Factor Analysis of Data Matrices*. Holt, Rinehart and Winston, New York.

[6] Hotelling, H. (1936). *Biometrika*, **28**, 321–377.

[7] Kettenring, J. R. (1971). *Biometrika*, **58**, 433–451. (Detailed study of construction of GCVs by five methods. Also gives interesting factor analytic interpretations.)

[8] Lee, S. Y. (1978). *Psychometrika*, **43**, 427–431. (Errata 1979, *Psychometrika*, **44**, 131.)

[9] McDonald, R. P. (1968). *Psychometrika*, **33**, 341–381.

[10] McKeon, J. J. (1966). *Psychometric Monogr.*, **13**. (Discusses applications of CVs.)

[11] Sen Gupta, A. (1980). Generalized Correlations in the Singular Case. *Tech. Rep. No. 46* (under Contract N00014-75-C-0442), Dept. of Statistics, Stanford University, Stanford, Calif.

[12] Sen Gupta, A. (1982). On the Problems of Construction and Statistical Inference Associated with a Generalization of Canonical Variables. *Tech. Rep. No. 168* (under NSF Grant MCS 78-07736), Dept. of Statistics, Stanford University, Stanford, Calif. (Introduces and formulates aspects of statistical inference related to GCVs and to new GCVs in particular.)

[13] Sen Gupta, A. (1982). On Tests for the Equicorrelation Coefficient and the Generalized Variance of Symmetric Standard Multivariate Normal Distribution. *Tech. Rep. No.* 55 (under Contract N00014-75-C-0442), Dept. of Statistics, Stanford University, Stanford, Calif.

[14] Steel, R. G. D. (1951). *Ann. Math. Statist.*, **22**, 456–460.

[15] Thurstone, L. L. and Thurstone, T. G. (1941). *Psychometric Monogr.*, **2**.

[16] Timm, N. H. (1975). *Multivariate Analysis with Applications in Education and Psychology*. Brooks/Cole, Monterey, Calif. (Discusses conditional GCCs.)

(CANONICAL ANALYSIS
GENERALIZED VARIANCE)

ASHIS SEN GUPTA

GENERALIZED HYPERGEOMETRIC DISTRIBUTIONS

Hypergeometric distributions* are defined by the probabilities

$$p(x) = \binom{M}{x}\binom{N-M}{n-x}\Big/\binom{N}{n} = \binom{n}{x}\binom{N-n}{M-x}\Big/\binom{N}{M} \qquad (1)$$

for $\max(0, n + M - N) \leqslant x \leqslant \min(n, M)$, where N, M, and n are positive integers. If $N \geqslant M + n$, this is written

$$p(x) = \frac{(N-M)!\,(N-n)!}{N!\,(N-M-n)!} \times \frac{(-M)^{[x]}(-n)^{[x]}}{(N-M-n+1)^{[x]}x!},$$

and the factor depending on x is a term of

the Gauss hypergeometric series

$$ {}_2F_1(\alpha, \beta; \gamma; s) = \sum_{x=0}^{\infty} \frac{\alpha^{[x]}\beta^{[x]}}{\gamma^{[x]}x!} s^x, \quad (2) $$

having $s = 1$ and suitable parameters. If α or β is a negative integer $-k$, say, and the other parameters have suitable values, then the series is defined to be finite with $x \leqslant k$. The infinite series with real parameters and $s = 1$ is convergent if and only if $\alpha + \beta < \gamma$. By using (2), the probability generating function* (PGF) of (1) is written in the form

$$ {}_2F_1(\alpha, \beta; \gamma; s)/{}_2F_1(\alpha, \beta; \gamma; 1). \quad (3) $$

We remark that, if $p(x)$ has a PGF of this general form, then

$$ p(x+1)/p(x) = (\alpha + x)(\beta + x)/\{(\gamma + x)(1 + x)\}. \quad (4) $$

Since $p(x)$ of (1) corresponds to a special case of (3), it is natural to ask how it can be generalized; Karl Pearson* and his school studied the generalization. Pearson-type curves* are based on a differential equation which is a continuous analog of the difference equation of $(p(x+1) - p(x))/p(x)$ obtained from (4) (see, e.g., Karl Pearson [18, 19], Camp [1], and Davies [4, 5]).

BASIC TYPES OF GENERALIZED HYPERGEOMETRIC DISTRIBUTIONS

One way of generalization is to find when (3) is actually PGF of a distribution on $[0, n]$ or $[0, \infty)$. It is shown that there are five basic types as shown in Table 1 (see Davies [5], Shimizu [21], and Kemp [13]).

Another approach to generalize $p(x)$ is to allow M, N, and n in (1) to be real and negative by extending the definition of binomial coefficient. In this way, Kemp and Kemp [15] classified eight types, cited by Johnson and Kotz [10]; they are compared with the classification in Table 1.

The classification was criticized by Sarkadi [20] and Shimizu [21], since $p(x)$ of (1) is a distribution on positive integers if $n + M > N$. The criticism was answered by Sibuya and Shimizu [23] by defining the function form of distributions in terms of gamma functions, and by setting up some conventional rules on negative integer arguments of the gamma functions and on distribution range. It turns out that all possible new types are distributions of $\pm X \pm m$, where X is a variable of one of types in Table 1 and m is an integer.

Type A1, A2, and B3 distributions of Table 1 are generated by many kinds of chance mechanisms, of which three are typical. First, if X and Y are independent binomial* (negative binomial*) variables with the same probability parameter, then the conditional distribution of X given $X + Y = s$ is type A1 (A2). Second, if the probability parameter of a binomial (negative binomial) distribution is a beta* variable, then the compound distribution is type A2 (B3). Third, in Pólya's urn model*, type A1 (A2) is generated in the negative (positive) contagion case, and in the inverse sampling of Pólya's urn, type A2 (B3) is generated in the negative (positive)

Table 1 Basic Generalized Hypergeometric Distributions

Range	Type	$(\alpha, \beta; \gamma)$[a]	Restriction	Kemp and Kemp Types
$[0, n]$	A1	$(-\xi, -n; \zeta)$	$\xi > n-1$	IA(i), (ii)
	A2	$(\xi, -n; -\zeta)$	$\zeta > n-1$	IIA, IIIA
$[0, \infty)$	B1	$(-n+\epsilon, -n+\delta; \zeta)$	—	IB
	B2	$(\epsilon, -n+\delta; -n+\rho)$	$\rho > \epsilon + \delta$	IIB, IIIB
	B3	$(\xi, \eta; \zeta)$	$\zeta > \xi + \eta$	IV

[a] ξ, η, and ζ are positive real numbers; n is a positive integer; and ϵ, δ, and ρ are real numbers between zero and 1.

contagion case (see, e.g., Johnson and Kotz [11]).

Because of these models and their various expressions, distributions of these three types are called by many names. Type A1 distributions are "positive hypergeometric," including the "ordinary hypergeometric" of (1), and best defined by

$$\binom{a}{x}\binom{b}{n-x}\Big/\binom{a+b}{n}$$
$$=\binom{n}{x}\frac{a^{(x)}b^{(n-x)}}{(a+b)^{(n)}}$$
$$=\frac{b^{(n)}}{(a+b)^{(n)}}\frac{(-a)^{[x]}(-n)^{[x]}}{(b-n+1)^{[x]}x!}, \tag{5}$$

where $0 \leqslant x \leqslant n$, $a,b > n-1$, and $x \leqslant a$ $(n-b \leqslant x)$ if $a(b)$ is an integer.

Type A2 distributions are called "negative hypergeometric," "binomial beta," "beta-binomial," "Pólya," "Markov–Pólya," and "Pólya–Eggenberger." The latter names are due to the fact that Pólya's urn was actually studied by Markov [16] and Eggenberger and Pólya [6]. Patil and Joshi [17] called type A2 "inverse hypergeometric." It is defined by

$$\binom{-a}{x}\binom{-b}{n-x}\Big/\binom{-a-b}{n}$$
$$=\binom{a+x-1}{x}\binom{b+n-x-1}{n-x}\Big/\binom{a+b+n-1}{n}$$
$$=\binom{n}{x}\frac{a^{[x]}b^{[n-x]}}{(a+b)^{[n]}}$$
$$=\frac{b^{[n]}}{(a+b)^{[n]}}\frac{a^{[x]}(-n)^{[x]}}{(-b-n+1)^{[x]}x!}, \tag{6}$$

where $0 \leqslant x \leqslant n$ and $a,b > 0$; if both a and b are integers, this is expressed as

$$x^{[a-1]}(n-x)^{[b-1]}/B(a,b)n^{[a+b-1]}, \tag{6a}$$

where $B(a,b)$ is the beta function.

Type B3 distributions are called "inverse Pólya," "inverse Pólya-Eggenberger," "negative binomial beta," "beta-Pascal," and "generalized Waring." The last name, by Irwin [7], is based on the fact that the probabilities are of a generalized form of Waring's inverse factorial series. The probabilities are

$$\frac{B(a+c,b+c)}{B(a+b+c,c)}\frac{a^{[x]}b^{[x]}}{(a+b+c)^{[x]}x!}$$
$$=\frac{\Gamma(a+c)\Gamma(b+c)\Gamma(a+x)\Gamma(b+x)}{\Gamma(a)\Gamma(b)\Gamma(c)\Gamma(a+b+c+x)x!}, \tag{7}$$

where $x = 0,1,2,\ldots$ and $a,b,c > 0$. The case $a = b = 1$ is called the "Yule distribution," and the case $b = c = 1$ the "Mizutani distribution" or "compound Poisson* distribution." When zero is deleted and β approaches zero, the limit is the digamma distribution, a compound logarithmic series distribution* (see Sibuya [22]). Sometimes types A2 and B3 are defined by their shifted forms.

Type B1 has unimodal or monotone-decreasing probabilities and its mean is greater than its variance if $n \geqslant 2$. Type B2 can be unimodal or bimodal and does not have a finite mean.

MULTIVARIATE GENERALIZED HYPERGEOMETRIC DISTRIBUTION*

Multivariate extensions were also studied by Karl Pearson [19] and his school. But a unified theory was developed more recently by Steyn [24, 25], Janardan and Patil [9], Janardan [8], and others. Let the p-variate hypergeometric series be defined by

$$F(\boldsymbol{\alpha},\beta;\gamma;\mathbf{s})$$
$$=\sum_{x=0}^{\infty}\left(\sum_{\sum x_i=x}\prod_{i=1}^{p}\alpha_i^{[x_i]}s_i^{x_i}/x_i!\right)\beta^{[x]}/\gamma^{[x]},$$

where $\boldsymbol{\alpha} = (\alpha_1,\ldots,\alpha_p)$, $\mathbf{s} = (s_1,\ldots,s_p)$. A PGF of the form $F(\boldsymbol{\alpha},\beta;\gamma;\mathbf{s})/F(\boldsymbol{\alpha},\beta;\gamma;\mathbf{1})$ is a natural extension of (3). There are not many types in this extension since if $(X_1,\ldots,X_p)$ is a variable of these distributions, the distributions of $X_1+\cdots+X_p$ and all the marginal distributions of X_i's must be univariate generalized hypergeometric, and the conditional distribution of $(X_1, \ldots, X_p)$ given $X_1 + \cdots + X_p = x$ must be a singular p-variate positive or negative hypergeometric distribution in the following. Complete classification, including distributions on negative regions, was done in Sibuya and Shimizu [23].

There are four types of p-variate distributions with finite nonnegative ranges. Two of them are extensions of type A1, and one is best expressed by its $(p+1)$-variate singular form

$$\prod_{i=0}^{p}\binom{a_i}{x_i}\bigg/\binom{a}{n} = \binom{n}{\mathbf{x}}\left(\prod_{i=0}^{p} a_i^{(x_i)}\right)\bigg/ a^{(n)} = \frac{a_0^{(n)}}{(a_0+a)^{(n)}}\frac{(-n)^{[x]}}{(a_0-n+1)^{[x]}}\prod_{i=1}^{p}\frac{(-a_i)^{[x_i]}}{x_i!}, \tag{8}$$

where $\mathbf{x} = (x_0, x_1, \ldots, x_p)$; n is a positive integer; $\binom{n}{\mathbf{x}}$ is the multinomial coefficient; $a_i > n-1$, $a = \sum_{i=1}^{p} a_i$; the x_i's are nonnegative integers; $x = \sum_{i=1}^{p} x_i$, $x_0 + x = n$; and $x_i \leqslant a_i$ if a_i is an integer less than n $(i = 0, 1, \ldots, p)$.

Another extension of type A1 is

$$\left\{\prod_{i=1}^{p}\binom{n_i}{x_i}\right\}\frac{a^{(x)}b^{(n-x)}}{(a+b)^{(n)}} = \left[\binom{a}{x}\binom{b}{n-x}\bigg/\binom{a+b}{n}\right]\left[\prod_{i=1}^{p}\binom{n_i}{x_i}\bigg/\binom{n}{x}\right] = \frac{b^{(n)}}{(a+b)^{(n)}}\frac{(-a)^{[x]}}{(b-n+1)^{[x]}}\prod_{i=1}^{p}\frac{(-n_i)^{[x_i]}}{x_i!}, \tag{9}$$

where n_i's are positive integers and $\sum n_i = n$; $0 \leqslant x_i \leqslant n_i$ $(i = 1, \ldots, p)$; $\sum x_i = x$; $a, b > 0$; and $x \leqslant a$ $(n - x \leqslant b)$ if $a(b)$ is an integer less than n. The distribution ranges of (8) are simplex, while those of (9) are rectangular, but if all the parameters in (8) and (9) are integers, both types are identical and "multivariate ordinary hypergeometric."

Two other finite types are extensions of A2, and one corresponds to (8):

$$\prod_{i=0}^{p}\binom{-a_i}{x_i}\bigg/\binom{-a}{n} = \prod_{i=0}^{p}\binom{a+x_i-1}{x_i}\bigg/\binom{a+x-1}{n} = \binom{n}{x}\prod_{i=0}^{p}\left(a_i^{[x_i]}\right)\bigg/ a^{[n]} = \frac{a_0^{[n]}}{(a_0+a)^{[n]}}\frac{(-n)^{[x]}}{(-a_0-n+1)^{[x]}}\prod_{i=1}^{p}\frac{a_i^{[x_i]}}{x_i!}, \tag{10}$$

where notation and ranges are the same as in (8), except that the range does not change for any values of a_i's. Another type corresponds to (9):

$$\left\{\prod_{i=1}^{p}\binom{n_i}{x_i}\right\}\frac{a^{[x]}b^{[n-x]}}{(a+b)^{[n]}} = \left[\binom{-a}{x}\binom{-b}{n-x}\bigg/\binom{-a-b}{n}\right] \times\left[\prod_{i=1}^{p}\binom{n_i}{x_i}\bigg/\binom{n}{x}\right] = \frac{b^{[n]}}{(a+b)^{[n]}}\frac{a^{[x]}}{(-b-n+1)^{[x]}}\prod_{i=1}^{p}\frac{(-n_i)^{[x_i]}}{x_i!}, \tag{11}$$

where notation and range are the same as (10). Types (10) and (11) are different; the ranges especially are different. When chance mechanisms generating A2 are extended to the multivariate case, some generate (10) and others (11), and this splitting does not look systematic [23].

There are a couple of types of distributions with infinite nonnegative range, but the typical one is an extension of B3:

$$\frac{B(a+c, b+c)}{B(a+b+c, c)}\frac{b^{[x]}}{(a+b+c)^{[x]}}\prod_{i=1}^{p}\frac{a_i^{[x_i]}}{x_i!} = \frac{\Gamma(a+c)\Gamma(b+c)\Gamma(b+x)}{\Gamma(b)\Gamma(c)\Gamma(a+b+c+x)}\prod_{i=1}^{p}\frac{\Gamma(a_i+x_i)}{x_i!}, \tag{12}$$

where all a_i's, b, and c are positive real

numbers, $a = \sum a_i$, x_i's are nonnegative integers, and $x = \sum x_i$.

Janardan and Patil [9] named (10) "multivariate negative hypergeometric," (11) "multivariate inverse hypergeometric," and (12) "multivariate negative inverse hypergeometric." They also named (8) and (10) "multivariate Pólya" and (11) and (12) "multivariate inverse Pólya."

OTHER GENERALIZATIONS

The function (3) suggests a wider class of univariate discrete distributions on $[0, n]$ or $[0, \infty)$ having PGF

$$\frac{{}_pF_q(a_1, \ldots, a_p; b_1, \ldots, b_q; \theta s)}{{}_pF_q(a_1, \ldots, a_p; b_1, \ldots, b_q; \theta)}, \quad (13)$$

which is a broad subclass of power series distribution*. The condition under which (13) defines a PGF was studied by Kemp [13] and Dacey [3] and the family thus defined is called the "Kemp family*." Dacey [3] lists PGFs having $p + q \leqslant 3$. This covers the binomial*, Poisson*, negative binomial*, hyper-Poisson [2], Katz* family [12], and others. Some limits of the Kemp family belong to the same family. Truncation, shift, compounding, and generalization of distributions of the family have PGFs expressed in terms of hypergeometric series.

Kemp and Kemp [14] and Tripathi and Gurland [26] studied distributions with PGF of the form ${}_pF_q(a_1, \ldots, a_p; b_1, \ldots, b_q; \theta(s-1))$.

References

[1] Camp, G. H. (1925). *Biometrika*, **17**, 61–67.

[2] Crow, E. L. and Bardwell, G. E. (1965). *Proc. Int. Symp. Discrete Distributions*, G. P. Patil, ed. Pergamon Press, Oxford, England, pp. 127–140.

[3] Dacey, M. F. (1972). *Sankhyā B*, **34**, 243–250.

[4] Davies, O. L. (1933). *Biometrika*, **25**, 295–322.

[5] Davies, O. L. (1934). *Biometrika*, **26**, 59–107.

[6] Eggenberger, F. and Pólya, G. (1923). *Zeit. Angew. Math. Mech.*, **3**, 279–289.

[7] Irwin, J. O. (1975). *J. R. Statist. Soc. A*, **138**, 18–31, 204–227, 374–384.

[8] Janardan, K. G. (1973). *Sankhyā A*, **35**, 465–478.

[9] Janardan, K. G. and Patil, G. P. (1972). *Sankhyā A*, **34**, 363–376.

[10] Johnson, N. L. and Kotz, S. (1969). *Distribution in Statistics: Discrete Distributions*. Wiley, New York.

[11] Johnson, N. L. and Kotz, S. (1978). *Urn Models and Their Application*. Wiley, New York.

[12] Katz, L. (1965). *Proc. Int. Symp. Discrete Distributions*, G. P. Patil, ed. Pergamon Press, Oxford, England, pp. 175–182.

[13] Kemp, A. W. (1968). *Sankhyā A*, **30**, 401–410.

[14] Kemp, A. W. and Kemp, C. D. (1974). *Commun. Statist.*, **3**, 1187–1196.

[15] Kemp, C. D. and Kemp, A. W. (1956). *J. R. Statist. Soc. B*, **18**, 202–211.

[16] Markov, A. A. (1917). *Izv. Imp. Akad. Nauk, Ser. VI*, **11**(3), 177–186 (in Russian). (*Selected Works*, Izd. AN SSSR, Moscow, 1951.)

[17] Patil, G. P. and Joshi, S. W. (1968). *A Dictionary and Bibliography of Discrete Distributions*. Oliver & Boyd, Edinburgh.

[18] Pearson, K. (1895). *Philos. Trans. R. Soc. Lond. A*, **186**, 343–414. (Early statistical papers, Cambridge University Press, Cambridge, 1948.)

[19] Pearson, K. (1924). *Biometrika*, **16**, 272–288.

[20] Sarkadi, K. (1957). *Publ. Math. Inst. Hung. Acad. Sci.*, **2**, 59–69.

[21] Shimizu, R. (1968). *Proc. Inst. Statist. Math.*, **16**, 147–165 (in Japanese).

[22] Sibuya, M. (1979). *Ann. Inst. Statist. Math.*, **31**, 373–390.

[23] Sibuya, M. and Shimizu, R. (1981). *Ann. Inst. Statist. Math.*, **33**, 177–190.

[24] Steyn, H. S. (1951). *Proc. Kon. Ned. Akad. Wet. A*, **54**, 23–30.

[25] Steyn, H. S. (1955). *Proc. Kon. Ned. Akad. Wet. A*, **58**, 588–595.

[26] Tripathi, R. C. and Gurland, J. (1979). *Commun. Statist. A*, **8**, 855–869.

(HYPERGEOMETRIC DISTRIBUTION
PEARSON SYSTEM OF CURVES
POLYA'S URN
URN MODELS)

M. SIBUYA

GENERALIZED INVERSES

The $n \times m$ matrix $\mathbf{G}$ is a *generalized inverse* of the $m \times n$ complex matrix $\mathbf{A}$ whenever

$$\mathbf{A} = \mathbf{AGA}. \quad (1)$$

A principal motivation for the definition (1) is that the consistent system of nonhomogeneous *simultaneous linear equations**

$$\mathbf{Ax} = \mathbf{b}$$

has as a solution $\mathbf{x} = \mathbf{Gb}$ if and only if $\mathbf{G}$ satisfies (1). The matrix $\mathbf{G}$ that satisfies (1) is unique if and only if $m = n$ and $\mathbf{A}$ is nonsingular. The nature of the nonuniqueness of generalized inverses $\mathbf{G}$ that satisfy (1) is perhaps best illustrated by considering a diagonal form of $\mathbf{A}$. Using the *singular value decomposition** there exist unitary matrices $\mathbf{U}$ $m \times m$ and $\mathbf{V}$ $n \times n$ such that

$$\mathbf{A} = \mathbf{U}\begin{pmatrix} \mathbf{D} & \mathbf{0} \\ \mathbf{0} & \mathbf{0} \end{pmatrix}\mathbf{V}^*,$$

where $\mathbf{D}$ is an $r \times r$ diagonal positive-definite matrix, with $r = \text{rank}(\mathbf{A})$. The superscript * on a matrix denotes conjugate transpose. Then every generalized inverse $\mathbf{G}$ satisfying (1) has the form

$$\mathbf{G} = \mathbf{V}\begin{pmatrix} \mathbf{D}^{-1} & \mathbf{X} \\ \mathbf{Y} & \mathbf{Z} \end{pmatrix}\mathbf{U}^*,$$

with $\mathbf{X}$, $\mathbf{Y}$, and $\mathbf{Z}$ arbitrary, and so the generalized inverses $\mathbf{G}$ form a translated linear space of dimension $mn - r^2$.

We will write $\mathbf{A}^-$ for a generalized inverse $\mathbf{G}$ satisfying (1). This notation is now widespread, but other notations, including $\mathbf{A}^c$, $\mathbf{A}^{(1)}$, and $\mathbf{A}^g$, are also current. The name "generalized inverse" for such a matrix $\mathbf{A}^-$ seems to have become very popular, but the terms "conditional inverse," "weak inverse," "*g*-inverse," "{1}-inverse," and "g_1-inverse" are also in use.

It seems that Fredholm [9] was the first author to consider the concept of generalized inverse with his 1903 paper on the "pseudo-inverse" of an integral operator. It was, however, not till much later that the generalized inverse of a *matrix* was first mentioned in print: Eliakim Hastings Moore defined a unique inverse or "general reciprocal" for every finite, square or rectangular, matrix in his book *General Analysis, Part I* [11], which was published posthumously in 1935 (the results were announced in 1920 in an abstract). Moore's work appears to have been largely unnoticed until a resurgence of interest in the early 1950s centered on the use of generalized inverses in *least-squares** problems, which were not considered by Moore. In 1955 Penrose [14], extending work published in 1951 by Bjerhammar [3], showed that Moore's generalized inverse is the unique matrix $\mathbf{G}$ that satisfies the four equations

$$\mathbf{A} = \mathbf{AGA}, \tag{1}$$

$$\mathbf{G} = \mathbf{GAG}, \tag{2}$$

$$(\mathbf{AG})^* = \mathbf{AG}, \tag{3}$$

$$(\mathbf{GA})^* = \mathbf{GA}. \tag{4}$$

This unique generalized inverse is now almost universally called the Moore–Penrose inverse, and denoted $\mathbf{A}^+$ or $\mathbf{A}^\dagger$ (almost equally by statisticians and by mathematicians, respectively). Unfortunately, the notation $\mathbf{A}^-$ is also in use for the Moore–Penrose inverse, but is much less popular than $\mathbf{A}^+$ or $\mathbf{A}^\dagger$. We will use $\mathbf{A}^+$.

Generalized inverses $\mathbf{G}$ satisfying (1) and some but not all of the other three conditions are also of interest. For *least-squares problems* the equation

$$\mathbf{A}^*\mathbf{AG} = \mathbf{A}^* \tag{5}$$

is necessary and sufficient for the vector $\hat{\mathbf{x}} = \mathbf{Gb}$ to satisfy the *normal equations** $\mathbf{A}^*\mathbf{A}\hat{\mathbf{x}} = \mathbf{A}^*\mathbf{b}$. It follows that a matrix $\mathbf{G}$ satisfies (5) if and only if it satisfies (1) and (3). Such a generalized inverse may be called a least-squares generalized inverse and denoted $\mathbf{A}_{\text{ls}}^-$; other notations include $\mathbf{A}^{(1,3)}$ and $\mathbf{A}^{g_{13}}$.

Another generalized inverse that has received considerable attention is the reflexive generalized inverse $\mathbf{G}$, which satisfies both (1) and (2). This is equivalent to $\mathbf{G}$ satisfying (1) and having the same rank as that of $\mathbf{A}$; a generalized inverse $\mathbf{G}$ satisfying only (1) has rank at least equal to the rank of $\mathbf{A}$. We denote a reflexive generalized inverse by $\mathbf{A}_{\text{r}}^-$; the notations $\mathbf{A}^{(1,2)}$ and $\mathbf{A}^{g_{12}}$ are also in use.

In terms of the expansion of a generalized inverse $\mathbf{G}$ following the singular value decomposition of $\mathbf{A}$ given above, we see that $\mathbf{G}$ is a least-squares generalized inverse if and only if $\mathbf{X} = \mathbf{0}$, that $\mathbf{G}$ is a reflexive generalized inverse if and only if $\mathbf{Z} = \mathbf{YDX}$, and that $\mathbf{G}$ is the Moore–Penrose inverse if and only if $\mathbf{X} = \mathbf{0}$, $\mathbf{Y} = \mathbf{0}$, and $\mathbf{Z} = \mathbf{0}$.

Several statistical subroutine packages compute generalized inverses (especially for regression* analysis); e.g., the SAS system [17, p. 287] uses the singular value decomposition to compute the Moore–Penrose inverse $\mathbf{X}^+$ of the design matrix $\mathbf{X}$ in the usual linear model $\mathscr{E}(\mathbf{y}) = \mathbf{X}\boldsymbol{\beta}$. An algorithm using *Householder transformations* is described by Seber [18, Sec. 11.5.3] to compute a reflexive least-squares generalized inverse of $\mathbf{X}$. *See also* STATISTICAL SOFTWARE.

In many applications we need only use a generalized inverse $\mathbf{G} = \mathbf{A}^-$ which satisfies (1). Noting that the square matrices $\mathbf{AA}^-$ and $\mathbf{A}^-\mathbf{A}$ are both idempotent* (although not necessarily Hermitian), it follows that

$$\mathscr{C}(\mathbf{A}) = \mathscr{N}(\mathbf{I} - \mathbf{AA}^-)$$

and

$$\mathscr{N}(\mathbf{A}) = \mathscr{C}(\mathbf{I} - \mathbf{A}^-\mathbf{A}),$$

where $\mathscr{C}$ denotes column space (range) and $\mathscr{N}$ denotes null space. Hence the system of nonhomogeneous simultaneous linear equations

$$\mathbf{Ax} = \mathbf{b} \text{ is consistent} \Leftrightarrow (\mathbf{I} - \mathbf{AA}^-)\mathbf{b} = \mathbf{0},$$

and then the general solution is given by

$$\mathbf{x} = \mathbf{A}^-\mathbf{b} + (\mathbf{I} - \mathbf{A}^-\mathbf{A})\mathbf{z},$$

where $\mathbf{z}$ is arbitrary. The normal equations*

$$\mathbf{A}^*\mathbf{A}\hat{\mathbf{x}} = \mathbf{A}^*\mathbf{b}$$

are always consistent, since

$$\left[\mathbf{I} - \mathbf{A}^*\mathbf{A}(\mathbf{A}^*\mathbf{A})^-\right]\mathbf{A}^* = \mathbf{0},$$

and hence

$$\hat{\mathbf{x}} = (\mathbf{A}^*\mathbf{A})^-\mathbf{A}^*\mathbf{b} + \left[\mathbf{I} - (\mathbf{A}^*\mathbf{A})^-\mathbf{A}^*\mathbf{A}\right]\mathbf{z}$$

is the general solution, where $\mathbf{z}$ is arbitrary. The matrix $(\mathbf{A}^*\mathbf{A})^-\mathbf{A}^*$ is a reflexive least-squares generalized inverse of $\mathbf{A}$ satisfying conditions (1), (2) and (3). The solution $\hat{\mathbf{x}}$ with shortest length $(\hat{\mathbf{x}}^*\hat{\mathbf{x}})^{1/2}$ is

$$\hat{\mathbf{x}}_0 = \mathbf{A}^+\mathbf{b},$$

where $\mathbf{A}^+$ is the Moore–Penrose inverse of $\mathbf{A}$.

The primary usefulness of generalized inverses in statistics is that many otherwise unduly complicated expressions can now be written in closed form. In the usual linear model $\mathscr{E}(\mathbf{y}) = \mathbf{X}\boldsymbol{\beta}$, for example, the vector $\mathbf{A}\boldsymbol{\beta}$ is *estimable* (possesses a linear unbiased estimator depending on $\mathbf{y}$) if and only if $\mathbf{AX}^-\mathbf{X} = \mathbf{A}$ (cf. ref. 18 [Sec. 3.8.2]). The least-squares predictor for $\mathbf{y}$ is $\mathbf{y} = \mathbf{XX}^+\mathbf{y}$; if $\mathbf{y}$ has covariance matrix proportional to $\mathbf{V}$ (possibly singular), then $\mathbf{X}(\mathbf{X}^*\mathbf{V}^+\mathbf{X})^+\mathbf{X}^*\mathbf{V}^+\mathbf{y}$ is the BLUE (best linear unbiased estimator) of $\mathbf{X}\boldsymbol{\beta}$ when the column space $\mathscr{C}(\mathbf{X})$ is contained in the column space $\mathscr{C}(\mathbf{V})$ (*see* GENERAL LINEAR MODEL). Other applications to statistics are provided by Rao and Mitra [16, Chaps. 10 and 11] to the distribution of quadratic forms* in normal variables, to variance components*, and to discriminant* functions in multivariate analysis*.

Another kind of generalized inverse has found considerable application in the study of *Markov chains* (cf. ref. 7). Let $\mathbf{A}$ be a square matrix and consider a reflexive generalized inverse $\mathbf{G}$ which also satisfies the condition

$$\mathbf{AG} = \mathbf{GA}.$$

Such a matrix $\mathbf{G}$ exists if and only if $\mathbf{A}$ is nonsingular or has index 1, i.e.,

$$\operatorname{rank}(\mathbf{A}^2) = \operatorname{rank}(\mathbf{A}),$$

and then $\mathbf{G}$ is unique and is called the group inverse of $\mathbf{A}$ and is often denoted $\mathbf{A}^{\#}$. The condition that $\mathbf{A}$ have index 1 is equivalent to the algebraic multiplicity of 0 as an eigenvalue of $\mathbf{A}$ being equal to the dimension of the null space of $\mathbf{A}$ (also called the geometric multiplicity of 0 as an eigenvalue of $\mathbf{A}$); the eigenvalue 0 is then called a regular eigenvalue of $\mathbf{A}$. If a finite *Markov chain* is governed by the row-stochastic matrix $\mathbf{P}$ of stationary transition probabilities, then the matrix $\mathbf{I} - \mathbf{P}$ has nullity 1 and $\mathbf{I} - \mathbf{P} + \mathbf{eu}^*$ is nonsingular for all vectors $\mathbf{u}$ satisfying $\mathbf{u}^*\mathbf{e} \neq 0$, where $\mathbf{e}$ is the column vector of 1's. Then $(\mathbf{I} - \mathbf{P} + \mathbf{eu}^*)^{-1}$ is a nonsingular generalized inverse of $\mathbf{I} - \mathbf{P}$, and the stationary distribution of the Markov chain $\boldsymbol{\ell}^* = \boldsymbol{\ell}^*\mathbf{P} = \mathbf{u}^*(\mathbf{I} - \mathbf{P} + \mathbf{eu}^*)^{-1}$. The group inverse $(\mathbf{I} - \mathbf{P})^{\#} = (\mathbf{I} - \mathbf{P} + \mathbf{e}\boldsymbol{\ell}^*)^{-1} - \mathbf{e}\boldsymbol{\ell}^*$, while the

Moore–Penrose inverse $(\mathbf{I}-\mathbf{P})^{+}=(\mathbf{I}-\mathbf{P}+\alpha\boldsymbol{\ell}\mathbf{e}^{*})^{-1}-\alpha\mathbf{e}\boldsymbol{\ell}^{*}$, where $\alpha=(n\boldsymbol{\ell}^{*}\mathbf{l})^{-1/2}$.

There are at least 11 books [1, 2, 4–8, 10, 12, 15, 16] on generalized inverses, including one in German [10]; almost all of these books include most of the definitions and properties given in this article. The most general and comprehensive book on generalized inverses must surely be by Ben-Israel and Greville [2], which also contains a wealth of material on general matrix theory, as well as an extensive bibliography; a much longer and extremely well annotated bibliography is by Nashed and Rall [13], who list 1775 references. This bibliography is published in the proceedings [12] of a conference on generalized inverses held in Madison, Wisconsin, in 1973; the proceedings [5] of an earlier conference on generalized inverses held in Lubbock, Texas, in 1968, have also been published. The more "statistical" of the books cited are by Albert [1], Bjerhammar [4], Pringle and Rayner [15], and Rao and Mitra [16]; these four books concentrate on the use of generalized inverses in regression analysis and linear models. The book by Campbell and Meyer [7], however, has quite a lot of material on Markov chains. Among the other books on generalized inverses is an introduction accessible to undergraduate mathematics majors [8] and a small monograph by Boullion and Odell [6], following up on the conference they organized in Lubbock in 1968.

References

"hb" indicates hard cover; "pb" soft cover.

[1] Albert, A. (1972). *Regression and the Moore–Penrose Inverse*. Academic Press, New York. (hb)

[2] Ben-Israel, A. and Greville, T. N. E. (1974). *Generalized Inverses: Theory and Applications*. Wiley, New York. (hb) (Reprinted with corrections 1980, Robert E. Krieger, Huntington, New York.)

[3] Bjerhammar, A. (1951). *Bull. Géod.*, **20**, 188–220.

[4] Bjerhammar, A. (1973). *Theory of Errors and Generalized Matrix Inverses*. Elsevier, Amsterdam. (hb)

[5] Boullion, T. L. and Odell, P. L., eds. (1968). *Proc. Symp. Theory Appl. Generalized Inverses Matrices*, Math. Ser. No. 4. Texas Tech Press, Lubbock, Tex. (pb)

[6] Boullion, T. L. and Odell, P. L. (1971). *Generalized Inverse Matrices*. Wiley, New York. (hb)

[7] Campbell, S. L. and Meyer, C. D., Jr. (1979). *Generalized Inverses of Linear Transformations*. Pitman, London. (hb) (Contains material on Markov chains.)

[8] Cline, R. E. (1979). *Elements of the Theory of Generalized Inverses for Matrices*. Modules and Monographs in Undergraduate Mathematics and Its Applications Project (UMAP), Educational Development Center, Newton, Mass. (pb)

[9] Fredholm, I. (1903). *Acta Math.*, **27**, 365–390. (First mention of generalized inverses.)

[10] Kuhnert, F. (1976). *Pseudoinverse Matrizen und die Methode der Regularisierung*. B. G. Teubner, Leipzig (in German). (pb)

[11] Moore, E. H. (1935). *General Analysis, Part I*. Mem. Amer. Philos. Soc., Philadelphia [see esp. pp. 147–209. Abstract: *Bull. Amer. Math. Soc.*, **26**, 394–395 (1920)].

[12] Nashed, M. Z., ed. (1976). *Generalized Inverses and Applications: Proceedings of an Advanced Seminar* (University of Wisconsin–Madison, Oct. 1973). Academic Press, New York. (hb)

[13] Nashed, M. Z. and Rall, L. B. (1976). In *Generalized Inverses and Applications: Proceedings of an Advanced Seminar* (University of Wisconsin–Madison, Oct. 1973), M. Z. Nashed, ed. Academic Press, New York, pp. 771–1054. (hb)

[14] Penrose, R. A. (1955). *Proc. Camb. Philos. Soc.*, **51**, 406–413. (Seminal paper on generalized inverses.)

[15] Pringle, R. M. and Rayner, A. A. (1971). *Generalized Inverse Matrices with Applications to Statistics*. Charles Griffin, London. (pb)

[16] Rao, C. R. and Mitra, S. K. (1971). *Generalized Inverse of Matrices and Its Applications*. Wiley, New York. (hb)

[17] *SAS User's Guide, 1979 Edition*. SAS Institute, Raleigh, N.C. (Statistical subroutine package manual.)

[18] Seber, G. A. F. (1977). *Linear Regression Analysis*. Wiley, New York.

Bibliography

Campbell, S. L., ed. (1982). *Recent Applications of Generalized Inverses*. Pitman, London. (pb)

(GENERAL LINEAR MODEL
LEAST SQUARES
MARKOV PROCESSES
NORMAL EQUATIONS
SIMULTANEOUS LINEAR EQUATIONS
SINGULAR VALUE DECOMPOSITION)

GEORGE P. H. STYAN

GENERALIZED LIKELIHOOD RATIO TESTS

Suppose that **X** is a vector of random variables to be observed, and the joint distribution of the elements of **X** depends on m unknown parameters $\theta_1, \ldots, \theta_m$. Let $f_{\mathbf{X}}(x; \theta_1, \ldots, \theta_m)$ denote the joint probability density function of the elements of **X** if **X** is continuous, and the probability that $\mathbf{X} = \mathbf{x}$ when **X** is discrete; Ω denotes the subset of m-dimensional space consisting of all possible parameter vectors $(\theta_1, \ldots, \theta_m)$. The problem is to test the null hypothesis H_0 that $(\theta_1, \ldots, \theta_m)$ is in a specified subset of Ω against the alternative hypothesis H_1 that $(\theta_1, \ldots, \theta_m)$ is in another specified subset of Ω. For convenience, the symbol H_0 will denote both the subset of Ω specified by the null hypothesis and the statement that $(\theta_1, \ldots, \theta_m)$ lies in that subset, with a similar dual use of the symbol H_1; no confusion will result from this dual use of symbols. In all cases to be considered, $H_0 \cap H_1$ will be empty.

If both H_0 and H_1 are simple hypotheses, i.e., if H_0 consists of a single point $(\theta_1(0), \ldots, \theta_m(0))$ and H_1 of a single point $(\theta_1(1), \ldots, \theta_m(1))$, the familiar Neyman–Pearson lemma* states that the most powerful test of level of significance α is based on the "likelihood ratio"

$$\frac{f_X(X; \theta_1(0), \ldots, \theta_m(0))}{f_X(X; \theta_1(1), \ldots, \theta_m(1))};$$

H_0 is accepted if the likelihood ratio is above a preassigned critical value $c(\alpha)$, and H_1 is accepted if the likelihood ratio is below $c(\alpha)$, where $c(\alpha)$ is chosen to give the desired level of significance α. Unless the distribution of the likelihood ratio is continuous, randomization* may have to be used when the likelihood ratio is equal to $c(\alpha)$, in order to make the level of significance exactly equal to α. The test just described is the *likelihood ratio test**.

A *generalized likelihood ratio test* (GLRT) is a generalization to the problem where at least one of H_0 or H_1 is a composite hypothesis (i.e., contains more than one point of Ω).

For any subset S of Ω, let $\text{lub}_S f(X)$ denote the least upper bound of $f_X(X; \theta_1, \ldots, \theta_m)$ as $(\theta_1, \ldots, \theta_m)$ varies over S. Define $\lambda(1)$ as $\text{lub}_{H_0} f(X)/\text{lub}_{\Omega} f(X)$, $\lambda(2)$ as $\text{lub}_{H_0} f(X)/\text{lub}_{H_1} f(X)$, and $\lambda(3)$ as $\text{lub}_{H_0} f(X)/\text{lub}_{H_0 \cup H_1} f(X)$. Each of these quantities has been considered as a generalization of the likelihood ratio. A GLRT based on $\lambda(i)$ is to accept H_0 if $\lambda(i) > c(\alpha, i)$ and to accept H_1 if $\lambda(i) < c(\alpha, i)$, where $c(\alpha, i)$ is chosen to give the desired level of significance α. Many authors call such a test a "likelihood ratio test" rather than a "generalized likelihood ratio test."

In general, $\lambda(1)$, $\lambda(2)$, and $\lambda(3)$ will have different values. One important case where they all have the same value is when $H_0 \cup H_1 = \Omega$, and $P[(\hat{\theta}_1, \ldots, \hat{\theta}_m) \text{ in } H_0] = 0$, where $\hat{\theta}_i$ denotes the maximum likelihood* estimator of θ_i. Also, it is easily seen that $\lambda(3) = \min(\lambda(2), 1)$, so that $\lambda(2)$ is more informative than $\lambda(3)$, in the sense that given $\lambda(2)$ we know the value of $\lambda(3)$, but not vice versa. Thus the GLRT based on $\lambda(3)$ can always be stated in terms of $\lambda(2)$, so there is no point in further discussion of $\lambda(3)$. An example illustrating the drawback of $\lambda(3)$ is given by Solomon [9].

The GLRT based on $\lambda(1)$ is intuitively appealing, because $\hat{\theta}_i$ usually has a high probability of being close to θ_i. Then if H_0 is true, $(\hat{\theta}_1, \ldots, \hat{\theta}_m)$ will probably be close to a point in H_0, making $\lambda(1)$ relatively large, so that H_0 will be accepted.

One case occurs so often that it deserves a special name. Suppose that $H_0 \cup H_1 = \Omega$ and that either H_0 consists of all vectors in Ω with $(m - s)$ specified coordinates having given values, or H_0 can be brought into this form by a one-to-one transformation to a new set of parameters (a process called *reparametrization*). We will call this case the *hyperplane case*, because H_0 consists of a hyperplane (a point, line, or plane are special cases of hyperplanes). There is no loss of generality in assuming that H_0 specifies values of the last $(m - s)$ coordinates of $(\theta_1, \ldots, \theta_m)$, so that H_0 consists of all vectors in Ω of the form $(\theta_1, \ldots, \theta_s, \theta_{s+1}(0), \ldots, \theta_m(0))$, where $\theta_i(0)$ is a specified value and $0 \leqslant s < m$. The parameters

$\theta_1, \ldots, \theta_s$, which are unspecified by H_0, are *nuisance parameters**. If $s = 0$, there are no nuisance parameters, and H_0 is simple. In many hyperplane cases, $P[(\hat{\theta}_1, \ldots, \hat{\theta}_m)$ in $H_0] = 0$, so that $\lambda(1) = \lambda(2)$ with probability 1.

Note that, since reparametrization is one to one, the values of $\lambda(1)$ and $\lambda(2)$ are unaffected by it. In practice, $\lambda(1)$ and $\lambda(2)$ are usually computed using the original parameters. However, it is easier to describe the properties of the GLRT in terms of the new parameters, and this will be done below.

A large number of commonly used tests are either equivalent to a GLRT or asymptotically equivalent to a GLRT. (Here "asymptotic" means in the limit as it becomes possible to estimate the parameters more precisely.) We give two examples of exact equivalence.

Example 1. $\mathbf{X} = (X_1, \ldots, X_n)$, where $X_1, \ldots, X_n$ are independent and identically distributed, each with a normal distribution with unknown standard deviation θ_1 and unknown mean θ_2. Ω consists of all vectors (θ_1, θ_2) with $\theta_1 > 0$. H_0 consists of all vectors $(\theta_1, \theta_2(0))$ with $\theta_2(0)$ a given value and $\theta_1 > 0$. H_1 consists of all vectors (θ_1, θ_2) with $\theta_1 > 0$ and $\theta_2 > \theta_2(0)$. This is not a hyperplane case, because $H_0 \cup H_1$ is not Ω. Define $\bar{X}$ as $(1/n)(X_1 + \cdots + X_n)$ and T as

$$\frac{\sqrt{n(n-1)}\left(\bar{X} - \theta_2(0)\right)}{\sqrt{\sum_{i=1}^{n}(X_i - \bar{X})^2}}.$$

The usual test of level of significance α accepts H_0 if T is less than the $(1-\alpha)$th quantile* of Student's t-distribution* with $(n-1)$ degrees of freedom. A simple calculation shows that $\lambda(2) = 1$ if $T \leqslant 0$, $\lambda(2) = [1 + T^2/(n-1)]^{-n/2}$ if $T \geqslant 0$. Thus $\lambda(2)$ is large when T is small, so the GLRT based on $\lambda(2)$ is exactly equivalent to the usual test. On the other hand, $\lambda(1) = \{1 + [T^2/(n-1)]\}^{-n/2}$ for all T, so $\lambda(1)$ is large when $|T|$ is small. This illustrates the fact that $\lambda(2)$ is more suitable for one-sided tests than $\lambda(1)$, which is obvious from the definitions of $\lambda(1)$ and $\lambda(2)$.

Example 2. $\mathbf{X} = (X_1, \ldots, X_n)$, where $X_1, \ldots, X_n$ are independent, each with a normal distribution with common unknown standard deviation σ, and $E(X_i) = \beta_1 z_{1i} + \beta_2 z_{2i} + \cdots + \beta_r z_{ri}$ for $i = 1, \ldots, n$, where (z_{ij}) are known and nonrandom, $r < n$, the matrix $\{z_{ij}\}$ has rank r, and $\beta_1, \ldots, \beta_r$ are unknown parameters. Then Ω consists of all vectors $(\sigma, \beta_1, \ldots, \beta_r)$ with $\sigma > 0$, H_0 consists of all such vectors with the parameters $\beta_1, \ldots, \beta_r$ satisfying p given independent linear equations, and $H_0 \cup H_1 = \Omega$. This is a hyperplane case, the familiar problem of testing a linear hypothesis in the normal linear regression model (*see* GENERAL LINEAR MODEL). Let Z denote the usual analysis-of-variance* F-ratio used to test this hypothesis. Then $\lambda(1) = \lambda(2) = [1 + pZ/(n-r)]^{-n/2}$; (see Graybill [4]). Thus $\lambda(i)$ is large when Z is small, so the GLRT is exactly equivalent to the usual test.

Before discussing an example of the asymptotic equivalence of a GLRT and a commonly used test, we introduce the additional notation n for an index with positive integral values such that the larger the value of n, the more information we have about the unknown parameters. Then the vector of observations and its joint distribution both depend on n: we write them as $\mathbf{X}(n)$, $f_{n,\mathbf{X}(n)}(\mathbf{x}(n); \theta_1, \ldots, \theta_m)$, respectively. We also replace $\lambda(i)$ by the symbol $\lambda_n(i)$ to emphasize the dependence on n, and $\Lambda_n(i)$ denotes $-2\log_e \lambda_n(i)$. Thus $\Lambda_n(i)$ is small if $\lambda_n(i)$ is large.

Example 3. $\mathbf{X}(n) = (X_1(n), \ldots, X_{m+1}(n))$, where the joint distribution of $X_1(n), \ldots, X_{m+1}(n)$ is multinomial, with n trials and unknown probabilities $\theta_1, \ldots, \theta_{m+1}$. That is,

$$f_{n,\mathbf{X}(n)}(\mathbf{x}(n); \theta_1, \ldots, \theta_{m+1}) = n! \prod_{i=1}^{m+1} \left\{ \frac{\theta_i^{x_i(n)}}{(x_i(n))!} \right\}$$

if $\{x_i(n)\}$ are nonnegative integers with $\sum_{i=1}^{m+1} x_i(n) = n$. Since $\sum_{i=1}^{m+1} \theta_i = 1$, there are only m "genuine" parameters and one "artificial" parameter. We assign the role of artificial parameter to θ_{m+1}; thus Ω consists of all vectors $(\theta_1, \ldots, \theta_m)$ with nonnegative elements such that $\sum_{i=1}^{m}\theta_i \leqslant 1$. H_0 is the simple hypothesis that $\theta_i = \theta_i(0)$ for $i = 1, \ldots, m$, where $\{\theta_i(0)\}$ are given positive values with $\sum_{i=1}^{m}\theta_i(0) < 1$, $H_0 \cup H_1 = \Omega$. Thus this is a hyperplane case with no nuisance parameters. Suppose that for index n, θ_i is actually equal to $\theta_i(0) + c_i(n)/\sqrt{n}$ for $i = 1, \ldots, m$. Then if $|c_i(n)| < B < \infty$ for all n and for $i = 1, \ldots, m$, a straightforward expansion shows that we can write $\Lambda_n(1)$ as $S_n + \Delta_n$, where

$$S_n = \sum_{i=1}^{m+1} \frac{\left[X_i(n) - n\theta_i(0)\right]^2}{n\theta_i(0)},$$

and Δ_n converges stochastically to zero as n increases. The commonly used test for this problem accepts H_0 if S_n is less than the $(1-\alpha)$th quantile* of the central chi-square distribution* with m degrees of freedom, where α is the desired level of significance; S_n is called a "chi-square statistic" because as n increases the distribution of S_n approaches a chi-square distribution*. Thus the desired level of significance α is achieved only approximately for large n. The fact that $\Lambda_n(1) = S_n + \Delta_n$ shows that the GLRT based on $\Lambda_n(1)$ is asymptotically equivalent to the test based on S_n, the latter test being commonly known as the "chi-square test*."

The chi-square statistic of Example 3 is just one of a large class of such statistics. Cochran [2] gives a good survey of this class. The term "chi-square statistic" applied to S_n is somewhat misleading, because it seems to imply that there is something special about Example 3 that leads to an asymptotic chi-square distribution for $\Lambda_n(1)$. As we will see, $\Lambda_n(1)$ has an asymptotic chi-square distribution in a wide variety of cases.

In Example 3, S_n and $\Lambda_n(1)$ are asymptotically equivalent, so for large n they give approximately the same results. But is one of these statistics better than the other for small n? The answer to this question is not known. Cochran mentions some comparisons that have been made, but for small n the computations are too complicated to allow definitive results. A different sort of comparison has been made by Hoeffding [5]. The asymptotic equivalence of the tests based on S_n and $\Lambda_n(1)$ holds for a fixed level of significance α. Hoeffding compares these tests when the level of significance α_n depends on the index n, and α_n approaches zero as n increases. Then the test based on $\Lambda_n(1)$ is asymptotically more powerful than the test based on S_n.

Returning to the general discussion of the GLRT, how good is such a test? For fixed n, examples exist where the GLRT is a poor test (see Lehmann [7]). But in some special cases, the GLRT has been shown to have desirable properties for fixed n. Thus the GLRT of Example 2 has been shown by Wolfowitz [14] to have a certain optimal property, described below. As n increases, there is a fairly general answer: In a wide variety of cases, the GLRT is asymptotically optimal in a certain natural sense. Details are given in the following section.

ASYMPTOTIC THEORY FOR THE ERGODIC HYPERPLANE CASE

Throughout this section we deal with a hyperplane case, reparametrized if necessary, so that H_0 consists of all vectors in Ω of the form $(\theta_1, \ldots, \theta_s, \theta_{s+1}(0), \ldots, \theta_m(0))$, where $\theta_i(0)$ is specified and $0 \leqslant s < m$. We also assume that for any vector $\bar{\boldsymbol{\theta}} \equiv (\bar{\theta}_1, \ldots, \bar{\theta}_m)$ in the interior of Ω,

$$\frac{\partial^2}{\partial\theta_i\,\partial\theta_j}\log_e f_{n,X(n)}(X(n); \theta_1, \ldots, \theta_m)\big]_{\bar{\boldsymbol{\theta}}} \equiv D_n(i, j, \bar{\boldsymbol{\theta}}),$$

say, exists for all i, j, and that there are $2m$ sequences of nonrandom positive quantities $\{K_1(n), \ldots, K_m(n)\}$, $\{M_1(n), \ldots, M_m(n)\}$, with $\lim_{n\to\infty} K_i(n) = \infty$, $\lim_{n\to\infty} M_i(n) = \infty$, $\lim_{n\to\infty}[M_i(n)/K_i(n)] = 0$ for $i = 1, \ldots, m$, such that:

1. $-[1/\{K_i(n)K_j(n)\}]D_n(i, j, \bar{\theta})$ converges stochastically, as n increases, to a nonrandom quantity $B_{ij}(\bar{\boldsymbol{\theta}})$, assuming that the parameters are actually $\bar{\theta}_1, \ldots, \bar{\theta}_m$. $B_{ij}(\bar{\boldsymbol{\theta}})$ is a continuous function of $\bar{\boldsymbol{\theta}}$, and

the $m \times m$ matrix $\mathbf{B}(\bar{\boldsymbol{\theta}})$ with $B_{ij}(\bar{\boldsymbol{\theta}})$ in row i and column j is positive definite.

2. $M_i(n)M_j(n)| - [1/\{K_i(n)K_j(n)\}] D_n(i,j,\bar{\boldsymbol{\theta}}) - B_{ij}(\bar{\boldsymbol{\theta}})|$ converges stochastically to zero as n increases.
3. Roughly speaking, the convergence in **2** is uniform for all vectors $\bar{\boldsymbol{\theta}}$ whose ith coordinates are no farther apart than $M_i(n)/K_i(n)$ for $i = 1, \ldots, m$. ($N_n(\bar{\boldsymbol{\theta}})$ denotes the set of vectors with ith coordinate within $M_i(n)/K_i(n)$ of the ith coordinate of $\bar{\theta}$, for all i.)

If all the assumptions above are satisfied, we say we are dealing with an *ergodic hyperplane case*. Many commonly encountered problems are of this type. For example, if $\mathbf{X}(n)$ consists of n independent and identically distributed components $X_1, \ldots, X_n$, so that $f_{n,\mathbf{X}(n)}(\mathbf{X}(n); \theta_1, \ldots, \theta_m) = \prod_{i=1}^n f_{\mathbf{X}}(X_i; \theta_1, \ldots, \theta_m)$, and if for all i and j,

$$\frac{\partial^2}{\partial\theta_i\,\partial\theta_j}\log_e f_{\mathbf{X}}(X_1;\theta_1,\ldots,\theta_m)$$

exists, has an expected value which is a continuous function of $(\theta_1, \ldots, \theta_m)$ and a finite variance, and if there are no artificial parameters, then the assumptions above are all satisfied with $K_i(n) = \sqrt{n}$ and with $M_i(n) = n^{(1/6)-\delta}$, where $0 < \delta < \frac{1}{6}$. Thus the following is an ergodic hyperplane case.

Example 4. Let $\mathbf{X}(n) = (X_1, \ldots, X_n)$, where $X_1, \ldots, X_n$ are independent and identically distributed, with common marginal probability density function

$$\frac{1}{2}\left[\frac{1}{\theta_1\sqrt{2\pi}}\exp\left\{-\frac{(x-\theta_2)^2}{2\theta_1^2}\right\} + \frac{1}{\sqrt{2\pi}}\exp\left\{-\frac{(x-\theta_2)^2}{2}\right\}\right];$$

Ω consists of all vectors (θ_1, θ_2) with $\theta_1 > 0$, and H_0 of the single vector $(\theta_1(0), \theta_2(0))$, with $\theta_1(0), \theta_2(0)$ as specified values; $H_0 \cup H_1 = \Omega$. Kiefer and Wolfowitz [6] showed that in this case $\text{lub}_\Omega f(X)$ and $\text{lub}_{H_1} f(X)$ are both infinite, so that $\lambda_n(1)$ and $\lambda_n(2)$ are both equal to zero.

In order to handle the problem of Example 4 and similar problems, we must modify our definition of the generalized likelihood ratio slightly. Suppose that we have available estimators $\tilde{\theta}_1(n), \ldots, \tilde{\theta}_m(n)$ of $\theta_1, \ldots, \theta_m$ such that for any vector $\bar{\boldsymbol{\theta}} \equiv (\bar{\theta}_1, \ldots, \bar{\theta}_m)$ in the interior of Ω, and any sequence $\{\theta(n)\} \equiv (\theta_1(n), \ldots, \theta_m(n))$ with $\theta(n)$ in $N_n(\bar{\boldsymbol{\theta}}) \cap \Omega$ for all n,

$$\lim_{n\to\infty} P_{\theta(n)}\left[\bigcap_{i=1}^m \{K_i(n)|\tilde{\theta}_i(n) - \theta_i(n)| \leqslant \tfrac{1}{4}M_i(n)\}\right] = 1,$$

where $P_{\theta(n)}$ denotes probability computed assuming that $(\theta_1(n), \ldots, \theta_m(n))$ are the true parameters at index n. [In Example 4, if $\bar{X}(n)$, $V(n)$ denote the mean and variance, respectively, of the sample $X_1, \ldots, X_n$, we can use $\bar{X}(n)$ as $\tilde{\theta}_2(n)$ and $\sqrt{2V(n)-1}$ as $\tilde{\theta}_1(n)$.] Then we define $S_1(n)$ as the set of all vectors $(\theta_1, \ldots, \theta_s, \theta_{s+1}(0), \ldots, \theta_m(0))$ with $K_i(n)|\theta_i - \tilde{\theta}_i(n)| \leqslant M_i(n)/4$ for $i = 1, \ldots, s$, and we define $S_2(n)$ as the set of all vectors $(\theta_1, \ldots, \theta_m)$ with

$$K_i(n)|\theta_i - \tilde{\theta}_i(n)| \leqslant M_i(n)/4$$

for $i = 1, \ldots, m$. Now define the generalized likelihood ratio $\lambda_n(4)$ as

$$\text{lub}_{S_1(n)\cap\Omega} f(X)/\text{lub}_{S_2(n)\cap\Omega} f(X).$$

Define $\Lambda_n(4)$ as $-2\log_e\lambda_n(4)$. Suppose that for each index n the actual parameter vector is

$$\left(\bar{\theta}_1, \ldots, \bar{\theta}_s, \theta_{s+1}(0) + \frac{c_{s+1}}{K_{s+1}(n)}, \ldots, \theta_m(0) + \frac{c_m}{K_m(n)}\right),$$

where $(\bar{\theta}_1, \ldots, \bar{\theta}_s, \theta_{s+1}(0), \ldots, \theta_m(0)) \equiv \boldsymbol{\theta}^*$, say, is in the interior of Ω; then, as n increases, the distribution of $\Lambda_n(4)$ approaches a noncentral chi-square distribution* with $(m - s)$ degrees of freedom and noncentrality parameter given by the following computation. Partition the matrix $\mathbf{B}(\theta^*)$ as

$$\begin{pmatrix} \mathbf{B}(1,1;\theta^*) & \mathbf{B}(1,2;\theta^*) \\ \mathbf{B}(2,1;\theta^*) & \mathbf{B}(2,2;\theta^*) \end{pmatrix},$$

where $\mathbf{B}(1,1;\theta^*)$ is $s \times s$. Let $\mathbf{c}$ denote the $1 \times (m-s)$ vector $(c_{s+1}, \ldots, c_m)$. Then the noncentrality parameter is $[\mathbf{c}\{\mathbf{B}(2,2;\theta^*) - \mathbf{B}(2,1;\theta^*)\mathbf{B}^{-1}(1,1;\theta^*)\mathbf{B}(1,2;\theta^*)\}\mathbf{c}^T]^{1/2}$, where $\mathbf{c}^T$ denotes the transpose of $\mathbf{c}$.

It follows from these results that the test which rejects H_0 when $\Lambda_n(4)$ is greater than the $(1-\alpha)$th quantile of the central chi-square distribution with $(m-s)$ degrees of freedom has level of significance asymptotically equal to α. It can be shown that this GLRT based on $\Lambda_n(4)$ has the following optimal property. Suppose that for each index n we have some competing test T_n, and suppose that the asymptotic level of significance of T_n is no greater than α. Then for each given positive Δ, there is some vector $\mathbf{c}(\Delta) \equiv (c_{s+1}(\Delta), \ldots, c_m(\Delta))$ with corresponding noncentrality parameter equal to Δ such that the asymptotic power of T_n against the alternative

$$\left(\bar{\theta}_1, \ldots, \bar{\theta}_s, \theta_{s+1}(0) + \frac{c_{s+1}(\Delta)}{K_{s+1}(n)}, \ldots, \theta_m(0) + \frac{c_m(\Delta)}{K_m(n)}\right)$$

is no greater than the asymptotic power of the GLRT against this same alternative.

The optimal property of the GLRT of Example 2, mentioned above, is that the asymptotic property just described holds exactly for each fixed n.

OTHER HYPERPLANE CASES

A unified asymptotic theory does not yet exist for hyperplane cases not satisfying the conditions of the preceding section. We give two examples to illustrate some of the asymptotic distributions that can occur.

Example 5. Let $\mathbf{X}(n) = (X_1, \ldots, X_n)$, where $X_1, \ldots, X_n$ are independent and identically distributed, with common marginal probability density function $e^{-(x-\theta)}$ if $x \geqslant \theta$ and zero if $x < \theta$; Ω is the interval $(-\infty, \infty)$, H_0 consists of the single value $\theta(0)$, and $H_0 \cup H_1 = \Omega$. In this case, we cannot differentiate the density function with respect to θ, so $D_n(1,1,\bar{\theta})$ of the preceding section does not exist. If the actual value of θ is $\theta(0) + c/n$, where $c \geqslant 0$, the distribution of $\Lambda_n(1)$ is the distribution of the random variable $(Z + 2c)$, where Z has a central chi-square distribution with two degrees of freedom.

Example 6. Let $X(n)$ consist of one element, with probability density function $\sqrt{n}/[\pi + \pi n(x-\theta)^2]$; Ω consists of the interval $(-\infty, \infty)$. H_0 consists of the single value $\theta(0)$, and $H_1 \cup H_0 = \Omega$. In this case, $D_n(1,1,\bar{\theta})$ exists, but there is no sequence $\{K_1(n)\}$ of nonrandom quantities such that $-D_n(1,1,\bar{\theta})/K_1^2(n)$ converges stochastically to a positive constant as n increases; this would be required in order to apply the theory of the preceding section. If H_0 is true, $P[\Lambda_n(1) \leqslant c] = (2/\pi)\{\arctan(\sqrt{e^{c/2} - 1})\}$ for all $c \geqslant 0$.

Basawa and Prakasa Rao [1] give the asymptotic distribution of $\Lambda_n(1)$ for the problem of testing a hypothesis about the parameters of a nonergodic stochastic process.

NONHYPERPLANE CASES

For the case where $X(n)$ consists of n independent and identically distributed components, and certain regularity conditions are satisfied, Feder [3] gives some general results on the asymptotic distribution of $\Lambda_n(2)$. The asymptotic distribution is more complicated than the chi-square distribution. We illustrate with one example.

Example 7. Let $X(n)$ consist of n pairs $(Y_1, Z_1), (Y_2, Z_2), \ldots, (Y_n, Z_n)$, which are independent, with identical bivariate distributions. The joint distribution of (Y_i, Z_i) is normal, with $E(Y_i) = \theta_1$, $E(Z_i) = \theta_2$, variances of Y_i and Z_i both equal to 1, and zero covariance; Ω consists of all vectors (θ_1, θ_2), H_0 consists of all vectors (θ_1, θ_2) with both elements nonnegative and at least one element equal to zero, and $H_0 \cup H_1 = \Omega$. Define $\overline{Y}$ as $(1/n)(Y_1 + Y_2 + \cdots + Y_n)$, and $\overline{Z}$ as $(1/n)(Z_1 + \cdots + Z_n)$. Then $\Lambda_n(1) = \Lambda_n(2) = n(\overline{Y}^2 + \overline{Z}^2)$ if $\overline{Y} < 0$ and $\overline{Z} < 0$; $\Lambda_n(1) = \Lambda_n(2) = n[\min(\overline{Y}, \overline{Z})]^2$ if either $\overline{Y} \geqslant 0$ or $\overline{Z} \geqslant 0$. If the true parameter vector is $(0, 0)$, the asymptotic distribution of $\Lambda_n(1)$ is not a chi-square distribution.

No general results on asymptotic optimality for nonhyperplane cases seem to be available.

HISTORICAL NOTE

The first authors to propose a GLRT were Neyman and Pearson in ref. 8. (These authors called their test a "likelihood ratio test," not a "generalized likelihood ratio test.") For a class of hyperplane cases where $X(n)$ consists of n independent and identically distributed components, and where $\lambda_n(1)$, $\lambda_n(2)$, and $\lambda_n(3)$ are asymptotically equivalent, Wilks [13] derived the asymptotic distribution of $\Lambda_n(1)$ when the hypothesis is true, and Wald [10] derived the asymptotic distribution in general and proved the asymptotic optimality of the GLRT. The results for the more general ergodic hyperplane case are due to Weiss [11; 12, Chap. 7].

References

[1] Basawa, I. V. and Prakasa Rao, B. L. S. (1980). *Statistical Inference for Stochastic Processes*. Academic Press, New York.

[2] Cochran, W. G. (1952). *Ann. Math. Statist.*, **23**, 315–345.

[3] Feder, P. I. (1968). *Ann. Math. Statist.*, **39**, 2044–2055.

[4] Graybill, F. A. (1961). *An Introduction to Linear Models*, Vol. 1. McGraw-Hill, New York.

[5] Hoeffding, W. (1965). *Ann. Math. Statist.*, **36**, 369–401.

[6] Kiefer, J. and Wolfowitz, J. (1956). *Ann. Math. Statist.*, **27**, 887–906.

[7] Lehmann, E. L. (1950). *Ann. Math. Statist.*, **21**, 1–26.

[8] Neyman, J. and Pearson, E. S. (1928). *Biometrika*, **20A**, 175–240, 264–294.

[9] Solomon, D. L. (1975). *Amer. Statist.*, **29**, 101–102.

[10] Wald, A. (1943). *Trans. Amer. Math. Soc.*, **54**, 426–482.

[11] Weiss, L. (1975). *J. Amer. Statist. Ass.*, **70**, 204–208.

[12] Weiss, L. and Wolfowitz, J. (1974). *Maximum Probability Estimators and Related Topics*. Springer-Verlag, New York.

[13] Wilks, S. S. (1938). *Ann. Math. Statist.*, **9**, 60–62.

[14] Wolfowitz, J. (1949). *Ann. Math. Statist.*, **20**, 540–551.

(CHI-SQUARE TESTS
LIKELIHOOD RATIO TESTS)

LIONEL WEISS

GENERALIZED LINEAR MODELS

A statistical model is the specification of a probability distribution. For example, the model implicit in much of regression analysis is that the observations have a normal distribution*, the means being linearly related to the covariate values. Similarly, a log-linear model for counts can be thought of as specifying that the counts are Poisson* variables, whose means are multiplicatively related to the explanatory factors.

In these examples, and indeed quite generally for any univariate statistical model, the model consists of one random variable, a set of explanatory variables, and the specification of a probability distribution for the random variable in terms of the explanatory ones. What makes the examples above special, however, is that the specification of the distribution takes a particularly simple form; this simplifies and unifies the theoretical results that can be derived and the subsequent statistical analysis and computation.

The models having this special form are known as *generalized linear models* (GLMs); they include many of those mentioned above. They were first defined by Nelder and Wedderburn [2], who also pointed out the advantages of the resulting theoretical unification.

(A popular by-product of the 1972 paper has been the computer package GLIM*, designed for the statistical analysis of GLMs, and discussed in a separate article (*see* GLIM).)

Subsequent sections describe models and analyses (e.g., analysis of variance, regression analysis, log-linear models) described elsewhere in this encyclopedia. We aim to point out the underlying similarity of these models behind the superficial differences, and to explain how a single methodology for estimation*, goodness of fit* and prediction*, and a single algorithm may replace the many individual theories and methods.

DEFINITION OF A GENERALIZED LINEAR MODEL

A probability distribution for an observed random variable vector **y** given explanatory vectors $\mathbf{x}_1, \ldots, \mathbf{x}_p$ is a GLM if it satisfies the following three conditions:

(a) The distribution of each element y_i of **y** $(i = 1, \ldots, n)$, given $x_{i1}, \ldots, x_{ip}$, belongs to an exponential family*, in the sense that the probability density function (PDF) for each y_i has the form

$$\exp\left[(y_i\theta_i - a(\theta_i))/\phi + b(y_i, \phi)\right]$$

where θ is a function of $x_{i1}, \ldots, x_{ip}$ that involves unknown parameters; the y_i are jointly independent.

Note:

1. θ_i is the *natural parameter*; ϕ, which is constant for all i, is the *scale parameter*; it can be shown that

$$\mu_i = E(y_i) = a'(\theta_i),$$
$$\mathrm{var}(y_i) = \phi a''(\theta_i).$$

2. Different choices of $a(\cdot)$ and ϕ give rise to different PDFs. For example:

$$a(\theta) = \theta^2/2 \Rightarrow \text{Normal}(\theta, \phi),$$
$$a(\theta) = e^{\theta}, \phi \equiv 1 \Rightarrow \text{Poisson with mean } e^{\theta},$$
$$a(\theta) = -\sqrt{-2\theta} \Rightarrow \text{inverse Gaussian with mean } 1/\sqrt{-2\theta}, \text{ variance} = \phi\mu^3.$$

3. We can sometimes relax the assumption of joint independence. The conditional distribution of otherwise independent observations on some marginal total will contain nonzero covariances, but the general algorithm can often be extended to allow for this (see the section "Multinomial Models").

4. If we postulate only the mean and variance [i.e., the first two derivatives of $a(\cdot)$] without asking to which PDF it corresponds, we effectively use the method of quasi-likelihood* [5].

(b) The explanatory variables enter only as a linear sum of their effects, the *linear predictor* $\boldsymbol{\eta}$; hence for each i,

$$\eta_i = \sum_{j=1}^{p} x_{ij}\beta_j$$

or $\boldsymbol{\eta} = \mathbf{X}\boldsymbol{\beta}$, where $\mathbf{X} = [\mathbf{x}_1 \ldots \mathbf{x}_p]$; the β_j effects are the *linear parameters*.

Note:

1. There is a variety of terminology in use. A useful distinction is between a *factor* and a *variate*, the former taking values in a finite set (such as the positive integers less than or equal to some m—its *number of levels*) and used to classify the observed values into disjoint subsets, the latter representing covariate measurements and taking values on the real line. A factor with m levels will appear in **X** (often termed the design matrix, but better called the *linear structure matrix*) as a set of m columns of 1's or zeros; a variate appears as a column vector of its values.

2. Although these concepts were developed for use in analysis of variance* (AOV) and regression*, they apply quite generally throughout the class of GLMs. Similarly, such extensions are interactions* and polynomials in factors and variates have obvious applications in the wider class.

3. Wilkinson and Rogers [6] developed a most useful (*linear*) *structure formula* no-

tation that can describe, for all commonly used designs, not only the makeup of the linear structure matrix but also the structure of the design and the set of covariates that the matrix represents. Extensions to complex polynomials and interactions are straightforward.

(c) The final condition for a GLM is that the expected value of each observation can be expressed as some known function of its linear predictor; i.e.,

$$E(y_i) = \mu_i(= a'(\theta_i)) = g(\eta_i);$$

$g(\cdot)$ is known as the *link function*, although sometimes it is g^{-1} that is so called.

Note:

1. A simple extension would allow the function g to vary with i.
2. A more general extension allows an expected value to be a function of more than one linear predictor, i.e.,

$$\mu_i = h_i(\boldsymbol{\eta}).$$

The most useful form of this occurs when $h(\cdot)$ is the composition of an ordinary link function and a linear transformation

$$\boldsymbol{\mu} = \mathbf{A}g(\boldsymbol{\eta}),$$

where $\mathbf{A}$ is a (not necessarily square) matrix. Such a form arises when considering grouped data* from an arbitrary distribution, $g(\cdot)$ being its cumulative distribution function (CDF) and $\mathbf{A}$ representing the differencing of successive values (see ref. 4).

3. Sometimes a model will fail to be a GLM because the link function contains one or more unknown parameters, (e.g., probit analysis* with unknown natural mortality). The methods given below for the standard GLM can sometimes be extended to this (see ref. 3).
4. The case where $g(\cdot) \equiv a'(\cdot)$ is of particular interest for each distribution, g then being known as the *canonical link* for the distribution. Only in this case do sufficient statistics* exist for the linear parameters $\boldsymbol{\beta}$.

EXAMPLES

ONE-WAY ANOVA*. This is one of the simplest GLMs, yet one that illustrates many important features of the class. The usual specification would be that $\mathbf{y}$ is distributed normally with mean $\boldsymbol{\mu}$ and (co)variance $\mathbf{I}\sigma^2$. We can rephrase this so that it is obvious that the model is a GLM. The probability distribution of the observations is normal and they are independent, so that condition (a) is satisfied. The only explanatory variable in the model is a factor A (say) and its contribution can be written as the linear sum $\boldsymbol{\eta} = \mathbf{X}\boldsymbol{\beta}$, so condition (b) is satisfied. Finally, if we set the identity function as the link function, so that $\mu_i = \eta_i$, we see that condition (c) is also fulfilled.

We call such a GLM, with normal probability and identity link, a *classical* GLM.

MULTIPLE REGRESSION*. If for the classical GLM we allow only variates in the linear predictor, we obtain the usual multiple regression model. Since such models in general are unbalanced, in the sense that the sum of squares for a variate in the linear predictor is not independent of which other variates are in the linear predictor, most of the features present in more complex GLMs are exhibited here.

LOG-LINEAR MODELS. The log-linear models usually proposed for contingency-table* data with random margins are GLMs with the following specification:

1. The Poisson probability distribution* applies for the counts in the table.
2. The linear predictor, as in AOV, has the form $\boldsymbol{\eta} = \mathbf{X}\boldsymbol{\beta}$.
3. The (inverse of the) link function is the natural logarithm $\eta_i = \log \mu_i$.

MULTINOMIAL MODELS. Contingency tables with all margins random do not occur as often in practice as those in which one set of margins is fixed, with the others random. Under such a model the entries are distributed multinomially, conditional on the given fixed margin. However, it is well known that we shall arrive at the same conclusions if we treat the counts as Poisson, but add a term representing the fixed margins into the linear predictor of the log-linear model.

LOGIT* AND PROBIT* ANALYSIS. These models are GLMs with

1. A binomial distribution for the counts based on sample sizes n_i
2. As always, a linear predictor of the form $\boldsymbol{\eta} = \mathbf{X}\boldsymbol{\beta}$
3. Either a logit link given by

$$\eta_i = \log(\mu_i/(n_i - \mu_i))$$

or a probit link given by

$$\eta_i = \Phi^{-1}(\mu_i/n_i),$$

where $\Phi(\cdot)$ is the normal probability integral

Further examples may be found in ref. 1.

Terminology

The term *general linear model** should be distinguished from the term "generalized linear model," as illustrated above. Searle's article defines a general linear model for a random vector $\mathbf{y}$ by $\mathbf{y} = \mathbf{x}\boldsymbol{\beta} + \mathbf{e}$, where

$$E(\mathbf{e}) = \mathbf{0}, \qquad \text{var}(\mathbf{e}) = \mathbf{V},$$

for $\mathbf{V}$ nonnegative definite.

Such models can be subdivided into those with a single error stratum (one component of variation) and those (such as split plots*) with multiple error strata. The former are written as $E(\mathbf{y}) = \mathbf{x}\boldsymbol{\beta}$, $\text{var}(\mathbf{y}) = \mathbf{I}\sigma^2$, and if we add a normal distribution for $\mathbf{y}$ we obtain the classical linear model defined above, a special case of a generalized linear model. General linear models can also be classified as to whether they are "generally balanced" or not (*see* GENERAL BALANCE). Almost all common designs, including the so-called orthogonal designs, are generally balanced; this property allows a particularly simple form of analysis of variance, as is implemented in the ANOVA facility of Genstat

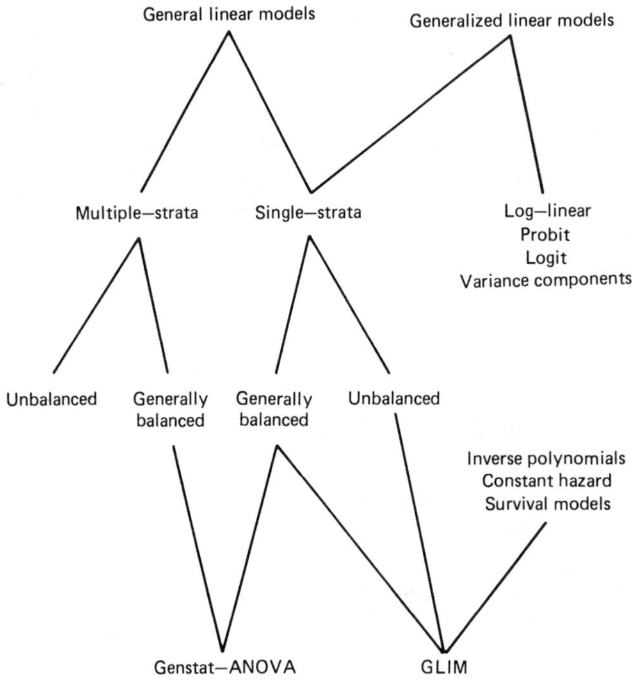

Figure 1

(*see* STATISTICAL SOFTWARE). This classification is illustrated in Fig. 1.

ESTIMATION

We deal with maximum likelihood estimation* under the standard GLM as defined in the section "Examples," pointing out extensions as necessary. The log-likelihood (or support, s) can be written as

$$s(\theta) = \sum_i [y_i\theta - a_i(\theta)]/\phi.$$

Differentiating with respect to $\boldsymbol{\beta}$, we get

$$\mathbf{c}(\boldsymbol{\beta}) = \mathbf{X}^t(\mathbf{HAV})^{-1}(\mathbf{y} - \boldsymbol{\mu}),$$

where $\mathbf{H} = \text{diag}\{\partial\boldsymbol{\eta}/\partial\boldsymbol{\mu}\}$, $\mathbf{A} = \text{diag}\{\phi\}$, and $\mathbf{V} = \text{diag}\{a''(\theta_i)\}$.

Differentiating again, and putting $\mathbf{t}(\boldsymbol{\beta})$ for the matrix of second derivatives, we find, after taking expectations and negating,

$$\mathbf{i}_e(\boldsymbol{\beta}) = -E[\mathbf{t}(\boldsymbol{\beta})] = \mathbf{X}^t(\mathbf{HVAH})^{-1}\mathbf{X}\phi^{-1}.$$

We solve the equation $\mathbf{c}(\boldsymbol{\beta}) = \mathbf{0}$ by the Newton–Raphson technique of repeatedly solving the approximation

$$\mathbf{0} = \mathbf{c}(\boldsymbol{\beta}_0) - \mathbf{t}(\boldsymbol{\beta}_0)(\boldsymbol{\beta} - \boldsymbol{\beta}_0),$$

where $\boldsymbol{\beta}_0$ is the value of $\boldsymbol{\beta}$ obtained at a previous cycle. Using Fisher's scoring technique we replace $\mathbf{t}(\cdot)$ by $-\mathbf{i}_e(\cdot)$ and on rearrangement obtain

$$\mathbf{i}_e(\boldsymbol{\beta}_0)\boldsymbol{\beta} = \mathbf{i}_e(\boldsymbol{\beta}_0)\boldsymbol{\beta}_0 - \mathbf{c}(\boldsymbol{\beta}_0),$$

which on substitution becomes

$$\mathbf{X}^t(\mathbf{H}_0\mathbf{V}_0\mathbf{H}_0)^{-1}\mathbf{X}\boldsymbol{\beta}$$
$$= \mathbf{X}^t(\mathbf{H}_0\mathbf{V}_0\mathbf{H}_0)^{-1}[\boldsymbol{\eta}_0 + \mathbf{H}_0(\mathbf{y} - \boldsymbol{\mu}_0)].$$

If we set $\mathbf{z}_0 = \boldsymbol{\eta}_0 + \mathbf{H}_0(\mathbf{y} - \boldsymbol{\mu}_0)$ and $\mathbf{W} = (\mathbf{H}_0\mathbf{V}_0\mathbf{H}_0)^{-1}$, the equation takes the form encountered in the least-squares* regression of $\mathbf{z}_0$ on $\mathbf{X}$ with weight $\mathbf{W}$, and immediately suggests a computational algorithm.

Note that $\mathbf{X}^t(\mathbf{H}_0\mathbf{V}_0\mathbf{H}_0)^{-1}\mathbf{X}$ will often be singular, in which case we take a g_2-inverse which implicitly supplies the necessary constraints on $\boldsymbol{\beta}$. Note also that the asymptotic (co-)variance matrix of such an estimate is given by $(\mathbf{X}^t(\mathbf{HVAH})^{-1}\mathbf{X})^{-}\phi^{-1}$.

GOODNESS-OF-FIT TESTS

Given a GLM with two possible linear predictors

$$\boldsymbol{\eta}_1 = \mathbf{X}\boldsymbol{\beta}_1 \quad \text{and} \quad \boldsymbol{\eta}_2 = \mathbf{X}\boldsymbol{\beta}_2 + \mathbf{Z}\boldsymbol{\alpha},$$

we can test the usefulness of the $\boldsymbol{\alpha}$ parameters through

$$S(1,2) = -2\log(l_1/l_2),$$

where $l_m(m = 1, 2)$ is the maximum value of the likelihood under the model with linear predictor η_m. S is termed the *scaled deviance*. It is well known that $S(1,2)$ is distributed asymptotically as χ_r^2, where r is the difference in ranks between $\mathbf{X}$ and $[\mathbf{X}\,\vdots\,\mathbf{Z}]$.

Note that S contains the scale parameter ϕ, and should ϕ be unknown, we can only calculate the *deviance* $D = \phi S$.

Given the sequence of models obtained from linear structure matrices $[\mathbf{X}_1]$, $[\mathbf{X}_1\,\vdots\,\mathbf{X}_2], \dots, [\mathbf{X}_1\,\vdots\,\dots\,\vdots\,\mathbf{X}_d]$, we can form the sequence of deviances

$$S(m-1, m) = -2\log(l_{m-1}/l_m),$$

where $l_m(m = 1, \dots, d)$ is the maximum value of the likelihood under the model with linear structure matrix $[X_1|\dots|X_d]$. Alternatively, we can note that $S(m-1, m) = S(m-1, 0) - S(m, 0)$, where l_0 is the likelihood value obtained by putting $\boldsymbol{\mu} = \mathbf{y}$, so that $S(m-1, m)$ can be formed as the difference of deviances each calculated on a single hypothesis. Any hypothesis expressible in terms of the sequence of models can then be tested by taking differences of such deviances; alternatively, the differences, together with their degrees of freedom and the names of the hypotheses, can be accumulated into an "analysis of deviance" table similar to that obtained in the analysis of variance.

In the classical balanced AOV model, using the main effect and interaction terms for X and Z above, we would obtain for D the sums of squares to be inserted into the AOV table. In the more general unbalanced case we can still construct an AOD (analysis of deviance) table, but the values obtained will depend on the order of fitting.

PREDICTION

When the fitting process is completed and a satisfactory model has been found, we are then interested in evaluating certain population values, which go under various names (predicted values, projections, population estimates, extrapolations, etc.), but they have in common that they are computed as functions of the estimated parameters in the final model.

The calculation and presentation of such values usually forms the last stage of the analysis of a body of data. In Release 4 of GLIM*, which has been developed to aid the analysis of GLMs, facilities exist to cope with all these stages.

References

[1] Baker, R. J. and Lane, P. W. (1978). *Compstat 1978, Proc. Comp. Statist.* Physica-Verlag-Wien, Leiden, pp. 31–36.

[2] Nelder, J. A. and Wedderburn, R. W. M. (1972). *J. R. Statist. Soc. A*, **135**, 370–384.

[3] Pregibon, D. (1980). *Appl. Statist.*, **29**, 15–24.

[4] Thompson, R. and Baker, R. J. (1981). *Appl. Statist.*, **30**, 125–131.

[5] Wedderburn, R. W. M. (1974). *Biometrika*, **64**, 439–447.

[6] Wilkinson, G. N. and Rogers, C. E. (1973). *Appl. Statist.*, **22**, 392–399.

(ANALYSIS OF COVARIANCE
ANALYSIS OF VARIANCE
CONTINGENCY TABLES
GENERAL BALANCE
GENERAL LINEAR MODEL
GLIM
LEAST SQUARES)

J. A. NELDER
R. J. BAKER

GENERALIZED MAXIMUM LIKELIHOOD ESTIMATION

Under certain regularity conditions, maximum likelihood* estimators are known to be asymptotically unbiased and asymptotically normally distributed, and to have the smallest possible asymptotic variance among a class of estimators satisfying certain assumptions (*see* ASYMPTOTIC NORMALITY). These results leave two important questions:

1. If the regularity conditions are not satisfied, can maximum likelihood estimators be shown to have desirable asymptotic properties?
2. Even if the regularity conditions are satisfied, does the fact that the maximum likelihood estimator has the smallest possible asymptotic variance necessarily imply that the maximum likelihood estimator (MLE) is better than competitors which may not be asymptotically normal, since a comparison of variances between distributions of different shape may not be meaningful?

Generalized MLEs were developed by Weiss and Wolfowitz [3] to help answer both of these questions. Generalized MLEs have optimal asymptotic properties even when their asymptotic distribution is not normal, and when compared to a very wide class of competing estimators. In many cases, generalized MLEs and MLEs are identical, but generalized maximum likelihood estimators exist and have optimal properties in certain cases where maximum likelihood estimators do not even exist.

THE CASE OF ONE UNKNOWN PARAMETER

For simplicity of exposition, we start with the case of a single unknown parameter. Since we are interested in asymptotic theory, we introduce an index n, with positive integral values, such that as n increases, we can estimate the unknown parameter more precisely. Sometimes n will represent a sample size, but this will not always be the case. We will be interested in what happens as n increases. Let $X(n)$ represent the vector of random variables to be observed when the index is n. There are not necessarily n components of $X(n)$, nor are the components necessarily independent or identically distributed. The joint distribution of the components of $X(n)$ depends on a single unknown parameter θ. Let $f_n(x(n);\theta)$ denote the joint probability density function of the components of $X(n)$ if $X(n)$ is continuous, and $P(X(n)=x(n)\mid\theta)$ if $X(n)$ is discrete, where $x(n)$ is a vector of the same dimension as $X(n)$, and $P(E\mid\theta)$ represents the probability of the event E when the parameter is equal to θ.

Suppose that for each n and for a fixed positive value r, there exists a function $\bar\theta(X(n),n,r)$ of $X(n)$, n, and r, and a positive nonrandom value $K(n)$ with $\lim_{n\to\infty}K(n)=\infty$ such that for all θ,

$$\lim_{n\to\infty} P\Big[K(n)\big(\bar\theta(X(n),n,r)-\theta\big)\leqslant y\mid\theta\Big] = L(y;\theta),$$

$$\lim_{n\to\infty} P\Big[K(n)\Big(\bar\theta(X(n),n,r)-\theta-\frac{r}{K(n)}\leqslant y\mid\theta+\frac{r}{K(n)}\Big] = L(y;\theta),$$

where $L(y;\theta)$ is a continuous cumulative distribution function, and such that

$$\frac{f_n(X(n);\theta+r/K(n))}{f_n(X(n);\theta)}\leqslant 1$$

implies that

$$\bar\theta(X(n),n,r) < \theta+\frac{r}{2K(n)}+\frac{A(X(n),n,r,\theta)}{K(n)},$$

$$\frac{f_n(X(n);\theta+r/K(n))}{f_n(X(n);\theta)}>1$$

implies that

$$\bar\theta(X(n),n,r) > \theta+\frac{r}{2K(n)}+\frac{A'(X(n),n,r,\theta)}{K(n)},$$

where $A(X(n),n,r,\theta)$ converges stochastically to zero as n increases when the true parameter is θ, and $A'(X(n),n,r,\theta)$ converges stochastically to zero as n increases when the true parameter is $\theta+r/K(n)$. [In all the above, we can ignore sets in $X(n)$-space whose probabilities approach zero.] We call $\bar\theta(X(n),n,r)$ a "generalized maximum likelihood estimator with respect to r." Then $\bar\theta(X(n),n,r)$ has the following asymptotic optimality property. Suppose that for each n, $T_n(X(n))$ is any estimator of θ satisfying the condition

$$\lim_{n\to\infty}\Big\{P\big[K(n)(T_n(X(n))-\theta)\leqslant -\tfrac{1}{2}r\mid\theta\big] - P\big[K(n)(T_n(X(n))-\theta-r/K(n)) \leqslant -\tfrac{1}{2}r\mid\theta+r/K(n)\big] = 0$$

for every θ. Then for every θ,

$$\lim_{n\to\infty} P\Big[-\frac{r}{2}<K(n)\big(\bar\theta(X(n),n,r)-\theta\big)<\frac{r}{2}\mid\theta\Big] \geqslant \lim_{n\to\infty}\sup P\Big[-\frac{r}{2}<K(n)(T_n(X(n))-\theta) \leqslant\frac{r}{2}\mid\theta\Big].$$

The proof is given in Weiss and Wolfowitz [3].

We now give some examples of generalized MLEs.

Example 1. $X(n)=(X_1,\dots,X_n)$, where $X_1,\dots,X_n$ are independent and identically distributed, each with probability density function equal to $e^{-(x-\theta)}$ if $x\geqslant\theta$, and equal to zero if $x<\theta$. Here we can take $K(n)$ equal to n. ($K(n)$ could also be taken as cn for a fixed positive c). In this case $\bar\theta(X(n),n,r)=\min(X_1,\dots,X_n)-r/2n$; $L(y;\theta)=0$ if $y\leqslant -r/2$, and $L(y;\theta)=1-e^{-(y+r/2)}$ if $y\geqslant -r/2$. Denoting the MLE by $\hat\theta_n$, we note that $\hat\theta_n=\min(X_1,\dots,X_n)$, which is equal to $\bar\theta(X(n),n,0)$ but not to $\bar\theta(X(n),n,r)$ for $r>0$.

Example 2. $X(n)=(X_1,\dots,X_n)$, where $X_1,\dots,X_n$ are independent and identically distributed, each with probability density function equal to $1/\theta$ if $0\leqslant x\leqslant\theta$, and equal to zero for other values of x. Here we can take $K(n)$ equal to n. In this case,

$$\bar\theta(X(n),n,r)=\max(X_1,\dots,X_n)+r/(2n),$$

$$L(y;\theta)=\exp\Big\{\frac{y-r/2}{\theta}\Big\}$$

if $y \leqslant r/2$; $L(y;\theta)=1$ if $y \geqslant r/2$. Once again, $\hat{\theta}_n = \bar{\theta}(n, X(n), 0)$.

Example 3. $X(n) = (X_1, \ldots, X_n)$, where $X_1, \ldots, X_n$ are independent and identically distributed, each with probability density function equal to 1 if $\theta \leqslant x \leqslant \theta + 1$, and equal to zero otherwise. Here we can take $K(n)$ equal to n. For any value h in the closed interval [0, 1], $h\{\min(X_1, \ldots, X_n) - r/(2n)\} + (1-h)\{\max(X_1, \ldots, X_n) - 1 + r/(2n)\}$ can be taken as $\bar{\theta}(X(n), n, r)$. $L(y; \theta)$ then depends on h and is given as follows. When h is in the open interval (0, 1), then

$$L(y;\theta) = (1-h)\exp\left\{\frac{y-(r/2)(1-2h)}{1-h}\right\} \quad \text{if } y \leqslant \tfrac{1}{2}r(1-2h);$$

$$L(y;\theta) = 1 - h\exp\left\{\frac{(r/2)(1-2h)-y}{h}\right\} \quad \text{if } y \geqslant \tfrac{1}{2}r(1-2h).$$

When $h=0$, then $L(y;\theta) = \exp\{y - r/2\}$ if $y \leqslant r/2$, $L(y;\theta)=1$ if $y \geqslant r/2$. When $h=1$, then $L(y;\theta)=0$ if $y \leqslant -r/2$, $L(y;\theta) = 1 - \exp\{-y - r/2\}$ if $y \geqslant -r/2$. We note that for any h in the closed interval [0, 1], $L(r/2;\theta) - L(-r/2;\theta) = 1 - e^{-r}$. Also, if $h = \frac{1}{2}$, $\bar{\theta}(X(n), n, r)$ does not depend on r, so that then $\frac{1}{2}\{\max(X_1, \ldots, X_n) + \min(X_1, \ldots, X_n) - 1\}$ is the generalized MLE with respect to positive r, which is a strong asymptotic optimality property. In Examples 1 and 2, no generalized MLE exists with respect to all positive r.

The three examples above are from Weiss and Wolfowitz [3, 4], where many other similar examples can be found.

Example 4. $X(n) = (X_1(n), X_2(n))$, and the joint probability density function is $(n/\pi)[1 + n(x_1-\theta)^2 + n(x_2-\theta)^2]^{-2}$. Here we can take $K(n) = \sqrt{n}$. Denote $\frac{1}{2}(X_1(n) + X_2(n))$ by $\bar{X}(n)$. It is easy to show that $\bar{X}(n)$ is a generalized MLE with respect to all positive r, and is also the MLE.

$$L(y;\theta) = \frac{1}{2}\left[1 + \frac{\sqrt{2}\,y}{\sqrt{1+2y^2}}\right].$$

This example is interesting, because $\bar{X}(n)$ is a consistent, unbiased estimator of θ with an infinite variance, and $X_1(n)$ is also a consistent, unbiased estimator of θ with an infinite variance. The standard technique of comparing consistent unbiased estimators by comparing their variances thus breaks down in this case. However,

$$P\left[-\tfrac{1}{2}r < \sqrt{n}\left(\bar{X}(n) - \theta\right) < \tfrac{1}{2}r \mid \theta\right] = \frac{r}{\sqrt{2}}\left(1 + \frac{r^2}{2}\right)^{-1/2},$$

while

$$P\left[-\tfrac{1}{2}r < \sqrt{n}\left(X_1(n) - \theta\right) < \tfrac{1}{2}r \mid \theta\right] = \frac{r}{2}\left(1 + \frac{r^2}{4}\right)^{-1/2} < \frac{r}{\sqrt{2}}\left(1 + \frac{r^2}{2}\right)^{-1/2}$$

for all $r > 0$, thus demonstrating the superiority of $\bar{X}(n)$ compared to $X_1(n)$ as an estimator of θ.

Now we discuss a class of what may be called "regular" cases. Suppose that the possible values of θ form an open interval Ω (perhaps infinite or semi-infinite). If θ^* is any value in the interior of Ω, we assume that $(\partial^2/\partial\theta^2)\log_e f_n(X(n);\theta)]_{\theta^*}$ exists, and denote this second derivative by $D_n(\theta^*)$. We denote $(1/K(n))(\partial/\partial\theta)\log_e f_n(X(n);\theta)]_{\theta^*}$ by $A_n(\theta^*)$. We assume that if θ^* is the true value of the parameter, then as n increases $-[1/K^2(n)]D_n(\theta^*)$ converges stochastically to a nonrandom positive value $B(\theta^*)$, which is a continuous function of θ^*. We also assume that roughly speaking, this stochastic convergence is uniform in θ^*: the exact condition is given in Weiss [2].

The assumptions in the preceding paragraph are satisfied in the familiar case where $X(n)$ consists of n independent and identically distributed components $X_1, \ldots, X_n$, each with probability density function $f_X(x;\theta)$, and $(\partial^2/\partial\theta^2)\log_e f_X(X_1;\theta)$ exists, has an expected value which is a continuous function of θ, and a finite variance; then the assumptions are satisfied with $K(n) = \sqrt{n}$. The assumptions are also satisfied in many other cases: see Weiss [1]. Whenever the assumptions are satisfied, we have the following results. Suppose that $\tilde{\theta}_n$ is any estimator of θ such that $\lim_{n\to\infty} P[K(n)|\tilde{\theta}_n - \theta| < L(n) | \theta] = 1$ for every sequence $\{L(n)\}$ of nonrandom positive quantities with $\lim_{n\to\infty} L(n) = \infty$. That is, $\tilde{\theta}_n$ can be a relatively crude estimator of θ. Define $\tilde{\theta}_n^*$ as $\tilde{\theta}_n + A_n(\tilde{\theta}_n)[K(n)B(\tilde{\theta}_n)]^{-1}$. Then $\tilde{\theta}_n^*$ is a generalized MLE with respect to all positive r. Also, $L(y;\theta)$ is the normal cumulative distribution function with mean zero and variance $1/B(\theta)$. These results follow from Weiss and Wolfowitz [3], supplemented by Weiss [1, 2].

THE CASE OF MORE THAN ONE UNKNOWN PARAMETER

For typographical simplicity, we consider two unknown parameters. It will be obvious how to extend the discussion to more than two unknown parameters. Then the distribution of the components of $X(n)$ is given by $f_n(x(n);\theta_1,\theta_2)$. Suppose that for each n and for fixed positive values r_1, r_2, there exist functions $\bar{\theta}_1(X(n),n,r_1,r_2)$ and $\bar{\theta}_2(X(n),n,r_1,r_2)$, which we abbreviate to $\bar{\theta}_1(n), \bar{\theta}_2(n)$, respectively, and positive nonrandom quantities $K_1(n)$, $K_2(n)$, with $\lim_{n\to\infty} K_1(n) = \infty$, $\lim_{n\to\infty} K_2(n) = \infty$, satisfying conditions given precisely in Weiss and Wolfowitz (1966), which will be stated somewhat imprecisely but more simply here. First, as n increases, $P[\bigcap_{i=1}^{2}\{K_i(n)(\bar{\theta}_i(n) - \theta_i) \leqslant y_i\} | \theta_1,\theta_2]$ approaches $L(y_1, y_2;\theta_1,\theta_2)$, where $L(y_1,y_2;\theta_1,\theta_2)$ is a continuous bivariate cumulative distribution function. Second, in the limit as n increases, which of the four quantities

$$f_n(X(n);\theta_1,\theta_2), \quad f_n\left(X(n);\theta_1 + \frac{r_1}{K_1(n)},\theta_2\right),$$

$$f_n\left(X(n);\theta_1,\theta_2 + \frac{r_2}{K_2(n)}\right),$$

$$f_n\left(X(n);\theta_1 + \frac{r_1}{K_1(n)},\theta_2 + \frac{r_2}{K_2(n)}\right)$$

is greatest is determined by $\bar{\theta}_1(n)$ and $\bar{\theta}_2(n)$: the first is greatest if $\bar{\theta}_1(n) < \theta_1 + \frac{1}{2}r_1/K_1(n)$ and $\bar{\theta}_2(n) < \theta_2 + \frac{1}{2}r_2/K_2(n)$; the second is greatest if $\bar{\theta}_1(n) > \theta_1 + \frac{1}{2}r_1/K_1(n)$ and $\bar{\theta}_2(n) < \theta_2 + \frac{1}{2}r_2/K_2(n)$; the third is greatest if $\bar{\theta}_1(n) < \theta_1 + \frac{1}{2}r_1/K_1(n)$ and $\bar{\theta}_2(n) > \theta_2 + \frac{1}{2}r_2/K_2(n)$; the fourth is greatest if $\bar{\theta}_1(n) > \theta_1 + \frac{1}{2}r_1/K_1(n)$ and $\bar{\theta}_2(n) > \theta_2 + \frac{1}{2}r_2/K_2(n)$. Then if $T_1(n)$, $T_2(n)$ are any estimators with $P[\bigcap_{i=1}^{2}\{-r_i/2 < K_i(n)(T_i(n) - \theta_i) < r_i/2\} | \theta_1,\theta_2]$ varying continuously with θ_1,θ_2, the maximum possible asymptotic value of this probability is given by $T_1(n) = \bar{\theta}_1(n)$ and $T_2(n) = \bar{\theta}_2(n)$. $\bar{\theta}_1(n)$, $\bar{\theta}_2(n)$ are called "generalized maximum likelihood estimators with respect to r_1, r_2."

Example 5. $X(n) = (X_0, X_1, \ldots, X_n)$, where X_i is the state at time i of a two-state stationary Markov chain with transition matrix

$$\begin{pmatrix} \theta_1 & 1-\theta_1 \\ \theta_2 & 1-\theta_2 \end{pmatrix};$$

$0 < \theta_1 < 1$, $0 < \theta_2 < 1$ (*see* MARKOV PROCESSES). Suppose that the two possible states are labeled 1, 2, and that $X_0 = 1$. Define N_{ij} as the number of transitions from state i to state j in the sequence $X_0, X_1, \ldots, X_n$, for $i = 1,2$ and $j = 1,2$. Here we can take $K_1(n) = K_2(n) = \sqrt{n}$. Let $\bar{\theta}_1(n) = N_{11}/(N_{11} + N_{12})$ and $\bar{\theta}_2(n) = N_{21}/(N_{21} + N_{22})$; $\bar{\theta}_1(n), \bar{\theta}_2(n)$ are generalized MLEs with respect to all positive r_1, r_2. The bivariate cumulative distribution function $L(y_1, y_2 | \theta_1,\theta_2)$ makes the two random variables independent and normal, the first having mean 0 and variance $\theta_1(1-\theta_1)(1-\theta_1+\theta_2)\theta_2^{-1}$, the second with mean 0 and variance $\theta_2(1-\theta_2)(1-\theta_1+\theta_2)(1-\theta_1)^{-1}$.

Example 6. $X(n) = (X_1, \ldots, X_n)$, where $X_1, \ldots, X_n$ are independent and identically distributed, each with probability density function equal to $\frac{1}{2}\exp\{-(x-\theta_1)\} + (1/(2\theta_2))\exp\{-(x-\theta_1)/\theta_2\}$ if $x \geqslant \theta_1$, and equal to zero if $x < \theta_1$. Here θ_1, θ_2 are unknown parameters, with $\theta_2 > 0$. In this example, the MLEs do not exist, since if we set the estimate of θ_1 equal to $\min(X_1, \ldots, X_n)$, and let the estimate of θ_2 approach zero, the likelihood approaches infinity. But in Weiss and Wolfowitz [3] generalized MLEs were constructed for this problem.

In both Examples 5 and 6, the estimators of the two parameters are asymptotically independent. That is what makes it possible to apply the method in these examples, since in each case we can maximize the asymptotic probability that $(\bar{\theta}_1(n), \bar{\theta}_2(n))$ falls in the appropriate rectangle by maximizing separate probabilities for $\bar{\theta}_1(n)$ and $\bar{\theta}_2(n)$. When the estimators are not asymptotically independent, this can no longer be done. Maximum probability estimators, described in Weiss and Wolfowitz [5], were developed to overcome this difficulty (*see* MAXIMUM PROBABILITY ESTIMATION).

HISTORICAL NOTE

There have been many attempts to explain why the maximum likelihood estimator works so well in so many cases. The approach described above originated in Wolfowitz [6], where the condition imposed on the estimator $T_n(X(n))$ was proposed as reasonable.

References

[1] Weiss, L. (1971). *J. Amer. Statist. Ass.*, **66**, 345–350.

[2] Weiss, L. (1973). *J. Amer. Statist. Ass.*, **68**, 428–430.

[3] Weiss, L. and Wolfowitz, J. (1966). *Teoriya Vyeroyatn.*, **11**, 68–93.

[4] Weiss, L. and Wolfowitz, J. (1968). *Teoriya Vyeroyatn.*, **13**, 657–662.

[5] Weiss, L. and Wolfowitz, J. (1974). *Maximum Probability Estimators and Related Topics*. Springer-Verlag, New York.

[6] Wolfowitz, J. (1965). *Teoriya Vyeroyatn.*, **10**, 267–281.

(MAXIMUM LIKELIHOOD ESTIMATION
MAXIMUM PROBABILITY ESTIMATION)

LIONEL WEISS

GENERALIZED MULTINOMIAL DISTRIBUTION *See* MULTINOMIAL DISTRIBUTIONS

GENERALIZED SEQUENTIAL PROBABILITY RATIO TESTS

TESTING ONE SIMPLE HYPOTHESIS AGAINST ANOTHER

Suppose that $X_1, X_2, \ldots$ are independent and identically distributed random variables, possibly multivariate. The distribution of X_i depends on an unknown parameter θ, which could be a vector, but we will assume it to be a scalar, with no real loss of generality. If X_i is discrete, let $f_X(x_i;\theta)$ denote $P(X_i = x_i; \theta)$, where $P(E;\theta)$ denotes the probability of the event E when the parameter is equal to θ. If X_i is continuous, let $f_X(x_i;\theta)$ denote the probability density function for X_i, when the parameter is equal to θ.

We are given two values, θ_0 and θ_1, and are asked to test the simple hypothesis H_0: $\theta = \theta_0$ against the simple alternative H_1: $\theta = \theta_1$. We are allowed to observe the X's sequentially; i.e., for each positive integer m, if we have observed $X_1, \ldots, X_m$, we may stop sampling and choose either H_0 or H_1, or we can decide to observe X_{m+1}. This makes the total number of X's we will observe a random variable, which we denote by N.

A sequential probability ratio test* (abbreviated SPRT) is defined as follows. Two values A, B are chosen, with $B \leqslant A$; X_1 is surely observed. For any positive integer m, if $X_1, \ldots, X_m$ have been observed, the SPRT acts as follows:

1. If $\prod_{i=1}^{m}\{f_X(X_i;\theta_1)/f_X(X_i;\theta_0)\} < B$, X_{m+1} is not observed and H_0 is accepted.
2. If $\prod_{i=1}^{m}\{f_X(X_i;\theta_1)/f_X(X_i;\theta_0)\} > A$, X_{m+1} is not observed and H_1 is accepted.
3. If $B < \prod_{i=1}^{m}\{f_X(X_i;\theta_1)/f_X(X_i;\theta_0)\} < A$, X_{m+1} is observed.
4. If $\prod_{i=1}^{m}\{f_X(X_i;\theta_1)/f_X(X_i;\theta_0)\} = B$, we can randomize in any way between observing X_{m+1} or stopping and choosing H_0.

5. If $\prod_{i=1}^{m}\{f_X(X_i;\theta_1)/f_X(X_i;\theta_0)\} = A$, we can randomize in any way between observing X_{m+1} or stopping and choosing H_1.

A *generalized sequential probability ratio test* (abbreviated GSPRT) differs from a SPRT only in that after $X_1, \dots, X_m$ have been observed, $\prod_{i=1}^{m}\{f_X(X_i;\theta_1)/f_X(X_i;\theta_0)\}$ is compared to B_m, A_m, respectively, instead of to B, A, where $\{B_1, B_2, \dots\}$ and $\{A_1, A_2, \dots\}$ are preassigned values with $B_m \leqslant A_m$ for all positive integers m. Thus, if $B_m = B$ and $A_m = A$ for all m, a GSPRT becomes a SPRT.

To describe the properties of these tests, we introduce the following notation. If E is any event and T is any test procedure, $P(E; \theta; T)$ denotes the probability of the event E when the parameter is equal to θ and the test procedure T is used. If Z is any random variable, $E\{Z;\theta;T\}$ denotes the expected value of Z when the parameter is equal to θ and the test procedure T is used.

Theorem 1 was proved by Wald and Wolfowitz [11].

Theorem 1. If T is any test procedure that surely observes X_1, there is a SPRT $S(T)$ with $P(H_i \text{ accepted}; \theta_i; S(T)) \geqslant P(H_i \text{ accepted}; \theta_i; T)$ and $E\{N;\theta_i;S(T)\} \leqslant E\{N; \theta_i; T\}$ for $i = 1, 2$.

From this theorem it also follows that if T' is any test procedure that assigns positive probabilities to observing X_1 and to not observing X_1, there is a test procedure $S(T')$ that proceeds like a SPRT once X_1 is observed, with $P(H_i \text{ accepted}; \theta_i; S(T')) \geqslant P(H_i \text{ accepted}; \theta_i; T')$ and where $E\{N;\theta_i;S(T')\} \leqslant E\{N;\theta_i;T'\}$, $i = 1, 2$.

Theorem 2 was proved by Weiss [12]; later a shorter proof was given by Le Cam [8].

Theorem 2. If T is any test procedure that surely observes X_1, there is a GSPRT $G(T)$ with $P(H_i \text{ accepted}; \theta_i; G(T)) \geqslant P(H_i \text{ accepted}; \theta_i; T)$ and $P(N \leqslant n; \theta_i; G(T)) \geqslant P(N \leqslant n; \theta_i; T)$ for $i = 1, 2$ and all positive integers n.

It follows from this that if T' is any test procedure that assigns positive probabilities to observing X_1 and to not observing X_1, there is a test procedure $G(T')$ that proceeds like a GSPRT once X_1 is observed, with $P(H_i \text{ accepted}; \theta_i; G(T')) \geqslant P(H_i \text{ accepted}; \theta_i; T')$ and $P(N \leqslant n; \theta_i; G(T')) \geqslant P(N \leqslant n; \theta_i; T')$ for $i = 1, 2$ and all positive integers n.

TESTING A ONE-SIDED HYPOTHESIS

$X_1, X_2, \dots$, and $f_X(X_i;\theta)$ are as above, but θ must be a scalar. Values θ_0, θ_1 are given, with $\theta_0 < \theta_1$; H_0 is now the hypothesis that $\theta \leqslant \theta_0$, and H_1 is now the hypothesis that $\theta \geqslant \theta_1$. For any test procedure T, $P(H_0 \text{ accepted}; \theta; T)$ is a function of θ, the *operating characteristic* (OC) *function* corresponding to T (*see* POWER), and $E\{N;\theta;T\}$ is a function of θ, the *average sample number** (ASN) function corresponding to T. Small positive values α and β are also given, and in order to be considered for use, a test procedure T must satisfy the following conditions (called OC conditions): $P(H_0 \text{ accepted}; \theta; T) \geqslant 1 - \alpha$ for all $\theta \leqslant \theta_0$, and $P(H_0 \text{ accepted}; \theta; T) \leqslant \beta$ for all $\theta \geqslant \theta_1$. That is, if $\theta \leqslant \theta_0$ we would like to accept H_0, and if $\theta \geqslant \theta_1$ we would like to reject H_0; the θ-interval (θ_0, θ_1) is the *indifference zone*, where it does not matter much whether we accept or reject H_0.

For commonly encountered distributions of X_i, it is shown in ref. 9 that if we use the SPRT of θ_0 against θ_1, with $B = \beta/(1-\alpha)$ and $A = (1-\beta)/\alpha$, we satisfy the OC conditions approximately. From the results above, it follows that among all test procedures approximately satisfying the OC conditions, this SPRT approximately minimizes the ASN function at $\theta = \theta_0$ and at $\theta = \theta_1$. But Bechhofer [2] has shown that if α and β are small, the maximum value (as θ varies) of the ASN function of the SPRT is far above the sample size required by the fixed sample size test satisfying the OC conditions. Since there is no such thing as an "indifference zone" for the ASN function, this is a definite disadvantage of the SPRT.

Suppose that T^* is a test procedure which satisfies the OC conditions, and that

$$\max_\theta E\{N;\theta;T^*\} \leqslant \max_\theta E\{N;\theta;T\}$$

for every procedure T satisfying the OC conditions. Then T^* is a "minimax* ASN test." Kiefer and Weiss [7] showed that for many of the commonly encountered distributions of X_i, a minimax ASN test is a GSPRT with the following structure: There exists a finite positive integer n such that $B_1 \leqslant B_2 \leqslant \cdots \leqslant B_n = A_n \leqslant A_{n-1} \leqslant \cdots \leqslant A_1$. This structure was applied to the case where X_i has a Bernoulli distribution* (i.e., X_i can only be 0 or 1, with respective probabilities $1-\theta, \theta$) to construct minimax ASN tests explicitly in refs. 4, 6, and 13. The reduction in the maximum expected sample size is illustrated in the following example.

Example: Bernoulli. Suppose that $\alpha = \beta = 0.01$, $\theta_0 = 0.3$, $\theta_1 = 0.4$. Let S denote the SPRT for this case, T^* the minimax ASN test for this case, and T the fixed sample size test for this case. Then $E\{N;\theta;T\} \simeq 476$ for all θ: $E\{N;\theta_0;T^*\} = 231.5$, $E\{N;\theta_1;T^*\} = 221.6$, $\max_\theta E\{N;\theta;T^*\} = 363$; $E\{N;\theta_0;S\} \simeq 208.4$, $E\{N;\theta_1;S\} \simeq 199.3$, $\max_\theta E\{N;\theta;S\} \simeq 475.8$.

Anderson [1] has constructed approximately minimax ASN tests for the case where X_i has a normal distribution with unknown mean θ. DeGroot [3] and Hall [5] have shown that GSPRTs should still be used when loss is measured more generally than by error probabilities and sampling cost is measured more generally than by sample size. Weiss [14] has reduced a class of problems to a problem of testing a one-sided hypothesis about the parameter of a Bernoulli distribution, and then used a minimax ASN test for the latter problem.

HISTORICAL NOTE

The SPRT was developed by Wald [9]. A history of this development is given in refs. 9 and 10. The term "generalized sequential probability ratio test" was first used by Weiss [12], but GSPRTs that were not also SPRTs had been used earlier: the truncated sequential test described in ref. 10 is an example.

References

[1] Anderson, T. W. (1960). *Ann. Math. Statist.*, **31**, 165–197.

[2] Bechhofer, R. E. (1960). *J. Amer. Statist. Ass.*, **55**, 660–663.

[3] DeGroot, M. H. (1961). *Ann. Math. Statist.*, **32**, 602–605.

[4] Freeman, D. and Weiss, L. (1964). *J. Amer. Statist. Ass.*, **59**, 67–88.

[5] Hall, W. J. (1980). In *Asymptotic Theory of Statistical Tests and Estimation*. Academic Press, New York, pp. 325–350.

[6] Higgins, J. (1969). *Bernoulli Sampling Plans Which Approximately Minimize the Maximum Expected Sample Size*. M. S. thesis, Cornell University.

[7] Kiefer, J. and Weiss, L. (1957). *Ann. Math. Statist.*, **28**, 57–74.

[8] Le Cam, L. (1954). *Ann. Math. Statist.*, **25**, 791–794.

[9] Wald, A. (1945). *Ann. Math. Statist.*, **16**, 117–186.

[10] Wald, A. (1945). *Sequential Analysis*. Wiley, New York.

[11] Wald, A. and Wolfowitz, J. (1948). *Ann. Math. Statist.*, **19**, 326–339.

[12] Weiss, L. (1953). *Ann. Math. Statist.*, **24**, 273–281.

[13] Weiss, L. (1962). *J. Amer. Statist. Ass.*, **57**, 551–566.

[14] Weiss, L. (1979). *Trabajos Estadíst.*, **30**, 79–91.

(HYPOTHESIS TESTING
SEQUENTIAL ANALYSIS
SEQUENTIAL PROBABILITY RATIO TEST
STOPPING RULES)

LIONEL WEISS

GENERALIZED VARIANCE

A measure of spread in the univariate distribution is the variance σ^2. In the multidimensional case, an obvious analog is the covariance matrix $\mathbf{\Sigma}$. For this case Wilks [31] introduced $|\mathbf{\Sigma}|$ as a scalar measure of dispersion. He called it the *generalized variance*. Frisch

[7] referred to the determinant of $\mathbf{R}$, the correlation matrix, as the scatter coefficient. For the sample, the generalized variance is taken to be the determinant of the sample covariance matrix

$$\mathbf{S} = \textstyle\sum_{\alpha=1}^{N}(\mathbf{X}_\alpha - \overline{\mathbf{X}})(\mathbf{X}_\alpha - \overline{\mathbf{X}})'/(N-1),$$

based on the random sample $\mathbf{X}_1, \mathbf{X}_2, \dots, \mathbf{X}_N$. Many likelihood ratio criteria* in multivariate analysis* are functions of the sample generalized variance (see ref. 1).

A more general concept of multidimensional scatter can be defined from a geometric interpretation of its measure. Let $\mathbf{X}_0$ be an arbitrary point and $\mathbf{X}_1, \dots, \mathbf{X}_N$ the sample points in the p-dimensional space. Consider the parallelotopes formed by taking as the principal edges the vectors obtained by using any p of these N points and $\mathbf{X}_0$ as the end points. The sum of the squared volumes of the $\binom{N}{p}$ parallelotopes that can be so formed is called the *multidimensional scatter* of the N points around the pivotal point $\mathbf{X}_0$ [32]. This scatter is minimized when $\mathbf{X}_0$ is chosen to be the sample mean $\overline{\mathbf{X}}$. Anderson [1, p. 170] has shown the value of the minimal scatter to be $(N-1)^p|\mathbf{S}|$. The minimal scatter of the N points $\mathbf{X}_1, \dots, \mathbf{X}_N$ is called their *internal scatter*. An alternative construction that can be placed on the internal scatter is given by the following theorem due to Anderson [1, p. 167]:

Theorem. Let the rows of the $p \times N$ matrix $(\mathbf{X}_1 - \overline{\mathbf{X}}, \mathbf{X}_2 - \overline{\mathbf{X}}, \dots, \mathbf{X}_N - \overline{\mathbf{X}})$ be represented by the columns of $\mathbf{Y}$ $(N \times p)$. The square of the p-dimensional volume of the parallelotope with $\mathbf{Y}_1, \dots, \mathbf{Y}_p$ as its principal edges is proportional to the generalized variance.

Under general conditions of the independence and identical distribution of $\mathbf{X}_1, \dots, \mathbf{X}_N$, it can be shown that the expected value of $|\mathbf{A}|$ is $\{\Gamma(N)/\Gamma(N-p)\}|\boldsymbol{\Sigma}|$. Here $\mathbf{A} = (N-1)\mathbf{S}$ is the sample sum of squares and products matrix [30].

Most of the developments in connection with the distribution and applications of $|\mathbf{S}|$ have been restricted to multinormal* (MN) populations.

DISTRIBUTION AND MOMENTS

Since $|\mathbf{S}|$ and $|\mathbf{A}|$ are simply related, the results for the latter are equally applicable to the former. If $\mathbf{X}_\alpha \sim \mathrm{MN}(\mathbf{0}, \boldsymbol{\Sigma})$, it can be shown that $\mathbf{A} = \sum_{\alpha=1}^{\nu}\mathbf{X}_\alpha\mathbf{X}_\alpha'$ has the central Wishart distribution* in p dimensions with ν degrees of freedom and variance matrix $\boldsymbol{\Sigma}$. Notationally, we say that $\mathbf{A} \sim W_p[\nu\,|\,\boldsymbol{\Sigma}]$. Using the Bartlett decomposition theorem it can be shown that $|\mathbf{A}|/|\boldsymbol{\Sigma}|$ is distributed as the product of the p independent χ^2 variates $\chi_\nu^2, \chi_{\nu-1}^2, \dots, \chi_{\nu-p+1}^2$ [16, p. 47]. Hence $E[|\mathbf{A}|^t] = |2\boldsymbol{\Sigma}|^t\Gamma_p[t+\nu/2]/\Gamma_p[\nu/2]$, where $\Gamma_p[k]$ is the p-variate gamma function* [28].

When A has the noncentral Wishart distribution*, with the noncentrality parameter $\boldsymbol{\Omega}$ and $n = N-1$ degrees of freedom, then $E[|\mathbf{A}|^t] = |2\boldsymbol{\Sigma}|^t\{\Gamma_p[t+n/2]/\Gamma_p[n/2]\}\, e^{(-\operatorname{tr}\boldsymbol{\Omega})} \cdot {}_1F_1[t+n/2; n/2; \boldsymbol{\Omega}]$, where ${}_1F_1[a; b; \mathbf{X}]$ is the confluent hypergeometric function* of matrix argument [4]. The probability density function of $|\mathbf{A}|$ can be obtained from this moment by the inverse Mellin transform technique (*see* INTEGRAL TRANSFORMS) as

$$f(x) = x^{-1}\left\{\pi^{p(p-1)/4}\Big/\Gamma_p\left(\frac{n}{2}\right)\right\}e^{(-\operatorname{tr}\boldsymbol{\Omega})}$$

$$\cdot \sum_k \sum_\kappa \left\{C_\kappa(\boldsymbol{\Omega})\Big/\left(k!\left(\frac{n}{2}\right)_\kappa\right)\right\}$$

$$\times G_{0,p}^{p,0}\left[x \,\Big|\, \frac{n}{2}+k_1, \frac{n-1}{2}+k_2, \dots, \frac{n-p+1}{2}+k_p\right], \quad x > 0.$$

Here G is the Meijer's G-function of matrix argument; $C_\kappa(Z)$ is the zonal polynomial* of degree k [19]. Special cases of the distribution have been examined by various authors for a variety of assumptions regarding the rank of Ω (see refs. 3, 5, and 20, for example). Setting $\boldsymbol{\Omega} = \mathbf{0}$, one obtains the central distribution of $|\mathbf{A}|$. Complex analogs of these results appear in refs. 10 and 14.

APPROXIMATIONS

Although the probability density function (PDF) of $|\mathbf{A}|$ can be cast in a computable form, approximations are useful. Hoel [11] suggested for $V = \{|\mathbf{A}|/|\mathbf{\Sigma}|\}^{1/p}$ the two-parameter gamma approximation with $c = (p/2)[1 - (p-1)(p-2)/(2N)]^{1/p}$ and $p(N-p)/2$ as its parameters. This result is exact when $p = 1, 2$. Steyn [27] has modified this result for the central and the noncentral cases by requiring that the approximating distribution have its first two moments equal to those of V. An application of the multivariate central limit theorem* gives the asymptotic distribution of $\sqrt{n}\,[|\mathbf{S}|/|\mathbf{\Sigma}| - 1]$ as $N(0, 2p)$ [1, p. 173]. From the representation of $|\mathbf{A}|/|\mathbf{\Sigma}|$ as the product of χ^2 variates, it is possible to write $\ln\{|\mathbf{A}|/|\mathbf{\Sigma}|\} = \sum_{j=1}^{p} \ln X_j$, where the X's are independent and X_j is χ^2 with $(n - j + 1)$ degrees of freedom. Thus the normal approximation for $\ln X_j$ yields the approximate distribution of $\ln\{|\mathbf{A}|/|\mathbf{\Sigma}|\}$ as $N[p \ln 2 + \sum_{j=1}^{p} \psi\{(n-j+1)/2\}, \sum_{j=1}^{p} \psi'\{(n-j+1)/2\}]$, where ψ and ψ' are the digamma and trigamma functions, respectively. Although no detailed assessment of these various approximations has been made, it is generally agreed that Hoel's procedure is good for $p \leqslant 3$, and the latter for $p > 3$ (see ref. 12, p. 198). In the noncentral case Fujikoshi [8] and Sugiura and Nagao [29] have given results on the asymptotic expansion for the distribution of $|\mathbf{S}|$ when Ω is $0(1)$, $0(\sqrt{n})$, and $0(n)$.

ESTIMATION

An unbiased estimator for $|\mathbf{\Sigma}|$ is seen to be $\{\Gamma(N-p)/\Gamma(N)\}|\mathbf{A}|$. The maximum likelihood estimator* is $|\hat{\mathbf{\Sigma}}|$ or $|\mathbf{A}|/N^p$. Selliah [22] has presented the minimax, best affine equivariant estimator* under constant risk as $\{(n-p+2)!/(n+2)!\}|\mathbf{A}|$. Using a quadratic loss function* Shorrack and Zidek [24] have developed an estimator that dominates Selliah's. An interval estimate of $|\mathbf{\Sigma}|$ can be given by using Anderson's approximation for the distribution of $|\mathbf{S}|/|\mathbf{\Sigma}|$ (see also Madansky and Olkin [17]).

TESTS OF HYPOTHESES

Although several test criteria involve the distribution of $|\mathbf{S}|$ or its functions, no direct tests of hypotheses regarding $|\mathbf{\Sigma}|$ have been considered until recently. Sen Gupta [23] has constructed likelihood ratio tests* for the standardized generalized variance, i.e., for $|\mathbf{\Sigma}_i|^{1/p_i} = \Delta_i^2$. The problems discussed are:

1. $H_{01} : \Delta_1^2 = \Delta_{10}^2$
2. $H_{02} : \Delta_1^2 = \Delta_2^2$ (the two-sample problem)
3. $H_{03} : \Delta_1^2 = \Delta_2^2 = \cdots = \Delta_k^2$ (the k-sample problem)

The criteria developed are functions of the sample values d_i^2 of Δ_i^2. The null and the nonnull distributions of the criteria are discussed by Sen Gupta. An extension of Hartley's $F_{\max}$ criterion* can be used for H_{03}, in the case when $p_i = p$, $N_i = N$. The criterion is $F_{p\,\max} = d_{\max}^2 / d_{\min}^2$.

APPLICATIONS

Eaton [6], Gnanadesikan and Gupta [9], and Regier [21] have used the generalized variance for ranking and selection* problems based on multiple characters. Optimal design construction by minimizing the determinant of the covariance matrix (or the inverse of the information matrix of the design) has been presented by Kiefer and Studden [15]. Stratification using multiple characters is achieved through minimizing the generalized variance by Aravantis and Afonja [2]. Steel [26] and Kettenring [13] have extended canonical correlation analysis* to several sets of variates by minimizing the generalized variance of the resulting vectors. Examination of efficiency of estimators in the multiparameter case has been done through their generalized variances.

Example. The problem of talker identification based on a five-dimensional summary of speech is attacked by Gnanadesikan and Gupta [9] by the use of generalized variance. The data consist of seven replicates of utterances of 10 words by the particular speaker. The word with the least variation is selected by comparing the generalized variances for each of the 10 words uttered.

References

[1] Anderson, T. W. (1958). *Introduction to Multivariate Statistical Analysis*. Wiley, New York. (An excellent introduction to mathematical aspects of multivariate analysis.)

[2] Aravantis, L. G. and Afonja, B. (1971). *Biometrics*, **27**, 119–127.

[3] Bagai, O. P. (1965). *Ann. Math. Statist.*, **36**, 120–130.

[4] Constantine, A. G. (1963). *Ann. Math. Statist.*, **34**, 1270–1285.

[5] Consul, P. C. (1964). *Math. Nachr.*, **28**, 169–179.

[6] Eaton, M. L. (1967). *Ann. Math. Statist.*, **38**, 941–943.

[7] Frisch, R. (1929). *Nord. Statist. J.*, **8**, 36–102.

[8] Fujikoshi, Y. (1968). *J. Sci. Hiroshima Univ. Ser. A-1*, **32**, 293–299.

[9] Gnanadesikan, M. and Gupta, S. S. (1970). *Technometrics*, **12**, 103–117.

[10] Goodman, N. R. (1963). *Ann. Math. Statist.*, **34**, 178–180. (Introduces the complex Wishart distribution.)

[11] Hoel, P. G. (1937). *Ann. Math. Statist.*, **8**, 149–158.

[12] Johnson, N. L. and Kotz, S. (1972). *Distributions in Statistics: Continuous Multivariate Distributions*. Wiley, New York.

[13] Kettenring, J. R. (1971). *Biometrika*, **58**, 433–451.

[14] Khatri, C. G. (1965). *Ann. Math. Statist.*, **36**, 98–114.

[15] Kiefer, J. and Studden, W. G. (1976). *Ann. Statist.*, **4**, 1113–1123.

[16] Kshirsagar, A. M. (1972). *Multivariate Analysis*. Marcel Dekker, New York. (Good discussion of Bartlett decomposition and random orthogonal transformations.)

[17] Madansky, A. and Olkin, I. (1969). In *Multivariate Analysis II*, P. R. Krishnaiah, ed. Academic Press, New York, pp. 261–288.

[18] Mathai, A. M. (1972). *Sankhyā A*, **34**, 161–170.

[19] Mathai, A. M. (1972). *Ann. Inst. Statist. Math.*, **24**, 53–63.

[20] Mathai, A. M. and Rathie, P. N. (1971). *Sankhyā A*, **33**, 45–60.

[21] Regier, M. H. (1976). *Technometrics*, **18**, 483–489.

[22] Selliah, J. B. (1964). Estimation and Testing Problems in a Wishart Distribution. *Tech. Rep. No. 10*, Dept. of Statistics, Stanford University, Stanford, Calif.

[23] Sen Gupta, A. (1981). Tests for Standardized Generalized Variances of Multivariate Normal Populations of Possibly Different Dimensions. *Tech. Rep. No. 50*, Dept. of Statistics, Stanford University, Stanford, Calif. (To appear 1983, *J. Multivariate Anal.*)

[24] Shorrock, R. W. and Zidek, J. V. (1976). *Ann. Statist.*, **4**, 629–638.

[25] Sinha, B. K. (1976). *J. Multivariate Anal.*, **6**, 617–625.

[26] Steel, R. G. D. (1951). *Ann. Math. Statist.*, **22**, 456–460.

[27] Steyn, H. S. (1978). *J. Amer. Statist. Ass.*, **73**, 670–675.

[28] Subrahmaniam, K. (1976). *Sankhyā A*, **38**, 221–258. (A good review of the zonal polynomials literature.)

[29] Sugiura, N. and Nagao, H. (1971). *Ann. Inst. Statist. Math.*, **23**, 469–475.

[30] van der Vaart, H. R. (1965). *Ann. Math. Statist.*, **36**, 1308–1312.

[31] Wilks, S. S. (1932). *Biometrika*, **24**, 471–494.

[32] Wilks, S. S. (1960). In *Contributions to Probability and Statistics: Essays in Honor of H. Hotelling*, Olkin et al., eds. Stanford University Press, Stanford, Calif., pp. 486–503. (Discusses multivariate analysis through multidimensional statistical scatter.)

(MULTIVARIATE ANALYSIS
WISHART DISTRIBUTION)

S. Kocherlakota
K. Kocherlakota

GENERAL LINEAR HYPOTHESIS *See* GENERAL LINEAR MODEL

GENERAL LINEAR MODEL[1]

DEFINITION

Many situations motivate representing a random variable Y as a function of other values $x_1, x_2, \ldots, x_p$; or representing the expected value* of Y as $E(Y) = f(\mathbf{x})$, where $f(\mathbf{x})$ is a function of $x_1, x_2, \ldots, x_p$, represented by the vector $\mathbf{x}$. If y is a realized value of Y, the difference $y - f(\mathbf{x})$ is taken to be random and is referred to as residual* or as error; $e = y - E(y) = y - f(\mathbf{x})$, so $y = f(\mathbf{x}) + e$.

In general, $f(\mathbf{x})$ can be any function of the x's, the most common being a linear function of unknown parameters $\beta_1, \ldots, \beta_k$, namely $\beta_1 x_1 + \beta_2 x_2 + \cdots + \beta_k x_k$. A parameter β_0 corresponding to no x-variable can also be included, so that

$$y = \beta_0 + \beta_1 x_1 + \beta_2 x_2 + \cdots + \beta_k x_k + e. \tag{1}$$

β_0 is called the "intercept" and $\beta_1, \ldots, \beta_k$ are called "slopes" corresponding, when $k = 1$, to β_1 being the slope of the straight line $y = \beta_0 + \beta_1 x_1$ in the Cartesian plane.

Equation (1) applies to each set of observations $y_i, x_{i1}, x_{i2}, \ldots, x_{ik}$ on Y and the k x's, so

$$y_i = \beta_0 + \beta_1 x_{i1} + \beta_2 x_{i2} + \cdots + \beta_k x_{ik} + e_i, \tag{2}$$

for $i = 1, 2, \ldots, N$. On defining the vectors and matrix

$$\mathbf{y} = \begin{bmatrix} y_1 \\ y_2 \\ \vdots \\ y_N \end{bmatrix},$$

$$\mathbf{X} = \begin{bmatrix} 1 & x_{11} & x_{12} & \cdots & x_{1k} \\ 1 & x_{21} & x_{22} & \cdots & x_{2k} \\ \vdots & \vdots & \vdots & \cdots & \vdots \\ 1 & x_{N1} & x_{N2} & \cdots & x_{Nk} \end{bmatrix}, \tag{3}$$

$$\boldsymbol{\beta} = \begin{bmatrix} \beta_0 \\ \beta_1 \\ \vdots \\ \beta_k \end{bmatrix}, \qquad \mathbf{e} = \begin{bmatrix} e_1 \\ e_2 \\ \vdots \\ e_N \end{bmatrix},$$

(2) is written as

$$\mathbf{y} = \mathbf{X}\boldsymbol{\beta} + \mathbf{e}. \tag{4}$$

This is the *model equation* of the general linear model.

The general linear model consists of the model equation and of statements about the stochastic nature of the random vector $\mathbf{e}$ and hence of $\mathbf{y}$. Because, by definition, $e_i = y_i - E(y_i)$,

$$\mathbf{e} = \mathbf{y} - E(\mathbf{y}), \qquad E(\mathbf{e}) = \mathbf{0}$$

and

$$E(\mathbf{y}) = \mathbf{X}\boldsymbol{\beta}. \tag{5}$$

The variance–covariance (dispersion) matrix* of $\mathbf{e}$ (and hence of $\mathbf{y}$) is nonnegative definite, to be denoted $\mathbf{V}$:

$$\begin{aligned} \text{var}(\mathbf{y}) = \text{var}(\mathbf{e}) &= E[\mathbf{y} - E(\mathbf{y})][\mathbf{y} - E(\mathbf{y})]' \\ &= E(\mathbf{ee}') = \mathbf{V}. \end{aligned} \tag{6}$$

Every e_i in $\mathbf{e}$ is commonly assumed to have the same variance σ^2, with every pair of (different) e_i's having zero covariance, so that $\mathbf{V} = \sigma^2\mathbf{I}$.

Equations (4), (5), and (6) constitute the general linear model, often with $\mathbf{V} = \sigma^2\mathbf{I}$ also. The stochastic properties attributed to $\mathbf{y}$ concern only first and second moments. For estimating $\boldsymbol{\beta}$, no particular form of probability distribution need be assumed; but when confidence intervals and/or hypothesis tests are required, normality* assumptions are customarily made. Some writers confine the "general" in general linear model to mean only those models having error terms normally distributed, i.e., $\mathbf{e} \sim N(\mathbf{O}, \mathbf{V})$; *generalized linear model** then means models in which error terms may or may not be normally distributed. Despite this distinction, general linear model remains the usual term for (4), (5), and (6). A special, widely used case is $\mathbf{V} = \sigma^2\mathbf{I}$.

APPLICATIONS

A few of the many applications of the general linear model are briefly noted.

Linear Regression

The most elementary application is simple linear regression*, when $k = 1$ and the model equation is $y_i = \beta_0 + \beta_1 x_{i1} + e_i$. Multiple linear regression* is where there are two or more x-variables: $y_i = \beta_0 + \beta_1 x_{i1} + \beta_2 x_{i2} + \beta_3 x_{i3} + e_i$.

Polynomial Regression

The "linear" in "linear model" pertains to the occurrence of the β's in the model equation. They occur linearly: $E(y_i)$ is a linear function* of the β's. This does not preclude

x's occurring in nonlinear ways; for example, if y varies as a polynomial function (cubic, say) of time, measured as t from some base point, then the model equation is

$$y_i = \beta_0 + \beta_1 t_i + \beta_2 t_i^2 + \beta_3 t_i^3 + e_i . \quad (7)$$

This is, (2) with $k = 3$ and $x_{i1} = t_i$, $x_{i2} = t_i^2$ and $x_{i3} = t_i^3$.

One rewriting of polynomial regression models like (7) leads to orthogonal polynomials*. Define $q_j(t)$ as a polynomial function* of t of order j. Then rewrite (7) as $y_i = \alpha_0 + \alpha_1 q_1(t_i) + \alpha_2 q_2(t_i) + \alpha_3 q_3(t_i) + e_i$: the α's are linear functions of the β's. Choosing the q-functions so that

$$\sum_{i=1}^{N} q_j(t_i) q_{j'}(t_i) = 0 \quad \text{and} \quad \sum_{i=1}^{N} \left[q_j(t_i) \right]^2 = 1$$

for $j \neq j' = 1, 2, 3$ defines them as orthogonal polynomials. This simplifies estimation of the α's. Generalization to $j, j' = 1, \ldots, k$ for any integer k is straightforward.

Nonpolynomial Functions of x's

The linear model can represent a variety of nonlinear functions of the x's. One example is $p(y_i) = \beta_0 + \beta \log x_i + e_i$, where x_i is dose rate and $p(y_i)$ is the probit* of y_i, the cumulative death (or survival) rate corresponding to dose x_i. Another is the reduction of $y_i = \beta_0 \epsilon_i \exp(\beta_1 x_{i1} + \beta_2 x_{i2})$ to $\log_e y_i = \beta_{00} + \beta_1 x_{i1} + \beta_2 x_{i2} + e_i$, where $\beta_{00} = \log_e \beta_0$ and $e_i = \log \epsilon_i$. The error term ϵ_i occurs multiplicatively with $E(y_i)$, whereas e_i occurs additively with $E(\log_e y_i)$.

A third example is the analysis of cell frequencies in contingency tables*. If f_{ij} is the observed cell frequency in categories i and j of a two-variable table, an appropriate model equation if $f_{ij} = p_0 p_{1(i)} p_{2(j)} \epsilon_{ij}$, where p_0 is some constant and $p_{1(i)}$ and $p_{2(j)}$ are the relative frequencies of categories i and j in variables 1 and 2, respectively, and ϵ_{ij} is a multiplicative error term. Defining $u_0 = \log p_0$, $u_{1(i)} = \log p_{1(i)}$, $u_{2(j)} = \log p_{2(j)}$ and $e_{ij} = \log \epsilon_{ij}$ gives a log-linear model (*see* CONTINGENCY TABLES)

$$\log f_{ij} = u_0 + u_{1(i)} + u_{2(j)} + e_{ij} . \quad (8)$$

Bishop et al. [1] have extensive discussion of these models, with many examples.

Dummy Variables

Classification variables such as sex, religion, and geographic location, can be accommodated in linear models in two ways. Suppose that a study is made of annual incomes for people aged 40 who had three different levels of education: (a) did not finish high school, (b) finished high school, and (c) attended college. This ordinal variable (*see* CATEGORICAL DATA) can be part of a linear model by defining an x-variable having values 1, 2, or 3 for people in classes (a), (b) or (c), respectively. If y_i is the ith person's income at age 40, the linear model would be that of simple linear regression, $y_i = \beta_0 + \beta_1 x_i + e_i$, as in the section "Linear Regression." The difficulty with this is that the codes 1, 2, and 3 are used as quantification of the amount of education inherent in the three classes—and as such, neither they nor any other set of three codes are unequivocally acceptable.

A preferred way of handling ordinal and nominal variables is, for each such variable, to define several x-variables. For example, for the three education classes three x-variables are defined in the following way: for class (a), $x_{i1} = 1$, $x_{i2} = 0$, $x_{i3} = 0$; for class (b), $x_{i1} = 0$, $x_{i2} = 1$, $x_{i3} = 0$; and for class (c), $x_{i1} = 0$, $x_{i2} = 0$, $x_{i3} = 1$. The model equation is then

$$y_i = \beta_0 + \beta_1 x_{i1} + \beta_2 x_{i2} + \beta_3 x_{i3} + e_i . \quad (9)$$

Thus, if in a small "survey" there were 2, 2, and 4 people, respectively, in the three classes, the model equation (4) would be

$$\mathbf{y} = \begin{bmatrix} 1 & 1 & \cdot & \cdot \\ 1 & 1 & \cdot & \cdot \\ 1 & \cdot & 1 & \cdot \\ 1 & \cdot & 1 & \cdot \\ 1 & \cdot & \cdot & 1 \\ 1 & \cdot & \cdot & 1 \\ 1 & \cdot & \cdot & 1 \\ 1 & \cdot & \cdot & 1 \end{bmatrix} \begin{bmatrix} \beta_0 \\ \beta_1 \\ \beta_2 \\ \beta_3 \end{bmatrix} + e \quad (10)$$

with dots in a matrix representing zeros. The x-variables defined in this manner are often called "dummy variables," although "indicator variables" is more descriptive. The unity value in each case represents the incidence of the β's in the y_i's, and thus the **X** matrix, of 0's and 1's, is an *incidence matrix**.

Linear models can also be combinations of the special cases described here. An important example of this is combining regression and dummy variables, which leads to the analysis of covariance* (see the section so named).

Experimental Design Models

Dummy variables are also the basis of model equations for data from designed experiments. For example, for a randomized complete blocks* experiment of 3 treatments and 4 blocks, the model equation is

$$\mathbf{y} = \begin{bmatrix} 1 & 1 & . & . & 1 & . & . & . \\ 1 & 1 & . & . & . & 1 & . & . \\ 1 & 1 & . & . & . & . & 1 & . \\ 1 & 1 & . & . & . & . & . & 1 \\ 1 & . & 1 & . & 1 & . & . & . \\ 1 & . & 1 & . & . & 1 & . & . \\ 1 & . & 1 & . & . & . & 1 & . \\ 1 & . & 1 & . & . & . & . & 1 \\ 1 & . & . & 1 & 1 & . & . & . \\ 1 & . & . & 1 & . & 1 & . & . \\ 1 & . & . & 1 & . & . & 1 & . \\ 1 & . & . & 1 & . & . & . & 1 \end{bmatrix} \begin{bmatrix} \mu \\ \tau_1 \\ \tau_2 \\ \tau_3 \\ \rho_1 \\ \rho_2 \\ \rho_3 \\ \rho_4 \end{bmatrix} + \mathbf{e}. \tag{11}$$

The pattern of 0's and 1's in the incidence matrix, especially in its submatrices, is very evident and is a consequence of the experimental design. The incidence matrix is accordingly called a *design matrix*. The name "model matrix" is also used, as an all-inclusive alternative to both incidence and design matrix. Very often, but not always, the term "linear model" refers to cases like (10) and (11), where **X** is just an incidence matrix.

Fixed-, Random-, and Mixed-Effects Models*

Model equations (11) are equivalent to

$$y_{ij} = \mu + \tau_i + \rho_j + e_{ij}, \tag{12}$$

where y_{ij} is the observation on treatment i in block j, for $i = 1,2,3$ and $j = 1,2,3,4$. In some contexts, interest centers on estimating μ, the τ_i's, and the ρ_j's, whereupon these parameters are called fixed effects, and the model is a *fixed-effects** model. In other contexts, one or more sets of effects [e.g., the ρ_j's in (12)] are considered as random variables having zero mean and some assumed second-moment properties, usually that of homoscedastic variance, σ_ρ^2 say, and zero covariances. In this case the ρ's are called random effects*. When all effects in a model [except for μ, a general mean, as in (12)] are random effects it is called a random-effects* or *variance components** model. The object then is to estimate the variance components, such as σ_ρ^2 and σ_e^2. And when a model has a mixture of fixed and random effects, it is called a mixed-effects model or, simply, a *mixed model**. General linear model theory is usually concerned only with estimating fixed effects; although it does embrace variance components estimation*, this difficult topic, which has wide application, is usually treated separately. (Harville [3] and Searle [10, 11] give reviews, details, and references.)

Survey Data

Suppose that the eight people of illustration (10), classified there into three different education classes, also come from four geographical regions, as shown in Table 1. A possible model equation for studying annual income in relation to education and region is

$$\mathbf{y} = \begin{bmatrix} 1 & 1 & . & . & . & 1 & . & . \\ 1 & 1 & . & . & . & . & 1 & . \\ 1 & . & 1 & . & . & 1 & . & . \\ 1 & . & 1 & . & . & . & . & 1 \\ 1 & . & . & 1 & 1 & . & . & . \\ 1 & . & . & 1 & . & . & 1 & . \\ 1 & . & . & 1 & . & . & . & 1 \\ 1 & . & . & 1 & . & . & . & 1 \end{bmatrix} \begin{bmatrix} \mu \\ \tau_1 \\ \tau_2 \\ \tau_3 \\ \rho_1 \\ \rho_2 \\ \rho_3 \\ \rho_4 \end{bmatrix} + \mathbf{e}. \tag{13}$$

This handles only main effects*; interactions* could also be included.

Table 1 Number of People

	Region of Country				
Education	N	S	E	W	Total
(a)	—	1	1	—	2
(b)	—	1	—	1	2
(c)	1	—	1	2	4
Total	1	2	2	3	8

Model equations (11) and (13) are both of the form $\mathbf{y} = \mathbf{X}\boldsymbol{\beta} + \mathbf{e}$. They illustrate how **X** matrices from designed experiments are well structured and have very particular patterns of 0's and 1's, as in (11), whereas those of survey data are usually much less structured, as in (13).

Balanced and Unbalanced Data

The big difference between (11) and (13) is that in (11) each subclass of the data (as defined by treatments and blocks) has the same number of observations whereas in (13) the subclasses (defined by education and region) have different numbers of observations, with some having none at all. These exemplify a dichotomy of data that is important in linear model analysis: balanced data (or equal-subclass-numbers data), wherein every innermost (or submost) subclass has the same number of observations; and unbalanced* (or unequal-subclass-numbers) data, wherein subclasses do *not* have all the same number of observations, including situations where some have none at all.

Well-designed and well-executed experiments yield data that are either balanced, or unbalanced in a carefully designed manner that could be called planned unbalancedness; e.g., a Latin square* of order n yields $(1/n)$th part of the n^3 subclasses defined by its factors. The analysis of such data (adapted by missing value* techniques when appropriate) is relatively straightforward, and can usually be expressed in terms of means and summation notation. In contrast, analysis of unbalanced data is more complicated and interpretation more difficult. At all times, balanced data are just special cases of unbalanced data. (For other uses of balance, *see* GENERAL BALANCE and BALANCE IN EXPERIMENTAL DESIGN; *also see* UNBALANCED DATA.)

ESTIMATION

Methods

The parameter vector $\boldsymbol{\beta}$ of the general linear model (4), (5), and (6) is usually estimated by generalized or ordinary least squares*, abbreviated GLS and OLS, respectively. GLS yields estimation equations

$$\mathbf{X}'\mathbf{V}^{-1}\mathbf{X}\hat{\boldsymbol{\beta}} = \mathbf{X}'\mathbf{V}^{-1}\mathbf{y} \tag{14}$$

where **V** is nonsingular; and OLS gives

$$\mathbf{X}'\mathbf{X}\hat{\boldsymbol{\beta}} = \mathbf{X}'\mathbf{y}. \tag{15}$$

These equations require no specific form for the distribution of **e**. When **e** has a multivariate normal distribution*, the GLS equations (14) [and hence (15) when $\mathbf{V} = \sigma^2\mathbf{I}$] are also the maximum likelihood* equations for estimating $\boldsymbol{\beta}$. When **V** is singular, (14) takes the more general form

$$\mathbf{X}'\mathbf{V}^{-}\mathbf{X}\hat{\boldsymbol{\beta}} = \mathbf{X}'\mathbf{V}^{-}\mathbf{y}, \tag{16}$$

where $\mathbf{V}^{-}$ is a generalized inverse* of **V** satisfying $\mathbf{V}\mathbf{V}^{-}\mathbf{V} = \mathbf{V}$: Rao [8], Searle [10], and Zyskind and Martin [22] give details.

Since $\mathbf{V}^{-} = \mathbf{V}^{-1}$ for nonsingular **V**, (16) includes (14). Provided that a symmetric $\mathbf{V}^{-}$ is used (and it always exists, e.g., $\mathbf{V}^{-}\mathbf{V}\mathbf{V}^{-\prime}$ for any nonsymmetric $\mathbf{V}^{-}$), it will be nonnegative definite because **V** is, so $\mathbf{V}^{-} = \mathbf{L}'\mathbf{L}$ for some real, full-row-rank matrix **L**. Then, with $\mathbf{W} = \mathbf{L}\mathbf{X}$ and $\mathbf{z} = \mathbf{L}\mathbf{y}$, (14) becomes $\mathbf{W}'\mathbf{W}\hat{\boldsymbol{\beta}} = \mathbf{W}'\mathbf{z}$, the same form as (15). Despite this equivalence in form, the practical difficulty is that **V** is seldom known, and some estimate must be used in its place, thus requiring estimation of variance components, as discussed in the section "Fixed-, Random-, and Mixed-Effects Models." To avoid this problem, or to circumvent it by using something in lieu of **W** and **z**, attention is usually confined to (15), as is now done.

Normal Equations with Many Solutions

Equations (15) are called the normal equations. Whenever $\mathbf{X}'\mathbf{X}$ is nonsingular, as is usual in regression*, the solution is $\hat{\boldsymbol{\beta}} = (\mathbf{X}'\mathbf{X})^{-1}\mathbf{X}'\mathbf{y}$. When $\mathbf{X}'\mathbf{X}$ is singular, there are many solutions, namely $\mathbf{G}\mathbf{X}'\mathbf{y} + (\mathbf{I} - \mathbf{G}\mathbf{X}'\mathbf{X})\mathbf{t}$ for arbitrary $\mathbf{t}$ of appropriate order, where $\mathbf{G}$ is any generalized inverse of $\mathbf{X}'\mathbf{X}$. (Properties of these solutions are discussed in Rao [8], Searle [10], and Seber [17].) To emphasize the existence of many solutions, we use the symbol $\boldsymbol{\beta}^0$, rewrite the normal equations as

$$\mathbf{X}'\mathbf{X}\boldsymbol{\beta}^0 = \mathbf{X}'\mathbf{y}, \qquad (17)$$

and consider a solution

$$\boldsymbol{\beta}^0 = \mathbf{G}\mathbf{X}'\mathbf{y} \qquad (18)$$

for any $\mathbf{G}$ satisfying $\mathbf{X}'\mathbf{X}\mathbf{G}\mathbf{X}'\mathbf{X} = \mathbf{X}'\mathbf{X}$. Clearly, $\boldsymbol{\beta}^0 = \hat{\boldsymbol{\beta}} = (\mathbf{X}'\mathbf{X})^{-1}\mathbf{X}'\mathbf{y}$ from (15) when $\mathbf{X}'\mathbf{X}$ is nonsingular.

Constraints on Solutions

The solution (18) requires a generalized inverse, although in many applications (17) can be solved by imposing constraints on elements of the solution vector so as to yield, in combination with (17), a solution to (17). From this, the corresponding generalized inverse can be derived for use as needed; e.g., for the variance of $\boldsymbol{\lambda}'\boldsymbol{\beta}^0$, as in **2** from the section "Estimable Functions." General discussion of such constraints and an algorithm for easy application are available in Searle [10, Sec. 5.7].

Solutions and Estimators

$\hat{\boldsymbol{\beta}} = (\mathbf{X}'\mathbf{X})^{-1}\mathbf{X}'\mathbf{y}$ is the only solution to the normal equations (15) when $\mathbf{X}'\mathbf{X}$ is nonsingular and it is an unbiased estimator* of $\boldsymbol{\beta}$. But none of the solutions $\boldsymbol{\beta}^0 = \mathbf{G}\mathbf{X}'\mathbf{y}$ to (17) is an unbiased estimator of $\boldsymbol{\beta}$ when $\mathbf{X}'\mathbf{X}$ is singular. Nevertheless, numerous functions of $\boldsymbol{\beta}^0$ are not only invariant with respect to $\boldsymbol{\beta}^0$, but are also unbiased estimators of functions of $\boldsymbol{\beta}$. Thus any one of the solutions $\boldsymbol{\beta}^0 = \mathbf{G}\mathbf{X}'\mathbf{y}$ is the basis of whatever unbiased estimation is available and the symbol ^ is reserved for this purpose. Furthermore, since $\boldsymbol{\beta}^0 = \hat{\boldsymbol{\beta}}$ when $\mathbf{X}'\mathbf{X}$ is nonsingular and $\mathbf{G} = (\mathbf{X}'\mathbf{X})^{-1}$, everything that follows in terms of $\boldsymbol{\beta}^0$ and singular $\mathbf{X}'\mathbf{X}$ also holds true for $\hat{\boldsymbol{\beta}}$ and nonsingular $\mathbf{X}'\mathbf{X}$.

CONSEQUENCES OF A SOLUTION

Properties of Generalized Inverses of X′X

Numerous properties of generalized inverses* are available in Rao [8], Searle [10], Seber [17], and elsewhere. Those that are especially pertinent to linear model theory for $\mathbf{G}$ of $\mathbf{X}'\mathbf{X}\mathbf{G}\mathbf{X}'\mathbf{X} = \mathbf{X}'\mathbf{X}$ are: (a) $\mathbf{G}$ and $\mathbf{G}'$ are both generalized inverses of $\mathbf{X}'\mathbf{X}$; (b) $\mathbf{X}\mathbf{G}\mathbf{X}'\mathbf{X} = \mathbf{X}$; (c) $\mathbf{X}\mathbf{G}\mathbf{X}'$ is symmetric and invariant with respect to $\mathbf{G}$; (d) $\mathbf{X}\mathbf{G}\mathbf{X}' = \mathbf{X}\mathbf{X}^+$, for $\mathbf{X}^+$ being the Moore–Penrose inverse of $\mathbf{X}$; and (e) even though $\mathbf{G}$ need not be symmetric, $\dot{\mathbf{G}} = \mathbf{G}\mathbf{X}'\mathbf{X}\mathbf{G}'$ is symmetric and is a reflexive generalized inverse of $\mathbf{X}'\mathbf{X}$, meaning that $\dot{\mathbf{G}}\mathbf{X}'\mathbf{X}\dot{\mathbf{G}} = \dot{\mathbf{G}}$, as well as $\mathbf{X}'\mathbf{X}\dot{\mathbf{G}}\mathbf{X}'\mathbf{X} = \mathbf{X}'\mathbf{X}$.

First and Second Moments

The solution vector $\boldsymbol{\beta}^0$ has expected value $E(\boldsymbol{\beta}^0) = \mathbf{G}\mathbf{X}'\mathbf{X}\boldsymbol{\beta}$, which is not invariant with respect to $\mathbf{G}$. Furthermore, $E(\boldsymbol{\beta}^0) \neq \boldsymbol{\beta}$, so that $\boldsymbol{\beta}^0$ is not an unbiased estimator of $\boldsymbol{\beta}$. Unbiasedness of $\boldsymbol{\beta}^0$, i.e., $E(\boldsymbol{\beta}^0) = \boldsymbol{\beta}$, occurs only when $\mathbf{X}'\mathbf{X}$ is nonsingular.

The sampling dispersion matrix of $\boldsymbol{\beta}^0$ is $\operatorname{var}(\boldsymbol{\beta}^0) = E(\boldsymbol{\beta}^0 - \mathbf{G}\mathbf{X}'\mathbf{X}\boldsymbol{\beta})(\boldsymbol{\beta}^0 - \mathbf{G}\mathbf{X}'\mathbf{X}\boldsymbol{\beta})' = \mathbf{G}\mathbf{X}'\mathbf{X}\mathbf{G}'\sigma^2 = \dot{\mathbf{G}}\sigma^2$, which is also not invariant with respect to $\mathbf{G}$.

Estimable Functions

The many solutions $\boldsymbol{\beta}^0$, and their differing first and second moments, mean that $\boldsymbol{\beta}^0$ is not a satisfactory estimator of $\boldsymbol{\beta}$. The confusion of multitudinous solutions $\boldsymbol{\beta}^0$ is avoided by concentrating on certain scalar, linear functions $\boldsymbol{\lambda}'\boldsymbol{\beta}$ of the elements of $\boldsymbol{\beta}$. Whenever $\boldsymbol{\lambda}'$ has the form $\boldsymbol{\lambda}' = \mathbf{t}'\mathbf{X}$ for some $\mathbf{t}'$, the function $\boldsymbol{\lambda}'\boldsymbol{\beta}$ is said to be an *estimable func-*

tion. The corresponding function $\boldsymbol{\lambda}'\boldsymbol{\beta}^0$ of elements of $\boldsymbol{\beta}^0$ then has three important properties, as follows. For $\boldsymbol{\lambda} = \mathbf{t}'\mathbf{X}$:

1. $\boldsymbol{\lambda}'\boldsymbol{\beta}^0$ is invariant with respect to $\boldsymbol{\beta}^0$.
2. $\boldsymbol{\lambda}'\boldsymbol{\beta}^0$ has variance $v(\boldsymbol{\lambda}'\boldsymbol{\beta}^0) = \boldsymbol{\lambda}'\mathbf{G}\boldsymbol{\lambda}\sigma^2$ and is invariant with respect to $\mathbf{G}$.
3. $\boldsymbol{\lambda}'\boldsymbol{\beta}^0$ is the best linear unbiased estimator (BLUE) (*see* GAUSS–MARKOV THEOREM) of $\boldsymbol{\lambda}'\boldsymbol{\beta}$. Using $\widehat{\boldsymbol{\lambda}'\boldsymbol{\beta}}$ to denote the BLUE of $\boldsymbol{\lambda}'\boldsymbol{\beta}$ gives

$$\widehat{\boldsymbol{\lambda}'\boldsymbol{\beta}} = \boldsymbol{\lambda}'\boldsymbol{\beta}^0. \tag{19}$$

The role of the symbol $\boldsymbol{\beta}^0$ is evident: when $\boldsymbol{\lambda}'\boldsymbol{\beta}$ is an estimable function of parameters, $\widehat{\boldsymbol{\lambda}'\boldsymbol{\beta}}$ is its BLUE, $\boldsymbol{\beta}^0$ is any solution to the normal equations, and from (19), a calculation formula for $\widehat{\boldsymbol{\lambda}'\boldsymbol{\beta}}$ is $\boldsymbol{\lambda}'\boldsymbol{\beta}^0$.

Important properties of estimable functions include the following:

1. The expected value of any observation is an estimable function.
2. Linear combinations of estimable functions are estimable.
3. $\boldsymbol{\lambda}'\boldsymbol{\beta}$ is estimable if $\boldsymbol{\lambda}'\mathbf{G}\mathbf{X}'\mathbf{X} = \boldsymbol{\lambda}'$ or, equivalently, if $\boldsymbol{\lambda} = \mathbf{G}\mathbf{X}'\mathbf{X}\mathbf{w}$ for any vector $\mathbf{w}$.

Predicted y or Estimated $E(\mathbf{y})$

Corresponding to the vector of observed values $\mathbf{y}$ is the vector of predicted values

$$\hat{\mathbf{y}} = \widehat{E(\mathbf{y})} = \widehat{\mathbf{X}\boldsymbol{\beta}} = \mathbf{X}\boldsymbol{\beta}^0 = \mathbf{X}\mathbf{G}\mathbf{X}'\mathbf{y} = \mathbf{X}\mathbf{X}^+\mathbf{y}, \tag{20}$$

which is invariant with respect to the solution vector $\boldsymbol{\beta}^0$ and to $\mathbf{G}$.

Estimating the Residual Variance

The residual sum of squares is the sum of squares of the deviations of each observed y from its corresponding predicted value in $\hat{\mathbf{y}}$, and has various equivalent forms:

$$\begin{aligned}\text{SSE} &= \sum_i (y_i - \hat{y}_i)^2 = (\mathbf{y} - \hat{\mathbf{y}})'(\mathbf{y} - \hat{\mathbf{y}}) \\ &= \mathbf{y}'(\mathbf{I} - \mathbf{X}\mathbf{X}^+)\mathbf{y} = \mathbf{y}'\mathbf{y} - \boldsymbol{\beta}^{0\prime}\mathbf{X}'\mathbf{y}. \end{aligned} \tag{21}$$

When the model includes the customary $\mathbf{V} = \sigma^2\mathbf{I}$, the expected value of SSE of (21) is $E(\text{SSE}) = (N - r)\sigma^2$, where r is the rank* of $\mathbf{X}$. Hence $\hat{\sigma}^2 = \text{SSE}/(N - r)$ is an unbiased estimator of the residual error variance σ^2.

Partitioning the Sum of Squares

The total sum of squares is $\text{SST} = \mathbf{y}'\mathbf{y} = \sum_{i=1}^N y_i^2$. The reduction in sum of squares due to fitting the model $E(\mathbf{y}) = \mathbf{X}\boldsymbol{\beta}$ is therefore

$$\begin{aligned} R(\boldsymbol{\beta}) &= \text{SST} - \text{SSE} = \boldsymbol{\beta}^{0\prime}\mathbf{X}'\mathbf{y} \\ &= \mathbf{y}'\mathbf{X}\mathbf{G}\mathbf{X}'\mathbf{y} = \mathbf{y}'\mathbf{X}\mathbf{X}^+\mathbf{y}. \end{aligned} \tag{22}$$

The equality $R(\boldsymbol{\beta}) = \boldsymbol{\beta}^{0\prime}\mathbf{X}'\mathbf{y}$ embedded in (22) indicates that the reduction in sum of squares due to fitting $E(\mathbf{y}) = \mathbf{X}\boldsymbol{\beta}$, namely $R(\boldsymbol{\beta})$, can be calculated as $\boldsymbol{\beta}^{0\prime}(\mathbf{X}'\mathbf{y})$, i.e., as the inner product* of the solution vector $\boldsymbol{\beta}^0$ and the vector $\mathbf{X}'\mathbf{y}$ of right-hand sides of the normal equations (17). Then in (21), $R(\boldsymbol{\beta})$ is simply subtracted from $\mathbf{y}'\mathbf{y}$ to get SSE.

The correction for the mean is $N\bar{y}_{\cdot}^2$, also called $R(\mu)$, i.e., $R(\mu) = N\bar{y}_{\cdot}^2$. Using this and $R(\boldsymbol{\beta})$, the total sum of squares can be partitioned as in Table 2. The three partitionings shown there are the basis of traditional analysis of variance* for the general linear model (see the section "Confidence Intervals"). $R(\boldsymbol{\beta})_m$ in Table 2 is the sum of squares due to fitting the model $E(\mathbf{y}) = \mathbf{X}\mathbf{b}$,

Table 2 Partitioning Sums of Squares

	$R(\mu) = N\bar{y}_{\cdot}^2$	
$R(\boldsymbol{\beta}) = \mathbf{y}'\mathbf{X}\mathbf{X}^+\mathbf{y}$	$R(\boldsymbol{\beta})_m = \mathbf{y}'\mathbf{X}\mathbf{X}^+\mathbf{y} - N\bar{y}_{\cdot}^2$	$R(\boldsymbol{\beta})_m = \mathbf{y}'\mathbf{X}\mathbf{X}^+\mathbf{y} - N\bar{y}_{\cdot}^2$
$\text{SSE} = \mathbf{y}'(\mathbf{I} - \mathbf{X}\mathbf{X}^+)\mathbf{y}$	$\text{SSE} = \mathbf{y}'(\mathbf{I} - \mathbf{X}\mathbf{X}^+)\mathbf{y}$	$\text{SSE} = \mathbf{y}'(\mathbf{I} - \mathbf{X}\mathbf{X}^+)\mathbf{y}$
$\text{SST} = \mathbf{y}'\mathbf{y}$	$\text{SST} = \mathbf{y}'\mathbf{y}$	$\text{SST}_m = \mathbf{y}'\mathbf{y} - N\bar{y}_{\cdot}^2$

adjusted (or corrected) for the mean; and SST_m is the total sum of squares corrected for the mean: $SST_m = \mathbf{y}'\mathbf{y} - N\bar{y}_{\cdot}^2 = (y_i - \bar{y}_{\cdot})^2$ for $\bar{y}_{\cdot} = \sum_{i=1}^{N} y_i/N$.

A statistic sometimes used as a measure of concordance of the data with the model is the *coefficient of determination**. It is the square of the product-moment correlation*, R, between observed y's and corresponding predicted y's (elements of $\hat{\mathbf{y}}$); and $R^2 = R(\boldsymbol{\beta})_m/SST_m$.

DISTRIBUTIONAL PROPERTIES

The only distributional properties attributed to the linear model $\mathbf{y} = \mathbf{X}\boldsymbol{\beta} + \mathbf{e}$ in the preceding sections are that $E(\mathbf{e}) = \mathbf{0}$ and $\text{var}(\mathbf{e}) = \mathbf{V}$, with $\mathbf{V}$ usually taken as $\mathbf{V} = \sigma^2\mathbf{I}$. But for confidence intervals* and hypothesis testing*, more precision about the form of the distribution of the elements of $\mathbf{e}$ is needed.

Normality

The assumption usually made is that $\mathbf{e}$ is multivariate normal*, with mean $E(\mathbf{e}) = \mathbf{0}$ and variance–covariance matrix $\text{var}(\mathbf{e}) = \sigma^2\mathbf{I}$. We write this as $\mathbf{e} \sim N(\mathbf{0}, \sigma^2\mathbf{I})$. Then $\mathbf{y}$ and $\boldsymbol{\beta}^0$ also have multivariate normal distributions: $\mathbf{y} \sim N(\mathbf{X}\boldsymbol{\beta}, \sigma^2\mathbf{I})$ and $\boldsymbol{\beta}^0 \sim N(\mathbf{G}\mathbf{X}'\mathbf{X}\boldsymbol{\beta}, \mathbf{G}\mathbf{X}'\mathbf{X}\mathbf{G}'\sigma^2)$. The BLUE of an estimable function $\boldsymbol{\lambda}'\boldsymbol{\beta}$ for $\boldsymbol{\lambda}' = \mathbf{t}'\mathbf{X}$, namely $\widehat{\boldsymbol{\lambda}'\boldsymbol{\beta}} = \boldsymbol{\lambda}'\boldsymbol{\beta}^0$, has a univariate normal distribution: $\widehat{\boldsymbol{\lambda}'\boldsymbol{\beta}} \sim N(\boldsymbol{\lambda}'\boldsymbol{\beta}, \boldsymbol{\lambda}'\mathbf{G}\boldsymbol{\lambda}\sigma^2)$. (See **2** and **3** in the section "Estimable Functions.")

Quadratic Forms*

Sums of squares like those of Table 2 are quadratic forms in the vector $\mathbf{y}$. Distributional and independence properties of sums of squares when $\mathbf{y}$ is normally distributed are determined by the following general theorems. The first concerns the noncentral χ^2-distribution*, denoted as $\chi^{2\prime}(n, \lambda)$, with n degrees of freedom* and noncentrality parameter λ*; and the second and third concern independence.

Theorem 1. When $\mathbf{y} \sim N(\boldsymbol{\mu}, \mathbf{V})$ for nonsingular $\mathbf{V}$, then $\mathbf{y}'\mathbf{A}\mathbf{y}$ has a noncentral χ^2-distribution if and only if $\mathbf{AV}$ is idempotent*; and that distribution is $\chi^{2\prime}(\rho, \frac{1}{2}\boldsymbol{\mu}'\mathbf{A}\boldsymbol{\mu})$ for $\rho =$ rank of $\mathbf{A}$.

Theorem 2. When $\mathbf{y} \sim N(\boldsymbol{\mu}, \mathbf{V})$ for nonsingular $\mathbf{V}$, then $\mathbf{y}'\mathbf{A}\mathbf{y}$ and $\mathbf{B}\mathbf{y}$ are distributed independently if and only if $\mathbf{BVA} = \mathbf{0}$.

Theorem 3. When $\mathbf{y} \sim N(\boldsymbol{\mu}, \mathbf{V})$ for nonsingular $\mathbf{V}$, then $\mathbf{y}'\mathbf{A}\mathbf{y}$ and $\mathbf{y}'\mathbf{B}\mathbf{y}$ are distributed independently if and only if $\mathbf{AVB} = \mathbf{0}$ (or, equivalently, $\mathbf{BVA} = \mathbf{0}$).

Proof, discussion, and corollaries of these theorems appear, for example, in Rao [8], Searle [10] and Seber [17]. There is also a comprehensive theorem adapting Theorems 1 and 3 to a sum of quadratic forms, being a broad extension of Cochran's theorem*. These theorems have counterparts for singular $\mathbf{V}$ and for nonhomogeneous quadratic forms such as $\mathbf{y}'\mathbf{A}\mathbf{y} + \mathbf{h}'\mathbf{y} + m$.

If $\mathbf{y} \sim N(\mathbf{X}\boldsymbol{\beta}, \sigma^2\mathbf{I})$ as in the preceding section, then Theorem 1 applied to the sums of squares in Table 2 gives

$$R(\boldsymbol{\beta})/\sigma^2 \sim \chi^{2\prime}\left(r, \boldsymbol{\beta}'\mathbf{X}'\mathbf{X}\boldsymbol{\beta}/(2\sigma^2)\right)$$

and

$$SSE/\sigma^2 \sim \chi^2_{N-r}, \tag{23}$$

where $r =$ rank of $\mathbf{X}$, and where χ^2_n represents a central χ^2-distribution* with n degrees of freedom. Also,

$$\begin{aligned} R(\mu)/\sigma^2 &\sim \chi^{2\prime}\left[1, (\mathbf{1}'\mathbf{X}\boldsymbol{\beta})^2/(2\sigma^2)\right], \\ R(\boldsymbol{\beta})_m/\sigma^2 &\sim \chi^{2\prime}\left[r-1, \theta\right], \\ \theta &= \boldsymbol{\beta}'\mathbf{X}'(\mathbf{I} - \bar{\mathbf{J}})\mathbf{X}\boldsymbol{\beta}/(2\sigma^2), \end{aligned} \tag{24}$$

where $\mathbf{1}'$ is a row vector of the 1's and $\bar{\mathbf{J}}$ is a square matrix with $1/N$ for every element. Theorem 2 establishes independence of $\boldsymbol{\beta}^0$ and $\hat{\sigma}^2 = SSE/(N-r)$; Theorem 3 provides independence of $R(\boldsymbol{\beta})$ and SSE of (23), and the pairwise independence (and hence in this case the mutual independence) of $R(\mu)$, $R(\boldsymbol{\beta})_m$, and SSE.

F-Statistics

The properties just stated result in the following noncentral F-distributions*, denoted as $F'(n_1, n_2, \lambda)$, where n_1, n_2 are the degrees of freedom of the numerator and denominator, respectively, and λ is the noncentrality parameter. With $\hat{\sigma}^2 = \text{SSE}/(N-r)$,

$$R(\boldsymbol{\beta})/r\hat{\sigma}^2 \sim F'[r, N-r, \boldsymbol{\beta}'\mathbf{X}'\mathbf{X}\boldsymbol{\beta}/2\sigma^2],$$

$$R(\mu)/\hat{\sigma}^2 \sim F'\left[1, N-r, (\mathbf{1}'\mathbf{X}\boldsymbol{\beta})^2/2\sigma^2\right],$$

$$R(\boldsymbol{\beta})_m/(r-1)\hat{\sigma}^2 \sim F'[r-1, N-r, \theta].$$

Under an appropriate hypothesis H, each of these noncentral F'-distributions becomes a central F-distribution*; so the corresponding statistic may be tested against tabulated values of the central $F(n_1, n_2)$-distribution with (n_1, n_2) degrees of freedom. Hence

$$F(\boldsymbol{\beta}) = R(\boldsymbol{\beta})/(r\hat{\sigma}^2) \qquad (25)$$

compared to $F(r, n-r)$ tests $H: \mathbf{X}\boldsymbol{\beta} = \mathbf{0}$. Similarly,

$$F(\mu) = R(\mu)/\hat{\sigma}^2 \qquad (26)$$

compared to $F(1, N-r)$ tests $H: \mathbf{1}'\mathbf{X}\boldsymbol{\beta} = \mathbf{0}$. This hypothesis is equivalent to $H: E(\bar{y}) = 0$, where $\bar{y}$ is the average data value; its F-statistic is the square of a t-statistic* because $F(\mu) = N\bar{y}_.^2/\hat{\sigma}^2 = [\bar{y}/(\hat{\sigma}/N)]^2$. Finally,

$$F(\boldsymbol{\beta}_m) = R(\boldsymbol{\beta})_m/\left((r-1)\hat{\sigma}^2\right) \qquad (27)$$

compared to $F(r-1, N-r)$ tests $H: \mathbf{X}\boldsymbol{\beta} - [E(\bar{y})]\mathbf{1} = \mathbf{0}$.

Further uses of F-statistics are considered in the sections "F-Statistic" and "Partitioning a Linear Model."

Analysis of Variance

Calculation of the preceding F-statistics is summarized in Table 3. Its first section corresponds to the first column of Table 2 and shows the $F(\boldsymbol{\beta})$ of (25). The second section corresponds to the second column of Table 2 and shows the F-statistics of (26) and (27). A third section could also be set out, of the terms in the third column of Table 2 and showing just $F(\boldsymbol{\beta})_m$ of (27).

Confidence Intervals

The BLUE of the estimable function $\boldsymbol{\lambda}'\boldsymbol{\beta}$ is $\widehat{\boldsymbol{\lambda}'\boldsymbol{\beta}} = \boldsymbol{\lambda}'\boldsymbol{\beta}^0 \sim N(\boldsymbol{\lambda}'\boldsymbol{\beta}, \boldsymbol{\lambda}'\mathbf{G}\boldsymbol{\lambda}\sigma^2)$, as in the sections "Estimable Functions" and "Normality." Hence, $(\boldsymbol{\lambda}'\boldsymbol{\beta}^0 - \boldsymbol{\lambda}'\boldsymbol{\beta})/\sqrt{\boldsymbol{\lambda}'\mathbf{G}\boldsymbol{\lambda}\hat{\sigma}^2} \sim t_{N-r}$, where t_{N-r} represents the t-distribution* with $N-r$ degrees of freedom. This provides a mechanism for establishing confidence intervals* for $\boldsymbol{\lambda}'\boldsymbol{\beta}$.

Table 3 Analyses of Variance

Source of Variation	d.f.	Sum of Squares	Mean Square	F-Statistic
		Fitting the model $E(\mathbf{y}) = \mathbf{X}\boldsymbol{\beta}$		
Model	r	$R(\boldsymbol{\beta}) = \boldsymbol{\beta}^{0\prime}\mathbf{X}'\mathbf{y}$	$M(\boldsymbol{\beta}) = R(\boldsymbol{\beta})/r$	$F(\boldsymbol{\beta}) = M(\boldsymbol{\beta})/\hat{\sigma}^2$
Residual	$N-r$	$\text{SSE} = \mathbf{y}'\mathbf{y} - \boldsymbol{\beta}^{0\prime}\mathbf{X}'\mathbf{y}$	$\hat{\sigma}^2 = \text{SSE}/(N-r)$	
Total	N	$\text{SST} = \mathbf{y}'\mathbf{y}$		
		Fitting the model $E(\mathbf{y}) = \mathbf{X}\boldsymbol{\beta}$ and adjusting for the mean		
Mean	1	$R(\mu) = N\bar{y}_.^2$	$M(\mu) = R(\mu)/1$	$F(\mu) = M(\mu)/\hat{\sigma}^2$
Model, a.f.m.[a]	$r-1$	$R(\boldsymbol{\beta})_m = \boldsymbol{\beta}^{0\prime}\mathbf{X}'\mathbf{y} - N\bar{y}_.^2$	$M(\boldsymbol{\beta})_m = R(\boldsymbol{\beta})_m/(r-1)$	$F(\boldsymbol{\beta})_m = M(\boldsymbol{\beta})_m/\hat{\sigma}^2$
Residual	$N-r$	$\text{SSE} = \mathbf{y}'\mathbf{y} - \boldsymbol{\beta}^{0\prime}\mathbf{X}'\mathbf{y}$	$\hat{\sigma}^2 = \text{SSE}/(N-r)$	
Total	N	$\text{SST} = \mathbf{y}'\mathbf{y}$		

[a] a.f.m., adjusted for the mean.

THE GENERAL LINEAR HYPOTHESIS

Formulation

A hypothesis about linear functions of parameters is a *general linear hypothesis*. Its usual test statistic is based on BLUEs of estimable functions. The general linear hypothesis in the linear model of (5) and (6) is stated as

$$H : \mathbf{K}'\boldsymbol{\beta} = \mathbf{m}, \tag{28}$$

where $\mathbf{m}$ is a vector of desired constants, in many cases null.

All linear hypotheses can be expressed in the form (28). Some can be tested and some cannot; those that can must satisfy two conditions.

(a) Each element of $\mathbf{K}'\boldsymbol{\beta}$ must be estimable; i.e.,

$$\mathbf{K}' = \mathbf{T}'\mathbf{X} \text{ for some } \mathbf{T}'. \tag{29}$$

(b) Elements of $\mathbf{K}'\boldsymbol{\beta}$ cannot be linear combinations of each other; i.e.,

$$r \geqslant \text{rank of } \mathbf{K}' = \text{number of rows in } \mathbf{K}' \quad \text{(full row rank)}. \tag{30}$$

These conditions are important both in practice and in theory; they are satisfied by a wide range of linear functions $\mathbf{K}'\boldsymbol{\beta}$. (a) ensures that $\mathbf{K}'\boldsymbol{\beta}^0$, which estimates $\mathbf{K}'\boldsymbol{\beta}$, is invariant with respect to $\boldsymbol{\beta}^0$; and (b) precludes formulating a hypothesis that includes redundant and/or inconsistent statements. For example, statements $\tau_2 - \tau_3 = 1$ and $\tau_2 - \tau_3 = 5$ when used in combination with $H : [\tau_1 - \tau_2 = 3$, and $\tau_1 - \tau_3 = 4]$ are, respectively, redundant and inconsistent: they do not satisfy (b).

F-Statistic

The F-statistic for testing $H : \mathbf{K}'\boldsymbol{\beta} = \mathbf{m}$, where $\mathbf{K}'$ satisfies conditions (a) and (b), is

$$F(H) = Q/s\hat{\sigma}^2 \quad \text{with}$$

$$Q = (\mathbf{K}'\boldsymbol{\beta}^0 - \mathbf{m})'(\mathbf{K}'\mathbf{G}\mathbf{K})^{-1}(\mathbf{K}'\boldsymbol{\beta}^0 - \mathbf{m}), \tag{31}$$

where s is the (row) rank of $\mathbf{K}'$ and $\hat{\sigma}^2$ is the estimated residual variance $\text{SSE}/(N - r)$. The statistic $F(H)$ is distributed as $F'(s, N - r, \lambda_F)$, noncentral F*, with

$$\lambda_F = (\mathbf{K}'\boldsymbol{\beta} - \mathbf{m})'(\mathbf{K}'\mathbf{G}\mathbf{K})^{-1}(\mathbf{K}'\boldsymbol{\beta} - \mathbf{m})/(2\sigma^2).$$

(Some writers define λ_F without the factor 2 in the denominator. *Ed.*) Hence, under the hypothesis, the statistic $F(H)$ is distributed as $F(s, N - r)$, a central F, and can be compared with tabulated values thereof as a basis for inference.

The rationale for this test is the likelihood ratio test*. Under normality assumptions $\mathbf{y} = \mathbf{X}\boldsymbol{\beta} + \mathbf{e} \sim N(\mathbf{X}\boldsymbol{\beta}, \sigma^2\mathbf{I})$, and the test statistic $F(H)$ is a single-valued monotonic function of the likelihood ratio*. Furthermore, $Q/(s\sigma^2)$ has a noncentral χ^2-distribution, independent of the central χ^2 density of $\hat{\sigma}^2/\sigma^2$, as may be shown using Theorems 1 and 3 from the section "Quadratic Forms."

When $H : \mathbf{K}'\boldsymbol{\beta} = \mathbf{m}$ is not rejected, one can estimate $\boldsymbol{\beta}$ under that hypothesis. A solution vector is

$$\boldsymbol{\beta}_H^0 = \boldsymbol{\beta}^0 - \mathbf{G}\mathbf{K}(\mathbf{K}'\mathbf{G}\mathbf{K})^{-1}(\mathbf{K}'\boldsymbol{\beta}^0 - \mathbf{m}). \tag{32}$$

The associated residual sum of squares is $\text{SSE}_H = (\mathbf{y} - \mathbf{X}\boldsymbol{\beta}_H^0)'(\mathbf{y} - \mathbf{X}\boldsymbol{\beta}_H^0) = \text{SSE} + Q$; see (31). Hence $Q = \text{SSE}_H - \text{SSE}$.

Nontestable Hypotheses

Condition (b) in the section "Formulation," that $\mathbf{K}'$ have full row rank, is both necessary and sufficient for $(\mathbf{K}'\mathbf{G}\mathbf{K})^{-1}$ to exist in Q of (31); and condition (a), that $\mathbf{K}' = \mathbf{T}'\mathbf{X}$, ensures that $H : \mathbf{K}'\boldsymbol{\beta} = \mathbf{m}$ is testable. But it is not a necessary condition for the existence of $(\mathbf{K}'\mathbf{G}\mathbf{K})^{-1}$. It is therefore possible to have a $\mathbf{K}'$ of full row rank with $\mathbf{K}' \neq \mathbf{T}'\mathbf{X}$, such that Q and the F-statistic of (31) can be calculated. The hypothesis then being tested by $F = Q/(s\hat{\sigma}^2)$ is $H : \mathbf{K}'\mathbf{G}\mathbf{X}'\mathbf{X}\boldsymbol{\beta} = \mathbf{m}$ and not $H : \mathbf{K}'\boldsymbol{\beta} = \mathbf{m}$, because $\mathbf{K}' \neq \mathbf{T}'\mathbf{X}$ implies that $\mathbf{K}'\mathbf{B}$ is not estimable, so condition (a) is not satisfied. If $\mathbf{K}'\boldsymbol{\beta}$ can be partitioned so that $\mathbf{K}_1'\boldsymbol{\beta}$ is estimable and $\mathbf{K}_2'\boldsymbol{\beta}$ is not, the hypothesis being tested is

$$H : \begin{bmatrix} \mathbf{K}_1'\boldsymbol{\beta} \\ \mathbf{K}_2'\mathbf{G}\mathbf{X}'\mathbf{X}\boldsymbol{\beta} \end{bmatrix} = \begin{bmatrix} \mathbf{m}_1 \\ \mathbf{m}_2 \end{bmatrix}.$$

(Details are shown in Searle [10, Sec. 5.5d], corrected in Searle [12].)

Independent and Orthogonal Contrasts

Suppose that $H:\mathbf{K}'\boldsymbol{\beta}=\mathbf{0}$ is expressed as H_i: $\mathbf{k}_i'\boldsymbol{\beta}=\mathbf{0}$ for $i=1,\ldots,s$, where $\mathbf{k}_i'$ is the ith row of $\mathbf{K}'$. For testing the composite hypothesis* $H:\mathbf{K}'\boldsymbol{\beta}=\mathbf{0}$, the numerator of (31) becomes $Q=\boldsymbol{\beta}^{0\prime}\mathbf{K}(\mathbf{K}'\mathbf{G}\mathbf{K})^{-1}\mathbf{K}'\boldsymbol{\beta}^0$. For testing the simple hypothesis* $H_i:\mathbf{k}_i'\boldsymbol{\beta}=\mathbf{0}$, the numerator is $q_i=(\mathbf{k}_i'\boldsymbol{\beta}^0)^2/\mathbf{k}_i'\mathbf{G}\mathbf{k}_i$. The following situation then pertains.

Theorem 4. The q_i are distributed independently if and only if $\mathbf{k}_i'\mathbf{G}\mathbf{k}_j=\mathbf{0}$ for $i\neq j=1,\ldots,s$.

Definition. $\mathbf{k}_i'\boldsymbol{\beta}$ and $\mathbf{k}_j'\boldsymbol{\beta}$ are said to be *orthogonal* when $\mathbf{k}_i'\mathbf{G}\mathbf{k}_j=0$.

Remark. With balanced data, $\mathbf{G}$ is often a scalar matrix*, or else it and the $\mathbf{k}_i$'s of interest partition so that $\mathbf{k}_i'\mathbf{G}\mathbf{k}_j=0$ reduces to $\mathbf{h}_i'\mathbf{G}\mathbf{h}_j=0$, where $\mathbf{h}_i$ is a subvector of $\mathbf{k}_i$ and $\mathbf{k}_i'\boldsymbol{\beta}\equiv\mathbf{h}_i'\boldsymbol{\beta}_1$, $\boldsymbol{\beta}_1$ being a subvector of $\boldsymbol{\beta}$ (*see* GENERAL BALANCE and the section "Balanced and Unbalanced Data").

Theorem 5. If, when $\mathbf{K}$ has the same rank as $\mathbf{X}$ (i.e., $s=r$), the q_i are independent, then $Q=\sum_{i=1}^{i=r}q_i$.

Remark. This theorem gives a sufficient condition for the numerator sums of squares of the simple hypotheses $H_i:\mathbf{k}_i'\boldsymbol{\beta}=\mathbf{0}$ to add up to that for the composite hypothesis H: $\mathbf{K}'\boldsymbol{\beta}=\mathbf{0}$. But it is not a necessary condition; i.e., with $r=s$, having $\sum q_i=Q$ does not necessarily imply that the q_i are independent.

Partitioning a Linear Model

It is sometimes convenient to write the model equation $E(\mathbf{y})=\mathbf{X}\boldsymbol{\beta}$ as

$$E(\mathbf{y})=\mathbf{X}_1\boldsymbol{\beta}_1+\mathbf{X}_2\boldsymbol{\beta}_2=[\mathbf{X}_1\,|\,\mathbf{X}_2]\begin{bmatrix}\boldsymbol{\beta}_1\\ \boldsymbol{\beta}_2\end{bmatrix}, \tag{33}$$

where $\boldsymbol{\beta}$ is partitioned into subvectors $\boldsymbol{\beta}_1$ and $\boldsymbol{\beta}_2$. Similar to (22) we then have $R(\boldsymbol{\beta}_1)=\mathbf{y}'\mathbf{X}_1\mathbf{X}_1^+\mathbf{y}$, and $R(\boldsymbol{\beta}_2)=\mathbf{y}'\mathbf{X}_2\mathbf{X}_2^+\mathbf{y}$ from fitting $E(\mathbf{y})=\mathbf{X}_1\boldsymbol{\beta}_1$ and $E(\mathbf{y})=\mathbf{X}_2\boldsymbol{\beta}_2$, respectively. Furthermore, from fitting (33) there is $R(\boldsymbol{\beta})\equiv R(\boldsymbol{\beta}_1,\boldsymbol{\beta}_2)=\mathbf{y}'\mathbf{X}\mathbf{X}^+\mathbf{y}$ for $\mathbf{X}=[\mathbf{X}_1\,|\,\mathbf{X}_2]$ of (33), so that one can also consider

$$R(\boldsymbol{\beta}_2\,|\,\boldsymbol{\beta}_1)=R(\boldsymbol{\beta}_1,\boldsymbol{\beta}_2)-R(\boldsymbol{\beta}_1), \tag{34}$$

$$=\mathbf{y}'\mathbf{M}_1\mathbf{X}_2'(\mathbf{X}_2'\mathbf{M}_1\mathbf{X}_2)^-\mathbf{X}_2'\mathbf{M}_1\mathbf{y}, \tag{35}$$

where $\mathbf{M}_1=\mathbf{I}-\mathbf{X}_1\mathbf{X}_1^+$ is symmetric and idempotent.

The definition in (34) makes it clear that $R(\boldsymbol{\beta}_2\,|\,\boldsymbol{\beta}_1)$ is that portion of the sum of squares due to fitting $E(\mathbf{y})=\mathbf{X}_1\boldsymbol{\beta}_1+\mathbf{X}_2\boldsymbol{\beta}_2$ which exceeds that due to fitting $E(\mathbf{y})=\mathbf{X}_1\boldsymbol{\beta}_1$. This is referred to variously as the sum of squares due to $\boldsymbol{\beta}_2$ over and above $\boldsymbol{\beta}_1$, or due to $\boldsymbol{\beta}_2$ after $\boldsymbol{\beta}_1$, or due to $\boldsymbol{\beta}_2$ adjusted for $\boldsymbol{\beta}_1$.

On extending the partitioning in (33) to three terms, $E(\mathbf{y})=\mathbf{X}_1\boldsymbol{\beta}_1+\mathbf{X}_2\boldsymbol{\beta}_2+\mathbf{X}_3\boldsymbol{\beta}_3$, Table 4 summarizes the hypotheses tested by the four possible kinds of $R(\cdot)$ terms:

1. $R(\boldsymbol{\beta}_1)$ when $\boldsymbol{\beta}_1$ is the only term in the model (line 1 of the table)
2. $R(\boldsymbol{\beta}_1)$ when $\boldsymbol{\beta}_1$ is part of the model (lines 2 and 4)
3. $R(\boldsymbol{\beta}_2\,|\,\boldsymbol{\beta}_1)$ when $\boldsymbol{\beta}_1$ and $\boldsymbol{\beta}_2$ constitute the whole model (line 3 and, equivalently, line 6 for $R(\boldsymbol{\beta}_3\,|\,\boldsymbol{\beta}_1,\boldsymbol{\beta}_2)$
4. $R(\boldsymbol{\beta}_2\,|\,\boldsymbol{\beta}_1)$ when $\boldsymbol{\beta}_1$ and $\boldsymbol{\beta}_2$ are only part of the model (line 5)

Example. The no-interaction model for row-by-column data exemplified in (13) is $E(y_{ij})=\mu+\alpha_i+\beta_j$ for $i=1,\ldots,a$, and $j=1,\ldots,b$ with $n_{ij}=0$ or 1 observations in the cell defined by row i and column j. The equivalent model for Table 4 is $E(\mathbf{y})=\mu\mathbf{X}_1+\mathbf{X}_2\boldsymbol{\alpha}+\mathbf{X}_3\boldsymbol{\beta}$, where $\mathbf{X}_1=\mathbf{1}_N$, a vector of N ones, and $\boldsymbol{\alpha}$ and $\boldsymbol{\beta}$ are vectors of the α_i's and β_j's, respectively; $\mathbf{X}_2=\bigoplus_{i=1}^{a}\mathbf{1}_{n_{i.}}$, the *direct sum* of vectors $\mathbf{1}_{n_{i.}}$, and $\mathbf{X}_3$ is the incidence matrix of the observations among the columns in each successive row. For unbalanced data $\mathbf{X}_3$ has no universal form, whereas for balanced data with all $n_{ij}=1$,

Table 4 F-Statistics in Partitionings of a Linear Model, and the Hypotheses They Test

	Model[a] for $E(\mathbf{y})$	F-Statistic[b]	Hypothesis Tested[c]
1	$\mathbf{X}_1\boldsymbol{\beta}_1$	$R(\boldsymbol{\beta}_1)/\hat{\sigma}^2 r_1$	H: $\mathbf{X}_1\boldsymbol{\beta}_1 = \mathbf{0}$
2	$\mathbf{X}_1\boldsymbol{\beta}_1 + \mathbf{X}_2\boldsymbol{\beta}_2$	$R(\boldsymbol{\beta}_1)/\hat{\sigma}^2 r_1$	H: $\mathbf{X}_1\boldsymbol{\beta}_1 + \mathbf{X}_1\mathbf{X}_1^+\mathbf{X}_2\boldsymbol{\beta}_2 = \mathbf{0}$
3		$R(\boldsymbol{\beta}_2 \mid \boldsymbol{\beta}_1)/(\hat{\sigma}^2(r_{12} - r_1))$	H: $\mathbf{M}_1\mathbf{X}_2\boldsymbol{\beta}_2 = \mathbf{0}$
4	$\mathbf{X}_1\boldsymbol{\beta}_1 + \mathbf{X}_2\boldsymbol{\beta}_2 + \mathbf{X}_3\boldsymbol{\beta}_3$	$R(\boldsymbol{\beta}_1)/\hat{\sigma}^2 r_1$	H: $\mathbf{X}_1\boldsymbol{\beta}_1 + \mathbf{X}_1\mathbf{X}_1^+(\mathbf{X}_2\boldsymbol{\beta}_2 + \mathbf{X}_3\boldsymbol{\beta}_3) = \mathbf{0}$
5		$R(\boldsymbol{\beta}_2 \mid \boldsymbol{\beta}_1)/(\hat{\sigma}^2(r_{12} - r_1))$	H: $\mathbf{M}_1\mathbf{X}_2\boldsymbol{\beta}_2 + \mathbf{M}_1\mathbf{X}_2(\mathbf{M}_1\mathbf{X}_2)^+\mathbf{X}_3\boldsymbol{\beta}_3 = \mathbf{0}$
6		$R(\boldsymbol{\beta}_3 \mid \boldsymbol{\beta}_1, \boldsymbol{\beta}_2)/(\hat{\sigma}^2(r_{123} - r_{12}))$	H: $\mathbf{M}_{12}\mathbf{X}_3\boldsymbol{\beta}_3 = \mathbf{0}$

[a]In each model, $\hat{\sigma}^2 = \text{SSE}/(N - r)$, where r is r_1, r_{12}, or r_{123}, the rank of $\mathbf{X}_1$, $(\mathbf{X}_1 \mid \mathbf{X}_2)$, and $(\mathbf{X}_1 \mid \mathbf{X}_2 \mid \mathbf{X}_3)$, respectively.
[b]In each case, degrees of freedom are the coefficient of the denominator $\hat{\sigma}^2$ in F, and $N - r$.
[c]$\mathbf{M}_1 = \mathbf{I} - \mathbf{X}_1\mathbf{X}_1^+$ and $\mathbf{M}_{12} = \mathbf{I} - (\mathbf{X}_1 \mid \mathbf{X}_2)(\mathbf{X}_1 \mid \mathbf{X}_2)^+ = \mathbf{M}_1 - \mathbf{M}_1\mathbf{X}_2(\mathbf{M}_1\mathbf{X}_2)^+$.

$\mathbf{X}_3 = \mathbf{1}_a \otimes \mathbf{I}_b$, the *direct product* of $\mathbf{1}_a$ and $\mathbf{I}_b$. [The direct sum of two matrices $\mathbf{A}$ and $\mathbf{B}$ is $\begin{bmatrix}\mathbf{A} & \mathbf{0}\\ \mathbf{0} & \mathbf{B}\end{bmatrix}$; and the direct product is the matrix $\{a_{ij}\mathbf{B}\}$ (see Searle [9]).]

Two different partitionings of sums of squares available for this model are shown in Table 5. In both parts of the table $R(\mu)$, SSE and SST are the same. But the partitioning $R(\boldsymbol{\alpha} \mid \mu)$ and $R(\boldsymbol{\beta} \mid \mu, \boldsymbol{\alpha})$ in (a) is not the same as $R(\boldsymbol{\alpha} \mid \mu, \boldsymbol{\beta})$ and $R(\boldsymbol{\beta} \mid \mu)$ in (b). Although each partitioning adds to the same thing, $R(\boldsymbol{\alpha} \mid \mu) + R(\boldsymbol{\beta} \mid \mu, \boldsymbol{\alpha}) = R(\boldsymbol{\alpha}, \boldsymbol{\beta} \mid \mu) = R(\boldsymbol{\beta} \mid \mu) + R(\boldsymbol{\alpha} \mid \mu, \boldsymbol{\beta})$, the individual terms differ and, very importantly, do not test the same hypotheses. Of particular importance is that in (a), whereas $R(\boldsymbol{\beta} \mid \mu, \boldsymbol{\alpha})$ can be used to test $H: \beta_j$'s all equal, $R(\boldsymbol{\alpha} \mid \mu)$ cannot be used with unbalanced data for testing $H: \alpha_i$'s all equal. This can be done using $R(\boldsymbol{\alpha} \mid \mu, \boldsymbol{\beta})$ of (b). In contrast, for balanced data, $R(\boldsymbol{\alpha} \mid \mu)$ and $R(\boldsymbol{\alpha} \mid \mu, \boldsymbol{\beta})$ are identical, equal to the familiar row sum of squares $b\sum(\bar{y}_{i\cdot} - \bar{y}_{\cdot\cdot})^2$. Details of establish-

Table 5 Analyses of Variance of Row-by-Column Data for a No-Interaction Model

Source of Variation	Degrees of Freedom[a]	Sum of Squares[b]	Hypothesis Tested Using Sum of Squares as Q in $F = Q/(f\hat{\sigma}^2)$
	(a) Fitting rows before columns		
Mean	1	$R(\mu)$	H: $E(\bar{y}) = 0$
Rows	$a - 1$	$R(\boldsymbol{\alpha} \mid \mu)$	H: all $(\alpha_i + \sum_j n_{ij}\beta_j/n_{i\cdot})$ equal
Columns after rows	$b - 1$	$R(\boldsymbol{\beta} \mid \mu, \boldsymbol{\alpha})$	H: β_j's all equal
Residual	N'	SSE	
Total	N	$\text{SST} = \mathbf{y}'\mathbf{y}$	
	(b) Fitting columns before rows		
Mean	1	$R(\mu)$	H: $E(\bar{y}) = 0$
Columns	$b - 1$	$R(\boldsymbol{\beta} \mid \mu)$	H: all $(\beta_j + \sum_i n_{ij}\alpha_i/n_{\cdot j})$ equal
Rows after columns	$a - 1$	$R(\boldsymbol{\alpha} \mid \mu, \boldsymbol{\beta})$	H: α_i's all equal
Residual	N'	SSE	
Total	N	$\text{SST} = \mathbf{y}'\mathbf{y}$	

[a]$N' = N - a - b + 1$.
[b]$\text{SSE} = \mathbf{y}'\mathbf{y} - R(\mu, \boldsymbol{\alpha}, \boldsymbol{\beta}) = N'\hat{\sigma}^2$.

ing Table 5, including computing formulae, derivation of F-statistics and of hypotheses tested, and also of the interaction model, together with extensions, appear in Searle [10, Chaps. 7, 8].

Tables 4 and 5 carry no implication that hypothesis testing is universally appropriate in linear model analysis. Even when it is appropriate, the hypotheses derived in this way may not always be useful for inferences. Tables 4 and 5 show the hypotheses tested by using certain traditional sums of squares as numerators of F-statistics. The purpose of the tables is simply to show what kinds of hypotheses they are, not necessarily to promote their use; indeed, if anything, it is to promote the nonuse of some of them, such as that associated with $R(\boldsymbol{\alpha} \mid \mu)$, for example. This is not useful because first, it is not a hypothesis about simple and universally interesting functions of the parameters, and in particular it is not the hypothesis that α_i's are equal. Second, the hypothesis is based on the data, because it is expressed in terms of the n_{ij}'s. Third, it is derived in a back-to-front manner as far as the logic of hypothesis testing is concerned. We have calculated the sum of squares $R(\boldsymbol{\alpha} \mid \mu)$ without any stated reason for doing so, insofar as hypothesis testing is concerned, and have then derived the associated hypothesis, namely $H: \alpha_i + \sum_j n_{ij}\beta_j / n_{i\cdot}$ equal for all i. The correct logic is to formulate a hypothesis $\mathbf{K}'\boldsymbol{\beta} = \mathbf{m}$ of interest, collect data in a manner that makes $\mathbf{K}'\boldsymbol{\beta}$ estimable, and then use (31) to test it.

SPECIAL CASES

Regression*

When $\mathbf{X}'\mathbf{X}$ is nonsingular (i.e., has full rank), all the preceding results hold true with $(\mathbf{X}'\mathbf{X})^- = \mathbf{G}$ replaced by $(\mathbf{X}'\mathbf{X})^{-1}$. Consequently, the solution vector $\boldsymbol{\beta}^0$ is then $\hat{\boldsymbol{\beta}}$, the BLUE of $\boldsymbol{\beta}$; every linear function of elements of $\boldsymbol{\beta}$ is estimable and any linear hypothesis $\mathbf{K}'\boldsymbol{\beta} = \mathbf{m}$ is testable as long as $\mathbf{K}'$ satisfies just condition (b) from the section "Formulation." With $E(y_i) = \beta_0 + \beta_1 x_{i1} + \cdots + \beta_k x_{ik}$, four special cases are of interest.

For testing

$H: \boldsymbol{\beta} = \mathbf{0}$, use $F = \hat{\boldsymbol{\beta}}'\mathbf{X}'\mathbf{X}\hat{\boldsymbol{\beta}}/((k+1)\hat{\sigma}^2)$;

$H: \boldsymbol{\beta} = \boldsymbol{\beta}_0$, use $F = (\hat{\boldsymbol{\beta}} - \boldsymbol{\beta}_0)'\mathbf{X}'\mathbf{X}(\hat{\boldsymbol{\beta}} - \boldsymbol{\beta}_0)/((k+1)\hat{\sigma}^2)$;

$H: \boldsymbol{\lambda}'\boldsymbol{\beta} = \mathbf{m}$, use $F = (\boldsymbol{\lambda}'\hat{\boldsymbol{\beta}} - \mathbf{m})^2/(\boldsymbol{\lambda}'(\mathbf{X}'\mathbf{X})^-\boldsymbol{\lambda}\hat{\sigma}^2;)$

$H: \boldsymbol{\beta}_q = \mathbf{0}$, where $\boldsymbol{\beta}_q$ is a subvector of order q of $\boldsymbol{\beta}$ and $\mathbf{T}_{qq}$ is the corresponding submatrix of order q of $(\mathbf{X}'\mathbf{X})^{-1}$, use $F = \hat{\boldsymbol{\beta}}_q'\mathbf{T}_{qq}^{-1}\hat{\boldsymbol{\beta}}_q/(q\hat{\sigma}^2)$ (see Searle et al. [16]).

Analysis of Covariance

Analysis of covariance* involves a model equation $E(\mathbf{y}) = \mathbf{X}\boldsymbol{\beta}$, some columns of $\mathbf{X}$ being dummy variables with values 0 and 1 (see the sections "Dummy Variables," "Experimental Design Models," and "Survey Data") and other columns being observed values of covariates*, as in regression. It is useful to specify these two kinds of columns separately and write the model equation as

$$\mathbf{y} = [\mathbf{X} \mid \mathbf{Z}]\begin{bmatrix}\boldsymbol{\alpha} \\ \boldsymbol{\beta}\end{bmatrix} + \mathbf{e} = \mathbf{X}\boldsymbol{\alpha} + \mathbf{Z}\boldsymbol{\beta} + \mathbf{e}, \quad (36)$$

where $\boldsymbol{\alpha}$ is the vector of effects for factors and interactions (including an overall mean, μ), $\mathbf{X}$ is the corresponding incidence matrix, $\mathbf{Z}$ is the matrix having columns that are values of the covariates, and $\boldsymbol{\beta}$ is the vector of "slopes" or coefficients corresponding to those covariates.

Estimators, confidence intervals, and hypothesis tests based on (36) are all derived from applying to it the principles described in the sections "Estimation," "Consequences of a Solution," "Distributional Properties," and "The General Linear Hypothesis." This results, for example, in the solution vectors

$$\boldsymbol{\alpha}^0 = (\mathbf{X}'\mathbf{X})^-\mathbf{X}'(\mathbf{y} - \mathbf{Z}\hat{\boldsymbol{\beta}}) \quad \text{and} \quad \hat{\boldsymbol{\beta}} = (\mathbf{R}'\mathbf{R})^{-1}\mathbf{R}\mathbf{y}, \quad (37)$$

where $\mathbf{R} = \mathbf{M}\mathbf{Z} = (\mathbf{I} - \mathbf{X}\mathbf{X}^+)\mathbf{Z}$ is a matrix with jth column $\mathbf{z}_j - \hat{\mathbf{z}}_j$, where $\mathbf{z}_j$ is the jth column of $\mathbf{Z}$ and $\hat{\mathbf{z}}_j = \mathbf{X}\mathbf{X}^+\mathbf{z}_j$, a "predicted" value of $\mathbf{z}_j$ in the manner of (20). (For further details, see Searle [10, 14].)

Restricted Models

Linear models are sometimes defined with restrictions on elements of $\boldsymbol{\beta}$, usually of the form $\mathbf{P}'\boldsymbol{\beta} = \boldsymbol{\delta}$. If $\mathbf{P}'\boldsymbol{\beta}$ is estimable, a solution vector for the model $E(\mathbf{y}) = \mathbf{X}\boldsymbol{\beta}$ restricted by $\mathbf{P}'\boldsymbol{\beta} = \boldsymbol{\delta}$ is

$$\boldsymbol{\beta}_r^0 = \boldsymbol{\beta}^0 - \mathbf{GP}(\mathbf{P}'\mathbf{GP})^{-1}(\mathbf{P}'\boldsymbol{\beta}^0 - \boldsymbol{\delta}),$$

akin to (32). The estimated residual variance is $\hat{\sigma}^2 = \text{SSE}_r/(N - r_{\mathbf{X}} + r_{\mathbf{P}})$, where $\text{SSE}_r = \text{SSE} + Q$, Q corresponding to testing $H: \mathbf{P}'\boldsymbol{\beta} = \boldsymbol{\delta}$; $r_{\mathbf{X}}$ and $r_{\mathbf{P}}$ are the ranks of $\mathbf{X}$ and $\mathbf{P}$, respectively. When $\mathbf{P}'\boldsymbol{\beta}$ is nonestimable the solution vector in the restricted model is $\boldsymbol{\beta}_r^0 = \boldsymbol{\beta}^0 + (\mathbf{I} - \mathbf{GX}'\mathbf{X})\mathbf{z}$, where $\mathbf{z}$ satisfies $\mathbf{P}'(\mathbf{I} - \mathbf{GX}'\mathbf{X})\mathbf{z} = \mathbf{P}'\boldsymbol{\beta}^0 - \boldsymbol{\delta}$. The estimator $\hat{\sigma}^2 = \text{SSE}/(N - r_{\mathbf{X}})$ is the same as in the unrestricted model.

The important consequence of restrictions is in their effect on estimable functions and testable hypotheses. For example, in the randomized complete blocks case of (11), the F-statistic based on $R(\mu)$ in Table 5 tests $H: E(\bar{y}) = 0$, which is $H: \mu + \frac{1}{4}(\tau_1 + \tau_2 + \tau_3 + \tau_4) + \frac{1}{3}(\rho_1 + \rho_2 + \rho_3) = 0$. If a model with restrictions $\sum \tau_i = 0$ and $\sum \rho_j = 0$ (called the $\sum$-restrictions) is used, this hypothesis reduces to $H: \mu = 0$.

A general discussion of restricted models is available in Searle [10, Sec. 5.6, with examples in Secs. 6.2g, 6.4g, 7.1h, and 7.2g]. Further examples are to be found in Hocking and Speed [6], Hocking et al. [7], Speed and Hocking [19], and Speed et al. [20]. The effect of restrictions on computational algorithms is considered in Searle et al. [16].

The Cell Means Model

The need for estimable functions in models using dummy (0, 1) variables arises from the implicit overparameterization. For example, in (10) there are four parameters but only three cell means available for estimating them. Using restricted models to negate this overparameterization works well for balanced data, with $\sum$-restrictions, but does not work well with unbalanced data, as illustrated by Searle et al. [16]. A viable alternative, which avoids both overparameterization and restrictions, is the model $E(\mathbf{y}) = \mathbf{X}\boldsymbol{\mu}$, where $\mathbf{X}$ is the direct sum of vectors $\mathbf{1}_t$, of order n_t, and where μ_t is the tth element of $\boldsymbol{\mu}$ and is the population mean of the tth subclass, containing data having n_t observations.

This is the *cell means model*, or μ_{ij}-model, corresponding to its use with row-by-column data in the form $E(y_{ijk}) = \mu_{ij}$. In this model the BLUE of μ_t for a cell containing data is $\hat{\mu}_t = \bar{y}_t$, the mean of the observations in the tth cell, and its sampling variance is σ^2/n_t. Each linear combination $\sum \lambda_t \mu_t$ of such μ_t's is estimable, with BLUE $\sum \lambda_t \bar{y}_t$ and sampling variance $\sigma^2(\sum \lambda_t^2/n_t)$; and all linear hypotheses about these μ_t's are testable.

The nature of the cell means model implies that interactions between all main effects implicit in the model are also part of the model. If desired, some or all interactions can be excluded by imposing restrictions on $\boldsymbol{\mu}$, say $\mathbf{P}'\boldsymbol{\mu} = \boldsymbol{\delta}$, and then using

$$\hat{\boldsymbol{\mu}}_r = \bar{\mathbf{y}} - \mathbf{DP}(\mathbf{P}'\mathbf{DP})^{-1}(\mathbf{P}'\bar{\mathbf{y}} - \boldsymbol{\delta}),$$

adapted from $\boldsymbol{\beta}_r^0$ of the preceding section, or its equivalent form,

$$\hat{\boldsymbol{\mu}}_r = \left[\mathbf{D} - \mathbf{DP}(\mathbf{P}'\mathbf{DP})^{-1}\mathbf{P}'\mathbf{D}\right]\mathbf{y} \qquad \text{when} \quad \boldsymbol{\delta} \equiv \mathbf{0},$$

where $\mathbf{D}$ is the diagonal matrix of elements $1/n_t$.

Cell means models are espoused by Hocking and Speed [6], Searle [10, Secs. 7.5, 8.1f], Speed [18], and Urquhart and Weeks [21]. They are extremely useful for data in which there are empty cells. Some easily overcome difficulties of estimation when restrictions are used are discussed by Searle and Speed [15].

The Multivariate Linear Model

The general linear model $E(\mathbf{y}) = \mathbf{X}\boldsymbol{\beta}$ with $\text{var}(\mathbf{y}) = \mathbf{V}$ ($= \sigma^2\mathbf{I}$ customarily) is for univariate data. Representing multivariate data by a matrix $\mathbf{Y}$ of order $N \times p$, for N observations on each of p variables, the corresponding multivariate linear model is

$$E(\mathbf{Y}) = \mathbf{X}\mathscr{B} \qquad \operatorname{cov}(\mathbf{y}_i, \mathbf{y}_j') = \sigma_{ij}\mathbf{I}_N, \quad (38)$$

where $\mathscr{B}$ is a matrix of parameters and, for $i, j = 1, \ldots, p$, the vectors $\mathbf{y}_i$ and $\mathbf{y}_j$ are columns of $\mathbf{Y}$.

This model can be put in a univariate framework, so as to permit using the general linear model for univariate data. This is achieved by defining $\mathbf{\Sigma} = \{\sigma_{ij}\}$ for $i, j = 1, \ldots, p$ and using the vec operator $\operatorname{vec}\mathbf{Y} = [\mathbf{y}_1', \mathbf{y}_2', \ldots, \mathbf{y}_p']'$, a vector of order $Np \times 1$ consisting of columns of $\mathbf{Y}$ stacked one beneath the other (see Henderson and Searle [4, 5]). Then (38) can be written in univariate form as

$$E(\operatorname{vec}\mathbf{Y}) = (\mathbf{I} \otimes \mathbf{X})\operatorname{vec}\mathscr{B},$$

$$\operatorname{var}(\operatorname{vec}\mathbf{Y}) = \mathbf{\Sigma} \otimes \mathbf{I}_N. \quad (39)$$

Application of the GLS equations (14) to (39) yields $(\mathbf{X}'\mathbf{X})\mathscr{B}^0 = \mathbf{X}'\mathbf{Y}$ and $\mathscr{B}^0 = (\mathbf{X}'\mathbf{X})^-\mathbf{X}'\mathbf{Y}$. The latter is symbolically akin to (18); however, in the univariate form $\operatorname{vec}\mathscr{B}^0 = [\mathbf{I} \otimes (\mathbf{X}'\mathbf{X})^{-1}\mathbf{X}]\operatorname{vec}\mathbf{Y}$, general linear model theory for univariate data is available for the multivariate linear model (38); see Searle [13] and Eaton [2].

Note

1. Paper No. BU-387 in the Biometrics Unit, Cornell University, Ithaca, N.Y.

References

[1] Bishop, Y. M. M., Fienberg, S. E., and Holland, P. (1975). *Discrete Multivariate Analysis*. MIT Press, Cambridge, Mass. (A very complete treatment of its subject.)

[2] Eaton, M. L. (1970). *Ann. Math. Statist.*, **41**, 528–538. (Estimation for multivariate linear models.)

[3] Harville, D. A. (1977). *J. Amer. Statist. Ass.*, **72**, 320–340. (A review of maximum likelihood estimation of variance components.)

[4] Henderson, H. V. and Searle, S. R. (1979). *Canad. J. Statist.*, **7**, 65–81. (Review of the vec operator.)

[5] Henderson, H. V. and Searle, S. R. (1981). *Linear and Multilinear Algebra*, **9**, 271–288. (Extensions of the vec operator.)

[6] Hocking, R. R. and Speed, F. M. (1975). *J. Amer. Statist. Ass.*, **70**, 706–712. (The cell means model.)

[7] Hocking, R. R., Hackney, O. P., and Speed, F. M. (1978). In *Contributions to Survey Sampling and Applied Statistics—Papers in Honor of H. O. Hartley*, H. A. David, ed. Academic Press, New York, pp. 133–150. (The cell means model.)

[8] Rao, C. R. (1973). *Linear Statistical Inference and Its Applications*. Wiley, New York. (Mathematically thorough, rigorous, and extensive.)

[9] Searle, S. R. (1982). *Matrix Algebra Useful for Statistics*. Wiley, New York. (Basic matrix algebra.)

[10] Searle, S. R. (1971). *Linear Models*. Wiley, New York. (Emphasis on unbalanced data, with detailed chapters on variance components.)

[11] Searle, S. R. (1971). *Biometrics*, **27**, 1–76. (A review of variance component estimation.)

[12] Searle, S. R. (1974). *Paper No. BU-501-M*, Biometrics Unit, Cornell University, Ithaca, N.Y. (Corrections to testing nontestable hypotheses.)

[13] Searle, S. R. (1978). In *Contributions to Survey Sampling and Applied Statistics—Papers in Honor of H. O. Hartley*, H. A. David, ed. Academic Press, New York, pp. 181–189. (The multivariate linear model.)

[14] Searle, S. R. (1979). *Commun. Statist. A*, **8**, 799–818. (Alternative covariance models.)

[15] Searle, S. R. and Speed, F. M. (1981). *Paper No. BU-730-M*, Biometrics Unit, Cornell University, Ithaca, N.Y. (Estimability in cell means models.)

[16] Searle, S. R., Speed, F. M., and Henderson, H. V. (1981). *Amer. Statist.*, **35**, 16–33. (Computational methods.)

[17] Seber, G. A. F. (1977). *Linear Regression Analysis*. Wiley, New York. (A broad compendium of mathematical results, largely concerned with regression.)

[18] Speed, F. M. (1969). *NASA Tech. Memo. TX 58030*, National Aeronautics and Space Administration, Houston, Tex. (Cell means models.)

[19] Speed, F. M. and Hocking, R. R. (1976). *Amer. Statist.*, **30**, 30–33. (Cell means models.)

[20] Speed, F. M., Hocking, R. R., and Hackney, O. P. (1978). *J. Amer. Statist. Ass.*, **73**, 105–112. (Cell means models.)

[21] Urquhart, N. S. and Weeks, D. L. (1978). *Biometrics*, **34**, 696–705. (Data analysis with the cell means model.)

[22] Zyskind, G. and Martin, F. B. (1969). *SIAM J. Appl. Math.*, **17**, 1190–1202. (General estimation theory.)

Bibliography

See the following works, as well as the references just cited, for more information on the topic of the general linear model.

Arnold, S. F. (1981). *The Theory of Linear Models and Multivariate Analysis*. Wiley, New York. (Theoretical.)

Graybill, F. A. (1976). *Theory and Application of the Linear Model*. Duxbury Press, North Scituate, Mass. (An extensive and very theorematic presentation.)

Lewis, T. O. and Odell, P. L. (1971). *Estimation in Linear Models*. Prentice-Hall, Engelwood Cliffs, N.J. (A succinct mathematical account, with special cases.)

Mendenhall, W. (1968). *Introduction to Linear Models and the Design and Analysis of Experiments*. Wadsworth, Belmont, Calif. (Elementary presentation.)

Neter, J. and Wasserman, W. (1974). *Applied Linear Statistical Models*. Richard D. Irwin, Homewood, Ill. (Intermediate treatment, mainly of balanced data.)

Neter, J., Wasserman, W. and Kutner, M. H. (1982). *Applied Linear Regression Models*. Richard D. Irwin, Homewood, Ill. (Revises part of the preceding reference.)

Seber, G. A. F. (1965). *The Linear Hypothesis: A General Theory*. Charles Griffin, London. (A concise monograph of the mathematical development.)

(ANALYSIS OF COVARIANCE
ANALYSIS OF VARIANCE
BALANCE IN EXPERIMENTAL DESIGN
ESTIMABILITY
FIXED-, RANDOM-, AND MIXED-EFFECTS MODELS
GENERAL BALANCE
GENERALIZED LINEAR MODELS
LEAST SQUARES
REGRESSION
VARIANCE COMPONENTS)

S. R. SEARLE

GENERATING FUNCTIONS

Two kinds of generating function are of particular interest in statistical theory: *moment generating functions* and *probability generating functions*.

PROBABILITY GENERATING FUNCTIONS

Let X be a random variable whose possible values are restricted to the nonnegative integers, and write $p_n = \Pr\{X = n\}$, for $n = 0, 1, 2, \ldots$. The probability generating function (PGF) is then defined as

$$\Pi_X(z) = \sum_{n=0}^{\infty} p_n z^n, \tag{1}$$

where z is a dummy variable (usually complex) which must be restricted to a region in which the power series in (1) is convergent. The power series always converges if $|z| \leqslant 1$, as is easily seen, and sometimes it converges for a more extensive range of z-values.

For example, if we have a *geometric distribution** with $p_n = qp^n$ for $n = 0, 1, 2, \ldots$, with $p + q = 1$, $p \geqslant 0$, $q \geqslant 0$, then $\Pi_X(z) = q/(1 - pz)$; in this case the series is absolutely convergent for all $|z| < p^{-1}$.

As a second example, for n a fixed positive integer, consider the *binomial distribution**, for which $p_r = \binom{n}{r} p^r q^{n-r}$, for $r = 0, 1, \ldots, n$, with p and q as before; here $\Pi_X(z) = (q + pz)^n$ and since this is a polynomial, the "series" converges absolutely for all finite values of z.

As a third example we cite the *Poisson distribution**, for which $p_n = e^{-\mu}\mu^n/n!$, for $n = 0, 1, 2, \ldots$, where $\mu > 0$ is some constant. In this case $\Pi_X(z) = e^{\mu(z-1)}$, which also gives a series absolutely convergent for all finite values of z.

These three examples bring out one obvious and attractive feature of the PGF: it summarizes a probability distribution on the integers succinctly as a function of the dummy variable z.

An alternative definition of $\Pi_X(z)$ is

$$\Pi_X(z) = \mathscr{E} z^X. \tag{2}$$

This is plainly equivalent to definition (1) but helps bring out other useful attributes of the PGF. For instance, it is possible to justify repeated differentiation of (2), with respect to z, in the region $|z| < 1$. As an illustration, if we twice differentiate (2) we get

$$\Pi_X''(z) = \mathscr{E} X(X-1) z^{X-2}.$$

On letting z increase to 1, we find that

$$\Pi_X''(1) = \mathscr{E} X(X-1) = \sum_{n=2}^{\infty} n(n-1) p_n.$$

Thus $\Pi_X''(1)$ gives the *second factorial moment** of X, which may or may not be finite. [Strictly speaking, we should take the limit of $\Pi_X''(z)$ as z increases to unity from below, but for simplicity we shall write $\Pi_X''(1)$.] Plainly the procedure we have adopted is quite general, and for any positive integer k

we will have

$$\Pi_X^{(k)}(1) = \sum_{n=k}^{\infty} n(n-1)(n-2)\cdots \times (n-k+1)p_n = \mu_{(k)}, \quad (3)$$

say, the kth factorial moment*. Further, it follows from Taylor's theorem that when, for some $\rho > 1$, the PGF $\Pi_X(z)$ is analytic in the circle $|z| = \rho$, then all the factorial moments are finite and (for $|z-1| < \rho - 1$):

$$\Pi_X(z) = \sum_{n=0}^{\infty} \frac{(z-1)^n}{n!} \mu_{[n]}.$$

This alternative expansion of $\Pi_X(z)$ can sometimes be helpful.

Equation (2) also brings out another important property. Suppose that X and Y are independent random variables with PGFs $\Pi_X(z)$ and $\Pi_Y(z)$, respectively. Then, by the independence,

$$\mathscr{E} z^{X+Y} = (\mathscr{E} z^X)(\mathscr{E} z^Y),$$

so that

$$\Pi_{X+Y}(z) = \Pi_X(z)\Pi_Y(z). \quad (4)$$

Thus the PGF provides a very useful tool for studying the sum of independent random variables.

Also by Taylor's theorem, the PGF uniquely determines the corresponding probability distribution. (It is plain that a given probability distribution on 0, 1, 2, . . . , uniquely determines the PGF.) This (1-1) correspondence between PGFs and probability distributions is crucial, and is what makes the PGF supremely useful, especially in conjunction with (4).

One can thus, for simple instances, very easily show that the sum of two independent binomial variables is itself a binomial variable, that a similar result holds for Poisson variables, and that the sum of several independent geometric variables gives a negative binomial* variable.

The PGF is also useful in studying the effect of *randomizing a parameter*. By way of an easy example, suppose that X has a Poisson distribution with parameter $Y > 0$, where Y has a PDF: $g(y) = e^{-\mu y} y^{\nu - 1}/\Gamma(\nu)$ on $y > 0$, with constants $\nu > 0$ and $\mu > 0$. Then, conditional upon Y, the PGF of X is simply $\exp\{Y(z-1)\}$. Thus (for $|z| < 1$) the *unconditional* PGF of X is

$$\int_0^{\infty} e^{y(z-1)} g(y)\,dy = \int_0^{\infty} e^{y(z-1)-\mu y}\left(\frac{\mu^{\nu} y^{\nu-1}}{\Gamma(\nu)}\right)dy,$$

$$= \left(\frac{\mu}{1+\mu-z}\right)^{\nu}.$$

The latter function is a negative binomial PGF, associated with probabilities $q = \mu(1+\mu)$ and $p = 1/(1+\mu)$. Thus we have painlessly established that if a random variable X has a Poisson distribution whose parameter has a gamma* type of distribution, then X has, unconditionally, a negative binomial distribution.

A further use for the PGF is in studying certain kinds of stochastic processes, especially "birth and death processes*." As a simple example, let $t > 0$ represent time, let $X(t)$ be the number of particles in a closed population at time t, and suppose that $X(t)$ is incremented by one whenever a random event, with intensity $\lambda > 0$ occurs (we are thinking of random events as in a Poisson process*). If one writes $p_n(t)$ for the probability that $X(t) = n$, then the following infinite system of differential equations is easily established ($n = 0, 1, \ldots,$):

$$\frac{d}{dt} p_n(t) = \lambda p_{n-1}(t) - \lambda p_n(t), \quad (5)$$

where $p_{-1}(t) \equiv 0$. If $\Pi(z,t)$ be the PGF of $X(t)$, then the infinity of equations (5) can be summarized as a single differential equation:

$$\frac{\partial}{\partial t}\Pi(z,t) = \lambda(z-1)\Pi(z,t). \quad (6)$$

This is easily solved; we obtain:

$$\Pi(z,t) = e^{\lambda t(z-1)}\Pi(z,0).$$

Thus it appears that $X(t)$ is the sum of two independent random variables: the initial population size $X(0)$ and a Poisson variable with parameter λt. This almost trivially easy illustration should give some inkling of the usefulness of the PGF as an instrument with which certain complicated stochastic process* problems may be successfully tackled.

Suppose now that we have an infinite sequence of PGFs, which we shall denote, with an abuse of notation: $\Pi_1(z), \Pi_2(z), \ldots$, where $\Pi_n(z) = \sum_{m=0}^{\infty} p_m^{(n)} z^m$. Suppose that at an infinity of points $z_1, z_2, \ldots$, all lying in the open disk $|z| < 1$, the sequence $\{\Pi_n(z)\}$ is convergent. That is, for every $j = 1, 2, 3, \ldots$, as $n \to \infty$, $\Pi_n(z_j)$ tends to a limit. Then there is a function $\Pi_\infty(z)$ which has a Taylor expansion $\sum p_m^{(\infty)} z^m$, say, in $|z| < 1$, and $\Pi_n(z) \to \Pi_\infty(z)$, as $n \to \infty$ for every $|z| \leqslant 1$. Moreover, for every fixed integer m, $p_m^{(n)} \to p_m^{(\infty)}$, as $n \to \infty$. Thus $p_m^{(\infty)} \geqslant 0$ for every m and $\Pi_\infty(z)$ has all the attributes of a PGF except for the fact that, maybe, $\sum_{m=0}^{\infty} p_m^{(\infty)} < 1$. In the latter case, $\Pi_\infty(z)$ is said to be the PGF of a *defective* probability distribution. Whether $\Pi_\infty(z)$ refers to the defective case or not can be discovered by calculating $\Pi_\infty(1)$. The important principle, that convergence of PGFs implies (and is implied by) convergence of the corresponding probability distributions, is referred to as the *continuity theorem* for PGFs. We cannot pursue this matter any further here and must beg the reader's forgiveness for interpolating a somewhat heuristic comment; it is possible to show that if $\Pi_X(z)$ and $\Pi_Y(z)$ do not differ by very much anywhere in the circle $|z| \leqslant 1$, then probabilities like $P\{X \in A\}$ and $P\{Y \in A\}$, where A is any set of nonnegative integers, will not differ by much either.

By way of illustration, suppose that X is the sum of the number of successes in n (very large) independent and identical trials, for each of which the probability p of a success is very small, say $p \simeq \tilde{\omega}/n$, $\tilde{\omega}$ being of "moderate" size. Then $\Pi_X(z) = (q + pz)^n$, the familiar PGF of the binomial distribution. But it is clear that

$$\Pi_X(z) \simeq \left[1 + (\tilde{\omega}/n)(z-1)\right]^n \simeq e^{\tilde{\omega}(z-1)},$$

the PGF of the Poisson distribution. Thus is suggested to us the result, which can be rigorously established: that X has in such a case a distribution close to the Poisson; if n is quite large, we shall not be far wrong, in a practical situation, in replacing the "correct," but awkward, binomial distribution with the more tractable Poisson.

It has been intimated that a PGF $\Pi_X(z)$ will be an analytic function of complex z in the open disk $|z| < 1$. The series (1) always converges and defines $\Pi_X(z)$ in the closed disk $|z| \leqslant 1$. In various delicate research investigations the analytic nature of $\Pi_X(z)$ on the circumference $|z| = 1$ is critical. Unfortunately, no general statements are possible; a wide range of possibilities exist. One result is generally true, however, and occasionally useful: If ρ be the radius of convergence of the power series in (1), then the point $z = \rho$ is *always* a singularity of $\Pi_X(z)$. Of course, there may well be additional singularities on the circumference $|z| = \rho$.

We close this section with the following. For $n = 0, 1, 2, \ldots$, let $r_n = p_{n+1} + p_{n+2} + \cdots$ be the "tail" probabilities. Then it can be shown that

$$\sum_{n=0}^{\infty} r_n = \mu_1 = \mathscr{E}X,$$

which we assume to be finite. Thus the sequence $\{r_n/\mu_1\}$, $n = 0, 1, \ldots$ constitutes another probability distribution on the integers; its PGF is given by the formula

$$\frac{1 - \Pi_X(z)}{\mu_1(1-z)},$$

which arises in various contexts.

MOMENT GENERATING FUNCTIONS

In this section the random variable X is no longer restricted in any way and we write $F(x)$ for its distribution function (df). We consider a dummy (possibly complex) variable T and define the (ordinary) moment generating function (MGF) by the equation

$$M_X(T) = \mathscr{E}e^{TX} = \int_{-\infty}^{+\infty} e^{Tx} F(dx). \quad (7)$$

Unfortunately, without careful restrictions on the range allowed the dummy variable T, the expectation (or integral) in (7) will not necessarily converge. If we restrict T to be purely imaginary, there is no difficulty of this sort; but if $T = i\theta$, for real θ, then the MGF is more usually called the *characteristic function**, discussed in a separate article.

In the present treatment we therefore assume the existence of real numbers $\sigma_1 > 0$

and $\sigma_2 > 0$ such that

$$\int_0^{\infty} e^{\sigma_1 x} F(dx) < \infty \quad \text{and} \quad \int_{-\infty}^{0} e^{\sigma_2 |x|} F(dx) < \infty. \tag{8}$$

In this case (7) defines the MGF for all complex values of $T = u + iv$, say, such that $-\sigma_2 < u < \sigma_1$; in other words, $M_x(T)$ is defined for all T in a vertical *strip* in the complex plane. Indeed, given the existence of the positive numbers σ_1 and σ_2, it will follow [from (8)] that $M_x(T)$ is an analytic function of the complex T at all interior points of the strip. Furthermore, by a result similar to one already described for the PGF, it can be shown that if σ_1^* be the supremum of all possible values for σ_1, and σ_2^* similarly for all possible values of σ_2, then $T = \sigma_1^*$ and $T = -\sigma_2^*$ are singularities of $M_X(T)$. (Should σ_1^* or σ_2^* be infinite, this remark needs obvious changes.)

For the rest of this section the discussion is formal, assuming that T is always within some suitable strip of analyticity of $M_X(T)$, and paying scant attention to questions of rigor.

In view of (8) all the ordinary moments of X (and, indeed, of $|X|$) are finite, and we shall write

$$\mu_n = \mathscr{E}X^n, \qquad n = 0, 1, 2, \dots .$$

Repeated differentiation of (8) is then possible, and yields, for $n = 1, 2, \dots,$

$$\left[\frac{d^n}{dT^n} M_X(T) \right]_{T=0} = \mu_n . \tag{9}$$

Thus, if we form the Taylor expansion

$$M_X(T) = \sum_{n=0}^{\infty} \frac{T^n}{n!} \left[\frac{d^n}{dT^n} M_X(T) \right]_{T=0}, \tag{10}$$

the moments of X can be obtained from the resulting coefficients.

For example, let X be normal with zero mean and variance σ^2. Then

$$M_X(T) = e^{(1/2)\sigma^2 T^2} = \sum_{n=0}^{\infty} \frac{1}{n!} \left(\frac{\sigma^2 T^2}{2} \right)^n .$$

Hence odd moments of X are all zero (as was to be expected) and for $n = 1, 2, \dots,$

$$\mu_{2n} = \frac{2n!}{n!} \left(\frac{\sigma^2}{2} \right)^n .$$

It is clear from (7) that, for real T such that $-\sigma_2^* < T < \sigma_1^*$,

$$\frac{d^2}{dT^2} M_x(T) \geqslant 0.$$

Thus if $M_X(T)$ be plotted against (real) T in the allowable range, the resulting graph is convex. Evidently, $M_X(0) = 1$, but in many cases, because of the convexity, there will be another real number $\tau \neq 0$ such that $M_X(\tau) = 1$ also. It is easy to see that if $\mu_1 < 0$, then $\tau > 0$, while if $\mu_1 > 0$, then $\tau < 0$. We ignore the case $\mu_1 = 0$. The existence of such a τ is valuable in sequential analysis*, especially in connection with Wald's fundamental identity*.

Let X and Y be *independent* random variables. Then (7) shows that

$$M_{X+Y}(T) = \mathscr{E}e^{T(X+Y)} = (\mathscr{E}e^{TX})(\mathscr{E}e^{TY}) = M_X(T) M_Y(T).$$

Thus the MGF is, like the PGF, a most convenient tool for the study of sums of independent random variables.

It is known that the MGF uniquely determines the df, and conversely. A continuity theorem is also true. If, abusing notation again, $\{M_n(T)\}$, for $n = 1, 2, 3, \dots,$ is a sequence of MGFs, each one analytic in a fixed strip of analyticity $\mathscr{S}$, say, and if a bounded subset of $\mathscr{S}$ contains an infinite sequence of points $T_1, T_2, \dots,$ at each of which, for $j = 1, 2, \dots,$

$$\lim_{n \to \infty} M_n(T_j) \tag{11}$$

exists, then the corresponding df's converge weakly to a limiting df, and, for all T in $\mathscr{S}$, $M_n(T)$ tends to some function $M_\infty(T)$, analytic in $\mathscr{S}$. In other words, if for $n = 1, 2, \dots,$ the MGF $M_n(T)$ corresponds to the df $F_n(x)$, then there is a limit df $F_\infty(x)$, say, such that $F_n(x) \to F_\infty(x)$, as $n \to \infty$, for all points x at which F_∞ is continuous. The converse result also holds, namely: The weak convergence of the sequence $\{F_n\}$ of df's, to a (possibly defective) limit df F_∞, implies the pointwise convergence of the corresponding MGFs $\{M_n\}$ to the MGF $M_\infty(t)$ of the limit df F_∞, in any open region $\mathscr{R}$ in which the $\{M_n\}$ are uniformly bounded. However, as has been hinted, the limit F_∞ may be "defective," corresponding to an "improper" random variable. It can be proved that F_∞ is proper if and only if $M_\infty(0) = 1$. The continuity theorem is most usually used when all the points $\{T_j\}$ are real.

Various other generating functions have occurred from time to time in the literature. These arise for special purposes and enjoy nothing like the importance of the two cases discussed in this article. However, mention should be made of the *factorial moment generating function*. This is the function of the dummy variable z defined by

$$\mathscr{E}\{(1+z)^X\}.$$

This expectation, when it exists for z in a small circle $|z| < \rho$, can be expanded as a Taylor series in powers of z and yields in the successive coefficients the factorial moments of X. Except when X is a nonnegative integer-valued random variable, it is little used.

LITERATURE

A great deal of information on probability generating functions is contained in the two volumes by Feller [1].

Apart from the most elementary aspects of the subject, little is given in textbooks about moment generating functions. The book by Lukács [2] contains some useful information. A valuable but difficult reference containing much profound material applicable to moment generating functions is the book by Paley and Wiener [3].

[*Editor's note.* Seal [4] gives an informative history of the use of probability generating functions in probability theory, tracing the subject back to de Moivre* and including a generalization to continuous distributions by Lagrange.]

References

[1] Feller, W. (1968, 1971). *An Introduction to Probability Theory and Its Applications*. Vols. 1, 2. Wiley, New York.

[2] Lukács, E. (1960). *Characteristic Functions*. Charles Griffin, London.

[3] Paley, R. E. A. C. and Wiener, N. (1934). *Fourier Transforms in the Complex Domain*. American Mathematical Society, Providence, R.I.

[4] Seal, H. L. (1949). *Bull. Ass. Actuaires Suisses*, **49**, 209–228 (reprinted in *Studies in the History of Statistics and Probability*, Vol. 2, M. G. Kendall and R. L. Plackett, eds. Macmillan, New York, 1977, pp. 67–86.)

(CHARACTERISTIC FUNCTIONS
CONVERGENCE OF SEQUENCES OF RANDOM VARIABLES
LIMIT THEOREMS
TAYLOR SERIES)

W. L. Smith

GENERATION OF RANDOM VARIABLES

The ability to synthesize observations from a wide variety of probability distributions is an essential part of statistical practice. Applications range from Monte Carlo* studies of analytically intractable problems in statistical theory (*see* EDITING STATISTICAL DATA) to simulation* studies of complex stochastic systems in operations research*.

Empirical synthesis of random variables (RVs) has a long history, dating back at least as far as "Student" [16], who discovered the sampling distributions of the t-statistic and the correlation coefficient* from simulations based on height and finger measurements of criminals. More recently, digital computers* have become invaluable aids in simulation and Monte Carlo work, so that now almost all synthetic observations or random numbers are generated by a computer when required, instead of being taken from a table.

The process begins by generating observations from the uniform distribution* on the interval (0, 1), which can be done quickly with the computer's arithmetic operations. If the application requires observations from some nonuniform distribution, either continuous or discrete, a variety of techniques are available. Some of these are standard methods, applicable to many distributions; others are clever special-purpose devices, invented for a single distribution. This article describes the basic techniques for generating uniform random numbers, several common techniques for continuous distributions, and some techniques for discrete distributions. It also examines the journal literature, where new methods continue to advance the state of the art.

UNIFORM RANDOM NUMBERS

Almost all algorithms for generating nonuniform random numbers rely on a sequence $\{U_n\}$ of independent observations from the uniform distribution on the interval (0, 1). In practice, the uniform "random numbers" that play the role of $\{U_n\}$ are not precisely random. Instead, *pseudo-random numbers* are obtained by some mechanism that, it is hoped, produces an adequate approximation to randomness. Some early computers used electronic circuits or electromechanical devices; they could deliver close approximations to randomness, but required continuing surveillance to ensure that they were functioning properly. Also, because their output was not reproducible, debugging a program that used them was rather uncertain. To avoid these practical difficulties, John von Neumann and others suggested using the computer's arithmetic operations to produce sequences of numbers that, while being entirely deterministic, had the appearance of randomness. Von Neumann's "middle-square" method, however, turned out to be quite unsatisfactory (see Knuth [7]).

For reasons of reproducibility and convenience, most sources of uniform pseudo-random numbers now use a technique suggested by D. H. Lehmer in 1948. This is the *linear congruential generator*, which constructs a sequence of integers through simple arithmetic, multiplying the previous integer by a constant, adding another constant, and taking the remainder from division by m:

$$X_{n+1} = aX_n + c(\text{mod } m). \qquad (1)$$

The corresponding uniform deviates are $U_n = X_n/m$. The *modulus*, m, is usually a large integer, related to the largest integer that the computer can hold in one word of storage. The *multiplier*, a, is an integer between 1 and m, chosen so that when a term X_n in the sequence is multiplied by a, increased by the *increment* c, and then reduced to its remainder modulo m, the result, X_{n+1}, appears to be unrelated to X_n. The *starting value* or *seed*, X_0, provides a simple way of reproducing any portion of the sequence.

Many generators in common use follow a special case ($c = 0$) of (1) known as the *multiplicative congruential generator*:

$$X_{n+1} = aX_n(\text{mod } m). \qquad (2)$$

Both in this form and in the linear congruential form of (1), the relationship between a and m determines how well the generated sequence approximates a truly random one. The modulus m controls how coarse the discrete sequence $\{U_n\}$ is, compared with a continuous uniform sequence, because the U_n are all integer multiples of $1/m$. It is customary to use a large value of m and to neglect the discreteness of the generated numbers, especially when consecutive numbers from the interval (0, 1) that can be exactly represented in the computer differ by $1/m$.

Another reason for using a fairly large modulus m is that the longest sequence generated by (1) or (2) is cyclic. That is, it eventually starts over at X_0. Because the remainders modulo m are $0, 1, \dots, m-1$, the length of the cycle cannot exceed m. In the linear congruential generator one can always choose a and c so that this maximum is achieved; but the best possible result in the multiplicative congruential generator depends on m, as well as on a suitable choice of a. For example, when m is a power of 2, the maximum cycle length is $m/4$, a must leave a remainder of either 3 or 5 modulo 8, and X_0 must be odd. Or, when m is a prime number, the maximum cycle length is $m - 1$, and a must, in the language of number theory, be a "primitive root" of m. These conditions (and analogous ones for other cases) are not particularly restrictive; when m is large, a great many potential values of a satisfy them. Not surprisingly, such a value of a ensures that the sequence will not run out prematurely, but it need not provide an adequate approximation to randomness*. That involves more complicated criteria and generally requires a search among potential a-values satisfying the basic conditions.

QUALITY OF CONGRUENTIAL GENERATORS

Unavoidably, congruential generators, multiplicative and linear alike, are far from perfect. As Marsaglia demonstrated in 1968 [8], all the n-tuples of consecutive numbers which they are capable of generating must, when viewed as points in n-dimensional space, lie on at most a certain number of parallel hyperplanes. The number of such hyperplanes is relatively small, and a poor choice of a can make it quite small indeed. Figure 1 illustrates this crystalline structure

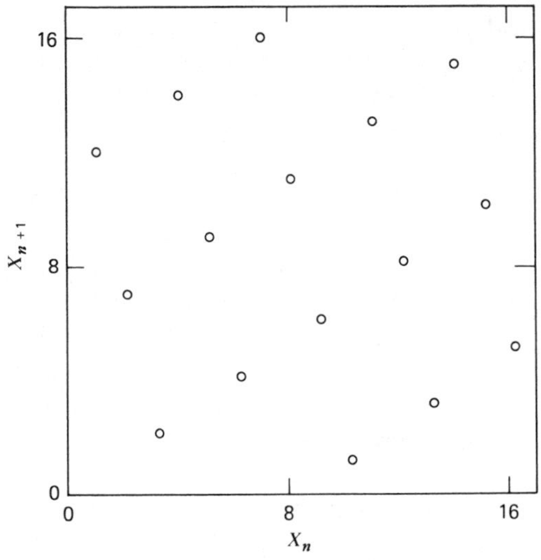

Figure 1 Structure of the pairs generated by the congruential sequence $X_{n+1} = 12X_n (\text{mod } 17)$.

in the two-dimensional output of the simple generator $X_{n+1} = 12X_n (\text{mod } 17)$. Given the limitations of the extremely small modulus, 17, this lattice of pairs is actually quite regular. A paradigm of poor behavior, the generator $X_{n+1} = 65539X_n (\text{mod } 2^{31})$, often known as RANDU [4], produces triples that lie on only 15 parallel planes in 3-space, defined by the relation $X_{n+2} - 6X_{n+1} + 9X_n \equiv 0 \ (\text{mod } 2^{31})$.

Two theoretical tests have been devised to measure the n-dimensional regularity of congruential generators for $n \geqslant 2$. The *lattice test* [9] finds the sides of a fundamental cell in the lattice of n-tuples and uses the length ratio of the longest side to the shortest side as a measure of irregularity. A perfectly cubic lattice would yield a ratio of 1; values below 2 are desirable.

Among the families of parallel hyperplanes in the n-dimensional lattice produced by a congruential generator, the *spectral test** [3, 7] finds the most widely separated family and calculates the distance between adjacent hyperplanes. (The name "spectral test" comes from its derivation by Coveyou and MacPherson [3] in terms of the frequencies of n-dimensional waves.) In the 0–1 scale (to facilitate comparisons among generators with different values of m) denote the maximum interhyperplane distance by d_n. Knuth [7] works with the n-dimensional "accuracy," $\nu_n = 1/d_n$, and rates generators according to

$$\mu_n = \frac{\pi^{n/2}\nu_n^n}{(n/2)!\, m},$$

where $(n/2)! = (n/2)[(n/2) - 1] \cdots (\frac{1}{2})\sqrt{\pi}$ when n is odd. A good generator should have $\mu_n \geqslant 1$ for $2 \leqslant n \leqslant 6$.

The spectral and lattice tests are theoretical in the sense that they work with the full period of the congruential sequence. Over the years, a variety of empirical tests for randomness* have been applied to actual generated sequences of random numbers. Unfortunately, many of those tests did not expose generators that later were found to be poor. The spectral test, however, has proved to be much more effective. Knuth points out [7, p. 89] that "not only do all good generators pass this test, all generators now known to be bad actually *fail* it." Atkinson [2] discusses the use of these theoretical tests and their comparison with certain empirical tests.

For technical reasons the spectral test and lattice test apply directly to only three classes of congruential generators: the linear congruential ones of (1), with $c \neq 0$ and suitable a; the multiplicative congruential ones of (2) with $m = 2^e$, $e \geqslant 4$, and $a \bmod 8 = 5$; and (2) with m prime and a a primitive root of m. This limitation causes no practical difficulty, however, because these three classes contain many generators that are adequate for most applications.

The discussions of methods for generating nonuniform continuous and discrete random variables (RVs) below assume a perfect source of uniform random numbers.

NONUNIFORM CONTINUOUS DISTRIBUTIONS

A few general techniques are available for transforming a sequence of uniform random numbers $\{U_n\}$ into a sequence of random observations from a nonuniform continuous distribution. The chief methods use the inverse cumulative distribution function, rejection, mixtures, and the ratio of uniform deviates. Often, however, an algorithm for a particular distribution employs some special property of that distribution (perhaps in conjunction with one or more of the basic general techniques). The literature abounds with imaginative approaches of this kind.

For the most common distributions, research has produced several competing algorithms. The dominant goal is speed, but some algorithms introduce a trade-off between speed and storage by using bulky tables. Generally, these algorithms are exact; i.e., they generate the desired distribution as accurately as the finite-precision arithmetic of the computer permits. Exact methods are available for all the most commonly used distributions, including normal, exponential, chi-squared, gamma, and Student's t. For example, one need never resort to approximating the normal distribution* by adding uniform deviates and appealing to the central limit theorem*. The algorithm based on $(U_1 + \cdots + U_{12}) - 6$ is both inexact and far slower than the best exact methods.

Inverse-CDF Method

Given a RV X with cumulative distribution function (CDF) F, a straightforward probability argument establishes that transforming X by F yields a uniform RV on $(0, 1)$ [i.e., $F(X) = U$]. Thus, when it is available in closed form, the inverse CDF, F^{-1}, provides a convenient link between a source of uniform random numbers and random observations from the desired nonuniform distribution: $X = F^{-1}(U)$. Two examples are the exponential distribution*, where $F(x) = 1 - e^{-x}$ and $F^{-1}(u) = -\log_e(1 - u)$, applied as $X = -\log_e U$ more conveniently, and the logistic distribution*, which has $F(x) = e^x/(1 + e^x)$ and

$$F^{-1}(u) = \log_e(u) - \log_e(1 - u)$$
$$= \log_e[u/(1 - u)].$$

Use of the inverse CDF is generally slower than other, seemingly more complicated methods; the exponential distribution provides a notable example [1]. When F^{-1} can only be approximated, other methods are preferable. For example, in addition to being inexact, an approximate inverse CDF for the normal distribution is quite slow. Still, the inverse CDF is an important basic technique, sometimes the only approach available.

Rejection Methods

One approach to generating random observations from a desired distribution begins with a simpler or more convenient distribution and, in a sense, cuts it to fit. Most

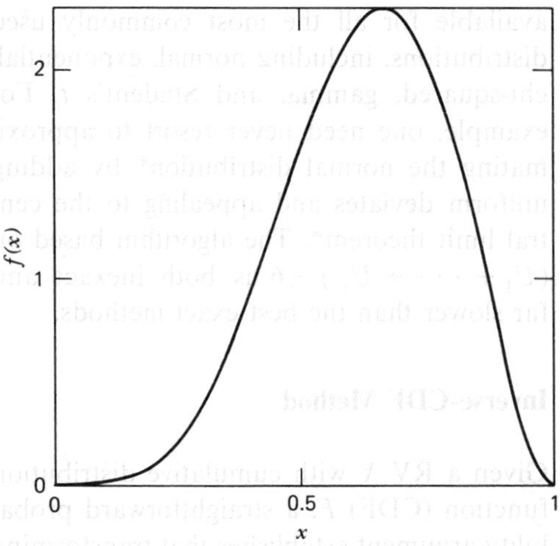

Figure 2 Rejection method with a uniform envelope for a beta density, $f(x) = 105x^4(1-x)^2$. The maximum value of f is $f(\frac{2}{3}) = 2.305$.

commonly, the procedure works with the probability density functions, f for the desired distribution and g for the more convenient one. Both densities are assumed to be nonzero throughout the same finite or infinite interval I. The basic idea is to use g as an envelope for f, choosing a constant, k, such that $kg(x) \geqslant f(x)$ for all $x \in I$. Ordinarily, then, $k = \max\{f(x)/g(x) \mid x \in I\}$. Figure 2 illustrates the relationship between f and kg in a simple situation: f is the density of the beta distribution* with parameters 5 and 3, $f(x) = 105x^4(1-x)^2$, $x \in (0,1)$, and g is uniform on (0, 1); $k = f(2/3) = 2.305$.

The actual sampling process begins by generating a random observation y from g. It next generates a uniform random number u. If $u \leqslant f(y)/kg(y)$, it accepts y as a random observation from f; otherwise, it rejects y and starts over. In this form the rejection method initially generates points uniformly over the area beneath kg and then rejects those not beneath f. Because g is a density, $1/k$ measures the efficiency of the rejection scheme; it is the fraction of observations generated from g that are accepted as observations from f. Thus the example in Fig. 2 is rather inefficient: $1/k = 1/2.305 = 0.434$.

A number of refinements in the rejection method can improve efficiency. For example, an envelope based on two triangular densities would come much closer to f in Fig. 2. And, when f/g is difficult to evaluate in the rejection step, it is often possible to find a fairly simple function h such that $h(x) \leqslant f(x)/g(x)$ throughout I (or a subinterval). Only when $u > h(y)$ is it necessary to calculate $f(y)/g(y)$ in determining acceptance or rejection.

Mixture Methods*

The random process for a mixture first selects a component distribution and then samples an observation from that distribution. If the m components have mixing probabilities $p_1, \ldots, p_m$ and CDFs $F_1, \ldots, F_m$, then the CDF for the mixture distribution is $F = p_1F_1 + \cdots + p_mF_m$. Similarly, the density function is

$$f = p_1 f_1 + \cdots + p_m f_m. \tag{3}$$

As an example, the occurrence of outliers* in data may follow a two-component mixture distribution which yields an observation from the normal distribution $N(\mu, \sigma^2)$ with probability $1-\gamma$ and otherwise $(p_2 = \gamma)$ yields an observation from $N(\mu, 9\sigma^2)$.

To generate random numbers from a desired distribution, one can begin with its density as f in (3) and construct components, f_i, and mixing probabilities, p_i, that yield a fast, exact algorithm [11]. Imaginative application of this approach can involve 30 or more components, few of which bear any resemblance to f.

Ratio of Uniform Deviates

Kinderman and Monahan [5] showed that it is possible to construct efficient algorithms for generating many distributions by sampling a point uniformly from a certain region in the plane and then taking the ratio of the coordinate values of the point. For the density f, the region is

$$C_f = \left\{(u,v) \mid 0 \leqslant u \leqslant \left[f(v/u)\right]^{1/2}\right\}. \tag{4}$$

If (U, V) is distributed uniformly over C_f,

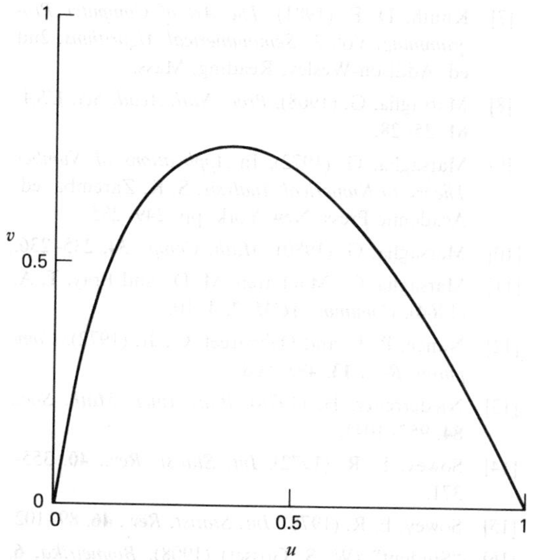

Figure 3 Acceptance region, C_f, for generating exponential random variables by the ratio-of-uniforms method. The curved boundary has the equation $v = -2u \log_e u$.

then $X = V/U$ has the desired density, f. Many densities permit an explicit solution of the equation $u = [f(v/u)]^{1/2}$, which serves as one boundary of C_f in (4). For example, the exponential density, $f(x) = e^{-x}$, $x \geqslant 0$, leads to C_f defined by $0 \leqslant u \leqslant 1$ and $0 \leqslant v \leqslant -2u \log_e u$, as shown in Fig. 3.

Example: Student's t-Distributions*

A selection of the algorithms that have been developed to generate random observations from Student's t with n degrees of freedom serves to emphasize the diversity of approaches. The direct method generates Z from $N(0, 1)$ and, independently, V from chi-squared* with n degrees of freedom and returns $X = Z/\sqrt{V/n}$. Kinderman et al. [6] discuss several algorithms based on rejection and on mixtures. Kinderman and Monahan use the ratio of uniform deviates. For any $n > 2$, Marsaglia [10] begins with RNOR and REXP, normal and exponential random variables generated by very fast procedures, and proceeds as follows: *Generate* $A = \text{RNOR}$, $B = A^2/(n-2)$, *and* $C = \text{REXP}/(\frac{1}{2}n - 1)$ *until* $e^{-B-C} \leqslant 1 - B$; *then exit with* $T = A[(1 - 2/n)(1 - B)]^{-1/2}$. Marsaglia's algorithm exploits a particular property of Student's t-distributions and has efficiency greater than $1 - 1/n$, even if n changes from one use to the next.

NONUNIFORM DISCRETE DISTRIBUTIONS

Algorithms for generating random variables from common discrete distributions, such as the binomial* and the Poisson*, generally rely on particular properties of the distribution. Among the few general techniques for discrete distributions, the main ones are table lookup and the method of two-point distributions. Both apply to random variables X that have a finite number of outcomes, $x_1, \ldots, x_k$, with corresponding probabilities $p_1, \ldots, p_k$ such that $p_1 + \cdots + p_k = 1$.

Table Lookup

The most straightforward approach generates a uniform random number U on the interval $(0, 1)$; successively checks whether it is less than $p_1, p_1 + p_2, \ldots, p_1 + \cdots + p_{k-1}$ until the comparison succeeds; and returns the corresponding x_i as the value of X. Thus, if $U < p_1$, the result is $X = x_1$; and if $U \geqslant p_1 + \cdots + p_{k-1}$, the method yields $X = x_k$.

Two-Point Distributions

At the heart of this method, devised by Walker [17], is the fact that any discrete distribution with k possible outcomes can be represented as an equiprobable mixture of k two-point distributions. Thus a preparatory step determines the alternative outcomes, $a_1, \ldots, a_k$, and the probabilities, $f_1, \ldots, f_k$, for a set of two-point distributions. The generation algorithm uses two random numbers: V discrete uniform on $1, \ldots, k$ and U uniform on $(0, 1)$. If $U < f_V$, it will return $X = x_V$; otherwise, it returns $X = a_V$. The representation in terms of two-point distributions thus eliminates all but one of the comparisons required in table lookup. As an example, for the binomial distribution with $n = 3$ and $p = 0.3$, the method of two-point distributions uses the a_i and f_i shown below.

i	x_i	p_i	a_i	f_i
1	0	0.343	0	1.0
2	1	0.441	0	0.872
3	2	0.189	0	0.756
4	3	0.027	1	0.108

LITERATURE

Generation of random numbers has a long history and a far-flung literature. Bibliographies by Sowey [14, 15] and by Nance and Overstreet [12] provide comprehensive coverage up through 1976. Starting with 1975, most articles can be found through *Current Index to Statistics**. The core journals are *ACM Transactions on Mathematical Software*; *Applied Statistics*; *Computing*; *Journal of the American Statistical Association**; *Journal of the Royal Statistical Society**, *Series A*; and *Mathematics of Computation*.

The best single account of theory and techniques in the generation of random numbers appears in the book by Knuth [7]. The survey article by Niederreiter [13] includes a mathematical discussion of tests on pseudo-random numbers and the theory of linear congruential and related generators.

References

[1] Ahrens, J. H. and Dieter, U. (1972). *Commun. ACM*, **15**, 873–882.

[2] Atkinson, A. C. (1980). *Appl. Statist.*, **29**, 164–171.

[3] Coveyou, R. R. and MacPherson, R. D. (1967). *J. ACM*, **14**, 100–119.

[4] IBM Corporation (1968). *System/360 Scientific Subroutine Package* (360A-CM-03X) *Version III, Programmer's Manual* (H20-0205-3).

[5] Kinderman, A. J. and Monahan, J. F. (1977). *ACM Trans. Math. Software*, **3**, 257–260.

[6] Kinderman, A. J., Monahan, J. F., and Ramage, J. G. (1977). *Math. Comp.*, **31**, 1009–1018.

[7] Knuth, D. E. (1981). *The Art of Computer Programming*, Vol. 2: *Seminumerical Algorithms*, 2nd ed. Addison-Wesley, Reading, Mass.

[8] Marsaglia, G. (1968). *Proc. Natl. Acad. Sci. USA*, **61**, 25–28.

[9] Marsaglia, G. (1972). In *Applications of Number Theory to Numerical Analysis*, S. K. Zaremba, ed. Academic Press, New York, pp. 249–285.

[10] Marsaglia, G. (1980). *Math. Comp.*, **34**, 235–236.

[11] Marsaglia, G., MacLaren, M. D., and Bray, T. A. (1964). *Commun. ACM*, **7**, 4–10.

[12] Nance, R. E. and Overstreet, C., Jr. (1972). *Computing Rev.*, **13**, 495–508.

[13] Niederreiter, H. (1978). *Bull. Amer. Math. Soc.*, **84**, 957–1041.

[14] Sowey, E. R. (1972). *Int. Statist. Rev.*, **40**, 355–371.

[15] Sowey, E. R. (1978). *Int. Statist. Rev.*, **46**, 89–102.

[16] "Student" (W. S. Gosset) (1908). *Biometrika*, **6**, 1–25, 302–310.

[17] Walker, A. J. (1977). *ACM Trans. Math. Software*, **3**, 253–256.

(COMPUTERS IN STATISTICS
RANDOM NUMBERS
SIMULATION
SPECTRAL TEST
UNIFORM DISTRIBUTION)

DAVID C. HOAGLIN

GENERATOR *See* DIFFUSION PROCESSES

GENETICS, STATISTICS IN

Statistical concepts and methods, together with those of stochastic process* theory, arise in genetics in three broad areas. First, standard methods of estimation* and hypothesis testing* are used to estimate genetic parameters and to test hypotheses about these parameters. Second, several central themes of plant and animal breeding theory and of biological evolution are phrased in statistical terms and make substantial use of statistical techniques, and finally stochastic process* theory is used to investigate properties of the evolution of a population subject to random changes in gene frequency. In fact, as noted below, the introduction of a number of important statistical concepts took place in a genetic or evolutionary setting.

We first introduce the genetical terminology used throughout. In most higher organisms chromosomes occur in pairs, one member of each pair being contributed by each parent of the individuals. Each chromosome may be considered as consisting of a sequence of gene loci with the genes at a given locus controlling, in whole or in part, the expression of one or several characters in the individual. In the simplest possible case it may be assumed, first, that a gene locus A controls one character, with no contribution from genes at other loci, and second that only two gene types, denoted A_1 and A_2, are possible at the locus. The latter assumption implies that any individual has genotype A_1A_1, A_1A_2 ($\equiv A_2A_1$) or A_2A_2 for this locus. Assuming that the character is measurable and ignoring any environmental contribution, it may be further assumed that every A_1A_1 individual has measurement m_{11} for this locus, all A_1A_2 individuals have measurement m_{12}, and all A_2A_2 individuals have measurement m_{22}. We make this assumption for the moment.

Under random mating, the frequencies of A_1A_1, A_1A_2, and A_2A_2 in the population are in (Hardy–Weinberg) form x^2, $2x(1-x)$, $(1-x)^2$, where x is the (unknown) population frequency of A_1. In this case the mean μ and variance σ^2 of the measurement in the population are

$$\begin{aligned} \mu &= m_{11}x^2 + 2m_{12}x(1-x) + m_{22}(1-x)^2, \\ \sigma^2 &= m_{11}{}^2x^2 + 2m_{12}{}^2x(1-x) \\ &\quad + m_{22}{}^2(1-x)^2 - \mu^2. \end{aligned} \tag{1}$$

The most elementary application of statistical methods arises in genetic counseling. Here prospective parents, aware of the histories of a genetic disease in both families, seek the probability that a child is free of the disease. The calculations usually involve conditional probabilities, the condition being the known family history of the disease, and use known transmission rules for the gene controlling the disease. As the simplest example, suppose that A_1A_1 individuals have a disease, whereas A_1A_2 and A_2A_2 individuals do not. The first child of two parents not having the disease has the disease and the parents seek the probability that a second child will have the disease. The data show that both parents transmitted an A_1 gene to the first child, and since neither parent has the disease, both must be A_1A_2. Thus the probability required is that each parent again transmits the A_1 gene, namely $\frac{1}{2} \times \frac{1}{2} = \frac{1}{4}$. In this example elementary logical arguments suffice, but in more complex cases detailed conditional probabilities are required.

A further elementary application concerns the estimation of the frequency x of A_1 if a random sample of n reveals n_{11} individuals of type A_1A_1, n_{12} of type A_1A_2, and n_{22} of type A_2A_2. Assuming random mating, the likelihood of this sample, apart from a constant, is $x^{2n_{11}+n_{12}}(1-x)^{n_{12}+2n_{22}}$ and maximum likelihood* methods yield $\hat{x} = (n_{11} + \frac{1}{2}n_{12})/n$. This value can now be used to test, via chi-square*, the assumption that the frequencies are indeed in the ratios $x^2 : 2x(1-x) : (1-x)^2$.

The two examples above do not involve any measurement on the individual. Suppose now that some measurement is taken and that the measurement for an individual of genotype A_iA_j is not fixed at m_{ij} but is rather a random variable from a distribution $f_{ij}(m)$ having mean m_{ij}. The density function of a single observation is then

$$\begin{aligned} f(m) &= x^2f_{11}(m) + 2x(1-x)f_{12}(m) \\ &\quad + (1-x)^2f_{22}(m). \end{aligned}$$

In this density function the frequency x and the parameters of $f_{ij}(m)$ are usually unknown and are estimated by maximum likelihood. Because of the form of $f(x)$, this usually involves computer and numerical methods. Goodness of fit* can again be tested by chi-square. Often several individuals in the sample come from the same family, in which case pedigree analysis (the calculations for which are similar to those of the conditional probabilities used in genetic counseling) can be used to further test the model.

Another class of statistical procedures of great importance in genetics concerns the estimation and testing of linkage values. Complete chromosomes are not necessarily passed on from parent to offspring: several "crossing-over" events may occur whereby parts of one parental chromosome together with the complementary parts of the other are transmitted. If we focus attention on two specific loci, let R be the probability of an odd number of crossings-over between these two loci; then if the loci are closely placed on the same chromosome, $R \simeq 0$, since there is a high probability that no crossover occurs between the loci and the genes from one or other parental chromosome will be transmitted together. On the other hand, for loci far apart on the same chromosome $R \simeq \frac{1}{2}$, and for loci on different chromosomes $R = \frac{1}{2}$. The statistical problems concern the estimation of R (using the genetic makeup of offspring of parents of carefully chosen genotypes) and tests of the hypothesis $R = \frac{1}{2}$. The latter procedure is used eventually to locate a given gene locus on one or other chromosome, and the former procedure leads to a map of the various gene loci along one given chromosome, both undertakings being of great genetical importance. Normally, maximum likelihood methods are used for the estimation process, although some workers prefer to incorporate, by Bayesian methods, prior views into their analysis (*see* BAYESIAN INFERENCE).

A statistical problem arising in particular in genetics concerns ascertainment bias. Suppose that some individuals suffer a certain genetic defect whereas others do not. It is often the practice to accumulate data on this defect by examining the entire family of an individual found with it. Suppose that it is required to estimate the probability p that a child of normal parents who can have defective children is in fact defective, and suppose that the data consist of families with normal parents ascertained through a defective child. The conditional probability that a family of s children has i defective children is proportional to

$$p^i(1-p)^{s-i}/\{1-(1-p)^s\}.$$

For small p, ignoring the "ascertainment bias" denominator term leads to severe bias in the estimation of p.

We turn now to the second broad area of the interaction of statistics and genetics: the influence of statistical concepts on population genetics and evolutionary theory (and, just as important, the influence of genetics in the initial development of important statistical ideas). The group of biometricians centered around Karl Pearson* in the later years of the nineteenth century (*see* ENGLISH SCHOOL OF BIOMETRY) introduced a variety of now standard statistical techniques, including those of correlation* and regression*, which were used to quantify aspects of the biological evolutionary process. The theories used required knowledge of the correlation between relatives for various characters (e.g., height), so empirical estimates of these correlations were obtained and were used in the analysis. Since Mendelism was not rediscovered until 1900, these theories were not cast in a genetical framework and are thus largely of historical interest only. Fortunately, however, the concept of the correlation between relatives has turned out to be essential, in a Mendelian context, for devising efficient plant and animal breeding programs and, to a lesser extent, for analyzing genetic diseases of importance to human beings.

To illustrate this briefly, we consider a measurable characteristic whose value is fixed at m_{ij} for an individual of genotype A_iA_j. Consider first the father–son correlation in this measurement. If the father is A_1A_1 and the population frequency of A_1 is x, the son will be A_1A_1 (with probability x) or A_1A_2 (with probability $1 - x$), depending on whether he gets an A_1 or A_2 gene from his mother. Considering all possible combinations, it is found that the correlation is

$$\tfrac{1}{2}\sigma_A^2/\sigma^2, \qquad (2)$$

where σ^2 is defined in (1) and

$$\sigma_A^2 = 2x(1-x)\{m_{11}x + m_{12}(1-2x) - m_{22}(1-x)\}^2.$$

It is natural to seek an interpretation for σ_A^2. From a statistical point of view, the most natural interpretation arises from a weighted regression of the values m_{11}, m_{12}, and m_{22} on the number (0, 1, or 2) of A_2 genes in the correspondong genotypes, with weights x^2, $2x(1-x)$, $(1-x)^2$. It is found that σ_A^2 is the regression sum of squares and for this reason is called the *additive genetic variance*. From a genetic point of view, we may construct a notional measurement for the A_1 gene in the following way. An A_1 gene combines, with probability x, with another A_1 gene to form a genotype whose measurement is m_{11}; with probability $1-x$ it combines with an A_2 gene to form a genotype whose measurement is m_{12}. This leads to a notional measurement m_1 for the A_1 gene defined by $m_1 = xm_{11} + (1-x)m_{12}$. Similarly, the notional measurement of A_2 is $m_2 = xm_{12} + (1-x)m_{22}$. A random variable taking the value m_1 with probability x and m_2 with probability $(1-x)$ has variance σ_A^2, thus supplying a genetic interpretation for this variance.

Correlation formulae parallel to (2) hold for other relationships, but we do not enter here into their mathematical form.

Suppose in the above that $m_1 = m_2$, so that the notional measurements for A_1 and A_2 are the same. It then follows that $\sigma_A^2 = 0$, and hence the father–son correlation is also zero. Furthermore, the overall mean m of the measurement in question, defined by $m = xm_1 + (1-x)m_2$, is at a turning point with respect to x. (This may be seen through elementary computation of dm/dx.) If this turning point is a maximum, there is no scope for plant or animal breeders to increase m by changing the genetic makeup of their stock (i.e., by varying x). Hence zero father–son correlation implies no capacity for improvement in stock, and more generally an increased correlation implies an increased capacity for improvement. This observation, quantified through the concept of heritability and used in practice in conjunction with other correlations as well as that between father and son, is at the basis of many plant and animal breeding programs, and our interest in it here is that it derives largely from the statistical concept of correlation.

The analyses above assume no environmental component. When environmental effects exist the analysis is far more complex and to make "nature–nurture" allocations, and even to define such an allocation, requires several assumptions on genetic–environment interaction covariance. The conclusions so far reached, especially those of social importance to human beings, are quite controversial.

A second use of statistics of broad significance in the quantitative description of evolution concerns the attempt by Fisher*, in his fundamental theorem of natural selection, to cast in Mendelian terms the essential features of the Darwinian theory, namely that evolution cannot take place without the existence of variation (and that the rate of evolution increases as the degree of variation), and that evolution leads to improvement in the individuals in the population. Suppose that the measurement m_{ij} above is the fitness of the genotype A_iA_j, i.e., the probability that an individual of this genotype will survive from the time of conception to the age of reproduction. (Note that we do not include a fertility component in fitness since this properly relates to a mating pair and the analysis becomes extremely complex.) Then if at the time of conception the frequencies of A_1A_1, A_1A_2, and A_2A_2 are x^2, $2x(1-x)$, and $(1-x)^2$, the mean fitness μ is as given in (1). At maturity the three genotypes will occur with relative frequencies $m_{11}x^2 : 2m_{12}x(1-x) : m_{22}(1-x)^2$, from which the frequency of A_1 at the time of conception of the daughter generation can be found. This leads to a mean fitness μ' in the daughter generation. Ignoring small-order terms, it is found that the difference

$\mu' - \mu$ is equal to the additive variance σ_A^2, and is thus nonnegative. This quantifies the Darwinian concept ($\mu' \geqslant \mu$) of improvement of the population through natural selection and also the view that the rate of improvement is proportional to the current variation (in fitness). Note that it is not the total variance that is relevant, only the additive component, so that the statistical concept of the analysis of variance*, with specific significance to each component in the analysis, has a great evolutionary importance. (It is more than likely that the concept of the analysis of variance occurred to Fisher through his derivation of the evolutionary result just described.)

Finally, we consider applications of stochastic process* theory in the study of genetic evolution. The unit evolutionary process is the replacement of an "inferior" gene in a population by a "superior" gene. In a finite population properties of such a replacement procedure involve stochastic process theory, since random events must be considered; it is possible that by chance the superior gene will die out and not replace the inferior one. The investigation of such processes led to some of the first practical applications of Markov chain, diffusion*, branching process*, and random walk* theory, and further, much of the early theory of these processes was introduced in a genetic context. We consider now some specific cases.

Suppose that a superior gene arises in a population by a single mutational event. It is possible that the individual carrying the mutation leaves no offspring or, if he or she does, that the superior gene is not passed on to any of them. To find the rate of evolution through the incorporation of such genes it is important to find the probability of eventual loss from the population of a favored mutant. This probability can be found from branching process theory, and indeed the first practical application of this theory was Fisher's derivation of this incorporation probability.

Evolution through natural selection cannot occur without genetic variation; it is therefore natural, as a second application of the stochastic theory, to investigate the rate of the random loss of genetic variation in natural populations. We do this here in the case where selection does not occur by comparing the rate of loss with that obtaining under one hypothetical non-Mendelian form of heredity, the "blending" theory, which assumes that the measurements of any character for an offspring tends to be the average of those for its two parents (and which, in pre-Mendelian times, was often taken as a reasonable hereditary mechanism). The simplest possible model of the formation of the genetic constitution of one generation from the preceding one is to suppose that the genes in the daughter generation are obtained by sampling with replacement from the genes of the parent generation. Assuming a population of fixed size N ($2N$ genes at the locus in question) and that there are i A_1 genes in the parent generation, the probability that there are j A_1 genes in the daughter generation is, under this model,

$$p_{ij} = \binom{2N}{j}\left(\frac{i}{2N}\right)^{j}\left(\frac{2N-i}{2N}\right)^{2N-j}. \quad (3)$$

This equation defines a Markov chain with transition matrix $\{p_{ij}\}$ (*see* MARKOV PROCESSES). Clearly, 0 and $2N$ are absorbing states, one or other of which must eventually be entered, leading to permanent loss of genetic variation. We seek a measure of the speed with which this loss occurs. The most immediate such measure is the mean time for loss of either A_1 or A_2, given the initial number of A_1 genes. The form of (3) is such that a simple explicit form for this mean time does not exist: (we mention below an approximation to this mean time). A second approach is to find the eigenvalues* of $\{p_{ij}\}$ and to argue that if the leading nonunit eigenvalue is close to unity, rather slow loss of genetic variation through random effects can be expected. The leading nonunit eigenvalue is found to be $1 - (2N)^{-1}$, so that for large populations, slow random loss of genetic variation is indeed expected. This calculation can be generalized to cover more realistic cases where two sexes exist, where the population size fluctuates, and so on, but conclusions similar to the above are reached except in certain extreme cases. This rate of loss of variation is far slower than that under any "blending" theory of heredity (e.g., under random "blending," the rate of loss of variation is 50% per generation) and shows the importance of the Mendelian scheme to an evolutionary process brought about by variation and natural selection.

In an attempt to reach further properties of the evolutionary process defined by the Markov chain (3), the chain can be approximated by a suitably chosen diffusion process*. For the diffusion process it is found, for example, that if the initial number of A_1 genes is moderate, the mean time until loss of A_1 or A_2 is of order N generations. This is a very long time in all but extremely small populations and agrees with the conclusion on the rate of loss of genetic variation reached by considering the eigenvalues of $\{p_{ij}\}$. A parallel conclusion arises for models more general and realistic than (3).

The model (3) does not allow for the possibility of mutation. When mutation from A_1 to A_2 and from A_2 to A_1 exists, there is no concept of permanent loss of A_1 or A_2; rather, a stationary distribution exists for the frequency of A_1. The form of this distribution is known for models generalizing (3) that do not involve selection and also for models that do. These distributions can be generalized to the case where several types of genes (not just two) can occur at a given locus, and this has led to the beginnings of a statistical theory of testing, given observed gene frequencies, whether or not natural selection is operating at the locus in question.

A vast literature on Markov chains, branching processes, random walks, and diffusion theory applied in genetics and evolution now exists: some references (and references to other problems discussed earlier) are given below. A significant proportion of this literature first arose through the need to show deductively, in part using probability theory, that evolution as a Mendelian process could and would occur. A rather complete deductive theory of genetic evolution, cast largely in probabilistic and statistical terms, now exists. Present research is now turning to a consideration of inductive techniques, based on this deductive theory and on the large amount of information now available on the genetic computation of real populations, on the likely course that evolution has taken so far. In this endeavor statistical methods can again be expected to play a significant part.

Bibliography

Bodmer, W. F. and Cavalli-Sforza, L. L. (1976). *Genetics, Evolution, and Man*. W. H. Freeman, San Francisco.

Crow, J. F. and Kimura, M. (1970). *An Introduction to Population Genetics Theory*. Harper & Row, New York.

Elandt-Johnson, R. C. (1971). *Probability Models and Statistical Methods in Genetics*. Wiley, New York.

Ewens, W. J. (1979). *Mathematical Population Genetics*. Springer-Verlag, Berlin.

Fisher, R. A. (1958). *The Genetical Theory of Natural Selection*. Dover, New York.

Kempthorne, O. (1957). *An Introduction to Genetic Statistics*. Wiley, New York.

Wright, S. (1969). *Evolution and the Genetics of Populations*, Vol. 2. University of Chicago Press, Chicago.

(FISHER, RONALD AYLMER
HUMAN GENETICS, STATISTICS IN)

W. J. EWENS

GENSTAT *See* STATISTICAL SOFTWARE

GEOGRAPHY, STATISTICS IN

Modern geographic research has enthusiastically, if not excessively, embraced probability and statistics. Not only have new applications been found for the common core of statistical methodology, but the special characteristics of some geographic problems have fostered interesting new developments in this methodology. For purposes of organization this article is divided into the subtopics *spatial probability models*, *interpolation and smoothing of geographic patterns*, and *multivariate analysis of geographic data*. Undoubtedly, many important statistical applications in geography have been omitted, but a fair idea of scope of applications should be conveyed here. A broad view is taken of the subject matter of geography to include aspects of cartography, demography*, ecology, hydrology*, and geology*. (*See also* ECOLOGICAL STATISTICS; GEOLOGY, STATISTICS IN; and GEOSTATISTICS.)

SPATIAL PROBABILITY MODELS

Spatial probability models in geographic research have been used for various ends. They may be used to provide a convenient summary description of a geographic pattern and for focusing comparisons of two or more patterns. When so used such models serve a purpose similar to common fitting of probability distributions to repetitive observations. It can be argued, however, that a few well-chosen "model-free" summary statistics would suffice and that generative models are not needed—by analogy with the computation of sample moments, say. For example, a spatial autocovariance function (or its Fourier transform) serves to summarize many features of a spatial pattern which appears to be spatially stationary or from which major trends have been removed.

Occasionally, probability models are developed which are consistent with certain simplified physical or cultural processes. Then the fitting of such models may assist in the understanding of how changes in the driving process parameters would affect the geographic pattern. (References for discussions of spatial probability models and their properties are found in refs. 3, 8, 11, and 13.) This objective is especially relevant for demographic and ecological patterns which have historical time scales of change.

Once a spatial probability model has been specified one could infer, in principle, what new models would result by forming local moving averages* of the original process. This is especially useful when the observational units have a nonnegligible areal extent and in particular when there is some choice in the size of observational unit. Probability modelling then provides a mechanism for relating pattern characteristics to size of observational unit. In practice this can be mathematically complicated, although the effect of local averaging on autocovariances is well understood (see refs. 9 and 17).

The range of probability models in the literature is quite vast. One sees Poisson point processes* in the plane with constant or smoothly varying rate parameters as well as contagious or clustered point processes as descriptions of the dispersal of human settlements or locations of establishments, occurrences of disease, and the like. Particularly in ecological work (*see* ECOLOGICAL STATISTICS) one sees probabilistic models for partitioning of geographic regions into domains; such models have been described in refs. 11 and 13. Finally, probability models for quantities that vary continuously over a geographic area, such as spatial Gaussian processes*, have been used especially in mineralogic, geochemical, hydrologic, and atmospheric problems. As well, there are specially developed models such as the one given in ref. 6 for drainage networks. One also has probability models defined on two-dimensional lattices which are clever generalizations of time-series* models (see, e.g., refs. 1 and 2). However, in geographic applications one is usually dealing with a plane rather than a plane lattice. *See also* SPATIAL PROCESSES.

The fitting of models to observed patterns and the estimation of model parameters is an area where there is much to be done. The simplest case of a spatially homogeneous Poisson point process (i.e., a completely random distribution of points) has had considerable attention. The usual objective has been to provide formal tests of significance* to guard against inferring spurious structure in an observed spatial distribution of points (e.g., trees, towns, etc.). Test statistics, which also serve as descriptive measures of pattern, are often based on the distribution of interpoint distances (see refs. 4 and 14). One real complication is that the shape of the effective planar region which can receive points may be very complicated due to the presence of lakes, parks, etc. This complication may seriously affect the distribution of interpoint distances and should be explicitly accounted for.

INTERPOLATION AND SMOOTHING OF GEOGRAPHIC PATTERNS

Particularly in physical geography one frequently encounters the problem of spatial interpolation using a data network of essentially punctual observations of a smoothly varying quantity. This is essentially the problem of drawing maps or determining level contours or boundaries from isolated observations. Various interpolation* algorithms are incorporated into computer mapping routines such as refs. 7 and 15. Commonly, the interpolated value at an unobserved location is taken to be a weighted average of the nearest observed values. The most fruitful weighting schemes seem to be those derived from treating the observed and unobserved values as a collection of correlated random variables. In this context, minimum mean squared error* interpolation yields weighted averages whose optimized coefficients depend on the point-to-point correlations. The correlations are usually estimated from the data by having them be a simple parametric function of interpoint distance. Such interpolation procedures together with their various refinements often go under the name of Kriging*. (Good general sources are refs. 5 and 12; *see also* GEOSTATISTICS.)

Kriging has been applied extensively, particularly in hydrologic and natural resource mapping problems. Simpler weighting schemes such as those with data weights inversely proportional to distance from the interpolation point will often give maps almost indistinguishable from kriged maps except when the geographic configuration of the data points is clustered. Another possible advantage of the Kriging (i.e., correlation) approach is that it gives estimates of the interpolation error which will be larger as one gets further away from the data. How useful or realistic these error estimates are may be debated; one possible check is to interpolate the data points themselves as though they were absent.

While a map represents the spatial variation of a single variable, one often can and should use available information on other variables. For example, average rainfall maps interpolated from a network of rain gauges should make use of the highly detailed topographic information. This use of covariates for map construction is largely unresearched but would seem to be important (see ref. 18).

The interpolation problem for a spatially distributed nominal or qualitative variable differs in that averaging of nearby observations no longer has any meaning. Instead, one might interpolate a presumed most probable value given the surrounding data according to some adopted probabilistic model. More simply, one could interpolate the value of the nearest datum. The map resulting from the nearest-neighbor interpolation algorithm will have physically implausible polygonal boundaries between colors. However, the degree of granularity of the map reflects the density of the available data, so such maps are honest if somewhat naive. A certain fraction of the map area will inevitably be incorrectly mapped. This fraction can be estimated using a model in which the probability that two geographic points have different values is a simple parametric function, say, of the distance between the points (see ref. 16). An application of this kind of error rate estimation is in the choice of a digitizing interval for computer storage and processing of multicolor maps. An acceptable error rate could then be related to a specific sampling rate. See Figures 1 and 2; the digitized map (Fig. 2) has a 5.2% actual error compared with the true map (Fig. 1). A calculated error of 5.2% was also obtained using only the data of the digitized map and the model-fitting procedure described in ref. 16. As an alternative to interpolating a single multicolor map for a geographical qualitative variable, one

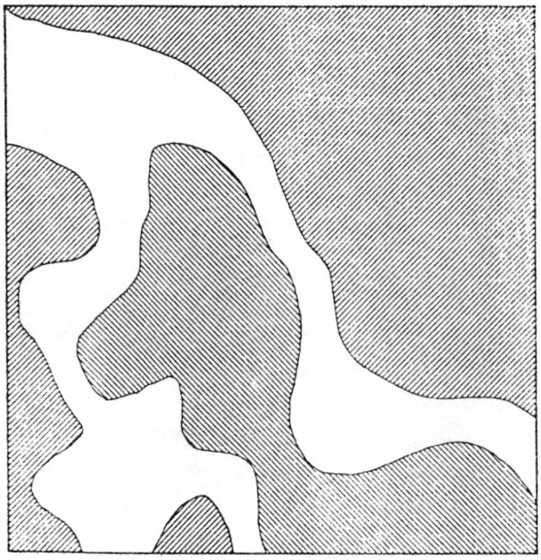

Figure 1 Portion of a pre-Pennsylvanian geologic map of Kansas (true map). (From Switzer [16].)

might try to produce a separate shade print map for each color where the darkness of shading at a point is proportional to the presumed probability of the given color at the given point. Such an undertaking would have the disadvantage of requiring substantial model input. A discussion of probability maps is given in ref. 23.

Typically, interpolated maps of all kinds show somewhat less variability than the "true" maps would. This has been shown both mathematically and empirically. However, for many purposes, one may wish to deliberately suppress a substantial part of the variability, in particular the local or high-frequency component. Such smoothing (*see* GRADUATION) will lead to less accurate interpolation but may produce maps in which the themes playing on a larger geographic scale are more visually apparent. It is also often easier to compare regions with one another on the basis of smoothed maps.

A widely used smoothing method for quantitative variables, called *trend surface analysis*, fits a simple parametric function of the geographic coordinates to the data, usually a low-order polynomial. The fitting is done by least squares* or, better, by generalized least squares in order to mitigate the effects of any clustering in the data. Examples of the use of trend surface analyses are given in ref. 22. Smoothing is commonly done in a stepwise manner until the data residuals* no longer appear to manifest regional patterns. There has been frequent

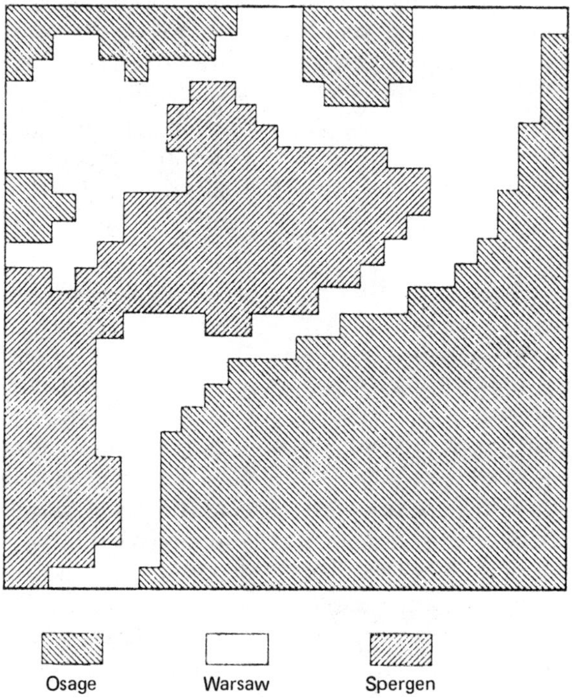

Figure 2 Estimated map formed by sampling a 25×25 grid. (From Switzer [16].)

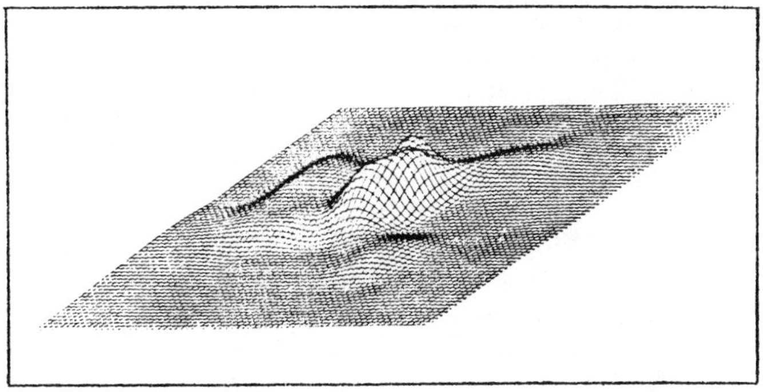

Figure 3 1970 population density of Ann Arbor, Michigan, based on census tract data.

criticism of trend surface analysis, but this criticism usually confuses the interpolation objective with the smoothing objective.

A special problem in smoothing maps of a demographic or socioeconomic variable arises when the available data are already aggregated over bureaucratic geographic units of substantial size relative to the total map area. The objective of smoothing is to make the map continuous, in particular along the bureaucratic boundaries. Reasonable algorithms to accomplish this goal are not simple, especially under the constraint that the smoothed map preserve the aggregated data values. Recent progress on this problem has been made in ref. 21. An example of the result of smoothing census tract data is shown in Fig. 3, where the smoothing methods of ref. 21 have been used.

MULTIVARIATE ANALYSIS OF GEOGRAPHIC DATA

Questions regarding the joint behavior of two or more variables on the same collection of geographic entities has led to the application of a variety of multivariate statistical methods. A list of references for such applications can be found in ref. 10. The objectives of the multivariate analysis* of geographic data are varied. The methods used are typically standard and rarely adapted to or particularly cognizant of the geographic support of the data. It is usual for general-purpose geographic data files to be multivariate, offering multiple opportunities for analysis.

In the analysis of demographic or socioeconomic data the observational units are often sufficiently large that it may be unreasonable to impose conditions of geographic smoothness on the analysis. This would be the case, for example, when the units are school districts and one wishes to relate per pupil expenditure to the size of the tax base and the proportion of minority pupils. The usual calculation of partial correlation coefficients might provide an adequate summary. However, physical data such as weather or air pollution measurements are measured at points and explicit use of the geographic configuration of the measurement stations in the analysis of the data could be useful. For example, a regression analysis* of mean annual ozone concentrations on mean annual hours of sunshine should take account of the correlation* of residuals between closely spaced stations. A suggestion for an explicitly geographic correlation coefficient has been given in ref. 20.

The potential shortcomings of routine multivariate analyses performed without regard to the geographic character of the data will sometimes be apparent when the outputs are finally mapped. For example, contour maps of principal components scores or discriminant function values may show spurious high-frequency components superimposed on the sought after geographic pattern. Or one may see strong geographic patterns in the residuals from predicted values in a multiple regression analysis.

The availability of earth satellite data for geographic classification has opened up new statistical vistas. These data have the form of energy spectra averaged over a small number of selected wavelength bands each averaged over a spatial resolution element usually called a pixel. For LANDSAT data pixels are about an acre in size and four wavelength bands are used. In a common type of analysis, pixels corresponding to known surface terrain categories are used to estimate discriminant functions, which in turn are used to classify pixels whose category is not known. Often this will be a standard four-variate linear discriminant analysis*, but variants of the standard analysis are discussed in ref. 19 in which account is taken of the likely similarity of adjacent pixels and the possibility that a pixel may be a mixture of more than one category.

References

[1] Bartlett, M. S. (1975). *The Statistical Analysis of Spatial Patterns.* Chapman & Hall, London. (Particular emphasis on plane lattice models.)

[2] Besag, J. E. (1974). *J. R. Statist. Soc. B*, **36**, 192–236.

[3] Cliff, A. D. and Ord, J. K. (1973). *Spatial Autocorrelation.* Pion, London. (A survey with emphasis on geographic problems.)

[4] Dacey, M. F. and Tung, T. H. (1962). *J. Reg. Sci.*, 83–96.

[5] Delfiner, P. and Delhomme, J. P. (1975). In *Display and Analysis of Spatial Data*, J. C. Davis and M. J. McCullogh, eds. Wiley, London, pp. 96–114.

[6] Dunkerley, D. L. (1977). *J. Geol.*, **85**, 459–470.

[7] Gaits, G. M. (1975). In *Display and Analysis of Spatial Data*, J. C. Davis and M. J. McCullogh, eds. Wiley, London, pp. 267–281.

[8] Gettis, A. and Boots, B. (1978). *Models of Spatial Processes.* Cambridge University Press, Cambridge. (Human geography.)

[9] Hsu, M. L. (1975). In *Display and Analysis of Spatial Data*, J. C. Davis and M. J. McCullogh, eds. Wiley, London, pp. 115–129.

[10] Johnston, R. J. (1978). *Multivariate Statistical Analyses in Geography*. Longman, London.

[11] Matérn, B. (1960). *Spatial Variation.* Royal Swedish Forestry Institute, Stockholm. (Probability models, sampling design problems for estimating areal averages.)

[12] Mathéron, G. (1971). *The Theory of Regionalized Variables and Its Applications.* Centre de Morphologie Mathématique, Fontainebleau, France. (An exposition of Kriging methods.)

[13] Pielou, E. C. (1969). *An Introduction to Mathematical Ecology*. Wiley, New York. (Models and applications.)

[14] Ripley, B. D. (1977). *J. R. Statist. Soc. B*, **39**, 172–212. (A survey with some geographic and ecological applications.)

[15] Sampson, R. J. (1975). In *Display and Analysis of Spatial Data*, J. C. Davis and M. J. McCullogh, eds. Wiley, London, pp. 244–266.

[16] Switzer, P. (1975). In *Display and Analysis of Spatial Data*, J. C. Davis and M. J. McCullogh, eds. Wiley, London, pp. 1–13.

[17] Switzer, P. (1976). *J. Appl. Prob.*, **13**, 86–95.

[18] Switzer, P. (1979). *Water Resour. Res.*, **15**, 1712–1716.

[19] Switzer, P. (1980). *Math. Geol.*, **12**, 417–423.

[20] Tjøstheim, D. (1978). *Biometrika*, **65**, 109–114.

[21] Tobler, W. R. (1979). *J. Amer. Statist. Ass.*, **74**, 519–536.

[22] Whitten, E. H. T. (1975). In *Display and Analysis of Spatial Data*, J. C. Davis and M. J. McCullogh, eds. Wiley, London, pp. 282–297.

[23] Wrigley, N. (1977). *Probability Surface Mapping*, Geo Abstract. University of East Anglia, Norwich, England.

Bibliography

See the following works, as well as the references just cited, for more information on the topic of statistics in geography.

Berry, B. J. L. (1971). *J. Amer. Statist. Ass.*, **66**, 510–523.

Gudgin, G. and Thornes, J. B. (1974). *The Statistician*, **23**, 157–177.

Mather, P. and Openshaw, S. (1974). *The Statistician*, **23**, 283–307.

(DEMOGRAPHY
ECOLOGICAL STATISTICS
GEOLOGY, STATISTICS IN
GEOSTATISTICS
INTERPOLATION
KRIGING
SPATIAL DATA ANALYSIS)

PAUL SWITZER

GEOLOGY, STATISTICS IN

Geology seeks to describe and understand the processes that have acted in the past, and are acting now, to form the continents and oceans with their mountains and valleys and which have lead to the varied sequences of rocks of differing compositions and structures. At the time of Darwin's 1832–1836 voyage on the *Beagle*, geology was closely linked with biology as the study of "natural history," and both then made great leaps forward. In fact, Lyell's [23] book on stratigraphy (epochs are defined statistically by fossil contents) was Darwin's inspiration. It is a curious historical fact that although the intense application of the physical sciences led the subjects to diverge, both had their next revolution at about the same time—molecular biology in the 1950s and plate tectonics in the 1960s.

Mathematics entered geology when the physics of the earth was studied—gravity and the figure of the earth, tides in the oceans, air, and solid earth, the cooling of the earth, earthquakes, and the propagation of waves around the earth must be a matter of inference, because it is necessarily indirect—only Jules Verne could imagine a "Voyage to the Center of the Earth." Also, geological field measurements are subject to greater errors than laboratory work in chemistry and physics, and it is often not possible to take "random samples." It is something of a coincidence that a little after Sir Ronald Fisher's development of statistics largely for biologists, an eminent geophysicist, Sir Harold Jeffreys, should develop his own theory of statistics [15]. Jeffreys' logical predisposition led him to a mathematical rule for deriving priors rather than to use a purely subjective origin. For his geophysical achievements and classical mathematical geophysics, readers should also consult his classic text, *The Earth* [16].

The earth sciences may also claim to have initiated several areas of statistical theory and practice. There are so many periodic or pseudo-periodic earth phenomena that Sir George Stokes' 1879 introduction of the Fourier transform of data and its development by Schuster (*see*, e.g., Brillinger [4] and INTEGRAL TRANSFORMS) was natural. The most advanced applications of *time-series* analysis* are still to be found in geophysics; a comprehensive bibliography has been given by Tukey [30]. The orientation of pebbles [22] and the direction of magnetism of rocks [12] led to the development methods for *directional data* analysis*; a survey with many references to papers in this area was given by Watson [32]. More recently, economic geology and efficient mining have led to *geostatistics**, an extensive application of random function theory by Matheron [24]. Chemical petrologists study the proportions of substances making up rocks, so their data add to unity. The study of the compositional data raises special problems that have occupied geologists more than other scientists (see, e.g., Chayes [7]). The study of their sections (e.g., Chayes [6]) has led to stereological and geometrical probability* problems, as has exploration geology. Geologists have always needed maps and photographs of regions. Now the computer is being used heavily to produce and process such information; Matheron [25] provides a theoretical background (*see also* MATHEMATICAL MORPHOLOGY).

There has been a rapid growth in the use of computers*, mathematics, and statistics in geology in the 1970s. This literature is fairly easy to enter. The American Geological Institute publishes a bibliography and index in which most of the relevant articles appear under the main heading "Automatic Data Processing," although some appear under "Mathematical Geology." Two journals, *Mathematical Geology* and *Computers and Geosciences*, specializing in these topics, began in the 1970s. There are a number of general texts (e.g., Agterberg [2] and Davis [11]) and a number devoted to specific topics to be mentioned below. Merriam has edited many symposia (e.g., ref. 26). As these quantitative methods become a recognized part of all subdivisions of all geology, the specialized journals (*Sedimentology*, to give just one example) all carry articles of statistical interest.

The following sections are chosen to show the methods and problems of special interest which are to be found currently in geology. The references given will lead the reader further. Exploration and resource estimation and exploitation is ignored here but covered partially in GEOSTATISTICS.

DATA BANKS

Efforts are being made to computerize data, so it can be accessed easily (see, e.g., Chayes [8] for igneous petrology). Many interesting data are unavailable because of their economic value to those who possess them.

STOCHASTIC MODELS

This field is very wide indeed. Earth movements lead to an interest in the growth of cracks or fractures (see, e.g., Vere-Jones [31]). The occurrence and strength of earthquakes and volcanic eruptions have been the subject of much point processes* modeling (see, e.g., Adamopoulos [1]). Erosion and sedimentation require a knowledge of particle size distributions*. Models for forming sands and powders often lead to the lognormal* (e.g., Kolmogoroff [20]) and Weibull* (see, e.g., Kittleman [19]) distributions. Considerations of the transport and deposition of sand (see, e.g., Bagnold [3] and Sen Gupta [28]) lead to other distributions and stochastic processes. Kolmogoroff [21] first modeled the deposition and subsequent erosion of sediments. His model was studied further by Hattori [14]. This theory is distinct from the literature that tries to fit Markoff chains (see below) to the succession of beds according to their composition rather than thicknesses, although Hattori deals with both approaches. The present writer regards the Markoff approach as more data analysis than modeling. Grenander [13] has provided a stationary stochastic model on the circle (which is easily generalized to the sphere) for the height of the land surface. Erosion is modeled by diffusion, which always smooths, and inequalities are maintained by uplifts at random times described by random independent functions.

The study of streams in drainage basins, their lengths, and topology is fascinating (see, e.g., Dacey and Krumbein [9]).

The distribution of elements has a large literature, but statistical models to explain them are few. Kawabe [18] gives a model and a literature list, including references to papers by Ahrens, who felt log-normal distributions of elements were a law of nature (*see also* PARTICLE SIZE DISTRIBUTIONS).

DATA ANALYSIS

Nowadays, all the common statistical procedures are used widely. Most data are observational*. One collects rocks where they happen to be exposed and accessible, so the problems of "nonrandom samples" are very serious. The earth is a sample of one. The list below gives leads to areas of particular interest.

CLUSTERING METHODS. Dendrograms* and other methods are often used to relate fossils, rocks, etc., to help explain their evolution. Petrofabric and other studies yield orientations plotted on a sphere. Deciding whether the points fall in groups or clusters is a common problem, but may be attacked differently.

FACTOR ANALYSIS. Factor analysis* is widely used in palaeontology and elsewhere (see, e.g., Jöreskog et al. [17]). Temple [29] gives a very critical review. In the analysis of data on the sedimentary composition of a closed basin the factors might be the few inputs to the basin. In the fossil content of oceanic cores, they might be the depositorial climates—tropical, polar, etc.

MARKOV CHAINS (*see* MARKOV PROCESSES). In studying the succession of different lithologies, often a small set (e.g., sand, silt, and clay) recurs in a partially cyclic way. It might be that the failure to be strictly cyclic is due to the complete erosion of some parts of the record (see Casshyap [5] and Miall [27]).

COMPOSITIONAL DATA. It is natural to study such data (e.g., the proportions $p_1, p_2, \ldots, p_k$ of the k constituents of a rock) to see if the relative amounts of substances 1 and 2 are associated. The facts that

$p_1 + p_2 < 1$, $\sum_1^k p_i = 1$ make the usual methods invalid (see Chayes [7] and Darroch and Ratcliff [10] for later work).

ORIENTATION DATA. Normals to bedding planes, cracks, and joints provide examples of axial data and the flow of glaciers and directions of magnetization provide examples of directional data analysis*.

TIME-SERIES ANALYSIS. Time-series analysis* is basic to seismological data processing.

References

[1] Adamopoulos, L. (1976). *Int. Ass. Math. Geol.*, **8**, 463–475.

[2] Agterberg, F. P. (1974). *Geomathematics*. Elsevier, Amsterdam. (A course on calculus, matrix algebra, and probability statistics—sound but classical—followed by two chapters on spatial statistics, time series, Markov chains, stochastic petrology, directional data, and multivariate analysis of spatial data.)

[3] Bagnold, R. A. (1954). *The Physics of Blown Sands and Desert Dunes*. Chapman & Hall, London.

[4] Brillinger, D. R. (1975). *Time Series, Data Analysis and Theory*. Holt, Rinehart and Winston, New York. (Best modern account of spectrum analysis; combines numerical insight and examples with the requisite formulas and theorems.)

[5] Casshyap, S. M. (1975). *Sedimentology*, **22**, 237–255.

[6] Chayes, F. (1956). *Petrofabric Modal Analysis*. Wiley, New York. (Now out of date, but shows the origins of the topic.)

[7] Chayes, F. (1971). *Ratio Correlation*. University of Chicago Press, Chicago. (For recent advances, see Darroch and Ratcliff [10].)

[8] Chayes, F., McCammon, R., Trochimczyk, J. and Velde, D. (1977). Rock Information System RKNFSYS: Carnegie Institution of Washington, Year Book 76, pp. 635–637.

[9] Dacey, M. F. and Krumbein, W. C. (1976). *Int. Ass. Math. Geol.*, **8**, 429–461.

[10] Darroch, J. H. and Ratcliff, D. (1978). *Int. Ass. Math. Geol.*, **10**, 361–368.

[11] Davis, J. C. (1973). *Statistics and Data Analysis in Geology*. Wiley, New York.

[12] Fisher, R. A. (1953). *Proc. R. Soc. Lond.*, A, **217**, 295–305.

[13] Grenander, U. (1975). *J. Math. Geol.*, **7**, 267–278.

[14] Hattori, I. (1973). *Sedimentology*, **20**, 331–345.

[15] Jeffreys, H. (1939). *Theory of Probability*. Clarendon Press, Oxford. (Author's version of Bayesian statistics, with many novel geophysical applications.)

[16] Jeffreys, H. (1961). *The Earth*, 4th ed. Cambridge University Press, Cambridge. (Classic text on geophysics.)

[17] Jöreskog, K. G., Klovan, J. E., and Reyment, R. A. (1976). *Geological Factor Analysis*. Elsevier, New York.

[18] Kawake, I. (1977). *Int. Ass. Math Geol.*, **9**, 39–54.

[19] Kittleman, L. R. (1964). *J. Sediment. Petrol.*, **34**, 483–502.

[20] Kolmogoroff, A. N. (1941). *Dokl. Akad. Nauk SSSR*, **31**, 99–101.

[21] Kolmogoroff, A. N. (1949). *Amer. Math. Soc. Trans.*, **53**.

[22] Krumbein, W. C. (1939). *J. Geol.*, **47**, 673–706.

[23] Lyell, C. (1800, 1832, 1833). *Principles of Geology*, Vols. 1, 2, 3 (republished in 1969 by the Johnson Reprint Corp., New York). (Darwin took Vol. 1 and received Vol. 2 in Montevideo in 1832.)

[24] Matheron, G. (1965). *Les Variables regionalisées et une estimation*. Masson & Cie, Paris. (A difficult book now outmoded by Huisbreghts and Journel; *see* GEOSTATISTICS.)

[25] Matheron, G. (1967). *Éléments pour une théorie des milieux poreux*. Masson & Cie, Paris. (In the first part, random sets are used to model sediments and tie descriptive theory to observations made by an image analyzer. In Part 2, the flow through porous media is studied. Outdated.)

[26] Merriam, D. F., ed. (1978). *Recent Advances in Geomathematics*. Pergamon Press, New York.

[27] Miall, A. D. (1973). *Sedimentology*, **20**, 347–364.

[28] Sen Gupta, S. (1975). *Sedimentology*, **22**, 257–273.

[29] Temple, J. T. (1978). *J. Math. Geol.*, **10**, 379–387.

[30] Tukey, J. W. (1965). *Proc. 35th Sess., Int. Statist. Inst.*, Belgrade, Book 1, pp. 267–305. (Much scientific detail; large bibliography.)

[31] Vere-Jones, D. (1977). *Int. Ass. Math. Geol.*, **9**, 455–482.

[32] Watson, G. S. (1970). *Bull. Geol. Inst. Uppsala N. S.*, **2**:**9**, 73–89.

Bibliography

See the following works, as well as the references just cited, for more information on the topic of statistics in geology.

Ahrens, L. H. (1963). *Geochim. Cosmochim. Acta*, **27**, 333–343. (Gives references to three earlier and many other leads to this literature.)

Aitchison, J. (1982). *J. R. Statist. Soc.*, *B*, **44**, 139–177.

Cubitt, J. M. and Henley, S. (1978). *Statistical Analysis in Geology*. Benchmark Papers in *Geology*, Vol. 37. Dowden, Hutchinson & Ross, Stroudsburg, Pa. (A collection of 23 papers on statistical ideas important in geology rather than techniques.)

Griffiths, J. C. (1967). *Scientific Method in the Analysis of Sediments*. McGraw-Hill, New York. (Rather old-fashioned statistically; largely sedimentology.)

Koch, G. S. and Link, R. F. (1971). *Statistical Analysis of Geological Data*, Vols. 1, 2. Wiley, New York. (More of interest to economic geologists than to statisticians.)

Mardia, K. V. (1972). *Statistics of Directional Data*. Academic Press, New York. (The only text on the topic; primarily for statisticians.)

McElhinny, M. W. (1973). *Palaeomagnetism and Plate Tectonics*. Cambridge University Press, Cambridge. (Most up to date on paleomagnetism and its implications, with many illustrations of statistical procedures.)

Ramsey, J. G. (1967). *Folding and Fracturing of Rocks*. McGraw-Hill, New York.

Reyment, R. A. (1971). *Introduction to Quantitative Paleoecology*. Elsevier, Amsterdam.

(DENDROCHRONOLOGY
DIRECTIONAL DATA ANALYSIS
GEOSTATISTICS
OBSERVATIONAL STUDIES)

G. S. Watson

GEOMETRIC DISTRIBUTION

Let there be a sequence of independent trials and p be the probability of success at each trial. Let X be the number of trials before the first success, and the probability function of the random variable X be given by

$$f(x) = pq^x, \qquad x = 0, 1, 2, \ldots; \quad q = 1 - p. \tag{1}$$

This is known as the *geometric distribution* because successive terms in the probability function above form a geometric progression. It is also known as the Pascal or Furry distribution. The sum of n independent, identically distributed geometric random variables has a negative binomial distribution* and (1) is thus treated as a special case of the negative binomial distribution. It belongs to the class of generalized power series distributions* [24].

Mean: p/q

Variance: p/q^2

Mean deviation: $2lq^l$, where $l = [\mu] + 1$ is the smallest integer greater than $[\mu]$

Coefficient of variation: $q^{-1/2}$

Coefficient of skewness: $(1+q)q^{-1/2}$

Coefficient of excess: $(6q + p^2)q^{-1}$

Moments about zero:

$$\mu'_{r+1} = q\left[\frac{\partial u'_r}{\partial q} + \frac{1}{p}\mu'_r\right]$$

Central moments:

$$\mu_{r+1} = q\left[\frac{\partial \mu_r}{\partial q} + \left(\frac{r}{p^2}\right)\mu_{r-1}\right]$$

(Ascending) factorial moments:

$$\mu'_{[r]} = r!\,\frac{q^r}{p^r}.$$

Cumulants:

$$\kappa_{r+1} = q\left[\frac{d\kappa_r}{dp}\right], \qquad r \geqslant 1$$

Moment generating function: $p(1 - qe^t)^{-1}$

Characteristic function: $p(1 - qe^{it})^{-1}$

Probability generating function: $p(1 - qt)^{-1}$

For the properties above, see Johnson and Kotz [13] and Patel et al. [23].

The geometric distribution is the only discrete distribution which enjoys the lack of memory property [9]; i.e.,

$$\Pr(X > a + b \mid X > a) = \Pr(X > [a + b] - [a]).$$

Hence some of its properties run parallel to those associated with the exponential distribution*, which is of the continuous type. The usefulness of the geometric distribution in applied stochastic processes*, queueing theory*, and reliability* is described in Bhat [3, pp. 130, 131, 132, 137, 210] and in Mann et al. [19, pp. 294, 395].

There are several characterizations* of this distribution [1, 2, 5, 7, 10, 14, 18, 26, 30–33].

For moments of order statistics*, see Margolin and Winokur [20].

For an empirical Bayes estimator, see Maritz [21]. For parameter estimation in the truncated case, see Kapadia and Thompson [15] and Thompson and Kapadia [34]. For confidence limits, see Clemans [6]. For unbiased estimation* and tests of hypotheses*, see Klotz [16], and for testing for homogeneity, see Vit [35]. For estimation of the parameters of a mixture of geometric distributions, see Krolikowska [17].

For applications of the geometric distribution, see Jagers [12], Phatarford [27], Holgate [11], Pielou [28, 29], Yang [36], and Chen [4]. For a mixture of geometric distributions and its applications to busy period distributions in equilibrium queueing systems, see Daniels [8].

For an application of a generalized geometric distribution, see Pandit and Sriwastav [22].

References

[1] Arnold, B. C. (1980). *J. Appl. Prob.*, **17**, 570–573.

[2] Arnold, B. C. and Ghosh, M. (1976). *Scand. Actuarial J.*, 232–234.

[3] Bhat, U. N. (1972). *Elements of Applied Stochastic Processes*. Wiley, New York.

[4] Chen, R. (1978). *J. Amer. Statist. Ass.*, **73**, 323–327.

[5] Chong, K.-M. (1977). *J. Amer. Statist. Ass.*, **72**, 160–161.

[6] Clemans, K. G. (1959). *Biometrika*, **46**, 260–264.

[7] Dallas, A. C. (1974). *J. Appl. Prob.*, **11**, 609–611.

[8] Daniels, H. E. (1961). *J. R. Statist. Soc. B*, **13**, 409–413.

[9] Feller, W. (1962). *An Introduction to Probability Theory and Its Applications*. Wiley, New York.

[10] Ferguson, T. S. (1965). *Amer. Math. Monthly*, **72**, 256–280.

[11] Holgate, P. (1966). *Biometrics*, **22**, 925–936.

[12] Jagers, P. (1973). *J. Amer. Statist. Ass.*, **68**, 801–804.

[13] Johnson, N. L. and Kotz, S. (1969). *Discrete Distributions*. Houghton Mifflin, Boston.

[14] Kagan, A., Linnik, Y. V., and Rao, C. R. (1973). *Characterization Problems in Mathematical Statistics*. Wiley, New York.

[15] Kapadia, C. H. and Thompson, R. L. (1975). *Ann. Inst. Statist. Math.*, **27**, 269–272.

[16] Klotz, J. (1970). *Ann. Math. Statist.*, **41**, 1078–1082.

[17] Krolikowska, K. (1976). *Demonstratio Math.*, **9**, 573–582.

[18] Lukács, E. (1965). *Proc. 3rd Berkeley Symp. Math. Statist. Probl.*, Vol. 2. University of California Press, Berkeley, Calif., pp. 195–214.

[19] Mann, N. R., Schafer, R. E., and Singapurwalla, N. C. (1974). *Methods for Statistical Analysis of Reliability and Life Data*. Wiley, New York.

[20] Margolin, B. H. and Winokur, H. S., Jr. (1967). *J. Amer. Statist. Ass.*, **62**, 915–925.

[21] Maritz, J. S. (1966). *Biometrika*, **53**, 417–429.

[22] Pandit, S. S. and Sriwastav, G. L. (1976). *Sankhyā B*, **38**, 68–71.

[23] Patel, J. K., Kapadia, C. H., and Owen, D. B. (1976). *Handbook of Statistical Distributions*. Marcel Dekker, New York.

[24] Patil, G. P. (1962). *Ann. Inst. Statist. Math., Tokyo*, **14**, 179–182.

[25] Patil, G. P. and Joshi, S. W. (1968). *A Dictionary and Bibliography of Discrete Distributions*. Oliver & Boyd, Edinburgh.

[26] Paulson, A. S. and Uppuluri, V. R. R. (1972). *Sankhyā A*, **34**, 297–300.

[27] Phatarford, R. M. (1963). *Ann. Math. Statist.*, **34**, 1588–1592.

[28] Pielou, E. C. (1962). *Biometrics*, **18**, 579–593.

[29] Pielou, E. C. (1963). *Biometrics*, **19**, 603–615.

[30] Puri, P. S. (1973). *Sankhyā A*, **35**, 61–78.

[31] Shaked, M. (1974). *A Modern Course on Statistical Distributions, Scientific Work*, Vol. 1. D. Riedel, Dordrecht, Holland.

[32] Shanbhag, D. N. (1974). *J. Amer. Statist. Ass.*, **69**, 1256–1259.

[33] Srivastava, R. C. (1974). *J. Amer. Statist. Ass.*, **69**, 267–269.

[34] Thompson, R. L. and Kapadia, C. H. (1968). *Ann. Inst. Statist. Math.*, **29**, 519–523.

[35] Vit, P. (1974). *Biometrika*, **61**, 565–568.

[36] Yang, M. C. K. (1975). *J. Appl. Prob.*, **12**, 148–156.

(NEGATIVE BINOMIAL DISTRIBUTION)

C. H. Kapadia

GEOMETRIC MEAN

The geometric mean of n positive numbers $X_1, X_2, \ldots, X_n$ is the positive nth root of their product. If any X_i are zero, the geometric mean is defined to be zero. If some X_i are negative, and none are zero, the geometric mean is not defined.

Various notations for the geometric mean include $\tilde{X}$, G, $G(X_i, \ldots, X_n)$, $\mathrm{GM}(X_1, \ldots, X_n)$, sometimes with a subscript indicating sample size. For a random variable X, a parameter analogous to the sample geometric mean is $\mathrm{GM}(X) = \exp(E[\ln(X)])$, when $E[|\ln(X)|] < \infty$. In the following, μ will denote $E(\ln(X))$, so that $\mathrm{GM}(X) = \exp \mu$.

The geometric mean of $X_1, X_2, \ldots, X_n$ can be calculated by

$$\tilde{X}_n = \left(\prod_{i=1}^{n} X_i \right)^{1/n}, \tag{1}$$

by

$$\tilde{X}_n = \exp\left(\frac{1}{n} \sum_{i=1}^{n} \ln X_i \right) \tag{2}$$

if all X_i are positive, or iteratively by

$$\tilde{X}_n = \tilde{X}_{n-1} \times (X_n / \tilde{X}_{n-1})^{1/n}. \tag{3}$$

The geometric mean of a sample or population of values can also be estimated from the corresponding arithmetic mean*, variance*, and higher-order moments. Yound and Trent [10], for example, investigated the approximations

$$\mathrm{GM}(X) \simeq \left(E(X)^2 - \mathrm{var}(X) \right)^{1/2}$$
$$\simeq E(X) - \mathrm{var}(X)/[2E(X)],$$

the Latane and Johnson approximations, respectively, and reported that approximations involving third- and fourth-order moments as well failed to improve substantially on these two approximations.

THE GEOMETRIC MEAN AS A MEASURE OF LOCATION

Although less obvious as a measure of location than the arithmetic mean, the geometric mean does arise naturally as a measure of location in at least three circumstances: when observations X_i possess a certain relation between their conditional means and variances, when observed values are thought to be the results of many minor multiplicative (rather than additive) random effects, and when products of moderate to large numbers of random variables are of interest. Such products can arise in economics (see, e.g., Latane [7] or Samuelson [8]) and in population genetics* [3].

The first circumstance arises if X_i is to be chosen for a randomly selected subpopulation, if subpopulations differ both in mean and in standard deviation, with the standard deviation of a subpopulation being proportional to the mean of that subpopulation. This might occur, for example, if X_i were the estimated yield of field i, the weight of animal i, the insect count at location $i, \ldots$. When the conditional standard deviation is proportional to the conditional mean, an appropriate variance stabilizing transformation* is the logarithm. The exponential of the arithmetic mean of $\ln(X_1), \ln(X_2), \ldots, \ln(X_n)$ is, of course, the geometric mean.

The second circumstance when the geometric mean is relevant occurs when X is the cumulative result of many minor influences which combine in a multiplicative way, so that the same influence has a greater absolute effect on a larger result than on a smaller one. Since $\ln(X)$ is thus the sum of a great many small random effects, the central limit theorem* suggests that X may well be close to log-normal* in distribution, even if the contributing influences are not all independent. We note that X is said to be log-normal $\Lambda(\mu, \sigma^2)$ (see Aitchison and Brown [1]) if $\ln(X)$ is normal (μ, σ^2). As noted by Kirkwood [6], σ provides a measure of dispersion for such X if interpreted in terms of the ratio of X to e^{μ}. With 0.95 probability, for example, X/e^{μ} will lie between $\exp(-1.96\sigma)$ and $\exp(1.96\sigma)$. The name *geometric standard deviation* (GSD) was proposed by Kirkwood for the quantity $\exp(\sigma)$.

The geometric mean is also relevant if products of moderate to large numbers of independent identically distributed random variables are of interest. By the Kolmogorov strong law of large numbers*, if $Y_N = \prod_{i=1}^{N} X_i$, the Nth root of Y_N converges almost surely to $\mathrm{GM}(X)$. Moreover, if the variance σ^2 of $\ln(X)$ is finite, $\ln(Y_N)$

is asymptotically normal with mean $N\ln(\text{GM}(X))$ and variance $N\sigma^2$. This implies that $\text{GM}(X)^N$ is asymptotically the median* of Y_N, and that ranges for Y_N of the form

$$\text{GM}(X)^N\exp\left(-z_{\alpha/2}\sigma\sqrt{N}\right) \leqslant Y_N \leqslant \text{GM}(X)^N\exp\left(z_{\alpha/2}\sigma\sqrt{N}\right) \quad (4)$$

asymptotically have probability $1-\alpha$ of being correct (*see* ASYMPTOTIC NORMALITY; CONVERGENCE OF SEQUENCES OF RANDOM VARIABLES).

DISTRIBUTION

Let $X_1, X_2, \ldots, X_n$ be independent, identically distributed random variables corresponding to n sampled values of a random variable X. As before, let μ and σ^2 denote the expected value and variance of $\ln(X)$, when they exist. (See discussions of the mean and of the variance for conditions of existence, and Springer [9] for a discussion of various particular distributions of the X_i, and for further references.)

Assume in the following that X is a nondegenerate random variable, i.e., that the probability distribution of X is not concentrated at a single value. By Jensen's inequality*

$$E(X) > \text{GM}(X).$$

More generally, Hölder's inequality* implies that, if $1 < n < m$,

$$E(X) = E(\tilde{X}_1) \geqslant E(\tilde{X}_n) \geqslant E(\tilde{X}_m) \geqslant \text{GM}(X)$$

with all inequalities strict if $\infty > E(X)$ and $\text{GM}(X) > 0$. Hence for all n, $\tilde{X}_n$ is a biased estimator of $\text{GM}(X)$.

Even though $\tilde{X}_n$ is a biased estimator of $\text{GM}(X)$, if μ exists, then by the Kolmogorov strong law of large numbers*, $\tilde{X}_n$ tends almost surely to $\text{GM}(X)$ as $n \to \infty$, so that $\tilde{X}_n$ is a consistent estimator* of $\text{GM}(X)$. If both μ and σ^2 exist, the central limit theorem applied to $\ln(X_1), \ln(X_2), \ldots, \ln(X_n)$ as $n \to \infty$ implies that $\tilde{X}_n$ is asymptotically lognormal $\Lambda(\mu, n^{-1}\sigma^2)$. The asymptotic distribution is exact if X is $\Lambda(\mu, \sigma^2)$. Hypothesis testing concerning, or confidence intervals for, $\text{GM}(X)$ can therefore be carried out essentially by working with $\ln(X_1)$, $\ln(X_m), \ldots, \ln(X_n)$, and then translating conclusions about μ into conclusions about $\text{GM}(X)$.

The observations $X_1, X_2, \ldots, X_n$ need not be independent for $\tilde{X}_n$ to be a consistent estimator of $\text{GM}(X)$. If the X_i are generated by an ergodic process (*see* ERGODIC THEOREMS), the convergence of $\tilde{X}_n$ to $\text{GM}(X)$ is again almost certain, but at a rate generally different from those described previously. Determination of this rate, or of the distribution of $\tilde{X}_n$ in such cases, requires the use of the theory of stochastic processes*.

COMPARISON OF THE GEOMETRIC AND ARITHMETIC MEANS AS MEASURES OF LOCATION

As noted previously, $\text{GM}(X)^N$ is close to the median of $Y_N = (\prod_{i=1}^N X_i)$ for large N, so that $\text{GM}(X)$ is in some sense a "typical" value of X when products are of interest. Since $E(Y_N) = [E(X)]^N$, $E(X)$ would also appear to be a useful "typical" value. In fact, for two-sided ranges for Y_N such as (4), if X is nondegenerate, the expected value of Y_N falls outside this range for sufficiently large N. If $\sigma < \infty$, for example, Jensen's inequality implies that for some $c > 0$,

$$\ln(E(X)) - \ln(\text{GM}(X)) = c\sigma > 0,$$

so that

$$\begin{aligned} E(Y_N) &= \left[E(X)\right]^N \\ &= \text{GM}(X)^N\exp(c\sigma N) \\ &> \text{GM}(X)^N\exp\left(z_{\alpha/2}\sigma\sqrt{N}\right) \end{aligned}$$

if $N > z_{\alpha/2}^2/c^2$. Thus, for large N, the expected value of the product of $X_1, X_2, \ldots, X_n$ reflects, not the typical behavior of the product, but rather the influence of the rare but extremely large values. The choice of whether $E(X)$ or $\text{GM}(X)$ is a suitable measure of the location of X when studying the behavior of products should therefore be based on whether the rare but very large values, or the more typical smaller values, are of primary relevance and interest.

RELATION TO OTHER MEANS

The geometric mean is the inverse transformation of the arithmetic mean of a certain transformation (the natural logarithm) of data $X_1, X_2, \ldots, X_N$, of a random variable. As the logarithmic transformation is one of the family

$$X^{(\lambda)} = \begin{cases} X^{\lambda} & \text{if } \lambda \neq 0 \\ \ln(X) & \text{if } \lambda = 0, \end{cases}$$

a family that also includes the identity, square root, and reciprocal transformation, the geometric mean can be regarded as one mean of many in a family including the arithmetic mean ($\lambda = 1$) and the harmonic mean* ($\lambda = -1$).

A well-known result connecting the arithmetic, geometric, and harmonic means of positive random variables is that the geometric mean is never greater than the arithmetic mean and never less than the harmonic mean [4, 5]. Cartwright and Field [2] have established that for a positive random variable X with probability concentrated on $[a, b]$ where $a > 0$, if the arithmetic mean, variance, and geometric mean of X are μ, σ^2, and GM(X), respectively, then

$$\sigma^2/2b \leqslant \mu - \text{GM}(X) \leqslant \sigma^2/2a.$$

EFFECTS OF SIMPLE TRANSFORMATIONS

Scaling and power law transformations have simple effects on the geometric mean since

$$\left(\prod_{i=1}^{n} cX_i\right)^{1/n} = c\tilde{X}_n$$

for $c \geqslant 0$, and

$$\left(\prod_{i=1}^{n} X_n^p\right)^{1/n} = \tilde{X}_n^p.$$

No such simplicity occurs for changes of origin (i.e., for adding a constant c to all data), since the geometric mean of $X_1 + c$, $X_2 + c, \ldots, X_n + c$ is not a function of c and X_n alone, and is not even defined if $c < -\min(X_1, X_2, \ldots, X_n)$. Nonetheless, for c for which the resulting geometric mean is defined, the geometric mean $\tilde{X}_n^{(+c)}$, of $X_1 + c, X_2 + c, \ldots, X_n + c$, or a variation of this, $\tilde{X}_n^{(+c)} - c$, is of practical interest as providing a class of measures of location. This class includes the geometric mean (for $c = 0$) and the arithmetic mean, since $\tilde{X}_n^{(+c)} - c \to \overline{X}_n$, the arithmetic mean, as $c \to \infty$. Hence $\tilde{X}_n^{(+c)}$ may be considered as providing various compromises between these two measures of location. Particularly common when some X_i are zero (as in count data) is the practice of setting $c = \frac{1}{2}$ or 1. When $\tilde{X}_n^{(+c)}$, or $\tilde{X}_n^{(+c)} - c$, is used for comparing different populations, the same value of c should of course be used throughout.

References

[1] Aitchison, J. and Brown, J. A. C. (1957). *The Lognormal Distribution*. Cambridge University Press, Cambridge.

[2] Cartwright, D. I. and Field, M. J. (1978). *Proc. Amer. Math. Soc.*, **71**, 36–38.

[3] Crow, J. F. and Kimura, M. (1970). *An Introduction to Population Genetics Theory*. Harper & Row, New York.

[4] Hardy, G. H., Littlewood, J. E., and Polya, G. (1964). *Inequalities*. Cambridge University Press, Cambridge.

[5] Kendall, M. G. and Stuart, A. (1969). *The Advanced Theory of Statistics*, Vol. 1. Charles Griffin, London.

[6] Kirkwood, T. B. L. (1979). *Biometrics*, **35**, 908–909.

[7] Latane, H. A. (1969). *J. Polit. Econ.*, **62**, 144–155.

[8] Samuelson, P. A. (1971). *Proc. Natl. Acad. Sci. USA*, **68**, 2493–2496.

[9] Springer, M. D. (1979). *The Algebra of Random Variables*. Wiley, New York.

[10] Yound, W. E. and Trent, R. H. (1969). *J. Financ. Quant. Anal.*, **4**, 179–199.

(ARITHMETIC MEAN
HARMONIC MEAN
LOGNORMAL DISTRIBUTION)

W. G. S. HINES

GEOMETRIC MOVING AVERAGE

The geometric moving average (GMA) of the discrete parameter time series $\{X_t\}$ is a related time series* $\{G_t\}$, with

$$G_t = (1-\lambda)\sum_{\tau \leqslant t} \lambda^{\tau - t} X_t.$$

The parameter λ, satisfying $0 < \lambda < 1$, is referred to as a smoothing coefficient, with the larger values of λ resulting in greater smoothing as a result of effectively averaging over more past values of the process (*see* GRADUATION). If $E(X_t) = \mu_t$ is constant, then $E(G_t) = \mu_t$. If μ_t is not constant but is so slowly varying that $\mu_\tau \simeq \mu_t$ for $\tau \leqslant t$ such that $\lambda^{\tau - t}$ is not negligible, then $E(G_t) \simeq \mu_t$. Hence, for processes with constant or slowly varying mean, G_t provides a simple estimator of the current mean value of the $\{X_t\}$ process. The geometric moving average is also referred to as an exponential or an *exponentially weighted moving average* [3], and its application to a time series as *simple exponential smoothing* [4, 10].

The geometric moving average is made more attractive as an estimator of $E(X_t)$ because of the simple recursive formula

$$G_t = (1-\lambda)X_t + \lambda G_{t-1}.$$

Since newly acquired values of X_t do not need to be retained after their initial incorporation into the geometric moving average, the memory requirements of this estimator are small.

Figure 1 demonstrates the effect of various choices of λ on the ability of a geometric moving average to estimate the current mean of a time series. With $G_0 = 0$, and X_t normal with mean μ_t and unit variance, the geometric moving average with larger λ tends to estimate μ_t more reliably when μ_t is constant ($\lambda = 0.8$, $\mu_t = 3$, $0 \leqslant t \leqslant 15$), but is relatively slower at following changes in μ_t: e.g., a sudden change of μ_t to $\mu_t = 5$ for $t \geqslant 16$.

Various methods exist for selecting the value of λ appropriate for a given time series. Values suggested by some researchers include the range 0.75 to 0.95 [2], close to 0.80 [5], and 0.70 to 0.99 [10]. Montgomery and Johnson [10] also suggest using the first several values of the $\{X_t\}$ process available to decide on various suitable λ, with the resulting forecasting errors for subsequent

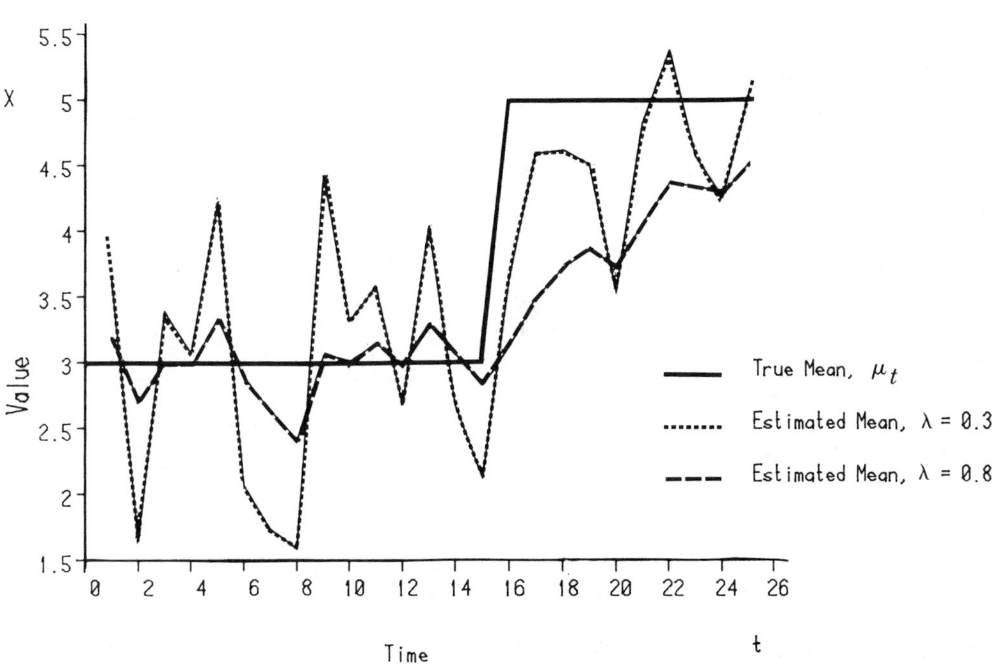

Figure 1 True mean values μ_t of a sequence of $N(\mu_t, 1)$ random variables, and two geometric-moving-average estimates of μ_t, with $G_t = 0.0$.

values then being determined and used to choose from among the different values of λ. They also suggest using initially small values of λ when very limited historical data are available, and increasing λ as more data are acquired. Another consideration when choosing λ is that the mean age of observations in the geometric moving average is $(1 - \lambda) \times (0 + \lambda + 2\lambda^2 + \cdots)$, or $\lambda/(1 - \lambda)$. This mean age should be acceptably small relative to the time scale on which detected changes of mean are of interest. Other, more sophisticated criteria for choosing λ arise from a consideration of average run lengths*, or from regarding the geometric moving average as one simple form of linear filter (*see* STATIONARY PROCESSES).

AVERAGE RUN LENGTHS

The average run length* (ARL) of a procedure for monitoring a time series in order to detect apparent failures of some hypothesis H_0 is the mean time after some initial event until the procedure rejects H_0. When the geometric moving average is used to detect apparent changes in the mean, $E(X_t)$, two average run lengths are of particular importance: the mean times to rejection of H_0: $E(X_t) =$ constant, in the cases when H_0 is true and when it is false. Ideally, the ARL in the former case should be infinite and in the latter should be zero, at least for departures from H_0 of interest to the user of the monitor.

In one procedure for detecting the failure of H_0, a one- or two-sided interval for G_t is decided on, depending on the alternative of interest, and H_0 is rejected if and when G_t moves outside this interval. ARLs for this procedure have been investigated: first by Roberts [12, 13] using simulation* procedures, and then by Hines [6, 7] using an approximation valid for $1 - \lambda$ close to zero, and by Robinson and Ho [14] using a numerical procedure. Hines obtained analytic expressions for one-sided ARLs, while Robinson and Ho produced tables of ARLs for both one- and two-sided tables and for a variety of intervals, values of λ, and shifts in mean.

THE GEOMETRIC MOVING AVERAGE AS A LINEAR FILTER

The geometric moving average can be regarded as one simple form of a linear filter* and its properties assessed by standard techniques. Particularly important in such an assessment are the gain and phase functions, obtainable by regarding the action of the geometric moving average on the time series $\{e^{i\omega t}\}$, for $0 \leqslant \omega \leqslant \pi$ (*see* STATIONARY PROCESSES). These functions are plotted in Fig. 2 for $\lambda = 0.1(0.2)0.9$. The plot of the gain function shows that rapid variations in an $\{X_t\}$ series, corresponding to ω much different from zero, are reduced, or smoothed, in a new series $\{G_t\}$. The smoothing is more severe, and affects a wider range of frequencies, as λ approaches unity. The effect of the geometric moving average on the slower variations, corresponding to small ω, can be inferred from the phase function plot. The near-linear nature of the phases near

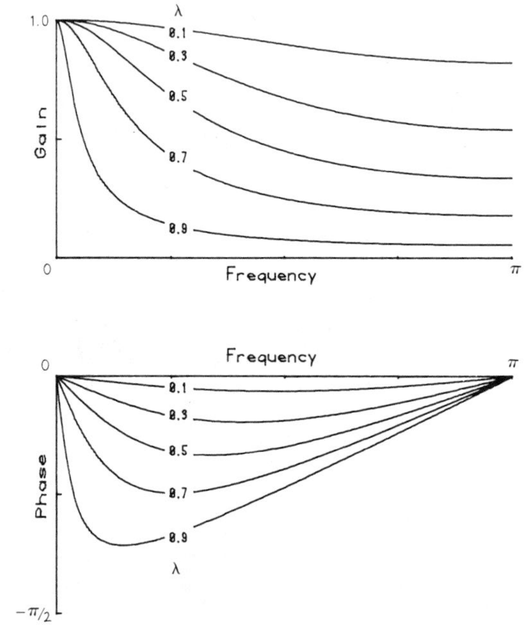

Figure 2 Gain and phase plots for the geometric moving average; $G_t = (1 - \lambda)X_t + \lambda G_{t-1}$.

$\omega = 0$ is characteristic of induced time delays which are frequency independent. For a particular value of λ, the time delay, which results from the averaging of past values with more recent ones, is found to be $\lambda/(1-\lambda)$, and can be considerable if λ is taken overly close to unity. Hence care must be taken when choosing λ to ensure that unacceptable time delays do not result.

COMPARISON WITH OTHER ESTIMATORS

The geometric moving average is one of a number of possible estimators for use with time series. Extensions of it, to cope for example with trend or seasonality, have been investigated by such researchers as Holt [9], Winters [15], Brown [2], Harrison [5], and others (see books by Brown [2], Montgomery and Johnson [10], and Granger and Newbold [4]). An extensive class of estimators and methods for choosing among them, now conventionally referred to as Box–Jenkins* methodology, have been discussed at length by Box and Jenkins [1].

While the geometric moving average is only one possible estimator of many, it can perform well. Of all estimators based on data up to time t, say $\tilde{X}_t$, of $E(X_t)$, G_t is the minimum (discounted) sum of squared deviations estimator. That is, for an estimator $\tilde{X}_t$, $\sum \lambda^{\tau-t}(X_\tau - \tilde{X}_t)^2$, summed over $\tau \leqslant t$, is minimized for $\tilde{X}_t = G_t$.

Muth [11] established that G_t is an optimal predictor of $X_{\tau+t}$, based on $\dots, X_{t-1}$, X_t, if and only if $\{X_t\}$ was generated by

$$X_t = X_{t-1} + \epsilon_t - \lambda\epsilon_{t-1}$$

for a sequence $\{\epsilon_t\}$ of independent, identically distributed (i.i.d.) random variables with zero mean and constant variance. Cox [3] compared G_t as a predictor of X_{t+1} or of X_{t+2} to the optimal linear predictor based on data up to time t when $\{X_t\}$ is a Markov* series, possibly with superimposed error. When strong positive correlation between adjacent values of the process existed, the geometric moving average was found to be little inferior to the optimal estimators for one- and two-step predictors. Hines [8] investigated the use of G_t for estimating $E(X_t) = V_t$ for the process generated by

$$V_t = \rho V_{t-1} + \eta_t,$$

$$X_t = V_t + \epsilon_t,$$

for $\{\eta_t\}$ and $\{\epsilon_t\}$ independent time series, each with i.i.d. successive values. In numeric studies of the case $\rho = 1$, he found that $\{G_t\}$ was practically as efficient an estimator of $\{V_t\}$ as the optimal estimator, based on $\dots, X_{t-1}, X_t$, at least after a slight initial inefficiency.

THE DISTRIBUTION OF $\{G_t\}$

For $\{X_t\}$ a stationary time series*, with constant mean μ_t and covariance function $\gamma_X(z)$ (see Granger and Newbold [4]), where

$$\gamma_X(z) = \sum_{\tau=-\infty}^{\infty} \operatorname{cov}(X_t, X_{t+\tau}) z^\tau,$$

the mean and covariance function of $\{G_t\}$ are, respectively, μ_t and

$$\gamma_G(z) = \gamma_X(z)/\left[(1-\lambda z)(1-\lambda z^{-1})\right].$$

For the time series $X_1, X_2, \dots, X_j, \dots$ with the X_j as i.i.d. random variables with common mean μ and cumulant* function $\phi_X(s)$ and $G_0 = \mu$, G_t does not in general have a normal or a limiting normal distribution as $t \to \infty$. The cumulant function of G_t, $\phi_G(s;t)$, is given by

$$\phi_G(s,t) = \sum_{\nu=1}^{\infty} \left[\frac{c_\nu(is)^\nu(1-\lambda)(1-\lambda^{\nu t})}{1-\lambda^\nu} \right];$$

$$\phi_G(s,\infty) = \sum_{\nu=1}^{\infty} c_\nu(is)(1-\lambda)/(1-\lambda^\nu)$$

as $t \to \infty$, where $\phi_X(s) = \sum c_\nu(is)^\nu$. The cumulant expansion was used by Robinson and Ho [14] to obtain an Edgeworth series expansion for the stationary distribution of $\{G_t\}$, leading to the average-run-length tables cited previously. Hines [6, 7] used the asymptotic normality* of the stationary distribution and the limiting correlation function of $\{G_t\}$, as $\lambda \to 1$, to obtain an approximation to $\{G_t\}$ in terms of the Ornstein–Uhlenbeck process (*see* GAUSSIAN PROCESSES), thus obtaining analytic expressions for average run lengths in three circumstances: when H_0 remains true, when H_0 has been erroneously rejected and a new (incorrect) H_0 postulated, and when H_0 has ceased to be true.

References

[1] Box, G. E. P. and Jenkins, G. M. (1970). *Time Series Analysis, Forecasting and Control*. Holden-Day, San Francisco.

[2] Brown, R. G. (1962). *Smoothing, Forecasting and Prediction of Discrete Time Series*. Prentice Hall, Englewood Cliffs, N.J.

[3] Cox, D. R. (1961). *J. R. Statist. Soc. B*, **23**, 414–422.

[4] Granger, C. J. W. and Newbold, P. (1977). *Forecasting Economic Time Series*. Academic Press, New York.

[5] Harrison, P. J. (1965). *Appl. Statist.*, **14**, 102–139.

[6] Hines, W. G. S. (1976). *IEEE Trans. Inf. Theory*, **IT-22**, 210–216.

[7] Hines, W. G. S. (1976). *IEEE Trans. Inf. Theory*, **IT-22**, 496–499.

[8] Hines, W. G. S. (1977). *Technometrics*, **19**, 313–318.

[9] Holt, C. C. (1957). Forecasting Seasonals and Trends by Exponentially Weighted Moving Averages. Carnegie Institute of Technology, Pittsburgh, Pa.

[10] Montgomery, D. C. and Johnson, L. A. (1976). *Forecasting and Time Series Analysis*. McGraw-Hill, New York.

[11] Muth, J. F. (1960). *J. Amer. Statist. Ass.*, **55**, 299–306.

[12] Roberts, S. W. (1959). *Technometrics*, **1**, 239–250.

[13] Roberts, S. W. (1966). *Technometrics*, **8**, 411–430.

[14] Robinson, P. B. and Ho, T. Y. (1978). *Technometrics*, **20**, 85–93.

[15] Winters, P. R. (1960). *Manag. Sci.*, **6**, 324–342.

Bibliography

A number of books on time series analysis are now available, with a variety of emphasis, styles, and levels of sophistication. Suitable introductory books include those given in the references, and the following.

Bowerman, B. L. and O'Connell, R. T. (1979). *Time Series and Forecasting: An Applied Approach*. Duxbury Press, North Scituate, Mass.

Fuller, W. A. (1976). *Introduction to Statistical Time Series*. Wiley, New York.

Hamming, R. W. (1977). *Digital Filters*. Prentice-Hall, Englewood Cliffs, N.J.

Kendall, M. G. (1973). *Time Series*. Hafner Press, New York.

Nelson, C. R. (1973). *Applied Time Series Analysis for Managerial Forecasting*. Holden-Day, San Francisco.

(GRADUATION
QUALITY CONTROL
SPECTRAL ANALYSIS
STATIONARY PROCESSES
TIME SERIES)

W. G. S. HINES

GEOMETRIC PROBABILITY THEORY

This subject covers problems that arise when we ascribe probability distributions to geometric objects such as points, lines, and planes (usually in Euclidean spaces), or to geometric operations such as rotations or projections. Its applications are extremely diverse and may be exemplified by the study of spatial pattern, the areal sampling of plants, the estimation of the amounts of overlap in areas of damage due to individual bombs, the estimation of the probability that the shading effect of antibodies will stop a virus from being infective, the modeling of crystals in metals, and the estimation of three-dimensional structures from two-dimensional cross sections.

Historically, the first such problem appears to be that of Buffon* in 1777 [2] (see Kendall and Moran [10]). Suppose that a needle of length L is thrown "at random" on a plane ruled with parallel straight lines at unit distance apart. Buffon easily found the probability that the needle would intersect at least one of these lines interpreting the phrase "at random" to mean that the angle θ made by the needle is uniformly distributed over $(0, 2\pi)$, and that the center of the needle is independently uniformly distributed over a unit interval on a line perpendicular to the parallel lines. For the early history of the subject, see Miles and Serra [14].

If geometric objects or operations are determined by coordinates or parameters, we usually want to apply to the parameter space a probability measure which is to be invariant when the Euclidean groups of translations, rotations, and reflections are applied to the spaces in which the geometric objects lie. Such operations then induce a transformation in the parameter space under which the probability measure is to be invariant. It is easy to see that for points, lines, and planes in Euclidean spaces this cannot be done directly, for the measure cannot be both invariant and also have unit value for the whole parameter space.

There are two ways out of this dilemma. In the first we define a measure for the whole parameter space which is invariant under the appropriate group of induced transformations. In most cases this measure is not itself a probability measure. If E is a measurable set in the probability space, write $M(E)$ for its measure and suppose that this is σ-finite. Then if E_1 is a set contained in a set E_2, we can regard $M(E_1)/M(E_2)$ as the probability that a geometric object corresponds to a point in E_1 when it is known to correspond to a point in E_2. In this way, for example, we can define the probability that a random line in the plane will hit a bounded convex figure K_1 if it is already known to hit another bounded convex figure K_2.

The other approach is to consider a Poisson field of geometric objects whose parameter points in the parameter space themselves form a Poisson field. This is defined to be a distribution of points in the parameter space such that the number lying in any measurable set E has a Poisson distribution* whose mean is equal to some specified measure $M(E)$. The definition of a Poisson field also requires the condition that the number of points lying in any set E be distributed independently of the number lying in any other set E_1 disjoint with E. Furthermore, in most problems the measure $M(E)$ will be chosen to be invariant under transformations in the parameter space induced by any Euclidean transformations of the space in which the geometric elements lie. $M(E)$ is usually not a probability measure but in the case of rotations and projections the parameter space is bounded and can easily be given a probability measure which is invariant under all rotations.

For points in Euclidean space the appropriate choice for $M(E)$ is Lebesgue measure. The one-dimensional case is of less directly *geometrical* interest but has many statistical applications. Thus if n points are independently and uniformly distributed in the unit interval $(0, 1)$, we can study the distribution of various functions of the $n + 1$ intervals that result. This is the basis of the extensive theory of spacings* and the theory of tests of goodness of fit* based on empirical distribution functions*. A more geometric problem (the "parking" problem*) is to find the distribution of the number of intervals of length $l < 1$ which can be put sequentially into a unit interval subject to the condition that they do not overlap previous intervals and that their left-hand (say) end points are otherwise uniformly distributed.

In two or more dimensions, problems concerning random points take on a more genuinely geometrical character. For example, it is not difficult to find the probability distribution of the distance between two independent points randomly and uniformly placed inside a circle or sphere, and there are a variety of similar problems. One of longstanding historical (but not practical) interest is Sylvester's problem of finding the probability that four points randomly placed inside a finite convex domain form a convex quadrilateral [20]. For a circle this probability is $1 - 35(12\pi^2)^{-1}$. The solutions of such problems are much simplified by using the remarkable theorems of Morgan Crofton [4, 5]; these relate such probabilities by a differential equation to similar probabilities conditional on one of the points lying on the boundary of the domain, and thus provide a general method for dealing with this part of the subject.

An extensive theory exists for Poisson fields of points in the plane. One can also construct more elaborate models in which the points are constrained, for example, to satisfy the condition that no two points are closer than a prescribed distance. These theories have extensive applications in the theory of plane sampling, and the analysis of spatial pattern*. Another problem, with a practical application in the design of communication networks, arises if we consider two points independently distributed with the same distribution in the plane [8]. It is then of interest to study the distribution of the distance between the two points conditional on their distances from the origin or, say, on the sum of the latter.

The appropriate measures for lines in R^2 and R^3 are easy to derive. If we parametrize lines in the plane by the polar equation $x\cos\theta + y\sin\theta = p$, the coordinates of the line can be taken as (p, θ), where $-\infty < p < \infty$, $0 \leqslant \theta < \pi$. The element of invariant measure is then $dp\, d\theta$. If we represent lines in three dimensions by the equations $x = az + p$, $y = bz + q$ (excluding lines parallel to the plane $z = 0$, which are to have zero measure), the appropriate element of invariant measure is $(1 + a^2 + b^2)^{-2}\, da\, db\, dp\, dq$. Much of the resulting theory is concerned with the intersection of such random lines with convex figures in R^2, and convex bodies in R^3, and thus links up closely with integral geometry. It is easy to show [3] that the mean length of the projection of a convex plane figure on a randomly chosen line is π^{-1} times the perimeter, while the mean area of the projection of a convex body on a random plane is one-fourth of its surface area. Crofton has given a number of other more elaborate theorems on such mean values; for example, the third moment of the length of a random chord of a convex figure is $3A^2/L$, where A is the area and L is the perimeter.

If we define planes in R^3 by their polar equation

$$x\sin\theta\cos\phi + y\sin\theta\sin\phi + z\cos\theta = p,$$

the appropriate element of invariant measure is $\sin\theta\, d\theta\, d\rho\, dp$ $(0 \leqslant \theta \leqslant \frac{1}{2}\pi,\ 0 \leqslant \phi < 2\pi,\ -\infty < p < \infty)$, and then the measure of the set of planes in R^3 that intersect a linear segment of length L is πL. Generalizing this, if we have a twisted curve of length L in R^3, and the number of intersections with a random plane with parameters (θ, ϕ, p) is $N(\theta, \phi, p)$, then the integral of the latter with respect to the invariant measure is again πL. Similarly, it can be shown that the average length of the curve that is the projection of a twisted curve of length L in R^3, on a random plane, is $nL/4$. A similar result in R^2 is the following. Suppose that C is an arbitrary rectifiable curve of length L in the plane, and $L(\theta)$ the total length of its projections on a line in the direction θ, multiple points being counted multiply. Then the mean of $L(\theta)$ over all directions is $2L/\pi$; this result is useful for estimating the lengths of empirically observed lines.

Minkowski [15] has shown that the measure of all planes meeting a convex body, K, with a differentiable surface, is equal to the integral of $\frac{1}{2}(\rho_1^{-1} + \rho_2^{-1})$ over the surface, where ρ_1, ρ_2 are the two principal radii of curvature. Write this integral as M; $M/2\pi$ is the *mean caliper diameter*. It then follows that the mean area of intersection of K with a random plane is $2\pi V/M$, where V is the volume of K. Similarly, the mean perimeter of the intersection is $\frac{1}{2}\pi^2 SM^{-1}$, where S is the surface area of K. It is also clear that if a convex body K_1 lies inside another convex body K_2, and the corresponding Minkowski integrals are M_1 and M_2, the probability that a random plane hitting K_2 also hits K_1 is $M_1 M_2^{-1}$. There are a number of other elegant similar results, and generalizations to higher dimensions.

A rotation in three-dimensional space is determined by three independent parameters lying in a bounded region, so a probability distribution can be defined which is invariant under rotations. Several different methods of parametrization are possible. One is to use the Euler angles; another obvious one is to give the direction cosines (α, β, γ, say, with $\alpha^2 + \beta^2 + \gamma^2 = 1$) of the axis of rotation together with the angle of rotation θ. For the latter the invariant probability ascribes a uniform measure to all directions of the axis and, independently of this, a probability density $\pi^{-1}\sin^2(\frac{1}{2}\theta)$ $(0 \leqslant \theta < 2\pi)$ for the angle of rotation (contrary to the intuitively plausible idea that θ should be uniformly distributed).

Another method of parametrization is to use quaternions. If (u_1, u_2, u_3, u_4) is a quaternion with a unit tensor ($\sum u^2 = 1$), an orthogonal 3×3 matrix representing a rotation can be defined with elements that are quadratic functions of the u's. The invariant probability measure is then found by imposing a uniform probability density on the three-dimensional surface of the four-dimensional sphere, $\sum u^2 = 1$, provided that opposite points are identified. This leads to one simple method for the statistical simulation of random rotations. Problems of estimating rotations occur in geophysics and in crystallography.

Many problems of practical importance are concerned with coverage* and clustering. As an example, consider the problem of determining the distribution of the number of points of a lattice which are covered by a randomly placed geometric figure (usually convex). This has practical applications in mensuration, and theoretical interest in that it led Kendall [9] to a probabilistic solution of a version of the classical unsolved problem of determining the exact order of the error term in the lattice point problem for a circle. For a general survey, *see* COVERAGE.

Another interesting field is clumping. Suppose, for example, that we have a Poisson field of points which are the centers of unit circles. We may study the clumps which these form (i.e., sets of circles which are topologically connected). Let the expected number of centers falling in unit area be λ and choose one circle at random. Let $E[N]$ be the expected number of circles which are topologically connected with the chosen one. Then it can be shown, using percolation theory, that there is a finite nonzero number λ_0 such that if $\lambda < \lambda_0$, $E[N]$ is finite, while if $\lambda > \lambda_0$, $E[N]$ is infinite. The exact value of λ_0 is not known but there are estimates obtained by Monte Carlo methods*. A similar result holds in R^3. Other problems of clumping have useful applications in microscopy.

There is an extensive body of theory (mainly due to Miles [12, 13]) concerned with the random division of Euclidean spaces. The simplest model is obtained by considering random lines in a plane; suppose that we have a Poisson field of lines in a plane with a density $\lambda\, dp\, d\theta$, where $-\infty < p < \infty$, $0 \leqslant \theta < \pi$. These divide the plane into polygonal regions. It is not difficult to show that their average area is $(\pi\lambda^2)^{-1}$. Similarly, the average number of sides is four. Writing N, S, and A for the number of vertices, the perimeter, and the area of the polygons, respectively, all the first- and second-order moments of N, S, and A, and some of the third-order moments, are known. Similar results are known for the random polyhedra formed by a Poisson field of random planes in R^3.

Another useful model for the random division of Euclidean space is formed by random Voronoi tessellations (and polyhedra) which can be defined as follows. Suppose that we have a homogeneous Poisson distribution of points in R^2 or R^3; for any point P of this process, define a corresponding polygon (polyhedron) to consist of all the points of the space which are closer to P than to any other point of the process. The first- and some of the second-order moments of the elementary geometrical measures of this polygon (polyhedron) and its edges, (faces), and vertices are known. Other, more elaborate models for the division of space have also been studied.

Stereology* is a subject based on geometrical probability with many applications. Suppose that a three-dimensional solid body is heterogeneous and consists of two or more different phases. For example, particles of one substance might be distributed in a matrix of a different material. We could explore the density and geometrical character of the inclusion by examining a random two-dimensional section of the body, or sometimes a one-dimensional line across the latter. It is then possible to make probabilistic inferences about the three-dimensional structure. This subject has practical applications in anatomy, metallurgy, and other sciences.

The literature on geometric probability theory is widely scattered because of the wide variety of applications. Introductory accounts are given in Kendall and Moran [10] and Solomon [19]. The bibliography in the first of these has been supplemented by four survey articles: Moran [16, 17], Little [11], and Baddeley [1]. Santaló [18] gives a comprehensive account of the theory, especially in its relation to integral geometry.

References

[1] Baddeley, A. (1977). *Adv. Appl. Prob.*, **9**, 824–860.

[2] Buffon, G. (1777). Essai d'arithmétique morale. Supplément à *l'Histoire naturelle*, Vol. 4.

[3] Cauchy, A. (1850). *Mém. Acad. Sci. Paris*, **22**, 3–13.

[4] Crofton, M. W. (1869). *Philos. Trans. R. Soc.*, **158**, 181–199.

[5] Crofton, M. W. (1877). *Proc. Lond. Math. Soc.*, **8**, 304–309.

[6] Crofton, M. W. (1885). Probability. In *Encyclopedia Britannica*, 9th ed., pp. 768–788.

[7] Davy, P. J. and Miles, R. E. (1977). *J. R. Statist. Soc. B*, **39**, 56–65.

[8] Gilbert, E. N. (1977). *J. Appl. Prob.*, **14**, 260–271.

[9] Kendall, D. G. (1948). *Quart. J. Math.*, **19**(2), 1–26.

[10] Kendall, M. G. and Moran, P. A. P. (1963). *Geometrical Probability*. Charles Griffin, London.

[11] Little, D. V. (1974). *Adv. Appl. Prob.*, **6**, 103–130.

[12] Miles, R. E. (1972). Suppl., *Adv. Appl. Prob.*, 243–266.

[13] Miles, R. E. (1973). *Adv. Math.*, **10**, 256–290.

[14] Miles, R. E. and Serra, J. (1978). Geometric Probability and Biological Structures; Buffon's 200th Anniversary. *Lec. Notes Biomath.*, **23**.

[15] Minkowski, H. (1903). *Math. Ann.*, **57**, 447–495.

[16] Moran, P. A. P. (1966). *J. Appl. Prob.*, **3**, 453–463.

[17] Moran, P. A. P. (1969). *Adv. Appl. Prob.*, **1**, 73–89.

[18] Santaló, L. A. (1976). Integral Geometry and Geometric Probability. In *Encyclopedia of Mathematics and Its Applications*, Vol. 1. Addison-Wesley, Reading, Mass.

[19] Solomon, H. (1978). *Geometric Probability*. SIAM, Philadelphia.

[20] Sylvester, J. J. (1891). *Acta Math.*, **14**, 185–205.

(COVERAGE
GEOMETRY IN STATISTICS
SPATIAL PATTERN ANALYSIS)

P. A. P. MORAN

GEOMETRY IN STATISTICS

The geometry of Euclidean n-space illustrates much of statistics. Because the relationships between points, lines, and planes as realized in Euclidean 3-space are familiar to many, geometry provides a means of picturing much of the mathematics of statistics. Because distributions, parameters, statistics, designs, and a great deal of the other mathematical paraphernalia of statistics are concerned with real numbers, sketches of Euclidean 1-, 2-, or 3-space are natural accompaniments to the study of statistical ideas. Because to a large degree statistics is concerned with extracting information from a set of n real numbers, the connection between statistics and Euclidean n-space is inherent and thus the geometry of n-space is inherently a part of statistics.

This illustrative role of geometry seems by far its most pervasive use in statistics. The role of geometry as an illustrator or illuminator of statistics is not confined to providing sketches in Euclidean 1-, 2-, or 3-space, however. Geometric ideas that elicit some knowing response from a reader can sometimes be connected with a difficult-to-understand statistical idea and the association thus formed can provide a deeper understanding of the statistical idea. For example, in the study of the invariance* structure of a statistical methodology the word "orbit" is used to define the equivalence class of all data equivalent under a group of transformations. "Orbit" provides a more suggestive word picture than "equivalence class," especially when the transformations are rotations. Or a geometric idea can serve as a connection between seemingly unrelated statistical ideas. An example of this is provided by Efron [8] in relating nearness of a distribution to the family of one-parameter exponential distributions and the second-order efficiency of Fisher [10] and Rao [22] via the curvature of the distribution.

Perhaps the prime example of the use of illustrative geometry in statistics is in the study of linear models. Illustrative geometry in vector space theory provides pictures, insights, and connectors for this study. Some examples that will be discussed subsequently are the picture which makes clear the solution of the problem of least-squares* estimation, the insights into conditional inverses provided by the geometry, and hypothesis testing* in linear models. A formal, more deductive role of geometry in statistics is provided by the work of Bose [4] wherein a finite, incidence geometry called a partial geometry is invented for the purpose of studying association schemes of partially balanced, incomplete block designs*.

The balance of this article will be devoted to four topics: a brief history of the use of geometry in statistics, an example of the use of formal geometry in statistics, several examples of the illustrative use of geometry in linear models, and two illustrative examples outside the field of linear models.

A BRIEF HISTORY OF GEOMETRY IN STATISTICS

Since R. A. Fisher* is widely recognized as using geometry in his work, specifically in the derivation of the distribution of the correlation* coefficient [9], he seems to provide a benchmark on the use of geometry in statistics. However, it is very difficult to say that any particular statistician did not use geometry. For example, Francis Galton*, who first plotted bivariate data, noticed that contours of equal intensity appeared to be ellipses. This visual aspect of geometry led him, with the help of a mathematician Hamilton Dixon, to the bivariate normal distribution and a measure of what Galton called "co-relation." He used the picture of an ellipse in his work on regression to explain and illustrate the joint variability of mid-parent height and adult offspring height [11] (*see* CORRELATION).

In his paper, Fisher is concerned with the distribution of the correlation coefficient of a sample of n pairs from a bivariate normal distribution. He does not help the reader very much with formulas, as can be seen in this excerpt from his discussion of transforming the two n-variate normal density of sample pairs (x_i, y_i) to a density in terms of the quantities $\bar{x}$, $\mu_1 = [n^{-1}\sum(x-\bar{x})^2]$, $\bar{y}$, $\mu_2 = [n^{-1}\sum(y-\bar{y})^2]$, and the sample correlation coefficient r.

> An element of volume in this n dimensional space may now without difficulty be specified in terms of $\bar{x}$ and μ_1; for given $\bar{x}$ and μ_1, P (the point $(x_1, x_2, \ldots, x_n)$) must lie on a sphere in $n-1$ dimensions, lying at right angles to the line OM (the line through the origin in the direction $(1, 1, \ldots, 1)$), and the element of volume is $C\mu_1^{n-2}\,d\mu_1\,d\bar{x}$, where C is some constant, which need not be determined.

If you see it, it is elegant; if you do not, there is little to help improve your vision. It seems the kind of discussion that inspires the reader to honor the genius that produced it, but does not inspire him or her to emulate the approach. In Fisher's case, the illustration becomes the work. Mahalanobis* in his biographical sketch of Fisher suggests why.

> On account of his (Fisher's) eyes he was forbidden to work by artificial light, and when he would go to work with Mr. Roseveare in the evenings the instruction was given purely by ear without the use of paper and pencil or any other visual aid. To this early training may be attributed Fisher's tendency to use hypergeometrical representation, and his great power in penetrating into problems requiring geometrical intuition. It is, perhaps, a consequence of this that throughout life his solutions have been singularly independent of symbolism. [18]

In any case Fisher's use of geometry does not seem to have inspired a popular turn toward geometry by statisticians in the first half of this century.

Some notable exceptions were Bose, Roy, and Mahalanobis. Bose [3] finds the exact distribution of the Mahalanobis distance* D^2 for normal data. His method is similar to Fisher's [9], although it is done in more detail and is thus easier to follow. Mahalanobis et al. [19] define the $p(p+1)/2$ rectangular coordinates which can be associated with a sample of size n from a p-variate normal distribution by using the Gram–Schmidt orthogonalization* process to convert the p n-tuples to mutually orthogonal unit vectors. They use a geometric point of view with appropriate pictures for $p = 3$, but do not appear to know they are using the Gram–Schmidt process.

At about the same time Bartlett [1] was thinking of a sample of size n as an n-dimensional vector. His approach might be termed analytic geometric in that each geometric idea is represented by analytic formulas. In the sense that the geometry was central rather than just illustrative, Bartlett's work was similar to that of Fisher, Bose, et al., and different from Galton's. An example is provided by Bartlett's discussion of the analysis of the row vector **S** of observations from a Latin square* design.

> We have a classification in rows, columns, and treatments. We write
>
> $$\mathbf{S} = \mathbf{R} + \mathbf{C} + \mathbf{T} + \mathbf{E} + \mathbf{M},$$
>
> where $\mathbf{R} = (\bar{x}_r - \bar{x})$ is the vector representing the differences of row means from the general mean, $\mathbf{C} = (\bar{x}_c - \bar{x})$ similarly for columns, $\mathbf{T} = (\bar{x}_t - \bar{x})$ for treatments, $\mathbf{M} = (\bar{x})$ as before, and
>
> $$\mathbf{E} = (x - \bar{x}_r - \bar{x}_c - \bar{x}_t + 2\bar{x})$$
>
> is the residual error term. From the algebraic relations
>
> $$\mathbf{RC}' = \mathbf{RT}' = \cdots = \mathbf{EM}' = 0,$$
>
> we have, analogously to the first relation
>
> $$\mathbf{S}^2 = \mathbf{R}^2 + \mathbf{C}^2 + \mathbf{T}^2 + \mathbf{E}^2 + \mathbf{M}^2.$$

He had previously mentioned that the algebraic relations meant the corresponding vectors were perpendicular and that $\mathbf{S}^2$ would be used for the squared length of **S**. Although the notation is rather lean, the ideas and approach have the fullness of the modern approach of Kruskal, for example.

In their study of the geometry of estimation, Durbin and Kendall [6] seem to revert to the purer geometric approach of Fisher. As an example, consider their discussion of finding the minimum variance, linear unbiased estimator of the common mean of a sample $x_1, x_2, \ldots, x_n$ of values of independent, identically distributed random variables with common variance σ^2. The estimate is of the form $\sum \lambda_j x_j$ with the unbiasedness* restriction $\sum \lambda_j = 1$.

> Consider now a Euclidean [n] space with co-ordinates $\lambda_1, \ldots, \lambda_n$, which we call the estimator space. The hyperplane $\sum \lambda_j = 1$ corresponds to the range of values of λ giving unbiased estimators and any point P in it determines just one estimator. Now the variance of the estimator is $\sigma^2 \sum \lambda_j^2$ and hence is $\sigma^2 OP^2$, where O is the origin. It follows that this is a minimum when P is the foot of the perpendicular from O on to the hyperplane. Symmetry alone is enough to show that the values of the λ's are then all equal.

A solution more in the spirit of Bartlett might run as follows. Suppose that $\boldsymbol{\lambda}$ is the n-tuple of λ values and $\mathbf{j}$ the n-tuple of 1's. Every such $\boldsymbol{\lambda}$ can be decomposed into a component in the direction of $\mathbf{j}$ and a component perpendicular to that direction, e.g., $\boldsymbol{\lambda} = \alpha\mathbf{j} + \mathbf{z}$ for α real and $\mathbf{z}$ perpendicular to $\mathbf{j}$. The condition $\sum \lambda_i = 1$ is $\boldsymbol{\lambda}\mathbf{j}' = 1$, using row vectors as Bartlett did. Since $\boldsymbol{\lambda}\mathbf{j}' = \alpha n + \mathbf{z}\mathbf{j}' = \alpha n$, it follows that $\alpha = n^{-1}$. Pythagoras's theorem gives $\|\boldsymbol{\lambda}\|^2 = n^{-2}\|\mathbf{j}\|^2 + \|\mathbf{z}\|^2$ for $\|\cdot\|^2$ as squared length. Thus the minimum length $\boldsymbol{\lambda}$ is $n^{-1}\mathbf{j}$.

With Kruskal [15] the phrase *coordinate-free* may have been introduced into the statistical vocabulary, although he almost certainly got it from Halmos [12]. *Coordinate-free* means that an analytic geometric approach is being considered in the spirit of Bartlett [1]. The reason for this name is to emphasize that an abstract vector-space point of view is being adopted and no particular coordinate system is being used. Thus a discussion involving n-tuples is not coordinate-free, since the writing down of n-tuples involves the use of the "usual" coordinate system in Euclidean n-space. A somewhat pithier explanation is that coordinate-free means free to use whatever coordinate system seems most appropriate at the moment. Kruskal [15] notes that "it is curious the coordinate-free approach to Gauss–Markov estimation, although known to statisticians, has infrequently been discussed in the literature on least squares and analysis of variance." He indicates that there are two major motivations for emphasizing the coordinate-free approach.

> First, it permits a simpler, more general, more elegant, and more direct treatment of the general theory of linear estimation than do its notational competitors, the matrix and scalar approaches. Second, it is useful as an introduction to infinite-dimensional spaces, which are important, for example, in the consideration of stochastic processes.

Kruskal credits L. J. Savage with introducing him to the coordinate-free approach. The second section of Kruskal [15] provides a succinct primer on the coordinate-free approach. It seems that Kruskal hoped his paper would encourage more statisticians to adopt this approach to linear models. It does not appear that this hope was realized during the next 10 years or so.

In stating conditions under which simple least-squares estimators are also best linear unbiased estimators, Zyskind [29] refers to "r orthogonal eigen-vectors" of the dispersion matrix forming "a basis for the column space" of the design matrix. This is the precursor of the more geometric work that Zyskind did with his student Justus Seely [25]. It is interesting to note that Kruskal [15] is not referenced in Zyskind's paper.

Following on Zyskind [29], Watson [28] illustrates Kruskal's contention that coordinate-free linear models are closely related to stochastic processes, for Watson makes extensive and effective use of the spectral decomposition of the dispersion matrix to study the error vectors in least-squares regression*. This spectral decomposition identifies n mutually orthogonal eigenvectors of the dispersion matrix; these n vectors can serve as a basis or coordinate system for n-space. This use of a convenient basis rather than one fixed at the outset is an excellent illustration that "coordinate-free" does not mean freedom from coordinates as much as freedom to choose appropriate coordinates for the task at hand. That Watson is thinking geometrically is illustrated by the following solution to least-squares regression.

> If a perpendicular is dropped from the point in n-space with position $\mathbf{y}$ onto the regression space, the foot of the perpendicular is $\mathbf{Xb}$, where $\mathbf{b}$ is the least squares estimate of $\boldsymbol{\beta}$.

Watson is aware of Kruskal's work in as much as he foretells the existence of Kruskal [16], but he still does not reference Kruskal [15].

In Kruskal [16], the question of equality of simple least squares and best linear unbiased estimates, which was considered in Zyskind [29] and Watson [28], is treated using a coordinate-free approach. Comparison of the parts of the three papers dealing with this question is very instructive; the simplicity and beauty of the coordinate-free approach is clearly demonstrated.

In Kruskal [17] a geometric view of generalized inverses* is presented with such skill and grace that the paper ought to be required reading for anyone who might be tempted to deal with generalized inverses. A more detailed look at this work will appear in the section on linear models.

In this brief history only a few papers have been discussed. They do not, of course, tell the whole story. As has been mentioned, L. J. Savage evidently was instrumental in getting Kruskal interested. R. C. Bose, whose notes on linear models were used for years by graduate students at Chapel Hill, has, through these notes, acquainted a large segment of the statistical profession with the comprehensiveness of the geometric approach to linear models, although he did not stress the geometry in his lectures. Seber [24] obviously appreciates the geometric ideas inherent in linear models. The book by Scheffé [23] is a classic in which the geometric ideas appear as asides, as though Scheffé appreciated the elegance of the geometry but did not believe the book would be accepted if it were all done geometrically. The dust jacket of this book features **the** picture for illustrating the geometry of hypothesis testing in a linear model.

At present there seems to be more interest in using geometry in statistics, especially in linear models, than prior to 1970. This is only an impression and may be an artifact of the explosion in publication generally during this period (see the bibliography for the present article).

AN EXAMPLE OF A FORMAL GEOMETRY IN STATISTICS

Bose [4] introduces the *partial geometry*, which serves to unify and generalize certain theorems on embedding of nets and the uniqueness of association schemes of partially balanced designs. A partial geometry is a rather restricted type of incidence geometry. Incidence geometries traditionally have been used to study graph theory* and Bose uses graph theoretic methods for studying association schemes of partially balanced designs. The partial geometry he invents is labeled by the triple (r, k, t); it is defined as a system of undefined points and lines and an undefined relation incidence, satisfying the following four axioms.

A1. Any two points are incident with not more than one line.

A2. Each point is incident with r lines.

A3. Each line is incident with k points.

A4. If the point p is not incident with the line l, there pass through p exactly t lines ($t \geqslant 1$) intersecting l.

He shows that the number v of points and the number b of lines in the partial geometry (r, k, t) are given by

$$v = k[(r-1)(k-1)+t]/t$$

$$b = r[(r-1)(k-1)+t]/t.$$

Then the graph G of a partial geometry (r, k, t) is defined as follows. The vertices of G are the points of the partial geometry. Two vertices are joined if the corresponding points of the geometry are incident with the same line. Two vertices of G are unjoined if the corresponding points of the partial geometry are not incident with the same line. He then shows that the graph of a partial geometry is strongly regular. Given a strongly regular graph G, Bose identifies its v vertices with v treatments. Then a partially balanced incomplete blocks (PBIB) design* is an arrangement of the v treatments into b sets (called blocks) with the following properties.

1. Each treatment is contained in exactly r blocks.
2. Each block contains k distinct treatments.
3. Any two treatments which are first associates (joined in G) occur together in exactly λ_1 blocks. Any two treatments which are second associates (unjoined in G) occur together in λ_2 blocks.

Such a design may be called a PBIB design $(r, k, \lambda_1, \lambda_2)$ based on the strongly regular graph G. Thus given a partial geometry (r, k, t) with graph G, it is clearly a PBIB design $(r, k, 1, 0)$.

Using these ideas Bose is able to prove several results, including the following theorem.

Theorem. Consider a PBIB $(r, k, \lambda_2, \lambda_1)$ design, $\lambda_1 > \lambda_2$, based on a strongly regular graph G with $d = k - r + 1$ and parameters

$$n_1 = (d-1)(k-1)(k-t)/t,$$

$$n_2 = d(k-1),$$

$$p_{11}^1 = [(d-1)(k-1)(k-t) - d(k-t-1) - t]/t,$$

$$p_{11}^2 = (d-1)(k-t)(k-t-1)/t.$$

The design can be extended by adding new blocks, containing the same treatments, in such a way that the extended design is a balanced incomplete block design* with $r_0 = r + d(\lambda_1 - \lambda_2)$ replications, block size k, and in which every pair of treatments occurs together in λ_1 blocks provided that

$$k > \tfrac{1}{2}[d(d-1) + t(d+1)(d^2 - 2d + 2)].$$

(*See* BALANCING IN EXPERIMENTAL DESIGN: GENERAL BALANCE.)

ILLUSTRATIVE GEOMETRY IN LINEAR MODELS

Least-Squares Estimation

Let $\mathbf{y} = (y_1, y_2, \ldots, y_n)'$ be a vector of data that lies in Euclidean n-space, R^n. Assume that there is a parameter vector $\boldsymbol{\beta} = (\beta_1, \beta_2, \ldots, \beta_k)'$ in R^k, $k \leqslant n$ and an $n \times k$ matrix $\mathbf{X}$ so that $\mathbf{y} = \mathbf{X}\boldsymbol{\beta} + \mathbf{error}$. The problem is to estimate $\boldsymbol{\beta}$. Using least squares* as a criterion, choose as estimates of β_i those values $\hat{\beta}_i$ which minimize

$$Q = \sum_{i=1}^{n} \left(y_i - \sum_{j=1}^{k} x_{ij}\beta_j \right)^2,$$

where x_{ij} is the (i, j)th element of $\mathbf{X}$. A more useful way of writing Q is

$$Q = \|\mathbf{y} - \mathbf{X}\boldsymbol{\beta}\|^2,$$

where $\| \cdot \|$ is the distance function in Euclidean n-space. In this format the geometry will make the solution clear (see Fig. 1). Notice that $\mathbf{X}\boldsymbol{\beta}$ is an n vector, constrained to be a linear combination of the columns of $\mathbf{X}$. Thus $\mathbf{X}\boldsymbol{\beta}$ is constrained to lie in the subspace of R^n spanned by the columns of $\mathbf{X}$. Denote this subspace by $[\mathbf{X}]$; Q is the squared distance from $\mathbf{y}$ to the point $\mathbf{X}\boldsymbol{\beta}$ in $[\mathbf{X}]$. Thus the problem is to find the point in $[\mathbf{X}]$ closest to $\mathbf{y}$. Having stated the problem geometrically, the answer is clear, the point in $[\mathbf{X}]$ directly below $\mathbf{y}$, i.e., the perpendicular projection of $\mathbf{y}$ on $[\mathbf{X}]$. If $\mathbf{X}\hat{\boldsymbol{\beta}}$ denotes the perpendicular projection of $\mathbf{y}$ on $[\mathbf{X}]$, then $\mathbf{X}\hat{\boldsymbol{\beta}}$ is unique and satisfies

$$\mathbf{y} = \mathbf{X}\hat{\boldsymbol{\beta}} + \mathbf{z}, \qquad \mathbf{z} \text{ perpendicular to } [\mathbf{X}].$$

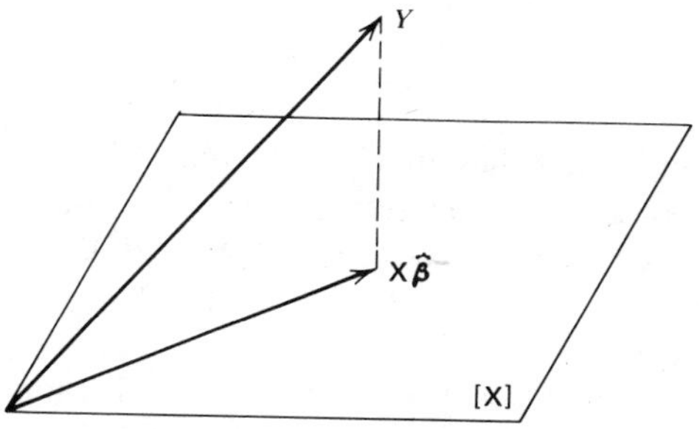

Figure 1 Least-squares estimate.

Since $\mathbf{z}$ perpendicular to $[\mathbf{X}]$ means that $\mathbf{z}$ is perpendicular to the columns of $\mathbf{X}$, it follows that $\mathbf{X}'\mathbf{z} = \mathbf{0}$. Thus $\hat{\boldsymbol{\beta}}$ must satisfy $\mathbf{X}'\mathbf{y} = \mathbf{X}'\mathbf{X}\hat{\boldsymbol{\beta}}$, which are referred to as the normal equations*. Notice that by the Pythagorean theorem,

$$\|\mathbf{y} - \mathbf{X}\boldsymbol{\beta}\|^2 = \|\mathbf{y} - \mathbf{X}\hat{\boldsymbol{\beta}}\|^2 + \|\mathbf{X}\boldsymbol{\beta} - \mathbf{X}\hat{\boldsymbol{\beta}}\|^2$$

for any $\boldsymbol{\beta}$ in R^k. Thus

$$\|\mathbf{y} - \mathbf{X}\hat{\boldsymbol{\beta}}\|^2 \leqslant \|\mathbf{y} - \mathbf{X}\boldsymbol{\beta}\|^2 \quad \text{for all } \boldsymbol{\beta} \in R^k,$$

which is the analytic version of "the point in $[\mathbf{X}]$ directly below $\mathbf{y}$ is the closest point."

Not only did geometry illustrate this problem but it pointed the way toward the solution as well. But that is not all. The geometry can also help isolate the critical elements of the solution so that the problem and its solution can be generalized. The critical elements in this case appear to be a vector space, a subspace, a distance, and a notion of perpendicularity (orthogonality), which together with the distance preserve the Pythagorean Theorem. Here is a possible generalization. Let V be a finite-dimensional vector space with an inner product $\langle \cdot, \cdot \rangle$ and a distance $\| \cdot \|$ so that $\langle \mathbf{v}, \mathbf{v} \rangle = \|\mathbf{v}\|^2$. For any two vectors $\mathbf{v}_1$ and $\mathbf{v}_2$ in V, define $\mathbf{v}_1 \perp \mathbf{v}_2$ ($\mathbf{v}_1$ perpendicular to $\mathbf{v}_2$) iff $\langle \mathbf{v}_1, \mathbf{v}_2 \rangle = 0$.

Let U be a subspace of V. Let $\mathbf{y}$ belong to V. The problem is to find the vector $\mathbf{u}$ in U which is closest to $\mathbf{y}$. If $\mathbf{X}$ is a linear transformation from a vector space W into V with range U, then the problem can be stated as before: Find $\boldsymbol{\beta}$ in W so that $\|\mathbf{y} - \mathbf{X}\boldsymbol{\beta}\|^2$ is minimum.

Now the vector in U closest to $\mathbf{y}$ ought to be the point in U "directly below" $\mathbf{y}$, i.e., $\mathbf{P}(\mathbf{y}; U)$ the perpendicular projection of $\mathbf{y}$ on U. $\mathbf{P}(\mathbf{y}; U)$ is unique and

$$\mathbf{y} = \mathbf{P}(\mathbf{y}; U) + \mathbf{z}, \qquad \mathbf{z} \perp U.$$

Notice that $\mathbf{y} - \mathbf{P}(\mathbf{y}; U) \perp U$, so that

$$\langle \mathbf{y} - \mathbf{P}(\mathbf{y}; U), \mathbf{u} \rangle = 0 \qquad \text{for all } \mathbf{u} \text{ in } U.$$

Thus

$$\begin{aligned} \|\mathbf{y} - \mathbf{u}\|^2 &= \|\mathbf{y} - \mathbf{P}(\mathbf{y}; U) + \mathbf{P}(\mathbf{y}; U) - \mathbf{u}\|^2 \\ &= \|\mathbf{y} - \mathbf{P}(\mathbf{y}; U)\|^2 + \|\mathbf{P}(\mathbf{y}; U) - \mathbf{u}\|^2, \end{aligned}$$

since $\mathbf{P}(\mathbf{y}; U) - \mathbf{u}$ is in U. The Pythagorean theorem holds, and as before

$$\|\mathbf{y} - \mathbf{P}(\mathbf{y}; U)\|^2 \leqslant \|\mathbf{y} - \mathbf{u}\|^2 \quad \text{for all } \mathbf{u} \in U.$$

Hence in this more general setting the least-squares estimate of $\mathbf{y}$ in U is $\mathbf{P}(\mathbf{y}; U)$.

If U is the range of a transformation $\mathbf{X}$ from another inner product space W, then $\mathbf{P}(\mathbf{y}; U) = \mathbf{X}\hat{\boldsymbol{\beta}}$ for some $\hat{\boldsymbol{\beta}}$ in W. The normal equations for $\hat{\boldsymbol{\beta}}$ are formally the same,

$$\mathbf{X}'\mathbf{y} = \mathbf{X}'\mathbf{X}\hat{\boldsymbol{\beta}},$$

but now $\mathbf{X}'$ is the adjoint of the transformation $\mathbf{X}$ with respect to the inner products on V and W. If $\{\mathbf{w} : \mathbf{X}\mathbf{w} = \mathbf{0}\} = \{\mathbf{0}\}$ (the null space of $\mathbf{X}$ consists of $\mathbf{0}$), then $\hat{\boldsymbol{\beta}} = (\mathbf{X}'\mathbf{X})^{-1}\mathbf{X}'\mathbf{y}$.

An application of this more general problem is in finding the best (minimum variance) linear unbiased estimate of $\boldsymbol{\beta}$ when the "error" in the original problem has expected value zero and dispersion matrix $\boldsymbol{\Gamma}$. Let $V = R^n$, $U = [\mathbf{X}]$, $W = R^k$, and $\langle \mathbf{v}_1, \mathbf{v}_2 \rangle = \mathbf{v}_1'\boldsymbol{\Gamma}^{-1}\mathbf{v}_2$. Then $\hat{\boldsymbol{\beta}} = (\mathbf{X}'\boldsymbol{\Gamma}^{-1}\mathbf{X})^{-1}\mathbf{X}'\boldsymbol{\Gamma}^{-1}\mathbf{y}$, where $\mathbf{X}'$ is the transpose of $\mathbf{X}$, is the generalized least-squares estimate and the best linear unbiased estimate, courtesy of the Gauss–Markov theorem*. Kruskal's [16] investigation of when the least squares and the generalized least-squares estimates are the same is particularly elegant.

Testing Linear Hypotheses

Let $\mathbf{y} = (y_1, y_2, \ldots, y_n)'$ be a vector of data in Euclidean n-space, R^n. Assume that there is a parameter vector $\boldsymbol{\beta} = (\beta_1, \beta_2, \ldots, \beta_k)'$ which lies in R^k, $k \leqslant n$, and an $n \times k$ matrix $\mathbf{X}$ so that $\mathbf{y} = \mathbf{X}\boldsymbol{\beta} + \textbf{error}$. The problem here is to test linear hypotheses about the β_i's.

To create an appropriate linear model assume that $\mathbf{Y}$ is a random vector whose observed value is $\mathbf{y}$ and further assume that $\mathbf{e} = \mathbf{Y} - \mathbf{X}\boldsymbol{\beta}$ is a multivariate normal* random vector with mean zero and dispersion matrix $\sigma^2\mathbf{I}$, $\sigma^2 > 0$. Thus the linear model is

$$\mathbf{Y} = \mathbf{X}\boldsymbol{\beta} + \mathbf{e}, \qquad \mathbf{e} \sim N(\mathbf{0}, \sigma^2\mathbf{I}).$$

For a specific example assume a one-way analysis-of-variance* model. Further, assume a cell mean model, i.e., β_i is the mean of the ith population. Then in partitioned form $\mathbf{X}$ is given by

$$\mathbf{X} = \begin{bmatrix} \mathbf{j}_{n_1} & \cdots & \mathbf{0} & \cdots & \mathbf{0} \\ \vdots & & & & \vdots \\ \mathbf{0} & \cdots & \mathbf{j}_{n_2} & \cdots & \mathbf{0} \\ \vdots & & & \ddots & \vdots \\ \mathbf{0} & \cdots & \mathbf{0} & \cdots & \mathbf{j}_{n_k} \end{bmatrix}$$

for $\mathbf{j}_m$ an $m \times 1$ vector of 1's and n_i the size of the sample from the ith population; $n = n_1 + n_2 + \cdots + n_k$. Thus $[\mathbf{X}]$, the linear subspace spanned by the columns of $\mathbf{X}$, is a k-dimensional subspace of R^n.

If $\mathbf{M}$ is any $p \times q$ matrix, let $[\mathbf{M}]$ denote the linear subspace spanned by the columns of $\mathbf{M}$. Suppose that A and B are linear subspaces of R^n. Then $A + B = \{\mathbf{a} + \mathbf{b} : \mathbf{a} \in A, \mathbf{b} \in B\}$ is another, often larger linear subspace of R^n. The set of vectors in $A + B$ which are orthogonal to B is denoted by $A \mid B$ and is read A *adjusted for* B. The subspace $A \mid B$ is the orthogonal complement of B in $A + B$. In some cases $B \subset A$, so that $A + B = A$; this is so with the subspaces $[\mathbf{X}]$ and $J = [\mathbf{j}_k]$, i.e., $J \subset [\mathbf{X}]$, so $[\mathbf{X}] + J = [\mathbf{X}]$. Finally, for any linear subspace A, let $\mathbf{P}(\mathbf{v} : A)$ denote the perpendicular projection of $\mathbf{v}$ on the subspace A.

Let $J_k = [\mathbf{j}_k]$ and $J = [\mathbf{j}_n]$. Then $[\mathbf{X}] \mid J$ is the subspace of all vectors in the estimation space $[\mathbf{X}]$ which are orthogonal to the mean line J. Note that the mean vector $E\mathbf{Y} = \mathbf{X}\boldsymbol{\beta}$. Consider now the usual hypothesis H, which, placed in a geometric setting, has the following equivalent statements:

$$H : \beta_1 = \beta_2 = \cdots = \beta_k;$$
$$H : \boldsymbol{\beta} \in J_k;$$
$$H : E\mathbf{Y} \in J;$$
$$H : E\mathbf{Y} \perp [\mathbf{X}] \mid J.$$

Now whether or not H is true, the mean vector $E\mathbf{Y}$ lies in $[\mathbf{X}]$. That is, $E\mathbf{Y}$ corresponds to a direction in $[\mathbf{X}]$. The question posed by the hypothesis is whether that direction is in J or not. Put differently, if H is false, then $E\mathbf{Y}$ is not in J and must have a component in $[\mathbf{X}] \mid J$, the orthogonal complement of J in $[\mathbf{X}]$ ($\mathbf{P}(E\mathbf{Y}; [\mathbf{X}] \mid J) \neq 0$), and conversely. Thus the directions in $[\mathbf{X}]|J$ correspond to violations of H in the sense that if $E\mathbf{Y}$ has a component in any one of those directions, the hypothesis is false. In this example $G = [\mathbf{X}]|J$ will be called the subspace corresponding to violations of the hypothesis H.

Consider now the data $\mathbf{y}$. The vector $\mathbf{y}$ can be decomposed (analyzed) into a component in J and one orthogonal to J, i.e., one in $J^{\perp} = R^n|J$ (see Fig. 2). The component in $J^{\perp}$, $\mathbf{P}(\mathbf{y}; J^{\perp})$, measures the variability of $\mathbf{y}$ about J in the sense that $\|\mathbf{P}(\mathbf{y}; J^{\perp})\|^2$ is the squared distance of $\mathbf{y}$ from J. This variability can be further analyzed as

$$\mathbf{P}(\mathbf{y}; J^{\perp}) = \mathbf{P}(\mathbf{y}; [\mathbf{X}]|J) + \mathbf{P}(\mathbf{y}; R^n|[\mathbf{X}]).$$

Since $[\mathbf{X}]|J \subset [\mathbf{X}]$ and $[\mathbf{X}]$ is perpendicular to $R^n|[\mathbf{X}]$, the Pythagorean theorem yields

$$\|\mathbf{P}(\mathbf{y}; J^{\perp})\|^2 = \|\mathbf{P}(\mathbf{y}; [\mathbf{X}]|J)\|^2 + \|\mathbf{P}(\mathbf{y}; R^n|[\mathbf{X}])\|^2,$$

which can be considered as a one-way analysis of the variability of $\mathbf{y}$. The magnitude of $\|\mathbf{P}(\mathbf{y}; [\mathbf{X}]|J)\|^2$ can be due to random error or to the direction of the mean vector $E\mathbf{Y}$ or both, whereas the size of $\|\mathbf{P}(\mathbf{y}; R^n|[\mathbf{X}])\|^2$ can be due only to random error. Since the spaces $[\mathbf{X}]|J$ and $R^n|[\mathbf{X}]$ are generally of different dimensions, $k-1$ and $n-k$, respectively, the corresponding variation should be compared on a per dimension basis. If a ratio comparison is made, large values of

$$\frac{\|\mathbf{P}(\mathbf{y}; [\mathbf{X}]|J)\|^2/(k-1)}{\|\mathbf{P}(\mathbf{y}; R^n|[\mathbf{X}])\|^2/(n-k)}$$

would indicate that there is rather more variability measured by the numerator than can reasonably be accounted for by random error.

The quantity $\|\mathbf{P}(\mathbf{y}; [\mathbf{X}]|J)\|^2$ is the usual sum of squares $\sum_i n_i(y_i - \bar{y})^2$ between groups, where $\bar{y}_i$ is the mean of the observations from the ith population and $\bar{y}$ is the overall mean. The quantity $\|\mathbf{P}(\mathbf{y}; R^n|[\mathbf{X}]\|^2$ is the within-groups or error sum of squares. The ratio above, of course, is the usual F-statistic for testing H.

It is worth noting that this F-statistic depends only on $[\mathbf{X}]$ and n. Thus however one chooses to parameterize the linear model, as long as $[\mathbf{X}]$ remains unchanged, the analysis does not change. That is, whatever the β_i's stand for and however many there are, as long as the corresponding transformation (matrix) has the linear span of its columns equal to $[\mathbf{X}]$, the analysis will be the same.

It is well known that this F-statistic can be derived from a likelihood ratio test* of H. A geometric approach to the attendant minimization problems provides an elegant exposition of that test.

The geometry provides a general point of view for testing linear hypotheses in linear models. Let $\mathbf{Y}$ be a random vector whose values, $\mathbf{y}$, lie in an inner product space $\mathbf{V}$. Suppose that $E\mathbf{Y}$ is restricted by the model

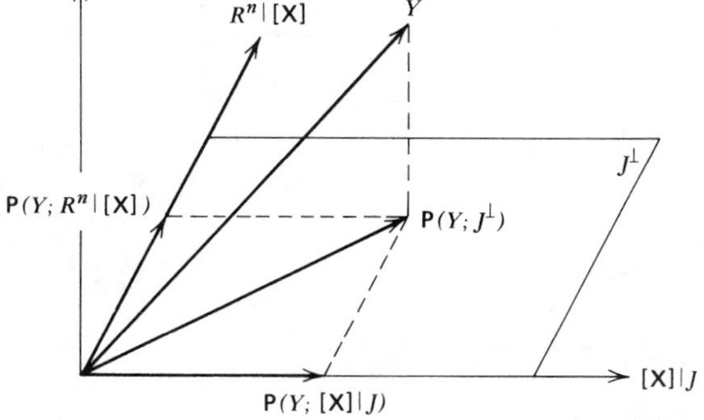

Figure 2 Analysis of the variability.

to lie in a subspace U of V. A linear hypothesis is of the form $E\mathbf{Y} \in W$, a subspace of U. The subspace $G = U \mid W$ corresponds to violations of the hypothesis and the test statistic is

$$\frac{\|\mathbf{P}(\mathbf{y}; G)\|^2/\dim G}{\|\mathbf{P}(\mathbf{y}; V \mid U)\|^2/\dim(V \mid U)}$$

for a suitably chosen distance function $\|\cdot\|^2$ on V. If $\mathbf{Y}$ is multivariate normal with dispersion matrix $\sigma^2\mathbf{I}$, this ratio is an F-statistic. If the dispersion matrix is more generally $\boldsymbol{\Gamma}$, the eigenstructure of $\boldsymbol{\Gamma}$ and how it fits the subspaces G and $V \mid U$ must be taken into account.

Generalized Inverses

The normal equations provide one example of a system of equations $\mathbf{Ax} = \mathbf{y}$ where the matrix $\mathbf{A}$ may not be of full rank. The *generalized inverse** $\mathbf{A}^-$ appears in trying to solve such equations. Kruskal [17] has provided a particularly nice geometric treatment of the structure of generalized inverses. A summary is included here because the geometry is especially informative.

Let V_1 and V_2 be finite-dimensional inner product spaces of dimension n_1 and n_2, respectively. Suppose that $\mathbf{A}$ defines a linear transformation from V_1 to V_2, and that the null space of $\mathbf{A}$, $N = \{\mathbf{x} : \mathbf{Ax} = \mathbf{0}\}$, has dimension ν_1. Let $R = \{\mathbf{y} : \mathbf{y} = \mathbf{Ax}\}$ be the range of $\mathbf{A}$ and $\dim R = \rho$. Then $\nu_1 = n_1 - \rho$. Let $\nu_2 = n_2 - \rho$; note that if $\nu_1 = \nu_2 = 0$, $\mathbf{A}$ is nonsingular and establishes a one-to-one relationship between V_1 and V_2. In this case $\mathbf{A}$ has a unique inverse $\mathbf{A}^{-1}$. However, if either ν_1 or ν_2 is positive, no such inverse exists and the problem is to construct some sort of inverse-like transformation $\mathbf{A}^-$, i.e., a generalized inverse. The problem with this problem is the number of different ways to go about its solution. The geometry of the situation will help codify these choices.

Two choices involve complements of N and R. Let M denote a complement of N (so that $V_1 = M + N$) and S a complement of R (so that $V_2 = R + S$). There are many ways to choose M and S; suppose for the moment that they have been chosen. The dimension of M is ρ and $\mathbf{A}$ is a one-to-one linear transformation from M to R. Thus whatever else ought to be true of $\mathbf{A}^-$ it seems reasonable that $\mathbf{A}^-$ be a one-to-one linear transformation from R back to M so that $\mathbf{A}^-\mathbf{y} = \mathbf{x}$ whenever $\mathbf{y} = \mathbf{Ax}$ and $\mathbf{x} \in M$. Then $\mathbf{A}^-\mathbf{A}$ is the identity on M and $\mathbf{AA}^-$ is the identity on R. That is, $\mathbf{A}^-$ reverses or inverts the action of $\mathbf{A}$ as long as $\mathbf{A}$ acts only on M. Thus if $\mathbf{x}$ is known to be in M, $\mathbf{A}^-\mathbf{y}$ is a solution of $\mathbf{Ax} = \mathbf{y}$. For any such $\mathbf{A}^-$ defined in terms of M, it follows that

$$\mathbf{AA}^-\mathbf{A} = \mathbf{A}. \tag{1}$$

Conversely, any linear transformation $\mathbf{A}^-$ from V_2 to V_1 which satisfies (1) is a one-to-one transformation from R to some M so that $\mathbf{A}^-\mathbf{y} = \mathbf{x}$ whenever $\mathbf{y} = \mathbf{Ax}$ and $\mathbf{x}$ is in M. Thus (1) seems the very least to require of a generalized inverse. Notice that $\mathbf{A}^-$ has not been defined on all of V_2. There are many ways to extend an $\mathbf{A}^-$ from R to all of V_2, i.e., to define $\mathbf{A}^-$ on S. One way that has a certain egalitarian spirit, to use Kruskal's words, is to define $\mathbf{A}^-$ on S so that $\mathbf{A}^-\mathbf{y} = \mathbf{0}$ for $\mathbf{y}$ in S. Then M becomes the range of $\mathbf{A}^-$, S becomes the null space of $\mathbf{A}^-$, and $\mathbf{A}^-$ has the same rank as $\mathbf{A}$. This leads to the condition that $\mathbf{A}^-\mathbf{A}$ is the identity on the range of $\mathbf{A}^-$, or

$$\mathbf{A}^-\mathbf{AA}^- = \mathbf{A}^-. \tag{2}$$

This is a convenient but not essential condition.

Now by (1), $\mathbf{A}^-\mathbf{A}$ is idempotent* and thus a projection on M. If M is chosen orthogonal to N, then $\mathbf{A}^-\mathbf{A}$ is an orthogonal projection and thus symmetric. Since the converse is true, choosing M orthogonal to N is equivalent to

$$\mathbf{A}^-\mathbf{A} = (\mathbf{A}^-\mathbf{A})'. \tag{3}$$

Finally, given (2), $\mathbf{AA}^-$ is idempotent and thus a projection on R. If S is chosen orthogonal to R, $\mathbf{AA}^-$ is an orthogonal projection on R and $\mathbf{AA}^-$ is symmetric. Again, the equivalence of the symmetry and the orthogonality of the projection yield the equivalent condition

$$\mathbf{AA}^- = (\mathbf{AA}^-)'. \tag{4}$$

If (1) is taken as essential, there are eight general categories of generalized inverses depending on whether or not each of (2), (3), or (4) are required. If all four conditions are required, $\mathbf{A}^-$ is uniquely defined and is referred to as the *Moore–Penrose inverse*.

In the case of the normal equations, $\mathbf{X}'\mathbf{X}\boldsymbol{\beta} = \mathbf{X}'\mathbf{y}$, for a linear model $\mathbf{y} = \mathbf{X}\boldsymbol{\beta} + \mathbf{error}$. The transformation is $\mathbf{A} = \mathbf{X}'\mathbf{X}$ and since the range of $\mathbf{X}'\mathbf{X}$ is the same as the range of $\mathbf{X}'$ there is always a solution. In this case $V_1 = V_2$, $\mathbf{A}$ is symmetric, and $N \perp R$. To get a unique solution it is only necessary to choose a particular M. An obvious choice is $M = R$, accomplished by requiring that $\boldsymbol{\beta}$ be orthogonal to N. In the literature this orthogonality is obtained by choosing a basis for N and requiring that $\boldsymbol{\beta}$ be orthogonal to these vectors. Vectors in N are sometimes referred to as *generating inestimable linear functions* of the β_i's. Thus the condition that $\boldsymbol{\beta}$ be orthogonal to N is often stated as an additional number of linearly independent linear equations (the number equal to the dimension of N) whose right-hand sides are zero.

Mixed Models

Let $\mathbf{y} = (y_1, y_2, \ldots, y_n)'$ be a vector of data in Euclidean n-space, R^n. To create a model for $\mathbf{y}$, assume that $\mathbf{Y}$ is a multivariate normal random vector whose observed value is $\mathbf{y}$. Suppose that $E\mathbf{Y}$ is restricted to lie in a subspace U of the inner product space $V = R^n$ and that the dispersion matrix is $\boldsymbol{\Gamma}$. If $\boldsymbol{\Gamma} = \sigma^2\mathbf{I}$ for some positive σ, the model is the usual fixed-effects model*. If $\boldsymbol{\Gamma}$ is assumed to have a more elaborate structure, the model includes random-effects models and mixed models. To test a hypothesis in this more general model a ratio of the form

$$\frac{\|\mathbf{P}(\mathbf{y}; G)\|^2/\dim G}{\|\mathbf{P}(\mathbf{y}; S)\|^2/\dim S}$$

is used. The geometry of the subspaces G and S together with the geometry of the eigenstructure of $\boldsymbol{\Gamma}$ determines whether this is an appropriate test statistic. It is assumed here that the eigenvectors of $\boldsymbol{\Gamma}$ are known but that the eigenvalues are not.

Consider the subspace S and suppose for simplicity that $\dim S = 2$. Let $\mathbf{s}_1$ and $\mathbf{s}_2$ be orthonormal (orthogonal with length 1) vectors in S. Then $\mathbf{P}(\mathbf{y}; S) = (\mathbf{s}_1'\mathbf{y})\mathbf{s}_1 + (\mathbf{s}_2'\mathbf{y})\mathbf{s}_2$. Here $\mathbf{s}_1'\mathbf{y}$ and $\mathbf{s}_2'\mathbf{y}$ are the coordinates of $\mathbf{y}$ with respect to the basis $\{\mathbf{s}_1, \mathbf{s}_2\}$. The bilinear form $\mathbf{v}_1'\mathbf{v}_2$ is the usual inner product on R^n and $\|\mathbf{v}\| = (\mathbf{v}'\mathbf{v})^{1/2}$ the usual Euclidean distance. Now $\mathbf{s}'\mathbf{Y}$ is a univariate normal random variable with mean $\mathbf{s}'E\mathbf{Y}$ and variance $\mathbf{s}'\boldsymbol{\Gamma}\mathbf{s}$. Thus $(\mathbf{s}'\mathbf{Y} - \mathbf{s}'E\mathbf{Y})^2/\mathbf{s}'\boldsymbol{\Gamma}\mathbf{s}$ is a chi-square* random variable with 1 degree of freedom. Note that

$$\|\mathbf{P}(\mathbf{Y}; S)\|^2 = (\mathbf{s}_1'\mathbf{Y})^2 + (\mathbf{s}_2'\mathbf{Y})^2.$$

If $\mathbf{s}_1$ and $\mathbf{s}_2$ can be chosen as eigenvectors of $\boldsymbol{\Gamma}$ corresponding to a common eigenvalue λ and if $\mathbf{s}_1'E\mathbf{Y} = \mathbf{s}_2'E\mathbf{Y} = 0$, then $\|\mathbf{P}(\mathbf{Y}; S)\|^2/\lambda$ is a chi-square random variable with 2 degrees of freedom. This follows because the variance of $\mathbf{s}_i'\mathbf{Y}$ is $\mathbf{s}_i'\boldsymbol{\Gamma}\mathbf{s}_i = \mathbf{s}_i'\lambda\mathbf{s}_i = \lambda\|\mathbf{s}_i\|^2 = \lambda$ and the covariance of $\mathbf{s}_1'\mathbf{Y}$ and $\mathbf{s}_2'\mathbf{Y}$ is $\mathbf{s}_1'\boldsymbol{\Gamma}\mathbf{s}_2 = \mathbf{s}_1'\lambda\mathbf{s}_2 = \lambda\mathbf{s}_1'\mathbf{s}_2 = 0$ under the assumed conditions. These conditions amount to assuming that S is a subspace of an eigenspace of $\boldsymbol{\Gamma}$ and that S is orthogonal to U.

If, under the null hypothesis, G satisfies the same conditions with respect to the same eigenvalue λ, then $\|\mathbf{P}(\mathbf{Y}; G)\|^2/\lambda$ will be a chi-square random variable with dim G degrees of freedom. Note here that the hypothesis may specify a further restriction of $E\mathbf{Y}$ (a fixed-effect hypothesis) or some restriction on the eigenvalues of $\boldsymbol{\Gamma}$ (a variance component hypothesis) or both. Thus if, under the null hypothesis, G is a subspace of the eigenspace of $\boldsymbol{\Gamma}$ corresponding to eigenvalue λ, $E\mathbf{Y}$ is orthogonal to G, and G is orthogonal to S, then

$$\frac{\|\mathbf{P}(\mathbf{Y}; G)\|^2/\dim G}{\|\mathbf{P}(\mathbf{Y}; S)\|^2/\dim S}$$

is distributed as a central F-statistic under the null hypothesis. The orthogonality of G and S is equivalent to independence of $\|\mathbf{P}(\mathbf{Y}; G)\|^2$ and $\|\mathbf{P}(\mathbf{Y}; S)\|^2$ under the assumption of normality.

This discussion could be carried out in the setting of a general inner product space without materially affecting the results. The geometry leads the way. Let $\langle \cdot, \cdot \rangle$ be an inner product on a general inner product space V and $\|\mathbf{v}\|^2 = \langle \mathbf{v}, \mathbf{v} \rangle$ be the squared length of $\mathbf{v}$. Then by defining a random vector $\mathbf{Y}$ with values in V to be normal if and only if for every $\mathbf{v} \in V$ $\langle \mathbf{v}, \mathbf{Y} \rangle$ is a univariate normal random variable with mean $\langle \mathbf{v}, E\mathbf{Y} \rangle$ and variance $\langle \mathbf{v}, \mathbf{\Gamma v} \rangle$ for some $\mathbf{\Gamma}$, the previous results follow with no formal change.

ILLUSTRATIVE GEOMETRY IN OTHER AREAS

Because statistical theory based on the assumption of normally distributed data plays such a central role, statistical theory for nonnormal data are often studied, at least in part, by examining how different it is from normal theory. The intimate relationship between normal theory and Euclidean geometry makes geometric investigations especially appealing. Two such investigations are mentioned here.

Efron [7, 8] has begun the study of the geometry of exponential families* of distributions. In ref. 8 his main purpose is to develop pictures to illustrate how maximum likelihood estimation (MLE) in one-parameter curved exponential families generalizes MLE in one-parameter exponential families. The pictures illustrate important contributions by Fisher [10], Rao [22], and Hoeffding [14] with respect to MLE. The curvature of a curved exponential family is introduced in ref. 7 and used in both refs. 7 and 8 to indicate how nearly exponential curved exponential families are, and how well statistical procedures which are in some sense optimal in exponential families work in curved exponential families. One-parameter exponential families can be viewed as straight lines in a class of multivariate exponential families and have curvature zero, while one-parameter curved exponential families will have a positive curvature. The question is: How does this curvature affect certain statistical procedures such as MLE, and how is the curvature related to other ideas in statistics such as Fisher information*?

Bates and Watts [2] consider the problem of nonlinear least squares in a multivariate setting. The solution locus for their problem is thus a p-dimensional surface in an n-dimensional space. They attempt to quantify the degree of nonlinearity of this surface so as to be able to compare the nonlinearity of different problems. Toward this end they consider the normal curvature in a given direction, which they label the intrinsic curvature because it depends only on the surface, not its parameterization. They combine the tangential curvatures into a measure they call the parameter-effects curvative because it is affected by the particular parameterization of the model. In using a linear approximation for a nonlinear problem two different assumptions are involved: first, that the solution locus is a hyperplane (the planar assumption), and second, that the curved parameter lines on the approximating tangent plane can be replaced by a grid of straight, parallel, equispaced lines (the uniform coordinate assumption). The intrinsic curvature is a measure of how well the planar assumption works and the parameter-effects curvature is a measure of how well the uniform coordinate assumption works. These ideas provide an illustrative as well as numerical means for examining the problem of nonlinearity.

References

[1] Bartlett, M. S. (1933–1934). *Proc. Camb. Philos. Soc.*, **30**, 327–340.

[2] Bates, D. M. and Watts, D. G. (1980). *J. R. Statist. Soc. B*, **42**, 1–25. (This paper with discussion sheds quite a bit of light on the use of geometry in statistics.)

[3] Bose, R. C. (1935–1936). *Sankhyā*, **2**, 143–154.

[4] Bose, R. C. (1963). *Pacific J. Math.*, **13**, 389–419.

[5] Burdick, D. S., Herr, D. G., O'Fallon, W. M., and O'Neill, B. V. (1974). *Commun. Statist.*, **3**, 581–595. (An analysis of unbalanced, two-way analysis of variance from a geometric point of view.)

[6] Durbin, J. and Kendall, M. G. (1951). *Biometrika*, **38**, 150–158.

[7] Efron, B. (1975). *Ann. Statist.*, **3**, 1189–1242.
[8] Efron, B. (1978). *Ann. Statist.*, **6**, 362–376.
[9] Fisher, R. A. (1915). *Biometrika*, **10**, 507–521.
[10] Fisher, R. A. (1925). *Proc. Camb. Philos. Soc.*, **122**, 700–725.
[11] Galton, F. (1886). *J. Anthropol. Inst. Gt. Brit. Ireland*, **15**, 246–263.
[12] Halmos, P. R. (1958). *Finite-Dimensional Vector Spaces*, 2nd ed. D. Van Nostrand, Princeton, N.J.
[13] Herr, D. G. (1980). *Amer. Statist.*, **34**, 43–47. (There is a more extensive bibliography of the history of the use of geometry in statistics, especially linear models, in this paper.)
[14] Hoeffding, W. (1965). *Ann. Math. Statist.*, **36**, 369–408.
[15] Kruskal, W. (1961). *Proc. 4th Berkeley Symp. Math. Statist. Prob.*, Vol. 1. University of California Press, Berkeley, Calif., pp. 435–451. (Contains an elegant primer on the coordinate-free approach to linear models.)
[16] Kruskal, W. (1968). *Ann. Math. Statist.*, **39**, 70–75.
[17] Kruskal, W. (1975). *J. R. Statist. Soc. B.*, **37**, 272–283. (An excellent treatment of the varieties of generalized inverses.)
[18] Mahalanobis, P. C. (1938). *Sankhyā*, **4**, 265–272. (An interesting discussion of R. A. Fisher.)
[19] Mahalanobis, P. C., Bose, R. C., and Roy, S. N. (1937). *Sankhyā*, **3**, 1–40.
[20] Pearson, E. S. (1956). *J. R. Statist. Soc. A.*, **119**, 125–149.
[21] Pukelsheim, F. (1978). *Sixth Int. Conf. Math. Statist.*, pp. 1–18. (Unbiased quadratic estimates of variance components are considered from a geometric point of view.)
[22] Rao, C. R. (1963). *Sankhyā*, **25**, 189–206.
[23] Scheffé, H. (1959). *The Analysis of Variance*. Wiley, New York.
[24] Seber, G. A. F. (1977). *Linear Regression Analysis*. Wiley, New York.
[25] Seely, J. and Zyskind, G. (1971). *Ann. Math. Statist.*, **42**, 691–703.
[26] Timm, N. H. (1975). *Multivariate Analysis with Applications in Education and Psychology*. Brooks/Cole, Belmont, Calif. (Perhaps the only textbook in applied statistics which makes a serious attempt to use a geometric approach.)
[27] Timm, N. H. and Carlson, J. E. (1975). *Analysis of Variance through Full Rank Models*. Multivariate Behavioral Research Monographs published by the Society of Multivariate Experimental Psychology. Texas Christian University Press, Fort Worth, Tex. (The appendix has a discussion of the geometry of the two-way analysis-of-variance design.)
[28] Watson, G. S. (1967). *Ann. Math. Statist.*, **38**, 1679–1699.
[29] Zyskind, G. (1967). *Ann. Math. Statist.*, **38**, 1092–1109.

(ANALYSIS OF VARIANCE
BALANCING IN EXPERIMENTAL DESIGN
ESTIMABILITY
GENERAL BALANCE
GENERALIZED INVERSE
GENERAL LINEAR MODEL
LEAST SQUARES
LIKELIHOOD RATIO TESTS
MAXIMUM LIKELIHOOD ESTIMATION
MULTIVARIATE NORMAL DISTRIBUTION)

DAVID G. HERR

GEOMETRY IN STATISTICS: CONVEXITY

The theory of convexity provides analytical tools which are used extensively in probability and statistics as well as in other areas of mathematics. Every mathematical statistician and probabilist should be aware of the most important results, which are surveyed in the following. A sample of applications is given.

CONVEX SETS

A subset C of R^n is *convex* if $\alpha x + (1-\alpha)y \in C$ whenever $x \in C$, $y \in C$, and $\alpha \in [0, 1]$. If $x_1, x_2, \ldots, x_m$ are m points in R^n, the vector sum $\alpha_1 x_1 + \alpha_2 x_2 + \cdots + \alpha_m x_m$ is a *convex combination* of $x_1, x_2, \ldots, x_m$ whenever the coefficients α_i are all nonnegative and $\alpha_1 + \alpha_2 + \cdots + \alpha_m = 1$. Clearly, *if a subset of R^n contains all the convex combinations of its elements, it is convex*. It is not difficult to show the converse: that *every convex set contains all the convex combinations of its elements*. A proof of this result and other elementary properties of convex sets can be found in Rockafellar [21, pp. 10–12].

The following are examples of convex sets in R^n: the closed unit cube $\{(x_1, \ldots, x_n): |x_i| \leqslant 1, i = 1, \ldots, n\}$, the closed unit ball $\{(x_1, \ldots, x_n): x_1^2 + \cdots + x_n^2 \leqslant 1\}$, the nonnegative orthant $\{(x_1, \ldots, x_n) : x_i \geqslant 0, i = 1, \ldots, n\}$ and the positive orthant $\{(x_1, \ldots, x_n) : x_i > 0, i = 1, \ldots, n\}$.

Let S be a collection of points in R^n. The *convex hull* of S is the set of all convex combinations of elements in S. Clearly, every convex hull is a convex set. For example, the closed unit cube is the convex hull of $\{x: |x_i| \leqslant 1, i = 1, \ldots, n$ and $|x_j| = 1$ for some $j = 1, \ldots, n\}$. It is also the convex hull of $\{x : x_i = \pm 1, i = 1, \ldots, n\}$. Actually, the latter is the collection of the extreme points of the closed unit cube (see definition below). Another example is the closed unit ball which is the convex hull of $\{x : x_1^2 + \cdots + x_n^2 = 1\}$.

From the first example it is seen that, given a convex set C, there may exist various point sets S such that C is the convex hull of S. It is of interest and use to find a small or a special S. A smallest S actually exists in many cases; it is the set of all extreme points, defined next.

A point x is an *extreme point* of the convex set C if $x \in C$ and there is no way to express x as a convex combination $\alpha y + (1 - \alpha)z$ such that $y \in C$, $z \in C$, $\alpha \in (0, 1)$, unless $y = z = x$. For example, every point $(x_1, \ldots, x_n)$ where each x_i is 1 or -1 is an extreme point of the closed unit cube and each point $(x_1, \ldots, x_n)$ such that $x_1^2 + \cdots + x_n^2 = 1$ is an extreme point of the closed unit ball. The only extreme point of the nonnegative orthant is $(0, \ldots, 0)$ and the positive orthant has no extreme points. A further discussion about extreme points and the geometry of convex sets in R^n is given in Rockafellar [21, Sec. 18].

One of the most important results of convexity theory is the Krein–Milman theorem and, its generalization, Choquet's theorem. Roughly, these say that many convex sets, C, are the convex hulls of the set of their extreme points. For such a statement to be true, C must have extreme points in the first place. It is not possible to give here all the background needed for the conditions of the most general statement of Choquet's theorem. In R^n the theorem says that *every compact convex set is the convex hull of its extreme points*. See Phelps [20] and references there for more general statements. Related results can be found in Stoer and Witzgall [26, Sec. 3.6].

Representation of points of a compact convex set as barycenters (= "convex combination" of as many as an uncountable number of elements) of the extreme points is a common tool in probability and statistics. The following is a sample of applications.

1. De Finetti's [4] theorem, a basic result of Bayesian statistics, say that every n-dimensional exchangeable distribution (*see* EXCHANGEABILITY) is a mixture (= barycenter) of distributions, each being the joint distribution of independent identically distributed random variables* (see also Shaked [22]). An elementary proof of a special case of de Finetti's theorem is given in Feller [7, Sec. VII.4].
2. Bochner's theorem says that a complex function is a characteristic function* if and only if it is a positive definite function. Edwards [5] proves this theorem using extreme points methods. Related results are discussed in Kendall [15] and Johansen [10]. These authors obtained the Lévy-Khinchin formula* for infinitely divisible* distributions by finding the extreme points of a certain convex set formed by the logarithms of the characteristic functions of infinitely divisible distributions, and then using the Krein-Milman theorem.
3. Langberg et al. [16–18], and references there, find the extreme points of various convex classes of distributions which are of importance in reliability theory*. Thus it is possible to represent every member of these classes as a mixture of the distributions which are the extreme points. The authors plan to use this information to obtain bounds, inequalities, and optimal values of convex functionals of distributions in those classes.

In addition to the Krein–Milman theorem, the other basic results that deal with convex sets are the *supporting hyperplane theorem* and the *separating hyperplane theorem*. The first theorem roughly says that if C is a convex set in R^n and x_0 is a boundary point of C, there exists a hyperplane, tangent to C at x_0, such that all points of C are on one side of it. The second theorem roughly says that if C_1 and C_2 are disjoint convex sets in R^n then there exists a hyperplane which separates C_1 and C_2, i.e., such that all the points of C_1 are on one side of the hyperplane and all the points of C_2 are on the other side. A discussion of these results in the context of statistical decision theory* appears in Ferguson [8, Sec. 2.7]. A more delicate analysis is given in Rockafellar [21, Sec. 11]. Further references are Berge and Ghouila-Houri [1, Chap. 3]; Valentine [28, Part II]; Stoer and Witzgall [26, Chap. 3]; Eggleston [6, Sec. 1.7] and Kelly and Weiss [14, Chap. 4].

Usage of the separating hyperplane theorem is illustrated in Ferguson [8, Chap. 2]. He uses it to prove the fundamental theorem of game theory*—the minimax* theorem and the important decision theoretic result—the complete class theorem.

CONVEX FUNCTIONS

A real function g defined on a convex subset of R^n is *convex* if

$$g(\alpha x + (1-\alpha)y) \leqslant \alpha g(x) + (1-\alpha) g(y) \tag{1}$$

whenever $\alpha \in [0, 1]$ and x and y are in the domain of g. The function g is *concave* if $-g$ is convex. Geometrically, g is convex if and only if the $(n+1)$-dimensional set $\{(x_1, \ldots, x_n, x_{n+1}) : x_{n+1} \geqslant g(x_1, \ldots, x_n), (x_1, \ldots, x_n)$ in the domain of $g\}$ (this set is the *epigraph* of g) is convex. In particular, if g is a continuous univariate convex function, then, corresponding to every point $(x, g(x))$ on the graph of g, there exists at least one straight line which passes through $(x, g(x))$ such that the curve $y = g(x)$ lies wholly above it (see, e.g., Hardy et al. [9, p. 94]).

In most instances convex functions are smooth. Actually, every univariate convex function is either very regular (i.e., continuous and possesses right and left derivatives at each point) or very irregular (i.e., oscillates in every interval range from $-\infty$ to ∞) (see Hardy et al. [9, p. 91]). In most applications g is continuous and then a necessary and sufficient condition for its convexity is

$$g\left(\frac{x+y}{2}\right) \leqslant \frac{g(x)+g(y)}{2} \tag{2}$$

whenever x and y are in the domain of g. Geometrically, condition (2) says that the midpoint of every line segment connecting two points on the graph of g lies above or on the graph. In comparison, condition (1) says that the whole line segment lies above or on the graph of g. Note that condition (2) is seemingly weaker than condition (1); however, if g is continuous, then (1) and (2) are equivalent (see Hardy et al. [9, p. 70]).

There are some useful sufficient conditions for convexity. If g is a univariate differentiable function, it is convex if its first derivative g' is monotone nondecreasing. If, in addition, g is twice differentiable, it is convex if its second derivative is nonnegative on the domain of g. An n-variate function that has all second-order derivatives, and which is defined on a convex domain C in R^n, is convex if and only if the Hessian matrix

$$Q_x = \{q_{ij}(x)\}_{i,j=1}^{n}, \qquad q_{ij} = \frac{\partial^2 g(x)}{\partial x_i \, \partial x_j}$$

is positive semidefinite for every $x \in C$. Rigorous proofs of these results and a list of additional properties of convex functions can be found in Rockafellar [21, Sec. 4]. (See also Feller [7, Sec. V.8] and Marshall and Olkin [19, Sec. 16.B].)

A basic result associated with convex functions is the *Jensen inequality**; roughly, it says that if X is an n-dimensional random vector and if g is an n-dimensional convex function, then

$$g(EX) \leqslant Eg(X). \tag{3}$$

More rigorous statements and proofs, as well as a converse and conditions for equality in (3), can be found, for example, in Ferguson [8, Sec. 2.8] or in Marshall and Olkin [19,

Sec. 16.C], where its use to obtain inequalities of importance in various fields of mathematics is also illustrated.

Convex and concave functions are encountered in numerous studies in probability and statistics. Scanning over any issue of a statistical journal the reader is bound, almost surely, to find some instances. In the following some properties will be given and a sample of usage will be described.

One important property is that if a convex function g is defined on a compact domain and is not constant on any nontrivial subset of its domain (interval, if g is univariate), it has one local minimum which is also the global minimum. Under analogous conditions a concave function has one local maximum which is also the global maximum. This fact is very useful when the existence and uniqueness of a maximum likelihood* estimator is to be shown. One just has to show that the likelihood function is concave and satisfies the aforementioned conditions to obtain existence and uniqueness. For an illustration, see Shaked [23].

An important partial ordering of probability measures is that of dilations, first used in a statistical context by Blackwell [2]. The probability measure μ is a *dilation* (or dilatation) of the probability measure λ, both with support Ω, if

$$\int_{\Omega} g(x)\lambda(dx) \leqslant \int_{\Omega} g(x)\mu(dx) \qquad (4)$$

for every convex function g. This relation is denoted as $\lambda \prec \mu$. If F and G are the distribution functions associated with λ and μ, respectively, we write $F \prec G$. Blackwell [2] uses it to define "comparison of experiments." Various conditions on λ and μ (or on F and G) which are equivalent to $\lambda \prec \mu$ have been derived by Blackwell [2] and Strassen [27]. (See Shaked [24] for a more detailed discussion.) When λ and μ are probability measures on the real line, (4) can be written as

$$Eg(X) \leqslant Eg(Y) \qquad (5)$$

for every convex function g, where X and Y are distributed, respectively, according to λ and μ. Note that (5) implies that $EX = EY$ because $g_1(u) = u$ and $g_2(u) = -u$ are both convex functions. Thus, substituting $g(u) = u^2$ in (5), it follows that, if $F \prec G$, the associated random variables X and Y satisfy $\text{var}(X) \leqslant \text{var}(Y)$. Similarly, other useful inequalities can be obtained from (5). Shaked [24] identifies several pairs of distributions F and G such that $F \prec G$. Based on the resulting inequalities, he is able to explain some statistical experimental phenomena, for example, the unexpected fact that the observed distribution of the number of males in sibships of a fixed size is not well approximated by the binomial distribution*.

Concavity arises in the study of *log-concave densities* (i.e., densities whose logarithm is concave), also called *Pólya frequency densities of order 2* (PF_2). The normal*, the gamma*, the beta*, the Weibull*, and many other densities are PF_2 (see Marshall and Olkin [19, p. 493] and Karlin [11, Chap. 7]). All PF_2 densities are unimodal*. Useful moment inequalities that are satisfied by log-concave and log-convex densities have been obtained (see Marshall and Olkin [19, p. 493]).

The importance of log-concave and log-convex densities follows from their properties and the fact that they are often easy to identify. For a weaker condition, $\log(1 - F)$ is concave, where F is a distribution function (see Karlin [11, p. 128] or Marshall and Olkin [19, p. 493]). The class of these distributions is of importance in reliability* theory. They are called the increasing hazard rate (IHR) distributions. The reason is that if F has a density f, then $\log(1 - F)$ is concave if and only if the hazard rate function, $r = f/(1 - F)$, is monotone nondecreasing (*see* HAZARD RATE CLASSIFICATION OF DISTRIBUTIONS). The important class of distributions F such that $\log(1 - F)$ is convex is that of decreasing hazard rate* (DHR) distributions.

We end the list of applications with the following theorem: *Every Laplace transform of a measure on the nonnegative half-line is log-convex* (*see* INTEGRAL TRANSFORMS). This well-known fact follows, for example, from Theorem 16.B.8 of Marshall and Olkin [19].

The Laplace transform of a probability measure on $(0, \infty)$ is a survival function because it can be thought of as a mixture of exponential survival functions. The associated densities are called *completely monotone*. A thorough discussion of this class of densities can be found in Feller [7, Sec. XIII.4] and Keilson [13, Chap. 5] (see also Brown [3] and Shaked [25]). Convexity and concavity play a major role in the development of any mathematical analysis of this class.

The idea of convexity can be generalized using its geometric interpretation. Some work has been done in this direction (see Karlin [11, Chap. 6] and Karlin and Studden [12, Chap. 1 and refs. there]). Another important analytical notion with applications in probability and statistics is Schur convexity (*see* MAJORIZATION AND SCHUR CONVEXITY).

References

[1] Berge, C. and Ghouila-Houri, A. (1965). *Programming, Games and Transportation Networks*. Methuen, New York. (A textbook on convexity theory with applications.)

[2] Blackwell, D. (1951). *Proc. 2nd Berkeley Symp. Math. Statist. Prob.*, University of California Press, Berkeley, Calif., pp. 93–102.

[3] Brown, M. (1981). *Ann. Prob.*, **9**, 891–895.

[4] DeFinetti, B. (1937). *Ann. Inst. H. Poincaré*, **7**, 1–68. (English transl. in *Studies in Subjective Probability*, H. E. Kyburg and H. E. Smokler, eds. Wiley, New York, 1964.)

[5] Edwards, R. E. (1965). *Functional Analysis, Theory and Applications*. Holt, Rinehart and Winston, New York.

[6] Eggleston, H. G. (1958). *Convexity*. Cambridge University Press, Cambridge. (A well-written text on convexity.)

[7] Feller, W. (1971). *An Introduction to Probability Theory and Its Applications*, Vol. 2, (2nd ed. Wiley, New York. (Contains a short and clear discussion on convex functions in Sec. V.8.)

[8] Ferguson, T. S. (1969). *Mathematical Statistics, a Decision Theoretic Approach*. Academic Press, New York. (Contains a clear discussion on convex sets in Chap. 2.)

[9] Hardy, G. H., Littlwood, J. E., and Pólya, G. (1952). *Inequalities*. Cambridge University Press, Cambridge. (Contains beautifully written proofs of some basic convexity results.)

[10] Johansen, S. (1966). *Z. Wahrscheinlichkeitsth. verw. Geb.*, **5**, 304–316.

[11] Karlin, S. (1968). *Total Positivity*, Vol. 1. Stanford University Press, Stanford, Calif.

[12] Karlin, S. and Studden, W. (1966). *Tchebycheff Systems: With Applications in Analysis and Statistics*. Wiley, New York.

[13] Keilson, J. (1979). *Markov Chains Models—Rarity and Exponentiality*. Springer-Verlag, New York.

[14] Kelly, P. J. and Weiss, M. L. (1979). *Geometry and Convexity*. Wiley, New York. (A good textbook on convexity theory.)

[15] Kendall, D. G. (1963). *Z. Wahrscheinlichkeitsth. verw. Geb.*, **1**, 295–300.

[16] Langberg, N. A., León, R. V., Lynch, J., and Proschan, F. (1979). Extreme Points of Certain Convex Subsets of Logconvex Functions. *Tech. Rep.* Dept. of Statistics, Florida State University, Tallahasee, Fla.

[17] Langberg, N. A., León, R. V., Lynch, J., and Proschan, F. (1980). *Math. Operat. Res.*, **5**, 35–42.

[18] Langberg, N. A., León, R. V., Lynch, J., and Proschan, F. (1980). Extreme Points of the Class of Discrete Decreasing Failure Rate Average Life Distributions. *Tech. Rep.*, Dept. of Statistics, Florida State University, Tallahassee, Fla.

[19] Marshall, A. W. and Olkin, I. (1979). *Inequalities: Theory of Majorization and Its Applications*. Academic Press, New York. (Contains a clear discussion on convex functions in Chap. 16.)

[20] Phelps, R. R. (1966). *Lectures on Choquet's Theorem*. Van Nostrand Reinhold, New York. (An advanced text on Choquet's Theorem.)

[21] Rockafellar, R. T. (1970). *Convex Analysis*. Princeton, N.J. (A modern "standard reference" for convexity in R^n. Somewhat confusing symbols.)

[22] Shaked, M. (1977). *Ann. Statist.*, **5**, 505–515.

[23] Shaked, M. (1978). *Commun. Statist. A*, **6**, 1323–1339.

[24] Shaked, M. (1980). *J. R. Statist. Soc. B*, **42**, 192–198.

[25] Shaked, M. (1981). *J. Appl. Prob.*, **18**, 853–863.

[26] Stoer, J. and Witzgall, C. (1970). *Convexity and Optimization in Finite Dimensions*, Vol. 1. Springer-Verlag, New York. (A somewhat advanced text on convexity.)

[27] Strassen, V. (1965). *Ann. Math. Statist.*, **36**, 423–439.

[28] Valentine, F. A. (1964). *Convex Sets*. McGraw-Hill, New York. (An advanced text on convex sets.)

(BAYESIAN INFERENCE
CHARACTERISTIC FUNCTIONS
DECISION THEORY

HAZARD RATE CLASSIFICATION OF DISTRIBUTIONS
INFINITE DIVISIBILITY
INTEGRAL TRANSFORMS
JENSEN'S INEQUALITY
MAJORIZATION AND SCHUR CONVEXITY
MAXIMUM LIKELIHOOD
RELIABILITY)

MOSHE SHAKED

GEOSTATISTICS

Etymologically, "geostatistics" should denote the application of statistics to problems in the earth sciences (geosciences). The accepted definition (at least among practitioners in the mining and extractive industry) is much narrower: "Geostatistics is the application of the formalism of random functions to the reconnaissance and estimation of natural phenomena" [5]. Indeed, the term "geostatistics" was made after the model "geophysics," and geophysics does not cover all the physics of earth sciences.

REGIONALIZED VARIABLES AND RANDOM FUNCTIONS

A natural phenomenon can often be characterized by the distribution in space of a certain number of measurable quantities called *regionalized variables*. Examples are the distribution of meteorological variables and (or) pollutant concentrations in a four-dimensional space (Cartesian space plus time), the distribution of ore grades in three-dimensional Cartesian space for a mineralization, the distribution of the densities of species of trees in the horizontal two-dimensional space in forestry, etc.

Mathematically, a regionalized variable is a function $z(\mathbf{x})$ of a point with coordinates $\mathbf{x}$ (in a n-dimensional space, $\mathbf{x}$ represents the n coordinates $u_1, u_2, \ldots, u_n$). For example, $z(\mathbf{x})$ may represent the grade measured from a core centered at point $\mathbf{x}$. The characteristic of regionalized variables in earth sciences is that their variation in space is locally chaotic, which precludes a direct deterministic study of the function $z(\mathbf{x})$ and calls for the measurement $z(\mathbf{x})$ to be interpreted as one realization of a *random variable* $Z(\mathbf{x})$. However, under a local chaotic appearance, a certain structure in the spatial variability of $z(\mathbf{x})$ is usually perceptible which calls for a pattern of correlation* between the various random variables $Z(\mathbf{x})$, $Z(\mathbf{x}')$.

The geostatistical approach interprets the regionalized variable $z(\mathbf{x})$ at hand as a particular realization of a *random function* $Z(\mathbf{x})$. A random function (RF) can be seen as a set of autocorrelated random variables $\{Z(\mathbf{x})$, for $\mathbf{x}$ taking any position within the field of study$\}$.

Similarly, if the natural phenomenon under study is characterized by the joint distribution of several regionalized variables $z_k(\mathbf{x})$, $k = 1$ to K, the geostatistical approach interprets these regionalized variables as a particular realization of a set of K inter- and autocorrelated RFs $\{Z_k(\mathbf{x}), k = 1 \text{ to } K\}$.

STATISTICAL INFERENCE

A RF $Z(x)$ is defined by its finite multidimensional distribution, i.e., by the set of all joint cumulative distribution functions (CDFs). It is clearly and always impossible to infer all these CDFs from a finite number of data. But this complete information is never needed; for the usual practical purpose of estimation the minimum needed is the knowledge of the *variogram* function, or moment of order 2 of the increment:

$$2\gamma(x,h) = E\left\{\left[Z(x+h) - Z(x)\right]^2\right\}.$$

The semivariogram function $\gamma(x,h)$ appears as a function of the interdistance vector h (modulus and directions) and also of the location x. Its practical inference thus requires a weak stationarity* hypothesis. The "intrinsic hypothesis" states that the moment of order 2, $E\{[Z(x+h) - Z(x)]^2\}$, is independent of the location x for limited interdistances $|h| < b$. It is then possible to infer without bias the variogram from the sample squared differences:

$$2\gamma^*(h) = \frac{1}{N(h)} \sum_{i=1}^{N(h)} \left[z(x_i + h) - z(x_i) \right]^2, \qquad \forall\ |h| < b.$$

Most often in practice, this intrinsic hypothesis can even be limited to a moving window, i.e., the function $\gamma(x,h)$ is expressed as the product of two terms:

$$\gamma(x,h) = f(x)\gamma_0(h), \qquad \forall\ |h| < b,$$

where $\gamma_0(h)$ characterizes an intrinsic variability constant over the field of study, and $f(x)$ characterizes an intensity of variability particular to each moving window centered at point x. The factor $f(x)$, called the "proportional effect", is usually estimated by a regression curve $\sigma^{2*} = f(m^*)$ fitted on the scattergram of sample means and variances.

Remarks. In the case of a natural phenomenon characterized by the joint distribution of K variables and modeled by a set of K RFs $\{Z_k(x),\ k = 1 \text{ to } K\}$, the joint variability is modeled by a matrix of semivariograms $\{\gamma_{kk'}(h),\ \forall\ k, k' = 1 \text{ to } K\}$. The diagonal terms $\gamma_{kk}(h)$, or direct semivariograms, characterize the spatial autocorrelation of each variable, and the rectangle terms $\gamma_{kk'}(h)$, $k \neq k'$, or cross-semivariograms, characterize the spatial intercorrelation between variables. In practice, data are seldom enough to allow a fair inference of a whole matrix of variograms for $K > 3$.

STRUCTURAL ANALYSIS

The structural analysis of a regionalized phenomenon is the process of data analysis which leads to a variogram model. The latter acts as a quantified summary of all available information on the structure of variability of the phenomenon. This summary of information is then channeled throughout all subsequent steps of reconnaissance and estimation. Variogram modeling and interpreting requires more than computer programs, essentially a good knowledge about the physics of the phenomenon (geology*, forestry*, bathymetry, etc., depending on the field of study) and also a good "craft" in data gathering and analysis.

As with a covariance model, a variogram model cannot be any function; it must ensure positiveness of all calculated variances; i.e., the covariance model $C(h)$ must be a positive-definite function and correspondingly the variogram model $\gamma(h)$ must be a "conditional positive definite" function (see Journel and Huijbregts [3, p. 35]). In practice this amounts to looking for a model which is a positive linear combination of basic functions known to be positive definite. For example, in three-dimensional Cartesian space a general model is

$$\gamma(h_1, h_2, h_3) = \sum_{i=1}^{n} C_i \gamma_i(|\mathbf{h}^{(i)}|),$$

where:

1. (h_1, h_2, h_3) are the three coordinates of the interdistance vector $\mathbf{h}$.
2. $|\mathbf{h}^{(i)}| = \sqrt{(h_1^{(i)})^2 + (h_2^{(i)})^2 + (h_3^{(i)})^2}$ is the modulus of a vector $\mathbf{h}^{(i)}$ whose three coordinates are linear transforms of the original coordinates (h_1, h_2, h_3), (i.e., in matrix notations $[\mathbf{h}^{(i)}] = [\mathbf{A}_i][\mathbf{h}]$.
3. The C_i's are positive constants.
4. The $\gamma_i(h)$'s are basic semivariogram models known to be conditional positive definite in any space of dimension $\leqslant 3$. The most currently used models are the spherical model of range a,

$$\gamma(h) = \begin{cases} \frac{3}{2}(h/a) - \frac{1}{2}(h/a)^3, & \forall\ h \in [0, a], \\ 1, & \forall\ h \geqslant a; \end{cases}$$

the exponential* model of parameter a,

$$\gamma(h) = 1 - e^{-h/a};$$

and the power model,

$$\gamma(h) = h^{\omega}, \qquad \text{with} \quad \omega \in]0, 2[.$$

Remarks. A linear combination of basic variogram schemes is particularly convenient as a model, since most other geostatistical operators appear as linear functionals

$\mathscr{L}(\gamma)$ of the variogram model; hence

$$\mathscr{L}(\gamma) = \sum_i C_i \mathscr{L}(\gamma_i).$$

A spherical model with a range a which is very small with regard to all distances $|h|$ of observation characterizes a phenomenon with no spatial correlation, also called the *pure nugget effect* phenomenon by extension of gold mining terminology. Note that all results from statistics of uncorrelated data are found by making γ a pure nugget effect model in the various geostatistical operators.

ESTIMATION VARIANCE

When estimating an unknown value $z(x)$ by some combination $z^*(x)$ of the available data, an error is made:

$$e(x) = [z(x) - z^*(x)].$$

Interpreted as a random variable, this error is defined by a distribution function whose two main characteristics are the mean and the variance. A good estimator $Z^*(\mathbf{x})$ should ensure "nonbias," i.e., $E\{Z(x) - Z^*(x)\} = 0$, and minimize the *estimation variance* or variance $E\{[Z(x) - Z^*(x)]^2\}$ of the unbiased error. For the very important class of linear estimators, the estimate being a linear combination of the available data within the local neighborhood defined by the intrinsic hypothesis $z^*(x) = \sum_{\alpha=1}^{n} \lambda_\alpha z(x_\alpha)$:

1. Unbiasedness is ensured when the sum of weights equals 1: $\sum_{\alpha=1}^{n} \lambda_\alpha = 1$.
2. The estimation variance can be expressed as a *linear* function of the semivariogram model $\gamma(h)$:

$$\sigma_E^2 = E\{[Z(x) - Z^*(x)]^2\}$$
$$= 2\sum_\alpha \lambda_\alpha \gamma(x - x_\alpha)$$
$$- \sum_\alpha \sum_\beta \lambda_\alpha \lambda_\beta \gamma(x_\alpha - x_\beta).$$

More generally, consider the estimation of the unknown average value $z_V(x) = (1/V)\int_{V(x)} z(u)\,du$ over a volume $V(x)$ of size V by *any* linear combination of n data,

$$z_{v_\alpha}(x_\alpha) = \frac{1}{v_\alpha}\int_{v_\alpha(x_\alpha)} z(u)\,du,$$

which are themselves mean values over volumes $v_\alpha(x_\alpha)$ of different sizes v_α. In the extractive industry, this is the classical problem of estimating the mean grade of a panel from data taken from cores or samples of variable sizes. Under the intrinsic hypothesis,

1. unbiasedness is ensured by equating the sum of weights to 1:

$$Z_V^*(x) = \sum_{\alpha=1}^{n} \lambda_\alpha Z_{V_\alpha}(x_\alpha); \qquad \sum_{\alpha=1}^{n} \lambda_\alpha = 1;$$

2. the estimation variance is then given by

$$E\{[Z_V(x) - Z_V^*(x)]^2\}$$
$$= 2\sum_\alpha \lambda_\alpha \bar{\gamma}(V, v_\alpha) - \bar{\gamma}(V, V)$$
$$- \sum_\alpha \sum_\beta \lambda_\alpha \lambda_\beta \bar{\gamma}(v_\alpha, v_\beta),$$

where

$$\bar{\gamma}(v_\alpha, v_\beta) = \frac{1}{v_\alpha v_\beta}\int_{v_\alpha(x_\alpha)} du \int_{v_\beta(x_\beta)} \gamma(u - u')\,du'$$

is the mean value of the semivariogram function $\gamma(h)$ when the two extremities of the interdistance vector $\mathbf{h}$ describe independently the two volumes $v_\alpha(x_\alpha)$ and $v_\beta(x_\beta)$.

Remarks. The last estimation variance formula is exact (no approximation is involved in its derivation) and completely general, whatever the size and relative locations of the volumes $V(x), v_\alpha(x_\alpha)$; it expresses the four essential and intuitive notions which condition all estimations; i.e., the quality of estimation depends on:

(a) The size and geometry of the volume to be estimated: term $\bar{\gamma}(V, V)$

(b) The interdistances between the unknown and the data: terms $\bar{\gamma}(V, v_\alpha)$

(c) The number but also the *configuration* of the data available: terms $\bar{\gamma}(v_\alpha, v_\beta)$

(d) Last and certainly not least, the degree of continuity (or variability) of the underlying phenomenon: utilization of the structural model $\gamma(h)$

Knowledge of the mean and the variance is usually not enough to characterize the distribution of the error of estimation and therefore to define the confidence intervals* of the estimates, unless a normal* or some other two-parameter distribution, is assumed. When dealing with earth sciences data, practice has shown that in most well-behaved cases the distributions of experimental errors, although not normal, tend to be symmetrical around a zero mean (unbiasedness) with more small errors and more very large errors than what would be predicted from the normal distribution with mean 0 and variance σ_E^2. Fortunately, the interval $[0 \pm 2\sigma_E]$ does contain approximately 95% of the observed errors.

The expression for the estimation variance only calls for the geometry of the information pattern and not for the data themselves; hence estimation variance can be forecasted before any information is actually taken and this allows for optimal design of sampling campaigns.

THE KRIGING TECHNIQUE

*Kriging** is a local estimation technique which provides the best linear unbiased estimator (BLUE) of any unknown value $z(x)$ or mean value $z_V(x)$. The information used are the n available data $z(x_\alpha)$ or $z_{V_\alpha}(x_\alpha)$ in the neighborhood of the value to be estimated. The problem is to derive the set of optimal weights $\{\lambda_\alpha, \alpha = 1 \text{ to } n\}$ such that the estimation variance attached to the linear estimator $Z_V^*(x) = \sum_\alpha \lambda_\alpha Z_{v_\alpha}(x_\alpha)$ is minimum under the unbiasedness condition $\sum_\alpha \lambda_\alpha = 1$.

The solutions $\{\lambda_\alpha\}$ together with a Lagrangian parameter μ appear as the unique solution of a system of $(n + 1)$ linear equations, called the *kriging system*:

$$\begin{cases} \sum_\beta \lambda_\beta \bar{\gamma}(v_\alpha, v_\beta) + \mu = \bar{\gamma}(v_\alpha, V), \\ \sum_\beta \lambda_\beta = 1. \qquad \forall \alpha = 1 \text{ to } n, \end{cases}$$

The corresponding minimum estimation variance, or *kriging variance*, is then

$$\sigma_K^2 = \sum_\alpha \lambda_\alpha \bar{\gamma}(v_\alpha, V) + \mu - \bar{\gamma}(V, V).$$

Remarks. The unbiased kriging estimator is also an exact interpolator; i.e., if the volume V to be estimated coincides with any of the data supports v_α, the kriging system provides a BLUE identical to that data value, i.e., $Z_V^*(x) \equiv Z_{v_\alpha}(x_\alpha)$, and a corresponding zero kriging variance.

The previous kriging procedure, also called *ordinary kriging*, can be generalized in various senses:

1. To allow for the linear coestimation of one element $z_k(x)$ using data $z_k(x_\alpha)$ on this element but also data from other intercorrelated elements $z_{k'}(x_\beta)$, $k' \neq k$ (see Journel and Huijbregts [3, pp. 324–326, 335–343]).
2. To allow for linear estimation of a nonstationary phenomenon $Z(x) = m(x) + Y(x)$. The RF $Z(x)$ is the sum of a deterministic component $m(x)$, called *trend* or *drift*, and a stationary (or locally) intrinsic fluctuation $Y(x)$ with its own pattern of spatial variability characterized by a semivariogram $\gamma_Y(h)$ (see Delfiner in Guarascio et al. [1, pp. 49–68]). Note that when $\gamma_Y(h)$ is made a pure nugget effect model, the fluctuations $Y(x)$ being uncorrelated, kriging the unknown trend $m(x)$ amounts to fitting it by least squares*. But such a total absence of correlation* is usually highly unrealistic when dealing with earth sciences data.

3. To allow for the derivation of best unbiased *non*linear estimators, such as conditional expectations and "disjunctive kriging" estimators (see Matheron in Guarascio et al. [1, pp. 221–236]).

It should be stressed that kriging *sensu lato* has fundamental connections with spline* theory [4] and the theory of projectors [2]. In this respect the name "kriging," although now widely accepted among practioners, is certainly misleading.

DISPERSION VARIANCE

In the reconnaissance of a natural phenomenon it is often necessary to have a measure of the in situ variability at various scales of its characteristic variable(s) $z(x)$.

Let V be an area within which stationarity of order 2 or the intrinsic hypothesis holds. The area (or volume) V is the union of N equally sized units $v(x_i)$: $V = \sum_{i=1}^{N} v(x_i)$. Let $z_v(x_i) = (1/v)\int_{v(x_i)} z(u)\,du$ be the average characteristic over each of the units $v(x_i)$.

A measure of the variability of unit values $z_v(x_i)$ within the area V is the mean squared difference:

$$s^2 = \frac{1}{N}\sum_{i=1}^{N}\left[z_v(x_i) - z_V(x)\right]^2.$$

This measure s^2 is usually unknown, since the values $z_v(x_i)$ are usually unknown. But once s^2 is interpreted as a realization of a random variable S^2, the expected value $E\{S^2\}$ can be calculated. By definition, this expected value is the *dispersion variance* $D^2(v \mid V)$ of unit v values within area V. It can be shown that this is a *linear* function of the semivariogram model $\gamma(h)$:

$$E\{S^2\} = D^2(v \mid V) = \bar{\gamma}(V,V) - \bar{\gamma}(v,v).$$

Remarks. $D^2(v \mid V)$, the dispersion variance characteristic of the *in situ* variability of unit v values at the scale of area V, should be clearly distinguished from the estimation variance σ_E^2, characteristic of the quality of an estimation.

When $v = 0$, i.e., in practice the unit size reduces to the size of the samples, $D^2(0 \mid V) = \bar{\gamma}(V,V) - \gamma(0) = \bar{\gamma}(V,V)$. Now if $V \to \infty$, in practice if the area V comprises all the field of study and if the stationarity hypothesis holds true over that field, the mean value $\bar{\gamma}(V,V) = C(0) - C(V,V) \to C(0)$, i.e., $D^2(0 \mid \infty) = C(0) = \operatorname{var}\{z(x)\}$. Similarly, $D^2(v \mid \infty) = \operatorname{var}\{z_v(x)\}$. The a priori variances, when they can be defined, are none other than the limits of dispersion variances within large areas (large with regard to the range of the semivariogram or covariance model).

It would be erroneous to try to estimate directly the previous variability measures s^2 by replacing the unknown values $z_v(x_i)$ by their estimates $z_v^*(x_i)$. Indeed, the *in situ* variability of the estimators $Z_v^*(x)$ is usually much smoother than the *in situ* variability of the true values $Z_v(x)$. Kriging estimators $Z_v^*(x)$ are no exception, but in their case the *smoothing relation* between the dispersion variances of kriged and true values can be established:

$$D^2(v \mid V) = D^2(v_K^* \mid V_K^*) + \overline{\sigma_{Kv}^2} - \sigma_{KV}^2$$

where

$$D^2(v_K^* \mid V_K^*) = E\left\{\frac{1}{N}\sum_{i=1}^{N}\left[Z_v^*(x_i) - Z_V^*(x)\right]^2\right\},$$

$$\overline{\sigma_{Kv}^2} = \frac{1}{N}\sum_{i=1}^{N} E\left\{\left[Z_v(x_i) - Z_v^*(x_i)\right]^2\right\},$$

$$\sigma_{KV}^2 = E\left\{\left[Z_V(x) - Z_V^*(x)\right]^2\right\},$$

these two estimation variances being provided directly by the kriging procedure.

In most practical cases, when $V \gg v$, the term σ_{KV}^2 can be neglected with regard to $\overline{\sigma_{Kv}^2}$.

Note that the smoothing effect $D^2(v \mid V) - D^2(v^*_K \mid V_K^*)$ is directly proportional to the average kriging variance $\overline{\sigma_{Kv}^2}$. The fewer data available, the greater is $\overline{\sigma_{Kv}^2}$ and the greater is the smoothing effect. For a given amount of information, the smaller the unit v, the greater is $\overline{\sigma_{Kv}^2}$ and consequently the smoothing effect. It is illusory to try to estimate too small units with too few data; the resulting estimates, although optimal, would give an unrealistic smoothed image of reality.

CONDITIONAL SIMULATIONS

The kriged image of reality, like any other estimated image, is smoothed; the idea is to produce another image of that reality which would identify at all scales the various dispersion variances. The goal is no longer to minimize the estimation variances but to identify the dispersion variances.

The regionalized variable $z(x)$ has been interpreted as a realization of a RF $Z(x)$. This class of RFs is characterized by a variogram model, and in the stationary case by a distribution function inferred from the histogram of data. The idea of *conditional simulations* is to draw other realizations $z_s(x)$ from this class of RFs, and to retain only those realizations $z_{cs}(x)$ which meet the experimental values at data locations. Simulated realizations $z_{cs}(x)$, except for data locations x_α, will differ point to point from the reality $z(x)$, but statistically will show the same structure of variability at all scales; $z_{cs}(x)$ and $z(x)$, since they are derived from the same variogram model, will have the same dispersion variances $D^2(v \mid V)$, for all v and V. The simulated realization, however, has the advantage of being known at almost all points x and not only at data points x_α.

Using this considerably extended data base $z_{cs}(x)$, the various intended procedures of survey, production, blending, etc., can be tested in great detail. In a way, the conditional simulation(s) represent(s) for the reconnaissance and production planning of a natural phenomenon what a flight simulator is for the design and testing of a new aircraft.

The Turning Band Technique

Many methods are available, and are currently used, to provide one-dimensional realizations of a stationary stochastic process* with a given covariance. However, when these procedures are extended to three-dimensional space with an anisotropic covariance, they prove to be inextricable and prohibitive in terms of computer time; the average simulation of a three-dimensional mineral deposit may involve over 1 million simulated values $z_{cs}(x)$.

The originality of the "turning band" technique, proposed by Matheron, and extensively described in Journel and Huijbregts [3, pp. 491–554], is to reduce any three or more generally n-dimensional simulations to several independent one-dimensional simulations along lines which are then rotated in n-dimensional space to generate the required n-dimensional simulation.

Remarks. A simulation does not replace good estimation. Indeed, if statistically the simulation $z_{cs}(x)$ reproduces the *in situ* variability (up to order 2) of the reality $z(x)$, then locally at any given point $x \neq x_\alpha$ the simulated value $z_{cs}(x)$ is a poor estimator of the unknown true value $z(x)$. It can be shown that the corresponding estimation variance is twice as big as the minimum estimation variance provided by kriging:

$$E\left\{\left[Z(x) - Z_{cs}(x)\right]^2\right\} = 2E\left\{\left[Z(x) - Z_K^*(x)\right]^2\right\}.$$

For problems of estimation and selection, kriged estimates should be used; for problems of evaluation of *in situ* variability, simulated values should be used.

Neither simulation nor kriging replaces data. The more and better the information available, the better would be the RF model and the greater would be the conditioning effect of data. In the limit, if an infinite amount of information about $z(x)$ is available, the conditioning effect $z_{cs}(x_\alpha) = z(x_\alpha)$ will identify all conditional simulations to the almost perfectly known reality $z(x)$ and also to the kriged image, kriging being an exact interpolator; $z_K^*(x_\alpha) = z(x_\alpha)$.

SOME HISTORY AND RECENT TRENDS

Until recently, geostatistical theory has been developed essentially by a single group of mine and geology oriented people headed by G. Matheron in France. Until the breakthrough of the "Geostat 1975" NATO Conference, this small group worked almost independently of other researchers in applied statistics. This resulted in unfortunate duplicated work and the use of idiomatic French notations and miner jargon in geostatistical publications. Fortunately, however, geostatistics was developed by practice-oriented people more interested in the solution than in the properties of the probabilistic model used; this resulted in extremely robust techniques such as linear kriging, which are in the trend of modern nonparametric statistics.

Since 1975 an explosion of information of unequal quality has appeared in the literature of both the theory and applications of geostatistics. Several trends can be detected:

1. The practice is no longer limited to the mining field. Case studies are now available in hydrology*, pluviometry, bathymetry, meteorology*, forest survey, geography*, and cartography.
2. In theory, research is aimed toward greater robustness of the established tools (variogram-kriging) with regard to stationarity in particular, and also toward an enrichment of the geostatistical model. Nonlinear geostatistics is no longer distribution-free, and the inference of a distribution function from a single realization of a random function poses severe problems of robustness*, although these are not peculiar to geostatistics. A positive trend in recently published case studies is the systematic use of cross-validation techniques, such as the jackknife*, to infer the various model parameters.

References

[1] Guarascio, M., David, M., and Huijbregts, C. (1975). *Proc. NATO ASI Geostat 75* (Rome, October 1975). D. Reidel, Dordrecht, Holland. (A must for the basic papers of P. Delfiner and G. Matheron.)

[2] Journel, A. G. (1977). *Math. Geol.*, **9**(6), 563–586.

[3] Journel, A. G. and Huijbregts, C. J. (1978). *Mining Geostatistics*. Academic Press, London. (The most complete and updated reference for methodology and practice.)

[4] Kimeldorf, G. and Wahba, G. (1971). *J. Math. Anal. Appl.*, **33**(1), 82–95.

[5] Matheron, G. (1962). *Traité de géostatistique appliquée*, Vol. 1 (1962), Vol. 2 (1963). Technip, Paris. (Most of linear geostatistics is already there. An historical landmark.)

Bibliography

See the following works, as well as the references just cited, for more information on the topic of geostatistics.

Alldredge, R. and Alldredge, N. (1978). *Int. Statist. Rev.*, **46**, 77–88. (A bibliography of geostatistics.)

Clark, I. (1979). *Practical Geostatistics*. Applied Science, London. (A good starter for practicioners.)

Delhomme, J. P. (1976). *Application de la théorie des variables regionalisées dans les sciences de l'eau*. Doctoral thesis, École Nationale Supérieure des Mines, Paris.

Krige, D. G. (1951). A Statistical Approach to Some Mine Valuation and Applied Problems at the Witwatersrand. Thesis, University of the Witwatersrand. (The historical starting point.)

Matern, B. (1960). Spatial variation. In *Almaenna Foerlaget*. Stockholm. (Another historical landmark with applications in forestry.)

Matheron, G. (1965). *Les Variables regionalisées et leur estimation*. Masson, Paris. (The mathematical statistics aspect of geostatistics.)

Matheron, G. (1970). *La Théorie des variables regionalisées et ses applications*. École Nationale Supérieure des Mines, Paris. (The basic course book of the French school.)

Matheron, G. (1973). *Adv. Appl. Prob.*, **5**, 439–468. (The assumption of stationarity is almost completely relieved for linear geostatistics.)

Pauncz, I. (1978). *Math. Geol.*, **10**(2), 253–260. (English language publications of the French school of geostatistics.)

Rendu, J. M. (1978). *An Introduction to Geostatistical Methods of Mineral Evaluation*. South African Institute of Mining and Metallurgy, Johannesburg.

Serra, J. (1967). *Mineralium Deposita*, **3**, 135–154.

(GEOGRAPHY, STATISTICS IN
GEOLOGY, STATISTICS IN
HYDROLOGY, STATISTICS IN

KRIGING
NONPARAMETRIC METHODS
REGRESSION
STOCHASTIC PROCESSES)

A. G. JOURNEL

GERMAN STATISTICAL ASSOCIATION *See* STATISTISCHE GESELLSCHAFT, DEUTSCHE

GERT *See* FLOWGRAPH ANALYSIS

GIBBS DISTRIBUTIONS I

Consider an open thermodynamical system in contact with a heat reservoir of temperature T. According to statistical thermodynamics, the system's energy is a random variable with the Gibbs distribution

$$\text{Prob}(u < U \leqslant u + du)$$
$$= dG(u)\exp(-u/kT)Z^{-1}(T),$$

where the function $dG(u)$ is called the *density of states*, and

$$Z(T) = \int \exp(-u/kT)\,dG(u)$$

is the partition function. The same distribution is known to the statistician as a distribution of exponential type*. Its main property is sufficiency*.

In refs. 1 and 2, sufficiency is related to the notion of thermal equilibrium (zeroth principle of thermodynamics). This leads to a new axiomatic foundation of thermodynamics. A previously shaky notion, "temperature of an isolated system," is interpreted as being a statistical estimate, hence as being intrinsically indeterminate. The most widely used variant, due to Boltzmann*, involves a maximum likelihood* estimate.

References

[1] Mandelbrot, B. B. (1962). *Ann. Math. Statist.*, **33**, 1021–1038.

[2] Mandelbrot, B. B. (1969). *J. Math. Phys.* **5**, 164–171.

(EXPONENTIAL FAMILIES
GIBBS DISTRIBUTIONS II
SUFFICIENCY)

BENOIT B. MANDELBROT

GIBBS DISTRIBUTIONS II

A Gibbs distribution is a probability distribution for a *spatial process** having a given system of conditional probability distributions. These arise in the study of systems with a large number of mutually interacting components; examples may be found in physics, plant ecology, cybernetics, etc. They are also of interest in probability theory because they offer generalizations of standard ideas such as *Markov chains* to multidimensional "time" (*see* MARKOV PROCESSES).

Consider a set S (referred to as the *lattice*), whose members will be called *sites*. In most applications, S will be a regular array of points in space. For each site $i \in S$ there is a random variable X_i, which we will assume takes values in a finite set W. The generalization to continuous, but bounded, ranges is quite straightforward, but unbounded ranges can lead to complications (see Lebowitz and Presutti [12] and Spitzer [18]). A *configuration* $x = (x_i)_{i \in S}$ is a function from S to W; for any subset A of S, we let $x_A = (x_i)_{i \in A}$ be the restriction of x to A. The σ-field generated by all the X_i will be denoted by $\mathscr{B}$, and the sub-σ-field generated by $\{X_i : i \in A\}$ will be called $\mathscr{B}_A$. It is traditional in mathematical physics to identify the X_i as functions on the space W^S of all configurations, and their distribution as a probability measure (often called a *state*) on $(W^S, \mathscr{B})$ (*see also* LATTICE SYSTEMS).

For example, S might be an orchard, with $X_i = 1$ or -1 depending on whether tree i suffers from a certain disease. In such a situation, it may be reasonable to expect that each tree is influenced rather strongly by the health of its near neighbors, but not directly affected by distant trees. Thus one might suppose that the conditional probability* $\Pr[X_i = 1 \mid \text{all other } X_j\text{'s}]$ is some function, say, of only the X_j's for sites j which are nearest neighbors of i. The choice of this function, however, is not completely arbitrary; some consistency conditions are necessary for the existence of a distribution with these conditional probabilities. Perhaps the best way to ensure that these consistency conditions are satisfied is to write down, not just the conditional probabilities for one site, but all the conditional probabilities $\Pr[X_A = x_A \mid \text{all } X_j \text{ for } j \notin A]$ for all finite subsets A of S, where by $X_A = x_A$ we mean $X_i = x_i$ for all $i \in A$. These can all be determined from the one-site conditional probabilities, assuming none of the latter are 0 or 1. Such a system of conditional probabilities is called a *specification*, and a distribution having these conditional probabilities is called a *Gibbs distribution*.

More formally, a specification is a system $\pi = \{\pi_A : A \subset S \text{ finite}\}$ of functions $\pi_A(x, y)$ on $W^A \times W^{S \setminus A}$ satisfying:

(a) $\pi_A(x, y) \geqslant 0$ for all x, y.

(b) $\sum_{x \in W^A} \pi_A(x, y) = 1$ for all y.

(c) $\pi_A(x, \cdot)$ is measurable in $\mathscr{B}_{A^c}$ for each x.

(d) For any $B \subset A$, $x \in W^A$ and $y \in W^{S \setminus A}$,

$$\pi_A(x, y) = \pi_B(x_B, x_{A \setminus B} \times y) \times \sum_{z \in W^B} \pi_A(z \times x_{A \setminus B}, y).$$

Condition (d) is a consistency condition, ensuring that

$$\Pr[X_A = x_A] = \Pr[X_B = x_B \mid X_{A \setminus B} = x_{A \setminus B}] \times \Pr[X_{A \setminus B} = x_{A \setminus B}].$$

A probability measure P is a Gibbs distribution for the specification π if

$$P[X_A = x_A \mid \mathscr{B}_{A^c}] = \pi_A(x_A, \cdot) \qquad P - \text{a.s.}$$

This is a very general definition: in fact [6], any probability measure on W^S is a Gibbs distribution for some specification. A more manageable class of specifications, to which we will restrict our attention in this article, arises from *interactions*. The prototype, and still the most important example, of these is the *Ising model*, proposed by Lenz in 1920 as a simple model of a magnet. The lattice S represents a crystal lattice of atoms, each with a *spin* that may be either "up" ($X_i = +1$) or "down" ($X_i = -1$). Each pair of nearest-neighbor spins X_i, X_j interacts with an energy $-JX_iX_j$, where J is a constant (positive for a *ferromagnet* or negative for an *antiferromagnet*). There may also be a contribution of $-h\sum_{i \in S} X_i$ to the energy from an external magnetic field h. If $H(X)$ is the total energy of the system, then according to the *canonical ensemble* of J. W. Gibbs, in equilibrium at temperature T the probability of a configuration x should be proportional to $e^{-H(x)/(kT)}$, where k is Boltzmann's constant. We will follow the common practice of rescaling H so that $kT = 1$. The specification for this distribution is

$$\pi_A(x, y) = \left(\sum_{x' \in W^A} e^{-H_A(x', y)} \right)^{-1} e^{-H_A(x, y)}, \tag{1}$$

where

$$H_A(x, y) = -J \sum^{(1)} x_i x_j - J \sum^{(2)} x_i y_j - h \sum_{i \in A} x_i. \tag{2}$$

Here $\sum^{(1)}$ is a sum over all pairs of nearest-neighbor sites i, j in A, and $\sum^{(2)}$ is a sum over all nearest-neighbor pairs with $i \in A$ and $j \notin A$. $H_A(x, y)$ is to be thought of as the energy of the system with configuration x in A and "boundary condition" y. The specification (1) may be interpreted as saying that if the "boundary condition" is fixed, then the system inside A is in equilibrium according to Gibbs. Note that, while the canonical ensemble is defined only for finite S [otherwise, the total energy $H(x)$ is a divergent sum], the specification (1) may also be used for infinite S. The definition of a Gibbs distribution using conditional probabilities was developed by Dobrushin [3] and Lanford and Ruelle [11] to deal with systems on infinite lattices.

The basic Ising model may be generalized to include interactions between more distant spins, or among several spins (e.g., terms such as $x_i x_j x_k$ in the energy). This leads to the following definitions. An *interaction* is a system $\Phi = \{\Phi_A : A \subset S \text{ finite}\}$, where each Φ_A is a real-valued function on W^A. The specification for the interaction Φ is given by (1), with

$$H_A(x, y) = \sum_{B \cap A \neq \phi} \Phi_B(x_{B \cap A} \times y_{B \setminus A}). \quad (3)$$

Some restrictions on Φ must be imposed to ensure that these sums converge: e.g., we may assume that each site is contained in only finitely many B for which Φ_B is not identically zero.

Outside of physics, the main interest in interactions comes from theorems which state that specifications arising from interactions are very general (see, e.g., Sullivan [19]) and that certain types of specifications come from particular types of interactions. For example, in our model of the orchard we wanted the conditional probability $\pi_{\{i\}}(1, y)$ to depend only on the restriction of y to the nearest neighbors of i. Assuming $\pi_{\{i\}}(1, y)$ to be never 0 or 1 and S to be a rectangular lattice, this implies that the specification comes from an interaction of the form

$$\Phi_{\{i\}}(x_i) = h_i x_i,$$

$$\Phi_{\{i,j\}}(x_i, x_j) = J_{ij} x_i x_j \quad (4)$$

$$\text{if } i \text{ and } j \text{ are nearest neighbors,}$$

$$\Phi_A = 0 \qquad \text{otherwise,}$$

which is the Ising model with h and J varying over the lattice (of course, a natural additional assumption is that all the h_i and all the J_{ij} are equal). A more general result holds for any W and S, and any definition of "neighbor" that is symmetric (so i is a neighbor of j if j is a neighbor of i) and such that each site has only finitely many neighbors: suppose $\pi_{\{i\}}(x, y)$ is never 0 and depends only on the restriction of y to the neighbors of i. Then π arises from an interaction Φ such that $\Phi_A = 0$ unless either A is a single site, or all sites in A are neighbors of each other. This result is sometimes called the *Hammersley–Clifford theorem**; it has been proved many times in varying degrees of generality, but the first appearance in print of the version above seems to be Grimmett [8].

The property that $\pi_{\{i\}}(x, y)$ depends only on the restriction of y to the neighbors of i implies that

$$\Pr[E \mid \mathscr{B}_{A^c}] = \Pr[E \mid \mathscr{B}_{\partial A}]$$

$$\text{for } A \text{ finite and } E \in \mathscr{B}_A, \quad (5)$$

where ∂A contains those sites outside A which are neighbors of sites in A. This is a type of *Markov property*, where A is thought of as the "future," ∂A as the "present," and $(A \cup \partial A)^c$ as the "past"; a process with this property is sometimes called a *Markov random field**. The question of whether (5) extends to infinite A (*global Markov property*) is more difficult (see Goldstein [7]).

Most of the important results of the theory of Gibbs distributions are for infinite S; these can be regarded as statements about the limit as $|S| \to \infty$. The most basic question is to describe the set $G(\Phi)$ of Gibbs distributions for a given interaction Φ. The existence of a Gibbs distribution is easy; more surprising is the fact that it is not always unique. Thus for the Ising model on an infinite rectangular lattice, if $h = 0$ and J is sufficiently large, there is a stationary Gibbs distribution P_+ for which $E[X_i] > 0$ (this is called *spontaneous magnetization*), and by the symmetry $X_i \leftrightarrow -X_i$ a second Gibbs distribution P_- with $E[X_i] < 0$. This is an example of *spontaneous breaking of symmetry*, where a symmetry of the specification is lacking in the Gibbs distribution. P_+ (respectively, P_-) can be obtained as limits of distributions on W^A for finite A tending to S, with probability mass functions $\Pr[X_A = x] = \pi_A(x, y)$ where all $y_i = +1$ (respectively, -1). The interpretation is that the influence of the boundary condition y is strong, even when the boundary is far away. Conversely, when the Gibbs distribution is unique, as for small J or for $h \neq 0$ and any $J > 0$, the limits above are equal for all y; this means that the influence of the boundary disappears as its distance increases.

Other types of symmetry, such as translation invariance, may also be spontaneously broken. For example, in the Ising model for $J < -J_c$ (where J_c is the "critical" J at which spontaneous magnetization begins) and $|h|$ sufficiently small there are nonstationary Gibbs distributions: if the sites are colored "black" and "white" in a checkerboard pattern, then $E[X_i]$ is positive for black sites and negative for white, or vice versa.

The set $G(\Phi)$ of Gibbs distributions for an interaction Φ is convex; i.e., if P_1 and P_2 are Gibbs distributions, so is $\alpha P_1 + (1-\alpha)P_2$ for $0 \leqslant \alpha \leqslant 1$. Moreover, it is a *Choquet simplex*, which means every Gibbs distribution has a unique representation as a weighted average of extreme points of $G(\Phi)$ (thus a triangle is a Choquet simplex, whereas a square is not). A Gibbs distribution is an *extreme point* of $G(\Phi)$ (i.e., is not an average of two other Gibbs distributions) if and only if its *tail field* is trivial, i.e., for any $E \in \mathscr{B}_\infty = \bigcap_{A \text{ finite}} \mathscr{B}_{A^c}$, $\Pr[E]$ is either 0 or 1 (*see* ZERO–ONE LAWS). Equivalently [11], the distribution has *short-range correlations*: given any event $E \in \mathscr{B}$ and $\delta > 0$, there is a finite set $A \subset S$ such that

$$|\Pr[E \cap F] - \Pr[E]\Pr[F]| < \delta \quad \text{for all} \quad F \in \mathscr{B}_{A^c}. \quad (6)$$

If S is a regular lattice such as $\mathbb{Z}^d$ and the interaction (and thus the specification) is translation invariant, we can consider the stationary Gibbs distributions. The configuration space W^S with a stationary Gibbs distribution and the action of lattice translations fits into the framework of *ergodic theory* (*see* ERGODIC THEOREMS), except that instead of a single transformation T and its iterates, we have a d-dimensional Abelian group of transformations; the basic results carry over with no change. The stationary Gibbs distributions for Φ themselves form a Choquet simplex. The extreme points of this simplex are called *ergodic* (an ergodic theorist would say the action of the group, not the measure, is ergodic) and are characterized by a *law of large numbers**: for any $\mathscr{B}$-measurable random variable Y with $E[|Y|] < \infty$,

$$\lim_{n\to\infty} (2n+1)^{-d} \sum_{j_1=-n}^{n} \cdots \sum_{j_d=-n}^{n} \tau_j(Y) = E[Y] \quad \text{a.s.,} \quad (7)$$

where $\tau_j(Y)$ is the translation of Y by j (this is the *Birkhoff ergodic theorem*). Ergodic Gibbs distributions need not have trivial tail fields and thus may be averages of nonstationary Gibbs distributions.

More specific results on $G(\Phi)$ include sufficient conditions for the Gibbs distribution to be unique, e.g., if the interaction is "small" in a suitable sense [4], or nonunique (the original proof by Peierls [16] of spontaneous magnetization in the two-dimensional Ising model is the ancestor of most of these). The results on the two-dimensional ferromagnetic Ising model are quite detailed: the Gibbs distribution is unique if $h \neq 0$ or $J < J_c = \frac{1}{2}\log(1+\sqrt{2})$, while for $h = 0$ and $J > J_c$ there are only the two extreme Gibbs distributions, P_+ and P_- ([1]; also proved independently by Higuchi). This model for $h = 0$ is unusual in that exact closed-form expressions for the moments are available, starting with the determination of $E[X_iX_j]$ for nearest neighbors by Onsager [15].

The behavior of the Ising model and its relatives near the critical point $h = 0$, $J = J_c$ is of particular interest. It is believed (and partly proven) that away from the critical point an extreme Gibbs distribution has covariance decaying exponentially with distance, while at the critical point the decay is only as a power of the distance. Moreover, away from the critical point an extreme Gibbs state should satisfy a central limit theorem*: in fact, Newman [14] showed that if $J > 0$ and $V \equiv \sum_{j\in S} \operatorname{cov}(X_0, X_j) < \infty$, then the *block spins*

$$S_j = L^{-d/2} \sum_{i \in C_j} (X_i - E[X_i])$$

(where $\mathbb{Z}^d$ is partitioned into d-dimensional cubes C_j of side L) tend in joint distribution to independent normal random variables of mean 0 and variance V as $L \to \infty$. At the critical point V is expected to be infinite, but the block spins (with $L^{-d/2}$ replaced by some other power) might have some other limiting distribution. The study of such matters (generally known as the *renormalization group*) is currently a very active topic.

So far *time* has not entered the picture, although (especially in physics) the Gibbs distribution is generally thought of as describing the equilibrium of a dynamical process. As yet, the Gibbs distribution cannot be derived from a realistic physical dynamics; however, certain processes can be constructed for which the Gibbs distributions are stationary* (in time). Suppose, for example, that $W = \{1, -1\}$, and imagine that the spin at each site i changes sign at random (continuous) times according to a rate $c_i(x)$ depending on the configuration x. Liggett showed [13] that under suitable conditions there is a standard *Markov process** $X(t)$ whose state space is W^S that corresponds to this description (the result is far from obvious because of the infinite number of sites). Gibbs distributions will be stationary for this process if the flip rates c_i satisfy

$$c_i(1 \times y)e^{-H_{\{i\}}(1,y)} = c_i(-1 \times y)e^{-H_{\{i\}}(-1,y)} \qquad (8)$$

for all $i \in S$ and $y \in W^{S\setminus\{i\}}$. For example, a popular choice is $c_i(x) = e^{H_{\{i\}}(x_i, x_{S\setminus\{i\}})}$. It is believed, but not yet proven in full generality, that all stationary (in time) distributions for the process are Gibbs distributions; this is known for one- and two-dimensional lattices [9]. These processes are related to a large number of other interacting Markov processes (both in discrete and continuous time) which have been studied.

Literature

A standard reference work for the physics of Ising and other models is Domb and Green [5]. A recent survey from the point of view of probability is Kindermann and Snell [10]. A very general treatment (including some models where the lattice is replaced by continuous space) is Preston [17]. Statistical applications are illustrated in Bartlett [2].

References

[1] Aizenman, M. (1980). *Commun. Math. Phys.*, **73**, 83–94.

[2] Bartlett, M. S. (1975). *The Statistical Analysis of Spatial Pattern*. Chapman & Hall, London.

[3] Dobrushin, R. L. (1968). *Theory Prob. Appl.*, **13**, 197–224.

[4] Dobrushin, R. L. (1968). *Funct. Anal. Appl.*, **2**, 302–312.

[5] Domb, C. and Green, M. S., eds. (1972, 1972, 1974, 1976, 1976, 1977). *Phase Transitions and Critical Phenomena*, Vols. 1–6. Academic Press, New York.

[6] Goldstein, S. (1978). *Z. Wahrscheinlichkeitsth. verw. Geb.*, **46**, 45–51.

[7] Goldstein, S. (1980). *Commun. Math. Phys.*, **74**, 223–234.

[8] Grimmett, G. R. (1973). *Bull. Lond. Math. Soc.*, **5**, 81–84.

[9] Holley, R. and Stroock, D. W. (1977). *Commun. Math. Phys.*, **55**, 37–45.

[10] Kindermann, R. and Snell, J. L. (1980). *Markov Random Fields and Their Applications*, Vol. 1: *Contemporary Mathematics*. American Mathematical Society, Providence, R.I.

[11] Lanford, O. E. and Ruelle, D. (1969). *Commun. Math. Phys.*, **13**, 194–215.

[12] Lebowitz, J. L. and Presutti, E. (1976). *Commun. Math. Phys.*, **50**, 195–218.

[13] Liggett, T. M. (1972). *Trans. Amer. Math. Soc.*, **165**, 471–481.

[14] Newman, C. M. (1980). *Commun. Math. Phys.*, **74**, 119–128.

[15] Onsager, L. (1944). *Phys. Rev.*, **65**, 117–149.

[16] Peierls, R. (1936). *Proc. Camb. Philos. Soc.*, **32**, 477–481.

[17] Preston, C. (1976). Random Fields, *Lect. Notes Math.*, **534**.

[18] Spitzer, F. (1975). *J. Funct. Anal.*, **20**, 240–255.

[19] Sullivan, W. (1973). *Commun. Math. Phys.*, **33**, 61–74.

(CRITICAL PHENOMENA
GIBBS DISTRIBUTIONS I
LATTICE SYSTEMS
MARKOV RANDOM FIELDS
SPATIAL PROCESSES)

R. B. ISRAEL

GINI INDEX *See* INCOME INEQUALITY MEASURES: LORENZ CURVE

GINI–SIMPSON INDEX *See* DIVERSITY, INDICES OF

GINI'S MEAN DIFFERENCE

In a long and influential paper on "variability and mutability" Gini [5] introduces what he terms the *mean difference* of the n quantities $x_1, \dots, x_n$, i.e.,

$$g = \frac{1}{n(n-1)} \sum_{i,j=1}^{n} |x_i - x_j|. \qquad (1)$$

Following common practice we use the symbol g and call it *Gini's mean difference*, although essentially the same function was studied much earlier in *Astronomische Nachrichten* (e.g., Jordan [10]; von Andrae [15]; Helmert [9]). In fact, Gini [5] states that he became aware of this work after completing his own. However, the points of view are different. The early writers took the x's to be observations of an unknown true value and succeeded in finding the mean and probable error of the mean difference for a normal parent population. Gini, on the other hand, regarded the x's themselves as constituting the population of interest for which g provides an index of variability.

For purposes of computation as well as comparison with other measures of variability it is helpful to express g in terms of the order statistics* $x_1', x_2', \dots, x_n'$ (i.e., the x's arranged in nondecreasing order of magnitude). Since

$$\frac{1}{2}\sum_{i,j}^{n} |x_i - x_j| = \sum_{i>j}^{n} (x_i' - x_j')$$
$$= \sum_{i=1}^{n} (2i - n - 1)x_i',$$

it follows that

$$g = \binom{n}{2}^{-1} \sum_{i=1}^{n} (2i - n - 1)x_i'. \qquad (2)$$

Easily derived variants are [5, 15]

$$g = \binom{n}{2}^{-1} \sum_{i=1}^{[(1/2)n]} (n - 2i + 1)(x_{n+1-i}' - x_i'), \qquad (3)$$

where the upper limit of summation is the integral part of $\frac{1}{2}n$, and [5]

$$g = \binom{n}{2}^{-1} \sum_{i=1}^{n} (n - 2i + 1)|x_i' - m|, \qquad (4)$$

where m is the sample median*; also [4],

$$g = \binom{n}{2}^{-1} \sum_{i=1}^{n-1} i(n-i)(x_{i+1}' - x_i'). \qquad (5)$$

Let G denote the statistic obtained on replacing the x's in g by random variables $X_1, X_2, \dots, X_n$. The mean and variance of G were derived by Nair [14], with an improvement by Lomnicki [13], when the X's are independent and have common cumulative distribution function (CDF) $F(x)$. The mean Δ, itself sometimes called the (coefficient of) mean difference, is from (1) simply $E|X_1 - X_2|$ and has other claims to interest (see, e.g., David [3]). Glasser [6] finds the variance of G when sampling is without replacement from a finite population.

If the X's are normal $N(\mu, \sigma^2)$, then $\Delta = 2\sigma/\sqrt{\pi}$, so that an unbiased estimator of σ is given by

$$\sigma^* = \sqrt{\pi} \sum_{i=1}^{n} (2i - n - 1)X_i'/[n(n-1)]. \qquad (6)$$

The first four moments of G have been obtained in closed form by Kamat [11] and Barnett et al. [1]. Either a χ or a χ^2 approximation gives a good fit to the distribution of G, which is asymptotically normal for *any* parent distribution with finite σ^2.

As an estimator of σ in normal samples σ^* has high efficiency (relative to the root-mean-square estimator S), decreasing to 97.8% asymptotically. For $n = 10$ the efficiency is 98.1% asymptotically. For $n = 10$ the efficiency is 98.1%, which may be compared with 99.0% for the best linear unbiased estimator σ^{**}. Clearly, σ^* has the advantage over σ^{**} of not requiring special tables for its calculation. It is also slightly less sensitive to the presence of outliers* than either S or σ^{**}. Although necessarily entailing a considerable loss in efficiency under normality, a symmetrically censored version of σ^* has been put forward by Healy

[8] as a simple robust estimator of σ. For k observations remaining, this takes the form

$$c_{k,n} \sum_{i=1}^{k} (2i - k - 1)X_i',$$

where $c_{k,n}$ is a constant depending on k and n.

Barnett et al. [1] consider the substitute t statistic

$$\sqrt{n}\,(\bar{X} - \mu)/\sigma^*$$

and show that it leads to little loss in power for normal samples; moreover, the power function is expressible approximately in terms of the normal CDF. The ratio of σ^* to S, calculated from the same sample, has been proposed by D'Agostino [2] as a test of normality (*see* DEPARTURES FROM NORMALITY, TESTS FOR).

Literature

A brief review of Gini's mean difference is given in David [3]. Its early (pre-1950) development may be traced from over 60 entries in Harter [7]. See also the index of Kendall and Stuart [12] under "mean difference".

References

[1] Barnett, F. C., Mullen, K., and Saw, J. G. (1967). *Biometrika*, **54**, 551–554.

[2] D'Agostino, R. B. (1972). *Biometrika*, **59**, 219–221.

[3] David, H. A. (1968). *Biometrika*, **55**, 573–575.

[4] De Finetti, B. and Paciello, U. (1930). *Metron*, **8**(3), 89–94.

[5] Gini, C. (1912). *Studi Econ.-Giuridici R. Univ. Cagliari*, **3**(2), i–iii, 3–159.

[6] Glasser, G. J. (1962). *J. Amer. Statist. Ass.*, **57**, 648–654.

[7] Harter, H. L. (1978). *A Chronological Annotated Bibliography on Order Statistics*, Vol. 1: *Pre-1950*. U.S. Government Printing Office, Washington, D.C.

[8] Healy, M. J. R. (1978). *Biometrika*, **65**, 643–645.

[9] Helmert, F. R. (1876). *Astron. Nachr.*, **88**, 127–132.

[10] Jordan, W. (1869). *Astron. Nachr.*, **74**, 209–226.

[11] Kamat, A. R. (1961). *Metron*, **21**, 170–175.

[12] Kendall, M. G. and Stuart, A. (1977). *The Advanced Theory of Statistics*, Vol. 1: *Distribution Theory*, 4th ed. Charles Griffin, London; Hafner, New York.

[13] Lomnicki, Z. A. (1952). *Ann. Math. Statist.*, **23**, 635–637.

[14] Nair, U. S. (1936). *Biometrika*, **28**, 428–436.

[15] von Andrae, C. C. G. (1872). *Astron. Nachr.*, **79**, 257–272.

(VARIABILITY, MEASURES OF)

H. A. DAVID

GIRDLE DISTRIBUTION *See* DIRECTIONAL DISTRIBUTIONS

GLAHN AND HOOPER CORRELATION COEFFICIENTS

Let $\mathbf{Y} = [\mathbf{y}_1, \ldots, \mathbf{y}_M]$ be the $T \times M$ matrix of T observations on M jointly determined variables; and $\mathbf{X} = [\mathbf{x}_1, \ldots, \mathbf{x}_K]$ be the $T \times K$ matrix of observations on K predetermined variables for a system of M linear stochastic equations with reduced form

$$\mathbf{Y} = \mathbf{X\Pi} + \mathbf{V}.$$

Here $\mathbf{\Pi}$ is a $K \times M$ matrix of unknown parameters and $\mathbf{V} = [\mathbf{v}_1, \ldots, \mathbf{v}_M]$ is a $T \times K$ matrix of disturbances. The corresponding least-squares estimators of $\mathbf{\Pi}$ and $\mathbf{V}$ are $\hat{\mathbf{\Pi}}$ and $\hat{\mathbf{V}}$; $\hat{\mathbf{Y}} = \mathbf{X}\hat{\mathbf{\Pi}}$, so that $\mathbf{Y} = \mathbf{X}\hat{\mathbf{\Pi}} + \hat{\mathbf{V}} \equiv \hat{\mathbf{Y}} + \hat{\mathbf{V}}$. All variables are measured as deviations from their respective means. Since $\hat{\mathbf{\Pi}}$ is chosen so that $\mathbf{X}'\hat{\mathbf{V}} = 0$, then

$$\mathbf{Y}'\mathbf{Y} = \hat{\mathbf{Y}}'\hat{\mathbf{Y}} + \hat{\mathbf{V}}'\hat{\mathbf{V}}, \tag{1}$$

or

$$\mathbf{I}_T = (\mathbf{Y}'\mathbf{Y})^{-1}\tilde{\mathbf{Y}}'\tilde{\mathbf{Y}} + (\mathbf{Y}'\mathbf{Y})^{-1}\hat{\mathbf{V}}'\hat{\mathbf{V}}. \tag{2}$$

In the special case of $M = 1$, equation (1) can be interpreted as resolving the total variation in the $\mathbf{y}$'s, $(\mathbf{y}_1'\mathbf{y}_1)$, into an explained part $(\hat{\mathbf{y}}_1'\hat{\mathbf{y}}_1)$, and an unexplained part, $(\hat{\mathbf{v}}_1'\hat{\mathbf{v}}_1)$; consequently, the single equation squared *multiple correlation coefficient**, defined as

$$R^2_{\mathbf{y}_1 \cdot \mathbf{x}_1, \mathbf{x}_2, \ldots, \mathbf{x}_K} = \hat{\mathbf{y}}_1'\hat{\mathbf{y}}_1 / \mathbf{y}_1'\mathbf{y}_1,$$

measures the fraction of variation in the $\mathbf{y}$'s explained by the $\mathbf{x}$'s.

Both Hooper and Glahn generalized this notion for the case of more than one equation. For $M \geqslant 1$, Hooper [2] defined the square *trace correlation* as

$$\bar{r}^2 = (1/M)\operatorname{tr}\left[(\mathbf{Y}'\mathbf{Y})^{-1}\hat{\mathbf{Y}}'\hat{\mathbf{Y}}\right]$$

$$= (1/M)\operatorname{tr}\left[\mathbf{I} - (\mathbf{Y}'\mathbf{Y})^{-1}\hat{\mathbf{V}}'\hat{\mathbf{V}}\right], \quad (3)$$

which could be "naturally interpreted as that part of the total variance of the jointly dependent variables that is accounted for by the systematic part of the reduced form, and $1 - \bar{r}^2$ as the unexplained part." Further, $\bar{r}^2$ shares the following properties with $R^2_{\mathbf{y}_1 \cdot \mathbf{x}_1, \mathbf{x}_2, \ldots, \mathbf{x}_K}$: $\bar{r}^2 + (1 - \bar{r}^2) \equiv 1$; $0 \leqslant \bar{r}^2 \leqslant 1$; and $\bar{r}^2$ is invariant with respect to the units in which the **x**'s and **y**'s are measured. Hooper [2] gives the asymptotic distributions of $\bar{r}^2$ and $1 - \bar{r}^2$. Using canonical correlation* theory, Hooper shows that Hotelling's *vector alienation coefficient** [3], $\sqrt{z}$ (which measures the independence of the **x**'s and the **y**'s), is a different scalar function of the same matrix, namely the positive square root of

$$z = |(\mathbf{Y}'\mathbf{Y})^{-1}\hat{\mathbf{Y}}'\hat{\mathbf{Y}}| = \prod_{i=1}^{M} (1 - \lambda_i^2),$$

where $\lambda_1, \lambda_2, \ldots$ are the *canonical correlations* of the **y**'s with the **x**'s.

Let $\tilde{\mathbf{Y}}$ be the predicted **y**'s obtained by predicting—in a least-squares* sense—the canonical dependent variables as functions of the canonical independent variables and then transforming back to the original variables. Canonical correlation theory ensures that $\tilde{\mathbf{Y}}'\mathbf{E} = \mathbf{0}$, where $\mathbf{E} = \mathbf{Y} - \tilde{\mathbf{Y}}$. Thus Glahn [1] obtains a second decomposition of the total sample covariance matrix of the $\mathbf{y}_i$'s,

$$(1/T)\mathbf{Y}'\mathbf{Y} = \tilde{\mathbf{Y}}'\tilde{\mathbf{Y}} + (1/T)\mathbf{E}'\mathbf{E}, \quad (4)$$

and bases his measure on (4). However, since $T\hat{\mathbf{Y}}'\hat{\mathbf{Y}} = \tilde{\mathbf{Y}}'\tilde{\mathbf{Y}}$ and $\mathbf{E}'\mathbf{E} = \hat{\mathbf{V}}'\hat{\mathbf{V}}$, (4) is equivalent to (2) and Glahn's [1] squared *composite correlation coefficient* $R^2_{\mathbf{y} \cdot \mathbf{x}}$ can be given as the ratio of the traces of the two square matrices in the first term on the right-hand side of (2),

$$R^2_{\mathbf{y} \cdot \mathbf{x}} = \frac{\operatorname{tr}(\hat{\mathbf{Y}}'\hat{\mathbf{Y}})}{\operatorname{tr}(\mathbf{Y}'\mathbf{Y})}$$

$$= \frac{\sum_{i=1}^{M} \sigma_i^2 R^2_{\mathbf{y}_i \cdot \mathbf{x}_1, \mathbf{x}_2, \ldots, \mathbf{x}_K}}{\sum_{i=1}^{M} \sigma_i^2},$$

where σ_i^2 is the sample variance of y_i and where $R^2_{\mathbf{y}_i \cdot \mathbf{x}_1, \mathbf{x}_2, \ldots, \mathbf{x}_K}$ is, for the ith equation, the single equation squared multiple correlation coefficient. Glahn [1] proposed that his measure supplants Hooper's. He interprets $R^2_{\mathbf{y} \cdot \mathbf{x}}$ as "the fractional part of the total variance of the dependent variables that is accounted for by the predictors," and $1 - R^2_{\mathbf{y} \cdot \mathbf{x}}$ as the unexplained part. Its properties are $0 \leqslant R^2_{\mathbf{y} \cdot \mathbf{x}} \leqslant 1$, $R^2_{\mathbf{y} \cdot \mathbf{x}} + (1 - R^2_{\mathbf{y} \cdot \mathbf{x}}) \equiv 1$; but $R^2_{\mathbf{y} \cdot \mathbf{x}}$ is not invariant with respect to scale transformations of the **y**'s.

Glahn's $R^2_{\mathbf{y} \cdot \mathbf{x}}$ is a special case of, and Hooper's $\bar{r}^2$ is closely related to, McElroy's R_z^2 [4]. Let $E[\mathbf{V}_i \mathbf{V}_j'] = \sigma_{ij}\mathbf{I}$, so that $\mathbf{\Sigma} = [\sigma_{ij}]$ is the $M \times M$ contemporaneous covariance matrix of the **V**'s. Let **S** be a consistent estimator for $\mathbf{\Sigma}$. Let q elements of $\mathbf{\Pi}$ be restricted to zero; then

$$R_z^2 = \frac{\operatorname{tr}\left[\mathbf{S}^{-1}\tilde{\mathbf{Y}}'\tilde{\mathbf{Y}}\right]}{\operatorname{tr}\left[\mathbf{S}^{-1}\mathbf{Y}'\mathbf{Y}\right]},$$

where $\tilde{Y}$ is predicted using Zellner's efficient technique [5] for seemingly unrelated regressions (or *joint generalized least squares*). R_z^2 has all of the analogous properties of the single-equation multiple correlation coefficient: $0 \leqslant R_z^2 \leqslant 1$; R_z^2 equals the correlation of weighted $\tilde{y}$'s with weighted **y**'s; it measures the fraction of the variation in the weighted **y**'s accounted for by the weighted **x**'s; R_z^2 is invariant with respect to linear transformations of both **x**'s and **y**'s; asymptotically,

$$u = \frac{R_z^2}{1 - R_z^2} \, \frac{M(T - K + 1) + q}{MK - q}$$

has an F distribution* with $MK - q$ and $M(T - K + 1) - q$ degrees of freedom. Glahn's $R^2_{\mathbf{y} \cdot \mathbf{x}}$ is the special case of McElroy's

R_z^2 where (a) the equations are contemporaneously uncorrelated and have a common variance (i.e., $\mathbf{\Sigma} = \sigma^2\mathbf{I}$), and (b) no elements of the $\mathbf{\Pi}$ matrix are restricted to be zero so that $q = 0$ and thus $\tilde{\mathbf{Y}} = \hat{\mathbf{Y}}$. Finally, in the special case when $M = 1$, all five measures coincide:

$$R^2_{\mathbf{y}_1 \cdot \mathbf{x}_1, \mathbf{x}_2, \ldots, \mathbf{x}_K} = \bar{r}^2 = \sqrt{z} = R^2_{\mathbf{y} \cdot \mathbf{x}} = R_z^2 .$$

For the example due to Zellner [5], where $T = 20$ observations on $M = 2$ dependent variables and on $K = 4$ independent variables (K *excludes* the intercept for each equation), we use the F-distribution with 4 and 34 degrees of freedom to test the hypothesis that the $q = K = 4$ coefficients on the independent variables are zero. We obtain $u = 14.65$ and the probability that all four coefficients (excluding the two intercepts) are zero is less than 0.005. The corresponding value of $R_z^2 = 0.632$. Also for these data, $\bar{r}^2 = 0.672$ and $R^2_{y \cdot z} = 0.736$.

References

[1] Glahn, H. (1969). *Econometrica*, **37**, 252–256.

[2] Hooper, J. W. (1959). *Econometrica*, **27**, 245–256.

[3] Hotelling, H. (1951). *Proc. 2nd Berkeley Symp. Math. Statist. Prob.*, J. Neyman, ed. University of California Press, Berkeley, Calif., pp. 23–41.

[4] McElroy, M. B. (1977). *J. Econometrics*, **6**, 381–387.

[5] Zellner, A. (1962). *J. Amer. Statist. Ass.*, **57**, 348–368.

(CANONICAL ANALYSIS
CORRELATION)

MARJORIE B. MCELROY

GLIM

GLIM is a computer program developed by the Working Party on Statistical Computing of the Royal Statistical Society*. It provides a framework for the fitting of generalized linear models to data, although its uses are considerably wider than this.

BACKGROUND

In 1972, Nelder and Wedderburn [2] published their paper on generalized linear models, (GLMs). This class of statistical models includes many commonly occurring models, such as those found in the analysis of variance*, multiple regression*, log-linear models, etc. A model within the class can be specified by means of three components: the probability distribution of the observations, a *linear predictor* containing the effects of the explanatory variables on the model, and a function linking the linear predictor for an observation to its expected value. A single algorithm can be used to find maximum likelihood estimates* and perform tests of significance and to produce predicted values from the fitted model. (*See* GENERALIZED LINEAR MODELS for further details.)

Although the algorithm is simple in principle, the computations are laborious and a program was required to perform the calculations. In early 1972 a proposal to this effect was put before the Working Party by J. A. Nelder, and by 1974 the work of M. R. B. Clarke, R. W. M. Wedderburn, and J. A. Nelder had produced the first release of GLIM, which stands for Generalized Linear Interactive Modeling. Release 2 followed in 1975 and, with R. J. Baker, Release 3 in 1978. A fourth Release is planned for 1983.

The principal motivation behind the development of GLIM was the provision of an algorithm for fitting GLMs, but other considerations had an important effect on the eventual design of the package.

First, GLIM was to be interactive. Although this requirement is more common today, most statistical packages in the early 1970s were intended for batch use only. The interactive nature of GLIM has been greatly appreciated and has undoubtedly proved its worth. Second, the program was designed for maximum flexibility. Whereas most statistical packages offer a range of detailed model options (e.g., an option for Latin squares*, one for a logit linear regression*, another for a two-factor log-linear model, etc.), GLIM instead provides a small set of building blocks from which these and many other models may be constructed. Such a setup does not constrain the user within the limits of a preconceived analysis but allows the preceding analysis to indicate how the succeeding step should be taken. Third, the program was to free the user of many of the "housekeeping" chores associated with the storage of data structures: such structures could be named, and the program would maintain the directories, handle space allocation and retrieval, and keep track of program control. These aspects of the program, plus the novelty of GLMs, have distinguished GLIM from the mainstream of statistical packages.

Approximately 50 copies of GLIM-1 were distributed and about 130 of GLIM-2. By summer 1982 over 550 copies of GLIM-3 had been distributed on about 35 machine ranges.

This article concerns mainly the facilities available in Release 3. See also the GLIM-3 manual [1] for a complete program description.

MODEL-FITTING FACILITIES

Instructions to the program take the form of sequences of *directives*.

In order to specify a GLM we need to know three things: (1) the probability distribution of the observations, (2) the link function, and (3) the makeup of the linear predictor. GLIM has three directives to specify these parts.

The ERROR directive specifies the distribution of the observations as in

$ERROR NORMAL.

The LINK directive specifies the link function, as in

$LINK LOG

which specifies that a linear predictor is the log of its fitted value.

Greater variety is possible in the makeup of the linear predictor, so the FIT directive, which is used to specify the linear predictor, is correspondingly more complex. The directive takes the form

$FIT *structure-formula*

and serves two purposes. The "structure-formula" is taken as the specification of the linear predictor as described below. Then, with the declaration of the model completed, the program performs a *fit*. That is, assuming the current model, the program evaluates the maximum likelihood estimates of the parameters implied by the model and prints the resulting goodness-of-fit (deviance) statistic for the fitted model.

The structure-formula describes the makeup of the linear predictor as follows. (See ref. 3 for the original paper on the subject.) The main effect of a factor A is specified as 'A' and the presence of a covariate X by 'X'. The interaction between factors and/or variates is denoted by the dot product so that the interaction between factors A, B, and C is denoted '$A.B.C$' and that between A and X is denoted '$A.X$', the latter specifying a set of slopes for X, one for each level of A. An identifier or the dot product of identifiers is known as a *term*. The presence of several terms in a structure formula is specified by concatenating them with '+'s, so that '$A + B$' stands for the main effects of A and B and the interaction can be included by writing '$A + B + A.B$'.

A structure-formula is defined to be any such sum of terms. Lengthy formulae can be shortened by use of special operators. For example

$$A*B \text{ is defined as } A + B + A.B$$

$$A*B*C \text{ as } A + B + C + A.B + A.C + B.C + A.B.C.$$

and similarly for higher-order products. Other operators handle nesting and deletion.

As an example, consider fitting the model of independence applied to a two-way contingency table*. Such a model implies main effects only on a log-linear scale. Assuming that the data have been read in, etc., and that the two factors are A and B, we need only specify

$ERROR POISSON $LINK LOG
$FIT A + B

The printed deviance gives the goodness of fit* for the table under independence, while the fitted values, residuals, etc., may be obtained by using the DISPLAY directive.

NUMERICAL METHODS

The algorithm used is that described in GENERALIZED LINEAR MODELS; we use the same terminology. The method, except for the classical linear model, is iterative. At each cycle the program forms, on the basis of current estimates, an adjusted dependent variate (z) and a weight vector, and performs a weighted regression of z on the linear structure matrix. Only the lower triangle of the (corrected) sums-of-squares-and-products (SSP) matrix is stored and the Gauss–Jordan* method is used to solve for β. The SSP matrix may be singular when the standard method of producing a g-inverse is employed.

By an initial inspection of the linear structure matrix the program is able to detect all columns that are intrinsically aliased (i.e., certain to be redundant) and will form the SSP matrix without them. This can lead to a very large saving in space.

Note that for the classical linear model many algorithms for computing β-parameter estimates impose constraints on the estimates, so that the values produced also serve as estimates of meaningful population values, such as expected marginal means or differences of them. In the general case, however, it becomes difficult, if not impossible, to combine these functions. It is simple enough to provide a parameterization that spans the required space, but its relationship to meaningful population parameters becomes complex. So, instead of trying to combine the two functions, GLIM separates them. A parameterization of the η-space is chosen to simplify the choice of β parameters, and any estimates of population values required are evaluated from the computed η and μ. Since η and μ are estimable quantities, this method also has the advantage that only estimable quantities can be computed from them.

OTHER FACILITIES

GLIM has many other facilities besides those specifically designed for model fitting. The ability to branch and loop, with macros as blocks, together with the provision of argument substitution, provides the user with a programming language of considerable generality. Secondary input files may be used and these may be divided into subfiles for individual access. The current program state may be dumped for later recovery. The CALCULATE directive has many facilities, including pseudo-random number generation, the standard arithmetic and relational functions, and indexing; sorting is also available. The PLOT directive produces scatter plots of up to nine variates simultaneously with automatic scaling of axes.

FUTURE DEVELOPMENTS

GLIM-4 is due for release in late 1983. Simultaneously, an analysis-of-variance package, AOV, will be released. This is based on a GLIM syntax and uses the same housekeeping routines as GLIM, but has the ANOVA routines of the statistical package Genstat as its algorithmic core; *see* STATISTICAL SOFTWARE.

References

[1] Baker, R. J. and Nelder, J. A. (1978). The GLIM System; Release 3. Numerical Algorithms Group, Oxford.

[2] Nelder, J. A. and Wedderburn, R. W. M. (1972). *J. R. Statist. Soc., A*, **135**, 370–384.

[3] Wilkinson, G. N. and Rogers, C. E. (1973). *Appl. Statist.*, **22**, 393–399.

(GENERALIZED LINEAR MODELS
STATISTICAL SOFTWARE)

R. J. BAKER
J. A. NELDER

GLIVENKO–CANTELLI THEOREMS

Let $X_1, X_2, \ldots$ be independent identically distributed (i.i.d.) random variables with common distribution function F, $F(x) = P(X \leqslant x)$ for $-\infty < x < \infty$, and let $\mathbb{F}_n$ denote the *empirical distribution function* of the first n X's (*see* EDF STATISTICS) defined for $-\infty < x < \infty$ by

$$n\,\mathbb{F}_n(x) = [\text{number of } i \leqslant n \text{ with } X_i \leqslant x] = \sum_{i=1}^{n} 1_{(-\infty, x]}(X_i).$$

For fixed x, $n\mathbb{F}_n(x)$ has a binomial distribution* with parameters n and $F(x)$, and hence, using the *weak law of large numbers** for (3), and the classical de Moivre–Laplace *central limit theorem** for (4),

$$E\,\mathbb{F}_n(x) = F(x), \tag{1}$$

$$\operatorname{var}(\mathbb{F}_n(x)) = F(x)(1 - F(x))/n, \tag{2}$$

$$\mathbb{F}_n(x) \underset{p}{\rightarrow} F(x) \quad \text{as } n \rightarrow \infty, \tag{3}$$

$$n^{1/2}(\mathbb{F}_n(x) - F(x)) \underset{d}{\rightarrow} N(0, F(x)(1 - F(x))) \quad \text{as } n \rightarrow \infty; \tag{4}$$

where E denotes expected value, "var" denotes the variance, "$\underset{p}{\rightarrow}$" denotes convergence in probability, and "$\underset{d}{\rightarrow}$" denotes convergence in law or in distribution (*see* CONVERGENCE OF SEQUENCES OF RANDOM VARIABLES).

The property of $\mathbb{F}_n$ that concerns us here strengthens (3) in two important ways: to uniform convergence (in x), and to convergence with probability 1 (w.p. 1) or almost sure convergence.

Theorem 1 [1, 8]:

$$P\Big(\lim_{n\rightarrow\infty} \sup_{-\infty<x<\infty} |\mathbb{F}_n(x) - F(x)| = 0\Big) = 1,$$

or, equivalently,

$$\lim_{n\rightarrow\infty} \|\mathbb{F}_n - F\| \equiv \lim_{n\rightarrow\infty} \sup_x |\mathbb{F}_n(x) - F(x)| = 0 \quad \text{w.p. } 1.$$

Theorem 1 was proved by Glivenko [8] for continuous distributions F, and by Cantelli [1] for general F (see, e.g., Loève [13] for a proof). It asserts that the empirical distribution function $\mathbb{F}_n$ estimates F to any desired degree of precision uniformly in x for sufficiently large sample size n. The true distribution function F can be "rediscovered from the data"; or the empirical distribution function $\mathbb{F}_n$ "looks like" the true distribution F for large n. The Glivenko–Cantelli theorem has been called the "central statistical theorem" by Loève [13] and the "fundamental statistical theorem" by Renyi [15].

The Glivenko–Cantelli theorem is of constant use in establishing the *consistency* of many different statistical tests and estimates. Two examples illustrate these types of applications.

Example 1: Consistency of the Kolmogorov Test. Consider testing the simple null hypotheses $H_0: F = F_0$, where F_0 is completely specified. Kolmogorov [11] suggested that H_0 be rejected when

$$D_n \equiv \sup_x |\mathbb{F}_n(x) - F_0(x)| \equiv \|\mathbb{F}_n - F_0\|$$

is large; *see* KOLMOGOROV–SMIRNOV TYPE TESTS OF FIT. When F_0 is the true distribution function, the Glivenko–Cantelli theorem asserts that

$$P_{F_0}\Big(\lim_{n\rightarrow\infty} D_n = 0\Big) = 1. \tag{5}$$

Kolmogorov [11] showed in fact that the distribution of D_n does not depend on F_0 if F_0 is continuous, and that

$$\lim_{n\rightarrow\infty} P_{F_0}(n^{1/2} D_n \geqslant \lambda) = 2\sum_{k=1}^{\infty} (-1)^{k+1} \exp(-2k^2\lambda^2) \equiv K(\lambda)$$

for all $0 \leqslant \lambda < \infty$. Thus if $K(\lambda_\alpha) = \alpha$ and $P_{F_0}(n^{1/2}D_n \geqslant \lambda_{n,\alpha}) \equiv \alpha$, $0 < \alpha < 1$, then

$$\lim_{n\rightarrow\infty} \lambda_{n,\alpha} = \lambda_\alpha. \tag{6}$$

If, however, some $F \neq F_0$ is the true distribution function, the Glivenko–Cantelli theorem implies that

$$P_F\left(\lim_{n\to\infty} D_n = d\right) = 1, \qquad (7)$$

where $d \equiv \sup_x |F(x) - F_0(x)| = \|F - F_0\| > 0$. Hence when $F \neq F_0$ is true, (6) and (7) imply that

$$\lim_{n\to\infty} P_F\left(n^{1/2} D_n \geqslant \lambda_{n,\alpha}\right) = 1. \qquad (8)$$

In other words, the probability of rejecting the null hypothesis $F = F_0$ when $F \neq F_0$ is the true distribution increases to 1 as the sample size becomes large. The Kolmogorov test is *consistent*.

Example 2: Consistency of the Mann–Whitney Estimator of $P(X \leqslant Y)$ in a Two-Sample Problem. Suppose that $Y_1, Y_2, \ldots$ are i.i.d. with common distribution function G, independent of the X's above, and let $\mathbb{G}_n$ denote the empirical distribution function of the first n Y's. Consider estimating $P(X \leqslant Y) = \int F\,dG$ based on the first m X's and first n Y's. The Mann–Whitney estimator of this probability is

$$W_{mn} \equiv \frac{1}{mn} \sum_{i=1}^{m} \sum_{j=1}^{n} 1_{[X_i \leqslant Y_j]} = \int \mathbb{F}_m \, d\mathbb{G}_n .$$

To show that $W_{mn} \to P(X \leqslant Y) = \int F\,dG$ w.p. 1, add and subtract $\int F\,d\mathbb{G}_n$ and integrate the second term by parts to obtain

$$\begin{aligned} &\left|\int \mathbb{F}_m \, d\mathbb{G}_n - \int F\,dG\right| \\ &= \left|\int (\mathbb{F}_m - F)\, d\mathbb{G}_n + \int F\, d(\mathbb{G}_n - G)\right| \\ &\leqslant \|\mathbb{F}_m - F\| + \left|\int (\mathbb{G}_n - G)\, dF\right| \\ &\leqslant \|\mathbb{F}_m - F\| + \|\mathbb{G}_n - G\| \\ &\to 0 + 0 = 0 \qquad \text{w.p. } 1 \end{aligned}$$

as $m \to \infty$, $n \to \infty$, by the Glivenko–Cantelli theorem. Thus W_{mn} is a (strongly) consistent estimator of $P(X \leqslant Y)$. (*See also* MANN–WHITNEY–WILCOXON TEST for further information concerning W_{mn}. Another proof of the consistency of W_{mn} is based on the fact that W_{mn} is a *U-statistic** and hence a reverse martingale*.)

Before leaving the classical case, two important related results should be mentioned: an exponential inequality for the random variable $\|\mathbb{F}_n - F\|$, and a law of the iterated logarithm*.

The *inequality* of Dvoretzky et al. [4] asserts that

$$P(\|\mathbb{F}_n - F\| \geqslant \lambda) \leqslant C \exp(-2n\lambda^2) \qquad (9)$$

for all $\lambda > 0$ where C is an absolute constant. [$C = 58$ works; the smallest C for which (9) holds is still unknown.] The factor of 2 appearing in this inequality is best possible; note that the lead term in the distribution $K(\lambda)$ is $2\exp(-2\lambda^2)$. For example,

$$P(\|\mathbb{F}_n - F\| \geqslant 0.04) \leqslant 0.10$$

if $n \geqslant \frac{1}{2} \cdot 625 \cdot \log(580) \cong 1989$.

The *iterated logarithm law* of Smirnov [17] and Chung [2] gives a rate of convergence for the Glivenko–Cantelli theorem: it asserts that

$$\limsup_{n\to\infty} \frac{n^{1/2}\|\mathbb{F}_n - F\|}{(2\log\log n)^{1/2}} = \sup_x\left[F(x)\{1 - F(x)\}\right]^{1/2} \leqslant \tfrac{1}{2} \qquad \text{w.p. } 1. \qquad (10)$$

Thus

$$\|\mathbb{F}_n - F\| = O\left(n^{-1/2}(\log\log n)^{1/2}\right) \qquad \text{w.p. } 1;$$

the supremum distance between $\mathbb{F}_n$ and F goes to zero only a little more slowly than $n^{-1/2}$ w.p. 1.

Since 1960 the Glivenko–Cantelli theorem has been extended and generalized in several directions: to random vectors and to observations X with values in more general metric spaces; to empirical probability measures indexed by families of sets; to observations that may be dependent or nonidentically distributed; and to metrics other than the supremum metric. Here we briefly summarize some of this work. More detailed information and further references can be found in the survey by Gaenssler and Stute [7].

Let $X_1, X_2, \ldots$ be i.i.d. random variables with values in a (measurable) space $(\mathbb{X}, \mathscr{B})$

and common probability measure P on $\mathbb{X}$; for many important applications in statistics $(\mathbb{X}, \mathscr{B}) = (R^k, \mathscr{B}^k)$, k-dimensional Euclidean space with its usual Borel sigma field. The *empirical measure* $\mathbb{P}_n$ of the first n X's is the probability measure that puts mass $1/n$ at each of $X_1, \dots, X_n$:

$$\mathbb{P}_n = (\delta_{X_1} + \cdots + \delta_{X_n})/n, \tag{11}$$

where $\delta_x(A) = 1$ if $x \in A$; 0 if $x \notin A$, for $A \in \mathscr{B}$.

Many of the generalizations referred to above assert that, in some sense, "$\mathbb{P}_n$ looks like P" for large n. It has become common practice to refer to any such theorem as a "Glivenko–Cantelli theorem."

For $(\mathbb{X}, d)$ a separable metric space, the convergence of $\mathbb{P}_n$ to P was first investigated by Fortet and Mourier [6] and Varadarajan [22], who proved that $\beta(\mathbb{P}_n, P) \to 0$ w.p. 1, where β is the dual-bounded-Lipschitz metric (see Dudley [3]) and $\mathbb{P}_n \to P$ weakly w.p. 1, respectively.

Let $\mathscr{C} \subset \mathscr{B}$ be some specified subclass of sets and set

$$D_n(\mathscr{C}, P) = \sup_{C \in \mathscr{C}} |\mathbb{P}_n(C) - P(C)|. \tag{12}$$

A number of results assert that $D_n(\mathscr{C}, P) \to 0$ w.p. 1 for specific spaces $\mathbb{X}$ and classes of sets $\mathscr{C}$. For example, when $\mathbb{X} = \mathbb{R}^k$ and $\mathscr{C}$ = all intervals in $\mathbb{R}^k$, or all half-spaces in $\mathbb{R}^k$, or all closed balls in $\mathbb{R}^k$, then $D_n(\mathscr{C}, P) \to 0$ for any probability measure P [5, 6, 10]. For a general class of sets $\mathscr{C}$, however, some restriction on P may be necessary: If $\mathbb{X} = \mathbb{R}^k$ and $\mathscr{C}$ = all convex sets in $\mathbb{R}^k$, then $D_n(\mathscr{C}, P) \to 0$ w.p. 1 if $P_c(\partial C) = 0$ for all $C \in \mathscr{C}$ where P_c is the nonatomic part of P [14]. For a discussion of more results of this type and further references, see Gaenssler and Stute [7].

In the just stated results the classes $\mathscr{C}$ were formed by subsets of $\mathbb{R}^k$ which have a common geometric structure; the methods of proof of the corresponding Glivenko–Cantelli theorems rely heavily on this fact. For arbitrary sample spaces $(\mathbb{X}, \mathscr{B})$ where geometrical arguments are not available, the most appealing approach to obtain Glivenko–Cantelli theorems for classes $\mathscr{C} \subset \mathscr{B}$ was given by Vapnik and Chervonenkis [21]. Based on combinatorial arguments they showed that given a class $\mathscr{C} \subset \mathscr{B}$ such that for some finite n, "$\mathscr{C}$ does not cut all subsets of any $E \subset \mathbb{X}$ with $\text{card}(E) = n$" [i.e., for any $E \subset \mathbb{X}$ with $\text{card}(E) = n$ there is a subset of E which is not of the form $E \cap C$ for some $C \in \mathscr{C}$], then (under some measurability assumptions) $D_n(\mathscr{C}, P) \to 0$ w.p. 1 for any probability measure P.

DEPENDENT OBSERVATIONS. When $\mathbb{X} = \mathbb{R}^1$, $\mathscr{C} = \{(-\infty, x] : x \in \mathbb{R}^1\}$, and

$$\mathbb{F}_n(x) = \mathbb{P}_n(-\infty, x],$$

Tucker [20] generalized the classical Glivenko–Cantelli theorem to *strictly stationary** sequences:

$$\|\mathbb{F}_n - F_\omega\| \to 0 \qquad \text{w.p. } 1, \tag{13}$$

where F_ω is a (possibly random) distribution function; when the X's are also *ergodic**, F_ω is simply the common one-dimensional marginal law of the X's. Tucker's Glivenko–Cantelli theorem applies to sequences of random variables satisfying a wide range of *mixing conditions*; it has been generalized to higher-dimensional spaces and more general index sets by Stute and Schumann [19] (see also Steele [18] and Kazakos and Gray [9]).

NONIDENTICALLY DISTRIBUTED OBSERVATIONS. If the X's are independent but not identically distributed, there is no common probability measure P to be recovered from the data. Nevertheless, letting P_i denote the probability law of X_i, $i = 1, 2, \dots,$ we still have

$$E\mathbb{P}_n(C) = n^{-1}(P_1 + \cdots + P_n)(C)$$
$$\equiv \overline{P}_n(C).$$

Thus it is still reasonable to expect that the empirical measure $\mathbb{P}_n$ "looks like" the average measure $\overline{P}_n$. When $\mathbb{X} = \mathbb{R}^1$, $\mathscr{C} = \{(-\infty, x]; x \in \mathbb{R}^1\}$, $\mathbb{F}_n(x) = \mathbb{P}_n(-\infty, x]$, and $\overline{F}_n(x) = \overline{P}_n(-\infty, x]$, Koul [12] and Shorack [16] have shown that

$$\|\mathbb{F}_n - \bar{F}_n\| \equiv \sup_x |\mathbb{F}_n(x) - \bar{F}_n(x)| \to 0 \qquad \text{w.p. } 1$$

always. When $(\mathbb{X}, d)$ is a separable metric space, Wellner [23] has shown that if $\{\bar{P}_n\}$ is tight, then $\beta(\mathbb{P}_n, \bar{P}_n) \to 0$ and $\rho(\mathbb{P}_n, \bar{P}_n) \to 0$ w.p. 1, where β and ρ are the dual-bounded Lipschitz and Prohorov metrics, respectively.

References

[1] Cantelli, F. P. (1933). *G. Ist. Ital. Attuari*, **4**, 421–424. (One of the original works; Glivenko's result for continuous distribution functions is extended to arbitrary df's.)

[2] Chung, K. L. (1949). *Trans. Amer. Math. Soc.*, **67**, 36–50. [Contains a proof of the law of the iterated logarithm (10) for $\|\mathbb{F}_n - F\|$.]

[3] Dudley, R. M. (1969). *Ann. Math. Statist.*, **40**, 40–50. (Bounds for expected Prohorov and dual-bounded-Lipschitz distances between the empirical measure $\mathbb{P}_n$ and true measure P are given using metric entropy methods.)

[4] Dvoretzky, A., Kiefer, J., and Wolfowitz, J. (1956). *Ann. Math. Statist.*, **27**, 642–669. [The exponential bound (9) is proved and used in a study of the asymptotic minimax properties of the empirical distribution function $\mathbb{F}_n$ as an estimator of F.]

[5] Elker, J., Pollard, D., and Stute, W. (1979). *Adv. Appl. Prob.*, **11**, 820–833. (Contains Glivenko–Cantelli theorems for the empirical measure $\mathbb{P}_n$ indexed by convex sets in k-dimensional Euclidean space.)

[6] Fortet, R. M. and Mourier, E. (1953). *Ann. Sci. École Norm. Sup.*, **70**, 266–285. (Convergence of the dual-bounded-Lipschitz distance from the empirical measure $\mathbb{P}_n$ to the true measure P is established.)

[7] Gaenssler, P. and Stute W. (1979). *Ann. Prob.*, **7**, 193–243. (A survey of results for empirical distribution functions and empirical processes with an extensive bibliography.)

[8] Glivenko, V. (1933). *G. Ist. Ital. Attuari*, **4**, 92–99. (Here Theorem 1 was first established for continuous distribution functions F.)

[9] Kazakos, P. P. and Gray, R. M. (1979). *Ann. Prob.*, **7**, 989–1002. (Glivenko–Cantelli theorems for finite-dimensional distributions of stationary processes.)

[10] Kiefer, J. and Wolfowitz, J. (1958). *Trans. Amer. Math. Soc.*, **87**, 173–186. [The Glivenko–Cantelli Theorem 1 and the law of the iterated logarithm (10) are extended to empirical distribution functions of random vectors in k-dimensional Euclidean space.]

[11] Kolmogorov, A. N. (1933). *G. Ist. Ital. Attuari*, **4**, 83–91. (One of the original and most important papers.)

[12] Koul, H. L. (1970). *Ann. Math. Statist.*, **41**, 1768–1773. (Contains a Glivenko–Cantelli theorem for the case of independent nonidentically distributed random variables.)

[13] Loève, M. (1977). *Probability Theory*, 4th ed. Springer, New York. (Graduate-level textbook; excellent reference for basic probability theory.)

[14] Ranga Rao, R. (1962). *Ann. Math. Statist.*, **33**, 659–680.

[15] Renyi, A. (1962). *Wahrscheinlichkeitsrechnung*, VEB, Deutscher Verlag der Wissenschaften, Berlin.

[16] Shorack, G. R. (1979). *Statist. Neerlandica*, **33**, 169–189. (The Glivenko–Cantelli theorem for independent but nonidentically distributed random variables is proved, and weak convergence of empirical processes of such variables studied.)

[17] Smirnov, N. V. (1944). *Uspehi Mat. Nauk*, **10**, 179–206. [The law of the iterated logarithm (10) for $\|\mathbb{F}_n - F\|$ is proved.]

[18] Steele, M. (1978). *Ann. Prob.*, **6**, 118–127. (General Glivenko–Cantelli theorems for empirical measures indexed by sets are obtained by using the combinatorial methods of Vapnik and Chervonenkis in combination with subadditive ergodic theory.)

[19] Stute, W. and Schumann, G. (1980). *Scand. J. Statist.*, **7**, 102–104. (Generalization of the result of Tucker [20] for stationary processes to empirical measures indexed by sets.)

[20] Tucker, H. G. (1959). *Ann. Math. Statist.*, **30**, 828–830. (Here the Glivenko–Cantelli theorem is proved for strictly stationary processes.)

[21] Vapnik, V. N. and Chervonenkis, A. Ya. (1971). *Theory Prob. Appl.*, **16**, 264–280. (Combinatorial methods are introduced and used to prove Glivenko–Cantelli theorems for general sample spaces and classes of index sets.)

[22] Varadarajan, V. S. (1958). *Sankhyā*, **19**, 23–26. (Classical paper showing that the empirical measure $\mathbb{P}_n$ converges weakly to P with probability 1 when the sample space is a separable metric space.)

[23] Wellner, J. A. (1981). *Stoch. Proc. Appl.*, **11**, 309–312.

(CONVERGENCE OF SEQUENCES OF RANDOM VARIABLES
EDF STATISTICS
LAWS OF LARGE NUMBERS
LAWS OF THE ITERATED LOGARITHM)

PETER GAENSSLER
JON A. WELLNER

GLOTTOCHRONOLOGY *See* LEXICOSTATISTICS

GOLDBERG CAUSAL VOTING MODEL *See* POLITICAL SCIENCE, STATISTICS IN

GOMORY CUT *See* INTEGER PROGRAMMING

GOMPERTZ DISTRIBUTION

In 1825, Benjamin Gompertz [4] introduced a distribution to fit mortality tables, with cumulative distribution function* (CDF)

$$F(x; k, c) = 1 - \exp\{-(k/c)(e^{cx} - 1)\},$$
$$x > 0; \quad k > 0, \quad c > 0. \quad (1)$$

The probability density function (PDF) is given by

$$f(x; k, c) = ke^{cx}\exp\{-(k/c)(e^{cx} - 1)\},$$
$$x > 0.$$

The distribution has the property that the survival function

$$G(x; k, c) = 1 - F(x; k, c)$$

yields the same form under a change of origin $x' = x - x_0$; in fact [3],

$$G(x; k, c) = G(x_0; k, c)$$
$$\times \exp\{-(k'/c)(e^{cx'} - 1)\},$$
$$x' > 0, \quad (2)$$

where $k' = ke^{cx_0}$. In the terminology of actuarial* science, k' is the force of mortality* at x_0; in the terminology of conditional distributions, if X is the random variable with CDF (1),

$$\Pr(X - x_0 > x'; k, c \mid X > x_0)$$
$$= G(x'; k, c)/G(x_0; k, c)$$
$$= G(x'; k', c). \quad (3)$$

This conditional distribution thus has PDF $f(x'; k', c)$. See Garg et al. [3], who derive likelihood equations for iterative derivation of maximum likelihood* estimates of k and of c.

Prentice and El Shaarawi [6] develop procedures to test the fit of the Gompertz model. They apply these to observed death rates in Ontario in 1964–1968; the fit was "very bad" in an unrestricted age range for both sexes, but was not significant for most causes of death in the age range 30 to 70 years. The Gompertz model is also used as a marginal survival distribution in competing risk* theory (see David and Moeschberger [2]).

An analogous form of the Gompertz CDF (1) over the real line is given by

$$F_1(x) = \exp(-ae^{-x/b}),$$
$$|x| < \infty, \quad a > 0, \quad b > 0. \quad (4)$$

This is an extreme value distribution* of Type I (see Johnson and Kotz [5, p. 271]). Ahuja and Nash [1] develop a generalized class based on (4) and on Verhulst distributions*, by introducing a further parameter. These are related to Pearson* Type III, VI, and I curves.

References

[1] Ahuja, J. C. and Nash, S. W. (1967). *Sankhyā A*, **29**, 141–156.

[2] David, H. A. and Moeschberger, M. L. (1978). *The Theory of Competing Risks*, Griffin's Statist. Monogr. No. 39. Macmillan, New York.

[3] Garg, M. L., Rao, B. R., and Redmond, C. K. (1970). *Appl. Statist.*, **19**, 152–159.

[4] Gompertz, B. (1825). *Philos. Trans. R. Soc. A*, **115**, 513–580.

[5] Johnson, N. L. and Kotz, S. (1970). *Distributions in Statistics. Continuous Univariate Distributions*, Vol. 2. Wiley, New York.

[6] Prentice, R. L. and El Shaarawi, A. (1973). *Appl. Statist.*, **22**, 301–314.

(ACTUARIAL STATISTICS—LIFE
COMPETING RISKS
MORBIDITY AND MORTALITY)

CAMPBELL B. READ

GOODMAN–KRUSKAL TAU AND GAMMA

Goodman and Kruskal's tau (τ_b) and gamma (γ) are proportional-reduction-in-error* (P-R-E) measures of association* for two-way cross classifications [3, 8–11]. Specifically, τ_b is an asymmetric P-R-E measure of association for nominal* scales and $|\gamma|$ is a symmetric P-R-E measure of association for ordinal* scales.

With respect to notation for both τ_b and γ, let n_{ij} be the observed cell frequency of events in the ith row and jth column ($i = 1, \dots, \alpha$; $j = 1, \dots, \beta$). Also, $n_{i.}$ denotes the observed ith row total, $n_{.j}$ denotes the observed jth column total, and $n_{..}$ denotes the observed overall total.

Suppose that a dependent nominal variable is represented by columns and that an independent nominal variable is represented by rows. Then the estimate of τ_b is characterized as the P-R-E measure of association for nominal scales given by

$$t_b = (E_1 - E_2)/E_1,$$

where E_1 is the expected number of classification errors of observed column totals given the observed overall total under the random allocation of events and E_2 is the expected number of classification errors of observed cell frequencies given the observed row totals under the random allocation of events. In particular,

$$E_1 = \sum_{j=1}^{\beta} \frac{n_{..} - n_{.j}}{n_{..}} n_{.j}$$

$$= n_{..} - \left(\sum_{j=1}^{\beta} n_{.j}^2 \right) / n_{..}$$

$$E_2 = \sum_{i=1}^{\alpha} \sum_{j=1}^{\beta} \frac{n_{i.} - n_{ij}}{n_{i.}} n_{ij}$$

$$= n_{..} - \sum_{i=1}^{\alpha} \left[\left(\sum_{j=1}^{\beta} n_{ij}^2 \right) / n_{i.} \right].$$

Both the large sample normality of t_b is established and the large sample variance of t_b is given by Goodman and Kruskal [10, 11]. In addition, a programmed computer algorithm based on these results yields large-sample confidence intervals* for τ_b (see Berry and Mielke [1]). If the role of columns and rows is reversed relative to representing the dependent and independent variables, the resulting P-R-E measure of association for nominal scales is termed τ_a and its estimated value designated by t_a is seldom equal to t_b (i.e., this is an asymmetric P-R-E measure). Specifically, consider the following 2×3 frequency table, where the first and second rows represent Democrats and Republicans, respectively, and the first, second, and third columns represent Jewish, Catholic, and Protestant religions, respectively.

15	20	20
5	10	20

The point estimate of $\tau_a(\tau_b)$ for this example is $t_a = 0.045$ ($t_b = 0.024$) and the large-sample 95% confidence interval for $\tau_a(\tau_b)$ is

$$0 \leqslant \tau_a \leqslant 0.131 \; (0 \leqslant \tau_b \leqslant 0.072).$$

In particular, Goodman and Kruskal's tau is asymmetric for tables larger than 2×2. In the 2×2 case, tau is equivalent to several of the older chi-square* measures: Pearson's ϕ^2 [14], Tschuprow's statistic [17], and Cramér's statistic [4]. Its only real competitor among P-R-E measures is Goodman and Kruskal's lambda [8], which, unfortunately, may yield null results even when the variables are not statistically independent. Margolin and Light [13] provide a technique for testing $\tau_a(\tau_b) = 0$ versus $\tau_a(\tau_b) > 0$.

Next suppose that one ordinal variable is represented by columns and another ordinal variable by rows. Then the estimator of $|\gamma|$ is characterized as the P-R-E measure of association for ordinal scales given by

$$|G| = (E_1 - E_2)/E_1,$$

where E_1 is the expected number of errors in predicting a pair's ordering for one variable with no knowledge of the pair's ordering for the other variable and E_2 is the expected number of errors in predicting a pair's ordering for one variable while knowing the pair's ordering for the other variable. In this instance,

$$E_1 = (C + D)/2,$$

$$E_2 = \min(C, D),$$

where C is the number of concordant pairs (all pairs such that the orderings of both variables are in the same direction for each pair), D is the number of discordant pairs (all pairs such that the orderings of both variables are in reversed directions for each pair), and $C + D$ is the number of pairs without ties for either variable. The value of $|G|$ is unchanged if the role of columns and rows is reversed (as anticipated since this is a symmetric P-R-E measure). In addition, the estimate of γ is given by

$$G = (C - D)/(C + D),$$

where $-1 \leqslant G \leqslant 1$ (the statistic $|G|$ is the previously defined P-R-E measure of association for ordinal scales). Again both the large-sample normality of G is established and the large-sample variance of G is given by Goodman and Kruskal [10, 11]. A programmed computer algorithm based on these results yields large-sample confidence intervals and testing procedures associated with γ (see Berry et al. [2]). Considering the same 2×3 frequency table used in the discussion of Goodman and Kruskal's tau, let the first and second rows now represent high and low political conservatism, respectively, and let the first, second, and third columns now represent upper-, middle-, and lower-income levels, respectively. The point estimate of γ for this example is $G = 0.360$ and the large-sample 95% confidence interval estimate for γ is $0.037 \leqslant \gamma \leqslant 0.683$. To illustrate a distributional concern, consider the following 2×2 table, where the letters merely designate cell locations.

a	b
c	d

Suppose that the true cell proportions are $p_a = p_b = \frac{1}{2} - \epsilon$, $p_c = 3\epsilon/2$, and $p_d = \epsilon/2$ where $0 < \epsilon < \frac{1}{2}$. For very small ϵ, the large-sample approximate skewness* of G for a sample of size n under the exact multinomial distribution* is $[2/(3\epsilon n)]^{1/2}$. This result clearly indicates that the convergence of the exact distribution of G to normality may be very slow in some instances (a similar concern holds for t_a or t_b). Gans and Robertson [7] discuss this concern in detail. Goodman and Kruskal's gamma is equivalent to Yule's Q statistic [18] in the 2×2 case. When there are no tied pairs, gamma is equivalent to Kendall's tau* statistic [12]. Gamma has several competitors among P-R-E measures; most notable of these is Kendall's tau statistic, which is limited (unlike gamma) to measuring only strong monotonicity. Considerable work has been done on partial gamma statistics [6]. Somers [15, 16] has extended gamma to asymmetric measures, d_{yx} and d_{xy}, which are equivalent to percentage differences in the 2×2 case, and Dabrowska et al. [5] have introduced a multivariate analog of gamma.

References

[1] Berry, K. J. and Mielke, P. W. (1976). *Educ. Psychol. Meas.*, **36**, 747–751. (Describes FORTRAN algorithms.)

[2] Berry, K. J., Mielke, P. W., and Jacobsen, R. B. (1977). *Educ. Psychol. Meas.*, **37**, 791–794. (Describes FORTRAN algorithms.)

[3] Costner, H. L. (1965). *Amer. Sociol. Rev.*, **30**, 341–353. (Develops P-R-E interpretation.)

[4] Cramér, H. (1946). *Mathematical Methods of Statistics*. Princeton University Press, Princeton, N.J. (See pp. 280–283.) (Extends ϕ^2 to asymmetric tables.)

[5] Dabrowska, D., Pleszczynska, E., and Szczesny, W. (1981). *Commun. Statist. A*, **10**, 2435–2445. (Proposes a multivariate analog of gamma.)

[6] Davis, J. A. (1967). *J. Amer. Statist. Ass.*, **62**, 189–193. (Partial gamma is developed.)

[7] Gans, L. P. and Robertson, C. A. (1981). *J. Amer. Statist. Ass.*, **76**, 942–946. (Concerned with the approximate distribution of gamma.)

[8] Goodman, L. A. and Kruskal, W. H. (1954). *J. Amer. Statist. Ass.*, **49**, 732–764. (Introduced tau and gamma.)

[9] Goodman, L. A. and Kruskal, W. H. (1959). *J. Amer. Statist. Ass.*, **54**, 123–163. (Relates work to previous literature.)

[10] Goodman, L. A. and Kruskal, W. H. (1963). *J. Amer. Statist. Ass.*, **58**, 310–364. (Initial variance development.)

[11] Goodman, L. A. and Kruskal, W. H. (1972). *J. Amer. Statist. Ass.*, **67**, 415–421. (Refined asymptotic variances.)

[12] Kendall, M. G. (1948). *Rank Correlation Methods*. Charles Griffin, London. (Develops various ordinal measures.)

[13] Margolin, B. H. and Light, R. J. (1974). *J. Amer. Statist. Ass.*, **69**, 755–764. (Provides a technique for testing $\tau_a(\tau_b) = 0$ versus $\tau_a(\tau_b) > 0$.)

[14] Pearson, K. (1904). *Drapers' Company Research Memoirs, Biometric Series*, Vol. 1. Cambridge University Press, Cambridge, pp. 1–35. (Develops ϕ^2.)

[15] Somers, R. H. (1962). *Amer. Sociol. Rev.*, **27**, 799–811. (Develops asymmetric counterparts of gamma.)

[16] Somers, R. H. (1962). *J. Amer. Statist. Ass.*, **57**, 804–812. (Relates various measures of association.)

[17] Tschuprow, A. A. (1939). *Principles of the Mathematical Theory of Correlation*. W. Hodge, London (translated by M. Kantorowitsch). (See pp. 50–57.) (Extends ϕ^2 to larger symmetric tables.)

[18] Yule, G. U. (1912). *J. R. Statist. Soc.*, **75**, 579–642. (Introduces one-way measure of association.)

(ASSOCIATION, MEASURES OF
CATEGORICAL DATA
CHI-SQUARE TESTS
FREQUENCY TABLE
LARGE-SAMPLE THEORY
LOGISTIC REGRESSION)

PAUL W. MIELKE, JR.

GOODMAN'S Y^2

Goodman's Y^2 [2] provides a statistical procedure for detecting differences between several multinomial* populations, and for constructing simultaneous confidence intervals* to investigate the nature of any differences. Suppose that observations are drawn at random from a population and are classified into one of several mutually exclusive categories. The original observations may be numerical, such as income which has been divided into several brackets, or they may represent nonnumerical quantities, such as marital status or race. In either case, the sample data can be described by a multinomial probability model in which the parameters of interest are often the probabilities of obtaining an observation from the various categories.

To compare several multinomial populations with common classification categories, initially a test for homogeneity* of the populations can be performed. For r populations with c categories, let p_{ij} $(i = 1, \dots, r;\ j = 1, \dots, c)$ represent the probability that an observation from population i is classified in category j. The hypothesis to be tested is $H_0 : p_{ij} = p_{kj}$ for all i, j, and k, versus $H_1 : p_{ij} \neq p_{kj}$ for some i, j, and k. The most commonly used test for homogeneity is based on Pearson's χ^2 statistic (*see* CHI-SQUARE TESTS). If n_i denotes the number of observations taken from population i, n_{ij} is the number of observations in this sample classified in category j, $n_{\cdot j} = \sum_{i=1}^{r} n_{ij}$ and $n = \sum_{i=1}^{r} n_i$, then Pearson's test rejects H_0 whenever

$$\chi^2 = n\left\{\left[\sum_{j=1}^{c}\sum_{i=1}^{r}\left[n_{ij}^2/n_i n_{\cdot j}\right]\right] - 1\right\}$$

exceeds $\chi^2_{(r-1)(c-1),1-\alpha}$, the $100(1-\alpha)$ percentile of the chi-square distribution* with $(r-1)(c-1)$ degrees of freedom. However, when a difference is found with this test, there is no easy procedure based on this test statistic for isolating the differences.

An alternative procedure is based on Goodman's Y^2 statistic. The test rejects H_0 whenever

$$Y^2 = n\left\{\left[\sum_{j=1}^{c}\bar{p}_j\right]^{-1} - 1\right\}$$

exceeds $\chi^2_{(r-1)(c-1),1-\alpha}$, where

$$\bar{p}_j = n\Big/\left[\sum_{i=1}^{r} n_i^2/n_{ij}\right].$$

The advantage of Goodman's Y^2 is the associated large-sample confidence intervals for all contrasts among the populations. Unfortunately, the χ^2 approximation to the distribution of Y^2 appears to set in rather slowly, and the procedure should only be used for truly large-sample sizes. An empirical study pointing out this possible problem appears in Knoke [4].

A *contrast* θ is a linear function of the p_{ij}, $\theta = \sum_{j=1}^{c}\sum_{i=1}^{r} c_{ij}p_{ij}$, where the c_{ij} are known constants for which $\sum_{i=1}^{r} c_{ij} = 0$ for all j. The maximum likelihood estimator of θ is $\hat{\theta} = \sum_{j=1}^{c}\sum_{i=1}^{r} c_{ij}\hat{p}_{ij}$, where $\hat{p}_{ij} = n_{ij}/n_i$, and the estimated variance of $\hat{\theta}$ is

$$s^2(\hat{\theta}) = \sum_{i=1}^{r} n_i^{-1} \left[\sum_{j=1}^{c} c_{ij}^2 \hat{p}_{ij} - \left(\sum_{j=1}^{c} c_{ij} \hat{p}_{ij} \right)^2 \right].$$

The correct procedure for obtaining simultaneous confidence intervals for several contrasts depends on whether the contrasts of interest, namely the c_{ij}, have been specified a priori or are decided upon after examining the data (*post hoc*).

Post hoc simultaneous confidence intervals for contrasts $\theta_1, \dots, \theta_k$ are obtained as follows. The probability that the k intervals $(\hat{\theta}_i - s(\hat{\theta}_i)L,\ \hat{\theta}_i + s(\hat{\theta}_i)L)$ simultaneously include $\theta_1, \dots, \theta_k$, respectively, is at least $1 - \alpha$ for large samples, where

$$L = +\left[\chi^2_{(r-1)(c-1),1-\alpha}\right]^{1/2}.$$

The hypothesis of homogeneity will be rejected with Goodman's Y^2 at level α if and only if there is at least one contrast for which the associated interval does not include zero.

When the contrasts are planned a priori, shorter intervals can often be obtained. For k planned contrasts, the probability that the intervals $(\hat{\theta}_i - s(\hat{\theta}_i)Z,\ \hat{\theta}_i + S(\hat{\theta}_i)Z)$ simultaneously include $\theta_1, \dots, \theta_k$, respectively, is at least $1 - \alpha$, where $Z = Z_{1-\alpha/(2k)}$ is the $100(1 - \alpha/(2k))$ percentile of the standard normal distribution. This result is obtained by using the facts that $(\hat{\theta}_i - \theta_i)/s(\hat{\theta}_i)$ has an approximate standard normal distribution for each i, in combination with the Bonferroni inequalities*. For some values of k and α, $L \leqslant Z$, in which case Z should be replaced by L even for planned comparisons.

Either of the preceding sets of simultaneous confidence intervals can be used to determine which contrasts are significantly different from zero, and in what directions. For each contrast, assert $\theta > 0$ if the computed interval is completely to the right of zero, and $\theta < 0$ if the interval is to the left of zero. The probability that all assertions are correct is at least $1 - \alpha$ as the sample sizes become large, regardless of the values of the p_{ij}.

Example. Parsons and Peterle [5] designed an experiment to study the effect of DDT on the cells of the parathyroid gland of raptors. Five birds (a subset of the data) were selected, two serving as controls while the remaining three were treated with DDT. The cells, randomly chosen from a tissue sample from the parathyroid gland of each bird, were classified into one of four phases: (1) inactive, (2) protein synthesis, (3) protein packaging, and (4) hormone release. The data appear in Table 1. $Y^2 = 86.324$, which indicates at level $\alpha = 0.001$ that there is a difference in the phase distribution of the cells of the *five* birds. Note that this test procedure will detect any differences between the birds, not just differences between control and treated birds.

Four planned comparisons contrasting the average phase distribution between the control and treated birds were of primary interest. They are

$$\hat{\theta}_i = (p_{1i} + p_{2i})/2 - (p_{3i} + p_{4i} + p_{5i})/3, \qquad i = 1, 2, 3, 4.$$

The computations yield

$$\hat{\theta}_1 = -0.219, \qquad \hat{\theta}_2 = 0.146,$$
$$\hat{\theta}_3 = 0.071, \qquad \hat{\theta}_4 = 0.002$$

with

$$s^2(\hat{\theta}_1) = 0.0012, \qquad s^2(\hat{\theta}_2) = 0.0011,$$
$$s^2(\hat{\theta}_3) = 0.0009, \qquad s^2(\hat{\theta}_4) = 0.0002.$$

Simultaneous 95% confidence intervals using $Z = 2.5$, the $100[1 - (0.05/8)]$ percentile of the standard normal distribution, yield

$$\theta_1 \in (-0.306,\ -0.132),$$
$$\theta_2 \in (\ 0.063,\ \ 0.229),$$
$$\theta_3 \in (-0.004,\ \ 0.146),$$
$$\theta_4 \in (-0.033,\ \ 0.037).$$

It can be concluded that $\theta_1 < 0$ and $\theta_2 > 0$. Since each cell continuously passes through all four phases, this could be interpreted as an indication that cells for the control birds are spending less time in the inactive phase and more time in the protein synthesis phase

Table 1

	Phase 1	2	3	4	
Control 1	36	42	24	4	106
Control 2	66	84	60	16	226
DDT 1	114	46	44	12	216
DDT 2	102	26	20	12	160
DDT 3	104	80	46	6	236

than those for the treated birds. Using a slightly lower confidence coefficient, it could also be concluded that $\theta_3 > 0$.

Some further applications of Goodman's Y^2 can be found in Hornstein et al. [3] and Payne [6]. A set of simultaneous confidence intervals for all linear combinations of the p_{ij}, not just contrasts, appear in Gold [1].

References

[1] Gold, R. (1963). *Ann. Math. Statist.*, **34**, 56–74.

[2] Goodman, L. (1964). *Ann. Math. Statist.*, **35**, 716–725.

[3] Hornstein, H., Masor, H., Sole, K., and Herlman, M. (1971). *J. Personality Social Psychol.*, **17**, 107–112.

[4] Knoke, J. (1976). *J. Amer. Statist. Ass.*, **71**, 849–853.

[5] Parsons, A. and Peterle, T. (1976). Unpublished data collected at The Ohio State University, Dept. of Zoology. Research supported by Patuxent Research Center, U. S. Dept. of the Interior.

[6] Payne, R. (1979). *Animal Behav.*, **27**, 997–1013.

(CHI-SQUARE TESTS
CONFIDENCE INTERVALS AND REGIONS
MULTIPLE COMPARISONS
SIMULTANEOUS STATISTICAL INFERENCE)

MICHAEL A. FLIGNER

GOODNESS OF FIT

In general, the term "goodness of fit" is associated with the statistical testing of hypothetical models with data. Examples of such tests abound and are to be found in most discussions on inference*, least-squares* theory, and multivariate analysis.* This article concentrates on those tests that examine certain features of a random sample to determine if it was generated by a particular member of a class of cumulative distribution functions* (CDFs). Such exercises fall under the broad heading of *hypothesis testing**. However, the feature that tends to characterize these "goodness-of-fit tests" is their preoccupation with the sample CDF, the population CDF, and estimates of it.

More specifically, let $X_1, X_2, \ldots, X_n$ be a random sample generated by CDF $G_X(x)$. It is required to test

$$H_0 : G_X(x) = F_X(x, \boldsymbol{\theta}), \qquad \boldsymbol{\theta} \in \Omega,$$

where $\boldsymbol{\theta}$ is a q-dimensional vector of parameters belonging to the parameter space Ω. If $\boldsymbol{\theta}$ is fixed at some value $\boldsymbol{\theta}_0$, say, then $F_X(x, \boldsymbol{\theta}_0) = F_X(x)$ is fully specified and H_0 is simple. Otherwise, the hypothesis states that $G_X(x)$ is some unspecified member of a family of CDFs and is composite*.

As an example, consider the normal* family $N_X(x; \boldsymbol{\theta})$, $\boldsymbol{\theta}' = (\theta_1, \theta_2)$, where θ_1 is the mean and θ_2 the variance of N_X. In this case $\Omega = \{(-\infty, \infty) \times (0, \infty)\}$ and it might be required to test whether or not a sample was generated by $N_X(x; \boldsymbol{\theta})$ for some unknown $\boldsymbol{\theta} \in \Omega$. Intuitively, and in fact, this is an intrinsically more difficult problem than testing whether the sample was generated by a particular normal CDF with known mean and variance. The latter case can always be reduced to the standard situation of testing $G_X(x) = N(x; \boldsymbol{\theta}_0)$, $\boldsymbol{\theta}'_0 = (0, 1)$.

Most useful tests are parameter-free; i.e., the distribution of the test statistics does not depend on $\boldsymbol{\theta}$. Among such tests are found both parametric and nonparametric tests which are either distribution specific or distribution-free. Since tests may require $F(x, \boldsymbol{\theta})$ to be continuous or a step function, later discussion will deal with continuous and discrete X_i separately.

There has been a recent resurgence of interest in the theory of goodness-of-fit tests. Technical advances have been made with some of the older tests, while new tests have been proposed and their power properties examined. This progress can be attributed in part to the availability of mathematical development in the theory of probability* and stochastic processes*. However, it is also in large measure due to the advent of the high-speed computer (*see* COMPUTERS IN STATISTICS), the associated numerical technology, and the increased demand for statistical services. This article can only summarize some of the available results and refer the reader to special sources for further detail.

Many statistical texts have introductory chapters on goodness-of-fit testing. For example, Kendall and Stuart [17, Vol. 2] and Lindgren [20] contain pertinent material lucidly presented. Pearson and Hartley [24] also contains accounts of specific tests illustrated by numerical examples.

The following general notation will be adopted, additional special symbols being introduced as required:

Probability density function* (PDF) corresponding to $F_X(x,\theta)$ and $G_X(x)$ (when they exist): $f_X(x,\theta)$, $g_X(x)$.

Order statistics*: $X_i' \leqslant X_2' \leqslant \cdots \leqslant X_n'$.

Expected values of order statistics: $E[X_i'] = \eta_i$.

Sample CDF: $G_n(x) = [\text{no. of } X_i \leqslant x]/n$.

A chi-square random variable with d degrees of freedom: $\chi^2(d)$.

$100(1-\alpha)$ percentile of the chi-square distribution* with d degrees of freedom: $\chi^2_{1-\alpha}(d)$.

The uniform density on [0, 1]: $U[0,1]$.

If X_n is a sequence of random variables, then $X_n \xrightarrow{L} \chi^2(d)$ will indicate convergence in law to a chi-square distribution with d degrees of freedom; if X_n is a sequence of random vectors, then $X_n \xrightarrow{L} N(\boldsymbol{\mu}, \boldsymbol{\Sigma})$ will indicate convergence* in law to a normal distribution with mean vector $\boldsymbol{\mu}$ and covariance matrix* $\boldsymbol{\Sigma}$ (*see* CONVERGENCE OF SEQUENCES OF RANDOM VARIABLES).

DISCRETE RANDOM VARIABLES

Simple H_0

Suppose that X is a discrete random variable, $\Pr\{X = x\} = f_X(x,\theta)$ for $\theta \in \Omega$ and $x \in \mathscr{X}$, where $\mathscr{X}$ is a finite or countable set of real numbers. By suitable definition, categorical data* with no evident numerical structure can be brought within this framework.

The simplest case is where $\mathscr{X} = \{1, 2, \ldots, k\}$, $f_X(j,\theta_0) = f_j$ is fully specified, and it is required to test $H_0 : g_X(j) = f_j$. Let N_j be the number of X_i in the sample such that $X_i = j$; then the probability under H_0 of obtaining the particular outcome vector, $\mathbf{n}' = \{n_1, n_2, \ldots, n_k\}$, where $\sum_j n_j = n$, is

$$P(\mathbf{n}) = \frac{n!}{\prod_j n_j!} \prod_j f_j^{n_j}. \tag{1}$$

An exact test of H_0 can, in principle, be constructed as follows:

1. Calculate $P(\mathbf{m})$ for all possible outcome vectors, $\mathbf{m}$.
2. Order the $P(\mathbf{m})$.
3. Sum all $P(\mathbf{m})$ which are less than or equal to $P(\mathbf{n})$.
4. Reject H_0 at level α if the cumulative probability of **3** is less than or equal to α.

This is known as the *multinomial test of goodness of fit*, and the necessary calculations must be carried out on a computer. Even then, this is only practicable if n and k are small.

Fortunately, for large n, there are ways around the distributional problem. Likelihood ratio theory can be invoked; the likelihood ratio test* of $H_0 : g_X(j) = f_j$ against the alternative $H_1 : g_X(j) \neq f_j$ is formed from the ratio $\Lambda = \prod_{j=1}^k (nf_j/N_j)^{N_j}$. It is known that

$$-2\ln\Lambda = -2\sum_{j=1}^{k} N_j\left[\ln nf_j - \ln N_j\right] \xrightarrow{L} \chi^2(k-1). \tag{2}$$

The null hypothesis is rejected at level α if the calculated value of $-2\ln\Lambda$ exceeds $\chi_{1-\alpha}(k-1)$.

A very old test dating back to the beginning of this century is based on

$$X^2 = \sum_{j=1}^{k} (N_j - nf_j)^2/(nf_j). \qquad (3)$$

This is known as Pearson's chi-square (*see* CHI-SQUARE TESTS) and has the same limiting distribution as (2). Since N_j is the observed number of X_i in the sample with $X_i = j$, (O_j), and $E[N_j] = nf_j = (E_j)$, (3) is sometimes written in the form of a mnemonic,

$$X^2 = \sum_{j=1}^{k} (O_j - E_j)^2/E_j.$$

Not only do (2) and (3) share the same limiting central distribution, but they are also asymptotically equivalent in probability. However, since X^2 is a direct measure of agreement between observation and expectation under H_0, it has some intuitive appeal not shared by (2).

Both (2) and (3) give asymptotic tests which tend to break down if the nf_j are too small. A common rule is that all nf_j should be greater than 1 and that 80% of them should be greater than or equal to 5. These conditions are sometimes hard to meet in practice. For a general discussion and further references, see Horn [15].

Radlow and Alf [28] point out that a direct comparison of X^2 with the multinomial test may be unjustified. The latter test orders experimental outcomes, **m**, in terms of $P(\mathbf{m})$ instead of ordering them in terms of discrepancies from H_0. It is suggested that X^2 should be compared with the following exact procedure:

1. Calculate $P(\mathbf{m})$ for all possible outcomes **m**.
2. Calculate X^2 for each **m** based on H_0, $X^2(\mathbf{m})$.
3. Sum $P(\mathbf{m})$ for which $X^2(\mathbf{m}) \geqslant X^2(n)$, the observed X^2 value.
4. Reject H_0 at level α if this sum exceeds α.

Numerical comparisons of this exact test with X^2 showed that the two agreed remarkably well, even for small n. The agreement of the exact multinomial test with X^2, on the other hand, was poor.

Thus care must be exercised in the assessment of the performances of large-sample tests using small n. Appropriate baseline exact tests must be used for comparisons.

Another procedure is the discrete Kolmogorov–Smirnov* goodness-of-fit test [4]. Let

$$D^- = \max_x(F_X(x) - G_n(x)),$$
$$D^+ = \max_x(G_n(x) - F_X(x)),$$
$$D = \max(D^-, D^+);$$

then D^-, D^+, and D test, respectively, $H_1: G_X(x) \leqslant F_X(x)$, $H_1: G_X(x) \geqslant F_X(x)$, and $H_1: G_X(x) \neq F_X(x)$. A discussion of the application of these statistics is given in Horn [15], where their efficiency relative to X^2 is discussed and a numerical example given. For more recent asymptotic results, see Wood and Altavela [38].

Composite H_0

When $H_0: G_X(x) = F_X(x, \boldsymbol{\theta})$, $\boldsymbol{\theta} \in \Omega$, is to be tested, $\boldsymbol{\theta}$ must be estimated and the theory becomes more elaborate. However, provided that asymptotically efficient estimators $\hat{\boldsymbol{\theta}}_n$ are used, tests (2) and (3) extend in a natural way and continue to be equivalent in probability.

More specifically, since X_i is assumed discrete, put $H_0: g_X(j) = f_j(\boldsymbol{\theta})$ and $H_1: g_X(j) \neq f_j(\boldsymbol{\theta})$, $\boldsymbol{\theta} \in \Omega$. Let $\hat{\boldsymbol{\theta}}_n$ be as above; e.g., $\hat{\boldsymbol{\theta}}_n$ could be the maximum likelihood estimator* (MLE) for $\boldsymbol{\theta}$ under H_0.

Then under H_0,

$$-2\ln\Lambda(\hat{\boldsymbol{\theta}}_n) = -2\sum_{j=1}^{k} N_j[\ln nf_j(\hat{\boldsymbol{\theta}}_n) - \ln N_j] \xrightarrow{L} \chi^2(k-q-1), \qquad (4)$$

$$X^2(\hat{\boldsymbol{\theta}}_n) = \sum_{j=1}^{k} (N_j - nf_j(\hat{\boldsymbol{\theta}}_n))^2/(nf_j(\hat{\boldsymbol{\theta}}_n)) \xrightarrow{L} \chi^2(k-q-1). \qquad (5)$$

The philosophy adopted in the preceding subsection for rejecting H_0 is used here, the critical level being $\chi^2_{1-\alpha}(k-q-1)$.

Although (4) and (5) are the standard tests recommended in textbooks and have received the most attention by practitioners and theoreticians, there are others. For example, a general class of goodness-of-fit tests can be based on quadratic form* theory for multinormally distributed random variables. Under sufficient regularity conditions the following results can be established by routine methods of probability calculus.

Put

$$\mathscr{N}' = (\mathscr{N}_1, \mathscr{N}_2, \ldots, \mathscr{N}_k),$$
$$\mathscr{N}_i = N_i/n,$$
$$(\mathbf{f}(\boldsymbol{\theta}))' = (f_1(\boldsymbol{\theta}), f_2(\boldsymbol{\theta}), \ldots, f_k(\boldsymbol{\theta}));$$

then $\sqrt{n}\,(\mathscr{N} - \boldsymbol{f}(\boldsymbol{\theta})) \xrightarrow{\text{L}} N(\mathbf{0}, \mathbf{V})$, where

$$\mathbf{V} = [v_{ij}],$$
$$v_{ii} = f_i(\boldsymbol{\theta})(1 - f_i(\boldsymbol{\theta})),$$
$$v_{ij} = -f_i(\boldsymbol{\theta})f_j(\boldsymbol{\theta}), \quad i \neq j,$$

and rank $(\mathbf{V}) = k - 1$. Now suppose that $\boldsymbol{\theta}_n^*$ is any estimator for $\boldsymbol{\theta}$ which can be expressed in the locally, suitably regular functional form $\boldsymbol{\theta}_n^* = g(\mathscr{N})$. Let

$$\mathbf{D} = [d_{ij}], \quad d_{ij} = \partial g_i(\mathscr{N})/\partial \mathscr{N}_j,$$
$$i = 1, 2, \ldots, q; \quad j = 1, 2, \ldots, k,$$
$$\mathbf{Q} = [q_{rs}], \quad q_{rs} = \partial f_r(\boldsymbol{\theta})/\partial \theta_s,$$
$$r = 1, 2, \ldots, k; \quad s = 1, 2, \ldots, q.$$

Then

$$\sqrt{n}\,(\mathscr{N} - \boldsymbol{f}(\boldsymbol{\theta}_n^*)) \xrightarrow{\text{P}} \sqrt{n}\,(\mathbf{I} - \mathbf{QD})(\mathscr{N} - \boldsymbol{f}(\boldsymbol{\theta}))$$
$$\xrightarrow{\text{L}} N(\mathbf{0}, \boldsymbol{\Sigma}),$$

where $\boldsymbol{\Sigma} = (\mathbf{I} - \mathbf{QD})\mathbf{V}(\mathbf{I} - \mathbf{QD})'$. If $\boldsymbol{\Sigma}^g$ is any generalized inverse* of $\boldsymbol{\Sigma}$ (i.e., $\boldsymbol{\Sigma}\boldsymbol{\Sigma}^g\boldsymbol{\Sigma} = \boldsymbol{\Sigma}$), it then follows that, under H_0,

$$Q_n(\boldsymbol{\theta}_n^*) = n(\mathscr{N} - \boldsymbol{f}(\boldsymbol{\theta}_n^*))'\boldsymbol{\Sigma}^g(\mathscr{N} - \boldsymbol{f}(\boldsymbol{\theta}_n^*))$$
$$\xrightarrow{\text{L}} \chi^2(k - q - 1). \tag{6}$$

The power of tests such as (6), which include (5) as a special case, can also be examined using sequences of local alternatives, similar arguments, and the noncentral chi-square* distribution. However, such studies are of limited use, as tests of fit should detect broad alternatives, a performance feature that can be checked only by computer simulations.

Note that $\boldsymbol{\Sigma}^g$ may be a function of $\boldsymbol{\theta}$, which can be replaced by $\boldsymbol{\theta}_n^*$ or any other consistent estimator for $\boldsymbol{\theta}$ without affecting the asymptotic distribution of $Q_n(\boldsymbol{\theta}_n^*)$.

For an early, rigorous account of the theory of chi-square tests*, see Cramér [5]. A modern and comprehensive treatment of this theory with many ramifications is given by Moore and Spruill [23]. Their paper is technical and covers cases where the $\mathbf{X}_i$ are random vectors and $\boldsymbol{\theta}$ is estimated by a variety of methods. An easy-to-read overview of this work is given by Moore [22].

CONTINUOUS RANDOM VARIABLES

The testing of goodness-of-fit hypotheses when $f_X(x, \boldsymbol{\theta})$ is a continuous function of x introduces features not exhibited by the discrete tests discussed in the section "Discrete Random Variables." However, by suitable constructions some of the latter tests remain useful. These points will be expanded on below.

Simple H_0

In order to apply results of the section "Discrete Random Variables," partition the real line into $k \geqslant 2$ sets:

$$I_1 = (-\infty, a_1], \qquad I_2 = (a_1, a_2], \ldots,$$
$$I_k = (a_{k-1}, \infty).$$

To test $H_0: g_X(x) = f_X(x)$, let N_j be the number of $X_i \in I_j$ in the sample and put $p_j = \int_{I_j} f_X(x)dx$. Then, under H_0, the N_j have a multinomial distribution* with parameters p_j, $n = \sum_{j=1}^k N_j$ and any of the tests from the discussion of simple H_0 in the section "Discrete Random Variables" can be applied.

Clearly, making a situation discrete which is essentially continuous leads to a loss of precision. The actual values of the X_i are suppressed and only their relationship with the I_j is used in the tests. The k classes are usually chosen to keep the np_j acceptably high. In order to achieve some standardization, it seems reasonable to use $p_j = k^{-1}$ and to determine the a_i by the equations $F_X(a_1) = k^{-1}$, $F_X(a_2) - F_X(a_1) = k^{-1}$, etc. (see Kendall and Stuart [17]). Nevertheless, there remains an essential nonuniqueness aspect to the tests. Given the same set of data, different statisticians can reach different conclusions using the same general procedures.

In fact, these tests condense the data and examine whether or not $g_X(x)$ is a member of the particular class of density functions with given content p_j for I_j. Despite these drawbacks, the approach outlined above has enjoyed wide support and is most commonly used in practice.

The method of condensation of data presented above when X is a continuous random variable may also have to be practiced when X is discrete. In this case subsets of $\mathscr{X}$ are used in place of individual elements to achieve cell expectations sufficiently large to render the asymptotic distribution theory valid.

A useful way of visually checking the adequacy of H_0 is to examine the order statistics $X_1', X_2', \ldots, X_n'$. Since $f_X(x)$ is fully specified, $E[X_i'] = \eta_i$ can be calculated and plotted against X_i'. If H_0 holds, this plot should be roughly linear.

There are analytical counterparts to the simple order statistics plots. Let $0 < \lambda_1 < \lambda_2 < \cdots < \lambda_k < 1$, $n_i = [n\lambda_i] + 1$, where $[x]$ is the greatest integer less than or equal to x, and consider X_{n_i}', $i = 1, 2, \ldots, k$. Under suitable regularity conditions on $f_X(x)$,

$$Y^2 = n \sum_{i=1}^{k} \left\{ \left[F_X(X_{n_i}') - F_X(X_{n_{i-1}}') \right] - p_i \right\}^2 p_i^{-1}$$
$$\xrightarrow{L} \chi^2(k-1), \tag{7}$$

where $p_i = \lambda_i - \lambda_{i-1}$ [3]. This bypasses the problem of constructing intervals I_j and uses part of the natural ordering of the sample.

A number of tests make specific use of the sample CDF, $G_n(x)$, for testing $H_0: G_X(x) = F_X(x)$. Some of these are now listed.

KOLMOGOROV–SMIRNOV* STATISTICS, D_n^-, D_n^+, AND D_n. Define

$$D_n^- = \sup_x \left[F_X(x) - G_n(x) \right],$$
$$D_n^+ = \sup_x \left[G_n(x) - F_X(x) \right], \tag{8}$$
$$D_n = \max(D_n^-, D_n^+).$$

Then D_n^- and D_n^+ can be used to test H_0 against the one-sided alternatives $H_1: G_X(x) \leqslant F_X(x)$ and $H_1: G_X(x) \geqslant F_X(x)$, respectively, while D_n tests $H_1: G_X(x) \neq F_X(x)$.

The CDFs of the three statistics are known exactly and are independent of $F_X(x)$ [7]. To see this, let $U = F_X(X)$; then $D_n^+ = \sup_{0 \leqslant u \leqslant 1}\{G_n(u) - u\}$, etc. The most useful set of tables is given by Pearson and Hartley [24], who also include some numerical examples.

A derivation of the asymptotic distributions of D_n^+ and D_n can be based on the stochastic process

$$y_n(t) = \sqrt{n}\,(G_n(t) - t), \qquad 0 \leqslant t \leqslant 1,$$

which has zero mean and

$$C[y(s), y(t)] = \min(s,t) - st,$$
$$0 \leqslant s, t \leqslant 1.$$

The central limit theorem* ensures that $[y_n(t_1), y_n(t_2), \ldots, y_n(t_k)]$ is asymptotically multinormal with null mean vector and the above covariance structure. Thus the finite-dimensional distributions of $Y_n(t)$ converge to those of $y(t)$, tied-down Brownian motion*.

Intuitively, the distributions of $\sup_t y_n(t)$ and $\sup_t |y_n(t)|$ will tend to those of $\sup_t y(t)$ and $\sup_t |y(t)|$. This can be verified using the theory of weak convergence. The two crossing problems thus generated can be solved to yield the desired limiting CDFs [7]. For a different approach, see Feller [12]. It is interesting that these investigations show $4n(D_n^+)^2$ to be asymptotically distributed as $\chi^2(2)$.

CRAMÉR–VON MISES TEST*. Let

$$W_n^2 = n\int_{-\infty}^{\infty}\left[G_n(X) - F_X(x)\right]^2 dx; \quad (9)$$

then W_n^2 is a measure of the agreement between $G_n(x)$ and $F_X(x)$ for all x and is known as the Cramér–von Mises statistic. By means of the probability transformation $U = F_X(X)$, (9) can be written

$$W_n^2 = n\int_0^1\left[G_n(u) - u\right]^2 du,$$

emphasizing that this test is also distribution-free.

The CDF of W_n^2 is not known for all n but has been approximated; the asymptotic distribution is derived in Durbin [7]. For easy-to-use tables, see Pearson and Hartley [24].

TESTS RELATED TO THE CRAMÉR–VON MISES TEST. Various modifications of W_n^2 are used for specific purposes. For instance, a weight function $\psi(t)$ can be introduced to give $\int_0^1[G_n(t) - t]^2\psi(t)\,dt$ as a test statistic. When $\psi(t) = [t(1-t)]^{-1}$, the resulting statistic is called the Anderson–Darling statistic, A_n^2, and leads to the Anderson–Darling test*. Since $E[n[G_n(t) - t]^2] = [t(1-t)]$, this weights discrepancies by the reciprocal of their standard deviations and puts more weight in the tails of the distribution, a feature that may be important. The same remarks made for W_n^2 apply to A_n^2.

A number of scientific investigations yield data in the form of directions and it may be required to test the hypothesis that these are orientated at random. Since each direction is represented by an angle measured from a fixed position P, such data can be represented as points on a unit circle. The test then concerns the randomness of the distribution of the points on this circle.

Watson [37] introduced the statistic

$$U_n^2 = n\int_0^1\left[G_n(t) - t - \overline{G_n(t) - t}\,\right]^2 dt, \quad (10)$$

where $\int_0^1[G_n(t) - t]\,dt = \overline{G_n(t) - t}$. It can be shown that U_n^2 is independent of the choice of P. The asymptotic distribution of U_n^2 is known [7] and appropriate tables may be found in Pearson and Hartley [24].

Under $H_0: G_X(x) = F_X(x)$, the variables $U_i = F_X(X_i)$ are distributed as $U[0,1]$. Hence $\overline{G}_n = \int_0^1 G_n(u)\,du = 1 - \overline{U}$, has expectation $\frac{1}{2}$, variance $(12n)^{-1}$ and tends rapidly to normality. This provides a direct large-sample test of H_0, although exact significance points are available [33].

Tests related to D_n^-, D_n^+, D_n, and W_n^2 have been proposed by Riedwyl [30]. He defines the ith discrepancy as $d_i = F(X_i') - F_n(X_i')$ and examines tests based on $\sum_1^n d_i^2$, $\sum_1^n |d_i|$, $\sum_1^n d_i$, $\max_i d_i$, $\max_i |d_i|$, etc. Some pertinent exact and asymptotic results are given.

Hegazy and Green [14] considered tests based on the forms $T_1 = n^{-1}\sum_1^n |X_i' - \nu_i|$ and $T_2 = n^{-1}\sum_1^n (X_i' - \nu_i)^2$, where $\nu_i = \eta_i$ and $\nu_i = \xi_i$, the mode of X_i'. Tests of the hypothesis $H_0: G_X(x) = F_X(x)$ can be reduced as shown above to testing whether or not $U_i = F_X(X_i)$ is distributed $U[0,1]$. Thus $\eta_i = i/(n+1)$ and $\xi_i = (i-1)/(n-1)$.

The powers of these T tests were examined against normal*, Laplace*, exponential*, and Cauchy* alternatives and compared with the powers of other tests. The conclusion was that T_1 and T_2 have similar performances and that it is slightly better to use ξ_i than η_i. These T statistics generally compare favorably with the tests just described, or minor modifications of them.

Hegazy and Green [14] provide an extensive bibliography of other studies of power of goodness-of-fit tests.

Composite H_0

The most common hypothesis that requires testing is $H_0: G_X(x) = F_X(x, \theta)$ for some $\theta \in \Omega$. The introduction of nuisance parameters* creates new technical difficulties which can only be touched on briefly here. In general, however, the same form of tests as those just presented are used, with modifications.

In order to make use of the results in the discussion of composite H_0 in the section "Discrete Random Variables," k intervals are introduced as in the preceding subsection. The interval contents are functions of θ, $p_j(\theta) = \int_{I_j} f(x,\theta)\,dx$ and if N_j is the number of X_i in I_j, a multinomial system is generated, the parameters being functions of the unknown θ. The whole problem may now be treated by the methods of the section on discrete variables, and the same comment concerning loss of information and nonuniqueness due to grouping applies.

A number of special points need emphasis. The estimation of θ must be made from the data in the grouped state if the distribution theory of the section on discrete variables is to hold. For instance, θ estimated from the X_i and $f(x,\theta)$ should not be used in the $X^2(\theta)$ statistic. Doing so results in a limiting CDF which depends on θ and a conservative test if the $\chi^2_{1-\alpha}(k-q-1)$ significance level is used.

Since θ is not known, there is some difficulty defining the intervals I_j. In general, the boundaries of the intervals are functions of θ; Moore and Spruill [23] have shown that, provided that consistent estimators of the boundary values are used, the asymptotic results (4), (5), and (6) remain valid if the random intervals are used as if they were the true ones.

For example, reconsider the problem of testing $H_0: G_X(x) = N_X(x,\boldsymbol{\theta})$. Consistent estimators of $\theta_1 = \mu$ and $\theta_2 = \sigma^2$ are $\bar{X} = \sum_1^n X_i/n$ and $S^2 = \sum_1^n (X_i - \bar{X})^2/(n-1)$ and it is appropriate that the I_j be constructed with $\bar{X}$ and S in place of μ and σ to ensure approximate contents of k^{-1}. Using these estimated intervals, the procedure requires that μ and σ^2 be estimated efficiently, by maximum likelihood, for instance, and the tests applied in the usual way.

A test developed by Moore [21] and suggested by Rao and Robson [29] has interesting flexibility and power potential. Let $\mathbf{V}_n(\boldsymbol{\theta})$ be a k-vector with ith component $(N_i - nf_i(\boldsymbol{\theta}))/\sqrt{nf_i(\boldsymbol{\theta})}$, $\mathbf{B}(\boldsymbol{\theta})$ a $k \times q$ matrix with elements $p_i(\theta)^{-1/2}\partial p_i(\boldsymbol{\theta})/\partial\theta_j$, and $\mathbf{J}(\boldsymbol{\theta})$ the usual information matrix for $F_X(x,\boldsymbol{\theta})$ (*see* FISHER INFORMATION). Define the statistic

$$T_n(\hat{\boldsymbol{\theta}}_n) = \mathbf{V}_n'(\hat{\boldsymbol{\theta}}_n) \times [\mathbf{I} - \mathbf{B}(\hat{\boldsymbol{\theta}}_n)\mathbf{J}^{-1}(\hat{\boldsymbol{\theta}}_n)\{\mathbf{B}(\hat{\boldsymbol{\theta}}_n)\}']^{-1}\mathbf{V}_n(\hat{\boldsymbol{\theta}}_n), \tag{11}$$

where $\hat{\boldsymbol{\theta}}_n$ is the ungrouped MLE for θ; then $T_n(\hat{\boldsymbol{\theta}}_n) \xrightarrow{L} \chi^2(k-1)$.

The problem of estimating intervals can be bypassed by the use of quantile* statistics. Define λ_i and p_i as for (7), and the statistic

$$Y_n^2(\boldsymbol{\theta}) = n\sum_1^k \{[F_X(X'_{n_i},\boldsymbol{\theta}) - F_X(X'_{n_{i-1}},\boldsymbol{\theta})] - p_i\}^2 p_i^{-1}. \tag{12}$$

If $\boldsymbol{\theta} = \tilde{\boldsymbol{\theta}}_n$ minimizes $Y_n^2(\boldsymbol{\theta})$, then $Y^2(\tilde{\boldsymbol{\theta}}_n) \xrightarrow{L} \chi^2(k-q-1)$.

Alternatively, the following test is available. Put $X_{n_i}/n = \mathscr{N}_i^*$ and let $\mathscr{N}^*$ be the $(k \times 1)$ vector of the $\mathscr{N}_i^*$; then it is well known that $\sqrt{n}\,(\mathscr{N}^* - \boldsymbol{\nu}) \xrightarrow{L} N(\mathbf{0},\mathbf{V})$, where ν_i is defined by $F_X(\nu_i) = \lambda_i$ and $v_{ij} = \lambda_i(1-\lambda_j)[f_X(\nu_i)f_X(\nu_j)]^{-1}$, $i \leqslant j$. In general, both $\boldsymbol{\nu}$ and $\mathbf{V}$ are functions of unknown $\boldsymbol{\theta}$, so define

$$A_n(\boldsymbol{\theta}) = n(\mathscr{N}^* - \boldsymbol{\nu}(\boldsymbol{\theta}))'\mathbf{V}^{-1}(\boldsymbol{\theta}) \times (\mathscr{N}^* - \boldsymbol{\nu}(\boldsymbol{\theta})) \tag{13}$$

and choose $\boldsymbol{\theta} = \boldsymbol{\theta}_n^*$ to minimize $A_n(\boldsymbol{\theta})$. Then $A_n(\boldsymbol{\theta}_n^*) \xrightarrow{L} \chi^2(k-q)$, $k > q$. If $q = 2$, θ_1 and θ_2 are location and scale parameters, respectively, and an explicit expression exists for θ_n^*. The matrix $\mathbf{V}$ for the standardized variable $(X - \theta_1)/\theta_2$ can be used in (13) and a single matrix inversion is needed to complete the test [36].

The tests described in the discussion of simple H_0 do not extend readily to composite hypotheses. In general, for the cases considered and reported in the literature to date, the resulting tests are not distribution-free but depend on $F(x,\boldsymbol{\theta})$ and on the method used to estimate $\boldsymbol{\theta}, \hat{\boldsymbol{\theta}}$. This is because the

CDF has a different limiting distribution when the parameters are estimated to that which results when the null hypothesis is simple [8]. Hence tables of critical values constructed for simple hypothesis cases cannot be used for testing composite hypotheses. In fact, different critical values are needed for each hypothesis tested; the tests are carried out replacing $F_X(x)$ by $F(x, \hat{\boldsymbol{\theta}})$ in the expressions of the preceding section.

KOLMOGOROV–SMIRNOV STATISTICS D_n^-, D_n^+, D_n. A technique for obtaining exact critical values was developed by Durbin [9], who applied it to obtain values for testing the composite hypothesis of exponentiality

$$H_0: f(x, \theta) = \theta^{-1}\exp(-x/\theta),$$
$$0 < x, \theta \in (0, \infty).$$

The technique is complicated, however, and has not been applied to other cases.

By a variety of techniques, including Monte Carlo methods*, Stephens [34] has given procedures for finding accurate critical values for testing composite hypotheses involving the normal and the exponential distributions. These procedures are also described by Pearson and Hartley [24]. For a treatment of this problem using sufficient statistics*, see Kumar and Pathak [19].

CRAMÉR–VON MISES STATISTIC* W_n^2. No technique is yet available for obtaining exact critical values of W_n^2 for testing composite hypotheses. The first accurate calculations of asymptotic significance points for testing exponentiality and normality were made by Durbin et al. [11]. Further extensions and related results were given by Stephens [35]. Again, methods of obtaining good approximations to finite-sample critical values for tests of exponentiality and normality are given by Stephens [34].

TESTS RELATED TO THE CRAMÉR–VON MISES TESTS. Similar treatments to those of W_n^2 are given to A_n^2 and U_n^2 by Stephens [34, 35] and Durbin et al. [11] for testing exponentiality and normality.

In summary, then, the development of tests of fit for composite hypotheses using the sample CDF has centered largely on the exponential and the normal distributions. The most useful reference for the practitioner is Stephens [34], where tables cater for most of the common tests when the hypothesis is simple, and for the composite hypothesis cases mentioned above. If the data are censored*, see Pettitt and Stephens [26] and Dufour and Maag [6].

FURTHER TESTS AND CONSIDERATIONS

In this final section a few other special tests and some additional ideas impinging on goodness of fit will be mentioned.

Two Sample Tests

Let $X_1, X_2, \ldots, X_m$ and $Y_1, Y_2, \ldots, Y_n$, $m \leqslant n$, be random samples from two different populations with continuous CDFs $F_X(x)$ and $F_Y(y)$ and sample CDFs $F_X^m(t)$ and $F_Y^n(t)$.

In analogy with the discussion on the Kolmogorov–Smirnov statistics D_n^-, D_n^+, and D_n pertaining to simple H_0 in the section "Continuous Random Variables," the hypothesis $H_0: F_X(t) = F_Y(t)$ can be tested against the alternatives $H_1: F_X(t) \leqslant F_Y(t)$, $H_1: F_X(t) \geqslant F_Y(t)$, and $H_1: F_X(t) \neq F_Y(t)$ by the respective statistics

$$D_{mn}^- = \sup_t \left[F_Y^n(t) - F_X^m(t) \right],$$
$$D_{mn}^+ = \sup_t \left[F_X^m(t) - F_Y^n(t) \right],$$
$$D_{mn} = \max\left[D_{mn}^-, D_{mn}^+ \right].$$

The exact distributions of these statistics are known; for finite sample critical points of D_{mn}, see Pearson and Hartley [24]. For further references to tabulations, see Steck [32]. If the statistics above are multiplied by $[mn(m+n)^{-1}]^{1/2}$ limiting distributions exist which are the same as those for $\sqrt{n}\, D_n^-$, $\sqrt{n}\, D_n^+$, and $\sqrt{n}\, D_n$.

Similar modifications can be made to the Cramér–von Mises statistic to cater for two sample tests; again, see Durbin [7].

Tests of Departure from Normality*

In view of the central role of the normal distribution in statistical theory and practice, a great deal of effort has been spent in developing tests of normality (see the first section of this article). Some of these tests have been dealt with in previous sections; only special tests tailored for the normal distribution will be covered here.

Let $m_r = n^{-1}\sum_1^n(X_i - \bar{X})^r$ and $S^2 = nm_2(n-1)^{-1}$; then statistics $\sqrt{b_1} = m_3/S^3$ and $b_2 = m_4/S^4$ measure skewness and kurtosis in the sample. If the population from which the sample is drawn is normal, $\sqrt{b_1}$ and b_2 should be near 0 and 3, respectively, and departure from these values is evidence to the contrary. Both $\sqrt{b_1}$ and b_2 have been examined separately, jointly, and as a linear combination (see Pearson and Hartley [24] and Pearson et al. [25]). Both were compared with other tests for power; a variety of skewed and leptokurtic distributions were used as alternatives; *see* MOMENT RATIOS.

The picture is somewhat confused, due in part to the wide spectrum of alternative distributions used and to the use of small numbers of Monte Carlo trials to establish the power properties. Nevertheless, a general pattern emerged; singly, the two statistics are useful for detecting departures from specific types of alternatives, and in combination they are reasonably robust against a large variety of alternatives.

Wilks–Francia Test*

There are fruitful extensions to the technique of plotting order statistics* against expectation as introduced in the discussion of simple H_0 in the section "Continuous Random Variables." Let the hypothesis be

$$H_0 : G(x,\theta) = F((x-\theta_1)/\theta_2),$$

i.e., G is determined up to a location and scale parameter. Then a plot of X_i' against the expectation of the standardized order statistics, η_i, should lie near the line $\theta_1 + \theta_2\eta_i$ under H_0.

Now, the unweighted least-squares estimator for θ_2 is $\bar{\theta}_2 = \sum_1^n X_i'\eta_i/\sum_1^n\eta_i^2$ and the residual sum of squares is

$$\begin{aligned} R_n^2 &= \sum_1^n \left(X_i' - \bar{X} - \bar{\theta}_2\eta_i\right)^2 \\ &= \sum_1^n \left(X_i' - \bar{X}\right)^2 - \left(\sum_1^n b_i X_i'\right)^2, \end{aligned}$$

where $b_i = \eta_i/(\sum_i^n \eta_i^2)^{1/2}$. Dividing both sides by $\sum_1^n (X_i' - \bar{X})^2$ to remove the scale effect yields

$$\begin{aligned} &\left(\sum_1^n \left(X_i' - \bar{X}\right)^2\right)^{-1} R_n^2 \\ &\quad = 1 - \left(\sum_1^n b_i X_i'\right)^2 \Big/ \sum_1^n \left(X_i' - \bar{X}\right)^2 \\ &\quad = 1 - W_n'. \end{aligned} \qquad (14)$$

Then W_n' is the *Wilks–Francia test statistic* and measures the departure of the order statistics from their expectations; it has been used to test normality specifically, but it clearly enjoys a wider application.

To carry out the test, tables of η_i are required as well as critical points; reference is made to the original paper by Shapiro and Francia [31]. Note that small values of W_n' are significant and that the test has been shown to be consistent. An asymptotic distribution for the test has been established (see Durbin [10], where further tests of fit using order statistics are discussed).

For completeness, it is pointed out that there exist goodness-of-fit procedures using the differences, or spacings*, between successive order statistics. Some of these tests are reviewed by Pyke [27], who developed the limiting distributions of functions of spacings and certain general limit theorems. More recent work in this area is reported by Kale [16] and Kirmani and Alam [18].

FINAL REMARKS

A few general considerations are worth raising. The single sample procedures outlined in previous sections deal with the problem of testing $H_0: G_X(x) = F_X(x,\theta)$, where θ is either fixed or is specified only up to a set Ω. If θ is not fixed, an estimator $\hat{\theta}$ is substituted for it in F and the concordance of this estimated model with the data assessed.

It is important in carrying out tests of fit not to lose sight of the fundamental purpose of the exercise. For example, tests of normality are often required as an intermediate step to further analyses. Alternatively, the performance of specific statistical processes, such as a random number generator, may need to be checked against specification. In these instances, the philosophy of using F, or a good estimate of it, to test against available data seems entirely reasonable.

A different situation is generated if *predictions** are required. In this case an estimate of F is to be used to predict future outcomes of the random variable X.

It is possible that $F_X(x,\hat{\theta})$ may allow satisfactory predictions to be made, especially if the model was appropriate and $\hat{\theta}$ based on a large sample. But there may be other candidates which would do a better job of prediction than $F_X(x,\hat{\theta})$, such as to set up a measure of divergence of one PDF from another (see Ali and Silvey [2]) and then to try to find that PDF, based on the data, which comes closest to the estimated PDF. This treatment may need Bayesian arguments to construct predictive densities [1].

More specifically, let $f_X(x,\theta)$ be the density which is to be estimated and introduce the weight function $p(\theta \mid z)$ on Ω based on data z. Put

$$q_X(x \mid z) = \int_\Omega p(\theta \mid z) f_X(x,\theta)\, d\theta; \qquad (15)$$

then q_X is called a *predictive density* for f_X. On the other hand, for any estimator for θ based on z, $\hat{\theta}(z)$, $f_X(x,\hat{\theta}(z))$ is called an *estimative density*.

Using the Kullback–Leibler directed measure of divergence, Aitchison [1] showed that $q_X(x \mid z)$ is optimal in the sense that it is closer to f_X than any other competing density, in particular $f_X(x,\hat{\theta}(z))$. Although this result may depend on the divergence measure used, it shows that $f_X(x,\hat{\theta}(z))$ may not always be the appropriate estimator for $f_X(x,\theta)$.

A chi-squared type of goodness-of-fit test for the predictive density has been developed by Guteman [13].

References

References are classified as follows: (A), applied; (E), expository; (R), review; (T), theoretical.

[1] Aitchison, J. (1975). *Biometrika*, **62**, 547–554. (T)

[2] Ali, S. M. and Silvey, S. D. (1966). *J. R. Statist. Soc. B*, **28**, 131–142. (T)

[3] Bofinger, E. (1973). *J. R. Statist. Soc. B*, **35**, 277–284. (T)

[4] Conover, W. J. (1972). *J. Amer. Statist. Ass.*, **67**, 591–596. (T)

[5] Cramér, H. (1945). *Mathematical Methods of Statistics*. Princeton University Press, Princeton, N.J. (E)

[6] Dufour, R. and Maag, U. R. (1978). *Technometrics*, **20**, 29–32. (A)

[7] Durbin, J. (1973). *Distribution Theory for Tests Based on the Sample Distribution Function*. Reg. Conf. Ser. Appl. Math. SIAM, Philadelphia. (T, R)

[8] Durbin, J. (1973). *Ann. Statist.*, **1**, 279–290. (T)

[9] Durbin, J. (1975). *Biometrika*, **62**, 5–22. (T)

[10] Durbin, J. (1977). Goodness-of-fit tests based on the order statistics. *Trans. 7th Prague Conf. Inf. Theory, Statist. Decision Functions, Random Processes/1974 Eur. Meet. Statist.*, Prague, 1974, Vol. A, 109–118. (R)

[11] Durbin, J., Knott, M., and Taylor, C. C. (1975). *J. R. Statist. Soc. B*, **37**, 216–237. (T)

[12] Feller, W. (1948). *Ann. Math. Statist.*, **19**, 177. (T)

[13] Guteman, I. (1967). *J. R. Statist. Soc. B*, **29**, 83–100. (T)

[14] Hegazy, Y. A. S. and Green, J. R. (1975). *Appl. Statist.*, **24**, 299–308. (A)

[15] Horn, S. D. (1977). *Biometrics*, **33**, 237–248. (A, R)

[16] Kale, B. K. (1969). *Sankhyā A*, **31**, 43–48. (T)

[17] Kendall, M. G. and Stuart, A. (1973). *The Advanced Theory of Statistics*, Vol. 2: *Inference and Relationship*, 3rd ed. Hafner Press, New York. (E)

[18] Kirmani, S. N. U. A. and Alam, S. N. (1974). *Sankhyā A*, **36**, 197–203. (T)

[19] Kumar, A. and Pathak, P. K. (1977). *Scand. Statist. J.*, **4**, 39–43. (T)

[20] Lindgren, B. W. (1976). *Statistical Theory*, 3rd ed. Macmillan, New York. (E)

[21] Moore, D. S. (1977). *J. Amer. Statist. Ass.*, **72**, 131–137. (T)

[22] Moore, D. S. (1979). In *Studies in Statistics*, R. V. Hogg, ed. Mathematical Association of America, Washington, D.C., pp. 66–106. (T, E)

[23] Moore, D. S. and Spruill, M. C. (1975). *Ann. Statist.*, **3**, 599–616. (T)

[24] Pearson, E. S. and Hartley, H. O. (1972). *Biometrika Tables for Statisticians*, Vol. 2. Cambridge University Press, Cambridge. (E)

[25] Pearson, E. S., D'Agostino, R. B., and Bowman, K. O. (1977). *Biometrika*, **64**, 231–246. (A)

[26] Pettitt, A. N. and Stephens, M. A. (1976). *Biometrika*, **63**, 291–298. (T)

[27] Pyke, R. (1965). *J. R. Statist. Soc. B*, **27**, 395–436. (T)

[28] Radlow, R. and Alf, E. F. (1975). *J. Amer. Statist. Ass.*, **70**, 811–813. (A)

[29] Rao, K. C. and Robson, D. S. (1974). *Commun. Statist.*, **3**, 1139–1153. (T)

[30] Riedwyl, H. (1967). *J. Amer. Statist. Ass.*, **62**, 390–398. (A)

[31] Shapiro, S. S. and Francia, R. S. (1972). *J. Amer. Statist. Ass.*, **67**, 215–216. (A)

[32] Steck, G. P. (1969). *Ann. Math. Statist.*, **40**, 1449–1466. (T)

[33] Stephens, M. A. (1966). *Biometrika*, **53**, 235–240. (T)

[34] Stephens, M. A. (1974). *J. Amer. Statist. Soc.*, **69**, 730–743. (A, E)

[35] Stephens, M. A. (1976). *Ann. Statist.*, **4**, 357–369. (T)

[36] Tallis, G. M. and Chesson, P. (1976). *Austr. J. Statist.*, **18**, 53–61. (T)

[37] Watson, G. S. (1961). *Biometrika*, **48**, 109–114. (T)

[38] Wood, C. L. and Altavela, M. M. (1978). *Biometrika*, **65**, 235–239. (T)

(CHI-SQUARE TESTS
DISTRIBUTION-FREE METHODS
HYPOTHESIS TESTING
KOLMOGOROV–SMIRNOV TESTS
NORMALITY, TESTS OF)

G. M. TALLIS

GOODNESS OF FIT: CENSORED DATA

To test whether a random sample with cumulative distribution function (CDF) $F(x)$ has specified CDF $G(x)$ or not, when some of the data are censored (as they may be in clinical trials*, for example), several authors have developed test procedures. Comparisons of the procedures and references are given in Hollander and Proschan [1]; procedures involving randomly censored grouped data are also cited.

Reference

[1] Hollander, M. and Proschan, F. (1979). *Biometrics*, **35**, 393–401.

(CENSORED DATA
GOODNESS OF FIT)

GOSSET, WILLIAM SEALY ("STUDENT")

Born: June 13, 1876, in Canterbury, England.

Died: October 16, 1937, in Beaconsfield, England.

Contributed to: sampling distributions, design of experiments, statistics in agriculture.

W. S. Gosset entered the service of the Guinness brewery business in Dublin, Ireland, in 1899, after obtaining a degree in chemistry at Oxford, with a minor in mathematics. He was asked to investigate what relations existed between the quality of materials such as barley and hops, production conditions, and the finished product. These practical problems led him to seek exact error probabilities of statistics from small samples, a hitherto unresearched area. His firm, which would later require him to write under the pseudonym "Student", arranged for him to spend the academic year 1906–1907 studying under Karl Pearson* at the Biometric Laboratory in University College, London. From Pearson, Gosset learned the theory of correlation and the Pearson system* of frequency curves, both of which helped him in developing statistical techniques to analyze small-sample data.

"The study of the exact distributions of statistics commences in 1908 with Student's paper *The Probable Error of a Mean*," wrote R. A. Fisher* [5, p. 22]. In this famous paper [11; 15, pp. 11–34], which came from his period at the Biometric Laboratory:

1. Gosset conjectured correctly the distribution of the sample variance s^2 in a normal sample.
2. Unaware of work by Abbe* and Helmert*, he proved that s^2 and the mean $\bar{x}$ of such a sample are uncorrelated and conjectured correctly that they are independent.
3. In essence he derived the t-distribution* named after him, as a Pearson type VII frequency curve*.

Gosset used some data which had appeared in the first volume of *Biometrika** to test the fit of the data to the theoretical distribution of $z = (\bar{x} - \mu)/s$. With the height and left-middle-finger measurements of 3000 criminals, he adopted a Monte Carlo* type of method to generate two sets of 750 "samples" of size 4, and obtained satisfactory fits for both sets of data. The frequency curves for both sets deviated slightly from normality, but Gosset noted that "This, however, appears to make very little difference to the distribution of z."

In Gosset's notation, $s^2 = \sum(x_i - \bar{x})^2/n$ for a sample of size n. It was Fisher who realized later that sampling distributions of regression coefficients could be related to z when based on a more general definition incorporating the notion of degrees of freedom*, and made the change from z to what is now called t (see Eisenhart [3]). In 1908, Gosset also correctly surmised the distribution of the sample correlation* coefficient in a bivariate normal sample when the population correlation coefficient is zero [12; 15, pp. 35–42].

Gosset did not establish his results rigorously, but his achievements initiated a period of intense research into the distributions of other sampling statistics. Fisher paid tribute to his lead in this direction [4] and acknowledged the "logical revolution" in thinking effected by his approach to research, because he sought to reason inductively from observations of data to the inferences to be made from them [1, p. 240]. But as frequently happens in scientific discovery, Gosset's work was largely unrecognized for several years; his tables of percent points of the t-distribution were used by few researchers other than those at Guinness and Rothamsted (*see* FISHER, R. A.) until the late 1920s. E. S. Pearson [8; 10, pp. 349–350] lists possible reasons for this lack of interest.

The collaboration of Gosset and Fisher in compiling, correcting, and publishing these tables lasted from 1921 to 1925, and is well documented in a collection of letters from Gosset to Fisher (see Box [2] and Gosset [6]), which also give many personal insights into the human side of Gosset's character. For example, when he was invited to referee the Master's thesis of one of Fisher's assistants, he wrote:

> I suppose they appointed me because the Thesis was about barley, so of course a brewer was required, otherwise it seems to me rather irregular. I fear that some of Miss Mackenzie's mathematics may be too 'obvious' for me. [6, letter no. 40].

The last refers to Fisher's use in mathematical work of the word "evidently," which for Gosset meant "two hours hard work before I can see why" [6, letter no. 6].

Gosset's work with the brewery led him to agricultural experiments. His work with a Dublin maltster, E. S. Beaven, and others led him to favor balanced designs. By comparing pairs of treatments in neighboring plots, he said, the correlation between them would be maximized and the error in treatment differences would be minimized. During 1936–1937 Gosset defended balanced designs such as those based on Beaven's half-drill strip method against the randomized designs of Fisher (*see* DESIGN AND ANALYSIS OF EXPERIMENTS and refs. 13, 14, and 15 [pp. 192–215]). The models over which they argued may have needed randomization* with some balance, but the practicality for ordinary farmers of performing and then analyzing balanced experiments was as important to Gosset as any theoretical consideration. His death in 1937 prevented them from resolving their differences.

Articles about "Student" by Fisher [4], McMullen and E. S. Pearson [7], and Pearson and Kendall [10, pp. 354–403] include a bibliography of his publications. In the latter Pearson quotes a letter of 1926 in which Gosset introduced the concept of "alternative hypothesis" in hypothesis testing* for the first time, and questioned the effect of nonnormality on sampling distributions; Gosset was frequently credited with being one step ahead of his contemporaries. A number of his letters also appear in Pearson [8, 9; 10, pp. 348–351, 405–408] (see also Box [1, 2] and Eisenhart [3]).

References

[1] Box, J. F. (1978). *R. A. Fisher. The Life of a Scientist*. Wiley, New York. (This biography contains insights into Gosset's influence on and friendship with Fisher.)

[2] Box, J. F. (1981). *Amer. Statist.*, **35**, 61–66. (A fascinating discussion of Gosset's correspondence and friendship with Fisher.)

[3] Eisenhart, C. (1979). *Amer. Statist.*, **33**, 6–10. (A mathematical discussion of the change from z to t.)

[4] Fisher, R. A. (1939). *Ann. Eugen.*, **9**, 1–19. (An appreciation of "Student" and the t-test.)

[5] Fisher, R. A. (1950). *Statistical Methods for Research Workers*, 11th ed. Oliver & Boyd, Edinburgh.

[6] Gosset, W. S. (1970). *Letters from W. S. Gosset to R. A. Fisher 1915–1936*. Arthur Guinness, Dublin. (Private circulation.)

[7] McMullen, L. and Pearson, E. S. (1939). *Biometrika*, **30**, 205–250. (Pearson's penetrating critique of "Student's" achievements is as relevant 40 years later as it was in 1939).

[8] Pearson, E. S. (1967). *Biometrika*, **54**, 341–355.

[9] Pearson, E. S. (1968). *Biometrika*, **55**, 445–457.

[10] Pearson, E. S. and Kendall, M. G., eds. (1970). *Studies in the History of Statistics and Probability*. Hafner, New York. (This collection of *Biometrika* reprints includes the preceding three references.)

[11] "Student" (1908). *Biometrika*, **6**, 1–25. (The classic paper introducing the t-distribution*.)

[12] "Student" (1908). *Biometrika*, **7**, 302–309.

[13] "Student" (1936). *J. R. Statist. Soc.* (Suppl.), **3**, 115–136. (Includes an illuminating discussion of balanced vs. randomized designs by Fisher, Wishart, Yates, and Beaven.)

[14] "Student" (1938). *Biometrika*, **29**, 363–379.

[15] "Student" (1942). *Collected Papers*. Biometrika, London.

(ENGLISH SCHOOL OF BIOMETRY
FISHER, RONALD AYLMER
PEARSON, KARL
t-DISTRIBUTION
t-TESTS)

CAMPBELL B. READ

GRADUATION

Graduation may be defined as the process of obtaining from an irregular set of observed values of a dependent variable a corresponding smooth set of values consistent in a general way with the observed values. Only the case of equally spaced arguments is considered here. The irregularities that it is desired to remove might include *random noise** and/or more or less systematic fluctuations such as seasonal variation*.

A time-honored and much used method of graduation is the *graphic* method, in which the observed data are plotted on graph paper and a smooth curve is drawn among them, perhaps with the help of mechanical aids. However, the discussion here will be limited to those methods in which the graduated values are obtained by some arithmetical procedure.

Graduation methods may be classified in several ways. There are *discrete* methods, in which it is desired merely to replace each observed value by a corresponding adjusted value, and *continuous* methods, in which a smooth curve (often represented mathematically by a piecewise function) is fitted to the observed data, so that the procedure yields a graduated value for every argument within a prescribed range. Such continuous methods might perhaps better be described as interpolation* with graduation. From another point of view, there are *local* methods, in which the graduated value for a given argument depends only on observed values for arguments within a stipulated distance from the given argument, and *global* methods, in which each graduated value depends on all the observed data.

These two dichotomies (discrete vs. continuous and local vs. global) give rise to four classes of graduation methods, three of which will be discussed in some detail. Continuous global methods amount to fitting an appropriate mathematical function to the data, and come under the head of curve fitting*. In this connection an important technique is the fitting of spline functions* (commonly called splines); an excellent reference is de Boor [1]. Useful sources concerning the fitting of parametric functions to demographic data are refs. 3, 9–12, and 20.

MOVING WEIGHTED AVERAGES

The most obvious discrete, local graduation method is the use of moving weighted averages. This was one of the earliest methods and is still widely used. It employs an adjustment formula of the form

$$u_x = \sum_{j=-m}^{q} c_j y_{x-j}, \tag{1}$$

where y_x is the observed value corresponding to the argument x, u_x is the corresponding graduated value, and the coefficients c_j satisfy the condition $\sum_{j=-m}^{q} c_j = 1$. Here m and q are positive integers. Usually, the formula is symmetrical, which implies that $q = m$ and $c_{-j} = c_j$ for all j. Hereafter we shall consider only symmetrical averages, so that (1) becomes

$$u_x = \sum_{j=-m}^{m} c_j y_{x-j}, \qquad c_{-j} = c_j. \tag{2}$$

It is customary to ensure a measure of fidelity to the observed data by imposing the condition that (2) be exact for polynomials of a certain degree. We say that (2) is *exact for the degree r* if the coefficients c_j are such that, for every polynomial $P(x)$ of degree r or less,

$$\sum_{j=-m}^{m} c_j P(x-j) = P(x) \tag{3}$$

for all x, but (3) is not true in general for polynomials of degree greater than r. Because of the symmetry of the coefficients, r must be odd. Usually, r is much smaller than m; in practice, $r = 3$ nearly always, while m is commonly 7 to 13. It is not difficult to see that if $r > 1$, some of the coefficients c_j must be negative. Usually, the negative coefficients are small in absolute value and occur in the "tails" (i.e., for values of j close to $\pm m$).

Since r is much smaller than the number of terms in the formula $(2m+1)$, the constraint (3) leaves a number of degrees of freedom in the choice of the coefficients c_j. These are utilized to make the formula an effective smoothing agent. One approach is to minimize some suitable measure of the "roughness" of the sequence of coefficients c_j. Such a measure [21] is R_k, the *smoothing coefficient of order* k of the formula (2), given by

$$R_k^2 = \sum_{j=-m-k}^{m} \left(\Delta^k c_j\right)^2 \Big/ \binom{2k}{k}, \tag{4}$$

where it is understood that $c_j = 0$ for $|j| > m$. For given m and r with $r < 2m$, there is, for each k, a unique set of coefficients c_j for which R_k is smallest. The graduation formulas (2) for various values of m with $r = 3$ that minimize R_3 are sometimes called [16] "Henderson's ideal" formulas. In British actuarial literature the reciprocal of R_3 is sometimes called the *power* of the formula.

Many moving-weighted-average graduation formulas have been derived on an ad hoc basis (mostly by British actuaries) and are known by the names of their originators. For example, one that has been used extensively by economic statisticians is Spencer's 15-term average, given by

$$\begin{aligned} u_x = \frac{1}{320}(&-3y_{x-7} - 6y_{x-6} - 5y_{x-5} \\ &+ 3y_{x-4} + 21y_{x-3} + 46y_{x-2} \\ &+ 67y_{x-1} + 74y_x + 67y_{x+1} \\ &+ 46y_{x+2} + 21y_{x+3} + 3y_{x+4} \\ &- 5y_{x+5} - 6y_{x+6} - 3y_{x+7}). \end{aligned} \tag{5}$$

In general, the objective has been to make R_3 small (but not minimal) and to make the coefficients c_j rational fractions with a relatively small common denominator. Other things being equal, the greater the number of terms in the formula, the more drastic is the smoothing effected by it.

Some writers (e.g., refs. 17 and 21) have considered the observed value y_x to be the sum of a "true" value U_x and an error e_x. If it is postulated that the errors are independently randomly distributed with zero mean and the same variance σ^2 for all x, the smoothed errors have the variance $R_0^2\sigma^2$, where $R_0^2 = \sum_{j=-m}^{m} c_j^2$. A good graduation formula should reduce the error variance, and therefore R_0^2 should be much less than 1. This is the case for all moving averages* used in practice. Some British actuarial writers call the reciprocal of R_0^2 the *weight* of the formula. By way of illustration, for Spencer's 15-term average R_3 is 0.0166 and R_0 is 0.4389, while for the 15-term Henderson's ideal formula R_3 is 0.0134 and R_0 is 0.4234.

I. J. Schoenberg [18] has defined the *characteristic function* of the average (1) as

$$\phi(t) = \sum_{j=-m}^{q} c_j e^{ijt},$$

where i is the imaginary unit. For a symmetrical average (2) this is a real function of the real variable t, and can be expressed in the alternative form

$$\phi(t) = c_0 + 2\sum_{j=1}^{m} c_j \cos jt.$$

He has argued persuasively that an average (1) can properly be called a "smoothing formula" only if its characteristic function satisfies $|\phi(t)| \leqslant 1$ for all t.

A serious shortcoming of the method of graduation by moving weighted averages is the fact that a symmetrical average of $2m + 1$ terms does not produce graduated values corresponding to the first m and the last m observations unless additional data are available beyond the extremities of the original data. E. L. DeForest, one of the earliest writers on the subject (see ref. 21), was acutely aware of this problem (see ref. 4) and made a suggestion for extending the graduation to the ends of the data. In the century since he wrote, other suggestions have been made (e.g., ref. 17 [p. 330]), but none has won general acceptance. Greville [4] has proposed an entirely new technique, based on the mathematical properties of the weighted average, that seems promising. Briefly, assume that the average is symmetrical, and write $f(z) = \phi(t)$, where $z = e^{it}$. Then, if the formula is exact for the degree $r = 2s - 1$, it is found that

$$f(z) = 1 - (-1)^s (z^{1/2} - z^{-1/2})^{2s} q(z)$$

for some $q(z) = \sum_{j=-m+s}^{m-s} q_j z^j$, with $q_{-j} = q_j$. We assume that $q(z)$ has no zeros on the unit circle of the complex plane (not an important restriction), so that it has $m - s$ zeros inside the unit circle and the same number outside (reciprocals of those inside). Let $p(z)$ be the polynomial of degree $m - s$ with leading coefficient unity whose zeros are the zeros of $q(z)$ that are inside the unit circle, and let

$$a(z) = (z-1)^s p(z) = z^m - \sum_{j=1}^{m} a_j z^{m-j}. \tag{6}$$

Then, if the observed values y_x range from $x = A$ to $x = B$, we calculate "extrapolated" values (which need not be realistic) for $x = A - 1$ to $A - m$ successively by the formula $y_x = \sum_{j=1}^{m} a_j y_{x+j}$ and for $x = B + 1$ to $B + m$ by the formula $y_x = \sum_{j=1}^{m} a_j y_{x-j}$. Application of the original moving weighted average to the composite sequence of observed and "extrapolated" values yields graduated values for $x = A, A + 1, \ldots, B$.

For example, in the case of Spencer's 15-term average (5),

$$f(z) = 1 - (z^{1/2} - z^{-1/2})^4 q(z)$$

with

$$q(z) = \tfrac{1}{320}(3z^{-5} + 18z^{-4} + 59z^{-3} + 137z^{-2} + 242z^{-1} + 318 + 242z + 137z^2 + 59z^3 + 18z^4 + 3z^5),$$

and, making use of (6), we find that "extrapolated" values y_x for $x = A - 1$ to

$A - 7$, inclusive, are obtained sequentially by

$$y_x = 0.961572y_{x+1} + 0.372752y_{x+2} + 0.015904y_{x+3} - 0.123488y_{x+4} - 0.125229y_{x+5} - 0.075887y_{x+6} - 0.025624y_{x+7}.$$

"Extrapolated" values for $x = B + 1$ to $B + 7$ are obtained by the identical formula except that the plus signs in the subscripts are changed to minus signs.

When this procedure is expressed in matrix-vector notation (see ref. 4), rather than in terms of "extrapolated" values, it becomes clear that the treatment of the values near the ends of the data is an integral part of a single overall operation, and not something extra grafted on at the ends.

For further information about graduation by moving weighted averages, refs. 2, 4, 7, 16–18, and 21 may be consulted; they contain further references.

DISCRETE GLOBAL METHODS

The Whittaker method* of graduation is the principal discrete, global method. It is known also as the Whittaker–Henderson method. It is based on minimization of a loss function that is a blend of a measure of departure of the graduated values from the observed data and a measure of roughness of the sequence of graduated values. This is

$$S = \sum_{x=A}^{B} W_x(u_x - y_x)^2 + g \sum_{x=A}^{B-s} (\Delta^s u_x)^2,$$

where W_x is a positive weight assigned to the particular observation and g is a positive constant chosen by the user to reflect the relative importance attached to smoothness and fidelity. Commonly, s is 2 or 3. The minimization problem becomes more tractable when S is expressed in matrix-vector notation. Thus,

$$S = (\mathbf{u} - \mathbf{y})'\mathbf{W}(\mathbf{u} - \mathbf{y}) + g(\mathbf{Ku})'\mathbf{Ku},$$

where $\mathbf{y}$ is the vector of observed values, $\mathbf{u}$ is the vector of graduated values, the "prime" denotes transposition, $\mathbf{W}$ is the diagonal matrix whose successive diagonal elements are the weights W_x, and $\mathbf{K}$ is the differencing matrix of $N = B - A + 1$ columns and $N - s$ rows that transforms a column of numbers into the column of their sth finite differences. Thus, if $\mathbf{K} = (k_{ij})$, then $k_{ij} = (-1)^{s+i-j}\binom{s}{j-i}$ if it is understood that $\binom{s}{l} = 0$ for $l < 0$ and for $l > s$. It is easily shown [4] that S is smallest when $\mathbf{Au} = \mathbf{Wy}$, where $\mathbf{A} = \mathbf{W} + g\mathbf{K}'\mathbf{K}$ is positive definite. Since $\mathbf{A}$, $\mathbf{W}$, and $\mathbf{y}$ are given, there is a unique solution for $\mathbf{u}$, the only unknown. In the "equally weighted" case, each $W_x = 1$, so $\mathbf{W} = \mathbf{I}$.

In the Whittaker method the graduation extends automatically to the extremities of the data. If, as is commonly done, the weight W_x is taken as the reciprocal of an estimate of the variance of the observation y_x, the graduated values are constrained toward the observations where these are reliable, and toward the form of a polynomial of degree $s - 1$ where the observations are dubious. Through the choice of the constant g, the user can regulate closely the degree of smoothness demanded (at the expense of fidelity).

In a generalized form of the Whittaker method, $\mathbf{W}$ is positive definite (not necessarily diagonal), $\mathbf{R}$ is positive semidefinite, and

$$S = (\mathbf{u} - \mathbf{y})'\mathbf{W}(\mathbf{u} - \mathbf{y}) + \mathbf{u}'\mathbf{Ru},$$

which is minimal for $(\mathbf{W} + \mathbf{R})\mathbf{u} = \mathbf{Wy}$. Schoenberg and Reinsch (see ref. 1) have adapted the Whittaker loss function to the fitting of a continuous curve, which turns out to be a spline*.

A somewhat related discrete, global method is the *Bayesian graduation* introduced by Kimeldorf and Jones [15], in which

$$\mathbf{u} = \mathbf{m} + (\mathbf{I} + \mathbf{BA}^{-1})^{-1}(\mathbf{y} - \mathbf{m}),$$

where $\mathbf{u}$ and $\mathbf{y}$ are as previously defined, $\mathbf{m}$ is an a priori estimate of $\mathbf{u}$, $\mathbf{A}$ is the postulated covariance matrix of the random variables whose means are the components of $\mathbf{m}$, and $\mathbf{B}$ (usually a positive definite diagonal matrix) is the "covariance matrix . . . for the graduator's model of the experiment." For further information, consult refs. 8 and 15.

CONTINUOUS LOCAL METHODS

Assuming without loss of generality that the arguments for which observations are available are consecutive integers, a graduation method of this class is based on a formula for u_{n+s} (where $0 \leqslant s \leqslant 1$) as a linear combination of $y_n, y_{n-1}, \ldots, y_{n-m+1}, y_{n+1}, y_{n+2}, \ldots, y_{n+m}$. The coefficients in the linear combination are functions of s only and do not depend on n. Usually, m is 2 or 3. Since one moves along the data with the formula from one unit interval to the next, there is a strong analogy to the moving-weighted-average method. The coefficients are chosen so that the (piecewise) curve of interpolated (graduated) values is not only continuous but has continuous first and sometimes second derivatives. The procedure has been described as "smooth-junction interpolation." Sometimes it is loosely termed "osculatory interpolation" (even when continuity is of the first order only).

An example is the formula of W. A. Jenkins [13]

$$\begin{aligned} u_{n+s} &= s\left(y_{n+1} - \tfrac{1}{36}\delta^4 y_{n+1}\right) \\ &\quad - \tfrac{1}{6}s(1-s^2)\left(\delta^2 y_{n+1} - \tfrac{1}{6}\delta^4 y_{n+1}\right) \\ &\quad + s'\left(y_n - \tfrac{1}{36}\delta^4 y_n\right) \\ &\quad - \tfrac{1}{6}s'(1-s'^2)\left(\delta^2 y_n - \tfrac{1}{6}\delta^4 y_n\right), \qquad (7) \end{aligned}$$

where $s' = 1 - s$ and δ denotes the finite difference taken centrally. The curve of interpolated values is a piecewise cubic with second-order continuity, or, in other words, a cubic spline*. This formula, published in 1927, was one of the earliest practical applications of spline functions. It is exact for cubics. Taking $s = 0$ in (7) gives $u_n = y_n - \frac{1}{36}\delta^4 y_n$ as the graduation formula (2) for the observations.

Schoenberg [18] has pointed out that an interpolation formula of the type described can be expressed in the form

$$u_x = \sum_{n=-\infty}^{\infty} L(x-n)y_n,$$

where the piecewise function $L(x)$, called the *basic function* of the interpolation formula, has its support contained in the interval $(-m, m)$. The order of continuity of the interpolating curve is that of the basic function. For a given interval of support and degree of exactness, there is (if the class of such formulas is nonempty) a unique formula for which $\int_{-m}^{m}[L^{(k)}(x)]^2\,dx$ is smallest, sometimes called the *minimized derivative formula of order k* of the class in question [6].

Consider, for example, the case $m = r = 3$, order of continuity 2. Such formulas are of the form

$$\begin{aligned} u_{n+s} &= sy_{n+1} + \tfrac{1}{6}s(s^2-1)\delta^2 y_{n+1} \\ &\quad + C(s)\delta^4 y_{n+1} + s'y_n \\ &\quad + \tfrac{1}{6}s'(s'^2-1)\delta^2 y_n + C(s')\delta^4 y_n, \end{aligned}$$

where $C(s)$ is a continuous function of S with continuous first and second derivatives in $[0, 1]$ satisfying the conditions $C(0) = C'(0) = C''(0) = 0$ and $C'(1) = -\frac{1}{12}$ (see ref. 14). If $C(s) = -\frac{1}{36}s^3$, we have Jenkins' formula (7). For the minimized derivative formula of order 3, $C(s) = \frac{1}{36}s^3(3s-5)$.

Greville and Vaughan [5] have used the theory of distributions of L. Schwartz to develop a general theory of smooth-junction interpolation formulas. For further information about such formulas, the reader may consult refs. 6, 14, and 17.

TESTS OF A SATISFACTORY GRADUATION

Two characteristics of a graduation need to be examined: smoothness and fidelity to the observed data. To a considerable extent they are antithetical: one can be improved only at the expense of the other. The degree of smoothness required is somewhat subjective and depends on the use to be made of the results. If a mathematical function involving few parameters (not a piecewise function) is fitted to the data, it may be presumed to be inherently smooth. In the case of a piecewise function, it will be useful to plot the resulting curve on graph paper, perhaps along with its first and/or second derivative. Marked undulations in the curve should be examined, unless these are a normal property of the kind of data under investigation. Also suspicious are numerically large values of the second derivative. The smoothness of graduations produced by discrete methods may be judged by tabulating first, second, and third differences of the graduated values. Second differences of large absolute value should be examined. The two quantities most commonly used to assess the relative smoothness of different graduations of the same data are $\sum(\Delta^3 u_x)^2$ and $\sum|\Delta^3 u_x|$; *see* FINITE DIFFERENCES, CALCULUS OF.

Fidelity to the observations is judged by comparing the graduated values at the data points to the corresponding observations. The deviations of graduated from observed values should be listed with their algebraic signs. If the observations can be considered statistically independent, the sum of the deviations and their first moment should be approximately zero. If this is not the case, a final adjustment of the graduation can be made by a linear transformation of the form $u'_x = au_x + b$, where u'_x is the final adjusted graduated value [17, p. 10]. Here a and b are chosen so that $\sum(u'_x - y_x) = \sum x(u'_x - y_x) = 0$. In some cases, a transformation of the data (e.g., taking logarithms if they are exponential in character) might be desirable before applying this test or adjustment. The most obvious method of testing fidelity to the data is the chi-squared test* of goodness of fit*. The mechanics of applying it in a particular case would depend on the nature and source of the data. Since the deviations are squared in applying this test, it does not take their signs into account, and these need to be examined. If there are N observations and they can be considered independent, the number of changes of sign in the deviations should be approximately $\frac{1}{2}(N-1)$. Long runs of deviations of the same sign should be questioned. Measures used to compare graduations of the same data are $\sum(u_x - y_x)^2$ and $\sum|u_x - y_x|$. Weights may be introduced, as in the Whittaker method.

Published discussions of tests of a satisfactory graduation focus primarily on graduations of (human) mortality rates by age. The interested reader may consult refs. 17 and 19.

References

[1] de Boor, C. (1978). *A Practical Guide to Splines*. Springer-Verlag, New York. (A very practical, computationally oriented treatment.)

[2] Borgan, Ø. (1979). *Scand. Actuarial J.*, 83–105. (This statistically oriented paper breaks new ground in the study of moving-weighted-average graduation.)

[3] Forsén, L. (1979). *Scand. Actuarial J.*, 167–178.

[4] Greville, T. N. E. (1981). *Scand. Actuarial J.*, 39–55, 66–81; *J. Approximation Theory*, **33**, 43–58.

[5] Greville, T. N. E. and Vaughan, H. (1954). *Soc. Actuaries Trans.*, **6**, 413–476.

[6] Greville, T. N. E. and Vaughan, H. (1982). *Utilitas Math.*, **22**.

[7] Henderson, R. (1938). *Mathematical Theory of Graduation*. Actuarial Society of America, New York. (This predecessor of ref. 17 gives more details on moving-average methods, much less on interpolation methods.)

[8] Hickman, J. C. and Miller, R. B. (1977). *Soc. Actuaries Trans.*, **29**, 7–21. (With discussion, pp. 23–49.)

[9] Hoem, J. M. (1973). *Proc. 6th Berkeley Symp. Math. Statist. Prob.*, University of California Press, Berkeley, Calif., pp. 569–600.

[10] Hoem, J. M. (1976). *Scand. J. Statist.*, **3**, 89–92.

[11] Hoem, J. M. and Berge, E. (1974). *Proc. 8th Biometrics Conf.*, Constanta, Romania, Aug., pp. 365–371.

[12] Hoem, J. M. and Berge, E. (1975). *Scand. Actuarial J.*, 129–144.

[13] Jenkins, W. A. (1927). *Trans. Actuarial Soc. Amer.*, **28**, 198–215.

[14] Kellison, S. G. (1975). *Fundamentals of Numerical Analysis*. Richard D. Irwin, Homewood, Ill. (No systematic coverage but contains useful exercises and examples.)

[15] Kimeldorf, G. S. and Jones, D. A. (1967). *Soc. Actuaries Trans.*, **19**, 66–112. (With discussion, pp. 113–127.)

[16] Macaulay, F. R. (1931). *The Smoothing of Time Series*. National Bureau of Economic Research, New York. (The classic text on smoothing econimic time series by moving averages.)

[17] Miller, M. D. (1946). *Elements of Graduation*. Actuarial Society of America and American Institute of Actuaries, New York (Society of Actuaries, Chicago, successor). (The standard text on graduation for North American actuaries.)

[18] Schoenberg, I. J. (1946). *Quart. Appl. Math.*, **4**, 45–99, 112–141. (The paper in which spline functions were first named and clearly defined.)

[19] Seal, H. L. (1943). *J. Inst. Actuaries*, **71**, 5–47. (With discussion, pp. 48–67.)

[20] Taylor, G. C. and Worcester, P. A. (1978). *J. Inst. Actuaries, Student Soc.*, **22**, 217–244.

[21] Wolfenden, H. H. (1925). *Trans. Actuarial Soc. Amer.*, **26**, 81–121. (The paper that first drew attention to De Forest's monumental but forgotten work on moving averages.)

(CURVE FITTING
FINITE DIFFERENCES, CALCULUS OF
INTERPOLATION
SPLINE FUNCTIONS
WHITTAKER METHOD)

T. N. E. GREVILLE

GRAECO-LATIN SQUARES

(This article is a continuation of D. A. Preece's article LATIN SQUARES with which it is closely linked).

Two Latin squares $A = (a_{ij})$ and $B = (b_{ij})$, $i, j = 1, 2, \ldots, n$, defined on the symbol set $\{0, 1, 2, \ldots, n-1\}$, are said to be (mutually) *orthogonal* if each of the n^2 possible ordered pairs of the symbols occurs just once among the pairs (a_{ij}, b_{ij}). For example, each pair of the squares in Fig. 2 is an orthogonal pair.

In an article published in 1782 [7], L. Euler used Roman letters for the symbols of the square A and Greek letters for the symbols of the square B. In consequence, a pair of orthogonal Latin squares is still sometimes called a *Graeco-Latin square* (sometimes also an *Eulerian square*; see Fig. 1 for an example). (When two squares are placed as in Fig. 1, they are said to be *juxtaposed* or *superimposed*).

a_α	b_β	c_γ	d_δ	e_ϵ
b_γ	c_δ	d_ϵ	e_α	a_β
c_ϵ	d_α	e_β	a_γ	b_δ
d_β	e_γ	a_δ	b_ϵ	c_α
e_δ	a_ϵ	b_α	c_β	d_γ

Figure 1

0	1	2	3
1	0	3	2
2	3	0	1
3	2	1	0

0	1	2	3
2	3	0	1
3	2	1	0
1	0	3	2

0	1	2	3
3	2	1	0
1	0	3	2
2	3	0	1

Figure 2

In the same paper, Euler showed that Graeco-Latin squares exist for all odd values of the order n (the *order* n denoting the number of symbols in each row and column of the square) and also for values of n which are multiples of 4. He was unable to construct a Graeco-Latin square of order 6 and conjectured that, for $n = 4m + 2$ ($m = 1, 2, \ldots$), no Graeco-Latin square existed. In 1900 G. Tarry proved by an enumerative method that Euler's conjecture is correct for $m = 1$; subsequently, shorter proofs of this fact have been obtained (see Dénes and Keedwell [5] for more details). In 1960, Bose et al. [2] proved by means of constructions using balanced incomplete designs* that Euler's conjecture is false for all $m > 1$. Thus Graeco-Latin squares exist for all orders n except 1, 2, and 6. Shorter and more direct proofs of this have been published, the shortest and most elegant of which the writer is aware being that of Zhu Lie [34].

Graeco-Latin squares (and sometimes sets of more than two pairwise orthogonal Latin squares) are used as experimental designs when more than the three orthogonal factors of a single Latin square (as explained in the article LATIN SQUARES) are to be taken into account. Each additional factor requires one further orthogonal Latin square. An article by Perry et al. [23] provides a fairly typical example. A table of Graeco-Latin squares for statistical purposes (of orders 3 to 12 excluding order 6) appears in Fisher and Yates [10]. If an $n \times n$ Graeco-Latin square is used as a row-and-column design* for two noninteracting sets of n treatments, the form of the analysis of variance* is as in Table 2 of LATIN SQUARES, except that there must now be a row of the table for each set of treatments, each of these rows having $n - 1$ degrees of freedom; consequently, the number of degrees of freedom for error must now be $(n - 1)(n - 3)$. Whether or not a Graeco-Latin square is used as a row-and-column design, the standard analysis assumes additivity of the effects of all four factors. Discussions of when and whether analysis of the design is possible if some observations are missing will be found in Yates [32], Nair [21], and Dodge and Shah [6].

A number of subsidiary problems present themselves. First, how many effectively different Graeco-Latin squares of a given order exist? Second, do sets of more than two pairwise orthogonal Latin squares of a given order n exist? [The maximum number of Latin squares of order n which can exist in a pairwise orthogonal set has been denoted by $N(n)$. For example, $N(4) = 3$, illustrated in Fig. 2.] Third, do pairs of orthogonal Latin squares exist which shall have some extra property? For instance, can we obtain a pair of orthogonal row-complete Latin squares? (*See* LATIN SQUARES and CHANGEOVER DESIGNS for an explanation of this concept and for the use of such squares as crossover designs or balanced changeover designs.)

As regards the first and second problems, for enumerative purposes two Latin squares are *equivalent* if one can be obtained from the other by rearranging the order of the rows, rearranging the order of the columns, or permuting the symbols (i.e., if they are in the same isotopy class; *see* LATIN SQUARES). Any single Latin square defined on the symbols $0, 1, \ldots, n - 1$ is equivalent to one (sometimes to several) in so-called *standard form*, in which the symbols which occur in the first row and first column are in natural order, like the first square in Fig. 2. If this square is orthogonal to one or more other squares, then, by permuting the symbols of each of the latter, we can replace them by squares for which the symbols of the first row (or first column, but not both) are in natural order. (Permuting the symbols of the second of a pair of orthogonal squares does not affect their orthogonality since the property that, when the squares are juxtaposed, each symbol of the first square is followed exactly once by each of the symbols of the second square, is unaffected.) The set of pairwise orthogonal Latin squares so obtained is said to be a *standardized set* (Fig. 2 provides an example). If, in Fig. 1, the symbols a, b, c, d, e are replaced by $0, 1, 2, 3, 4$, respectively, and the symbols $\alpha, \beta, \gamma, \delta, \epsilon$ likewise, then the Graeco-Latin square so obtained is in standardized form. As far as the author is aware, the problem of enumerating such standardized Graeco-Latin squares has not been much considered (see the discussion of enumeration in the article LATIN SQUARES). However, a complete enumeration of the 5×5 Graeco-Latin squares is given in Finney [8], [as orthogonal partitions of type (1^5)] and of the 7×7 Graeco-Latin squares in Norton [22] (see also the introductory section of Fisher and Yates [10]).

When the Latin squares of a pairwise orthogonal set have been arranged as a standardized set, the pairs of symbols (i, i), $i = 0, 1, \ldots, n - 1$, all occur in the cells of the first row when any pair of the set are juxtaposed. Consequently, the symbol in the second cell of the first column must be different for each square of the set and not equal to 0, because 0 is the entry in the first cell of the first column (see, e.g., Fig. 2). It follows that at most $n - 1$ Latin squares can occur in a pairwise orthogonal set of Latin squares. When this number is attained, the squares are said to form a *complete set* of pairwise orthogonal latin squares or sometimes a *hyper-Graeco-Latin square* as in Bose [1]. The squares in Fig. 2 form a complete set of pairwise orthogonal Latin squares of order 4. Evidently, such complete sets cannot exist for all orders n since, for example, there does not exist a Graeco-Latin square of order 6 (or 2). However, complete sets exist for all orders n which are prime powers and, for each such order, one complete set can be constructed in a very simple way from the Galois field* of that order. This is shown in Dénes and Keedwell [5, p. 167], and also in the earlier references given there: Bose [1] and Stevens [27]. Note that a complete set of pairwise orthogonal Latin squares of order n is equivalent to a finite projective plane with $n + 1$ points on every line or to a symmetric balanced incomplete block design* (*see* BALANCE IN EXPERIMENTAL DESIGN *and* GENERAL BALANCE) with parameters $b = v = n^2 + n + 1$, $r = k = n + 1$, and $\lambda = 1$. For values of n for which isomorphically distinct projective planes (with $n + 1$ points per line) exist, there will be complete sets of pairwise orthogonal Latin squares of more than one type. The smallest such value is $n = 9$.

It is not known for which (if any) non-prime power values of n complete sets of pairwise orthogonal Latin squares exist. However, Bruck and Ryser proved in 1949 that they do not exist if n is congruent to 1 or 2 modulo 4 and if the square-free part of the prime decomposition of n contains at least one prime factor p of the form $4m + 3$. In particular, they do not exist if $n = 6$, 14, 21, 22, or 30. If $N(n)$ denotes the maximum number of Latin squares in a pairwise orthogonal set, then $N(6) = 1$, $N(10) \geqslant 2$, $N(12) \geqslant 5$, (see Dénes and Keedwell [5, pp. 479–481]), $N(14) \geqslant 2$, $N(15) \geqslant 4$ (see Schellenburg et al. [26]). Also $N(n) \geqslant 3$ for all $n > 14$ [30]. Much work has been done on finding lower bounds for $N(n)$ for larger values of n and, although not of much statistical interest, we mention briefly some of the "best" results obtained so far. These are $N(n) \geqslant 4$ for all $n > 52$ (R. Guerin), $\geqslant 5$ for all $n > 62$ (H. Hanani), $\geqslant 6$ for all $n > 76$ (M. Wojkas), $\geqslant 7$ for all $n > 780$ (A. E. Brouwer). Of possibly more interest is the fact that, although $N(6) = 1$, there exists a pair of 6×6 Latin squares which almost form a Graeco-Latin square (see Fig. 3). The pairs of symbols $(1, 4)$ and $(3, 5)$ are missing and the pairs $(1, 5)$ and $(3, 4)$ occur twice. It is easy to prove that this result (obtained by

0_0	1_1	2_2	3_3	4_4	5_5
1_5	2_0	3_4	5_2	0_3	4_1
2_3	5_4	0_1	4_5	1_2	3_0
3_4	4_3	1_5	2_1	5_0	0_2
4_2	0_5	5_3	1_0	3_1	2_4
5_1	3_2	4_0	0_4	2_5	1_3

Figure 3

Euler) is the best possible. Also, although it is not known whether $N(10) > 2$, a triad of pairwise almost orthogonal 10×10 Latin squares has been constructed. Each two of the squares have four pairs of symbols repeated and an equal number missing (see Keedwell [20]).

A. E. Brouwer (unpublished work) has derived a quadruple of almost orthogonal 10×10 Latin squares from a so-called group divisible* partially balanced* incomplete block design on 48 elements. Each two of his squares have just two pairs of symbols repeated and two pairs missing.

It is important to note that, even though Graeco-Latin squares exist for all orders n except 2 and 6, it does not follow that all Latin squares of these orders admit orthogonal mates. In particular, if n is even, the cyclic square (*see* LATIN SQUARES) has no orthogonal mate. Freeman [11] has discussed the problem of adding a further set of treatments to a square which does not have an orthogonal mate so as to retain as much balance as possible. He has shown that the analysis of the resulting design may sometimes be simplest if $n + 1$ treatments are added.

As regards the third problem, of constructing pairs of orthogonal Latin squares with some extra property, a comprehensive account of results obtained up to 1974 appears in Dénes and Keedwell [5]. We confine ourselves here to more recent results, especially those likely to be of interest to statisticians.

A Latin square that is orthogonal to its own transpose (obtained by interchange of rows and columns) has been mistakenly called *self-orthogonal*; such squares exist for all orders n except 2, 3, and 6 (see Brayton et al. [3, 4]). Hedayat [14] has discussed the applications of such squares in designing experiments. He points out that self-orthogonal Latin squares can be used to construct experimental designs of types O : OT : TOO and O : OO : SSS (details of the construction will be found in Hedayat et al. [18]) and that they can also serve as partially replicated Latin squares designs (introduced by Youden and Hunter [33]). In Hedayat [14, 15] one self-orthogonal Latin square is listed for every order up to 20. Note that Weisner [31] was the first to exhibit a self-orthogonal Latin square of order 10, in contradiction to the statement of Hedayat [14, p. 394]. We may also ask whether a self-orthogonal Latin square of order n exists with a self-orthogonal subsquare of order k. Heinrich [19] has shown that a necessary condition is $n \geqslant 3k + 1$. She has also proved that if $n = 3k + 1$, such squares exist for all k except 2, 3, and 6; if $n = 3k + 2$, such squares exist for all odd $k \neq 3$; if $n = 3k + 3$, such squares exist for $4 \leqslant k \leqslant 21$, $k \neq 6$. Drake and Lenz have recently proved that the condition $n \geqslant 4k + 3$ is sufficient for all $k \geqslant 304$.

Row complete Latin squares are defined in LATIN SQUARES, and their use as crossover designs mentioned. Recent work of Heinrich and Keedwell has shown that row complete Graeco-Latin squares exist. Precisely, if $n = pq$, where p and q are distinct primes such that $q = 2ph + 1$ for some integer h and such that 2 is a primitive root of the prime p, then there exist $p - 1$ pairwise orthogonal row complete Latin squares of order n. In particular, there is a row complete Graeco-Latin square of order 21, the smallest order to which the result is applicable.

Also, in LATIN SQUARES, *Knut Vik designs* (sometimes called *totally diagonal Latin squares*) are defined. The name Knut Vik design for a Latin square of order n in which each of the n symbols occurs exactly once in each of the n left and right (broken) diagonals was introduced by Hedayat and Federer [17] because Vik [29] had proposed the use of such a design for a field experiment in order that the different replicates for the same treatment should be removed as far from one another as possible. In fact, Vik had proposed the use of a knight's move Latin square of order five for this purpose but for that order the concepts of knight's move Latin square and totally diagonal Latin square coincide. (The squares which define a knight's tour form what Hedayat [16] has called a *super diagonal*.) Hedayat [16] and Hedayat and Federer [17] have shown that i) Knut Vik designs exist if and only if n is not divisible by 2 or 3; ii) there exist at most $n - 3$ pairwise orthogonal Knut Vik designs of order n, and, if n is a prime, this number of pairwise orthogonal designs can be attained; iii) for every order n, for which Knut Vik designs exist, at least one pair of orthogonal designs (i.e., one totally diagonal Graeco-Latin square) can be constructed. An example of a pair of orthogonal Knut Vik designs of order 5 is given in Fig. 4.

Finally, we mention *symmetric Latin squares*. These are left unaffected if the roles of rows and columns are interchanged. Consequently, if the order of such a square is an odd integer, the symbols of its main left-to-right diagonal must be all different. If it is an even integer, each element that occurs on the main diagonal at all must occur on it an even number of times (see Dénes and Keedwell [5, p. 31]). Two symmetric Latin squares cannot be orthogonal. However, two such squares are said to be *perpendicular* if, when they are juxtaposed, no ordered pair of the symbols occurs more than once on or above the main diagonal and if a total of $\frac{1}{2}n(n + 1)$ ordered pairs of distinct symbols occur all together. Such pairs of perpendicular symmetric Latin squares exist for all odd orders n except 3 and 5.

0_0	1_1	2_2	3_3	4_4
2_3	3_4	4_0	0_1	1_2
4_1	0_2	1_3	2_4	3_0
1_4	2_0	3_1	4_2	0_3
3_2	4_3	0_4	1_0	2_1

Figure 4

Let $\nu(n)$ denote the maximum number of pairwise perpendicular symmetric Latin squares of the odd order n. A table of lower bounds for $\nu(n)$ has been given by Gross et al. [12]. In particular, it is known that $\nu(7) \geqslant 3$ and that $\nu(11) \geqslant 5$.

Latin cubes of the first and second orders are defined in LATIN SQUARES. It is also possible to define *orthogonal latin cubes* (Graeco–Latin cubes); for details see Dénes and Keedwell [5, pp. 187–189]. Orthogonal Latin cubes have been proposed for use as experimental designs. A bibliography on this topic has been prepared by Preece [24, 25]. However, we mention especially Fisher [9] and a series of papers by E. V. Markova.

LITERATURE

For the sources of matters mentioned above without an explicit reference, see Dénes and Keedwell [5]. Useful alternative references are Hall [13] and Vajda [28].

References

[1] Bose, R. C. (1938). *Sankhyā*, **3**, 323–338. (Of historical significance. This paper and, independently, W. L. Stevens [27] showed that complete sets of mutually orthogonal Latin squares exist for all prime power orders and gave a simple construction for such sets.)

[2] Bose, R. C., Shrikhande, S. S., and Parker, E. T. (1960). *Canad. J. Math.*, **12**, 189–203. (Contains the mathematical proof that Graeco-Latin squares exist for all orders except 2 and 6.)

[3] Brayton, R. K., Coppersmith, D., and Hoffman, A. J. (1974). *Bull. Amer. Math. Soc.*, **80**, 116–118.

[4] Brayton, R. K., Coppersmith, D., and Hoffman, A. J. (1976). Colloq. Int. Teorie Comb., Rome, 1973; *Atti dei Convegni Lincei*, No. 17, Vol. II, Roma, pp. 509–517.

(The second paper by Brayton et al. contains proofs of the results announced in the first: namely, self-orthogonal Latin squares exist of all orders except 2, 3, and 6.)

[5] Dénes, J. and Keedwell, A. D. (1974). *Latin Squares and Their Applications*. Akadémiai Kiadó, Budapest/English Universities Press, London-/Academic Press, New York. (This is a main reference source of information on the topic of the title and contains a comprehensive bibliography both of mathematical and statistical papers published up to 1974.)

[6] Dodge, Y. and Shah, K. R. (1977). *Commun. Statist. A*, **6**(15), 1465–1472. (Discusses estimation of parameters in Latin squares and Graeco-Latin squares with missing observations.)

[7] Euler, L. (1782). *Verh. Zeeuwsch Genootsch. Wet. Vlissengen*, **9**, 85–239. (Modern reprint: *Leonhardi Euleri Opera Omnia, Série 1*, **7**, 291–392, 1923.) (The earliest substantive work on Latin squares and still interesting.)

[8] Finney, D. J. (1946). *Ann. Eugen.*, **13**, 1–3.

[9] Fisher, R. A. (1945). *Ann. Eugen.*, **12**, 283–290. (Proposes orthogonal Latin cubes as appropriate designs for certain kinds of investigation.)

[10] Fisher, R. A. and Yates, F. (1963). *Statistical Tables for Biological, Agricultural and Medical Research*, 6th ed. Oliver & Boyd, Edinburgh (also reprinted by Longmans, London). (A standard work which contains detailed explanations of how to use the tables when designing an experiment.)

[11] Freeman, G. H. (1964). *Biometrics*, **20**, 713–729.

[12] Gross, K. B., Mullin, R. C., and Wallis, W. D. (1973). *Utilitas Math.*, **4**, 239–251. (Of mainly mathematical interest.)

[13] Hall, M. (1967). *Combinatorial Theory*. Blaisdell, Toronto. (This is a useful and well-written reference source for information on all aspects of its title suitable for more mathematically inclined readers.)

[14] Hedayat, A. (1973). *Biometrics*, **29**, 393–396.

[15] Hedayat, A. (1975). *Biometrics*, **31**, 755–759.

(The first of the above two papers discusses the existence of self-orthogonal Latin squares and their value as experimental designs. The second summarizes the results of Brayton, Coppersmith, and Hoffman, [3] and [4], and gives tables of self-orthogonal Latin squares of orders up to 20.)

[16] Hedayat, A. (1977). *J. Comb. Theory A*, **22**, 331–337.

[17] Hedayat, A. and Federer, W. T. (1975). *Ann. Statist.*, **3**, 445–447.

(Taken together, the two papers above provide a complete solution to the existence and nonexistence of Knut Vik designs and orthogonal Knut Vik designs.)

[18] Hedayat, A., Parker, E. T., and Federer, W. T. (1970). *Biometrika*, **57**, 351–355. (Concerns experimental designs suitable for conducting a sequence of experiments on the same material but with changing treatments.)

[19] Heinrich, K. (1977). *Ars Comb.*, **3**, 251–266.

[20] Keedwell, A. D. (1980). *Ars Comb.*, **9**, 3–10.

(The two papers above are of mainly mathematical interest.)

[21] Nair, K. R. (1940). *Sankhyā*, **4**, 581–588.

[22] Norton, H. W. (1939). *Ann. Eugen.*, **9**, 269–307. (A much-quoted paper which enumerates the Latin squares and Graeco-Latin squares of order 7.)

[23] Perry, J. N., Wall, C., and Greenway, A. R. (1980). *Ecol. Entomol.*, **5**, 385–396. (Discusses in detail the application of Latin square and Graeco-Latin square designs in experiments for comparing different insect sex attractants. Also discusses the advantages and disadvantages of such designs.)

[24] Preece, D. A. (1975). *Aust. J. Statist.*, **17**, 51–55.

[25] Preece, D. A. (1979). *Aust. J. Statist.*, **21**, 170–172.

(The two papers above provide a comprehensive bibliography of designs for experiments in three dimensions.)

[26] Schellenburg, P. J., van Rees, G. H. T., and Vanstone, S. A. (1978). *Ars Comb.*, **6**, 141–150. (Of mainly mathematical interest.)

[27] Stevens, W. L. (1939). *Ann. Eugen.*, **9**, 82–93. (See R. C. Bose [1].)

[28] Vajda, S. (1967). *The Mathematics of Experimental Design. Incomplete Block Designs and Latin Squares*. Griffin Statist. Monogr. Charles Griffin, London. (A useful and easy-to-read reference book.)

[29] Vik, K. (1924). *Meldinger fra Norges Landbrukshøiskole*, **4**, 129–181. (The original paper proposing use of so-called Knut Vik squares.)

[30] Wang, S. P. and Wilson, R. M. (1978). *Proc. 9th S. E. Conf. Combinatorics, Graph Theory and Computing*, p. 688.

[31] Weisner, L. (1963). *Canad. Math. Bull.*, **6**, 61–63.

(The two papers above are of mainly mathematical interest.)

[32] Yates, F. (1933). *Emp. J. Exper. Agric.*, **1**, 129–142. (Discusses the analysis of replicated experiments when the field results are incomplete.)

[33] Youden, W. J. and Hunter, J. P. (1955). *Biometrics*, **11**, 399–405. (Concerns the analysis of partially replicated Latin squares.)

[34] Zhu, L. (1977). *Acta Math. Appl. Sin.*, No. 3, 56–61 (in Chinese. Edited English translation, 1982, in *Ars Combinatoria*.)

(BALANCED INCOMPLETE BLOCK DESIGN
CHANGEOVER DESIGNS
KNUT VIK DESIGN
LATIN SQUARES
RANDOMIZATION
RANDOMIZED BLOCK DESIGN)

A. D. KEEDWELL

GRAM–CHARLIER SERIES

The Swedish astronomer Charlier [4] coined the terms Type A and Type B to denote certain orthogonal expansions* of frequency functions

$$f_X(x) = \phi(x) \sum_{r=0}^{\infty} c_r p_r(x)/r! \qquad (1)$$

based on the fundamental normal* and Poisson* limit laws, respectively. In the Type A series for a standardized variate X,

$$\phi(x) = (2\pi)^{-1/2} e^{-x^2/2} \qquad (-\infty < x < \infty),$$

$$p_r(x) = (-1)^r \frac{d^r\phi(x)}{dx^r} \Big/ \phi(x) = H_r(x), \qquad (2)$$

while for Type B, with Poisson parameter λ,

$$\phi(x) = e^{-\lambda}\lambda^x/x! \qquad (x = 0, 1, 2, \dots)$$

$$p_r(x) = \frac{\partial^r\phi(x)}{\partial\lambda^r} \Big/ \phi(x) = G_r(\lambda, x). \qquad (3)$$

Charlier [7] also defined a Type C expansion,

$$f_X(x) = \exp\left\{ \sum_{r=0}^{\infty} a_r H_r(x) \right\}. \qquad (4)$$

EXPANSIONS OF CHEBYSHEV AND GRAM

The problem of developing an arbitrary $f(x)$ in an orthogonal expansion (1) was formulated by Chebyshev* [8], in connection with fitting an approximation function to a given set of points by weighted least squares*. The appropriate orthogonal polynomials in any situation appeared as denominators in the expansion of a continued fraction. Chebyshev [9] derived the Type A form as a special case. His use of the $H_n(x)$ thus preceded Hermite's [24], whence Gnedenko and Kolmogorov [22] term them Chebyshev–Hermite polynomials*. However, they were certainly known to Laplace* [26]. The Charlier polynomials $G_r(\lambda, x)$ in (3) are also implicit in Chebyshev's work.

Gram [23] essentially reformulated Chebyshev's theory, giving the orthogonal polynomials in determinantal form. There appears to be little justification for associating his name with the series.

The Type A expansion may also be approached by means of the characteristic function* $C(t)$ of $f_X(x)$. If $\kappa_1 = \mu$, $\kappa_2 = \sigma^2$, $\kappa_3, \kappa_4, \dots$ are the cumulants* of X, and $\gamma(t)$ denotes the characteristic function of the normal $N(\mu, \sigma^2)$ distribution, then

$$C(t) = \gamma(t)\exp\left\{\sum_{r=3}^{\infty} \kappa_r (it)^r/r!\right\}$$

$$= \gamma(t)\sum_{r=0}^{\infty} c_r (it)^r/r!, \quad i = \sqrt{-1}, \quad (5)$$

and the Type A series follows by inversion. The c_r are thus obtained by setting $\kappa_1 = \kappa_2 = 0$ in the formulae for raw moments in terms of cumulants,

$$c_0 = 1, \quad c_1 = c_2 = 0, \quad c_3 = \kappa_3, \quad c_4 = \kappa_4,$$

$$c_5 = \kappa_5, \quad c_6 = \kappa_6 + 10\kappa_3^2, \dots . \quad (6)$$

Chebyshev [10] used the foregoing approach to derive the Type A series for X, where $X = n^{-1/2}\sum_{j=1}^{n} U_j$ and the U_j are independent variates with zero means. If the cumulants of U_j are $\{\kappa_r^{(j)}\}$, then X has cumulants

$$\mu = 0, \sigma^2 = n^{-1}\sum_{j=1}^{n} \kappa_2^{(j)},$$

$$\kappa_r = \alpha_r / n^{(1/2)r-1} \quad (r = 3, 4, \dots), \quad (7)$$

where $\alpha_r = n^{-1}\sum_{j=1}^{n} \kappa_r^{(j)}$. Substitution in (6) yields Chebyshev's c_r.

CURVE FITTING: CHARLIER AND EDGEWORTH

Thiele [30] and Bruns [2] proposed the Type A series as a generalization of the normal curve for representing variation in statistical populations. However, detailed statistical application of the expansion commenced in 1905 with Charlier and Edgeworth. Both appealed to the "hypothesis of elementary errors," which suggested that observed variation results from small additive and independent elementary errors. Their mathematical treatment thus roughly followed Chebyshev [10]. Charlier's [3] original presentation was faulty (see footnote in Charlier [5]), but he obtained correctly

$$f_X(x) = \phi(x) - \frac{c_3}{3!}\phi^{(3)}(x) + \frac{c_4}{4!}\phi^{(4)}(x)\dots, \quad (8)$$

where $\phi(x)$ here denotes the $N(\mu, \sigma^2)$ density. From (6), c_3 and c_4 are related to the skewness* and excess, respectively.

The Type B expansion represented Charlier's [4] initial attempt to deal with highly skew curves, for which Type A is unsuited. His approach was based on a variant of the "hypothesis of elementary errors" in which each error could take only two values 0 and α, with probabilities near 1 and 0, respectively. His expansion follows heuristically on the lines of (5). Let $f_X(x)$ denote a frequency function over $x = 0, 1, 2, \dots$ and $C(t)$ its characteristic function; then

$$f_X(x) = (2\pi)^{-1}\int_{-\pi}^{\pi} C(t)e^{-ixt}\,dt.$$

If $Q(s)$ denotes the descending factorial moment generating function*, and $\kappa_{(1)} = \lambda$, $\kappa_{(2)}, \dots$ are the factorial cumulants, then

$$C(t) = Q(e^{it} - 1)$$

$$= \exp\{\lambda(e^{it} - 1)\}\exp\left\{\sum_{r=2}^{\infty} \kappa_{(r)}(e^{it} - 1)^r/r!\right\}$$

$$= \exp\{\lambda(e^{it} - 1)\}\sum_{r=0}^{\infty} b_r(e^{it} - 1)^r/r!,$$

say. Hence

$$f_X(x) = \sum_{r=0}^{\infty} b_r(-1)^r \nabla^r \phi(x)/r!, \quad (9)$$

where

$$\nabla g(x) = g(x) - g(x-1),$$

$$\phi(x) = (2\pi)^{-1}\int_{-\pi}^{\pi} \exp\{\lambda(e^{it} - 1) - ixt\}\,dt. \quad (10)$$

Clearly, the b_r are obtained by setting $\kappa_{(1)} = 0$ in the formulae for factorial moments in terms of the $\kappa_{(r)}$. When x is an integer, $\phi(x) = e^{-\lambda}\lambda^x/x!$ and (9) reduces to (1) with (3) substituted. But Charlier also attempted to extend (10) to all $x \geqslant 0$, obtaining

$$\phi(x) = e^{-\lambda}\frac{\sin \pi x}{\pi}\sum_{r=0}^{\infty}\frac{(-\lambda)^r}{r!(x-r)}.$$

However, it is not clear under what conditions this represents a nonnegative density.

In subsequent papers, Charlier described practical procedures for fitting Type A and B series to real data.

Edgeworth [15] took the important step of rearranging the Type A series for the sum of elementary errors in its correct form as an asymptotic expansion. Pointing out that the c's given by (6) and (7) are not of decreasing order in $n^{1/2}$, he grouped terms of like order; to order n^{-1},

$$f_X(x) = \phi(x) - \frac{c_3}{3!}\phi^{(3)}(x) + \left[\frac{c^4}{4!}\phi^{(4)}(x) + \frac{1}{2}\left(\frac{c_3}{3!}\right)^2\phi^{(6)}(x)\right], \qquad (11)$$

(compare with (8); *see* CORNISH–FISHER /EDGEWORTH EXPANSIONS). Gnedenko and Kolmogorov [22] attribute Edgeworth's expansion to Chebyshev [10], but as indicated above, Chebyshev gave his series in Type A form without explicitly recommending the rearrangement. The usual association of Edgeworth with (11) thus appears justified.

The Charlier and Edgeworth expansions quickly found critics and supporters. Karl Pearson* ([27]; footnote to Elderton [16]) maintained that the "hypothesis of elementary errors" was inappropriate as a model for biological variation. He insisted that the elementary errors were correlated, rather than independent, and that his own system of skew curves reflected this requirement (*see* FREQUENCY CURVES, SYSTEMS OF). Pearson also warned of the risks involved in using sample moments beyond the fourth. Elderton [16] noted some of the now familiar difficulties in Charlier's approach, including the possibility of negative frequencies with Type A series. Further discussion and examples are given by Elderton and Johnson [17]. Edgeworth [15] himself noted that the expansions may give poor results at the "tails." Barton and Dennis [1] discussed the skewness and kurtosis regions in which (8) and (11) give unimodal and/or positive curves. A number of Scandinavian statisticians defended Charlier's methods, notably A. Fisher [18]. Wicksell [31] sought to develop the "hypothesis of elementary errors" to meet objections. Charlier [7] proposed his Type C series (4) to obtain a positive density in which the a_r appear to be of decreasing order in $n^{1/2}$, following a_2.

Using a generalization of (5), Charlier [6] gave the straightforward extension of the Type A series to the multivariate case.

PROBLEMS OF CONVERGENCE

The mathematical problem of the convergence of Type A expansions appears to have been discussed as early as 1905–1906 (see Hille [25] for references). Cramér showed that if $f(x)$ is of bounded variation in $(-\infty, \infty)$, and $e^{x^2/4}f(x)$ has a convergent integral, then the series converges to $f(x)$ at every continuity point. However, Cramér emphasized that the statistically important problem is whether the early terms give a satisfactory approximation and in 1928 [12] he gave conditions for the asymptotic validity of the Edgeworth and Type A series.

APPLICATIONS IN ROBUSTNESS STUDIES

Edgeworth's expansion has proved more useful in statistics than Charlier's version. However, the Type A form has advantages as an intermediate analytical step, for example, in the theoretical study of robustness*. Expansions for basic statistics have been presented for Type A and Edgeworth populations by Quensel [28] and Gayen [19–21] in particular. Davis [14] has given a formal approach to these expansions. Heuristically, (5) yields a decomposition $X = U + Z$, where U has the $N(\mu, \sigma^2)$ density $\phi(x \mid \mu, \sigma^2)$, and Z is an independent pseudo-variate with mean and variance zero, and the same higher-order cumulants as X. Hence

$$f_X(x) = E_Z\phi(x \mid \mu + Z, \sigma^2), \qquad (12)$$

which yields the Type A expansion when developed as a Taylor series in Z, since $E_Z(Z^r) = c_r$. The distribution of a statistic in Type A populations may thus be formally constructed from an appropriate normal-theory noncentral distribution by "averaging" over the Z's for a sample.

For further historical aspects of Gram–Charlier expansions, see Särndal [29] and Cramér [13].

References

[1] Barton, D. E. and Dennis, K. E. R. (1952). *Biometrika*, **39**, 425–427.

[2] Bruns, H. (1898). *Philos. Studien*, **14**, 339–375.

[3] Charlier, C. V. L. (1905). *Ark. Mat. Astron. Fys.*, **2**(8).

[4] Charlier, C. V. L. (1905). *Ark. Mat. Astron. Fys.*, **2**(15).

[5] Charlier, C. V. L. (1914). *Ark. Mat. Astron. Fys.*, **9**(25).

[6] Charlier, C. V. L. (1914). *Ark. Mat. Astron. Fys.*, **9**(26).

[7] Charlier, C. V. L. (1928). *Lunds Univ. Arsskr., N.F. Afd. 2*, **24**(8).

[8] Chebyshev, P. L. (1858). *J. Math. Pures Appl.*, **3**, 289–323.

[9] Chebyshev, P. L. (1860). *Bull. Acad. Imp. Sci. St. Petersbourg*, **1**(3), 193–200.

[10] Chebyshev, P. L. (1890). *Acta Math.*, **14**, 305–315.

[11] Cramér, H. (1925). *Trans. 6th Congr. Scand. Math.*, 399–425.

[12] Cramér, H. (1928). *Skand. Aktuarietidskr.*, **11**, 13–74, 141–180.

[13] Cramér, H. (1972). *Biometrika*, **59**, 205–207.

[14] Davis, A. W. (1976). *Biometrika*, **63**, 661–670.

[15] Edgeworth, F. Y. (1905). *Trans. Camb. Philos. Soc.*, **20**, 36–65, 113–141.

[16] Elderton, W. P. (1906). *Biometrika*, **5**, 206–210.

[17] Elderton, W. P. and Johnson, N. L. (1969). *Systems of Frequency Curves*. Cambridge University Press, Cambridge, England.

[18] Fisher, A. (1922). *Frequency Curves*. Macmillan, New York.

[19] Gayen, A. K. (1949). *Biometrika*, **36**, 353–369.

[20] Gayen, A. K. (1950). *Biometrika*, **37**, 236–255.

[21] Gayen, A. K. (1951). *Biometrika*, **38**, 219–247.

[22] Gnedenko, B. V. and Kolmogorov, A. N. (1968). *Limit Distributions for Sums of Independent Random Variables*, 2nd ed. Addison-Wesley, Reading, Mass.

[23] Gram, J. P. (1883). *J. Reine angew. Math.*, **94**, 41–73.

[24] Hermite, C. (1864). *C. R. Acad. Sci. Paris*, **58**, 93–100, 266–273.

[25] Hille, E. (1926). *Ann. Math.*, **27**, 427–464.

[26] Laplace, P. S. (1812). *Théorie analytique des probabilités*. Paris.

[27] Pearson, K. (1905). *Biometrika*, **4**, 169–212.

[28] Quensel, C. E. (1938). *Lunds Univ. Arsskr. N. F. Afd. 2*, **34**(4).

[29] Särndal, C.-E. (1971). *Biometrika*, **58**, 375–391.

[30] Thiele, T. N. (1889). *Forelaesninger over almindeling iaktlagelseslaere*. Gads, Copenhagen.

[31] Wicksell, S. D. (1917). *Ark. Mat. Astron. Fys.*, **12**(20).

(CORNISH–FISHER/EDGEWORTH EXPANSIONS
ORTHOGONAL EXPANSIONS)

A. W. DAVIS

GRAMIAN DETERMINANTS

Given a sequence of n-dimensional vectors

$$\boldsymbol{\alpha}_i = (a_{i1}, a_{i2}, \ldots, a_{in}); \qquad i = 1, \ldots, n,$$

let $(\boldsymbol{\alpha}_i, \boldsymbol{\alpha}_j)$ denote the inner product of the vectors $\boldsymbol{\alpha}_i$ and $\boldsymbol{\alpha}_j$. The determinant

$$\begin{vmatrix} (\boldsymbol{\alpha}_1, \boldsymbol{\alpha}_1) & (\boldsymbol{\alpha}_1, \boldsymbol{\alpha}_2) & \cdots & (\boldsymbol{\alpha}_1, \boldsymbol{\alpha}_n) \\ (\boldsymbol{\alpha}_2, \boldsymbol{\alpha}_1) & (\boldsymbol{\alpha}_2, \boldsymbol{\alpha}_2) & \cdots & (\boldsymbol{\alpha}_2, \boldsymbol{\alpha}_n) \\ & & \cdots & \\ (\boldsymbol{\alpha}_n, \boldsymbol{\alpha}_1) & (\alpha_n, \alpha_2) & \cdots & (\alpha_n, \alpha_n) \end{vmatrix}$$

$$= \begin{vmatrix} a_{11} & a_{12} & \cdots & a_{1n} \\ a_{21} & a_{22} & \cdots & a_{2n} \\ & & \cdots & \\ a_{n1} & a_{n2} & \cdots & a_{nn} \end{vmatrix}^2$$

is called the *Gramian* determinant (or simply Gramian) of the vectors α_i.

In statistical methodology Gramians are used for computational techniques for fitting multiple regressions* (see, e.g., Seber [22, Chap. 11]).

Gramian determinants of a special kind—used in distribution theory in connection with singular distributions*—are of the form

$$G(f_1, \ldots, f_n) = \begin{vmatrix} (1,1) & \cdots & (1,n) \\ (2,1) & \cdots & (2,n) \\ \vdots & & \vdots \\ (n,1) & \cdots & (n,n) \end{vmatrix}$$

where $(j,k) = \int_a^b f_j(x) f_k(x)\,dx$; j, $k = 1$,

$\ldots, n$, and $f_i(x)$ $(i = 1, \ldots, n)$ are n continuous functions defined on an interval $[a, b]$ in R^1. The two main properties of $G(f_1, \ldots, f_n)$ are:

(a) $G(f_1, \ldots, f_n) \geqslant 0$.

(b) $G(f_1, \ldots, f_n) = 0$ if and only if $f_1, \ldots, f_n$ are linearly dependent.

The latter property is used in characterizing singular distributions (see also ref. 1).

References

[1] Cramér, H. (1946). *Mathematical Methods of Statistics*. Princeton University Press, Princeton, N.J.

[2] Seber, G. A. F. (1977). *Linear Regression Analysis*. Wiley, New York.

GRAM–SCHMIDT ORTHOGONALIZATION

The Gram–Schmidt orthogonalization process is designed to determine an orthonormal* set of vectors $\mathbf{V}_i$ $(i = 1, \ldots, p)$ which forms a basis* for the space spanned by the columns $\mathbf{x}_i$ of an $n \times p$ matrix $\mathbf{X}$. In other words, constants a_{ij} $(i = 1, \ldots, p,\ j = 1, \ldots, p)$ are determined such that

$$\mathbf{v}_1 = \frac{1}{a_{11}}\mathbf{x}_1,$$

$$\mathbf{v}_2 = \frac{1}{a_{22}}\mathbf{x}_2 - \frac{a_{12}}{a_{22}}\mathbf{v}_1,$$

$$\vdots \qquad\qquad \vdots$$

$$\mathbf{v}_p = \frac{1}{a_{pp}}\mathbf{x}_p - \frac{a_{1p}}{a_{pp}}\mathbf{v}_1 - \cdots - \frac{a_{p-1,p}}{a_{pp}}\mathbf{v}_{p-1} \qquad (a_{ii} > 0).$$

Alternatively,

$$\mathbf{x}_1 = a_{11}\mathbf{v}_1,$$

$$\mathbf{x}_2 = a_{12}\mathbf{v}_1 + a_{22}\mathbf{v}_2,$$

$$\vdots \qquad\qquad \vdots$$

$$\mathbf{x}_p = a_{1p}\mathbf{v}_1 + a_{2p}\mathbf{v}_2 + \cdots + a_{pp}\mathbf{v}_p.$$

Thus $\mathbf{X}$ can be expressed as $\mathbf{X} = \mathbf{V}_p\mathbf{A}$ where $\mathbf{A}$ is a $p \times p$ upper triangular matrix, and $\mathbf{V}_p$ is a $n \times p$ matrix with *orthonormal* columns. Moreover,

$$\mathbf{X} = \mathbf{V}_p\mathbf{A} = \mathbf{V}_p\mathbf{D}\mathbf{A} = \mathbf{R}_p\mathbf{A},$$

where $\mathbf{D}$ is the diagonal matrix with the diagonal entries $a_{11} \cdots a_{pp}$; $\mathbf{A}$ is an upper triangular matrix with diagonal elements and $\mathbf{R}_p$ is a $n \times p$ matrix with orthonormal columns satisfying $\mathbf{R}_p'\mathbf{R}_p = \mathbf{D}$.

The Gram–Schmidt procedure is a useful auxiliary computational tool for fitting linear multiple regression* models. (See, e.g., Seber [4, Chap. 11].) Two basic algorithms for transforming $\mathbf{X}$ to $\mathbf{R}_p$ are known in statistical literature: the classical Gram–Schmidt algorithm (CGSA) and the modified Gram–Schmidt algorithm (MGSA), described in detail in Golub [3] and Farebrother [2], among others. Theoretical investigations by Björck [1] which compare accuracy and stability of these algorithms yield the conclusion that the second method is preferable computationally, especially if $\mathbf{X}$ is ill conditioned*.

References

[1] Björck, A. (1967). *BIT*, **7**, 1–21.

[2] Farebrother, R. W. (1974). *Appl. Statist.*, **23**, 470–476.

[3] Golub, G. H. (1969). In *Statistical Computation*, R. C. Milton and J. A. Nedler, eds. Academic Press, New York, pp. 365–397.

[4] Seber, G. A. F. (1977). *Linear Regression Analysis*. Wiley, New York.

(DOOLITTLE METHOD
GEOMETRY OF STATISTICS
LINEAR ALGEBRA, COMPUTATIONAL
MATRIX DECOMPOSITION
REGRESSION ANALYSIS)

GRAPHICAL REPRESENTATION, COMPUTER AIDED

Computer-drawn graphs or pictures are used in many situations not primarily related to mathematics or statistics. A cartoonist can produce an animated movie from pictures drawn by a computer. An engineer can design a car, building, or electrical circuit while sitting at a computer graphics terminal. An artist can create a "painting" at a color terminal. A pilot can fly a pseudo-airplane while watching a computer-drawn scenario. A ticket agency employee can sell theater tickets, referring to a map that is being dynamically updated by a computer to show seats currently available. A scientist can view a computer-drawn model of a galaxy or a molecule.

Computer graphics is also being utilized effectively by statisticians. Computer graphs to be discussed are those of a statistical and/or mathematical nature, that is, graphs that display statistical data or some function of the data (such as histograms*), and graphs representing a theoretical concept (such as a plot of a probability density function).

Computers are used to draw graphs for the same reasons that they are used numerically to analyze data; they are fast and accurate (*see* COMPUTERS AND STATISTICS). A graph may be physically complicated to draw, it may display a large amount of data, or the mathematical definition of the graph may contain complicated formulas. Large numbers of graphs may be required during the exploration of data when many different techniques or slight variations of a given technique are used, or during the analysis of many sets of data. Graphs may be needed quickly, in real time, while the user waits to make a decision. Computers can help in all these cases, examples of which are given below. Following these, some specialized applications of computer graphics are described, and hardware and software for computer graphics are discussed briefly.

For surveys of historical and recent activities in graphics, including computer graphics, see Feinberg and Franklin [17], Beniger and Robyn [8], Fienberg [19], and Izenman [29] (*see also* GRAPHICAL REPRESENTATION OF DATA).

EXAMPLES

Chernoff faces* are physically complicated graphs to draw. An example in CHERNOFF FACES shows 24 cartoon faces, each depicting measurements of 18 variables. Facial features such as the shape of the head, eye size, curvature of the mouth, etc., are functions of the measurements. An artist would be slow at producing these faces and would find it extremely difficult to show variations due to the data without introducing variations not due to the data.

Figure 1 is a graph of data from the U.S. draft lottery of 1970. The y-coordinates of the asterisks on the plot are birth dates (values between 1 and 366), supposedly drawn randomly from a box to determine an order for drafting people. The x-coordinates are the number of the draw on which the corresponding birth date was picked. Superimposed are three jagged lines, each a scatter plot* of 366 points connected by straight lines; using a complicated smoothing technique (see Cleveland and Kleiner [13] and Gentleman [23]), the upper, middle, and lower portions of the distribution of birth dates have been described as a function of a set of 366 smoothed x-coordinates. Thus Fig. 1 consists of four superimposed scatter plots; it contains $4 \times 366 = 1464$ points defined by $6 \times 366 = 2196$ coordinate values. The smoothing of the original scatter plot facilitates perception of nonrandomness in the selection of birth dates; days late in the year tended to be picked at an early draw (see Fienberg [18] and Gentleman [22]). The computer can perform the necessary calculations and draw the 1464 points in a few seconds. The smoothing technique (*see* GRADUATION) itself depends on a parameter (the number of points to be smoothed at a time, here 50), so in an exploratory mode, the computer might be called on to draw the plot several times while the user varied the parameter and watched for a satisfactory degree of smoothing to be achieved.

Figure 1 is not an extreme example of the computer facilitating the plotting of large amounts of data. The *Washington Post* Magazine, in a feature article "Computer Graphics" [32], says that "in many cases a picture on a computer graphics terminal can contain more information—information readily identifiable and understandable—than

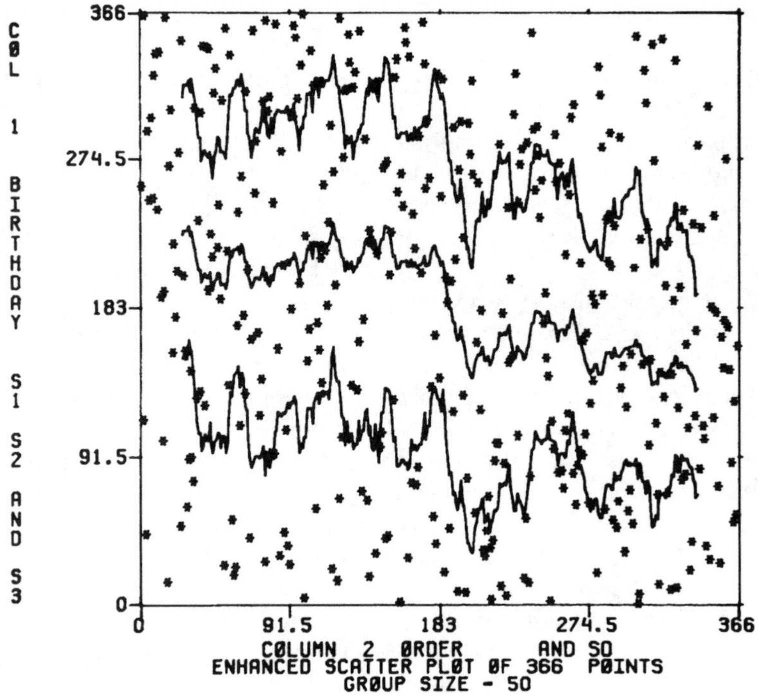

Figure 1 Smoothed scatter plot.

10,000 numbers." Clearly, the old adage that a picture is worth a thousand words is a considerable underestimate!

Figures 2, 3, and 4 show plots involving small amounts of data, which nevertheless require complicated calculations in order to draw them. Figure 2 displays the empirical cumulative distribution function* (ECDF) of 30 independent observations, superimposed on which is the theoretical cumulative distribution function (CDF) for the normal distribution, the mean and variance of which were estimated from the data (the usual $\bar{x}$ and s^2). The normal CDF involves a complicated integral which cannot be evaluated explicitly. After plotting the ECDF (which is a scatter plot of sample cumulative proportions versus ordered observations), the computer used a numerical approximation to the normal (0, 1) CDF [27], adjusting it to have the desired estimated parameters, and plotted the CDF at enough points, connecting them with straight lines, to make the function look smooth. Figure 2 shows that the data are not normal, and that they come from a distribution (such as the exponential) with a shorter left and longer right tail. For other examples, see Wilk and Gnanadesikan [45].

Superimposition of a probability density function (PDF) on a histogram is similarly useful; a numerical approximation to the desired PDF is sometimes needed. Construction of probability plots* can require a numerical approximation to the inverse CDF of a distribution. Formulas and/or programs for some useful numerical approximations* involving probability distributions may be found in Abramowitz and Stegun [1], International Mathematical and Statistical Libraries [28], Kennedy and Gentle [30], and the Statistical Algorithms section of issues of *Applied Statistics* [4].

Figure 3 shows a third-degree polynomial regression. Only 14 data points are plotted, but considerable computation was necessary to produce the fitted curve (solid line), the previously fitted second-degree polynomial (small dashes), and a 95% confidence band (large dashes). Plotting the two fitted curves provides a visual interpretation of the usual F-tests*, and the confidence band provides,

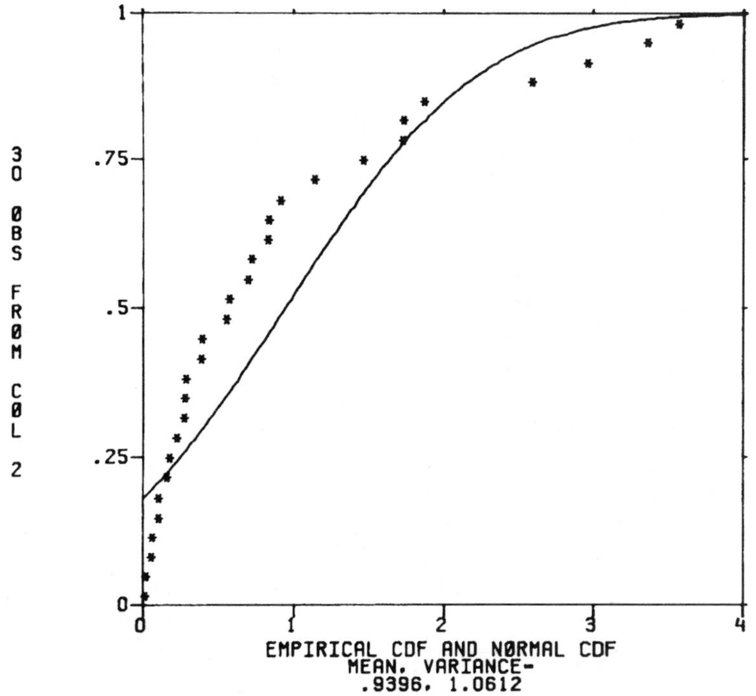

Figure 2 Plot of ECDF and CDF.

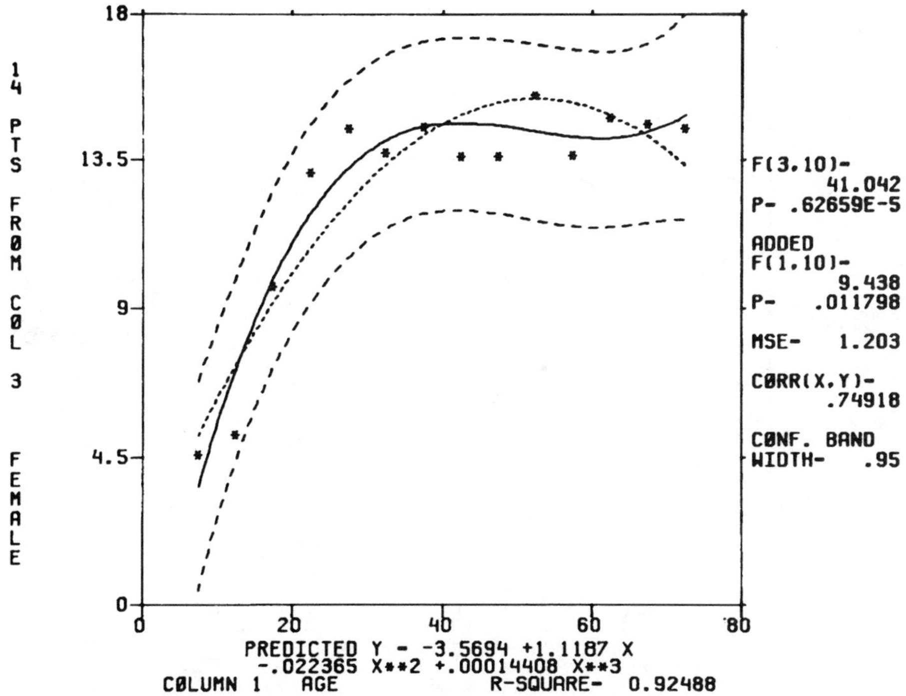

Figure 3 Polynomial regression plot.

at each x-coordinate, a 95% confidence interval for a future observation of the dependent variable. The ability of the computer to draw plots such as these based on complicated calculations, and to redraw them in an interactive mode in response to the demands of the user, is a powerful aid to data analysis.

Figure 4 is a cumulative sum (cusum) chart* (a scatter plot of 22 points, usually a time series), with a V-shaped "mask" superimposed (see Barnard [5]). For a description, see Davies and Goldsmith [14]. The rightmost point is necessarily inside the V-mask; if any points are outside, as they are in Fig. 4, the system generating the data is said to be *out of control*. Given the data and the values of several parameters which define the position and size of the V-mask, a computer can assist greatly in producing these plots on a regular basis in a real-time situation, such as in a factory.

Some special categories of graphs which rely heavily on the computer are graphs of multivariate data, maps, color graphs, movies, and videotapes.

A plot on a two-dimensional surface of multivariate data having more than two dimensions usually entails a severe loss of information. Complicated methods have been developed to retain as much information as possible. Most often, multivariate data are plotted using various types of black-and-white still graphs, examples of which are the Chernoff faces mentioned earlier, so-called "polygons," "stars," "glyphs," "metroglyphs," "weathervanes," and the graph in Fig. 5. The last (see Andrews [3]) contains one curve, a function of six tooth measurements, for each of 10 subjects. Two subjects are known to be human, six to be apes, and two are of unknown species. At an x-coordinate slightly below zero, the curves for apes form one tight cluster; the curves for men and for the two unknowns form another, suggesting that the unknown teeth came from human beings.

Sometimes additional dimensions of information can be conveyed, e.g., by the use of color or by parallax resulting from motion. An example of the latter appears in the PRIM-9 movie [20] and the PRIMH video-

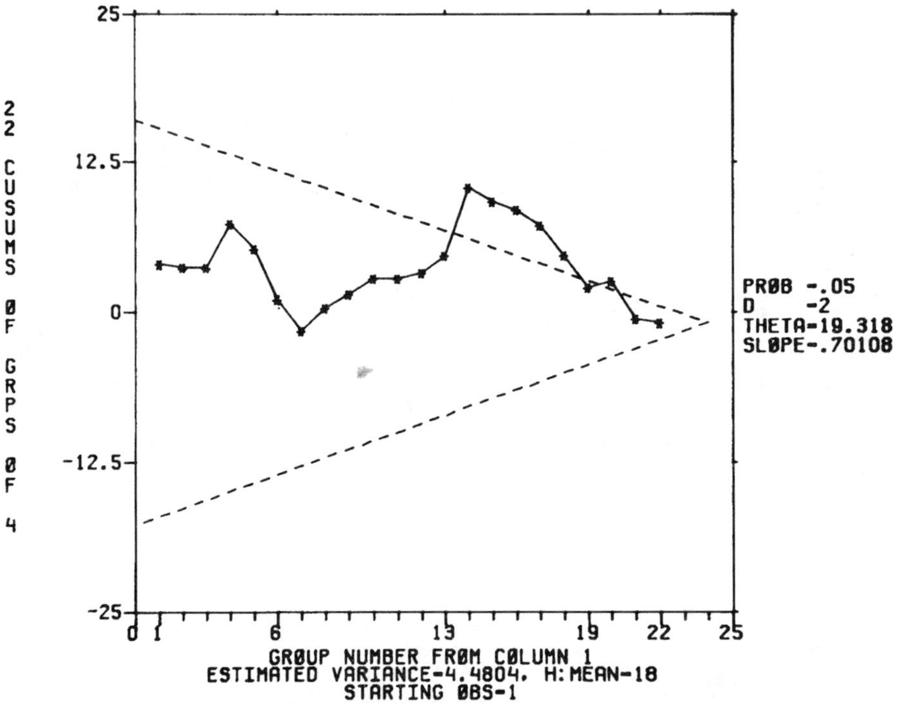

Figure 4 Cusum chart with *V*-mask.

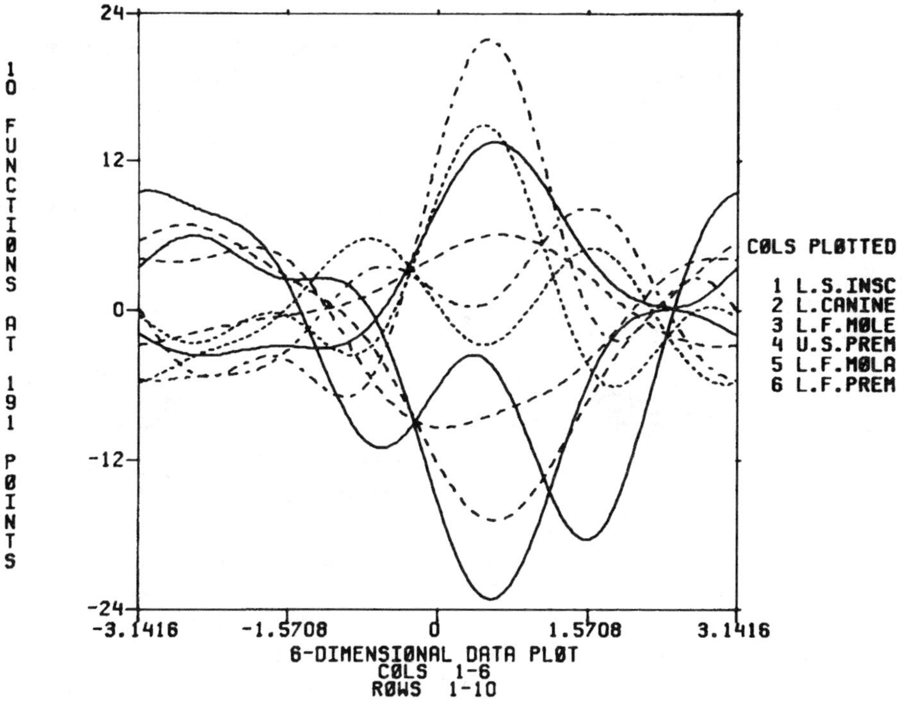

Figure 5 Multidimensional data plot.

tape [15], in which two-dimensional projections (scatter plots) of data points are plotted in sequence as the viewer would see them while moving through three-dimensional space. As long as the points appear to be moving, parallax enables the viewer to perceive the third dimension.

Three-dimensional theoretical surfaces can be plotted as if seen from selected perspectives. Hidden-line-removal algorithms can prevent the plotting of portions of the surface that the viewer would not be able to see from a hypothetical viewpoint in three-dimensional space. Figure 6, for example, is a plot of the surface defined by the function $f(x, y) = \exp(-0.08r)\cos(r)$, where $r = (x^2 + y^2)^{1/2}$. To obtain Fig. 6, $f(x, y)$ was evaluated over a 46×46 grid of (x, y) values, for $-9 \leqslant x \leqslant 9$ and $-9 \leqslant y \leqslant 9$, and a line of sight was selected by locating the viewer's eye at point (20, 16, 8) in three-dimensional space, looking toward point (0, 0, 0). (The supporting plotting program was the NCAR Graphics System [36].)

Holograms can be used to produce graphs of three-dimensional data. Holography is a form of photography in which coherent light is used with ordinary photographic plates to produce images that appear to be three-dimensional, and which can be viewed without the need for special optical equipment. For example, the Laboratory for Computer

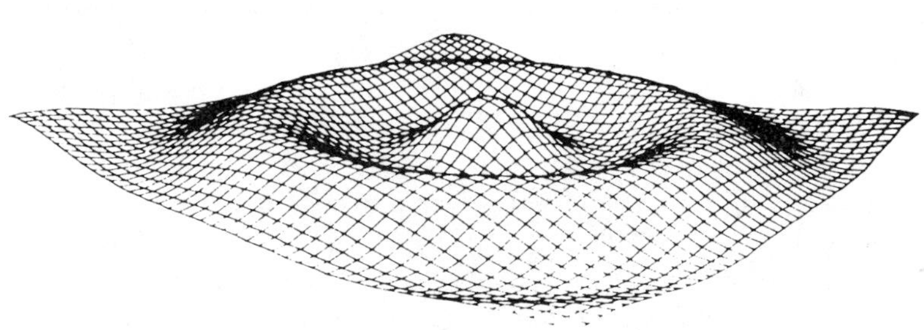

Figure 6 Plot of three-dimensional surface with hidden line removal.

Graphics and Spatial Analysis [25] produced 181 computer-generated maps of the United States showing its population (plotted as peaks above the map) and occupied land area from 1790 to 1970. The maps are displayed in time sequence on a cylindrical hologram which the viewer perceives as a three-dimensional movie (and which the viewer can walk around while it is showing).

For examples and descriptions of plotting techniques for multivariate data, see Fienberg [19], Gnanadesikan [24], Tukey and Tukey [41], Friedman and Rafsky [21], Chambers and Kleiner [11], Chambers et al. [12], and GRAPHICAL REPRESENTATION OF DATA.

Computer mapping is widely used. The U.S. Bureau of the Census* regularly produces both black-and-white maps and attractive color maps of the United States, representing geographical statistics (see Meyer et al. [35], Fienberg [19], and Broome and Witiuk [9]). Space satellites can transmit digital measurements which are converted by computer into maps depicting various desired characteristics. For information on computer mapping and its applications, see the six volumes of the Harvard Library of Computer Graphics (Harvard University [26]; *see also* GEOGRAPHY, STATISTICS IN).

A simple example of how color in graphs can enhance data perception is given in HISTOGRAMS. Showing two superimposed black-and-white histograms, the article points out that the comparison of the two samples would be greatly facilitated by using different colors for each histogram, with a blend of the two colors where they overlap. The graph of multivariate data in Fig. 5 would be much more effective if curves for human beings were in one color, curves for apes in another, and curves for the unknown species in a third color. Color helps distinguish among different portions of a graph, and can increase the effective number of dimensions. (See Beatty [6] for a useful tutorial on color graphics.)

Figures 7 and 8 are examples of both computer mapping and color graphics. Each of these two-variable color maps uses colors from a 4×4 grid to represent bivariate data. Figure 7 (previously shown and discussed in Meyer et al. [35]) shows, simultaneously, the education (percent of high school graduates) and income (per capita) for each county in the continental United States. High positive correlation* between these two variables is indicated by the prevalence of yellows and purples (colors from the grid's reverse diagonal) rather than greens and reds (colors from the main diagonal). The tendency of yellow-colored counties to cluster raises questions about the socioeconomic reasons for low income and low education occurring together in certain geographical areas. Information about this type of spatial correlation cannot be conveyed by the numerical value of a correlation coefficient.

The two maps in Fig. 8 use a different 4×4 grid of colors, as proposed by Trumbo [39]. The upper map in Fig. 8 plots education (median school years of persons age 25 or more) and income (median family income in dollars), again for U.S. counties. The four education categories, labeled A, B, C, D on the plot, are < 9.8, $[9.8, 11.3)$, $[11.3, 12.2)$, and $\geqslant 12.2$, respectively. The four income categories, labeled 1, 2, 3, 4 are < 6140, $[6140, 7440)$, $[7440, 8650)$, and $\geqslant 8650$, respectively. In this map, positive correlation (about 0.7) is indicated by the presence of mostly low-saturation colors (blacks, grays, whites) from the reverse diagonal of the color grid. The lower map in Fig. 8 shows two new variables which are linear combinations of education and income. The transformation reduces the correlation between the two plotted variables to approximately 0. This is apparent from the presence in the lower plot of a mixture of both low-saturation colors and high-saturation ones (blues, greens, reds) from the main diagonal. The proponents of this type of color grid feel that its use improves the ability to perceive underlying patterns in the data; the data analyst can initially try to get a general impression of the distribution of only four types of colors: light, dark, "warm," and "cool" colors, representing the four corners of the grid. Research into the use of such maps would be prohibitively difficult without the use of the computer.

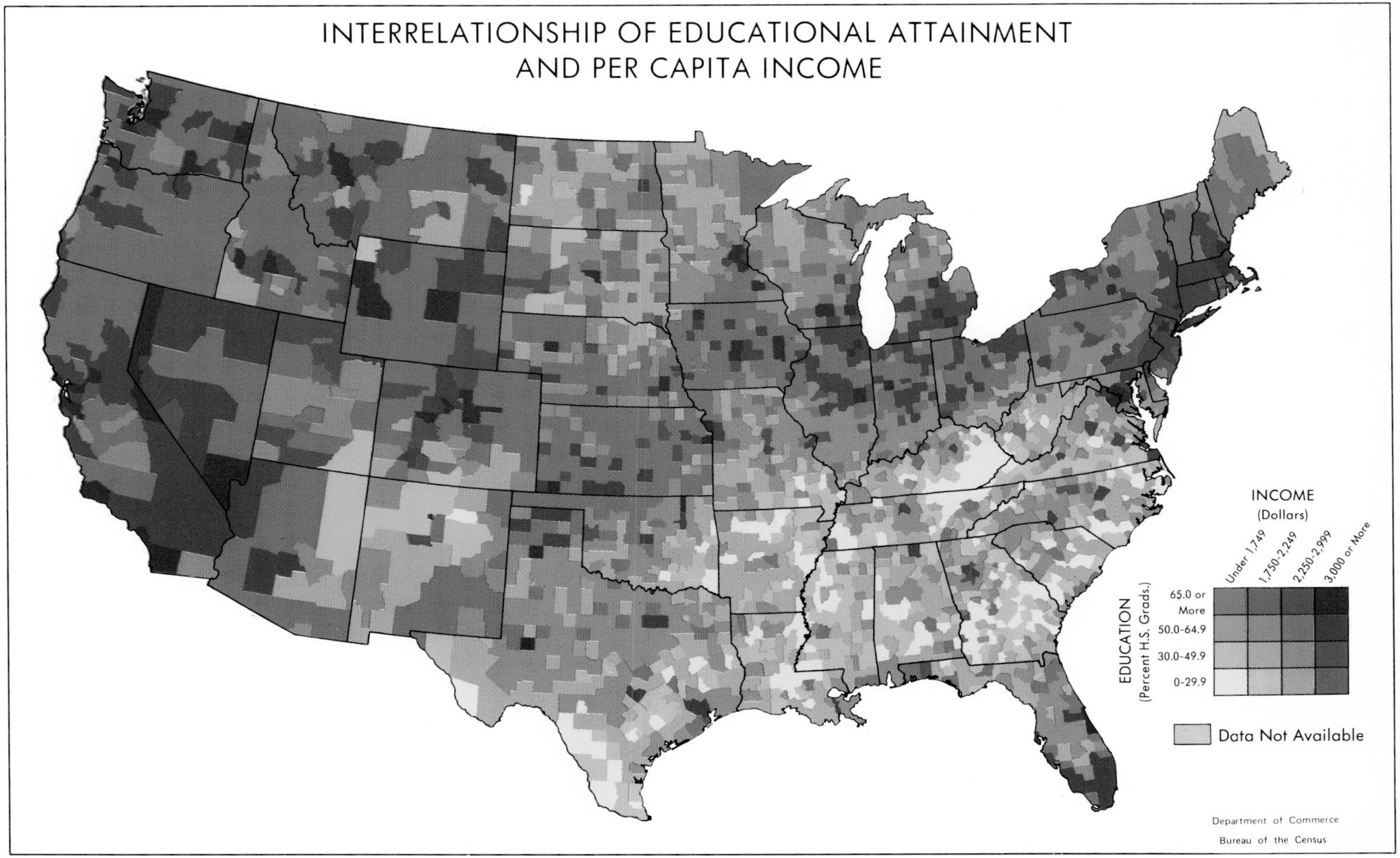

Figure 7 Two-variable color map. (Courtesy of the Geography Division of the U.S. Bureau of the Census*.)

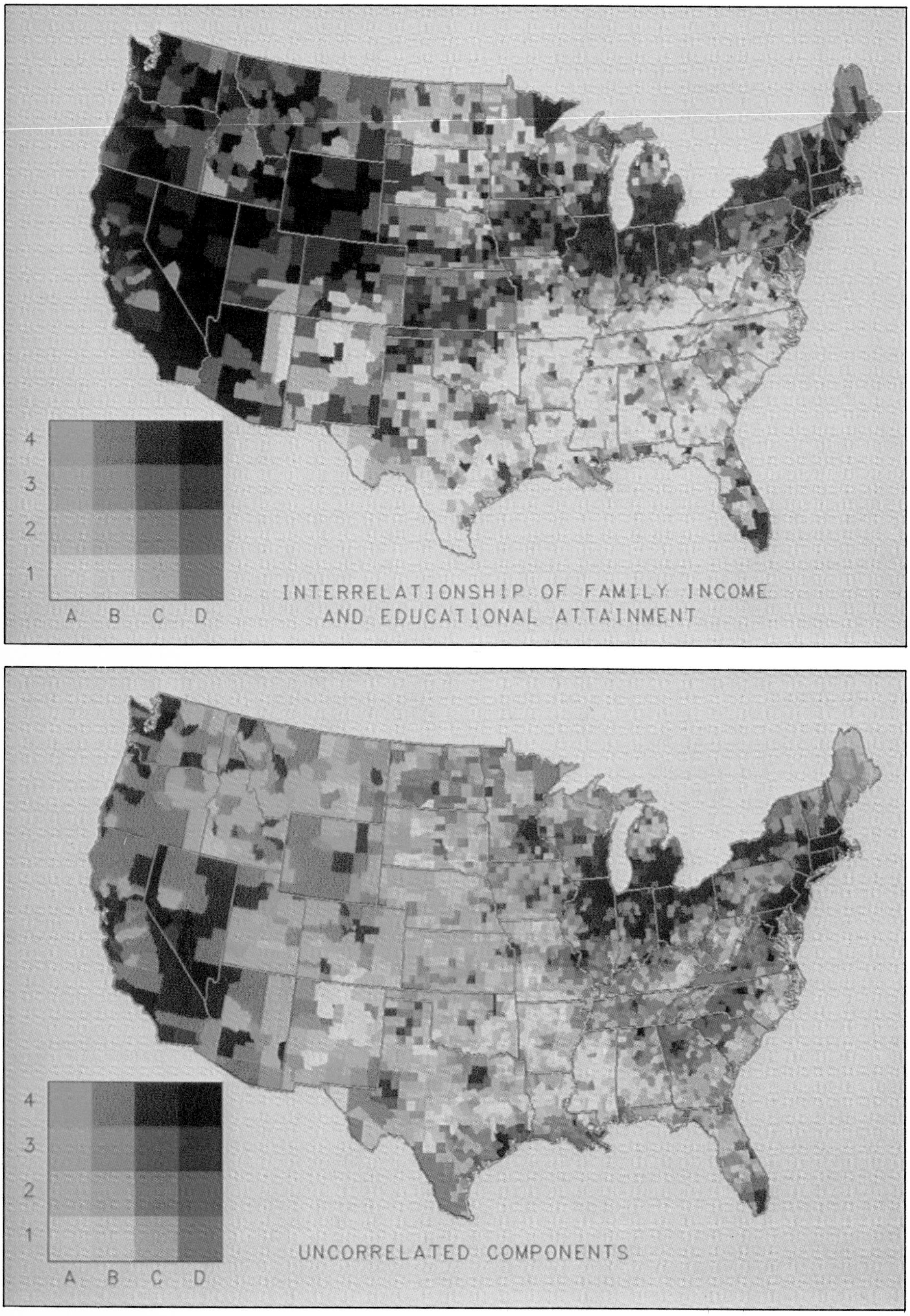

Figure 8 Two-variable color maps. (Produced by Albert Yen and William E. Johnston, with the cooperation of Lawrence Berkeley Laboratory, SEEDIS Project, sponsored by the U.S. Departments of Energy and Labor.)

A movie requires many frames, some nearly identical, to be shown in each second. A computer graphics terminal can be attached to a camera and left for many hours, if necessary, while a running program draws pictures and activates the camera, which photographs the terminal screen. Or, more conveniently, a computer program can transmit the graphical information directly to a film recorder. Max [31], for example, produced a computer-made film which demonstrates the results of a mathematical theorem; it shows how to turn a sphere topologically inside out "by passing the surface through itself without making any holes or creases." The PRIM-9 movie and PRIMH videotape mentioned earlier show how data perception can be enhanced by using these media.

HARDWARE

The simplest type of computer-drawn graph is formed from characters. The computer prints the same characters used for text in appropriate positions to form a plot. This can be done quite effectively if the available grid size is large enough (e.g., on a 132-character-wide printer). With a narrow printing device or terminal screen, the results are cruder, but can still be effective, depending on the graph being drawn. One particularly simple and effective contour plotting technique is to evaluate a function $f(x, y)$ of two variables over a rectangular grid having the same number of (x, y) coordinates as there are character-printing positions available, assign different characters to different possible values of $f(x, y)$, and print the resulting grid of characters. Figure 9 shows such a plot of contours of a bivariate normal PDF with means 0 and 0, variances 0.5 and 1, and correlation -0.5. The character "o" is used to plot values of $f(x, y)$ which are less than 1% of the maximum $f(x, y)$, blank space is used for values between 1 and 10% and between 40 and 99%, "x" for values between 10 and 40%, and "*" for values over 99%. The grid size is 20×60, and the entire plot, including labels, will fit on the screen of most terminals. Some useful programs for producing character plots are given in McNeil [33] and Velleman and Hoaglin [42].

For higher-quality permanent plots, computers can be equipped with film recorders, pen plotters, or electrostatic raster plotters. Film recorders produce the plot on film rather than paper, receiving the plot data from a program. Pen plotters draw on paper in response to programmed commands, moving from point to point, with the pen tip on or above the paper, as instructed. Some pen plotters offer a selection of a few colored pen tips; the pen draws some designated lines, moves to the side of the plotter, changes its own pen tip, and continues plotting. Electrostatic plotters form a plot by darkening, or not darkening, dots on a grid of paper. Electrostatic plotters thus excel at shading areas; pen plotters are best at drawing graphs composed of lines. To make a permanent record of a graphics terminal screen image, various types of hard-copy units and printers are available as directly attached accessories. Some "intelligent terminals" and desktop minicomputers and microcomputers have their own local graphics capability and some also have built-in hard-copy units.

Three types of graphics terminal available are vector refresh terminals, storage scopes (storage tubes), and the newer raster scan terminals. *Vector refresh terminals* maintain a "display list" containing end-point information for lines to be drawn. The image on the screen displaying these lines is refreshed (redrawn) 30 times a second, which gives excellent dynamics. However, these terminals are relatively expensive. The most commonly used *storage scopes* can display smooth lines with high precision, but cannot vary the intensity of the lines (i.e., use a gray scale) or use color. These terminals are not capable of "selective erasure" (of only a portion of the picture), and clearing of the entire screen is accompanied by a flash. Thus they can achieve motion only by adding lines (e.g., filling a glass of water by adding one line at a time). They cannot provide motion achieved by repositioning (e.g., drawing a pitcher, erasing it, and

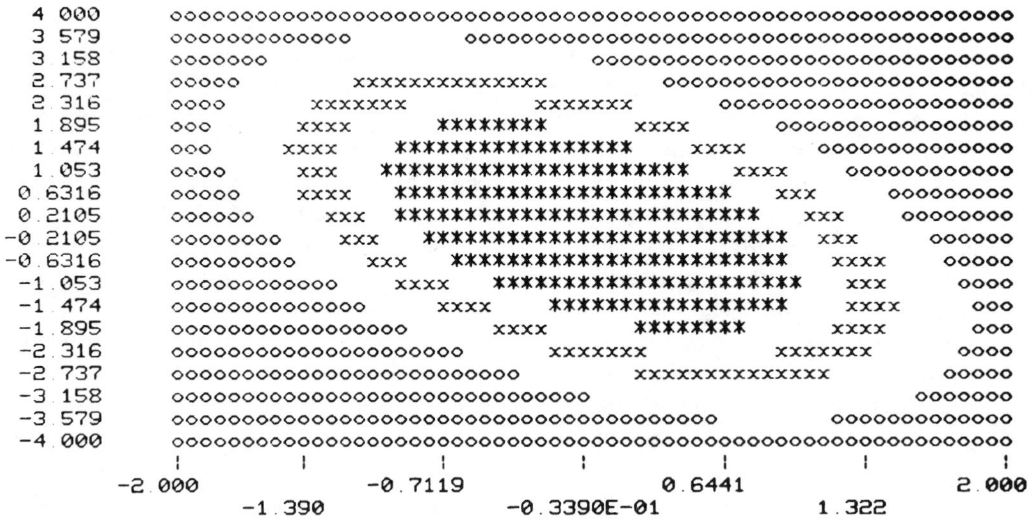

Figure 9 Character contour plot.

quickly drawing it again at a different angle, etc., to make it appear to be tilting as it pours into a glass). The seven black-and-white figures of this article were produced on such a storage terminal attached to a hard-copy unit. Some storage scopes have a limited refresh capability; in the terminal's "write through" mode, a small amount of plotted information can be displayed dynamically. New technology is improving this capability. *Raster terminals* use a rectangular grid of dots to represent graphs, and can provide gray scale, color, and selective erasure.

For more information on graphics hardware, see Newman and Sproull [37], Chambers [10], and Beatty [6].

SOFTWARE

Computer hardware technology is usually far ahead of applications software. Technology is advancing rapidly, and software takes a considerable amount of time to produce. Writing a program for statistical graphics requires understanding of the statistical/graphical technique being used, development of an algorithm for implementing it, and the ability to use the programming routines that cause the graphs to be drawn. Most programs for statistical graphics are specific to a particular stage of technological development and, often, to a particular brand and model of terminal. [For example, the program package (see Dunn and Gentleman [16]) that generated the first five figures of this article utilizes plotting routines obtained from the terminal manufacturer]. A graphical program package that attempts to be relatively portable and device independent is the Graphical Algorithms portion of Bell Labs' *S* System [7]. It is often hard to persuade users, who suffer from inertia and are inundated with technological changes, to switch to a newer mode of computation. Also, applications of new technology have to be invented before programs can be written to implement them. Color graphics is an example of this; statisticians are reviewing the literature on human color perception and are investigating methods of exploiting the capabilities of color for analyzing data. For an example, see the debate about two-variable color maps (such as those in Figs. 7 and 8) in Abt [2], Fienberg [19], Meyer [34], Trumbo [39], Tukey [40], Wainer [43], and Wainer and Francolini [44].

THE FUTURE

In 1973, the Japanese Cabinet declared October 18 to be National Statistics Day in Japan. Since then, annual observances there of this day have increased so much that all of October has become a statistical observance season [38]. By 1977, there were 29,836 entries in the National Statistics Day statistical-graph-drawing contest. Judges screened the entries three times and gave the prize to a pictogram (*see* HISTOGRAMS) which used children's faces as symbols showing the frequency with which mothers play with their children. The winning plot was submitted by five 7-year-olds, who would have been strong competition for any computer graphics program.

The appreciation and use of graphics, including computer graphics, can be expected to spread. With advances in technology, reduction in the cost of computers, and the acceptance of personal microcomputers, computer graphics can be expected to move not just into more offices, factories, and classrooms, but also into the home. In fact, computer games, which are a form of computer graphics, have already moved into the home. Computer networks will make vast supplies of data readily available, and graphs, as always, will be a useful way to make sense out of the data.

References

[1] Abramowitz, M. and Stegun, I. A. (1964). *Handbook of Mathematical Functions*, Dover, New York.

[2] Abt, C. C. (1981). *Amer. Statist.*, **35**, 57.

[3] Andrews, D. F. (1972). *Biometrics*, **28**, 125–136.

[4] *Applied Statistics*. Statistical Algorithms section: published programs appearing in each issue.

[5] Barnard, G. A. (1959). *J. R. Statist. Soc. B*, **21**, 239–271.

[6] Beatty, J. C. (1983). *Amer. Statist.*, **37** (in press).

[7] Becker, R. A. and Chambers, J. M. (1980). *S. A Language and System for Data Analysis*. Bell Laboratories, Murray Hill, N.J.

[8] Beniger, J. R. and Robyn, D. L. (1978). *Amer. Statist.*, **32**, 1–11.

[9] Broome, F. R. and Witiuk, S. W. (1980). In *The Computer in Contemporary Cartography*, D. R. F. Taylor, ed. Wiley, New York, Chap. 9.

[10] Chambers, J. M. (1977). *Computational Methods for Data Analysis*, Wiley, New York.

[11] Chambers, J. M. and Kleiner, B. (1981). In *Handbook of Statistics*, Vol. 2, P. R. Krishnaiah, ed. North-Holland, Amsterdam (in press).

[12] Chambers, J. M., Cleveland, W. S., Tukey, P. A., and Kleiner, B. (1983). *Graphical Methods in Statistics*. Wiley, New York (in press).

[13] Cleveland, W. C. and Kleiner, B. (1975). *Technometrics*, **17**, 447–454.

[14] Davies, O. L. and Goldsmith, P. L. (1972). *Statistical Methods in Research and Production*, 4th ed. rev. Oliver & Boyd, Edinburgh (for Imperial Chemical Industries Ltd.).

[15] Donoho, D., Huber, P. J., and Thoma, M. (1981). Interactive Graphical Analysis of Multidimensional Data. Videotape, Harvard University, Cambridge, Mass.

[16] Dunn, R. M. and Gentleman, J. F. (1976). *Proc. 7th Ontario Univ. Comput. Conf.*, Waterloo, pp. 306–317.

[17] Feinberg, B. M. and Franklin, C. A. (1975). *Social Graphics Bibliography*. Bureau of Social Science Research, Washington, D.C.

[18] Fienberg, S. E. (1971). *Science*, **171**, 255–261.

[19] Fienberg, S. E. (1979). *Amer. Statist.*, **33**, 165–178.

[20] Fisherkeller, M. A., Friedman, J. H., and Tukey, J. W. (1973). PRIM-9. 16-mm color film, Bin 88 Productions, Stanford Linear Accelerator Center, Stanford, Calif.

[21] Friedman, J. H. and Rafsky, L. C. (1981). *J. Amer. Statist. Ass.*, **76**, 277–287.

[22] Gentleman, J. F. (1977). *Amer. Statist.*, **31**, 166–175.

[23] Gentleman, J. F. (1978). *Appl. Statist.*, **27**, 354–358.

[24] Gnanadesikan, R. (1977). *Methods for Statistical Data Analysis of Multivariate Observations*. Wiley, New York.

[25] Harvard University (1978). American Graph Fleeting. Flyer describing holographic map animation, Laboratory for Computer Graphics and Spatial Analysis, Cambridge, Mass.

[26] Harvard University (1979). *The Harvard Library of Computer Graphics*. Laboratory for Computer Graphics and Spatial Analysis, Cambridge, Mass.

[27] Hastings, C., Jr. (1955). *Approximations for Digital Computers*. Princeton University Press, Princeton, N.J.

[28] International Mathematical and Statistical Libraries (1980). *IMSL Library, Reference Manual*, 8th ed. Houston, Tex.

[29] Izenman, A. J. (1980). *Proc. First Gen. Conf. Social Graphics*, Oct. 22–24, 1978, Leesburg, Va., pp. 51–79.

[30] Kennedy, W. J. and Gentle, J. E. (1980). *Statistical Computing*, Marcel Dekker, New York.

[31] Max, N. L. (1978). Turning a Sphere Inside Out. 16-mm color film distributed by International Film Bureau Inc., Chicago.

[32] Mclellan, J. (1980). Computer Graphics, *The Washington Post Magazine*, Oct. 6, 12–17.

[33] McNeil, D. R. (1977). *Interactive Data Analysis—A Practical Primer*. Wiley, New York.

[34] Meyer, M. A. (1981). *Amer. Statist.*, **35**, 56–57.

[35] Meyer, M. A., Broome, F. R., and Schweitzer, R. H., Jr. (1975). *Amer. Cartogr.*, **2**, 100–117.

[36] NCAR Graphics System (1977). National Center for Atmospheric Research, Boulder, Colo.

[37] Newman, W. M. and Sproull, R. F. (1979). *Principles of Interactive Computer Graphics*, 2nd ed. McGraw-Hill, New York.

[38] *New York Times* (1977). Data-Loving Japanese Rejoice on Statistics Day. Oct. 28, p. A1.

[39] Trumbo, B. E. (1981). *Amer. Statist.*, **35**, 220–226.

[40] Tukey, J. W. (1979). *J. Amer. Statist. Ass.*, **74**, 786–793. (See p. 792.)

[41] Tukey, P. A. and Tukey, J. W. (1981). In *Interpreting Multivariate Data*, V. Barnett, ed. Wiley, London, Chaps. 10–12, pp. 187–275.

[42] Velleman, P. F. and Hoaglin, D. C. (1981). *Applications, Basics, and Computing of Exploratory Data Analysis*. Duxbury Press, Boston.

[43] Wainer, H. (1981). *Amer. Statist.*, **35**, 57–58.

[44] Wainer, H. and Francolini, C. M. (1980). *Amer. Statist.*, **34**, 81–93.

[45] Wilk, M. B. and Gnanadesikan, R. (1968). *Biometrika*, **55**, 1–17.

(COMPUTERS AND STATISTICS
GRAPHICAL REPRESENTATION OF DATA)

JANE F. GENTLEMAN

GRAPHICAL REPRESENTATION OF DATA

Graphs, charts, and diagrams offer effective display and enable easy comprehension of complex, multifaceted relationships. Gnanadesikan and Wilk [29] point out that "man is a geometrical animal and seems to need and want pictures for parsimony and to stimulate insight." Various forms of statistical graphs have been in use for over 200 years.

Beniger and Robyn [7] cite the following as first or near-first uses of statistical graphs: Playfair's [60] use of the bar chart to display Scotland's 1781 imports and exports for 17 countries; Fourier's [25] cumulative distribution of population age in Paris in 1817; Lalanne's [41] contour plot of temperature by hour and month; and Perozzo's [59] stereogram* display of Sweden's population for the period 1750–1875 by age groupings. Fienberg [24] notes that Lorenz [45] made the first use of the P-P plot in 1905. Each of these plots is discussed subsequently. Beniger and Robyn [7], Cox [14], and Fienberg [24] provide additional historical accounts and insights on the evolution of graphics.

Today, graphical methods play an important role in all aspects of a statistical investigation—from the beginning exploratory plots, through various stages of analysis, to the final communication and display of results. Many persons consider graphical displays as the single most effective, robust statistical tool.

Not only are graphical procedures helpful, but in many cases essential. Tukey [72] claims that "the greatest value of a picture is when it forces us to notice what we never expected to see." This is no better exemplified than by an example of Anscombe [4], where plots of four equal-size data sets (Fig. 1) reveal large differences among the sets even though all sets produce the same linear regression* summaries. Mahon [47] maintains that statisticians' responsibilities include communication of their findings to decision makers, who frequently are statistically naive, and the best way to accomplish this is through the power of the picture.

Good graphs should be simple, self-explanatory, and not deceiving. Cox [14] offers the following guidelines:

1. The axes should be clearly labeled with the names of the variables and the units of measurement.
2. Scale breaks should be used for false origins.
3. Comparison of related diagrams should be easy, for example, by using identical scales of measurement and placing diagrams side by side.
4. Scales should be arranged so that systematic and approximately linear relations are plotted at roughly 45° to the x-axis.

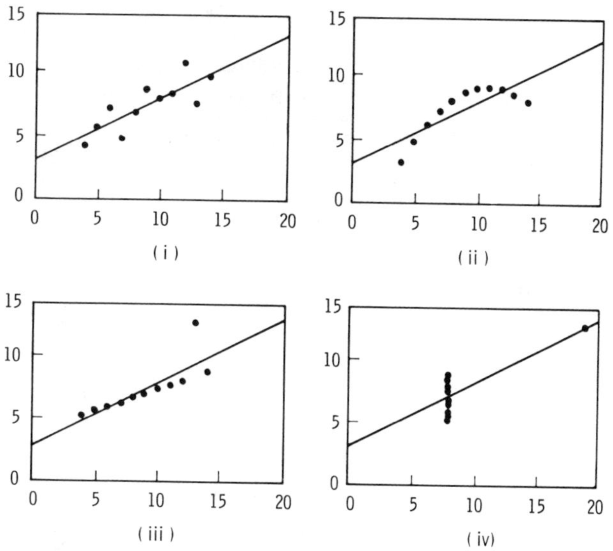

Figure 1 Anscombe's [4] plots of four equal-size data sets, all of which yield the same regression summaries.

5. Legends should make diagrams as nearly self-explanatory (i.e., independent of the text) as is feasible.
6. Interpretation should not be prejudiced by the technique of presentation.

Most of the graphs discussed here which involve spatial relationships are implicitly or explicitly on Cartesian or rectangular coordinate grids, with axes that meet at right angles. The horizontal axis is the *abscissa* or x-axis and the vertical axis is the *ordinate* or y-axis. Each point on the grid is uniquely specified by an x- and a y-value, denoted by the ordered pair (x, y). Ordinary graph paper utilizes linear scales for both axes. Other scales commonly used are logarithmic (see Fig. 8) and inverse distribution function (see Fig. 6). Craver [15] includes a discussion of plotting techniques together with over 200 graph papers that may be copied without permission of the publisher.

This discussion includes old and new graphical forms that have broad application or are specialized but commonly used. Taxonomies based on the uses of graphs have been addressed by several authors, including Tukey [71], Fienberg [24], and Schmid and Schmid [63]. This discussion includes references to more than 50 different graphical displays (Table 1) and is organized according to the principal functions of graphical techniques:

Exploration
Analysis
Communication and display of results
Graphical aids

Some graphical displays are used in a variety of ways; however, each display here is discussed only in the context of its widest use.

EXPLORATORY GRAPHS

Exploratory graphs are used to help diagnose characteristics of the data and to suggest appropriate statistical analyses and models. They usually do not require assumptions about the behavior of the data or the system or mechanism that generated the data.

Data Condensation

A listing or tabulation of data can be very difficult to comprehend, even for relatively

Table 1 Graphical Displays Used in the Analysis and Interpretation of Data

Exploratory Plots		
	Relationship Among Variables	
Data Condensation	Two Variables	Three or More Variables
Histogram	Scatter plot	Labeled scatter plot
Dot-array diagram	Sequence plot	Glyphs and metroglyphs
Stem and leaf diagram	Autocorrelation plot	Weathervane plot
Frequency polygon	Cross-correlation plot	Biplot
Ogive		Face plots
Box and whisker plot		Fourier plot
		Cluster trees
		Similarity and preference maps
		Multidimensional scaling displays

Graphs Used in the Analysis of Data		
Distribution Assessment	Model Adequacy and Assumption Verification	Decision Making
Probability plot	Average versus standard deviation	Control chart
Q-Q plot	Residual plots	Cusum chart
P-P plot	Partial-residual plot	Youden plot
Hanging histogram	Component-plus-residual plot	Half-normal plot
Rootogram		C_p plot
Poissonness plot		Ridge trace

Communication and Display of Results		
Quantitative Graphics	Summary of Statistical Analyses	Graphical Aids
Bar chart	Means plots	Power curves
Pictogram	Sliding reference distribution	Sample-size curves
Pie chart	Notched box plot	Confidence limits
Contour plot	Factor space/response	Nomographs
Stereogram	Interaction plot	Graph paper
Color map	Contour plot	Trilinear coordinates
	Predicted response plot	
	Confidence region plot	

small data sets. Data condensation techniques, discussed in most elementary statistics texts, include several types of *frequency distributions* (see, e.g., Freund [26] and Johnson and Leone [37]). These associate the frequency of occurrence with each distinct value or distinct group of values in a data set. Ordinarily, data from a continuous variable will first be grouped into intervals, preferably of equal length, which completely cover without overlap the range of the data. The number or length of these intervals is usually best determined from the size of the data set, with larger sets able effectively to support more intervals. Table 2 presents four commonly used forms: frequency, relative frequency, cumulative frequency, and cumulative relative frequency. Here carbon monoxide emissions (grams per mile) of 794 cars are grouped into intervals of length 24, where the upper limit is included (denoted by the square upper interval bracket) and the lower limit is not (denoted by the lower open parenthesis). Columns 4 and 6 depict

Table 2 Frequency Distributions of Carbon Monoxide Data

(1) Interval	(2) Interval Midpoint	(3) Frequency	(4) Relative Frequency	(5) Cumulative Frequency	(6) Cumulative Relative Frequency
1. [0–24]	12	13	0.016	13	0.016
2. (24–48][a]	36	98	0.123	111	0.140
3. (48–72]	60	161	0.203	272	0.343
4. (72–96]	84	189	0.238	461	0.581
5. (96–120]	108	148	0.186	609	0.767
6. (120–144]	132	85	0.107	694	0.874
7. (144–168]	156	45	0.057	739	0.931
8. (168–192]	180	30	0.038	769	0.969
9. (192–216]	204	10	0.013	779	0.981
10. (216–240]	228	5	0.006	784	0.987
11. (240–264]	252	5	0.006	789	0.994
12. (264–288]	276	1	0.001	790	0.995
13. (288–312]	300	2	0.003	792	0.997
14. (312–336]	324	1	0.001	793	0.999
15. (336–360]	348	1	0.001	794	1.000

[a]Notation designates inclusion of all values greater than 24 and less than or equal to 48.

relative frequencies which are scaled versions (divided by 794) of columns 3 and 5, respectively.

The four distributions tabulated in Table 2 are useful data summaries; however, plots of them can help the data analyst develop an even better understanding of the data. A *histogram** is a bar graph associating frequencies or relative frequencies with data intervals. The histogram for carbon monoxide data shown in Fig. 2 clearly shows a positively skew, unimodal distribution with modal interval (72–96]. Other forms of histograms use symbols such as dots (*dot-array*

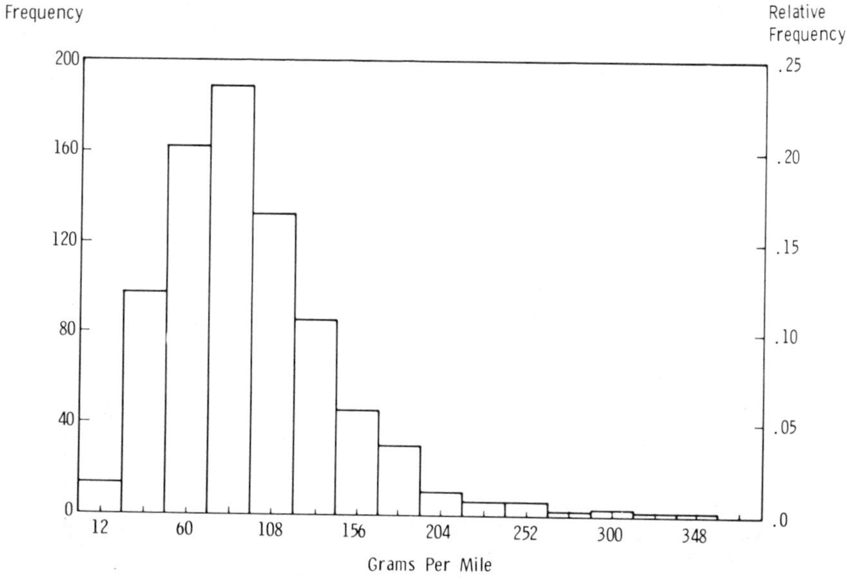

Figure 2 Histogram of Environmental Protection Agency surveillance data (1957–1967) on carbon monoxide emissions from 794 cars.

diagram) or asterisks in place of bars, with each symbol representing a designated number of counts.

A *frequency polygon** is similar to a histogram. Points are plotted at coordinates representing interval midpoints and the associated frequency; consecutive points are connected with straight lines (e.g., in Table 2, plot column 3 vs. column 2). The form of this graph is analogous to that of a probability density function.

A disadvantage of a grouped-data* histogram is that individual data points cannot be identified since all the data falling in a given interval are indistinguishable. A display that circumvents this difficulty is the *stem and leaf diagram**, a modified histogram with "stems" corresponding to interval groups and "leaves" corresponding to bars. Tukey [72] gives a thorough discussion of stem and leaf and its variations. For examples, *see* HISTOGRAMS *and* EXPLORATORY DATA ANALYSIS.

An *ogive** is a graph of the cumulative frequencies (or cumulative relative frequencies) against the upper limits of the intervals (e.g., from Table 2, plot column 5 versus the upper limit of each interval in column 1) where straight lines connect consecutive points. An ogive is a grouped data analog of a graph of the empirical cumulative distribution function and is especially useful in graphically estimating percentiles (quantiles*), which are data values associated with specified cumulative percents. Figure 3 shows the ogive for the carbon monoxide data and how it is used to obtain the 25th percentile (i.e., lower quartile).

Another display, which highlights five important characteristics of a data set, is a *box and whisker* or box plot. The box, usually aligned vertically, encloses the interquartile range*, with the lower line identifying the 25th percentile (lower quartile) and the upper the 75th (upper quartile). A line sectioning the box displays the 50th percentile (median*) and its relative position within the interquartile range. The whiskers at either end may extend to the extreme values, or, for large data sets, to the 10th/90th or 5th/95th percentiles. These plots are especially convenient for comparing two or more data sets, as in Fig. 4 for winter snowfalls of Buffalo and Rochester, New York. (See Tukey [72] for further discussion and McGill et al. [51] for some variations.)

Relationships between Two Variables

Often of interest is the relationship, if any, between x and y, or developing a model to predict y given the value of x. Ordinarily, an

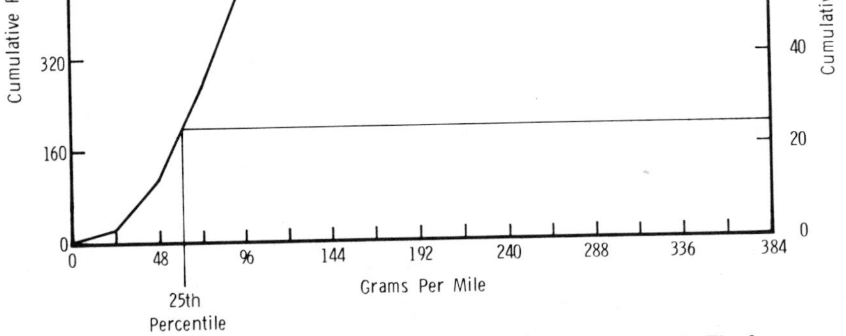

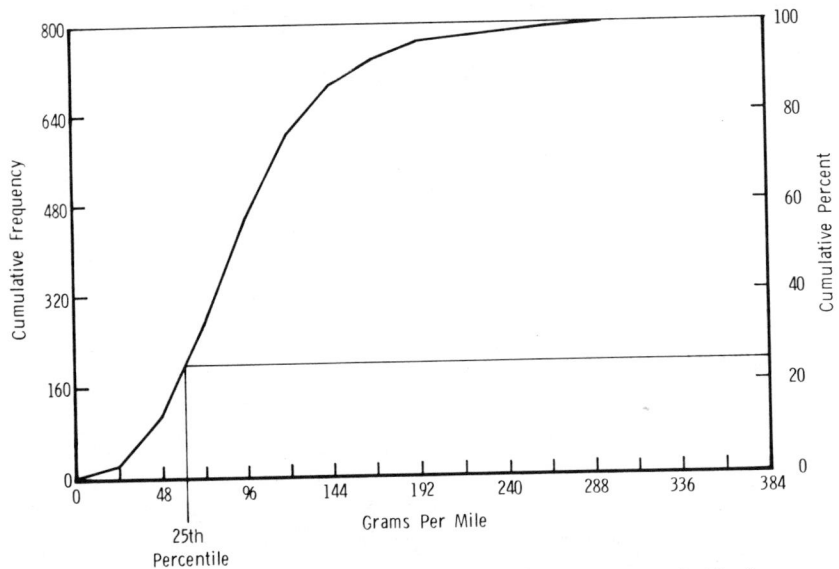

Figure 3 Ogive of auto carbon monoxide emissions data shown in Fig. 2.

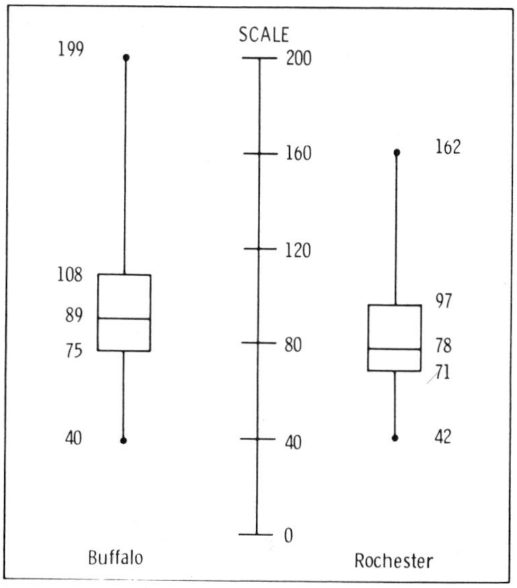

Figure 4 Box and whisker plots comparing winter snowfalls of Buffalo and Rochester, New York, (1939–1940 to 1977–1978) and demonstrating little distributional difference (contrary to popular belief). (Based on local climatological data gathered by the National Oceanic and Atmospheric Administration, National Climatic Center, Asheville, N.C.)

(x, y) measurement pair is obtained from the same experimental unit, such as the following:

Unit	x	y
Object	Diameter	Weight
Person	Age	Height
Product	Raw material purity	Quality

A usual first step is to construct a *scatter plot**, a collection of plotted points representing the measurement pairs (x_i, y_i), $i = 1, \ldots, n$. The importance of scatter plots was seen in Fig. 1. These four very different sets of data yield the same regression* line (drawn on the plots) and associated statistics [4]. Consequently, the numerical results of an analysis, without the benefit of a look at plots of the data, could result in invalid conclusions.

The objective of regression analyses is to develop a mathematical relationship between a measured response or dependent variable, y, and two or more predictor or independent variables, $x_1, x_2, \ldots, x_p$. A usual initial step is to plot y versus each of the x variables individually and to plot each x_i versus each of the other x's. This results in a total of $p + p(p-1)/2$ plots. Plots of y versus x_i enable one to identify the x_i-variables which appear to have large effects, to assess the form of a relationship between the y- and x_i-variables, and to determine whether any unusual data points are present. Plots of x_i vs. x_j, $i \neq j$, help to identify strong correlations that may exist among the predictor variables. It is important to recognize such correlations because least-squares* regression techniques work best when these correlations are small [35].

In many instances data are collected sequentially in time and a plot of the data versus sequence of collection can help identify sources of important effects. Figure 5 shows a *sequence plot* of gasoline mileage of an automobile versus the sequence of gasoline fill-ups. The large seasonal effect (summer mileage is higher than winter mileage) and the increase in gasoline mileage due to a major tune-up are clearly evident.

Observations collected sequentially in time, such as the gasoline mileage data plotted in Fig. 5, form a time series*. Statistical modeling of a time series is largely accomplished by studying the correlation* between observations separated by $1, 2, \ldots, n-1$ units in time. For example, the lag 1 autocorrelation is the correlation coefficient between observations collected at time i and time $i+1$, $i = 1, 2, \ldots, n-1$; it measures the linear relationship among all pairs of consecutive observations. An *autocorrelation plot* may be used to study the "correlation structure" in the data where the lag j autocorrelation coefficient computed between observations i and $i+j$ is plotted versus lag j. A *cross-correlation plot* between two time series is developed similarly. The lag j correlation coefficient, computed between observations i in one series and observations $i+j$ in the other series, is plotted versus the lag j. Box and Jenkins [9] discuss how to construct and interpret these plots, in addition to presenting examples on the use of sequence plots and the modeling of time series.

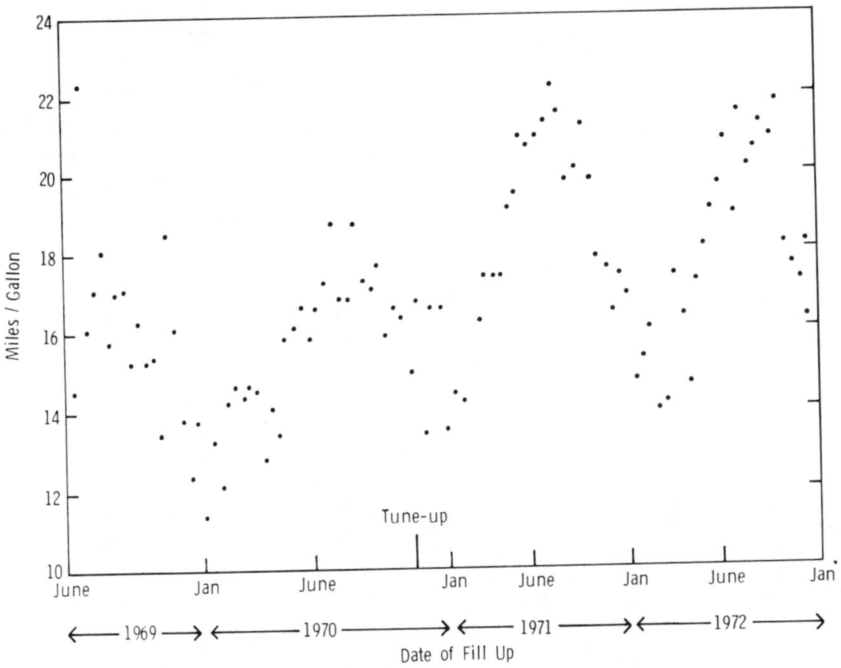

Figure 5 Sequence plot of gasoline mileage data. Note the seasonal variation and the increased average value and decreased variation in mileage after tune-up.

Relationships among More Than Two Variables

Scatter plots directly display relationships between two variables. Values of a third variable can be incorporated in a *labeled scatter plot*, in which each plotted point (whose location designates the values of two variables) is labeled by a symbol designating a level of the third variable. Anderson [1] extended these to "pictorialized" scatter plots, called *glyphs* and *metroglyphs*, where each coordinate point is plotted as a circle and has two or more rays emanating from it; the length of each ray is indicative of the value of the variable associated with that ray. Bruntz et al. [11] developed a variation of the glyph for four variables, called the *weathervane plot*, where values of two of the variables are again indicated by the plotted coordinates and the other two by using variable-size plotting symbols and variable-length arrows attached to the symbols.

Tukey [72], Gabriel's [27] *biplot**, and Mandel [48] provide innovative methods for displaying two-way tables of data. All three approaches involve fitting a model to the table of data and then constructing various plots of the coefficients in the fitted model to study the relationships among the variables.

Chernoff [12] used facial characteristics (*see* CHERNOFF FACES) to display values of up to 18 variables through *face plots*. Each face represents a multivariate datum point, and each variable is represented by a different facial characteristic, such as size of eyes or shape of mouth. Experience has shown that the interpretation of these plots can be affected by how the variables are assigned to facial characteristics.

Exploratory plots of raw data are usually less effective with measurements on four or more variables (e.g., height, weight, age, sex, race, et., of a subject). The usual approach then is to reduce the dimensionality of the data by grouping variables with common properties or identifying and eliminating unimportant variables. The variables plotted may be functions of the original variables as specified by a statistical model. Exploratory analysis* of the data is then conducted with plots of the "reduced" data. Andrews' *Fourier plots* [2], *data clustering* [33], *similarity and preference mapping* [32], and geometrical representations of multidimensional scaling* analyses [44] are examples of such procedures. One often attempts to determine whether there are two or more groups (i.e., clusters) of observations within the data set. When several different groups are identified, the next step is usually to determine why the groups are different. These methods are sophisticated and require computer programs to implement on a routine basis. (See Gnanadesikan [28], Everitt [22], Fienberg [24], and GRAPHICAL REPRESENTATION, COMPUTER-AIDED.)

GRAPHS USED IN THE ANALYSIS OF DATA

The graphical methods discussed next generally depend on assumptions of the analysis. Decisions made from these displays may be either subjective in nature, such as a visual assessment of an underlying distribution, or objective, such as an out-of-control signal from a control chart*.

Distribution Assessment and Probability Plots

The *probability plot* is a widely used graphical procedure for data analysis. Since other graphical techniques discussed in this article require a basic understanding of it, a brief discussion follows (*see also* PROBABILITY PLOTTING).

A probability plot on linear rectangular coordinates is a collection of two-dimensional points specifying corresponding quantiles from two distributions. Typically, one distribution is empirical and the other is a hypothesized theoretical one. The primary purpose is to determine visually if the data could have arisen from the given theoretical distribution. If the empirical distribution is similar to the theoretical one, the expected shape of the plot is approximately a straight line; conversely, large departures from linearity suggest different distributions and may indicate how the distributions differ.

Imagine a sample of size n in which the data $y_1, \dots, y_n$ are independent observations on a random variable Y having some continuous distribution function (df). The ordered data $y_{(i)}, y_{(1)} \leqslant \cdots \leqslant y_{(n)}, i = 1, \dots, n$, represent sample quantiles* and are plotted against theoretical quantiles, $x_i = F^{-1}(p_i)$, where F^{-1} denotes the inverse of F, the hypothesized df of Y. Moreover, $F(y)$ may involve unknown location (ν) and scale (δ) parameters (not necessarily the mean and the standard deviation) as long as $F((y - \nu)/\delta)$ is completely specified. If Y has df F, then $p_i = F(x_i) = F((y_{(i)} - \nu)/\delta)$; x_i is called the reduced $y_{(i)}$-variate and is a function of the unknown parameters.

Selection of p_i for use in plotting has been much discussed. Suggested choices have been $i/(n+1)$, $(i - 1/2)/n$, $(2i-1)/2n$, and $(i - 3/8)/(n + 1/4)$. Kimball [39] discusses some p_i-choices in the context of probability plots.

Now $F^{-1}(p_i)$ is not expressible in closed form for most commonly encountered distributions and thus provides an obstacle to easy evaluations of x_i. An equivalent procedure often employed to avoid this difficulty is to plot $y_{(i)}$ against p_i on *probability paper*, which is rectangular graph paper with an F^{-1} scale for the p-axis. The p_i-entries on the p-axis are commonly called plotting positions. Naturally, a different type of probability paper is needed for each family of distributions, F.

Normal probability paper, with F^{-1} based on the normal distribution, is most common, although many others have been developed (see, e.g., King [40]). Normal probability paper is available in two versions: arithmetic probability paper, where the data axis has a linear scale, and logarithmic probability paper, where the data axis has a natural logarithmic scale. The latter version is used to check for a lognormal distribution*.

To illustrate the procedure, two data sets of size $n = 15$ have been plotted on normal (arithmetic) probability paper shown in Fig.

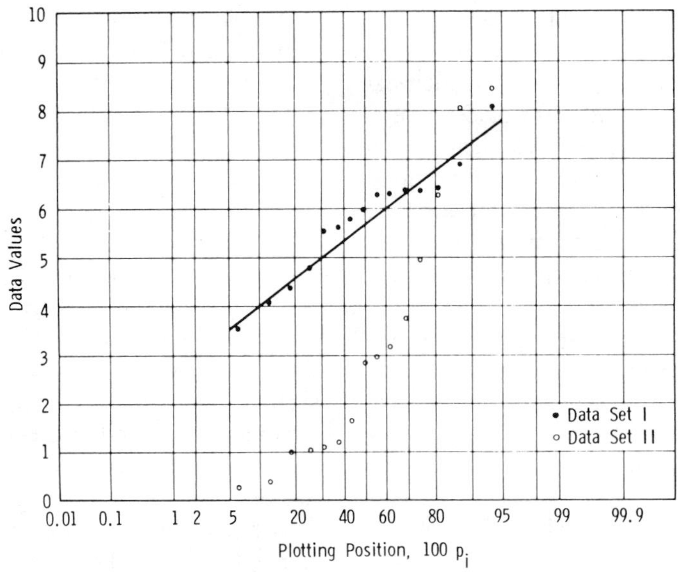

Figure 6 Comparison of two data sets plotted on normal probability paper. Set I can be adequately approximated by a normal distribution, whereas set II cannot.

6, using $p_i = i/(n + 1)$. Here the horizontal axis is labeled in percent; $100p_i$ has been plotted against the ith smallest observation in each set. The plotted points of set I appear to cluster around the straight line drawn through them, visually supportive evidence that these data come from a population which can be adequately approximated by a normal distribution. The points for set II, however, bend upward at the higher percents, suggesting that the data come from a distribution with a longer upper tail (i.e., larger upper quantiles) than the normal distribution.

If set I data are viewed as sufficiently normal, graphical estimates of the mean (μ) and standard deviation (σ) are easily obtained by noting that the 50th percentile of the normal distribution corresponds to the mean and that the difference between the 84th and 50th percentiles corresponds to one standard deviation. The respective graphical estimates from the line fitted by eye for μ and σ are 5.7 and $7.0 - 5.7 = 1.3$, respectively.

The conclusions from Fig. 6 were expected because the data from set I are, in fact, random normal deviates with $\mu = 6$ and $\sigma = 1$. The data in set II are random log-normal deviates with $\mu = 0.6$ and $\sigma = 1$. A plot of set II data on logarithmic probability paper produces a more nearly linear collection of points.

Visual inference, such as determining here whether the collection of points forms a straight line, is fairly easy, but should be used with theoretical understanding to enhance its reliability. A curvilinear pattern of points based on a large sample offers more evidence against the hypothesized distribution than does the same pattern based on a smaller sample. For example, the plot of set I data exhibits some asymmetric irregularities which are due to random fluctuations; however, a similar pattern of irregularity based on a much larger sample would be much more unlikely from a normal distribution. Daniel [16] and Daniel and Wood [17] give excellent discussions on the behavior of normal probability plots.

The probability plot is a special case of the *Q-Q plot* [74], which is a quantile–quantile comparison of two distributions, either or both of which may be empirical or theoretical, whereas a probability plot is typically a display of sample data on probability paper (i.e., empirical versus theoretical). *Q-Q* plots are particularly useful because a straight line will result when comparing the distributions of X and Y, whenever one variable can be expressed as a linear function of the other. *Q-Q* plots are relatively more discriminating in low-density or low-frequency regions (usually the tails) of a distribution than near high-density regions, since in low-density regions quantiles are rapidly changing functions of p. In the plot this translates into comparatively larger distances between consecutive quantiles in low-density areas. The quantiles in Fig. 6 illustrate this, especially the larger empirical ones of set II.

A related plot considered by Wilk and Gnanadesikan [74] is the *P-P* (probability–probability) *plot*. Here, for varying x_i, $p_{i1} = F_1(x_i)$ is plotted against $p_{i2} = F_2(x_i)$, where F_j, $j = 1, 2$, denotes the df (empirical or theoretical). If $F_1 = F_2$ for all x_i, the resulting plot is a straight line with unit slope through the origin. This plot is especially discriminating near high-density regions, since here the probabilities are more rapidly changing functions of x_i than in low-density regions. The *P-P* plot is not as widely used as the *Q-Q* plot since it does not remain linear if either variable is transformed linearly (e.g., by a location or scale change).

For large data sets, an obvious approach for comparison of data with a probability model is a graph of a fitted theoretical density (parameters estimated from data), with the appropriate scale adjustment, superimposed on a histogram. Gross differences between the ordinates of the distributions are easily detected. A translation of the differences to a reference line (instead of a reference curve) to facilitate visual discrimination is easily accomplished by hanging the bars of the histogram from the density curve [73]. Figure 7 illustrates the *hanging histogram*, where the histogram for the carbon monoxide data (Fig. 2) is hung from a lognormal distribution. Slight, systematic variation about the reference line suggests that the data are slightly more skewed to the right in the high-density area than in the lognormal distribution.

Further improvements in detecting systematic variation may be achieved by rootograms [73]. A *hanging rootogram* is analogous to a hanging histogram except that the square roots of the ordinate values are

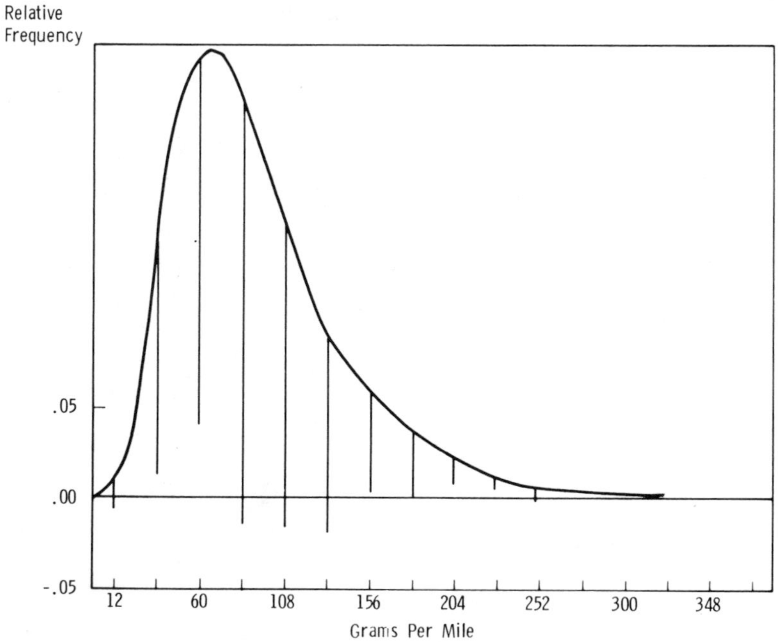

Figure 7 Hanging carbon monoxide histogram from fitted lognormal distribution.

graphed. The *suspended rootogram* is an upside-down graph of the residuals about the baseline of the hanging rootogram.

Graphical assessments for discrete distributions can also be made by comparing the histogram of the data to the fitted probability density, $p(x)$. However, as with continuous distributions, curvilinear discrimination may be difficult and linearizing procedures are helpful. A general approach is to determine a function of $p(x)$, say, $r(x) = r(p(x))$, which is linearly related to a function of x, say, $s(x)$. Then using sample data one calculates relative frequencies to estimate $p(x)$, evaluates $r(x)$, and plots $r(x)$ against $s(x)$. The absence of systematic departures from linearity offers some evidence that the data could arise from density $p(x)$. The slope and intercept will be functions of the parameters and can be used to estimate the parameters graphically. A suitable $r(x)$ may be obtained by simply transforming $p(x)$; for example, taking logarithms of the density of the discrete Pareto distribution*, where $p(x) \propto x^{\lambda}$, gives $r(x) = \log p(x)$ and $s(x) = \log x$. In other cases ratios of consecutive probabilities [e.g., $p(x+1)/p(x)$] are linear functions of $s(x)$ [21]. Table 3 summarizes these ratios for three commonly encountered discrete distributions.

Ord [54] expands on the foregoing ideas by defining a class of discrete distributions where $r(x) = xp(x)/(p(x-1))$ is a linear function of x [i.e., $s(x) = x$)], thereby keeping the same abscissa scale. Distributions in this class are the binomial, negative binomial, Poisson, logarithmic, and uniform.

These graphical tests for discrete distributions may be difficult to interpret because the sample relative frequencies have nonhomogeneous variances. This difficulty may be compounded when using ratios of the relative frequencies as functions of $s(x)$. These procedures are therefore recommended more as exploratory than confirmatory.

Another graphical technique for the Poisson distribution is the *Poissonness plot* [34], similar in spirit to probability plotting. It can also be applied to truncated Poisson data or any one parameter exponential family* of discrete distributions such as the binomial.

For further insights, see Parzen [57].

Model Adequacy and Assumption Verification

Any statistical analysis is based on certain assumptions. Those usually associated with least-squares regression analysis are that experimental errors are independent, have a homogeneous variance, and have a normal (Gaussian) distribution. It is standard practice to check these assumptions as part of the analysis. These checks, most often done graphically, have the desirable by-product of forcing the analyst to look at the data critically. This can be effectively accomplished by graphical analysis of both the raw data and residuals from the fitted model. In addition to assumption verification, this evalua-

Table 3

Distribution	$p(x)$	$r(x)$	=	Intercept	+	Slope	×	$s(x)$
Binomial	$\binom{n}{x}\pi^x(1-\pi)^{n-x}$, $x = 0, \ldots, n$	$\dfrac{p(x+1)}{p(x)}$		$-\dfrac{\pi}{1-\pi}$		$\dfrac{(n+1)\pi}{1-\pi}$		$\dfrac{1}{x+1}$
Poisson	$\dfrac{e^{-\lambda}\lambda^x}{x!}$, $x = 0, 1, 2, \ldots$	$\dfrac{p(x)}{p(x+1)}$		$\dfrac{1}{\lambda}$		$\dfrac{1}{\lambda}$		x
Pascal	$\binom{x-1}{k-1}\pi^k(1-\pi)^{x-k}$, $x = k, k+1, \ldots$	$\dfrac{p(x)}{p(x+1)}$		$\dfrac{1}{1-\pi}$		$\dfrac{k-1}{1-\pi}$		$\dfrac{1}{x}$

tion frequently results in the discovery of unusual observations or unsuspected relationships. Most of the plots discussed below are applications of graphical forms previously discussed.

If repeat observations have been obtained for each of k groups representing different situations or conditions being studied, a scatter plot of the group standard deviation, s_i, vs. the group mean, $\bar{y}_i$, $i = 1, \ldots, k$, will appear random and show little correlation when the homogeneous variance assumption is satisfied. Box et al. [10] point out that if these assumptions are not satisfied, this plot can be used to determine a transformed measurement scale on which the assumptions will be more nearly satisfied (Fig. 8).

The normal distribution assumption can be checked in this situation from a histogram of the residuals*, $r_{ij} = y_{ij} - \bar{y}_i$, between

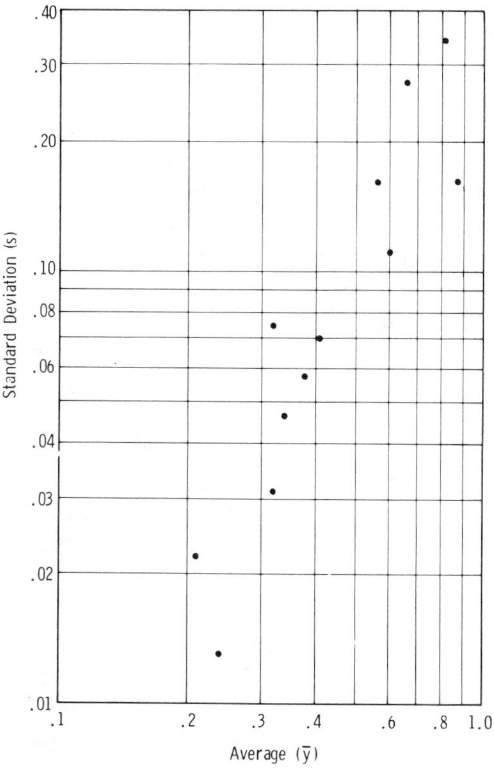

Figure 8 Toxic agent data [10, Table 7.11]. Linear log-log relationship suggests that a power transformation will produce homogeneous variances. The slope of the line indicates the necessary power.

the observations in each group (y_{ij}) and the average of the group ($\bar{y}_i$). This histogram will tend to be bell-shaped if the normal distribution and homogeneous variance assumptions are satisfied. Alternatively, especially for small data sets, the r_{ij}'s may be plotted on normal probability paper. The expected shape is a straight line if these assumptions are appropriate.

Replicate observations often are not available. The analysis assumptions and adequacy of the form of the model can be still checked, however, by constructing plots of the residuals or standardized residuals [18], from the fitted model. The residual associated with observation y_i is $r_i = y_i - \hat{y}_i$, where $\hat{y}_i$ is the value of y_i predicted by the model fitted to the data. Four types of *residual plots* are routinely constructed [19]: plots on normal probability paper, residuals (r_i) vs. predicted values ($\hat{y}_i$), sequence plot of residuals, and residuals (r_i) vs. predictor variables (x_j). Note that although the residuals are not mutually independent, the effect of the correlation structure on the utility of these plots is negligible [5].

The *plot of residuals on normal probability paper* provides a check on the normal distribution assumption. Substantive deviations from linearity may be due to a nonnormal distribution of experimental errors, the presence of atypical (i.e., outlying) data points, or an inadequate model (Fig. 9).

The *residuals* (r_i) *vs. the predicted values* ($\hat{y}_i$) *plot* will show a random distribution of points (no trends, shifts, or peculiar points) if all assumptions are satisfied. Any curvilinear relationship indicates that the model is inadequate (Fig. 10). Nonhomogeneous variance is indicated if the spread in r_i changes with $\hat{y}_i$. When the spread increases linearly with $\hat{y}_i$, a log transformation (i.e., replace y in the analysis by $y' = \log y$) will often produce a response scale on which the homogeneous variance assumption will be satisfied (Fig. 10). Plots of the residuals (r_i) vs. raw observations (y_i) are of little value because they will always show a linear correlation whose value is $(1 - R^2)^{1/2}$, where R^2 is the

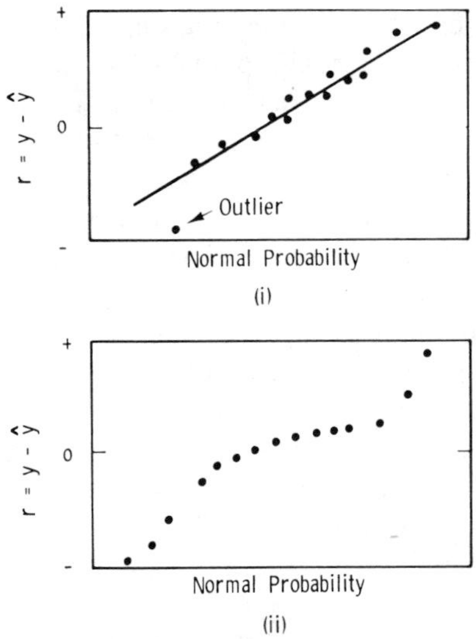

Figure 9 Normal probability plot of residuals. Plot (i) shows an outlying data point and plot (ii) shows a set of residuals not normally distributed.

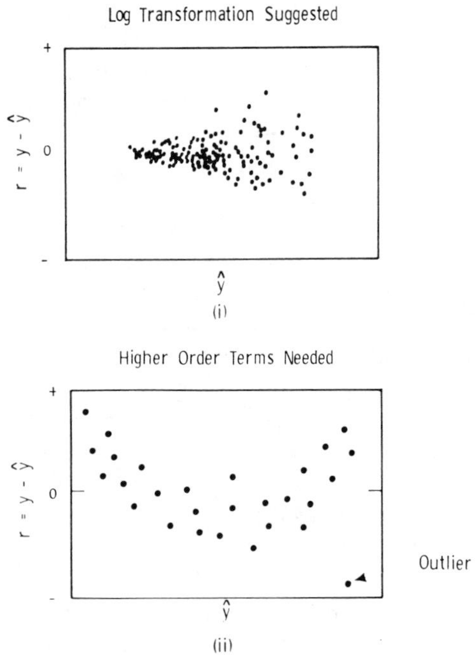

Figure 10 Residuals versus fitted values. Plot (i) shows increased residual variability with increasing fitted values and plot (ii) shows a curvilinear relationship.

coefficient of determination of the fitted model [36].

If all analysis assumptions are satisfied, the *sequence plot of the residuals* (r_i) can be expected to show a random distribution of

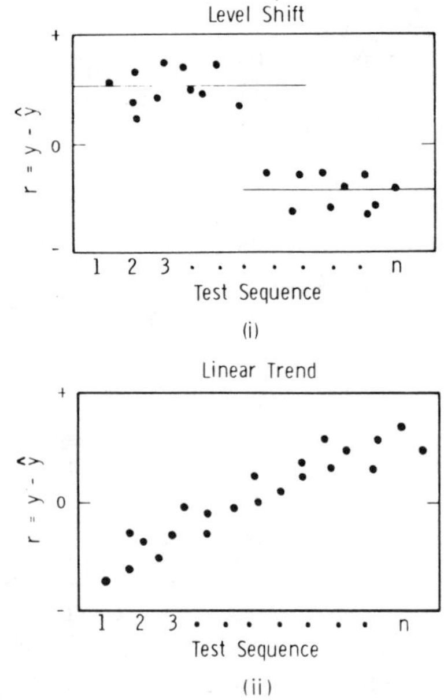

Figure 11 Sequence plot of residuals. Plot (i) shows an abrupt change and plot (ii) shows a gradual change due to factors not accounted for by the model.

points and contain no trends, shifts, or atypical points. Any trends or shifts here suggest that one or more variables not included in the model may have changed during the collection of the data (Fig. 11). This plot may show cycles* in the residuals and other dependencies, indicating that the assumption of independence of experimental errors is not appropriate. This assumption can also be checked by constructing an *autocorrelation plot* of the residuals (see earlier discussion).

The *residuals* (r_i) *vs. the predictor variables* (x_i) *plot* should also show a random distribution of points. Any smooth patterns or trends suggests that the form of the model may not be appropriate (Fig. 12).

The scatter plots of residuals discussed above are not independent of each other. Peculiarities and trends observed in one plot usually show up in one or more of the others. Collectively, these plots provide a good check on data behavior and the model construction process.

With a large number of predictor variables it is sometimes hard to see relationships between y and x_i in scatter plots. Lar-

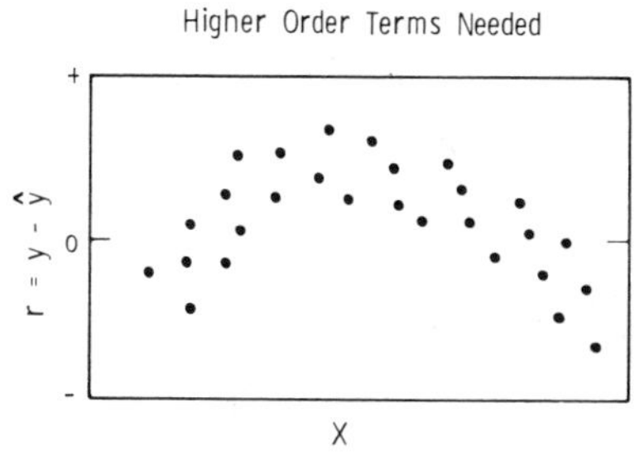

Figure 12 Residuals versus a predictor variable.

sen and McCleary [42] developed the *partial-residual plot* to overcome this problem. Wood [75] and Daniel and Wood [17] refer to these as *component-plus-residual plots*. A regression model must be fitted to the data before the plot can be constructed. In effect, the relationships of all the other variables are removed and the plot of the component-plus-residual vs. x_i displays only the computed relationship between y and x_i and the residual variation in the data.

Plots for Decision Making

At various points in a statistical analysis, decisions concerning the effects of variables and differences among groups of data are made. Statisticians and other scientists have developed a variety of statistical techniques which use graphical displays to make these decisions. In some instances (e.g., control charts, Youden plots) these contain both the data in raw or reduced form and a measure of their uncertainty. The user, in effect, makes decisions from the plot rather than calculating a test statistic. In other situations (e.g., half-normal plot*, ridge trace*) a measure of uncertainty is not available but the analyst makes decisions concerning the magnitude of an effect or the appropriateness of a model by assessing deviations from expected or desired appearance.

The *control chart** is widely used to control industrial production and analytical measurement processes [31]. It is a *sequence plot* of a measurement or statistic (average, range, etc.) vs. time sequence together with limits to reflect the expected random variation in the plotted points. For example, on a plot of sample averages, limits of ± 3 standard deviations are typically shown about the process average to reflect the uncertainty in the averages. The process is considered to be out of control if a plotted average falls outside the limits. This suggests that a process shift has occurred and a search for an assignable cause should be made. The *cumulative sum control chart** (Cusum chart) is another popular process control technique, particularly useful in detecting small process shifts [6, 46].

Ott [56] used the control chart concept to develop his analysis-of-means plotting procedure for the interpretation of data that would ordinarily be analyzed by analysis-of-variance techniques. Schilling [62] systematized Ott's procedure and extended it past the cross-classification designs to incomplete block experiments and studies involving random effects. The analysis-of-means procedure enables those familiar with control chart concepts and technology to develop quickly an ability to analyze the results from experimental designs.

The *Youden plot** [76] was developed to study the ability of laboratories to perform a test procedure. Samples of similar materials, A and B, are sent to a number of laboratories participating in a collaborative test. Each laboratory runs a predetermined number of replicate tests on each sample for a number of different characteristics. A Youden plot is constructed for each measured characteristic. Each point represents a different laboratory, where the average of the replicate results on material A is plotted versus the average results on material B (Fig. 13). Differences along a 45° line reflect between-lab variation. Differences in the direction perpendicular to the 45° line reflect within-lab and lab-by-material interaction* variation. An uncertainty ellipse can be used to identify problem laboratories. Any point outside this ellipse is an indication that the associated laboratory's results are significantly different from those of the laboratories within the ellipse. Mandel and Lashof [49] generalized and extended the construction and interpretation of Youden's plot. Ott

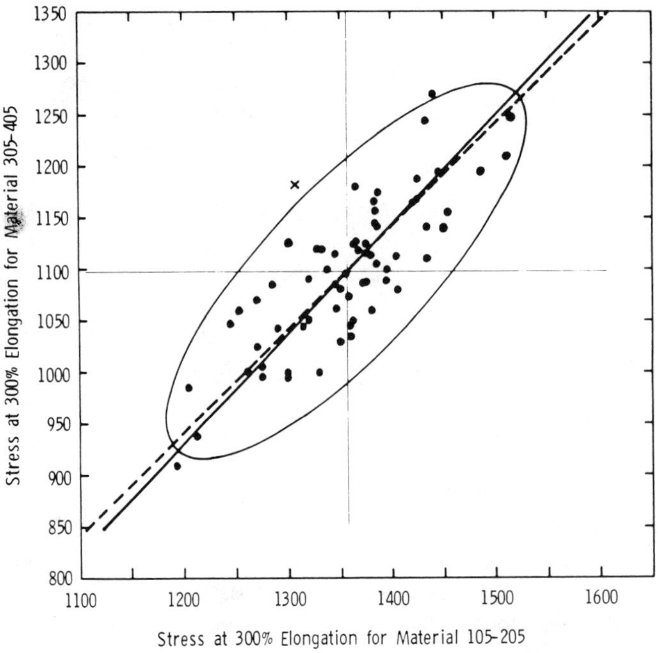

Figure 13 Youden plot [49].

[55] showed how this concept can be used to study paired measurements. For example, "before" and "after" measurements are frequently collected to evaluate a process change or a manufacturing stage of an industrial process.

Daniel [16] developed the half-normal plot* to interpret two-level factorial and fractional factorial experiments*. It displays the absolute value of the $n - 1$ contrasts (i.e., main effects and interactions) from an n-run experiment, versus the probability

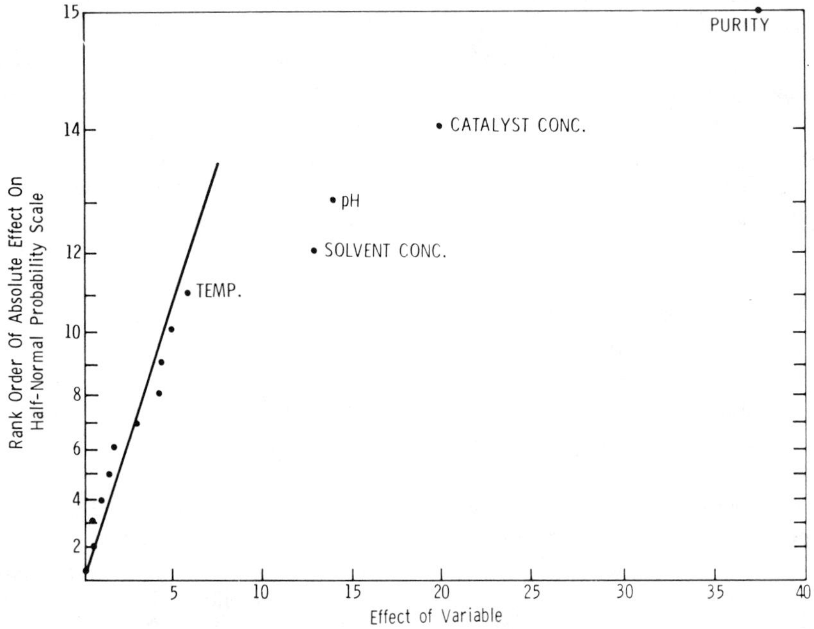

Figure 14 Half-normal plot from a 2^{5-1} factorial experiment showing four important effects on a color response.

scale of the half-normal distribution* (Fig. 14). Identification of large effects (positive or negative) is enhanced by plotting the absolute value of the contrast. Alternatively, to preserve the sign of the contrast, the contrast value may be plotted on normal probability paper [10]. With no significant effects, either plot will appear as a straight line. Significant effects are indicated by the associated contrast falling off the line. Although this plot is usually assessed visually, uncertainty limits and decision guides have been developed by Zahn [77, 78]. The half-normal plot is not restricted to two-level experiments and can be used in the interpretation of any experiment for which the effects can be described by independent one-degree-of-freedom contrasts.

The half-normal plot was significant in marking the beginning of extensive research on probability plotting methods in the early 1960s. During this time the statistical community became convinced of the usefulness and effectiveness of graphical techniques; many of these developments were discussed earlier.

Research in the 1960s and 1970s also focused on regression analysis and the fitting of equations to data when the predictor variables (x's) are correlated. Two displays, the *C_p plot* and the *ridge trace**, were developed as graphical aids in these studies.

The *C_p plot*, suggested by C. L. Mallows and popularized by Gorman and Toman [30] and Daniel and Wood [17], is used to determine which variables should be included in the regression equation. It (Fig. 15) is constructed by plotting, for each equation considered, C_p vs. p, where $C_p = \mathrm{RSS}_p/s^2 - (n - 2p)$, p is the number of

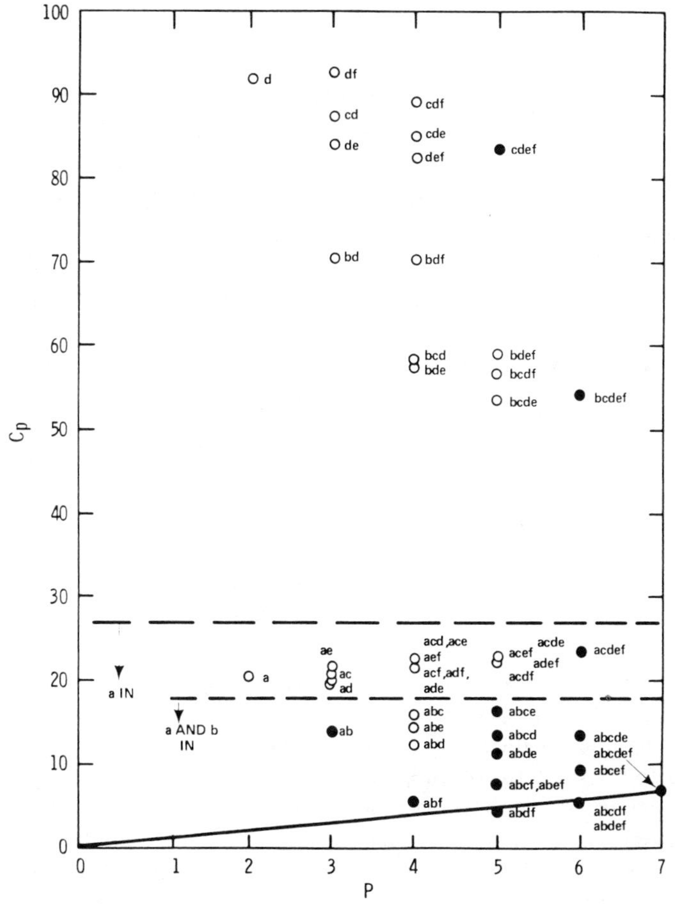

Figure 15 C_p plot [30]. The six variables in the equation are noted by a, b, c, d, e, and f. The number of terms in the equation is denoted by P.

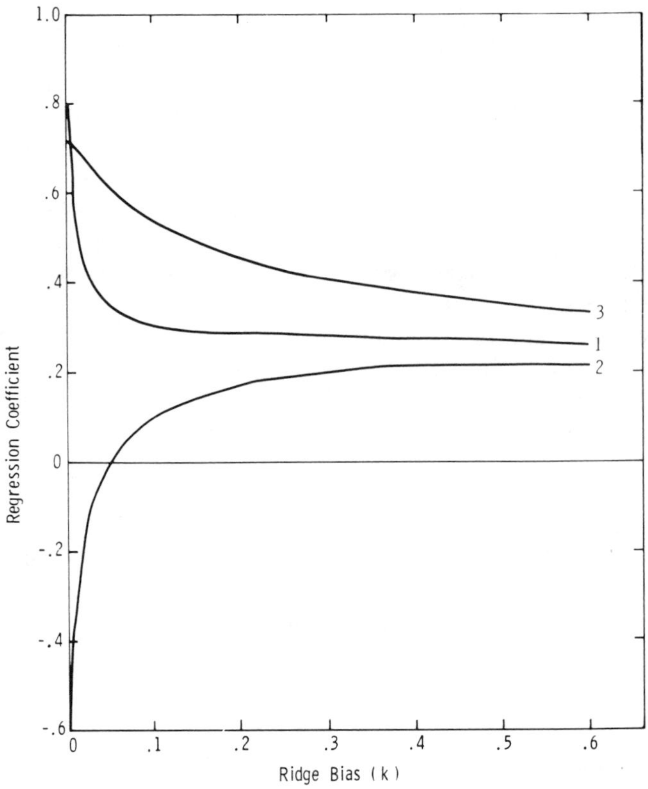

Figure 16 Ridge trace showing instability of regression coefficients 1 and 2.

terms in the equation, RSS_p is the residual sum of squares for the p-term equation of interest, n is the total number of observations, and s^2 is the residual mean square obtained when all the variables are included in the equation. If all the important terms are in the equation, $C_p = p$. The line $C_p = p$ is included on the plot and one looks for the equations that have points falling near this line. Points above the line indicate that significant terms have not been included. The objective is to find the equation with the smallest number of terms for which $C_p \doteq p$.

The *ridge trace*, developed by Hoerl and Kennard [35] and discussed by Marquardt and Snee [50], identifies which regression coefficients are poorly estimated because of correlations among the predictor variables (x's). This is accomplished by plotting the regression coefficients vs. the bias parameter, k, which is used to calculate the ridge regression coefficients, $\hat{\beta} = (\mathbf{X}'\mathbf{X} + k\mathbf{I})^{-1}\mathbf{X}'\mathbf{Y}$ (Fig. 16). Coefficients whose sign and magnitude are affected by correlations among the predictor variables (x's) will change rapidly as k increases. Hoerl and Kennard recommend that an appropriate value for k is the smallest for which the coefficients "stabilize" or change very little as k increases. If the coefficients remain nearly constant for $k > 0$, then $k = 0$ is suggested; this indicates that the least-squares coefficients should be used.

COMMUNICATION AND DISPLAY OF RESULTS

Quantitative Graphics

Quantitative graphics encompass general methods of presenting and summarizing numerical information, usually for easy comprehension by the layman. There can be no complete catalog of these graphics since their form is limited only by one's ingenuity.

Beniger and Robyn [7] trace the historical development since the seventeenth century of quantitative graphics and offer several pictorial examples illustrating the improved sophistication of graphs. Schmid and Schmid [63] in a comprehensive handbook discuss and illustrate many quantitative graphical forms. A few of the more common forms follow briefly.

A *bar chart** is similar in appearance to a histogram (Fig. 2) but more general in application. It is frequently used to make quantitative comparisons of qualitative variables, as a company might do to compare revenues among departments. Comparisons within and between companies are easily made by superimposing a similar graph of another comparable company and identifying the bars.

A *pictogram* is similar to the bar chart except that "bars" consist of objects related to the response tallied. For example, figures of people might be used in a population comparison where each figure represents a specified number of people [38]. When partial objects are used (e.g., the bottom half of a person), it should be stated whether the height or volume of the figure is proportional to frequency.

A *pie chart** is a useful display for comparing attributes on a relative basis, usually a percentage. The angle formed by each slice of the pie is an indication of that attribute's relative worth. Governments use this device to illustrate the number of pennies of a taxpayer's dollar going into each major budget category.

A *contour plot* may effectively depict a relationship among three variables by one or more contours. Each contour is a locus of values of two variables associated with a constant value of the third. A relief map that shows latitudinal and longitudinal locations of constant altitude by contours or isolines is a familiar example. Contours displaying a response surface, $y = f(x_1, x_2)$, are discussed subsequently (Fig. 21). A *stereogram* is another form for displaying spatial relationship of three variables in two dimensions, similar to a draftsman's three-dimensional perspective drawing in which the viewing angle is not perpendicular to any of the object's surfaces (planes).

The field of *computer graphics* has rapidly expanded, offering many new types of graphics. Feinberg and Franklin [23] present a bibliography for these developments. (*See* GRAPHICAL REPRESENTATION, COMPUTER-AIDED, where a discussion and illustration of *statistical color maps* are also presented.)

Summary of Statistical Analysis

Many of the displays previously discussed are useful in summarizing and communicating the results of a study. For example, histograms or dot*-array diagrams are used to display the distribution of data. Box plots are useful (see Fig. 4) for large data sets or when several groups of data are compared.

The results of many analyses may be displayed by a *means plot* of the group means ($\bar{y}$) together with a measure of their uncertainty, such as standard error ($\bar{y} \pm$ SE), confidence limits ($\bar{y} \pm t$SE), or least significant interval limits (LSI $= \bar{y} \pm$ LSD/2). The use of the LSI [3] is particularly advantageous because the intervals provide a useful decision tool (Fig. 17). Any two averages are significantly different at the assigned probability level if and only if their LSIs do not overlap. Intervals based on the standard error and confidence limits for the mean do not have this straightforward interpretation. Similar intervals can be developed for other multiple comparison* procedures, such as those developed by Tukey (honest significant difference), Dunnett, or Scheffé.

Box et al. [10] use the *sliding reference distribution* to display and compare means graphically. The distribution width is determined by the standard error of the means. Any two means not within the bounds of the distribution are judged to be significantly different. They also use this display in the interpretation of factor effects in two-level factorial experiments*.

The *notched-box plot* of McGill et al. [51] is the nonparametric analog of the LSI-type plots discussed above. The median* of the

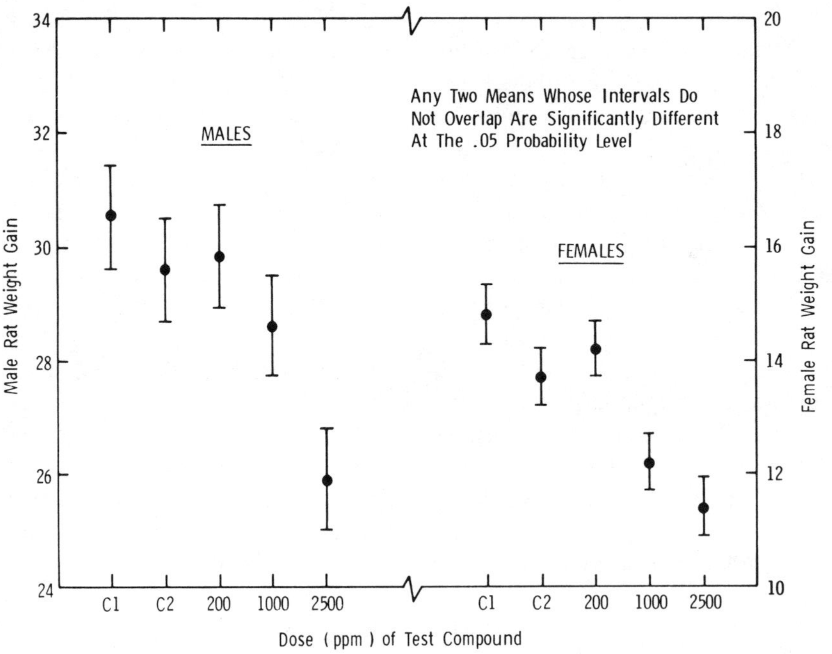

Figure 17 Least significant interval plot [70].

sample and confidence limits for the difference between two medians are displayed in this plot; any two medians whose intervals do not overlap are significantly different. McGill et al. [51] discuss other variations and applications of the box plot.

Factorial experimental designs* are widely used in science and engineering. In many instances the effects of two and three variables are displayed on *factor space/response plots* with squares and cubes similar to those shown in Fig. 18. The numbers in these

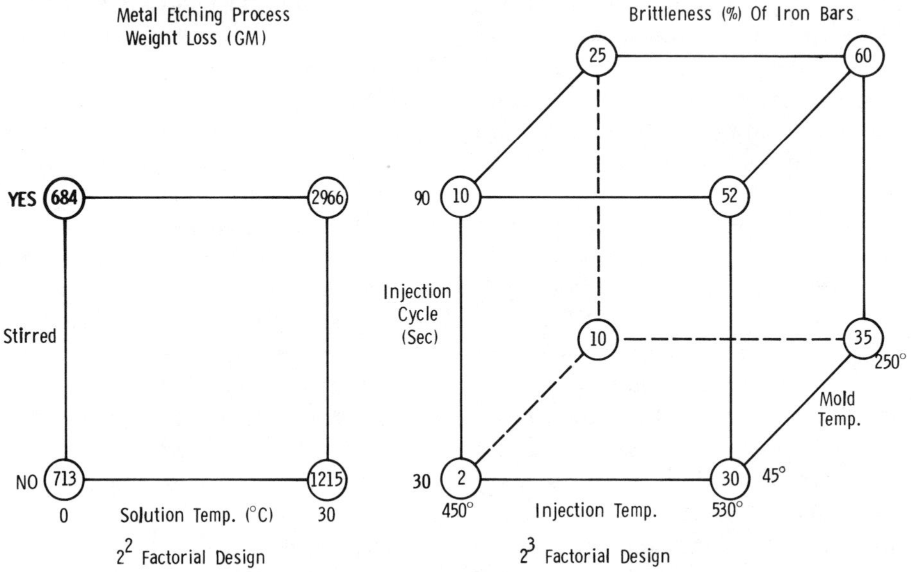

Figure 18 Factor space/response plots showing results of 2^2 and 2^3 experiments. The average responses are shown at the corners of the figures [8].

figures are the mean responses obtained at designated values of the independent variables studied. These figures show the region, or factor space, over which the experiments were conducted, and aid the investigator in determining how the response changes as one moves around in the experimental region.

An *interaction plot* is used to study the nature and magnitude of the interaction effect of two variables from a designed experiment by plotting the response (averaged over all replicates) versus the level of one of the independent variables while holding the other independent variable(s) fixed at a given level. In Fig. 19 it is seen by the nonparallel lines that the size of the effect of solution temperature on weight loss depends on whether or not stirring occurred. Solution temperature and stirring are said to *interact*, another way of saying that their effects are not additive. Monlezun [52] discusses ways of plotting and interpreting three-factor interactions.

The objective of many experiments is to develop a prediction model for the response (y) of the system as a function of experimental (predictor) variables. A typical two-predictor variable model, based on a second-order Taylor series, is

$$y = b_0 + b_1X_1 + b_2X_2 + b_{12}X_1X_2 + b_{11}X_1^2 + b_{22}X_2^2,$$

where the b's are coefficients estimated by regression analysis techniques. One of the best ways to understand all the effects described by this equation is by a *contour plot*, which gives the loci of X_1 and X_2 values associated with a fixed value of the response. By constructing contours on a rectangular X_1–X_2 grid for a series of fixed values of y, one obtains a global picture of the response surface (Fig. 20a). The response surface of a mixture system such as a gasoline or paint can be displayed by a contour plot on a triangular grid [13, 68]. Another use of trilinear coordinates is shown in Fig. 22.

Predicted response plots help in interpreting interaction terms in regression equations

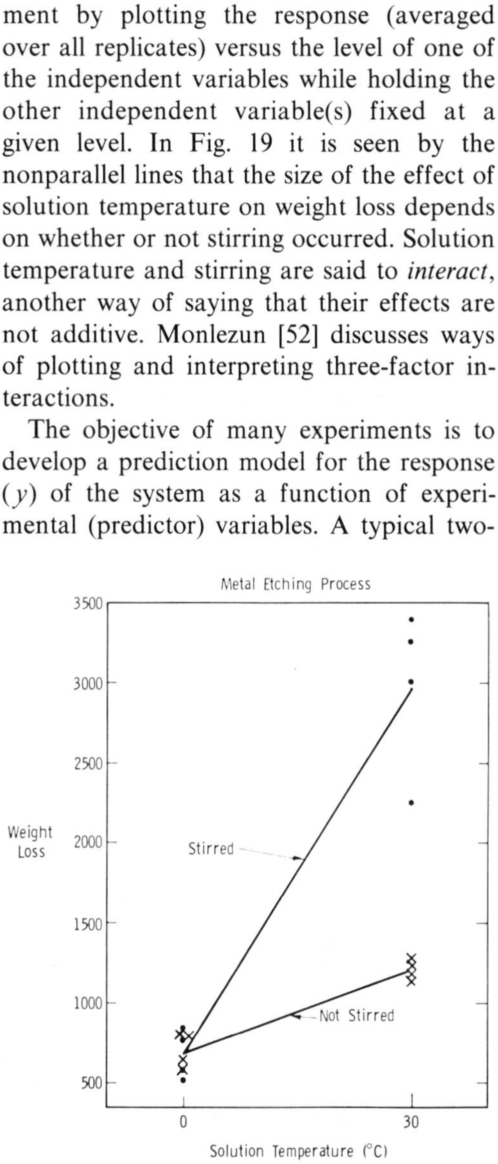

Figure 19 Nonparallelism of lines suggests an interaction between stirring and solution temperature. Stirring has a larger effect at 30°C than at 0°C.

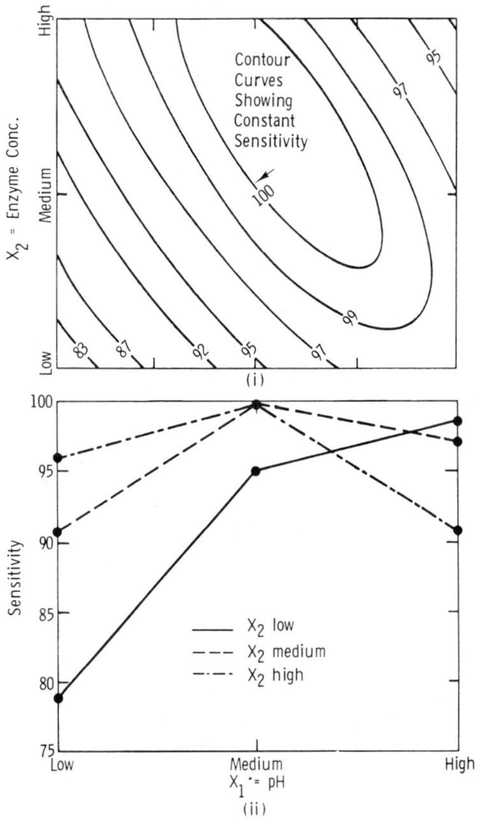

Figure 20 Contour plot of (1) quadratic response surface and (ii) X_1X_2-interaction plot [61].

[64]. The interaction between X_i and X_j is studied by fixing the other variables in the equation at some level (e.g., set $X_k = \overline{X}_k$, $k \neq i, j$) and constructing an interaction plot of the values predicted by the equation for different values of X_i and X_j (Fig. 20b). Similar plots are also useful in interpreting response surface models for mixture systems [67] and in the analysis of the results of mixture screening experiments [69].

A contour plot is also the basis for a *confidence region* plot*, used to display a plausible region of simultaneous hypothesized values of two or more parameters. For example, Fig. 21 [65] shows a contour representing the joint confidence region for regression parameters β_1 and β_2 associated with each of four coal classes. The figure clearly shows that these four groups do not all have the same values of regression parameters.

Confidence region plots on trilinear coordinates [66] can also be used to display and identify variable dependence in two-way contingency tables* where one variable is partitioned into three categories (see also Draper et al. [20]). A two-dimensional plot results (Fig. 22) from the constraint that the sum of the three probabilities, π_1, π_2, and π_3, is unity. It is also possible to display three-dimensional response surfaces or confidence regions via a series of two-dimensional slices through three-dimensional space or with special computer graphics systems that utilize cathode ray tubes (i.e., TV screens) to display the regions.

GRAPHICAL AIDS

Statisticians, like chemists, physicists, and other scientists, rely on graphical devices to help them do their job more effectively. The following graphs are used as "tools of the trade" in offering a parsimonious representation for complex functional relationships among two or more variables.

Graphs of *power* functions* or *operating characteristics* (e.g., Natrella [53]) are used for the evaluation of error probabilities of hypothesis tests (*see* HYPOTHESIS TESTING) when expressed as a function of the unknown parameter(s). Test procedures are easily compared by superimposing two power curves on the same set of axes. *Sample-size curves* are constructed from a family

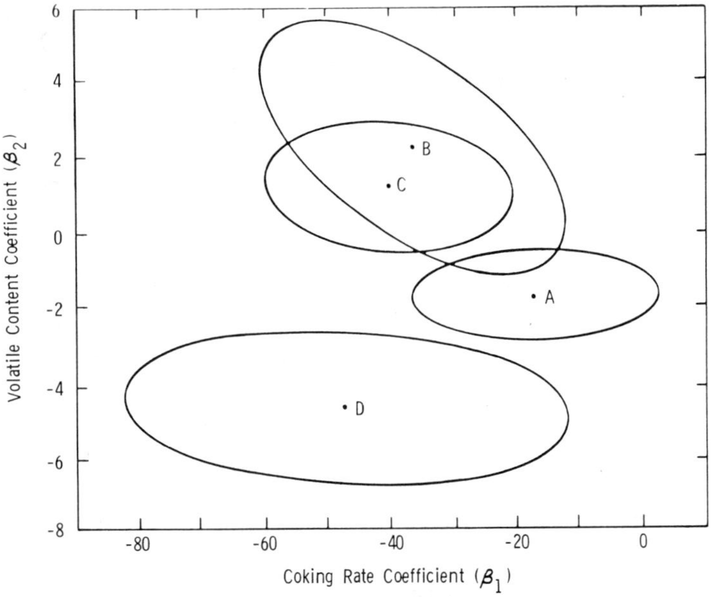

Figure 21 Joint 95% β_1–β_2 confidence regions for coal classes A, B, C, and D [65].

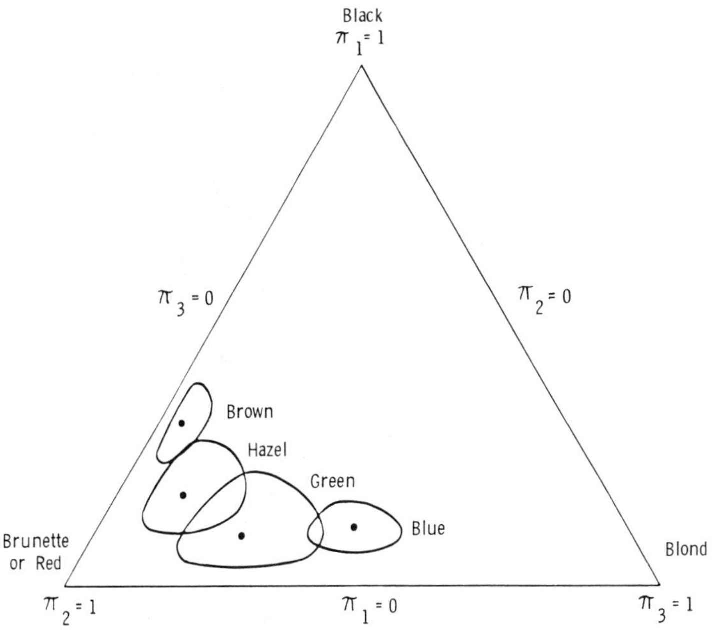

Figure 22 Joint 95% confidence region plots for hair color probabilities for brown, hazel, green, and blue eye colors [66].

of operating characteristic curves, each associated with a different sample size. These curves are useful in planning the size of an experiment to control the chances of wrong decisions. Natrella [53] offers a number of these for some common testing situations.

Similar contour graphics using families of curves indexed by sample size are also useful for determining *confidence limits*. These are especially convenient when the end points are not expressible in closed form, such as those for the correlation coefficient or the success probability in a Bernoulli sample [58].

Nomographs are graphical representations of mathematical relationships, frequently involving more than three variables. Unlike most graphics, they do not offer a picture of the relationship, but only enable the determination of the value of (usually) any one variable from the specification of the others. Levens [43] discusses techniques for straight-line and curved scale nomograph construction. Other statistical nomographs have appeared in the *Journal of Quality Technology**; *see also* NOMOGRAMS.

References

[1] Anderson, E. (1960). *Technometrics*, **2**, 387–391.

[2] Andrews, D. F. (1972). *Biometrics*, **28**, 125–136.

[3] Andrews, H. P., Snee, R. D., and Sarner, M. H. (1980). *Amer. Statist.*, **34**, 195–199.

[4] Anscombe, F. J. (1973). *Amer. Statist.*, **27**, 17–21.

[5] Anscombe, F. J. and Tukey, J. W. (1963). *Technometrics*, **5**, 141–160.

[6] Barnard, G. A. (1959). *J. R. Statist. Soc. B*, **21**, 239–271.

[7] Beniger, J. R. and Robyn, D. L. (1978). *Amer. Statist.* **32**, 1–11. (The history of graphics from the seventeenth to the twentieth centuries is traced with special attention to uses in the display of spatial organization, continuous distributions, discrete comparisons, and multivariate distributions and correlations. Extensive references.)

[8] Bennett, C. A. and Franklin, N. L. (1954). *Statistical Analysis in Chemistry and the Chemical Industries*. Wiley, New York.

[9] Box, G. E. P., and Jenkins, G. M. (1970). *Time Series Analysis: Forecasting and Control*. Holden-Day, San Francisco.

[10] Box, G. E. P., Hunter, W. G., and Hunter, J. S. (1978). *Statistics for Experimenters*. Wiley-Interscience, New York. (Several graphical displays used in the analysis and interpretation of designed experiments are discussed.)

[11] Bruntz, S. M., Cleveland, W. S., Kleiner, B., and Warner, J. L. (1974). *Proc. Symp. Atmos. Diffus. Air Pollut., Amer. Meteorol. Soc.*, pp. 125–128.

[12] Chernoff, H. (1973). *J. Amer. Statist. Ass.*, **68**, 361–368.

[13] Cornell, J. A. (1981). *Experiments with Mixtures*, Wiley-Interscience, New York.

[14] Cox, D. R. (1978). *Appl. Statist.*, **27**, 4–9.

[15] Craver, J. S. (1980). *Graph Paper from Your Copier*, H. P. Books, Tucson, Ariz. (Over 200 graph papers that may be copied without permission of the publisher. Elementary introduction to plotting techniques.)

[16] Daniel, C. (1959). *Technometrics*, **1**, 311–341. (The half-normal plot marked the beginning of the extensive research activity on probability plotting methods.)

[17] Daniel, C. and Wood, F. S. (1980). *Fitting Equations to Data*, 2nd ed. Wiley, New York.

[18] Draper, N. R. and Behnken, D. W. (1972). *Technometrics*, **14**, 101–111.

[19] Draper, N. R. and Smith, H. (1981). *Applied Regression Analysis*, 2nd ed. Wiley, New York. (Graphical displays for regression interpretation and diagnostics.)

[20] Draper, N. R., Hunter, W. G., and Tierney, D. E. (1969). *Technometrics*, **11**, 309–320.

[21] Dubey, S. D. (1966). *Amer. Statist.*, **20**, 23–24.

[22] Everitt, B. S. (1978). *Graphical Techniques for Multivariate Data*. North-Holland, Amsterdam. (Broad, medium-level discussions and illustrations of graphical displays for multivariate data.)

[23] Feinberg, B. M. and Franklin, C. A. (1975). *Social Graphics Bibliography*. Bureau of Social Science Research, Washington, D.C.

[24] Fienberg, S. E. (1979). *Amer. Statist.*, **33**, 165–178. (A review of some of the highlights in the historical development of statistical graphics and a taxonomy of the current use of graphical methods. Extensive references.)

[25] Fourier, J. B. J. (1821). *Recherches statistiques sur la Ville de Paris et le Department de la Seine*, Vol. 1, pp. 1–70.

[26] Freund, J. E. (1976). *Statistics: A First Course*, 2nd ed. Prentice-Hall, Englewood Cliffs, N.J.

[27] Gabriel, K. R. (1971). *Biometrika*, **58**, 453–467.

[28] Gnanadesikan, R. (1977). *Methods for Statistical Data Analysis of Multivariate Observations*. Wiley, New York.

[29] Gnanadesikan, R. and Wilk, M. B. (1969). In *Multivariate Analysis*, Vol. 2, P. R. Krishnaiah, ed. Academic Press, New York, pp. 593–637.

[30] Gorman, J. W. and Toman, R. J. (1966). *Technometrics*, **8**, 27–51.

[31] Grant, E. L. and Leavenworth, R. S. (1980). *Statistical Quality Control*, 5th ed. McGraw-Hill, New York.

[32] Green, P. E. and Carmone, F. J. (1970). *Multidimensional Scaling and Related Techniques in Marketing Analysis*. Allyn and Bacon, Boston.

[33] Hartigan, J. A. (1975). *Clustering Algorithms*. Wiley-Interscience, New York.

[34] Hoaglin, D. C. (1980). *Amer. Statist.*, **34**, 146–149.

[35] Hoerl, A. E. and Kennard, R. W. (1970). *Technometrics*, **12**, 55–70.

[36] Jackson, J. E. and Lawton, W. H. (1967). *Technometrics*, **9**, 339–341.

[37] Johnson, N. L. and Leone, F. C. (1977). *Statistics and Experimental Design in Engineering and the Physical Sciences*, 2nd ed., Vol. 1. Wiley, New York.

[38] Joiner, B. L. (1975). *Int. Statist. Rev.*, **43**, 339–340.

[39] Kimball, B. F. (1960). *J. Amer. Statist. Ass.*, **55**, 546–560.

[40] King, J. R. (1971). *Probability Charts for Decision Making*. Industrial Press, New York. (The use of a wide variety of probability plotting papers as problem-solving tools is discussed and illustrated.)

[41] Lalanne, L. (1845). Appendix to *Cours complet de météorologie de L. F. Kaemtz*, translated and annotated by C. Martins, Paris.

[42] Larsen, W. and McCleary, S. (1972). *Technometrics*, **14**, 781–790.

[43] Levens, A. S. (1959). *Nomography*. Wiley, New York.

[44] Lingoes, J. C., Roskam, E. E., and Borg, I. (1979). *Geometrical Representations of Relational Data—Readings in Multidimensional Scaling*. Mathesis Press, Ann Arbor, Mich. [Extensive discussion on the emergence and development of a great variety of computerized procedures (graphical and analytical) for multidimensional scaling.]

[45] Lorenz, M. O. (1905). *J. Amer. Statist. Ass.*, **9**, 209–219.

[46] Lucas, J. M. (1976). *J. Quality Tech.*, **8**, 1–12.

[47] Mahon, B. H. (1977). *J. R. Statist. Soc. A*, **140**, 298–307.

[48] Mandel, J. (1971). *Technometrics*, **13**, 1–18.

[49] Mandel, J. and Lashof, T. W. (1974). *J. Quality Tech.*, **6**, 22–36.

[50] Marquardt, D. W. and Snee, R. D. (1975). *Amer. Statist.*, **29**, 3–20.

[51] McGill, R., Tukey, J. W., and Larsen, W. A. (1978). *Amer. Statist.*, **32**, 12–16.

[52] Monlezun, C. J. (1979). *Amer. Statist.*, **33**, 63.

[53] Natrella, M. G. (1963). *Experimental Statistics, National Bureau of Standards Handbook 91*. U.S. Government Printing Office, Washington, D.C.

[54] Ord, J. K. (1967). *J. R. Statist. Soc. A*, **13**, 232–238.

[55] Ott, E. R. (1957). *Ind. Quality Control*, **13**, 1–4.

[56] Ott, E. R. (1967). *Ind. Quality Control*, **24**, 101–109. (A control chart procedure for the analysis of cross-classified data, typically analyzed by analysis-of-variance techniques, is developed.)

[57] Parzen, E. (1979). *J. Amer. Statist. Ass.*, **74**, 105–121.

[58] Pearson, E. S. and Hartley, H. O. (1970). *Biometrika Tables for Statisticians*, Vol. 1. Cambridge University Press, Cambridge.

[59] Perozzo, L. (1880). *Ann. Statist.*, **12**, 1–16.

[60] Playfair, W. (1786). *The Commercial and Political Atlas*. London.

[61] Rautela, G. S., Snee, R. D., and Miller, W. K. (1979). *Clin. Chem.*, **25**, 1954–1964.

[62] Schilling, E. G. (1973). *J. Quality Tech.*, Part 1, **5**, 93–108; *ibid.*, Parts 2 and 3, **5**, 147–159.

[63] Schmid, C. F. and Schmid, S. E. (1979). *Handbook of Graphic Presentation*, 2nd ed. Wiley, New York. (Excellent dissertation on quantitative graphical forms for statistical data. Numerous illustrations.)

[64] Snee, R. D. (1973). *J. Quality Tech.*, **5**, 67–79.

[65] Snee, R. D. (1973). *J. Quality Tech.*, **5**, 109–122.

[66] Snee, R. D. (1974). *Amer. Statist.*, **28**, 9–12.

[67] Snee, R. D. (1975). *Technometrics*, **17**, 425–430.

[68] Snee, R. D. (1979). *Chemtech*, **9**, 702–710.

[69] Snee, R. D. and Marquardt, D. W. (1976). *Technometrics*, **18**, 19–29.

[70] Snee, R. D., Acuff, S. K., and Gibson, J. R. (1979). *Biometrics*, **35**, 835–848.

[71] Tukey, J. W. (1972). In *Statistical Papers in Honor of George W. Snedecor*, T. A. Bancroft, ed. Iowa State University Press, Ames, Iowa.

[72] Tukey, J. W. (1977). *Exploratory Data Analysis*. Addison-Wesley, Reading, Mass. (The philosophy and tools of exploratory data analysis are developed and discussed by their originator, J. W. Tukey.)

[73] Wainer, H. (1974). *Amer. Statist.*, **28**, 143–145.

[74] Wilk, M. B. and Gnanadesikan, R. (1968). *Biometrika*, **55**, 1–17. (Classic reference on graphical techniques using the empirical cumulative distribution function, *Q-Q* plots, *P-P* plots, and hybrids of these. Medium level.)

[75] Wood, F. S. (1973). *Technometrics*, **15**, 677–695.

[76] Youden, W. J. (1959). *Ind. Quality Control*, **15**, 133–137. (Original paper on the Youden plot for comparing test methods between and within laboratories.)

[77] Zahn, D. A. (1975). *Technometrics*, **17**, 189–200.

[78] Zahn, D. A. (1975). *Technometrics*, **17**, 201–212.

(BIPLOT
CHERNOFF FACES
EXPLORATORY DATA ANALYSIS
EYE ESTIMATE
GRAPHICAL REPRESENTATION, COMPUTER-AIDED
HALF-NORMAL PLOTS
HISTOGRAMS
PROBABILITY PLOTTING
RESIDUALS
RIDGE TRACE
SCATTER PLOTS
STEREOGRAMS
YOUDEN PLOTS)

RONALD D. SNEE
CHARLES G. PFEIFER

GRAPHS, RANDOM *See* RANDOM GRAPHS

GRAPH-THEORETIC CLUSTER ANALYSIS

A *cluster* is a maximal collection of suitably similar objects drawn from a larger collection of objects. Those cluster analysis procedures that determine a partition of the objects into clusters or determine a hierarchy of cluster partitions (*see* HIERARCHICAL CLUSTER ANALYSIS; CLASSIFICATION; DENDROGRAMS) are often considered as classification procedures. The entries CLASSIFICATION and HIERARCHICAL CLUSTER ANALYSIS treat extensively several cluster analysis methods where the association data for the objects is assumed available in the form of an *objects* × *variables* attribute matrix and/or an *objects* × *objects* similarity (or distance) matrix. In contrast to these numeric-matrix oriented cluster analysis procedures, graph-theoretic cluster analysis provides a simpler combinatorial cluster analysis model. This is most appropriate where either the raw data are in the form of a similarity relation or where the number of objects is too large for distance matrix methods to be computationally tractable. The entries CLASSIFICATION and HIERARCHICAL CLUSTER

ANALYSIS should be consulted in conjunction with this entry for comparative features of cluster analysis models.

In the graph-theoretic model of cluster analysis, objects are represented by vertices and those pairs of objects satisfying a particular similarity relation are termed adjacent and constitute the edges of the graph. Clusters are then characterized by appropriately defined subgraphs (*see also* GRAPH THEORY).

We first summarize several instances where graph-theoretic cluster analysis is most appropriate. We then identify some computational advantages of the graph theoretic model. Graph-theoretic characterizations of clusters are then reviewed with regard to properties such clusters must obtain. The graph-theoretic methods are compared with analogous distance matrix methods in most cases.

WHEN IS GRAPH-THEORETIC CLUSTER ANALYSIS APPROPRIATE?

Several instances where the graph-theoretic model of cluster analysis is most appropriate are:

1. **Relational Association Data.** The association data between objects is a single (algebraic) relation on the objects.

Example. A sociological study of employee work habits where the data are the algebraic relation indicating each pair of employees that work well together.

2. **Sparse* Association Data.** For each object there are data given only on those relatively few objects that are most similar to the object.

Example. In a study of economic activities, each firm provides information on those other firms that are felt to be significant competitors.

3. **Computationally Intractable Distance Matrix.** Suppose that the number of objects is very large (e.g., $n = 5000$), so that the full distance matrix is too large ($n^2 = 25{,}000{,}000$) to compute and/or store efficiently. In this case it may be possible to determine only a limited number of pairwise distances corresponding to all distances below some threshold. If a procedure allows the resulting number m of sufficiently similar object pairs to be computed in time $O(m)$ and satisfy $m \ll n^2$, computational tractability may be achievable.

Example. Suppose that the object data can be considered as points in k-dimensional space. Then distribute the points into k-dimensional cells bounded in size related to the threshold, and determine all objects at distance less than the threshold for each given object by only investigating other objects in the same or appropriate neighboring cells.

4. **Ordinal Pairwise Association Data.** A ranking of all pairs of objects in their order of similarity is available either fully or to some threshold level. The data may then be considered as a hierarchy of graphs or as in Matula [6] by a *proximity graph*, which is a graph where the edge set satisfies an order relation.

Example. Suppose that the raw *objects* × *variables* data in a taxonomy application contains considerable nonnumeric data, such as color, shape, or other nominal data*. Computation of any meaningful real-valued distance function between each pair of objects may be considered too subjective and therefore unreliable for the application of distance matrix cluster analysis methods. However, ranking of all object pairs that are sufficiently close by some acceptably objective criteria may be possible, and thus provide the basis for application of graph-theoretic cluster determination methods on the resulting proximity graph.

COMPUTATIONAL ADVANTAGES OF THE GRAPH-THEORETIC MODEL

Storage space and computation time are the critical requirements that eventually render a cluster determination algorithm intractable as the size of the application grows. Graph-theoretic cluster determination algorithms can incorporate efficiencies in storage space and execution time that allow problems with a relatively large number of objects to be investigated.

Storage Space

Data structures available for efficient representation of graphs are:

1. **Packed Adjacency Matrix.** For a computer with a 32-bit word size, the $n \times n$ $\{0, 1\}$-adjacency matrix can be stored in packed binary form in $n^2/32$ words. For 1000 objects we then need less than 33,000 words for the packed adjacency matrix, whereas the distance matrix on the same 1000 objects would require 1,000,000 words.
2. **Sequential Adjacency List Structure.** For each object a sequential list of the adjacent objects is stored. These sequential lists are concatenated into one long list with pointers indicating the start and end of the objects adjacent to each given object. For 5000 objects with the threshold for the similarity relation set so that an average of about 50 objects are adjacent to each object, the sequential adjacency list structure requires only about 255,000 words of storage. The full distance matrix on 5000 objects would require 25 million words of storage.
3. **Linked Adjacency List Structure.** The adjacent pairs are linked together, with pointer fields associated with every object pair allowing determination of the objects adjacent to any given object by a linked-list traversal. This linked structure can handle the storage of a proximity graph where the edges form an ordered set (e.g., ranked similarities) simply by prescribing the sequential order of the pairs as the rank order of the edges, as noted by Matula [6]. Storage space for the pointer fields increases the storage requirement by a factor of 2 to 3 over the sequential adjacency structure, but still provides a great advantage over the full distance matrix representation when the number of adjacent pairs m satisfies $m \ll n^2$.

Execution Time

The design of cluster determination algorithms can incorporate appropriate graph data structures to realize efficiencies in execution time. For the packed adjacency matrix, logical operations on words provide a convenient level of parallelism to speed up some of the combinatorial subprocedures of certain cluster algorithms. For a cluster analysis problem on n objects with m adjacent object pairs in the relation, note that the adjacency list data structure requires only $O(m)$ as opposed to $O(n^2)$ storage. Furthermore, algorithm design techniques incorporating these data structures can result in certain subprocedures or possibly the whole cluster determination algorithm possessing an execution time of $O(m)$ rather than $O(n^2)$. Since $m \ll n^2$ might often be associated with applications for large values of n, these dual savings in space and time are significant features of the graph theoretic model. General techniques for integration of graph data structures and algorithm design are considered by Aho et al. [1].

GRAPH-THEORETIC CHARACTERIZATION OF CLUSTERS

A major advantage of the graph-theoretic model of cluster analysis is that the characterization of clusters can be based on rigorous graph-theoretic properties which are *independent* of the method or algorithm for determination of the clusters. Given the characterization of clusters as specific kinds of subgraphs, the question of algorithm and data structure design to determine these subgraphs can then be approached separately. The following cluster characterizations are discussed first in terms of their graph-theoretic characterization and properties, and separately in terms of their computational tractability. Correspondences with established distance matrix methods are noted. Questions regarding "divisive" and/or "agglomerative" cluster methods which pervade the literature on distance-based cluster methods (*see* HIERARCHICAL CLUSTER ANALYSIS) are relegated to the area of algorithmic techniques in the graph model. Heuristic distance-based cluster methods which might yield different clusters due to accidental features such as the ordering of rows and columns or how "ties" in sequential choice criteria are resolved are inherently avoided in the graph model.

Components/Single Linkage

For a specific graph, the components (i.e., the maximal connected subgraphs) are the clusters. The objects are then uniquely partitioned into clusters. A common problem with this characterization is that some clusters may be very weakly connected, appearing "long and stringy." For the proximity graph, where the edges are ordered and implicitly introduced in "levels," the hierarchy of cluster partitions provided by the components of the graph at each level forms the dendrogram* that is referred to as the "nearest neighbor*" or "single linkage" cluster method in the distance matrix cluster model. For a graph with n vertices and m edges, the components can be found in "linear" time $O(n + m)$ and linear space $O(n + m)$. For the proximity graph where the m edges are ordered and sufficient in number to connect the graph, the dendrogram for the single linkage method can be computed in $O(n + m)$ time if the edges are provided in order, and $O(n + m \log m)$ time if the edges must be sorted. Reference to Aho et al. [1] for the minimum-weight spanning tree problem suggests the appropriate algorithmic approaches.

Cliques/Complete Linkage

For a specific graph, the cliques (i.e., the maximal complete subgraphs) are the clusters. Each cluster is then completely connected, but the number of distinct cliques in an n-vertex graph can be as large as $n^{n/3}$. Experience shows practically that enumeration of all cliques is too costly in time even for relatively small graphs, and even then the cliques may overlap too widely for meaningful interpretation.

Enumeration of cliques should be compared with the complete linkage method in the distance matrix cluster model. The complete linkage method effects a partition into cliques at every threshold level, avoiding the problem of enumerating all cliques. However, the complete linkage method is well known to be very sensitive to the order of the edges. A single transposition in the edge order in the proximity graph can allow a widely different cluster partition over *most* of the threshold levels by complete linkage, and the method is not even well defined if two edges "tie" in rank order. See Matula [6] for an illustrative example.

Min Degree/k-Linkage

For a specific graph, the subgraphs that are connected and maximal with respect to their minimum degree are the clusters. For any given k, the subgraphs that are maximal connected with minimum degree at least k, together with the individual vertices that are in no subgraph of minimum degree as high as k, form a partition of the vertices. Considering values of k as levels of similarity, the min degree method then determines a hierarchical clustering from a single graph. This method provides the opportunity of determining a hierarchical classification* even though the data may be simply a single

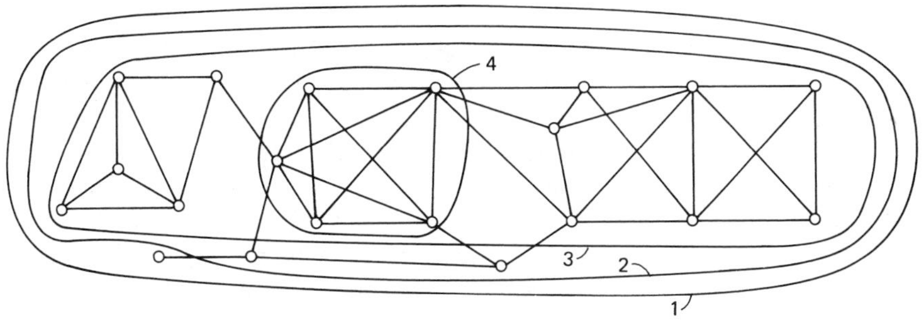

Figure 1 Min degree cluster hierarchy for a graph. The subgraphs that are maximal connected of minimum degree k are shown with corresponding values of k.

similarity relation on the objects, as seen in Fig. 1.

A problem with the min degree method is that two otherwise disjoint maximal connected subgraphs of minimum degree k must be joined in the presence of a single edge between them. The min degree cluster hierarchy is discussed by Matula [6], who notes that the hierarchy can be computed in time linear with the number m of object pair similarities (i.e., the number of edges in the graph).

A related distance matrix cluster method is the k-linkage [or (k, r)-cluster] procedure introduced by Ling [4]. For any *fixed* k, the proximity graph interpretation of k-linkage obtains for each level (in the edge rank order) a partition into maximal connected subgraphs of minimum degree at least k. For a proximity graph with m edges the k-linkage cluster hierarchy can be computed in time $O(m \log m)$ if the edges need to be sorted, and in time $O(m)$ if the edges are given in sorted order. Both the k-linkage and the min degree procedures are tractable for applications with a large number of objects.

k-Components/Strong k-Linkage

For a specific graph, the subgraphs that are maximal with respect to their edge connectivity are the clusters. (*See* GRAPH THEORY or Harary [2] for definitions of graph-theoretic terms.) For any given k, the maximal k-edge connected subgraphs are termed k-components and form a partition of the vertices (see Matula [5]). The k-component cluster hierarchy is shown in Fig. 2 for the graph of Fig. 1. The k-component hierarchy requires that two otherwise disjoint maximal k-edge-connected subgraphs must be joined by no less than k edges before they need be considered to be in the same cluster. This criterion avoids the weak linkage problem of the min degree hierarchy at the expense of considerably greater computing time.

A corresponding distance matrix method is provided by the strong k-linkage proce-

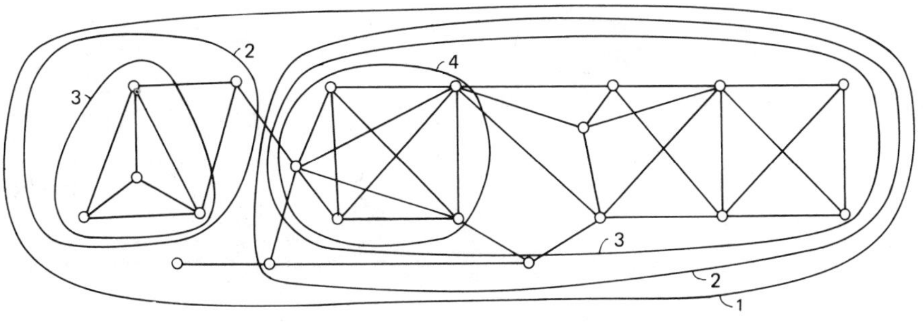

Figure 2 k-Component cluster hierarchy for a graph. The subgraphs that are maximal of edge connectivity k are shown with corresponding values of k.

dure. For any *fixed* k, the proximity graph interpretation of strong k-linkage obtains for each edge rank order level a partition into k-components.

A particular strength of these methods is derived from the following fundamental graph-theoretic result.

Theorem (Menger; see Harary [2, p. 49]). For any two vertices of a graph, the maximum number of edge-disjoint paths joining them equals the minimum number of edges whose removal separates them.

Thus if a cluster is characterized by being a k-component (i.e., a maximal k-edge-connected subgraph), it is *equivalently* characterized by either of the following properties:

1. The cluster is maximal in the sense that it cannot be separated by the removal of fewer than k edges (i.e., the disallowal of fewer than k object pair similarities).
2. The cluster is maximal in the sense that any two objects of the cluster must be joined by at least k edge-disjoint paths within the cluster.

k-Blocks / B_k

For a specific graph, the subgraphs that are maximal with respect to their vertex connectivity are the clusters. For any given k, the maximal k-vertex connected subgraphs are termed *k-blocks* and any two k-blocks can overlap in no more than $k - 1$ vertices. The k-block stratified clustering for the graph of Fig. 1 is shown in Fig. 3.

A closely associated distance matrix method is the B_k method of Jardine and Sibson [3] (see Matula [6] for a further discussion of the correspondence). The vertex connectivity variation of Menger's fundamental theorem [2, p. 47] allows the following observations.

Suppose that a cluster is characterized by being a k-block (i.e., a maximal k-vertex connected subgraph). Then it is *equivalently* characterized by either of the following properties:

1. The cluster is maximal in the sense that it cannot be separated by the removal of fewer than k vertices.
2. The cluster is maximal in the sense that any two objects of the cluster must be joined by at least k vertex-disjoint paths within the cluster.

References

[1] Aho, A. V., Hopcroft, J. E., and Ullman, J. D. (1974). *The Design and Analysis of Computer Algorithms*. Addison-Wesley, Reading, Mass.

[2] Harary, F. (1969). *Graph Theory*. Addison-Wesley, Reading, Mass.

[3] Jardine, N. and Sibson, R. (1971). *Mathematical Taxonomy*. Wiley, London.

[4] Ling, R. F. (1972). *Computer J.*, **15**, 326–332.

[5] Matula, D. W. (1972). *SIAM J. Appl. Math.*, **22**, 459–480.

[6] Matula, D. W. (1977). In *Classification and Clustering*, J. Van Ryzin, ed. Academic Press, New York, pp. 95–129.

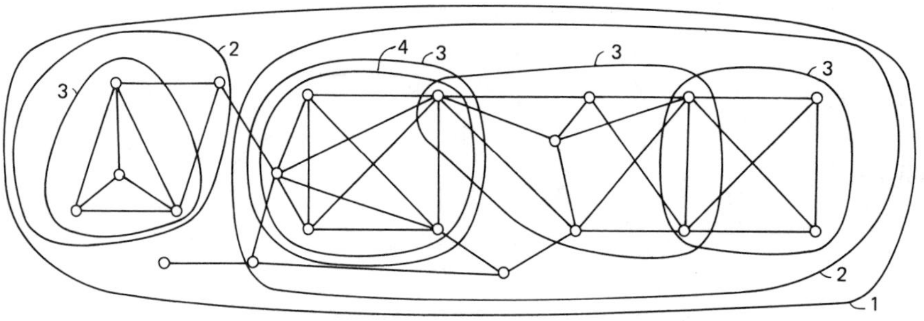

Figure 3 k-Block cluster hierarchy for a graph. The subgraphs that are maximal of vertex connectivity k are shown with corresponding values of k.

(CLASSIFICATION
DENDROGRAMS
GRAPH THEORY
HIERARCHICAL CLASSIFICATION
HIERARCHICAL CLUSTER ANALYSIS
RANDOM GRAPHS)

DAVID W. MATULA

GRAPH THEORY

Perhaps no topic in mathematics has enjoyed such explosive growth in recent years as graph theory. This stepchild of combinatorics* and topology has emerged as a valuable tool in applied mathematics and as a fascinating topic for research in its own right. In this article we outline briefly some major results and their applications.

A *graph* G is a finite nonempty set $V(G)$ together with a (possibly empty) set $E(G)$ of two-element subsets of distinct elements of $V(G)$. The elements of $V(G)$ are the *vertices* of G and the elements of $E(G)$ are the *edges* of G.

A graph can be conveniently pictured as a diagram where the vertices appear as small circular dots and the edges are indicated with line segments joining two appropriate dots. Two drawings of the graph G with vertex set $V(G) = \{v_1, v_2, v_3, v_4\}$ and edge set $E(G) = \{v_1v_2, v_2v_3, v_3v_1, v_3v_4\}$ are given in Fig. 1.

The origins of graph theory are obscure. The famous eighteenth-century Swiss mathematician Leonhard Euler was perhaps the first to solve a problem using graphs when he was asked to consider the problem of the Königsberg bridges. In the 1730s the flow of the river Pregel through the city of Königsberg was interrupted by two islands. Seven bridges connected the islands with each other and with the opposing banks, as shown in Fig. 2*a*. A popular puzzle of the day was to try and plan a walking route through the city that would start and stop at the same spot and which would include crossing each bridge exactly once.

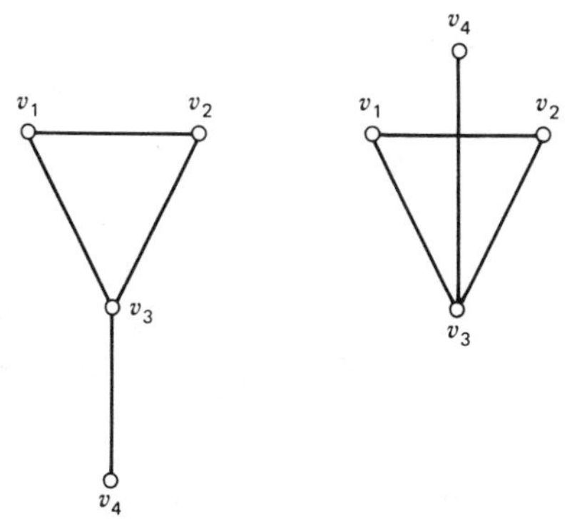

(a) (b)

Figure 1

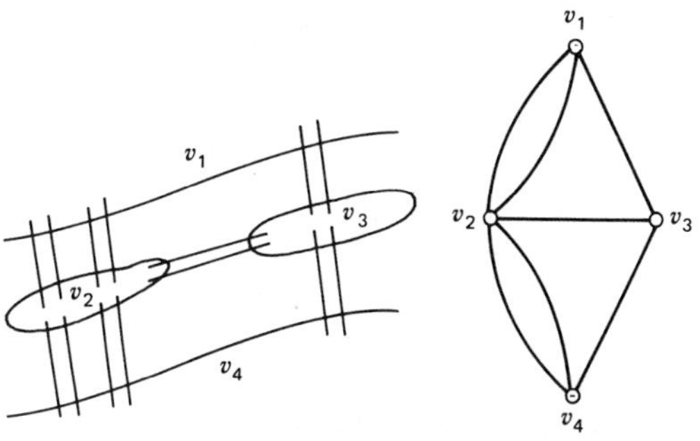

(a) (b)

Figure 2

Euler [5] proved that no such route was possible. The pattern of bridges and landmasses can be modeled with a "multigraph" as in Fig. 2*b*, where vertices v_1, v_4, v_2, and v_3 represent the two banks and the two islands, respectively, and where the edges represent the bridges. (Technically, Fig. 2*b* is not a graph because two vertices in a graph can be connected with no more than one edge.) Euler also characterized the class of graphs that admit such a walk. We present his famous result with the aid of technical terms consistent with Behzad et al. [2].

Let u and v be vertices in a graph G. A u–v *walk* in the graph G is a finite, alternating sequence of vertices and edges of G, beginning with u and ending with v, such that every edge in the walk is immediately preceded and followed by the two vertices that form it. A u–v walk is *closed* or *open* according as $u = v$ or $u \neq v$. A u–v *trail* is a u–v walk in which no edge is repeated, and a u–v *path* is a u–v walk in which no vertex is repeated. A graph G is *connected* if for each two vertices u and v of G, there exists a u–v walk in G.

An *Eulerian circuit* of a graph G is a closed trail which contains all the edges of G. Euler's characterization theorem relies on the notion of the *degree* of a vertex in a graph, defined to be the number of edges that contain the vertex as an element. For the graph in Fig. 1, vertices v_1 and v_2 have degree 2, vertex v_3 has degree 3, and vertex v_4 has degree 1.

Theorem 1: Euler. Let G be a connected graph containing more than one vertex. Then G possesses an Eulerian circuit if and only if the degree of every vertex of G is even.

As one practical application of this theorem, consider the problem of efficient use of snowplows after a snowstorm in a small community. Because of the high cost of fuel, it is important that each snowplow be removing snow all the time it is in operation. We assume here that all the snow on a road can be removed by one pass of the plow. Thus we want to avoid having the plow travel down roads more than once, if possible. We also would like each plow to follow a path around town that starts and stops at the same spot: its storage building. The network of roads in the town can be modeled by a graph where the vertices represent road intersections and the edges represent roads. Then optimal usage of the plows can occur in the sense that each plow will travel its assigned roads once and only once, starting and stopping at its storage building if and only if each plow follows an Eulerian circuit in the graph modeling its assigned roads. Theorem 1 would allow a city official to tell at a glance whether a given graph contains an Eulerian circuit, and hence whether such an optimal route is possible.

Another application of graph theory is to the theory of electrical circuits [3], in which an electrical circuit is a collection of circuit wires interconnected in a specified way. We form the graph of an electrical circuit by drawing an edge for each circuit wire, and a vertex for each electrical connection between wires.

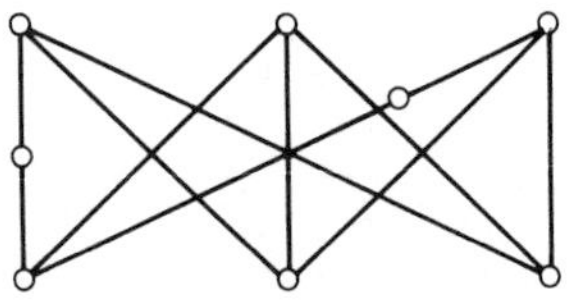

Figure 3 A nonplanar graph.

Suppose that the graph drawn twice in Fig. 1 represents two possible layouts of an electrical circuit on a circuit board. Clearly, Fig. 1*a* is superior because, in that drawing, two edges intersect only at a vertex. In Fig. 1*b*, however, the two edges v_3v_4 and v_1v_2 intersect at a point even though the circuit wires v_1v_2 and v_3v_4 have no direct electrical connection. If the drawing in Fig. 1*b* were to be implemented on a circuit board, it would be necessary to pass the wire v_3v_4 above or below the wire v_1v_2, causing unnecessary additional expense. These considerations suggest the following concept.

A graph G is *planar* if G can be drawn in the plane so that the edges of G intersect only at a vertex mutually common to them. Unfortunately, not all graphs are planar (see Fig. 3). It would clearly be of interest to electricians to be able to decide, for a given electrical circuit, whether or not the graph for the circuit is planar. With the aid of the following terminology, this question can be answered.

Let G be a graph with vertex set $V(G)$ and edge set $E(G)$. A graph H is a *subgraph* of G if $V(H) \subseteq V(G)$ and $E(H) \subseteq E(G)$. Let G be a graph and let $e = uv$ be an edge of G. An *elementary subdivision* of G is a new graph H obtained from G by deleting the edge uv and adding a new vertex w and new edges uw and vw. We say that a graph I is *homeomorphic from* G if I can be obtained from G by a finite (possibly empty) sequence of elementary subdivisions. For example, the graph I is homeomorphic from the graph G in Fig. 4. Two important classes of graphs need to be identified for our upcoming characterization of planar graphs. Let n be a positive integer. The *complete graph of order* n, denoted $K(n)$, is that graph with n verti-

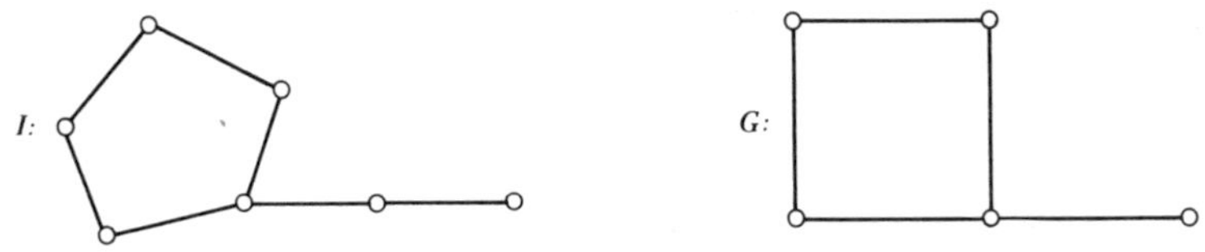

Figure 4

ces, every pair of which forms an edge. If m and n are positive integers, the *complete bipartite graph* $K(n, m)$ is that graph whose vertex set can be partitioned into two subsets V_1 and V_2 with n vertices in V_1 and m vertices in V_2, where uv is an edge in $K(n, m)$, if and only if $u \in V_i$ and $v \in V_j$ for $i \neq j$. The graphs $K(5)$ and $K(3, 3)$ are drawn in Fig. 5.

The following characterization of planar graphs is due to Kuratowski [11].

Theorem 2. A graph is planar if and only if it contains no subgraph homeomorphic with $K(5)$ or $K(3, 3)$.

It may be necessary to construct a circuit board whose underlying graph is not planar. Of course, any drawing of such a graph must include at least one extraneous edge crossing. It would be of interest to minimize the number of such crossings. Given a graph G, the *crossing number of* G, denoted by $\nu(G)$, is the minimum number of crossings of edges of G among the drawings of G in the plane. Relatively little is known about this difficult parameter. Guy [6, 7] has shown that

$$\nu(K(p)) \leqslant \frac{1}{4}\left[\frac{p}{2}\right]\left[\frac{p-1}{2}\right] \times \left[\frac{p-2}{2}\right]\left[\frac{p-3}{2}\right],$$

with equality holding for $1 \leqslant p \leqslant 10$, where $[x]$ is the greatest integer not exceeding the real number x. Zarankiewicz [14] established that

$$\nu(K(m, n)) \leqslant \left[\frac{m}{2}\right]\left[\frac{m-1}{2}\right]\left[\frac{n}{2}\right]\left[\frac{n-1}{2}\right];$$

Kleitman [10] proved equality in the above for $1 \leqslant \min\{m, n\} \leqslant 6$.

Kuratowski's work with planar graphs has been extended in another way. A nonplanar graph cannot be drawn in the plane without extraneous edge crossings, but it may be drawn on other topological surfaces without such crossings. For example, $K(3, 3)$ is not planar, but it can be drawn on the surface of a torus with no improper edge crossings. This is accomplished in Fig. 6, where the torus is formed by identifying opposite sides of the rectangle.

The torus can be thought of topologically as a sphere on which has been placed a "handle." In general, for any graph G there exists a nonnegative integer n such that G can be drawn on a sphere to which has been attached n handles in such a way that edges of G intersect only at common vertices. For a given graph G, the least nonnegative integer n for which this is possible is called the *genus of* G, denoted by $\gamma(G)$.

Since drawing graphs on spheres and planes are equivalent, a graph has genus 0 if and only if it is planar. It also follows immediately that a graph has genus 1 if and only if it can be drawn on the torus without improper edge crossings, and it is not planar.

Exact results have been obtained, after much labor, for the genus of any complete

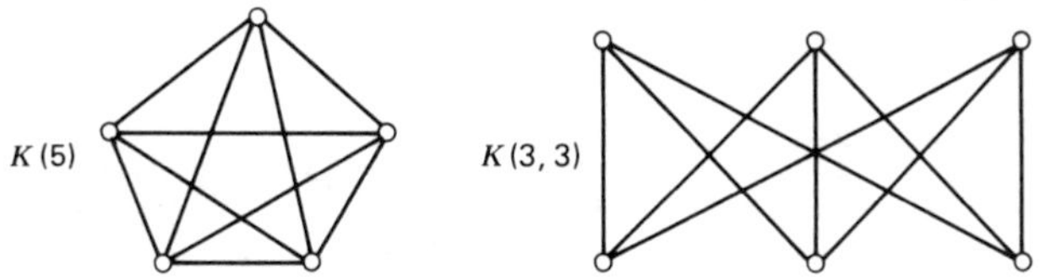

Figure 5

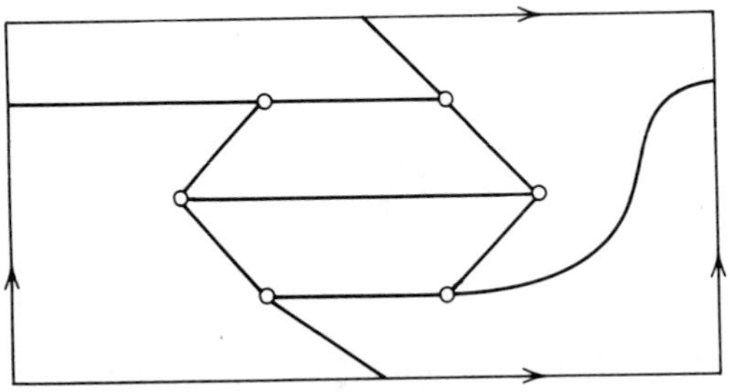

Figure 6

graph and any complete bipartite graph. Both results are due to Ringel, and appear in refs. 12 and 13 respectively.

Theorem 3. For $p \geqslant 3$,

$$\gamma(K(p)) = \left\{ \frac{(p-3)(p-4)}{12} \right\}.$$

Theorem 4. For m and $n \geqslant 2$,

$$\gamma(K(m,n)) = \left\{ \frac{(m-2)(n-2)}{4} \right\}.$$

One reason that graph theory is applicable to such a wide range of academic pursuits is that a graph can be used to model any finite set upon which is defined a binary relation.

In sociology, for example, conclusions can be reached based on theorems in a branch of graph theory called *Ramsey theory*. Consider a finite collection of people. For the moment, let us assume that for any two people in the collection, exactly one of two possible relationships exist; either these two individuals are friends or they are not friends. We will show, graphically, that if six or more people are in the group, there will always exist in the group three mutual friends or three mutual nonfriends. In other words, for any group of six or more people, there will exist in the group three people every pair of which are friends, or there will exist in the group three people every pair of which are not friends.

Let G be a graph with vertex set $V(G)$ and edge set $E(G)$. The *complement of* G, denoted $\overline{G}$, is that graph with vertex set $V(G)$, where $uv \in E(\overline{G})$ if and only if $uv \notin E(G)$. A graph and its complement are drawn in Fig. 7.

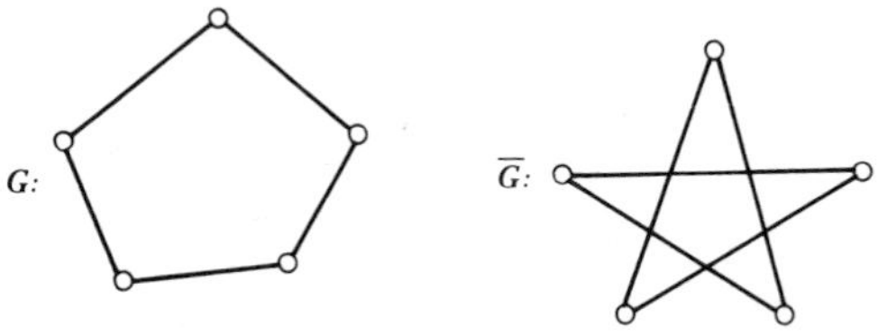

Figure 7

For positive integers m and n, the Ramsey number $r(m,n)$ is the least positive integer t such that for every graph G with t vertices, either G contains $K(m)$ as a subgraph, or $\overline{G}$ contains $K(n)$ as a subgraph.

We use a graph to model our sociological situation as follows. Let each person in our collection of people be represented by a vertex. Let two vertices be joined by an edge if and only if the two corresponding people are friends. The assertion that in any group of six or more people, there exist three mutual friends or three mutual nonfriends is equivalent to establishing the Ramsey theory result $r(3,3) = 6$.

We outline proof of this result because of its unusual appeal. For details, see Behzad et al. [2, p. 284].

Since neither graph in Fig. 7 contains $K(3)$ as a subgraph, $r(3,3) \geqslant 6$. Let G be any graph of order 6, and let v be a vertex of G. Because the degree of v in G plus the degree of v in $\overline{G}$ is 5, one of these two numbers must be at least 3. Without loss of generality we assume that vv_1, vv_2, and vv_3 are edges of G. If any of v_1v_2, v_1v_3, or v_2v_3 is an edge of G, then G contains $K(3)$ as a subgraph. The only other possibility is that $v_1v_2v_3$ form a $K(3)$ subgraph in $\overline{G}$, completing the proof.

The difficulty in determining $r(m,n)$ for other values of m and n is greater. Indeed, only eight results are currently known [2]:

$$\begin{array}{ll} r(1,n) = 1 & r(2,n) = n \quad (\text{for } n \geqslant 1) \\ r(3,3) = 6 & r(3,6) = 18 \\ r(3,4) = 9 & r(3,7) = 23 \\ r(3,5) = 14 & r(4,4) = 18. \end{array}$$

The *four-color theorem*, one of the most famous results in mathematics, can be easily

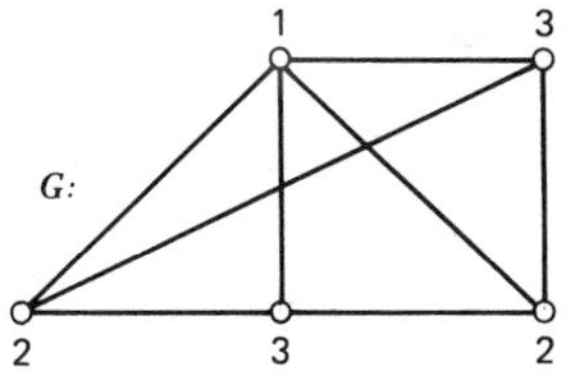

Figure 8

formulated in terms of another graph theoretic parameter called the chromatic number.

Let G be a graph. The *chromatic number* of G, denoted $X(G)$, is the least number of colors (colors being elements of some set) that need be assigned to the vertices of G, one color to each vertex, so that vertices connected by an edge are assigned different colors. The graph G in Fig. 8 has chromatic number 3, and an appropriate coloring of the vertices of G using colors 1, 2, and 3 is given. It is easy to observe that for positive integers m and n, $X(K(n)) = n$ and also $X(K(m,n)) = 2$.

It is said that mapmakers have "known" for hundreds of years that four colors were sufficient to color the countries of any map so that any two countries which share a common boundary line are colored differently. (Here a "country" must be connected.) It was not until approximately 1860 that this conjecture appeared in the mathematical literature, and until 1977 that it was proved by Appel et al. [1] using a lengthy mathematical argument and about 1200 hours of computer time. Letting a vertex represent each country and connecting two vertices with an edge if and only if the corresponding countries share a boundary line, we can state the theorem simply.

Theorem 5. If G is a planar graph, then $X(G) \leqslant 4$.

The determination of $X(G)$ where G is not necessarily planar has also been of interest. An important result for arbitrary graphs uses the following definition.

Let $C(n)$ be the graph with vertices v_1, $v_2, \ldots, v_n$ and edges $v_i v_{i+1}$ for $i = 1, n$ (modulo n). $C(n)$ is called an *even* (*odd*) *cycle* according as n is even (odd). A valuable theorem due to Brooks [4] states that if G is a connected graph that is neither an odd cycle nor a complete graph, then $X(G) \leqslant \Delta(G)$, where $\Delta(G)$ is the largest degree that occurs among the vertices of G.

We close with a brief description of an application of graph theory to cluster analysis [9]. Given a set $S = \{o(1), o(2), \ldots, o(p)\}$ of objects, one can define a nonnegative real-valued "proximity" function F on $S \times S$ where $F(o(i), o(j))$ for $1 \leqslant i, j \leqslant p$, is a number measuring the "similarity" of the two objects $o(i)$ and $o(j)$. Usually, $F(o(i), o(j)) < F(o(i), o(k))$ has the meaning that the objects $o(i)$ and $o(j)$ are more similar than the objects $o(i)$ and $o(k)$. Given such a function F and a number s, a graph $G(s)$ can be defined where $V(G(s)) = S$ and $o(i)o(j) \in E(G(s))$ if and only if $F(o(i), o(j)) \leqslant s$. The probability that a given set T has similarity structure indicated in the graph $G(s)$ can then be investigated by comparing the graph $G(s)$ with all other graphs having the same number of vertices and edges as $G(s)$.

References

[1] Appel, K., Haken, W., and Koch, J. (1977). *Ill. J. Math.*, **21**, 429–567. (A high-level outline of the proof of the four-color theorem. Knowledge of graph theory is assumed.)

[2] Behzad, M., Chartrand, G., and Lesniak-Foster, L. (1979). *Graphs & Diagraphs*. Prindle, Weber & Schmidt, Boston. (An excellent introductory text. Complete, compact, and very readable.)

[3] Beineke, L. and Wilson, R., eds. (1979). *Applications of Graph Theory*. Academic Press, New York. (A collection of in-depth articles; knowledge of graph theory assumed.)

[4] Brooks, R. (1941). *Proc. Camb. Philos. Soc.*, **37**, 194–197. (A central result on chromatic numbers is established in this classic paper. Very readable.)

[5] Euler, L. (1736). Solutio problematis ad geometriam situs pertinentis, *Comment. Acad. Sci. I. Petropolitanae*, **8**, 128–140. [This is the first paper in graph theory. The main result is proved (in English) in ref. 2 above.]

[6] Guy, R. (1960). *Bull. Malay. Math. Soc.*, **7**, 68–72. (A powerful attack on the difficult crossing-number parameter.)

[7] Guy, R. (1972). In *Graph Theory and Applications*. Springer-Verlag, New York. (A continuation of earlier work on crossing numbers of graphs. Knowledge of graph theory is assumed.)

[8] Harary, F. (1969). *Graph Theory*. Addison-Wesley, Reading, Mass. (A classic early text in graph theory, written in a compact style.)

[9] Killough, G. and Ling, R. (1976). *J. Amer. Statist. Ass.*, **71**, 293–300. (An advanced article on the interactions of graph theory and cluster analysis.)

[10] Kleitman, D. (1970). *J. Comb. Theory*, **9**, 315–323. [An advanced article that computes the crossing number of the graphs $K(5, n)$.]

[11] Kuratowski, K. (1930). *Fund. Math.*, **16**, 271–283. [A classic article (in French) characterizing planar graphs. A shorter proof of this result appears in ref. 2 above.]

[12] Ringel, G. (1965). *Abhand. Math. Semin. Univ. Hamburg*, **28**, 139–150 (in German). (This article establishes the genus of the complete graphs.)

[13] Ringel, G. (1974). *Map Color Theorem*. Springer-Verlag, New York. (This text contains the clever arguments needed to compute the genus of any complete bipartite graph.)

[14] Zarankiewicz, K. (1954). *Fund. Math.*, **41**, 137–145. [An early article (containing some flaws) on crossing numbers.]

DAVID BURNS

(GRAPH-THEORETIC CLUSTER ANALYSIS
RANDOM GRAPHS)

GRAUNT, JOHN *See* BIOSTATISTICS

GRAY AND WILLIAMS MEASURE

This is a variant for partial association of the Goodman–Kruskal τ* measure of association. It was introduced by Gray and Williams [2]; its asymptotic variance is discussed by Anderson and Landis [1].

References

[1] Anderson, R. J. and Landis, J. R. (1979). *Commun. Statist. A*, **8**, 1303–1314.

[2] Gray, L. N. and Williams, J. S. (1975). *Amer. Statist. Ass., Proc. Soc. Statist. Sec.*, pp. 444–448.

(ASSOCIATION, MEASURES OF
GOODMAN–KRUSKAL TAU AND GAMMA)

GREENWOOD'S FORMULA

An approximate formula for the product of a number of independent binomial proportions. If $X_1, X_2, \ldots, X_m$ are independent variables with binomial* distributions with parameters $(n_1, p_1), (n_2, p_2), \ldots, (n_m, p_m)$, respectively, then $\{\hat{p}_i = X_i/n_i\}$ are unbiased estimators of $\{p_i\}$, and

$$\hat{p}_1 \hat{p}_2 \cdots \hat{p}_m$$

is an unbiased estimator* of $P_m = p_1 p_2 \cdots p_m$ and

$$\operatorname{var}(\hat{p}_1 \hat{p}_2 \cdots \hat{p}_m) \doteqdot P_m^2 \sum_{j=1}^{m} (n_j p_j)^{-1}(1 - p_j).$$

The formula is used in survival analysis*, with p_i denoting the probability of surviving over an interval t_{i-1} to t_i and P_m the probability of surviving from t_0 to t_m. When using the formula it is necessary to check that the assumptions that $\hat{p}_1, \ldots, \hat{p}_m$ are mutually independent are reasonable.

(KAPLAN–MEIER ESTIMATOR
SURVIVAL ANALYSIS)

GREENWOOD'S STATISTIC

If $X_1, X_2, \ldots, X_n$ are n independent random variables, each having a common distribution on a unit interval $[a, a+1]$, and if

$$D_i = X_i - X_{i-1}; \qquad i = 2, 3, \ldots, n,$$
$$D_1 = X_1 - a,$$
$$D_{n+1} = a + 1 - X_n,$$

so that $D_1, D_2, \ldots, D_{n+1}$ are the *spacings** of the observations, then the statistic

$$G = \sum_{i=1}^{n+1} D_i^2$$

provides a test of the hypothesis H_0 that the data come from a uniform distribution*. Pyke [5] lists G as one of six statistics used to provide tests based on spacings; the original context of each was to test the randomness of events in time. This was the aim of the British epidemiologist Greenwood [3] in a statistical study of infectious diseases, when the use of G was first proposed (*see* EPIDEMIOLOGICAL STATISTICS).

Under the hypothesis H_0, the first four moments of G are given (see Moran [4]) by

$$E(X) = 2/(n+2),$$

$$\mathrm{var}(X) = \mu_2 = \frac{4n}{(n+2)^2(n+3)(n+4)},$$

$$\mu_3 = \frac{16n(5n-2)}{(n+2)^3(n+3)(n+4)(n+5)(n+6)},$$

$$\mu_4 = \frac{16n(3n^3+303n^2+42n-24)}{(n+2)^3(n+3)(n+4)\cdots(n+8)}.$$

Moran also shows that G has an asymptotic normal* distribution, but the convergence is very slow; when $n = 100$, for example, $\mu_3/\sigma^3 = 0.86$ and $\mu_4/\sigma^4 = 5.03$. Moran gives some lower percent points of G. More detailed tables were published in 1979 and 1981 (see Burrows [1], Currie [2], and Stephens [6]). These cover a wide range of probability values for $n \leqslant 20$, and for $n = 25, 30(10) 60, 80, 100, 200, 500$. See Stephens [6] for a brief discussion of the role of Greenwood's statistic in tests for exponentiality and for further references.

References

[1] Burrows, P. M. (1979). *J. R. Statist. Soc. A*, **142**, 256–258.

[2] Currie, I. D. (1981). *J. R. Statist. Soc. A*, **144**, 360–363.

[3] Greenwood, M. (1946). *J. R. Statist. Soc. A*, **109**, 85–109.

[4] Moran, P. A. P. (1947). *J. R. Statist. Soc. B*, **9**, 92–98. Corrigendum. *J. R. Statist. Soc. A*, **144**, 388 (1981).

[5] Pyke, R. (1965). *J. R. Statist. Soc. B*, **27**, 395–449.

[6] Stephens, M. A. (1981). *J. R. Statist. Soc. A*, **144**, 364–366.

(SPACINGS
UNIFORM DISTRIBUTION)

CAMPBELL B. READ

GREENWOOD VARIANCE ESTIMATOR

See KAPLAN–MEIER ESTIMATOR

GROSS NATIONAL PRODUCT DEFLATOR

The gross national product (GNP) deflator is a price index for the whole economy. The GNP is the market value of the whole range of goods and services produced in the economy for final use during a year, and the GNP deflator is an index number* of the prices of all these products.

For a run of years the GNP is a time series* of *nominal* money aggregates called "GNP at current prices," where "current" refers to the year in which the transactions took place. The GNP at current prices is by itself not a very illuminating statistic because, for example, it can increase when either prices or quantities increase. What is required is an estimate of the *real* GNP, which is defined as the value of all GNP products expressed in the prices of an arbitrarily chosen base year. The revalued GNP is called "GNP at constant prices," where "constant prices" refers to the prices of the base year. The price index obtained as the ratio of the GNP at current prices to the GNP at constant prices is called the "GNP deflator" or the "GNP implicit deflator." When this kind of price index refers to the gross domestic product (GDP), it is called the GDP deflator.

THE GNP AT CONSTANT PRICES

The GNP deflator is derived from the GNP at constant prices. The practical problem, therefore, is how to estimate the GNP at constant prices. In principle the GNP at constant prices can be measured from data on income, output, or expenditure. While national practices in this area vary widely and are often based on numerous ad hoc and unsystematic estimation procedures, the two basic approaches use output and expenditure data.

The *output approach* revalues net output in the current year item by item and industry by industry at the prices of the base year. The notion of "net output" or "value added" is introduced to avoid double counting, and is defined as the value of gross output less the value of intermediate inputs. For some GNP components, data limitations might dictate instead the use of a Laspeyres-type quantity index for value added which, when multiplied by the value added GNP at base year prices, yields the net output GNP at constant prices. Alternatively, the method of *double deflation* might be used. In this method net output GNP at constant prices is obtained as the difference between gross output at current prices deflated by a gross output Paasche-type price index and intermediate input at current prices deflated by an input Paasche-type price index.

The *expenditure approach* focuses on expenditures on final goods and services covered by the GNP. If p_{it} stands for the price of commodity i in the current year t, p_{i0} the price of the same commodity in the base year 0, and q_{it} and q_{i0} are the corresponding quantities, then the GNP at current and constant prices is given, respectively, by $\sum_i p_{it}q_{it}$ and $\sum_i p_{i0}q_{it}$, where summation extends over all the final goods and services. If complete information is available, then $\sum_i p_{i0}q_{it}$ is identically equal to (1) and (2):

$$\sum_i p_{it}q_{it} \Big/ \Big(\sum_i p_{it}q_{it} \Big/ \sum_i p_{i0}q_{it}\Big), \qquad (1)$$

$$\sum_i p_{i0}q_{i0}\Big(\sum_i p_{i0}q_{it} \Big/ \sum_i p_{i0}q_{i0}\Big). \qquad (2)$$

But this identity generally does not hold because available data on final goods and services are incomplete. For the same reason $\sum_i p_{i0}q_{it}$ cannot be obtained for all i directly by multiplying current year quantities by base year prices. Instead, some GNP components are expressed at constant prices as in expression (1) by deflating current expenditures on final goods and services by appropriate Paasche-type price indexes. The value of some components of the GNP at constant prices may also be obtained because of data constraints from expression (2) by extrapolating base year values of those components with Laspeyres-type quantity indexes.

THE GNP DEFLATOR

The GNP deflator is not constructed directly and independently as a general price index for the whole economy. Rather, components of the GNP are expressed in various ways at base year prices and summed to give an overall figure of the GNP at constant prices. When these constant price aggregates are divided into the corresponding current price aggregates and multiplied by 100, average price relationships or "implicit" price deflators emerge. Thus in many countries implicit price deflators are published for the total GNP as well as for various groups and components of expenditures.

The GNP deflator is one of three major measures of inflation, the other two being the consumer price index* and the wholesale or producer price index*. For a given period these three measures usually show different rates of inflation; the differences could be significant in some cases. Many of these differences between alternative measures of inflation are due to differences in methodology, coverage, weights, definitions, and valuation.

NATIONAL OUTPUT DEFLATOR

The GNP deflator is a statistical construct which can be interpreted as an estimate of the "true" national output deflator, which measures the relative values of outputs produced in two periods with a fixed level of technology and resource usage. The analysis is valid for binary comparisons and for an output-optimizing economy with a fixed production possibility map. Denote the current and base years by 1 and 0; the output market prices in these years by p_{i1} and p_{i0}, where i refers to the ith output; the quantities produced by q_{i1} and q_{i0}; and the two levels of factors usage in the current and base years by U_1 and U_0. The economic problem then is to maximize the value of output in year j ($j = 0, 1$) subject to factors usage U_j, which permits particular efficient combinations of outputs.

Given p_{i1} and U_0, optimal output is the unobserved q_{i0}^*, which is greater than q_{i0} because the latter cannot be optimal when current year prices are considered. Thus $\sum_i p_{i1}q_{i0}^*$ is greater than $\sum_i p_{i1}q_{i0}$, and the true national output deflator in the current year relative to base year prices and base year factors usage $P_{01}(U_0)$ is greater than the Laspeyres-type GNP deflator $P_{01}(q_0)$; that is,

$$P_{01}(U_0) = \frac{\sum_i p_{i1}q_{i0}^*}{\sum_i p_{i0}q_{i0}} > \frac{\sum_i p_{i1}q_{i0}}{\sum_i p_{i0}q_{i0}} = P_{01}(q_0).$$

The fact that $\sum_i p_{i1}q_{i0}^* > \sum_i p_{i1}q_{i0}$ also implies that the true real national output index is smaller than the Paasche-type real output index.

Similarly, given p_{i0} and U_1, optimal output is the unobserved q_{i1}^*, which is greater than q_{i1}. Thus $\sum_i p_{i0}q_{i1}^*$ is greater than $\sum_i p_{i0}q_{i1}$, which implies, first, that the true national output deflator in the current year relative to base year prices and current year factors usage is smaller than the Paasche-type GNP deflator; and second, that the real national output index in the current year relative to base year output and current year factors usage is greater than the Laspeyres-type real output index.

Bibliography

Allen, R. G. D. (1975). *Index Numbers in Theory and Practice*. Macmillan, London. (Discusses British practice; economic interpretation; extensive references.)

Backman, J. and Gainsbrugh, M. R. (1966). *Inflation and the Price Indexes*. U. S. Congress, Joint Economic Committee, Subcommittee on Economic Statistics, 89th Congress, 2nd Session, Washington, D.C. (Elementary; Appendix C by G. Kipnis contains a detailed account of U.S. practice.)

Emery, B. J. and Garston, G. J. (1969). *Rev. Income Wealth*, **15**, 1–32. (Elementary; few references.)

Fisher, F. M. and Shell, K. (1972). *The Economic Theory of Price Indices*. Academic Press, New York. (Intermediate; economic theory; mathematical; few references.)

Maunder, W. F., ed. (1970). *Bibliography of Index Numbers*. Athlone Press, London. (Exhaustive from 1707 up to the first half of 1968.)

Sen, A. (1979). *J. Econ. Lit.*, **17**, 1–45. (Elementary; economic theory; critical survey; references.)

United Nations (1979). *Manual on National Accounts at Constant Prices*. Statistical Papers, Ser. M, No. 64 (Elementary; useful general discussion; few references.)

Usher, D. (1980). *The Measurement of Economic Growth*. Blackwell, Oxford. (Intermediate; theory and applications; references.)

(CONSUMER PRICE INDEX
INDEX NUMBERS
PRODUCER PRICE INDEX)

NURI T. JAZAIRI

GROUP-DIVISIBLE DESIGNS

Group-divisible designs are a class of partially balanced incomplete block designs*, which were introduced by Bose and Nair [2]. They are designs for t treatments in b blocks of k plots each with every treatment appearing in exactly r blocks.

In balanced incomplete block designs (*see* BLOCKS, BALANCED INCOMPLETE) there is the additional requirement that every pair of treatments appears together in the same number of blocks. The five parameters t, b, r, k, and λ satisfy two equalities:

$$rt = bk = \text{total number of plots}, \quad (1)$$

$$r(k-1) = \lambda(t-1). \quad (2)$$

Equation (2), coupled with the requirement that λ be an integer, restricts the number of balanced designs, especially when $t-1$ is a prime number (*see* GENERAL BALANCE). For example, if $t = 8$ and $k = 3$, (1) requires that r be divisible by 3, and (2) requires that r be divisible by 7. Thus the smallest balanced design has $r = 21$, corresponding to the set of all triples from the eight treatments. This is too large a design for many applications.

In group-divisible designs the requirement of balance is relaxed in the following way. Let $t = mn$, where m and n are integers. The treatments are divided into m groups of n treatments each. In the design each treatment appears λ_1 times with every treatment in its own group, and λ_2 times with every treatment from different groups.

The division of the treatments into groups is the *association scheme*; two treatments are *first associates* of one another if they are in the same group, and *second associates* if they are in different groups. In the following example there are eight treatments, denoted by 1, 2, 3, 4, 5, 6, 7, 8; they are divided into four groups: 1 2, 3 4, 5 6, and 7 8.

Example 1. A design with $t = b = 8$, $r = k = 3$, $\lambda_1 = 0$, $\lambda_2 = 1$.

137, 352, 574, 726,
248, 461, 683, 815.

As in the case of the balanced incomplete block design, we express the model as

$$y_{ij} = \mu + \tau_i + \beta_j + e_{ij},$$

where y_{ij} is an observation on the ith treatment in the jth block, μ is an overall mean, τ_i and β_j are the effects (contributions) of the ith treatment and the jth block, and e_{ij} is the random error. If $\lambda_2 > 0$, the treatment effects are estimated, subject to the side condition $\sum \tau_i = 0$.

Let T_i be the total of all the observations on the ith treatment and B_j be the total for the jth block. The adjusted treatment total is

$$Q_i = T_i - \sum_j n_{ij} B_j / k,$$

where n_{ij} is the number of times that the ith treatment appears in the jth block. Then the estimate of τ_i is given by

$$[r(k-1) + \lambda_1]\hat{\tau}_i$$
$$= kQ_i - k(\lambda_2 - \lambda_1)[Q_i - S_1(Q_i)]/(t\lambda_2),$$

where $S_1(Q_i)$ is the sum $\sum_s Q_s$ taken over all the first associates of the ith treatment.

The variance of the estimate of a simple treatment comparison takes two values:

$$[r(k-1) + \lambda_1]V(\hat{\tau}_h - \hat{\tau}_i)$$
$$= \begin{cases} 2k\sigma^2 & \text{if they are first associates,} \\ 2k\sigma^2[1 - (\lambda_2 - \lambda_1)]/(\lambda_2 t), & \\ \quad \text{if they are second associates.} & \end{cases}$$

The estimates given above are the usual (intrablock) estimates. For large experiments there are procedures for obtaining interblock estimates, as in the case of the balanced designs.

The normal equations* have the same form as those for the balanced designs. In obtaining the intrablock estimates above, the block effects are eliminated from the normal equations, which then reduce to

$$\mathbf{Q} = (r\mathbf{I} - \mathbf{nn}'/k)\hat{\boldsymbol{\tau}} = \mathbf{C}\boldsymbol{\tau},$$

where $\mathbf{Q}$ is the vector of adjusted treatment totals and $\boldsymbol{\tau}$ is the vector of estimated treatment effects; $\mathbf{n} = (n_{ij})$ is the *incidence matrix** of the design. The matrix $\mathbf{C}$ is singular, since $\mathbf{C1} = \mathbf{0}$ ($\mathbf{1}$ is a vector of unit elements).

If $\lambda_2 > 0$, the rank of $\mathbf{C}$ is $t - 1$, and all the simple comparisons $(\tau_h - \tau_i)$ are estimable; the design is then said to be *connected*. If, however, $\lambda_2 = 0$, the design degenerates into a collection of balanced subdesigns in the individual groups of treatments, and the comparisons $(\tau_h - \tau_i)$ are estimable only if the two treatments are in the same group; the rank of $\mathbf{C}$ is then $t - m$, and the design is *disconnected*.

Example 2. A disconnected design with $t = b = 8$, $r = k = 3$, $m = 2$, $n = 4$, $\lambda_1 = 2$, $\lambda_2 = 0$. The two groups are 1, 2, 3, 4 and 5, 6, 7, 8. The design is:

123, 234, 341, 412,
567, 678, 785, 856.

The matrix $\mathbf{nn}'$ is the *concordance matrix* of the design. Its diagonal elements are r; off the diagonal, the element in the hth row and ith column is equal to the number of times that the hth and ith treatments concur (i.e., appear together in the same block). Its latent roots are $\theta_0 = rk$ with multiplicity 1, $\theta_1 = r - \lambda_1$ with multiplicity $m(n-1)$, and $\theta_2 = rk - \lambda_2 t$ with multiplicity $m - 1$. The corresponding latent roots of $\mathbf{C}$ are given by $\psi_i = (rk - \theta_i)/k$.

Group-divisible designs have been divided by Bose and Connor [1] into three classes. Designs with $\theta_1 = 0$ are called *singular*. They may all be obtained by taking a balanced design for m treatments, and replacing each treatment by a group.

Example 3. A singular design with $t = 8$, $b = 4$, $k = 6$, $r = \lambda_1 = 3$, $\lambda_2 = 2$. It is constructed from the balanced design *ABC*, *BCD*, *CDA*, *DAB*; *A* is replaced by 1 and 2, *B* by 3 and 4, and so on, giving the design

1 2 3 4 5 6, 3 4 5 6 7 8,
5 6 7 8 1 2, 6 7 1 2 3 4.

Designs with $\theta_2 = 0$ are called *semiregular*, designs with both $\theta_1 > 0$ and $\theta_2 > 0$ are called *regular*. For semiregular designs k is a multiple of m, and each block contains exactly k/m treatments from each group.

Example 4. A semiregular design with $t = 8$, $k = 4$, $m = 4$, $\lambda_1 = 0$, $\lambda_2 = 2$:

1 3 5 7, 1 3 6 8, 1 4 5 8, 1 4 6 7,
2 3 5 7, 2 3 6 8, 2 4 5 8, 2 4 6 7.

For group-divisible designs, (2) is replaced by

$$r(k-1) = (n-1)\lambda_1 + n(m-1)\lambda_2. \quad (3)$$

Conditions (1) and (3) are necessary, but not sufficient, for the existence of a design. The concordance matrix **nn**′ is nonnegative definite. Consider, for example, the set of parameters $t = 8$, $r = k = 4$, $m = 4$, $n = 2$, $\lambda_1 = 0$, $\lambda_2 = 3$. Equations (1) and (3) are satisfied, but such a design would be impossible since $\theta_2 = 16 - 24 < 0$.

Details of the mathematical derivations and further information may be found in John [4, 5].

References

[1] Bose, R. C. and Connor, W. S. (1952). *Ann. Math. Statist.*, **23**, 367–383.

[2] Bose, R. C. and Nair, K. R. (1939). *Sankhyā*, **4**, 337–372.

[3] Clatworthy, W. H. (1973). Tables of two-associate-class partially balanced incomplete block designs, *Nat. Bur. Stand. (U.S.) Appl. Math. Ser. 63*.

[4] John, P. W. M. (1971). *Statistical Design and Analysis of Experiments*. Wiley, New York.

[5] John, P. W. M. (1980). *Incomplete Block Designs*. Marcel Dekker, New York.

(BALANCE IN EXPERIMENTAL DESIGN
BLOCKS, BALANCED INCOMPLETE
DESIGN AND ANALYSIS OF
EXPERIMENTS
GENERAL BALANCE
PARTIALLY BALANCED DESIGNS)

Peter W. M. John

GROUPED DATA

One of the primary reasons for grouping data is descriptive, as for graphical* and tabular presentations of data. Assessment of empirical distributions, insight into the behavior of variables, and the choice of the statistical technique to be used for inference are examples of how a statistician can benefit from such presentations. Data-displaying techniques such as frequency polygons*, histograms*, or tables require specification of intervals, grouping the data into these intervals, and often condensing the grouped data into a point. Condensing is needed for the construction of frequency polygons and for most grouped data analyses.

Another important reason for grouping data arises in studies involving variables of a medical, socioeconomic, or demographic nature. These data sets are often obtained from confidential sources, and thus must be presented and analyzed so as to maintain the privacy of the individual's record.

Third, when data sets are large, grouping is often desirable in order to economize on computation, data recording, transmission, and storage.

Fourth, it is often difficult or even impossible to obtain accurate measurements. Because of measuring instrument limitations or because of difficulties in data handling, the data are measured on a discrete, rather than a continuous scale and then grouped by the appropriate ranges. A case in point is life testing*, where due to physical or economic restrictions, the system is inspected for failure only at certain time intervals. Thus the precise time of failure of each object is unknown and the only available information may be that failure occurred between the two inspection times. Hence the data are classified according to the inspection points.

The vast literature on grouped data can be divided into four broad categories. The first involves relationships between moments or other parameters calculated from the same data before and after grouping. Another, not necessarily related question, is that of making inferences from samples of grouped data (in short "grouped samples") about the ungrouped population. The issue of optimal grouping is reviewed next. This is important where the researcher has control over the grouping of the original data. Finally, we discuss situations where estimation methods require grouping. Related topics, such as order statistics*, quantile* estimation, censored* samples, rounding errors, aggregation*, and partial grouping (i.e., mixture of grouped and nongrouped data) will not be discussed here.

Before discussing these four aspects, we need a few definitions.

DEFINITIONS AND NOTATION

Data grouping is the process by which any variable X with a given distribution function $F(x)$ (continuous or discrete) is *condensed* into a *discrete distribution function*, i.e.,

$$p_i = \int_{c_{i-1}}^{c_i} dF(x), \qquad i = 1, \dots, k,$$

where $X \in [c_0, c_k]$ is partitioned by $c_0 < c < \cdots < c_k$ into k disjoint and exhaustive groups. The c_i's, which are set in advance, are termed the *interval limits* or *boundaries*, and (c_{i-1}, c_i) the ith *interval* or *group*. Hence grouping essentially transforms one distribution function, continuous or discrete, into a multinomial distribution* function. The number of cases falling into the ith group will be denoted by n_i, $\sum_{i=1}^{k} n_i = n$, and will be called the ith *group frequency*. The condensation is usually into the interval midpoints m_i $[m_i = (c_{i-1} + c_i)/2]$, but quite often it may be into the interval means or centroids $\bar{x}_i$, defined as the conditional expectation of X given $c_{i-1} \leqslant X < c_i$, i.e., $\bar{x}_i = \int_{c_{i-1}}^{c_i} dF(x)/p_i$, or any other point in the ith interval. The intervals may be of different widths $c_i - c_{i-1}$. In the special case of equal width, termed *equispaced* or *equidistanced*, h denotes the common value $c_i - c_{i-1}$, $i = 1, \dots, k$.

MOMENT CORRECTIONS

Most of the early statistical literature on grouped data dealt with the derivation of relationships between moments calculated from data before and after grouping. These are relationships between population values; they are merely expressions for the moments of one population, the parent population, in terms of the moments of another population which was derived from it, that of grouped data; inference is not an issue.

Attempts to arrive at these relationships proceeded along three main lines:

1. Attempts to relate directly the same moments calculated from the parent and grouped populations, respectively. Exact relationships are only possible for known distributions *and* fixed groupings.
2. Recognizing the limitations of the first approach, the second approach characterizes a family of distribution functions by one member of this family and uses the relationships between the moments of the grouped and ungrouped "representative" distribution to correct the moments of the other related distributions.
3. The third approach smooths a histogram* constructed from the grouped data in such a way that the area of each column under the smoothed curve will be identical to the corresponding area in the histogram. The approximate distribution function so constructed can be used either to compute various characteristics of the parent distribution or to derive relationships between moments of grouped and ungrouped data.

The widest known representative of the first type is *Sheppard's corrections* (SC) (see Sheppard [52] and CORRECTION FOR GROUP-

ING). These corrections express approximately the moments calculated from populations of midpoints of data grouped into *equispaced* intervals of width h in relation to the ungrouped parent population. For instance,

$$\mu_1' = \bar{\mu}_1', \quad \mu_2' = \bar{\mu}_2' - \tfrac{1}{12}h^2, \quad \mu_3' = \bar{\mu}_3' - \tfrac{1}{4}\,\bar{\mu}_1' h^2,$$

$$\mu_4' = \bar{\mu}_4' - \tfrac{1}{2}\,\bar{\mu}_2' h^2 + \tfrac{7}{240}h^4,$$

or there is the Wold [56] representation for the rth moment:

$$\mu_r' = \sum_{j=0}^{r} \binom{r}{j}(2^{1-j} - 1)B_j h^j\,\bar{\mu}_{r-j}',$$

where μ_i' and $\bar{\mu}_i'$ are the ith moments about zero of the ungrouped and grouped distributions, respectively, and B_j is the Bernoulli number* obtained as the coefficient of $t^j/j!$ in $t/(e^t - 1) = \sum_{j=0}^{\infty} B_j t^j/j!$. The grouped moments are calculated with the midpoints weighted by the corresponding group frequencies.

It is important to recognize the limitations of SC. Sufficient conditions for their validity are that the range of the distribution function is finite, that there is high-order contact at the terminals of the range (i.e., that the first and last frequencies are relatively small), and that the reminder term in the Euler–Maclaurin expansion* used in deriving the corrections is negligible. For a more explicit statement of the assumptions, see Kendall [36]. The requirement of high-order contact at the terminals of the range precludes the application of SC to most truncated and J-shaped distributions. As for the U-shaped distributions, a widely neglected fact is that SC can be used, but with the signs reversed. For instance, the correction for the second moment of a U-shaped distribution is $+h^2/12$.

Analytical solutions to the general problem of moment computations for grouped distributions under varying assumptions about distributional form, including distribution functions that have infinite ordinates and any slopes at either end of the terminals (e.g., J- and U-shaped distributions), were first introduced by Karl Pearson* [49]. Subsequent literature on the subject, mainly by Pairman and Pearson [46], Pearse [48], Martin [42], Lewis [39], Davies and Bruner [16], and Aigner [3], elaborated on it. The last three follow the third approach mentioned above, while the others follow the first approach. Elderton [20], who used the second approach, suggested corrections to J-shaped distributions by borrowing the corrections from the exponential density. All these corrections require equispaced intervals.

On the whole Pearse's method is most successful in bringing the moments of a great variety of condensed distribution functions closest to the true parent moments. However, her method requires laborious computations. The SC, with the sign reversal for J- or U-shaped distributions, should suffice for most practical purposes, at least for second and higher moments (see Table 1 for an illustration using a "twisted" J-shaped distribution).

Until now it has been assumed that the group limits were constant and fixed relative to the location of the distribution. If this assumption is relaxed and, instead, one lets

Table 1 Moment Adjustments of a Twisted J-Shaped Distribution

Parameter	True Values	Unadjusted Values	Pearse's Adjustments	Martin's Adjustments	Elderton's Adjustments	Sheppard's[a] Corrections
Means	2.5	2.583500	2.499970	2.529440	2.5491670	2.583500
Standard deviation	2.5	2.434071	2.499079	2.472198	2.4511750	2.451129
β_1	1	1.155141	0.994249	1.048328	0.9887035	1.017339
β_2	3	3.124304	2.987057	3.033454	2.9578665	3.119492

Source. K. Pearson [50].
[a]Computed by author.

c_1 vary randomly in the equispaced case, thereby determining the remaining c_i, SC should be used to correct the grouped moments so as to bring them *on the average* closest to the true moments (see Abernethy [1]). This demonstrates the real essence of SC: it eliminates the *systematic* errors due to grouping, but not the possible *accidental* errors due to the locations of the group intervals.

Bounds in terms of grouped moments, between which the real moments must lie, were derived by Norton [45]. This is a somewhat trivial result because it was derived under the two extreme cases in which the densities within each group were condensed to its two limits. The same boundaries can be obtained, of course, by direct calculation. A more useful result is that of Gastwirth and Krieger [24] and Krieger [37], who, for decreasing and unimodal* densities, constructed "upper" and "lower" decreasing and unimodal densities whose moments bound the real ones. Hartley [29] obtains measures of accuracy for SC applied to any distribution function, as well as detailing procedures to deal with lack of high-order contact at both ends of the distribution.

Corrections for the following cases are also available:

1. **Moments about the *Calculated* Mean from the Grouped Distribution.** Craig [13] includes a term to account for possible discrepancies between the calculated and the real means.
2. **Discrete Distributions Grouped into Intervals So That Each Contains Several Values of the Discrete Variable.** A further adjustment is to be added to the SC if the resulting moments should *on the average* be close to the true ones [11] or a correction due to Pierce [51] for any single grouping.
3. **Bivariate (Product) Moments.** No corrections are necessary for the covariance and the first product moment. Higher product moments do require Sheppard-like corrections (See Baten [6] and Wold [57]). Corrections for bivariate second-order moments in the presence of autocorrelations appear in Daniels [15].
4. **SC for Factorial Moments and Cumulants.** (See Wold [56].)

ESTIMATION WITH MIDPOINTS

The relationships between grouped and ungrouped moments discussed in the preceding section are applicable only for estimates when allied to the method of moments (which lays down that the moments of the sample are to be identified with those of the population). Although the method of moments* is easy to apply and it assures consistency, a more widely acceptable principle of estimation is that of maximum likelihood* (MLE). If we now replace the population value p_i by n_i/n and reinterpret m_i and $\bar{x}_i$ to be sample values, $\hat{\boldsymbol{\theta}}$, the MLE of the q-component parameter vector $\boldsymbol{\theta}$, is that value of $\boldsymbol{\theta} \in \Omega$, which provides the absolute maximum of the likelihood function $L = \prod_{i=1}^{k} p_i^{n_i}$, where $p_i = p_i(c_{i-1}, c_i; \boldsymbol{\theta})$ a function of the class limits and the parameters of the distribution, and Ω is the parameter space. That is, under suitable regularity conditions, the MLEs are the roots of each of the equations

$$\sum_i n_i \frac{\partial \log p_i}{\partial \theta_j} = 0 \qquad (j = 1, \ldots, q)$$

with asymptotic covariance matrix

$$\text{as var}(\boldsymbol{\theta}) = \left(-n \sum_{i=1}^{k} p_i \frac{\partial^2 \log p_i}{\partial \theta_l \, \partial \theta_j} \right)^{-1} \qquad (l, j = 1, \ldots, q)$$

A set of sufficient conditions for the existence and uniqueness for such an absolute maximum is that the derivative $\partial \log L / \partial \boldsymbol{\theta}$ is continuous for every $\boldsymbol{\theta} \in \Omega$, its sign is different at the two extremes of the parameter space, and $\partial^2 \log L / \partial \boldsymbol{\theta}^2$ is negative for any such root (see Kulldorff [38]). Using these conditions, MLEs were derived for the parameters of the following distributions:

1. Location, scale, and both simultaneously for the exponential* and truncated exponential [38].
2. Location and scale for the normal* [26, 38] and the Weibull* [43].
3. The scale parameter for the compound exponential [8] and Pareto* [5]; the latter uses an ingenious nonlinear regression approach.

There exist also numerical methods for MLEs of location and scale parameters from grouped samples, where the parent distribution is assumed known. Computer programs for such MLEs were written by Benn and Sidebottom [7] and by Swan [53], who specializes to the normal distribution case, and can include other cases such as probit* analysis. A more recent procedure for providing MLEs from grouped data is the EM (expectation maximization) algorithm by Dempster et al., [17], which was programmed by Wolynetz [58, 59] for grouped data.

For equispaced grouping, Lindley [40] obtained maximum likelihood equations based on Taylor's expansion, which were solved by Newton's method to produce Sheppard-like corrections to the estimated parameter θ:

$$-\frac{1}{24}h^2\left\{\sum_{i=1}^{k}(\partial/\partial\theta)(f''/f)\Big/\sum_{i=1}^{k}(\partial^2/\partial\theta^2)\log f\right\} + O(h^3),$$

where $f = F'(x)$ and a dash denotes differentiation with respect to the random variable X. If $F(x)$ is the univariate normal distribution, SC for both estimated mean and variance are obtained, thus advancing a third interpretation to Sheppard's corrections. Lindley's results were extended by Tallis [54] to the multivariate distribution under equispaced grouping and to the univariate distribution under unequal grouping. Tallis derived corrections to the five parameters of the bivariate normal distribution* which agreed with their population counterparts derived by Wold [57]. He also derived the grouped sampling covariance matrix of these five estimates. Similar corrections should be applied to multivariate normal correlation and regression coefficients. No correction is needed for the covariances. Fryer and Pethybridge [21], extend Lindley's analysis to a second term approximation, thereby obtaining more accurate corrections for simple regression coefficients. They also considered the cases where either the regressor or the regressand is grouped, but not both. Approximate MLEs for multivariate regression when both or either regressors and regressand are grouped are derived in Indrayan and Rustagi [32].

Another principle of estimation used in grouped data estimation was the minimum chi-square* criterion, which shares with MLE the property of best asymptotic normality*. It was suggested by Hughes [31] for the estimation of the variance of the normal distribution, by Kulldorff [38] as an alternative to the MLE of the parameters of the exponential distribution, and by Aigner and Goldberger [5] as an alternative to the MLE of the scale parameter of the Pareto distribution. Another consistent estimator for equidistance grouped data was offered by McNeil [44], which is valid also for coarsely grouped data. It is computationally simpler than the MLE and is consistent even when the MLE is not, while the loss of efficiency is trifling. However, it is extremely difficult to apply in multivariate cases.

ESTIMATION WITH CENTROIDS

Only in the early 1940s did statisticians begin to appreciate the additional information that the interval centroids convey in the calculation of moments. The combined use of interval centroids, midpoints, and frequencies in the calculation of the second [33] or higher moments (in an instrumental variable* fashion by Dwyer [19]) was shown to improve appreciably their precision. Much later, Cox [12] justified the reporting of the interval means as the points of condensation in a general-purpose data grouping (e.g., for convenience of exposition) on the grounds that means minimize the quadratic loss due to grouping. Still the fact remains that there is no known correction for second or higher moments computed from interval means. For instance, there is no general way of correcting for the within-group variation (in the analysis of variance*) which is lost when the data are grouped.

Probably the widest use of centroids in statistical analysis is in estimating regression models. If the regression model in the original observations is homoscedastic, the regression error becomes heteroscedastic* after the data are combined in groups of different sizes. Best linear unbiased estimates (BLUEs)—or MLEs when normality of the error term is assumed—are easily obtained if the class centroids used in the estimation are weighted by their corresponding class frequencies. For the linear multiple regression model, these can be obtained only if all the regressors need to be cross-classified into cells and the centroids and frequencies for the independent and dependent variables must be known in each cell. For instance, in a regression of Y on two variables X_1 and X_2, let the original n observations be grouped into $m_1 \times m_2$ cells by subdividing X_1 and X_2 into m_1 and m_2 classes, respectively. In each cell the means and frequencies for Y, X_1 and X_2 need to be reported. The BLUEs are obtained by repeating each cell's means for all three variables as often as the cell's frequency. (Note that the number of degrees of freedom* for estimating the error variance is $m_1 m_2 - 3$). However, often one does not have all this information. Sometimes only the marginal means, rather than the cell means, are available together with the cell frequencies. In other situations, even the cell frequencies are not available but only the marginal frequencies are given. This is the common situation where survey data are summarized in one-way classification* tables. Generalized least-squares* estimates for the former were produced by Houthakker and reported in Haitovsky [28], while the latter case was treated by Haitovsky [28], who offered two basic approaches:

1. **An Unbiased Method.** To compute the ith row in the normal equations from the table classified by the ith explanatory variable
2. **An Efficient Method.** To combine the marginal means from the one-way classification tables efficiently (i.e., with appropriate weights derived from the marginal frequencies) to produce the regression estimates

The latter requires a two-stage estimation. Two simplifications of this approach are also available and illustrated on real data (see Table 2). Note that, although the information about the original data is substantially reduced in the consecutive models, the estimates of the regression coefficients remain remarkably close to the ungrouped estimates as long as the information from both tables is used, either cross-classified or not. Only when the regression is estimated from each of the one-way classification tables are the estimates, although unbiased, far from their ungrouped counterparts; all are highly nonsignificant. An interesting analysis of the relative efficiency of the regression coefficient and of the behavior of the coefficient of determination when data are grouped as a function of the method of grouping is found in Cramer [14].

Another widespread use of the interval centroids in statistical inference came about by a recent resurgence of interest in measuring income inequality*, usually by the Lorenz concentration ratio and by the Gini coefficient. Family income data are customarily available only in grouped form. For instance, the usual source of such data in the United States is the Current Population Report Series published by the Census Bureau: *The Statistical Abstract of the U.S.A.* Here we cite Gastwirth [22], who uses the available interval centroids to obtain lower and upper bounds on the Gini index. This method is then extended to construct lower and upper bounds for the Lorenz curve*. Kakwani and Podder [34] specify a Lorenz curve equation that would fit the actual data reasonably well, and in [35] propose four

Table 2 Effect of Different Groupings on the Estimation of Regression Equations

Model	Intercept	Coefficient of X_1	Coefficient of X_2	R^2	σ^2
Ungrouped ($n = 1218$)	17.10 (7.30)[a]	0.7578 (0.1398)	− 0.1778 (0.0367)	0.0347	5250
Complete cross classification ($n = 56$)	16.47 (10.32)	0.7473 (0.1203)	− 0.1624 (0.0323)	0.4969	3914
X_1-table ($n = 7$)	10.86 (5.78)	0.5505 (1.6139)	0.0382 (1.8752)	0.7284	9027
X_2-table ($n = 8$)	73.74 (11.26)	− 0.6532 (2.5391)	− 0.0931 (0.1572)	0.9098	1345
Houthakker's method	18.08 (5.87)	0.7263 (0.1259)	− 0.1719 (0.0338)	0.8139	4285
Haitovsky: method 1	18.03 (5.62)	0.7271 (0.1033)	− 0.1718 (0.0282)	0.7705	4337
Haitovsky: method 2	16.70 (8.80)	0.7560 (0.2007)	− 0.1725 (0.0187)		
Haitovsky: method 2 simplified	15.58	0.7749 (0.1211)	− 0.1682 (0.0333)		
Pooled regression	19.62	0.7133 (0.1320)	− 0.1698 (0.0355)		

Source. Y. Haitovsky [28].

[a] Figures in parentheses are standard errors of the coefficients above them.

alternative methods to estimate the proposed Lorenz curve from grouped data, using a regression framework very similar to Aigner and Goldberger's [5] mentioned above; Gastwirth and Glauberman [23] use Hermite interpolations to estimate the Lorenz curve and Gini index.

OPTIMAL GROUPING

The notion of optimal grouping has been treated in two contexts: grouping for data presentation and estimating parameters from grouped data. The former was mentioned in the preceding section and is typified by the early work of Dwyer [19] and Cox [12]. Cox uses a loss function approach to arrive at optimal grouping. He specifies a standardized quadratic loss function to be minimized for a given number of groups k, resulting in a minimum within-group variance for that k. Since within-group variance is minimized by the group centroids, they are the points of condensation, while the optimal choice of k is determined from the trade-off between the additional information gained and cost incurred when k is increased. Table 3 exhibits the information lost by grouping for a normal distribution, as measured by the above-mentioned loss function when the data are grouped optimally, in contrast to equiprobability groups, equidistanced groups within 3 standard deviations of the mean,

Table 3 Comparison of the Information Lost by Four Grouping Schemes of the Normal Distribution (%)

Number of Groups	(1) Optimal Grouping	(2) Equal Probability Grouping	(3) Equidistanced Grouping	(4) Equidistanced with the Two Extreme Groups Open
2	36.34	36.34	36.34	36.34
3	19.02	20.68	26.17	19.02
4	11.75	13.87	16.29	11.75
5	7.99	10.31	11.03	8.04
6	5.80	8.06	7.93	5.88
7	4.40	6.57	5.97	4.51
8	3.45	5.51	4.64	3.58
9	2.79	4.71	3.71	2.92
10	2.29	4.10	3.03	2.44

Source. Column 1: Cox [12, Table 1], Kulldorff [38, Table 8.1] and Bofinger [9, Table III].
Column 2: For $2 \leqslant k \leqslant 6$ Cox [12, Table 1] and the rest computed by the author.
Column 3: Computed by the author.
Column 4: Kulldorff [38, Table 8.3].

and equidistanced groups where the two extreme groups are open. The last performs almost as well as the optimal, while the second is close to the optimal for $k > 6$ and computationally simpler.

Regarding grouping for estimation, the criterion most commonly used for optimal grouping is the relative efficiency* (REF) of the ungrouped estimator relative to the corresponding grouped estimator. An approximation to this ratio for MLEs from midpoints of equidistanced groups and the corresponding ungrouped estimator of the parameter θ is given by Lindley [40]:

$$1 + \frac{h^2}{24}\,\frac{E\left\{\dfrac{\partial^2}{\partial\theta^2}\left(\dfrac{f''}{f}\right) + \dfrac{f''}{f}\dfrac{\partial^2}{\partial\theta^2}(\log f)\right\}}{E\left\{\dfrac{\partial^2}{\partial\theta^2}(\log f)\right\}}.$$

Thus the REFs in estimating the mean and variance of a normal population using midpoints are $1 - h^2/(12\sigma^2)$ and $1 - h^2/(6\sigma^2)$, respectively, from which h, and hence k, can be chosen to satisfy a prescribed level of REF for given σ^2. This result was extended to the bivariate normal by Haitovsky [28], where it is shown, for instance, that the regression coefficients can be most efficiently estimated for a given number of cells by cross-classifying the two normally distributed variables into an equal number of groups (see also Don [18] for multivariate distributions).

The REF can be improved if the distribution is grouped so as to maximize Fisher's information* on θ in the grouped distribution, or equivalently, by minimizing the asymptotic variance of the MLE of θ. Such a grouping scheme will be called *optimal*, and can always be attained by choosing intervals to satisfy $\xi(c_i) = (\bar{\xi}_i + \bar{\xi}_{i+1})/2$ for $i = 1, \ldots, k-1$, where $\xi(x) = (\partial/\partial\theta)\log f$ is the score function and $\bar{\xi}_i$ is the conditional expectation $E\{\xi(x)|c_{i-1} \leqslant X \leqslant c_i\}$, provided that $\xi(x)$ is strictly monotonic and continuous in x. If the score function is *linear* in x, the $\bar{\xi}_i$'s become the interval centroids and hence $c_i = (\bar{x}_i + \bar{x}_{i+1})/2$. For instance, for the exponential family of distributions $B(\boldsymbol{\theta})h(x)\exp[\sum_i Q_i(\boldsymbol{\theta})R_i(x)]$, the score function is $\xi(x)$, where

$$\xi_j(x) = [B(\boldsymbol{\theta})]^{-1}\partial B(\boldsymbol{\theta})/\partial\theta_j + \sum_i R_i(x)\partial Q_i(\boldsymbol{\theta})/\partial\theta_j,$$

which is linear in x when $R_i(x) = x$ for all

$\partial Q_i(q)/\partial\theta_j \neq 0$. Thus, for estimating the mean of the normal distribution, the optimal group limits are obtained by solving $c_i = (\bar{x}_i + \bar{x}_{i+1})/2$ iteratively and are given by, e.g., Kulldorff [38, Table 8.1] for $k = 2, \dots, 10$. The loss in efficiency incurred by grouping is the same as given in Table 3 for the various grouping schemes. Examples of optimal groupings for estimating other parameters include the scale of the normal and exponential distributions* [38], the scale and location of the logistic distribution*, and the scale and shape of the Weibull distribution* [43].

The problem is that for most applications $\xi(x)$ will be a function of the parameter(s) we wish to estimate. The use of $\xi(x)$ to determine the optimal grouping is therefore impossible. From the author's experience, if $k > 5$, one can safely use equidistanced grouping with only a small loss in efficiency.

For linear score functions, the optimal groups for estimating with grouped data will be the same as the optimal strata boundaries in stratified sampling* and the optimal spacings* of order statistics* (see Adatia and Chan [2] and Bofinger [10]).

GROUPING REQUIRED FOR ESTIMATION

There are situations in which individual observations are available, but grouping is performed, either to improve an estimator's properties or as an estimation device.

For instance, in fitting a regression to two variates, both subject to measurement error, Wald [55] proposed the method of *grouped averages* to ensure consistency. It is done by dividing the data by the median observations on the regressor into two groups and fitting a line through the group means. The resulting estimators, although widely advocated, achieve consistency in very exceptional cases only (see Madansky [41]), because the dividing point between the groups is affected by the error in the regressor. In fact, when the regressor and its error are jointly normal, the asymptotic bias of the slope estimator is identical to that of least squares (see Pakes [47]). Hence, the Wald estimator, which is less efficient than the least squares, has no asymptotic bias advantages in the normal case either. No similar results are available for nonnormal cases where extra care must be exercised before applying Wald's method.

Another case is logit* and probit* analyses. It is common to categorize the observations in order to facilitate the transformations necessary for analyzing these models in a linear regression framework. However, such a practice is not necessary, since the superior MLE can be used instead, thereby avoiding the problem arising in weighted regression analysis (see Haberman [27]).

A third situation is grouping data for contingency tables*. A common practice is to regroup sparse contingency tables in coarse category level, but one should be aware of the fact that as the number of categories is reduced, the chi-square statistics used for testing for independence will systematically decrease [25], risking a type II error.

In conclusion, obtaining data in grouped form often reduces substantially the cost of collection, transmission, and storage. It is apparent from the discussion above that methods exist for drawing inferences from grouped data that sacrifice little in terms of precision. Thus the gain in precision using raw data is frequently outweighed by the increase in cost.

References

[1] Abernethy, J. R. (1933). *Ann. Math. Statist.*, **4**, 263–277.

[2] Adatia, A. and Chan, L. K. (1981). *Scand. Actuar. J.*, 193–202.

[3] Aigner, D. J. (1968). *Technometrics*, **10**, 793–809.

[4] Aigner, D. J. (1970). *Rev. Int. Statist. Inst.*, **38**, 210–219.

[5] Aigner, D. J. and Goldberger, A. S. (1970). *J. Amer. Statist. Ass.*, **65**, 712–723.

[6] Baten, W. D. (1931). *Ann. Math. Statist.*, **2**, 309–319.

[7] Benn, R. T. and Sidebottom, S. (1976). *Appl. Statist.*, **25**, 88–93.

[8] Boardman, T. J. (1973). *Technometrics*, **15**, 271–273.

[9] Bofinger, E. (1970). *J. Amer. Statist. Ass.*, **65**, 1632–1638.

[10] Bofinger, E. (1975). *J. Amer. Statist. Ass.*, **70**, 151–154.

[11] Carver, H. C. (1930). *Ann. Math. Statist.*, **1**, 101–121.

[12] Cox, D. (1957). *J. Amer. Statist. Ass.*, **52**, 543–547.

[13] Craig, C. C. (1941). *Ann. Math. Statist.*, **12**, 339–345.

[14] Cramer, J. S. (1964). *J. Amer. Statist. Ass.*, **59**, 233–250.

[15] Daniels, H. E. (1947). *J. R. Statist. Ass. B*, **9**, 245–249.

[16] Davies, G. R. and Bruner, N. (1943). *J. Amer. Statist. Ass.*, **38**, 63–68.

[17] Dempster, A. P., Laird, N. M., and Rubin, D. B. (1977). *J. R. Statist. Soc. B*, **39**, 1–22.

[18] Don, F. J. H. (1981). *J. Multivariate Anal.*, **11**, 452–458.

[19] Dwyer, P. S. (1942). *Ann. Math. Statist.*, **13**, 138–155.

[20] Elderton, W. P. (1933). *Biometrika*, **25**, 179–180.

[21] Fryer, J. G. and Pethybridge, R. J. (1972). *Appl. Statist.*, **21**, 142–154.

[22] Gastwirth, J. L. (1972). *Rev. Econ. Statist.*, **54**, 306–316.

[23] Gastwirth, J. L. and Glauberman, M. (1976). *Econometrica*, **44**, 479–484.

[24] Gastwirth, J. L. and Krieger, A. K. (1975). *J. Amer. Statist. Ass.*, **70**, 468–471.

[25] Gilula, Z. (1983). *J. Amer. Statist. Ass.*, **78**, No. 1.

[26] Gjeddebaek, N. F. (1970). *On Grouped Observations*. Egeltorlag/Ferroson, Søborg, Denmark.

[27] Haberman, S. J. (1974). *The Analysis of Frequency Data*. University of Chicago Press, Chicago.

[28] Haitovsky, Y. (1973). *Regression Estimation from Grouped Observations*. Charles Griffin, London/Hafner, New York.

[29] Hartley, H. O. (1950). *Biometrika*, **37**, 145–148.

[30] Hassanein, K. M. and Sebaugh, J. L. (1973). *Skand. Aktuarietidskr.*, 1–10.

[31] Hughes, H. M. (1949). *Univ. Calif. Publ. Statist.*, **1**, 37–52.

[32] Indrayan, I. and Rustagi, J. S. (1979). In *Optimizing Methods in Statistics*, J. S. Rustagi, ed. Academic Press, New York, pp. 301–319.

[33] Jones, H. L. (1941). *J. Amer. Statist. Ass.*, **36**, 525–529.

[34] Kakwani, N. C. and Podder, N. (1973). *Int. Econ. Rev.*, **14**, 278–291.

[35] Kakwani, N. C. and Podder, N. (1976). *Econometrica*, **44**, 137–148.

[36] Kendall, M. G. (1938). *J. R. Statist. Soc.*, **101**, 592–605.

[37] Krieger, A. M. (1979). *J. Amer. Statist. Ass.*, **74**, 375–378.

[38] Kulldorff, G. (1961). *Contributions to the Theory of Estimation from Grouped Samples*. Wiley, New York.

[39] Lewis, W. T. (1935). *Ann. Math. Statist.*, **6**, 11–20.

[40] Lindley, D. V. (1950). *Proc. Camb. Philos. Soc.*, **46**, 106–110.

[41] Madansky, A. (1959). *J. Amer. Statist. Ass.*, **54**, 173–205.

[42] Martin, E. S. (1934). *Biometrika*, **26**, 12–58.

[43] Marymont, I. (1975). *Corrections and Inference for Grouped Samples*. M. A. thesis, Tel-Aviv University.

[44] McNeil, D. R. (1966). *Biometrika*, **53**, 545–557.

[45] Norton, K. A. (1938). *Sankhyā*, **3**, 265–272.

[46] Pairman, E. and Pearson, K. (1919). *Biometrika*, **12**, 231–250.

[47] Pakes, A. (1982). *Int. Econ. Rev.*, **23**, 491–497.

[48] Pearse, G. E. (1928). *Biometrika*, **20**, 314–355.

[49] Pearson, K. (1902). *Biometrika*, **1**, 265–303.

[50] Pearson, K. (1933). *Biometrika*, **25**, 181.

[51] Pierce, J. A. (1943). *J. Amer. Statist. Ass.*, **38**, 57–62.

[52] Sheppard, W. F. (1886). *Proc. Lond. Math. Soc.*, **29**, 353–380.

[53] Swan, A. V. (1969). *Appl. Statist.*, **18**, 110–114.

[54] Tallis, A. M. (1967). *Technometrics*, **9**, 599–608.

[55] Wald, A. (1940). *Ann. Math. Statist.*, **11**, 284–300.

[56] Wold, H. (1934). *G. Ist. Ital. Att.*, **4**, 304.

[57] Wold, H. (1934). *Skand. Aktuarietidskr.*, **17**, 248–255.

[58] Wolynetz, M. S. (1979). *Appl. Statist.*, **28**, 185–195.

[59] Wolynetz, M. S. (1979). *Appl. Statist.*, **28**, 195–206.

(CORRECTION FOR GROUPING
MAXIMUM LIKELIHOOD ESTIMATION
REGRESSION)

Y. HAITOVSKY

GROUP TESTING

Group tests are customarily defined as tests designed for simultaneous administration to a group of examinees. Useful group testing is standardized with respect to administration of the test, wording and order of the questions, and scoring of the answers. In short, group testing has as its goal the efficient, simultaneous measurement of a large group of people under standardized conditions.

We distinguish between tests versus questionnaires or inventories. The correct answers to test questions are examiner-defined, whereas the answers to questionnaires or inventories are examinee-defined, and are therefore not classified by the examiner as correct or incorrect. A test measures an ability or an achievement; a questionnaire or inventory does not; for example, we do not classify an attitude questionnaire as a test. Some experts in testing do not make this distinction, although it does exist.

The first extensive use of standardized group testing was undertaken by the U.S. Army during World War I. Group intelligence tests were administered to almost 2 million army recruits for purposes of selection and classification. The methods were deemed a success and the fundamental procedures developed during that period have served as a standard model for group testing.

TEST THEORY

The construction of a group test is guided by rigorous statistical theory of mental testing. A formal rigorous presentation of the major results can be found in Lord and Novick's *Statistical Theories of Mental Test Scores* [2], which supersedes Gulliksen's classic *Theory of Mental Tests* [1]. Lord and Novick derived the assumptions of classical test theory from a more basic theory (*see* PSYCHOLOGICAL TESTING).

Classical test theory is based on a rather small number of basic concepts and assumptions. One begins with a countable set P of persons and a countable set Q of test forms. Let X_i be the observed-score random variable, T_i the true-score random variable, and E_i the error-of-measurement random variable for test form i. The fundamental assumption is that

$$X_i = T_i + E_i. \qquad (1)$$

The following additional assumptions are assumed to hold for every nonnull subset of P:

$$\mathbf{E}[E_i] = 0, \qquad (2)$$

where **E** denotes expectation;

$$\rho(T_i, E_i) = 0; \qquad (3)$$

$$\rho(E_i, E_j) = 0, \qquad i \neq j; \qquad (4)$$

$$\rho(E_i, T_j) = 0, \qquad i \neq j, \qquad (5)$$

where ρ denotes correlation*.

The first derived result is that the expected true score for a test form i equals the expected observed score:

$$\mathbf{E}[T_i] = \mathbf{E}[X_i - E_i] = \mathbf{E}[X_i].$$

RELIABILITY AND VALIDITY

Two important psychometric properties of a test are *reliability* and *validity*. Assessments of reliability are used to determine the consistency of test measurements and validity to determine the value of test measurements.

The reliability of a test is the squared correlation between the observed score (X) and the true score (T), denoted as ρ_{XT}^2. It follows from the assumptions of classical test theory that

$$\rho_{XT}^2 = \sigma_T^2 / \sigma_X^2 = 1 - (\sigma_E^2 / \sigma_X^2),$$

where σ_X^2 and σ_T^2 are the variances of X and T, respectively. The term σ_E is the *standard error of measurement*. If the standard error of measurement is zero, the reliability is 1, its maximum value.

Two test measurements X_1 and X_2 that satisfy (a) $X_1 - E_1 = X_2 - E_2 = T$ and (b) $\sigma^2(X_1) = \sigma^2(X_2)$ are said to be *parallel test measurements*. It follows that

$$\rho(X_1, X_2) = \sigma_T^2 / \sigma_X^2 = \rho_{XT}^2.$$

Hence, the correlation between two parallel test measurements equals the common reliability of the parallel tests from which the measurements are obtained.

A test measurement X is usually defined as the sum of a set of n item or component measurements (Y_i, $1 \leqslant i \leqslant n$), so that $X = Y_1 + Y_2 + \cdots + Y_n$. An important lower bound for the reliability of such a measurement X is *Cronbach's coefficient* alpha (α), where

$$\alpha = \frac{n}{n-1}\left[1 - \frac{\sum_{i=1}^{n}\sigma^2(Y_i)}{\sigma_X^2}\right] \leqslant \rho_{XT}^2;$$

a necessary and sufficient condition for equality to hold is that for $Y_i = T_i + E_i$, $T_i = T_j + a_{ij}$ $(1 \leqslant i, j \leqslant n)$ for all i and j, where a_{ij} is a constant. Measurements with this property are said to be *essentially tau-equivalent*.

If the Y_i are also parallel measurements, then for all i and j $(i \neq j)$,

$$\alpha = \rho_{XT}^2 = \frac{n\rho(Y_i, Y_j)}{1 + (n-1)\rho(Y_i, Y_j)}.$$

The term on the right-hand side of this equation is the *generalized Spearman–Brown formula*, which shows the relation between test length (k parallel subtests) and reliability.

In practice, the reliability coefficient has been estimated by three general methods: test–retest, parallel forms, and internal consistency. Reliability is estimated from administration of the same test form on two separate occasions with the *test–retest* method, from administration of different but parallel test forms on two separate occasions with the *parallel-forms* method, and from administration of a single test form with the *internal-consistency* method. Since reliability is a rather general term in practical applications, the methods used in estimating it should always be specified.

The *validity* of a test measurement X with respect to a second measure Y may be defined by the *validity coefficient* $|\rho(X, Y)|$, where

$$|\rho(X, Y)| = \sigma_{xy}/\sigma_x\sigma_y.$$

Given pairs of parallel test scores (X, X') and (Y, Y'), it can be shown that

$$\rho(T_x, T_y) = \frac{\rho(X, Y)}{\sqrt{\rho(X, X')\rho(Y, Y')}},$$

where T_x and T_y are the true-score random variables for X and Y, respectively. Note that the square root of the reliability sets an upper bound to the validity:

$$\rho(X, Y) \leqslant \rho(X, T_x) = \sqrt{\rho(X, X')}$$

$$\rho(X, Y) \leqslant \rho(Y, T_y) = \sqrt{\rho(Y, Y')}.$$

Validity coefficients are used to evaluate group tests with respect to three broad kinds of validity: predictive, concurrent, and construct validity. One evaluates the *predictive validity* of a test score X by its correlation with a measure Y (criterion variable) taken later. One evaluates the *concurrent validity* of a test score by its correlation with a measure taken concurrently.

If the relation between X and Y is predicted from a psychological theory in which T_x (true score for X) represents a measure of a property (construct) of a psychological process, then $\rho(X, Y)$ is a measure of *construct validity*.

TEST SCORES

Raw scores (X) are sometimes converted to percentile scores, but more commonly to standard scores (Y), as follows:

$$Y = \mu + \sigma\left(\frac{X - \bar{X}}{S}\right),$$

where $(\bar{X}, S)$ and (μ, σ) are the mean and standard deviation for X and Y, respectively. Hence one can transform X to any desired mean (μ) and standard deviation (σ). Test scores are often standardized to a mean of 50 and a standard deviation of 10.

This article has emphasized classical test theory as a guide to group testing. Conventional group tests have difficulty measuring the extremes of a heterogeneous group. Modern developments in test theory have led to the construction of latent* trait models of examinees' performance. Such models assume that the score on each test item is a function of some k-dimensional vector of psychological characteristics called *traits*, which are not directly observable and are therefore termed *latent*. Those new advances have particular application to testing tailored to the individual. Computerized and individualized tailored testing may provide an economically feasible alternative to standard paper-and-pencil group testing in the not-so-distant future.

References

[1] Gulliksen, H. (1950). *Theory of Mental Tests*. Wiley, New York. (A classic introduction to classical test theory.)

[2] Lord, F. M. and Novick, M. R. (1968). *Statistical Theories of Mental Test Scores*. Addison-Wesley, Reading, Mass. (The authoritative introduction to classical test theory with some latent trait theory.)

Bibliography

See the following works, as well as the references just cited, for more information on the topic of group testing.

American Psychological Association (1974). *Standards for Educational and Psychological Tests*. Americal Psychological Association, Washington, D.C. (A valuable guide for test developers and test constructors.)

Hambleton, R. K., Swaminathan, H., Cook, L. L., Eignor, D. R., and Gifford, J. A. (1978). *Rev. Educ. Res.*, **48**, 467–510. (A nontechnical introduction to latent trait theory.)

Lord, F. M. (1977). *J. Educ. Meas.*, **14**, 117–138. (An interesting survey of applications of latent trait theory.)

Weiss, D. J. and Davison, M. L. (1981). *Ann. Rev. Psychol.*, **32**, 629–658. (A provocative evaluation of classical test theory and modern developments.)

(FACTOR ANALYSIS
LATENT STRUCTURE ANALYSIS
PSYCHOLOGICAL TESTING
PSYCHOMETRICS)

DONALD D. DORFMAN

GROWTH CURVES

Although the study of growth is an important topic in many biological sciences, the term *growth curve* has a special meaning in statistics. Growth curve analysis applies to data consisting of repeated measurements* over time of some characteristic, obtained from each member of a group of individuals.

Elston and Grizzle [4] give a simple example in which the ramus height of 20 boys was measured at 8, 8.5, 9, and 9.5 years. The data show a steady growth with age, and we would like to fit a straight line or some other simple model to describe the pattern of growth. However, the distributional assumptions must take account of the statistical dependence of repeated measurements on the same individual. Growth curve analysis includes a variety of techniques for handling this problem, some related to multivariate analysis of variance* and some to random-effects models (*see* FIXED-, RANDOM-, AND MIXED-EFFECTS MODELS).

One approach, first suggested by Rao [11], begins with the multivariate distribution of the vector of observations for each individual. If growth is linear in time and we assume a multivariate normal* distribution for the 4×1 vector $\mathbf{Y}_i$ of observations from the ith individual, then we assume that

$$\mathbf{Y}_i \sim N(\mathbf{X}\boldsymbol{\beta}, \boldsymbol{\Sigma});$$

that is, $\mathbf{Y}_i$ has a multivariate normal distribution with mean $\mathbf{X}\boldsymbol{\beta}$ and covariance matrix $\boldsymbol{\Sigma}$. To fit a linear growth model to the data on ramus height, the matrix $\mathbf{X}$ is 4×2 and can be written

$$\mathbf{X}' = \begin{bmatrix} 1 & 1 & 1 & 1 \\ 8 & 8.5 & 9 & 9.5 \end{bmatrix},$$

and $\boldsymbol{\beta}$ is the 2×1 vector of regression parameters. Some writers favor using orthogonal polynomials as the columns of $\mathbf{X}$, but this can create difficulties when the times of measurement differ among individuals.

Rao [11] gives a lucid and thorough discussion of this model. He shows that the minimum variance unbiased* estimator of $\boldsymbol{\beta}$ is

$$(\mathbf{X}'\mathbf{S}^{-1}\mathbf{X})^{-1}\mathbf{X}'\mathbf{S}^{-1}\overline{\mathbf{Y}},$$

where $\mathbf{S}$ is the sample covariance matrix,

$$\mathbf{S} = \sum_{i=1}^{N} (\mathbf{Y}_i - \overline{\mathbf{Y}})(\mathbf{Y}_i - \overline{\mathbf{Y}})'/N,$$

of the N observations and

$$\overline{\mathbf{Y}} = \sum_{i=1}^{N} \mathbf{Y}_i/N.$$

Rao also constructs and obtains the distribution theory for a statistic testing the adequacy of the linear model.

Elston and Grizzle found that a linear growth model gave a satisfactory fit to the ramus height data, but the method applies directly to all polynomial growth curves as well as any other model for which the population mean vector is linear in the parameters. For example, Ware and Bowden [13] describe the application of this methodology to the analysis of sinusoidal fluctuation arising from circadian rhythm. The concept of growth curve analysis applies also to growth functions that are not linear in the parameters. Bock and Thissen [1] consider triple logistic functions for growth in stature from infancy to adult life. However, nonlinear models require iterative techniques for parameter estimation in place of the closed-form solutions afforded by the multivariate linear model (*see* GENERAL LINEAR MODEL).

The multivariate model provides a degree of generality for the covariance structure of $\mathbf{Y}_i$ that is not always needed. In addition, this model cannot be applied when each individual has a different design matrix*, for example, when ramus height is obtained at regular dental visits scheduled in a nonuniform way. Since this situation is typical, we need a more flexible model.

Suppose that each individual has a set of growth parameters $\boldsymbol{\beta}_i$, and that the conditional distribution of $\mathbf{Y}_i$ given $\boldsymbol{\beta}_i$ is

$$\mathbf{Y}_i \mid \boldsymbol{\beta}_i \sim N(\mathbf{X}\boldsymbol{\beta}_i, \sigma^2\mathbf{I}).$$

Suppose also that $\boldsymbol{\beta}_i$ has a normal distribution

$$\boldsymbol{\beta}_i \sim N(\boldsymbol{\beta}, \boldsymbol{\Lambda}).$$

Then the marginal distribution of $\mathbf{Y}_i$ is

$$\mathbf{Y}_i \sim N(\mathbf{X}\boldsymbol{\beta}, \mathbf{X}\boldsymbol{\Lambda}\mathbf{X}' + \sigma^2\mathbf{I}).$$

Fearn [5] and others call this a two-stage model; its advantages derive from the special form assumed for $\boldsymbol{\Sigma}$ and its applicability for unbalanced designs.

Rao [12] gives an informative discussion of this model. For an arbitrary vector $\mathbf{p}$ and with $\boldsymbol{\beta}$, $\boldsymbol{\Lambda}$, and σ^2 assumed known, he shows that the minimum mean square error* estimate of $\mathbf{p}'\boldsymbol{\beta}_i$ is $\mathbf{p}'\boldsymbol{\beta}_i^{(b)}$, where if $\mathbf{U} = \sigma^2(\mathbf{X}'\mathbf{X})^{-1}$,

$$\boldsymbol{\beta}_i^{(b)} = \boldsymbol{\beta}_i^{(l)} - \sigma^2\mathbf{U}(\boldsymbol{\Lambda} + \sigma^2\mathbf{U})^{-1}(\boldsymbol{\beta}_i^{(l)} - \boldsymbol{\beta})$$

and $\boldsymbol{\beta}_i^{(l)}$ is the least-squares* estimate of $\boldsymbol{\beta}_i$. Rao calls this the Bayes estimate of $\boldsymbol{\beta}_i$ and shows its relationship to the ridge regression* estimate of $\boldsymbol{\beta}_i$.

When σ^2, $\boldsymbol{\Lambda}$, and $\boldsymbol{\beta}$ are unknown, Rao gives unbiased estimates, including

$$\boldsymbol{\beta}^* = \sum_{i=1}^{N} \boldsymbol{\beta}_i^{(l)}/N.$$

When these estimates are substituted into the expression for $\boldsymbol{\beta}_i^{(b)}$, Rao obtains the empirical Bayes* estimate. This paper clearly shows the relationships among the least-squares, empirical Bayes, and ridge estimates* of $\boldsymbol{\beta}_i$. Rao discusses the case of unequal design matrices, but later work discussed below has further clarified that situation.

Fearn [5] gives a clear introduction to the growth curve problem and develops a Bayesian approach to growth curves based on the general linear model developed by Lindley and Smith [10]. Fearn assumes first that σ^2 and $\boldsymbol{\Lambda}$ are known and that $\boldsymbol{\beta}$ has a vague prior distribution, achieved by assuming that

$$\boldsymbol{\beta} \sim N(\mathbf{0}, \mathbf{A})$$

and letting $\mathbf{A}^{-1}$ go to zero. His results are equivalent to Rao's in this setting, although he assumes a unique design matrix, $\mathbf{X}_i$, for each individual.

When σ^2 and $\boldsymbol{\Lambda}$ are unknown, Fearn assumes a chi-square* hyperprior for $(\sigma^2)^{-1}$ and a Wishart* hyperprior for $\boldsymbol{\Lambda}^{-1}$. However, the posterior distributions cannot be obtained analytically, so he recommends using the expressions obtained when the variances are known and substituting estimates of σ^2 and $\boldsymbol{\Lambda}$. With this strategem, the empirical Bayes solution proposed by Rao and the three-stage Bayes model developed by Fearn differ very little.

Fearn uses the two-stage model to reanalyze the ramus height data of Elston and

Grizzle, and confirms the adequacy of the linear growth model. Although the functional form for the population mean growth curve is identical in the multivariate and two-stage models, the two-stage model can provide superior precision for growth curve parameters when it applies. A good example in the context of circadian rhythm analysis is provided by Zerbe and Jones [15].

Geisser [7], Lee and Geisser [9], Rao [12], and Fearn [6] consider the problem of prediction* from the growth curve model. To predict a new vector of observations conditional on the data, Geisser [7] and Fearn [6] obtain the (normal) predictive distribution for the new observations. In the partial prediction problem, part of the new vector is observed and the remainder is to be predicted. Rao [12], Lee and Geisser [9], and Fearn [6] obtain linear predictors for the unobserved values. The linear predictor can again be viewed as the mean of the predictive distribution.

To this point we have ignored the possibility that the population has identifiable subgroups with different mean growth curves or that the growth parameters depend on individual characteristics. Grizzle and Allen [8] treat this problem in the context of a general covariance matrix by assuming that

$$E(\mathbf{Y}_i \mid \boldsymbol{\beta}_i) = \mathbf{X}\boldsymbol{\beta}_i$$

as before, but

$$E(\boldsymbol{\beta}_i) = \gamma_{k\times q} a^{(i)}_{q\times 1}.$$

Here k is the dimension of $\boldsymbol{\beta}_i$ and $\mathbf{a}^{(i)}$ is a vector of individual characteristics, possibly including indicator variables for group membership. The matrix $\gamma_{k\times q}$ allows each element of $\boldsymbol{\beta}_i$ to have a different regression on $\mathbf{a}^{(i)}$. Then the marginal distribution of $\mathbf{Y}_{p\times N}$, the matrix whose columns are the individual observation vectors, is

$$\mathbf{Y}_{p\times N} \sim N(\mathbf{X}\gamma\mathbf{A}, \boldsymbol{\Sigma}\times\mathbf{I}),$$

where $\mathbf{A}$ is the $q\times N$ matrix with columns $\mathbf{a}^{(i)}$, and $\boldsymbol{\Sigma}\times\mathbf{I}$ is the $pN\times pN$ matrix with $\boldsymbol{\Sigma}$ appearing in the block diagonals and $\mathbf{0}$ elsewhere. Grizzle and Allen develop estimates of γ and hypothesis testing* procedures for this model.

Dempster et al. [3] treat all random-effects models for measured response, including growth curves as a special case. Their formulation generalizes Grizzle and Allen's model, and allows an integrated approach to parameter estimation and likelihood ratio testing*, based on the EM method of maximum likelihood* estimation described by Dempster et al. [2].

Dempster et al. begin with the representation

$$\mathbf{Y} = \mathbf{X}\boldsymbol{\beta} + \mathbf{e},$$

where $\mathbf{Y}$ is the $n\times 1$ response vector of *all* observations (n would be Np for earlier examples), $\mathbf{X}$ is a fixed $n\times r$ design matrix, and $\boldsymbol{\beta}$ is an $r\times 1$ vector of linear effects. In our earlier examples, $\boldsymbol{\beta}_i$ for each individual had length k. When $E(\boldsymbol{\beta}_i) = \boldsymbol{\beta}$, the method of Dempster et al. requires $r = 2kN$. Finally, $\mathbf{e}$ is an $n\times 1$ vector of errors with independent $N(0,\sigma^2)$ distributions.

By noting that classical estimates of fixed effects correspond to the limits of posterior means as the relevant variances go to infinity, Dempster et al. treat both fixed and random effects in this framework. At the second stage, $\boldsymbol{\beta}$ is partitioned into fixed effects $\boldsymbol{\beta}_1$, and random effects $\boldsymbol{\beta}_2$. Then

$$\boldsymbol{\beta} = \begin{bmatrix} \boldsymbol{\beta}_1 \\ \boldsymbol{\beta}_2 \end{bmatrix} \sim N\left(\mathbf{0}, \left[\begin{array}{c|c} \boldsymbol{\Sigma}_{11} & \boldsymbol{\Sigma}_{12} \\ \hline \boldsymbol{\Sigma}_{21} & \boldsymbol{\Sigma}_{22} \end{array}\right]\right)$$

and $\boldsymbol{\Sigma}_{11}$ goes to infinity in such a way that

$$\boldsymbol{\Sigma}^{-1} \to \left[\begin{array}{c|c} \mathbf{0} & \mathbf{0} \\ \hline \mathbf{0} & \boldsymbol{\Sigma}_{22}^{-1} \end{array}\right].$$

For further discussion of the model, see Dempster et al. [3].

By choosing maximum likelihood as the estimation criterion, these writers allow the use of the EM method as the numerical algorithm. The EM algorithm is a general technique for maximum likelihood estimation with incomplete observations. It is especially useful with random-effects models, as the random-effects parameters can be treated as missing data. This leads to easy representation of the marginal distribution of **Y**, $f(\mathbf{Y})$, as

$$f(\mathbf{Y}) = f(\mathbf{Y} \mid \boldsymbol{\beta}) f(\boldsymbol{\beta}).$$

Implementation of the EM algorithm is not always simple, but the programming is straightforward. Further research is needed to compare it to other numerical techniques for parameter estimation in this setting.

The reader interested in an overview of the growth curve literature can begin with the classic paper by Wishart [14] and continue with Rao [11, 12] and Fearn [5, 6]. Grizzle and Allen [8] introduce the more general model allowing heterogeneity between individuals; Dempster et al. [3] provide a unified approach to modeling and computation of the random-effects growth curve models.

References

[1] Bock, R. D. and Thissen, D. M. (1976). *Proc. Ninth Int. Biometrics Conf., Vol. 1.* International Biometric Society, Raleigh, NC, pp. 431–442.

[2] Dempster, A. P., Laird, N., and Rubin, D. B. (1977). *J. R. Statist. Soc. B*, **39**, 1–38.

[3] Dempster, A. P., Rubin, D. B., and Tsutakawa, R. K. (1979). *J. Amer. Statist. Ass.*, **76**, 341–353.

[4] Elston, R. C. and Grizzle, J. E. (1962). *Biometrics*, **18**, 148–159.

[5] Fearn, T. (1975). *Biometrika*, **62**, 89–100.

[6] Fearn, T. (1977). *Biometrika*, **64**, 141–143.

[7] Geisser, S. (1970). *Sankhyā A*, **32**, 53–64.

[8] Grizzle, J. E. and Allen, D. M. (1969). *Biometrics*, **25**, 357–382.

[9] Lee, J. C. and Geisser, S. (1972). *Sankhyā A*, **34**, 393–412.

[10] Lindley, D. V. and Smith, A. F. M. (1972). *J. R. Statist. Soc. B*, **34**, 1–41.

[11] Rao, C. R. (1959). *Biometrika*, **46**, 49–58.

[12] Rao, C. R. (1975). *Biometrics*, **31**, 545–554.

[13] Ware, J. H. and Bowden, R. (1977). *Biometrics*, **33**, 566–572.

[14] Wishart, J. (1938). *Biometrika*, **30**, 16–28.

[15] Zerbe, G. O. and Jones, R. H. (1980). *J. Amer. Statist. Ass.*, **75**, 507–509.

(EMPIRICAL BAYES THEORY
FIXED-, RANDOM-, AND MIXED-EFFECTS MODELS
MULTIVARIATE ANALYSIS
REPEATED MEASURES)

JAMES H. WARE

GRUBBS' ESTIMATORS (PRECISION AND ACCURACY OF MEASUREMENT)

BACKGROUND

An observation or measurement with an instrument consists of the true unknown value of the item or characteristic being measured plus an error of measurement. These components of the observation are confounded, but require study to determine the precision* and accuracy* of the instrument system or measurement process. The measurement errors* should be small and of limited variation in order that unquestionable judgments can be made concerning the items or product measured. The error of measurement itself will usually consist of a random component, following a normal distribution, $N(0, \sigma_e^2)$, plus a bias or a systematic error of the instrument, the latter perhaps due to improper calibration*. The standard deviation of the random errors of measurement (SDEM) is often called the *imprecision of measurement*, and is needed to determine the significance of the instrumental bias or systematic error. In applications, instrumental biases for a series of measurements may either be constant or possess a trend, and so must be modeled properly. A special variance components* analysis, initiated by Grubbs [4] and involving two or more instruments taking the same set of measurements, isolates and estimates the imprecisions of measurement of the instruments. These in turn are used to study the significance of the instrumental biases and hence the overall accuracy of the instrument or measurement process.

Analytical techniques for estimating the standard deviations of the errors of measurement result in quantities known as *Grubbs' estimators*.

In practice, estimates of imprecision and bias* are used to make precision and accuracy statements about the instruments or measurement process. In some applications, such as interlaboratory or "round-robin" testing, reference samples with known amounts of chemicals or materials with near-zero variability may be tested at different laboratories, so that errors of measurement may be studied more directly. However, Grubbs' estimators will often aid in the check of assumptions, in comparing operators at a laboratory, or even in judging laboratory capabilities. For many applications, the modeling of the biases or the problem of picking the best calibrated instrument can often be troublesome. As with other components of variance analyses, negative estimates of variance frequently arise, calling for close scrutiny of the data.

Since the original 1948 paper, this area of investigation has grown considerably (see Bradley and Brindley [1], Draper and Guttman [2], Grubbs [5, 6], Hahn and Nelson [7], Hanumara [8], Hanumara and Thompson [9], Jaech [10, 11], Maloney and Rastogi [12], Russell and Bradley [16], Shukla [17, 18], and Thompson [19, 20, 21]).

Under appropriate assumptions, only two instruments are required to estimate the variances in errors of measurement of the instruments and the product variability. Unfortunately, the variances of the estimators for the two-instrument case involve the product variance, which ordinarily should be many times the variances in the errors of measurement. The use of three instruments to take the same set of measurements seems to be ideal, for then there exists a separation of the errors of measurement from the level or value of the characteristic measured (product variability), and the imprecision standard deviations and biases may be studied more directly. Moreover, the variances of efficiencies of the imprecision estimators will not depend on the product variability [4]. With more than three instruments, the efficiency of the imprecision estimators may be enhanced considerably.

EXAMPLE (TWO INSTRUMENTS; RANDOM PRODUCT VARIABILITY)

A simple example for the two-instrument case will illustrate the principles. In military acceptance testing, a sample of n mechanical time fuses (here $n = 24$) is drawn at random from a production lot; the fuses are assembled to live projectiles and complete rounds, and fired from a gun at night. The fuses are set for a projectile explosion time of 5.00 seconds, and the elapsed time is determined by electric clocks (or chronographs) started by a muzzle switch and stopped by two independent observers at "flash" time. The gun firing test destroys the sample rounds and is expensive; the precision and accuracy of measurement is questionable. It is of interest, therefore, to see what a special statistical analysis might contribute to such a problem. Typical data are given in Table 1.

The data indicate that the total variation is confined to a relatively short interval, that the true running times of the fuses represent random variation, that the lag of the second operator in stopping his or her clock may involve a distinct bias, and that there could be random errors of measurement. Thus one might model the data approximately as

$$Y_{ij} = X_i + \beta_j + e_{ij}, \qquad (1)$$

$$i = 1, \ldots, n; \quad j = 1, 2; \quad \text{where}$$

Y_{ij} = observed or measured fuses time for the ith fuses with the jth instrument,

X_i = true running time of the ith fuse (with variance σ_X^2),

β_j = bias or systematic error of the jth instrument,

e_{ij} = random error of measurement for the ith fuse time as measured by instrument j, $N(0, \sigma_{ej}^2)$.

Let s_1^2 and s_2^2 be the observed sample variances for the first and second instruments, s_{12} the sample covariance, s_{1+2}^2 the variance

Table 1 Running Times of 20 Mechanical Time Fuses Measured by Operators Stopping Two Independent Clocks

Measurements by First Instrument (sec)		Measurements by Second Instrument (sec)	
4.85		5.09	
4.93		5.04	
4.75		4.95	
4.77		5.02	
4.67		4.90	
4.87		5.05	
4.67		4.90	
4.94		5.15	
4.85		5.08	
4.75		4.98	
4.83		5.04	
4.92		5.12	
4.74		4.95	
4.99		5.23	
4.88		5.07	
4.95		5.23	
4.95		5.16	
4.93		5.11;	$\bar{y}_2 = 5.063$
4.92;	$\bar{y}_1 = 4.853$	5.11;	$s_2 = 0.092$
4.89;	$s_1 = 0.096$	5.08;	$s_{1-2} = 0.035$

of the sum of the paired readings, and s_{1-2}^2 the variance of the difference of paired readings of the two instruments. Then for practically verifiable independence assumptions, Grubbs [4] shows that the product variance σ_X^2, and the variances in the errors of measurement of the first and second instruments, σ_{e1}^2 and σ_{e2}^2, may be estimated from

$$\hat{\sigma}_X^2 = s_{12} = \left(s_{1+2}^2 - s_{1-2}^2\right)/4 \quad (= 0.008529), \quad (2)$$

$$\hat{\sigma}_{e1}^2 = s_1^2 - s_{12} = \left(s_1^2 - s_2^2 + s_{1-2}^2\right)/2 \quad (= 0.000775), \quad (3)$$

$$\hat{\sigma}_{e2}^2 = s_2^2 - s_{12} = \left(s_2^2 - s_1^2 + s_{1-2}^2\right)/2 \quad (= 0.000441). \quad (4)$$

The standard deviation of the true fuse running times, and the standard deviation of the errors of measurement (SDEM), or imprecisions, of the first and second instruments are then estimated to be

$$\hat{\sigma}_X = 0.092 \text{ sec}, \quad \hat{\sigma}_{e1} = 0.028 \text{ sec},$$

$$\hat{\sigma}_{e2} = 0.021 \text{ sec}.$$

Thanks to the Pitman–Morgan study of normal correlation, [13, 15] a t-test* of whether $\sigma_{e1} = \sigma_{e2}$ is available [12]. In fact, if $r(\text{sd})$ is the sample correlation coefficient calculated from the sums and differences of the paired instrument readings, the statistic to test whether $\sigma_{e1} = \sigma_{e2}$ is

$$t(n-2, \sigma_{e1} = \sigma_{e2}) = r(\text{sd})\sqrt{n-2}\,/\left[1 - r^2(\text{sd})\right]^{1/2}. \quad (5)$$

Since the observed Student's t is about 0.31, we conclude that the SDEMs of the two instruments are equal. Consequently, we pool them and obtain an average SDEM of about 0.025 second. This figure, divided by the product variability ($0.025/0.092 = 0.27$), is only slightly more than one-fourth of the fuse-time variation, and may be acceptable in validating the measuring system, especially since $[(0.096)^2 - (0.025)^2]^{1/2} \simeq 0.093$ still.

There is, however, a definite problem with the two instrumental biases or systematic errors. The estimated difference in biases is about 0.21, with instrument 2 lagging, and a straightforward t-test of the difference is highly significant. Ordinarily, instruments are calibrated and put into service; the problem arising here is detected only in service use. From experience it is known that the true fuse running times will be somewhat different from the setting, so that recalibration of both instrument systems may be in order, but the measured times of instrument 2 should be decreased by 0.21 second, due to the bias or lag of the second operator.

For a simple precision* and accuracy statement of the instrumental systems here, it can be said:

> *The random errors of measurement of the instruments may be described with a standard deviation or imprecision of about* 0.025 *second, and the biases or systematic errors (shifts in location of random error of measurement populations) may be* 0.21 *second—and very significant. With the estimated imprecision of measurement of the instruments, a difference or shift in biases equal to about*
>
> $$|\beta_1 - \beta_2| \simeq s_{1-2} t_{0.975}(n-1)/\sqrt{n} \qquad (6)$$
>
> (= about 0.016 sec here)
>
> *could be detected with* 95% *assurance.*

ESTIMATION OF IMPRECISIONS OF MEASUREMENT WITH THREE OR MORE INSTRUMENTS (RANDOM PRODUCT VARIABILITY; CONSTANT INSTRUMENTAL BIASES)

When three or more instrument systems can take the same series or simultaneous measurements of random observations, one can work with the differences only in errors of measurement to study imprecision and inaccuracy, since the value of the product level may be eliminated altogether. Let there be $k \geqslant 3$ instruments, and take the sample variance of the differences in errors of measurement for the ith and jth instruments to be S_{i-j}^2. For the first designated instrument Grubbs [4] recommends

$$\hat{\sigma}_{e1}^2 = \frac{1}{k-1}\left[\sum_{j=2}^{k} S_{1-j}^2 - \frac{1}{k-2}\sum_{2 \leqslant i < j}^{k} S_{i-j}^2\right]. \qquad (7)$$

The variances in the measurement errors for all the other instruments may be obtained from an obvious rotation of the subscripts, or renumbering the instruments.

The variance of the estimate given by (7) is free of σ_X^2 and is

$$\operatorname{var}(\hat{\sigma}_{e1}^2) = \frac{2}{n-1}\sigma_{e1}^4 + \frac{1}{n-1}\left[\frac{4}{(k-1)^2}\sum_{j=2}^{k}\sigma_{e1}^2\sigma_{ej}^2 + \frac{4}{(k-1)^2(k-2)^2}\sum_{2 \leqslant i < j}^{j=k}\sigma_{ei}^2\sigma_{ej}^2\right]. \qquad (8)$$

Model (1) still applies, and thus a unique and practical way of analyzing only the errors of measurement of instrument systems is available. It seems natural to contrast this statistical technique with the treatment of residual error in experimental designs (see Russell and Bradley [16]).

For the third instrument, Hahn and Nelson [7] used the second of two instruments to obtain a repeated set of measurements; some special estimation procedures and significance tests concerning parameters arise. Shukla [17, 18] studied the overall problem, gave some exact tests of hypothesis concerning the parameters and suggested confidence intervals for various variance ratios. Thompson [19, 20, 21] studied the problem of negative variance estimates for the two-instrument case, and developed some useful tests of hypotheses concerning ratios of variances arising from product variability and instrumental measurement errors. See also Hanumara and Thompson [9] and Hanumara [8]; the confidence interval procedures of Thompson and Hanumara are especially useful in applications. For a rather complete account of hypothesis-testing procedures for population parameters and confidence interval estimation with many examples, see Grubbs [5, 6 (Chaps. 2 and 6]. For the Bayesian approach to the analysis of simultaneous measurement procedures, see Draper and Guttman [2]; they give a three-instrument example for which the estimates of the imprecision parameters and product variability check well with estimates of Grubbs [4].

Jaech [11] develops a variety of useful significance tests of equality of the imprecision parameters, using likelihood ratio*-type statistics involving least-squares* solutions to obtain estimates of the parameters developed by Grubbs [4] for the case of four or more instruments; Jaech also gives a very informative example for a six-instrument case.

Despite increased costs of experiments, experience has shown that the use of three or more instrumental systems will remove many doubts concerning the precision and accuracy of measurement. Moreover, since in applications the product variability is expected to be several times that of the imprecision measurement errors, the use of three or more instruments leads to much less trouble with negative variances estimates of the errors of measurement.

EXTENSIONS TO MORE COMPLEX MODELS (LEVEL-DEPENDENT BIASES AND RANDOM ERRORS OF MEASUREMENT)

In many applied areas of investigation, it becomes starkly clear that the instrumental biases will exhibit trends and the standard errors of measurement will vary in size with the level of the quantity or characteristic studied. For two-instrument applications there exists an analogy with the linear regression problem involving errors in both variables. It is well known that, for a linear relationship between the true values of the dependent and independent variables, all five parameters of interest—the true intercept, slope, the variance of the quantity measured, and the two variances of the errors—cannot be satisfactorily estimated without the use of ancillary information or assumptions. Nevertheless, for measuring system problems, progress may be achieved through the use of redundant instrumentation.

In a study of tests performed on a reactor fuel element prior to charging in a Hanford reactor, Jaech [10] recommended a model for the analysis of measurement errors. He uses several "instruments" to measure the same characteristic. The "instruments" may refer to different measuring devices, or possibly to different observers or operators of the same device, different laboratories, etc. The data analyzed to evaluate measurement errors consist of recorded measurements of n items on each of k instruments, where r_j "runs" are made on instrument j $(j = 1, 2, \ldots, k)$. The total number of data points is

$$nR = n\sum_{j=1}^{k} r_j \qquad \left(R = \sum_{j=1}^{k} r_j\right). \tag{9}$$

Since the reactor fuel item measurement might depend linearly on magnitude of the true quantity measured, Jaech [10] used the following model:

$$Y_{ip} = \beta_p + \alpha_p x_i + e_{ip}, \tag{10}$$

where

Y_{ip} = observed value for item i, run p,

x_i = "true" value for item i,

$i = 1, 2, \ldots, n$;

$p = 1, 2, \ldots, R$.

Note that the parameters α_p and β_p for run p are joint for the measurement bias, which depends on the level x_i. If $\beta_p = 0$ and $\alpha_p = 1$, there is no bias, but if $\alpha_p = 1$ and $\beta_p \neq 0$, there is a constant bias as in (1). Otherwise, the total bias or systematic error depends linearly on the level of x measured, and perhaps on the run made. The quantity x_i may be a "mathematical" variable or a random variable over a wide range, so that y may have a trend with a random element.

As before, the random errors of measurement e_{ip} are normally distributed with zero expectation [i.e., $N(0, \sigma_{ep}^2)$].

All unknown parameters of the model can be estimated, using sample covariances S_{jp} and variances S_p^2 as in Jaech [10], and are

$$\hat{\alpha}_p = \left(\underset{j \neq 1,p}{\overset{R}{\pi}} S_{jp}/S_{1j} \right)^{1/(R-2)}, \quad p \neq 1, \quad (11)$$

$$\hat{\sigma}_x^2 = \left(\underset{\substack{p=2 \\ p<q}}{\overset{R}{\pi}} S_{1p}S_{1q}/S_{pq} \right)^{2/(R-1)(R-2)}, \quad (12)$$

$$\hat{\sigma}_{e1}^2 = S_1^2 - \hat{\sigma}_x^2, \quad (13)$$

$$\hat{\sigma}_{ep}^2 = S_k^2 - \hat{\alpha}_p^2\hat{\sigma}_p^2, \qquad p \neq 1, \quad (14)$$

$$\hat{\beta}_p = \bar{y}_p - \hat{\alpha}_p \bar{y}_1; \qquad p \neq 1, \quad (15)$$

$$\hat{\mu}_x = \bar{y}_1 = \text{estimate of the mean } x, \quad (16)$$

$\bar{y}_1$ = mean of readings on run 1,

$\bar{y}_p$ = mean of readings on the pth run.

Jaech [10] indicates that the "run" designated as 1 is to be chosen as the base run, and therefore $\hat{\alpha}_k$ actually estimates α_p/α_1 and $\hat{\beta}_p$ actually estimates $\beta_p - \alpha_p\beta_1/\alpha_1$. Thus it would be desirable for the first run to be made the "best" (most precise and accurate) or on a standard instrument, if identifiable. Jaech [10] also gives estimates of the variances of the imprecisions, and analyzes the between-instrument and the between runs/within instruments variation.

Later, Jaech [11] develops large-sample tests of various hypotheses on instrumental precisions of measurement for multiple instruments in general, with assumptions similar to those of Grubbs [4, 5].

In Jaech's problem, the response of each instrument, y_{ip} of (10), is modeled by a linear part depending on the level of the characteristic measured plus a random error of measurement. The intercept and slope may well depend on the instrument and run—hence the need for redundancy in runs and instruments. However, for many linear regression* problems with measurement errors in both variables, the use of two or more redundant instruments measuring either the dependent or the independent variable will lead to sufficient overdetermination, so that the major parameters of interest can be estimated in many important applications.

Precision and accuracy of instrumentation problems in applied fields can become complex. Experiments can be costly; redundancy of instrumentation may be a prime requirement for determining imprecisions of measurement, and hence for being able to evaluate and model biases or systematic errors and the trends in them. Since one must usually concentrate on differences in measurements, which often vary with the levels of measurement, it becomes important to use the concept of a standard or reference instrument having fairly well known capabilities. Frequent recalibration may be called for. In all applications, the experimenter should be on the lookout for varying errors of measurement and the sizes of their standard errors over ranges of interest. In fact, the relative sizes of instrumental biases, standard deviations of measurement errors, and their trends, over experimental ranges may become very troublesome to model. An example is the determination of the concentration of stratospheric ozone with instruments aboard rockets fired into the atmosphere described by Grubbs [6, Chap. 6]. Redundancy of instrumentation was recommended for this study; both the sizes of the standard errors of measurement, or imprecisions, and the paired differences in instrumental biases varied with altitude, so that nonlinear models had to be considered. The basic model, of the form

$$Y_{ij}(h) = \beta_j(h) + \alpha_j(h)\omega_i(h) + e_{ij}(h), \quad (17)$$

where

$Y_{ij}(h) = Y_{ij}(h_i)$, the measurement of ozone concentration at altitude h_i with instrument j,

$\omega_i(h)$ = true concentration of ozone at altitude h_i,

and the intercept instrumental bias, slope, and random errors of measurement may vary with altitude. In this analysis, differences in paired instrumental biases could be modeled with orthogonal least squares, and the residual variances of differences in paired random instrument errors determined. The latter variance is the sum of the variances of random instrumental errors taken two at a time, and could be checked independently with a technique of Morse and Grubbs [14] for estimating residual dispersion from any data displaying trends, thereby helping with model validation [6]. Moreover, trends of differences in biases could be modeled and improved accuracy of measurements effected during the analysis; recalibration also results.

Although these statistical procedures appear promising for describing the precision and accuracy capability of instrumentation, the need for the joint use of appropriate physical and statistical models seems evident also.

ADDITIONAL REMARKS

As is evident from (1) and from the work of Russell and Bradley [16], the Grubbs estimators technique is closely tied in with the two-way analysis of variance*. In a private communication, Ralph Bradley and Dennis Brindley [1] of Florida State University indicate some striking accomplishments for the three-instrument case. They use (1) with $\sum_j \beta_j = 0$ and $\operatorname{var}(e_{ij}) = \sigma_{ej}^2$. Then if

$$S_j = \sum_{i=1}^{n} \left(Y_{ij} - \bar{Y}_{i.} - \bar{Y}_{.j} + \bar{Y}_{..} \right)^2, \quad (18)$$

(where dots denote a sum on that subscript and bars indicate average values), and if

$$Q_j = kS_j/(n-1)(k-2) - \sum_{j=1}^{k} S_j/(n-1)(k-1)(k-2), \quad (19)$$

$$E(Q_j) = \sigma_{ej}^2. \quad (20)$$

For the case $k = 3$, Bradley and Brindley indicate that they have found the joint probability density of Q_1, Q_2, and Q_3, and have established the likelihood ratio test* of the null hypothesis

$$H_0: \quad \sigma_{e1}^2 = \sigma_{e2}^2 = \sigma_{e3}^2 = \sigma_e^2 \quad (21)$$

versus the alternative

$$H_a: \quad \text{some } \sigma_{ej}^2 \neq \sigma_{eq}^2, \quad j \neq q. \quad (22)$$

The likelihood ratio statistic is

$$\hat{\lambda} = 3(Q_1 Q_2 + Q_1 Q_3 + Q_2 Q_3)/(Q_1 + Q_2 + Q_3)^2; \quad (23)$$

under H_0 the density of $\hat{\lambda}$ is

$$f(\lambda) = \frac{n-2}{2} \lambda^{(n-4)/2}, \qquad 0 \leqslant \lambda \leqslant 1. \quad (24)$$

Any probability level α of $\hat{\lambda}$ is given by

$$\lambda_\alpha = \alpha^{2/(n-2)}; \qquad \Pr(\hat{\lambda} \leqslant \lambda_\alpha) = \alpha. \quad (25)$$

Brindley and Bradley also established the power* function for this test. For other work in this area see Russell and Bradley [16].

References

[1] Bradley, R. A. and Brindley, D. (1980). Private communication. (Gives hypothesis test of equal imprecisions for the three-instrument case.)

[2] Draper, N. and Guttman, I. (1976). *J. Amer. Statist. Ass.*, **71**, 605–607. (Gives a Bayesian approach to Grubbs' estimators.)

[3] Ellenberg, J. H. (1977). *J. Amer. Statist. Ass.*, **72**, 407–411.

[4] Grubbs, F. E. (1948). *J. Amer. Statist. Ass.*, **43**, 243–264. (A basic paper on estimating imprecision of measurements.)

[5] Grubbs, F. E. (1973). *Technometrics*, **15**, 53–66. (Gives significance tests, confidence intervals, and applications.)

[6] Grubbs, F. E. (1983). *Selected Topics in Experimental Statistics with Army Applications*, U.S. Army Materiel Development and Readiness Command Pamphlet DARCOM-P 706-103 (in press). National Technical Information Service, Department of Commerce, Springfield, VA 22151. (Includes a rather complete account of imprecision and inaccuracy of measurement, with examples.)

[7] Hahn, J. H. and Nelson, W. (1970). *Technometrics*, **12**, 95–102. (Uses two instruments, with one taking a duplicate set of measurements.)

[8] Hanumara, R. C. (1975). *Technometrics*, **17**, 299–302.

[9] Hanumara, R. C. and Thompson, W. A. (1968). *Biometrika*, **55**, 505–512. (Has tables to determine confidence intervals on the imprecision parameters.)

[10] Jaech, J. L. (1964). *Technometrics*, **6**, 293–300. (Accounts for a linear model of the biases or systematic errors.)

[11] Jaech, J. L. (1976). *Technometrics*, **18**, 127–133. (Contains hypothesis tests about the imprecision parameters.)

[12] Maloney, C. J. and Rastogi, S. C. (1970). *Biomet-*

rics, **26**, 671–676. (Significance tests for two instruments.)

[13] Morgan, W. A. (1939). *Biometrika*, **31**, 13–19.

[14] Morse, A. P. and Grubbs, F. E. (1947). *Ann. Math. Statist.*, **18**, 194–214.

[15] Pitman, E. J. G. (1939). *Biometrika*, **31**, 9–12.

[16] Russell, T. and Bradley, R. A. (1958). *Biometrika*, **45**, 111–129. (Gives tie-in of the estimates of imprecision with two-way ANOVA).

[17] Shukla, G. K. (1973). *Biometrics*, **29**, 373–377.

[18] Shukla, G. K. (1976). *Amer. Statist.*, **30**, 151–152.

[19] Thompson, W. A., Jr. (1962). *Ann. Math. Statist.*, **33**, 273–289. (Discusses handling of negative estimates of imprecision variance.)

[20] Thompson, W. A., Jr. (1962). *J. Res. Natl. Bur. Stand. 3: Math. Math. Statist.*, **66B**(4), 161–164.

[21] Thompson, W. A., Jr. (1963). *J. Amer. Statist. Ass.*, **58**, 474–479.

(ACCURACY
ANALYSIS OF VARIANCE
CALIBRATION
MEASUREMENT ERRORS
PRECISION
QUALITY CONTROL, STATISTICAL)

FRANK E. GRUBBS

G-SPECTRAL ESTIMATOR, THE

The *G-transformation* was introduced by H. L. Gray et al. [2] for the purpose of increasing the rate of convergence of sequences. If $F(t) = \int_a^t f(x)\,dx$, the transformation is defined by

$$G[F(t); h, n] = \begin{vmatrix} F(t) & F(t+h) & \cdots & F(t+nh) \\ f(t) & f(t+h) & \cdots & f(t+nh) \\ \vdots & & & \vdots \\ f(t+(n-1)h) & & \cdots & f(t+(2n-1)h) \end{vmatrix} \div \begin{vmatrix} 1 & 1 & \cdots & 1 \\ f(t) & f(t+h) & \cdots & f(t+nh) \\ \vdots & & & \vdots \\ f(t+(n-1)h) & & \cdots & f(t+(2n-1)h) \end{vmatrix}; \tag{1}$$

it has been shown to be of value in evaluating improper integrals and to be especially effective in cases where the integrand behaves like a linear combination of exponentials with real or complex arguments. These ideas were used by Gray and later by Gray and Foster [1] to define the *G-spectral estimator*. For continuous index stationary time series*, $[X(t)]$, their definition is as follows. Let $\hat{R}(\tau; T)$ be an estimate, based on a record of length T, for the autocorrelation of $X(t)$ and let G denote the G-transformation. Then the original G-spectral estimator $\hat{S}(w; T)$ is defined by

$$\hat{S}(w, T) = G\left[2\int_0^{T_0} \hat{R}(\tau, T)\cos 2\pi w\tau\, d\tau; h, n\right], \tag{2}$$

where T_0, h, and n are parameters to be determined; normally, $h = 1$. The problem of selecting n and T_0 is almost the problem of finding p and q in the ARMA (p, q) process (*see* AUTOREGRESSIVE-MOVING AVERAGE MODELS). When the data are discrete, the G-spectral estimator is defined by (2) with the integral typically replaced by its trapezoid approximation.

In a rather lengthy paper, Gray et al. [3] studied the properties of (1) to some extent for discrete index set time series (primarily ARMA processes). They concluded at that time that, although for the proper choice of n and T_0 the estimator has a number of nice theoretical properties, the problem of determining T_0 and n may make the estimator impractical except for a very knowledgeable user.

In his Ph.D. thesis Morton [5] modified the original definition of the G-spectral estimator for discrete time series in order to offset some of the shortcomings noted by Gray. Morton's definition results from replacing the cosine function in (1) by the complex exponential and adjusting the lower limit in (1). Since his study is for discrete index sets, the integrals in (1) and (2) are replaced by sums. The resulting modified G-spectral estimator, $G_{n,m}$, is defined as

$$G_{n,m}(w) = 1 + 2\left[\text{Real } G(F_{m-n+1}; 1, n) - F_0\right], \tag{3}$$

where

$$F_m = \sum_{j=-m}^{m} \hat{\rho}(j)\exp(2\pi i w j)$$

and $\hat{\rho}(m)$ is the sample autocorrelation. The parameters n and m are estimated from the data as the order of the ARMA (n, m) process which best fits the data. The form in (3) can be simplified to

$$G_{n,m}(w) = 2\,\text{Real}(U), \qquad (4)$$

where

$$U = \frac{Z_{m+1} - \alpha_1 Z_m - \cdots - \alpha_n Z_{m-n+1}}{1 - \alpha_1 - \alpha_2 - \cdots - \alpha_n},$$

$$Z_j = F_j - F_0 + \tfrac{1}{2},$$

$$\alpha_j = \hat{\phi}_j \exp(2iwj),$$

and the $\hat{\phi}$ are the Yule–Walker estimates for the autoregressive parameters in the associated ARMA (n, m) process.

It can be shown that $G_{n,m}(w)$ is the method of moments* ARMA (n, m) spectral estimator. This is of particular interest since one would expect that the method of moments estimates for the moving-average parameters would be required to obtain such an estimator. Equation (3) or (4) shows that this is not the case, which is valuable, since method-of-moments moving-average parameter estimates cannot be obtained in closed form.

Several applications of (3) are possible. Morton [5] established its value for estimating the moving-average parameters of an underlying ARMA process by making use of inverse autocorrelations. Hart [4] has used the result to generalize the notion of autoregressive density estimator to ARMA density estimators while showing how such estimators are related to Fourier series density estimators.

The simplicity of (3), or equivalently (4), together with its practical and theoretical value, are no doubt virtues which will allow it to supersede the original definition in (2). Thus the G-spectral estimator is currently correctly defined by (3). No apparent loss, theoretical or practical, occurs in this modification of the definition. Further studies and applications of (3) or (4) are apt to be fruitful. The following example demonstrates that the G-spectral estimator can be used satisfactorily on processes which are not ARMA.

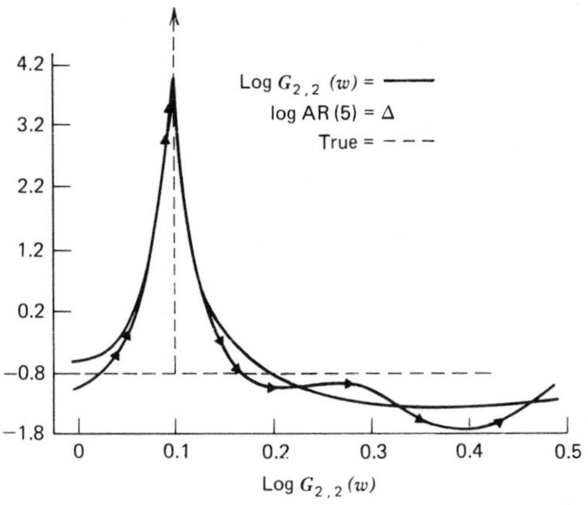

Figure 1

Example. A realization of length 30 was generated from the signal-plus-noise model $\{X_t\}$, where

$$X_t = \sqrt{20}\cos(0.2\pi t + \phi) + a_t,$$

a_t is a white noise* process with unit variance, and ϕ is uniform* on $(-\pi, \pi)$. Figure 1 shows the logarithm of the true spectrum, $G_{2,2}(w)$ and the autoregressive spectral estimator, which is an AR(5) as determined by Akaike's* FPE or Parzen's CAT criteria.

References

[1] Gray, H. L. and Foster, M. R. (1975). In *The Search for Oil: Some Statistical Methods and Techniques*, D. B. Owen, ed. Marcel Dekker, New York, pp. 169–189.

[2] Gray, H. L., Atchison, T. Q., and McWilliams, G. V. (1971). *SIAM J. Numer. Anal.*, **8**(2), 365–438.

[3] Gray, H. L., Houston, A. G., and Morgan, F. W. (1978). *Proc. 1976 Tulsa Symposium on Applied Time Series*. Academic Press, New York, pp. 39–138.

[4] Hart, J. D. (1981). *On Complex S-Arrays and Their Application to Density Estimation*. Ph.D. thesis, Southern Methodist University.

[5] Morton, M. J. (1981). *The Modified G-Spectral Estimator*. Ph.D. thesis, Southern Methodist University.

(AUTOREGRESSIVE-MOVING AVERAGE MODELS)

HENRY L. GRAY

GUESSING MODELS

Guessing models are statistical structures which aim to provide insight into the normative and optimal behaviors of people who must make choices in the apparent absence of data. Such models are related to aspects of *Bayesian* statistics* and the *Delphi method*[1], but they have a flavor and a theory of their own.

This is most easily illustrated by the anecdote of two statisticians, Bob and Mike, who engaged in a contest to guess the weights of people at a party. Bob agreed to always guess first, and on person 1 Bob guessed 142 pounds. Mike then guessed 142.1 pounds, and the subject declared Mike the winner. The contest continued; and, when final tallies were made, Bob found he had lost almost three-fourths of the time.

To model this scenario, consider a system of four p vectors

Target values:	$(\theta_1, \theta_2, \ldots, \theta_p) = \boldsymbol{\theta}$
First guess:	$(X_1, X_2, \ldots, X_p) = \mathbf{X}$
Second guesser's hunch:	$(Y_1, Y_2, \ldots, Y_p) = \mathbf{Y}$
Second guess:	$(G_1, G_2, \ldots, G_p) = \mathbf{G}$

The θ_i denote the real values to be guessed, so, for example, θ_2 would denote the weight of the second person considered by Bob and Mike. The X_i are the guesses made by the first guesser, and the Y_i represent the second guesser's best estimate of θ_i. Finally, the G_i are the guesses that are announced by the second guesser. The first problem in this theory is to determine how $\mathbf{G}$ should be determined by $\mathbf{X}$ and $\mathbf{Y}$.

As the anecdote suggests, each player wishes to come closer to each θ_i than his opponent, so we consider

$$V(\mathbf{G}, \boldsymbol{\theta}) = \sum_{j=1}^{p} V_j(\mathbf{G}, \boldsymbol{\theta})$$

where

$$V_j(G, \theta) = \begin{cases} 1 & \text{if } |G_j - \theta_j| < |X_j - \theta_j| \\ 0 & \text{otherwise.} \end{cases}$$

With the objective of maximizing $V(\mathbf{G}, \boldsymbol{\theta})$, it is intuitive that the second guesser should always guess just a bit higher or a bit lower than the first guesser. This can be proved to be true under very general circumstances. Specifically, if $\boldsymbol{\theta}$, $\mathbf{X}$, and $\mathbf{Y}$ are assumed to have a joint distribution that is continuous and if $\nu_i(\mathbf{X}, \mathbf{Y})$ denotes the conditional median of θ_i given $\mathbf{X}$ and $\mathbf{Y}$, a key role is played by the strategies

$$G_i^{\epsilon} = \begin{cases} X_i + \epsilon & \text{if } X_i < \nu_i(\mathbf{X}, \mathbf{Y}) \\ X_i - \epsilon & \text{otherwise.} \end{cases}$$

These guesses $\mathbf{G}^{\epsilon}$ are called *Hotelling strategies* and the first result in the theory of guessing models is the following:

Theorem 1. The Hotelling strategies are ϵ optimal, i.e.,

$$\lim_{\epsilon \to 0} EV(\mathbf{G}^{\epsilon}, \theta) = \sup_{\mathbf{G}} EV(\mathbf{G}, \boldsymbol{\theta}).$$

Although this result reassures intuitive feelings, it just makes the first step in telling the second guesser how to guess. Considerable ingenuity may be required to ferret out those models in which $\nu_i(\mathbf{X}, \mathbf{Y})$ can be calculated, and much of the theory of guessing strategies resides in the calculation of suitable approximations.

Consider, for example, the strategies

$$\tilde{G}_i = \begin{cases} X_i + \epsilon & \text{if } X_i < Y_i \\ X_i - \epsilon & \text{if } X_i > Y_i. \end{cases}$$

These are the "hunch" guided strategies and they model the reasonable actions of Mike in the anecdote. In some cases the hunch guided strategies are, in fact, Hotelling strategies, but even when these strategies are not ϵ optimal they have surprising power.

Theorem 2. If for each $1 \leqslant i \leqslant p$, X_i and Y_i are independent and identically distributed with a distribution that is symmetric about θ_i, then the hunch guided strategy $\tilde{G}_i$ has a $\frac{3}{4}$ probability of winning the ith contest as $\epsilon \to 0$.

The fact that one shrinks the first guess Y_i toward one's hunch X_i anticipates that the fact that other types of *shrinkers* are relevant in guessing models. In particular, when $p \geqslant 3$ and all the Y_i's are available to the second guesser, a very powerful strategy for the second guesser can be based on the *James–Stein estimator**.

NOTE

1. *Editor's note*: The Delphi method referred to above is the subject of a book [1], and is defined there as follows: "Delphi may be characterized as a method for structuring a group communication process so that the process is effective in allowing a group of individuals, as a whole, to deal with a complex problem." [1, p. 3]

(*See also* PUBLIC ADMINISTRATION, STATISTICS IN, *and* SOCIOLOGY, STATISTICS IN.)

Reference

[1] Linstone, H. A. and Turoff, M., eds. (1975). *The Delphi Method.* Addison-Wesley, Reading, Mass.

Bibliography

Hwang, J. T. and Zidek, J. V. (1981). *J. Appl. Prob.*, **19**, 321–331. (Investigates the strategy $G_i = X_i \pm \epsilon$ according as $X_i < \overline{X}$ or not.)

Pittinger, A. O. (1980). *J. Appl. Prob.*, **17**, 1133–1137. (Studies the problem of how the $\frac{3}{4}$ theorem extends to n-dimensional guessing.)

Steele, J. M. and Zidek, J. V. (1980). *J. Amer. Stat. Ass.*, **75**, 596–601. (Sets the foundation for the theory of second guessing, and provides the basis for this exposition. Contains much more information on Stein guided guessing as well as results of computer simulations.)

J. MICHAEL STEELE

GUMBEL DISTRIBUTION

The theory of statistical extremes has a short effective history. Beginning, essentially, with a paper by Dodd [10] on the distribution of the extremes (maxima and minima) of a univariate independent and identically distributed (i.i.d.) sample, the basic results regarding special properties and asymptotic behavior are contained in Fréchet [12], Fisher and Tippett [11], Gumbel [17], and von Mises [46], culminating in the fundamental paper by Gnedenko [14] (*see* EXTREME-VALUE DISTRIBUTIONS). The initial results concerning the law of large numbers* for extremes—not dealt with here—can be found in de Finetti [9]. The basic bibliography for statistical problems is still Gumbel [19]: many results and examples can be found in this fundamental reference. A large block of references can be found in Johnson and Kotz [22] and Harter [20]. A modern and essential reference for probabilistic results is Galambos [13]. Extensions and applications can be found at the end of the entry.

BASIC RESULTS

Consider a sample of k i.i.d. random variables $(Y_1, \ldots, Y_k)$ with cumulative distribution function (CDF) $F(x) = \Pr[Y \leqslant x]$. Then the CDF of $\max(Y_1, \ldots, Y_k)$ is

$$\Pr\left[\max(Y_1, \ldots, Y_k) \leqslant x\right] = \prod_1^k \Pr\left[X_i \leqslant x\right] = F^k(x);$$

in the same way,

$$\Pr\left[\min(Y_1, \ldots, Y_k) \leqslant x\right] = 1 - (1 - F(x))^k.$$

In general, to deal with samples of maxima (or minima), all obtained under the same conditions (with the same k and F), it would be necessary to know the form (and parameters) of F. But if k is large, we can try to use asymptotic distributions in statistical analysis. In fact, it can be shown that, in many cases, coefficients α_k and $\beta_k (> 0)$ exist such that $\Pr[(\max(X_1, \ldots, X_k) - \alpha_k)/\beta_k \leqslant x] = F^k(\alpha_k + \beta_k x)$ has a nondegenerate limit CDF; note that (α_k, β_k) are not uniquely defined. There are only three such limit forms (Weibull*, Gumbel, and Fréchet) which can be integrated in a condensed von Mises [46]–Jenkinson [21] form.

The Gumbel distribution plays a central role, sometimes being called *the* distribution of extremes; it is also named the type I distribution. The method of statistical choice between the three forms, for large samples, can be found in Tiago de Oliveira [42].

We will say that X has a *Gumbel distribution* (or is a Gumbel random variable), with location parameter λ and dispersion parameter $\delta(>0)$ if

$$\Pr[X \leqslant x] = \Pr[(X-\lambda)/\delta \leqslant (x-\lambda)/\delta]$$
$$= \Lambda((x-\lambda)/\delta), \quad -\infty < x < \infty,$$

where $\Lambda(z) = \exp(-e^{-z})$; we say that $Z = (X-\lambda)/\delta$ (location parameter 0 and dispersion parameter 1) is a *standard Gumbel random variable*: its CDF is $\Pr[Z \leqslant z] = \Lambda(z)$. The probability density function (PDF) of X is

$$\frac{d\Lambda((x-\lambda)/\delta)}{dx} = \frac{1}{\delta}\exp\left(-\frac{x-\lambda}{\delta}\right)\exp\left(-e^{-(x-\lambda)/\delta}\right)$$
$$= \frac{1}{\delta}\exp\left(-\frac{x-\lambda}{\delta}\right)\Lambda((x-\lambda)/\delta),$$
$$-\infty < x < \infty.$$

The graph for the standard form $\Lambda'(z)$ is given in Fig. 1.

Minima can be dealt with analogously or using the fact that $\min(X_1, \ldots, X_k) = -\max(-X_1, \ldots, -X_k)$; the Gumbel CDF for minima is $1 - \exp(-e^{(x-\lambda)/\delta})$. Note that, if X has this CDF, then $X' = -X$ has the CDF $\exp(-e^{-(x+\lambda)/\delta}) = \Lambda((x-\lambda')/\delta')$ with $\lambda' = -\lambda$, $\delta' = \delta$. Thus we need only deal with Gumbel distributions for maxima.

There are various complicated conditions on the parent CDF F for maxima to have an asymptotic Gumbel distribution. The most useful, necessary, and sufficient condition is (Mejzler [29]):

$$\frac{U(tx)-U(t)}{U(ty)-U(t)} \to \frac{\log x}{\log y}$$

as $t\downarrow 0$ for all $x, y > 0$ and $y \neq 1$, where $U(x) = \inf\{y \mid 1 - F(y) \leqslant x\}$; note that, with a translation, the condition that $F(0) < 1$ is immediate; for other formulations, and for characterizations, *see* EXTREME-VALUE DISTRIBUTIONS and Galambos [13, Chap. 2].

If X has a Gumbel distribution, then $\exp(-(X-\lambda)/\delta)$ has the standard exponential distribution*; this is useful when δ is known. More generally, e^{-X} has a Weibull distribution*.

Recall that, since $\Lambda(\cdot)$ is a continuous CDF, the convergence of $F^k(\alpha_k + \beta_k x)$ to $\Lambda(x)$ is uniform. This justifies the use of $\Lambda(x)$ as an approximation to $F^k(\alpha_k + \beta_k x)$. In fact, in statistical analysis $\Lambda((x-\lambda)/\delta)$ is used as the (approximate) CDF of maxima obtained from equal-size samples of i.i.d. random variables. Note also that $\Lambda^c(z) = \Lambda(z - \log c)$, the *stability* property. For more moderate samples, a better approximation can be obtained using a convenient Weibull distribution, as remarked by Fisher and Tippett [11] and recently confirmed by Gomes [16].

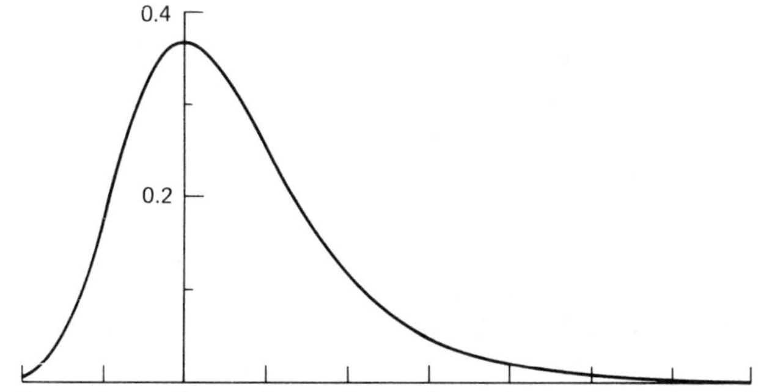

Figure 1 Graph of $e^{-z} \cdot e^{-e^{-z}}$.

Other references are David [8] and Bury [4].

IMPORTANT CHARACTERISTICS

The most important parameters are

1. Moments:

   mean: $\mu = \lambda + \gamma\delta$, $\gamma = 0.57722\ldots$ (Euler's constant)

   variance: $\mu_2 = \sigma^2 = \pi^2\delta^2/6$

   $= 1.64493\ldots\delta^2;$

   $\sigma = 1.28255\ldots\delta$

   $\mu_3 = 2.40411\ldots\delta^3$

   $\mu_4 = 3\pi^4\delta^4/20$

   $= 14.61136\ldots\delta^4$

   so that the associated coefficients are:

   coefficient of skewness: $\sqrt{\beta_1} = \mu_3/\sigma^3 = 1.13651\ldots$

   coefficient of excess: $\beta_2 = \mu_4/\sigma^4 = 5.4$

   The characteristic function is given by $E(e^{itX}) = e^{i\lambda t}\Gamma(1 - i\delta t)$.

2. Quantiles*, defined by $\Lambda(\psi_p) = p$, so that $\psi_p = \lambda - \log(-\log p)\delta$.

   median*: $\psi_{1/2} = \lambda + 0.36651\ldots\delta$

   first quartile: $\psi_{1/4} = \lambda - 0.32663\ldots\delta$

   third quartile: $\psi_{3/4} = \lambda + 1.24590\ldots\delta$

3. Mode: $\lambda = \psi_{1/e}$.

Quantiles are easy to compute even with a pocket calculator. Note that $-2 < Z < 7$ with probability 0.998.

It is very easy to construct probability paper for the Gumbel distribution: the x-axis is graduated in the usual arithmetic scale and the y-axis in the functional scale $-\log(-\log p)$, but marked p. For details, see Gumbel [19, Secs. 1.2.5–2.28] and Tiago de Oliveira [36].

ESTIMATION OF PARAMETERS

If $(x_1, \ldots, x_n)$ is a sample of i.i.d. random variables from a Gumbel distribution (λ, δ unknown), the maximum likelihood* estimators (λ, δ) are given by the equations

$$\hat{\delta} = \bar{x} - \frac{\sum_{i=1}^n x_i e^{-x_i/\hat{\delta}}}{\sum_1^n e^{-x_i/\hat{\delta}}}$$

$$\hat{\lambda} = -\hat{\delta}\log\left(\frac{\sum_{i=1}^n e^{-x_i/\hat{\delta}}}{n}\right),$$

where $\bar{x} = \sum_1^n x_i/n$.

The second equation gives $\hat{\lambda}$ immediately, once $\hat{\delta}$ is known. The first equation can be solved for $\hat{\delta}$ by the Newton–Raphson method, with the initial value obtained by the method of moments (i.e., $\delta_1 = \sqrt{6}\,s/\pi = 0.77970s$), where $s^2 = (\sum_{i=1}^n x_i^2/n) - \bar{x}^2$; numerical convergence is attained after half-a-dozen steps.

Linear combinations of order statistics* can be used as estimators of the quantiles, with efficiencies of 80% and above; they are not worthwhile if we have access to computers. The method of moments* is poor: its (joint) efficiency (Cramér [5] sense) is 48% (see Tiago de Oliveira [35]).

As maximum likelihood* estimators are regular, it can be shown that the random pair $(\hat{\lambda}, \hat{\delta})$ is asymptotically bivariate normal* (binormal) with expected values (λ, δ), variances

$$\left[1 + \frac{6(1-\gamma)^2}{\pi^2}\right]\frac{\delta^2}{n} = 1.10866\frac{\delta^2}{n}$$

and

$$\frac{6}{\pi^2}\frac{\delta^2}{n} = 0.60793\frac{\delta^2}{n},$$

respectively, covariance

$$\frac{6(1-\gamma)}{\pi^2}\frac{\delta^2}{n} = 0.25702\frac{\delta^2}{n},$$

and correlation coefficient

$$\rho = \left[1 + \frac{\pi^2}{6(1-\gamma)^2}\right]^{-1/2} = 0.31307.$$

The asymptotic confidence region* for (λ, δ) with confidence coefficient $1-\alpha$ is given by the ellipse

$$\left(\frac{\hat{\lambda}-\lambda}{\delta}\right)^2 - \frac{2(1-\gamma)(\hat{\lambda}-\lambda)(\hat{\delta}-\delta)}{\delta^2} + \left[\frac{\pi^2}{6}+(1-\gamma)^2\right]\left(\frac{\hat{\delta}-\delta}{\delta}\right)^2 < -\frac{2}{n}\log\alpha,$$

where $2(1-\gamma) = 0.84556$ and where $\frac{1}{6}\pi^2 + (1-\gamma)^2 = 1.82368$.

This result follows because the quadratic form on the left-hand side has a χ^2 distribution with 2 degrees of freedom, i.e., an exponential distribution. An asymptotic test of the hypothesis H_0: $(\lambda, \delta) = (\lambda_0, \delta_0)$ with significance level α is immediate: we will accept H_0 if (λ_0, δ_0) lies inside the given ellipse, that is, if the left-hand side is smaller than $-(2/n)\log\alpha$.

It is useful to refer also to quantile estimation. The maximum likelihood estimator of the quantile $\lambda + \psi\delta$ is evidently $\hat{\lambda} + \psi\hat{\delta}$. Since $(\hat{\lambda}, \hat{\delta})$ is asymptotically binormal, $\hat{\lambda} + \psi\hat{\delta}$ is asymptotically normal with expected value $\lambda + \psi\delta$ and variance

$$\left[1 + \frac{6}{\pi^2}(1-\gamma+\psi)^2\right]\frac{\delta^2}{n}.$$

For more information, see Gumbel [19] and Tiago de Oliveira [36] and references therein; see also papers by David, Gumbel, Epstein, and Lieblein in Sarhan and Greenberg [34] and White [48].

OTHER STATISTICAL PROBLEMS

Forecasting

Let $(x_1, \ldots, x_n)$ be the (observed) sample and consider a second sample $(X_{n+1}, \ldots, X_{n+m})$, to be observed. Taking $\hat{\lambda} + k\hat{\delta}$ as the natural forecaster of $\max(X_{n+1}, \ldots, X_{n+m})$, we see that the mean square error* of $(\hat{\lambda} + k\hat{\delta}) - \max(X_{n+1}, \ldots, X_{n+m})$ is minimized for

$$k = \frac{\gamma + \log m - 6(1-\gamma)/(n\pi^2)}{1 + 6/(n\pi^2)},$$

as $\lambda + (\gamma + \log m)\delta$ is the expected value of $\max(X_{n+1}, \ldots, X_{n+m})$. In practice we take $k = \gamma + \log m$ and the asymptotic variance of the deviation between the forecaster and $\max(X_{n+1}, \ldots, X_{n+m})$ is

$$\frac{\pi^2}{6}\delta^2 + \left[1 + \frac{6}{\pi^2}(1-\gamma+\log m)^2\right]\frac{\delta^2}{n},$$

the sum of the variances of $\max(X_{n+1}, \ldots, X_{n+m})$ and of the quantile estimator $\hat{\lambda} + (\gamma + \log m)\hat{\delta}$, used as forecaster.

The one-sided prediction* interval is defined by

$$\Pr\left[\hat{\lambda} + c\hat{\delta} \leqslant \max(X_{n+1}, \ldots, X_{n+m})\right] = 1 - \omega,$$

where ω is the prediction level; it is given by

$$c = c' + \log m + \frac{1-e^{-c'}}{2n}\left[1 + \frac{6}{\pi^2}(1-\gamma+c'+\log m)^2\right],$$

where $\Lambda(c') = \omega$ or $c' = -\log(-\log\omega)$.

The shortest two-sided prediction interval for $\max(X_{n+1}, \ldots, X_{n+m})$ with prediction level ω is given by $[\hat{\lambda} + a\hat{\delta}, \hat{\lambda} + b\hat{\delta}]$ $(a < b)$, where $a = a' + \log m$, $b' = b + \log m$, apart from terms of order n^{-1}, and a' and b' satisfy the equations

$$a' + e^{-a'} = b' + e^{-b'},$$

$$\Lambda(b') - \Lambda(a') = 1 - \omega.$$

For more details on forecasting for Gumbel distributions, see Tiago de Oliveira [36] and Tiago de Oliveira and Littauer [44].

Discrimination

Suppose that we have two Gumbel populations, one corresponding to the sample $(X_1, \ldots, X_n)$ and parameters (λ, δ) and the other to the sample $(x'_1, \ldots, x'_{n'})$ and parameters (λ', δ'), with $\delta' = \delta$; λ, λ' and δ are unknown. We want to decide if a new random variable y belongs to the first or the second population. If we impose the asymptotic equality of misclassification errors, i.e., of probabilities of deciding (wrongly) that y belongs to the first population if it belongs to the second, and vice versa, the decision rule is as follows: Compute $\hat{\lambda}$, $\hat{\lambda}'$, and $\hat{\delta}$ by

$$\hat{\lambda} = -\hat{\delta}\log\left(\sum_1^n e^{-x_i/\hat{\delta}}/n\right),$$

$$\hat{\lambda}' = -\hat{\delta}\log\left(\sum_1^{n'} e^{-x_i/\hat{\delta}}/n'\right),$$

$$n\left(\frac{\bar{x}}{\hat{\delta}} - 1 - \frac{\sum_1^n x_i e^{-x_i/\delta}}{\sum_1^n e^{-x_i/\hat{\delta}}}\right)$$

$$+n'\left(\frac{\bar{x}'}{\hat{\delta}} - 1 - \frac{\sum_1^n x_i' e^{-x_i'/\delta}}{\sum_1^n e^{-x_i'/\hat{\delta}}}\right) = 0.$$

If $\hat{\lambda} < \hat{\lambda}'$ we shall allocate y to the first population if $y < w(\hat{\lambda}, \hat{\lambda}', \hat{\delta})$ and to the second if $y > w(\hat{\lambda}, \hat{\lambda}', \hat{\delta})$; w is the solution of

$$\Lambda\left(\frac{w-\hat{\lambda}}{\hat{\delta}}\right) + \Lambda\left(\frac{w-\hat{\lambda}'}{\hat{\delta}}\right) = 1;$$

the misclassification error is, approximately, $\Lambda((w-\hat{\lambda})/\hat{\delta})$; if $\hat{\lambda}' < \hat{\lambda}$, the decision rule is reversed. For details, see Tiago de Oliveira [37].

Quick Estimation Methods

In some cases, it may be useful to use a first (inefficient) estimate of (λ, δ) using few quantiles. If we denote by q' and q'' the sample quantiles and $\lambda + \nu'\delta$ and $\lambda + \nu''\delta$ are the quantiles [of probabilities $\Lambda(\nu)$ and $\Lambda(\nu')$], then the estimators to be used are

$$\lambda^* = \frac{\nu'' q' - \nu' q''}{\nu'' - \nu'}, \qquad \delta^* = \frac{q'' - q'}{\nu'' - \nu'}.$$

The maximum efficiency is attained for ν' and ν'' close to the first and third quartiles and its efficiency is about 25%. Also (λ^*, δ^*) is asymptotically binormal. If we take the quartiles [i.e., $\nu' = -\log(-\log 0.25) = -0.32663$, $\nu'' - \log(-\log(0.75)) = 1.24590$], then (λ^*, δ^*) is asymptotically binormal with expected values (λ, δ), having variances $1.42876(\delta^2/n)$ and $0.95130(\delta^2/n)$, covariance $0.05622(\delta^2/n)$ and correlation coefficient $\rho = 0.04822$. For details, see Tiago de Oliveira [36]; also Sarhan and Greenberg [34], Kubat [24], and Kubat and Epstein [25].

Samples of Different Size

It can be shown that if the initial CDF $F(x)$ is such that, for convenient α and β (>0), $e^{\alpha+\beta x}(1-F(x)) \to 1$ as $x \to \infty$, then for maxima of sizes k and k' we can take $\alpha_{k'} = \alpha_k + \beta_k \log(k'/k)$ and $\beta_{k'} = \beta_k$. Then when dealing with maxima of samples of sizes $(n_1, \dots, n_j, \dots, n_p)$ their asymptotic CDF can be taken as

$$\Lambda\left(\frac{x-\lambda}{\delta} + \log\frac{n_j}{n_i}\right)$$

and, consequently, one can estimate (λ, δ) by the maximum likelihood method. For details, see Ramachandran [33] and Tiago de Oliveira [39].

TABLES AND RANDOM NUMBERS

Gumbel [18] contains tables of $\Lambda(z)$ and $\Lambda'(z)$, to seven decimal places for $z = -3(0.1) - 2.4(0.05)0.00(0.1)4.0(0.2)8.0(0.5)17.0$, its inverse (the quantiles) $z = -\log(-\log \Lambda)$ to five decimal places for $\Lambda = 0.0001(0.0001)0.0050(0.001)0.988(0.0001)0.9994(0.00001)0.99999$, between other tables associated with the Gumbel distribution, as for instance the PDF expressed in terms of the CDF [i.e., $p(\Lambda) = -\Lambda\log\Lambda$].

Owen [31] contains a table to four decimal places of $z = -\log(-\log \Lambda)$ for Λ varying from 0.0001 to 0.99999995. Tables for the expected values and variances of all order statistics for sample sizes $n = 1(1)50(5)100$, to seven decimal places, are given by White [49], extending previous tables. Goldstein [15] gives a table of 500 random numbers from the Gumbel distribution, to three decimal places.

EXTENSIONS

Many applications are based not only on previous results, connected with samples of maxima from i.i.d. random variables, but on various types of extensions in which those restrictions are partially removed.

Juncosa [23] shows that we can substitute the "identical distributions" condition for a weaker one. Watson [47], Barndorff-Nielsen [1], Newell [30], Loynes [28], Cramér [6], Leadbetter [26], Berman [2, 3], Pickands [32], and others have weakened the independence condition, either assuming m-dependence (i.e., that in a random sequence X_i and X_j are independent if $|i-j|>m$) or some conditions on the correlation coefficient for Gaussian processes*, etc. For details, see Cramér and Leadbetter [7, Secs. 12.3 and 13.5] and the report by Leadbetter et al. [27].

More generally, Tiago de Oliveira [40] has shown that both i.i.d. conditions can be weakened by imposing a restriction on the behavior of the right tail of the distribution of the maxima. Note that some independence at large must always be assumed.

Tiago de Oliveira [38] has shown that if we add to an i.i.d. sequence a periodic disturbance, the asymptotic behavior of maxima is the same; Turkman [45] extended this and previous results.

Bivariate and multivariate extensions of the Gumbel distribution have also been developed (see Tiago de Oliveira [41, 43]).

APPLICATIONS

The use of the Gumbel distribution in dealing with maxima obtained from large samples of i.i.d. random variables is immediately justified.

The first application of interest was to the study of yearly floods (or droughts, for minima): although not independent, daily maxima of river discharges are independent at large and the result was successful in dam planning (see Gumbel [19]). Essentially, it consisted in the estimation of a quantile of large probability (i.e., with a small probability of being overpassed).

Largest gusts, largest waves, etc., use the asymptotic distribution, justified by the results briefly described in the previous section —essentially the idea of m-dependence or weak association.

Other applications are to the breaking strength of materials; the "weakest link theory"—i.e., that a piece breaks when the tension is larger than the tolerance of the weakest section (link)—justifies the use of a distribution of minima, and consequently, applications to building codes, geology, etc.

In fire and earthquake insurance, when we have maxima obtained from samples of unequal size, we can apply the results briefly described above.

References

[1] Barndorff-Nielsen, O. (1963). *Ann. Math. Statist.*, **34**, 992–1002.

[2] Berman, S. M. (1962). *Ann. Math. Statist.*, **33**, 894–908.

[3] Berman, S. M. (1964). *Ann. Math. Statist.*, **35**, 502–516.

[4] Bury, K. V. (1975). *Statistical Models in Applied Science*. Wiley, New York.

[5] Cramér, H. (1946). *Mathematical Methods of Statistics*. Princeton University Press, Princeton, N.J.

[6] Cramér, H. (1965). *Theory Prob. Appl.*, **10**, 126–128.

[7] Cramér, H. and Leadbetter, M. R. (1967). *Stationary and Related Stochastic Processes*. Wiley, New York.

[8] David, H. A. (1970). *Order Statistics*. Wiley, New York.

[9] de Finetti, B. (1932). *Metron*, **9**, 127–138.

[10] Dodd, E. L. (1923). *Trans. Amer. Math. Soc.*, **25**, 525–539.

[11] Fisher, R. A. and Tippett, L. H. C. (1928). *Proc. Camb. Philos. Soc.*, **24**, 419–438.

[12] Fréchet, M. (1927). *Ann. Soc. Pol. Math.* (Cracow), **6**, 93–116.

[13] Galambos, J. (1978). *The Asymptotic Theory of Extreme Order Statistics*. Wiley, New York.

[14] Gnedenko, B. V. (1943). *Ann. Math.*, **44**, 423–453.

[15] Goldstein, N. (1963). *Publ. Inst. Statist. Univ. Paris*, **12**, 137–158.

[16] Gomes, M. I. (1978). Some Probabilistic and Statistical Problems in Extreme Value Theory. Ph.D. thesis, University of Sheffield.

[17] Gumbel, E. J. (1935). *Ann. Inst. H. Poincaré*, **4**, 115–168.

[18] Gumbel, E. J. (1953). *Probability Tables for the Analysis of Extreme Value Data, Natl. Bur. Stand. (U.S.) Appl. Math. Ser.*, **33**. U.S. Government Printing Office, Washington, D.C.

[19] Gumbel, E. J. (1958). *Statistics of Extremes*. Columbia University Press, New York.

[20] Harter, H. L. (1978). *Int. Statist. Rev.*, **46**, 278–306.

[21] Jenkinson, A. F. (1955). *Quart. J. Meteorol. Soc.*, **81**, 158–171.

[22] Johnson, N. L. and Kotz, S. (1970). *Continuous Univariate Distributions*, Vol. 1. Houghton and Mifflin, New York.

[23] Juncosa, M. L. (1949). *Duke Mat. J.*, **16**, 609–618.

[24] Kubat, P. (1975). Simplified Estimation Procedures Based on Order Statistics. Ph.D. thesis, Technion, Israel Institute of Technology.

[25] Kubat, P. and Epstein, B. (1980). *Technometrics*, **22**, 575–581.

[26] Leadbetter, M. R. (1974). *Z. Wahrscheinlichkeitsth. verwend. Geb.*, **28**, 289–303.

[27] Leadbetter, M. R., Lindgren, G., and Rootzén, H. (1979, 1980). Extremal and Related Properties of Stationary Processes. *Stat. Res. Rep.*, University of Umea, Umea, Sweden.

[28] Loynes, R. M. (1965). *Ann. Math. Statist.*, **36**, 993–999.

[29] Mejzler, D. G. (1949). *Sb. Tr. Inst. Mat. Akad. Nauk. Ukrain*, **12**, 31–35 (in Russian).

[30] Newell, G. F. (1964). *Ann. Math. Statist.*, **35**, 1322–1325.

[31] Owen, D. B. (1962). *Handbook of Statistical Tables*. Addison-Wesley, Reading, Mass.

[32] Pickands, J., III (1967). *Z. Wahrscheinlichkeitsth. verwend. Geb.*, **7**, 190–233.

[33] Ramachandran, G. (1975). In *Statistical Distributions in Scientific Work*, Vol. 2. D. Reidel, Dordrecht, Holland, pp. 355–367.

[34] Sarhan, A. E. and Greenberg, B. G. (1962). *Contributions to Order Statistics*. Wiley, New York.

[35] Tiago de Oliveira, J. (1963). *Trabajos Estadist. Inv. Operat.*, **14**, 61–81.

[36] Tiago de Oliveira, J. (1972). In *Structural Safety and Reliability*. Pergamon Press, New York, pp. 91–105.

[37] Tiago de Oliveira, J. (1973). In *Discriminant Analysis and Applications*. Academic Press, New York, pp. 291–309.

[38] Tiago de Oliveira, J. (1976). *Ann. Inst. Statist. Math.*, **28**, 19–23.

[39] Tiago de Oliveira, J. (1977). In *Recent Developments in Statistics*. North-Holland, Amsterdam, pp. 613–617.

[40] Tiago de Oliveira, J. (1979). *Metron*, **36**, 3–21.

[41] Tiago de Oliveira, J. (1979). In *Multivariate Analysis*, Vol. 5. North-Holland, Amsterdam, pp. 349–366.

[42] Tiago de Oliveira, J. (1981). In *Statistical Distributions in Scientific Work*, Vol. 6. D. Reidel, Dordrecht, Holland, pp. 367–387.

[43] Tiago de Oliveira, J. (1982). In *Some Recent Advances in Statistics*. Academic Press, New York, pp. 101–110.

[44] Tiago de Oliveira, J. and Littauer, S. B. (1976). *Naval Res. Logist. Quart.*, **23**, 487–511.

[45] Turkman, K. F. (1980). Limiting Distributions of Maxima of Certain Types of Non-stationary Stochastic Processes. Ph.D. thesis, University of Sheffield.

[46] von Mises, R. (1936). In *Selected Papers*, Vol. 2. American Mathematical Society, Providence, R.I., pp. 271–294.

[47] Watson, G. S. (1954). *Ann. Math. Statist.*, **25**, 798–800.

[48] White, J. S. (1964). *J. Ind. Math.*, **14**, 21–60.

[49] White, J. S. (1969). *Technometrics*, **11**, 373–386.

(DISCRIMINANT ANALYSIS
EXTREME-VALUE DISTRIBUTIONS
MULTIVARIATE EXTREME-VALUE DISTRIBUTIONS
WEIBULL DISTRIBUTION)

J. TIAGO DE OLIVEIRA

GUTTMAN SCALING *See* CORRESPONDENCE ANALYSIS

H

HAAR DISTRIBUTIONS

Haar distributions are generalizations of the uniform distribution* to sample spaces more complicated than the unit interval. They exist when a sample space has an operation defined on it which makes it into a mathematical group as long as that group has a topology (*see* TOPOLOGICAL CONCEPTS IN STATISTICS) under which it is compact. Haar probability distributions are special cases of Haar measures, which exist on groups which are only locally compact. They are named after A. Haar, who gave an almost general existence proof in 1933 [2], although such measures on orthogonal groups and other matrix groups date back to 1897 work of A. Hurwitz [3].

The distinguishing property which makes a measure Haar is group invariance*; the group operation does not change the measure of sets. In rigorous terms, a Haar measure is defined to be any Borel measure on a locally compact topological group invariant under the action of the group. These properties will be explained in succeeding paragraphs. The important examples in statistics are random matrices*, which form groups under matrix multiplication. For instance, if a multivariate normal* sample of dimension p has zero-mean vector, full rank, and the identity matrix for its expected covariance matrix, then the rotation matrix which defines its sample principal components* belongs to the group of $p \times p$ orthogonal matrices and has a Haar distribution.

The most familiar Haar measure is Lebesgue measure on the real line, when the real line is endowed with the usual Euclidean topology and regarded as a group under addition. The interval $\{x : a < x < b\}$ has the same length and so the same measure as the shifted interval $\{x + y : a < y < b\}$, whatever real number y we choose. This invariance of measure when we form a new set $\{x + y\}$ from a set $\{x\}$ by applying the group operation + holds for all Lebesgue-measurable sets and makes this measure a Haar measure.

A second example is the positive real line regarded as a group under multiplication. Group invariance requires the measure of the interval $\{x : t < x < t + h\}$ of length h to equal the measure of the interval $\{xy : ty < xy < ty + th\}$ of length th. The ratio of measure to length goes down the farther to the right an interval lies. That suggests, correctly, that a Haar measure on the multiplicative reals must have a density with respect to Lebesgue measure proportional to $1/t$.

Any constant multiple of $1/t$ yields an invariant measure, but no other densities do.

Both of these Haar measures assign infinite mass to the whole real line, so they cannot be normalized to give the whole group mass 1 and become probability measures. The simplest example of a Haar probability measure is the uniform distribution* on the unit interval $[0, 1]$ under the group operation of addition modulo unity. The uniform distribution is invariant under this group operation, inasmuch as it assigns any interval $[a, b]$ the same probability as the shifted interval $[a + y, b + y]$ or as the sets $[a + y, 1] \cup [0, b + y - 1]$ or $[a + y - 1, b + y - 1]$, whichever fits inside $[0, 1]$. Another statement of this invariance is that the expectation of any continuous function g of x in $[0, 1]$ is the same as the expectation of $g(x + y$ modulo 1) for any y in $[0, 1]$.

These examples are all commutative groups: $x + y = y + x$, etc. An example of a noncommutative group is the lower-triangular nonsingular $p \times p$ matrices under matrix multiplication. A set of matrices $\{T\}$ transforms into different sets $\{VT\}$ or $\{TV\}$ depending on whether V acts on the left or the right. The measures invariant here are different, one called left Haar measure, the other right Haar measure. Right Haar measure has a density proportional to $T_{11}^{-p}T_{22}^{1-p} \cdots T_{pp}^{-1}$. whereas left Haar measure has $T_{11}^{-1}T_{22}^{-2} \cdots T_{pp}^{-p}$. The latter density may be checked by showing that $f(T)$ equals $f(VT)$ times the Jacobian* $|dVT/dT|$ of the transformation from T to VT, the condition needed to make the expectations of $g(VT)$ and $g(T)$ equal for any lower-triangular matrix V and continuous function g.

Compactness is the property of the unit interval not shared by the real line or triangular matrices, which gives the unit interval a Haar probability measure. Compactness is the necessary and sufficient condition for a group with a topology to carry a Haar probability measure. Thus the group of $p \times p$ rotation matrices, being compact, carries a Haar probability measure, while the group of $p \times p$ lower-triangular nonsingular matrices only carries Haar measures which cannot be normalized to be probabilities. Any (left or right) Haar measure is unique up to constant multiples, like the density $1/t$ for the real line under multiplication. Since only one constant multiple can make the mass of the whole group equal unity, a Haar probability distribution on a group with a given topology is strictly unique. Furthermore, for compact groups, left and right Haar probability measures are the same.

Haar measure depends on the sample space, the group operation on it, and also the topology: the topology must make the group operation continuous and make the group locally compact, that is, must give every element an open neighborhood whose closure is compact. This condition rules out groups that are "too big," that is, whose points are mostly not close enough to each other in the sense of closeness defined by the topology. The groups of interest to statisticians that are small enough to carry Haar measures are predominantly groups like matrix groups that can be viewed as subsets of vector spaces of finite or countable dimensionality with the usual topology defined by a Euclidean norm.

Unfortunately, interesting topologies often make groups too big for Haar measures. The topology of weak convergence fails to make the set of all univariate probability distributions locally compact. These distributions do form a subset of the interesting group (under addition) whose elements are continuous linear functionals defined on all bounded continuous functions. From a naive point of view, a Haar measure on this group might seem likely to lead to a Bayesian prior* expressing total ignorance as to a choice of univariate probability models. But the failure of the group to be locally compact dashes any such proposal. For matrix groups, on the other hand, which are locally compact, Haar measures do often form reasonable Bayesian priors.

The proof of uniqueness of Haar measure is very easy, especially for compact groups, and reliance on uniqueness leads to short illuminating proofs of many results in multivariate* statistics. For instance, samples with rotationally symmetric distributions (like centered samples of independent, identically distributed normal variables) have sample covariance matrices (like Wishart matrices) whose eigenvectors are independent of their eigenvalues* (*see* WISHART DISTRIBUTION). This independence can be proved by arranging the eigenvectors as columns of an orthogonal matrix* whose distribution conditional on the eigenvalues, being rotationally symmetric like that of the sample, must be Haar. But this Haar distribution, being unique, cannot depend on what values the eigenvalues take, and so must be independent of them.

Such Haar uniqueness arguments are often alternatives to calculations with Jacobians of matrix transformations. Another case in point is the derivation of Bartlett's decomposition of a central standard Wishart matrix* into the product of a lower-triangular matrix of independent normal and chi-distributed variables times its transpose. Conversely, Jacobian calculations yield explicit expressions for densities of Haar measures.

Care with Haar densities is required, for the matrix elements themselves are generally not all functionally independent of each other. The group must be "parametrized" by expressing all matrix elements as functions of some special elements or other variables. The expressions

$$\begin{bmatrix} t & +(1-t^2)^{1/2} \\ -(1-t^2)^{1/2} & t \end{bmatrix} \quad \text{and}$$

$$\begin{pmatrix} \sin\theta & \cos\theta \\ -\cos\theta & \sin\theta \end{pmatrix}$$

both represent orthogonal matrices. A Haar density for t would be proportional to $(1-t^2)^{-1/2}$ on $[-1,1]$, whereas a Haar density for θ would be uniform on $[0, 2\pi)$. Neither expression, however, applies to all orthogonal matrices. Those with negative determinants require other parametrizations, and the densities must be pasted together. Such technicalities, bound up with the theory of Lie groups and the Jacobian formulas for Lie algebras, make Haar density calculations messy, in contrast to abstract derivations exploiting the uniqueness property.

Often sample spaces arise which are not themselves groups but are acted on by groups of transformations. The space of p-dimensional vectors is acted on by the rotation group. Haar techniques apply to such spaces if the spaces are split into subsets on which the group acts transitively, that is, on which any point can be transformed into any other point by some group element. Each origin-centered sphere in p-space is a transitive subset of p-space under the rotation group. Each sphere carries a unique probability measure invariant under the rotation group, called the uniform distribution on the sphere. In general, unique invariant probability distributions on transitive spaces can be derived from Haar probability measures using the theory of group quotient spaces, as treated succinctly in Loomis [5, pp. 130–133]. However, uniqueness only holds for transitive sets. There are many different rotation-invariant measures on the whole of p-space, assigning spheres of different radius different probabilities.

Alan James [4] pioneered multivariate sampling theory proofs relying on invariant measures on quotient spaces, an approach adopted and extended in the textbook by Dempster [1, Chap. 13], still the best extensive treatment of the subject. Replacing the ordinary orthogonal group by the group of transformations preserving an arbitrary inner product on p-space, the theory automatically embraces the elliptical distributions. James was able, by Haar methods, to rederive formulas like the joint density of canonical correlations (*see* CANONICAL ANALYSIS) under the standard null hypothesis because these canonical correlations are the singular values of a rectangular submatrix of a Haar-distributed orthogonal matrix. These Haar submatrix singular values have a simple asymptotic empirical distribution function, as proved in Wachter [8], with density proportional to

$$(x^2 - A^2)^{1/2}(B^2 - x^2)^{1/2}/((1 - x^2)x).$$

Broad classes of random orthogonal matrices share certain other limiting properties with Haar orthogonal matrices, leading Silverstein [7] to pursue a notion of "asymptotic Haar." With these exceptions, the asymptotic theory of Haar matrices is still largely unexplored.

On the whole, Haar measures have gained prominence in statistics with the realization that many consequences of multivariate normal assumptions do not depend on normality itself but only on rotational symmetries. For graphically based data analysis, symmetry assumptions are often preferable to parametric distributional assumptions such as normality. Thus, curiously enough, data analytic emphasis in multivariate statistics has promoted ties with the highly mathematical theory of Haar distributions.

References

[1] Dempster, A. (1969). *Elements of Continuous Multivariate Analysis*. Addison-Wesley, Reading, Mass. (Applications to multivariate sampling theory, highly recommended.)

[2] Haar, A. (1933). *Ann. Math.*, **34**, 147–169.

[3] Hurwitz, A. (1897). *Goettinger Nachr.*, 71–90.

[4] James, A. (1954). *Ann. Math. Statist.*, **25**, 40–75. (Initiated statistical applications, but assumes knowledge of differential forms.)

[5] Loomis, L. (1953). *Abstract Harmonic Analysis*. D. Van Nostrand, Princeton, N.J. (Concise and mathematical.)

[6] Nachbin, L. (1965). *The Haar Integral*. D. Van Nostrand, Princeton, N.J. (The most readable general treatment.)

[7] Silverstein, J. (1981). *SIAM J. Math. Anal.*, **12**, 274–281.

[8] Wachter, K. W. (1980). *Ann. Statist.*, **8**, 937–957. (No comprehensive elementary treatment of Haar measures for statisticians has yet appeared.)

(INVARIANCE CONCEPTS IN STATISTICS
MULTINORMAL DISTRIBUTION
RANDOM MATRICES)

KENNETH W. WACHTER

HADAMARD DERIVATIVE *See* STATISTICAL FUNCTIONALS

HADAMARD MATRICES

A square $n \times n$ matrix **H** is a Hadamard matrix if

(a) All the elements of **H** are $+1$ or -1.

(b) $\mathbf{H}'\mathbf{H} = n\mathbf{I}$, where **I** is the identity matrix.

Hadamard matrices exist only if $n = 2$ or if n is divisible by 4. The first row and column can be arranged so that their elements are all $+1$. Hadamard matrices up to $n = 100$ (excepting $n = 92$) were constructed cyclically by Plackett and Burman [2]; they are particularly useful in constructing d-optimal 2^t fractional factorial designs* and in forming optimum weighing designs*. A good discussion of the properties of Hadamard matrices and their use in the design of experiments* is to be found in Raghavarao [3, Secs. 15.5 and 17.4] and in Hedayat and Wallis [1].

References

[1] Hedayat, A. and Wallis, W. D. (1978). *Ann. Statist.*, **6**, 1184–1238.

[2] Plackett, R. L. and Burman, J. P. (1946). *Biometrika*, **33**, 305–325.

[3] Raghavarao, D. (1971). *Constructions and Combinatorial Problems in Design of Experiments*. Wiley, New York.

(FRACTIONAL FACTORIAL DESIGNS
WEIGHING DESIGNS)

HÁJEK–RENYI INEQUALITY

The *Hájek–Renyi inequality* is an extension of Kolmogorov's inequality*. Let $X_1, X_2, \ldots, X_n, \ldots, X_m$ $(n < m)$ be independent random variables with $E(X_j^2) < \infty$; $j = 1, \ldots, m$. Let

$$S_k = X_1 + \cdots + X_k \qquad (k = 1, \ldots, m).$$

Then, given an arbitrary sequence of positive numbers $\epsilon_n, \epsilon_{n+1}, \ldots, \epsilon_m$ such that $0 < \epsilon_n \leqslant \epsilon_{n+1} \leqslant \cdots \leqslant \epsilon_m$,

$$\Pr[|S_k - E(S_k)| \geqslant \epsilon_k \text{ for some integer } k \text{ such that } n \leqslant k \leqslant m]$$

$$= \Pr\left[\bigcup_{k=n}^{m} |S_k - E(S_k)| \geqslant \epsilon_k\right]$$

$$\leqslant \frac{\sum_{i=1}^{n}\operatorname{var}(X_i)}{\epsilon_n^2} + \sum_{k=n+1}^{m}\left(\frac{\operatorname{var}(X_k)}{\epsilon_k^2}\right). \quad (1)$$

When $n = 1$, inequality (1) becomes

$$\Pr\left[\bigcup_{k=1}^{m} |S_k - E(S_k)| \geqslant \epsilon_k\right] \leqslant \sum_{k=1}^{m}\left\{\operatorname{var}(X_k)/\epsilon_k^2\right\}.$$

When $\epsilon_n = \epsilon_{n+1} = \cdots = \epsilon_m$, this reduces to Kolmogorov's inequality.

The Hájek–Renyi inequality [3] yields a version of the strong law of large numbers* for nonidentically distributed random variables. The condition of independence for (1) to hold may be weakened to the martingale* property (see Chow and Teicher [1, pp. 243–244]). For more recent extensions and analogs with applications to convergence theorems, see Dunnage [2].

References

[1] Chow, Y. S. and Teicher, H. (1978). *Probability Theory*. Springer-Verlag, New York. (An advanced mathematical treatment.)

[2] Dunnage, J. E. A. (1975). *Quart. J. Math.* (2nd ser.), **26**, 361–376.

[3] Hájek, J. and Renyi, A. (1955). *Acta Math. Acad. Sci. Hung.*, **6**, 281–283.

(KOLMOGOROV'S INEQUALITY
LAWS OF LARGE NUMBERS)

HÁJEK'S PROJECTION LEMMA

The projection lemma of Jaroslav Hájek [1] gives an approximation of a function of independent random variables by a sum of independent random variables. This approximation is best in an extensive class of such sums, in a sense explained in the statement of the lemma. Since much is known about asymptotic distributions of sums of independent random variables, the lemma may enable us to find the asymptotic distribution of a more general function of independent random variables (see the Corollary below).

Hájek's Projection Lemma. *Let $X_1, \ldots, X_n$ be independent random variables and $S = s(X_1, \ldots, X_n)$ be a statistic such that $ES^2 < \infty$. Let*

$$\hat{S} = \sum_{i=1}^{n} E(S \mid X_i) - (n-1)ES.$$

Then $E\hat{S} = ES$ and

$$E(S - \hat{S})^2 = \operatorname{var} S - \operatorname{var} \hat{S}.$$

Moreover, if $L = \sum_{i=1}^{n} l_i(X_i)$, where $El_i^2(X_i) < \infty$, $i = 1, \ldots, n$, then

$$E(S - L)^2 = E(S - \hat{S})^2 + E(\hat{S} - L)^2.$$

Now for each $N = 1, 2, \ldots,$ let $(X_{N1}, \ldots, X_{Nn})$ be $n = n(N)$ independent random variables, let $S_N = s_N\ (X_{N1}, \ldots, X_{Nn})$, assume that $ES_N^2 < \infty$, and define $\hat{S}_N$ as $\hat{S}$, in terms of $S_N, X_{N1}, \ldots, X_{Nn}$. The lemma implies the following

Corollary. *If, as $N \to \infty$,*

(a) $$\frac{\operatorname{var}\hat{S}_N}{\operatorname{var} S_N} \to 1$$

and $(\hat{S}_N - E\hat{S}_N)/(\operatorname{var}\hat{S}_N)^{1/2}$ has a limit distribution, then $(S_N - ES_N)/(\operatorname{var} S_N)^{1/2}$ has the same limit distribution.

Some applications of the corollary—and an example where condition (a) is not satisfied—are mentioned in the entry ASYMPTOTIC NORMALITY.

Reference

[1] Hájek, J. (1968). *Ann. Math. Statist.*, **39**, 325–346.

W. HOEFFDING

HALDANE–SMITH TEST, THE

The Haldane–Smith nonparametric* test was invented in the first place to deal with the following problem. Several congenital abnormalities, for example, Down's syndrome ("mongolism") tend to occur in late-born children, conceivably through accumulated mutation in parents. It is therefore of interest to test whether late or early birth affects the probability of babies suffering other congenital malformations.

Haldane and Smith suggested a test based on the total birth ranks of affected children. Consider the data in Table 1 on nine families, in which some of the children are affected by phenylketonuria [4]. The children in each family are written in order of birth, with N denoting a normal child, A an affected one. In family [1] the affected children are numbers 5 and 8, with total birth rank $S = 13$. In a family of c children, of which a are affected, the expected sum of birth ranks will be $E(S) = a(c+1)/2$, and the variance of S is $a(c-a)(c+1)/12$. Hence we may analyze the data of Table 1 as in Table 2.

The difference (total observed S-total expected S) $= 50.0 - 48 = 2.0$ is divided by the standard error

$$(= \sqrt{\text{total variance}} = \sqrt{34.67} = 5.89).$$

The quotient $2.0/5.89 = 0.34$ is an approximate normal deviate (here showing no significant deviation from the null hypothesis that affected occur at random). In any appreciable body of data the normal approximation should be very good, since the distribution in large families approximates to normality, while with many small families we can appeal to the central limit theorem*.

Unclassified individuals are ignored in this calculation. If there are ties, the values of $E(S)$ and $\text{var}(S)$ must be found as a particular case of the following. Suppose that c children in a family have respective values $x_1, x_2, \ldots, x_c$. Find $T = \sum x_i$, $U = \sum x_i^2$, $V = (U - T^2/c)/(c-1)$. Suppose that a "affected" children are chosen at random out of the c children in the family. Let S be the sum of their x_i. Then

$$E(S) = Ta/c, \qquad \text{var}(S) = Va(c-a)/c.$$

The x_i will usually be the birth rank of child

Table 1

[1] NNNNANNA	[2] ANANNN	[3] NNAANANN
[4] AANAN	[5] NNNNNAN	[6] NANNN
[7] AN	[8] A	[9] ANN

Table 2 Test for Birth-Order Effect in Phenylketonuria Data of Table 1

Family Number	Number of Children, c	Number of Affected, a	Observed Total Rank Affected, S	$E(S)$	$\text{var}(S)$
1	8	2	13	9.0	108/12
2	6	2	4	7.0	56/12
3	8	3	13	13.5	135/12
4	5	3	7	9.0	36/12
5	7	1	6	4.0	48/12
6	5	1	2	3.0	24/12
7	2	1	1	1.5	3/12
8	1	1	1	1.0	0/12
9	3	1	1	2.0	6/12
Total	45	15	48	50.0	416/12 = 34.67

number i. Thus, with no ties, $x_i = 1$, and numerical values of $E(6S)$ and var$(6S)$ are given by Haldane and Smith [1].

If we are comparing birth orders of normal and affected in one single family, this test is identical with the Wilcoxon [6] and Mann–Whitney [3] tests* and equivalent to the Kruskal–Wallis [2] test*.

In applying this test it is important to be sure that the sampling is unbiased. Families will often come to the investigators' attention through having one or more affected members. If they come to attention soon after an affected member is born, and before the completion of the family, the result will be an excess of late born affected. If the disease is lethal, early born affected may die and the family with no living affected remaining may escape notice. Either way there will be a spurious effect.

If a real effect is found, it is possible that it may be due to mother's or father's age or birth order. Disentangling these can be difficult: for a simple but incomplete solution, see Smith [5].

References

[1] Haldane, J. B. S. and Smith, C. A. B. (1949). *Ann. Eugen.*, **14**, 117–124. (Statement of the Haldane–Smith test, with instructions for use.)

[2] Kruskal, W. H. and Wallis, W. A. (1952). *J. Amer. Statist. Ass.*, **47**, 583–621; *ibid.*, **48**, 910 (1953). (Generalization of the test, analogous to simple ANOVA.)

[3] Mann, H. B. and Whitney, D. R. (1947). *Ann. Math. Statist.*, **18**, 50–60. (Statement of the Mann–Whitney test, virtually equivalent to Wilcoxon and Haldane–Smith.)

[4] Munro, T. A. (1947). *Ann. Eugen.*, **14**, 60–88. (Data on birth order of individuals affected by phenylketonuria.)

[5] Smith, C. A. B. (1972). *Ann. Hum. Genet.*, **35**, 337–342. (A use of partial regression to discriminate between the effects of maternal age, paternal age, and birth order.)

[6] Wilcoxon, F. (1945). *Biometrics*, **1**, 80–83. (The first formulation of the test.)

(EPIDEMIOLOGICAL STATISTICS
HUMAN GENETICS, STATISTICS IN)

CEDRIC A. B. SMITH

HALF-NORMAL DISTRIBUTION *See* FOLDED DISTRIBUTIONS

HALF-NORMAL PLOTS

In 1954 a data analyst confronted with the results of a 2^8 factorial experiment* had only one published example [4, 5] to guide him in handling the mass of contrasts* produced by the standard computations. In these papers no account was given of any inspection of the data or of the resulting contrasts to spot bad values, misprints, or variable variance.

It seemed natural to plot the ordered absolute values of the contrasts on normal probability paper, together with the properly scaled theoretical cumulative distribution of the range for duplicate pairs as a reference distribution. Each unsigned contrast could be viewed as the range of a pair of normally distributed random variables, since the signs are partly arbitrary and since the central limit theorem* justified the assumption of normality. Probability plots* had long been used to judge the normality of simple samples of observations and to find *unwanted* outliers*. My purpose was rather to find *desired* outliers among the contrasts since these would correspond to real effects.

Allan Birnbaum showed me that the distribution of ranges of pairs from the normal distribution is the midfolded or half-normal (*see* FOLDED DISTRIBUTIONS). Its cumulative distribution function* can be linearized by using one-half of a normal grid. Cutting a sheet of "normal probability paper" in half at the 50% line, and then relabeling the P (percent) axis on the $P > 50\%$ section as $P' = 2P - 100$, we have a half-normal grid. Thus the former 84% point becomes the $P' = 68\%$ point. The other half of the sheet can be saved by substituting $P' = (100 - 2P)$ for P. The values of P' for N points are given with sufficient accuracy by the formula $P' = (i - \frac{1}{2})/N$, for $i = 1, 2, \ldots N$. For further details, see Zahn [8].

If the experiment is a null one, no factor having effect or interaction*, the whole collection of contrasts should fall near a

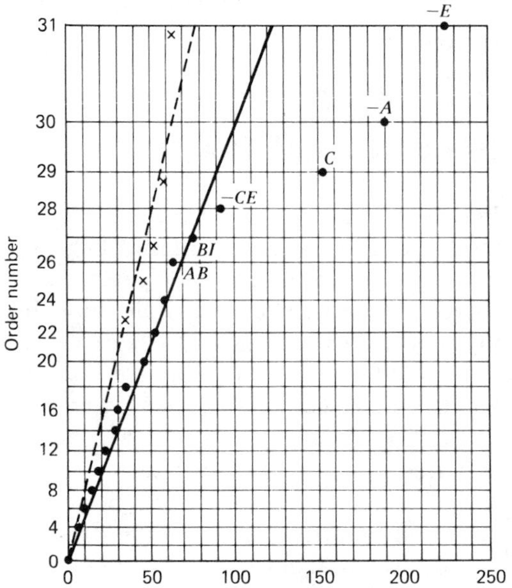

Figure 1 Half-normal plot of a 2^5 experiment. Dots and solid line, 31 contrasts; crosses and dashed line, 26 smallest contrasts.

straight line through the origin, whose 68.3% point (the theoretical value) would approximate the standard error of the contrasts plotted. If one or a few real effects are present, the corresponding contrasts should plot well off a straight line drawn through the origin and the remaining smaller contrasts.

The solid circles and the straight line of Fig. 1 show a typical plot. The experiment was an unreplicated 2^5 factorial on five factors thought to influence the yield of penicillin in surface culture, described in Davies [3, pp. 383–387, 416]. If the five largest contrasts are judged to be real, the remaining 26 are taken to be *error contrasts*. In order to estimate their standard error they are replotted, taking account of their number. This gives the dashed line in the figure and an estimated standard error of 33, instead of the original apparent 53.

It was naive to assign all contrasts either to significance or to error. An intermediate set likely to contain small real effects should have been excluded from the error set. As Wilk et al. [6] indicated analytically, and as Zahn [9, pp. 207ff.] confirmed by simulation* for the 2^4 case, better sensitivity is attained by this means.

By 1958 the results of several dozen 2^{p-q} experiments were available. (A 2^{-q} fraction of a 2^p factorial design is called a 2^{p-q}; *see* FRACTIONAL FACTORIAL DESIGNS.) While about half of these gave acceptable plots, many unexpected shapes emerged. Some turned out to be characteristic stigmata for common failures to meet the usual assumptions. A single observation out of the pattern of the rest of the data forces the plotted points—except the very smallest—to fall on a straight line with positive x-intercept. As a second example, inadvertent plot splitting on one factor shows up in a half-normal plot as a bend to the right at or above the middle order statistic. The two slopes correspond to the subplot error and to the total error, respectively. Numerical examples are given in Daniel [1].

Experience gained since 1959 has convinced me that some serious defects in experimental data may be concealed by half-normal plots. We should *always* examine the residuals*, even from an early biased fitting equation containing only the largest effects. Thus in the 2^5 experiment mentioned above there probably was a discrepant value at the treatment combination *abcd*, and the observed variance of the 16 observations at high E was 4.8 times that of the low E trials. This and other examples of defects not noticeable in half-normal plots are discussed in Daniel [1; 2, Chap. 7]. Both of the troubles just mentioned become obvious from residual plots.

Uncritical use of half-normal plots without detailed inspection of residuals can, then, be disastrous. The data analyst should not deliberately provide still another example of the besetting sin of analysts generally, the mistake of *premature aggregation**.

If the experimental conditions were precisely followed, *if* there are no defective values, *if* the error variance is nearly constant over all experimental conditions, and *if* there are only a few real effects (say fewer than $N/5$), *then* a half-normal plot often looks very nice and can be used to make decisions about effects with roughly known experiment-wise probability of type I error.

A serious mistake in Daniel [1] was uncovered by Zahn [7, 8, 9]. The "guard rails" originally published for judging when one or more contrasts were improbably large and hence real, are much too close to the null line, especially for the desired 0.05 probability error rate. Zahn deduced, I believe correctly, how the mistake must have been made. Any user of the guard rails will surely prefer Zahn's limits.

Zahn also made a detailed study for the 2^4 design of the operating characteristics of several methods (called by him versions) for deciding which effects are real, and for estimating error. His conclusions are clear and important. The method he calls "version S" seems consistently best, at least for 2^4 plans. This version requires estimation of the standard error of the contrasts, x_i, by

$$s_S = \sum_1^{11} x_i z_{i,15} \Big/ \sum_1^{11} z_{i,15}^2,$$

where the $z_{i,15}$ are the expected values of the normal order statistics for a sample of size 15. It is assumed, critically, that not more than *four* contrasts measure real effects. If it is likely, or if the data suggest, that more than four effects are real, one would naturally contract the range of i, perhaps to seven, corresponding to Zahn's "version R."

I have found the half-normal grid useful also for a somewhat different purpose. When several sets of 10 or more pairs of duplicate measurements are available, each set taken under different experimental conditions, a collection of half-normal plots of the ranges of the pairs, one for each set, provides an overview of the within-set error. Occasional maverick observations are easily seen; heterogeneity of error among experimental conditions is clearly visualized. Even if the observations are *counts*, then provided that 15 or more pairs are obtained, the ranges of duplicate counts follow the half-normal distribution quite closely.

The rather disappointing sensitivity of the 15-contrast half-normal plot documented by Zahn for many alternatives makes one wish that a corresponding study be made for the 31-contrast case. More than twice as many degrees of freedom* would usually be available for error estimation; the error variance of effects would be halved; plot splitting and other forms of replication degeneracy would be more easily spotted; fractional replication would be safer.

Sixty-four-run experiments are rather rare, so it does not seem likely that a satisfactory treatment of the 32-run case will provoke a demand for still another extension.

The earliest papers on the half-normal plot made the probability (or order number) the ordinate, and ticked off the contrasts on the abscissa axis. This conformed to almost a century of usage in plotting empirical and theoretical cumulative distributions. The reason given for interchanging axes (by Wilk et al. [6], followed by Zahn) was that it is traditional to plot the exact variable horizontally in linear regression*, versus the variable containing error vertically. I appeal to the older tradition and am tempted to ask somewhat fearfully if these authors propose to plot all empirical cumulative distributions with P as abscissa, and (for consistency perhaps) even histograms* on end.

In the null case the removal of the signs and the reordering of the absolute values of the $(2^p - 1)$ contrasts loses no information, *provided* that everything is "all right"—that is, provided that the key assumptions of the standard model are satisfied. But if one observation is widely discrepant by, say, δ units, every contrast is deflected by $\pm\delta$. The pattern of signs of the contrasts in standard order will usually reveal this. There is a different pattern for every treatment in the standard order. It is the complement of the corresponding effect vector, since the transformation matrix taking observations into effects is symmetrical. Thus if treatment combination *abcd* in a 2^5 experiment on factors A, B, C, D, and E, is disturbed, then the signs of all contrasts containing E will be moved in one direction, those without E being of course moved in the opposite direction. If *two* observations are disturbed by the amount δ, half the contrasts are displaced by

$\pm 2\delta$ and half are undisturbed. It makes sense, then, to inspect the vector of contrasts in standard order, responding to observed patterns, before going to the order statistics* and to their plot. In an obvious sense the order statistics are sensitive to, not robust to, real disturbances in one or two observations.

It appears safe to conclude that, with Zahn's corrections, the half-normal plot gives a convenient visualization of the results of 2^4 factorial experiments, provided that the data and the contrasts have been scrutinized as recommended above, and provided that there are not more than five or six real effects. Extension of Zahn's results to the 2^5 design would probably more than double the value of the corresponding plots.

References

[1] Daniel, C. (1959). *Technometrics*, **1**(4), 311–341.

[2] Daniel, C. (1976). *Applications of Statistics to Industrial Experimentation*. Wiley, New York.

[3] Davies, O. L., ed. (1971). *Design and Analysis of Industrial Experiments*. Hafner, New York.

[4] Kempthorne, O. (1952). *The Design and Analysis of Experiments*. Wiley, New York.

[5] Kempthorne, O. and Tischer, R. G. (1953). *Biometrics*, **9**(3), 302 ff.

[6] Wilk, M. B., Gnanadesikan, R., and Freeny, A. E. (1963). *J. Amer. Statist. Ass.*, **58**, 152–166.

[7] Zahn, D. A. (1969). An Empirical Study of the Half-Normal Plot. Ph.D. thesis, Harvard University.

[8] Zahn, D. A. (1975). *Technometrics*, **17**(2), 189–200.

[9] Zahn, D. A. (1975). *Technometrics*, **17**(2), 201–211.

(FOLDED DISTRIBUTIONS
FRACTIONAL FACTORIAL DESIGNS
GRAPHICAL REPRESENTATION OF DATA
PROBABILITY PLOTS)

CUTHBERT DANIEL

HALF-SAMPLE TECHNIQUES

Many modern sample surveys* are based on complex sample designs and estimation procedures, including stratification (*see* STRATIFIED SAMPLING), clustering (*see* CLUSTER SAMPLING), multistage selection of sample elements, poststratification, and ratio estimation*. As a result, exact analytical expressions for estimated variances of parameter estimates are often not obtainable. To overcome these difficulties a general class of methods, often referred to as *replication methods* and, in particular, half-sample techniques, has emerged over the last two decades. Half-sample techniques involve repeatedly computing estimates of the parameter of interest, each time using only half the data available in the sample. The sampling variance of the parameter estimate can be estimated from these calculated values.

To illustrate, consider a sample design of L strata of equal sizes, with two bivariate observations randomly selected from each.

Strata	Observations	
1	(X_{11}, Y_{11}),	(X_{12}, Y_{12})
2	(X_{21}, Y_{21}),	(X_{22}, Y_{22})
$\vdots$		$\vdots$
L	(X_{L1}, Y_{L1}),	(X_{L2}, Y_{L2})

Suppose that it is desired to estimate the population ratio $R = \sum Y / \sum X$. A half-sample is formed by selecting, from each stratum, one of the two available observations and estimating the parameter of interest from the L selected observations. That is, to estimate R, let

$$\hat{R} = \frac{1}{2L}\left(\sum_{i=1}^{L} \sum_{j=1}^{2} Y_{ij} \Big/ \sum_{i=1}^{L} \sum_{j=1}^{2} X_{ij} \right)$$

be the estimate of R based on all $2L$ observations, and let

$$\hat{R}_{(i)} = \frac{1}{L} \frac{y_{1u} + y_{2v} + \cdots + y_{Lw}}{x_{1u} + x_{2v} + \cdots + x_{Lw}},$$

$$u = 1 \text{ or } 2, \quad v = 1 \text{ or } 2, \quad w = 1 \text{ or } 2,$$

be the estimate of R based only on L observations (one from each stratum). Clearly, 2^L such half-sample estimates could be constructed.

An appropriate estimate of the variance of $\hat{R}$ computed from the 2^L possible half-samples would be

$$\hat{V}(\hat{R}) = \frac{1}{2^L} \sum_{i=1}^{2^L} (\hat{R}_{(i)} - \hat{R})^2.$$

Usually, 2^L is so large that it is impractical to compute every one of the possible $\hat{R}_{(i)}$'s. A solution to this problem was developed by McCarthy [5], who, through the use of matrices originally constructed by Plackett and Burman [6], showed that it is possible to select, in a rigidly prescribed manner, a small subset of the half-samples and still obtain satisfactory variance estimates. In fact, he showed that for linear estimates, use of the small subset gave precisely the same variance estimate as using all 2^L half-samples. Cross-product terms that arise as a result of between-strata contributions to variance are shown to cancel out when half-samples are selected in this "balanced" manner.

The design matrices* used are square matrices of dimension m with elements ± 1. These elements provide the rule for determining which sample elements are to appear in each half-sample. The matrices are orthogonal*, with all elements in the final column equal to -1. As a consequence, if δ_{ij} denotes the element in the ith row and jth column,

$$\sum_{i=1}^{m} \delta_{ij} = 0, \qquad j = 1, \ldots, m-1$$

[but $\sum_{i=1}^{m} \delta_{im} = \sum_{i=1}^{m}(-1) = -m$], and

$$\sum_{i=1}^{m} \delta_{ij}\delta_{ik} = 0, \qquad \text{all } j \neq k.$$

Plackett and Burman presented $m \times m$ matrices of dimension $m = \{4, 8, 12, \ldots, 100\}$ excluding $m = 92$. The dimension of the matrix to be used in a given problem would depend on the number of strata. McCarthy uses matrices in which m is a multiple of 4 and $L \leqslant m \leqslant L + 3$, where, in all cases with $L > 2$, $m \ll 2^L$. Once the $m \times m$ matrix is selected, all rows but only the first L columns are used.

As an example illustrating the process of creating half-samples, suppose that we have $L = 3$ strata with two observations each. We would use the first three columns of the 4×4 Plackett–Burman matrix. Each row of this matrix is a blueprint of the establishment of a half-sample. A half-sample is constructed in accordance with the value of δ_{ij} as follows:

$$\text{If } \delta_{ij} = \begin{cases} +1, & \text{use } x_{j1}, \\ -1, & \text{use } x_{j2}. \end{cases}$$

Hence the appropriate matrix is

		Strata 1	2	3
Half-samples	1	+1	+1	+1
	2	+1	−1	−1
	3	−1	−1	+1
	4	−1	+1	−1

The half-sample estimates of R are:

$$\hat{R}_{(1)} = \frac{y_{11} + y_{21} + y_{31}}{x_{11} + x_{21} + x_{31}};$$

$$\hat{R}_{(2)} = \frac{y_{11} + y_{22} + y_{32}}{x_{11} + x_{22} + x_{32}};$$

$$\hat{R}_{(3)} = \frac{y_{12} + y_{22} + y_{31}}{x_{12} + x_{22} + x_{31}};$$

$$\hat{R}_{(4)} = \frac{y_{12} + y_{21} + y_{32}}{x_{12} + x_{21} + x_{32}}.$$

Note that each element appears in an equal number of half-samples. This is due to the balanced nature of the matrix.

The balanced half-sample estimate of the variance of $\hat{R}$ may now be computed as

$$\hat{V}(\hat{R}) = \frac{1}{m} \sum_{i=1}^{m} (\hat{R}_{(i)} - \hat{R})^2.$$

There are a number of variations of this formula (including the substitution of $\overline{R} = \sum_{i=1}^{m} \hat{R}_{(i)}/m$ for $\hat{R}$), but the general idea behind the controlled selection of half-samples is common to all.

In actual survey situations, two primary sampling units (PSUs) may be selected from each of the L strata. From the ith PSU of stratum h, estimates x'_{hi} and y'_{hi} are obtained of the PSU totals for the two variables using the sampled individuals in that PSU. These x'_{hi} and y'_{hi} are then used as were the X_{hi} and Y_{hi} in the previous discussion.

It also should be noted that often surveys are designed in which there are $2L$ strata, and one PSU is selected from each. In order to apply the balanced half-sample method, the PSUs are paired to form L "pseudo-strata" and the estimation process goes on as previously described. Care is generally exercised in this pairing to keep the PSUs as similar as possible with respect to known demographic information.

The balanced half-sample method has also been useful for estimating the variance of other nonlinear estimates such as slopes and correlation coefficients. Research has demonstrated that the method works quite effectively under a range of situations studied (see Bean [1], Kish and Frankel [2], Lemeshow and Epp [3], and Lemeshow and Levy [4]). The method is currently used by the National Center for Health Statistics* in a number of large-scale, ongoing sample surveys such as the Health and Nutrition Examination Survey (HANES) and the Health Interview Survey (HIS).

Other methods available for the purpose of variance estimation include the linearization or Taylor series method* and the jackknife*.

References

[1] Bean, J. A. (1975). Distribution and Properties of Variance Estimators for Complex Multistage Probability Samples: An Empirical Distribution. *Data Evaluation and Methods Research*. NCHS, Ser. 2, No. 65. DHEW Publ. No. (HRA) 75-1339.

[2] Kish, L. and Frankel, M. (1974). *J. R. Statist. Soc. B*, **36**, 1–37.

[3] Lemeshow, S. and Epp, R. (1977). *Commun. Statist. A*, **6**(13), 1259–1274.

[4] Lemeshow, S. A. and Levy, P. S. (1978). *J. Statist. Comp. Simul.*, **8**, 191–205.

[5] McCarthy, P. J. (1966). Replication. An Approach to the Analysis of Data from Complex Surveys. *Vital and Health Statistics*. NCHS, Ser. 2, No. 14.

[6] Plackett, R. L. and Burman, J. P. (1946). *Biometrika*, **33**(Pt. IV), 305–325.

(CLUSTER SAMPLING
SAMPLE SURVEYS
SAMPLING
STRATIFIED SAMPLING)

STANLEY LEMESHOW

HAMMERSLEY–CLIFFORD THEOREM, THE

Given a graph, a state space for each node, and a definition of the neighbors of each, we can define a Markov property: that the probability of a node being in a particular state given the states of all other nodes should depend only on the states of the neighbors. The Hammersley–Clifford [4] theorem identifies the class of probability distributions on the states of the nodes which are consistent with the Markov property. The theorem has applications to statistical mechanics*, the graph representing, perhaps, a lattice and the states being presence or absence (*see* LATTICE SYSTEMS). Alternatively, the nodes might represent geographic regions, or fruit trees in an orchard; one can then define "neighbor" and formulate parametric models for the joint probability distribution.

Suppose, then, that we have a collection of n *sites*, at each of which a finite set of *states* is available. The state of the ith site is denoted by a real number x_i. We consider probability distributions over the set of realizations $\mathbf{x} = (x_1, \ldots, x_n)$. Denote the probability of $\mathbf{x}$ by $P(\mathbf{x})$. Site j is a *neighbor* of site i $(i \neq j)$ iff the functional form of $P(x_i \mid x_1, \ldots, x_i - 1, x_i + 1, \ldots, x_n)$ is dependent on the variable x_j. As the simplest example, suppose that $X_1, \ldots, X_n$ is a Markov chain (*see* MARKOV PROCESSES). Then it is easily shown that site i $(2 \leqslant i \leqslant n-1)$ has neighbors $i-1$, $i+1$, while sites 1 and n have the single neighbors x_2, x_{n-1}, respectively. A more interesting case is that of a rectangular lattice, where the neighbors of an interior site with coordinates (r,s) might be taken to be $\{(r-1,s),(r+1,s),(r,s-1),(r,s+1)\}$, or perhaps these four augmented by the quartet $\{(r \pm 1, s \pm 1)\}$. Any set of sites which either consists of a single site or else in which every site is a neighbor of every other site in the set is a *clique*.

The dependence of conditional probabilities only on the neighbors is clearly an extension of the Markov property. A system of n sites, with specified neighbors, generates a class of valid probability distributions called *Markov random fields**.

At each site, we attach the label "zero" to one of the states which has positive probability there. Thus $\mathbf{0}$ represents a realization of 0 at each node. Assume the *positivity condi-*

tion: $P(\mathbf{x}) > 0$ for all $\mathbf{x}$. That is, there are no forbidden states. It turns out to be convenient to deal with

$$Q(\mathbf{x}) \equiv \ln\{P(\mathbf{x})/P(\mathbf{0})\} \qquad (1)$$

rather than with $P(\mathbf{x})$. The Hammersley–Clifford theorem answers the following question: Given the neighbors of each site, what is the most general form $Q(\mathbf{x})$ may take in order to give a valid probability structure? Now algebraically, it may be seen that there is a unique expansion of $Q(\mathbf{x})$ of the form

$$\begin{aligned} Q(\mathbf{x}) = &\sum_{1 \leqslant i \leqslant n} x_i G_i(x_i) + \sum\sum_{1 \leqslant i < j \leqslant n} x_i x_j G_{i,j}(x_i, x_j) \\ &+ \sum\sum\sum_{1 \leqslant i < j < k \leqslant n} x_i x_j x_k G_{i,j,k}(x_i, x_j, x_k) + \cdots \\ &+ x_1 x_2 \ldots x_n G_{1,2,\ldots,n}(x_1, x_2, \ldots, x_n). \end{aligned} \qquad (2)$$

For example, we have $x_i G_i(x_i) \equiv Q(0, \ldots, 0, x_i, 0, \ldots, 0) - Q(\mathbf{0})$, with the analogous difference formulae for the higher-order G-functions. We might interpret $G_i(x_i)$ as indicating the relative likelihood of state x_i at site i, and so on. According to the theorem, the characterization of a Markov random field is that: *For any* $1 \leqslant i < j < \cdots < s \leqslant n$, *the function* $G_{i,j,\ldots,s}$ *in* (2) *may be nonnull if and only if the sites* $i, j, \ldots, s$ *form a clique. Subject to this restriction, the G-functions may be chosen arbitrarily.*

DISCUSSION

The foregoing formulation of the theorem follows Besag [2], who gives a simple proof. The original proof [4] did not proceed through (2), and was rather lengthy. A further proof was given by Grimmett [3]; see also Averintsev [1], Preston [8], and Sherman [10]. The positivity condition turns out to be necessary for the theorem to hold [5].

In statistical mechanics*, the *potential function* of a system of sites is $\Phi(x) = -\ln\{P(x)\}$; Φ is a *Gibbs potential* [9] if it is of the form

$$\Phi(x) = \sum \phi_{i,j,\ldots,k}(x_i, x_j, \ldots, x_k),$$

the sum being taken only over sets $\{i, j, \ldots, k\}$ which are cliques. Thus the theorem shows the equivalence of the class of Markov random fields and that of Gibbs systems. (See Moussouris [5], GIBBS DISTRIBUTIONS, and LATTICE SYSTEMS.)

An important case is the rectangular lattice with only two states; here the theorem had already been proved by Spitzer [11] and others. If the neighbors of a site are those sharing a common edge, and under an assumption of spatial homogeneity, we get the classical *Ising model* of statistical mechanics; in this case (2) defines an exponential family for $\{P(x)\}$. The G_{ij}'s are interaction parameters measuring the tendencies toward clustering or segregation; if all the G_{ij}'s are zero, the states of the sites are independent. The sufficient statistics* are the counts of the number of pairs of adjacent sites in the same state. Pickard [6, 7] gives central limit theorems* for these statistics. Strauss [12, 13] discusses the estimation of the clustering parameters for the two-state and multistate cases (see also Besag [2]).

References

[1] Averintsev, M. B. (1970). *Problemy Peredaci Inf.*, **6**, 100–109.

[2] Besag, J. E. (1974). *J. R. Statist. Soc. B*, **36**, 192–236. (A comprehensive statistical treatment of related lattice models, including a wealth of discussion, numerical examples, and references.)

[3] Grimmett, G. R. (1973). *Bull. Lond. Math. Soc.*, **5**, 81–84. (An early and rather complex proof of the main theorem.)

[4] Hammersley, J. M. and Clifford, P. (1971). Markov Fields on Finite Graphs and Lattices. Unpublished manuscript.

[5] Moussouris, J. (1974). *J. Statist. Physics*, **10**, 11–33. (An excellent account both of the theorem—and its equivalent forms—and the considerably more complex case when positivity fails.)

[6] Pickard, D. K. (1976). *J. Appl. Prob.*, **13**, 486–497.

[7] Pickard, D. K. (1977). *Adv. Appl. Prob.*, **9**, 476–501.

[8] Preston, C. J. (1973). *Adv. Appl. Prob.*, **5**, 242–261.

[9] Ruelle, D. (1969). *Statistical Mechanics*. W. A. Benjamin, New York.

[10] Sherman, S. (1973). *Isr. J. Math.*, **14**, 92–103.

[11] Spitzer, F. (1971). *Amer. Math. Monthly*, **78**, 142–154.

[12] Strauss, D. J. (1975). *J. Appl. Prob.*, **12**, 702–712.
[13] Strauss, D. J. (1977). *J. Appl. Prob.*, **14**, 135–143.

(GIBBS DISTRIBUTIONS
GRAPH THEORY
LATTICE SYSTEMS
MARKOV RANDOM FIELDS)

D. J. STRAUSS

HANNING

This is a smoothing technique in exploratory data analysis*. In a sequence of numbers, adjacent values are first averaged; the resulting sequence of averages is then treated in like manner, and the two extreme values of the data set are included in the smoothed data set.

Example.

Basic Data	5	6	11	9	12	21
First Averaging	$5\frac{1}{2}$	$8\frac{1}{2}$	10	$10\frac{1}{2}$	$16\frac{1}{2}$	
Smoothed Data	5	7	$9\frac{1}{4}$	$10\frac{1}{4}$	$13\frac{1}{2}$	21

Hanning is often applied to monotonic or nearly monotonic sequences of numbers, which may themselves result from smoothing techniques to remove peaks and valleys from more scattered data sets (see Tukey [1, Chap. 7]).

Reference

[1] Tukey, J. W. (1977). *Exploratory Data Analysis*. Addison-Wesley, Reading, Mass.

(EXPLORATORY DATA ANALYSIS
GRADUATION
WAVE-CUTTING FORMULAS)

HANSEN FREQUENCIES

A scoring technique frequently used in the study of animal behavior is to record whether a behavior of interest occurred at least once or not at all during some fixed interval of time, t. The resulting data are often referred to as *Hansen frequencies* from the work of E. W. Hansen [4] and his students, although such a scoring system was used as early as 1929 by Olson, and Fisher [3] considered a similar problem in dilution experiments.

If the behavior of interest has a temporal distribution governed by a Poisson process*, the Hansen frequencies then follow a binomial distribution*. Thus if p_0 is the probability of no occurrence of the behavior in a time interval of length t,

$$p_0 = e^{-\lambda t}, \tag{1}$$

where λ is the mean rate of occurrence of the behavior in a unit time interval. From (1) we get $\lambda = -t^{-1}\log p_0$. Since n_0, the number of intervals in which the behavior did not occur, is binomially distributed with parameter p_0 and N, the total number of intervals, the maximum likelihood estimate* of λ is given by

$$\lambda^* = t^{-1}\log(n_0/N). \tag{2}$$

Using the δ method [1, pp. 486–489], we can show that λ^* has an asymptotic normal distribution with mean λ and with variance $(e^{\lambda t} - 1)/(Nt^2)$.

The asymptotic relative precision of λ^* as compared to the usual maximum likelihood estimate, $\hat{\lambda}$, based on complete data (i.e., counts of the number of occurrences) is

$$\text{asymptotic relative precision} = \frac{\lambda t}{e^{\lambda t} - 1} < 1. \tag{3}$$

When the observed behavior no longer has a distribution governed by a Poisson process, λ^* of expression (2) is no longer a suitable estimate of the mean rate of behavior [2].

References

[1] Bishop, Y. M. M., Fienberg, S. E., and Holland, P. W. (1975). *Discrete Multivariate Analysis: Theory and Practice*. MIT Press, Cambridge, Mass.

[2] Fienberg, S. E. (1972). *Primates*, **13**, 323–326.

[3] Fisher, R. A. (1922). *Philos. Trans. R. Soc. Lond. A*, **222**, 309–368.

[4] Hansen, E. W. (1966). *Behavior*, **27**, 107–149.

(CATEGORICAL DATA
POISSON PROCESS
PROPAGATION OF ERRORS)

STEPHEN E. FIENBERG

HANSEN–HURWITZ METHOD FOR SUBSAMPLING NONRESPONDENTS

THE PROCEDURE

When nonresponse occurs in sample surveys*, the estimators for the population means, totals, and proportions based on the responding units are biased. Hansen and Hurwitz [3] suggest a method for subsampling the nonrespondents (see also Hansen et al. [4, pp. 473–475]).

Let y_i $(i = 1, 2, \ldots, N)$ denote the values of the characteristic of interest of the N population units. The population mean is $\overline{Y} = \sum_1^N y_i/N$ and its variance is $S^2 = \sum_1^N (y_i - \overline{Y})^2/(N-1)$. Let N_1 denote the size of the respondents of the population—the units that would respond to the initial call if they were drawn into the sample. Let $\overline{Y}_1$ and S_1^2 denote the population mean and variance of the respondents.

The initial random sample of size n, drawn without replacement, results in n_1 respondents and $n_2 = (n - n_1)$ nonrespondents. The sample mean $\bar{y}_1 = \sum_{i=1}^{n_1} y_i/n_1$ is unbiased for $\overline{Y}_1$, but it is biased for $\overline{Y}$. Hansen and Hurwitz [3] suggest drawing a random sample of size $m_2 = n_2/k$, where k $(k > 1)$ is fixed in advance, and obtaining response on all of them through intensive efforts. Let $\bar{y}_{2m}$ denote the mean of the m_2 units.

Now, the estimator for $\overline{Y}$ is

$$\hat{\overline{Y}} = \frac{n_1}{n}\bar{y}_1 + \frac{n_2}{n}\bar{y}_{2m}, \qquad (1)$$

which is unbiased for $\overline{Y}$ and has variance

$$V(\hat{\overline{Y}}) = \frac{N-n}{Nn}S^2 + W_2\frac{k-1}{n}S_2^2, \qquad (2)$$

where $W_2 = N_2/N$ is the proportion of the nonrespondents in the population. An estimate of the variance in (2) is given in Cochran [1, p. 333] and Rao [5]. The second term on the right of (2) is the addition to the variance due to subsampling.

OPTIMUM VALUES OF n AND k

The cost of sampling may be of the form

$$C' = C_0 n + C_1 n_1 + C_2 m_2, \qquad (3)$$

where C_0 is the initial cost for setting up the survey, C_1 is the cost per unit for obtaining the responses from the n_1 units and processing them, and C_2 is the cost for contacting the subsampled units and for obtaining and processing responses from them. From (3), the average cost is

$$C = \left(C_0 + C_1 W_1 + \frac{C_2 W_2}{k}\right)n, \qquad (4)$$

where $W_1 = N_1/N$ and $W_2 = N_2/N$.

For minimizing the average cost for a given value V of the variance in (2), the optimum values of k and n are given by

$$k_{\text{opt}}^2 = \frac{C_2(S^2 - W_2 S_2^2)}{S_2^2(C_0 + C_1 W_1)}, \qquad (5)$$

$$n_{\text{opt}} = n_0\left[1 + \frac{(k_{\text{opt}} - 1)W_2 S_2^2}{S^2}\right], \qquad (6)$$

where $n_0 = NS^2/(NV + S^2)$ is the sample size required when $W_2 = 0$. Similarly, for minimizing the variance in (2) for a given value of the average cost C in (4), the optimum value of k is the same as in (5), but the initial sample size is given by

$$n_{\text{opt}} = \frac{k_{\text{opt}} C}{k_{\text{opt}}(C_0 + C_1 W_1) + C_2 W_2}. \qquad (7)$$

As an illustration, consider the example of Hansen and Hurwitz [3] in which $C_0 = 0.1$, $C_1 = 0.4$, and $C_2 = 4.5$. The required precision is that given by a sample of size 1000 if there were no nonrespondents [i.e., if $n_0 = 1000$ and $V = (N - n_0)S^2/(Nn_0)$]. Suppose that it is assumed that S^2 and S_2^2 are equal. If the value of W_2 is thought to be equal to 0.4, $k_{\text{opt}} = 2.82$ from (5) and $n_{\text{opt}} = 1728$ from (6). Now, the "expected" size of the subsample is $m_{2\text{opt}} = n_{\text{opt}} W_2 / k_{\text{opt}} = 245$ (see Cochran [1, p. 373]).

ALTERNATIVE PROCEDURES FOR DETERMINING THE SAMPLE SIZES

Since the variance in (2) depends on the unknown W_2, Srinath [8] and Rao [5] suggest determining the subsample size as $m_2^* = n_2^2/(k^*n + n_2)$, where k^* is "fixed in advance." For this procedure, the expected variance takes the form

$$V^*(\hat{\bar{Y}}) = \frac{1-f}{n} S^2 + \frac{k^*}{n} S_2^2, \qquad (8)$$

which does not depend on W_2. However, the optimum values of n and k^* for minimizing the variance or the cost, as before, depend on W_2. Details of deriving these values and a discussion of the appropriateness of this procedure are given in Rao [7].

Cochran [1, p. 372] suggests a practical procedure. Find the value of n from (5) and (6) for a series of values of W_2 from zero to a "safe upper limit" and use the maximum of these values as the initial sample size n. The value of n_2 is obtained at the end of the initial survey. With these values of n and n_2, the value of k can be obtained by prescribing an upper limit V_0 to the conditional variance $\{w_2(k-1)/n\}S_2^2$, where $w_2 = n_2/n$. Thus, for each sample, this procedure will guarantee that $V(\hat{\bar{Y}})$ does not exceed $V = (N-n)S^2/n + V_0$. This procedure determines the subsample size from the observed value of (n_2/n).

EXTENSIONS

Cochran [1] presents the approximate mean square errors* of the ratio and regression type of estimators when the nonrespondents are subsampled (see also Rao [7]). Procedures for determining the initial sample and the subsampling fractions in the case of stratification are discussed by the author in Rao [6].

In some situations, it may not be possible to obtain responses from all the m_2 subsampled units; to cover such cases, El-Badry [2] extends the procedure of Hansen and Hurwitz to more than two calls.

References

[1] Cochran, W. G. (1977). *Sampling Techniques*, 3rd ed. Wiley, New York. (A widely used textbook on the theory and applications of sample surveys.)

[2] El-Badry, M. A. (1956). *J. Amer. Statist. Ass.*, **51**, 209–227. (Extends the procedure of Hansen and Hurwitz to more than two calls.)

[3] Hansen, M. H. and Hurwitz, W. N. (1946). *J. Amer. Statist. Ass.*, **41**, 517–529. (Suggest subsampling the nonrespondents and provide the methodology.)

[4] Hansen, M. H., Hurwitz, W. N., and Madow, W. G. (1953). *Sample Survey Methods and Theory*, Vol. 1. Wiley, New York. (A good textbook on sample surveys.)

[5] Rao, J. N. K. (1973). *Biometrika*, **60**, 125–133. (Gives a general procedure for selecting the sample sizes in the context of double sampling for stratification.)

[6] Rao, P. S. R. S. (1980). Nonresponse and Subsampling: Stratification and Optimum Sample Sizes. *Res. Rep.*, U.S. Bureau of the Census and the University of Rochester. (Different procedures for selecting the sizes of the initial sample and the subsample are suggested and the resulting variances of the estimators are compared.)

[7] Rao, P. S. R. S. (1983). Nonresponse and Double Sampling. Panel on Incomplete Data, National Research Council, Volume on Theory and Methods, Academic Press, New York (in press). (A comprehensive review of the Hansen and Hurwitz method is provided. Alternative procedures for selecting the sample sizes are compared.)

[8] Srinath, K. P. (1971). *J. Amer. Statist. Ass.*, **66**, 583–586. (Suggests a method for determining the size of the subsample for the Hansen and Hurwitz procedure; see also refs. 5 and 7 for further details.)

(SURVEY SAMPLING
SURVEYS, HOUSEHOLD)

PODURI S. R. S. RAO

HARDY–WEINBERG MODEL *See* GENETICS, STATISTICS IN; HUMAN GENETICS, STATISTICS IN

HARMONIC MEAN

The harmonic mean of n quantities $X_1, X_2, \ldots, X_n$ is the reciprocal of the arithmetic mean* of their reciprocals. Symbolically,

$$H = \frac{n}{\sum_{i=1}^{n}(1/X_i)}.$$

AN EXAMPLE

Alder and Roessler [1] present the following example showing how the harmonic mean can be used. Suppose that it is desired to find the average speed required for a car to cover a fixed distance d after the car has traveled the distance at speeds r_1, r_2, and r_3. The average speed is found by dividing the total distance traveled by the total time required:

$$\text{average speed} = \frac{d + d + d}{t_1 + t_2 + t_3}$$

However, since $d = r_i t_i$ or $t_i = d/r_i$, the average speed can be rewritten as

$$\frac{d + d + d}{d/r_1 + d/r_2 + d/r_3}$$

$$= \frac{3d}{d(1/r_1 + 1/r_2 + 1/r_3)}$$

$$= \frac{3}{1/r_1 + 1/r_2 + 1/r_3},$$

which is the harmonic mean of the three speeds.

As a numerical illustration of this example, consider an airplane flying 3 legs of 300 miles each at rates 300, 250, and 300 miles per hour. The harmonic mean gives the average speed

$$\frac{3}{\frac{1}{300} + \frac{1}{250} + \frac{1}{300}} = 281.25 \text{ miles per hour.}$$

If the arithmetic mean had been used, an incorrect value of $(300 + 250 + 300)/3 = 283.33$ miles per hour would have been obtained. However, the arithmetic mean would be appropriate if times were held constant rather than distance.

OTHER APPLICATIONS

A widely employed use of the harmonic mean in applied statistics is that of providing a compromise sample size when sample sizes are unequal and the statistical procedure under consideration requires equal sample sizes. An application of this type is given by Iman [5] for testing the equality of correlation* coefficients based on two sample correlation coefficients. Iman has provided graphs for quick 5% and 1% tests of hypotheses of equality of population correlation coefficients based on Fisher's z-transformation*. While Fisher's z-transformation accommodates unequal sample sizes, the graphs had to be restricted to equal sample sizes. Iman indicates that the harmonic mean provides a good compromise sample size to use with the graphs when sample sizes are unequal, but that its use creates a slightly nonconservative test. However, the size of the increase in the α level is quite small when the minimum sample size is at least 15.

In another application Miller [8] suggests use of the harmonic mean in the inverse formula of the Freeman–Tukey double arcsine transformation, which is used for stabilizing the variance and achieving approximate normality when data have been obtained from binomial distributions (*see* ANGULAR TRANSFORMATION). Miller suggests using the harmonic mean when the original proportions involve different totals. In rather a unique application Good [4] suggested that if the tail-area probabilities of several statistics calculated from the same data are $P_1, P_2, \ldots, P_n$, it is reasonable to summarize them using their harmonic mean or weighted harmonic mean. The harmonic mean has been recommended for use with multiple comparisons* by Bancroft [2]; however, Keselman et al. [7] have shown that inversely pairing unequal variances with unequal samples causes the observed type I error to exceed true α when using the harmonic mean with Tukey's multiple comparison statistic. The harmonic mean is also employed in the adjustment of mean square calculations when using an unweighted means analysis of a factorial experiment* with unequal numbers of observations per cell (see Snedecor [10]). The harmonic mean is also used occasionally in the area of business statistics. Freund and Williams [3] give one such application and references to other books containing business applications.

RELATIONSHIP TO OTHER SIMPLE MEANS

Several authors have given fundamental inequalities involving simple means. In particular, if the X_i represent a finite sequence of positive numbers and we let $A = \sum_{i=1}^{n} X_i/n$ and $G = (\Pi_{i=1}^{n} X_i)^{1/n}$ represent the arithmetic mean* and geometric mean*, respectively, then (Mitrinović [9])

$$\text{minimum } X_i \leqslant H \leqslant G \leqslant A \leqslant \text{maximum } X_i .$$

In addition, Mitrinović presents a number of inequalities involving means as well as references to related work. Kendall and Stuart [6] and Mitrinović both consider the quantity

$$M(t) = \left[\frac{1}{n} (X_1^t + X_2^t + \cdots + X_n^t) \right]^{1/t}$$

where the X's are again defined as above. These authors show that $M(t)$ is an increasing function of t and for $t = 1$, $M(t)$ is the arithmetic mean; when $t = -1$, $M(t)$ is the harmonic mean; and as t tends to zero, $M(t)$ is the geometric mean.

References

[1] Alder, H. L. and Roessler, E. B. (1964). *Introduction to Probability and Statistics*. W. H. Freeman, San Francisco.

[2] Bancroft, T. A. (1968). *Topics in Intermediate Statistical Methods*, Vol. 1. Iowa State University Press, Ames, Iowa.

[3] Freund, J. E. and Williams, F. J. (1969). *Modern Business Statistics*. Prentice-Hall, Englewood Cliffs, N.J.

[4] Good, I. J. (1958). *J. Amer. Statist. Ass.*, **53**, 799–813.

[5] Iman, R. L. (1977). *J. Quality Tech.*, **9**, 172–175.

[6] Kendall, M. G. and Stuart, A. (1969). *The Advanced Theory of Statistics*, Vol. 1, 3rd edition, Hafner, New York.

[7] Keselman, H. J., Toothaker, L. E., and Shooter, M. (1975). *J. Amer. Statist. Ass.*, **70**, 584–587.

[8] Miller, J. J. (1978). *Amer. Statist.*, **32**, 138–139.

[9] Mitrinović, D. S. (1970). *Analytic Inequalities*. Springer-Verlag, New York.

[10] Snedecor, G. W. (1956). *Statistical Methods*. Iowa State University Press, Ames, Iowa.

(ARITHMETIC MEAN
GEOMETRIC MEAN)

R. L. IMAN

HARRIS POLL *See* PUBLIC OPINION POLL

HARTER'S ADAPTIVE ROBUST METHOD

Harter's adaptive robust method, in its original form, involves the use of a criterion based on the sample kurtosis* (standardized fourth moment) K of deviations from the mean (or from a provisional least-squares* regression) to decide whether to use the maximum likelihood* (ML) estimates of mean, standard deviation, and regression coefficients for the uniform*, the normal*, or the double exponential distribution (*see* LAPLACE DISTRIBUTION). These ML estimates are those obtained by the method of least pth powers* (the L_p estimates—see Jackson [9]) for $p = \infty$, 2, and 1, respectively.

This method was first proposed in a 1972 technical report and repeated in an expanded and updated journal article by Harter [4], Part V. In the original version,

Table 1

Distribution	Regression Coefficients	Mean	Standard Deviation
Uniform (rectangular)	Minimax* estimates	Sample midrange	(Sample semirange)/$\sqrt{3}$
Normal	Least-squares estimates	Sample mean	$\sqrt{\text{Sample variance}}$ (with n, not $n-1$ in denominator)
Double exponential	Least absolute values* estimates	Sample median*	$\sqrt{2}$ (average deviation from sample median)

the error distribution is taken to be uniform (a typical platykurtic or short-tailed distribution) if $K < 2.2$, normal (a typical mesokurtic or medium-tailed distribution) if $2.2 \leqslant K \leqslant 3.8$, and double exponential (a typical leptokurtic or long-tailed distribution) if $K > 3.8$. Then the ML estimates for the chosen distribution are used, as shown in Table 1.

Several modifications of the original procedure have been proposed. To avoid undue influence of outlying observations, Hogg [8] suggested taking deviations from the median or from a provisional Brown–Mood* [2] regression. Studies by Jorgenson [10], Forth [3], Bourdon [1], and Rugg [11] demonstrated that the appropriate critical values K_L and K_U, for small to moderate samples, depend on the sample size. Harter et al. [6] determined the best values, for various sample sizes, of the critical values of K and of alternate criteria based on a statistic Q suggested by Hogg [7] and on the sample likelihoods, for classifying a sample as coming from a uniform, normal, or double exponential population and estimating location and scale parameters of symmetric populations. The same critical values of K, Q, and functions of likelihoods would be expected to be optimal (or nearly so) for estimating regression coefficients.

Example 1. Given the following random sample of size 24 from a population of unspecified type: -0.0561, 0.3437, -0.0100, 1.0304, 0.1152, -0.2727, 1.8107, 0.0202, -0.2235, -2.3779, 1.8607, -0.0799, -0.7472, -0.7363, 1.2053, -3.0925, -0.1559, -0.8488, 4.0254, -1.1312, 1.5517, -0.7744, -0.3060, 0.5614. Estimate the population mean μ and standard deviation σ.

The sample kurtosis is found to be $K = 4.404$. Since $K > 3.8$, the distribution is taken to be double exponential, and hence the adaptive robust estimates of μ and σ are the sample median, -0.0680, and $\sqrt{2}$ times the average deviation from the sample median, 1.3674, respectively.

Example 2. Let it be required to find the straight line that best fits the following 24 points: (1, 2.3394), (2, -0.4190), (3, 3.9360), (4, 5.7104), (5, 4.9803), (6, 5.0956), (7, 5.6350); (8, 5.7270), (9, 6.6854), (10, 6.5562), (11, 8.2599), (12, 7.7398), (13, 8.4688), (14, 9.1321), (15, 9.7863), (16, 9.8558), (17, 10.3997), (18, 11.2892), (19, 11.4872), (20, 11.8735), (21, 11.0908), (22, 13.5829), (23, 13.6063), (24, 13.7849).

The least-squares regression line is found to be $\hat{Y}_2 = 1.860 + 0.5065X$. The kurtosis of the vertical deviations of the given points from this provisional regression line is $K = 9.326$. Since $K > 3.8$, the error distribution is taken to be double exponential, and hence the adaptive robust estimates of the regression coefficients are the least absolute values estimates. These estimates are not always unique (see Harter [5]). In the problem under consideration, there are four limiting least absolute values regression lines:

$$\hat{Y}_{1,1} = 2.2212 + 0.48768X,$$

$$\hat{Y}_{1,2} = 2.1456 + 0.49166X,$$

$$\hat{Y}_{1,3} = 2.2758 + 0.47988X,$$

$$\hat{Y}_{1,4} = 2.1908 + 0.48414X.$$

Lines 1 and 4 intersect at the point $(-8.59, -1.97)$ and lines 2 and 3 intersect at the point $(11.05, 7.58)$. As recommended by Harter [5], the line joining these two points of intersection, $\hat{Y}_1 = 2.21 + 0.486X$, is taken as the compromise least absolute values regression line, and hence as the adaptive robust regression line.

References

[1] Bourdon, G. A. (1974). A Monte Carlo Sampling Study for Further Testing of the Robust Regression Procedure Based upon the Kurtosis of the Least Squares Residuals. M. S. thesis (GSA/MA/74D-1), Air Force Institute of Technology, Wright-Patterson Air Force Base, Ohio.

[2] Brown, G. W. and Mood, A. M. (1951). In *Proc. 2nd Berkeley Symp. Math. Statist. Prob.* University of California Press, Berkeley, Calif., pp. 159–166.

[3] Forth, C. R. (1974). Robust Estimation Techniques for Population Parameters and Regression Coefficients. M.S. thesis (GSA/MA/74-1), Air Force Institute of Technology, Wright-Patterson Air Force Base, Ohio. AD777865.

[4] Harter, H. L. (1974). *Int. Statist. Rev.*, **42**, 147–174, 235–264, 282; *ibid.*, **43**, 1–44, 125–190, 269–278 (1975); *ibid.*, **44**, 113–159 (1976).

[5] Harter, H. L. (1977). *Commun. Statist. A*, **6**, 829–838.

[6] Harter, H. L., Moore, A. H. and Curry, T. F. (1979). *Commun. Statist. A*, **8**, 1473–1491.

[7] Hogg, R. V. (1972). *J. Amer. Statist. Ass.*, **67**, 422–424.

[8] Hogg, R. V. (1974). *J. Amer. Statist. Ass.*, **69**, 909–923; discussion, 923–927.

[9] Jackson, D. (1924). *Ann. Math.*, (2), **25**, 185–192.

[10] Jorgenson, L. W. (1973). Robust Estimation of Location and Scale Parameters. M.S. thesis (GSA/MA/73-2), Air Force Institute of Technology, Wright-Patterson Air Force Base, Ohio. AD766882.

[11] Rugg, B. J. (1974). Adaptive Robust Estimation of Location and Scale Parameters Using Selected Discriminants. M.S. thesis (GSA/MA/74D-3), Air Force Institute of Technology, Wright-Patterson Air Force Base, Ohio.

(ADAPTIVE METHODS
LEAST SQUARES
LINEAR REGRESSION
MAXIMUM LIKELIHOOD
METHOD OF LEAST ABSOLUTE VALUES
METHOD OF LEAST pTH POWERS
MINIMAX METHOD)

H. LEON HARTER

HARTLEY, HERMAN OTTO

Born: April 13, 1912, in Berlin, Germany.

Died: December 30, 1980, in Durham, North Carolina.

Contributed to: analysis of variance, numerical analysis, sampling distributions and theory, statistical methods.

H. O. Hartley was one of the most prominent and influential statisticians of the twentieth century. He was born in Germany and in 1933 obtained his Ph.D. in mathematics in Berlin. In 1934 he emigrated to England, where his collaboration with Egon Pearson* after the war led to the extensive and well-known *Biometrika Tables for Statisticians* [11, 12].

In 1953, Hartley visited the United States, and stayed to become Research Professor at Iowa State College; from 1963 to 1977 he founded and administered the Institute of Statistics at Texas A&M University, moving in "retirement" in 1979 to a full-time position at Duke University, where he served until his death.

Hartley's versatile and penetrating mind produced significant contributions in many branches of statistics over a period of 40 years. These include data processing, numerical analysis* and tabulation (see [4–6] and his article COMPUTERS AND STATISTICS in Vol. 2 of this encyclopedia). They include innovative results in the analysis of variance* [2, 4, 8], estimation* [8, 9, 10], sampling theory [3, 7], the study of sampling distributions [1, 4], hypothesis testing* [2], variance components* [4], sample surveys* [9], and the development of statistical methods in fields of application such as carcinogenic experiments [10]. The references given here are only an illustrative cross section of his research publications.

Professor Hartley was a Fellow of the American Statistical Association* and in 1979 served as its President (see ref. 6). He also served as President of the Eastern North American Region of the Biometric Society* and was a Fellow of the Institute of Mathematical Statistics*; among many honors he was the recipient in 1973 of the S. S. Wilks* medal.

References

[1] David, H. A., Hartley, H. O., and Pearson, E. S. (1954). *Biometrika*, **41**, 482–493.

[2] Hartley, H. O. (1950). *Biometrika*, **37**, 308–512.

[3] Hartley, H. O. (1966). *J. Amer. Statist. Ass.*, **61**, 739–748.

[4] Hartley, H. O. (1967). *Biometrics*, **23**, 105–114.

[5] Hartley, H. O. (1976). In *On the History of Statistics and Probability*, D. B. Owen, ed. Marcel Dekker, New York, pp. 419–442.

[6] Hartley, H. O. (1980). *J. Amer. Statist. Ass.*, **75**, 1–7. (Hartley's presidential address to the American Statistical Association; a thought-provoking essay for anybody concerned with statistics as a science and as a profession.)

[7] Hartley, H. O. and Rao, J. N. K. (1962). *Ann. Math. Statist.*, **33**, 350–374.

[8] Hartley, H. O. and Rao, J. N. K. (1967). *Biometrika*, **54**, 93–108.

[9] Hartley, H. O. and Rao, J. N. K. (1968). *Biometrika*, **55**, 547–557.

[10] Hartley, H. O. and Sielken, R. L., Jr. (1977). *Biometrika*, **33**, 1–30.

[11] Pearson, E. S. and Hartley, H. O. (1966). *Biometrika Tables for Statisticians*, Vol. 1, 3rd ed. Cambridge University Press, Cambridge.

[12] Pearson, E. S. and Hartley, H. O. (1972). *Biometrika Tables for Statisticians*, Vol. 2. Cambridge University Press, Cambridge.

HARTLEY'S $F_{\max}$ TEST

A simple test for testing the hypothesis of homogeneity of variances,

$$\sigma_1^2 = \sigma_2^2 = \cdots = \sigma_r^2,$$

in a single-factor experiment. If the number of observations for each one of the treatments (populations) is the same, say n, the test statistic is given by

$$F_{\max} = \frac{\text{largest of the } k \text{ within-class variances}}{\text{smallest of the } k \text{ within-class variances}} = \frac{s^2 \text{ largest}}{s^2 \text{ smallest}}.$$

Assuming independent random samples from a normal population under the hypothesis $\sigma_i^2 = \sigma_j^2$, $i \neq j = 1, \ldots, r$, the critical values of this statistic at the 1% and 5% levels were tabulated by Hartley [2] and are reproduced, for example, in Table 31 of the *Biometrika Tables for Statisticians*, Vol. 1 (2nd ed.); see also ref. 1. The parameters of this distribution are r (the number of populations) and $n - 1$, which is referred to as degrees of freedom* associated with each population. If the number of observations in each of the classes is not constant, but relatively close to being equal, the largest $n_j - 1$ is used instead of $n - 1$ for the degrees of freedom, which leads to a positive bias (i.e., rejecting the alternative more often than should be the case). Alternatively, the average sample size could be used. As with the Bartlett-M test (*see* BARTLETT'S TEST), Hartley's test is quite sensitive to departures from normality assumptions (*see also* COCHRAN'S (TEST) STATISTIC).

References

[1] David, H. A. (1952). *Biometrika*, **39**, 422–424. (Presents upper 5 and 1% points of $F_{\max}$.)

[2] Hartley, H. O. (1950). *Biometrika*, **37**, 308–312 (original paper).

[3] Winer, B. J. (1971). In *Statistical Principles in Experimental Design*, 2nd ed. McGraw-Hill, New York, pp. 206–207. (Presents a numerical example.)

(BARTLETT'S TEST OF HOMOGENEITY OF VARIANCES
COCHRAN'S (TEST) STATISTIC
HETEROSCEDASTICITY)

HAT MATRIX

This is the matrix in the analysis of the general linear model* which converts n observed response or dependent variables $\mathbf{y}$ by ordinary least squares* into predicted or fitted values $\hat{\mathbf{y}}$, based on a set of given or control variables $\mathbf{X}$, where, say,

$$\mathbf{X} = (\mathbf{1}\mathbf{x}_1\mathbf{x}_2 \cdots \mathbf{x}_k);$$

$\mathbf{1}$ is a $n \times 1$ vector of 1's, and $\mathbf{x}_i$ a known $n \times 1$ vector; $i = 1, \dots, k$.

If the model is

$$\mathbf{y} = \mathbf{X}\boldsymbol{\beta} + \mathbf{Z}, \qquad E(\mathbf{Z}) = 0,$$

then the normal equations are

$$\mathbf{X}'\mathbf{X}\hat{\boldsymbol{\beta}} = \mathbf{X}'\mathbf{y}.$$

For nonsingular $\mathbf{X}'\mathbf{X}$ these solve to give

$$\hat{\boldsymbol{\beta}} = (\mathbf{X}'\mathbf{X})^{-1}\mathbf{X}'\mathbf{y};$$

$\mathbf{H}$ is the hat matrix, where the fitted values are

$$\hat{\mathbf{y}} = \mathbf{H}\mathbf{y} = \mathbf{X}\hat{\boldsymbol{\beta}} = \mathbf{X}(\mathbf{X}'\mathbf{X})^{-1}\mathbf{X}'\mathbf{y}.$$

Then $\mathbf{H}$ is a symmetric, idempotent* projection matrix*. The (i, j)th element h_{ij} measures the influence of the data element y_j on the fitted element $\hat{y}_i$, the diagonal elements h_{ii} most directly so; $i, j = 1, \dots, n$. Thus $\mathbf{H}$ is useful as a diagnostic tool for identifying multivariate outliers*. One can show that

$$h_{ii} = h_{ii}^2 + \sum_{i \neq j}\sum h_{ij}^2,$$

$$\sum_i h_{ii} = p = \text{rank of } \mathbf{X},$$

so that $0 \leqslant h_{ii} \leqslant 1$ and $h_{ij} = 0$ whenever $h_{ii} = 0$ or $h_{ii} = 1$. Hoaglin and Welsch [1] discuss these and other properties with several examples. If $\mathbf{H}$ is $n \times n$, a rule of thumb is to call h_{ii} "large" (indicating y_i as an exceptional data point) if $h_{ii} > 2p/n$.

Reference

[1] Hoaglin, D. C. and Welsch, R. E. (1978). *Amer. Statist.*, **32**, 17–22; correction, *ibid.*, 146.

(GENERAL LINEAR MODEL
REGRESSION DIAGNOSTICS)

HAUSDORFF DIMENSION

BACKGROUND AND DEFINITION

The notion of dimension is implicit in the most fundamental mathematics. A straight line has dimension one, a filled-in square has dimension two, a solid cube has dimension three, and so forth. There are situations, however, in which the limitation of using only integers to describe the dimension of some set is unnecessarily restrictive. Such situations arose initially (around 1900) in pure mathematics, then later in the study of stochastic processes*. More recently, following the establishment of this rather theoretical background, it has been noticed that such situations also seem to arise in nature itself, and so in the mathematical modeling of certain natural phenomena.

The interest of pure mathematicians in this area arose at the same time as their attention was drawn to objects such as plane-filling *Peano curves*. In 1890, Peano [7] showed that a single point, moving continuously over a square, could (in a finite time) reach any point within it. Since curves were generally thought of as one-dimensional objects, while squares are two-dimensional, such curves seemed to indicate the possibility of a "continuity" of dimension, as it passed from one integer to another. At about the same time mathematical interest became focused on *Cantor-type* sets, such as the classic Cantor ternary set. This is simply formed in stages, by taking the interval [0, 1] and as first step deleting the middle third

$(\frac{1}{3}, \frac{2}{3})$. Each of the remaining two pieces (i.e. $[0, \frac{1}{3}]$, and $[\frac{2}{3}, 1]$) then has its middle third deleted, leaving four pieces, of which the middle thirds are deleted, and so on, ad infinitum. The resulting set can easily be shown to have zero length (Lebesgue measure), while at the same time its points can be placed in one-to-one correspondence with the whole interval [0, 1] from which it was formed. Thus the Cantor ternary set seemed to have neither dimension zero (as a finite, or even infinite but countable, set of points would) nor dimension one, since its length was zero.

To cope with these and similar problems, Hausdorff [4], in 1919, introduced a notion of fractional dimension that later came to be named for him. To describe it, let A be a subset of some n-dimensional Euclidean space, $\mathbb{R}^n$. An open ball in $\mathbb{R}^n$ is a set of the form $\{\mathbf{x} \in \mathbb{R}^n : |\mathbf{x} - \mathbf{y}| < \epsilon\}$ for some $\mathbf{y}$ and ϵ, where $|\cdot|$ denotes Euclidean distance. The number 2ϵ is the *diameter* of the sphere. Consider the expression

$$S_\alpha(A) = \lim_{\epsilon \downarrow 0} \inf \sum (\text{diameter } B_i)^\alpha,$$

where α is any real number. For every $\epsilon > 0$ the infimum is taken over all collections of open balls $\{B_i\}$ whose union covers A and for which the diameter of each B_i is not greater than ϵ; the sum is taken over each particular collection $\{B_i\}$.

As α increases, $S_\alpha(A)$ decreases, and it can be shown that $S_\alpha(A)$ takes at most three values; zero, infinity, and for at most one value of α, a nonzero finite value. This one value of α, at which $S_\alpha(A)$ changes from being infinite to being finite, is called the *Hausdorff dimension* of A and is denoted by $\dim(A)$. More formally, $\dim(A) = \inf\{\alpha : S_\alpha(A) < \infty\} = \sup\{\alpha : S_\alpha(A) = 0\}$. When $\dim(A)$ turns out to be an integer, it yields what we normally think of as dimension. Furthermore, in this case, $S_\alpha(A)$ is, except for a constant, the α-dimensional Lebesgue measure of A [e.g., if A is a straight line, $S_1(A)$ = length of A; if A is a two-dimensional set, $S_2(A)$ is proportional to its area, etc.].

The power of this concept, however, lies in situations when $\dim(A)$ is not an integer. For example, the Cantor ternary set has dimension $\log 2/\log 3 = 0.63 \ldots$, confirming intuition, which seems to say that this set lies somewhere between an ordinary set of points and an interval. The graph of the space-filling Peano curve mentioned above turns out to have Hausdorff dimension two. Similarly constructed, but not space-filling curves exist with graphs of dimension anywhere between one and two. An example is given by the curve discovered by von Koch [9] in 1904, obtained by cutting pieces out of a triangle (ad infinitum) as depicted in Fig. 1. The depicted curve has dimension $\log 4/\log 3 = 1.26 \ldots$, although in general the dimension increases from 1 to 2 as θ decreases from π to $\pi/2$.

STOCHASTIC PROCESSES

Hausdorff dimension theory has found its primary application in statistics as a tool for describing various aspects of the sample functions of certain stochastic processes*. The most classic example is the case of the

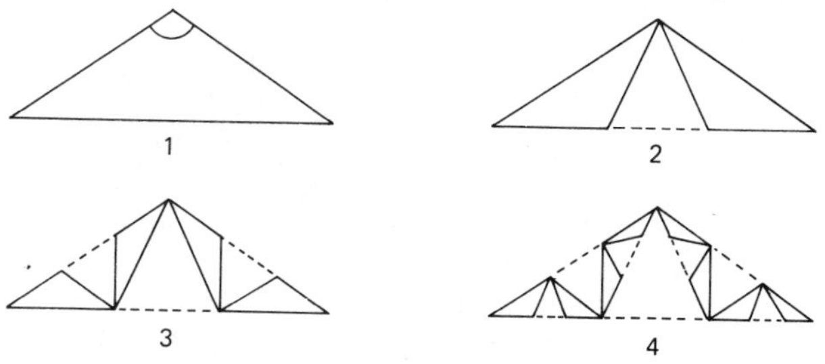

Figure 1 Constructing a von Koch curve.

Brownian motion*, $B(t)$, $t \geqslant 0$. The sample functions of a Brownian motion are of unbounded variation in every finite time interval, and so have infinite arc length; furthermore, its *zero set*, $\{t : 0 \leqslant t \leqslant 1, B(t) = 0\}$, is infinite. These facts seem to indicate that the graph of the Brownian motion has dimension exceeding one. In 1955, S. J. Taylor [8] showed that this was in fact the case, proving that the graph of B has dimension $1\frac{1}{2}$, while the dimension of its zero set was $\frac{1}{2}$. This result spawned an impressive literature of the Hausdorff dimension of sets such as the graph and zero set of stochastic processes, with much of the first decade and a half of this literature authored or coauthored by Taylor. In this earlier literature Taylor's result was extended to include many Gaussian processes and processes with independent increments.

Much of recent literature is concerned with *random fields** (i.e., stochastic processes for which "time" is multidimensional), often even lifting the level of generality to vector-valued fields.

The random field analog of the Brownian motion is a process called the *Brownian sheet* or *multiparameter Wiener process*. This field, which we denote by $B(t_1, \dots, t_n)$, is a Gaussian process* defined on the positive orthant of n-dimensional Euclidean space [i.e., on the set $\{(t_1, \dots, t_n): t_i \geqslant 0, i = 1, \dots, n\}$]. It has mean zero and covariance function

$$E\{X(s_1, \dots, s_n)X(t_1, \dots, t_n)\} = \prod_{i=1}^{n} \min(s_i, t_i).$$

Figure 2 shows an example of a Brownian sheet on the unit square in the plane; here the field has been "chopped off" at the level zero, and all negative values replaced by zero. This example shows a striking resemblance to a rough landmass jutting out of the sea, and, indeed, this resemblance has led to the formal modeling of such phenomena by spatial fields closely related to the Brownian sheet, as noted below. Let us consider the zero contour line $\{(s,t): B(s,t) = 0\}$ and a method of quantifying the "size" of this line. In particular, set $L(x) = \{(s,t): 0 \leqslant s, t \leqslant x, B(s,t) = 0\}$.

Figure 3*a* displays the zero-level contour for the example of Fig. 2; this is the set $L(1)$. The figure suggests that a possible means of quantifying the size of $L(1)$ would be to count the number of "islands" or "lakes" that it encloses. However, Fig. 3*b* indicates that this is a poorly defined concept, for here is depicted again the lower left-hand corner of Fig. 3*a*, i.e., $L(\frac{1}{2})$, magnified by a factor

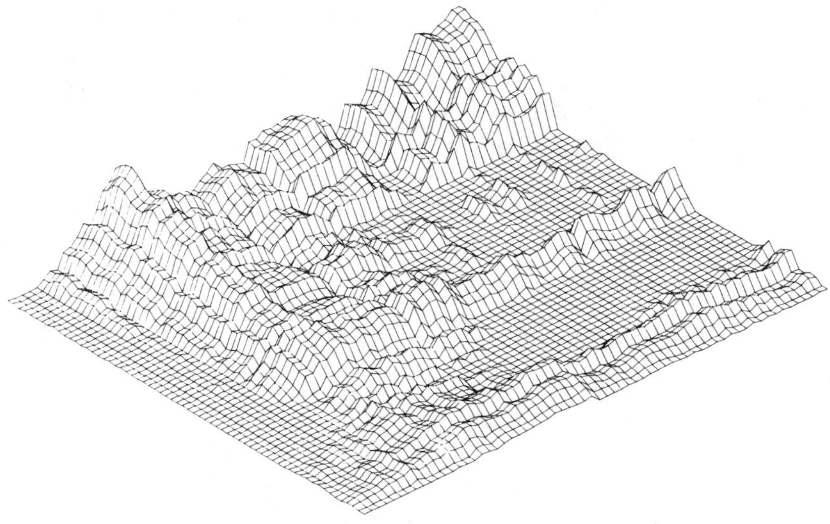

Figure 2 Brownian sheet on the unit square. (Reproduced by kind permission of the Applied Probability Trust.)

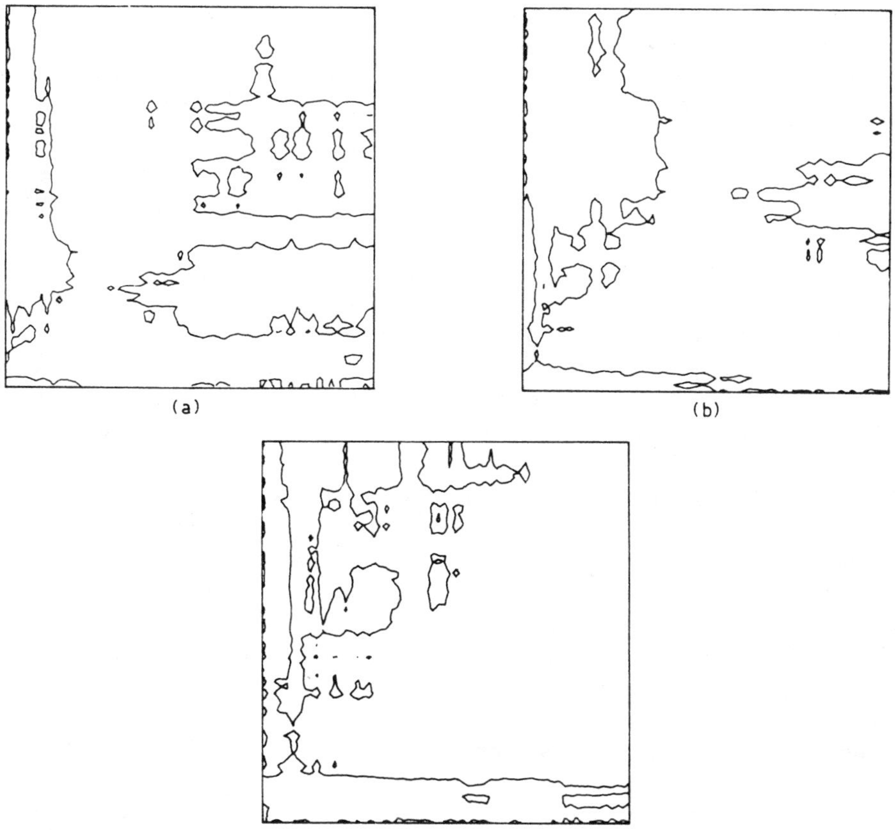

Figure 3 Zero-level curves of the Brownian sheet on (*a*) the unit square; (*b*) $(0,\frac{1}{2})\times(0,\frac{1}{2})$; (*c*) $(0,\frac{1}{4})\times(0,\frac{1}{4})$.

of 2. However, whereas $L(1)$ was obtained by observing the sheet on a square grid of points separated by a distance of $\frac{1}{64}$, the distance between points used to obtain Fig. 3*b* was only $\frac{1}{128}$. A brief glance shows that the magnification in Fig. 3*b* indicates the existence of further islands and lakes which were not noticeable in the original diagram. In Fig. 3*c*, $L(\frac{1}{4})$ is shown, again magnified, but this time the gap in the underlying grid is only $\frac{1}{256}$. The appearance of new islands and lakes is even more noticeable here. This process can be continued indefinitely and new islands and lakes would appear without end, thus making nonsense of any attempt to either count them, or measure their area or the length of their boundaries. Such phenomena are indicative of the fact that the Hausdorff dimension of these contour lines is greater than one. It has recently been shown that the zero set of an n-parameter Brownian sheet has dimension $n-\frac{1}{2}$. Similar dimension results are known to hold for other Gaussian fields, but there is not yet the completeness for fields of the aforementioned theory for processes on the line.

PROBABILITY

The concepts of Hausdorff dimension turn out to have a curious connection with the notion of entropy*, an important concept of both ergodic* and information* theory. Billingsley [2] has shown that the dimensions of sets of numbers in [0, 1] with particular types of continued fraction expansions are most easily calculated using entropy-type concepts. Although the results so obtained are more within the realm of pure mathematics than statistics, the methodology of the proofs as well as their relevance to information theory places them well within the interest of the more theoretical probabilist.

MODELING

Although mathematical models possessing interesting dimensional properties have long been studied (Brownian motion being the most obvious and important of these), it is only in the last 15 years that dimension, per se, has been considered an important consideration in such models. The most obvious application, via models of the type exemplified in Figs. 2 and 3, has been in modeling geographical phenomena. Such models seem particularly appropriate in this area, for the phenomenon of changing information with changing scale discussed above in relation to Fig. 2 has long been observed by cartographers when dealing with actual maps. Other geographical phenomena are also understandable only in terms of such models (*see* GEOGRAPHY, STATISTICS IN).

By far the most active worker in the area of applying Hausdorff dimension and similar concepts to describe real-life phenomena has been Benoit Mandelbrot [5, 6]. He convincingly argues that a wide variety of natural phenomena can only be understood in terms of fractals*, a term he has coined to describe objects possessing noninteger Hausdorff dimension. In a recent and fascinating book [6] he has considered applications in such widely diverse areas as geomorphology, turbulence, the distribution of stellar matter, and vascular geometry.

MATHEMATICAL METHODOLOGY

The principles underlying the derivation of the dimension of particular random sets are basically straightforward, but unfortunately the details of these derivations tend to be technically exacting. Many proofs of results of the type described above for stochastic processes rely on a theorem proved in 1935 by Frostman [3] about a relationship between capacities in potential theory and Hausdorff dimension. This result states, in essence, that if a set $E \subset \mathscr{R}^n$ has positive β-capacity, then $\dim(E)$ is at least β. [Positive β-capacity means that there exists a nonzero measure μ, supported by E, for which $\int_E \int_E |\mathbf{x} - \mathbf{y}|^{-\beta}\, d\mu(\mathbf{x})\, d\mu(\mathbf{y}) < \infty$.] Since it is generally not too difficult to determine when a stochastic process-generated random set does have finite β-capacity, this result provides a straightforward technique for obtaining lower bounds on dimensions. Corresponding upper bounds are then obtained from a principle which basically states that the smoother the sample paths of a process, the smaller will be the dimension of such objects as its graph, zero set, etc. This principle can be carefully formulated in terms of smoothness, or *Lipschitz conditions* on the sample functions. [A function $f(t)$ is said to satisfy a Lipschitz condition of order γ, $0 < \gamma < 1$ throughout some interval if $|f(t) - f(s)| \leqslant C|t - s|^{\gamma}$ for some finite C and all t, s belonging to the interval.]

Intimately involved in this methodology is the concept of the local time* of a stochastic process. A general principle, initially enunciated by Berman [1], is that smooth local times are associated with irregular sample paths and high Hausdorff dimensions.

References

[1] Berman, S. M. (1969). *Trans. Amer. Math. Soc.*, **137**, 277–299.

[2] Billingsley, P. (1965). *Ergodic Theory and Information*. Wiley, New York.

[3] Frostman, O. (1935). *Medd. Lunds Univ. Math. Semin.*, **3**, 1–118.

[4] Hausdorff, F. (1919). *Math. Ann.*, **79**, 157–179.

[5] Mandelbrot, B. (1977). *Fractals: Form, Chance, and Dimension*. W. H. Freeman, San Francisco. [This may well turn out to be one of the most influential works in stochastic modeling/probability of recent times. In what is a long (360 pages) essay, rather than a book, Mandelbrot convincingly argues that Hausdorff dimension is an extremely useful and totally natural concept. The work requires little formal background of its readers, but demands an ability to accept an immense diversity of new ideas at a dizzying rate.]

[6] Mandelbrot, B. (1982). *The Fractal Geometry of Nature*. W. H. Freeman, San Francisco.

[7] Peano, G. (1890). *Math. Ann.*, **36**, 157–160. (Translated in *Peano's Selected Works*, H. C. Kennedy, ed. Toronto University Press, Toronto, 1973.)

[8] Taylor, S. J. (1955). *Proc. Camb. Philos. Soc.*, **51**(Pt. II), 265–274.

[9] von Koch, H. (1904). *Ark. Mat. Astron. Fys.*, **1**, 681–704.

FURTHER READING

Adler, R. J. (1981). *The Geometry of Random Fields*. Wiley, New York. (Two chapters of this book are devoted to Hausdorff dimension, and related problems, for Gaussian random fields. Although the exposition is self-contained, the reader needs a solid grounding in probability and stochastic processes.)

Gardner, M. (1976). Mathematical Games. *Sci. Amer.*, Dec., pp. 124–133. (This is a popular article, requiring no background. As is usual with Gardner, the article is entertaining as well as informative.)

Rogers, C. A. (1970). *Hausdorff Measures*. Cambridge University Press, Cambridge. (A technical book, for the measure theorist only.)

(BROWNIAN MOTION
FRACTALS
RANDOM FIELDS)

R. J. ADLER

HAUSMAN SPECIFICATION TEST

There are several alternative definitions of a specification test procedure as opposed to conventional procedures which are used to test a set of parametric restrictions. Hausman [2] proposed a general form of specification test designed to test the implications of an hypothesis in terms of bias or inconsistency of an estimator.

As an introduction to Hausman's procedure, consider the regression* model $y = X\beta + \epsilon$. It is usually assumed that the conditional expectation $E(\epsilon \mid X)$ is zero. If this assumption is not satisfied, the ordinary least-squares* (OLS) estimator of β is inconsistent. On the other hand, one can find an instrumental variable* (IV) estimator of β which is consistent under both the hypotheses $E(\epsilon \mid X) \neq 0$ and $E(\epsilon \mid X) = 0$. Following Hausman [2], one may construct a test statistic for the problem of testing the null hypothesis $E(\epsilon \mid X) = 0$ against the alternative hypothesis* $E(\epsilon \mid X) \neq 0$ based on the following idea. Under the usual regularity assumptions, the OLS estimator of β is consistent, asymptotically normal, and asymptotically efficient (i.e., attains the asymptotic Cramér–Rao bound*). Under the alternative hypothesis, this estimator will be inconsistent. In contrast, the above-mentioned consistent IV estimator of β will not be asymptotically efficient under the null hypothesis. The specification test proposed by Hausman [2] is a chi-square test* based on the asymptotic distribution of the difference between the IV and OLS estimators of β. He also showed that the asymptotic variance of this difference is equal to the difference between the asymptotic variances of these two estimators. More generally, Hausman has suggested a general form of specification test, that is, a procedure for testing $E(\epsilon \mid X) = 0$ against $E(\epsilon \mid X) \neq 0$, based on the same idea of examining the difference between a consistent estimator (under both the null and alternative hypotheses) and an efficient estimator under the null hypothesis but otherwise inconsistent under the alternative hypothesis. It is worth noting that the theory presented by Hausman concentrates on the large-sample case where it is assumed that the asymptotic distribution of all the estimators under consideration is normal (*see* ASYMPTOTIC NORMALITY). Notice that the basic idea underlying this specification test procedure had also been put forward by Durbin [1].

More recently Holly [4] showed that Hausman's specification test could be derived from a maximum likelihood* approach within a general framework of nonlinear models. The advantage of this approach is that its asymptotic nonnull distribution can also be derived; this provides a way to compare it with more conventional test procedures in terms of asymptotic local power.

We summarize the work of Holly [4]. Some of the results given below appear also in Hausman and Taylor [3]. Suppose that we have a family of models giving for a sample of size n a log-likelihood $\mathscr{L}(\boldsymbol{\theta}, \boldsymbol{\gamma})$ depending on two vectors $\boldsymbol{\theta}$ and $\boldsymbol{\gamma}$ of p and q parameters, respectively; $\boldsymbol{\theta}$ denotes a vector of parameters of interest, and $\boldsymbol{\gamma}$ a vector of nuisance parameters. The null hypothesis is H_0: $\boldsymbol{\theta} = \boldsymbol{\theta}^0$. Holly [4] investigated the distribution of Hausman's test statistic on the sequence of alternative hypotheses $\boldsymbol{\theta}_n^0 = \boldsymbol{\theta}^0 + n^{-1/2}\boldsymbol{\beta}$, where $\boldsymbol{\beta}$ is a given vector. The "true value" of $\boldsymbol{\gamma}$ is denoted by $\boldsymbol{\gamma}^0$.

Corresponding to $\boldsymbol{\theta}$ and $\boldsymbol{\gamma}$, the information matrix $\mathscr{I}$ is partitioned as

$$\mathscr{I} = \begin{pmatrix} \mathscr{I}_{\theta\theta} & \mathscr{I}_{\theta\gamma} \\ \mathscr{I}_{\gamma\theta} & \mathscr{I}_{\gamma\gamma} \end{pmatrix}$$

Denote by $\hat{\boldsymbol{\gamma}}^0$ the estimator of $\boldsymbol{\gamma}$ under the restriction $\boldsymbol{\theta} = \boldsymbol{\theta}^0$ and by $\hat{\boldsymbol{\gamma}}$ the unconstrained estimator of $\boldsymbol{\gamma}$. We have the following general result (see Holly [4] and Hausman and Taylor [3]).

The distribution of Hausman's test statistic

$$n(\hat{\boldsymbol{\gamma}} - \hat{\boldsymbol{\gamma}}^0)'\left[\left(\mathscr{I}_{\gamma\gamma} - \mathscr{I}_{\gamma\theta}\mathscr{I}_{\theta\theta}^{-1}\mathscr{I}_{\theta\gamma}\right)^{-1} - \mathscr{I}_{\gamma\gamma}^{-1}\right]^{-}(\hat{\boldsymbol{\gamma}} - \hat{\boldsymbol{\gamma}}^0)$$

converges to the noncentral chi-square distribution with degrees of freedom equal to the rank of $\mathscr{I}_{\gamma\theta}$ and noncentrality parameter given by*

$$\lambda^2 = \boldsymbol{\beta}'\mathscr{I}_{\theta\gamma}\mathscr{I}_{\gamma\gamma}^{-1}\mathscr{A}^{-}\mathscr{I}_{\gamma\gamma}^{-1}\mathscr{I}_{\gamma\theta}\boldsymbol{\beta},$$

$$\mathscr{A} = \mathscr{I}_{\gamma\gamma}^{-1}\mathscr{I}_{\gamma\theta}\left(\mathscr{I}_{\theta\theta} - \mathscr{I}_{\theta\gamma}\mathscr{I}_{\gamma\gamma}^{-1}\mathscr{I}_{\gamma\theta}\right)^{-1}\mathscr{I}_{\theta\gamma}\mathscr{I}_{\gamma\gamma}^{-1}.$$

In addition, λ^2 is invariant for any choice of the g-inverse.

Notice that $(\mathscr{I}_{\gamma\gamma} - \mathscr{I}_{\gamma\theta}\mathscr{I}_{\theta\theta}^{-1}\mathscr{I}_{\theta\gamma})^{-1}$ and $\mathscr{I}_{\gamma\gamma}^{-1}$ are the asymptotic variance of $n^{1/2}(\hat{\boldsymbol{\gamma}}^0 - \boldsymbol{\gamma}^0)$ and $n^{1/2}(\hat{\boldsymbol{\gamma}} - \boldsymbol{\gamma}^0)$, respectively.

It is interesting to consider the case where the number of nuisance parameters is less or equal to the number of parameters under test. If $\mathscr{I}_{\gamma\theta}$ has rank q, the generalized inverses* which occur in the theorem are the inverses of the matrices.

Although the derivation of Hausman's test from the maximum likelihood approach has been obtained under a sequence of local alternative hypotheses, the null hypothesis actually tested by this procedure is H_0^*: $\mathscr{I}_{\gamma\gamma}^{-1}\mathscr{I}_{\gamma\theta}\boldsymbol{\beta} = \mathbf{0}$ against $H_1^*: \mathscr{I}_{\gamma\gamma}^{-1}\mathscr{I}_{\gamma\theta}\boldsymbol{\beta} \neq \mathbf{0}$. Notice that H_0^* reduces to H_0 whenever we have simultaneously $q \geqslant p$ and rank $(\mathscr{I}_{\gamma\theta}) = p$.

Now, suppose that H_0 and H_0^* are not equivalent, and that Hausman's procedure is used for the problem of testing the null hypothesis H_0. Two main results can be derived. First, if n is sufficiently large and $\boldsymbol{\theta}_n^0$ not near $\boldsymbol{\theta}^0$, the power of Hausman's test will not be near 1 in all the directions of the parameter space. Thus there is a strong possibility that the test might not be consistent. Second, there exist directions for which the test has a better power than conventional procedures such as the likelihood ratio test*.

Finally, Hausman [2] suggested an alternative procedure for the specification error testing problem. He pointed out that in many situations the null hypothesis of no specification error may be tested in an expanded regression framework.

References

[1] Durbin, H. (1954). *Rev. Int. Statist. Inst.*, **22**, 23–32.

[2] Hausman, J. A. (1978). *Econometrica*, **46**, 1251–1271.

[3] Hausman, J. A. and Taylor, W. E. (1981). *Economics Letters*, **8**, 239–245.

[4] Holly, A. (1982). *Econometrica*, **50**, 749–759.

(ECONOMETRICS)

ALBERTO HOLLY

HAZARD PLOTTING

The old Chinese proverb "one picture is worth a thousand words" is exemplified by the practitioners of graphical* methods in statistics. One particular graphical method that is used in the analysis of reliability and survival data is *hazard plotting*, first introduced by Nelson [5]. The principal purpose of hazard plotting is to determine graphically how well a particular probability distribution, when characterized by its cumulative hazard function, fits a given set of failure data. The procedure allows for censoring* of the data.

It is first necessary to review some basic concepts of survival/reliability theory*. The *survival function* $S_T(t)$ is defined as the probability that an individual survives at least time t ($t > 0$). That is,

$$S_T(t) = \Pr\{T > t\},$$

where T is the random variable that describes the length of life of an individual.

The *hazard function* $\lambda_T(t)$ characterizes the instantaneous failure rate when $T = t$, conditioned on survival to time t, and may be expressed as

$$\lambda_T(t) = -S_T'(t)/S_T(t),$$

where $S_T'(t) = dS_T(t)/dt$.

The *cumulative hazard function* $\Lambda_T(t)$ is defined as

$$\Lambda_T(t) = \int_0^t \lambda_T(u)\,du = -\ln S_T(t),$$

since ordinarily, $S_T(0) = 1$. The usual graphic procedure places t on the y-axis and $\Lambda_T(t)$ on the x-axis. Thus t written as a function of Λ is

$$t(\Lambda) = S^{-1}\{\exp(-\Lambda)\},$$

where $S^{-1}(\cdot)$ is the inverse survival function; i.e., if $S(y) = p$, then $S^{-1}(\cdot)$ is the upper $100p$th percentile of the distribution of Y.

The function $t(\Lambda)$ is exhibited next for some of the most common parametric survival distributions; (*see also* HAZARD RATE CLASSIFICATION OF DISTRIBUTIONS).

1. **Exponential***. $\Lambda_T(t) = \lambda t$. Thus
$$t = \Lambda/\lambda, \qquad \lambda > 0,$$
a straight line with positive slope that goes through the origin of the (Λ, t) plane. Ordinary arithmetic graph paper is used for plotting.
2. **Weibull***. $\Lambda_T(t) = \lambda t^{\gamma}$. In this case,
$$\ln t = (-\ln \lambda + \ln \Lambda)/\gamma,$$
$$\lambda > 0, \quad \gamma > 0, \quad \Lambda > 0.$$
It follows that the graph of t vs. Λ is a straight line when plotted on log-log graph paper.
3. **Type I Extreme-Value Distribution*** (see Hahn and Shapiro [3]). $\Lambda_T(t) = \exp\{(t - \mu)/\sigma\}$. One sees that
$$t = \mu + \sigma \ln \Lambda,$$
$$\sigma > 0, \quad -\infty < \mu < \infty, \quad \Lambda > 0.$$
Thus the graph of t vs. Λ is a straight line when plotted on semilogarithmic graph paper.
4. **Lognormal Distribution***
$$\Lambda_T(t) = -\ln\left(1 - \Phi\left(\frac{\log t - \mu}{\sigma}\right)\right),$$
where $t > 0$, $-\infty < \mu < \infty$, $\sigma > 0$, $\Lambda > 0$, $\Phi(\cdot)$ is the cumulative standard normal distribution and the logarithm of t is measured to some convenient base, usually base 10 or the base of natural logarithms. It now follows that
$$\log t = \mu + \sigma \Phi^{-1}\{1 - \exp(-\Lambda)\},$$
where $\Phi^{-1}(\cdot)$ is the inverse of the cumulative standard normal distribution; e.g., if $\Lambda = 1$, $\Phi^{-1}(1 - e^{-1}) = 0.337$. Lognormal hazard paper is available, so that a plot of $\log t$ vs. Λ can be accomplished.
5. **Linear Hazard Rate Distribution*** (see Kodlin [4] or Gross and Clark [2])
$$\Lambda_T(t) = \lambda_0 t + \lambda_1 t^2$$
Here
$$t = \left(\sqrt{\lambda_0^2 + 4\Lambda\lambda_1} - \lambda_0\right)/2\lambda_1,$$
$$\lambda_0 > 0, \quad \lambda_1 > 0, \quad \Lambda > 0$$
The graph of t vs. Λ is the branch of a parabola in the first quadrant of the (Λ, t) plane that passes through the origin.

In order to use hazard plotting as a tool in describing survival data, it is necessary to determine Λ empirically. Suppose that n items are subjected to a life-testing* procedure wherein the failure time or censored time for each item is recorded. It is then possible to denote ordered observation times of these items as $t_{(1)} \leqslant t_{(2)} \leqslant \cdots \leqslant t_{(n)}$, where $t_{(i)}$, the actually selected observation

Table 1 Hazard Plot Calculations for a Weibull Fit

Reverse Rank	Failure Time	Log Failure Time	Hazard (%)	Cumulative[a] Hazard (%)	Log Cumulative Hazard
20	0.001	− 6.91	5.00	5.00	1.61
19	0.030	− 3.51	5.26	10.26	2.33
18	0.071	− 2.65	5.56	15.82	2.76
17	0.185	− 1.69	5.88	21.70	3.08
16	0.345	− 1.06	6.25	27.95	3.33
15	0.435	− 0.83	6.67	34.62	3.54
14	0.469	− 0.76	7.14	41.76	3.73
13	0.470	− 0.76	7.69	49.45	3.90
12	0.505	− 0.68	8.33	57.78	4.06
11	0.664	− 0.41	9.09	66.87	4.20
10	0.806	− 0.22	10.00	76.87	4.34
9	0.970	− 0.03	11.11	87.98	4.48
8	1.033	0.03	12.50	100.48	4.61
7	1.550	0.44	14.29	114.77	4.74
6	1.550	0.44	16.67	131.44	4.88
5	2.046^{+}	0.72^{+}			
4	3.532	1.26	25.00	156.44	5.05
3	7.057	1.95	33.33	189.77	5.25
2	9.098^{+}	2.21^{+}			
1	57.628	4.05	100.00	289.77	5.67

[a]The reader should note that the cumulative hazard may exceed 100%.

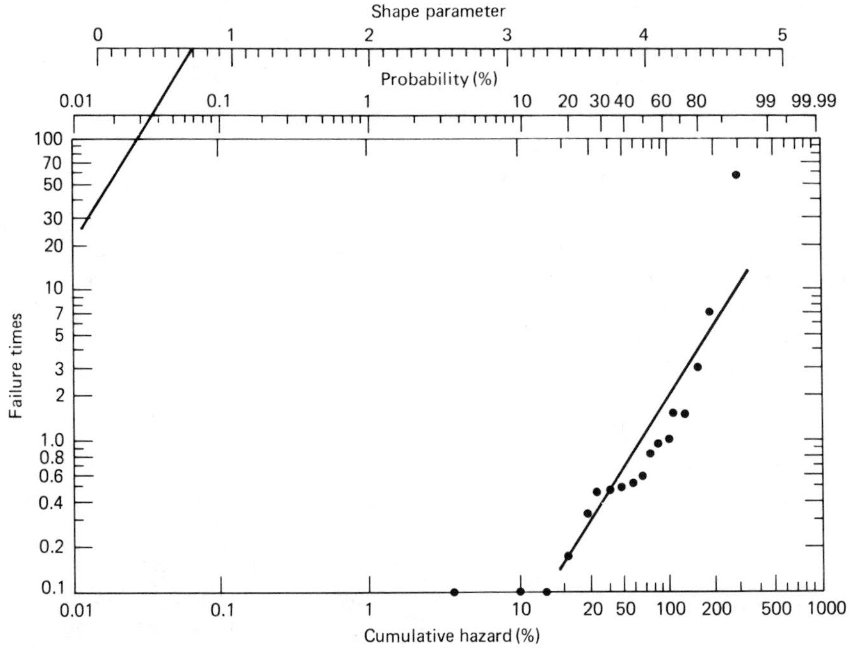

Figure 1 Hazard plot on Weibull paper. (From ref. 2.)

time of the ith item in the ordering is its time to failure if the item fails during the test or is its censoring time if it fails beyond the length of the test. Let d be the number of items that fail during the test and $t_{r(1)} \leqslant t_{r(2)} \leqslant \cdots \leqslant t_{r(d)}$ the corresponding observed failure times. Nelson [5] defines $r(1) > r(2) > \cdots > r(d)$ as the *reverse ranks*. As an illustration, suppose that a life test on electronic equipment yields $t_{(1)} = 600$ hours, $t_{(2)} = 805^+$ hours, $t_{(3)} = 818$ hours, $t_{(4)} = 902^+$ hours, and $t_{(5)} = 1031$ hours, where 805^+ hours and 902^+ hours are the test times of the two surviving items. In this case $n = 5$, $d = 3$, $t_{r(1)} = 600$ hours, $t_{r(2)} = 818$ hours, and $t_{r(3)} = 1031$ hours; further, $r(1) = 5$, $r(2) = 3$, $r(3) = 1$.

Since Λ generally involves one or more unknown parameter(s) a suitable substitute is required to obtain the requisite plotting positions. Nelson [5] indicates that when $t_{r(m)}$ is plotted on the y-axis, the corresponding plotting position for Λ on the x-axis is $\sum_{j=1}^{m}[r(j)]^{-1}$. Note that only those points are plotted for which a failure has occurred. The censored observations are accounted for in the determination of the reverse ranks $r(m)$, $m = 1, 2, \ldots, d$.

The steps of the plotting procedure are now summarized (see Gross and Clark [2, pp. 146–147]).

1. Decide on the theoretical survival distribution* to be fitted to the observed failure data.
2. Plot each failure time vertically against its corresponding cumulative hazard value on the horizontal axis on the suitably chosen hazard paper.
3. When special hazard paper is used (e.g., Weibull or lognormal), the fit of the theoretical distribution is adequate if the failure times reasonably lie on a straight line.
4. When ordinary arithmetic graph paper is used, the fit of the theoretical distribution is adequate if the failure times reasonably lie on the requisite curve (e.g., a parabola through the origin for the linear hazard rate distribution).
5. Although it is convenient to have the appropriate hazard paper, arithmetic graph paper can be used if a suitable transformation is made on the failure data.

As an example of hazard plotting, consider the following ordered sample to which a Weibull distribution is to be fitted: 0.001, 0.030, 0.071, 0.185, 0.345, 0.435, 0.469, 0.470, 0.505, 0.664, 0.806, 0.970, 1.033, 1.550, 1.550, 2.046^+, 3.532, 7.057, 9.098^+, 57.628. The values 2.046^+ and 9.098^+ are censored observations. These data first appeared in an article by Cohen [1] and were analyzed in a different context by Gross and Clark [2]. Table 1 describes the plotting procedure for these data and Fig. 1 shows their hazard plot.

To plot these data the cumulative hazard is plotted along the lower vertical axis and is read directly from Table 1 as is the corresponding failure time, which is plotted along the vertical axis. The circled dot appears at position 20 along the vertical axis (on a scale with upper limit 100) and 0 on the shape parameter scale. Finally, the empirical cumulative frequency distribution comprises the upper horizontal axis, recalling the relationship

$$F_T(t) = 1 - \exp\left[-\Lambda_T(t)\right].$$

To estimate the shape parameter, γ, of the fitted Weibull distribution, a straight line is drawn that is parallel to the fitted straight line so that it passes through the circled dot marked in the upper left-hand corner of the paper and through the shape parameter scale as shown in Fig. 1. The graphical estimate of γ is 0.7.

References

[1] Cohen, A. C. (1965). *Technometrics*, **7**, 579–588.

[2] Gross, A. J. and Clark, V. A. (1975). *Survival Distributions: Reliability Applications in the Biomedical Sciences*. Wiley, New York.

[3] Hahn, G. J. and Shapiro, S. S. (1967). *Statistical Models in Engineering*. Wiley, New York.

[4] Kodlin, D. (1967). *Biometrics*, **25**, 227–239.

[5] Nelson, W. B. (1972). *Technometrics*, **14**, 945–965.

(CENSORED DATA
GRAPHICAL REPRESENTATION OF DATA
HAZARD RATE CLASSIFICATION OF DISTRIBUTIONS
LIFE TESTING
RELIABILITY THEORY
SURVIVAL DISTRIBUTIONS)

ALAN J. GROSS

HAZARD RATE AND OTHER CLASSIFICATIONS OF DISTRIBUTIONS

The hazard rate (or the failure rate) concept is one mathematical way of describing aging. Hazard rates have been investigated extensively, particularly in reliability* studies and actuarial sciences*.

CONTINUOUS VARIABLES

If the lifetime X of a unit is a random variable (rv) with probability density function (PDF) $f(x)$ and cumulative distribution function (CDF) $F(x)$, then the reliability or survival function of the unit at time x is $S(x) = 1 - F(x)$. The *hazard rate function* (HRF) of the PDF $f(x)$ is defined by

$$\lambda_x(x) = \lambda(x) = \frac{f(x)}{1 - F(x)} = -\frac{S'(x)}{S(x)}$$

$$= -\frac{d[\ln S(x)]}{dx}.$$

The HRF is known as the *force of mortality** in actuarial work, and the *intensity function* in extreme value theory. In economics, its reciprocal for the half-normal distribution is known as "Mills' ratio*." A probabilistic interpretation of the HRF is that $\lambda(x)\,dx$ represents the conditional probability that a unit of age x will fail in the interval $(x, x + dx)$.

Several books (see refs. 2–4, 7, 8, 13, 14, 16) have discussed the hazard rate concept and its properties, including a generalization [4] to multivariate distributions. The PDFs and CDFs, which appear to have the same general shape, may have considerably different hazard rates and the probability distributions can be distinguished on this basis. Some typical hazard rates are the following:

An increasing hazard rate indicates that the unit of age x is more likely to fail in a given increment of time than it would be in the same increment of time at an earlier age. This would be the case if the unit is wearing out or deteriorating with age. Similarly, a decreasing HRF means that the unit is improving with age. A theoretical explanation of why a life distribution would under certain circumstances exhibit a decreasing hazard rate is given in Proschan [25]. Another typical hazard rate is a bathtub*-shaped or U-shaped function. This may arise due to three types of failures. Early failures are often related to errors during production or errors that have escaped quality control* measures. As items fail by these causes the HRF will tend to decrease. Chance failures occur more or less randomly and are often independent of product age. The HRF is often constant or nearly constant in this range. Wearout failures are due to the product simply wearing out (*see* FATIGUE MODELS). During this time the HRF tends to increase. The linear HRF is useful in studying survival times of certain human populations (*see* LINEAR HAZARD RATE DISTRIBUTION). Several examples of HRFs for mortality* survival times in biomedical situations are given in Gross and Clark [8, pp. 9–10].

The CDF $F(x)$ associated with the HRF $\lambda(x)$ can be determined by the relationship

$$F(x) = 1 - \exp\left\{-\int_{-\infty}^{x} \lambda(y)\,dy\right\}.$$

If the PDF $f(x)$ is truncated, so that $a < x \leqslant b$, then the HRF of the truncated distribution (see Elandt-Johnson and Johnson [7, pp. 54–55]) is

$$\lambda(x \mid a < X \leqslant b) = \frac{f(x)}{S(x) - S(b)}$$

$$= \lambda(x)\frac{S(x)}{S(x) - S(b)}.$$

If $b = \infty$, then $\lambda(x \mid X > a) = \lambda(x)$. The left-hand truncation does not change the HRF, whereas the right-hand truncation increases it.

Now consider series and parallel systems made up of m independent units. Let $\lambda_i(x)$, $\lambda_s(x)$, and $\lambda_p(x)$ be, respectively, the HRFs of an ith unit, the series system, and the parallel system. Then from Bain [2, pp. 114–117],

$$\lambda_s(x) = \sum_{i=1}^{m} \lambda_i(x),$$

and when $m = 2$ with $\lambda_1(x) = \lambda_2(x) = \lambda(x)$,

$$\lambda_p(x) = \lambda(x)\frac{2F(x)}{1 + F(x)},$$

where $F(x)$ is the CDF of a unit. Notice that $\lambda_p(x) < \lambda(x)$, but $\lambda_p(x) \to \lambda(x)$ as $x \to \infty$.

Now consider the case of mixed populations. Suppose that a certain kind of unit is received from k different suppliers, with a fraction p_i of units received from the ith supplier, where $\sum_{i=1}^{k} p_i = 1$. Suppose that the outgoing lots consist of random mixtures of these units. Then a unit selected at random from a lot has a life distribution with the HRF (see ref. 2 [pp. 117–118])

$$\lambda_m(x) = \frac{\sum_{i=1}^{k} \lambda_i(x) p_i S_i(x)}{\sum_{i=1}^{k} p_i S_i(x)}.$$

The function $\lambda_m(x)$ can be interpreted as the weighted average of the individual $\lambda_i(x)$, $i = 1, 2, \ldots, k$.

It is difficult to express a relatively simple CDF which will allow a bathtub-shape or a U-shaped hazard rate. One possibility is to determine the lower-order polynomial HRF which provides a good fit. Various models and a method of estimating the parameters are discussed in Bain [2, pp. 401–407]. A k-step piecewise constant HRF to approximate the bathtub shape to any desired degree of accuracy is given in ref. 6.

The distributions possessing increasing or decreasing HRFs play an important role in reliability studies. Barlow, Proschan, and others have investigated properties of these distributions. The following results are from refs. 3 (pp. 9–39) and 4 (52–126).

A CDF $F(x)$ is said to be an *increasing (decreasing) hazard rate* (IHR and DHR, respectively) *distribution* if its HRF $\lambda(x)$ is nondecreasing (nonincreasing) in x. Some properties are:

1. If rvs X_1 and X_2 are IHR with hazard rates $\lambda_1(x)$ and $\lambda_2(x)$, respectively, then $(X_1 + X_2)$ is also IHR with hazard rate
$$\lambda_h(x) \leqslant \min\{\lambda_1(x), \lambda_2(x)\}.$$
The DHR property is not preserved.
2. A mixture of DHR distributions is also DHR. This is not necessarily true for IHR distributions.
3. Parallel and series systems of identical IHR units are IHR. For the series systems the units do not have to be identical.
4. Order statistics* from an IHR distribution have IHR distributions. However, this is not true for spacings* from an IHR distribution. Order statistics from a DHR distribution are not necessarily DHR. However, spacings from a DHR are DHR.
5. The DHR $f(x)$ is a decreasing function. The IHR $f(x)$ need not be unimodal.

The exponential distribution*, with a constant hazard rate, is a natural boundary between IHR and DHR distributions. A number of bounds comparable to the well-known Chebyshev*-type bounds are available; a summary of these is given in Johnson and Kotz [13, pp. 284–287]. They are of the following type:

1. If F is IHR, and $F(\xi_p) = p$ (i.e., ξ_p is a $100p$th percentile), then
$$S(x) \geqslant \exp(-\alpha x), \quad x \leqslant \xi_p$$
$$\leqslant \exp(-\alpha x), \quad x \geqslant \xi_p$$
where $\alpha = -[\log(1-p)]/\xi_p$. The direction of the bound reverses when F is DHR.
2. If F is IHR with mean μ, then
$$S(x) \geqslant \exp(-x/\mu), \quad x < \mu$$
$$0, \quad x \geqslant \mu$$
$$S(x) \leqslant 1, \quad x \leqslant \mu$$
$$\exp(-wx), \quad x > \mu$$

where w depends on x and satisfies $1 - w\mu = \exp(-wx)$.

Moment inequalities, like the following, provide comparisons with the corresponding moments for the exponential distributions.

1. If F is IHR with rth moment μ_r, then

$$\begin{aligned} \mu_r &\leqslant \Gamma(r+1)\,\mu_1^r, & r &\geqslant 1 \\ &\geqslant \Gamma(r+1)\,\mu_1^r, & 0 &\leqslant r \leqslant 1. \end{aligned}$$

The direction of the inequalities reverses when F is DHR. A consequence, when $r = 2$, is that an IHR (DHR) distribution is more (less) peaked than the exponential distribution.

2. The mean life of a series system with IHR units whose means are μ_i $(i = 1, 2, \ldots, m)$ exceeds the mean life of a series system with exponential units with means μ_i. Just the reverse is true for a parallel system.

A broader class than the IHR (DHR) distributions is that of distributions with *increasing (decreasing) hazard rate on the average* (IHRA and DHRA, respectively). A CDF F is IHRA (DHRA) if $-(1/x)\log S(x)$ is nondecreasing (nonincreasing) in $x \geqslant 0$. Notice that $-\log S(x)$ represents the cumulative hazard rate $\int_0^x \lambda(y)\,dy$.

An IHRA distribution can arise in the following situation. Suppose that shocks occur according to a Poisson process* in time, each independently causing random damage to a device, the damages accumulating until a critical threshold is exceeded, at which time the device fails. This time of failure can be shown to be governed by an IHRA distribution.

Classes of distributions more general than the IHRA (DHRA) arise quite naturally in considering replacement policies. These are *new better (worse) than used* [i.e., NBU (NWU)] and *new better (worse) than used in expectation* [i.e., NBUE (NWUE)] (see Marshall and Proschan [18] and Barlow and Proschan [4, pp. 158–189]).

A distribution F is NBU if

$$S(x+y) \leqslant S(x)S(y) \quad \text{for } x \geqslant 0, y \geqslant 0.$$

The inequality for the function S is reversed if F is NWU. This is equivalent to stating that if F is NBU (NWU), then the conditional survival probability $S(x+y)/S(x)$ of a unit of age x is less (greater) than the corresponding survival probability $S(y)$ of a new unit. The equality holds if and only if F is the exponential distribution.

A distribution F is NBUE if

(a) F has finite mean μ.

(b) $\int_x^\infty S(y)\,dy \leqslant \mu S(x)$ for $x \geqslant 0$.

The inequality for the function S is reversed if F is NWUE. Now $\int_x^\infty \{S(y)/S(x)\}\,dy$ represents the conditional mean remaining life of a unit of age x. Hence, if F is NBUE (NWUE), it implies that a used unit of age x has smaller (larger) mean remaining life than a new unit.

Basic properties of IHRA (DHRA) and the NBU (NWU) or NBUE (NWUE) are described in ref. 4 (pp. 81–124, 158–189).

Another useful classification is based on the *mean residual life*, defined by

$$r_X(x) = r(x) = E[X - x \mid X > x], \quad x \geqslant 0;$$

the function is nonnegative and satisfies $r'(x) \geqslant -1$, $r(x) = o(x \ln x)$ as $x \to \infty$ [19]. It is related to the hazard rate $\lambda(x)$ by

$$\lambda(x) = \{1 + r'(x)\}/r(x).$$

See Muth [19], who argues that $r(x)$ provides a more descriptive measure of an aging or wear-out process than does $\lambda(x)$.

If $r(x)$ is nonincreasing (nondecreasing) in x, then X is said to have a *decreasing (increasing) mean residual life distribution* (DMRL and IMRL, respectively). The exponential distribution is again a natural boundary between DMRL and IMRL distributions, having a constant mean residual lifetime.

The classes of life distributions discussed above satisfy the following relations (see refs. 5 and 12):

IHR $\subset$ IHRA $\subset$ NBU $\subset$ NBUE,
DHR $\subset$ DHRA $\subset$ NWU $\subset$ NWUE,
IHR $\subset$ DMRL $\subset$ NBUE,
DHR $\subset$ IMRL $\subset$ NWUE.

INFERENCE

Maximum likelihood* estimators (MLEs) of the hazard rate for IHR (DHR) distributions have been given in Marshall and Proschan [17], and Johnson and Kotz [13, p. 288]. Contrary to the IHR case, the MLE in the DHR case is not unique.

Several tests are available for testing

H_0: F is exponential.
H_1: F has a monotone hazard rate and is not exponential.

The first such test for an IHR alternative used the normalized spacings of the order statistics as a statistic. A review of this and other tests is given in ref. 9 (pp. 581–584). Also available are a nonparametric test [1], a test using a Bayesian approach [15], and a test for testing whether new is better than used (see ref. [11] and HOLLANDER–PROSCHAN NBU TEST).

Some known results for exponential distributions have been generalized for the foregoing classes of distributions. Following is a partial list of some topics (additional sources on a topic may be found in the given references): tolerance* and confidence* limits (see refs. 21 and 22), prediction* limits [23], ranking* and selection* problems [10], life test* sampling plans, and inequalities on linear combinations of order statistics [9], and some characterizations of IHR (DHR) distributions [24]; *see also* TESTS FOR LIFE DISTRIBUTIONS).

DISCRETE VARIABLES

In fatigue* studies the time to failure is measured in the number of cycles* to failure and is, therefore, a discrete rv. Let $p(k)$ be the probability function (PF) of a discrete rv X; then its hazard rate $\lambda(k)$ is defined by

$$\lambda(k) = p(k) \Big/ \sum_{j=k}^{\infty} p(j).$$

A PF $p(k)$ is said to be IHR (DHR) distribution if $\lambda(k)$ is nondecreasing (nonincreasing) in k. Notice that $\lambda(k) \leqslant 1$. Some properties of $\lambda(k)$ have been summarized in ref. 3 (pp. 18–45). The MLE of $\lambda(k)$ is given in ref. 17.

In the following, some useful continuous and discrete distributions have been classified as having IHR (DHR). The CDFs are omitted in more commonly known distributions.

CONTINUOUS DISTRIBUTIONS

(1) Exponential*: IHR, DHR; (2) gamma*: IHR ($c \geqslant 1$), DHR ($c \leqslant 1$), c is the shape parameter; (3) Weibull*: IHR ($c \geqslant 1$), DHR ($c \leqslant 1$), c is the shape parameter; (4) half-normal (*see* FOLDED DISTRIBUTIONS): IHR; (5) modified extreme value*: IHR [20]; (6) Pareto*: DHR [20]; (7) lognormal*: the HRF is hump-shaped [2, p. 392]; (8) log-logistic: $F(x) = (1 + \theta x)^{-p}$, $\theta > 0$, $p > 0$; DHR ($p \leqslant 1$), if $p > 1$ the HRF resembles the HRF of the lognormal distribution [14, p. 28]; (9) Birnbaum and Saunders: the HRF on the average is nearly nondecreasing [16, p. 155]; (10) inverse Gaussian*: the HRF is not monotonic in general.

DISCRETE DISTRIBUTIONS

(1) Binomial*: IHR; (2) Poisson*: IHR; (3) geometric*: IHR, DHR; (4) negative binomial*: $p(k) = \binom{r+k-1}{r} p^r (1-p)^k$, IHR ($r \geqslant 1$), DHR ($r \leqslant 1$); (5) logarithmic series*: DHR [20]; (6) uniform*: IHR.

References

[1] Ahmad, I. A. (1975). *Commun. Statist. A*, **4**, 967–974; corrections and amendments, *ibid.*, **5**, 1549–1552 (1976).

[2] Bain, L. J. (1978). *Statistical Analysis of Reliability and Life-Testing Models*. Marcel Dekker, New York.

[3] Barlow, R. E. and Proschan, F. (1965). *Mathematical Theory of Reliability*. Wiley, New York.

[4] Barlow, R. E. and Proschan, F. (1975). *Statistical Analysis of Reliability and Life Testing*. Holt, Rinehart and Winston, New York.

[5] Bryson, M. C. and Siddiqui, M. M. (1969). *J. Amer. Statist. Ass.*, **64**, 1472–1483.

[6] Colvert, R. E. and Boardman, T. J. (1976). *Commun. Statist.*, *A*, **5**, 1013–1029.

[7] Elandt-Johnson, R. C. and Johnson, N. L. (1980). *Survival Models and Data Analysis*. Wiley, New York. (There is a good discussion of multiple types of failure.)

[8] Gross, A. J. and Clark, V. A. (1975). *Survival Distributions: Reliability Applications in the Biomedical Sciences*. Wiley, New York. [There is a good discussion of estimation of parameters of some IHR (DHR) distributions.]

[9] Gupta, S. S. and Panchapakesan, S. (1974). In *Reliability and Biometry, Statistical Analysis of Life Lengths*, F. Proschan and R. J. Serfling, eds. SIAM, Philadelphia, pp. 503–596.

[10] Gupta, S. S. and Panchapakesan, S. (1979). *Multiple Decision Procedures: Theory and Methodology of Selecting and Ranking Populations*. Wiley, New York.

[11] Hollander, M. and Proschan, F. (1972). *Ann. Math. Statist.*, **43**, 1136–1146.

[12] Hollander, M. and Proschan, F. (1975). *Biometrika*, **62**, 585–593.

[13] Johnson, N. L. and Kotz, S. (1970). *Continuous Univariate Distributions—2*. Wiley, New York.

[14] Kalbfleisch, J. D. and Prentice, R. L. (1980). *The Statistical Analysis of Failure Time Data*. Wiley, New York. (There is a good discussion of the proportional hazard model and its inference.)

[15] Lochner, R. H. and Basu, A. P. (1977). In *The Theory and Application of Reliability*, Vol. 1, C. P. Tsokos and I. N. Shimi, eds. Academic Press, New York, pp. 67–83.

[16] Mann, N. R., Schafer, R. E., and Singpurwalla, N. D. (1974). *Methods for Statistical Analysis of Reliability and Life Data*. Wiley, New York.

[17] Marshall, A. W. and Proschan, F. (1965). *Ann. Math. Statist.*, **36**, 69–79.

[18] Marshall, A. W. and Proschan, F. (1972). *Proc. 6th Berkeley Symp. Math. Statist. Prob.*, Vol. 1. University of California Press, Berkeley, Calif., pp. 395–415.

[19] Muth, E. J. (1977). In *Theory and Applications of Reliability with Emphasis on Bayesian and Nonparametric Methods*, Vol. 2, C. P. Tsokos and I. N. Shimi, eds. Academic Press, New York, pp. 401–435.

[20] Patel, J. K. (1973). *Commun. Statist.*, **1**, 281–284.

[21] Patel, J. K. (1976). *Technometrics*, **18**, 221–225.

[22] Patel, J. K. (1980). *IEEE Trans. Rel.*, **R-29**, 154–157.

[23] Patel, J. K. (1980). *IEEE Trans. Rel.*, **R-29**, 406–409.

[24] Patel, J. K. and Read, C. B. (1975). *J. Amer. Statist. Ass.*, **70**, 238–244.

[25] Proschan, F. (1963). *Technometrics*, **5**, 375–384.

(BATHTUB CURVE
LIFE TESTING
RELIABILITY
TESTS FOR LIFE DISTRIBUTIONS)

Jagdish K. Patel

HEALTH STATISTICS, NATIONAL CENTER FOR

The National Center for Health Statistics (NCHS) is the only federal agency established specifically to collect and disseminate data on health in the United States. The Center designs and maintains national data collection systems, conducts research in statistical and survey methodology, and cooperates with other agencies in the United States and in foreign countries in activities to increase the availability and usefulness of health data.

The National Center for Health Statistics was created in August 1960 as a result of a recommendation of the Study Group on Mission and Organization of the Public Health Service. The Center was formed by combining two existing units, the national Office of Vital Statistics and the National Health Survey Program. The latter program had been created four years earlier following the passage of the National Health Survey Act. The vital statistics* program had been in the Public Health Service since 1946, when it was transferred from the U.S. Bureau of the Census*. The compilation of national vital statistics by the Bureau of the Census dates back to the beginning of the twentieth century.

The mission of NCHS is, first and foremost, to develop and maintain systems capable of providing reliable, general-purpose, national, descriptive health statistics on a continuing basis and to publish these statistics for the use of the health industry and related industries, both public and private.

The words "Health Statistics" in its title are defined by the Center to include statistics concerning the health of people; the health services they receive; the personnel and facilities resources that provide the services; and certain basic demographic data, particularly in the area of fertility* and family formation and dissolution, which are closely linked to health problems and the population being served. Also covered under the heading of "Health Statistics" are statistics on health attitudes and practices and on payment for health services.

The terms "health statistics" and "health industry" in this statement are interpreted broadly. Some of the statistics are of particular interest to demographers and economists (e.g., marriage and divorce statistics and statistics of births and rates of natural increase) and to actuaries (e.g., construction and interpretation of life tables*).

Three other terms in the statement require exposition if one is to understand thoroughly how the Center views its mission. They are of particular importance because they have a limiting effect which helps to distinguish the Center's mission from that of other statistical programs of the Public Health Service (PHS) and, for example, the statistics collected in the medicare program. These three terms are *general purpose*, *national*, and *descriptive*.

By *general purpose* is meant that NCHS tries to design the data collection systems so that they will serve the needs of a variety of users and leave to others the ad hoc studies and also the highly program-related statistics. This distinction is not sharp, since the obvious attempt is to make data as relevant to current needs as possible. But the Center generally avoids, for example, what are known as service statistics—statistics on the services rendered in a single program. Also, the Center does not conduct strict epidemiological* research such as that designed to identify etiologic factors for a particular disease (although the findings of general-purpose statistics on mortality* and disease prevalence and incidence in various population subgroups do provide epidemiologists with hypotheses to be tested further or evidence for or against earlier hypotheses).

By *national* is meant that the statistics cover the entire nation and are not confined to lesser areas or to particular classes or subgroups of the population. However, national data are frequently presented for various subnational demographic groups and geographic areas, even to the extent of presenting small-area data when circumstances warrant it. An exception is made in methodological research which may be confined to a local area or special group that serves, in effect, as a research laboratory.

By *descriptive* is meant that the statistics show things as they are and not necessarily as they should be. They do not, for example, show needs in terms of numbers of services or resources (although they can be used by others to construct estimates of needs if certain assumptions are introduced). They differ markedly from the types of data obtained in experiments or epidemiological studies, which ordinarily are intended to answer a limited number of specific questions (*see* EPIDEMIOLOGICAL STATISTICS).

Three important subsidiary missions of the Center are (1) the conduct of applied research to improve methods in the field of health and demographic statistics, (2) contribution in all ways possible to improvement of the critical personnel situation in health statistics, including the conduct of a program of training in applied health statistics methods, and (3) the provision of technical assistance to others.

Methodological research in support of the Center's data collection systems has been an important component of its activities. Recent examples of such applied research include the estimation of sampling errors from complex sample surveys* using pseudo-replication methods, the application of network sampling theory to improve the precision of sample survey estimates for rare events in populations, and the development of procedures to improve response rates in telephone surveys*.

The Center maintains data collection systems in the following broad areas:

General Population Surveys: In the annual National Health Interview Survey, NCHS collects information on illness, injuries, and disability in the population and on the costs and use of health care. Its National Health and Nutrition Examination Survey is the source of data from special physical examinations on the prevalence of chronic diseases and related health care needs, nutritional status, and other health measurements.

Vital Statistics: From information on the records of births, deaths, marriages, and divorces maintained by the states, NCHS produces the nation's official vital statistics. Other data on fertility, family planning, and infant and maternal health are collected in the periodic National Survey of Family Growth.

Health Resources Utilization Surveys: In these surveys, NCHS obtains information from office-based physicians, hospitals, nursing homes, and family planning clinics which is used to estimate the volume and types of health care rendered in the United States and the characteristics and health problems of the people served.

Health Resources Surveys: NCHS periodically conducts inventories of inpatient health care facilities, family planning clinics, and selected health occupations to obtain data on the number, location, and characteristics of these resources.

Ad Hoc Surveys: Periodically, NCHS conducts surveys based on samples of vital records to obtain health data not available from those records. The periodic National Medical Care Utilization and Expenditures Survey provides detailed information on expenditures for various types of health care, sources of payment, health insurance coverage, and health and demographic characteristics of the population using health care.

Cooperative Health Statistics System: NCHS is authorized to develop, demonstrate, and partially fund the Cooperative Health Statistics System (CHSS), a shared system operating to meet data needs of public and private agencies at national, state, and local levels. The primary objective is timely, accurate, and comparable data relevant to high-priority needs. States participating in CHSS designate state focal agencies to coordinate data collection and analytic activities to meet the needs of planning, public health, utilization review, rate setting, and other state and local agencies.

The principal form of output of the Center's work is published statistical reports. These come out in several series and individual special reports. The major types are:

Vital and Health Statistics Series: This master series of publications includes over 600 reports presenting findings from the Center's data collection systems as well as background, methodological, and analytical studies. Some typical topics that have been covered include: height, weight, and other body measurements; dietary intake, food consumption, and nutrient levels; dental examination findings; trends in breast feeding in American mothers; geographic patterns in the risk of dying; and average length of stay in short-stay hospitals.

Advance Data from Vital and Health Statistics: These short summary reports provide a means of early release of selected findings from health and demographic surveys.

Vital Statistics of the United States: Annual volumes present final mortality, natality, and marriage and divorce statistics, with tabulations by demographic variables and for geographic areas.

Monthly Vital Statistics Report: Regular issues provide monthly provisional data on births, deaths, marriages, and divorces. *MVSR* supplements are used for final vital statistics. There is also an annual provisional summary.

Clearinghouse on Health Indexes: A bibliography of printed articles and research in the area of health indices, their construction and use, published quarterly with an annual summary.

Catalogs and Guides to Data: NCHS issues an annual catalog of publications, supplemented by the quarterly Publication Note. A catalog of public use data tapes is also available. In addition, separate bibliographies cover NCHS publications on the health of minorities, the elderly, women, and other specific groups or topics. An annual bibliography on statistical methodology covers published and unpublished papers, reports, and presentations.

The extent of data usage is indicated by the 10,000 direct requests for data received in the Center each month and the 400,000 copies of publications which go out annually to people who have requested them. In addition to publications, the Center has issued more than 150 public-use computer tapes containing detailed nonconfidential data from nearly all the surveys and data systems operated in the past 10 years. Several hundred tapes are sold annually, mostly to scientists who wish to make additional analyses of the data. Within the limits of staff and equipment, special tabulations of data are prepared, the costs being borne by the requester.

The Center engages in a variety of international activities, including programs of research, analysis, and technical assistance, that support its national mission and reflect its worldwide reputation.

One of the oldest international health statistics activities of the Center and its predecessor agencies is participation in the development, revision, and application of the International Classification of Diseases (ICD). The ICD, a statistical system that has been evolving for 200 years, provides an invaluable tool for the study of temporal and spatial distribution of disease; for the estimation of the effects of disease on populations, particularly with respect to morbidity and mortality by age, sex, and other demographic characteristics; and for the investigation of other epidemiologic aspects of disease in human populations.

In April 1976, the World Health Organization designated NCHS as the WHO Center for Classification of Diseases for North America. The work of the North American Center includes not only general support and assistance to users of the current revision of ICD, but also the organization and planning for input into future revisions from a wide variety of users in the United States, Canada, and other English-speaking countries in the Western Hemisphere.

Internationally, the Center has become a recognized source of expertise in the field of health statistics. A stream of visitors comes through the offices for briefing and training.

The Center is now widely recognized both within the government and outside as a major part of the U.S. statistical system. Its data are quoted in hundreds of scientific and popular journals each year and are used in congressional testimony, reports of national commissions, and the planning activities of many states and communities (often, incidentally, not only for their value as national data but also for want of comparable locally generated data of acceptable quality).

Increasingly, NCHS is being looked upon by state and local governments and universities as the focal point for national leadership in the development of coordinated health statistics programs to serve the needs of both public and private programs. This national leadership role had been well recognized in the vital statistics field before the Center was established and is becoming a reality in the fields of morbidity, disability, health services, and health resources statistics as well.

PUBLICATIONS

For copies of or information about publications, NCHS data, or programs, contact the Scientific and Technical Information Branch, Division of Data Services, NCHS, Room 1-57, 3700 East-West Highway, Hyattsville, MD 20782, (301)436-8500.

(FEDERAL STATISTICS)

ROBERT A. ISRAEL

HEAVY-TAILED DISTRIBUTIONS

A heavy-tailed distribution is one whose extreme probabilities approach zero relatively slowly: the survival distribution* function $1 - F(x)$ approaches zero slowly as $x \to \infty$, or, if the support includes the entire real line, the cumulative distribution function* $F(x)$ goes to zero slowly as $x \to -\infty$. The intent is to characterize absolute shape, not scale, and one distribution is not regarded as more heavy-tailed than another simply because its variance is larger. While there is no universally accepted cutoff point for heavy-tailedness, most statisticians would agree that normal distributions* can be described as light-tailed and Pareto distributions* as heavy-tailed. The exponential* and gamma* families seem to occupy a middle ground. Terminology is also far from universal: the terms "long-tailed," "thick-tailed," "large-tailed," and "fat-tailed" will be encountered with exactly the same meaning as "heavy-tailed." So commonly is the Pareto family used to typify heavy-tailedness that the adjective "Paretian" is sometimes used.

Early researchers such as Mills [10] often used the coefficient of kurtosis* as a shape-characterizing parameter, with large values ("leptokurtosis") believed to indicate both a heaviness in the tail and a large degree of central peakedness. This is now recognized as a gross oversimplification; moreover, kurtosis is infinite in enough cases of interest to render it an ineffective discriminator.

The nonexistence of finite variance is an important criterion for heavy-tailedness, as noted by Mandelbrot [7], who effectively treated distributions as heavy-tailed if and only if the variance is infinite. Finiteness of variance affects most of the common inferential procedures by placing a distribution in the domain of attraction of the normal distribution (*see* STABLE DISTRIBUTIONS) with respect to limiting behavior of the sample mean. Both Mandelbrot [7] and Fama [2] discussed the frequent occurrence of infinite variances in economic data and suggested that standard least-squares-based analytic procedures could be accordingly ineffective. Granger and Orr [4] provide a recent summary of such implications. Similar concern in the area of estimation has prompted a huge volume of recent work in robustness*.

When variance is infinite, a standardized sum analogous to the sample mean may have as its limiting distribution a nonnormal stable distribution* [3]. The characteristic exponent of the stable distribution ($0 < \alpha \leqslant 2$) can serve as one measure of heavy-tailedness, with the smaller α-values indicating heavier tails. The variance is finite only for $\alpha = 2$ (normal distribution) and the mean exists for $\alpha > 1$. Mandelbrot [7] noted the importance of the nonnormal stables, utilizing the term "Paretian stable" to describe them—although actually the lighter-tailed members of the Pareto family do have finite variance and hence fall into the normal's domain of attraction.

Bryson [1] suggested the less restrictive criterion of *increasing conditional mean exceedance*: a distribution $F_x(x)$ is called heavy-tailed if the expectation $E(X - x \mid X > x)$ is increasing for large x or if it does not exist. This definition places the exponential family (with constant conditional mean exceedance) at the borderline between light and heavy tails. A key role is similarly played by the exponential in the more refined taxonomy developed by Mantel [8]. His five-way tail classification was based on the behavior of Mills' ratio*, the ratio of a survival function $1 - F(x)$ to the corresponding density $f(x)$. The more heavy-tailed distributions (which, again, are those with heavier-than-exponential tails) are those for which Mills' ratio diverges for large x. Mantel's further refinement consisted of classifying exceptionally light- or heavy-tailed distributions for which Mills' ratio goes to zero or infinity at finite x.

Yet another classification identifies distributions as heavy-tailed depending on the limiting distribution of the maximum of a random sample. Gumbel [5] shows that for lighter-tailed distributions (including the exponential) the maximum has as its limiting distribution the type I extreme value distri-

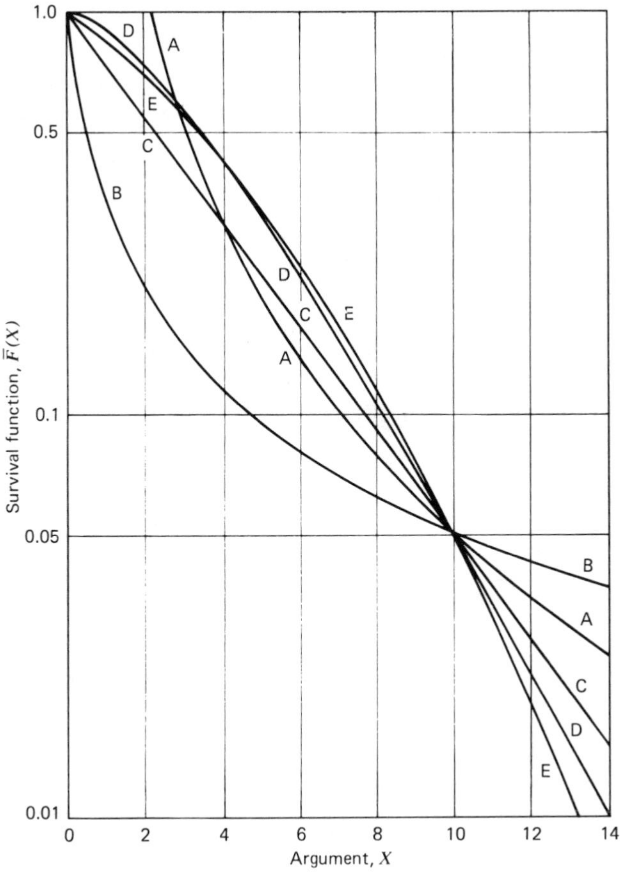

Figure 1 Comparative log-survival function graphs.

bution* with standardized cumulative distribution function (CDF)

$$F(x) = \exp[-\exp(-x)],$$

while maxima from heavier-tailed distributions such as the Pareto and log-gamma follow the limiting type II law,

$$F(x) = \exp[-x^{-k}], \quad x > 0; \quad k > 0.$$

These various criteria are not equivalent. The lognormal distribution*, for example, is heavy-tailed by the Mantel and Bryson criteria, but belongs to the type I extreme-value domain of attraction. Pareto distributions with shape parameter $k > 2$ belong to the type II extreme-value domain, but still have finite variance.

Practical implications of heavy-tailedness are many. If data are suitably modeled by an infinite variance distribution (which in finite samples may be the case even for physical phenomena that are theoretically bounded), routine use of the central limit theorem* or related methodology may be misleading. Mandelbrot [7] and Samuelson [12] are among those finding this important in market price analyses. Salas-La Cruz and Boes [11] use Paretian stable distributions in one of several possible explanations for the controversial "Hurst phenomenon*" of hydrologic sequence sums. Hsu [6] found that the conventionally used normal distribution was too light-tailed for modeling navigational errors. Shen et al. [13] found that flood maxima in the United States more commonly follow type II than type I extreme-value distributions. Mielke [9] found a Paretian-tailed family suitable for describing precipitation phenomena. Finally, Zipf's law* can be shown to imply a Paretian-

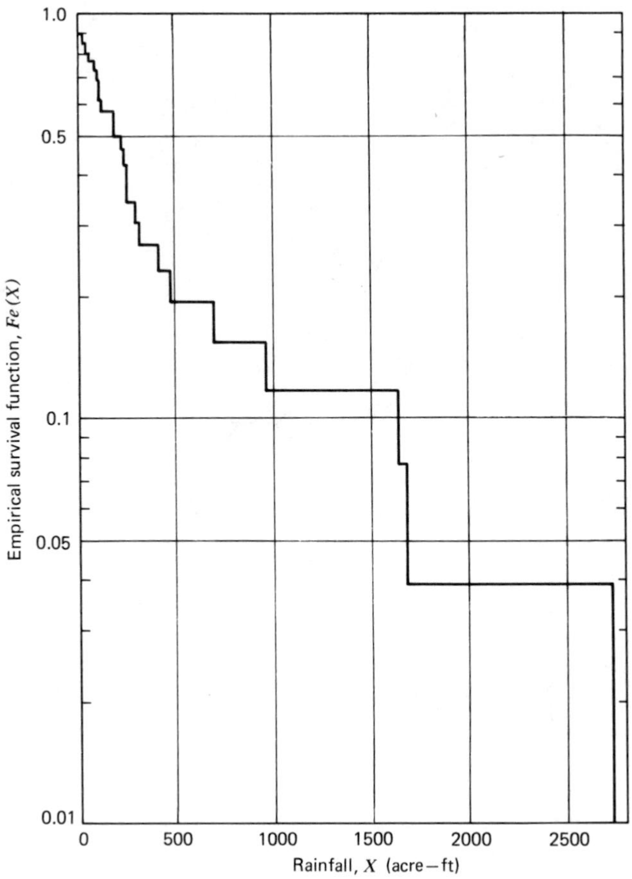

Figure 2 Example: Florida precipitation data.

tailed distribution for textual word frequencies and many other phenomena.

In working with empirical data, various tests for heavy-tailedness may be useful. Granger and Orr [4] suggest a test for infinite variance. After arranging the data in random order, plot the sample variance of the first n observations against n. The absence of any apparent convergence with increasing n is an indication of infinite population variance. Bryson [1] suggested and gave critical values for the test statistic

$$\frac{\bar{x}(x_{\max})}{\bar{x}_{\mathrm{GA}}^2}$$

where $\bar{x}$ is the sample mean, $x_{\max}$ the sample maximum, and $\bar{x}_{\mathrm{GA}}$ a modified geometric mean*

$$\bar{x}_{\mathrm{GA}} = \prod_{i=1}^{n} (x_i + A)^{1/n}$$

with A equal to $x_{\max}/(n-1)$. This test statistic is a minor modification of one that gives a most powerful test of the exponential distribution against the alternative of a heavy-tailed Pareto,

$$F(x) = 1 - (1 + \lambda x)^{-1}.$$

Graphically, a convenient procedure for comparing exponential against Pareto alternatives is to plot the log survival function, $\ln(1 - F(x))$, against x. Figure 1 illustrates such a plot for distributions having the following five density functions:

(A) $f(x) = 2A^2/x^3 \quad (x > A)$

(B) $f(x) = A/(A + x)^2 \quad (x > 0)$

(C) $f(x) = A^{-1}\exp(-x/A) \quad (x > 0)$

(D) $f(x) = A^{-2}x\exp(-x/A) \quad (x > 0)$

(E) $f(x) = 2(\sqrt{2\pi A})^{-1}\exp(-x^2/(2A^2))$ $\quad (x > 0)$

The scale parameter A was chosen in each case to force $F(10) = 0.95$. It is clear that the heavy-tailed Pareto distributions (A) and (B) exhibit noticeable concavity, while the light-tailed normal (E) is distinctly convex. The same kind of concavity, indicating a degree of heavy-tailedness, is apparent in Fig. 2 for the following set of 26 Florida precipitation data from Simpson [14]:

129.6	302.8	200.7	978.0	118.3	40.6
31.4	119.0	274.7	198.6	255.0	
2745.6	4.1	274.7	703.4	115.3	
489.1	92.4	7.7	1697.8	242.5	
430.0	17.5	1656.0	334.1	32.7	

In testing the exponential-tail hypothesis against the Pareto alternative, Bryson [1] found these data to be heavy-tailed at a significance level of about 0.02. The convergence test of sample variances seems to indicate failure of a finite-variance model as well: unless the random ordering of the data happens to locate the three largest observations early in the data set, sample variances do not appear to be convergent.

References

[1] Bryson, M. C. (1974). *Technometrics*, **16**, 61–67. (Largely heuristic, motivating a definition of heavy-tailedness and deriving a test statistic for one form of heavy-tailed behavior.)

[2] Fama, E. G. (1963). *J. Bus.*, **36**, 420–429. (Fairly nonmathematical discussion of some important data-analytic implications of Mandelbrot's [7] article.)

[3] Gnedenko, B. V. and Kolmogorov, A. N. (1954). *Limit Distributions for Sums of Independent Random Variables*, K. L. Chung, trans. Addison-Wesley, Reading, Mass. (The classical and definitive work on limiting distributions of sums; rigorously mathematical.)

[4] Granger, C. W. J. and Orr, D. (1972). *J. Amer. Statist. Ass.*, **67**, 275–285. (A readable and thorough discussion of the role of stable distributions in economic modeling, with nice examples and an appendix summarizing mathematical properties; motivated by, and extending, the work of Mandelbrot [7].)

[5] Gumbel, E. J. (1960). *Statistics of Extremes*. Columbia University Press, New York. (The classical reference for extreme-value distributions, not difficult but sometimes cumbersome mathematically.)

[6] Hsu, D. A. (1979). *Appl. Statist.*, **28**, 62–72. (An example of probability modeling in navigation.)

[7] Mandelbrot, B. (1963). *J. Bus.*, **36**, 394–419. (A truly seminal paper discussing the existence and importance of heavy-tailed distributions in commodity prices and introducing their farther-reaching implications; more heuristic than rigorous, and interesting if not easy to read.)

[8] Mantel, N. (1976). *Amer. Statist.*, **30**, 14–17. (A fairly nonmathematical discussion of how distributions may be placed into categories according to tail weight.)

[9] Mielke, P. W. (1973). *J. Appl. Meteorol.*, **12**, 275–280. (Introduction of the heuristically derived "kappa" distribution, a heavy-tailed family suitable for precipitation modeling.)

[10] Mills, F. C. (1927). *The Behavior of Prices*. National Bureau of Economic Research, New York. (Old but still interesting, with lengthy examples of classical probability modeling based mainly on perturbations of the normal distribution.)

[11] Salas-La Cruz, J. D. and Boes, D. C. (1974). *Water Resour. Res.*, **10**, 457–463. (Discusses the "Hurst phenomenon" in hydrology, and suggests possible use of stable distribution theory in explaining it; summarizes key mathematical results from other work.)

[12] Samuelson, P. A. (1976). *J. Financ. Quant. Anal.*, **11**, 485–503. (More implications from Mandelbrot's work, especially relating to portfolio management.)

[13] Shen, H. W., Bryson, M. C., and Ochoa, I. D. (1980). *Water Resour. Res.*, **16**, 361–364. (Emphasizing data analysis, with results indicating that streamflow maxima tend to follow type II rather than type I extreme-value distribution.)

[14] Simpson, J. (1972). *Monthly Weather Rev.*, **100**, 309–312. (Empirical curve-fitting using gamma distributions to model precipitation.)

(KURTOSIS
PARETO DISTRIBUTIONS
STABLE DISTRIBUTIONS)

M. C. BRYSON

HEDONIC INDEX NUMBERS

Hedonic index numbers are price and quantity index numbers adjusted for quality changes by regression* methods. Generally, price and quantity index numbers rest on the hypothesis that the relative change in money value can be factored out into a price index and a quantity index. This price index is supposed to be a "pure" price index in that it measures the relative change in the prices of the same products bought and sold in the same markets. Thus a variety of a product which is in some sense better than another should be counted as more quantity than the poorer one.

Hedonic index numbers improve on this system by accounting for the change in the quality of the product. The hedonic approach assumes that a commodity can be viewed as a bundle of physical and performance characteristics (such as the horsepower, weight, length, and power brakes of different makes and models of automobiles), that quality is a transform of those characteristics, that a change in the market price of a commodity can be factored out into a quality price component and a pure price component, and that the price of a product is correlated with its characteristics.

For example, suppose that the price of a certain type of milk in cents (P) is linearly related to the volume of milk in quarts (V) and the number of calories (C). Suppose further that the ordinary least-squares* estimate of this relation from cross-section data is $\hat{P} = 50V + 0.06C$. If at a future date the price of a quart of milk of this type rose from 80 cents to 90 cents and its calorie content increased from 500 calories to 600, we would attribute 6 cents of the price increase to quality change since the regression equation predicts that the price of an additional 100 calories is 6 cents (0.06×100), and the remaining 4 cents is pure price change. The observed price has increased by 12.5% [$(90/80) \times 100$], and the quality of milk has improved by 7.5% [$(86/80) \times 100$]. If it is assumed that the observed price increase of 12.5% is the sum of a pure price change and a quality change, then the hedonic price of milk has increased by 5%. This is one of many ways of constructing a quality-adjusted or pure price index. The observed quantity index is unchanged, while the quality-adjusted or hedonic quantity index has increased by 7.5%.

Pioneering work on hedonic indexes can be traced to Court [4], Houthakker [11], and Stone [15], while modern influential work on the subject has been done by Griliches [8], Adelman and Griliches [1], Burstein [3], Lancaster [12], and Muth [13].

HOFSTEN'S QUALITY INDEX

The hedonic approach can be viewed as a generalization of a method studied by Hofsten [10]. Hofsten's method is based on the proposition that, in competitive market equilibrium, relative prices, relative rates of substitution to the consumer, and relative marginal costs to the producer are all equal. Hofsten assumed that $p_b(t) = gp_a(t)$, where $p_b(t)$ and $p_a(t)$ are the prices of an old item a and a new item b at time t, and g is a *quality index*, which is assumed to be independent of all other prices and purchases. Thus the change in quality is proportional to the change in price: item b is g times as good as item a and g times the price; and $g > 1$ if the quality change is an improvement, and $g < 1$ if the change is a deterioration. There are many ways to estimate g. First, if the two items a and b are both sold at time t, then their price ratios before and after the introduction of the new item are *linked* or *spliced* to obtain the relative price between $t+1$ and $t-1$ via t as follows:

$$\frac{p_{t+1}}{p_{t-1}} = \frac{p_{a(t)}}{p_{a(t-1)}} \frac{p_{b(t+1)}}{p_{b(t)}} \qquad \left(= \frac{p_{b(t+1)}}{gp_{a(t-1)}} \right).$$

Second, if the two items do not overlap at time t, an estimate of the quality index g may be based on the marginal cost of item b to that of item a. Third, the hedonic technique can be used.

THE HEDONIC APPROACH

In the context of consumer theory, the hedonic approach is based, at least partially, on the more formal proposition that a transformation matrix $\mathbf{B}$ exists to transform the vector of the goods $\mathbf{y}$ purchased by the consumer into a vector of characteristics $\mathbf{x}$ (i.e., $\mathbf{x} = \mathbf{B}\mathbf{y}$). The consumer utility function $u = u(x_1, x_2, \dots, x_m)$ is then maximized subject to the two constraints $M = \mathbf{p}'\mathbf{y}$ and $\mathbf{x} = \mathbf{B}\mathbf{y}$, where M is income and $\mathbf{p}'$ is a price vector.

The shadow or implicit prices of the characteristics are interpreted as hedonic prices, and estimated by regressing the observed prices of different varieties of the same commodity on their measurable characteristics. The hedonic regression is of the form

$$p_{it} = \sum_j \beta_{jt} x_{jit} + \epsilon_{it},$$

where p_{it} is the price of the ith variety of a commodity at time t, x_{jit} is the level of the jth characteristic contained in the ith variety at time t, ϵ_{it} is a disturbance term, and the estimated coefficients are the implicit prices.

Hedonic regressions have been estimated from single years' cross-section data, and from pooled time series* and cross-section data for at least two years. In the latter case, the coefficients of the time dummy variables included in the regression equation are interpreted as estimates of the pure price change. In the case of single years' cross-section data, the estimated regression is used in two ways to account for quality change. One is to estimate the quality index g for period 1 relative to period 0 for the ith variety from a Laspeyres-type expression given by

$$\sum_j \hat{\beta}_{j0} x_{ji1} \Big/ \sum_j \hat{\beta}_{j0} x_{ji0}.$$

The pure price relative for variety i is the observed price relative divided by the quality index $[(p_{i1}/p_{i0})/g]$. In the other use, the implicit prices $\hat{\beta}_{j1}$ and $\hat{\beta}_{j0}$ are used to price some average bundle of characteristics in the two periods 0 and 1. For example, a Laspeyres-type pure price index for variety i is given by

$$\sum_j \hat{\beta}_{j1} x_{ji0} \Big/ \sum_j \hat{\beta}_{j0} x_{ji0}.$$

The quality or pure price indexes of the varieties can be aggregated for a set of commodities.

THE FUNCTIONAL FORM

Applications of the hedonic approach typically regress the logarithms of the prices of different varieties of a product on their characteristics. This semilogarithmic transformation is a special case of the Box–Cox [2] transformation* which has been used in hedonic regression. For this transformation, assume that there exists a scalar λ such that

$$\frac{(p_{it}^{\lambda} - 1)}{\lambda} = \sum_j \beta_{jt} x_{jit} + \epsilon_{it},$$

where the disturbance term ϵ_{it} is assumed to satisfy the standard assumptions of the regression model. For $\lambda = 1$, the transformation is linear, and as $\lambda \to 0$, it reduces to the semilogarithmic form (see Freeman [5] and Goodman [6]).

THE VINTAGE PRICE METHOD

Hall [9] proposed a "modified hedonic method" which combines data on the quality characteristics of a commodity with its different varieties and vintages in a single regression. The use of vintage prices as an alternative method to adjust for quality change was first proposed by Burstein [3]. The vintage price method assumes that differences in the prices of secondhand goods of different vintages can be accounted for by quality differences, depreciation, and an overall average price. Hall combines this method with the hedonic method by representing the quality differences by the measurable characteristics of the product (see Hall [9], Gordon [7], and Ohta and Griliches [14]).

References

[1] Adelman, I. and Griliches, Z. (1961). *J. Amer. Statist. Ass.*, **56**, 535–548. (Theory; examples.)

[2] Box, G. E. and Cox, D. R. (1964). *J. R. Statist. Soc. B*, **26**, 211–252.

[3] Burstein, M. L. (1961). *Manchester Sch. Econ. Soc. Stud.*, **29**, 267–279. (The first to use vintage prices.)

[4] Court, A. T. (1939). In *The Dynamics of Automobile Demand.* General Motors Corporation, New York, pp. 99–117. (The first to use hedonic regression.)

[5] Freeman, A. M. (1979). In *The Economics of Neighborhood*, D. Segal, ed. Academic Press, New York, pp. 193–218. (Discusses functional form; references.)

[6] Goodman, A. C. (1978). *J. Urban Econ.*, **5**, 471–484. (Uses the Box–Cox transformation.)

[7] Gordon, R. J. (1971). *Rev. Income Wealth*, **17**, 121–174. (Survey; producer durables; references.)

[8] Griliches, Z. (1961). In *The Price Statistics of the Federal Government*. Nat'l. Bur. Econ. Res. Gen. Ser., No. 73. National Bureau of Economic Research, New York, pp. 137–196. (Key paper reviving interest in hedonic index numbers.)

[9] Hall, R. E. (1971). In *Price Indexes and Quality Change*. Z. Griliches, ed. Harvard University Press, Cambridge, Mass., pp. 240–271.

[10] Hofsten, E. V. (1952). *Price Indexes and Quality Changes*. George Allen & Unwin, London. (Early systematic study of quality changes; references.)

[11] Houthakker, H. S. (1952). *Rev. Econ. Stud.*, **19**, 155–164. (Economic theory.)

[12] Lancaster, K. (1966). *J. Pol. Econ.*, **74**, 132–157. (Economic theory.)

[13] Muth, R. F. (1966). *Econometrica*, **34**, 699–708. (Economic theory.)

[14] Ohta, M. and Griliches, Z. (1975). In *Studies in Income and Wealth*, Vol. 40: *Household Production and Consumption*, N. E. Terleckyj, ed. pp. 325–390. (Investigates many of the questions still being discussed in the current literature on hedonic index numbers; references.)

[15] Stone, R. (1956). *Quantity and Price Indexes in National Accounts*. OEEC, Paris.

Bibliography

Griliches, Z., ed. (1971). *Price Indexes and Quality Change*. Harvard University Press, Cambridge, Mass. (Collection of important contributions; extensive references.)

Halvorsen, R. and Palmquist, R. (1980). *Amer. Econ. Rev.*, **70**, 474–475.

Maunder, W. F., ed. (1970). *Bibliography of Index Numbers*. Athlone Press, London. (Exhaustive from 1707 up to the first half of 1968.)

Triplett, J. E. (1975). In *Analysis of Inflation*, P. H. Earl, ed. Lexington Books, Lexington, Mass. (Elementary; good survey; references.)

(INDEX NUMBERS)

NURI T. JAZAIRI

HELLINGER DISTANCE

Let $\mathscr{X}$ be a sample and $\mathscr{A}$ a σ-algebra of events in $\mathscr{X}$; let $\{P_\theta, \theta \in \Theta\}$ be a family of distributions on $(\mathscr{X}, \mathscr{A})$ for all values of the parameter θ in Θ, with density $f(x;\theta)$ defined by $f(x;\theta) = dP_\theta/d\nu$, where ν is a σ-finite measure on $\mathscr{A}$. Then the *Hellinger distance* r_p between two measures $\int_A f(x;\theta)\,d\nu$ and $\int_A f(x;\theta')\,d\nu$ (in obvious measure-theoretic notation) for $A \in \mathscr{A}$ is defined as:

$$r_p(\theta,\theta') = \left\{ \int_{\mathscr{X}} \left| [f(x;\theta)]^{1/p} - [f(x;\theta')]^{1/p} \right|^p \nu(dx) \right\}^{1/p}, \quad p \geqslant 1.$$

The Hellinger distance satisfies the property that

$$0 \leqslant r_p(\theta,\theta') \leqslant 1;$$

$$r_p(\theta,\theta') = 1 \quad \text{if and only if} \quad \theta = \theta'.$$

The distances r_1 and r_2 have been used in problems in the theory of statistical estimation*, and for constructing consistent estimators (see, e.g., ref. 1).

Reference

[1] Ibragimov, I. A. and Has'minskii, R. Z. (1981). *Statistical Estimation (Asymptotic Theory), Appl. Math.*, **16**. Springer-Verlag, Berlin.

HELLY–BRAY THEOREMS

This is a group of useful theorems concerned with infinite sequences of distribution functions. Many authors refer to them as Helly theorems, dropping the reference to Bray, since it appears that Helly has the priority of discovery by a few years.

Let $\{F_n(x)\}$, for $n = 1, 2, 3, \ldots$, be a sequence of distribution functions, which may or may not be proper [i.e., we allow $F_n(\infty) - F_n(-\infty) < 1$].

We say $\{F_n(x)\}$ converges *weakly* to a distribution function $F(x)$, also possibly improper, if at every point x where $F(x)$ is continuous, $\{F_n(x)\}$ converges as $n \to \infty$ to $F(x)$ in the ordinary sense of convergence (*see* CONVERGENCE OF SEQUENCES OF RANDOM VARIABLES).

The first theorem of Helly–Bray asserts that every sequence $\{F_n(x)\}$ must contain a subsequence $F_{n_1}(x), F_{n_2}(x), \dots$, which converges weakly to some, possibly improper, distribution function $F(x)$.

The second theorem of Helly–Bray assumes $\{F_n(x)\}$ to be weakly convergent to $F(x)$ at all continuity points of $F(x)$ in a finite closed interval $[a,b]$. It also assumes $F(x)$ to be continuous at $x = a$ and $x = b$, and then asserts that, as $n \to \infty$,

$$\int_a^b g(x)\,dF_n(x) \to \int_a^b g(x)\,dF(x)$$

for every function $g(x)$ which is continuous throughout $[a,b]$.

Often one wants to use the second theorem for infinite intervals instead of finite ones. One can do so if it is further assumed that $g(x)$ is bounded and continuous on $(-\infty, +\infty)$, and that the limit distribution function $F(x)$ is proper.

These theorems are of considerable theoretical value and facilitate the derivation of many limit theorems*. They are discussed in most treatises on probability theory.

Bibliography

The following books contain detailed discussions of Helly–Bray theorems; many others doubtless also do so.

Gnedenko, B. V. (1963). *The Theory of Probability*. Chelsea, New York.

Loève, M. (1977, 1978). *Probability Theory*, Vols. 1, 2, 4th ed. Springer-Verlag, New York.

Moran, P. A. P. (1968). *An Introduction to Probability Theory*. Clarendon Press, Oxford.

(CONVERGENCE OF SEQUENCES OF RANDOM VARIABLES
LIMIT THEOREMS
LIMIT THEOREMS, CENTRAL)

W. L. SMITH

HELMERT, FRIEDRICH ROBERT

Born: July 31, 1843, in Freiberg (Saxony), Germany.

Died: June 15, 1917, in Potsdam (Prussia), Germany.

Contributed to: geodesy, sampling distributions, transformations.

Friedrich Robert Helmert was a mathematical physicist whose main research was in geodesy, although this led him to investigate several statistical problems. In his doctoral dissertation [2] he developed a theory of the "ellipse of error," and in 1872 [3] he used the method of least squares* in an examination of measuring instruments. In 1872 he was appointed Professor of Geodesy at the technical school in Aachen, and following the favorable reception of his work *Die mathematischen und physikalischen Theorien der höheren Geodäsie* (Part I in 1880 and Part II in 1884), he became Professor of Advanced Geodesy at the University of Berlin in 1887, and Director of the Prussian Geodetic Institute [1].

Until the 1960s, the derivation of the chi-square (χ^2) distribution* was frequently attributed in statistical literature to Helmert, although credit for this more properly belongs to Ernst Abbe* and, to a large extent to Irenée-Jules Bienaymé* (see Sheynin [7]). However, in 1876 Helmert [4] proved that $S = \sum_{i=1}^n (x_i - \bar{x})^2/\sigma^2$ has a χ^2_{n-1} distribution (with $n-1$ denoting degrees of freedom*) if $x_1, \dots, x_n$ is a random sample from a normal population having common variance σ^2 and an unknown common mean μ, and where $\bar{x} = \sum x_i/n$.

Helmert showed first that if $\lambda_i = x_i - \bar{x}$, $i = 1, \dots, n$, then the joint density of $\lambda_1, \dots, \lambda_{n-1}$ and $\bar{x}$ (with $\lambda_n = -\lambda_1 - \cdots - \lambda_{n-1}$) is proportional to

$$\exp\left\{-\tfrac{1}{2}\sigma^{-2}\left[\lambda_1^2 + \cdots + \lambda_n^2\right]\right\} \times \exp\left\{-\tfrac{1}{2}n\sigma^{-2}(\bar{x}-\mu)^2\right\}. \qquad (1)$$

He thus established the independence of $\bar{x}$ and any function of $x_1 - \bar{x}, \dots, x_n - \bar{x}$, including S.

In order to obtain the distribution of S, Helmert introduced the transformation (presented here exactly as he gave it)

$$\begin{aligned} t_1 &= \sqrt{2}\,(\lambda_1 + \tfrac{1}{2}\lambda_2 + \tfrac{1}{2}\lambda_3 + \tfrac{1}{2}\lambda_4 \cdots + \tfrac{1}{2}\lambda_{n-1}) \\ t_2 &= \sqrt{\tfrac{3}{2}}\,(\lambda_2 + \tfrac{1}{3}\lambda_3 + \tfrac{1}{3}\lambda_4 \cdots + \tfrac{1}{3}\lambda_{n-1}) \\ t_3 &= \sqrt{\tfrac{4}{3}}\,(\lambda_3 + \tfrac{1}{4}\lambda_4 \cdots + \tfrac{1}{4}\lambda_{n-1}) \\ &\ \vdots \\ t_{n-1} &= \sqrt{n/(n-1)}\,\lambda_{n-1}. \end{aligned} \tag{2}$$

This changes the joint density of $\lambda_1, \dots, \lambda_{n-1}$, namely

$$\sqrt{n}\left(\frac{h}{\sqrt{\pi}}\right)^{n-1} \exp\left\{-h^2(\lambda_1^2 + \cdots + \lambda_n^2)\right\}, \tag{3}$$

into that of $t_1, \dots, t_{n-1}$, i.e.,

$$\left(\frac{h}{\sqrt{\pi}}\right)^{n-1} \exp\left\{-h^2\left(t_1^2 + \cdots + t_{n-1}^2\right)\right\}, \tag{4}$$

where $h = 1/(\sqrt{2}\,\sigma)$ is the *modulus* of the distribution of each of $X_1, \dots, X_n$, or for that matter of $t_1, \dots, t_{n-1}$ (the term "modulus" and the notation being in general use during the nineteenth century). Since (4) is the joint distribution of $n - 1$ independent identical normal variates with zero means, and since

$$\sigma^{-2} \sum_{i=1}^{n-1} t_i^2 = \sigma^{-2} \sum_{j=1}^{n} \lambda_j^2 = S, \tag{5}$$

it follows that S has the χ^2_{n-1} distribution.

Lancaster [6] points out that the term "Helmert transformation" refers either to the combined transformation

$$\mathbf{t} = \mathbf{A}\mathbf{x} \tag{6}$$

where $\mathbf{x}' = (x_1 \cdots x_n)$ and $\mathbf{t}' = (t_1 \cdots t_{n-1})$, or to the inverse transformation. It could also describe the transformation

$$\begin{pmatrix} \sqrt{n}\,\bar{x} \\ \mathbf{t} \end{pmatrix} = \mathbf{B}\mathbf{x} \tag{7}$$

or its inverse, given by

$$\mathbf{x} = \mathbf{B}^{-1}\begin{pmatrix} \sqrt{n}\,\bar{x} \\ \mathbf{t} \end{pmatrix},$$

$$\mathbf{B}^{-1} = \begin{bmatrix} \frac{1}{\sqrt{n}} & \frac{1}{\sqrt{1.2}} & \frac{1}{\sqrt{2.3}} & \frac{1}{\sqrt{3.4}} & \cdots & \frac{1}{\sqrt{(n-1)n}} \\ \frac{1}{\sqrt{n}} & -\frac{1}{\sqrt{1.2}} & \frac{1}{\sqrt{2.3}} & \frac{1}{\sqrt{3.4}} & \cdots & \frac{1}{\sqrt{(n-1)n}} \\ \frac{1}{\sqrt{n}} & & -\frac{2}{\sqrt{2.3}} & \frac{1}{\sqrt{3.4}} & \cdots & \frac{1}{\sqrt{(n-1)n}} \\ \frac{1}{\sqrt{n}} & & & -\frac{3}{\sqrt{3.4}} & \cdots & \frac{1}{\sqrt{(n-1)n}} \\ \vdots & & & \vdots & & \vdots \\ \frac{1}{\sqrt{n}} & & & & \cdots & -\frac{n-1}{\sqrt{(n-1)n}} \end{bmatrix} = \mathbf{B}', \tag{8}$$

since $\mathbf{B}$ is orthogonal. When $\mathbf{x}$ is distributed as above, any orthogonal transformation $\mathbf{y} = \mathbf{C}\mathbf{x}$ preserves independence, and $\sum_{i=1}^n y_i^2 = \sum_{i=1}^n x_i^2$. The Helmert matrix, $\mathbf{B}$, however, is probably the simplest orthogonal matrix having $1/\sqrt{n}$ for each element in the first row; see Lancaster [5], who also defines and discusses a generalized Helmert transformation and applications.

References

[1] Fischer, F. (1972). In *Dictionary of Scientific Biography*, Vol. 6. Scribner's, New York, pp. 239–241.

[2] Helmert, F. R. (1868). *Studien über rationelle Vermessungen der höheren Geodäsie*. University of Leipzig, Leipzig.

[3] Helmert, F. R. (1872). *Die Ausgleichsrechnung nach der Methode der kleinsten Quadrate mit Anwendungen auf die Geodäsie und die Theorie der Messinstrumente*. Leipzig.

[4] Helmert, F. R. (1876). *Astron. Nachr.*, **88**, cols. 113–120.

[5] Lancaster, H. O. (1965). *Amer. Math. Monthly*, **72**, 4–12. (Discusses a general class of Helmert-type matrices.)

[6] Lancaster, H. O. (1966). *Aust. J. Statist.*, **8**, 117–126. (A historical account of precursors of the χ^2-distribution.)

[7] Sheynin, O. B. (1966). *Nature*, **211**, 1003–1004.

(ABBE, ERNST
CHI-SQUARE DISTRIBUTION
QUADRATIC FORMS)

CAMPBELL B. READ

HELMERT TRANSFORMATION *See* HELMERT, FRIEDRICH ROBERT

HENDERSON METHOD *See* VARIANCE COMPONENTS

HERITABILITY *See* GENETICS, STATISTICS IN

HERMITE DISTRIBUTIONS

Let x_1 and x_2 be two independent Poisson* variables with parameters a_1 and a_2. The probability distribution of the random variable $x_1 + 2x_2$ is the *Hermite distribution* with parameters a_1 and a_2 and is given by

$$P_n = \exp[-(a_1 + a_2)] \sum_{j=0}^{[n/2]} \frac{a_1^{n-2j} a_2^j}{(n-2j)!\, j!},$$
$$n = 0, 1, 2, \ldots, a_1, \quad a_2 \geqslant 0, \tag{1}$$

where $[t]$ denotes the integer part of t. The distribution has multimodes. The probability generating junction* (PGF) of (1) is

$$P(s) = \sum_{n=0}^{\infty} P_n s^n$$
$$= \exp[a_1(s-1) + a_2(s^2 - 1)]. \tag{2}$$

The distribution can be used as a substitute for the Poisson when the initial population comprises singlets and pairs. In a phagocytic experiment to be discussed later, McKendrick [14] observed that bacteria which were being ingested were not all discrete; some of them were united in pairs. Using moment estimates he fitted the data with the Hermite distribution and found the fit to be very satisfactory. Some other examples are: number of larvae per plant, cornborers per hill, number of accidents involving one and two cars, and mistakes in copying groups of random digits.

MODELS LEADING TO THE DISTRIBUTION

1. If x_1 and x_2 have the bivariate Poisson distribution with parameters a_1, a_2, and a_{12} (see Johnson and Kotz [10, p. 298]) then $x_1 + x_2$ has the Hermite distribution with parameters $a_1 + a_2$ and a_{12}.
2. The Hermite distribution with parameters $2\lambda p(1-p)$ and λp^2 is a special case of the Poisson binomial distribution with parameters λ, n, and p when $n = 2$ (see Johnson and Kotz [10, p. 190]).
3. Suppose that the parameter λ of the Poisson distribution is a normal random variable with mean μ and variance σ^2; then the PGF of the resulting distribution is

$$G(s)$$
$$= (2\pi\sigma^2)^{-1/2} \int_{-\infty}^{\infty} \exp\left[\lambda(s-1) - \tfrac{1}{2}(\lambda - \mu)^2/\sigma^2\right] d\lambda$$
$$= \exp\left[(\mu - \sigma^2)(s-1) + \tfrac{1}{2}\sigma^2(s^2 - 1)\right],$$

which is that of the Hermite distribution with parameters $\mu - \sigma^2$ and $\frac{1}{2}\sigma^2$, provided that $\mu - \sigma^2 \geqslant 0$.

The distribution derives its name from Hermite polynomials [8], also known as Chebyshev–Hermite polynomials* [3], as it can be expressed in terms of their modified form. Hermite polynomials of degree n are defined by

$$H_n(x) = \sum_{j=0}^{[n/2]} \frac{(-)^j n!\, x^{n-2j}}{(n-2j)!\, j!\, 2^j}, \tag{3}$$

and have the generating function

$$\exp\left(tx - \tfrac{1}{2}t^2\right) = \sum_{n=0}^{\infty} H_n(x) t^n / n!. \tag{4}$$

Their properties are listed in Jordan [11], Erdelyi [6], and Kendall and Stuart [13]. For example,

$$\frac{d}{dx}\left[H_n(x)\right] = nH_{n-1}(x)$$

(*see also* CHEBYSHEV–HERMITE POLYNOMIALS). The polynomials possess the following orthogonality property:

$$\int_{-\infty}^{\infty} H_n(x) H_m(x) e^{-x^2/2} dx = 0, \quad n \neq m,$$

$$\int_{-\infty}^{\infty} \left[H_n(x)\right]^2 e^{-x^2/2} dx = n!\sqrt{2\pi}.$$

If a function $f(x)$ and its first two derivatives are continuous and finite from $-\infty$ to

∞ and if $f(\pm\infty) = 0$, $(d/dx)[f(\pm\infty)] = 0$, $(d^2/dx)^2[f(\pm\infty)] = 0$, then $f(x)$ can be expressed in the following convergent series:

$$f(x) = \left[C_0 + C_1 H_1(x) + C_2 H_2(x) + \cdots\right] e^{-x^2/2}/\sqrt{2\pi}.$$

Using the orthogonality property of Hermite polynomials, we obtain

$$C_n = \frac{1}{n!}\int_{-\infty}^{\infty} H_n(x) f(x)\,dx = \sum_{j=0}^{[n/2]} \frac{(-)^j M_{n-2j}}{(n-2j)!\,j!\,2^j},$$

where $M_n = \int_{-\infty}^{\infty} x^n f(x)\,dx$. In particular,

$$C_0 = M_0,\quad C_1 = M_1,\quad C_2 = (M_2 - M_0)/2,$$
$$C_3 = (M_3 - 3M_1)/6,$$
$$C_4 = (M_4 - 6M_2 + 3M_0)/24.$$

These results have been used in expanding a continuous probability density function $f(x)$ and its cumulative function in Edgeworth and Gram–Charlier series* (see Kendall and Stuart [13]).

To express the probabilities in terms of modified Hermite polynomials, we write $a_1 = \alpha\beta$ and $a_2 = \frac{1}{2}\alpha^2$ in (1) and (2), so that

$$P(s) = \exp\left[-\left(\alpha\beta + \frac{\alpha^2}{2}\right)\right]\exp\left(\alpha\beta s + \frac{\alpha^2 s^2}{2}\right) = \exp\left[-\left(\alpha\beta + \frac{\alpha^2}{2}\right)\right]\sum_{n=0}^{\infty} H_n^*(\beta)\frac{\alpha^n s^n}{n!}$$

and

$$p_0 = \exp\left[-\left(\alpha\beta + \frac{\alpha^2}{2}\right)\right],$$
$$p_n = p_0 \alpha^n H_n^*(\beta)/n!,\quad n = 0, 1, 2, \ldots$$

where $H_n^*(\beta)$ are the *modified Hermite polynomials* [12] given by

$$H_n^*(\beta) = i^{-n} H_n(i\beta) = \sum_{j=0}^{[n/2]} \frac{n!\,\beta^{n-2j}}{(n-2j)!\,j!\,2^j}.$$

The properties of $H_n^*(x)$ are similar to those of $H_n(x)$. For example,

$$\frac{d}{dx}\left[H_n^*(x)\right] = nH_{n-1}^*(x),$$
$$H_{n+1}^*(x) = xH_n^*(x) + nH_{n-1}^*(x).$$

These modified polynomials have as coefficients the absolute values of the Hermite coefficients.

Expanding the polynomials gives the probabilities

$$p_1 = p_0\alpha\beta,\quad p_2 = p_0\alpha^2(\beta^2 + 1)/2!;$$

in general,

$$p_{n+1} = (\alpha\beta p_n + \alpha^2 p_{n-1})/(n+1), \tag{7}$$

a recurrence relation for calculating the probabilities.

MOMENTS

Using the moment generating function* $M(t)$ of the Hermite distribution and collecting the coefficient of $t^r/r!$ in the expansion of $\log M(t)$ gives the rth cumulant as

$$K_r = a_1 + 2^r a_2.$$

The mean and the succeeding three central moments about the mean are

$$\mu = \kappa_1 = a_1 + 2a_2 = \alpha(\alpha + \beta)$$
$$\mu_2 = \kappa_2 = a_1 + 4a_2 = \alpha(2\alpha + \beta)$$
$$\mu_3 = \kappa_3 = a_1 + 8a_2 = \alpha(4\alpha + \beta)$$
$$\mu_4 = \kappa_4 + 3\kappa_2^2 = a_1 + 16a_2 + 3(a_1 + 4a_2)^2 = \alpha(8\alpha + \beta) + 3\alpha^2(2\alpha + \beta)^2.$$

Expanding the factorial moment generating function* of the Hermite distribution the rth factorial moment can be obtained in terms of modified polynomials. This is given by

$$\mu'_{(r)} = E\left[x(x-1)\cdots(x-r+1)\right] = \alpha^r H_r^*(\alpha + \beta),$$

and a use of the recurrence relation for modified polynomials gives

$$\mu'_{(r+1)} = \alpha(\alpha + \beta)\mu'_{(r)} + \alpha^2 r\mu'_{(r-1)}.$$

ESTIMATES OF PARAMETERS

The moment estimates can be obtained by using the expressions for μ and μ_2 and replacing them by their sample estimates (*see* METHOD OF MOMENTS). The maximum likelihood* (ML) estimates can be obtained by solving the equations

$$S_\alpha = \sum_n f_n\left[\frac{n}{\alpha} - (\alpha + \beta)\right] = 0, \qquad (8)$$

giving

$$\bar{x} = \alpha(\alpha + \beta), \qquad (9)$$

and

$$S_\beta = \sum_n f_n \alpha\left(\frac{p_{n-1}}{p_n} - 1\right) = 0. \qquad (10)$$

To solve (8) and (10) for the maximum likelihood values $\hat{\alpha}$, $\hat{\beta}$, two methods were used by Kemp and Kemp [12]. In the method of false position, take two real values α_0 and α_1, determine the corresponding values β_0 and β_1 from (9), compute the p_n from (7) and calculate S_β from (10). The ML value $\hat{\alpha}$ is that for which $S_\beta = 0$. Using these values, interpolate for a new α_2 nearer $\hat{\alpha}$. Next, interpolate between α_2 and either α_0 or α_1, and so on. Usually, a number of cycles are required to have $S_\beta \simeq 0$.

In the Newton–Raphson method, the second derivatives of log L are replaced by their expected values:

$$E\left(\frac{\partial^2 \log L}{\partial \alpha^2}\right) = -N\left(1 + \frac{\mu}{\alpha^2}\right)$$
$$= N\left(1 + \frac{\bar{x}}{\alpha^2}\right) = -I_{\alpha\alpha},$$

$$E\left(\frac{\partial^2 \log L}{\partial \alpha \partial \beta}\right) = -N = -I_{\alpha\beta},$$

$$E\left(\frac{\partial^2 \log L}{\partial \beta^2}\right) = N\alpha^2 \sum_{n=0}^{\infty} p_n\left(\frac{p_{n-2}}{p_n} - \frac{p_{n-1}^2}{p_n^2}\right)$$
$$= -N\alpha^2 \sum_{n=0}^{\infty} p_n\left(\frac{p_{n-1}}{p_n} - 1\right)^2$$
$$= -I_{\beta\beta},$$

where $N = \sum_{n=0}^{k} f_n$, and using the fact that $\sum p_n = \sum p_{n-1} = \sum p_{n-2}$. The method is applied as follows. Take a trial value β_0 for β and choose a positive root α_0 by solving (8). Then

$$\beta_1 = \beta_0 + \frac{I_{\alpha\alpha} S_\beta}{I_{\alpha\alpha} I_{\beta\beta} - I_{\alpha\beta}^2}.$$

After the final cycle estimate of I's, α, and β, the estimates of the large-sample variances and covariances of $\hat{\alpha}$ and $\hat{\beta}$ can be obtained from the inverse matrix,

$$\begin{pmatrix} I_{\alpha\alpha} & I_{\alpha\beta} \\ I_{\alpha\beta} & I_{\beta\beta} \end{pmatrix}^{-1}.$$

Although each cycle takes a little longer than in the false position method, convergence seems to be more rapid [12].

TWO GENERALIZATIONS

Generalized Hermite Distribution [7]

Suppose that x_1 and x_2 are two independent Poisson variables with parameters α and β. The random variable $x_1 + mx_2$ then has a generalized hermite distribution. Its PGF is

$$P(s) = \exp[-(\alpha + \beta)]\exp(\alpha s + \beta s^m)$$
$$= \sum_{n=0}^{\infty} p_{n,m}(\alpha, \beta) s^n.$$

The probabilities are

$$p_{n,m}(\alpha, \beta) = p_0 H_{n,m}(\alpha, \beta)/n!, \qquad n = 0, 1, 2, \ldots,$$

where

$$p_0 = \exp[-(\alpha + \beta)],$$

and

$$H_{n,m}(\alpha, \beta) = \sum_{k=0}^{[n/m]} \frac{\beta^k}{k!} \frac{n!}{(n - mk)!} \alpha^{n-mk}$$

are the generalized polynomials. Expansion of the polynomials gives the probabilities

$$p_{j,m} = p_0 \alpha^j / j!, \quad j = 0, 1, 2, \ldots, m - 1,$$
$$p_{m,m} = p_0(\alpha^m + \beta m!)/m!$$
$$p_{r,m} = (\alpha p_{r-1,m} + \beta m p_{r-m,m})/r, \quad r \geqslant m,$$

where $p_{r,m} \equiv p_{r,m}(\alpha, \beta)$. The first four moments are found to be

$$\mu = \alpha + \beta m, \quad \mu_2 = \alpha + \beta m^2,$$
$$\mu_3 = \alpha + \beta m^3,$$
$$\mu_4 = \alpha + \beta m^4 + 3(\alpha + \beta m^2)^2.$$

Table 1

Number of Bacteria per Leucocyte	Observed Frequency	Poisson	Hermite McKendrick by Moments	Hermite Kemp and Kemp by ML	Borel–Hermite Type B by Moments
0	269	245.7	268	269.6	269.3
1	4	49.1	7	3.7	3.5
2	26	4.9	23	25.2	26.1
3	0	0.3	0.6	0.3	0.3
4	1	0.0	1.1	1.2	0.8
Total	300.00	300.00	299.7	300.00	300.00
χ^2		601.97	1.50	0.00	0.00

Borel–Hermite Distributions

Jain and Plunkett [9] considered the "random sum distribution" of an ordinary generalized Poisson variable and a generalized Poisson "doublet" variable. Suppose that an initial population is comprised of groups or clusters. If the groups are assumed to follow the Poisson distribution with mean m, while the group sizes are assumed to have the Borel distribution [1]

$$g_j = (\lambda j)^{j-1}e^{-\lambda j}/j!, \qquad j = 1, 2, \ldots$$

then the distribution of the final counts of the population is the generalized Poisson distribution [4, 9],

$$b_j = m\lambda^j(m+j)^{j-1}e^{-\lambda(m+j)}/j!, \qquad j = 0, 1, 2, \ldots$$

$$= \lambda_1(\lambda_1 + \lambda j)^{j-1}e^{-(\lambda_1+\lambda j)}/j!, \qquad j = 0, 1, 2, \ldots$$

where $\lambda_1 = m\lambda > 0$, $0 < \lambda < 1$.

Now assume further that the initial population consists of two independent sets; one of singlets and the other of pairs, and that each one is distributed according to the generalized Poisson distribution. The distribution of the total count of the population can be written as

$$p_n = \sum_{k=0}^{[n/2]} \frac{m_1(m_1+k)^{k-1}e^{-\lambda_1(m_1+k)}m(m+n-2k)^{n-2k-1}\lambda^{n-2k}}{k!(n-2k)!} \times \exp[-\lambda(m+n-2k)].$$

The first four moments of the Borel–Hermite distribution are

$$\mu = \frac{m\lambda}{1-\lambda} + \frac{2m_1\lambda_1}{1-\lambda_1},$$

$$\mu_2 = \frac{m\lambda}{(1-\lambda)^3} + \frac{4m_1\lambda_1}{(1-\lambda_1)^3},$$

$$\mu_3 = \frac{m(1+2\lambda)}{(1-\lambda)^5} + \frac{8m_1\lambda_1(1+2\lambda_1)}{(1-\lambda_1)^5},$$

$$\mu_4 = \frac{m\lambda(1+8\lambda+6\lambda^2)+3m^2\lambda^2(1-\lambda)}{(1-\lambda)^7} + \frac{16\left[m_1\lambda_1(1+8\lambda_1+6\lambda_1^2)+3m_1^2\lambda_1^2(1-\lambda_1)\right]}{(1-\lambda)^7} + \frac{24m\lambda m_1\lambda_1}{(1-\lambda)^3(1-\lambda_1)^3}.$$

The Borel–Hermite distribution has four parameters: m, λ, m_1, and λ_1. The following distributions with two and three parameters have been defined by Jain and Plunkett [9]:

1. Borel–Hermite type A, when $m\lambda = m_1\lambda_1$, $\lambda = \lambda_1$
2. Borel–Hermite type B, when $\lambda = \lambda_1$, $m \neq m_1$
3. Borel–Hermite type C, when $m\lambda = m_1\lambda_1$, $\lambda \neq \lambda_1$
4. Borel–Hermite type D, when $m = m_1$, $\lambda \neq \lambda_1$

GOODNESS OF FIT

Table 1 considers McKendrick's multimodal data [14] which he fitted with a Hermite distribution by using moment estimates. The maximum likelihood estimates were used by

Kemp and Kemp [12] which significantly improved the fit. Jain and Plunkett [9] fitted the data with a Borel–Hermite type B distribution by the method of moments. By a comparison of χ^2-values it is evident that Hermite and Borel–Hermite type B distributions provide the best fit.

References

[1] Borel, E. (1942). *C. R. Acad. Sci. Paris*, **214**, 452–456.

[2] Charlier, C. V. L. (1905). *Ark. Mat. Astron. Fys.*, **2**(20), 1–35.

[3] Chebyshev, P. L. (1860). *Bull. Acad. Imp. Sci. St. Petersbourg*, **1**(3), 193–200.

[4] Consul, P. C. and Jain, G. C. (1973). *Technometrics*, **15**, 791–799.

[5] Edgeworth, F. Y. (1907). *J. R. Statist. Soc. A*, **70**, 102–104.

[6] Erdelyi, A. (1953). *Higher Transcendental Functions*. McGraw-Hill, New York.

[7] Gupta, R. P. and Jain, G. C. (1974). *SIAM J. Appl. Math.*, **27**, 359–363.

[8] Hermite, C. (1864). *C. R. Acad. Sci. Paris*, **58**, 93–100, 266–273.

[9] Jain, G. C. and Plunkett, I. G. (1977). *Biom. J.*, **19**, 347–354.

[10] Johnson, N. L. and Kotz, S. (1969). *Distributions in Statistics. Discrete Distributions*. Wiley, New York.

[11] Jordan, C. (1947). *Calculus of Finite Differences*. Chelsea, New York.

[12] Kemp, C. D. and Kemp, A. W. (1965). *Biometrika*, **52**, 381–394.

[13] Kendall, M. G. and Stuart, A. (1977). *The Advanced Theory of Statistics*, Vol. 1: *Distribution Theory*, 4th ed. Macmillan, New York.

[14] McKendrick, A. G. (1926). *Proc. Edinb. Math. Soc.*, **44**, 98–130.

(CHEBYSHEV–HERMITE POLYNOMIALS
POISSON DISTRIBUTIONS)

GOPI C. JAIN

HERMITE FORM

An $n \times n$ matrix $\mathbf{A}$ is in upper Hermite form if and only if $\mathbf{A}$ is an upper-triangular matrix, the diagonal entries consist of zeros and ones only, every element is zero in any row having zero as its diagonal element, and every off-diagonal element is zero in any column having 1 as its diagonal element.

Hermite forms are useful in the study of conditional inverses (see Graybill [1, pp. 120–129]).

Reference

[1] Graybill, F. A. (1969). *Introduction to Matrices with Applications in Statistics*. Wadsworth, Belmont, Calif.

(GENERALIZED INVERSE)

HERMITE POLYNOMIALS *See* CHEBYSHEV–HERMITE POLYNOMIALS

HETEROGENEITY *See* HOMOGENEITY

HETEROSCEDASTICITY

If one has observations from several sources, say $X_{i1}, X_{i2}, \ldots$ (which are independent and identically distributed random variables) from source i $(i = 1, 2, \ldots, k)$, interest is often in the means $\mu_i = E(X_{i1})$ and variances $\sigma_i^2 = \text{var}(X_{i1})$. One talks of *homoscedasticity* if $\sigma_1^2 = \cdots = \sigma_k^2$, and of *heteroscedasticity* otherwise.

Until recently, the procedures available for these problems assumed normality and $\sigma_1^2 = \cdots = \sigma_k^2 = \sigma^2$ and provided performance characteristics (e.g., power* for a test, confidence coefficient or length for a confidence interval*, probability of correct selection for a ranking-and-selection* procedure) which depended on the unknown σ^2. The solutions given below for these problems do not assume equal variances, yet do allow full control of such performance characteristics as power*, confidence interval* length, and probability of correct selection. These solutions, and those of a large number of other problems involving heteroscedasticity, have as a cornerstone the solution of a simpler problem given below.

When the X_{ij} have nonnormal distributions,

$$\bar{X}_i = \sum_{j=1}^{n} X_{ij}/n \quad \text{and}$$

$$s_i^2 = \sum_{j=1}^{n} \left(X_{ij} - \bar{X}_i\right)^2/(n-1)$$

are still unbiased estimators of μ_i and σ_i^2, respectively. However, $\bar{X}_i$ and s_i^2 are no longer independent random variables and confidence intervals for μ_i (even with random length) are not generally available. Moreover, $\bar{X}_i$ and s_i^2 (while each asymptotically normal by the central limit theorem) are no longer jointly minimal sufficient statistics for (μ_i, σ_i^2) in this setting (*see* ASYMPTOTIC NORMALITY; SUFFICIENCY). Here transformations* are often used; i.e., one lets $Y_{ij} = \xi(X_{ij})$, $1 \leqslant j \leqslant n$, for some function $\xi(\cdot)$ such that $\xi(X_{ij})$ is normally distributed. Then $E(Y_{ij})$ and $\mathrm{var}(Y_{ij})$ can be estimated as before, and confidence intervals provided. Methods are available for relating these estimates and intervals back to μ_i and σ_i^2, the quantities of primary interest.

The procedures given require the experimenter to have design control, but generalize to any statistical problem via the heteroscedastic method (to be discussed in the section "Heteroscedastic Method"). Problems of nonnormality, comparison with the usual variance-stabilizing-transformation approach, and other comparisons and questions that arise in practical implementation are discussed throughout.

BASIC SAMPLING RULE $\mathscr{S}_B(n_0, w)$

If we are able to observe independent and identically distributed normal random variables $X_1, X_2, \ldots$ with mean μ and variance σ^2 (both unknown), and wish to make inferences about μ, recall that $(\bar{X} - \mu)/(s/\sqrt{n})$ has Student's t-distribution* with $n - 1$ degrees of freedom, which gives not only a point estimate, but also a $100(1 - \alpha)\%$ confidence interval for μ:

$$\bar{X} - t_{n-1,1-\alpha/2}\frac{s}{\sqrt{n}} \leqslant \mu \leqslant \bar{X} + t_{n-1,1-\alpha/2}\frac{s}{\sqrt{n}}.$$

But the interval length, $2t_{n-1,1-\alpha/2}s/\sqrt{n}$, is a random variable and cannot be controlled to be $\leqslant 2L$, for example, by choice of the sample size n. Instead, therefore, we use the following:

Sampling Rule $\mathscr{S}_B(n_0, w)$. Take an initial sample $X_1, \ldots, X_{n_0}$ of size $n_0 (\geqslant 2)$, and calculate

$$\bar{X}(n_0) = \sum_{j=1}^{n_0} X_j/n_0,$$

$$s^2 = \sum_{j=1}^{n_0}\left[X_j - \bar{X}(n_0)\right]^2/(n_0 - 1),$$

$$N = \max\left\{n_0 + 1, \left[(ws)^2\right]\right\},$$

where $w > 0$ (depends on the problem under consideration) and $[y]$ denotes the smallest integer $\geqslant y$ (e.g., $[5.1] = 6$ . . . introduced because sample sizes must be integers). Take $N - n_0$ additional observations $X_{n_0+1}, \ldots, X_N$ and calculate

$$\bar{X}(N - n_0) = \sum_{j=n_0+1}^{N} X_j/(N - n_0),$$

$$\tilde{\bar{X}} = b\bar{X}(n_0) + (1 - b)\bar{X}(N - n_0),$$

$$\bar{X} = \sum_{j=1}^{N} X_j/N,$$

where

$$b = \frac{n_0}{N}\left[1 - \sqrt{1 - \frac{N}{n_0}\left[1 - \frac{N - n_0}{(ws)^2}\right]}\right].$$

Stein showed in 1945 [22] that, for his sampling rule $\mathscr{S}$, $(\tilde{\bar{X}} - \mu)/(1/w) \sim t_{n_0-1}$ (*see* FIXED-WIDTH CONFIDENCE INTERVALS). Therefore,

$$\tilde{\bar{X}} - t_{n_0-1,1-\alpha/2}\frac{1}{w} \leqslant \mu \leqslant \tilde{\bar{X}} + t_{n_0-1,1-\alpha/2}\frac{1}{w}$$

is an exact $100(1 - \alpha)\%$ confidence interval for μ, and its half-length can be fully controlled to a preset number $L > 0$ by choosing w in $\mathscr{S}_B(n_0, w)$ such that

$$t_{n_0-1,1-\alpha/2}\frac{1}{w} = L, \quad \text{i.e., } w = \frac{t_{n_0-1,1-\alpha/2}}{L}.$$

Note that, since N is an increasing function of w, the total sample size N required is larger for small values of half-length L as well as for high confidence coefficients $1 - \alpha$. (While use of $\bar{X}$ in this setting would yield a slight uniform improvement over use of $\tilde{\bar{X}}$, it is noted in the section "Heteroscedasticity (Several Sources): Tests" that

Table 1 $w = t_{n_0-1,1-\alpha/2}/L$ **for** $L = 1$, $1-\alpha = 0.95$

n_0	2	3	4	5	6	7	8	9	10	11	12	∞
w	12.706	4.303	3.182	2.776	2.571	2.447	2.365	2.306	2.262	2.228	2.201	1.960

this is not usually the case when dealing with heteroscedasticity of several sources. Attempts to show the opposite retarded development of procedures for heteroscedasticity for many years.)

The procedure described above is valid for any preliminary sample size $n_0 \geqslant 2$. Since $t_{n_0-1,1-\alpha/2}$ decreases as n_0 increases, it is reasonable to make n_0 large if possible. The decrease is negligible after $n_0 \geqslant 12$ or so; hence it is reasonable to take $n_0 = 12$ (or larger, for example, if one is sure to take $n_0 \geqslant B$ for some positive integer $B \geqslant 12$).

The validity of the procedure for any n_0 (with no "optimal" choice of n_0 being obvious) bothered early workers in the field and led to disuse of these early procedures. The realization that $n_0 \geqslant 12$ is all that is required for good results in practice as far as n_0 is concerned is a factor leading to great current interest in these procedures and their extensions as will be discussed later. One may think of the situation as follows: one's total sample size N is approximately w^2s^2, and taking n_0 very small will force a large total sample size simply because of a poor initial estimate s^2 (see Table 1).

BASIC NONNORMALITY AND TRANSFORMATIONS

If X_1 in the preceding section is nonnormal, one often uses such transformations as

$$\xi_1(X_1) = \sqrt{X_1 - a}$$

$$\xi_2(X_1) = X_1^{1/3}$$

$$\xi_3(X_1) = \log_{10}(X_1)$$

$$\xi_4(X_1) = \arcsin\sqrt{X_1}$$

$$\xi_5(X_1) = \sinh^{-1}\sqrt{X_1}\,.$$

If one of these, say $\xi(X_1)$, is normally distributed, then the mean and variance of $Y_i = \xi(X_i)$ may be estimated by

$$\overline{Y} = \sum_{j=1}^{n} Y_j/n, \quad s_Y^2 = \sum_{j=1}^{n}\left(Y_j - \overline{Y}\right)^2/(n-1).$$

However, interest in many cases is not in $E\xi(X_1)$ and $\operatorname{var}\xi(X_1)$, but in the original problem units, EX_1 and $\operatorname{var}(X_1)$. Simply using the inverse transformation, for example to estimate EX_1 by $\overline{Y}^2 + a$ in the case of ξ_1, results in a biased estimate. However, Neyman* and Scott showed in 1960 that the unique minimum variance unbiased estimators* (MVUEs) of $E(X_1)$ are as shown in Table 2, assuming that $\xi(X_1)$ is normally distributed. General results for second-order entire functions were also given by Neyman and Scott.

In 1968, Hoyle provided the MVUEs of $\operatorname{var}(X_1)$ and, more important, of the variances of the estimators of EX_1 given in Table 2 (see Table 3). [Here it is also assumed that $\xi(X_1)$ is normally distributed.] The latter can be used to obtain approximate 95% confidence intervals for EX_1; for example, when using $\sqrt{X_1 - a}$,

$$\mu \in \overline{Y}^2 + a + \left(1 - \frac{1}{n}\right)s_Y^2 \pm 2\sqrt{\lambda}\,,$$

$$\lambda = \frac{4}{n}s_Y^2\overline{Y}^2 + s_Y^4\left\{\left(1-\frac{1}{n}\right)^2 - \frac{n-1}{n+1}\left[1 - 2\left(1-\frac{1}{n}\right)^2 + 3\left(1-\frac{1}{n}\right)^4\right]\right\}.$$

More recent work and comparisons with other methods are given by Land [18].

HETEROSCEDASTICITY (SEVERAL SOURCES): TESTS

Let $X_{i1}, X_{i2}, \ldots$ be independent and identically distributed normal random variables with mean μ_i and variance σ_i^2 ($i = 1, 2, \ldots, k$). Experimenters have often been cautioned that "the assumption of equal

Table 2 Transformations and MVUEs[a] of $E(X_1)$

$\xi(X_1)$	MVUE of EX_1
$\sqrt{X_1 - a}$	$\overline{Y}^2 + a + (1 - \frac{1}{n})s_Y^2$
$\log_{10}(X_1)$	$10^{\overline{Y}} S[(\ln 10)^2(1 - \frac{1}{n})(n-1)s_Y^2, n-1]$
$\arcsin \sqrt{X_1}$	$(\sin^2 \overline{Y} - 0.5)S[4(\frac{1}{n} - 1)(n-1)s_Y^2, n-1] + 0.5$
$\sinh^{-1} \sqrt{X_1}$	$(\sinh^2 \overline{Y} + 0.5)S[4(1 - \frac{1}{n})(n-1)s_Y^2, n-1] - 0.5$

[a] Here $S(a,b) = \sum_{i=0}^{\infty}[(1/i!)(\Gamma(b/2)/\Gamma(i + (b/2)))(a/4))^i]$; this series converges faster than the series for the exponential function.

Table 3 MVUEs[a] of Variances of Table 2 Estimators of EX_1

$\xi(X_1)$	MVUE of Variance of MVUE of EX_1
$\sqrt{X_1 - a}$	$\frac{4}{n}s_Y^2\overline{Y}^2 + s_Y^4\{(1 - \frac{1}{n})^2 - \frac{n-1}{n+1}[1 - 2(1 - \frac{1}{n})^2 + 3(1 - \frac{1}{n})^4]\}$
$\log_{10}(X_1)$	$10^{2\overline{Y}}\{S^2[(\ln 10)^2(1 - \frac{1}{n}), n-1] - S[2(\ln 10)^2(1 - \frac{2}{n}), n-1]\}$
$\arcsin \sqrt{X_1}$	$(\widehat{EX_1})^2 - \frac{1}{4} - \frac{1}{8}S(-8, n-1) + \frac{1}{2}\cos(2\overline{Y})S[-4(1 - \frac{1}{n}), n-1] - \frac{1}{8}\cos(4\overline{Y})S[-8(1 - \frac{2}{n}), n-1]$
$\sinh^{-1} \sqrt{X_1}$	$(\widehat{EX_1})^2 - \frac{1}{4} - \frac{1}{8}S(8, n-1) + \frac{1}{2}\cosh(2\overline{Y})S[4(1 - \frac{1}{n}), n-1] - \frac{1}{8}\cosh(4\overline{Y})S[8(1 - \frac{2}{n}), n-1]$

[a] For $S(a, b)$, see the footnote to Table 2.

variability should be investigated" (e.g., by Cochran and Cox in 1957 [6], by Juran et al. in 1974 [17]). For some tests for homoscedasticity, see, e.g., Harrison and McCabe [13] or Bickel [2]; typically such tests have low power, and may not detect even substantial heteroscedasticity. However, no exact statistical procedures have been available for dealing with cases where one finds that variabilities are unequal. [A variance-stabilizing transformation is commonly employed (e.g., arcsin for binomial data; *see* EQUALIZATION OF VARIANCES); however, if X_{ij} *is* normal, then $\xi(X_{ij})$ will be nonnormal. The transformation method has not been developed to handle this problem except in special cases, and even there one deals not with the parameters $\mu_1, \ldots, \mu_k$ of basic interest if one uses such a transformation, but rather with some transform whose meaning (i.e., interpretability) will not often be clear. We do not therefore regard transformations as of general use for $k \geqslant 2$ when $\mu_1, \ldots, \mu_k$ are parameters of natural interest (not arbitrary parametrizations).]

It was first developed in the 1970s by E. J. Dudewicz that, applying sampling procedure $\mathscr{S}_B(n_0, w)$ from the section "Basic Sampling Rule $\mathscr{S}_B(n_0, w)$" separately to each source of observations, one would obtain the ability to control fully the performance characteristics of statistical procedures even in the presence of heteroscedasticity. Let $\tilde{\tilde{X}}_i$ result from applying the sampling procedure to $X_{i1}, X_{i2}, \ldots$ $(i = 1, 2, \ldots, k)$. When $k = 1$, one can develop procedures (as in the preceding section) using $\tilde{\tilde{X}}$, but if one replaces this by $\overline{X}$ at the end, the procedure is still valid; it has slightly better performance characteristics (higher power), and is even simpler ($\overline{X}$

being simpler than $\tilde{\bar{X}}$, which is a random-weighted combination of the sample means of the first and second stages of sampling). However, this improvement is not large: approximately the amount that increasing sample size from N to $N+1$ will buy. This improvement of $\bar{X}$ over $\tilde{\bar{X}}$ has been shown *not* to hold generally when $k \geqslant 2$: in most such cases, if $\bar{X}_1, \ldots, \bar{X}_k$ are used to replace $\tilde{\bar{X}}_1, \ldots, \tilde{\bar{X}}_k$, the procedure no longer has the desired performance characteristics.

We describe the new analysis-of-variance* procedures for the one-way* layout; similar procedures are available for r-way layouts, $r > 1$. In the one-way layout, we might want to test the null hypothesis

$$H_0: \mu_1 = \mu_2 = \cdots = \mu_k .$$

Define

$$\tilde{F} = \sum_{i=1}^{k} w^2(\tilde{\bar{X}}_i - \tilde{\bar{X}}.)^2,$$

$$\tilde{\bar{X}}. = \frac{1}{k}\sum_{i=1}^{k}\tilde{\bar{X}}_i;$$

reject H_0 if and only if

$$\tilde{F} > \tilde{F}^{\alpha}_{k,n_0},$$

where $\tilde{F}^{\alpha}_{k,n_0}$ is the upper αth percent point of the null distribution of $\tilde{F}$. This distribution is also that of $Q = \sum_{i=1}^{k}(t_i - \bar{t}.)^2$, where the $\{t_i\}$ are independent identically distributed Student's-t variates with $n_0 - 1$ degrees of freedom and $\bar{t}. = (1/k)\sum_{i=1}^{k} t_i$.

Values of $\tilde{F}^{\alpha}_{k,n_0}$ obtained by a Monte Carlo* sampling experiment, together with the power attained at various alternatives measured by $\delta = \sum_{i=1}^{k}(\mu_i - \bar{\mu}.)^2$, for various given $1/w^2$ values, appear in Bishop and Dudewicz [3]. There is a need for approximations to the percentage points of the $\tilde{F}$ statistics under the null and alternative distributions. Such approximations are available in the general setting (see Dudewicz and Bishop [9]), and have been studied in special cases (see Bishop et al. [4]). Consider first the limiting distribution of $\tilde{F}$ as $n_0 \to \infty$. This is noncentral chi-square* with $k-1$ degrees of freedom and noncentrality parameter $\Delta = \sum_{i=1}^{k} w^2(\mu_i - \bar{\mu}.)^2$, denoted by $\chi^2_{k-1}(\Delta)$. However, numerical results indicate that for small n_0 the tails of this distribution are too light to give a good approximation. One therefore approximates by a random variable with a $[(n_0 - 1)/(n_0 - 3)]\chi^2_{k-1}(\Delta)$ distribution (in which case $\tilde{F}$ and its approximating distribution have the same expected value under H_0).

Example. Suppose that we wish to test the hypothesis that four different chemicals are equivalent in their effects. Suppoe that we decide to take initial samples of size 10 with each treatment, that we want only a 5% chance of rejecting H_0 if in fact H_0 is true, and an 85% chance of rejecting H_0 if the spread among μ_1, μ_2, μ_3, and μ_4 is at least 4.0 units. We then proceed, step by step, as follows.

Step 1: Problem Specification. With $k = 4$ sources of observations, we desire an $\alpha = 0.05$ level test of H_0: $\mu_1 = \mu_2 = \mu_3 = \mu_4$, and if the spread among μ_1, μ_2, μ_3, and μ_4 is $\delta = 4.0$ units or more, we desire power (probability of then rejecting the false hypothesis H_0) of at least $P^* = 0.85$.

Step 2: Choice of Procedure. Assuming we do not know that $\sigma_1^2 = \sigma_2^2 = \sigma_3^2 = \sigma_4^2$, only procedure $\mathscr{S}_B(n_0, w)$ can guarantee the specifications. It requires that we sample n_0 observations in our first stage, and recommends that n_0 be at least 12 (although any $n_0 \geqslant 2$ will work). Suppose that the experimenter only wants to invest 40 units in first-stage experimentation and sets $n_0 = 10$.

Step 3: First Stage. Draw $n_0 = 10$ independent observations from each source, with results as shown in Table 4.

Step 4: Analysis of First-Stage Data. Calculate the first-stage sample variances s_1^2, s_2^2, s_3^2, s_4^2, the total sample sizes N_1, N_2, N_3, N_4, needed from the four sources, and the factors b_1, b_2, b_3, b_4 to be used in the second-

Table 4 First-Stage Samples

Chemical 1	Chemical 2	Chemical 3	Chemical 4
77.199	80.522	79.417	78.001
74.466	79.306	78.017	78.358
82.746	81.914	81.596	77.544
76.208	80.346	80.802	77.364
82.876	78.385	80.626	77.554
76.224	81.838	79.011	75.911
78.061	82.785	80.549	78.043
76.391	80.900	78.479	78.947
76.155	79.185	81.798	77.146
78.045	80.620	80.923	77.386

stage analysis. These quantities appear in Table 5. The value of w is found as follows.

We desire power $P^* = 0.85$ (step 1 above) when

$$\Delta = \frac{w^2\delta^2}{4} = \frac{w^2(4.0)^2}{4} = 4.0w^2.$$

To set w for this requirement, we first need to know when we reject. We will later reject H_0 if $\tilde{F} > \tilde{F}_{4,10}^{0.05}$, where, approximately,

$$\tilde{F}_{4,10}^{0.05} = \frac{n_0 - 1}{n_0 - 3}(7.81) = 10.04,$$

and where a chi-square variable with $k - 1 = 3$ degrees of freedom exceeds 7.81 with probability $\alpha = 0.05$ (see standard tables, e.g., Pearson and Hartley [20, p. 137], Dudewicz [8, p. 459]). The power will be, approximately,

$$P\left[\chi_3^2(\Delta) > 7.81\right] = 0.85$$

if $\Delta = 12.301$ (see the tables in Haynam et al. [14, p. 53]), so $w^2 = 12.301/4.0 = 3.075$.

Step 5: Second Stage. Draw $N_i - n_0$ observations from source i ($i = 1, 2, 3, 4$), yielding Table 6.

Table 6 Second-Stage Samples

Chemical 1	Chemical 2	Chemical 3	Chemical 4
82.549	79.990	80.315	78.037
78.970			
78.496			
78.494			
80.971			
80.313			
76.556			
80.115			
78.659			
77.697			
80.590			
79.647			
82.733			
80.552			
79.098			

Step 6: Final Calculations. Calculate the $\tilde{\bar{X}}_i$ and $\tilde{F}$, and find

$$\tilde{\bar{X}}_1 = 79.079, \quad \tilde{\bar{X}}_2 = 80.688,$$
$$\tilde{\bar{X}}_3 = 80.197, \quad \tilde{\bar{X}}_4 = 77.597;$$
$$\tilde{\bar{X}}_. = 79.390, \quad \tilde{F} = 17.38.$$

Step 7: Final Decision. Since $\tilde{F} = 17.38$ exceeds $\tilde{F}_{4,10}^{0.05} = 10.04$, we reject the null hypothesis and decide that the chemicals differ.

HETEROSCEDASTICITY (SEVERAL SOURCES): CONFIDENCE INTERVALS

The case of a confidence interval for the mean when $k = 1$ was considered in the section "Basic Sampling Rule $\mathscr{S}_B(n_0, w)$". When $k = 2$, a two-sided confidence interval for the difference $\mu_1 - \mu_2$, of half-length $L > 0$ and with confidence coefficient $1 - \alpha$,

Table 5 Analysis of First Stage

	Chemical 1	Chemical 2	Chemical 3	Chemical 4
n_0	10	10	10	10
Sample mean	77.837	80.580	80.122	77.625
s_i^2	7.9605	1.8811	1.7174	0.6762
w	1.754	1.754	1.754	1.754
N_i	25	11	11	11
b_i	0.330	0.936	0.939	0.969

Table 7 $c_{1-\gamma}(n_0)$

n_0 / $1-\gamma$	10	11	12	13	14	15	20	25	30
0.75	1.03	1.02	1.02	1.01	1.01	1.00	0.99	0.98	0.98
0.80	1.29	1.28	1.27	1.26	1.26	1.25	1.24	1.23	1.22
0.85	1.60	1.59	1.57	1.56	1.56	1.55	1.53	1.51	1.51
0.90	2.00	1.98	1.96	1.95	1.94	1.93	1.90	1.88	1.87
0.95	2.61	2.58	2.56	2.53	2.52	2.50	2.45	2.42	2.41
0.975	3.18	3.13	3.09	3.06	3.04	3.02	2.95	2.91	2.88
0.99	3.89	3.82	3.76	3.71	3.67	3.64	3.54	3.48	3.45
0.995	4.41	4.31	4.24	4.18	4.13	4.09	3.96	3.89	3.85
0.999	5.61	5.45	5.32	5.22	5.14	5.07	4.86	4.74	4.67

is given by

$$\left(\tilde{\bar{X}}_1 - \tilde{\bar{X}}_2\right) - L \leqslant \mu_1 - \mu_2 \leqslant \left(\tilde{\bar{X}}_1 - \tilde{\bar{X}}_2\right) + L$$

if we choose [in $\mathscr{S}_B(n_0, w)$]

$$w = \frac{c_{1-\alpha/2}(n_0)}{L},$$

where c is as given in Table 7.

Note that the corresponding test solves the Behrens–Fisher problem* exactly in two stages, with controlled level and power. This was first noted by Chapman [5].

For $k > 2$, multiple-comparison* procedures are also available for many of the usual multiple-comparison confidence interval goals see, e.g., Dudewicz et al. [12]).

HETEROSCEDASTICITY (SEVERAL SOURCES): RANKING AND SELECTION*

Here $k \geqslant 2$ and, in the indifference-zone formulation of the problem, we wish to select that source having mean value $\max(\mu_1, \ldots, \mu_k)$. Let $\mu_{[1]} \leqslant \cdots \leqslant \mu_{[k]}$ denote the ordered values of $\mu_1, \ldots, \mu_k$; thus $\mu_{[k]}$ denotes $\max(\mu_1, \ldots, \mu_k)$, etc.

The performance characteristic of interest is the probability that we will make a correct selection (CS), i.e., that the population selected is the one that has mean $\mu_{[k]}$. Following Bechhofer [1], we require Pr(CS) to have at least a specified value P^* $(1/k < P^* < 1)$ whenever the largest mean is at least δ^* more than the next-to-largest mean; i.e. we require

$$\Pr(\mathrm{CS}) \geqslant P^*$$

whenever

$$\mu_{[k]} - \mu_{[k-1]} \geqslant \delta^* > 0$$

The procedure (see Dudewicz and Dalal [10]) is to select that source which yields the largest of $\tilde{\bar{X}}_1, \ldots, \tilde{\bar{X}}_k$; i.e.,

$$\text{select } \pi_i \text{ iff } \tilde{\bar{X}}_i = \max\left(\tilde{\bar{X}}_1, \ldots, \tilde{\bar{X}}_k\right).$$

In the sampling rule $\mathscr{S}_B(n_0, w)$ one chooses $w = c_{P^*}(n_0)/\delta^*$, where $c_{P^*}(n_0)$ for specified values of P^* and n_0 is given in Table 7 for $k = 2$, and in Dudewicz et al. [12] for $k > 2$. Approximations for $k > 25$ are given by Dudewicz and Dalal [10], as are procedures for the subset-selection formulation of the problem.

THE HETEROSCEDASTIC METHOD

The special-case solutions described above have been placed into a general theory with the heteroscedastic method of Dudewicz and Bishop [9]. In a general decision-theoretic setting, they show how to develop procedures like the one above in any problem. It is also shown that no single-stage procedure can solve most such problems.

Some questions one might ask about the procedures thus produced are as follows. First, how do they perform under violation of normality? Iglehart [15] has shown, in some computational settings, that replacing s^2 by a jackknife* estimator is sufficient to preserve the main properties of the procedures. Other recent work [11] shows asymptotic validity under asymptotic normality*.

Second, are they preferable to comparable sequential procedures? In most cases there are no "comparable" sequential procedures: those of Chow–Robbins type (*see* FIXED-WIDTH CONFIDENCE INTERVALS) which are usually mentioned only have asymptotic validity even under exact normality, while the $\mathcal{S}_B(n_0, w)$-based two-stage procedures have exact known properties. It is sometimes claimed that the sequential procedures are more efficient, but this is only as, for example, $\sigma_i^2 \to 0$. The so-called inefficiency of $\mathcal{S}_B(n_0, w)$ in this situation is because it then requires $N = n_0 + 1$ (since $N \geqslant n_0 + 1$ always) and in fact (as $\sigma_i^2 \to 0$) and $N \to 1$ will suffice. This appears to have little practical relevance, as one usually knows that trivial sample sizes will be insufficient for one's problems; it is rather a curiosity of mathematical interest only.

As a final note, we mention that while variance-stabilizing transformations and other approximate methods have existed for many years, most experimental situations are such that the problem is far from solved by these approximate methods. For example, such methods misallocate sample size by taking the same sample size from a treatment with relatively small variability, as from a treatment with relatively large variability, even though the need for observations on the latter is substantially greater and they have a greater beneficial effect on performance characteristics of the overall analysis. Also, procedures based on $\mathcal{S}_B(n_0, w)$ behave acceptably even if variances are equal; hence the equality-of-variances tests, which are known to be weak in power, can be skipped and these new procedures can be applied directly without regard to equality or inequality of variances.

References

[1] Bechhofer, R. E. (1954). *Ann. Math. Statist.*, **25**, 16–39. (The original paper on ranking and selection methods.)

[2] Bickel, P. J. (1978). *Ann. Math. Statist.*, **6**, 266–291. (A theoretical study of asymptotic power functions of tests for heteroscedasticity, especially in linear models under nonnormality.)

[3] Bishop, T. A. and Dudewicz, E. J. (1978). *Technometrics*, **20**, 419–430. (The original ANOVA procedures for heteroscedastic situations, with tables and approximations needed for implementation.)

[4] Bishop, T. A., Dudewicz, E. J., Juritz, J., and Stephens, M. A. (1978). *Biometrika*, **65**, 435–439. (Considers approximating the $\tilde{F}$ distribution.)

[5] Chapman, D. G. (1950). *Ann. Math. Statist.*, **21**, 601–606. [Considered the $k = 2$ test of H_0: $\mu_1 = \mu_2$ vs. H_1: $|\mu_1 - \mu_2| = d$, and also H_0: $\mu_1 = r\mu_2$. Tabled c for $n_0 = 2(2)12$ and $1 - \alpha = 0.975$, 0.995, correct to 0.1 (except for a gross error when $n_0 = 4$).]

[6] Cochran, W. G. and Cox, G. M. (1957). *Experimental Designs* 2nd ed. Wiley, New York. (On p. 91, notes the need to test for heteroscedasticity.)

[7] Dantzig, G. B. (1940). *Ann. Math. Statist.*, **11**, 186–191. (First to show that one-stage procedures could not solve many practical problems.)

[8] Dudewicz, E. J. (1976). *Introduction to Statistics and Probability*. American Sciences Press, Columbus, Ohio.

[9] Dudewicz, E. J. and Bishop, T. A. (1979). In *Optimizing Methods in Statistics*, J. S. Rustagi, ed. Academic Press, New York, pp. 183–203. (Develops the heteroscedastic method as a unifying procedure in a general setting, and shows how the procedures referred to here fit in as special cases.)

[10] Dudewicz, E. J. and Dalal, S. R. (1975). *Sankhyā B*, **37**, 28–78. (Solves the heteroscedastic ranking and selection problem in indifference-zone and subset-selection settings. Gives extensive tables and suggestions on solutions of other problems with similar methods.)

[11] Dudewicz, E. J. and van der Meulen, E. C. (1980). *Entropy-Based Statistical Inference, II: Selection-of-the-Best / Complete Ranking for Continuous Distributions on* (0, 1), *with Applications to Random Number Generators*, Communication No. 123. Mathematical Institute, Katholieke Universiteit, Leuven, Belgium. [New results on validity of $\mathcal{S}_B(n_0, w)$ under asymptotic (rather than exact) normality.]

[12] Dudewicz, E. J., Ramberg, J. S., and Chen, H. J. (1975). *Biom. Zeit.*, **17**, 13–26. [Gives procedures and theory for one-sided multiple comparisons with a control, plus extensive tables of $c_{1-\gamma}(n_0)$ useful in many problems.]

[13] Harrison, M. J. and McCabe, B. P. M. (1979). *J. Amer. Statist. Ass.*, **74**, 494–499. (Introduces and compares tests for heteroscedasticity, especially in linear regression models.)

[14] Haynam, G. E., Govindarajulu, Z., and Leone, F. C. (1970). *Selected Tables in Mathematical Statistics*, Vol. 1, H. L. Harter and D. B. Owen, eds.

Markham, Chicago, pp. 1–78. (Tables of the cumulative noncentral chi-square distribution.)

[15] Iglehart, D. L. (1977). *TIMS Stud. Manag. Sci.*, **7**, 37–49. [Suggested using a jackknife variance estimator with $\mathscr{S}_B(n_0, w)$, and indicated it solves nonnormality problems in his context.]

[16] Johnson, N. L. and Kotz, S. (1970). *Distributions in Statistics: Continuous Univariate Distributions*, Vol. 2. Wiley, New York.

[17] Juran, J. M., Gryna, F. M., Jr., and Bingham, R. S., Jr., eds. (1974). *Quality Control Handbook*, 3rd ed. McGraw-Hill, New York. (Recommend testing for heteroscedasticity on p. 46 of Section 27.)

[18] Land, C. E. (1974). *J. Amer. Statist. Ass.*, **69**, 795–802. (Considers and compares several methods for confidence interval estimation for original means after data transformations to normality, including the method considered in the section "Basic Nonnormality and Transformations.")

[19] Miller, R. G. (1974). *Biometrika*, **61**, 1–15. (Recent survey of jackknife methods, with an extensive bibliography.)

[20] Pearson, E. S. and Hartley, H. O., eds. (1970). *Biometrika Tables for Statisticians*, Vol. 1, 3rd ed. Cambridge University Press, Cambridge (reprinted with additions).

[21] Ruben, H. (1962). *Sankhyā A*, **24**, 157–180. (Looks at testing H_0: $\mu_1 = \mu_2$ when $k = 2$, concentrating attention on $\bar{X}_1$ and $\bar{X}_2$ . . . hence missing the generalizations found in the 1970s.)

[22] Stein, C. (1945). *Ann. Math. Statist.*, **16**, 243–258. [The original reference to $\mathscr{S}_B(n_0, w)$, but did not consider heteroscedasticity, perhaps due to emphasis on $\bar{X}$ as a replacement for $\tilde{\bar{X}}$.]

Acknowledgment

This research was supported by Office of Naval Research Contract N00014-78-C-0543.

(EQUALIZATION OF VARIANCE
FIXED-WIDTH CONFIDENCE INTERVALS
MULTIPLE COMPARISONS
RANKING AND SELECTION)

EDWARD J. DUDEWICZ

H-FUNCTION DISTRIBUTION

The *H*-function was introduced by C. Fox [4] in 1961 as a symmetric Fourier kernel to Meijer's *G*-function. While there may be slight variations and generalizations in the definitions of the *H*-function in the literature, it is usually defined [3, pp. 49–50] as

$$h(z) = H^{m,n}_{p,q}\left[z \middle| \begin{matrix} (a_1, \alpha_1), \dots, (a_p, \alpha_p) \\ (b_1, \beta_1), \dots, (b_q, \beta_q) \end{matrix} \right]$$

$$= \frac{1}{2\pi i} \int_C \frac{\prod_{j=1}^{m} \Gamma(b_j - \beta_j s) \prod_{j=1}^{n} \Gamma(1 - a_j + \alpha_j s)}{\prod_{j=m+1}^{q} \Gamma(1 - b_j + \beta_j s) \prod_{j=n+1}^{p} \Gamma(a_j - \alpha_j s)} z^s \, ds \tag{1}$$

where $i = \sqrt{-1}$; $0 \leqslant m \leqslant q$; $0 \leqslant n \leqslant p$; $\alpha_j > 0$, $j = 1, 2, \dots, p$; $\beta_j > 0$, $j = 1, 2, \dots, q$ and where a_j $(j = 1, 2, \dots, p)$ and b_j $(j = 1, 2, \dots, q)$ are complex numbers such that no pole of $\Gamma(b_j - \beta_j s)$ for $j = 1, 2, \dots, m$ coincides with any pole of $\Gamma(1 - a_j + \alpha_j s)$ for $j = 1, 2, \dots, n$. Further, C is a contour in the complex s-plane going from $c - i\infty$ to $c + i\infty$ (c real) such that points $s = (b_j + k)/\beta_j$, $j = 1, 2, \dots, m$; $k = 0, 1, \dots$ and points $s = (a_j - 1 - k)/\alpha_j$, $j = 1, 2, \dots$; $k = 0, 1, \dots$ lie to the right and left of C, respectively. In other words, the *H*-function is a Mellin–Barnes integral.[1] Structure-wise, it is the Mellin inversion integral

$$h(z) = \frac{1}{2\pi i} \int_{c-i\infty}^{c+i\infty} z^s M_s(h(z)) \, ds; \tag{2}$$

$$M_s(h(z)) = \frac{\prod_{j=1}^{m} \Gamma(b_j - \beta_j s) \prod_{j=1}^{n} \Gamma(1 - a_j + \alpha_j s)}{\prod_{j=m+1}^{q} \Gamma(1 - b_j + \beta_j s) \prod_{j=n+1}^{p} \Gamma(a_j - \alpha_j s)} \tag{3}$$

is the Mellin transform of $h(z)$ (*see* INTEGRAL TRANSFORMS). It defines a family of functions, including many special functions in applied mathematics [6, p. 196].

Since the *H*-function is a transcendental function, its integral over the range of the (nonnegative) variable is not in general unity. If k, c, and the values of the parameters a_j, α_j, b_j, β_j, in (1) are such that

$$\int_0^\infty k h(cz) \, dz = 1, \tag{4}$$

it is known as the *H*-function distribution [6,

p. 200]. It then defines a family of H-function random variables (rvs), i.e., rvs whose probability density functions (PDFs) are H-functions. This family includes the majority of the classical PDFs [6, pp. 202–208]: the half-normal, beta*, gamma* (including, of course, the exponential* and chi-square*), half-Cauchy, half-Student, F*, Weibull*, Maxwell, Rayleigh, and general hypergeometric*. For a specific set of parameters, the Mellin inversion integral (2), when normalized via (4), yields a unique PDF obtainable through application of the residue theorem [6, Chap. 7, App. F]. Products, quotients, and rational powers (but, in general, not sums and differences) of independent H-function rvs are themselves H-function rvs [6, Chap. 6]. Moreover, the parameters in the resulting H-function PDF are expressible in terms of the parameters of the component H-functions. Thus, the PDF $h(z)$ of the product $z = x_1 x_2$ of two independent beta rvs x_1, x_2 with PDFs

$$\begin{aligned} f(x_j) &= \frac{x_j^{\theta_j - 1}(1 - x_j)^{\phi_j - 1}}{B(\theta_j, \phi_j)} \\ &= \frac{\Gamma(\theta_j + \phi_j)}{\Gamma(\theta_j)} H_{1,1}^{1,0}\left[x_j \middle| \begin{matrix} (\theta_j + \phi_j - 1, 1) \\ (\theta_j - 1, 1) \end{matrix} \right], \\ &\qquad 0 < x_j < 1, \quad j = 1, 2 \quad \theta_j, \phi_j > 0 \\ &= 0, \qquad \text{otherwise} \end{aligned}$$

is given by

$$\begin{aligned} h(z) &= \prod_{j=1}^{2} \frac{\Gamma(\theta_j + \phi_j)}{\Gamma(\theta_j)} \\ &\quad \times H_{2,2}^{2,0}\left(z \middle| \begin{matrix} (\theta_1 + \phi_1 - 1, 1), (\theta_2 + \phi_2 - 1, 1) \\ (\theta_1 - 1, 1), (\theta_2 - 1, 1) \end{matrix} \right), \\ &\qquad\qquad 0 \leqslant z \leqslant 1 \\ &= 0, \qquad \text{otherwise.} \end{aligned}$$

For example, if $\theta_1 = 2$, $\phi_1 = 2$; $\theta_2 = 2$, and $\phi_2 = 1$,

$$\begin{aligned} h(z) &= \frac{\Gamma(4)\Gamma(3)}{(\Gamma(2))^2} H_{2,2}^{2,0}\left(z \middle| \begin{matrix} (3, 1), (2, 1) \\ (1, 1), (1, 1) \end{matrix} \right) \\ &= \frac{1}{2\pi i} \int_{c - i\infty}^{c + i\infty} \frac{12 z^{-s}}{(s + 2)(s + 1)^2} \, ds \\ &= 12z(z - \ln z - 1), \qquad 0 \leqslant z \leqslant 1 \\ &= 0, \qquad \text{otherwise.} \end{aligned}$$

The determination of the exact PDF $h(z)$ is considerably simplified by an algorithm (using recursive relationships) for which a computer program is operational [2]. Also, one may use an alternative procedure to obtain an approximation to $h(z)$ to any desired degree of accuracy by using a determinable number of moments in a method developed by Woods and Posten [6, pp. 284–300]. If one is primarily interested in the distribution function

$$H(z) = \int_0^z h(w) \, dw \tag{5}$$

of any H-function rv, it can be obtained without first deriving $h(z)$ by evaluating the inversion integral

$$H(z) = 1 - \frac{1}{2\pi i} \int_{c - i\infty}^{c + i\infty} \frac{z^s}{s} M_{s+1}(h(z)) \, ds. \tag{6}$$

It is noteworthy that the moments of the H-function distribution (2) are directly obtainable from the Mellin transform (3). That is, the mth moment μ'_m about the origin is precisely $M_s(h(z))|_{s = m+1}$, $m = 0, 1, 2, \ldots$. This is particularly significant since one often knows the moments of $h(z)$, even though $h(z)$ is itself unknown. For example, if x_1 and x_2 are two H-function independent random variables with known PDFs $f_1(x_1)$ and $f_2(x_2)$, respectively, then the mth moment of the PDF $h(z)$ of the product $z = x_1 x_2$ is immediately obtainable from the known Mellin transforms $M_s(f_1(x_1))$ and $M_s(f_2(x_2))$. Specifically,

$$\begin{aligned} &[M_s(h(z))]|_{s = m+1} \\ &\quad = [M_s(f_1(x_1)) M_s(f_2(x_2))]|_{s = m+1} \\ &\quad = \mu'_m, \qquad m = 0, 1, 2, \ldots \end{aligned}$$

While the appearance of the H-function distribution may suggest that it has little application to real-world problems, such is not the case. In fact, the H-function distribution has proven useful in system reliability analysis, in the detection of weak signals masked by noise, in the analysis of certain cash flow problems in economics, in the multiplicative processing of images, in queueing* problems, and in various military operations research* problems related to radar discrimination [2; 6, pp. 6–9; 8]. The H-function has also been utilized in the development of theoretical statistics, including the derivation of (a) a multivariate generalization of the analysis-of-variance* test [6, p. 311]; (b) the exact noncentral distribution of Votaw's criteria for testing the compound symmetry of a covariance matrix [6, p. 387]; (c) the exact distribution of the product of generalized F variables [6, p. 374]; and (d) the exact distribution of products of powers of Dirichlet* rvs [6, p. 377]. (See also [5, Chap. 4].)

As an example that brings the utility of the H-function into focus for multiplicative models, consider the problem of determining probability limits (sometimes referred to as Bayesian "confidence limits") for the reliability* $R = \prod_{j=1}^{n} R_j$ of a complex system consisting of n independent subsystems in series with unknown reliabilities $R_1, R_2, \ldots, R_n$. For these systems (such as space systems and nuclear power plants) it is not feasible to test entire systems per se to determine system reliability. A Bayesian probability model enables one to utilize test data for individual subsystems to derive the posterior PDF $h(R)$ and distribution function (DF) $H(R)$ of system reliability. For a series system composed of n subsystems having exponential life density functions, the posterior PDF and DF of system reliability are, respectively, the H-functions [7]

$$h(R) = \sum_{j=1}^{n} R^{b_j}\left[K_{ja_j}\frac{(\ln 1/R)^0}{0!} + \cdots + K_{j0}\frac{(\ln 1/R)^{a_j}}{a_j!}\right]$$

and

$$H(R) = \sum_{j=1}^{n} R^{b_j+1}\left[A_{ja_j}(\ln 1/R)^0 + \cdots + A_{j0}(\ln 1/R)^{a_j}\right],$$

where a_j, b_j, K_{ja_j}, and A_{ja_j}, $j = 1, 2, \ldots, n$, are constants obtained from the test data.

Note

1. These introductory statements are reprinted from *The Algebra of Random Variables*, by M. D. Springer, Wiley (1979), p. 195, with permission of the publisher.

References

[1] Carter, B. D. and Springer, M. D. (1978). *SIAM J. Appl. Math.*, **33**, 542–558.

[2] Eldred, B. S. (1978). The Application of Integral Transform Theory to the Analysis of Probabilistic Cash Flows. Doctoral dissertation, College of Engineering, University of Texas.

[3] Erdelyi, A. (1954). *Higher Transcendental Functions*, Vol. 1. McGraw-Hill, New York.

[4] Fox, C. (1961). *Trans. Amer. Math. Soc.*, **98**, 395–429.

[5] Mathai, A. M. and Saxena, R. K. (1978). *The H-Function with Applications in Statistics and Other Disciplines*. Wiley, New York. (Excellent bibliography on H-functions.)

[6] Springer, M. D. (1979). *The Algebra of Random Variables*. Wiley, New York.

[7] Springer, M. D. and Thompson, W. E. (1967). *IEEE Trans. Rel.*, **R-16**, 86–89.

[8] Webb, E. L. R. (1956). *Canad. J. Phys.*, **34**, 679–691.

(HYPERGEOMETRIC DISTRIBUTIONS
INTEGRAL TRANSFORMS)

MELVIN D. SPRINGER

HIERARCHICAL CLASSIFICATION

Given a set S containing n objects, a hierarchical classification (HC) refers to an ordered collection of subsets of S that can be used to represent similarity relationships between the objects. The objects themselves are arbitrary and could refer to people, words, tests, stimuli, items, plants, and so on. The similarity relationships that are summarized by the HC may reflect empirical data or describe some theoretical structure among the n objects (*see* CLASSIFICATION *and* HIERARCHICAL CLUSTER ANALYSIS for a discussion of various construction methods). Compared to an exhaustive listing of all the individual similarity relationships for the object pairs, an HC represents these relationships in a concise form that is easier to understand. In short, the purpose of an HC can be described as data simplification or reduction.

More formally, an HC can be defined by $K + 1$ ordered levels of subsets numbered from 0 to K with level k containing N_k subsets. The collection of sets as a whole satisfies four properties:

1. **Level 0 Triviality.** The n sets at level 0 each consist of a single object from $S(N_0 = n)$.
2. **Level K Triviality.** The single set at level K is inclusive and contains all objects from $S(N_K = 1)$.
3. **Subset Exclusivity.** No set at level k is a proper subset of another set at the same level.
4. **Hierarchy Principle.** Each set at level k is either present at level $k + 1$ or is a proper subset of a set at level $k + 1$.

Each HC defines a measure of relationship or proximity between each unordered pair of objects in S. The pair is merely assigned a proximity value equal to the minimum level at which the two objects appear together within a single subset. Once defined, this latter measure can be used to reconstruct the complete HC; thus it describes perfectly the numerical relationships between the objects represented by the HC. In addition, the measure constructed from an HC can be compared to an original data source from which the hierarchy was constructed to assess the loss of information, if any, in using the HC representation.

There are three special cases of an HC, defined by imposing further constraints on the subsets at each level that have particular significance in the literature:

1. **Complete Partition Hierarchy.** There are $n - k$ subsets at level k defining a partition of the n objects in S. Thus, $K = n - 1$ and the partition at level $k + 1$ is constructed by uniting a single pair of sets from level k. Most methods of hierarchical cluster analysis are concerned with this special case.
2. **Single Nontrivial Level.** Besides the two trivial partitions, the single nontrivial level contains N_1 subsets; thus $K = 3$. If the subsets actually partition S, the system contains a single nontrivial partition.
3. **Overlap Restriction.** The overlap for any pair of subsets at level k is less than or equal to some fixed constant. This condition is a basic property for an important class of hierarchical clustering procedures developed by Jardine and Sibson [3].

The most familiar examples of HC are the taxonomic structures of biology and botany. However, the concept is important in other disciplines as well. In psychology*, for instance, the HC notion can be used to discuss the phrase structure of sentences [5], the definition of a story grammar, now central to research on reading comprehension [4], and the organization and associated retrieval processes of human memory [2].

Example. We give an explicit illustration of a hierarchical structure for a cognitive theory. Following Royce [6], the set S contains 19 specific cognitive abilities: (1) Sensitivity to Problems, (2) Semantic Redefinition, (3) Verbal Comprehension, (4) Ideational Fluency, (5) Word Fluency, (6) Associational Fluency, (7) Expressed Fluency, (8) Associative Memory, (9) Memory Span, (10) Induction, (11) Deduction, (12) Syllogistic Reasoning, (13) Spontaneous Flexibility, (14) Spatial Scanning, (15) Visualization, (16) Spatial Relations, (17) Figural Adaptive Flexibility, (18) Flexibility of Closure and (19) Speed of Closure. Each of these 19 objects forms a separate subset at level 0; level 1 consists of five disjoint subsets: verbal—$\{1, 2, 3\}$, fluency—$\{4, 5, 6, 7\}$, memory—$\{8, 9\}$, reasoning—$\{10, 11, 12, 13\}$, visualization—$\{14, 15, 16, 17, 18, 19\}$. In turn, three level 2 sets with overlap are defined from the level 1 sets: symbolizing—{verbal, fluency, memory}, conceptualizing—{fluency, memory, reasoning}, perceiving—{memory, visualization}. Finally, level 3 has the single inclusive set containing all 19 cognitive abilities. If the original 19 cognitive abilities are interpreted as elemental, the successive groupings at levels 1 and 2 define higher-order cognitive abilities. Consequently, the latter must be evaluated through their individual components even though they may have separate conceptual identities in a general theory of cognitive functioning.

For a further discussion of the HC concept the reader is referred to the books by Bock [1], Jardine and Sibson [3], and Sneath and Sokal [7] and to the bibliographies each contains.

References

[1] Bock, H. H. (1974). *Automatische Klassifikation*. Vandenhoeck & Ruprecht, Gottingen. (A very comprehensive source on classification.)

[2] Bower, G. H., Clark, M. C., Lesgold, A. M., and Winzenz, D. (1969). *J. Verb. Learn. Verb. Behav.*, **8**, 323–343.

[3] Jardine, N. and Sibson, R. (1971). *Mathematical Taxonomy*. Wiley, New York.

[4] Mandler, J. M. and Johnson, N. S. (1977). *Cognit. Psychol.*, **9**, 111–151.

[5] Martin, E. (1970). *Psychol. Bull.*, **74**, 155–166.

[6] Royce, J. R. (1976). In *Nebraska Symposium on Motivation, 1975*, Vol. 23, W. J. Arnold, ed. University of Nebraska Press, Lincoln, Nebr., pp. 1–63.

[7] Sneath, P. H. A. and Sokal, R. R. (1973). *Numerical Taxonomy*. W. H. Freeman, San Francisco. (Excellent general reference with an extensive bibliography.)

(CLASSIFICATION
HIERARCHICAL CLUSTER ANALYSIS)

LAWRENCE J. HUBERT

HIERARCHICAL CLUSTER ANALYSIS

Hierarchical cluster analysis refers to the use of some data analysis procedure that generates a hierarchical classification for an object set S (*see* CLASSIFICATION *and* HIERARCHICAL CLASSIFICATION). The composition of the set S is unspecified and could contain people, stimuli, plants, and so on; the hierarchical classification is defined explicitly by subsets of S that are arranged at ordered levels and constructed from some type of empirical information, usually in the form of numerical similarity or proximity measures defined between the objects (for convenience, the term "proximity" will be used throughout to refer to any numerical index of relationship between objects). Supposedly, the levels at which subsets (or clusters) are formed, as well as what clusters are actually present, tell us something about the structure of the original object set.

The purpose of hierarchical cluster analysis is descriptive. We wish to represent the information in a proximity measure through a collection of hierarchically arranged subsets. At times, the hierarchy restriction may be motivated by theoretical considerations from a substantive area (e.g., plant taxonomies in biology, phrase structure grammars in psychology, and so on), but more often a hierarchy is a convenience in much the same way that regression functions are typically assumed to be linear. In nonhierarchical cluster analysis the proximity function being reconstructed is dichotomous since an object pair is either within the same set or not. Hierarchical classification, however, allows a more complex structure with the hope of a more complete representation of the information available in a proximity measure.

Our attention is restricted to the best known case of a hierarchical classification defined by an ordered sequence of *partitions*:

1. The subsets at each level from 0 to $n - 1$ form a partition of S in the sense that each object is contained within one and only one subset.
2. The partition at level $k + 1$ is formed by uniting a single pair of subsets at level k.
3. The partition at level 0 is trivial and defined by n single subsets each containing a single object.
4. The partition at level $n - 1$ consists of a single class that includes all the objects in S.

Table 1 Agglomerative Complete-Link Partition Hierarchy for the Kuennapas and Janson Data

Level	Partition
0	$\{\{O_1\}, \{O_2\}, \{O_3\}, \{O_4\}, \{O_5\}, \{O_6\}, \{O_7\}, \{O_8\}, \{O_9\}, \{O_{10}\}\}$
1	$\{\{O_9, O_{10}\}, \{O_1\}, \{O_2\}, \{O_3\}, \{O_4\}, \{O_5\}, \{O_6\}, \{O_7\}, \{O_8\}\}$
2	$\{\{O_9, O_{10}\}, \{O_2, O_4\}, \{O_1\}, \{O_3\}, \{O_5\}, \{O_6\}, \{O_7\}, \{O_8\}\}$
3	$\{\{O_9, O_{10}\}, \{O_2, O_4\}, \{O_3, O_5\}, \{O_1\}, \{O_6\}, \{O_7\}, \{O_8\}\}$
4	$\{\{O_9, O_{10}\}, \{O_2, O_4, O_8\}, \{O_3, O_5\}, \{O_1\}, \{O_6\}, \{O_7\}\}$
5	$\{\{O_9, O_{10}\}, \{O_2, O_4, O_8\}, \{O_1, O_3, O_5\}, \{O_6\}, \{O_7\}\}$
6	$\{\{O_6, O_9, O_{10}\}, \{O_2, O_4, O_8\}, \{O_1, O_3, O_5\}, \{O_7\}\}$
7	$\{\{O_6, O_9, O_{10}\}, \{O_2, O_4, O_7, O_8\}, \{O_1, O_3, O_5\}\}$
8	$\{\{O_6, O_9, O_{10}\}, \{O_1, O_2, O_3, O_4, O_5, O_7, O_8\}\}$
9	$\{\{O_1, O_2, O_3, O_4, O_5, O_6, O_7, O_8, O_9, O_{10}\}\}$

These specifications could be relaxed to allow overlap at a given level and/or a different number of levels (see Jardine and Sibson [28], Hubert [16], and Peay [37, 38]).

As a simple example, the partition hierarchy of Table 1 was constructed using a very common method in the behavioral science literature called the *complete-link procedure* (see Johnson [29]). The 10 objects O_1, $O_2, \ldots, O_{10}$ refer to the first 10 lowercase letters $a, b, \ldots, j$; the hierarchy was formed from empirical proximities between the object pairs given in Table 2 taken from Kuennapas and Janson [30]. Each entry in Table 2 represents the average over 57 human subjects of direct estimates of letter similarity. Specifically, all subjects were presented with each letter pair and asked to subjectively judge the similarity of the two letters along a predefined numerical scale. For our purposes it is only necessary to note that the entries are symmetric (the values for the pairs $\{O_i, O_j\}$ and $\{O_j, O_i\}$ are the same), and as a technical convenience, smaller proximities represent more similar letter pairs. Generalizations to asymmetric proximities are available (e.g., Hubert [14]).

Proximities may be constructed in many ways other than by direct judgment (*see* CLASSIFICATION and the book by Sneath and Sokal [42] for a more complete discussion). For the present, our concern is only with carrying out a hierarchical cluster analysis based on whatever proximities are given,

Table 2 Kuennapas and Janson Proximity Matrix for the First 10 Letters[a]

	Letter									
Letter	O_1	O_2	O_3	O_4	O_5	O_6	O_7	O_8	O_9	O_{10}
O_1	×	734	620	662	490	859	623	771	854	886
O_2		×	664	208	698	737	459	316	816	802
O_3			×	582	270	852	652	821	892	890
O_4				×	724	759	485	470	826	798
O_5					×	814	725	800	827	895
O_6						×	804	696	686	576
O_7							×	713	871	580
O_8								×	750	751
O_9									×	106
O_{10}										×

[a]The letters *a* through *j*, which appear in the text, correspond to the objects denoted here by O_1 through O_{10}.

irrespective of whether they are provided directly or constructed indirectly from a collection of measurements on the original objects. A discussion of hierarchical clustering on an object by variable data matrix without using intermediate proximities is given by Hartigan [10, 11].

The complete-link method used to construct Table 1 can be viewed as a prototypic hierarchical clustering strategy that proceeds agglomeratively from level 0 to level $n - 1$; e.g., two subsets at a given level are united to form a single new subset at the next level. To form the new set at level $k + 1$, an agglomerative method picks a pair of subsets to unite at level k based on some function. The complete-link criterion, for instance, chooses a pair of subsets to minimize the resultant diameter of the new set, where the diameter is defined as the maximum proximity between any pair of objects in the set. Thus all object pairs within a set have proximities less than or equal to the diameter; i.e., all pairs within a set are completely linked at a value equal to the diameter or less.

GENERAL AGGLOMERATIVE METHODS

Although the complete-link method has been used as a model, the same sequential process of forming new sets could be carried out with criteria other than the diameter. All that is needed is some way of picking a pair of subsets to unite at level k, presumably by choosing those two that are "closest" together in some well-defined sense, and then redefining this closeness measure between all pairs of subsets present at level $k + 1$. There is a basic ambiguity in characterizing what a cluster actually is; typically, the method itself is used to define in a rather vague and implicit manner what particular notion of a cluster is being pursued.

As one strategy suggested by Lance and Williams [31], suppose that the two subsets S_a and S_b at level k are united to form S_{ab} at level $k + 1$. If S_c represents some third cluster at level $k + 1$, the "closeness" of S_{ab} to S_c, say $d(S_{ab}, S_c)$, is defined as

$$d(S_{ab}, S_c) = \alpha_a d(S_a, S_c) + \alpha_b d(S_b, S_c) + \beta d(S_a, S_b) + \gamma |d(S_a, S_c) - d(S_b, S_c)|,$$

where α_a, α_b, β, and γ are parameters and $d(S_a, S_c)$, $d(S_b, S_c)$, and $d(S_a, S_b)$ have been calculated previously at level k. By varying the four parameters, a number of common agglomerative strategies are available as special cases. These are listed below together with references to the literature (n_a, n_b, and n_c refer to the number of objects in S_a, S_b, and S_c, respectively):

1. **Single-Link [29].** $\alpha_a = \alpha_b = \frac{1}{2}$, $\beta = 0$, $\gamma = -\frac{1}{2}$
2. **Complete-Link [29].** $\alpha_a = \alpha_b = \frac{1}{2}$, $\beta = 0$, $\gamma = +\frac{1}{2}$
3. **Median [9].** $\alpha_a = \alpha_b = \frac{1}{2}$, $\beta = -\frac{1}{4}$, $\gamma = 0$
4. **Group Average [31].** $\alpha_a = n_a/(n_a + n_b)$, $\alpha_b = n_b/(n_a + n_b)$, $\beta = 0$, $\gamma = 0$
5. **Centroid [44].** $\alpha_a = n_a/(n_a + n_b)$, $\alpha_b = n_b/(n_a + n_b)$, $\beta = -\alpha_a\alpha_b$, $\gamma = 0$
6. **Ward's Method [45].** $\alpha_a = (n_a + n_c)/(n_a + n_b + n_c)$, $\alpha_b = (n_b + n_c)/(n_a + n_b + n_c)$, $\beta = -n_c/(n_a + n_b + n_c)$, $\gamma = 0$

Other clustering possibilities may or may not be reducible to this simple linear form. For example, a general class of methods based on graph theory* could be constructed by defining the "closeness" between any two subsets S_a and S_b as the minimum proximity value h for which the set $S_a \cup S_b$ generates the subgraph with a particular property in the larger graph defined by the objects in S (see Hubert [16] and Ling [32]). In this larger graph a line is placed between any pair of objects with proximity less than or equal to h.

DIVISIVE METHODS

Agglomerative hierarchical clustering methods may be the best known in the literature, but divisive strategies that proceed in the opposite order from level $n-1$ to level 0 can also be defined. The complete object set S is subdivided into two groups to form the partition at level $n-2$; one of these two subsets is then chosen for resplitting to form two additional clusters at level $n-3$, and so on.

As one example, a divisive complete-link analog could be implemented, but it will not necessarily produce the same hierarchy as the agglomerative version. To form the two subsets S_a and S_b from S at level $n-2$, we would proceed by building up S_a and S_b sequentially. The members of the most dissimilar object pair are split to form the nuclei of what will eventually be S_a and S_b. The next most dissimilar object pair is found that has one of its constituent objects already allocated to what will become S_a or S_b and the other not. If the allocated object belongs to the subset that will be denoted by $S_a(S_b)$, the unallocated object becomes part of the future set $S_b(S_a)$. The process continues until all objects are allocated. Among the two subsets S_a and S_b at level $n-2$, the one with the largest diameter is chosen for resplitting; the same allocation process is implemented to form the two new subsets at level $n-3$, and so on. The process produces the partition hierarchy given in Table 3 using the Kuennapas and Janson data. All object pairs within any subset at level k are completely linked at a proximity value no larger than h, where h is the largest diameter among all subsets at level k (see Hubert [13, 15]).

The procedure just defined for dividing S into S_a and S_b produces an optimal two-group split in the sense that we minimize the maximum diameter over all two-group partitions (see Rao [39]). Specifically, if S_a^* and S_b^* define the subsets in any two-group partition, and if $D(S_c)$ gives the diameter of any arbitrary subset S_c, then $\max\{D(S_a^*), D(S_b^*)\} \geqslant \max\{D(S_a), D(S_b)\}$. This can be seen numerically in the two-group partition, $\{\{O_6, O_9, O_{10}\}, \{O_1, O_2, O_3, O_4, O_5, O_7, O_8\}\}$, from the agglomerative complete-link partition hierarchy, which supposedly keeps the diameters as small as possible in proceeding from level 0 to level $n-1$. The two-group partition, $\{\{O_6, O_8, O_9, O_{10}\}, \{O_1, O_2, O_3, O_4, O_5, O_7\}\}$, from the divisive method is "better" since

$$\begin{aligned} 821 &= \max\{D(\{O_6, O_9, O_{10}\}), \\ &\qquad D(\{O_1, O_2, O_3, O_4, O_5, O_7, O_8\})\} \\ &\geqslant \max\{D(\{O_6, O_8, O_9, O_{10}\}), \\ &\qquad D(\{O_1, O_2, O_3, O_4, O_5, O_7\})\} \\ &= 751. \end{aligned}$$

Unfortunately, no such optimality property is guaranteed for the divisive hierarchy beyond level $n-2$. As an example, the three-cluster partition from the agglomerative complete-link hierarchy is "better" than the three-cluster partition from the divisive hier-

Table 3 Divisive Complete-Link Partition Hierarchy for the Kuennapas and Janson Data

Level	Partition
0	$\{\{O_1\}, \{O_2\}, \{O_3\}, \{O_4\}, \{O_5\}, \{O_6\}, \{O_7\}, \{O_8\}, \{O_9\}, \{O_{10}\}\}$
1	$\{\{O_9, O_{10}\}, \{O_1\}, \{O_2\}, \{O_3\}, \{O_4\}, \{O_5\}, \{O_6\}, \{O_7\}, \{O_8\}\}$
2	$\{\{O_9, O_{10}\}, \{O_2, O_4\}, \{O_1\}, \{O_3\}, \{O_5\}, \{O_6\}, \{O_7\}, \{O_8\}\}$
3	$\{\{O_9, O_{10}\}, \{O_2, O_4\}, \{O_3, O_5\}, \{O_1\}, \{O_6\}, \{O_7\}, \{O_8\}\}$
4	$\{\{O_9, O_{10}\}, \{O_2, O_4, O_7\}, \{O_3, O_5\}, \{O_1\}, \{O_6\}, \{O_8\}\}$
5	$\{\{O_9, O_{10}\}, \{O_2, O_4, O_7\}, \{O_1, O_3, O_5\}, \{O_6\}, \{O_8\}\}$
6	$\{\{O_6, O_9, O_{10}\}, \{O_2, O_4, O_7\}, \{O_1, O_3, O_5\}, \{O_8\}\}$
7	$\{\{O_6, O_9, O_{10}\}, \{O_1, O_2, O_3, O_4, O_5, O_7\}, \{O_8\}\}$
8	$\{\{O_6, O_8, O_9, O_{10}\}, \{O_1, O_2, O_3, O_4, O_5, O_7\}\}$
9	$\{\{O_1, O_2, O_3, O_4, O_5, O_6, O_7, O_8, O_9, O_{10}\}\}$

archy, since

$$\begin{aligned} 713 = \max\{ & D(\{O_6, O_9, O_{10}\}), \\ & D(\{O_2, O_4, O_7, O_8\}), \\ & D(\{O_1, O_3, O_5\})\} \\ \leqslant \max\{ & D(\{O_6, O_9, O_{10}\}), \\ & D(\{O_1, O_2, O_3, O_4, O_5, O_7\}), \\ & D(\{O_8\})\} = 734. \end{aligned}$$

For further discussion see Hubert and Baker [23].

OPTIMALITY CONSIDERATIONS

Almost all of the hierarchical clustering techniques discussed in the literature are heuristic. There is no guarantee that some global adequacy measure is being optimized, and even when one can be found, as in the single-link strategy (see Bock [3]), it probably is not of great practical interest. For the same data set, it is possible to construct different partition hierarchies with alternative methods that may lead to differing substantive interpretations. Even when the same general type of local optimization is being carried out (e.g., by keeping diameters as small as possible in the agglomerative and divisive complete-link strategies), serious anomalies can occur. (For an extensive discussion of computational problems in cluster analysis, the reader is referred to Brucker [4].)

ULTRAMETRICS

Suppose that $s(\cdot,\cdot)$ denotes the proximity function used to construct a hierarchical classification and let $l_0, l_1, \ldots, l_{n-1}$ be the partitions defined at levels $0, 1, \ldots, n-1$. A function $d(\cdot,\cdot)$ can be constructed that numerically represents the partition hierarchy. Let

$$d(O_i, O_j) = \min\{k \mid O_i \text{ and } O_j \text{ are placed in a single subset in } l_k\};$$

then $d(\cdot,\cdot)$ satisfies four properties, the last of which is the most restrictive:

1. $d(O_i, O_j) \geqslant 0$
2. $d(O_i, O_j) = 0$ if and only if $O_i = O_j$
3. $d(O_i, O_j) = d(O_j, O_i)$
4. $d(O_i, O_j) \leqslant \max\{d(O_i, O_k), d(O_k, O_j)\}$

Any strictly monotone increasing function of $d(\cdot,\cdot)$ that maps zero to zero would also be an ultrametric (see Jardine and Sibson [28] and Johnson [29]). As a further generalization, the integers defining $d(\cdot,\cdot)$ could be replaced by values that specify the "fusing" levels at which a given partition is constructed. In the complete-link method this value could be the diameter of the newly formed subset at level k; in the single-link method it could be the minimum proximity between the two initial subsets. For both of these cases, the fusing numbers will be monotonically related to the levels 0 to $n - 1$.

If the original proximity function $s(\cdot,\cdot)$ is monotone with respect to $d(\cdot,\cdot)$, then $s(\cdot,\cdot)$ is itself an ultrametric and the sequence of partitions $l_0, l_1, \ldots, l_{n-1}$ is a perfect structure for $s(\cdot,\cdot)$. Most proximity functions will not satisfy an ultrametric condition exactly. Nevertheless, any hierarchical clustering method will still locate a "good" sequence of partitions that could be used to obtain an ultrametric function $d(\cdot,\cdot)$. The degree to which $d(\cdot,\cdot)$ is related to the function $s(\cdot,\cdot)$ could be formalized by an index that measures the extent to which the imperfect patterning of the proximity values matches the perfect ultrametric structure of $d(\cdot,\cdot)$ (see Baker and Hubert [1]).

Given an arbitrary proximity function $s(\cdot,\cdot)$ and the sequence of partitions constructed by some hierarchical clustering method, one rather well known graphic technique used in representing the analysis rearranges the proximity matrix in a very specific way (cf. BMDP program manual: Dixon [7, pp. 621–632]). If the values assigned by $d(\cdot,\cdot)$ are organized into the form of an $n \times n$ matrix $\mathbf{D} = \{d(O_i, O_j)\}$, then there exists an ordering (not unique) of the rows and simultaneously the columns of **D** that will give the reordered matrix **D*** the following properties:

1. **D*** can be partitioned as

$$\mathbf{D}^* = \begin{bmatrix} \mathbf{D}_{11}^* & \mathbf{D}_{12}^* \\ \mathbf{D}_{21}^* & \mathbf{D}_{22}^* \end{bmatrix},$$

   where all the elements of $\mathbf{D}_{12}^*$ and $\mathbf{D}_{21}^*$ are equal to the single largest elements of **D**.
2. The submatrices $\mathbf{D}_{11}^*$ and $\mathbf{D}_{22}^*$ are partitionable as in property 1.
3. The partitioning process can be repeated until all the resulting submatrices are of order 1.

Given the reorganized matrix **D***, the original proximity matrix, say $\mathbf{P} = \{s(O_i, O_j)\}$, can also be restructured by using the same row and column ordering. Denoting this reordered matrix by **P***, the degree to which **P*** does not have the exact same partition structure as did **D*** indicates the adequacy of the hierarchical clustering.

EVALUATION STRATEGIES

One of the most disconcerting aspects of using hierarchical clustering is the lack of comprehensive statistical inference methods for evaluating the results of such an analysis. Most of the available work has emphasized those methods that rely only on the rank ordering of the proximity measures (e.g., the single-link and complete-link techniques), since there is some hope of testing an overall measure of goodness of fit for a hierarchy using reference tables generated by simple randomization* (e.g., see Ling [32], Ling and Killough [33], Baker and Hubert [1, 2], Hubert and Baker [20, 22], Hubert [12, 17], Hubert and Schultz [24]). Even here the null hypothesis that can be dealt with conveniently is probably very unrealistic for most applications. The null assumption that all proximity measures have been assigned at random to the object pairs does not take into account any structural constraints on the original measures: for example, metric conditions, positive definiteness, and the like. In short, the null distributions that can be generated under complete randomness may not reflect a more appropriate null model that would respect the algebraic properties of the original proximities.

There are two major problems of statistical evaluation, however, that reoccur continually in the general field of cluster analysis and that can now be handled rather easily (see Rohlf [40] for some historical background):

1. Given a collection of objects and two different classification schemes for these objects obtained from *distinct* data sets, can the two classification schemes be considered similar?
2. Given a collection of objects and a symmetric measure of proximity defined for each objects pair, do these measures reflect a particular classification scheme obtained through some independent process, that is, without reference to the original proximities? For instance, the classification scheme may be hypothesized a priori from the literature, or possibly, found through some cluster analysis of a second data matrix.

For both of these problems the same general statistical procedure may be used that is based on correlating the corresponding off-diagonal entries in two symmetric square matrices. In problem 1, the matrices are constructed from the two given classification schemes using, for instance, the ultrametric representation defined earlier; in problem 2, one matrix defines the empirical proximity matrix and the second is reconstructed from the classification scheme being evaluated. Once these correlations are obtained (referred to as cophenetic or matrix correlations in the clustering literature), the additional problem of assessing statistical significance can be attacked directly through the use of Mantel's generalized permutation test strategy [34]. A more complete discussion is available in Hubert and Baker [21] and Sokal [43].

As one alternative to the difficult task of assessing global structure and whether a complete proximity matrix is well represented by some partition hierarchy, it is possible to evaluate variation in structure within an object set if the whole proximity matrix is used to define an operational population. Evaluating the degree to which a *subset* of S satisfies the conjectured property of being well represented by a partition hierarchy can be approached through a confirmatory inference strategy. A formal statistical test is based on a reference distribution for some adequacy measure generated over all equally likely subsets of the same size as the original subset being considered. For a more complete discussion of this topic, the reader is referred to Hubert [19].

For current work on cluster analysis, the reader is referred to the work on foundations by Janowitz [25, 26, 27], Day [5, 6], Matula [35, 36], and Hubert [18]; several useful reviews of the statistical aspects of the topic are available in Dubes and Jain [8] and Smith and Dubes [41].

References

[1] Baker, F. B. and Hubert, L. (1975). *J. Amer. Statist. Ass.*, **10**, 31–38.

[2] Baker, F. B. and Hubert, L. J. (1976). *J. Amer. Statist. Ass.*, **71**, 870–878.

[3] Bock, H. H. (1974). *Automatische Klassification*. Vandenhoeck & Ruprecht, Gottingen.

[4] Brucker, P. (1978). In *Optimization and Operations Research*, Lect. Notes Econ. Math. Syst., R. Henn, B. Korte, and W. Oletti, eds. Springer-Verlag, Berlin.

[5] Day, W. H. E. (1977). *Math. Biosci.*, **36**, 299–317.

[6] Day, W. H. E. (1981). *Math. Social Sciences*, **3**, 269–287.

[7] Dixon, W. J., ed. (1977). *Biomedical Computer Programs, P-Series*. University of California Press, Berkeley, Calif.

[8] Dubes, R. and Jain, A. K. (1979). *Pattern Recognition*, **11**, 235–254.

[9] Gower, J. C. (1967). *Biometrics*, **23**, 623–638.

[10] Hartigan, J. A. (1972). *J. Amer. Statist. Ass.*, **67**, 123–129.

[11] Hartigan, J. A. (1975). *Clustering Algorithms*. Wiley, New York.

[12] Hubert, L. J. (1972). *Psychometrika*, **37**, 261–274.

[13] Hubert, L. J. (1973). *Psychometrika*, **38**, 47–62.

[14] Hubert, L. J. (1973). *Psychometrika*, **38**, 63–72.

[15] Hubert, L. J. (1974). *Brit. J. Math. Statist. Psychol.*, **27**, 14–28.

[16] Hubert, L. J. (1974). *Psychometrika*, **39**, 283–309.

[17] Hubert, L. J. (1974). *J. Amer. Statist. Ass.*, **69**, 698–704.

[18] Hubert, L. J. (1977). *J. Math. Psychol.*, **15**, 70–88.

[19] Hubert, L. J. (1980). *J. Math. Psychol.*, **21**, 247–264.

[20] Hubert, L. J. and Baker, F. B. (1976). *J. Educ. Statist.*, **1**, 87–111.

[21] Hubert, L. J. and Baker, F. B. (1977). *J. Math. Psychol.*, **16**, 233–253.

[22] Hubert, L. J. and Baker, F. B. (1977). In *Classification and Clustering*, J. Van Ryzin, ed. Academic Press, New York, pp. 131–153.

[23] Hubert, L. J. and Baker, F. B. (1979). *J. Educ. Statist.*, **4**, 74–92.

[24] Hubert, L. J. and Schultz, J. V. (1975). *Brit. J. Math. Statist. Psychol.*, **28**, 121–133.

[25] Janowitz, M. F. (1978). *SIAM J. Appl. Math.*, **34**, 55–72.

[26] Janowitz, M. F. (1978). *Discrete Math.*, **21**, 47–60.

[27] Janowitz, M. F. (1979). *SIAM J. Appl. Math.*, **37**, 148–165.

[28] Jardine, N. and Sibson, R. (1971). *Mathematical Taxonomy*. Wiley, New York.

[29] Johnson, S. C. (1967). *Psychometrika*, **32**, 241–254.

[30] Kuennapas, T. and Janson, A. (1969). *Percept. Motor Skills*, **28**, 3–12.

[31] Lance, G. N. and Williams, W. T. (1967). *Computer J.*, **9**, 373–380.

[32] Ling, R. F. (1973). *J. Amer. Statist. Ass.*, **68**, 156–164.

[33] Ling, R. F. and Killough, G. G. (1976). *J. Amer. Statist. Ass.*, **71**, 293–300.

[34] Mantel, N. (1967). *Cancer Res.*, **27**, 209–220.

[35] Matula, D. W. (1977). In *Classification and Clustering*, J. Van Ryzin, ed. Academic Press, New York, pp. 96–129.

[36] Matula, D. W. (1978). *J. Comb. Theory B*, **24**, 1–13.

[37] Peay, E. R. (1974). *Sociometry*, **37**, 54–65.

[38] Peay, E. R. (1975). *Psychometrika*, **40**, 297–313.

[39] Rao, M. R. (1971). *J. Amer. Statist. Ass.*, **66**, 622–626.

[40] Rohlf, F. J. (1974). *Ann. Rev. Ecol. Systematics*, **5**, 101–113.

[41] Smith, S. P. and Dubes, R. (1980). *Pattern Recognition*, **12**, 177–187.

[42] Sneath, P. H. A. and Sokal, R. R. (1973). *Numerical Taxonomy*. W. H. Freeman, San Francisco.

[43] Sokal, R. R. (1979). *Syst. Zool.*, **28**, 227–232.

[44] Sokal, R. R. and Michener, C. D. (1958). *Univ. Kans. Sci. Bull.*, **38**, 1409–1438.

[45] Ward, J. H. (1963). *J. Amer. Statist. Ass.*, **58**, 236–244.

(CLASSIFICATION
GRAPH-THEORETIC CLUSTER ANALYSIS
HIERARCHICAL CLASSIFICATION)

LAWRENCE J. HUBERT

HIERARCHICAL KAPPA STATISTICS

The analysis of hierarchical kappa statistics is a methodological procedure for assessing the pattern of agreement among two or more classifications of some categorical response variable for each subject in a study for which observer reliability is of interest. Its formulation involves two considerations. The first is the use of *kappa statistics** as chance-corrected *measures of agreement** relative to judgmental criteria for interpreting different response categories as equivalent. The other is a hierarchical sequence of successively less stringent definitions for such criteria which yield respectively larger kappa statistics for correspondingly broader (or more liberal) views of agreement.

The respective kappa statistics express the extent to which the observed pattern of agreement is more apparent than would be expected from chance* (via the ratio of the observed vs. expected difference to the maximum possible such agreement). Their range is typically from 0 to 1, with 0 corresponding to chance agreement and 1 to perfect agreement, but negative values are possible. A hierarchical relationship among a set of kappa statistics provides a framework for investigating whether patterns of disagreement pertain primarily to interchanges among similar response categories as opposed to substantively important misclassifications. Such information is potentially useful for ascertaining whether the measurement process for a response variable has acceptable reliability.

The manner in which hierarchical kappa statistics are formulated can be seen by considering their application to a specific example. The data summarized in Table 1 are based on observed frequencies from a study to compare the standard version and a modified version of the same diagnostic procedure for a particular health status attribute, but they have been hypothetically modified to correspond to classifications of $n = 1000$ subjects. If $n_{jj'}$ denotes the number of subjects who are classified into the jth response category for the modified version diagnosis and the j'th response category for the standard version with $j, j' = 1, 2, 3, 4$, then one

Table 1 Frequencies for Cross-Classification of 1000 Subjects according to Standard Version and Modified Version of a Diagnostic Procedure

	Standard Version Diagnosis				
Modified Version Diagnosis	Strongly Negative	Moderately Negative	Moderately Positive	Strongly Positive	Total Number of Subjects
Strongly negative	287	70	3	2	362
Moderately negative	98	201	37	3	339
Moderately positive	7	29	37	20	93
Strongly Positive	2	2	17	185	206
Total number of subjects	394	302	94	210	1000

Table 2 Criterion Weights for Hierarchical Kappa Statistics[a]

	Modified Version × Standard Version Response (jj')															
Criterion (h)	11	12	13	14	21	22	23	24	31	32	33	34	41	42	43	44
1	1	0	0	0	0	1	0	0	0	0	1	0	0	0	0	1
2	1	1	0	0	1	1	0	0	0	0	1	0	0	0	0	1
3	1	1	0	0	1	1	0	0	0	0	1	1	0	0	1	1
4	1	1	0	0	1	1	1	0	0	1	1	1	0	0	1	1

[a]The entry in the hth row and jj'th column is the weight $w_{h,jj'}$ for the hth criterion.

set of hierarchical kappa statistics can be formulated as $\hat{\kappa}_h = (\hat{\lambda}_h - \hat{\gamma}_h)/(1 - \hat{\gamma}_h)$, where $\hat{\lambda}_h = \sum_{j=1}^{4}\sum_{j'=1}^{4} w_{h,jj'}(n_{jj'}/n)$ is the observed level of agreement relative to the hth criterion weights $\{w_{h,jj'}\}$ and where $\hat{\gamma}_h = \sum_{j=1}^{4}\sum_{j'=1}^{4} w_{h,jj'}(n_{j+}n_{+j'}/n^2)$ is the estimated expected agreement under independence of the two diagnoses; also $n_{j+} = \sum_{j'=1}^{4} n_{jj'}$ and $n_{+j'} = \sum_{j=1}^{4} n_{jj'}$. The weights $w_{h,jj'}$ are defined to satisfy $w_{h,jj} = 1$ and $0 \leqslant w_{h,jj'} \leqslant w_{h',jj'} \leqslant 1$ if $j \neq j'$ and $h < h'$. Four sets of criterion weights that are of interest for the data in Table 1 are given in Table 2. The first set of weights defines agreement as the occurrence of the same response category for the two diagnosis procedures; the second is less stringent by regarding categories 1 and 2 as equivalent; the third is additionally less stringent by regarding categories 1 and 2 as equivalent and categories 3 and 4 as equivalent; the fourth is the least stringent by regarding successive pairs of adjacent categories as equivalent. The kappa statistics corresponding to the criterion weights in Table 2 and their estimated standard errors (SEs) were computed by the matrix methods described in Landis and Koch [1]. They are as follows:

$$\hat{\kappa}_1 = 0.587 \qquad \hat{\kappa}_2 = 0.735$$
$$\hat{\kappa}_3 = 0.798 \qquad \hat{\kappa}_4 = 0.947$$
$$\mathrm{SE}(\hat{\kappa}_1) = 0.020 \quad \mathrm{SE}(\hat{\kappa}_2) = 0.021$$
$$\mathrm{SE}(\hat{\kappa}_3) = 0.021 \quad \mathrm{SE}(\hat{\kappa}_4) = 0.012$$

Since these kappa statistics have an approximate multivariate normal distribution* for large samples (e.g., $n \geqslant 100$), *chi-square tests** for linear hypotheses concerning them can be undertaken with Wald statistics. The use of linear models methods for this purpose is discussed in Landis and Koch [1]. For the data in Table 1, such tests indicated that the successive differences between $\hat{\kappa}_1$ and $\hat{\kappa}_2$, $\hat{\kappa}_2$ and $\hat{\kappa}_3$, and $\hat{\kappa}_3$ and $\hat{\kappa}_4$ are statistically significant, and that the agreement for the fourth criterion is "almost perfect" in the sense of exceeding 0.800. Thus the extent of disagreement can be interpreted as mainly pertaining to interchanges between adjacent categories.

Hierarchical kappa statistics can also be applied to situations where there are more than two observers; a specific example is discussed in Landis and Koch [2].

References

[1] Landis, J. R. and Koch, G. G. (1977). *Biometrics*, **33**, 159–174.

[2] Landis, J. R. and Koch, G. G. (1977). *Biometrics*, **33**, 363–374.

Acknowledgment

This research was partially supported by the U.S. Bureau of the Census through Joint Statistical Agreement JSA 80-19.

(KAPPA STATISTICS
MEASURES OF AGREEMENT)

GARY G. KOCH

HIERARCHY OF MODELS

A collection $H_1, H_2, \ldots, H_m, H$ of statistical models is a *hierarchy* if

$$H_1 \subset H_2 \subset \cdots \subset H_m \subset H. \qquad (1)$$

Usually, H_1 will be a model of interest, H is a general class of alternatives, and H_2 and other intermediate models describe restricted alternative models. If H_0 is a further model such that $H_0 \subset H$, then each of the hierarchies

$$H_0 \cap H_1 \subset H_0 \subset H,$$
$$H_0 \cap H_1 \subset H_1 \subset H_2 \subset \cdots \subset H$$

may be of interest. As an example, in a two-way analysis of variance*,

H_1: No main effects or interaction, having normally distributed errors,

H_2: No main effects, having normally distributed errors,

H: Normally distributed errors,

satisfies the hierarchical structure (1).

The hierarchical structure is important in likelihood ratio tests* (*see* PARTITION OF CHI-SQUARE). The same notation (H_0, H_1, etc.) can be used to denote the hypotheses being tested; H_0 can denote a specific model or the assertion that the model holds, without confusion. In the second case, (1) describes a *hierarchy of hypotheses*.

(LIKELIHOOD RATIO TESTS
PARTITION OF CHI-SQUARE)

HIGHSPREAD

In a set of data, the high spread is the difference (upper extreme value)–(median).

(FIVE-NUMBER SUMMARIES
LOW SPREAD)

HILDRETH–LU SCANNING METHOD

Consider the linear regression model

$$y_t = \beta_0 + \beta_1 X_{t1} + \cdots + \beta_{p-1} X_{t,p-1} + \epsilon_t, \quad t = 1, \ldots, N, \qquad (1)$$

where y_t is the tth observation on the dependent variable, X_{tj} is the tth observation on the jth nonstochastic independent variable, and ϵ_t is the tth observation on the error term. This can be written in matrix form as $\mathbf{y} = \mathbf{X}\boldsymbol{\beta} + \boldsymbol{\epsilon}$, where $\mathbf{y}$ is $(N \times 1)$ in dimension, $\mathbf{X}$ is $(N \times p)$, $\boldsymbol{\beta}$ is $(p \times 1)$, and $\boldsymbol{\epsilon}$ is $(N \times 1)$. The usual assumptions on the error vector $\boldsymbol{\epsilon}$ are that $E(\boldsymbol{\epsilon}) = \mathbf{0}$ and $E(\boldsymbol{\epsilon}\boldsymbol{\epsilon}') = \sigma^2\mathbf{I}$. In this case, the ordinary least-squares* estimator of $\boldsymbol{\beta}$, denoted by $\hat{\boldsymbol{\beta}}_0$, is given by $\hat{\boldsymbol{\beta}}_0 = (\mathbf{X}'\mathbf{X})^{-1}\mathbf{X}'\mathbf{y}$ (*see* GENERAL LINEAR MODEL).

The assumption that the error terms are uncorrelated often breaks down in time-series* studies and sometimes in cross-sectional studies, in which case we state that the error terms are autocorrelated or serially correlated. We denote this by writing $E(\boldsymbol{\epsilon}\boldsymbol{\epsilon}') = \boldsymbol{\Sigma}$. Mixed autoregressive-moving average* processes are used to describe this serial correlation (*see* AUTOREGRESSIVE-MOVING AVERAGE (ARMA) MODELS). Specifically,

$$\epsilon_t = \phi_1\epsilon_{t-1} + \cdots + \phi_p\epsilon_{t-p} + a_t - \theta_1 a_{t-1} - \cdots - \theta_q a_{t-q}, \qquad (2)$$

where $E(a_t) = 0$, $V(a_t) = \sigma_a^2$, and the a_t's are uncorrelated. To ensure stationarity, we require that the roots of $1 - \phi_1 x - \phi_2 x^2 - \cdots - \phi_p x^p = 0$ lie outside the unit circle. The case that has been considered most frequently in the econometrics* literature is when the error terms are AR(1). That is,

$$\epsilon_t = \phi_1\epsilon_{t-1} + a_t. \qquad (3)$$

In (3), it has become customary to replace ϕ_1 by ρ, where we require $|\rho| < 1$ for a stationary process*. We will focus our attention on the autocorrelation structure specified in (3).

For an AR(1) process, it is shown in Box and Jenkins [2] that $\sigma_\epsilon^2 = \sigma_a^2/(1-\rho^2)$ and $E(\epsilon_t\epsilon_{t-k}) = \rho^k$. Thus the covariance matrix $\boldsymbol{\Sigma}$ associated with $(\epsilon_1, \epsilon_2, \ldots, \epsilon_N)$ is

$$\boldsymbol{\Sigma} = \sigma_\epsilon^2\mathbf{A} = \sigma_\epsilon^2 \begin{bmatrix} 1 & \rho & \rho^2 & \cdots & \rho^{N-1} \\ \rho & 1 & \rho & \cdots & \rho^{N-2} \\ \rho^2 & \rho & 1 & \cdots & \rho^{N-3} \\ \vdots & \vdots & \vdots & & \vdots \\ \rho^{N-1} & \rho^{N-2} & \rho^{N-3} & \cdots & 1 \end{bmatrix}$$
$$= \sigma_a^2\mathbf{B}. \qquad (4)$$

When ρ *is known*, the generalized least-squares estimator of $\boldsymbol{\beta}$, denoted by $\hat{\boldsymbol{\beta}}_G$, is found by minimizing $(\mathbf{y} - \mathbf{X}\boldsymbol{\beta})'\mathbf{B}^{-1}(\mathbf{y} - \mathbf{X}\boldsymbol{\beta})$. Since $\mathbf{B}$ is positive definite, there is a nonsingular matrix $\mathbf{H}$ such that $\mathbf{B} = (\mathbf{H}'\mathbf{H})^{-1}$ and $\mathbf{B}^{-1} = \mathbf{H}'\mathbf{H}$. Thus minimizing $(\mathbf{y} - \mathbf{X}\boldsymbol{\beta})'\mathbf{B}^{-1}(\mathbf{y} - \mathbf{X}\boldsymbol{\beta})$ with respect to $\boldsymbol{\beta}$ is equivalent to minimizing $(\mathbf{y}^* - \mathbf{X}^*\boldsymbol{\beta})'(\mathbf{y}^* - \mathbf{X}^*\boldsymbol{\beta})$ via ordinary least-squares, where $\mathbf{y}^* = \mathbf{H}\mathbf{y}$ and $\mathbf{X}^* = \mathbf{H}\mathbf{X}$. It follows that

$$\begin{aligned}\hat{\boldsymbol{\beta}}_G &= (\mathbf{X}^{*\prime}\mathbf{X}^*)^{-1}\mathbf{X}^{*\prime}\mathbf{y}^* \\ &= (\mathbf{X}'\mathbf{B}^{-1}\mathbf{X})^{-1}\mathbf{X}'\mathbf{B}^{-1}\mathbf{y}. \qquad (5)\end{aligned}$$

For the AR(1) error structure, the transformation $\mathbf{H}$ that permits ordinary least-squares estimation is

$$\mathbf{H} = \begin{bmatrix} \sqrt{1-\rho^2} & 0 & 0 & \cdots & 0 & 0 \\ -\rho & 1 & 0 & \cdots & 0 & 0 \\ 0 & -\rho & 1 & \cdots & 0 & 0 \\ \vdots & \vdots & \vdots & & \vdots & \vdots \\ 0 & 0 & 0 & \cdots & -\rho & 1 \end{bmatrix}. \qquad (6)$$

In (5), one could have used $\mathbf{A}^{-1}$ or $\boldsymbol{\Sigma}^{-1}$ in place of $\mathbf{B}^{-1}$ since the scalars cancel out.

When ρ *is not known*, Judge et al. [7] point out that three procedures are available for parameter estimation: estimated generalized least-squares, nonlinear least-squares, and maximum likelihood*.

Let $\hat{\boldsymbol{\beta}}_E$ denote the estimated generalized least-squares estimator of $\boldsymbol{\beta}$. $\hat{\boldsymbol{\beta}}_E$ is obtained by using the estimator in (5) after estimating ρ. Thus these procedures are called two-step procedures. Several methods are available for estimating ρ. These include:

1. The Cochrane–Orcutt procedure [3], where $\hat{\rho}_1 = \sum_{t=2}^{N}\hat{\epsilon}_t\hat{\epsilon}_{t-1}/\sum_{t=1}^{N}\hat{\epsilon}_t^2$ and the $\hat{\epsilon}_t$'s are obtained by using ordinary least-squares on $\mathbf{y} = \mathbf{X}\boldsymbol{\beta} + \boldsymbol{\epsilon}$. More precisely, this is termed the Prais–Winsten [10] procedure, since all N elements of $\mathbf{y}$ and all N rows of $\mathbf{X}$ were affected by the $\mathbf{H}$ transformation in obtaining $\mathbf{y}^*$ and $\mathbf{X}^*$. Theil [11] proposed a modification of $\hat{\rho}_1$: namely, $(N-p)/(N-1)\hat{\rho}_1$.
2. The estimate of ρ obtained from the Durbin–Watson* statistic, D. Specifically, $\hat{\rho}_2 = 1 - D/2$. Theil and Nagar [12] give the following modification of $\hat{\rho}_2$: $(N^2\hat{\rho}_2 + p^2)/(N^2 - p^2)$.
3. The estimate of ρ obtained from the Durbin procedure [4]. Let $\mathbf{H}_0$ denote the $(N-1) \times N$ matrix obtained by deleting the first row of $\mathbf{H}$ in (6). Let $\hat{\rho}_3$ denote the estimated coefficient of y_{t-1} in the model: $\mathbf{H}_0\mathbf{y} = \mathbf{H}_0\mathbf{X}\boldsymbol{\beta} + \mathbf{H}_0\boldsymbol{\epsilon}$. For the simple linear regression* model, we have

$$\begin{aligned} y_t = \beta_0(1-\rho) &+ \rho y_{t-1} \\ &+ \beta_1 x_t - \beta_1\rho x_{t-1} + a_t, \\ &\qquad t = 2, \ldots, N. \end{aligned}$$

By using this method, Maddala [8] points out that one is ignoring the constraint that (coefficient of x_{t-1}) $= -$(coefficient of x_t) $\cdot$ (coefficient of y_{t-1}).

In the nonlinear least-squares procedure, one needs to find those estimates of $\boldsymbol{\beta}$ and ρ that simultaneously minimize $(\mathbf{y}^* - \mathbf{X}^*\boldsymbol{\beta})'(\mathbf{y}^* - \mathbf{X}^*\boldsymbol{\beta})$. Although nonlinear optimization algorithms can be used, Hildreth and Lu [6] suggested a search procedure. For values of ρ from -1.0 to 1.0 in increments of 0.1, calculate $\hat{\boldsymbol{\beta}}_G$ as stipulated in (5) and the corresponding sum of squares, $(\mathbf{y}^* - \mathbf{X}^*\hat{\boldsymbol{\beta}}_G)'(\mathbf{y}^* - \mathbf{X}^*\hat{\boldsymbol{\beta}}_G)$. Choose that value of ρ which minimizes this sum of squares. Higher decimal accuracy can be obtained by finding the sum of squares for several additional values of ρ near the minimizing value. Although the Hildreth–Lu method is not computationally efficient, the minimum sum of squares obtained should be global rather than local if some care is exercised in the search procedure. Obviously, the value of ρ so obtained need not equal any of the values used in the estimated generalized least-squares procedure.

Under the assumption that the a_t's are normally distributed, the maximum likelihood procedure can be used. Judge et al. [7] show that maximizing the concentrated likelihood function is equivalent to minimizing $(1-\rho^2)^{-1/N}(\mathbf{y}^* - \mathbf{X}^*\boldsymbol{\beta})'(\mathbf{y}^* - \mathbf{X}^*\boldsymbol{\beta})$; this

Table 1 Estimates of First-Order Autocorrelation Coefficient by Different Methods

	Method			
Firm	Cochrane–Orcutt	Durbin	Hildreth–Lu	Maximum Likelihood
GM	0.458	0.816	0.67	0.64
U.S. Steel	0.481	0.874	0.74	0.69
GE	0.461	1.061	0.50	0.47
Chrysler	− 0.020	− 0.346	− 0.05	− 0.04
Atlantic-Richfield	− 0.236	− 0.737	− 0.22	− 0.21
IBM	0.114	0.624	0.18	0.17
Union Oil	0.098	0.125	0.12	0.11
Westinghouse	0.241	0.297	0.30	0.28
Goodyear	0.246	0.706	0.39	0.36
Diamond Match	0.402	0.385	0.65	0.57

differs from the nonlinear least-squares procedure by the $(1 - \rho^2)^{-1/N}$ factor. Algorithms for maximizing the concentrated likelihood function are presented in Hildreth and Dent [5] and in Beach and MacKinnon [1], although a search procedure similar to the Hildreth–Lu method could be utilized.

Empirical results for some of the procedures discussed above are presented in Maddala [8]. For annual data from 1935 to 1954, and 10 different firms, Maddala regresses gross investment on two independent variables: value of the firm, and stock of plant and equipment. The results are presented in Table 1. Inspection of the entries in Table 1 reveals that the maximum likelihood and Hildreth–Lu estimates are always in the same neighborhood, with Durbin's estimates differing substantially from these two.

Pindyck and Rubinfeld [9] also present two numerical examples using the Hildreth–Lu scanning method.

Although the name Hildreth–Lu has been reserved to refer to the search procedure for first-order autoregressive error, a similar search procedure could be employed for any ARMA error structure, as discussed in Judge et al. [7].

References

[1] Beach, C. M. and MacKinnon, J. G. (1978). *Econometrica*, **46**, 51–58.

[2] Box, G. E. P. and Jenkins, G. M. (1970). *Time Series Analysis, Forecasting and Control.* Holden-Day, San Francisco.

[3] Cochrane, D. and Orcutt, G. H. (1949). *J. Amer. Statist. Ass.*, **44**, 32–61.

[4] Durbin, J. (1960). *J. R. Statist. Soc. B*, **22**, 139–153.

[5] Hildreth, C. and Dent, W. (1974). In *Econometrics and Economic Theory: Essays in Honor of Jan Tinbergen*, W. Sellekaert, ed. Macmillan, London, pp. 3–25.

[6] Hildreth, C. and Lu, J. Y. (1960). Demand Relations with Autocorrelated Disturbances. *Mich. State Univ. Agric. Exp. Stn. Tech. Bull. 276*, East Lansing, Mich.

[7] Judge, G. G., Griffiths, W. E., Hill, R. C., and Lee, T.-C. (1980). *The Theory and Practice of Econometrics.* Wiley, New York.

[8] Maddala, G. S. (1977). *Econometrics.* McGraw-Hill, New York.

[9] Pindyck, R. S. and Rubinfeld, D. L. (1976). *Econometric Models and Economic Forecasts.* McGraw-Hill, New York.

[10] Prais, S. J. and Winsten, C. B. (1954). Trend Estimators and Serial Correlation. *Cowles Comm. Discuss. Paper No. 383*, Chicago.

[11] Theil, H. (1971). *Principles of Econometrics.* Wiley, New York.

[12] Theil, H. and Nagar, A. L. (1961). *J. Amer. Statist. Ass.*, **56**, 793–806.

(AUTOREGRESSIVE-MOVING AVERAGE (ARMA) MODELS
DURBIN–WATSON TEST
LEAST SQUARES)

FRANK B. ALT

HINGES *See* FIVE-NUMBER SUMMARIES

HIRSCHMAN'S INDEX *See* INCOME INEQUALITY MEASURES

HISTOGRAMS

A histogram is a *graphical representation** of a *frequency distribution*, typically utilizing bars to exhibit the frequency or relative frequency of occurrence of each value or group of values in a data set.

Figure 1 illustrates the bar form of a histogram for winter snowfall data in Buffalo, New York. The data, which are identified in Fig. 6 and cover a 39-year period starting with the winter of 1939–1940, have been grouped into eight intervals, each 20 measurement units (inches) wide. The numbers on the horizontal axis in Fig. 1 denote the midpoints of the intervals. Data having values on the interval boundaries (e.g., 60, 80, . . .) belong to the interval preceding the boundary.

The term "histogram" appears to have been used first by Karl Pearson* in 1895 [10]. The use by Guerry of *bar charts** [5] to display crime frequencies by numerical characteristics such as age is cited by Beniger and Robyn [1] as the first use of graphical displays of empirical frequency distributions.

Discussions of histograms, and related distributional forms such as the frequency polygon* and ogive*, may be found in Freund [4], King [9], Johnson and Leone [6], and most elementary statistics textbooks (*see also* GRAPHICAL REPRESENTATION OF DATA).

USES

A histogram is used to:

1. Condense a set of data for easy visual comprehension of general characteristics

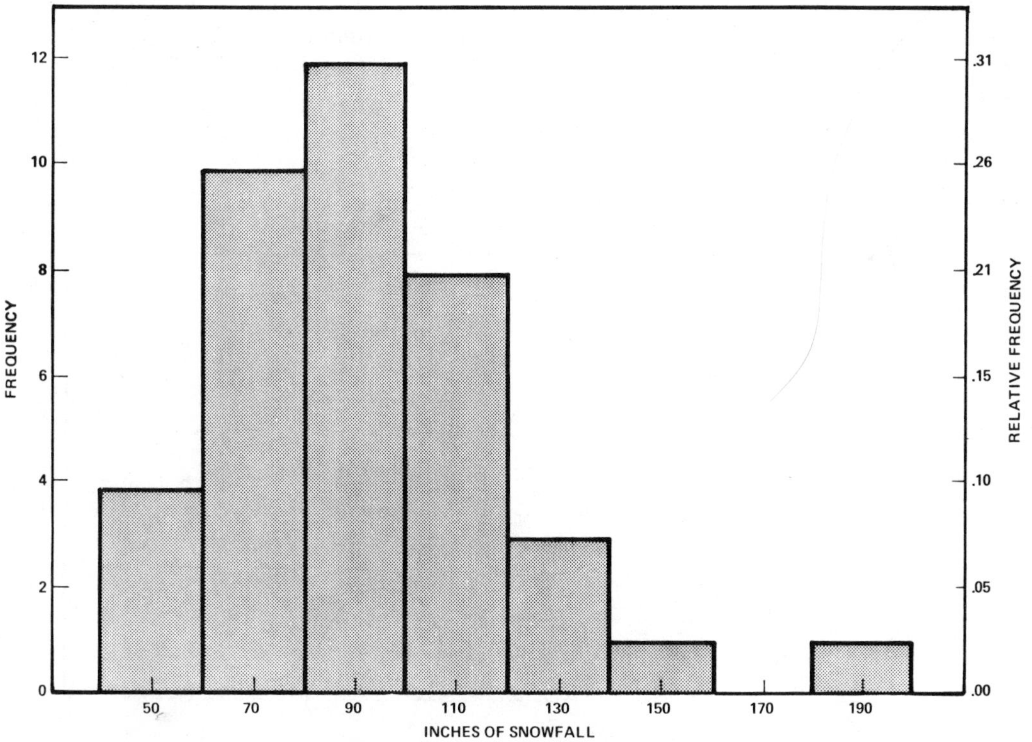

Figure 1 Histogram of Buffalo winter snowfall covering a 39-year period starting in 1939–1940. (Based on local climatological data gathered by the National Oceanic and Atmospheric Administration, National Climatic Center, Asheville, N.C.)

such as typical values, spread or variation, and shape
2. Suggest probability models or transformations for subsequent analysis
3. Detect unexpected behavior or unusual values in the data

The distribution of the snowfall data in Fig. 1 is seen to be unimodal and nearly symmetric, with the mean, median, and mode clearly associated with the interval 80–100. The mean can be thought of as the point on the horizontal axis at which a fulcrum should be placed to balance the histogram where the bar heights are proportional to weight. The data (as seen in Fig. 1) cover a range of up to 160 inches, with over three-fourths (30/39) of the data concentrated between 60 and 120. If the one extreme value is deleted, the distribution is somewhat normal-like (i.e., Gaussian); consequently, the standard deviation may be roughly approximated as one-fourth of the centered data range that includes about 95% of data. In this case the range of about 120 (= 160 − 40) gives a standard deviation of 30, which compares favorably with the calculated value of 33.

Histograms are frequently used as an exploratory tool prior to statistical analysis and modeling (*see* EXPLORATORY DATA ANALYSIS). The shape of a histogram may be suggestive of candidate probability models (e.g., normal, lognormal*, etc). Statistical procedures frequently assume that the data follow a normal distribution*. A non-normal shaped histogram may suggest a transformation to render the data more normal-like or an augmentation of a regression* or time-series* model to ensure a more normal-like error structure.

A histogram is a useful diagnostic tool for detecting outlying values such as the one seen in Fig. 1. Atypically shaped histograms often provide important clues to the nature of the system or process that generated the data. The double hump or bimodality exhibited in Fig. 2 led to the discovery that two machines were producing paint can ears of different average thickness. The unusually high frequency at multiples of 5°F in Fig. 3 suggests that the data recorder was biased toward increments of 5°F and that this bias was different at 20, 30, 40, and 50°F (19, 29, 39, and 49°F were rounded up 1°F) than at 25, 35, and 45°F (24, 34, and 44°F were rounded up while 26, 36, and 46°F were rounded down). The abrupt change in frequencies around the lower specification limit in Fig. 4 [3] identified that some out-of-specification readings were misrepresented by being recorded as being within specifications. It is important to note that the histograms depicted in Figs. 3 and 4 might

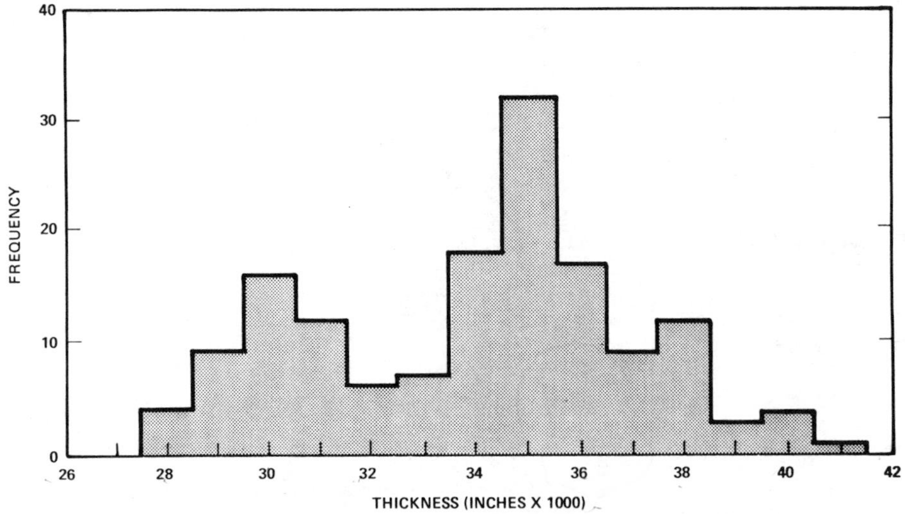

Figure 2 Histogram of the thickness of paint can ears, exhibiting bimodality and identifying the fact that the two machines were producing different average thicknesses.

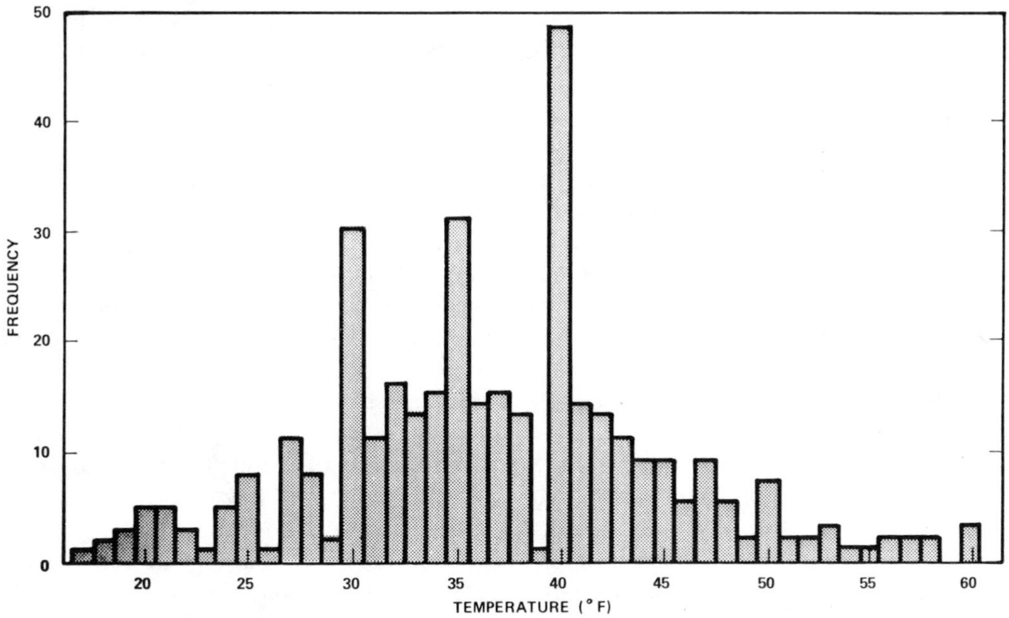

Figure 3 Histogram of maximum daily temperature during January (1950–1961) for a southern U.S. city, suggesting inconsistent rounding.

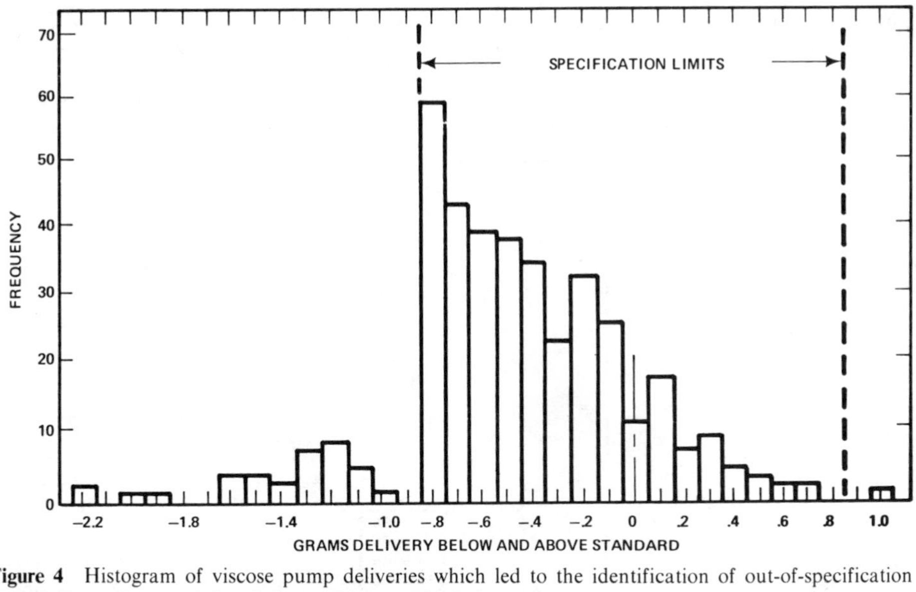

Figure 4 Histogram of viscose pump deliveries which led to the identification of out-of-specification data deliberately recorded as being within specifications.

not have been as effective diagnostically if fewer intervals had been used.

CONSTRUCTION

Data are often grouped into intervals (e.g., Fig. 1) in order to provide an informative display of the distribution. To ensure fairness of display, all intervals should be the same width, be mutually exclusive, and include all possible data values. The selection of the number of intervals or the interval width should depend at least on the number of data and the proposed use of the histogram.

The shape of the histogram is particularly important for uses **1** and **2** listed earlier, with too many intervals producing an undesirable erratic pattern of frequencies. Experience is generally the best guide to interval specification. Formulas that may be effectively used are $1 + 3.3 \log_{10}(n)$ for the number of intervals [12] and $3.49sn^{-1/3}$ for the interval width [11], where n denotes the sample size and s denotes the sample standard deviation. Both of these rules tend to produce too few intervals when the distribution is asymmetric or outlying values are present. For use **3** it is important that aberrations in the data are not masked by too few intervals (e.g., Figs. 3 and 4).

The construction process for bivariate data proceeds by grouping according to two variables forming a two-way table of cells (instead of intervals). The frequency of occurrence of each cell is then exhibited by "towers" (instead of bars) in a stereogram representation of the three dimensions [8].

To portray denseness when constructing a histogram, it is appropriate to display the bars without spaces between them when the data are from a continuous variable (e.g., snowfall data) and with spaces when the data are from a discrete variable.

SUBJECTIVITY AND MISUSE

Grouping data is a subjective process that may affect the appearance of the resulting histogram. For example, if the snowfall data is grouped into intervals of width 20 but with the first interval starting at 30, the resulting shape (Fig. 5) is more suggestive of a skew rather than a symmetric distribution. Consequently, it is usually warranted to use several groupings before conclusions are drawn (*see also* GROUPED DATA).

Unfortunately, a histogram can be misused to present a distorted picture of the distribution. Some examples include selectively choosing a grouping which best serves one's needs, using unequal interval widths without proper scaling adjustments, or starting the frequency scale above zero, thereby

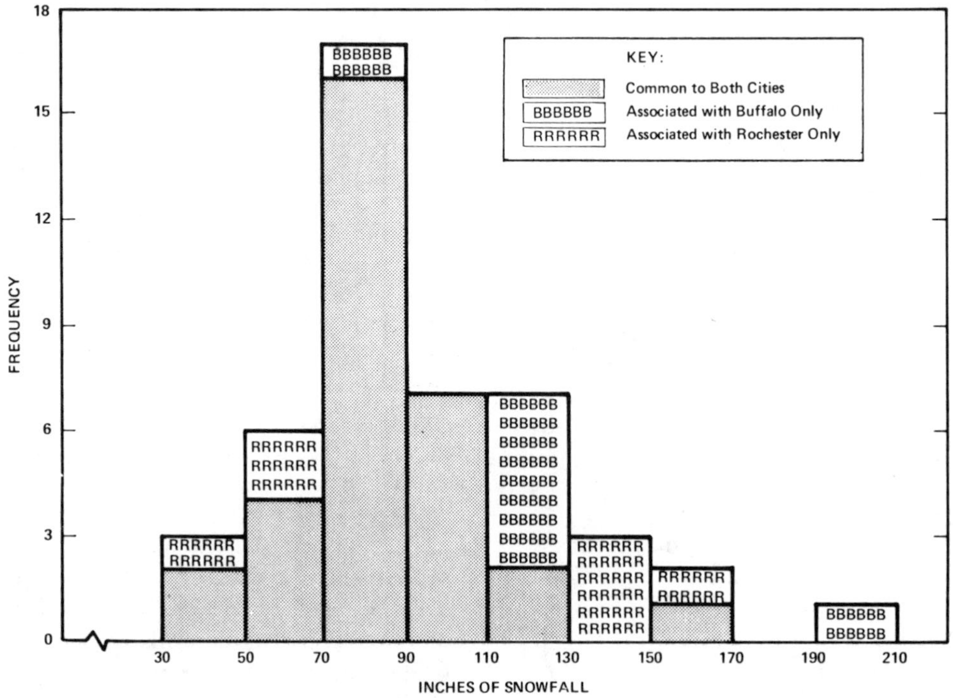

Figure 5 Overlayed histograms comparing Buffalo and Rochester snowfalls for a 39-year period starting in 1939–1940. (Based on local climatological data gathered by the National Oceanic and Atmospheric Administration, National Climatic Center, Asheville, N.C.)

visually distorting comparisons among frequencies.

COMPARISONS OF DISTRIBUTIONS

Common methods for comparing two (or more) empirical distributions include separate histograms displaced vertically over the same horizontal axis; back-to-back or dual histograms, where horizontal bars extend from a common vertical data axis [2]; or histograms overlayed on each other with bar portions not common to both identified. Figure 5 illustrates the overlay procedure comparing winter snowfalls of Buffalo, and Rochester, New York. Color overlays are particularly effective, when feasible, with the common portions of both distributions being shown as a blend of the two colors. Relative frequencies should be used for a fair comparison when overlaying different-size data sets.

A graphical comparison of an empirical and a theoretical distribution may be achieved by simply superimposing the theoretical density curve on a histogram of the data, where the density is appropriately scaled so that the total area under the curve equals the total area of the bars of the histogram. Visual discrimination between the empirical and fitted theoretical distributions may be further enhanced by hanging the histogram from the fitted curve [14].

VARIATIONS AND RELATED GRAPHICS

Some common variations of the histogram for quantitative data substitute symbols, such as asterisks, dots (*dot-array diagram*), or figures (*pictogram*) for the bars. A *stem and leaf* diagram* [13] is a histogram form which uses the numerical data values to build the bars (*see* EXPLORATORY DATA ANALYSIS). In Fig. 6 the stems identify the tens digit of the Buffalo snowfall and the leaves correspond to the last two digits of the snowfall. A pictogram using human figures is a familiar way to illustrate distributions of people. Joiner [7] exhibits "living histograms" by photographs of students arranged by height.

Bar (or symbol) *charts** can be used to display frequencies for qualitative (or nonnumerical) data and will have the appearance of histograms, although no unique ordering of the data is possible. For example, a bar chart relating frequency of annual cancer deaths to cancer type could be displayed using any ordering of cancer types desired. A *pie chart** is an alternative form which deemphasizes order. Each piece of pie is associated with a different value of nonnu-

TENS DIGIT OF INCHES OF SNOWFALL

STEM	LEAVES
4–5	58, 40, 40, 56
6–7	78, 79, 65, 71, 72, 71, 66, 72, 78, 79
8–9	90, 86, 89, 83, 90, 85, 90, 98, 97, 89, 96, 83
10–11	11, 05, 14, 15, 16, 02, 01, 10
12–13	21, 25, 21
14–15	54
16–17	
18–19	99

Figure 6 Stem and leaf diagram displaying the distribution of the Buffalo snowfall data and providing a tabulation of the data. (Based on local climatological data gathered by the National Oceanic and Atmospheric Administration, National Climatic Center, Asheville, N.C.)

merical data and the angle formed by the piece is proportional to the frequency of occurrence. Strictly speaking, these charts are not histograms, as they do not portray distribution characteristics. More accurately, they offer quantitative comparisons of qualitative data.

References

[1] Beniger, J. R. and Robyn, D. L. (1978). *Amer. Statist.*, **32**, 1–11. (Historical accounts of graphics with references.)

[2] Dallal, G. and Finseth, K. (1977). *Amer. Statist.*, **31**, 39–41.

[3] Deans, P. N. (1976). *Unforgettable Problems*. Presented at the Tennessee Quality Control Clinic, Knoxville, Tenn.

[4] Freund, J. E. (1976). *Statistics: A First Course*, 2nd ed. Prentice-Hall, N.J. (Elementary.)

[5] Guerry, A. M. (1833). *Essai sur la statistique morale de la France*. Paris.

[6] Johnson, N. L. and Leone, F. C. (1977). *Statistics and Experimental Design in Engineering and the Physical Sciences*, 2nd ed., Vol. 1. Wiley, New York. (Elementary.)

[7] Joiner, B. L. (1975). *Int. Statist. Rev.*, **43**, 339–340.

[8] Kendall, M. G. and Stuart, A. (1977). *The Advanced Theory of Statistics*, Vol. 1, 4th ed. Macmillan, New York. (Intermediate. Good discussions on distributions.)

[9] King, J. R. (1971). *Probability Charts for Decision Making*. Industrial Press, New York. (Detailed discussion of histogram construction.)

[10] Pearson, K. (1895). *Philos. Trans.*, **186**(Pt. 1), 343–414.

[11] Scott, D. W. (1979). *Biometrika*, **66**, 605–610.

[12] Sturges, H. A. (1926). *J. Amer. Statist. Ass.*, **21**.

[13] Tukey, J. W. (1977). *Exploratory Data Analysis*. Addison Wesley, Reading, Mass.

[14] Wainer, H. (1974). *Amer. Statist.*, **28**, 143–145.

(BAR CHART
DOT DIAGRAM
EXPLORATORY DATA ANALYSIS
FREQUENCY POLYGON
GRAPHICAL REPRESENTATION OF DATA
OGIVE
PIE CHART
STEM AND LEAF DIAGRAM)

RONALD D. SNEE
CHARLES G. PFEIFER

HISTORICAL CONTROLS

A comparative clinical trial* is a planned experiment on human patients involving two or more treatments, where the primary purpose is to evaluate the relative effectiveness of the treatments. Often, the comparison is between two treatments, a "standard" and a proposed new treatment. A large number of patients usually have been studied on the standard treatment and patient characteristics related to prognosis may be known. In many clinical trials, patients entering the study are randomized to the available treatments (arguments for this have been given by Byar et al. [3] and summarized in the article CLINICAL TRIALS), but there may be circumstances for preferring to select the group of patients on the standard (or control) treatment from a historical series. Such patients are generally termed *historical controls*; examples are patients chosen from articles reported in the literature or from a preceding clinical trial in a sequence of studies. Recent articles giving arguments for historical control groups are Gehan and Freireich [6], Gehan [8], and Freireich and Gehan [5].

The general arguments for utilizing historical control groups are: all knowledge is historical and modifications are made as evidence accumulates. In a nonrandomized clinical trial, results of a new treatment in consecutive patients are compared with those from a historical control group. This approach is consistent with the accumulation of knowledge using the principles of the scientific method. Predictions are based on the premise that the past is the best guide to knowledge of the future. Confirmation of results observed solidifies their acceptance. Proponents of randomized clinical trials must accept some historical data, namely their own studies; otherwise, the clinical trials would have no predictive value.

Studies utilizing historical controls require a much smaller number of patients and shorter time period than randomized studies designed for equivalent objectives. Further, a larger number of patients will generally be available, since some patients will not accept randomization* to treatment, whereas they would accept assignment to a new treatment. If an investigator is studying a new (A) versus standard (B) treatment in a nonrandomized study and sufficient data are available so that the response rate for the standard treatment may be assumed known (say p), then the number of patients required to compare A with B is only one-fourth that for a randomized study with equivalent statistical significance level* and power* [6]. When the response rate is not assumed known, but is estimated from a historical series of moderate size (say 75 patients or more), Makuch and Simon [10] give tables for the number of patients required. When the historical control series is moderate or large and it is desired to detect at least a 20% improvement for the new over the historical control treatment, the number of patients required on the new treatment is always less than that for the new treatment group alone in a prospective randomized study.

For a clinical investigator who does not randomize, there is no ethical dilemma either when advising patients about entry into study or when the study has been conducted for a period of time and some results are available. The ethical basis of the randomized clinical trial depends on there being no convincing evidence about the relative merits of the treatments. It is rare that the evidence favoring two or more treatments is equivalent prior to a study and, after initiation, the ethical basis for continuing may become less tenable. Interim results may suggest the new is better than the historical control treatment at some level of statistical significance, say $P = 0.20$. Such a circumstance could arise in a clinical trial designed to accrue a fixed number of patients or in a sequential trial where a boundary point had not quite been reached. It would be difficult to argue that the weight of the evidence favoring each of the treatments is identical. If it is accepted that all clinical investigators seek better treatments, it follows that no study should be started unless there is preliminary evidence suggesting that the new therapy is at least as good or possibly better than the standard. In the historical control group study, a clinical investigator would be entering all patients on the new therapy.

The outstanding criticism of historical control groups is that consciously or unconsciously, patients may be selected to receive the new treatment that are more favorable than patients receiving the standard. Hence the trial of the new treatment may yield a positive result merely because the group of patients and not the treatment was more favorable. When a large body of data is available on the standard treatment, techniques for determining prognostic factors are well known [1] and knowledge of these factors may be used to stratify patients or to adjust the comparison of the new vs. standard treatment by use of regression models.

If a regression model* is available relating the outcome of treatment to prognostic factors, the model may be used to test for treatment effects after adjustment for the prognostic features of the patients. An example of Cox's model* comparing disease-free survival and survival between treatment groups in a breast cancer study is given by Buzdar et al. [2]. Gehan et al. [9] give some approaches to validating regression models in making adjustments for prognostic factors.

Planners of randomized clinical trials can rely on randomization, stratification*, and regression procedures as techniques for adjustment in conducting and analyzing their studies, whereas those preferring a historical control group can only use the latter two procedures. Arguing that a historical control group might not be comparable with a new treatment group involves asserting that there was an unknown prognostic feature of major importance in addition to those already accounted for that was responsible for an observed treatment difference. It seems extremely unlikely that there could be a major prognostic characteristic that was unknown, balanced by randomization but not by time period, and was very important after accounting for other prognostic features. If all these unfavorable events did occur in a single study, the investigator who did not randomize would have to discover in a subsequent confirmation study that the new treatment was not as beneficial as expected, whereas the investigator who randomized would discover this within the trial.

Since clinical research has produced many efficacious new treatments, especially in the last 30 years, it is reasonable to ask: Which of the important advances in a particular clinical field in the last 30 years can be attributed to historical control group clinical studies and which to randomized clinical trials? At least in cancer research, the evidence is very strong that new treatment regimens for acute leukemia, choriocarcinoma, lymphoma, lung cancer, osteosarcoma, breast cancer, and sarcoma have come from nonrandomized studies [8]. Although it is true that randomized clinical trials have debunked false claims made for some new treatments [4], this was mainly because the historical control group studies were poorly conducted, rather than any special virtues of the subsequent randomized clinical studies.

References

[1] Armitage, P. and Gehan, E. A. (1974). *Int. J. Cancer*, **13**, 16–36. (Review article giving methods for detecting and utilizing prognostic factors.)

[2] Buzdar, A. U., Gutterman, J. U., Blumenschein, G. R., Hortobagyi, G. N., Tashima, C. K., Smith, T. L., Hersh, E. M., Freireich, E. J., and Gehan, E. A. (1978). *Cancer*, **41**(3), 1064–1075. (Example of the use of Cox's regression model in a historical control group study of breast cancer.)

[3] Byar, D. P., Simon, R. M., Freidewald, W. T., Schlesselman, J. J., DeMets, D. L., Ellenberg, J. H., Gail, M. H., and Ware, J. H. (1976). *N. Engl. J. Med.*, **295**, 74.

[4] Chalmers, T. C., Block, J. B., and Lee, S. (1972). *N. Engl. J. Med.*, **287**, 75. (Gives reasons for always randomizing patients and some examples of misleading historical control studies.)

[5] Freireich, E. J. and Gehan, E. A. (1979). In *Methods in Cancer Research*, Vol. 17, H. Busch and V. Devita, eds. Academic Press, New York, Chap. 8, pp. 277–310. (Discusses limitations of the randomized trial with respect to conduct of studies, tests of hypotheses, and drawing conclusions.)

[6] Gehan, E. A. (1978). *Biomedicine*, **28**, 13–19. (Example of use of historical controls in breast cancer and some techniques for validating use of regression models.)

[7] Gehan, E. A. and Freireich, E. J. (1974). *N. Engl. J. Med.*, **290**, 198–203. (Gives arguments for conduct of historical control group studies in cancer research.)

[8] Gehan, E. A. and Freireich, E. J. (1980). *Semin. Oncol.*, **8**, 430–436.

[9] Gehan, E. A., Smith, T. L., and Buzdar, A. U. (1980). *Cancer Treat. Rep.*, **64**, 2–3, 373–379.

[10] Makuch, R. and Simon, R. (1979). *J. Chronic Dis.*, **33**, 175–181.

(BIOSTATISTICS
CLINICAL TRIALS
RANDOMIZATION)

EDMUND A. GEHAN

HISTORICAL SERIES *See* TIME SERIES

HISTORY, STATISTICS IN *See* STATISTICS IN HISTORICAL STUDIES

HISTOSPLINES *See* SPLINE FUNCTIONS

HITTING TIME *See* RANDOM WALK

HODGES BIVARIATE SIGN TEST *See* SIGN TESTS

HODGES–LEHMANN ESTIMATORS

THE ESTIMATORS

Let $X_1, \ldots, X_m$, $Y_1, \ldots, Y_n$, be samples from continuous distributions F and G, respectively, which are assumed to satisfy

$$G(y) = F(y - \Delta), \tag{1}$$

so that G is obtained by shifting F by an amount Δ. The Hodges–Lehmann (HL) estimator $\hat{\Delta}$ of the shift parameter Δ is the median* of the mn differences $(Y_j - X_i)$. It is a *robust** competitor of the classical estimator $\bar{\Delta} = \bar{Y} - \bar{X}$, less strongly influenced by outlying observations (*see* ROBUSTNESS). The asymptotic relative efficiency* (ARE) of $\hat{\Delta}$ to Δ, if F has density f, is

$$e_{\hat{\Delta},\Delta}(F) = 12\sigma^2\left(\int_{-\infty}^{\infty} f^2(x)\,dx\right)^2 \tag{2}$$

where σ^2 is the variance of F. This efficiency is always $\geqslant 0.864$; it is $3/\pi$, or approximately 0.955, when F is normal, and can be arbitrarily large (even infinite) when f has sufficiently heavy tails.

There is a corresponding estimator of the center θ of a continuous symmetric distribution F based on a sample $Z_1, \ldots, Z_n$ from F. This one-sample HL estimator is the median $\hat{\theta}$ of the $n + \binom{n}{2}$ averages $\frac{1}{2}(Z_i + Z_j)$, $i \leqslant j$. Its ARE with respect to $\bar{Z}$ is again given by the right side of (2).

Convenient methods for calculating $\hat{\Delta}$ or $\hat{\theta}$ have been suggested by Høyland [5] and by Moses [8], who considers the closely related and somewhat more general problem of calculating the associated distribution-free confidence intervals* for Δ and θ. [See also Walker and Lev [14, Chap. 18] (contributed by Moses).] A fast algorithm is given by Boos and Monahan [1].

The estimators $\hat{\Delta}$ and $\hat{\theta}$ were proposed by Hodges and Lehmann [3] and by Sen [10]; a related estimator for regression coefficients was suggested earlier by Theil [13] and was generalized by Sen [11].

A GENERAL PRINCIPLE

The estimators of the preceding section can be derived from, and their properties related to, the two- and one-sample Wilcoxon tests*, respectively, by means of a general principle that converts test statistics into estimators. Suppose $T = T(X, \mu_H)$ is a statistic that uses data X to test the hypothesis that a parameter μ has the value μ_H. If T is sufficiently large, we would (say) be inclined to reject μ_H in favor of the alternatives $\mu > \mu_H$; if sufficiently small, to reject in favor of $\mu < \mu_H$. If $c(\mu_H)$ is in some reasonable sense the "central" value of T when $\mu = \mu_H$, then on observing $T = c(\mu_H)$ we would incline toward alternatives on neither side of μ_H; i.e., $c(\mu_H)$ is the value of T that gives the strongest support to the hypothesis $\mu = \mu_H$. Suppose, as frequently happens, that $c_0 = c(\mu_H)$ is independent of μ_H, and that $T(x, \mu_H)$ is continuous in μ_H and decreases as μ_H increases, so that there exists a unique value $\mu^*(x)$ of μ_H for which

$$T[x, \mu^*(x)] = c_0, \tag{3}$$

One may then reasonably think of $\mu^*(x)$ as the value of μ that is in best accord with the data, and use it as an estimator of μ. This natural way of converting test statistics into point estimators gains interest, because the resulting estimators will often share the attractive properties of the tests.

Example 1. Let $X_1, \ldots, X_n$ be independently distributed with common mean θ and variance σ^2 and consider the test statistics

$$t(X, \theta_H) = \sqrt{n(n-1)}\,(\bar{X} - \theta_H)\Big/\sqrt{\Sigma(X_i - \bar{X})^2}$$

and

$$S^2(X, \sigma_H^2) = \sum (X_i - \bar{X})^2 / \sigma_H^2 .$$

Under the hypotheses $\theta = \theta_H$ and $\sigma = \sigma_H$ we have, respectively, $E[t(X, \theta_H)] = 0$ and $E[S^2(X, \sigma_H^2)] = n - 1$, and if we take these expectations as the central values of the test statistics, we are led to $\theta^*(X) = \bar{X}$ and $\sigma^{*2}(X) = \sum(X_i - \bar{X})^2/(n-1)$ as unique solutions of (3).

If $T(x, \mu_H)$ is not continuous in μ_H, (3) may not have a solution. However, there will then exist a unique value μ^* such that $T(x, \mu_+^*) < c_0 < T(x, \mu_-^*)$, and this μ^* will be a natural estimator of μ. A further generalization occurs when T is assumed to be nonincreasing rather than strictly decreasing, so that the solution of (3) is not unique. The totality of solutions then forms an interval, the midpoint μ^* of which provides a reasonable value for μ.

Application of the conversion principle in this generalized form to the one- and two-sample Wilcoxon test leads to the HL estimators of the preceding Section.

Example 2. Let F be an unknown cumulative distribution function* for which there exists a unique value μ such that $F(\mu) = 1 - p$, with p given $(0 < p < 1)$. It is desired to estimate μ on the basis of a sample

$X_1, \ldots, X_n$ from F. A test of $\mu = \mu_H$ can be based on the number $T(x, \mu_H)$ of x's $> \mu_H$. Since T has a binomial distribution* with success probability p, $E(T) = np$ provides a reasonable value for t_0. If $X_{(1)} < \cdots < X_{(n)}$ denote the ordered X's, the resulting estimator is $X_{(i)}$ if $F(i/n) = 1 - p$ and is $\frac{1}{2}[X_{(i)} + X_{(i+1)}]$ if $F(i/n) < 1 - p < F[(i+1)/n]$.

GENERALIZATIONS OF THE HL ESTIMATORS

A large class of robust estimators, called the R-estimators* were obtained by Hodges and Lehmann [3] by applying the principle of the preceding section to general rank tests* or signed-rank tests. In the one-sample problem of the section "The Estimators" for instance, rank the differences $X_i - \theta_H$ by increasing absolute value. Each of the ranks $j = 1, \ldots, n$ is assigned a score $S(j)$, the scores being positive and increasing with j. The statistic T is defined as the sum of the scores attached to positive differences. If the X's are distributed symmetrically about θ_H (when $\theta = \theta_H$), the distribution of T is independent of θ_H and its expectation is a reasonable center. [If the scores $S(j)$ are symmetric, i.e., $S(j) + S(n + 1 - j) = 2c$ is independent of j, then T is symmetrically distributed about $t_0 = c$.] An R-estimator of particular interest is that based on *absolute normal scores**; the HL estimator corresponds to the scores $S(j) = j$. A detailed study of the efficiency of R-estimators (including the HL and absolute normal scores estimators) is provided by Hampel [2].

A different generalization of the HL estimator concerns the estimation of contrasts in the analysis of variance*. Since contrasts are functions of differences of all means, the differences can first be estimated by HL, and then combined into overall estimators of the contrasts. This method leads to inconsistencies, since the estimators of different contrasts will not satisfy the linear relationships existing among these contrasts. The difficulty can be removed by adjusting the method through least squares* or weighted least squares*. These adjustments do not affect the AREs of the estimators. (For details, see Lehmann [6] and Spjøtvoll [12].)

AN EXAMPLE

The following example illustrates the relationship of the median $\tilde{X}$, the mean $\bar{X}$, the one-sample HL estimator W, and the estimator N based on the absolute normal scores, for the estimation of θ in the one-sample problem. Let $n = 4$ and suppose that the four observations are 13, 18, 22, and 43, so that $\bar{X} = 24$.

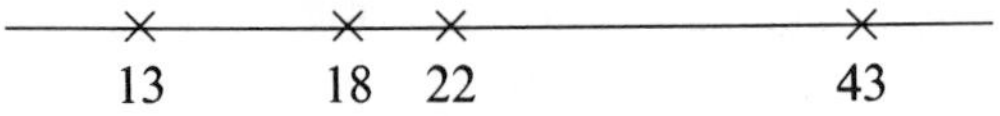

The estimators $\tilde{X}$, W, and N correspond to the scores $S(j) = 1$, $S(j) = j$, and $S(j)$ = absolute normal score, respectively, and their calculation from these scores is shown below.

1. If $S(j) = 1$, independent of j, then $S(j) + S(5 - j) = 2$ and we must find a value θ_H such that $\sum S(j) = 2$ when the summation is extended over the values j for which $X_j > \theta_H$. Here any value of θ_H will serve for which exactly two X's exceed θ_H, i.e., θ_H can be any value between 18 and 22, and the natural choice is the midpoint $\tilde{\theta} = 20$.
2. Let $S(j) = j$, so that each difference $|X_j - \theta_H|$ receives as score its rank. Then any θ_H between 20 and 22 will make both positive and negative scores add up to $5 = 2 + 3 = 1 + 4$. Thus the midpoint 21 is the HL estimator W.
3. If the ranks of the $|X_j - \theta_H|$ are assigned the absolute normal scores for $n = 4$: i.e., 0.26, 0.55, 0.91, and 1.46 (taken from Table 21 of *Biometrika Tables for Statisticians*, Vol. 2, Camb. Univ. Press, 1972), then a θ_H just to the left of 22 has positive score sum $(0.26 + 1.46 = 1.72)$ greater than the negative score sum $(0.55 + 0.91 = 1.46)$. But for any θ_H just to the right of 22, the inequality is re-

versed. The estimate N is thus equal to 22.

Note that $\tilde{X} < W < N < \bar{X}$, which reflects the increasing attention paid to the outlier* 43.

A discussion of the Hodges–Lehmann estimators can be found in most recent books on nonparametric methods, among them Hollander and Wolfe [4, Chap. 3, Sec. 2; Chap. 4, Sec. 2], Lehmann [7, Chap. 2, Sec. 5; Chap. 4, Sec. 4], and Randles and Wolfe [9, Chap. 7].

References

[1] Boos, D. D. and Monahan, J. (1982). In *Computer Science and Statistics: Proc. 4th Annual Symp. on the Interface*.

[2] Hampel, F. R. (1982). In *A Festschrift for Erich L. Lehmann*. Wadsworth, Belmont, Calif.

[3] Hodges, J. L., Jr. and Lehmann, E. L. (1963). *Ann. Math. Statist.*, **34**, 598–611.

[4] Hollander, M. and Wolfe, D. A. (1973). *Nonparametric Statistical Methods*. Wiley, New York.

[5] Høyland, A. (1964). *Norske Vid. Selsk. Fork.*, **37**, 42–47.

[6] Lehmann, E. L. (1963). *Ann. Math. Statist.*, **34**, 957–966.

[7] Lehmann, E. L. (1975). *Nonparametrics: Statistical Methods Based on Ranks*. Holden-Day, San Francisco.

[8] Moses, L. (1965). *Technometrics*, **7**, 257–260.

[9] Randles, R. H. and Wolfe, D. A. (1979). *Introduction to the Theory of Nonparametric Statistics*. Wiley, New York.

[10] Sen, P. K. (1963). *Biometrics*, **19**, 532–552.

[11] Sen, P. K. (1968). *J. Amer. Statist. Ass.*, **63**, 1379–1389.

[12] Spjøtvoll, E. (1968). *Ann. Math. Statist.*, **39**, 1486–1492.

[13] Theil, H. (1950). *Proc. Kon. Ned. Akad. Wet. A*, **53**, 1397–1412.

[14] Walker, H. M. and Lev, J. (1953). *Statistical Inference*. Holt, Rinehart and Winston, New York.

(DISTRIBUTION-FREE METHODS
NONPARAMETRIC METHODS
NORMAL SCORES TESTS
RANKING PROCEDURES
R-ESTIMATORS
ROBUSTNESS
WILCOXON TESTS)

J. L. HODGES, JR.
ERICH L. LEHMANN

HODGES SUPEREFFICIENCY

Let $X_1, X_2, \ldots$ be a sequence of independent and identically distributed observations with their common distribution determined by a real-valued parameter θ taking values in an open interval. Let $L(\theta, X_1)$ denote the natural logarithm of the likelihood* when the data consist of X_1, and let

$$I(\theta) = -E_\theta(L''(\theta, X_1)), \qquad (1)$$

where dashes denote partial differentiation with respect to θ, and E_θ denotes expected value when θ obtains. $I(\theta)$ is the *Fisher information** in X_1 for the estimation of θ. Under general regularity conditions we have

$$E_\theta(L'(\theta, X_1)) = 0, \quad E_\theta(L'(\theta, X_1))^2 = I(\theta), \quad 0 < I(\theta) < \infty.$$

For each integer $n = 1, 2, \ldots,$ let $T_n(X_1, \ldots, X_n)$ be a real-valued statistic, and suppose that T_n is a consistent and asymptotically normal estimate of θ with asymptotic variance $v(\theta)/n$, i.e.,

$$(T_n - \theta)/[v(\theta)/n]^{1/2} \to N(0, 1)$$

in distribution when θ obtains and $n \to \infty$. Then the asymptotic efficiency* of T_n, say ϕ, is

$$\phi(\theta) = \frac{1/I(\theta)}{v(\theta)} \qquad (2)$$

(*see* ASYMPTOTIC NORMALITY).

Definitions (1) and (2), which are due to R. A. Fisher* [4, 5], are based in part on the following considerations. (a) Under general regularity conditions there exist consistent and asymptotically normal estimates of θ with asymptotic variance $1/\{nI(\theta)\}$, e.g., the maximum likelihood* estimate based on $(X_1, \ldots, X_n)$. Let $\hat{T}_n$ be such an estimate. (b) For given θ, let m and n be large sample sizes such that $m/n = \phi(\theta)$. Then, by the present distributional assumptions, $\hat{T}_m$ and T_n are equally precise estimates of θ. Thus ϕ is a relative efficiency. (c) $1/nI(\theta)$ is a lower bound for the *asymptotic variance* of any consistent and asymptotically normal estimate of θ, so that

$$\phi(\theta) \leqslant 1 \tag{3}$$

and ϕ is in fact an absolute efficiency. This consideration is related to but not identical with the Cramér–Rao bound* for the actual variance of any unbiased estimate of θ.

It was discovered in 1951 by J. L. Hodges, Jr., that consideration (c) is not literally correct; there can exist estimates T_n which are superefficient, i.e., $\phi(\theta) > 1$, for certain values of θ. Suppose, for example, that the X_i are real-valued $N(\theta, 1)$ variables. Then $I(\theta) \equiv 1$. For each n, let $\bar{X}_n = \sum_1^n X_i/n$. Let α be a constant, $0 < \alpha < 1$, and

$$T_n = \begin{cases} \bar{X}_n & \text{if } |\bar{X}_n| > n^{-1/4}, \\ \alpha\bar{X}_n & \text{if } |\bar{X}_n| \leqslant n^{-1/4}. \end{cases}$$

Then T_n is consistent and asymptotically normal with asymptotic variance $v(\theta)/n$, where $v(\theta) = 1$ if $\theta \neq 0$ and $v(\theta) = \alpha^2$ if $\theta = 0$. In this example of Hodges, T_n is fully efficient if $\theta \neq 0$ and superefficient if $\theta = 0$.

It was shown by LeCam [7, 8] that consideration (c) is essentially correct in the following sense: In the general case, the set of all θ where (3) does not hold is of Lebesgue measure zero. A simple proof of LeCam's theorem is given in Bahadur [1]. It follows from the theorem that (3) holds at every continuity point of ϕ.

It has also been shown that superefficiency is a technical term rather than a recommendation; superefficient estimates always have certain undesirable properties. For example, if T_n is superefficient at a particular parameter value θ_0, if $\theta_n = \theta_0 + an^{-1/2}$ where $a > 0$, and if b_n is the probability that $T_n < \theta_n$ when θ_n obtains, then there exists $c > 0$ such that $b_n > \frac{1}{2} + c$ for all sufficiently large n [1]. Again, for each sufficiently large n, the maximum expected squared error of a superefficient estimate is larger than that of certain other estimates [6, 7].

Useful expositions of large sample* theories of estimation are given in Chernoff [3], Bahadur [2], and Hájek [6]. See Roussas [10] and LeCam [9] for certain generalizations and other technical developments.

References

[1] Bahadur, R. R. (1964). *Ann. Math. Statist.*, **35**, 1545–1552.

[2] Bahadur, R. R. (1967). *Ann. Math. Statist.*, **38**, 303–324.

[3] Chernoff, H. (1956). *Ann. Math. Statist.*, **27**, 1–22.

[4] Fisher, R. A. (1922). *Phil. Trans. R. Soc. A*, **222**, 309–368.

[5] Fisher, R. A. (1925). *Proc. Camb. Philos. Soc.*, **22**, 700–725.

[6] Hájek, J. (1972). *Proc. 6th Berkeley Symp. Math. Statist. Prob.*, Vol. 1. University of California Press, Berkeley, Calif., pp. 175–194.

[7] LeCam, L. (1953). *Univ. Calif. Publ. Statist.*, **1**, 277–300.

[8] LeCam, L. (1958). *Publ. Inst. Statist. Univ. Paris*, **7**, 17–35.

[9] LeCam, L. (1979). In *Contributions to Statistics* (Hájek Memorial Volume). Academia, Prague, pp. 119–135.

[10] Roussas, G. G. (1972). *Contiguity of Probability Measures: Some Applications in Statistics*. Cambridge University Press, Cambridge, England.

(ASYMPTOTIC NORMALITY
CRAMÉR–RAO LOWER BOUND
EFFICIENCY
MAXIMUM LIKELIHOOD ESTIMATION
NORMAL DISTRIBUTION)

R. R. BAHADUR

HOEFFDING INEQUALITY *See* PROBABILITY INEQUALITIES FOR SUMS OF BOUNDED RANDOM VARIABLES

HOEFFDING'S INDEPENDENCE TEST

Let the random vector (X, Y) have the cumulative distribution function* (CDF) $F(x, y)$. Let $\mathscr{F}$ be the class of all continuous bivariate CDFs, and $\mathscr{F}_0$ be the class of all $F \in \mathscr{F}$ such that $F(x, y) = F(x, \infty) F(\infty, y)$. Assume that $F \in \mathscr{F}$. The hypothesis H_0 that X and Y are independent is equivalent to the hypothesis that $F \in \mathscr{F}_0$.

Let $(X_1, Y_1), \ldots, (X_n, Y_n)$ be n independent observations of (X, Y). For $n \geqslant 5$ there is an unbiased estimator D_n of

$$\Delta(F) = \int \{F(x, y) - F(x, \infty)F(\infty, y)\}^2 \, dF(x, y),$$

which is symmetric in the n observations; it is unique under the assumption $F \in \mathscr{F}$. Hoeffding [3] proposed the test which rejects H_0 if D_n exceeds a constant.

An explicit expression for D_n is

$$D_n = \frac{A - 2(n-2)B + (n-2)(n-3)C}{n(n-1)(n-2)(n-3)(n-4)},$$

where

$$A = \sum_{i=1}^{n} a_i(a_i - 1)b_i(b_i - 1),$$

$$B = \sum_{i=1}^{n} (a_i - 1)(b_i - 1)c_i,$$

$$C = \sum_{i=1}^{n} c_i(c_i - 1),$$

$a_i + 1$ and $b_i + 1$ are the ranks of X_i among $X_1, \ldots, X_n$ and of Y_i among $Y_1, \ldots, Y_n$, respectively, and c_i is the number of observations (X_j, Y_j) such that both $X_j < X_i$ and $Y_j < Y_i$.

Under H_0 the distribution of D_n does not depend on the distribution of (X_i, Y_i). It has been tabulated for $n \leqslant 7$ in Hoeffding [3]. As $n \to \infty$, the distribution of nD_n under H_0 converges to a nondegenerate limit distribution, whose characteristic function* is given in Hoeffding's paper. For a fixed distribution $F \in \mathscr{F} - \mathscr{F}_0$, $n^{1/2}[D_n - \Delta(F)]$ has a nondegenerate normal limit distribution.

Blum et al. [1] considered a closely related test, based on the statistic

$$B_n = \int \{S_n(x, y) - S_n(x, \infty)S_n(\infty, y)\}^2 \, dS_n(x, y),$$

where $S_n(x, y)$ is the empirical CDF.

Assume for simplicity that $F(x, \infty) = F(\infty, x) = x$, $0 \leqslant x \leqslant 1$. The authors show that as $n \to \infty$, the distribution of nB_n under H_0 converges to that of

$$B = \int_0^1 \int_0^1 T^2(x, y) \, dx \, dy,$$

where $T(x, y)$ is the Gaussian process* on the unit square which satisfies $ET(x, y) = 0$,

$$ET(x, y)T(u, v) = \{min(x, u) - xu\}\{min(y, v) - yv\}.$$

The B_n-test is asymptotically equivalent to the D_n-test. The limit distribution of nB_n under H_0 is the same as that of $nD_n + \frac{1}{36}$. (The distribution of B is similar to the limit distribution of the Cramér–von Mises* goodness-of-fit* statistic in the null case.)

The authors show that the asymptotic power properties of the B_n-test are superior to those of a comparable chi-square test*. They tabulate the limit distribution of $\frac{1}{2}\pi^4 nB_n$ under H_0. Furthermore, they give an expression for B_n which is convenient for computations, and study the power of the test at alternatives close to the hypothesis. Finally, they consider statistics of a similar type for testing independence or partial independence in m-variate distributions with $m > 2$.

Csörgö [2] supplemented these results by establishing strong invariance principles* for the random functions

$$T_n(x, y) = S_n(x, y) - S_n(x, \infty)S_n(\infty, y)$$

under H_0. For example, he showed that Gaussian processes $T^{(n)}(x, y)$, $n = 1, 2, \ldots$, each distributed as the process $T(x, y)$ above, can be constructed such that

$$\sup_{x,y} |n^{1/2}T_n(x, y) - T^{(n)}(x, y)|$$

converges to zero almost surely at a specified rate as $n \to \infty$.

References

[1] Blum, J. R., Kiefer, J., and Rosenblatt, M. (1961). *Ann. Math. Statist.*, **32**, 485–498.

[2] Csörgö, M. (1979). *J. Multivariate Anal.*, **9**, 84–100.

[3] Hoeffding, W. (1948). *Ann. Math. Statist.*, **19**, 546–557.

(DEPENDENCE, TESTS FOR)

W. Hoeffding

HOEFFDING'S LEMMA

It is well known that if a random variable X has finite expectation, then

$$EX = \int_0^\infty [1 - F(x)]\,dx - \int_{-\infty}^0 F(x)\,dx.$$

Indeed, this is a special case of the general result, when $E|X|^n < \infty$, that

$$EX^n = \int_0^\infty x^{n-1}[1 - F(x)]\,dx - \int_{-\infty}^0 x^{n-1}F(x)\,dx.$$

Wassily Hoeffding gave a bivariate version of this identity [1], but the lemma did not become widely known until it was quoted by Lehmann [3]. Let F_{XY}, F_X, and F_Y denote the joint and marginal distribution functions for random variables X and Y, and suppose that $E|XY|$, $E|X|$, and $E|Y|$ are finite. Then

$$E(XY) - (EX)(EY) = \int_{-\infty}^{\infty}\int_{-\infty}^{\infty} [F_{X,Y}(x, y) - F_X(x)F_Y(y)]\,dx\,dy.$$

The proof uses Franklin's identity*. Let (X_1, Y_1) and (X_2, Y_2) be independent and identically distributed random variables with distribution function $F_{X,Y}$. Then

$$\begin{aligned}2[E(X_1Y_1) - (EX_1)(EY_1)] &= E[(X_1 - X_2)(Y_1 - Y_2)] \\ &= E\left\{\int_{-\infty}^{\infty} [I(u, X_1) - I(u, X_2)]\,du \times \int_{-\infty}^{\infty} [I(v, Y_1) - I(v, Y_2)]\,dv\right\},\end{aligned}$$

where $I(u, x) = 1$ if $u \leqslant x$ and 0 otherwise. This last expression equals

$$E\int_{-\infty}^{\infty}\int_{-\infty}^{\infty} [I(u, X_1)I(v, Y_1) - I(u, X_1)I(v, Y_2) - I(u, X_2)I(v, Y_1) + I(u, X_2)I(v, Y_2)]\,du\,dv.$$

Since $E|XY|$, $E|X|$, and $E|Y|$ are finite, this expression is absolutely integrable and we may interchange the order of taking expectation and integration. Then the last expression equals

$$2\left[\int_{-\infty}^{\infty}\int_{-\infty}^{\infty} [F_{X_1,Y_1}(u, v) - F_{X_1}(u)F_{Y_1}(v)]\,du\,dv\right],$$

giving the required identity.

Hoeffding used his lemma to identify the bivariate distributions with given marginal distribution functions F_X and F_Y that minimize and maximize the correlation* between X and Y. Note that

$$\begin{aligned}P[X \leqslant x] - P[Y > y] &\leqslant P[X \leqslant x, Y \leqslant y] \\ &\leqslant \min\{P[X \leqslant x], P[Y \leqslant y]\}.\end{aligned}$$

Then

$$\begin{aligned}\max\{F_X(x) + F_Y(y) - 1, 0\} &\leqslant F_{X,Y}(x, y) \\ &\leqslant \min\{F_X(x), F_Y(y)\};\end{aligned}$$

the lemma shows that $\max\{F_X + F_Y - 1, 0\}$ and $\min\{F_X, F_Y\}$ achieve the smallest and largest correlations, respectively, among all bivariate distributions with the given marginals. Whitt [4] presents other representations of these distributions useful in data analysis and Monte Carlo* studies.

Another direct consequence of Hoeffding's lemma is to present a rich class of bivariate distributions where uncorrelatedness implies independence. Lehmann [3] describes bivariate distributions $F_{X,Y}$ such that $F_{X,Y}(x, y) - F_X(x)F_Y(y)$ is nonnegative for all (x, y) in the support of $F_{X,Y}$ as the class of *positively quadrant dependent* (p.q.d.) *distributions*. That is, (X, Y) is p.q.d. if

$$\Pr[X \leqslant x, Y \leqslant y] \geqslant \Pr[X \leqslant x]\Pr[Y \leqslant y].$$

If the inequalities between probabilities are reversed, we say that (X, Y) is *negatively quadrant dependent* (n.q.d.); (*see* QUADRANT DEPENDENT DISTRIBUTIONS). Lehmann used Hoeffding's lemma and the right continuity of distribution functions to prove the following result. Suppose that $E(XY)$, $E(X)$, and $E(Y)$ are finite. If (X, Y) is p.q.d., then X and Y have nonnegative covariance; and if (X, Y) is n.q.d., the covariance is nonpositive; and if (X, Y) is p.q.d. or n.q.d. and uncorrelated, X and Y are independent. Examples of quadrant dependence include the bivariate normal distribution; other examples can be constructed by the methods given by Lehmann (*see also* DEPENDENCE, CONCEPTS OF).

Jogdeo [2] extends Lehmann's bivariate result to the multivariate case. In the three-dimensional case, he considers triples (X, Y, Z) of random variables with all third-order absolute moments finite. We say that (X, Y, Z) is *positive orthant dependent* (p.o.d.) if

$$\Pr[X \leqslant x, Y \leqslant y, Z \leqslant z] \geqslant \Pr[X \leqslant x]\Pr[Y \leqslant y]\Pr[Z \leqslant z].$$

Jogdeo uses Lehmann's result on p.q.d. bivariate random vectors to prove that if (X, Y, Z) is p.o.d., then they are independent if and only if (a) each pair is uncorrelated and (b) one of the pairs, say (XY), is conditionally uncorrelated given the third, Z; that is,

$$E[XY \mid Z] = E[X \mid Z]E[Y \mid Z].$$

Jogdeo next applied Hoeffding's method of proof to the independent and identically distributed triples (X_1, Y_1, Z_1) and $(-X_2, Y_2, Z_2)$ to show that if (X, Y, Z) are p.o.d., then X, Y, and Z are independent if and only if they are uncorrelated and $EXYZ = (EX)(EY)(EZ)$. Jogdeo notes that this result holds for all types of orthant dependent random vectors.

References

[1] Hoeffding, W. (1940). *Schr. Math. Inst. Univ. Berl.*, **5**, 181–233.

[2] Jogdeo, K. (1968). *Ann. Math. Statist.*, **39**, 433–441. (An extension of Lehmann's [3] article to several variables.)

[3] Lehmann, E. L. (1966). *Ann. Math. Statist.*, **37**, 1137–1153. (A study of bivariate quadrant dependence and the resultant unbiasedness of tests based on Pearson's *r*, Spearman's rho, Kendall's tau, etc.)

[4] Whitt, W. (1976). *Ann. Statist.*, **4**, 1280–1289. (A practical approach to bivariate distributions that have extreme correlations given fixed marginals.)

(CORRELATION
DEPENDENCE, CONCEPTS OF
FRANKLIN'S IDENTITY
QUADRANT DEPENDENT DISTRIBUTIONS)

GERALD A. SHEA

HÖLDER'S INEQUALITY

Hölder's inequality is a generalization of the Cauchy–Schwarz inequality*. Let $p > 1$ and $q > 1$ such that $p^{-1} + q^{-1} = 1$.

1. For sequences $\{a_n\}$ and $\{b_n\}$ of real numbers,

$$\left|\sum_{n=0}^{\infty} a_n b_n\right| \leqslant \left(\sum_{n=0}^{\infty} |a_n|^p\right)^{1/p} \left(\sum_{n=0}^{\infty} |b_n|^q\right)^{1/q},$$

   whenever the sums on the right converge.

2. For Lebesgue-integrable functions $f(x)$ and $g(x)$,

$$\left|\int_{-\infty}^{\infty} f(x)g(x)dx\right| \leqslant \left\{\int_{-\infty}^{\infty} |f(x)|^p dx\right\}^{1/p} \times \left\{\int_{-\infty}^{\infty} |g(x)|^q dx\right\}^{1/q}$$

   whenever the integrals on the right exist and are finite. The result holds if the range of integration is an interval I and f and g are integrable on I.

3. For random variables X and Y,

$$|E(XY)| \leqslant \{E|X^p|\}^{1/p}\{E|Y^q|\}^{1/q}$$

   whenever $E|X^p| < \infty$ and $E|Y^q| < \infty$.

The Cauchy–Schwarz inequality* follows when $p = q = 2$. As a further application, Chow and Teicher [1, p. 104], use Hölder's inequality to prove a result attributed to Liapunov*: if X is a random variable, and nonnegative almost surely, and if $E|X^p| < \infty$ for all $p > 0$, then $\ln[E(X^p)]$ is a convex function of p.

Reference

[1] Chow, Y. S. and Teicher, H. (1978). *Probability Theory*. Springer-Verlag, New York.

(CAUCHY–SCHWARZ INEQUALITY)

HOLLANDER BIVARIATE SYMMETRY TEST

INTRODUCTION

For a bivariate random vector (X, Y), with corresponding distribution function $F(x, y) = P(X \leqslant x; Y \leqslant y)$, the hypothesis of *bivariate symmetry* is

$$H_0: F(x, y) = F(y, x) \quad \text{for all } (x, y). \quad (1)$$

Thus H_0 asserts that the joint distribution of (X, Y) is the same as the joint distribution of (Y, X). That is, the joint distribution of (X, Y) is exchangeable (*see* EXCHANGEABILITY).

The hypothesis H_0 arises in a natural way in experiments for comparing a treatment with a control (or comparing treatment A with treatment B), where subjects serve as their own controls. If, say, we call the control response X, and the treatment response Y, then H_0 is the hypothesis of no treatment effect. For example, X could be a measure of depression taken on a patient prior to that patient receiving a tranquilizer, Y would be the measure taken a suitable period after administration of the tranquilizer, and then H_0 asserts that the tranquilizer has no effect (in that period, on the particular measure of depression).

Hollander [2] developed a conditionally distribution-free* test of H_0 based on a random sample from F. Let

$$\mathbf{Z} = \{(X_1, Y_1), \dots, (X_n, Y_n)\},$$

where $(X_1, Y_1), \dots, (X_n, Y_n)$ are assumed to be n mutually independent bivariate random vectors, each distributed according to the bivariate cumulative distribution function F. Note that we do *not* assume that within a pair X and Y are independent; in most applications X and Y are dependent.

Hollander's test of H_0 is based on the statistic H^2, where

$$H^2 = n \int\int \{F_n(x, y) - F_n(y, x)\}^2 dF_n(x, y),$$

$$F_n(x, y) = n^{-1} \sum_{j=1}^{n} \phi(X_j, x)\phi(Y_j, y), \quad (2)$$

and $\phi(a, b) = 1$ if $a \leqslant b$ and $= 0$ otherwise. Note that F_n is the empirical distribution function calculated from the random sample (*see* EDF STATISTICS).

Hollander's test is a permutation test* with respect to the group G of 2^n transformations, where

$$g_{(j_1, \dots, j_n)}(\mathbf{Z}) = \{(X_1, Y_1)^{(j_1)}, \dots, (X_n, Y_n)^{(j_n)}\},$$

where each j_i is either 0 or 1, $(X_i, Y_i)^{(0)} = (X_i, Y_i)$ and $(X_i, Y_i)^{(1)} = (Y_i, X_i)$. The hypothesis H_0 implies that for each $g \in G$, $g(\mathbf{Z})$ has the same distribution as $\mathbf{Z}$. This leads to the consideration of the conditional measures

$$P_c\{(X_i, Y_i) = (X_i, Y_i)^{(j_i)}, i = 1, \dots, n \,|\, \mathbf{Z}\} = 2^{-n} \quad \text{for each } (j_1, \dots, j_n). \quad (3)$$

The test, defined by (10) and (11) in the following section, is exact with respect to these conditional measures.

Hollander's test is an omnibus test in that it is consistent (i.e., has power* tending to 1 as n tends to ∞) against a very broad class of alternatives to H_0. Hollander [2] shows that the conditional test is consistent against all distributions F for which the population parameter $\int\int \{F(x, y) - F(y, x)\}^2 dF(x, y)$ is positive; in particular, the test is consistent against *all* absolutely continuous F's not satisfying H_0. Power values of the test based on H^2, against various alternatives, are given by Hollander [2] and Koziol [4].

THE TEST PROCEDURE

To simplify the computation of H^2, arrange the $n(X_i, Y_i)$ pairs so that the sequence $\{\min(X_i, Y_i)\}$ increases with i, and define the vector $\mathbf{r} = (r_1, \dots, r_n)$, where, for $i = 1, \dots, n$,

$$r_i = \begin{cases} 1 & \text{if } X_i < Y_i, \\ 0 & \text{if } X_i > Y_i. \end{cases} \quad (4)$$

(If $X_i = Y_i$, we arbitrarily set $r_i = 0$, since such a tied situation does not contribute to the value of H^2.) With these definitions, H^2

can be rewritten as

$$H^2 = n^{-2} \sum_{j=1}^{n} T_j^2; \tag{5}$$

$$T_j = \sum_{i=1}^{n} s_i d_{ij}, \tag{6}$$

$$s_i = 2r_i - 1, \tag{7}$$

and, for $i, j = 1, \ldots, n$,

$$d_{ij} = \begin{cases} 1 & \text{if } a_i < b_i \leqslant b_j \text{ and } a_i \leqslant a_j, \\ 0 & \text{otherwise,} \end{cases} \tag{8}$$

where $a_i = \min(X_i, Y_i)$, $b_i = \max(X_i, Y_i)$.

The distribution of H^2, with respect to the conditional measures P_c defined by (3), is obtained by computing H^2 for each of the possible 2^n **r**-configurations, where each r_i can be 0 or 1. [In performing these 2^n calculations, the d's defined by (8) remain the same.] Let

$$h^{(1)} \leqslant h^{(2)} \leqslant \cdots \leqslant h^{(2^n)}$$

denote the 2^n ordered values of H^2. We note that, since the h's depend on **Z** only through the ordering pattern of the $2n$ X's and Y's, it is possible, though extremely tedious for large n, to table the P_c distribution of H^2.

The conditional test, with probability of a type I error equal to α, is defined as follows. Set

$$m = 2^n - [2^n\alpha], \tag{9}$$

where $[2^n\alpha]$ is the greatest integer less than or equal to $2^n\alpha$. Let M_1 be the number of values $h^{(1)}, \ldots, h^{(2^n)}$ which are greater than $h^{(m)}$, and let M_2 be the number of values $h^{(1)}, \ldots, h^{(2^n)}$ which are equal to $h^{(m)}$. The α-level test is

$$\begin{array}{lll} \text{reject } H_0 & \text{if} & H^2 > h^{(m)}, \\ \text{accept } H_0 & \text{if} & H^2 < h^{(m)}, \end{array} \tag{10}$$

and if $H^2 = h^{(m)}$, make a randomized decision that rejects H_0 with probability p and accepts H_0 with probability $1 - p$, where

$$p = (2^n\alpha - M_1)/M_2. \tag{11}$$

In (10), H^2 is the value corresponding to our observed **r**-configuration.

The permutation test defined by (10) and (11) has the desirable feature that no adjustments for ties are necessary. The procedure is well defined when ties occur and treats ties automatically. It does have the disadvantage that the calculation of 2^n H^2 values is tedious. This difficulty can be circumvented by applying a large-sample approximation to the conditional distribution of the test statistic. To use the large-sample approximation, only the value of H^2 corresponding to the observed **r**-configuration is calculated.

Under the mild regularity conditions that $F(x, \infty)$ and $F(y \mid x)$ are continuous, respectively, in x and in y for each x, Koziol [4] has derived the asymptotic conditional distribution of H^2 under H_0. The large-sample approximation to the exact α-level conditional test is

$$\begin{array}{lll} \text{reject } H_0 & \text{if} & H^2 > h_{1-\alpha}, \\ \text{accept } H_0 & \text{if} & H^2 \leqslant h_{1-\alpha}, \end{array} \tag{12}$$

where $h_{1-\alpha}$ is the upper α (asymptotic) percentile point of H^2. Table 1 gives the asymptotic percentage points.

Table 1 Asymptotic Percentage Points of H^2

Cumulative Probability, α	Percentage Point, h_α	Cumulative Probability, α	Percentage Point, h_α
0.05	0.0581	0.65	0.1701
0.10	0.0681	0.70	0.1858
0.15	0.0765	0.75	0.2048
0.20	0.0842	0.80	0.2284
0.25	0.0918	0.85	0.2597
0.30	0.0994	0.90	0.3053
0.35	0.1073	0.95	0.3870
0.40	0.1156	0.975	0.4722
0.45	0.1245	0.99	0.5889
0.50	0.1341	0.995	0.6794
0.55	0.1447	0.999	0.8940
0.60	0.1566		

Source. Koziol [4].

AN EXAMPLE

Shelp et al. [6], in a study of renal transplants, considered the inulin clearance capacity of the donor and recipient after the transplant was performed. Table 2 gives inulin clearance values for seven recipients and their corresponding donors. The patient numbers in Table 2 are different from those

Table 2 Inulin Clearance of Living Donors and Recipients of Their Kidneys

	Inulin Clearance (ml/min)	
Patient	Recipient, X_i	Donor, Y_i
1	61.4	70.8
2	63.3	89.2
3	63.7	65.8
4	80.0	67.1
5	77.3	87.3
6	84.0	85.1
7	105.0	88.1

Source. Data from Shelp et al. [6].

in the Shelp et al. study, because we have renumbered so that $a_1 < a_2 < \cdots < a_7$.

From Table 2, and (4), we see that our observed **r**-configuration is $\mathbf{r} = (1, 1, 1, 0, 1, 1, 0)$. After calculating the $n^2 = 49$ values of d_{ij} defined by (8), we readily compute $T_1 = 1$, $T_2 = 2$, $T_3 = 1$, $T_4 = 0$, $T_5 = 0$, $T_6 = 1$, $T_7 = 0$, and from (5) we obtain $H^2 = 7/49$. Calculation of H^2 for each of the other $2^7 - 1 = 127$ **r**-configurations is straightforward but omitted here (details appear in Hollander and Wolfe [3, Sec. 10.3], an introductory text on nonparametric statistical methods). The 128 ordered values are

$$h^{(1)} = \cdots = h^{(8)} = \tfrac{3}{49},$$

$$h^{(9)} = \cdots = h^{(40)} = \tfrac{7}{49},$$

$$h^{(41)} = \cdots = h^{(88)} = \tfrac{11}{49},$$

$$h^{(89)} = \cdots = h^{(120)} = \tfrac{15}{49},$$

$$h^{(121)} = \cdots = h^{(128)} = \tfrac{19}{49}.$$

We illustrate the $\alpha = (8/128) = 0.0625$ level test defined by (10) and (11). From (9) we find $m = 120$ and thus $h^{(m)} = h^{(120)} = \frac{15}{49}$. We also find $M_1 = 8$, $M_2 = 32$, $p = 0$, so that the $\alpha = 0.0625$ test is to reject H_0 if $H^2 > \frac{15}{49}$, and accept H_0 otherwise. Since our observed value of H^2 is $\frac{7}{49}$, we accept H_0 at $\alpha = 0.0625$. Furthermore, since there are 120 **r**-configurations that yield an H^2-value greater than or equal to $\frac{7}{49}$, the P value (i.e., the lowest α-value at which we can, with a nonrandomized test, reject H_0 with our observed data) is $P = (120/128) = 0.9375$. Thus the data support H_0.

To obtain an approximate P value based on the large-sample approximation, we enter Table 1 with the value $H^2 = (7/49) = 0.143$ to find $P \approx 0.55$. Thus the large-sample approximation also shows that the data support H_0. That the approximate P value of 0.55 is not close to the exact P value of 0.9375 should not be too disturbing. The large-sample approximation will typically be better in the upper tail of the distribution. [Note that for our **Z**, $P_c(H^2 \leqslant 15/49) = 0.9375$, that is, 0.306 is the 93.75 percentage point. Table 1 gives the approximate 90 percentage point as 0.3053.] Furthermore, $n = 7$ is a very small sample size and for larger sample sizes the large-sample approximation will tend to give closer approximations.

See Sen [5] and Bell and Haller [1] for other tests of bivariate symmetry, and Smith [7] for a test of bivariate circular symmetry.

References

[1] Bell, C. B. and Haller, H. S. (1969). *Ann. Math. Statist.*, **40**, 259–269. (Technical paper; proposes various tests for bivariate symmetry; considers various formulations of symmetry.)

[2] Hollander, M. (1971). *Biometrika*, **58**, 203–212. (Technical paper; develops the bivariate symmetry test described in this article.)

[3] Hollander, M. and Wolfe, D. A. (1973). *Nonparametric Statistical Methods*. Wiley, New York. (General reference.)

[4] Koziol, J. A. (1979). *Commun. Statist. A*, **8**(3), 207–221. (Technical paper; derives and tables the asymptotic distribution of the Hollander bivariate symmetry test statistic.)

[5] Sen, P. K. (1967). *Sankhyā A*, **29**, 351–372. (Technical paper; uses the conditional approach to provide distribution-free bivariate symmetry tests.)

[6] Shelp, W. D., Bach, F. H., Kisken, W. A., Newton, M., Rieselbach, R. E., and Weinstein, A. B. (1970). *J. Amer. Med. Ass.*, **213**, 1143–1447. (Contains the data set used in this article.)

[7] Smith, P. J. (1977). *Commun. Statist. A*, **6**(3), 209–220. (Technical paper; provides a nonparametric test for bivariate circular symmetry.)

Acknowledgment

Research supported by the Air Force Office of Scientific Research AFSC, USAF, under Grant AFOSR78-3678.

(DISTRIBUTION-FREE METHODS
PERMUTATION TESTS)

MYLES HOLLANDER

HOLLANDER EXTREME TEST

INTRODUCTION

In certain two-sample experiments, concerned with comparing a control (or standard) group with a treated (or experimental) group, the effect of the experimental treatment may tend to increase the scores of some subjects and tend to decrease the scores of other subjects. The control group responses, however, are not expected to be extreme in either direction. Situations in which some treated subjects may react extemely in one direction and other treated subjects may react extremely in the opposite direction include:

1. **Psychological Studies of Defensive Behavior.** In certain association or recognition situations defensive behavior may manifest itself in either a rapid or a delayed response. Similar considerations arise in certain personality tests where the experimental group consists of hostile subjects trying to conceal a personality trait.
2. **Relief of Anxiety.** A drug (or operant conditioning technique) is designed to reduce anxiety. However, the drug may increase the anxiety of certain subjects while decreasing the anxiety of others.
3. **Technique to Increase Efficiency.** A plant manager institutes a program to increase the output of workers performing a repetitive industrial task. The program may motivate some workers to be more productive but lead others to decrease their efficiency deliberately.

Essentially, we wish to test whether two independent random samples can be viewed as a combined sample from a common population. Let $X_1, X_2, \ldots, X_n$ be a random sample from the "control" population Π_1 with continuous distribution function $F_1(x) = P(X \leqslant x)$, and let $Y_1, Y_2, \ldots, Y_n$ be an independent (of the X's) random sample from the "treatment" population Π_2 with continuous distribution function $F_2(x) = P(Y \leqslant x)$. We wish to test the null hypothesis

$$H_0 : \Pi_1 \equiv \Pi_2, \quad \text{or} \quad H_0 : F_1(x) = F_2(x) \quad \text{for all } x,$$

equivalently. Note that H_0 specifies that $F_1 = F_2$ but does not specify the form (shape, etc.) of the common distribution.

There are various ways to model alternatives to H_0 that would corresponds to the "extreme reactions" situation described previously. Some possibilities with the property that, even under the alternative both populations are centered around the same value, include the alternatives H_1 and H_2 defined below:

H_1: $F_2(x) = pF_1(x - \theta) + pF_1(x + \theta) + (1 - 2p)F_1(x)$, F_1 unspecified, $\theta > 0$, θ unspecified, $0 < p < \frac{1}{2}$, p unspecified;

H_2: $F_1(x) = H((x - \mu)/\sigma)$, $F_2(x) = H((x - \mu)/\tau)$, for some (unspecified) distribution H having mean 0, $-\infty < \mu < \infty$, μ unspecified, $\sigma > 0$, σ unspecified, $\tau > 0$, τ unspecified.

The H_1 alternatives represent situations where a randomly selected value Y from the treatment population can be viewed as follows: With probability p, $Y \overset{d}{=} X + \theta$; with probability p, $Y \overset{d}{=} X - \theta$; and with probability $1 - 2p$, $Y \overset{d}{=} X$, where X is a randomly selected value from the control population, and the symbol "$\overset{d}{=}$" is to be read "has the same distribution as." That is, with probability p a randomly selected Y is like a randomly selected value from the population which is obtained by shifting the X population to the right by θ, with probability p the randomly selected Y is like a randomly selected value from the population which is obtained by shifting the X population to the left by θ, and with probability $1-2p$ the randomly selected Y is like a randomly selected value from the (unshifted) X population.

The H_2 alternatives assert that $(X-\mu)/\sigma$ and $(Y-\mu)/\tau$ both have the same distribution function H. If the mean of H (assumed to exist) is 0, then the mean of X is μ and the mean of Y is μ, but X and Y have different "dispersions," τ/σ being the ratio of the "scale" parameters. If we assume further that the variance of H is finite, then $\gamma^2 \overset{\text{def.}}{=} \tau^2/\sigma^2 = \text{var}(Y)/\text{var}(X)$. If $\gamma^2 > 1$, the Y-population is more "spread out" than the X-population; if $\gamma^2 < 1$, the Y-population is less "spread out" than the X-population.

In the following section we describe a test, proposed by Hollander [3], of H_0 versus the alternative that population Π_2 is extreme in both directions relative to the population Π_1. In the third section we discuss the relationship of Hollander's test to a test for extreme reactions proposed by Moses [5] and to a test for dispersion alternatives proposed by Mood [4]. The fourth section contains an example illustrating the use of Hollander's test.

THE TEST PROCEDURE

To test H_0 versus the alternative that the population Π_2 tends to be extreme in both directions relative to population Π_1, first rank all $N = m + n$ X's and Y's jointly, from least to greatest. We denote the rank of X_i, in this joint ranking, by r_i. The test statistic proposed by Hollander [3] is

$$G = \sum_{i=1}^{m} (r_i - \bar{r})^2 = \left(\sum_{i=1}^{m} r_i^2\right) - m\bar{r}^2, \quad (1)$$

where $\bar{r} = \sum_{i=1}^{m} r_i/m$ is the sample mean of the X-ranks.

Note that $G/(m-1)$ is the sample variance of the X ranks. If the treatment group does exhibit extreme responses, the control observations will tend to be compressed relative to the treatment observations, and this will tend to yield small values of G. Thus Hollander's test rejects H_0 in favor of the alternative that Π_2 tends to be extreme in both directions relative to Π_1, if $G \leqslant C_\alpha$. The critical value C_α is chosen so that the type I error probability equals α, that is,

$$P(\text{rejecting } H_0 \mid H_0 \text{ is true}) = P(G \leqslant C_\alpha \mid H_0 \text{ is true}) = \alpha.$$

Critical values C_α are easily obtained using the fact that under H_0 (and the assumption that the common population is continuous) all $\binom{N}{m}$ possibilities for the m X-ranks are equally likely, each having probability $1/\binom{N}{m}$. Critical values C_α are given in Hollander [3] and are reproduced as Table A.10 of Daniel [1]. (Critical values are for $\alpha = 0.01, 0.05, 0.10$, and various m, N pairs with $m \leqslant 12$, $N \leqslant 20$).

To define the large-sample approximation to the exact α-level test, let

$$G^* = \{G - E_0(G)\}/\{\text{var}_0(G)\}^{1/2}, \quad (2)$$

where $E_0(G)$ and $\text{var}_0(G)$, the mean and variance, respectively, of G under H_0, are

$$E_0(G) = (m-1)(N^2+N)/12, \quad (3)$$

$$\text{var}_0(G) = E_0(G^2) - [E_0(G)]^2; \quad (4)$$

$$E_0(G^2) = \frac{(m-1)^2}{720}\left[-\frac{6(N^4+2N^3+N^2)}{m} + \frac{m+1}{m-1} \times (5N^4+6N^3-5N^2-6N)\right]. \quad (5)$$

The large-sample approximation to the exact test rejects H_0, in favor of the alternative that Π_2 is extreme with respect to Π_1, if $G^* \leqslant -z_\alpha$, and accepts H_0 otherwise. Here z_α is the upper α-percentile point of a standard normal distribution, i.e., if Φ denotes the standard normal cumulative distribution function, then $\Phi(z_\alpha) = 1 - \alpha$. From Tables 1, 2, and 3 of Hollander [3], the normal approximation is seen to be good when m and n are both at least as large as 7.

RELATED TESTS

Moses [5] proposed a distribution-free test that is sensitive to the alternative that Π_2 is extreme relative to Π_1. His test statistic, S_h, is the smallest number of consecutive ranks necessary to include all the X-ranks after exclusion of the h greatest and smallest ones. S_h is a measure of how compressed the controls are, and thus H_0 is rejected for small values of S_h.

The tests based on G and S_h are similar in character, except that the G test is based on the variance of X-ranks, whereas S_h is based on a "range" of X-ranks. For example, for $h = 0$, the statistic S_h reduces to

$$S_0 = \max[r_1, \ldots, r_m] - \min[r_1, \ldots, r_m] + 1.$$

There are, however, several disadvantages to the test based on S_h, including:

1. For small values of h, S_h may be determined by just a few observations; for large values S_h sacrifices too much of the information in the samples.
2. There is no rule for determining h.

Mood [4] proposed a distribution-free test, intended to detect dispersion alternatives to H_0, that is based on a statistic which is similar to G. Mood's statistic is

$$M = \sum_{i=1}^{m} \left(r_i - \frac{N+1}{2} \right)^2.$$

Thus, whereas G is a sum of squared deviations of X-ranks from their sample mean, M is a sum of squared deviations of X-ranks from the constant $(N+1)/2$; the latter being the expected value of each r_i under $H_0(E_0(r_i) = E_0(\bar{r}) = (N+1)/2)$.

Mood's test was advanced as a test of H_0 versus the alternative that Π_1, Π_2 differed wtih respect to dispersion. In particular, significantly small values of M are supposed to be indicative of the alternative $\mathrm{var}(Y) > \mathrm{var}(X)$. However, in order that Mood's test, or any rank test [i.e., a test based on the ranks of the X's (or Y's) in a joint ranking of all N observations] be interpretable as a test for dispersion alternatives, one *must* know or assume that Π_1, Π_2 have equal locations. This point has been made abundantly clear by Moses [6]. Mood's test is sensitive to one-sided dispersion alternatives but also to translation alternatives or "bilateral translation." The latter two possibilities made impossible the interpretation of a significant value of M as evidence of differences in dispersion, unless one knows a priori that locations are equal. If Π_1, Π_2 do have equal locations, then Hollander's test, Mood's test, and Moses' test can be viewed as rank tests of dispersion, and the tests based on G and M will have very similar properties. *See also* MOOD DISPERSION TEST.

For other references to rank tests of dispersion, see the survey article by Duran [2].

AN EXAMPLE

In a study concerned with defensive behavior and memory, a control group of eight subjects and a treatment group of eight subjects were exposed to an incident. Recollection of the incident was tested by posing a question and measuring the response time. The treatment group was given implausible information embedded in the question. This implausible information was not given to the control group. One point of interest was whether the blatantly misleading information would lead to defensive behavior in the treatment group. The response times, in seconds, for the control group (X's) and the treatment group (Y's) are as follows:

Control Observations	Treatment Observations
$X_1 = 184$	$Y_1 = 215$
$X_2 = 150$	$Y_2 = 401$
$X_3 = 118$	$Y_3 = 425$
$X_4 = 205$	$Y_4 = 85$
$X_5 = 381$	$Y_5 = 156$
$X_6 = 96$	$Y_6 = 88$
$X_7 = 194$	$Y_7 = 105$
$X_8 = 390$	$Y_8 = 256$

Arranging the observations in ascending order we obtain the sequence

1	2	3	4	5	6	7	8	9	10
Y	Y	X	Y	X	X	Y	X	X	X

11	12	13	14	15	16
Y	Y	X	X	Y	Y

with the X-ranks receiving the values 3, 5, 6, 8, 9, 10, 13, 14. Thus

$$\bar{r} = \tfrac{1}{8}(3 + 5 + 6 + 8 + 9 + 10 + 13 + 14)$$
$$= 8.5.$$

From (1) we obtain

$$G = (3)^2 + (5)^2 + (6)^2 + (8)^2 + (9)^2 + (10)^2 + (13)^2 + (14)^2 - 8(8.5)^2 = 102.$$

We find, from Table 3 of Hollander [3] entered at $m = 8$, $N = 16$, that $C_{0.10} = 103.5$. Since $G < 103.5$ we reject H_0, at $\alpha = 0.10$, in favor of the alternative that the treatment population is extreme relative to the control population.

To apply the large-sample approximation to the $\alpha = 0.10$ test we compute, from (2) to (5),

$$E_0(G) = 7\{(16)^2 + 16\}/12 = 158.67,$$

$$E_0(G^2) = \frac{(7)^2}{720}\left[-\frac{6\{(16)^4 + 2(16)^3 + (16)^2\}}{8} + \frac{9}{7} \times \{5(16)^4 + 6(16)^3 - 5(16)^2 - 6(16)\}\right]$$

$$= 26{,}925.73,$$

$$\text{var}_0(G) = 1749.56,$$

$$G^* = \frac{102 - 158.67}{\sqrt{1749.56}} = -1.35.$$

From tables of the standard normal distribution we find that the P-value corresponding to G^* is approximately 0.09, and thus in this example the normal approximation and the exact tests are in close agreement.

References

[1] Daniel, W. W. (1978). *Applied Nonparametric Statistics*. Houghton Mifflin, Boston. (General reference for descriptions of nonparametric tests.)

[2] Duran, B. S. (1976). *Commun. Statist. A*, **5**, 1287–1312. (Survey paper; reviews various rank tests of dispersion.)

[3] Hollander, M. (1963). *Psychometrika*, **28**, 395–403. (Technical paper; developes the test for extreme reactions described in this article.)

[4] Mood, A. M. (1954). *Ann. Math. Statist.*, **25**, 514–522. (Technical paper; develops a two-sample test for dispersion alternatives and calculates the asymptotic efficiencies of various nonparmetric two-sample tests.)

[5] Moses, L. E. (1952). *Psychometrika*, **17**, 239–247. (Technical paper; developes a test for extreme reactions which motivated the test described in this article.)

[6] Moses, L. E. (1963). *Ann. Math. Statist.*, **34**, 973–983. (Technical paper; shows that the two-sample rank tests of dispersion are inadequate unless the population locations are equal.)

Acknowledgment

Research supported by the Air Force Office of Scientific Research AFSC, USAF, under Grant AFOSR 76-3678.

(MOOD DISPERSION TEST
RANK TESTS)

Myles Hollander

HOLLANDER PARALLELISM TEST

INTRODUCTION

Consider the linear regression* model

$$Y_{ij} = \alpha_i + \beta_i x_{ij} + e_{ij}, \quad i = 1, 2, \quad j = 1, \ldots, N. \quad (1)$$

Here α_1, α_2, β_1, and β_2 are unknown parameters (α_i is the Y-intercept and β_i is the slope of line i, $i = 1, 2$), the x's are known nonrandom quantities (called independent variables, predictors, or regressors) that are typically determined by the experimenter, and the Y's are observable random variables (called dependent variables, predictands, or regressands). Think of Y_{ij} as the "response" at x_{ij}. The e's are mutually independent unobservable random variables with distribution functions $P(e_{ij} \leqslant t) = F_i(t)$. Hollander

[1] proposed a method for constructing a test of

$$H_0 : \beta_1 = \beta_2,$$

which is distribution-free under H_0 when $(F_1, F_2) \in \mathscr{F}$, where

$$\mathscr{F} = \{(F_1, F_2)[F_1, F_2 \text{ are continuous}\}, \quad (2)$$

thus providing an exact test of H_0 under very mild assumptions concerning the underlying distributions. The test statistic, defined in the following section, is the Wilcoxon one-sample signed-rank statistic (*see* WILCOXON TESTS) applied to $N/2$ independent random variables of the form

$$[(Y_{1s} - Y_{1s'})/(x_{1s} - x_{1s'})] - [(Y_{2t} - Y_{2t'})/(x_{2t} - x_{2t'})].$$

The need to test H_0 can arise (a) in a business context, when comparing two production rates, (b) in bioassay* when assessing the potency of an unknown parameter preparation relative to a standard preparation, (c) in experimental psychology* when comparing two food-intake rates in taste-aversion studies, and in numerous other situations.

THE TEST PROCEDURE

Assume, without loss of generality that $x_{i1} \leqslant x_{i2} \leqslant \cdots \leqslant x_{iN}$. (If this is not the case, relabel the Y's—with the corresponding x's —so that the ordering is satisfied.) Furthermore, assume that $N = 2n$, discarding at random an observation from both samples if necessary. For line 1, form n pairs by pairing x_{1j} with $x_{1,j+n}, j = 1, \ldots, n$, and compute n line-1 slope estimators of the form

$$u_{1j} = \frac{Y_{1,j+n} - Y_{1j}}{x_{1,j+n} - x_{1j}}, \qquad j = 1, \ldots, n. \quad (3)$$

Similarly, for line 2, form n pairs to obtain n line-2 slope estimators of the form

$$u_{2t} = \frac{Y_{2,t+n} - Y_{2t}}{x_{2,t+n} - x_{2t}}, \qquad t = 1, \ldots, n. \quad (4)$$

Randomly pair the u_{1j}'s with the u_{2t}'s, so that each u appears in one and only one pair, and compute n differences of the form

$$Z = u_{1j} - u_{2t}.$$

Call these differences $Z_1, Z_2, \ldots, Z_n$. The test statistic is the Wilcoxon one-sample signed-rank statistic applied to the Z's, i.e.,

$$T^+ = \sum_{i=1}^{n} R_i \psi_i, \quad (5)$$

where R_i is the rank of $|Z_i|$ in the joint ranking from least to greatest of $|Z_1|, \ldots, |Z_n|$, and $\psi_i = 1$ if $Z_i > 0$, 0 otherwise.

The one-sided test of H_0 against alternatives $\beta_1 > \beta_2$ (respectively, $\beta_1 < \beta_2$) rejects for significantly large (respectively, small) values of T^+. The two-sided test of H_0 against alternatives $\beta_1 \neq \beta_2$ rejects for significantly large values of $|T^+|$. Exact critical values of T^+ appear in Table A.4 of Hollander and Wolfe [2]. For large n (say $n \geqslant 10$), the distribution of

$$\frac{T^+ - \frac{1}{4}n(n+1)}{[n(n+1)(2n+1)/24]^{1/2}}$$

under H_0 is approximately standard normal. (For modifications to treat tied observations, see Hollander and Wolfe [2, Sec. 3.1].)

The random pairing of the u_{1j}'s with the u_{2t}'s, could be replaced by any pairing of the u_{1j}'s with the u_{2t}'s, as long as the pairing depends only on the x's (and not on the u_1 and u_2 values) and each u appears in one and only one pair. Random pairing here acts as a safeguard against the introduction of biases. We emphasize that the user is entitled to only one random pairing in a given analysis. It is incorrect to try different random pairings until one reaches a significant decision. Such action invalidates all inferences.

At first glance, the random pairing and the formation of only n slopes for each line (when one could in fact compute $N^{(i)}$ line-i sample slopes, where $N^{(i)}$ denotes the number of positive $x_{ij'} - x_{ij}$ differences) appear to be extremely wasteful of the information in the data. However, for the equally spaced model, where one takes $N/(2k)$ observations at each of the $2k$ points $\mathscr{C}_i + 2j$ on line

$i, j = 0, \ldots, 2k - 1$, the Pitman asymptotic relative efficiency* of the signed-rank test with respect to the normal theory t-test* of H_0, for the case $F_1 = F_2 = F$ (say) and F normal, is 0.955 for $k = 1$ and 0.716 as $k \to \infty$. The corresponding values for F uniform and F exponential are, respectively, (0.919, 0.689) and (1.172, 0.879). For the same model with F_1 uniform and F_2 exponential, these efficiencies are (1.437, 1.077).

The pairings and the restriction to equal numbers of observations for each line are disadvantages of this parallelism test. These defects can be viewed as the price of a reasonably efficient test that has the distribution-free property with respect to the very large class $\mathscr{F}$, given by (2).

Sen [4] tests for parallelism of k ($k \geqslant 2$) regression lines in a nonparametric context. Sen's procedures have good efficiency properties but require distributional assumptions that are more restrictive than those given in (2), are not distribution-free in finite samples but only asymptotically distribution-free (i.e., the nominal level of the test is achieved asymptotically), and depend on estimators that must be determined by trial-and-error methods. Potthoff [3] proposes a very conservative test for parallelism that is analogous to the two-sample Wilcoxon rank sum test. The procedures of Sen and Potthoff are not restricted to equal numbers of observations for each line.

AN EXAMPLE

In many bioassays, the question of parallelism is extremely important, since the concept of relative potency (of a test preparation with respect to a standard) depends on the assumption that the dose–response lines are parallel. The data in Table 1 are from an analysis of Wardlaw and van Belle [5] of the mouse hemidiaphragm method for assaying insulin (see Hollander and Wolfe [2, pp. 210–212]). We use these data to illustrate the parallelism test, associating line 1 with standard insulin and line 2 with sample 1 insulin.

Letting $d = \log(1.5) - \log(0.3) = 0.699$, and replacing $Y_{ij}(x_{ij})$ by $Y_{i,j}(x_{i,j})$ for notational clarity, we have

$$u_{11} = \frac{Y_{1,7} - Y_{1,1}}{x_{1,7} - x_{1,1}} = \frac{365 - 230}{d} = 193.1,$$

$$u_{12} = \frac{Y_{1,8} - Y_{1,2}}{x_{1,8} - x_{1,2}} = \frac{325 - 290}{d} = 50.1,$$

$$u_{13} = \frac{Y_{1,9} - Y_{1,3}}{x_{1,9} - x_{1,3}} = \frac{360 - 265}{d} = 135.9,$$

$$u_{14} = \frac{Y_{1,10} - Y_{1,4}}{x_{1,10} - x_{1,4}} = \frac{300 - 225}{d} = 107.3,$$

$$u_{15} = \frac{Y_{1,11} - Y_{1,5}}{x_{1,11} - x_{1,5}} = \frac{360 - 285}{d} = 107.3,$$

$$u_{16} = \frac{Y_{1,12} - Y_{1,6}}{x_{1,12} - x_{1,6}} = \frac{385 - 280}{d} = 150.2.$$

Similarly, for line 2, we obtain $u_{21} = 150.2$, $u_{22} = 157.4$, $u_{23} = 107.3$, $u_{24} = -28.6$, $u_{25} = 178.8$, and $u_{26} = 143.1$.

Randomly pairing the u_1's with the u_2's led to the six Z differences: $Z_1 = u_{11} - u_{21}$, $Z_2 = u_{12} - u_{22}$, $Z_3 = u_{13} - u_{24}$, $Z_4 = u_{14} - u_{25}$, $Z_5 = u_{15} - u_{26}$, $Z_6 = u_{16} - u_{23}$. We now illustrate, in tabular form, the computations used to evaluate the Wilcoxon signed-rank statistic T^+ applied to the Z's.

i	Z_i	$\lvert Z_i \rvert$	R_i	ψ_i	$R_i\psi_i$
1	42.9	42.9	2.5	1	2.5
2	−107.3	107.3	5	0	0
3	164.5	164.5	6	1	6
4	−71.5	71.5	4	0	0
5	−35.8	35.8	1	0	0
6	42.9	42.9	2.5	1	2.5

From (5) we obtain

$$T^+ = \sum_{i=1}^{6} R_i\psi_i = 11.$$

From Table A.4 of Hollander and Wolfe [2], we find that the two-sided test of H_0 versus $\beta_1 \neq \beta_2$, at the type I error probability level of $\alpha = 0.156$, rejects H_0 if $T^+ \geqslant 18$ or if $T^+ \leqslant 3$. Hence with the value of

Table 1. Glycogen Content of Hemidiaphragms Measures by Optical Density in the Anthrone Test × 1000

	Standard Insulin		Sample 1 Insulin	
j	x_{1j} (log dose)	Y_{1j} (glycogen content)	x_{2j} (log dose)	Y_{2j} (glycogen content)
1	log(0.3)	230	log(0.3)	310
2	log(0.3)	290	log(0.3)	265
3	log(0.3)	265	log(0.3)	300
4	log(0.3)	225	log(0.3)	295
5	log(0.3)	285	log(0.3)	255
6	log(0.3)	280	log(0.3)	280
7	log(1.5)	365	log(1.5)	415
8	log(1.5)	325	log(1.5)	375
9	log(1.5)	360	log(1.5)	375
10	log(1.5)	300	log(1.5)	275
11	log(1.5)	360	log(1.5)	380
12	log(1.5)	385	log(1.5)	380

Source. Data from Wardlaw and van Belle [5].

$T^+ = 11$ we accept the hypothesis of parallelism at the 0.156 level.

References

[1] Hollander, M. (1970). *J. Amer. Statist. Ass.*, **65**, 387–394. (Technical paper; proposes the parallelism test described in this article and compares it with various competitors.)

[2] Hollander, M. and Wolfe, D. A. (1973). *Nonparametric Statistical Methods*. Wiley, New York. (General reference.)

[3] Potthoff, R. F. (1974). *Ann. Statist.*, **2**, 295–310. (Technical paper; proposes a conservative test for parallelism of two regression lines; considers the one-sample problem of testing that the slope is equal to a specified value.)

[4] Sen, P. K. (1969). *Ann. Math. Statist.*, **40**, 1668–1683. [Technical paper; proposes tests for parallelism of $k(k \geqslant 2)$ regression lines; gives asymptotic relative efficiency results for tests and related estimators.]

[5] Wardlaw, A. C. and van Belle, G. (1964). *Diabetes*, **13**, 622–633. (Discusses and analyzes the mouse hemidiaphragm method for assaying insulin.)

Acknowledgment

Research supported by the Air Force Office of Scientific Research AFSC, USAF, under Grant AFOSR 78-3678.

(LINEAR REGRESSION
RANK TESTS)

MYLES HOLLANDER

HOLLANDER–PROSCHAN NEW-BETTER-THAN-USED TEST

THE NEW-BETTER-THAN-USED CLASS

Let X denote a nonnegative random variable. Thus X could denote the time to the occurrence of an end-point event such as pregnancy, failure of a component, death of a person, or relapse of a patient. The distribution function $F(x) = P(X \leqslant x)$, of the random variable X, is called a *life* distribution [i.e., a life distribution is a distribution for which $F(x) = 0$ for $x < 0$]. The *survival* function is $\bar{F}(x) = 1 - F(x)$.

A life distribution F is said to be *new better than used* (NBU) if

$$\bar{F}(x+y) \leqslant \bar{F}(x)\bar{F}(y) \quad \text{for all} \quad x, y \geqslant 0. \tag{1}$$

The corresponding concept of a *new worse than used* (NWU) distribution is defined by reversing the inequality in (1). The boundary members of the NBU class, obtained by insisting on equality in (1), are the exponential distributions, for which used items are no worse and no better than new items.

Note that property (1) has a direct physical interpretation in terms of wearout. Property (1) states that the chance $\bar{F}(x)$ that a new unit will survive to age x is greater than or equal to the chance $\bar{F}(x+y)/\bar{F}(y)$ that an unfailed unit of age y will survive an additional time x. That is, a new unit has stochastically greater life than a used one of any age.

By taking logarithms on both sides of inequality (1), it is seen that F is NBU if and only if $-\ln \bar{F}(x)$ is a *superadditive* function for $x > 0$; that is, for all $x, y \geqslant 0$, $-\ln \bar{F}(x+y) \geqslant -\ln \bar{F}(x) - \ln \bar{F}(y)$.

In performing studies of life lengths, it has also been found very useful to classify life distributions according to monotonicity properties of the failure rate. One frequently used class of life distributions is the increasing failure rate (IFR) class. The distribution F is said to be IFR if $-\ln \bar{F}(x)$ is convex. If F has a density f, this condition is equivalent to the condition that the failure rate $q(x) = f(x)/\bar{F}(x)$ is increasing in x [x such that $\bar{F}(x) > 0$]. The IFR class is contained in the NBU class and there are NBU distributions where the failure rate can fluctuate and in particular need not be increasing. The NBU test, described in the following section, can detect such NBU distributions where wear-out is occurring but where the wear-out need not be in the very restrictive sense of an increasing failure rate.

The NBU class plays a fundamental role in the study of replacement policies [8] and in the study of shock models [4]. Both of these references describe situations where it is important to know whether or not the underlying distribution is NBU.

We briefly mention two typical results, concerning replacement policies, from Marshall and Proschan [8]. Under an *age replacement policy*, a unit is replaced upon failure or upon reaching a specified age T, whichever comes first. Under a *block replacement policy*, a replacement is made whenever a failure occurs, and additionally at specified times $T, 2T, 3T, \ldots$. Marshall and Proschan show that a necessary and sufficient condition for failure-free intervals to be stochastically larger (smaller) under age replacement than under a policy of replacement at failure only is that the underlying distribution be NBU (NWU). Marshall and Proschan also show that a necessary and sufficient condition that the number of failures in a specified interval $[0, t]$ be stochastically smaller (larger) under age replacement than under a policy of replacement at failure only is that the underlying distribution be NBU (NWU). Similar comparisons hold using block replacement. These, and other related results, show that in reaching a decision as to whether to use an age (block) replacement policy or not, it is important to know whether or not the underlying distribution is NBU.

For more details on the NBU class, see Barlow and Proschan [1], a modern book on reliability concepts and models (*see also* HAZARD RATE AND OTHER CLASSIFICATIONS OF DISTRIBUTIONS).

THE HOLLANDER–PROSCHAN NBU TEST

We now describe a test of

$$H_0 : F(x) = 1 - \exp(-\lambda x),\quad x \geqslant 0,\ \lambda > 0\ (\lambda \text{ unspecified}), \tag{2}$$

versus

$$H_1 : F \text{ is NBU (and not exponential)}, \tag{3}$$

based on a random sample $X_1, \ldots, X_n$ from the life distribution F.

Let $X_{(1)} \leqslant \cdots \leqslant X_{(n)}$ denote the ordered X's. Compute

$$T = \sum_{i>j>k} \psi(X_{(i)}, X_{(j)} + X_{(k)}), \tag{4}$$

where $\psi(a,b) = 1$ if $a > b$, $\frac{1}{2}$ if $a = b$, 0 if $a < b$. Note that the summation in (4) is over the $n(n-1)(n-2)/6$ ordered triples (i, j, k) with $i > j > k$.

To test H_0 vs. H_1, with type I error probability α,

$$\begin{aligned} &\text{reject } H_0 \text{ if } T \leqslant t_1(\alpha, n),\\ &\text{accept } H_0 \text{ if } T > t_1(\alpha, n), \end{aligned} \tag{5}$$

where $t_1(\alpha, n)$, the lower α percentile point of the null distribution of T, satisfies $P\{T \leqslant t_1(\alpha, n) \mid H_0\} = \alpha$.

To test H_0 vs. H_1': F is NWU (and not exponential), with type I error probability α,

$$\begin{aligned}&\text{reject } H_0 \text{ if } T \geqslant t_2(\alpha, n),\\&\text{accept } H_0 \text{ if } T < t_2(\alpha, n),\end{aligned} \tag{6}$$

where $t_2(\alpha, n)$, the upper α percentile point of the null distribution of T, satisfies $P\{T \geqslant t_2(\alpha, n) | H_0\} = \alpha$. Table 4.1 of Hollander and Proschan [5] and Table A.27 of Hollander and Wolfe [6] (an introductory text on nonparametric statistical methods) contain approximate lower-tail critical values $t_1(\alpha, n)$ and upper-tail critical values $t_2(\alpha, n)$ for $n = 4(1)20(5)50$, and $\alpha = 0.01$, 0.025, 0.05, 0.075, and 0.10.

To define the large-sample approximations to procedures (5) and (6), set

$$T^* = \{T - E_0(T)\}/\{\text{var}_0(T)\}^{1/2}, \tag{7}$$

where $E_0(T)$, $\text{var}_0(T)$, the mean and variance, respectively, of T under H_0, are

$$E_0(T) = n(n-1)(n-2)/8, \tag{8}$$

$$\begin{aligned}\text{var}_0(T) &= \tfrac{3}{2}n(n-1)(n-2)\\&\quad\times\left[\tfrac{5}{2592}(n-3)(n-4)\right.\\&\quad\left.+ (n-3)\tfrac{7}{432} + \tfrac{1}{48}\right].\end{aligned} \tag{9}$$

The large-sample approximation to procedure (5) rejects H_0 in favor of NBU alternatives if $T^* \leqslant -z_\alpha$, and accepts H_0 otherwise. The large-sample approximation to procedure (6) rejects H_0 in favor of NWU alternatives if $T^* \geqslant z_\alpha$, and accepts H_0 otherwise. Here z_α is the upper α percentile point of a standard normal distribution.

The motivation for the test based on T is (briefly) as follows. The parameter

$$\gamma(F) = \int\int\{\bar{F}(x)\bar{F}(y) - \bar{F}(x+y)\}\,dF(x)\,dF(y)$$

can be viewed as a measure of the deviation of F from H_0, being 0 when F is exponential and positive when F is continuous, NBU, and not exponential. It can be shown (see Hollander and Proschan [5] for details) that the statistic $\frac{1}{4} - [2T/\{n(n-1)(n-2)\}]$ estimates $\gamma(F)$ and thus significantly small (large) values of T indicate NBU (NWU) alternatives.

AN EXAMPLE

A study discussed by Siddiqui and Gehan [9], and also considered by Bryson and Siddiqui [2], contains survival times, measured in days from the date of diagnosis, of 43 patients suffering from chronic granulocytic leukemia. For such studies, the IFR class may be too restrictive. Hopefully, the treatment, applied after diagnosis, will (at least for a period of time) decrease the failure rate.

The ordered survival times $X_{(1)}, \ldots, X_{(43)}$ are: 7, 47, 58, 74, 177, 232, 273, 285, 317, 429, 440, 445, 455, 468, 495, 497, 532, 571, 579, 581, 650, 702, 715, 779, 881, 900, 930, 968, 1077, 1109, 1314, 1334, 1367, 1534, 1712, 1784, 1877, 1886, 2045, 2056, 2260, 2429, 2509. We will test H_0 against NBU alternatives. For these data $T = 8327$ and $T^* = -1.46$. Thus the lowest value of α at which we can reject H_0 in favor of an NBU alternative, using the large-sample approximation to procedure (5), is $P = 0.07$. Thus there is evidence against H_0, in the NBU direction. See Koul [7] for a competitor of the Hollander–Proschan NBU test and see Chen et al. [3] for a generalization, of the Hollander–Proschan NBU test, to accommodate right-censored data.

References

[1] Barlow, R. E. and Proschan, F. (1975). *Statistical Theory of Reliability and Life Testing: Probability Models.* Holt, Rinehart and Winston, New York. (Excellent general reference for modern reliability theory.)

[2] Bryson, M. C. and Siddiqui, M. M. (1969). *J. Amer. Statist. Ass.*, **64**, 1472–1482. (Technical paper; considers various notions of aging.)

[3] Chen, Y. Y., Hollander, M., and Langberg, N. (1983). *Ann. Statist.*, **11** (in press). (Technical paper; generalizes the test described in this article so that it can accommodate censored data.)

[4] Esary, J. D., Marshall, A. W., and Proschan, F. (1973). *Ann. Prob.*, **1**, 627–649. [Technical paper;

considers shock models and wear processes (the NBU class plays a fundamental role).]

[5] Hollander, M. and Proschan, F. (1972). *Ann. Math. Statist.*, **43**, 1136–1146. (Technical paper; develops the test described in this article.)

[6] Hollander, M. and Wolfe, D. A. (1973). *Nonparametric Statistical Methods*. Wiley, New York. (General reference.)

[7] Koul, H. L. (1977). *Commun. Statist. A*, **6**, 563–573. (Technical paper; proposes a competitor of the test described in this article.)

[8] Marshall, A. W. Proschan, F. (1972). *Proc. 6th Berkeley Symp. Math. Statist. Prob.*, Vol. 1. University of California Press, Berkeley, Calif., pp. 395–415. [Technical paper; considers replacement policies (the NBU class plays a fundamental role).]

[9] Siddiqui, M. M. and Gehan, E. A. (1966). *Statistical Methodology for Survival Time Studies*. U.S. Government Printing Office, Washington, D.C. (A National Cancer Institute monograph.)

Acknowledgment

Research supported by the Air Force Office of Scientific Research AFSC, USAF, under Grant AFOSR 78-3678.

(HAZARD RATE AND OTHER CLASSIFICATIONS OF DISTRIBUTIONS
TESTS OF LIFE DISTRIBUTIONS)

MYLES HOLLANDER

HOMOGENEITY AND TESTS OF HOMOGENEITY

Homogeneity refers to sameness or similarity. In comparing several treatments, such as competing methods of nondestructive assay, interest may be focused on a particular population characteristic, for instance the mean error of measurement. At issue is whether the various means have the same value. In general, a collection of populations is said to be homogeneous with respect to a given characteristic or set of characteristics if the populations are identical with respect to this characteristic or set. Thus homogeneity of means holds if the population means are all equal, regardless of whether other population attributes are shared. Depending on the context, the degree of similarity implied by the term "homogeneity" may vary from the minimum of a single common attribute, as in the equality of means, to the extreme of total sameness, that is, equivalence of the populations.

The negation of homogeneity with respect to a certain characteristic is *heterogeneity* with respect to this characteristic. Means are heterogeneous, then, if they are not all equal. To decide statistically between homogeneity and heterogeneity, a test of homogeneity may be constructed and performed. The null hypothesis H_0 specifies a particular form of homogeneity; the alternative hypothesis H_a, an analogous form of heterogeneity (*see* HYPOTHESIS TESTING). In some investigations it is desirable to use as alternative hypothesis* not the logical negation of H_0 but rather some restricted version. To illustrate, in testing the homogeneity of means when one population is the "control" and all other populations are "treatments," H_a might assert that each treatment population mean exceeds the control population mean. In general, the appropriate form of test criterion for a test of homogeneity depends on such considerations as the parametric assumptions that are made, the type of heterogeneity specified in the alternative hypothesis, and the number of populations involved.

STANDARD PARAMETRIC TESTS OF HOMOGENEITY

A collection of populations modeled by distributions within a single specified parametric family may be tested for possessing a given form of homogeneity of interest. For instance, do fuel rods produced by five different assembly lines have the same mean enrichment, assuming that each line is modeled by gamma distributed measurements of enrichment? In general, assume that $k \geqslant 2$ populations are modeled by distributions in the parametric family $\mathscr{F}$ indexed by the parameter $\boldsymbol{\theta} = (\theta_1 \ldots \theta_p)$ of dimension $p \geqslant 1$. Denote by $\boldsymbol{\theta}^{(i)'}$ the true parameter value for population i; $i = 1, \ldots, k$. Suppose that interest is focused on the parametric function $g(\boldsymbol{\theta})$ of dimension r, where

$1 \leqslant r \leqslant p$. Homogeneity of $g(\boldsymbol{\theta})$ then implies that $g(\boldsymbol{\theta}^{(1)}) = \cdots = g(\boldsymbol{\theta}^{(k)})$. The statistical literature is replete with parametric tests of homogeneity for various families $\mathscr{F}$, parametric functions g, and schemes for sampling the populations. Traditionally, much attention is devoted to the parametric functions that characterize the population mean, the population variance (for which homogeneity historically is called *homoscedasticity*), and the individual components of $\boldsymbol{\theta}$. The special case $g(\boldsymbol{\theta}) = \boldsymbol{\theta}$, which addresses total homogeneity (i.e., equality of distributions) is also well researched for many important parametric families. The mathematical derivation used to formulate an optimal test criterion typically exploits peculiarities of the family postulated. As a dependable yet not necessarily optimal device for providing a result, the likelihood ratio* procedure (see [13, Chap. 24]) is frequently employed.

A great deal of study has been devoted to the family of normal distributions*. To highlight some of the effort, consider first two populations, where a random sample is drawn from each population, and the samples are independent. For testing equality of means with known variances, a normal distributed test statistic is optimum, whereas if the variances are unknown but assumed equal, a Student's t-test* statistic is appropriate [23, Chap. 9, Sec. 4.1]. For the case of unknown and not necessarily equal variances (the Behrens–Fisher* problem), several tests of homogeneity of means are arguably satisfactory [18]. On the other hand, for testing homogeneity of variances (i.e., homoscedasticity), an F statistic is recommended (*see* F-TESTS and [23, Chap. 9, Sec. 4.2]), and for testing homogeneity of both means and variances (i.e., equality of distributions), a few satisfactory tests are available [28]. Analogous results have been developed to handle situations in which independent random samples are taken from $k > 2$ normal populations. For homogeneity of means, under the assumption of known variances, a likelihood ratio test procedure may be used [13, Chap. 24]. Conversely, when assuming unknown but equal variances (the classical one-factor analysis-of-variance* context), an F statistic is appropriate [32, Sec. 3.1]. For the k-sample Behrens–Fisher problem, in which no restrictions are placed on the variances, at least one available test of homogeneity of means is considered adequate [38]. For the homoscedasticity problem, several test criteria have been devised, among them the methods of Bartlett* [3; 25, Sec. 15.6], Cochran* [4, Sec. 7.7; 7] and Hartley* [9; 25, Sec. 15.6]. For homogeneity of distributions, a test by Neyman* and Pearson* [26] is standard. Similar issues of homogeneity may be addressed relative to the family of multivariate normal distributions [1]. In addition, fundamental questions of homogeneity are central to the study of the general linear model*. For example, in the analysis of variance, F-tests of homogeneity are widely used in comparisons of main effects* and various interactions* [19, Chap. 7; 32]. Similarly, in regression* analysis, F-tests of homogeneity appear in the context of parallelism and coincidence of regression lines or surfaces [33, Sec. 7.5].

Selected tests of homogeneity developed from parametric assumptions other than normality include work by Potthoff and Whittinghill [29] and Wisniewski [40] for the binomial distribution*; Vit [37] for the geometric distribution*; Chi [6] and Kudo [16] for the negative binomial distribution*; Potthoff and Whittinghill [30] for the Poisson distribution*; Pearson [27], Potthoff and Whittinghill [29], and Madansky [22] for the multinomial distribution*; Sukhatme [36], Hogg and Tanis [10], and Nagarsenker [24] for the exponential distribution*; Lentner and Buehler [21] for the gamma distribution*; Barr [2] for the uniform distribution*; and Lawless and Mann [17] for the extreme value* and Weibull distributions*.

STANDARD NONPARAMETRIC TESTS OF HOMOGENEITY

Questions of homogeneity frequently arise in contexts where for reasons of prudence or ignorance it is not acceptable to use a parametric model. To illustrate, suppose that independent random samples are taken from $k \geqslant 2$ populations. To test whether the populations are identical, while making no assumptions about their respective probability distributions, one traditional approach is to impose a multinomial structure. A partition into a fixed number of categories is effected which allows a contingency table* accounting of how many observations fall into each category for each sample. From the table a "chi-square" statistic is then constructed to test homogeneity (see [23, Chap. 9, Sec. 5.3]). A large body of other nonparametric approaches to this problem concentrates on ranks. Let $X_{i1}, \dots, X_{in_i}$, $i = 1, \dots, k$, denote the respective random samples. By ordering from low to high the combined sample of $n = \sum_{i=1}^{k} n_i$ values, the ranks $R_{i1}, \dots, R_{in_i}$, $i = 1, \dots, k$, are generated, which identify relative positions in the ordering. Motivated by such considerations as form of the alternative hypothesis and suspected distributional flavor, various rank tests* have been proposed. (See, for example, Lehmann [20] and Hollander and Wolfe [11].) For the two-sample case (see [20, Chap. 2]), these include the Wilcoxon rank-sum* [39] and the normal scores* [8] tests for location shift alternatives (loosely analogous to a discrepancy in means in the parametric setting); the Siegel–Tukey* [34], the Capon* [5], and the Savage* [31] tests for scale shift alternatives (loosely analogous to variance discrepancies); and the Smirnov [35] test for general, or omnibus, alternatives. For the $k > 2$ sample case (see [11, Chap. 6; 20, Chap. 5]), there is the Kruskal–Wallis* [15] test for location shift alternatives (the nonparametric one factor analysis of variance analog), the Jonckheere* [12] test for ordered location shift alternatives, and the Kiefer* [14] test for omnibus alternatives.

GENERAL COMMENTS

No hypothesis testing procedure should be used without examining its power* function to assess probabilities of detecting (i.e., rejecting) various alternatives of interest. In tests of homogeneity, such examination is particularly crucial. Because the null hypothesis H_0 embraces equality, while departures from homogeneity, however minor, reside in H_a, it may be practically undesirable to reject H_0 when only a slight deviation from H_0 is the case. Yet if the test is so-called "consistent against all alternatives", as most widely used tests are, then any alternative, no matter how "close" to H_0, will be rejected with probability tending to unity as sample sizes tend to infinity. To illustrate, suppose that a medical treatment is to be compared with a control or placebo. The null hypothesis of strict homogeneity (i.e., complete absence of treatment effect) may be regarded as an implausible ideal. The implicit goal of the investigation is really to decide whether the treatment has essentially no effect (i.e., there is virtual, or approximate, homogeneity). However, if sample sizes are large enough, the hypothesis of homogeneity will be rejected with near certainty, even if in fact the treatment effect is nonzero but negligible. The potential problem of "too much" power is obviated by a careful determination of sample sizes. Alternatives "close" to H_0 are selected for which H_0 would preferably be accepted, and alternatives which represent important departures from H_0 are selected for which H_0 would preferably be rejected. Sample sizes are then chosen small enough such that power is suitably small for the former class of alternatives yet large enough that power is suitably large for the latter class.

Another consideration in tests of homogeneity is robustness*. It is often instructive, for example, to evaluate the sensitivity of the size and the power function to certain conceivable departures from underlying model assumptions. The effect of augmentation of H_a may also be examined. A final consideration is the typical need for a follow-up strategy of action in the event of rejection of the hypothesis of homogeneity. Confidence sets and rankings are among the usual devices to estimate the extent of heterogeneity.

References

[1] Anderson, T. W. (1958). *An Introduction to Multivariate Statistical Analysis*. Wiley, New York.

[2] Barr, D. R. (1966). *J. Amer. Statist. Ass.*, **61**, 856–864.

[3] Bartlett, M. S. (1937). *Proc. R. Soc. Lond. A*, **160**, 268–282.

[4] Bowker, A. and Lieberman, G. J. (1972). *Engineering Statistics*, 2nd ed. Printice-Hall, Englewood Cliffs, N.J.

[5] Capon, J. (1961). *Ann. Math. Statist.*, **32**, 88–100.

[6] Chi, P. Y. (1980). *Biometrika*, **67**, 252–254.

[7] Cochran, W. G. (1941). *Ann. Eugen.*, **11**, 47–52.

[8] Fisher, R. A. and Yates, F. (1938). *Statistical Tables for Biological, Agricultural and Medical Research*, Oliver & Boyd, Edinburgh.

[9] Hartley, H. O. (1950). *Biometrika*, **37**, 308–312.

[10] Hogg, R. V. and Tanis, E. A. (1963). *J. Amer. Statis. Ass.*, **58**, 435–443.

[11] Hollander, M. and Wolfe, D. A. (1973). *Nonparametric Statistical Methods*. Wiley, New York.

[12] Jonckheere, A. R. (1954). *Biometrika*, **41**, 133–145.

[13] Kendall, M. G. and Stuart, A. (1973). *The Advanced Theory of Statistics*, Vol. 2: *Inference and Relationship*, 3rd ed. Hafner, New York.

[14] Kiefer, J. (1959). *Ann. Math. Statist.*, **30**, 420–447.

[15] Kruskal, W. H. and Wallis, W. A. (1952). *J. Amer. Statist. Ass.*, **47**, 583–612.

[16] Kudo, A. (1978). *Commun. Statist. A*, **7**, 977–986.

[17] Lawless, J. F. and Mann, N. R. (1976). *Commun. Statist. A*, **5**, 389–405.

[18] Lee, A. F. S. and Gurland, J. (1975). *J. Amer. Statist. Ass.*, **70**, 933–941.

[19] Lehmann, E. L. (1959). *Testing Statistical Hypotheses*. Wiley, New York.

[20] Lehmann, E. L. (1975). *Nonparametrics: Statistical Methods Based on Ranks*. Holden-Day, San Francisco.

[21] Lentner, M. M. and Buehler, R. J. (1963). *J. Amer. Statist. Ass.*, **58**, 670–677.

[22] Madansky, A. (1963). *J. Amer. Statist. Ass.*, **58**, 97–119.

[23] Mood, A. M., Graybill, F. A., and Boes, D. C. (1974). *Introduction to the Theory of Statistics*, 3rd ed. McGraw-Hill, New York.

[24] Nagarsenker, P. B. (1980). *Biometrika*, **67**, 475–478.

[25] Neter, J. and Wasserman, W. (1974). *Applied Linear Statistical Models*. Richard D. Irwin, Homewood, Ill.

[26] Neyman, J. and Pearson, E. S. (1931). *Bull. Acad. Pol. Sci. Lett. A*, 460–481.

[27] Pearson, K. (1900). *Philos. Mag.*, **50**, 157–175.

[28] Perng, S. K. and Littell, R. C. (1976). *J. Amer. Statist. Ass.*, **71**, 968–971.

[29] Potthoff, R. F. and Whittinghill, M. (1966). *Biometrika*, **53**, 167–182.

[30] Potthoff, R. F. and Whittinghill, M. (1966). *Biometrika*, **53**, 183–190.

[31] Savage, I. R. (1956). *Ann. Math. Statist.*, **27**, 590–615.

[32] Scheffé, H. (1959). *The Analysis of Variance*. Wiley, New York.

[33] Seber, G. A. F. (1977). *Linear Regression Analysis*. Wiley, New York.

[34] Siegel, S. and Tukey, J. W. (1960). *J. Amer. Statist. Ass.*, **55**, 429–444.

[35] Smirnov, N. V. (1939). *Bull. Univ. Mosc.*, **2**, 3–14.

[36] Sukhatme, P. V. (1936). *Statist. Res. Mem.*, **1**, 94–112.

[37] Vit, P. (1974). *Biometrika*, **61**, 565–568.

[38] Welch, B. L. (1947). *Biometrika*, **34**, 28–35.

[39] Wilcoxon, F. (1945). *Biometrics*, **1**, 80–83.

[40] Wisniewski, T. K. M. (1968). *Biometrika*, **55**, 426–428.

(DISTRIBUTION-FREE METHODS
HETEROSCEDASTICITY
HYPOTHESIS TESTING)

R. E. GLASER

HONEST SIGNIFICANT DIFFERENCE *See* TUKEY'S SIMULTANEOUS COMPARISON PROCEDURE

HOOPER CORRELATION COEFFICIENT *See* GLAHN AND HOOPER CORRELATION COEFFICIENT

HORIZON *See* CLINICAL TRIALS

HORVITZ–THOMPSON ESTIMATOR

The objectives of this paper, Horvitz and Thompson [5] are to give an unbiased estimator of the population total when sampling with known unequal probabilities. The sampling variance of this estimator is also given, as well as an unbiased estimator of the sampling variance. It is well known that no unbiased estimator with minimum variance can be given, except by a fluke, because if the probability π_i is proportional to y_i (usually unknown) the estimator $\sum y_i/\pi_i$ is known to have zero variance when $\pi_i = y_i/Y_i$, where $Y_i = \sum y_i$ is the population total.

If n sampling units are drawn independently from a population with probabilities π_i, equal or unequal, the *Horvitz–Thompson unbiased estimator* of the population total is

$$\hat{Y} = \sum_{i=1}^{n} \frac{Y_i}{\pi_i},$$

where Y_i is the total on the ith unit, and π_i is assumed nonzero for any method of sampling. Horvitz and Thompson showed that the variance of the estimator is

$$V(\hat{Y}) = \sum_{i=1}^{N} \frac{1-\pi_i}{\pi_i} Y_i^2 + 2\sum_{i \neq j}^{N} \frac{\pi_{ij} - \pi_i\pi_j}{\pi_i\pi_j} Y_i Y_j,$$

where π_{ij} is the probability (assumed nonzero for any pair) that units i and j are both in the sample. They also showed that an unbiased sample estimator of variance is

$$\hat{V}(\hat{Y}) = \sum_{i=1}^{n} \frac{1-\pi_i}{\pi_i^2} Y_i^2 + 2\sum_{i=1}^{N}\sum_{j \neq i}^{N} \frac{\pi_{ij} - \pi_i\pi_j}{\pi_i\pi_j} Y_i Y_j.$$

An alternative sample estimate of variance, given by Yates and Grundy [12], is

$$V(Y) = \sum_{i=1}^{n}\sum_{j \neq i}^{n} \frac{\pi_{ij} - \pi_i\pi_j}{\pi_{ij}} \left(\frac{Y_i}{\pi_i} - \frac{Y_j}{\pi_j}\right)^2$$

taken over every pair of units in the sample, with the same restrictions that $\pi_{ij} \neq 0$ for any pair of units in the sample. Both of the estimators of variance can be negative and are therefore rather unstable quantities. Rao and Singh [9] compared the two variance estimators in samples with $n = 2$ from 34 small natural populations using a method of sample selection due to Brewer [1] which will be described later. The Yates–Grundy estimator of the variance was found to be considerably more stable.

In connection with the Yates–Grundy estimator of variance, Vijayan [11] noted that if $n = 2$ a necessary and sufficient condition that their estimate of variance be nonnegative is $(\pi_i\pi_j - \pi_{ij}) \geqslant 0$ for all i, j, but by a counter-example he pointed out that if $n > 2$ and a nonnegative estimator of variance exists, this condition is sufficient but not necessary for all estimators.

If the sample is drawn with equal probabilities, the HT estimator becomes the estimator $\hat{Y} = NY/n$ commonly used, since $\pi_i = n/N$. But as Horvitz and Thompson remark, there may be an advantage in accuracy in drawing the sample with unequal probabilities. As an example, they take a population consisting of 20 city blocks in Ames, Iowa, the variate Y_i being the estimated number of households in the block. For sampling, they divide the population into two strata by the number of houses and consider a sample with $n = 2$, one unit being drawn from each stratum. When units are drawn with equal probability the variance of the estimated total is 7873, but when units are drawn with probability proportional to Y_i, the variance of the estimated population total falls to 3934.

With $n > 1$ when sampling without replacement, one must be careful in order to attain the desired probability, since they always move toward equality if we are careless. The probability that the ith unit is drawn at either the first or the second draw, if we follow the natural method and draw the second unit with probability $\pi_i/(1 - \pi_j)$, where unit j was drawn first, is

$$\pi_i + \sum_j \frac{\pi_i\pi_j}{1-\pi_j} = \pi_i\left(1 + \sum_{j \neq i}^{N} \frac{\pi_j}{1-\pi_j}\right) = \pi_i\left(1 + A - \frac{\pi_i}{1-\pi_i}\right),$$

where

$$A = \sum_{j=1}^{N} \frac{\pi_i}{1-\pi_i}.$$

Dividing these values by 2, Yates and Grundy [12] found that with original probabilities $\pi_i = 0.4$, 0.3, 0.2, and 0.1, and $n = 2$, the average probabilities of drawing unit i on either the first or second draw become 0.1173, 0.2206, 0.3042, and 0.3579. Thus some care is necessary in order to attain any desired π's. For example, Brewer [1], Rao [8], and Durbin [2] all gave methods that have the desired π_i and π_{ij}. Brewer draws the first unit with probability proportional to $\pi_i(1 - \pi_i)/(1 - 2\pi_i)$ and the second with probability equal to $\pi_i/(1 - \pi_j)$, where unit j was the unit drawn first. The divisor needed to convert the terms for the probability of being drawn first into probabilities is of course their sum

$$D = \sum_{i=1}^{N} \frac{\pi_i(1 - \pi_i)}{1 - 2\pi_i}$$

$$= \frac{1}{2}\left(1 + \sum_i \frac{\pi_i}{1 - 2\pi_i}\right).$$

With $n = 2$ the probability that a unit was drawn is the sum of the probabilities that the unit was drawn first and that it was drawn second. Thus this probability is

$$\frac{\pi_i(1 - \pi_i)}{D(1 - 2\pi_i)} + \frac{1}{D} \sum_{j \neq i}^{N} \frac{\pi_j(1 - \pi_j)\pi_i}{(1 - 2\pi_j)(1 - \pi_j)}$$

$$= \frac{\pi_i}{D}\left(1 + \sum_{j=1}^{N} \frac{\pi_j}{1 - 2\pi_j}\right).$$

But this equals $2\pi_i$ by definition of D.

Durbin draws the first unit with probability π_i and the second unit with probability proportional to

$$\pi_j\left(\frac{1}{1 - 2\pi_i} + \frac{1}{1 - 2\pi_j}\right),$$

where unit j was drawn first. It turns out that the probability that the ith unit was drawn either first or second is $2\pi_i$. Thus Durbin's method has the property, which Brewer's does not, that π_i is the probability of being drawn at either the first or second draw, which is useful if one wants to replace a panel of people.

For systematic sampling Madow [6] has given a simple method which keeps $\pi_i \propto nX_i$, where the units are cluster units of size X_i. As usual with systematic sampling, no unbiased sample estimate of variance is available. Sometimes one wants to draw the sample with probability proportional to $\sum x_i$, where x_i is an auxiliary variable with an approximately constant ratio to the variable Y_i. If we draw the first unit with probability x_i, and subsequent units with equal probabilities, the probability of a sample is proportional to $\sum x_i$, as shown by Midzuno [7].

Later writers examined further properties of the estimator: Hanurav [3] investigated good sampling strategies with $\pi_i \propto X_i$ and the use of the Horvitz–Thompson estimator. He considered that $\pi_{ij}/(\pi_i\pi_j)$ should be positive, less than 1, but not too small, otherwise the Yates–Grundy estimator has a large variance. He produced a method of drawing the sample that has these properties. Hege [4], following Ajagaonkar, stated that if an estimator is unbiased and has the smallest variance of any unbiased estimator in a class, it is called the "necessary best estimator" in that class. He showed that the Horvitz–Thompson estimator is a necessary best estimator among the class of linear estimates of Y, with arbitrary probabilities of selection without replacement at each draw. T. J. Rao [10] examined several good sampling strategies, with use of the Horvitz–Thompson estimator and fixed cost of sampling.

Editors' note: This article is one of Professor Cochran's last publications; he died at Cape Cod, Massachusetts, on March 29, 1980.

References

[1] Brewer, K. W. R. (1963). *Aust. J. Statist.*, **5**, 5–13.

[2] Durbin, J. (1967). *Appl. Statist.*, **16**, 152–164.

[3] Hanurav, T. V. (1967). *J. R. Statist. Soc. B*, **29**, 374–391.

[4] Hege, V. S. (1967). *J. Amer. Statist. Ass.*, **62**, 1013–1017.

[5] Horvitz, D. G. and Thompson, D. J. (1952). *J. Amer. Statist. Ass.*, **47**, 663–685.

[6] Madow, W. G. (1949). *Ann. Math. Statist.*, **20**, 333–354.

[7] Midzuno, H. (1951). *Ann. Inst. Statist. Math.*, **2**, 99–108.

[8] Rao, J. N. K. (1965). *J. Ind. Statist. Ass.*, **3**, 173–180.

[9] Rao, J. N. K. and Singh, M. P. (1973). *Aust. J. Statist.*, **15**, 95–104.

[10] Rao, T. J. (1971). *J. Amer. Statist. Ass.*, **66**, 872–875.

[11] Vijayan, K. (1975). *J. Amer. Statist. Ass.*, **70**, 713–716.

[12] Yates, F. and Grundy, P. M. (1953). *J. R. Statist. Soc. B*, **15**, 253–261.

(ESTIMATION, POINT
UNBIASEDNESS)

WILLIAM G. COCHRAN

HOT DECK METHOD *See* EDITING STATISTICAL DATA

HOTELLING, HAROLD

Born: September 29th, 1895, Fulda, Minnesota.

Died: December 26th, 1973, Chapel Hill, North Carolina.

Contributed to: econometrics, multivariate analysis, statistical inference.

Harold Hotelling, who was responsible for so much pioneering theoretical work in both statistics and mathematical economics, and who did so much to encourage and improve the teaching of statistics at U.S. universities, was born on September 29, 1895, in Fulda, Minnesota. At an early age his father's business forced a move to Seattle, and so, not surprisingly, he attended the University of Washington.

It is interesting that he chose to study journalism as his major, and that he worked on various newspapers in Washington while a student, obtaining his degree in journalism in 1919. This doubtless accounts for the particular interest he always had, even when he was a distinguished leader of the group of theoretical statisticians at the University of North Carolina at Chapel Hill, in ensuring that the achievements of his colleagues received ample attention in the local (and in some cases the national) press.

While majoring in journalism he also took some mathematics courses from Eric Temple Bell, who recognized his analytical abilities and steered him toward mathematics. Thus it was that, now with mathematics as his major subject, he took a Master of Science degree at Washington in 1921, and a Doctorate of Philosophy at Princeton in 1924. His Ph.D. dissertation was in the field of topology, published in 1925 in the *Transactions of the American Mathematical Society*.

After obtaining his doctorate he began work at Stanford University, originally in the Food Research Institute, and his interest in probability and statistics began to take hold; he taught his first courses in statistical theory and practice (a great novelty at that time) and began to publish the first of a long series of scholarly articles. His earliest applications of mathematical ideas concerned journalism and political science*; from these he turned to population and food supply, and then to theoretical economics, in which he was one of the initiators of the modern theories of imperfect competition and welfare economics. At the same time he was producing a series of publications in theoretical statistics which were often of such originality and importance that they provoked a considerable amount of later research by many scholars in many lands. In 1931 he published in the *Annals of Mathematical Statistics* what is quite possibly his most important contribution to statistical theory, his paper, "The generalization of Student's ratio."

In 1931 he was appointed Professor of Economics at Columbia University, where he would stay for 15 years. It was while he was there that he was able to assist various refugee scholars from central Europe, including the late Abraham Wald*. During World War II he organized at Columbia University the famous Statistical Research Group*, which was engaged in statistical work of a military nature; the group included Wald, Wallis, and Wolfowitz, and one of its signal achievements was the theory of sequential procedures (*see* SEQUENTIAL ANALYSIS).

In 1946 came his final move, to the University of North Carolina at Chapel Hill, where he was almost given carte blanche to create a theoretical statistics department. He rapidly recruited many able scholars, including R. C. Bose, W. Hoeffding, W. G. Madow, H. E. Robbins, S. N. Roy, and P. L. Hsu.

In statistical theory he was a leader in multivariate analysis*, being responsible for some of the basic ideas and tools in the treatment of vector-valued random variables. His major contribution to this area has come to be called "Hotelling's generalized T^2*." He also played a major role in the development of the notions of principal components and of canonical correlations (*see* COMPONENT ANALYSIS *and* CANONICAL ANALYSIS).

As early as 1927 he studied differential equations subject to error, a topic of general current interest, and published one of the first papers in this field. His papers on rank correlation*, on statistical prediction*, and on the experimental determination of the maximum of a function, also stimulated much further research in succeeding decades.

In economic theory, his papers on demand theory, on the incidence of taxation and welfare economics, are already regarded as classics that form the basis for much further work that has been done since they were written. In demand theory he was one of a small number of pioneers to revolutionize the basis of that theory and to extend its applications. His work on the incidence of taxation is important still in the literature of public finance*.

In 1955 he was awarded an honorary LL.D. by the University of Chicago; in 1963 he was awarded an honory D.Sc. by the University of Rochester. He was an Honorary Fellow of the Royal Statistical Society* and a Distinguished Fellow of the American Economic Association. He served as President of the Econometric Society in 1936–1937 and of the Institute of Mathematical Statistics* in 1941. In 1970 he was elected to the National Academy of Sciences, in 1972 he received the North Carolina Award for Science, and in 1973 he was elected to a membership of The Accademia Nazionale dei Lincei, in Rome. It is sad to say that, by the time this last honor had come his way, he had already suffered a severe stroke, in May 1972, which led to his death on December 26, 1973.

(CANONICAL ANALYSIS
CORRELATION
ECONOMETRICS
HOTELLING'S T^2
HOTELLING'S TRACE
MULTIVARIATE ANALYSIS)

WALTER L. SMITH

HOTELLING–PABST TEST

The Hotelling–Pabst test is a form of the Spearman rank correlation test* of independence of two random variables.

Bibliography

Conover, W. J. (1980). *Practical Nonparametric Statistics*. Wiley, New York. (See Sections 5.4 and Table A11.)

Hotelling, H. and Pabst, M. R. (1936). *Ann. Math. Statist.*, **7**, 29–43.

HOTELLING'S T^2

DEFINITION

The statistic is defined as follows:

$$T^2 = N(\overline{\mathbf{X}} - \boldsymbol{\mu})'\mathbf{S}^{-1}(\overline{\mathbf{X}} - \boldsymbol{\mu}),$$

where $\overline{\mathbf{X}}$ is a p-dimensional column vector of means computed from a sample of size $N(> p)$ drawn from a p-variate normal population $N_p(\boldsymbol{\mu}, \boldsymbol{\Sigma})$ with mean vector $\boldsymbol{\mu}$ and covariance matrix $\boldsymbol{\Sigma}$, $\mathbf{S}^{-1}$ is the inverse of the sample covariance matrix with

$n = N - 1$ degrees of freedom, and $\mathbf{A}'$ denotes the transpose of the matrix $\mathbf{A}$.

It may be noted that T^2 generalizes the well-known Student's t^2-statistic from univariate normal theory (*see* t-TESTS). Hotelling [10] proposed T^2 actually in the two-sample context but since that case is known as Mahalanobis D^2, the present treatment concerns mainly the single-sample case.

DERIVATION

Two derivations are discussed, of which the first uses the p-variate normal density while the second starts with the univariate t^2.

1. **Likelihood Ratio Criterion.** Consider the test of the following null hypothesis:

$$H_0 : \boldsymbol{\mu} = \boldsymbol{\mu}_0 \text{ (specified) vs. } H_1 : \boldsymbol{\mu} \neq \boldsymbol{\mu}_0$$

where $\boldsymbol{\Sigma} > \mathbf{0}$, i.e., positive definite, and unknown in $N_p(\boldsymbol{\mu}, \boldsymbol{\Sigma})$. If the likelihood ratio* criterion for the test above is denoted by λ, then (see Anderson [1])

$$\lambda^{2/N} = \left[\sup_{\omega} L(\boldsymbol{\mu}_0, \boldsymbol{\Sigma}) / \sup_{\Omega} L(\boldsymbol{\mu}, \boldsymbol{\Sigma}) \right]$$

$$= \left(1 + T^2/n\right)^{-1},$$

where $\omega = \{(\boldsymbol{\mu}, \boldsymbol{\Sigma}) : \mu = \boldsymbol{\mu}_0 \text{ and } \boldsymbol{\Sigma} > 0\}$, $\Omega = \{(\boldsymbol{\mu}, \boldsymbol{\Sigma}) : -\infty < \mu_i < \infty,\ i = 1, \ldots, p, \text{ and } \boldsymbol{\Sigma} > \mathbf{0}\}$ and in T^2 above $\boldsymbol{\mu} = \boldsymbol{\mu}_0$. Since the critical region for a level α test is given by $\lambda < \lambda_\alpha$, it is equivalently given by $T^2 > T^2_{1-\alpha}$.

2. **Union–Intersection Principle*.** If $\mathbf{a}$, $p \times 1$, is a nonnull vector and if

$$\omega = \bigcap_{\mathbf{a}} (\mathbf{a}'\boldsymbol{\mu} = \mathbf{a}'\boldsymbol{\mu}_0, \mathbf{a}'\boldsymbol{\Sigma}\mathbf{a} > \mathbf{0})$$

$$= \bigcap_{\mathbf{a}} H_{0\mathbf{a}},$$

then for testing $H_{0\mathbf{a}}$ vs. $H_{1\mathbf{a}} : (\mathbf{a}'\boldsymbol{\mu} \neq \mathbf{a}'\boldsymbol{\mu}_0, \mathbf{a}'\boldsymbol{\Sigma}\mathbf{a} > \mathbf{0})$, from properties of univariate t^2, the critical region is of the form

$$\text{reject } H_{0\mathbf{a}} \text{ if } N\left[\mathbf{a}'\left(\overline{\mathbf{X}} - \boldsymbol{\mu}_0\right)\right]^2 / \mathbf{a}'\mathbf{S}\mathbf{a} > c,$$

where c depends on $\mathbf{a}$. The union of the foregoing critical regions over all $\mathbf{a}$ gives

reject H_0 if

$$\sup_{\mathbf{a}} \left\{ N\left[\mathbf{a}'\left(\overline{\mathbf{X}} - \boldsymbol{\mu}_0\right)\right]^2 / \mathbf{a}'\mathbf{S}\mathbf{a} \right\} > c,$$

which (see Roy [23], Kshirsager [15], Morrison [17], and Srivastava and Khatri [27]) reduces to

$$\text{reject } H_0 \text{ if } N\left(\overline{\mathbf{X}} - \boldsymbol{\mu}_0\right)'\mathbf{S}^{-1}\left(\overline{\mathbf{X}} - \boldsymbol{\mu}_0\right) > c,$$

$$c = T^2_{1-\alpha}.$$

DISTRIBUTION

Writing $T^2 = N\mathbf{d}'\mathbf{S}^{-1}\mathbf{d} = (\mathbf{d}'\mathbf{S}^{-1}\mathbf{d}/\mathbf{d}'\boldsymbol{\Sigma}^{-1}\mathbf{d}) N\mathbf{d}'\boldsymbol{\Sigma}^{-1}\mathbf{d}$, where $\mathbf{d} = \overline{\mathbf{X}} - \boldsymbol{\mu}_0$, for a given $\mathbf{d}$, $n\mathbf{d}'\boldsymbol{\Sigma}^{-1}\mathbf{d}/\mathbf{d}'\mathbf{S}^{-1}\mathbf{d}$ is distributed as chi-square* with $n - p + 1$ degrees of freedom, χ^2_{n-p+1}. The chi-square distribution does not involve $\mathbf{d}$ and hence $n\mathbf{d}'\boldsymbol{\Sigma}^{-1}\mathbf{d}/\mathbf{d}'\mathbf{S}^{-1}\mathbf{d}$ is distributed independently of $\mathbf{d}$. Since $\mathbf{d}$ is distributed $N_p(\boldsymbol{\mu} - \boldsymbol{\mu}_0, \boldsymbol{\Sigma}/N)$, $N\mathbf{d}'\boldsymbol{\Sigma}^{-1}\mathbf{d}$ is distributed as noncentral chi-square* with p degrees of freedom* (df) and noncentrality parameter $\tau^2 = N(\boldsymbol{\mu} - \boldsymbol{\mu}_0)'\boldsymbol{\Sigma}^{-1}(\boldsymbol{\mu} - \boldsymbol{\mu}_0)$. Hence $[(n - p + 1)/(pn)]T^2$ has a noncentral F distribution*, with p and $n - p + 1$ df and noncentrality parameter τ^2. If $\boldsymbol{\mu} = \boldsymbol{\mu}_0$, $(n - p + 1)T^2/(pn)$ is distributed as a central $F_{p,n-p+1}$. (See Rao [21] for details; Wijsman [30] and Bowker [4] for the representation of T^2 as the ratio of independent chi-squares; Hotelling [10] for the central distribution; Bose and Roy [3] and Hsu [11, 12] for the noncentral distribution.) Tang [29] has given tables of probability of type II error for various values of τ^2 for the levels of significance 0.05 and 0.01. (For other tables and charts, see references in Anderson [1] and Kshirsagar [15].)

OPTIMUM PROPERTIES

Given a sample of size N from $N_p(\boldsymbol{\mu}, \boldsymbol{\Sigma})$, of all tests of the hypothesis $\boldsymbol{\mu} = \mathbf{0}$ that are invariant under the group of nonsingular linear transformations, the T^2 test is uniformly most powerful. Simaika [26] has in fact shown that, given a sample as above, of all tests of $\boldsymbol{\mu} = \mathbf{0}$ with power depending only on $N\boldsymbol{\mu}'\boldsymbol{\Sigma}^{-1}\boldsymbol{\mu}$, the T^2 test is a uniformly most powerful* test (see Anderson [1]). Further, the T^2 test is admissible and minimax* (*see* HYPOTHESIS TESTING). Stein [28] has shown that the T^2 test is admissible for large deviations from H_0. Giri et al. [9] have proved that among all α-level tests the T^2 test maximizes the minimum power when $p = 2$, $N = 3$ and for each choice of deviation mean vector and α. Salaevski [24] has extended their result for general p and N establishing that the T^2 test is minimax in general. Further, Giri and Kiefer [8] have shown some local and asymptotic minimax properties of the T^2 test. In addition, Kiefer and Schwartz [14] have established the admissible Bayes character of T^2. Also, to test $\boldsymbol{\mu} = \mathbf{0}$ versus $\boldsymbol{\mu} \neq \mathbf{0}$, the T^2 test which uses the bivariate random sample (X_{1i}, X_{2i}), $i = 1, \ldots, M$, and discards the additional independent observations X_{1i}, $i = M + 1, \ldots, N$, is admissible. (*See* ADMISSIBILITY; also see Cohen [5]; for other references, see Kshirsagar [15], Giri [7], and Srivastava and Khatri [27].)

In regard to robustness* against nonnormality, Arnold [2] has shown by Monte Carlo methods* that the size of the T^2 test is not influenced if samples are drawn from bivariate distributions in which both marginal distributions are either of rectangular or of double exponential form. Again, Monte Carlo study on the negative exponential* population has shown that T^2 is more sensitive to the measure of skewness* than to the measure of kurtosis* (see Mardia [16]).

Further, Kariya [13] has studied a robustness property of T^2 in the following setup: Let $\mathbf{X}(N \times p) = (\mathbf{x}_1, \ldots, \mathbf{x}_N)'$, be a random matrix distributed in the form

$$|\boldsymbol{\Sigma}|^{-N/2} q\left(\sum_{i=1}^{N} (\mathbf{x}_i - \boldsymbol{\mu})'\boldsymbol{\Sigma}^{-1}(\mathbf{x}_i - \boldsymbol{\mu}) \right), \quad q \in Q,$$

where Q is a set of nonincreasing convex functions from $[0, \infty)$ into $[0, \infty)$ and $N > p$. For the test of $H_0 : \boldsymbol{\mu} = \mathbf{0}$ versus $H_1 : \boldsymbol{\mu} \neq \mathbf{0}$, $\boldsymbol{\Sigma} > \mathbf{0}$, Kariya shows that the T^2-test is uniformly most powerful invariant in this class of PDF's, the invariance of the distribution of $\mathbf{X}$ being under the transformation $\mathbf{X} \to \mathbf{HX}$ for $\mathbf{H} \in O(N)$, where $O(N)$ denotes the group of all real $N \times N$ orthogonal matrices.

The null distribution of T^2 under any member of the class is the same as that under normality and in this sense the T^2-test is robust against departures from normality (*see also* HOTELLING'S TRACE).

APPLICATIONS

1. The test of the hypothesis $H_0 : \boldsymbol{\mu} = \boldsymbol{\mu}_0$ vs. $H_1 : \boldsymbol{\mu} \neq \boldsymbol{\mu}_0$ in $N_p(\boldsymbol{\mu}, \boldsymbol{\Sigma})$ at level α discussed above in "Derivation" is an application of T^2; the test could be made using an F-statistic through the relation $T^2_{1-\alpha} = [np/(n - p + 1)]F_{p,n-p+1,1-\alpha}$.
2. A $100(1 - \alpha)\%$ confidence interval* statement for $\boldsymbol{\mu}$ using T^2 is given by

$$N(\overline{\mathbf{X}} - \boldsymbol{\mu})'\mathbf{S}^{-1}(\overline{\mathbf{X}} - \boldsymbol{\mu}) \leqslant T^2_{1-\alpha}.$$

Further, for all nonnull $\mathbf{a}(p \times 1)$ (see Roy [23]),

$$\mathbf{a}'\overline{\mathbf{X}} - \left[T^2_{1-\alpha}\mathbf{a}'\mathbf{S}\mathbf{a}/N \right]^{1/2} \leqslant \mathbf{a}'\boldsymbol{\mu} \leqslant \mathbf{a}'\overline{\mathbf{X}} + \left[T^2_{1-\alpha}\mathbf{a}'\mathbf{S}\mathbf{a}/N \right]^{1/2}$$

gives $100(1 - \alpha)\%$ simultaneous confidence bounds on all linear functions of $\boldsymbol{\mu}$.
3. Consider the $(p - 1) \times p$ contrast matrix $\mathbf{C}$ of rank $(p - 1)$ defined by $\mathbf{C}\boldsymbol{\epsilon} = \mathbf{0}$, where $\boldsymbol{\epsilon}' = (1, \ldots, 1)$. For testing H_0: $\mathbf{C}\boldsymbol{\mu} = \mathbf{0}$ vs. $H_1 : \mathbf{C}\boldsymbol{\mu} \neq \mathbf{0}$, if each of the N observation vectors undergoes a linear transformation with $\mathbf{C}$ as the matrix of transformation, the test statistic to be used is $T^2 = N(\mathbf{C}\overline{\mathbf{X}})'(\mathbf{CSC}')^{-1}(\mathbf{C}\overline{\mathbf{X}})$ with n df and $(p - 1)$ dimensions. This statistic is independent of the choice of $\mathbf{C}$, as it is invariant under any linear transformation in the $p - 1$ dimensions orthogonal to $\boldsymbol{\epsilon}$. For suitable choice of $\mathbf{C}$ the test

of $H_0: \mu_1 = \mu_2 = \cdots = \mu_p$ is a special case of the hypothesis $\mathbf{C}\boldsymbol{\mu} = \mathbf{0}$.

4. Consider a vector $\mathbf{Y}(p \times 1)$, distributed $N_p(\mathbf{A}\boldsymbol{\theta}, \boldsymbol{\Sigma})$, where the matrix $\mathbf{A}(p \times q)$ is of rank r and $\boldsymbol{\theta}(q \times 1)$ is a vector of unknown parameters. If a matrix $\mathbf{S}$ with n df independent of the vector $\mathbf{Y}$ is available such that $E\mathbf{S} = \boldsymbol{\Sigma}$ and $n\mathbf{S}$ has a central Wishart distribution*, then for the test of $H_0: E\mathbf{Y} = \mathbf{A}\boldsymbol{\theta}$ vs. $H_0: E\mathbf{Y} \neq \mathbf{A}\boldsymbol{\theta}$, the test statistic is

$$[(n-(p-r)+1)/((p-r)n)]T^2 = F_{p-r,n-(p-r)+1},$$

where T^2 is $(p-r)$-dimensional with n df. Here $T^2 = \min_\theta[\mathbf{Y} - \mathbf{A}\boldsymbol{\theta})'\mathbf{S}^{-1}(\mathbf{Y} - \mathbf{A}\boldsymbol{\theta})$.

For these and other applications see Anderson [1], Kshirsagar [15], Rao [21], and Morrison [17]. Also for the use of T^2 in step-down procedures [22, 27], see Mudholkar and Subbaiah [18, 19, 20]. They use the fact that $(1 + T^2/n)^{-1} = \prod_{i=1}^p B_i$, where B_i is the likelihood ratio statistic for testing the hypothesis concerning the additional information supplied by the ith variate (see Rao [21]). In addition, for the two-sample version of T^2 *see* MAHALANOBIS D^2. For a further generalization of T^2 to test the equality of p-dimensional mean vectors of l normal populations having an unknown common covariance matrix see Hotelling's T_0^2, defined in the HOTELLING'S TRACE entry. Also, for the complex analog of T^2, see Giri [6, 7] and Srivastava and Khatri [27].

EXAMPLE (of the test discussed in "Derivation")

Table 1 gives the mean vector and the covariance matrix computed from data on four physical measurements—height (inches), weight (pounds), chest (inches), and waist (inches)—of 60 male officers of ages 29 to 31 of the Philippine Army (see Sen [25]). The four variables are denoted respectively by X_1, X_2, X_3, and X_4. Now let us test H_0: $\boldsymbol{\mu} = \boldsymbol{\mu}_0$ vs. H_1: $\boldsymbol{\mu} \neq \boldsymbol{\mu}_0$, where $\boldsymbol{\mu}_0' = (64, 125; 32, 28)$. Using the data in Table 1, we get $T^2 = 35.8851$ and $F_{4,56} = 8.5151$, which is significant even at $\alpha = 0.001$.

Table 1 Sample Mean Vector and Sample Covariance Matrix from Four Physical Measurements of 60 Male Officers of the Philippine Army

		$\mathbf{S} = (s_{ij}),\ i, j = 1, 2, 3, 4$			
i	$\bar{\mathbf{X}}$	1	2	3	4
1	63.86	2.792	11.009	0.599	0.427
2	125.05		172.387	15.325	21.880
3	32.94			2.865	2.373
4	28.36				5.086

It may be seen from the example that although the univariate means are within one standard deviation of the respective hypothesized means, the T^2-test is highly significant and we reject the hypothesis that the sample comes from a four-variate normal population with the hypothesized mean vector.

References

[1] Anderson, T. W. (1958). *An Introduction to Multivariate Statistical Analysis*. Wiley, New York.

[2] Arnold, H. J. (1964). *Biometrika*, **51**, 65–70.

[3] Bose, R. C. and Roy, S. N. (1938). *Sankhyā*, **4**, 19–38.

[4] Bowker, A. H. (1960). In *Contributions to Probability and Statistics* (*Essays in Honor of Harold Hotelling*). Stanford University Press, Stanford, Calif., pp. 142–149.

[5] Cohen, A. (1977). *J. Multivariate Anal.*, **7**, 454–460.

[6] Giri, N. (1965). *Ann. Math. Statist.*, **36**, 664–670.

[7] Giri, N. (1977). *Multivariate Statistical Inference*. Academic Press, New York.

[8] Giri, N. and Kiefer, J. (1964). *Ann. Math. Statist.*, **35**, 21–35.

[9] Giri, N., Kiefer, J., and Stein, C. (1963). *Ann. Math. Statist.*, **34**, 1524–1535.

[10] Hotelling, H. (1931). *Ann. Math. Statist.*, **2**, 360–378.

[11] Hsu, P. L. (1938). *Ann. Math. Statist.*, **9**, 231–243.

[12] Hsu, P. L. (1945). *Ann. Math. Statist.*, **16**, 278–286.

[13] Kariya, T. (1981). *Ann. Statist.*, **9**, 211–214.

[14] Kiefer, J. and Schwartz, R. (1965). *Ann. Math. Statist.*, **36**, 747–770.

[15] Kshirsagar, A. M. (1972). *Multivariate Analysis*. Marcel Dekker, New York.
[16] Mardia, K. V. (1970). *Biometrika*, **57**, 519–530.
[17] Morrison, D. F. (1976). *Multivariate Statistical Methods*. McGraw-Hill, New York.
[18] Mudholkar, G. S. and Subbaiah, P. (1976). *J. Amer. Statist. Ass.*, **71**, 429–434.
[19] Mudholkar, G. S. and Subbaiah, P. (1978). *Biom. J.*, **20**, 15–24.
[20] Mudholkar, G. S. and Subbaiah, P. (1978). *J. Amer. Statist. Ass.*, **73**, 414–418.
[21] Rao, C. R. (1973). *Linear Statistical Inference and Its Applications*. Wiley, New York.
[22] Roy, J. (1958). *Ann. Math. Statist.*, **29**, 1177–1187.
[23] Roy, S. N. (1957). *Some Aspects of Multivariate Analysis*. Wiley, New York.
[24] Salaevski, O. V. (1968). *Sov. Math. Dokl.*, **9**, 733–735.
[25] Sen, P. (1957). *On a Multivariate Test Criterion and Its Applications*. Thesis, The Statistical Center, University of the Philippines.
[26] Simaika, J. B. (1941). *Biometrika*, **32**, 70–80.
[27] Srivastava, M. S. and Khatri, C. G. (1979). *An Introduction to Multivariate Statistics*. North-Holland, New York.
[28] Stein, C. (1956). *Ann. Math. Statist.*, **27**, 616–623.
[29] Tang, P. C. (1938). *Statist. Res. Mem.*, **2**, 126–157.
[30] Wijsman, R. A. (1957). *Ann. Math. Statist.*, **28**, 415–423.

(HOTELLING'S TRACE
HYPOTHESIS TESTING
MAHALANOBIS' D^2
NONCENTRAL CHI-SQUARE
NONCENTRAL F
t-TESTS)

K. C. S. PILLAI

HOTELLING'S TRACE

INTRODUCTION

It was stated elsewhere (*see* HOTELLING'S T^2) in the single-sample context that Hotelling proposed T^2 actually in the two-sample case. However, the two-sample T^2 is more often known as Mahalanobis D^2*. The statistic known as *Hotelling's trace* considered here is a constant times Hotelling's T_0^2 (see below for definition), where T_0^2 is an l-sample generalization of Hotelling's T^2 (or Mahalanobis D^2) for the test of equality of p-dimensional mean vectors of l p-variate normal populations having a common unknown covariance matrix. However, Hotelling's trace should be considered not only in the context of the test of this hypothesis, but also in the contexts of tests of two other hypotheses described below.

Consider the following three hypotheses [Pillai (1955, 1960 [8], 1976 [9])]:

(I) Equality of covariance matrices of two p-variate normal populations;
(II) Equality of p-dimensional mean vectors of l p-variate normal populations having a common unknown covariance matrix, known as MANOVA* (or general linear hypothesis);
(III) Independence between a p-set and a q-set ($p \leqslant q$) in a $(p+q)$-variate normal population.

Tests proposed for these three hypotheses are generally invariant tests [Giri (1977 [4]), Lehmann (1959), Roy (1957); *see* INVARIANCE PRINCIPLES IN STATISTICS], which, under the null hypotheses, depend only on the characteristic (ch.) roots of matrices based on samples. For example, in (I), the ch. roots are those of $\mathbf{S}_1\mathbf{S}_2^{-1}$, where $\mathbf{S}_1$ and $\mathbf{S}_2$ are sum-of-products (SP) matrices with n_1 and n_2 df. In (II), the matrix is $\mathbf{S}^*\mathbf{S}^{-1}$, where $\mathbf{S}^*$ is the between-SP matrix and $\mathbf{S}$ the within-SP matrix with $l-1$ and $N-l$ df, respectively, and N is the total of l sample sizes. In (III), the matrix is $\mathbf{S}_{12}\mathbf{S}_{22}^{-1}\mathbf{S}_{12}'(\mathbf{S}_{11} - \mathbf{S}_{12}\mathbf{S}_{22}^{-1}\mathbf{S}_{12}')^{-1}$, where $\mathbf{S}_{ij}(i, j = 1, 2)$ is the SP matrix of the ith set with the jth set (1 denoting p-set and 2, q-set). $\mathbf{S}_{12}\mathbf{S}_{22}^{-1}\mathbf{S}_{12}'$ and $\mathbf{S}_{11} - \mathbf{S}_{12}\mathbf{S}_{22}^{-1}\mathbf{S}_{12}'$ have q and $n' - 1 - q$ df respectively, where n' is the sample size.

In each of the three cases above, under the null hypothesis, the $s \leqslant p$ nonzero ch. roots, $0 < f_1 < f_2 < \cdots < f_s < \infty$, have the same form of joint density function—the well-known Fisher–Girshick–Hsu–Mood–Roy distribution [see Pillai (1960 [8], 1977

[10]) for references] given below:

$$f(f_1, \ldots, f_s) = C(s,m,n) \prod_{i=1}^{s} \left\{ \frac{f_i^m}{(1+f_i)^{m+n+s+1}} \right\} \prod_{i>j} (f_i - f_j).$$

[See Pillai (1960 [8], 1976 [9]) for $C(s,m,n)$.] Here s, m, and n are to be understood differently for different situations. For (I), if $p \leqslant n_1, n_2$, then $s = p$; $m = \frac{1}{2}(n_1 - p - 1)$; $n = \frac{1}{2}(n_2 - p - 1)$. For (II), $s = \min(p, l-1)$; $m = \frac{1}{2}(|l - 1 - p| - 1)$; $n = \frac{1}{2}(N - l - p - 1)$. For (III), if $p + q < n'$, $s = p$; $m = \frac{1}{2}(q - p - 1)$; $n = \frac{1}{2}(n' - 1 - q - p - 1)$. Alternatively, m and n could be written in terms of ν_1 and ν_2 df, where for (I) $\nu_1 = n_1$ and $\nu_2 = n_2$; for (II) $\nu_1 = l - 1$ and $\nu_2 = N - l$; and for (III) $\nu_1 = q$ and $\nu_2 = n' - 1 - q$. Now Hotelling's trace is $U^{(s)}$, where $U^{(s)} = \sum_{i=1}^{s} f_i$ [Pillai (1954, 1955)], [i.e., $U^{(s)}$ is the trace of the matrix giving ch. roots in I, II, or III], while $T_0^2 = \nu_2 U^{(s)}$, [Hotelling (1944, 1947, 1951)]. Hotelling's trace is also, although less frequently, known as the *Lawley–Hotelling trace*. T_0^2 was proposed by Hotelling in 1944 as a generalized T^2-test statistic and measure of multivariate dispersion; it was previously considered by several authors [Lawley (1938), in a generalization of Fisher's z-test; Bartlett (1939); and Hsu (1940)].

DISTRIBUTION

The null and nonnull distribution problems may be considered separately.

Null Distribution

Pillai (1954, 1956) studied the first four moments of $U^{(s)}$ and suggested an F approximation given by $F_{\gamma_1,\gamma_2} = (\gamma_2/\gamma_1)(U^{(s)}/s)$, where $\gamma_1 = s(2m + s + 1)$, $\gamma_2 = 2(sn + 1)$. The approximation is recommended for $n - m > 30$. Pillai and Samson (1959) obtained approximate upper 5 and 1% points for $s = 2$, 3, and 4 and various values of m and n using the moment quotients, and Pillai (1957, 1960) has given such approximate percentage points for $s = 2(1)8$. [See Pillai and Young (1971), Tiku (1971), and Hughes and Saw (1972) for other approximations.]

No general forms for the exact null distribution have yet been obtained. See Hsu (1940), Hotelling (1951) for the distribution of $U^{(2)}$; Pillai and Young (1971), Pillai and Sudjana (1974) for inversion of the Laplace transform (*see* INTEGRAL TRANSFORMS) of $U^{(s)}$; also Krishnaiah and Chang (1972) for the Laplace transform; Davis (1968) for the distribution of $U^{(s)}$ satisfying a differential equation. Davis (1970 [1]) has obtained upper 5 and 1% points of $U^{(s)}$ for $s = 3$, 4, 5 and later (1980 [2]) for $s = 6(1)10$. Pillai's approximate percentage points (1957, 1960 [8]) have been generally shown to have three-decimal-place accuracy except for small values of ν_2.

As for asymptotic distributions, Ito (1956) has obtained asymptotic expansions for Hotelling's T_0^2, both for the CDF and percentiles as a chi-square* series up to order ν_2^{-2}, the first term being $\chi^2_{p\nu_1}$. [See Davis (1968, 1970) for extension to order ν_2^{-3}; Muirhead (1970) for general asymptotic expansions for functions satisfying a system of partial differential equations.]

Nonnull Distribution

The nonnull distribution of $U^{(p)}$ for (II) has been obtained by Constantine (1966) through inverse Laplace transforms (*see* INTEGRAL TRANSFORMS):

$$C_1(p,\nu_1,\nu_2)(U^{(p)})^{(1/2)p\nu_1 - 1} \times \sum_{k=0}^{\infty} \left\{ (-U^{(p)})^k / (\tfrac{1}{2}p\nu_1)_k k! \right\} \sum_{\kappa} (\tfrac{1}{2}\nu)_\kappa L_\kappa^m(\boldsymbol{\Omega}),$$

[see Constantine (1966), Pillai (1977 [10]) for $C_1(p,\nu_1,\nu_2)$], $\nu = \nu_1 + \nu_2$; $\boldsymbol{\kappa} = (k_1, \ldots, k_p)$ is a partition of k into not more than p parts such that $k_1 \geqslant \cdots \geqslant k_p \geqslant 0$, $k_1 + \cdots + k_p = k$. The generalized Laguerre polynomial $L_\kappa^m(\boldsymbol{\Omega})$ is defined in Constantine (1966), and $(a)_\kappa = \prod_{i=1}^{p}(a - \frac{1}{2}(i-1))_{k_i}$ with $(a)_k = a(a+1)\cdots(a+k-1)$. Here $p \leqslant \nu_1, \nu_2$. $\boldsymbol{\Omega} = \boldsymbol{\Sigma}^{-1}\mathbf{MM}'$, where $\mathbf{M}(p \times \nu_1)$ is the mean matrix and $\boldsymbol{\Sigma}(p \times p)$ is the common covariance matrix.

The series is convergent for $|U^{(p)}| < 1$. The density of $U^{(p)}$ for $\nu_1 < p \leqslant \nu_2$ can be obtained from the density above by making the following substitutions: $(\nu_1, \nu_2, p) \to (p, \nu_1 + \nu_2 - p, \nu_1)$. The null distribution can be obtained by putting $\mathbf{\Omega} = \mathbf{0}$. Pillai (1973) [see also Pillai and Sudjana (1974)] used the density above to suggest the following form:

$$C_1(p, \nu_1, \nu_2)(U^{(p)})^{(1/2)p\nu_1 - 1} \times \sum_{k=0}^{\infty} \left\{ E_k(-U^{(p)}/p)^k / (1 + U^{(p)}/p)^{k+(1/2)p\nu} \right\},$$

where

$$E_k = \left\{ p^k / (\tfrac{1}{2} p\nu_1)_k k! \right\} \sum_{\kappa} (\tfrac{1}{2}\nu)_\kappa L_\kappa^m(\mathbf{\Omega}) - \sum_{r=0}^{k-1} \left\{ \prod_{j=r}^{k-1} (\tfrac{1}{2}\nu p + j)/(k-r)! \right\} E_r, \quad E_0 = 1.$$

This series is convergent for $0 < U^{(p)} < p/(p-2)$ [see Davis (1980 [2])] and while it may yield some useful results for $p \leqslant 6$ and small ν_1, for large p and ν_1 it will be useful only for larger ν_2. Davis (1970 [1]) employed analytic continuation on the Constantine series with $\mathbf{\Omega} = \mathbf{0}$ for computation of percentage points for $p = 3, 4, 5$, while for $p = 6(1)10$ he used the differential equation approach. For $p = 2$ and $\mathbf{\Omega} = \mathbf{0}$, Pillai's series reduces to the exact form given by Hotelling [Constantine (1966)]. See also Pillai and Jayachandran (1967, 1968) for the exact distribution of $U^{(2)}$ for (I), (II), and (III) using up to the sixth-degree zonal polynomial* [James (1964)]; Khatri (1967) for the Constantine-type series for (I); Pillai and Sudjana (1975) for a single distributional form for (I) and (II) of which Constantine's and Khatri's series are special cases; and Pillai and Hsu (1979 [11]) for (III).

Constantine (1966) derived the general moment of $U^{(p)}$ for (II) in the form

$$E(U^{(p)})^k = (-1)^k \sum_{\kappa} L_\kappa^m(-\mathbf{\Omega}) / (\tfrac{1}{2}(p+1-\nu_2))_\kappa, \qquad \nu_2 > 2k + p - 1.$$

For earlier work, see Hsu (1940) for the first two moments; Ghosh (1963) for the first moment and the variance for $p = 3$ and $p = 4$; Khatri and Pillai (1967) for the first four moments; and Pillai (1977 [10]) for other references.

In regard to asymptotic distributions, considerable work has been carried out. The nonnull distribution of T_0^2 was given by Siotani (1957) and later by Ito (1960) up to order ν_2^{-1}. Further, Siotani (1968, 1971) extended his result up to order ν_2^{-2}. The above authors generally used Taylor expansion and perturbation* techniques. [See Pillai (1977 [10]) for other work on asymptotic expansions for (I), (II), and (III) by Fujikoshi, Muirhead, Lee, Sugiura and Nagao, Chattopadhyay and Pillai, Pillai and Saweris, and others.]

OPTIMUM PROPERTIES

The union–intersection* character of Hotelling's trace has been demonstrated by Mudholkar, Davidson, and Subbaiah (1974) in the following manner. Consider the MANOVA or multivariate general linear model* $E\mathbf{X} = \mathbf{A}\boldsymbol{\xi}$, concerning a matrix $\mathbf{X}_{N \times p}$ of N independently normally distributed rows with a common covariance matrix $\mathbf{\Sigma}_{p \times p}$, where $\mathbf{A}_{N \times q}$ is a known design matrix of rank $r \leqslant N - p$ and $\boldsymbol{\xi}_{q \times p}$ is a matrix of unknown parameters. Let a testable hypothesis be $H_0 : \mathbf{\Delta} = \mathbf{C}\boldsymbol{\xi}\mathbf{U} = \mathbf{0}$, where $\mathbf{C}_{g \times q}$ and $\mathbf{U}_{p \times u}$ are given matrices with ranks $g(\leqslant r)$ and $u(\leqslant p)$, respectively. Here

$$\mathbf{S}^* = \hat{\mathbf{\Delta}}'\mathbf{W}^{-1}\hat{\mathbf{\Delta}} \quad \text{and}$$

$$\mathbf{S} = \mathbf{U}'\mathbf{X}'\left(\mathbf{I} - \mathbf{A}(\mathbf{A}'\mathbf{A})^{-}\mathbf{A}'\right)\mathbf{X}\mathbf{U},$$

where $\hat{\mathbf{\Delta}} = \mathbf{C}(\mathbf{A}'\mathbf{A})^{-}\mathbf{A}'\mathbf{X}\mathbf{U}$, $\mathbf{W} = \mathbf{C}(\mathbf{A}'\mathbf{A})^{-}\mathbf{C}'$, and $(\mathbf{A}'\mathbf{A})^{-}$ is a generalized inverse* of $\mathbf{A}'\mathbf{A}$. Now the matrix decomposition of the MANOVA hypothesis $H_0 : \mathbf{\Delta} = \mathbf{0}$ is

$$H_0 : \bigcap_{\mathbf{M} \in \mathscr{M}} \{H_0(\mathbf{M}) : \operatorname{tr} \mathbf{M}'\mathbf{\Delta} = \mathbf{0}\},$$

where $\mathscr{M}$ is the set of all matrices of order $(g \times u)$. Further, the union–intersection character of Hotelling's trace is observed from the result

$$\sup_{\mathbf{M}} \left\{ \operatorname{tr}(\mathbf{M}'\hat{\mathbf{\Delta}}) / \operatorname{tr}^{1/2}(\mathbf{M}'\mathbf{W}\mathbf{M}\mathbf{S}) \right\} = \operatorname{tr}^{1/2}(\mathbf{S}^*\mathbf{S}^{-1}).$$

For (I), let $\lambda_1, \ldots, \lambda_p$, be the ch. roots of $\Sigma_1\Sigma_2^{-1}$ ($\Sigma_h(p \times p)$ is the covariance matrix of the hth population, $h = 1, 2$). Similarly for (II), let $w_1, \ldots, w_p$, be the ch. roots of Ω and for (III) let $\rho_1^2, \ldots, \rho_p^2$, be the ch. roots of $\Sigma_{11}^{-1}\Sigma_{12}\Sigma_{22}^{-1}\Sigma_{12}'$, where Σ_{ij} is the covariance matrix of the ith set with the jth ($i, j = 1, 2$). It has been shown by several authors [Anderson and Das Gupta (1964a, 1964b) and Das Gupta, Anderson, and Mudholkar (1964) by set convexity arguments; Eaton and Perlman (1974) using Schur convexity; Mudholkar (1965) using symmetric gauge functions; Roy and Mikhail (1961); J. N. Srivastava (1964)] that the power of the test based on Hotelling's trace increases monotonically in each nonzero population ch. root, i.e., in each λ_i for (I), in each nonzero w_i for (II) and in each nonzero ρ_i^2 for (III), $i = 1, \ldots, p$. Again, the admissibility* of $U^{(p)}$ for (II) has been established by Ghosh (1964 [3]) for large values of the parameters in the alternative hypotheses, i.e., against unrestricted alternatives. Further, Pillai and Jayachandran (1967, 1968) have observed for (I), (II), and (III), through exact power tabulations, that $U^{(2)}$ has the largest power compared to those of other available tests when the two population roots are far apart with their sum constant.

As regards robustness* against nonnormality and the violation of the assumption of equality of covariance matrices, although Pillai's trace* test performs best among the MANOVA tests [see Mardia (1971 [6]), Olson (1974 [7]), Pillai and Sudjana (1975)] the Monte Carlo study of Olson (1974 [7]) also indicates that Hotelling's trace is reasonably robust against the kurtosis* aspect of nonnormality. Some large sample results [Ito (1969 [5]); Ito and Schull (1964)] indicate that $U^{(p)}$ behaves in almost the same way as a univariate F-statistic against nonnormality and covariance heterogeneity, including favorable results under equality of sample sizes. Further, exact robustness studies are available for (I), (II), and (III) using Pillai's distribution of the ch. roots under violations [see Pillai (1975); Pillai and Sudjana (1975); Pillai and Hsu (1979 [11])]. Based on the numerical values of the ratio $e = (p_1 - p_0)/(p_0 - \alpha)$, where p_1 = power* under violation of assumptions, p_0 = power without violation and $\alpha = 0.05$, Hotelling's trace seems to rank third in terms of robustness* behind Pillai's trace* and Wilks' criterion*, where robustness for (I) and (III) considered is against nonnormality and for (II) is as above.

APPLICATION

In order to illustrate the test procedure for (II) a numerical example is given below.

Example. A study was made [see Ventura (1957); Pillai (1960 [8])] for a test of hypothesis (II) with four variables based on measurements of (1) height (inches), (2) weight (pounds), (3) chest (inches), and (4) waist (inches) of male reserve officers in civilian status of the Armed Forces of the Philippines, hailing from six different regions of the Philippine Islands but all within the age interval 29 to 31. The sample contained 25 officers from each of the regions. (The assumption of equality of covariance matrices was found to be justified in view of earlier tests.)

The null hypothesis is H_0: $\boldsymbol{\mu}_1 = \cdots = \boldsymbol{\mu}_6$ vs. H_1: not all $\boldsymbol{\mu}_h$'s equal in $N_4(\boldsymbol{\mu}_h, \boldsymbol{\Sigma})$, $h = 1, \ldots, 6$.

The within-SP matrix is given by

$$\mathbf{S} = \begin{array}{c} 1 \\ 2 \\ 3 \\ 4 \end{array}\begin{bmatrix} 471.2 & 1{,}118.8 & 60.6 & 23.2 \\ & 24{,}919.0 & 2{,}053.1 & 3{,}048.8 \\ & & 435.8 & 305.5 \\ & & & 675.2 \end{bmatrix}$$
$$\begin{array}{cccc} 1 & 2 & 3 & 4 \end{array}$$

and the between SP matrix is given by

$$\mathbf{S}^* = \begin{bmatrix} 15.9 & 17.9 & -7.5 & -7.8 \\ & 554.0 & 117.2 & 72.1 \\ & & 41.6 & 27.9 \\ & & & 28.0 \end{bmatrix}.$$

$U^{(4)} = \operatorname{tr}(\mathbf{S}^*\mathbf{S}^{-1}) = 0.1953 < U_{0.95}^{(4)}$ ($m = \frac{1}{2}(|l - 1 - p| - 1) = 0$, $n = \frac{1}{2}(N - l - p - 1) = 69.5$ since $N = 150$) [Davis (1970 [1]), Pillai (1960 [8])]. Hence do not reject H_0.

References

[For references not listed here, see Pillai [10]).]

[1] Davis, A. W. (1970). *Biometrika*, **57**, 187–191.

[2] Davis, A. W. (1980). *Commun. Statist. B*, **9**, 321–336.

[3] Ghosh, M. N. (1964). *Ann. Math. Statist.*, **35**, 789–794.

[4] Giri, N. (1977). *Multivariate Statistical Inference*. Academic Press, New York.

[5] Ito, K. (1969). In *Multivariate Analysis II*, Academic Press, New York, pp. 87–120.

[6] Mardia, K. V. (1971). *Biometrika*, **58**, 105–121.

[7] Olson, C. L. (1974). *J. Amer. Statist. Ass.*, **69**, 894–908.

[8] Pillai, K. C. S. (1960). *Statistical Tables for Tests of Multivariate Hypotheses*. The Statistical Center, University of the Philippines, Manila.

[9] Pillai, K. C. S. (1976). *Canad. J. Statist.*, **4**, 157–183. [This paper attempts at a review of the work on distributions of characteristic roots (see Introduction) in real Gaussian multivariate analysis, surveying developments in the field from the start and covering about 50 years. The exact null distribution has been reviewed in this Part I, and subasymptotic and asymptotic expansions of the distributions, mostly for large-sample sizes studied by various authors, have been briefly discussed. Such distributional studies of four test criteria (Hotelling's trace, Pillai's trace $= \sum_{i=1}^{s} f_i/(1+f_i)$, Roy's largest (smallest) root, and Wilks' criterion $= \prod_{i=1}^{s}(1+f_i)^{-1}$) and a few less important ones which are functions of ch. roots have been discussed further in view of the power comparisons made in connection with tests of three multivariate hypotheses. The one-sample case has also been considered. The topics discussed in order are: null distributions; some test criteria; individual ch. roots giving the Roy–Pillai reduction formula and Mehta–Krishnaiah pfaffian method, Davis differential equation method, Pillai–Fukutomi–Sugiyama zonal polynomial series; the two traces; Wilks' criterion; and other criteria. The references are included in Part II.]

[10] Pillai, K. C. S. (1977). *Canad. J. Statist.*, **5**, 1–62. (This paper is Part II of a review paper dealing with the noncentral distributions of the ch. roots in connection with the three hypotheses but as special cases of Pillai's distribution of the ch. roots under violations [Pillai (1975), *Ann. Math. Statist.*, **3**, 773–779]. The main topics are somewhat similar to those of Part I but dealing with the noncentral cases. There are 409 references.)

[11] Pillai, K. C. S. and Hsu, Y. S. (1979). *Ann. Inst. Statist. Math.*, **31**, 85–101.

(GENERAL LINEAR MODEL
HOTELLING'S T^2
MULTIVARIATE ANALYSIS
PILLAI'S TRACE
WILKS' LAMBDA)

K. C. S. Pillai

HOTELLING STRATEGY *See* GUESSING MODELS

H-SPREAD *See* FIVE-NUMBER SUMMARIES

HUMAN GENETICS, ANNALS OF

The *Annals of Human Genetics* was one of the earliest and has remained one of the foremost journals concerned with research into genetics*; it is specifically concerned with human genetics*. The original title was *Annals of Eugenics*, subtitled "A Journal for the scientific study of racial problems." The word "Eugenics" had been coined by Sir Francis Galton*, who defined it in 1904 as "the science which deals with all the influences that improve the inborn qualities of a race; also those that develop them to the utmost advantage." In his foreword to the first volume of the *Eugenics Review* in 1909 he said that "the foundation of Eugenics is laid by applying mathematical statistical treatment to a large collection of facts." He was a man of wide scientific interests, which included stockbreeding, psychology, and the use of fingerprints for identification, and he was a cousin of Charles Darwin. He was also one of the early white explorers of Africa and a prolific writer on these and many other subjects.

The journal was founded in 1925 by Karl Pearson*, who was head of the Department of Applied Statistics and of the Galton Laboratory at University College; it was printed by the Cambridge University Press. The Galton Laboratory had been founded under Galton's will, and Karl Pearson was the first occupant of the Galton Chair of Eugenics. He had been an intimate friend and disciple of Galton and remained faithful to many of his ideas. The journal's aims were set out in a foreword to the first volume, where eugenics is defined as "the study of agencies under social control that may improve or impair the racial qualities of future generations either physically or mentally." The quotations from Galton and Darwin still retained on the cover of the *Annals* indicate its commitment to mathematical and statistical techniques. Until the arrival of the computer* and the resulting enormous increase in the amount of information collected, the journal emphasized the necessity of publishing data with papers; this is now most often deposited in record offices when it is extensive.

Karl Pearson retired in 1933 and was succeeded as editor in 1934 by Ronald A. Fisher*, who was also keenly interested in the development of statistical techniques, but was critical of some of Pearson's statistical methods. He became Galton Professor in 1934, and the editorship of the *Annals* went with the Chair then as now.

Fisher innovated many well-known statistical techniques and showed how they could be applied to genetical problems. He changed the subtitle of the *Annals* to "A Journal devoted to the genetic study of human populations." He coopted several members of the Eugenics Society on to the editorial board. This society had been founded independently in 1908 as the Eugenics Education Society; its purpose was to propagate Galton's ideas and the work of the Laboratory, and Galton had accepted the presidency. Ronald Fisher had been an active member from its early days. This partnership seems only to have lasted until the outbreak of war, when Fisher returned to Rothamsted Experimental Station. He stayed there until 1943, when he accepted the Chair of Genetics at Cambridge and his editorship ended. In the foreword to the first volume of the *Annals* which he edited, Vol. 6, he announces its policy to be consistent with the aims of its founder:

> The contents of the journal will continue to be representative of the researches of the Laboratory and of kindred work, contributing to the further study and elucidation of the genetic situation in man, which is attracting increasing attention from students elsewhere. The two primary disciplines which contribute to this study are genetics and mathematical studies.

In 1945, Lionel S. Penrose succeeded him as editor of the *Annals*. He was a distinguished medical man and alienist, and under him the journal became more medical in content. Some of his papers on Down's anomaly, a permanent interest of his, and other aspects of inherited mental illness appeared in the journal. A feature of it in his time was the printing of pedigrees of inherited diseases covering several generations. He was responsible for changing the title from *Annals of Eugenics* to *Annals of Human Genetics*, a change for which it was necessary for an act of Parliament to be passed. The subtitle was also changed again to "A Journal of Human Genetics." He retired in 1965, and so did M. N. Karn, who had been the assistant editor since prewar days. Penrose was editor for a longer period than either of his predecessors and under his guidance the journal broadened its coverage and drew its contributions from a wider field.

Harry Harris, also a medical man and biochemist, succeeded to the Galton Chair and the editorship in 1965, and coopted C. A. B. Smith, mathematician and biometrician, who had been on the editorial board since 1955, to be coeditor, which he still is. Professor Harris was also head of a Medical Research Council Unit of Biochemical Genetics which became associated with the Galton Laboratory. Reflecting the editors' interests, the contents of the *Annals* inevitably became more concerned with biochemical genetics and statistics; *Annals of Eugenics* was dropped from the title page.

In 1975, Harris accepted the Chair of Human Genetics at the University of Pennsylvania, and for the next two years Cedric Smith was virtually the sole editor, as the Galton Chair remained vacant. In 1978, Elizabeth B. Robson was appointed to it, and she and C. A. B. Smith are currently the editors.

The journal is a specialized one dealing only with human genetics, but has changed with the progress of research and changing methods. As a general rule papers on statistical and mathematical methods are included only when illustrated by application to genetical data. The *Annals of Human Genetics* is published by the Galton Laboratory under the ownership of University College London, and its purpose is to provide a medium for the publication of original research in the field of human genetics from any source as well as the work of the Laboratory.

(ENGLISH SCHOOL OF BIOMETRY
FISHER, RONALD AYLMER
GALTON, FRANCIS
HUMAN GENETICS, STATISTICS IN
PEARSON, KARL)

JEAN EDMISTON

HUMAN GENETICS, STATISTICS IN

For about 100 years the sciences of human genetics and statistics have been closely related. Sir Francis Galton* was very interested in human genetics, partly viewed as a means to eugenics, i.e., the improvement of humanity by selective breeding. He invented the regression* and correlation* coefficients as tools of statistical investigation in genetics (see Galton [11, 12]). These methods were further developed by his friend Karl Pearson*. In 1900, Mendel's work was rediscovered, and the classical tools for mathematical and statistical investigation of Mendelian genetics were forged by Sir Ronald Fisher*, Sewall Wright, and J. B. S. Haldane in the period 1920–1950.

APPLICATIONS

Present-day applications of statistical theory to human genetics could conveniently be classified under four headings.

THEORY. Theory refers to the precise specification of the probabilistic mechanism of inheritance. Many discrete characters have *simple Mendelian* inheritance. The ABO blood groups are an example. In a slightly simplified treatment they can be considered as determined by three genes, A, B, O, with the rule that any one individual carries exactly two of these, not necessarily different. Thus individuals have six possible pairs of genes, or *genotypes*, AA, AO, BB, BO, AB, OO. Each child gets at random one gene from its mother, and one from its father, so a child of an $AB \times OO$ mating can be AO or BO, each with probability $\frac{1}{2}$.

Continuously (or almost continuously) varying characters such as height or fingerprint ridge count are believed to be due to the combined action of many genes of small effect (*multifactorial* or *polygenic* characters) together with possible environmental effects. Some discrete characters, such as diseases, are regarded as *threshold* characters [8, 9]; there is a hypothetical multifactorial continuous character x such that when $x >$ some threshold x_T, the individual is affected by the disease or disability. Other more complicated types of inheritance are conceivable, together with infection, environmental and social inheritance, and combinations of these.

VERIFICATION. Verification is checking whether a supposed mode of inheritance is compatible with observed family data. Mendelian inheritance can often be checked by goodness of fit* χ^2, e.g., the testing of the $1AO:1\ BO$ ratio from an $AB \times OO$ mating. Multifactorial inheritance can be tested by the observed correlations between relatives, e.g., mother–daughter and brother–sister correlations. In the simplest situation, when all genes act perfectly additively and there is negligible environmental effect, this correlation is equal to the expected proportion of genes the relatives have in common. Thus, since a daughter inherits just one of each pair of genes carried by her mother, daughter and mother have half their genes in common, and hence correlation $\frac{1}{2}$. Only a slightly more complicated argument shows that brother and sister have correlation $\frac{1}{2}$, grandparent–grandchild $\frac{1}{4}$, etc. Fingerprint ridge count shows correlations approximating to these [16]. The theoretical correlations for more complicated situations were first found by Fisher [10] in a classical paper, whose conclusions have not been greatly modified by subsequent work.

ESTIMATION OF GENETIC PARAMETERS. Many genetic parameters are proportions*. Examples are the proportion of *AO* individuals from an $AB \times OO$ mating; the proportion of *AB* individuals in a population; the proportion of *A* genes in a population; the mutation rate, or frequency with which a gene changes into a different gene in the next generation; the viability of a genotype, or probability of surviving from birth to reproductive maturity; the proportion of marriages between cousins; the recombination fraction θ in genetic linkage expressing the frequency with which certain "crossover" events occur, which result in the joining of a part of one chromosome (a body carrying the genes) with a part of another chromosome.

In the simplest cases the observation of x occurrences of an event in n trials gives the obvious estimate x/n of the probability of occurrence, with binomial SE = $\sqrt{x(n-x)/n^3}$. In practice the situation is complicated by various factors, of which the most important is recessivity. Thus, in the ABO blood groups we have anti-*A* and anti-*B* sera, which will show the respective presence or absence of the *A* and *B* genes in an individual by reacting or not reacting with a drop of her or his blood. But there is no reliable direct test for the presence or absence of *O*, so that an individual whose blood reacts with anti-*A* but not with anti-*B* (so-called blood group A) can be either *AA* or *AO*. The gene *O* which has no direct test for its presence is *recessive*. Thus a mating between individuals of groups A and O can be genetically either $AA \times OO$ or $AO \times OO$, producing a mixed distribution of offspring, not directly amenable to testing by χ^2. A problem also occurs in the testing of segregation ratios in the case of a rare recessive condition like phenylketonuria. Two apparently normal parents, genetically $Ph.ph \times Ph.ph$, produce on the average 25% phenylketonuric offspring *ph.ph* (who will be mentally and physically stunted unless specially treated). But the sampling will be biased because families with $Ph.ph \times Ph.ph$ parents which happen not to have any phenylketonuric offspring will not be noticed. Many methods exist for coping with biases due to recessivity, including the fully efficient "counting method" [2, 28], and simple nearly efficient alternatives [13, 19].

As far as estimation of the "recombination fraction" θ in the genetic linkage investigations is concerned, it is generally most convenient to calculate the log-likelihood function numerically for specified values of θ, such as 0, 0.05, 0.1, 0.15, . . . , 0.5 [21, 22]. Inferences are thus drawn from the likelihood curve.

Quantitative characters give rise to two distinct estimation problems. The data will consist of measurements on individuals in families of varying sizes. To find the correlations we must first find the covariances between relatives and their variances. This is equivalent to estimating variance and covariance components*, a problem on which there is a multitude of papers (Sahai [27]; see also Smith [29]). Then these correlations or variance and covariance components have in turn to be interpreted as showing the magnitudes of environmental and genetical effects and their interactions. This can be done either by analysis-of-variance* techniques [20] or by path coefficients [26]. For example, Rao and Morton claim to have shown that in American children the "heritability" of IQ is about 70%, which can roughly be interpreted as the proportional contribution of genetic factors, but in adults it is only 35%, although the precision of these estimates is questioned by Goldberger [14].

PRACTICE. Galton envisaged the chief practical use of human genetics as *eugenics*, i.e., the creation of better individuals through the encouragement of suitably desirable parents. Nowadays the practical value is more modestly limited to genetic counseling, i.e., to calculating the risks of inherited disease and abnormality for the benefit of anxious potential parents. This is in principle no more than the calculation of a conditional probability of abnormality given the known data on the family. For Mendelian characters, this uses only the simplest laws of probability [25], although the calculations can be quite complicated in detail. For quantitative and threshold characters it is usual to assume for simplicity that the joint distributions in relatives are multivariate normal*. Even so, methods at present available involve formidable problems of numerical integration*.

Population Size

In addition to the four classifications given above, we can also classify genetic questions as relating either to (1) individuals and individual families, or (2) to populations. Examples of (1) are Mendelian probabilities, e.g., the fact that children of an $AB \times AB$ mating have probabilities $\frac{1}{4}AA + \frac{1}{2}AB + \frac{1}{4}BB$. The questions (2) concern the immense field of population genetics, with issues involving the proportions of genes and genotypes in a population, and the effects of inbreeding, migration, selection, random fluctuations, or "drift" in small populations, mutation (an alteration in type of genes passed from parent to child), the effects of population intermarriage, etc.

GENETIC MECHANISMS

The fundamental law of human population genetics is the Hardy–Weinberg law of population proportions in a large population mating at random [15, 32].

In its original form it stated that if some character was controlled by two genes, G, g, which occur with respective frequencies P and $Q = 1 - P$ in the population, then the frequencies of genotypes are

$$P^2.GG + 2PQ.Gg + Q^2.gg.$$

This is easily extended to more complicated situations; e.g., if the frequency of the three blood group genes are $p.A + q.B + r.O$ ($= 0.26.A + 0.05.B + 0.69.O$ in Britain), then the frequency of genotype AA is p^2 ($= 0.07$), that of AB is $2pq$ ($= 0.03$), and so on. In theory, the Hardy–Weinberg law holds only in quite restricted situations, but in practice it is almost always found to apply to a quite satisfactory approximation.

The most important exception to Hardy–Weinberg relates to the effect of inbreeding on the frequency of rare recessive characters. An individual (I, say) is *inbred* if his or her mother and father share a common ancestor (C) a few generations back. If so, there is the possibility that (as regards any one particular character, such as phenylketonuria) the two genes that I gets from his or her mother and father were both descended from one single gene in C; they are said to be *identical by descent*. The probability F of this happening is I's *coefficient of inbreeding*. Thus if I's parents are first cousins, $F = \frac{1}{16}$.

If the original gene in the common ancestor C happened to be *ph*, then I would have genotype *ph.ph*, and therefore be phenylketonuric. The probability of this happening is accordingly Fq, where q (approximately 0.006 in Britain) is the frequency of the *ph* gene. Thus Fq ($\simeq 400 \times 10^{-6}$) is much larger than q^2 ($\simeq 36 \times 10^{-6}$), the probability that a child of unrelated parents is phenylketonuric [24]; that is, inbred children have a raised probability of suffering from recessive abnormalities: in this case the probability is greater by a factor of about $\frac{400}{36} = 11$. Since in the general British population at the time of Munro's investigation roughly one marriage in 100 was between cousins, we would expect that among parents of phenylketonurics about 1 in 100/11, i.e., 1 in 9 would be pairs of cousins. In fact, Munro found 10 cousin pairs out of 104, in good agreement (especially considering the roughness of the approximations involved).

A study of consanguinity can assist in the analysis of more complicated situations. Thus Chung et al. [4] noted that if in Stevenson and Cheeseman's [31] investigation of congenital deafness in Northern Ireland we consider those families in which both parents are normal, then when there is only one ("isolated") congenitally deaf child, the mean inbreeding coefficient for the children is about 2.8×10^{-3}, whereas when more than one child is affected, it is 8.5×10^{-3}. Since in the general population they estimated the mean inbreeding coefficient to be 0.4×10^{-3}, and since where more than one child in a family is affected it is very plausible that a recessive gene is involved, we can estimate that a proportion $(2.8 - 0.4)/(8.5 - 0.4) = 0.3$ of isolated cases are due to recessives.

Data on the survival of inbred individuals also gives information about the frequency of deleterious recessive genes in the population. Morton et al. [23] note that in American families studied by G. L. B. Arner in 1908, a proportion 0.168 of children died before the age of 20 when the parents were first cousins (inbreeding coefficient $F = 0.0625$), but only 0.116 when the parents were unrelated ($F = 0$). The difference $0.168 - 0.116 = 0.052$ is evidently due to deleterious recessives and may be taken as a crude estimate of $0.0625 \sum q$, where $\sum q$ is the total frequency of these all recessive genes, together, and is therefore estimated as $0.052/0.0625 = 0.8$. This calculation assumes that all these recessives are invariably lethal when in homozygous form *gg*. If they kill in only a proportion of cases, their total frequency must be correspondingly increased, to produce the same number of juvenile deaths. Thus we conclude that although any one particular type of deleterious recessive (such as phenylketonuria) is rare, altogether they are sufficiently numerous to have a high total frequency. (The presentation here of inbreeding calculations is kept deliberately simple; for more careful treatment, see the original papers.)

If a character is damaging to an organism, so that it does not reproduce, the corresponding genes are lost from the population. It is usually plausible that the supply of such genes is maintained by mutation, whereby normal genes are changed into abnormal (usually deleterious) ones. If the new genes show their effect immediately, we can count them and thus "directly" determine the rate of mutation. Otherwise, we have to rely on the principle that in the long run the number of new mutants entering the population by mutation must equal the number lost by premature death. So, with suitable reservations, the mutation rate can be "indirectly" estimated from the number of deaths. In either case it generally turns out that most genes have a mutation rate of the order of once in each 10^5 generations.

LITERATURE

Further information on these and other topics in human genetical statistics can be obtained partly from Ewens's article GENETICS, STATISTICS IN, in this volume, and also from many textbooks, although these tend to concentrate on theoretical population genetics. The books by Cavalli-Sforza and Bodmer [1], Charlesworth [3], Crow and Kimura [5], Elandt-Johnson [6], Ewens [7], Kempthorne [17], Li [18], and Wright [33] seem particularly worth consulting. A good general guide to human genetics is provided by Stern [30].

References

[1] Cavalli-Sforza, L. L. and Bodmer, W. F. (1971). *The Genetics of Human Populations*. W. H. Freeman, San Francisco. (A very readable and informative textbook.)

[2] Ceppellini, R., Siniscalco, M., and Smith, C. A. B. (1955). *Ann. Hum. Genet.*, **20**, 97–115.

[3] Charlesworth, B. (1980). *Evolution in Age-Structured Populations*. Cambridge University Press, Cambridge. (The first book to deal with the effects of continuous reproduction.)

[4] Chung, C. S., Robison, O. W. and Morton, N. E. (1959). *Ann. Hum. Genet.*, **23**, 357–366.

[5] Crow, J. F. and Kimura, M. (1970). *An Introduction to Populations Genetics Theory*. Harper & Row, New York.

[6] Elandt-Johnson, R. C. (1971). *Probability Models and Statistical Methods in Genetics*. Wiley, New York. (An elementary introduction to the field, lucidly written.)

[7] Ewens, W. J. (1979). *Mathematical Population Genetics*. Springer-Verlag, Berlin.

[8] Falconer, D. S. (1960). *Introduction to Quantitative Genetics*. Oliver & Boyd, Edinburgh.

[9] Falconer, D. S. (1965). *Ann. Hum. Genet.*, **29**, 51–76. (A classic paper, showing how the methods of quantitative genetics can be applied to discontinuous characters, especially inherited diseases.)

[10] Fisher, R. A. (1918). *Trans. R. Soc. Edinb.*, **52**, 399–433. (A classic paper.)

[11] Galton, F. (1877). *Proc. R. Inst.*, **8**, 282-301. (Introduces regression.)

[12] Galton, F. (1885). *Proc. R. Soc.*, **45**, 135–145. (Introduces the correlation coefficient from a genetical point of view.)

[13] Gart, J. J. (1968). *Amer. Hum. Genet.*, **31**, 283–292. (Simple estimation of parameters in a truncated binomial.)

[14] Goldberger, A. S. (1978). In *Genetic Epidemiology*, N. E. Morton and C. S. Chung, eds. Academic Press, New York, pp. 195–222.

[15] Hardy, G. H. (1908). *Science*, **28**, 49–50. (A classical paper, introducing what is now called the Hardy–Weinberg law of population genetics.)

[16] Holt, S. B. (1968). *The Genetics of Dermal Ridges*. Charles C. Thomas, Springfield, Ill.

[17] Kempthorne, O. (1957). *An Introduction to Genetic Statistics*. Iowa State University Press, Ames, Iowa. (An excellent introduction to the subject.)

[18] Li, C. C. (1976). *First Course in Population Genetics*. Boxwood Press, Pacific Grove, Calif. (A clear, detailed introduction.)

[19] Li, C. C. and Mantel, N. (1968). *Amer. J. Hum. Genet.*, **31**, 283–292. (Simple estimation of parameters in a truncated binomial.)

[20] Mather, K. and Jinks, J. L. (1971). *Biometrical Genetics*, 2nd ed.

[21] Morton, N. E. (1955). *Amer. J. Hum. Genet.*, **7**, 277–318.

[22] Morton, N. E. (1957). *Amer. J. Hum. Genet.*, **9**, 55–75. (These last two papers show how likelihoods for linkage can be quickly evaluated.)

[23] Morton, N. E., Crow, J. F., and Muller, H. J. (1956). *Proc. Natl. Acad. Sci.*, **42**, 855–863. (A classical paper, showing how to estimate the average number of deleterious recessives carried by man.)

[24] Munro, T. A. (1939). *Proc. Seventh International Congress of Genetics*. Cambridge University Press, Cambridge. (Investigates the genetics of phenylketonuria.)

[25] Murphy, E. A. and Chase, G. A. (1975). *Principles of Genetic Counseling*. Year Book Medical Publishers, Chicago.

[26] Rao, D. C. and Morton, N. E. (1978). In *Genetic Epidemiology*, N. E. Morton and C. S. Chung, eds. Academic Press, New York, pp. 145–193.

[27] Sahai, H. (1979). *Int. Statist. Rev.*, **47**, 177–222. (A list of papers on variance components.)

[28] Smith, C. A. B. (1957). *Ann. Hum. Genet.*, **21**, 254–276. (A simple method for estimating most genetic parameters.)

[29] Smith, C. A. B. (1980). *Ann. Hum. Genet.*, **44**, 95–105. (A simple method for estimating correlations between relatives.)

[30] Stern, C. (1960). *Principles of Human Genetics*, 2nd ed. W. H. Freeman, San Francisco.

[31] Stevenson, A. C. and Cheeseman, E. A. (1956). *Ann. Hum. Genet.*, **20**, 177–207.

[32] Weinberg, W. (1908). *Jahresh. Ver. vaterl. Naturkd. Württemberg*, **64**, 368–382.

[33] Wright, S. (1968, 1969, 1977, 1978). *Evolution and the Genetics of Populations*, Vols. 1–4. University of Chicago Press, Chicago. (A summary of his life's work by America's leading population geneticist.)

(GALTON, FRANCIS
GENETICS, STATISTICS IN
HUMAN GENETICS, ANNALS OF)

CEDRIC A. B. SMITH

HUNGARIAN CONSTRUCTIONS OF EMPIRICAL PROCESSES

Let $\xi_1, \ldots, \xi_n$ be independent uniform* (0, 1) random variables (rvs) having empirical distribution function (df) $\mathbb{G}_n$. Letting 1_A denote the indicator function of the set A,

$$\mathbb{U}_n(t) \equiv n^{1/2}\left[\mathbb{G}_n(t) - t\right]$$
$$= n^{-1/2} \sum_{i=1}^{n} \left[1_{[0,t]}(\xi_i) - t\right],$$
$$0 \leqslant t \leqslant 1, \quad (1)$$

is the *uniform empirical process**. Let $\mathbb{U}$ denote a *Brownian bridge*; that is, $\mathbb{U}$ is a normal process with continuous sample paths on [0, 1], having

$$E\,\mathbb{U}(t) = 0,$$
$$\operatorname{cov}\left[\mathbb{U}(s), \mathbb{U}(t)\right] = s \wedge t - st$$
$$\text{for all } 0 \leqslant s, \quad t \leqslant 1; \quad (2)$$

here $s \wedge t$ denotes the minimum of s and t. Since $n\mathbb{G}_n(t)$ is a binomial* (n, t) rv [while

$n\mathbb{G}_n(s)$ and $n\mathbb{G}_n(t)$ are "cumulative multinomials"], it is clear that the finite-dimensional distributions of $\mathbb{U}_n$ converge to those of $\mathbb{U}$; we denote this by writing $\mathbb{U}_n \rightarrow_{\text{f.d.}} \mathbb{U}$. For functionals g that are continuous (we omit the necessarily lengthy discussion of an appropriate topology), we would like to conclude that $g(\mathbb{U}_n) \rightarrow_{\text{d}} g(\mathbb{U})$. [An example of such a functional is $g(\mathbb{U}_n) = \|\mathbb{U}_n\| \equiv \sup\{|\mathbb{U}_n(t)|: 0 \leqslant t \leqslant 1\}$.] Unfortunately, $\rightarrow_{\text{f.d.}}$ of $\mathbb{U}_n$ to $\mathbb{U}$ does not imply this. This raises a natural question. Does $\mathbb{U}_n$ converge to $\mathbb{U}$ in some convenient mode that carries with it the conclusion

$$g(\mathbb{U}_n) \rightarrow_{\text{d}} g(\mathbb{U}) \quad \text{for a large class of functionals } g? \tag{3}$$

We will mention four approaches: (a) weak convergence, (b) Skorokhod's construction, (c) Skorokhod's embedding, and (d) the Hungarian construction. This will lead to a discussion of rates of convergence, and a mention of Berry–Esseen* type theorems for $g(\mathbb{U}_n)$.

See also PROCESSES, EMPIRICAL.

THE SIMPLEST SPECIAL CONSTRUCTION

Let ξ denote a uniform (0, 1) rv, F an arbitrary df, and define $X \equiv F^{-1}(\xi)$, where $F^{-1}(t) \equiv \inf\{x: F(x) \geqslant t\}$. Then the rv x has df F. We call this the *inverse transformation*.

Now suppose that F_n denotes a sequence of df's such that $F_n \rightarrow_{\text{d}} F$ as $n \rightarrow \infty$. Define $X_n \equiv F_n^{-1}(\xi)$. Then not only does $X_n \rightarrow_{\text{d}} X$ as $n \rightarrow \infty$, but it also happens that $X_n \rightarrow_{\text{a.s.}} X$ as $n \rightarrow \infty$ (see Billingsley [3, p. 337]), or just draw a picture). Thus, starting with a sequence of rvs converging in distribution, rvs X_n having the same marginal df's F_n have been constructed that in fact converge in the stronger a.s. sense. We shall call the constructed X_n's *versions* of the original rvs. (We claim nothing for the joint distribution of X_n's.)

This is a useful "theorem-proving mechanism." For example, if g denotes a function continuous except on a set that the df F assigns measure 0, then it is clear that $g(X_n) \rightarrow_{\text{d}} g(x)$ as $n \rightarrow \infty$ (since $g(X_n) \rightarrow_{\text{a.s.}} g(X)$ as $n \rightarrow \infty$ is in fact immediate for the specially constructed rvs X_n).

We now turn to analogs of this in which special versions of the empirical process are constructed that converge a.s. In fact, a rate of convergence is established. This is important because it is a useful theorem-proving mechanism for functions of these random processes.

WEAK CONVERGENCE AND THE SKOROKHOD CONSTRUCTION

Weak Convergence

Let D denote the set of all right continuous functions on [0, 1] that have left-hand limits at each point. We let d denote a complete, separable metric (not defined here) on D having the property that the sup norm

$$\|f_n - f\| \equiv \sup\{|f_n(t) - f(t)|: t \in [0, 1]\} \rightarrow 0$$

implies that $d(f_n, f) \rightarrow 0$ for functions f_n, f in D, while $d(f_n, f) \rightarrow 0$ implies that $\|f_n - f\| \rightarrow 0$ if f is continuous. Then $\mathbb{U}_n$ is said to converge weakly to $\mathbb{U}$ on (D, d) provided that $Eg(\mathbb{U}_n) \rightarrow Eg(\mathbb{U})$ for all bounded, d-continuous functions g on D. The σ-field induced by the metric d is the same as that induced by the finite-dimensional distributions. See Billingsley [2, p. 105] for these and the lengthy details of the result that

$$\mathbb{U}_n \text{ converges weakly to } \mathbb{U} \text{ on } (D, d). \tag{4}$$

This result implies that

$$(3) \text{ holds true for all } d\text{-continuous functionals } g. \tag{5}$$

Skorokhod's Construction*

We know from the section "The Simplest Special Construction" that rvs that converge weakly can be replaced by versions that converge a.s. Skorokhod [8, p. 281] showed that this also holds for random processes converging weakly on a complete, separable metric space. Thus there exist versions, which we shall call *Skorokhod versions*, of $\mathbb{U}_n$ and $\mathbb{U}$ that satisfy $d(\mathbb{U}_n, \mathbb{U}) \to_{\text{a.s.}} 0$; and because $\mathbb{U}$ has continuous sample paths, the d-convergence implies that we also have

$$\|\mathbb{U}_n - \mathbb{U}\| \to_{\text{a.s.}} 0 \quad \text{for the Skorokhod versions.} \quad (6)$$

From the sample paths of the Skorokhod version of $\mathbb{U}_n$, one can define independent uniform (0, 1) rvs $\xi_{n1}, \ldots, \xi_{nn}$ whose empirical process is the Skorokhod version of $\mathbb{U}_n$. Thus it holds that a triangular array of row independent uniform (0, 1) rvs exists such that the empirical process $\mathbb{U}_n$ of the nth row satisfies (4). (As in the preceding section, we claim nothing for the joint distribution of the rvs in different rows. In the next section, we deal with this deficiency as well as establishing a rate of convergence.) Thus from (6) we can conclude that

We have (3) for all measurable (with respect to the finite-dimensional σ-field) functionals g that are $\|\cdot\|$-continuous. (7)

THE HUNGARIAN CONSTRUCTION

We define the *sequential uniform empirical process* $\mathbb{K}_n$ by

$$\mathbb{K}_n(s,t) = \frac{1}{n^{1/2}} \sum_{i=1}^{[ns]} \left(1_{[0,t]}(\xi_i) - t\right) = \left(\frac{[ns]}{n}\right)^{1/2} \mathbb{U}_{[ns]}(t), \quad 0 \leqslant s, t \leqslant 1; \quad (8)$$

thus along the line $s = k/n$ the process $\mathbb{K}_n$ equals the scale multiple $(k/n)^{1/2}$ times the empirical process $\mathbb{U}_k$ of the first k observations. It is trivial to show that $\mathbb{K}_n \to_{\text{f.d.}} \mathbb{K}$ where $\mathbb{K}$ is a *Kiefer process**; that is, $\mathbb{K}$ is a normal process with continuous sample paths on $[0, 1]^2$ having

$$E\mathbb{K}(s,t) = 0, \quad \operatorname{cov}\left[\mathbb{K}(s_1, t_1)\right] = (s_1 \wedge s_2)\left[t_1 \wedge t_2 - t_1 t_2\right] \quad (9)$$

for all points in the unit square. In fact, it can be shown that

$$\mathbb{K}_n \text{ converges weakly to } \mathbb{K} \quad \text{(on an appropriate metric space).} \quad (10)$$

[Although application of Skorokhod [8] to (7) is possible, its conclusion does not help us toward our stated purpose.]

Note that $\mathbb{K}(ns, t)/n^{1/2}$ is again a Kiefer process, while each $\mathbb{B}_n = \mathbb{K}(n, \cdot)/n^{1/2}$ is a Brownian bridge.

The results of Komlos et al. [7] show that there exists a Kiefer process and a sequence $\xi_1, \xi_2, \ldots$ of independent uniform (0, 1) rvs for which the sequential uniform empirical process $\mathbb{K}_n$ of the first n rvs satisfies

$$\limsup_{n\to\infty} \left(\frac{\max_{1 \leqslant k \leqslant n} \left[\|\mathbb{K}(k/n, \cdot) - \mathbb{K}(k, \cdot)/n^{1/2}\|\right]}{(\log n)^2/n^{1/2}} \right) \leqslant \text{some } M < \infty \text{ a.s.} \quad (11)$$

Of course, setting $k = n$ in (11) immediately implies

$$\limsup_{n\to\infty} \|\mathbb{U}_n - \mathbb{B}_n\| \Big/ \frac{(\log n)^2}{n^{1/2}} \leqslant M < \infty \text{ a.s.}; \quad (12)$$

this implies (7), and also gives a rate of convergence. The authors in fact prove for their construction that

$$P\Big(\max_{1 \leqslant k \leqslant n} \|n^{1/2}\mathbb{K}_n(k/n, \cdot) - \mathbb{K}(k, \cdot)\| > (c_1 \log n + x)\log n\Big) \leqslant c_2 e^{-c_3 x} \quad (13)$$

for all x and all $n \geqslant 1$, where the c_i denote positive absolute constants.

A different construction by these authors allows $\log n$ to replace $(\log n)^2$ in (12) and allows the second $\log n$ in (13) to be replaced by 1; it must be noted for this result, however, that $\mathbb{B}_n$ merely denotes some Brownian bridge (and no underlying Kiefer process $\mathbb{K}$ is present). [It would be of great value if someone could show that the extra factor of $\log n$ in (12) and (13) is superfluous.]

An application of these results to give a Berry–Esseen type of theorem for the Cramér–von Mises statistic* is given by Csörgö

[6]. An application to establishing a rate in the justification of "bootstrapping*" the empirical process is given by Bickel and Freedman [1]. At least one log n factor in (12) and (13) is known to be necessary; it can prove an obstacle to establishing best possible rates via this method.

COMMENTS

Skorokhod's Embedding

If $Y_1, \ldots, Y_{n+1}$ are independent exponential rvs each with mean 1, then the rvs $\xi_{n:i} \equiv \sum_1^i Y_j / \sum_1^{n+1} Y_j$, $1 \leqslant i \leqslant n$, are distributed as the order statistics of n independent uniform (0, 1) rvs. Since $Y - 1$ is a (0, 1) rv, Skorokhod's embedding method allows us to define a stopping time τ for a given Brownian motion* S so that $S(\tau)$ has the same distribution as does $Y - 1$. By taking such an embedded version of the partial sums of the Y_i's, it can be shown that the empirical process $\mathbb{U}_n$ of the resulting uniform order statistics and the Brownian bridges

$$\mathbb{B}_n(t) \equiv -\left[S(nt) - tS(n) \right] / n^{1/2}, 0 \leqslant t \leqslant 1,$$

satisfy [the rate compares adversely with (12)]

$$\limsup_{n\to\infty} \frac{\|\mathbb{U}_n - \mathbb{B}_n\|}{n^{-1/4}(\log n)^{1/2}(\log\log n)^{1/4}} < \infty \quad \text{for Skorokhod embedding;} \qquad (14)$$

(see Brillinger [4] for details); this also implies (7), with a weak rate.

Skorokhod embedding preceded the Hungarian construction. Work by Csörgö and Révész [5] broke the bottleneck of opinion that the $n^{1/4}$ rate of (14) was best possible for a construction.

The proofs of most of the results cited, especially (11), are complicated. However (6), (11), and (12) are very easy to comprehend and use.

If $X_1, \ldots, X_n$ are independent with arbitrary df F and empirical df $\mathbb{F}_n$, then the empirical process $n^{1/2}(\mathbb{F}_n - F)$ has the version $\mathbb{U}_n(F)$ for any of the special $\mathbb{U}_n$'s above. This is the bridge that allows results for $\mathbb{U}_n$ to be transformed into results for the general empirical process.

References

[1] Bickel, P. and Freedman, D. (1981). *Ann. Statist.*, **9**, 1196–1217.

[2] Billingsley, P. (1968). *Convergence of Probability Measures*. Wiley, New York.

[3] Billingsley, P. (1979). *Probability and Measure*. Wiley, New York.

[4] Brillinger, D. (1969). *Bull. Amer. Math. Soc.*, **75**, 545–547.

[5] Csörgö, M. and Révész, P. (1975). *Zeit. Wahrscheinlichkeitsth.*, **31**, 255–269.

[6] Csörgö, S. (1976). *Actu. Sci. Math.*, **38**, 45–67.

[7] Komlos, J., Major, P., and Tusnady, G. (1975). *Zeit. Wahrscheinlichkeitsth.*, **32**, 111–131; *ibid.*, **34**, 33–58.

[8] Skorokhod, A. (1956). *Theory Prob. Appl.*, **1**, 138–171.

Acknowledgment

This material is based on work supported by the National Science Foundation under Grant MCS-81-02568.

(BROWNIAN MOTION
SKOROKHOD CONSTRUCTIONS OF
EMPIRICAL PROCESSES)

GALEN R. SHORACK

HUNT–STEIN THEOREM

One of the unpleasant facts about statistical decision problems is that they are generally too big or too difficult to admit of practical solutions. Introducing invariance principles*, restricting attention to invariant statistical tests, allows us to consider a subclass of the class of all available statistical tests. A natural question thus arises: under what conditions an invariant test which is optimum in a certain sense among the class of all invariant tests is also optimum among the class of all available tests if such a result can, at all, be achieved? Toward the end of World War II Hunt and Stein [12] proved (in their unpublished yet famous work) that under certain conditions on the group of transformations $\mathscr{G}$ leaving the problem invariant, there exists an invariant test of level

α which is minimax*, that is, minimizes the maximum error of the second kind. Many proofs have now appeared in the literature. The version of the Hunt–Stein theorem, published by Lehmann [19], is probably close in spirit to that originally developed by Hunt and Stein.

Pitman [24] suggested on intuitive grounds the use of best invariant procedures in certain estimation* and hypothesis testing* problems concerning scale and location parameters. Wald* [27] in the same year had the idea that for certain nonsequential estimation problems concerning location parameters, under certain restrictions on the transformation group there exists an invariant estimator that is minimax. Peisakoff [23] pointed out that there seems to be a lacuna in Wald's proof. He gave a comprehensive and a fairly general development of the theory of minimax decision procedures by the method of invariance*. Girshick and Savage [10] proved similar results for the location parameter case with squared error or bounded loss functions*. Blackwell and Girshick [1] treated the discrete case for location and scale parameter problems. Kudo [18] extended the results of Blackwell and Girshick to certain nonsequential estimation problems (all the foregoing results deal with nonsequential decision problems). Peisakoff [23] only mentions that sequential decision problems can be treated by his approach if one restricts the set of decision rules to consist of procedures which take at least a first observation with probability one. Kiefer [15] proved an analog of the Hunt–Stein theorem for sequential decision problems involving both continuous- and discrete-time observations, and extended the Hunt–Stein approach to other decision problems. Wesler [28] generalized the Hunt–Stein theorem for modified minimax tests based on slices of the parameter space. Karlin [14] used a streamlined version of this theorem in the context of invariant games.

The original version of the Hunt–Stein theorem is in terms of most stringent tests* and is contained in Hunt and Stein [11]. Let χ be the sample space, $\mathscr{A}$ be the σ-algebra* of subsets of χ, $\{P_\theta, \theta \in \Omega\}$ be the family of distributions over $(\chi, \mathscr{A})$ dominated by a σ-finite measure μ. Denote by p_θ the Radon–Nikodym derivative* $dP_\theta/d\mu$, so that for A in $\mathscr{A}$, $P_\theta(A) = \int_A p_\theta(x)\,d\mu(x)$. Assume that $\Omega = \Omega_0 \cup \Omega_1$, a disjoint union of Ω_0 and Ω_1. Suppose that we want to test the null hypothesis H_0: $\theta \in \Omega_0$ against the alternative hypothesis H_1: $\theta \in \Omega_1$. Let $\mathscr{G}$ be the group of transformations from χ onto χ which leave the testing problem invariant. Let $\mathscr{B}$ be a σ-algebra of subsets of $\mathscr{G}$ such that for any $A \in \mathscr{A}$ the set of pairs (x, g) with $g(x) \in A$ is in $\mathscr{A} \times \mathscr{B}$ and for any B in $\mathscr{B}$ and g in $\mathscr{G}$ the set Bg is in $\mathscr{B}$.

Hunt–Stein Theorem. Let there exist, for the problem of testing H_0 against H_1, an asymptotically right invariant sequence of probability measures $\{\nu_n\}$ on $(\mathscr{G}, \mathscr{B})$. Then there exists an invariant test ϕ_0 among all tests ϕ for which $\int \phi(x) p_\theta(x)\,d\mu(x) < \alpha$ for all $\theta \in \Omega_0$ and which also maximizes $\inf \int_{\theta \in \Omega_1} \phi(x) p_\theta(x)\,d\mu(x)$. In other words, ϕ_0 is minimax.

Thus for any invariant problem if there exists an invariant test which is minimax among all tests, the uniformly most powerful invariant test, if it exists, would then automatically have the desired minimax property.

A measure ν on $(\mathscr{G}, \mathscr{B})$ is said to be *right invariant* under $\mathscr{G}$ if $\nu(Bg) = \nu(B)$ for every B in $\mathscr{B}$ and every g in $\mathscr{G}$. A sequence $\{\nu_n\}$ of probability measures on $(\mathscr{G}, \mathscr{B})$ is *asymptotically right invariant* if

$$\lim_{n\to\infty} \left[\nu_n(Bg) - \nu_n(B) \right] = 0$$

for every B in $\mathscr{B}$ and every g in $\mathscr{G}$.

Let $\mathscr{G}$ be the additive group of reals, $\mathscr{B}$ be the ordinary Borel sets. Then the Lebesgue measure* ν is right invariant. But no invariant probability measure exists for this group. Define for each B in $\mathscr{B}$,

$$\nu_n(B) = \frac{1}{2n} \nu(B \cap [-n, n]),$$

the conditional probability given that g belongs to the closed interval $[-n, n]$. The ν_n are all probability measures with $1/(2n)$ as their normalizing factor. Since, for g in $\mathscr{G}$,

$Bg \cap [-n,n] = B \cap [-n-g, n-g]$, we obtain

$$
\begin{aligned}
|\nu_n(Bg) - \nu_n(B)| \\
&= \frac{1}{2n}\,|\nu(B \cap [-n-g, n-g]) \\
&\quad - \nu(B \cap [-n,n])| \\
&\leqslant \frac{|g|}{2n} \to 0 \quad \text{as } n \to \infty.
\end{aligned}
$$

Thus $\lim_{n\to\infty} \sup_B |\nu_n(Bg) - \nu_n(B)| = 0$ for every g in $\mathscr{G}$.

A remarkable feature of the Hunt–Stein theorem is that its assumptions concern only the structure of group $\mathscr{G}$, but not the distribution P_θ. They assert that if the transformation group in question is a locally compact Abelian group or a compact group, the result holds. The topological requirements are relevant because the class of measurable sets is related to the topology. In addition, if the result holds for a normal subgroup and the corresponding quotient group, it holds for the full group. It thus holds for the group of rigid motions in a Euclidean space, and for the common solvable groups, in particular, the multiplicative group of nonsingular triangular matrices. Brillinger [2] treated the case of Lie groups for invariance. Lie groups appear to be natural groups for the statisticians to use in connection with invariance. They are locally compact and hence possess the Haar measure required for the Hunt–Stein theorem. However, Kiefer [17] pointed out that examples for Lie groups can be constructed which do not satisfy the conditions of the Hunt–Stein theorem. We now consider a statistical example due to Stein, as reported in Lehmann [19, p. 338] to show that the general linear group $\mathscr{G}_l(p)$ does not satisfy the conditions of the Hunt–Stein theorem.

Let $\mathbf{X} = (X_1, \ldots, X_p)'$, $\mathbf{Y} = (Y_1, \ldots, Y_p)'$ be independently normally distributed p-vectors with the same $\mathbf{0}$ mean and covariance matrices $E(\mathbf{XX}') = \mathbf{\Sigma}$, $E(\mathbf{YY}') = \delta\mathbf{\Sigma}$, where δ is an unknown scalar quantity and $\mathbf{\Sigma}$ an unknown positive-definite matrix. The problem of testing H_0: $\delta = 1$ against the alternatives H_1: $\delta > 1$ remains invariant under the full linear group $\mathscr{G}_l(p)$ operating as $(\mathbf{X}, \mathbf{Y}; \mathbf{\Sigma}) \to (g\mathbf{X}, g\mathbf{Y}; g\mathbf{\Sigma}g')$ for $g \in \mathscr{G}_l(p)$. Since $\mathscr{G}_l(p)$ is transitive with probability 1 over the sample space of values of $(\mathbf{X}, \mathbf{Y})$, the uniformly most powerful test of level α of H_0 against H_1 is the trivial test $\phi(x, y) = \alpha$, which rejects H_0 with constant probability α for all values $(\mathbf{x}, \mathbf{y})$ of $(\mathbf{X}, \mathbf{Y})$. Hence the maximum power that can be achieved by using an invariant test is α. On the other hand, the level α test with the rejection region $y_i^2/x_i^2 > c$ for any i has a strictly increasing power function* $\beta(\delta)$ whose minimum over the set of alternatives $\delta \geqslant \delta_1 > 1$ is $\beta(\delta_1)$, where $\beta(\delta_1) > \beta(1) = \alpha$.

In almost all standard hypothesis-testing problems in multivariate analysis*, in particular for multinormal variations, there are no meaningful nonasymptotic (in regard to sample size) optimum properties either for the classical tests or for any other tests. The property of being best invariant under the full linear group of transformations that leave the problem invariant is often unsatisfactory because the Hunt–Stein theorem does not hold. Although the theorem holds for the subgroup $\mathscr{G}_T$ of $p \times p$ nonsingular triangular matrices, the dimension of the maximal invariant in the sample space and that of the corresponding maximal invariant in the parametric space are considerably larger than those under the full linear group. This results, very often, in the nonexistence of the best invariant test under $\mathscr{G}_T$. However, this theorem has been successfully used by Giri et al. [9] and by Linnik et al. [22] to solve the long-time open problem of the minimax character of Hotelling's T^2 test and by Giri and Kiefer [8] to prove the minimax character of the R^2-test* in some special cases. It has been successfully utilized among others by Giri [3, 6], Schwartz [25], and Giri and Kiefer [7] to establish the locally and asymptotically minimax character of various classical multivariate tests.

To verify the admissibility* of statistical tests by means of the Hunt–Stein theorem, the situation is more complicated. Apart from the trivial case of compact groups, only the one-dimensional translation parameter case has been studied by Lehmann and Stein [21]. If $\mathscr{G}$ is a finite or a compact group the most powerful invariant test is admissible. For other groups the statistical structure of the problem plays an important role. For verifying other optimum properties such as Isaacson's [13] type D or type E properties of several multivariate tests, the Hunt–Stein theorem has been used successfully by, among others, Giri and Kiefer [7] and by Sinha and Giri [26].

The Hunt–Stein theorem was originally stated in the context of most stringent tests. For an invariant testing problem with respect to the group $\mathcal{G}$ satisfying the conditions of the Hunt–Stein theorem, if there exists a uniformly most powerful almost invariant test with respect to $\mathcal{G}$, then that test is most stringent (see Pitman [24], Lehmann [19, 20], Kiefer [16], and Giri and Kiefer [7]).

For additional information on the Hunt–Stein theorem, see Giri [4, 5].

References

[1] Blackwell, D. and Girshick, M. A. (1954). *Theory of Games and Statistical Decisions*. Wiley, New York. (Treats discrete problems.)

[2] Brillinger, D. R. (1963). *Ann. Math. Statist.*, **36**, 492–500.

[3] Giri, N. (1968). *Ann. Math. Statist.*, **39**, 171–178.

[4] Giri, N. (1975). *Invariance and Statistical Minimax Tests*. Hindusthan Publishing Corp., India.

[5] Giri, N. (1977). *Multivariate Statistical Inference*. Academic Press, New York. (It deals with invariance for multivariate statistical inference, a graduate-level textbook.)

[6] Giri, N. (1979). *Canad. Statist.*, **7**, 53–60.

[7] Giri, N. and Kiefer, J. (1964). *Ann. Math. Statist.*, **35**, 21–35.

[8] Giri, N. and Kiefer, J. (1964). *Ann. Math. Statist.*, **35**, 1475–1490.

[9] Giri, N., Kiefer, J., and Stein, C. (1963). *Ann. Math. Statist.*, **34**, 1524–1535.

[10] Girshick, M. A. and Savage, L. J. (1950). *Proc. 2nd Berkeley Symp. Math. Stat. Prob.*, University of California Press, Berkeley, Calif., pp. 53–73.

[11] Hunt, and Stein, C. M. (1946).

[12] Hunt, and Stein, C. M. (1948).

[13] Isaacson, S. L. (1951). *Ann. Math. Statist.*, **22**, 217–234.

[14] Karlin, S. (1953). *Ann. Math. Statist.*, **24**, 371–401.

[15] Kiefer, J. (1957). *Ann. Math. Statist.*, **28**, 573–601.

[16] Kiefer, J. (1958). *Ann. Math. Statist.*, **29**, 675–699.

[17] Kiefer, J. (1966). *Multivariate Optimality Results in Multivariate Analysis*, P. R. Krishnaiah, ed. Academic Press, New York, pp. 255–274.

[18] Kudo, H. (1955). *Natl. Sci. Rep., Ochanomizu Univ.*, **6**, 31–73.

[19] Lehmann, E. L. (1959). *Testing Statistical Hypotheses*. Wiley, New York. (It contains two chapters dealing with invariance and the Hunt–Stein theorem, a graduate-level textbook.)

[20] Lehmann, E. L. (1959). *Ann. Math. Statist.*, **30**, 881–884.

[21] Lehmann, E. L. and Stein, C. (1953). *Ann. Math. Statist.*, **24**, 473–479.

[22] Linnik, Ju. V., Pliss, V. A., and Salaevskii, O. V. (1966). *Sov. Math. Dokl.*, **7**, 719.

[23] Peisakoff, M. P. (1950). *Transformation Parameters*. Ph.D. thesis, Princeton University.

[24] Pitman, E. J. G. (1939). *Biometrika*, **31**, 200–215.

[25] Schwartz, D. (1964). *Ann. Math. Statist.*, **35**, 939.

[26] Sinha, B. K. and Giri, N. (1976). *Sankhyā*, **38**, 244–248.

[27] Wald, A. (1939). *Ann. Math. Statist.*, **10**, 299–326.

[28] Wesler, A. (1959). *Ann. Math. Statist.*, **30**, 1–20.

(ADMISSIBILITY
HOTELLING'S T^2
INVARIANCE CONCEPTS IN STATISTICS
MINIMAX METHOD
MOST STRINGENT TESTS)

N. C. GIRI

HURST COEFFICIENT (RESCALED RANGE ANALYSIS)

Rescaled range analysis, or R/S analysis, is a statistical technique introduced by Mandelbrot and Wallis [8], and given mathematical foundation in Mandelbrot [4]. It introduces the important distinction between "short run" and "very long run" forms of statistical dependence*. The constant J on which this distinction hinges is the *Hurst coefficient* or *R/S exponent*, which can lie anywhere between 0 and 1. Even before defining J, one can describe its significance. The special value $J = 0.5$ is characteristic of independent, Markov, and other short-run dependent random functions. Therefore, the absence of very long run nonperiodic statistical dependence in empirical records or in sample functions can be investigated by *test-*

ing whether the hypothesis* that $J = 0.5$ is statistically acceptable. If not, the intensity of very long run dependence is measured by $J - 0.5$, and can be *estimated* from the data. The exponent J is *robust* with respect to the marginal distribution; not only is it effective when the underlying data or random functions are near Gaussian, but also when $X(t)$ is so far from Gaussian that $EX^2(t)$ diverges. On the contrary the usual second-order techniques (lag correlation, spectral analysis*) are invalid in the latter case.

The assumptions that $EX^2 < \infty$ and that X is short-run dependent used to be a matter of course in practical statistics. However, Mandelbrot [2] showed that long-tailed empirical records are often best interpreted by allowing $EX^2 = \infty$ (*see also* HEAVY-TAILED DISTRIBUTIONS). The question of whether an empirical record is weakly (short run) or strongly (long run) dependent was first faced in ref. 3, where Mandelbrot injected long-run dependence to interpret an empirical finding in hydrology,* called the Hurst phenomenon [1]. The mixture of long-tailedness and very long run dependence might have been unmanageable, but R/S analysis makes it possible to disregard the distribution of $X(t)$ and to tackle its long-run dependence.

DEFINITION OF THE STATISTIC R/S

When $X(t)$ is a random function in continuous time t, define $X^*(t) = \int_0^1 X(u)\,du$, $X^{2*}(t) = \int_0^t X^2(u)\,du$, and $X^{*2} = (X^*)^2$. When $X(i)$ is a sequence in discrete time i, define $X^*(0) = 0$, $X^*(t) = \sum_{i=1}^{[t]} X(i)$, with $[t]$ the integer part of t. For every $d > 0$, called the *lag*, define the *adjusted range of* $X^*(t)$ in the time interval 0 to d, as

$$R(d) = \max_{0 \leqslant u \leqslant d} \left\{ X^*(u) - \frac{u}{d} X^*(d) \right\} - \min_{0 \leqslant u \leqslant d} \left\{ X^*(u) - \frac{u}{d} X^*(d) \right\}.$$

Then evaluate the *sample standard deviation of* $X(t)$ itself (not of $X^*(t)$),

$$S^2(d) = X^{2*}(d)/d - X^{*2}(d)/d^2.$$

The expression $Q(d) = R(d)/S(d)$ is the *R/S statistic*, or *self-rescaled self-adjusted range of* $X^*(t)$. Obviously, R/S does not change if $X(t)$ is replaced by $aX(t) + b$.

DEFINITION OF THE R/S EXPONENT J

Suppose that there exists a real number J such that, as $d \to \infty$, $(1/d^J)(R(d)/S(d))$ converges in distribution to a nondegenerate limit random variable. It is shown in ref. 4

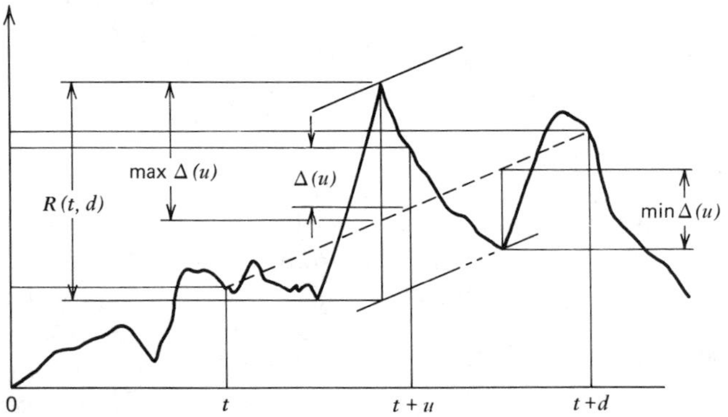

Figure 1 Construction of the sample range $R(t, d)$ [8]. The hydrologic interpretation of R and of S is of historical interest. Let $X(t)$ be the total discharge of a river into a reservoir during year t, and $X^*(u)$ the discharge in the first u years. Setting a time horizon of d years, suppose that the quantity of water average $X^*(d)/d$ can be withdrawn each year from the reservoir. At year u, $1 \leqslant u \leqslant d$, the amount of water in the reservoir exceeds the initial amount by $\Delta(u) = X^*(u) - (u/d)X^*(d)$. To avoid running out of water during these d years, the reservoir must initially contain a quantity at least equal to $-\min_{0 \leqslant u \leqslant d} \Delta(u)$. Suppose that it is initially filled with that minimal quantity. To avoid the reservoir's overflowing during these d years, its volume must be at least equal to the adjusted range $R(d)$—hence the use of $R(d)$ in an old method of design of water reservoirs, due to Rippl. Hurst added S as a natural normalizing factor, without noting the special virtues that make R/S so valuable.

that $0 \leqslant J \leqslant 1$. The function X is then said to have the R/S *exponent* J with a constant R/S prefactor.

Suppose, more generally, that $[1/d^J L(d)]$ $[R(d)/S(d)]$ converges in distribution to a nondegenerate random variable, where $L(d)$ denotes a slowly varying function at infinity, that is, a function that satisfies $L(td)/L(d) \to 1$ as $d \to \infty$ for all $t > 0$. The simplest example is $L(d) = \log d$. The function X is then said to have the R/S exponent J, and the R/S prefactor $L(d)$.

X i.i.d. Gaussian (White Noise), $J = 0.5$

One can assume that the X are $N(0, 1)$. Furthermore, $S(d) \to 1$ almost surely as $d \to \infty$; thus $R(d)/S(d) \sim R(d)$. $X^*(t) = B([t])$, where $B(t)$ is the standard Brownian motion*. As $d \to \infty$, weak convergence of random functions ensures that $R(d)/d^{0.5}$ tends toward the limit random variable $R_B(1)$ as $d \to \infty$, where

$$R_B(1) = \sup_{0 \leqslant v \leqslant 1} \{B(v) - vB(1)\} - \inf_{0 \leqslant v \leqslant 1} \{B(v) - vB(1)\}.$$

The same result holds for $d^{-0.5}R(d)/S(d)$. Thus X has the R/S exponent $J = 0.5$ with a constant R/S prefactor. More precisely, $e^{-\delta J}R(e^{\delta})/S(e^{\delta})$ is a stationary random function of $\delta = \log d$.

A Broader Class for Which $J = 0.5$

Let the random function $X(t)$ be such that $S(d) \Rightarrow \text{var}(X)$ and that the rescaled $a^{-0.5}X^*(at)$ converges weakly to Brownian motion $B(t)$ as $a \to \infty$. Then it happens that $d^{-0.5}R(d)/S(d)$ converges to $R_B(1)/\sqrt{\text{var}(X)}$. The convergence may be much slower than for white Gaussian noise*.

THE HURST PHENOMENON

Hurst [1] examined the records of annual water discharge through the Nile and other rivers, and other empirical records that are nonperiodic yet exhibit "cycles" of all periods, the slowest period being of the order of magnitude of the record or sample. Such cyclic behavior in hydrological records is described in the Biblical story of Joseph (Mandelbrot and Wallis call it the "Joseph effect" in ref. 6, hence the letter J). A naive application of the preceding two subsections would lead to the expectation that $R(d)/S(d)$ behaves like $d^{0.5}$. In fact, Hurst finds that $R(d)/S(d)$ *fails* to behave like $d^{0.5}$; instead, it fluctuates around d^J with $J > 0.5$; "typically" J is about 0.74. Additional evidence in ref. 7 (and in ref. 5, relative to diverse related contexts) suggests that the Hurst phenomenon does not reflect a transient effect. Rather, it is a widespread phenomenon to be handled on the basis of asymptotic theorems, relative to processes for which the foregoing definition of J is applicable.

There is a temptation to attribute $J \neq 0.5$ solely to the long-tailedness of X, more precisely to EX^2 being infinite. In a later section we show that this explanation is *not* valid in the absence of long-run dependence.

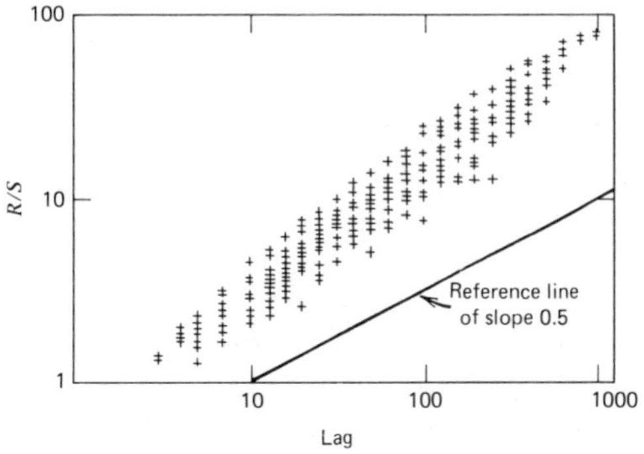

Figure 2 Graphical R/S analysis. A pox diagram of R/S for a record of annual tree ring indices for a bristlecone pine from White Mountains, California. Data provided by Laboratory for Tree Ring Research, University of Arizona, Tucson, Arizona. The diagram tightly clusters around a straight line, whose slope is an estimate of the exponent J. It obviously exceeds the slope 0.5 of the reference line. For the starting point t, $R(t, d)$ and $S(t, d)$ are defined by replacing $X^*(u)$ by $X^*(t + u) - X^*(t)$. Given a sample of T values $X(t)$, one subdivides it into k parts and computes $R(t_i, d)/S(t_i, d)$ for $t_1 = 1$, $t_2 = k + 1$, $t_3 = 2k + 1, \ldots,$ where all t_i are such that $(t_i - 1) + d \leqslant T$. One thus obtains T/k sample values of R/S for small lags and few values when d is close to T. One takes logarithmically spaced values of d. The present "pox diagram" is the plot of the sample values of $\log R(t_i, d)/S(t_i, d)$ vs. $\log d$.

ROLE OF SCALING

The random function $Z(t)$, with $Z(0) = 0$, is said to be *scaling* with the exponent H, if $Z(t)$ and $a^{-H}Z(at)$ have identical finite-dimensional distributions for all a. The function $a^{-H}Z(at)$ is deduced from $Z(t)$ by an affinity, hence this scaling expresses "self-affinity" (often wrongly called "self-similarity"). $Z(t)$ is *asymptotically scaling* with the exponent H and the slowly varying prefactor $L(a)$ if the limits for $a \to \infty$ of the finite-dimensional distributions of

$$a^{-H}L^{-1}(a)Z(at)$$

are identical to those of a scaling random function $W(t)$, called the *attractor* of $Z(t)$.

The foregoing derivation of $R(d)/S(d)$ for white Gaussian noise uses the fact that the Brownian motion $B(t)$ is scaling with the exponent taking on the standard value $H = 0.5$. It follows that $R(d)/S(d)$ satisfies a form of asymptotic scaling. Mandelbrot [3] interprets the Hurst phenomenon as involving scaling with a nonstandard exponent $H \neq 0.5$. The role of scaling in science is discussed in ref. 5; it is proving to be widespread and important.

$EX^2 = \infty$, X^* a Discrete Fractional Gaussian Noise with $H \neq 0.5$, $J = H$

The discrete fractional Gaussian noise is defined in FRACTIONAL BROWNIAN MOTION AND GAUSSIAN NOISES. It satisfies $S(d) \to 1$ almost surely, and $X^*(t) = B_H([t])$. For this noise, $J = H$.

Other Finite Variance Cases with $J \neq 0.5$

More generally, in order to obtain $J \neq 0.5$ with a constant prefactor, it suffices that $S(d) \to \text{var}(X)$ and that $X^*(t)$ be a random function attracted by $B_H(t)$ with $H = J$, and $EX^*(t) \sim t^{2H}$.

Still more generally, $J \neq 0.5$ with the prefactor $L(d)$ prevails if $S(d) \to \text{var}(X)$ almost surely, and $X^*(t)$ is attracted by $B_H(t)$ with $J = H$ and $EX^*(t) \sim t^{2H}L(t)$.

Finally, $J \neq 0.5$ when $S(d) \to \text{var}(X)$, and X^* is attracted by a non-Gaussian scaling random function of exponent $H = J$. Examples of such functions are given in ref. 9.

X a White Lévy Stable Noise, $J = 0.5$

White Lévy stable noise is the i.i.d. random function made of the increments of $\Lambda_\alpha(t)$, the Lévy stable random function* of exponent $\alpha \in]0, 2[$. Thus $X^*(t) = \Lambda_\alpha([t])$. It is known that $d^{-1/\alpha}\Lambda_\alpha(t)$ and $d^{-1/\alpha}R(d)$ tend to nondegenerate random variables as $d \to \infty$. Moreover, $S(d)$ *can no longer* be neglected in evaluating R/S, because $d^{-1/\alpha+0.5}S(d)$ does not tend to a nonrandom limit, but to a limit *random variable*. Combining $R(d) \sim d^{1/\alpha}$ and $S(d) \sim d^{1/\alpha - 0.5}$ suggests that $R(d)/S(d) \sim d^{0.5}$, and indeed it can be shown (Mandelbrot [4]) that $J = 0.5$.

If X Becomes Stationary When Differenced (or Differentiated), $J = 1$

An example is Brownian motion. More generally, suppose that $X(t)$ is scaling with parameter H in $]0, 1[$. Then X^* is scaling with parameter $H' = H + 1$, X^{2*} is scaling with parameter $H'' = 2H + 1$ and one proves [4] that $J = H' - 0.5H'' + 0.5 = 1$. Since this value of J is independent of $X(t)$, it tells nothing about $X(t)$; one should not perform R/S analysis over such functions, but over their differences (or derivatives), until a value of J below 1 is detected.

GENERAL MATHEMATICAL SETUP OF R/S ANALYSIS

Assume that the vector process of coordinates X and X^2 belongs to the domain of attraction of a vector scaling process, meaning that, as $d \to \infty$,

$$\left[X^*(td)/d^{H'}L'(d), X^{2*}(td)/d^{H''}L''(d)\right] \to \left[U(t), V(t)\right],$$

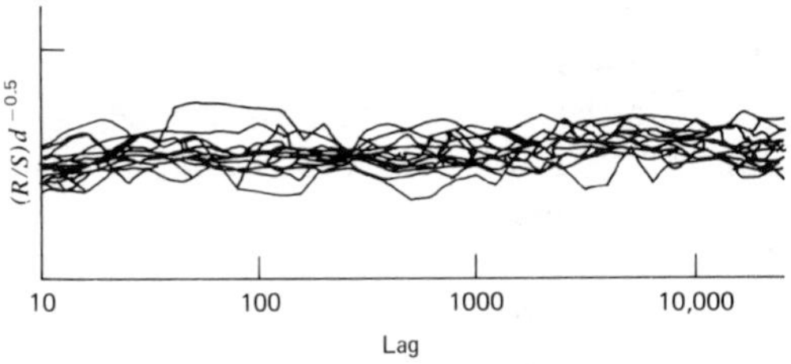

Figure 3 Behavior of $Q(d)d^{-0.5}$ for 15 independent samples, each made of 30,000 independent lognormal random variables. The trend of each of the 15 graphs is definitely horizontal and the fluctuations around this trend are small. We see that R/S testing for independence is blind to the extremely non-Gaussian character of the lognormal process. The kurtosis of the various samples ranged around 1000.

where the limit coordinate functions $U(t)$ and $V(t)$ are nondegenerate, are scaling with the respective exponents H' and H'', and are such that the sample functions are either almost surely continuous or right continuous with left limits, so that they belong to the function space $D(0, \infty)$. Under these assumptions (see Mandelbrot [4]) X has the R/S exponent $J = H' - 0.5H'' + 0.5$, and the R/S prefactor is the slowly varying function $L(d) = L'(d)/[L''(d)]^{0.5}$. This result includes all the preceding ones as special cases; for example, the result that, when the X's are i.i.d. with $EX^2 = \infty$, X being in the domain of attraction of a stable process with index $0 < \alpha < 2$, $J = 0.5$.

References

[1] Hurst, H. E. (1951). *Trans. Amer. Soc. Civil Eng.* **116**, 770–808. (A great seminal paper.)

[2] Mandelbrot, B. B. (1963). *J. Bus. Univ. Chicago*, **36**, 394–419. (Reprinted in *The Random Character of Stock Market Prices*, P. H. Cootner, ed. MIT Press, Cambridge, Mass., 1964, pp. 297–337.)

[3] Mandelbrot, B. B. (1965). *C. R. Acad. Sci. Paris*, **260**, 3274–3277.

[4] Mandelbrot, B. B. (1975). *Zeit. Wahrscheinlichkeitsth.*, **31**, 271–285. (Full mathematical treatment of R/S analysis.)

[5] Mandelbrot, B. B. (1982). *The Fractal Geometry of Nature*. W. H. Freeman, San Francisco. (For a capsule review, see HAUSDORFF DIMENSION.)

[6] Mandelbrot, B. B. and Wallis, J. R. (1968). *Water Resour. Res.*, **4**, 909–918. (An introduction to the Hurst phenomenon.)

[7] Mandelbrot, B. B. and Wallis, J. R. (1969). *Water Resour. Res.*, **5**, 321–340. (A collection of data confirming the Hurst phenomenon.)

[8] Mandelbrot, B. B. and Wallis, J. R. (1969). *Water Resour. Res.*, **5**, 967–988. (A purely empirical study of R/S by computer simulation.)

[9] Taqqu, M. S. (1975). *Zeit. Wahrscheinlichkeitsth.*, **31**, 287–302.

Bibliography

Kottegoda, N. T. (1980). *Stochastic Water Resources Technology*. Macmillan, London. (This textbook refers to the developments since the original papers in the references.)

Mandelbrot, B. B. (1972). *Ann. Econ. Social Meas.* **1**, 257–288. (Introduction addressed to sophisticated but practically minded statisticians.)

(FRACTIONAL BROWNIAN MOTION AND GAUSSIAN NOISES
HYDROLOGY, STOCHASTIC
STABLE PROCESSES)

BENOIT B. MANDELBROT

HUYGENS, CHRISTIAAN

Born: April 14, 1629, in The Hague, Netherlands.

Died: June 8, 1695, in The Hague, Netherlands.

Contributed to: probability theory, mathematics, physics, astronomy.

Huygens, who was descended from a politically and artistically prominent family, enjoyed a splendid education. He showed remarkable technical and scientific gifts at an early age. From 1645 to 1649 he studied mathematics and law in Leiden and Breda. In 1655 he stayed for a few months in Paris and took a doctor's degree at Angers. With this opportunity he learned about Pascal* and Fermat's achievements in probability. Back in Holland he wrote a small treatise on probability—the first in history—*Van Rekeningh in Spelen van Geluck* (calculation in hazard games). He sent the treatise to Van Schooten, who was glad to incorporate it into a work he was just preparing to be published in Latin and Dutch (1657 and 1660, respectively), in the Latin version under the title *De ratociniis in ludo aleae*. An anonymous work of 1692 *Of the laws of chance . . .*, probably by John Arbuthnot, contains a translation of Huygens' treatise. Pierre Rémond de Montmort (1708) and Abraham de Moivre* (1711) were certainly acquainted with Huygens' treatise. James Bernoulli* inserted it, with numerous comments, as a first part in an incompleted manuscript, *Ars Conjectandi*, which was posthumously published in 1713.

Although for half a century the only work on probability, Huygens' treatise, compared with his achievements in mathematics, physics, and astronomy, is only a minor work. Nevertheless, it shows some remarkable features.

The first chapters constitute an axiomatic introduction (as it were) to probability. Huygens founded probability on what is now called *expectation*. The term "expectatio" was introduced by Van Schooten in his Latin version, albeit in the sense of payoff table of a game; our "expectation" is indicated in Van Schooten's text by terms that translated would read "value of expectation."

Indeed, it is Huygens' fundamental question to ask for the value of the prospect of receiving payments $a_1, \ldots, a_n$, which are equally likely. (This formulation is of course ours.) The answer is obtained by a remarkable transformation: the payoff table is replaced with an equitable n-persons game with possible outputs $a_1, \ldots, a_n$; by definition, the value of the payoff table equals the stake required to participate in that n-persons game.

To explain Huygens' procedure, consider the case $n = 2$. The payoff table consists of the payments a and b, which are equally likely to be won. Instead, let two persons play an equitable game with the stake u while agreeing that the winner will pay the amount b to the loser (which is again an equitable stipulation). This means that the winner will actually earn $2u - b$, while the loser earns b. In order for the winner to get the required a, we shall put $u = (a + b)/2$, which is the stake of the game, and hence by definition the value of the payoff table.

In a similar way Huygens deals with the case $n = 3$, and then the case of a payoff table that grants p times the payment a and q times the payment b, with of course $(pa + gb)/(p + q)$ as its value.

After this introduction Huygens continues with a number of cases of "le problème des partis" (sharing the stakes if a sequence of games is interrupted prematurely) and an inductive rule how to solve the general case. Next in order, a large number of dice problems are proposed and solved. The treatise finishes with five problems for the reader, the last of which concerns a game of virtually infinite duration.

Huygens' probabilistic work does not contain mathematical statistical elements. The first such arguments are found in a note (1712) of John Arbuthnot (refuting the equiprobability of male and female births) and in James Bernoulli's *Ars Conjectandi* (estimating the ratio of black and white balls in a bag).

Bibliography

Bernoulli, Jacob. (1975). Die Werke, Vol. III. Basel.

Freudenthal, H. (1980). Huygens' Foundation of Probability. *Historia Math.*, **7**, 113–117.

Huygens, C. (1914). Oeuvres complètes, Vol. XIV. The Hague.

(BERNOULLIS, THE
DE MOIVRE, ABRAHAM
FERMAT, PIERRE DE
PASCAL, BLAISE)

HANS FREUDENTHAL

HYDROLOGY, STOCHASTIC

In a broad sense, the science of hydrology is concerned with the waters of the earth, their occurrence, circulation and distribution, physical and chemical properties, and interaction with the environment, including all forms of life. However, the notion of *stochas-*

tic hydrology* derives from the more common narrower concept of hydrology as a science dealing with the quantitative aspects of the processes governing the depletion and replenishment of water resources of the land areas of the earth. These processes comprise the hydrologic cycle and include precipitation, evaporation, evapotranspiration, infiltration and percolation, groundwater flow, and surface runoff, whose most important component is the streamflow. Their quantitative aspects involve mathematical descriptions of the behavior of the relevant variables, e.g., streamflow discharge rate Q (in cubic meters per second) or precipitation intensity P (in millimeters per hour), in time and/or space, and of their mutual relationships.

The prevailing view is that whenever some variables or parameters in mathematical formulations of hydrologic processes or relationships are defined as variates* (random variables*), the formulations belong under the label of stochastic hydrology. The usage, however, is not uniform. Some authors restrict its meaning to stochastic process* formulations or even further to a stochastic process representation of streamflow [8]. Some hold that it is merely a manipulation of statistical characteristics of hydrologic variables with the aim of quantification of risk or reliability associated with various water management and engineering projects (water supply schemes, hydropower plants, irrigation and flood protection systems, dams, etc.) for which hydrologic variables represent part of the inputs; others see it chiefly as a stochastic theory of the hydrologic cycle (discussed in [10]).

HISTORIC OUTLINE

The impetus for the emergence of stochastic hydrology was the conflict between the virtual unpredictability of streamflow for long periods of time ahead (months, years) and the importance of knowledge of future streamflow for the safe design of bridges across rivers, levees protecting low-lying areas against flooding, and for the planning of adequate water supply and hydropower facilities.

The first result was the introduction of probability distributions of hydrologic variables via frequency analysis, first applied to peak flow rates of floods (e.g., the concept of "100-year flood" is defined from the cumulative distribution function* of maximum annual flows as an annual maximum with a 1% probability of exceedance) and then to average flow rates and rainfall intensities per interval (usually hour, day, month, or year). Hydrological applications of frequency analysis and other statistical techniques that ignore the sequential order of observations of a given variable are often labeled statistical hydrology*. This term is also used as a synonym for stochastic hydrology.

The next step was the introduction of sequential order of observations into the analysis, first done with streamflow series used for the design of storage capacity of a reservoir that is to deliver a prescribed rate of continuous release ("safe draft," "target draft," etc.) over a given period of time. Rippl [16] observed as early as 1883 that the contemporary method of computing the required storage (as the deficit that would be incurred under the given target draft in the driest year of the record) was wrong, because a succession of several less dry years could lead to a larger cumulative deficit. This established the importance of sequential order of annual flows for storage design and, in 1914, led Hazen [7] (the inventor of probability paper*) to estimate a "probability of dry year" (a year in which a reservoir supplying a given target draft would become empty so that a failure in water supply would result; modern literature on storage theory employs the term *failure year* rather than dry year) from a long sequence of arbitrarily arranged standardized historic streamflow records from several rivers. This triggered two lines of development, both aimed at storage reservoir design: (1) stochastic modeling of streamflow series to be used as reservoir inflows, and (2) stochastic models of storage reservoirs.

The first line started in the 1920s with the generation of synthetic annual streamflows as a random series constructed by random sampling without replacement from a distribution of mean annual flows. The concept was received with suspicion by the engineering profession and stagnated for about three decades. Interest was revived in the 1950s with the advent of the digital computer; the introduction of series generation via sampling with replacement falls into this period. The second line started in the 1930s and by the end of the 1950s produced several procedures for the estimation of the probability distribution of storage via storage-state transition probabilities. The latter were derived from the probability distribution of annual flows (estimated from a historic flow record) and a chosen release rule on the assumption of independence of annual flows, and later for serially correlated annual flows. The classical example for the independent case is the Moran model for discrete approximation of stationary storage distribution in a finite reservoir by an m-dimensional vector

$$\mathbf{P} = \mathbf{p}\mathbf{P}$$

where $\mathbf{p}$ is an $m \times m$ matrix of transition probabilities [11, 13]. The stochastic theory of storage reservoirs has developed into an independent mathematical discipline with emphasis on explicit results rather than hydrological applications; *see also* DAM THEORY.

The 1960s brought a keen interest in the probabilistic structure of the sequential dependence of streamflow time series and with it time-series analysis* [9], namely the application of autocorrelation, spectral analysis*, and analysis of the range of cumulative departures from the mean (started by Hurst in the 1950s—see refs. 10 and 12). These techniques were applied mainly to mean annual, monthly, and daily flows and later extended to precipitation and other hydrologic variables [6, 17, 18]. Time-series analysis significantly improved stochastic models for the generation of synthetic series, in the sense that the former high degree of arbitrariness in the selection of their sequential structure was replaced by selection on the basis of significant statistics of the historic record. Preservation of these statistics in the generated sequences was the sole aim of the model and the only criterion of its adequacy. Models of this kind are *operational* (prescriptive, empirical) *stochastic hydrologic models*. Their only purpose is to represent the uncertainty inherent in hydrologic data in a form suitable for an approximate quantification of various hydrological risks (*see* DECISION THEORY). The stochastic structure of these models has no intrinsic hydrologic meaning; they are basically statistical interpolation* formulae of hydrologic data and extrapolation (e.g., for generating synthetic series much longer than the underlying historic series, or for estimation of probabilities of rare events far outside the range of observations) has no hydrologic justification.

Operational stochastic hydrologic models are usually divided into short-memory and long-memory models. An example of the former is the so-called Thomas–Fiering model for series of mean monthly (or seasonal) flows,

$$x_{i,j} = \mu_j + \rho_j \frac{\sigma_j}{\sigma_{j-1}} (x_{i-1,j-1} - \mu_{j-1}) + \epsilon_i \sigma_j \left(1 - \rho_j^2\right)^{1/2},$$

where i is the sequential number of flow x; j is number of the month (or season) within the annual cycle; μ_j and σ_j are the mean and standard deviation, respectively, of mean flow x_j in jth month (season); ρ_j is the correlation coefficient between flows x_j and x_{j-1}; and $\{\epsilon_i\}$ is a standardized independent sequence of given distribution [5, 12].

An example of a long-memory model (applicable to synthesis of annual series) is discrete fractional Gaussian noise* [12, 14], whose variates are defined as

$$x_t = (H - 0.5) \sum_{u=t-M}^{t-1} (t - u)^{H-1.5} \epsilon_u,$$

where M is a memory constant analogous to the order of a moving-average process, ϵ_u is a Gaussian variate with zero mean and unit variance, and H is the Hurst coefficient* (in hydrological series $0.5 < H < 1$).

Both short- and long-memory models were extended to multivariate models capable of generating synchronous synthetic series of several hydrologic variables (e.g., streamflows in several gauging stations along one or more rivers in a given area or precipitation and runoff series, etc.) while preserving the cross-correlation of the historic records [12, 15].

During the first half of the 1970s, the repertoire of operational stochastic hydrologic models was increased by the introduction of fast fractional Gaussian noise (*see* FRACTIONAL BROWNIAN MOTIONS), fractional non-Gaussian noise, broken-line models, and the new class of autoregressive-moving average* (ARMA) and autoregressive-integrated moving average* (ARIMA) models [1, 12]. The introduction of systems analysis* deserves special attention, and the concept of transfer function*, which made it possible to model sequentially the transformation of one hydrologic series into another (e.g., precipitation series into streamflow series). As an extension of regression it incorporates the dimension of time. This expanded the scope of stochastic hydrology from simulation* into real-time forecasting* of hydrological time series. While stochastic transfer function models have been used primarily to model dependence between causally related hydrologic variables, they are nevertheless operational models with no intrinsic hydrologic value.

The second half of the 1970s saw the introduction of many new mathematical techniques, e.g., intervention analysis*, Kalman filter*, kriging*, event-based modeling, ridge regression*, and cluster analysis. An important new development has been the introduction of stochastic space-time processes* and the theory of random fields*, which have been applied to the design and analysis of hydrological networks and to the modeling of areally distributed processes such as precipitation, groundwater levels, snow cover, etc.; these traditionally have been treated as point processes* at the respective gauging sites. Space–time analysis is a generalization of multi-site modeling.

The 1970s also witnessed the emergence of *physically based stochastic hydrology*, which strives to derive stochastic properties of a given hydrologic variable (e.g., streamflow) from stochastic properties of its causative factors (precipitation, evapotranspiration, etc.) and the physical mechanisms governing the transformation. An early prototype of such a model is the stochastic storage model, where the properties of the storage process (probability distribution, sequential structure) are derived from those of the inflow using a release rule representing the physical mechanism of reservoir operation. The model can serve, for example, as a crude representation of a river basin if the reservoir is identified with the basin, reservoir inflow with the (stochastic) precipitation process and the release rule with the dynamic equation(s) governing the movement of water through the basin. The output of the model is the basin runoff, whose stochastic properties are a function of those of the precipitation process, of the parameters of the basin, and of the release rule.

For example [10], if the precipitation (after subtraction of evaporation and other losses) is a random series x_t, $t = 1, 2, \ldots$, with mean $\mu(X)$, variance $\sigma^2(X)$ and coefficient of skewness $\gamma(X)$, and if the dynamic equation of the basin can be represented by the linear release rule $y_t = as_t$, where y and s are runoff and basin storage, respectively, and a is a constant $0 < a < 1$, then the runoff is the first-order autoregressive process with an autocorrelation function

$$\rho_{yy}(t) = \left(\frac{1}{a+1}\right)^t,$$

$$\text{mean } \mu(Y) = \mu(X),$$

$$\text{variance } \sigma^2(Y) = \frac{a}{a+2}\sigma^2(X),$$

and coefficient of skewness

$$\gamma(Y) = \frac{a^{1/2}(a+2)^{3/2}}{a^2+3a+3}\gamma(X).$$

Physically based stochastic hydrologic models of greater sophistication (based, for instance, on a kinematic wave approximation of the momentum equation, storage reservoirs with nonlinear release rules, cascades of reservoirs; see refs. 3, 4 and 10) have been suggested for the derivation of probability distributions of peak annual discharges and annual water yield, for a possible explanation of the Hurst phenomenon, negative skewness of probability distributions of streamflows observed in some rivers, and for other hydrological problems.

NUMERICAL EXAMPLE FOR THOMAS–FIERING MODEL

The following is adapted from Clarke [2]. Mean monthly flows, standard deviations of monthly flows, and correlation coefficients between each month's flow and that of the preceding month, at a gauging station on an Ethiopian river have been estimated from historic flow records as shown in Table 1.

From these data, estimates of regression coefficients, $\beta_j = \rho_j \sigma_j / \sigma_{j-1}$, $j = 1, 2, \ldots, 12$, are

$$\begin{array}{ll}
\beta_1 = 0.520 & \beta_2 = 0.865 \\
\beta_3 = 1.769 & \beta_4 = 0.405 \\
\beta_5 = 0.403 & \beta_6 = -0.022 \\
\beta_7 = 0.763 & \beta_8 = 0.999 \\
\beta_9 = 0.729 & \beta_{10} = 0.093 \\
\beta_{11} = 0.053 & \beta_{12} = 0.681
\end{array}$$

and those of the conditional standard deviations, $\sigma_{j|j-1} = \sigma_j(1 - \rho_j^2)^{1/2}$, $j = 1, 2, \ldots, 12$, are

$$\begin{array}{ll}
\sigma_{1|12} = 3.192 & \sigma_{2|1} = 2.089 \\
\sigma_{3|2} = 23.184 & \sigma_{4|3} = 5.994 \\
\sigma_{5|4} = 8.959 & \sigma_{6|5} = 23.939 \\
\sigma_{7|6} = 54.753 & \sigma_{8|7} = 103.617 \\
\sigma_{9|8} = 171.888 & \sigma_{10|9} = 66.982 \\
\sigma_{11|10} = 3.905 & \sigma_{12|11} = 3.546
\end{array}$$

so that the model consists of the following 12 equations:

$$\begin{aligned}
x_{i,1} &= 15.70 + 0.520(x_{i-1,12} - 18.01) \\
&\quad + 3.192\epsilon_i \\
x_{i+1,2} &= 13.62 + 0.865(x_{i,1} - 15.70) \\
&\quad + 2.089\epsilon_{i+1} \\
x_{i+2,3} &= 26.21 + 1.769(x_{i+1,2} - 13.62) \\
&\quad + 23.184\epsilon_{i+2} \\
x_{i+3,4} &= 22.25 + 0.405(x_{i+2,3} - 26.21) \\
&\quad + 5.994\epsilon_{i+3} \\
&\ \vdots \\
x_{i+11,12} &= 18.01 + 0.681(x_{i+10,11} - 20.84) \\
&\quad + 3.546\epsilon_{i+11}\,.
\end{aligned}$$

If, for instance, the generated December flow for the third year (i.e., the 36th term) of a synthetic series where $x_{36,12} = 18.01$, and the next four terms of the random deviates were $\epsilon_{37} = 2.289$, $\epsilon_{38} = -0.445$,

Table 1

Month	j	Mean ($m^3 \times 10^6$)	Standard Deviation ($m^3 \times 10^6$)	Correlation Coefficient
January	1	15.70	4.14	0.637
February	2	13.62	4.15	0.864
March	3	26.21	24.32	0.302
April	4	22.25	11.52	0.854
May	5	23.03	10.09	0.460
June	6	37.54	23.94	− 0.009
July	7	206.09	57.71	0.316
August	8	619.60	118.56	0.486
September	9	371.62	192.37	0.449
October	10	92.45	69.33	0.258
November	11	20.84	5.34	0.682
December	12	18.01	5.08	0.716

$\epsilon_{39} = -1.238$, $\epsilon_{40} = -0.841, \ldots,$ then the generated January, February, March, and April flows of the fourth year would be 23.0, 19.0, 7.0, and 9.4, respectively.

References

[1] Box, G. E. P. and Jenkins, G. M. (1970). *Time Series Analysis*. Holden-Day, San Francisco. (A classical basic textbook on ARMA, ARIMA, and transfer function models; very readable.)

[2] Clarke, R. T. (1973). *Mathematical Models in Hydrology*. Food and Agriculture Organization of the United Nations, Rome. [A clear, concise review of stochastic and deterministic hydrologic models (excluding storage models); contains simple solved numerical examples, FORTRAN program listings, and references.]

[3] Eagleson, P. S. (1972). *Water Resour. Res.* **8**, 878–897.

[4] Eagleson, P. S. (1978). *Water Resour. Res.* **14**, 705–776.

[5] Fiering, M. B. (1967). *Streamflow Synthesis*. Harvard University Press, Cambridge, Mass. (Gives the rationale of operational stochastic hydrologic modeling, a very readable elementary treatment of autoregressive models for streamflow series, and storage problem applications.)

[6] Freeze, R. A. (1979). *Water Resour. Res.*, **16**, 391–408.

[7] Hazen, A. (1914). *Trans. Amer. Soc. Civil Eng.*, **77**, 1539–1640.

[8] Kartvelishvili, N. A. (1976). *Soviet Hydrology: Selected Papers*, **15**, 110–113. (A brief nonmathematical outline of stochastic hydrology in the USSR. No references.)

[9] Kisiel, C. C. (1969). *Adv. Hydrosci.*, **5**, 1–119. [An excellent detailed (medium level) outline of time series analysis of hydrologic data, with emphasis on autocorrelation and spectral analysis. Extensive references.]

[10] Klemeš, V. (1978). *Adv. Hydrosci.*, **11**, 285–356. (A review of physically based stochastic hydrology. Extensive references.)

[11] Klemeš, V. (1981). *Adv. Hydrosci.*, **12**, 79–141. (A review of applied stochastic theory of storage. Extensive references.)

[12] Lawrence, A. J. and Kottegoda, N. T. (1977). *J. R. Statist. Soc. A*, **140** (Pt. 1), 1–47). (An excellent review of stochastic modeling of streamflow time series, with emphasis on the mathematical structure of the models. Extensive references.)

[13] Lloyd, E. H. (1967). *Adv. Hydrosci.* **4**, 281–339. (Gives a rigorous but relatively simple mathematical description of theoretical and numerical approaches to stochastic theory of storage. Many references.)

[14] Mandelbrot, B. (1977). *Fractals: Form, Chance and Dimension*. W. H. Freeman, San Francisco. (An excellent and highly readable exposition of the class of stochastic processes known as fractional motions and noises. Extensive references.)

[15] Matalas, N. C. (1967). *Water Resour. Res.*, **3**, 937–945.

[16] Rippl, W. (1883). *Proc. Inst. Civil Eng.*, **71**, 270–278.

[17] Todorovic, P. and Woolhiser, D. A. (1974). *J. Appl. Meteorol.*, **14**, 17–24.

[18] Todorovic, P. and Zelenasic, E. (1970). *Water Resour. Res.*, **6**, 1641–1648.

Bibliography

For recent developments in stochastic hydrology, consult the following journals: *Water Resources Research*, *Journal of Hydrology*, *Journal of Hydrological Sciences*, *Advances in Water Resources*.

Chin, C.-L., ed., (1978). *Applications of Kalman Filter to Hydrology, Hydraulics and Water Resources*. Department of Civil Engineering, University of Pittsburgh, Pittsburgh, Pa.

Chow, V. T., (1978). *Adv. Hydrosci.*, **11**, 1–93. (A review article on stochastic modeling of watershed systems.)

Fiering, M. B. and Jackson, B. B. (1971). *Synthetic Streamflows*. American Geophysical Union, Washington, D. C. [An elementary manual for design of operational autoregressive streamflow models for computer generation of time series (with numerical examples).]

Kartvelishvili, N. A. (1969). *Theory of Stochastic Processes in Hydrology and River Runoff Regulation*. Israel Program for Scientific Translations (available from U.S. Department of Commerce, Washington, D.C.). [A rigorous mathematical treatment; the theory of storage (referred to as river runoff regulation) is presented in terms of integral equations. Extensive references to Soviet sources.]

Kartvelishvili, N. A. (1975). *Stochastic Hydrology*. Gidrometeoizdat, Leningrad (in Russian). (A rigorous mathematical treatment of stochastic aspects of the streamflow process, with a detailed discussion of distributions of hydrologic variables, cyclicity, and maximum flows. Extensive references to Soviet sources.)

Klemeš, V. (1974). *Water Resour. Res.* **10**, 675–688. (A critique of long-memory stochastic hydrologic models.)

Kottegoda, N. T. (1980). *Stochastic Water Resources Technology*. Macmillan, London. (A comprehensive textbook on hydrologic time series analysis, statistical analysis; stochastic modeling; stochastic storage theory, systems analysis, and decision theory. Examples, extensive references.)

Mandelbrot, B. (1982). *The Fractal Geometry of Nature*. W. H. Freeman, San Francisco. (An expanded edition of Mandelbrot [14].)

Mandelbrot, B. and Wallis, J. R. (1968). *Water Resour. Res.* **4**, 909–918. (Exposition of the rationale for long-memory operational stochastic hydrologic models. For complete references and a review, see Lawrence and Kottegoda [12].)

Muzylev, S. V., Prival'skiy, V. E. and Ratkovich, D. Ya. (1982). *Stochastic Models in Engineering Hydrology*. Nauka, Moscow (in Russian). (A short account of stochastic hydrologic modeling techniques with emphasis on fluctuations of lake levels. Extensive references.)

Salas, J. D., Delleur, J. W., Yevjevich, V. and Lane, W. L. (1980). *Applied Modeling of Hydrologic Time Series*. Water Resources Publications, Littleton, Colo. (A comprehensive handbook containing the basic theory, numerical examples, and FORTRAN listings of many computer programs.)

Svanidze, G. G. (1980). *Mathematical Modeling of Hydrologic Series*. Water Resources Publications, Fort Collins, Colo. [Translation of a Russian original (published in 1977). A comprehensive summary of techniques used in the USSR, some little known in the West. Extensive references.]

Yevjevich, V. (1972). *Probability and Statistics in Hydrology*. Water Resources Publications, Fort Collins, Colo. (Elementary exposition of basic concepts; contains many engineering applications.)

Yevjevich, V. (1972). *Stochastic Processes in Hydrology*. Water Resources Publications, Fort Collins, Colo. (Elementary exposition of basic concepts, with many engineering applications. Draws heavily on research done under the author's supervision.)

(AUTOREGRESSIVE-INTEGRATED MOVING AVERAGE MODELS
DAM THEORY
FRACTIONAL BROWNIAN MOTIONS
GEOGRAPHY, STATISTICS IN
HURST COEFFICIENT
KRIGING
STORAGE THEORY
TIME SERIES
WEATHER MODIFICATION, STATISTICS IN)

V. KLEMEŠ

HYPERBOLIC DISTRIBUTIONS

The name *hyperbolic distribution* derives from the fact that for such a distribution the graph of the log-probability function is a hyperbola or, in several dimensions, a hyperboloid. In comparison, a normal log-probability function is representable by a parabola or a paraboloid. Therefore, the normal case is a limiting instance of the hyperbolic. Another limiting instance is that of a conical log-probability graph; in particular, the Laplace distribution* is obtainable as a limit of hyperbolic distributions.

Figure 1 shows a number of observed univariate distributions from various fields of investigation, together with the fitted hyperbolic distributions, determined by maximum likelihood*. In keeping with the geometric nature of the hyperbolic distributions, a logarithmic scale is used for the ordinate axes in this figure. Note that in ordinary nonlogarithmic plotting the tails of the hyperbolic distributions fall off exponentially.

The particle size distribution of a sand sample shown in Fig. 1*a* is typical of the size distributions found in aeolian sand deposits [2] and it originally motivated the study of the statistical properties of the one-dimensional hyperbolic distributions [4]. The extensions to several dimensions were briefly indicated in Barndorff-Nielsen [4, 5, 6] and followed up by Blaesild [13] (see also Blaesild and Jensen [15]) in connection with a reinvestigation of a classical example of a nonnormal bivariate distribution, namely Johannsen's measurements of the lengths and breadths of 9440 beans.

Actually, however, the isotropic version of the three-dimensional hyperbolic distribution turned up in statistical physics* much earlier, in 1911. One of the most basic derivations of the normal distribution is in the form of Maxwell's law for the joint distribution of the three components of the momentum (or velocity) vector for a single particle in an ideal gas, as established from the general Boltzmann's law and the formula for the energy of the particle in the gas. This derivation is based on Newtonian physics. But if the same calculation is made for Einstein's relativistic physics the resulting momentum distribution is three-dimensional hyperbolic (see Jüttner [25], Chandrasekhar [17], and Barndorff-Nielsen [8]). Incidentally, the Newtonian approximation to the relativistic solution illustrates the above-mentioned fact that the normal distribution occurs as a limit of hyperbolic distributions.

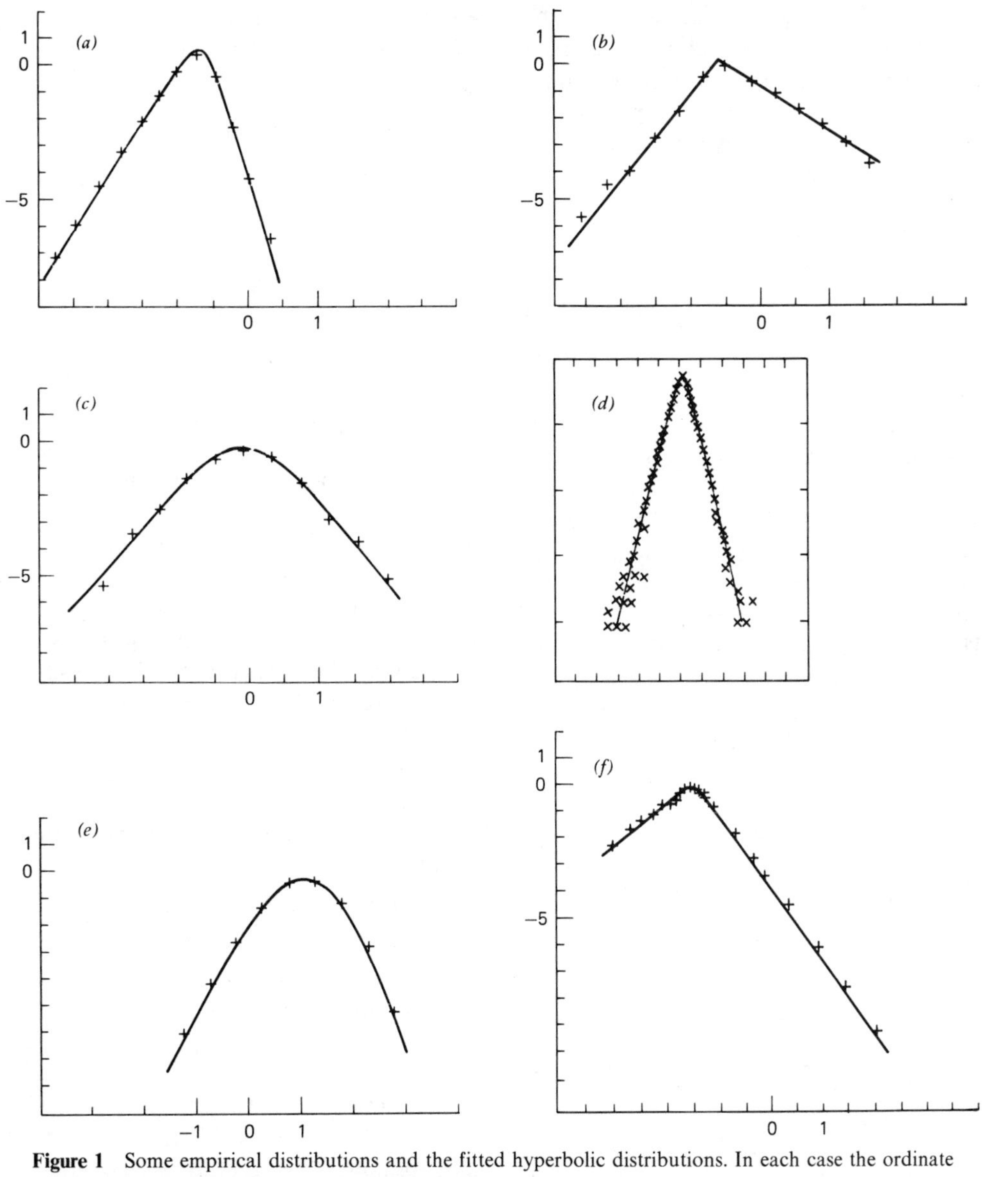

Figure 1 Some empirical distributions and the fitted hyperbolic distributions. In each case the ordinate scale is logarithmic, while the abscissa scale is logarithmic in cases where the basic measurements are positive [i.e., in (*a*), (*b*), and (*f*)]. The quantity n given in parentheses in the following indicates the sample size: (*a*) grain sizes, aeolian sand deposits ($n \simeq \infty$); (*b*) grain sizes, river bed sediment ($n \simeq \infty$); (*c*) differences between logarithms of duplicate determinations of content of gold per ore ($n = 1419$); (*d*) differences of streamwise velocity components in a turbulent atmospheric field of large Reynold number; distance between points of observation 42.8 cm ($n = 615.000$); (*e*) lengths of beans whose breadths lie in a fixed interval ($n = 2579$); (*f*) personal incomes in Australia, 1962–1963 ($n = 4.553.720$).

For applications of hyperbolic distributions to sedimentology and in turbulence, see Bagnold and Barndorff-Nielsen [3], Barndorff-Nielsen et al. [12], and Barndorff-Nielsen [4, 7].

A more technical discussion is given below, treating, in order, the family of one-dimensional hyperbolic distributions, the multivariate versions, and certain related distributions.

THE ONE-DIMENSIONAL HYPERBOLIC DISTRIBUTIONS

By the geometric interpretation of the hyperbolic distributions they fall off exponentially in the tails. They may be characterized by four parameters, two of which may be taken as a location parameter μ and a scale parameter δ. For $\mu = 0$ and $\delta = 1$ the probability (density) function of the hyperbolic distribution can be written

$$\left\{2\sqrt{1+\pi^2}\,K_1(\zeta)\right\}^{-1} \times \exp\left\{-\zeta\left[\sqrt{1+\pi^2}\sqrt{1+x^2}-\pi x\right]\right\}, \tag{1}$$

where $-\infty < x < \infty$, and ζ and π are parameters such that $0 < \zeta$ and $-\infty < \pi < \infty$. The standard Bessel function* notation K_ν is employed here and in the sequel. The parameter ζ is a measure of the distribution's peakedness; for fixed ζ the parameter π expresses the asymmetry of the distribution, symmetry occurring for $\pi = 0$.

The following three forms of the probability (density) function for the hyperbolic distribution with general location-scale parameter (μ, δ) are each of some interest:

$$\left\{2\delta\sqrt{1+\pi^2}\,K_1(\zeta)\right\}^{-1} \times \exp\left\{-\zeta\left[\sqrt{1+\pi^2}\sqrt{1+\left(\frac{x-\mu}{\delta}\right)^2}-\pi\frac{x-\mu}{\delta}\right]\right\}; \tag{2}$$

$$\left\{\sqrt{\alpha^2-\beta^2}\Big/\left[2\delta\alpha K_1\left(\delta\sqrt{\alpha^2-\beta^2}\right)\right]\right\} \times \exp\left\{-\alpha\sqrt{\delta^2+(x-\mu)^2}+\beta(x-\mu)\right\}; \tag{3}$$

$$\left\{\sqrt{\varphi\gamma}\Big/\left[\delta(\varphi+\gamma)K_1\left(\delta\sqrt{\varphi\gamma}\right)\right]\right\} \times \exp\left\{-\varphi/2\left[\sqrt{\delta^2+(x-\mu)^2}-(x-\mu)\right] -\gamma/2\left[\sqrt{\delta^2+(x-\mu)^2}+(x-\mu)\right]\right\}. \tag{4}$$

It appears from (3) and (4) that for any fixed values of the location-scale parameter (μ, δ) the hyperbolic distributions constitute a regular exponential family* with canonical parameter (α, β) or, alternatively, (φ, γ).

One consequence of this is that, since x is a component of the canonical statistic corresponding to (α, β), the mean, variance, and higher-order cumulants* of x are simply obtainable by differentiation with respect to β of the logarithm of the norming constant in (3). Thus

$$\begin{aligned} Ex &= \mu + \delta\pi R_1(\zeta), \\ Vx &= \delta^2\left\{\zeta^{-1}R_1(\zeta) + \pi^2 S_1(\zeta)\right\}; \\ R_\lambda(\zeta) &= K_{\lambda+1}(\zeta)/K_\lambda(\zeta), \\ S_\lambda(\zeta) &= \left\{K_{\lambda+2}(\zeta)K_\lambda(\zeta) - K_{\lambda+1}^2(\zeta)\right\}/K_\lambda^2(\zeta). \end{aligned} \tag{5}$$

The hyperbolic distributions form a subclass of the *generalized hyperbolic distributions* (see the section on related distributions below); a mathematical and numerical analysis in Barndorff-Nielsen and Blaesild [10] indicates that the domain of joint variation for the skewness* γ_1 and kurtosis* γ_2 of the generalized hyperbolic distributions is as indicated by the dashed curve in Fig. 2. This curve is given by $\gamma_2 = \frac{3}{2}\gamma_1^2$. Also shown in Fig. 2, by a solid curve, is the (γ_1, γ_2)-domain for the hyperbolic distributions themselves.

The connection between the mode point ν and the location μ of the hyperbolic distribution is given by

$$\nu = \mu + \delta\pi,$$

and the curvature of the log-probability function at the mode point is

$$\tau^2 = \delta^{-2}\zeta/(1+\pi^2).$$

This quantity is a measure of the peakedness of the distribution locally at ν. For a normal distribution the curvature would equal the inverse of the variance of the distribution, and if φ, γ, and δ all tend to infinity in such a way that τ^2 and $\varphi - \gamma$ remain fixed, then the hyperbolic distribution tends to the normal distribution with variance $\sigma^2 = \tau^{-2}$ and mean $\mu + (\varphi - \gamma)/(2\tau^2)$.

The discrepancy between the hyperbolic distribution and the normal distribution with the same mean and variance, in the case $\varphi = \gamma = 1$, $\mu = 0$, and $\delta = 1$, is shown in Fig. 3.

The parameters φ and γ [see (4)] are simply the slopes of the two linear asymptotes of the hyperbolic log-probability function.

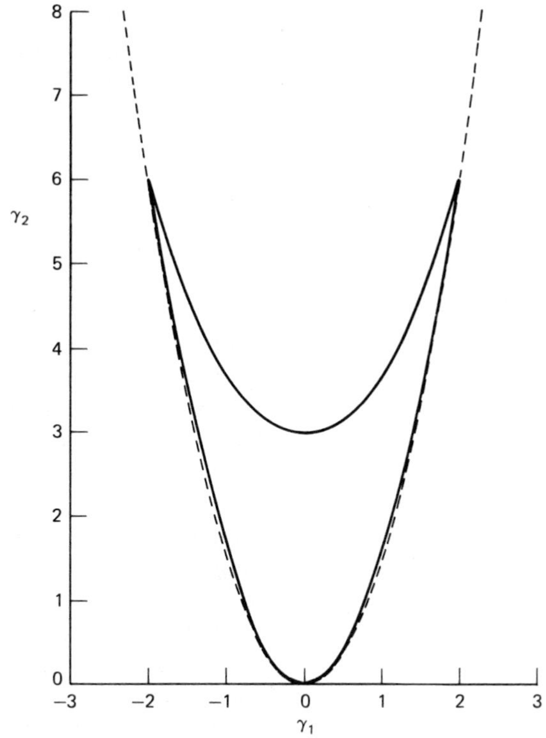

Figure 2 Domain of joint variation for the skewness γ_1 and kurtosis γ_2 of hyperbolic distributions (inside solid curve) and of the generalized hyperbolic distributions (inside dashed curve).

(Strictly speaking, the slope of the right-hand asymptote is $-\gamma$.) It may, furthermore, be noted that $\varphi = \alpha + \beta$, $\gamma = \alpha - \beta$, and

$$\beta/\alpha = \pi/\sqrt{1 + \pi^2}\ .$$

Figure 4 indicates geometrically the role of the various parameters.

The log-concavity of the probability function (1) or (2) means that the hyperbolic distribution is strongly unimodal. More important statistically, the positivity of the se-

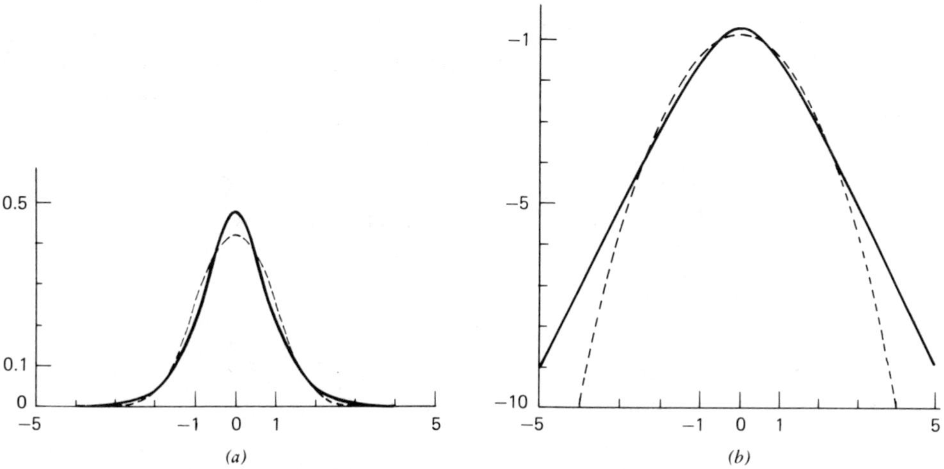

Figure 3 Probability functions (*a*) and log-probability functions (*b*) of the hyperbolic distribution (solid curve) with parameters $\mu = 0$, $\delta = 1$, $\varphi = \gamma = 1$, and the normal distribution (dashed curve) with the same mean and variance.

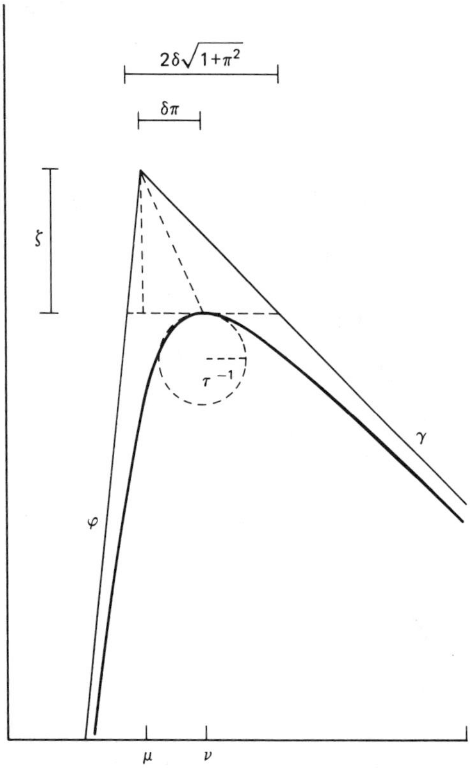

Figure 4 Log-probability function of a hyperbolic distribution with indication of the geometrical interpretation of the parameters. Note that $\zeta = \delta\sqrt{\varphi\gamma}$.

cond-order derivative of the log-probability function implies the existence and uniqueness of the maximum likelihood estimate in arbitrary linear models with a hyperbolic distribution (1) as error distribution, (see Barndorff-Nielsen and Blaesild [9] and Burridge [16]).

It has already been indicated how the normal distribution* and the Laplace distribution occur as limits of the hyperbolic distribution. The Laplace distribution with location parameter and scale parameter is obtained for $\pi = 0$ and $\delta \to 0$. For $\pi \neq 0$ and $\delta \to 0$ a "skew Laplace distribution" is found in the limit. Distributions of this kind have been considered by Hinkley and Revankar [19]. Yet another type of limiting distribution emerges if the slope φ of the left asymptote is allowed to tend to infinity while the right-hand slope γ, the mode ν, and the location μ are kept fixed. The resulting distribution is situated on the half-line (μ, ∞), and the distribution with $\mu = 0$ might be called the *positive hyperbolic distribution*. The probability (density) function of such a distribution can be written

$$\left\{2\sqrt{\chi/\psi}\,K_1\left(\sqrt{\chi\psi}\right)\right\}^{-1} \times \exp\left\{-\tfrac{1}{2}(\chi x^{-1} + \psi x)\right\}, \qquad (6)$$

the parameters χ and ψ being both positive. The distributions (6) belong to the wider class of generalized inverse Gaussian distributions* to be discussed briefly later.

It is a fact of some importance that the hyperbolic distribution is a mixture of normal distributions. Specifically, if x follows a normal distribution with variance w and mean of the form $\mu + \beta w$ and if w is endowed with the positive hyperbolic distribution (6) such that $\chi = \delta^2$ and $\psi = \alpha^2 - \beta^2$, the resulting distribution for x is given by (3). This mixture representation has been instrumental in proving that the hyperbolic distribution is self-decomposable (see Halgreen [18] and Shanbhag and Sreehari [28]). It is, moreover, possible to derive the hyperbolic distribution from the bivariate normal distribution by conditioning on a hyperbola (see Blaesild [14]). An efficient algorithm for simulating the hyperbolic distributions as well as a number of the other distributions mentioned in this article has been developed by Atkinson [1]. One general area of application of the hyperbolic distributions is in parametric robustness* studies. In this connection it seems of interest to note that the symmetric distribution corresponding to Huber's "most robust M-estimator*" (derived nonparametrically) is very nearly hyperbolic [20, 21].

MULTIVARIATE HYPERBOLIC DISTRIBUTIONS

Up to multivariate location-scale (or affine) transformations the probability function of the r-dimensional hyperbolic distribution may be expressed as

$$a_r(\zeta, \boldsymbol{\pi})\exp\left\{-\zeta\left[\sqrt{1+\boldsymbol{\pi}\cdot\boldsymbol{\pi}}\sqrt{1+\mathbf{x}\cdot\mathbf{x}} - \boldsymbol{\pi}\cdot\mathbf{x}\right]\right\}. \qquad (7)$$

The norming constant $a_r(\zeta, \pi)$ is given by

$$a_r(\zeta, \boldsymbol{\pi}) = \frac{\zeta^{(r-1)/2}}{(2\pi)^{(r-1)/2} 2\sqrt{1 + \boldsymbol{\pi} \cdot \boldsymbol{\pi}}\, K_{(r+1)/2}(\zeta)}, \tag{8}$$

ζ is a positive parameter, $\boldsymbol{\pi}$ is an arbitrary r-dimensional parameter vector, and $\cdot$ denotes the usual inner product. [The irrational number π in formula (8) should not be confused with the parameter $\boldsymbol{\pi}$.]

For even r an explicit expression is available for the Bessel function* $K_{(r+1)/2}$. In particular, the two-dimensional version of (8) is

$$a_2(\zeta, \boldsymbol{\pi}) = \left\{2\pi\sqrt{1 + \boldsymbol{\pi} \cdot \boldsymbol{\pi}}\right\}^{-1} \zeta^2 e^{\zeta}/(1 + \zeta).$$

The general form of the hyperbolic distribution (i.e., with arbitrary multivariate location-scale parameters introduced) is

$$|\boldsymbol{\Sigma}|^{-1/2} a_r(\zeta, \boldsymbol{\pi}) \exp\left\{-\zeta\left[\sqrt{1 + \boldsymbol{\pi} \cdot \boldsymbol{\pi}} \times \sqrt{1 + (\mathbf{x} - \boldsymbol{\mu})\boldsymbol{\Sigma}^{-1}(\mathbf{x} - \boldsymbol{\mu})'} - \boldsymbol{\pi}\boldsymbol{\Sigma}^{-1/2}(\mathbf{x} - \boldsymbol{\mu})'\right]\right\}; \tag{9}$$

or, writing δ^{2r} for the determinant $|\boldsymbol{\Sigma}|$ of the positive definite matrix $\boldsymbol{\Sigma}$ and setting $\boldsymbol{\Delta} = \delta^{-2}\boldsymbol{\Sigma}$, $\alpha = \delta^{-1}\zeta\sqrt{1 + \boldsymbol{\pi} \cdot \boldsymbol{\pi}}$, and $\boldsymbol{\beta} = \delta^{-1}\zeta\boldsymbol{\pi}\boldsymbol{\Delta}^{-1/2}$, the form is

$$\delta^{-r} a_r(\alpha, \boldsymbol{\beta}) \times \exp\left\{-\alpha\sqrt{\delta^2 + (\mathbf{x} - \boldsymbol{\mu})\boldsymbol{\Delta}^{-1}(\mathbf{x} - \boldsymbol{\mu})'} + \boldsymbol{\beta} \cdot (\mathbf{x} - \boldsymbol{\mu})\right\}, \tag{10}$$

where now

$$a_r(\alpha, \boldsymbol{\beta}) = \frac{(2\pi/\delta)^{-(r-1)/2}\left(\sqrt{\alpha^2 - \boldsymbol{\beta}\boldsymbol{\Delta}\boldsymbol{\beta}'}\right)^{(r+1)/2}}{\left\{2\alpha K_{(r+1)/2}\left(\delta\sqrt{\alpha^2 - \boldsymbol{\beta}\boldsymbol{\Delta}\boldsymbol{\beta}'}\right)\right\}}.$$

For fixed $\boldsymbol{\mu}$, δ, and $\boldsymbol{\Delta}$ the distributions (10) constitute a regular exponential family of order $r + 1$, and the cumulants of $\mathbf{x}$ may be derived by differentiating $\log a_r(\alpha, \boldsymbol{\beta})$ with respect to the coordinates of $\boldsymbol{\beta}$. Consequently, using the notations (5),

$$E\mathbf{x} = \boldsymbol{\mu} + \delta R_{(r+1)/2}(\zeta)\boldsymbol{\pi}\boldsymbol{\Delta}^{1/2}$$

$$V\mathbf{x} = \delta^2\left\{\zeta^{-1}R_{(r+1)/2}(\zeta)\boldsymbol{\Delta} + S_{(r+1)/2}(\zeta)(\boldsymbol{\pi}\boldsymbol{\Delta}^{1/2})'(\boldsymbol{\pi}\boldsymbol{\Delta}^{1/2})\right\}.$$

If $(\mathbf{x}^{(1)}, \mathbf{x}^{(2)})$ is a partitioning of $\mathbf{x}$ into vectors of dimension p and q, respectively, and if—in an obvious notation—$\boldsymbol{\Delta}_{12} = \mathbf{0}$ and either $\boldsymbol{\beta}^{(1)} = \mathbf{0}$ or $\boldsymbol{\beta}^{(2)} = \mathbf{0}$, then $\mathbf{x}^{(1)}$ and $\mathbf{x}^{(2)}$ are uncorrelated, but not independent.

Furthermore, one has the following relation between the location $\boldsymbol{\mu}$ and the mode $\boldsymbol{\nu}$ of (9):

$$\boldsymbol{\nu} = \boldsymbol{\mu} + \delta\boldsymbol{\pi}\boldsymbol{\Delta}^{1/2}.$$

Conditioning on a subset of the coordinates of $\mathbf{x}$ yields again a hyperbolic distribution. Thus the conditional distribution of $\mathbf{x}^{(2)}$ given $\mathbf{x}^{(1)}$ is hyperbolic, with parameters

$$\alpha_{2\cdot 1} = \alpha|\boldsymbol{\Delta}_{11}|^{1/(2q)}, \qquad \boldsymbol{\beta}_{2\cdot 1} = \boldsymbol{\beta}^{(2)},$$

$$\boldsymbol{\mu}_{2\cdot 1} = \boldsymbol{\mu}^{(2)}(\mathbf{x}^{(1)} - \boldsymbol{\mu}^{(1)})\boldsymbol{\Delta}_{11}^{-1}\boldsymbol{\Delta}_{12},$$

$$\delta^2_{2\cdot 1} = |\boldsymbol{\Delta}_{11}|^{-1/q}\left\{\delta^2 + (\mathbf{x}^{(1)} - \boldsymbol{\mu}^{(1)}) \times \boldsymbol{\Delta}_{11}^{-1}(\mathbf{x}^{(1)} - \boldsymbol{\mu}^{(1)})'\right\}$$

$$\boldsymbol{\Delta}_{2\cdot 1} = |\boldsymbol{\Delta}_{11}|^{1/q}\left\{\boldsymbol{\Delta}_{22} - \boldsymbol{\Delta}_{21}\boldsymbol{\Delta}_{11}^{-1}\boldsymbol{\Delta}_{12}\right\}.$$

Consequently, the conditional mode of $\mathbf{x}^{(2)}$ as a function of $\mathbf{x}^{(1)}$ traces a hyperbolic surface of dimension q in the r-dimensional space of $\mathbf{x}$; in other words, the mode regression of $\mathbf{x}^{(2)}$ on $\mathbf{x}^{(1)}$ is hyperbolic. A linear mode regression occurs if and only if $\boldsymbol{\beta}^{(2)} = \mathbf{0}$.

On the other hand, the marginal distribution of $\mathbf{x}^{(1)}$ is not hyperbolic, but it does belong to the class of generalized hyperbolic distributions discussed in the next section.

As in the one-dimensional case, the multivariate hyperbolic distribution is representable as a mixture of normal distributions, and the representation can be used to prove that the hyperbolic distribution is infinitely divisible in any number of dimensions (see the following section).

For more details on the multivariate hyperbolic distribution, see Blaesild and Jensen [15].

RELATED DISTRIBUTIONS

Consider the distribution on the positive half-line with probability (density) function

$$\left\{2(\chi/\psi)^{\lambda/2}K_\lambda\left(\sqrt{\chi\psi}\right)\right\}^{-1}x^{\lambda-1}$$
$$\times\exp\left\{-\tfrac{1}{2}(\chi x^{-1}+\psi x)\right\}. \quad (11)$$

The class of these distributions as the parameters λ, χ, and ψ vary over the domain given by

$$\chi \geqslant 0, \psi > 0 \quad \text{for} \quad \lambda > 0$$
$$\chi > 0, \psi > 0 \quad \text{for} \quad \lambda = 0$$
$$\chi > 0, \psi \geqslant 0 \quad \text{for} \quad \lambda < 0$$

is the family of *generalized inverse Gaussian distributions*. It includes the inverse Gaussian distributions*, obtained for $\lambda = -\frac{1}{2}$, the gamma distributions*, and the positive hyperbolic distributions (6). An extensive investigation of this family is available in Jørgensen [24]. In the present context interest in the family stems mainly from its occurrence in the following normal mixtures.

Given w, let $\mathbf{x}$ follow an r-dimensional normal distribution with mean vector $\boldsymbol{\mu} + w\boldsymbol{\beta}\boldsymbol{\Delta}$ and variance matrix $w\boldsymbol{\Delta}$. If w follows the generalized inverse Gaussian distribution (11) with $\chi = \delta^2$ and $\psi = \alpha^2 - \boldsymbol{\beta}\boldsymbol{\Delta}\boldsymbol{\beta}'$, then $\mathbf{x}$ has marginally a probability function which is expressible in terms of the Bessel functions K_ν and whose parameters are λ, α, $\boldsymbol{\beta}$, $\boldsymbol{\mu}$, δ, and $\boldsymbol{\Delta}$. The normal mixtures generated in this way are the r-dimensional *generalized hyperbolic distributions*. This class is a one-parameter extension of the hyperbolic distributions and it is closed under both conditioning and margining. The r-dimensional hyperbolic distributions obtain for $\lambda = (r+1)/2$. Another interesting class of r-dimensional distributions, the *hyperboloid distributions*, appears by setting $\lambda = (r-1)/2$. The probability (density) functions of these have an explicit form which for $(\boldsymbol{\mu}, \delta, \boldsymbol{\Delta}) = (\mathbf{0}, 1, \mathbf{I})$ is given by

$$a_r(\zeta)\sqrt{1+\mathbf{x}\cdot\mathbf{x}}^{\,-1}\exp\{-\zeta g(\mathbf{x};\boldsymbol{\pi})\}, \quad (12)$$

where

$$g(\mathbf{x};\boldsymbol{\pi}) = \sqrt{1+\boldsymbol{\pi}\cdot\boldsymbol{\pi}}\sqrt{1+\mathbf{x}\cdot\mathbf{x}} - \boldsymbol{\pi}\cdot\mathbf{x},$$
$$a_r(\zeta) = \frac{\zeta^{(r-1)/2}}{(2\pi)^{(r-1)/2}2K_{(r-1)/2}(\zeta)}.$$

The hyperboloid distributions (12) have properties very similar to, and in certain respects more tractable than, those of the von Mises–Fisher distributions* (see Jensen [22] and the references given there).

The generalized inverse Gaussian distributions are infinitely divisible; this implies that the same holds for the generalized hyperbolic distributions (see Barndorff-Nielsen and Halgreen [11]). The infinite divisibility of the Student distribution is a particular instance, since the class of one-dimensional generalized hyperbolic distributions includes the Student distributions (*see* t-DISTRIBUTIONS).

Specifically, the Student distributions result from the mixture construction described above by taking ψ in (11) equal to 0, i.e., by letting w have the distribution of the reciprocal of a gamma variate. If, instead, χ is set to 0, so that w is endowed with a gamma distribution, one obtains a class of distributions first investigated by McKay [26] (see the discussion in Johnson and Kotz [23, Chap. 12, Sec. 4.4]). Sichel [29] arrived at the same distributions by the normal mixture construction, in a paper concerned with the size distributions of diamonds excavated from marine deposits in South West Africa. Earlier, Rowland and Sichel [27] had applied the symmetric distributions from this class in setting up a statistical quality control* of underground gold ore valuation.

An appropriate alternative name for the generalized inverse Gaussian distributions would be *the positive generalized hyperbolic distributions*, for they are obtainable as limits of generalized hyperbolic distributions in a manner analogous to that mentioned previously for the hyperbolic case $\lambda = 1$.

The tail behavior of any one-dimensional generalized hyperbolic distribution is of the form $ax^be^{-c|x|}$, for some constants a, b, and c. Another type of generalization of the hyperbolic distributions, for which the tail behavior is as $ae^{-b|x|^c}$, has been considered in connection with turbulence by Barndorff-Nielsen [7].

References

[1] Atkinson, A. C. (1979). The Simulation of Generalised Inverse Gaussian, Generalised Hyperbolic, Gamma and Related Random Variables. *Res. Rep. 52*, Dept. Theoretical Statistics, Aarhus University, Aarhus, Denmark.

[2] Bagnold, R. A. (1941). *The Physics of Blown Sands and Desert Dunes*. Methuen, London.

[3] Bagnold, R. A. and Barndorff-Nielsen, O. (1980). *Sedimentology*, **27**, 199–207.

[4] Barndorff-Nielsen, O. (1977). *Proc. R. Soc. Lond. A*, **353**, 401–419.

[5] Barndorff-Nielsen, O. (1977). In *Scand. J. Statist.*, **4**. (A contribution to the discussion of D. R. Cox's paper on significance tests, pp. 49–70.)

[6] Barndorff-Nielsen, O. (1978). *Scand. J. Statist.*, **5**, 151–157.

[7] Barndorff-Nielsen, O. (1979). *Proc. R. Soc. Lond. A*, **368**, 501–520.

[8] Barndorff-Nielsen, O. (1982). *Scand. J. Statist.*, **9**, 43–46.

[9] Barndorff-Nielsen, O. and Blaesild, P. (1980). Global Maxima, and Likelihood in Linear Models. *Res. Rep. 57*, Dept. Theoretical Statistics, Aarhus University, Aarhus, Denmark.

[10] Barndorff-Nielsen, O. and Blaesild, P. (1981). In Taillie, C., Patil, G. P. and Baldessari, B. A. (Eds.), *Statistical Distributions in Scientific Work*, Vol. 4. D. Reidel, Dordrecht, Holland, pp. 19–44.

[11] Barndorff-Nielsen, O. and Halgreen, C. (1977). *Z. Wahrscheinlichkeitsth. verw. Geb.*, **38**, 309–312.

[12] Barndorff-Nielsen, O. Dalsgaard, K., Halgreen C., Kuhlman, H., Møller, J. T., and Schou, G. (1982). *Sedimentology*, **29**, 53–65.

[13] Blaesild, P. (1981). *Biometrika*, **68**, 251–263.

[14] Blaesild, P. (1979). *Ann. Statist.*, **7**, 659–670.

[15] Blaesild, P. and Jensen, J. L. (1981). In Taillie, C., Patil, G. P. and Baldessari, B. A. (Eds.), *Statistical Distributions in Scientific Work*, Vol. 4. D. Reidel, Dordrecht, Holland, pp. 45–66.

[16] Burridge, J. (1980). *J. R. Statist. Soc. B*, **43**, 41–45.

[17] Chandrasekhar, S. (1957). *An Introduction to the Study of Stellar Structure*. Dover, New York.

[18] Halgreen, C. (1979). *Z. Wahrscheinlichkeitsth. verw. Geb.*, **47**, 13–18.

[19] Hinkley, D. V. and Revankar, N. S. (1977). *J. Econometrics*, **5**, 1–11.

[20] Huber, P. J. (1964). *Ann. Math. Statist.*, **35**, 73–101.

[21] Huber, P. J. (1977). *Robust Statistical Procedures*. SIAM, Philadelphia.

[22] Jensen, J. L. (1981). *Scand. J. Statist.*, **8**, 193–206.

[23] Johnson, N. L. and Kotz, S. (1970). *Continuous Univariate Distributions*, Vol. 1. Wiley, New York.

[24] Jørgensen, B. (1982). Statistical Properties of the Generalized Inverse Gaussian Distribution. *Lecture Notes in Statistics*, **9**. Springer-Verlag, Heidelberg, W. Germany.

[25] Jüttner, F. (1911). *Ann. Phys.*, **34**, 856–882.

[26] McKay, A. T. (1932). *Biometrika*, **24**, 39–44.

[27] Rowland, R. St. J. and Sichel, H. S. (1960). *J. S. Afr. Inst. Min. Metall.*, **60**, 251–284.

[28] Shanbhag, D. N. and Sreehari, M. (1979). *Z. Wahrscheinlichkeitsth. verw. Gebiete*, **47**, 19–26.

[29] Sichel, H. S. (1973). *J. S. Afr. Inst. Min. Metall.*, **73**, 235–243.

(EXPONENTIAL FAMILIES
INFINITE DIVISIBILITY
INVERSE GAUSSIAN DISTRIBUTION
LAPLACE DISTRIBUTION
NORMAL DISTRIBUTION)

O. Barndorff-Nielsen
P. Blæsild

HYPERGEOMETRIC DISTRIBUTIONS

INTRODUCTION

Consider a situation for which the following conditions are satisfied:

(a) A set (lot, urn) contains N items.

(b) The N items are divided into two categories. Let k be the number having a characteristic so that $N - k$ do not have the characteristic.

(c) From the N items a simple random sample of size n is selected (sampling without replacement).

Let X be the number of items in the sample which have the characteristic. Using an elementary definition of probability and the fundamental counting principle, it is easy to show that

$$\Pr(X = x) = p(N, n, k, x) = \frac{\binom{k}{x}\binom{N-k}{n-x}}{\binom{N}{n}}, \quad a \leqslant x \leqslant b, \quad (1)$$

$a = \max[0, n - (N - k)]$, $b = \min[k, n]$. Conditions (a) to (c) are *hypergeometric conditions*, X is a *hypergeometric random variable*, and (1) is the hypergeometric probability (or density) function. One of the best known applications occurs in sampling inspection, with N being the number of items in a lot and k the number of defective items in that lot.

Applications usually require a sum of probabilities like

$$\Pr(X \leqslant r) = P(N, n, k, r) = \sum_{x=a}^{r} p(N, n, k, x). \quad (2)$$

Both (1) and (2) have been tabulated by Lieberman and Owen [5] (whose notation has been used) to six decimal places for $N = 2(1)50(10)100$ and all combinations of k, n. [A few probabilities are given for $N = 200(100)1000$.]

APPROXIMATIONS AND EXACT CALCULATION

Usually, the best way to evaluate (2) is to read it from the Lieberman and Owen table or to obtain it from a canned computer program. There are several approximations, some of which will now be discussed. Further information is given by Lieberman and Owen [5, pp. 16–22] and Johnson and Kotz [4, pp. 148–151].

Most of the simpler approximations are based on the binomial distribution*. Let

$$E(r; n, p) = \sum_{w=r}^{n} \binom{n}{w} p^{w}(1-p)^{n-w}. \quad (3)$$

Then, one approximation is obtained by replacing the terms of the hypergeometric by the corresponding terms of the binomial with $p = k/N$, getting

$$P(N, n, k, r) \simeq 1 - E(r+1; n, k/N), \quad (4)$$

which is reasonably good if $k \geqslant n$ and $n/N \leqslant 0.10$. It is easy to justify the reasonableness of (4) since, if sampling is done with replacement, the binomial would be appropriate. If $k < n$ and $k/N \leqslant 0.10$, then

$$P(N, n, k, r) \simeq 1 - E(r+1; k, n/N) \quad (5)$$

is usually better than (4). A more accurate approximation than either (4) or (5) is

$$P(N, n, k, r) \simeq 1 - E(r+1; n, p'), \quad (6)$$

where

$$p' = \frac{k - r/2}{M}, \qquad M = N - \frac{n-1}{2};$$

again, if $k < n$, interchange k and n.

The evaluation of the right-hand sides of (4), (5), and (6) may be done by using one of several well-known tables of the binomial. Unfortunately, k/N, n/N, p' may not be entries (the standard entries being multiples of 0.01) and interpolation* would be required. A better alternative is to use one of the modern desk or hand-held calculators which have binomial programs.

Other approximations based on the normal distribution* are discussed by Patel and Read [8]. Some of these appear to be quite accurate and several include bounds on the probability.

If a canned program for (2) is not available, it is relatively easy to prepare one. One procedure is to compute the first term and then to obtain successive terms recursively. If $a = 0$, $k < n$, for example, the first term is

$$\begin{aligned} p(N, n, k, 0) &= p(N, k, n, 0) \\ &= \frac{(N-n)\cdots(N-n-k+1)}{N\cdots(N-k+1)}, \end{aligned} \quad (7)$$

which can be evaluated by alternatively dividing and multiplying. Then successive terms can be found by using

$$p(N, n, k, x+1) = \frac{(n-x)(k-x)}{(x+1)(N-n-k+x+1)} p(N, n, k, x). \quad (8)$$

Algorithms for computing hypergeometric probabilities are given by Freeman [1] and Lund [6].

MOMENTS

Since all the probabilities in a probability distribution sum to 1, a special case of (2) is

$$P(N,n,k,b) = 1, \tag{9}$$

a result which, incidentally, provides a probability proof of the identity

$$\sum_{x=a}^{b} \binom{k}{x}\binom{N-k}{n-x} = \binom{N}{n}. \tag{10}$$

Use of (9) permits an easy method of obtaining moments. Thus

$$\begin{aligned} E(X) &= \sum_{x=a}^{b} xp(N,n,k,x) \\ &= n\frac{k}{N}P(N-1,n-1,k-1,b') \\ &= n\frac{k}{N}, \end{aligned} \tag{11}$$

where $b' = \min[k-1, n-1]$ (and a becomes $a' = \max[0, n-1-(N-1-k+1)]$). Next, $E(X^2) = E[X(X-1)] + E(X)$ and

$$\begin{aligned} E[X(X-1)] &= \frac{n(n-1)k(k-1)}{N(N-1)}P(N-2,n-2,k-2,b'') \\ &= \frac{n(n-1)k(k-1)}{N(N-1)}, \end{aligned} \tag{12}$$

where $b'' = \min[n-2, k-2]$. The obvious generalization of (12) yields

$$E[X(X-1)\cdots(X-r+1)] = E[X^{(r)}] = \frac{n^{(r)}k^{(r)}}{N^{(r)}} \tag{13}$$

(*see* FACTORIAL MOMENTS). Then (11) and (12) yield

$$\begin{aligned} \mathrm{var}(X) &= n\frac{k}{N}\frac{N-k}{N}\frac{N-n}{N-1} \\ &= np(1-p)\frac{N-n}{N-1}, \end{aligned} \tag{14}$$

where $p = k/N$.

POINT ESTIMATION

If N and n are known (as in the sampling inspection* setting mentioned earlier), it may be of interest to estimate k. From (11) it is obvious that

$$K = NX/n \tag{15}$$

is an unbiased estimator. (It is a minimum variance unbiased* estimator.) Since (15) may not be an integer, one may prefer to round it to the nearest integer. The maximum likelihood* estimator is also easily found by requiring that

$$p(N,n,k,x) \geqslant p(N,n,k-1,x)$$
$$p(N,n,k,x) \geqslant p(N,n,k+1,x).$$

These inequalities yield

$$k = \text{greatest integer in } \frac{x(N+1)}{n}. \tag{16}$$

If $x(N+1)/n$ is an integer, both this number and $x(N+1)/n - 1$ are maximum likelihood estimates.

Sometimes an estimate of N is sought when k and n are known. This situation may arise if N represents the number of creatures in a biological population (i.e., number of fish in a lake). In order to get an estimate, k creatures are caught, tagged, and then released. After waiting for the tagged creatures to disperse, a second sample of n creatures are observed of which x have tags. Assuming that the second sample is a simple random sample, the maximum likelihood estimate may be found by requiring that

$$p(N,n,k,x) \geqslant p(N-1,n,k,x),$$
$$p(N,n,k,x) \geqslant p(N+1,n,k,x).$$

The estimate is

$$N = \text{greatest integer in } \frac{kn}{x}, \qquad x \neq 0. \tag{17}$$

If kn/x is an integer, both this number and $kn/x - 1$ are maximum likelihood estimates. (*See* CAPTURE–RECAPTURE METHODS.)

CONFIDENCE INTERVALS

Suppose that N, n are known, x is observed, and a one-sided confidence interval* for k

of the type $(0, k_2)$ is sought having confidence coefficient at least $1 - \alpha$. Then the general method of obtaining a confidence interval (i.e., Mood et al. [7, pp. 389–393]) yields

$$k_2 = \min k \text{ satisfying } P(N, n, k, x) \leqslant \alpha. \quad (18)$$

Similarly, if an interval of the type (k_1, N) is desired, then

$$k_1 = \max k \text{ satisfying } P(N, n, k, x - 1) \geqslant 1 - \alpha. \quad (19)$$

Finally, if a two-sided interval (k_1, k_2) is to be found, use (18) and (19) with α replaced by $\alpha/2$.

If k, n are known, x is observed and a one-sided interval for N of the type $(0, N_2)$ is sought, (18) is replaced by

$$N_2 = \min N \text{ satisfying } P(N, n, k, x - 1) \geqslant 1 - \alpha, \quad (20)$$

and for an interval (N_1, ∞), replace (19) by

$$N_1 = \max N \text{ satisfying } P(N, n, k, x) \leqslant \alpha. \quad (21)$$

For a two-sided interval, use both (20) and (21) with α replaced by $\alpha/2$.

HYPOTHESIS TESTING FOR k

In acceptance sampling* it may be of interest to test

$$H_0 : k = k_0 \text{ against } H_1 : k > k_0, \quad (22)$$

subject to the conditions

$$\text{OC}\begin{cases} \geqslant 1 - \alpha_0 & \text{if } k = k_0, \\ \leqslant \beta_1 & \text{if } k = k_1 > k_0, \end{cases} \quad (23)$$

where OC = the operating characteristic* = 1 − power*. The acceptance region for (22) is $x \leqslant c$, and (23) can be written

$$\begin{aligned} P(N, n, k_0, c) &\geqslant 1 - \alpha_0 \\ P(N, n, k_1, c) &\leqslant \beta_1. \end{aligned} \quad (24)$$

Usually, the minimum n and accompanying c satisfying (24) will be desired. With a hypergeometric table this solution is quickly found by inspection using iteration on c.

A good approximation for solutions of (24) is obtained by computing a_1, a_2 for various choices of c, solutions existing for all n such that $a_2 \leqslant n \leqslant a_1$ makes sense. Here

$$a_1 = \frac{c(k_0 - \frac{1}{2}c) + \chi^2_{2c+2;\alpha_0}[N - \frac{1}{2}(k_0 - \frac{1}{2}c - 1)]}{2k_0 - c + \frac{1}{2}\chi^2_{2c+2;\alpha_0}}, \quad (25)$$

$$a_2 = \frac{c(k_1 - \frac{1}{2}c) + \chi^2_{2c+2;1-\beta_1}[N - \frac{1}{2}(k_1 - \frac{1}{2}c - 1)]}{2k_1 - c + \frac{1}{2}\chi^2_{2c+2;1-\beta_1}}, \quad (26)$$

and $\chi^2_{\nu;p}$ is the quantile* of order p for the chi-square distribution* with ν degrees of freedom. Again iterate on c until the minimum n is found.

Further discussion of hypothesis testing* with numerical examples can be found in the monograph by Guenther [3, pp. 15–24], which also contains information on hypothesis testing with curtailed sampling, double sampling*, and the sequential probability ratio test*.

THE INVERSE HYPERGEOMETRIC

Closely related to the hypergeometric is another distribution, known as the *inverse hypergeometric*, the negative hypergeometric, or the hypergeometric waiting-time distribution. This distribution appears in a number of places in the literature, but most researchers who encountered it failed to recognize it and consequently could not take advantage of its known properties.

Again assume that conditions (a) and (b) are satisfied, but replace (c) by:

(c′) Items are drawn one at a time in such a way that on each draw every remaining item has an equal chance of selection terminating when the cth item having the characteristic is obtained.

When the drawing produces an item having this characteristic, call the drawing a success. Now let X be the number of drawings required to obtain exactly c successes. Then

$$\begin{aligned}\Pr(X = x) = &\Pr(c-1 \text{ successes are obtained in the first } x-1 \text{ draws})\\ &\times \Pr(\text{a success occurs on the } x\text{th draw, given that } c-1 \text{ successes have already been obtained}).\end{aligned}$$

This is

$$p^*(N,k,c,x) = \frac{\binom{k}{c-1}\binom{N-k}{x-c}}{\binom{N}{x-1}} \frac{k-c+1}{N-x+1} \quad (27)$$

$$= \frac{\binom{x-1}{c-1}\binom{N-x}{k-c}}{\binom{N}{k}},$$

$$x = c, c+1, \ldots, N-k+c. \quad (28)$$

Letting $U = X - c$ = the number of failures obtained before the cth success occurs, the probability function of U is

$$p_1^*(N,k,c,u) = \frac{\binom{u+c-1}{c-1}\binom{N-u-c}{k-c}}{\binom{N}{k}},$$

$$u = 0, 1, \ldots, N-k, \quad (29)$$

a form frequently encountered in the literature.

Using the notation

$$P^*(N,k,c,r) = \sum_{x=c}^{r} p^*(N,k,c,x),$$

$$P_1^*(N,k,c,s) = \sum_{u=0}^{s} p_1^*(N,k,c,u),$$

it is easy to show with a brief probability argument that

$$P^*(N,k,c,r) = 1 - P(N,r,k,c-1), \quad (30)$$

$$P_1^*(N,k,c,s) = 1 - P(N,s+c,k,c-1). \quad (31)$$

Hence inverse hypergeometric probabilities can be obtained from hypergeometric probabilities.

Moments of the inverse hypergeometric can be found using the method that produced hypergeometric moments. It is easy to verify that

$$E(X) = c\frac{N+1}{k+1}. \quad (32)$$

Further information on the inverse hypergeometric is contained in a review paper by Guenther [2]. Some additional topics discussed are estimation of parameters, order statistics* for simple random sampling, distribution-free* prediction intervals, exceedances, and a multivariate hypergeometric distribution.

THE MULTIVARIATE HYPERGEOMETRIC DISTRIBUTION

Generalize the situation of condition (b) to read

(**b**′) The N items are divided into r categories containing $k_1, k_2, \ldots, k_r$ items respectively, where $k_1 + k_2 + \cdots + k_r = N$.

Then, let X_i be the number of items in the sample that have characteristic i, $i = 1, 2, \ldots, r$. The same argument which gave (1) yields

$$\Pr(X_1 = x_1, \ldots, X_r = x_r) = \frac{\binom{k_1}{x_1}\binom{k_2}{x_2}\cdots\binom{k_r}{x_r}}{\binom{N}{n}}, \quad (33)$$

where $x_i = 0, 1, \ldots, \min[n, k_i]$, $\sum_{i=1}^{n} x_i = n$. The fraction (33) is called the *multivariate hypergeometric probability function*. A few facts concerning this distribution are given by Patil and Joshi [9]; these authors also give a brief discussion of a multivariate "inverse hypergeometric" distribution and a multivariate "negative hypergeometric" distribution.

OTHER HYPERGEOMETRIC DISTRIBUTIONS

If in (1) the parameters N, n, k are allowed to take on any real values, then various cases of *generalized hypergeometric distributions** are obtained. Several other distributions containing the word "hypergeometric" are encountered in the literature; two of these are the "noncentral" hypergeometric and the "extended" hypergeometric. Some discussion and further references are given by Johnson and Kotz [4] and Patil and Joshi [9].

References

[1] Freeman, P. R. (1973). *Appl. Statist.* **22**, 130–133. (An algorithm for computing individual hypergeometric terms.)

[2] Guenther, W. C. (1975). *Statist. Neerlandica*, **29**, 129–144. (A review article containing a definition of the inverse hypergeometric, a number of properties of the distribution, and some practical applications.)

[3] Guenther, W. C. (1977). *Sampling Inspection in Statistical Quality Control*. Stat. Monogr. No. 37, Charles Griffin, London. (A quality control textbook which covers attribute and variables sampling plans, rectifying inspection, and tolerance intervals.)

[4] Johnson, N. L. and Kotz, S. (1969). *Distributions in Statistics, Discrete Distributions*. Wiley, New York. (A book covering the best known discrete distributions, their properties and uses, and many references.)

[5] Lieberman, G. J. and Owen, D. B. (1961). *Tables of the Hypergeometric Probability Distribution*. Stanford University Press, Stanford, Calif. (A table of individual and cumulative terms with an introduction containing useful information about the distribution.)

[6] Lund, R. E. (1980). *Appl. Statist.* **29**, 221–223. (An algorithm for computing cumulative hypergeometric sums.)

[7] Mood, A. M., Graybill, F. A., and Boes, D. C. (1974). *Introduction to the Theory of Statistics*. McGraw-Hill, New York. (A good intermediate-level text in mathematical statistics.)

[8] Patel, J. K. and Read, C. B. (1982). In *Handbook of the Normal Distribution*. Marcel Dekker, New York, Chap. 7. (A compendium of results relating to normal distributions, including normal approximations to the hypergeometric and other distributions.)

[9] Patil, G. P. and Joshi, S. W. (1968). *A Dictionary and Bibliography of Discrete Distributions*. Hafner, New York. (A book listing over 100 discrete distributions, a few facts about each, and some references to be used for more detailed information.)

(CAPTURE–RECAPTURE METHODS
ESTIMATION, POINT
FINITE POPULATIONS, SAMPLING FROM
GENERALIZED HYPERGEOMETRIC DISTRIBUTIONS
HYPOTHESIS TESTING
MULTIVARIATE HYPERGEOMETRIC DISTRIBUTIONS
URN MODELS)

WILLIAM C. GUENTHER

HYPERPARAMETER *See* CONJUGATE FAMILIES OF DISTRIBUTIONS

HYPOTHESIS TESTING

INTRODUCTION

Many problems lead to repetitions of an experiment with just two possible outcomes, e.g., (yes, no), (dead, alive), (response, no response), which reminds us of coin tossing, where we get either heads or tails. Our interest is usually the proportion p of responses. We may estimate p by the observed proportion in our experiments. Often, a theory or hypothesis says that p should have some specific value, and our purpose is to test this hypothesis. To fix our words, let us speak of coin tossing, where our hypothesis is that the coin is "fair" (i.e., that $p = \frac{1}{2}$). This is a *statistical hypothesis* since it governs the pattern of outcomes of a series of independent but identical experiments, Bernoulli trails*, in fact.

Common sense suggests that if x/n, the observed proportion of heads in n tosses or trials, is near $\frac{1}{2}$, this hypothesis gains support from our data, whereas if x/n is far from $\frac{1}{2}$, we will begin to doubt that $p = \frac{1}{2}$. Of course, random fluctuations can lead to values of x/n far from $\frac{1}{2}$ even if $p = \frac{1}{2}$. As n becomes larger, these become less likely, so we should then have a more powerful test. Probability theory enables us to sharpen these ideas. The number of heads, x, in n trials has a *Binomial* distribution*• which may be approximated by a *normal** or *Gaussian* distribution with mean np and variance npq, which equal $n/2$ and $n/4$, respectively, for the "null hypothesis" $p = \frac{1}{2}$. Thus, if the coin is fair, the chance that x will differ from $n/2$ by more than $1.96\sqrt{n/4}$ is fairly accurately $1/20$. For example, if $n = 100$, the probability that x is outside the interval $(50 - 1.96 \times 5, 50 + 1.96 \times 5)$ or roughly $(40, 60)$ when the coin is fair is about 5% or $\frac{1}{20}$. It has become commonplace in science to say that deviations of less than 2σ (2 is an approximation to 1.96 and here $\sigma = \sqrt{n/4} = 5$ is the standard deviation of the approximately normally distributed "statistic" x) are to be expected and so ignored, but that greater deviations are "significant" (i.e., worth looking into). There is nothing sacred about 5%. Scientists use this as a guide—not for making decisions, which depend upon many other factors.

If we have to reach a yes–no decision, where should we draw the line? Clearly, if we knew that p could never be less than $1/2$, we would only doubt the "$p = \frac{1}{2}$" hypothesis if x/n were suspiciously larger than $\frac{1}{2}$. A small x/n can only be due to random fluctuations. Thus our test must depend on the *alternative* hypothesis, as well as on the null hypothesis. Again, the (e.g., financial) consequences of saying erroneously that $p = \frac{1}{2}$ when $p \neq \frac{1}{2}$ might be much less than of saying erroneously that $p \neq \frac{1}{2}$ when in fact $p = \frac{1}{2}$. Then we would want to be very confident before we reject the null hypothesis. Hence the consequences of the decision must play a role. Finally, experiments cost time and money, so we want to make as few as are necessary to reach a "reliable" decision. With "reliable" clearly defined, how many observations are needed? It is clear that we can never be sure that our decision is right, but we can control the probability that is right. It is also evident that we must be careful if we look at the first data collected before deciding on n—one might be tempted to cheat.

The development of the ideas in the last paragraph comprises the subject matter of *hypothesis testing*, which is intimately connected with a large part of the theory of mathematical statistics, especially *estimation**. Most experiments have more than two outcomes; e.g., they may yield single or multiple measurements whose postulated probability distributions usually depend on several parameters, not just one as we had above. Null and alternative hypotheses may now be more complicated. It is then easy to devise various procedures for testing the same hypothesis. Hence we will need rules for selecting tests.

If one takes a Bayesian view of the general inference problem, hypothesis tests do not arise—all knowledge is summarized in (prior and posterior) probability distributions of the parameters (*see* BAYESIAN INFERENCE). Classical hypothesis testing seeks only the best control of the probabilities of the two kinds of erroneous decisions, illustrated above, when the null hypothesis and sample size are specified beforehand. In *decision theory**, explicit account is taken of the numerical consequences of decisions. *Sequential analysis** studies particularly efficient methods for terminating a sequence of experiments.

Nonparametric methods involve tests which are valid no matter what the distributions of the observations may be. *Multiple testing* theory began with ways of testing hypotheses suggested by the data rather than being given a priori. *Robustness** is concerned with tests that remain sensitive to departures from the null hypothesis when the distribution of the observations falls in some broad class, and so is a sharpened form of nonparametrics which tries to assume nothing.

EARLY IDEAS

As with so much statistics, the early examples are astronomical. An essay in 1734 by Daniel Bernoulli* won a prize at the French Academy, which set the following problem: Is the near coincidence of the orbital planes of the planets an accident or a consequence of their creation? Bernoulli invented a test statistic, i.e., a quantity that would be small if the planes were oriented randomly and independently of one another, and large if they were near coincidence. A modern improvement would use the length of the sum of the unit vectors which are perpendicular to each of the planes. Needless to say, in both cases the quantity is far larger than would arise often by chance if the orbits were in independent randomly oriented planes (*see* STATISTICS IN ASTRONOMY).

Although many tests were invented in the interim, the chi-square test* (Karl Pearson*, 1900 [15]), Student's t-test* (Gosset*, 1908a [16, Paper 2]), and Fisher's derivation of the distribution of the correlation* coefficient opened the modern era of statistics. Many important fields of application for statistics were found in the experimental sciences as well as the world of affairs, since in more and more areas, data were being consciously gathered to investigate specific hypotheses. Many of these experiments were small. Mathematical techniques became available to find the exact probability distributions of many quantities (i.e., statistics) computed from samples. It was then realized that there were many possible tests and estimators in every problem and so a problem of choice. To bring order out of chaos, it was necessary to seek general principles that would tell the statistician which method *should* be used in any given situation.

While Karl Pearson [14] in his book on the scientific method, *The Grammar of Science* (1892), had supported the use of Bayes' formula (the principle of inverse probability) to join previous and current data, he rarely used it. His chi-square statistic,

$$\chi^2 = \sum_{i=1}^{k} (x_i - np_i)^2/(np_i),$$

tests whether the numbers $x_1, \ldots, x_k$ of k types in a sample size n have been drawn from a population whose members are of k types in proportions $p_1, \ldots, p_k$. The expected value of x_i is np_i, so χ^2 will be large when at least some of the x_i's are far from their expectations when we might doubt the hypothesis. If the calculated value is χ_0^2, he suggested that one should look at P_0, the probability that samples of n from the multinomial $(p_1, \ldots, p_k)$ population would give values of χ^2 greater than χ_0^2. If we doubt the hypothesis given data yielding χ_0^2, we should also doubt it for data yielding any larger value of χ^2. He showed how to compute P_0, the size of which is a measure of the agreement between the data and the assumed distribution $(p_1, \ldots, p_k)$. If P_0 is small, we have two explanations—a rare event has happened, or the assumed distribution is wrong. This is the essence of the *significance test** argument. Not to reject the null hypothesis (that the proportions are $p_1, \ldots, p_k$) means only that it is accepted for the moment on a provisional basis. "Student" took the same approach for his t-statistic and for r, the correlation statistic (1908b [16, Paper 3]) with two important asides. First, he remarks that he would prefer to use inverse probability—but does not know how to set his priors. Second, he recognizes that actual samples may not come from normal distributions which he has had to assume, and gives a reason why his t-test* might yet be reliable for samples from some nonnormal populations.

By 1925 we find the subject, almost as one finds it today, in Fisher's highly influential book, *Statistical Methods for Research Workers* [7] (*see* FISHER, RONALD AYLMER). The hypothesis being tested, often the status quo (so to speak), is the *null hypothesis**. For the fair-coin example, this is $p = \frac{1}{2}$. If Pearson's chi-square were applied to testing that a six-sided die was fair, the null hypothesis would be that $p_1 = p_2 \cdots = p_6 = \frac{1}{6}$. Significance levels* of 5% and 1% are suggested. Thus, if the P_0 for chi-square is less than 0.05, it is said to be "significant." Fisher prepared tables to make these and many other significance tests easier to apply by computing the value of the statistic that would need to be exceeded in order that P_0 be less than 0.05, 0.01, etc. Such numbers are *significance points** of the relevant statistic. In his book, however, P-values have not entirely disappeared. It is said that Fisher tabulated significance points rather than P-values only because Karl Pearson held the copyrights to the P-value tables!

By 1925, Fisher had established his theory of point estimation, including the concepts of sufficiency*, consistency*, efficiency*, likelihood*, and maximum likelihood* estimation. But he never had a theory of significance test construction, although he invented or developed almost all the common tests, including some basic nonparametric tests (e.g., use of normal scores*, randomization*, and permutation tests*). All his life, Fisher objected in the strongest terms to the use of inverse probability, but clearly felt the need for something like it since he tried to develop his own substitute, fiducial probability*. Further, he always thought and wrote as a mathematician applying his skills in areas of basic natural science, but never, for example, in industry, so he was not attracted to decision theory*.

Theory, to assess and derive optimal significance tests, was provided in the next 10 years by Neyman and Egon Pearson. The story of their collaboration has been given by Pearson [12]. The key papers are conveniently collected in Neyman [10], Neyman and Pearson [11], and Pearson [13]. Pearson, like "Student," was interested in the use of statistical procedures to guide routine work in industry where rules of thumb were needed. Neyman brought a more philosophical and formal approach to the problem.

If, for example, the test for the fairness of a coin mentioned above is routinely applied (recall that the significance level chosen was 5%), and if "significant" means that we reject the null hypothesis $p = \frac{1}{2}$, then two kinds of errors may be made. Five percent of the time when p actually equals $\frac{1}{2}$, we will say it does not (i.e., reject the null hypothesis that $p = \frac{1}{2}$). However, when p is not equal to $\frac{1}{2}$ (e.g., $p = \frac{3}{4}$), we will sometimes get a number of heads x in n tosses which fall in the range

$$\left((n/2) - 1.96\sqrt{n/4}\right),$$

$$(n/2) + 1.96\sqrt{n/4}\ .$$

If this happens, we will not reject the hypothesis $p = \frac{1}{2}$ even though it is false. These are called, respectively, *errors of the first and second kinds*. Here, since we are setting up a method for routine application, we imagine that we will always make a definite assertion, or decision, every time. Hence we can talk about long-run rates of erroneous decisions. In nonroutine science, hypotheses do not stand or fall on the results of a statistical test, and this formulation is less appropriate. The only distinction between significance tests and hypothesis tests seems to be Fisher's insistence that there is one. Neither gives the weight of evidence for a hypothesis which is perhaps what one might like to have. Significance tests are usually derived or evaluated using the theory of hypothesis testing. It is somewhat ironic that, although Fisher was adamantly opposed to the latter, his tabulations of significant points and his later derivation of nonnull distributions set everything up for hypothesis-testing theory.

In their first paper (1928), Neyman and Pearson defined these errors and investigated an intuitive principle for test construction that yields likelihood ratio tests* [11]. The latter had appealed to them because Fisher had shown how to use likelihood to obtain "good" estimators and because it arises if one uses the Bayesian approach—which they were reluctant to do. The likelihood ratio principle (to be explained later) was extraordinarily fruitful; it enabled them to derive all the tests known, including Pearson's chi-square, from one rule.

But the nagging question—what is the *best* test?—remained. In 1928, they had implicitly formulated their criterion: Keep the rate of errors of the first kind constant, minimize the rate of errors of the second kind. By 1933, all the key definitions had been made to allow an orderly presentation whose centerpiece is the Neyman–Pearson lemma*. The subject called hypothesis testing has been developed on this basis ever since; the classic text is that of Lehmann [9]. Various other subdisciplines split off as the years went by. The concern that tests should not be dependent on, or insensitive to, distributional assumptions respectively led to distribution-free or nonparametric* and robust* methods. Efforts to minimize the cost or sample size in routine sampling led to sequential analysis*. The vagueness of the significance level can be resolved if one takes into account the costs of making errors; this led to decision theory. These last two developments are largely due to A. Wald* [17] and arose during and after World War II.

The theory of hypothesis testing gradually came to be interpreted rather more rigidly than its originators had in mind. This gave rise to many paradoxes. Practitioners tend to avoid these by less informal usage and ad hoc procedures. Some of the paradoxes are eliminated by recognizing that some problems are neither estimation nor testing but an incompletely formulated activity to do with the discovery of statistical regularities in data. Other paradoxes do not appear in the Bayesian formulation and so have led to its revival and development. However, the key ideas of hypothesis testing will always remain central to the theory of statistics and its applications. We will now define the main concepts needed in the theory in terms of examples since orderly developments are easy to find at a variety of levels of abstraction (e.g., Lehmann [9] and Bickel and Doksum [2]). The paradoxical aspects may be seen in Cox and Hinkley [5]. Barnett [1] gives a general discussion of statistical inference*.

THE NEYMAN–PEARSON LEMMA*

In the binomial and multinomial examples above, the data is a sample from a distribution fully specified on the null hypothesis; i.e., on the null hypothesis the probability of the sample is uniquely determined. However, there are many alternative hypotheses (e.g., $p \neq \frac{1}{2}$ for the binomial case). We say then that the null hypothesis is *simple* and the alternative is *composite*.

Had the data been a sample from a Gaussian distribution with mean μ and variance σ^2, a common null hypothesis is $\mu = 0$, with σ^2 unspecified. This null hypothesis is composite. To test $\mu = 0$, one instinctively thinks of rejecting the null hypothesis if $\bar{x}$ differs too much from zero. But use of the median* makes sense, too. Which is better? Also, here we can speak of the probability density rather than the probability of the data; let "probability function" or "likelihood" denote either.

The problem may be stated fairly generally now. Suppose that the probability function of the data $\mathbf{x}$ is $p(\mathbf{x}; \boldsymbol{\theta})$, where $\mathbf{x}$ is a point in the *sample space* $\mathscr{X}$ and $\boldsymbol{\theta}$ is a point in the parameter space Θ. Define the null hypothesis by $\boldsymbol{\theta} \in \Theta_0 \subset \Theta$ and the alternative hypothesis by $\boldsymbol{\theta} \in \Theta_1 \subset \Theta$, $(\Theta_0 \cap \Theta_1 = \varnothing)$. In the binomial case $\mathscr{X} = \{0, 1, \dots, n\}$ and $\boldsymbol{\theta} = p$, $\Theta = \{p \mid 0 \leqslant p \leqslant 1\}$, $\Theta_0 = \{\frac{1}{2}\}$, $\Theta_1 = \{p \mid p \neq \frac{1}{2}\}$. We will sometimes use H_0 and H_1 for the null and alternative hypothesis. In the examples, we rejected the null hypothesis when $\mathbf{x}$ fell in a set such that some function, the test statistic, say $t(\mathbf{x})$, of the data took too extreme a value. This means that any test divides $\mathscr{X}$ into two mutually exclusive regions; R, the rejection or critical region and its complement, the acceptance region. Choosing a test is choosing a test statistic $t(\mathbf{x})$ or equivalently a region R. For the test to have significance level α (e.g., 0.05), we require R such that

$$\Pr(\mathbf{x} \in R \mid \boldsymbol{\theta} \in \Theta_0) \leqslant \alpha. \tag{1}$$

This ensures that, although the left-hand side of (1) may vary as $\boldsymbol{\theta}$ takes different values in Θ_0, the probability of type 1 errors, or false alarms, never exceeds α. Now there will usually be many such regions R, so we want to find one that will maximize

$$\Pr(\mathbf{x} \in R \mid \boldsymbol{\theta} \in \Theta_1). \tag{2}$$

This probability, for a fixed R, is a function of $\boldsymbol{\theta}$, the *power* function* of the test R. It is the probability of rejecting the null hypothesis, which is what we want to do when $\boldsymbol{\theta} \in \Theta_1$. the alternative hypothesis. Hence maximizing (2) is the same as minimizing the probability of errors of the second kind.

To progress with this general formulation, we consider the simplest case when $\Theta_0 = \{\boldsymbol{\theta}_0\}$, $\Theta_1 = \{\boldsymbol{\theta}_1\}$ (i.e., both the null and alternative hypotheses are simple). Then the problem posed by (1) and (2) is: Find R such that

$$\int_R p(\mathbf{x}, \boldsymbol{\theta}_0)\, d\mu(\mathbf{x}) \leqslant \alpha, \tag{3a}$$

$$\int_R p(\mathbf{x}, \boldsymbol{\theta}_1)\, d\mu(\mathbf{x}) = \max. \tag{3b}$$

It is clearly advantageous to have equality in (3a). This is usually possible except for discrete distributions (see below). Assume that there are many regions R satisfying (3a) with equality. To achieve the maximum required in (3b) we would want to choose points for R where $p(\mathbf{x}, \boldsymbol{\theta}_1)$ is relatively large but where $p(\mathbf{x}, \boldsymbol{\theta}_0)$ is relatively small. This suggests that the optimal R might be R^*, the set of points $\mathbf{x}$ where the ratio of $p(\mathbf{x}, \boldsymbol{\theta}_1)$ to $p(\mathbf{x}, \boldsymbol{\theta}_0)$ is greatest, i.e.,

$$\frac{p(\mathbf{x}, \boldsymbol{\theta}_1)}{p(\mathbf{x}, \boldsymbol{\theta}_0)} \geqslant k, \tag{4}$$

where k is chosen so that (3a) is satisfied. Then R and R^* both satisfy (3a), but

$$\int_{R^*} p(\mathbf{x}, \boldsymbol{\theta}_1)\, d\mu(\mathbf{x}) \geqslant \int_R p(\mathbf{x}, \boldsymbol{\theta}_1)\, d\mu(\mathbf{x}),$$

since in the part of R^* not intersecting R, $p(\mathbf{x}, \boldsymbol{\theta}_1) \geqslant kp(\mathbf{x}, \boldsymbol{\theta}_0)$, while in the part of R not intersecting R^*, $p(\mathbf{x}, \boldsymbol{\theta}_1) \leqslant kp(\mathbf{x}, \boldsymbol{\theta}_0)$. This is the Neyman–Pearson lemma—the *most powerful test* of $\boldsymbol{\theta}_0$ vs. $\boldsymbol{\theta}_1$ is based on the likelihood ratio region (4).

Example 1. Let x be binomial (n, p), the null hypothesis $p = p_0$, and the alternative hypothesis $p = p_1 > p_0$. The likelihood ratio is

$$\frac{\binom{n}{x} p_1^x (1 - p_1)^{n-x}}{\binom{n}{x} p_0^x (1 - p_0)^{n-x}} = K\left(\frac{p_1/(1 - p_1)}{p_0/(1 - p_0)} \right)^x.$$

Because $p_1 > p_0$, this ratio increases with x so that the critical region is $\{c, c+1, \dots, n\}$ when we choose c to make

$$\sum_c^n \binom{n}{x} p_0^x (1 - p_0)^{n-x}$$

as near to α as possible. Had p_1 been less than p_0, the critical region would have been in the lower tail of the distribution. Both tests are as intuition would suggest; $p_1 > p_0$ is a one-sided alternative hypothesis and we have a one-tailed test.

To get α exactly here, we must resort to *randomization**. For example, if $n = 10$, $p_0 = \frac{1}{2}$, $\alpha = 0.05$, and $p_1 > \frac{1}{2}$, observe that Pr(6 heads) = 28/256, Pr(7 heads) = 8/256, and Pr(8 heads) = 1/256. Consider the rule:

> If we get 6 heads, reject with probability δ; if we get 7 or 8 heads, reject with probability 1.

Then

$$\Pr\left(\text{reject} \mid p = \frac{1}{2}\right) = \frac{28\delta + 8 + 1}{256} = \frac{1}{20}$$

if $\delta = 19/140$. For this reason, the general theory uses not a critical region but a function $\phi(\mathbf{x})$, the probability of rejection given sample $\mathbf{x}$. In the simplest case, $\phi = 1$ or 0 and is the indicator function of the critical region.

The example shows another special feature. When the alternative to $p = p_0$ is $p_1 > p_0$, the *same* test is also optimal for all $p_1 > p_0$, and is said to be *uniformly most powerful** (U.M.P.), clearly a most desirable property. However, had the alternative been $p \neq p_0$, a two-sided alternative, this argument fails. Clearly, we need to have two-tailed tests (i.e., some of each tail in the critical region), but we need a theory to tell us how much of each.

Finally, had the null hypothesis been $p \leqslant p_0$ (composite) and the alternative $p \geqslant p_1 > p_0$, the same test may be shown to be U.M.P.

Example 2. Let $X_1, \dots, X_n$ be independent and Gaussian (μ, σ^2) with *known* σ^2, and the null and alternative hypotheses be $\mu = \mu_0$, $\mu = \mu_1$, respectively. The likelihood ratio is

$$\frac{(\sqrt{2\pi}\,\sigma)^{-n} \exp\{-\sum(x_i - \mu_1)^2/2\sigma^2\}}{(\sqrt{2\pi}\,\sigma)^{-n} \exp\{-\sum(x_i - \mu_0)^2/2\sigma^2\}},$$

which reduces to

$$\exp\{(n/2\sigma^2)[(\mu_1 - \mu_0)\bar{x} - \mu_1^2 + \mu_0^2]\}.$$

Thus, if $\mu_1 > \mu_0$, the ratio is large for large positive $\bar{x}$, so the critical region is $\bar{x} > c$. To calculate c we observe that on the null hypothesis $\mu = \mu_0$, $(\bar{x} - \mu_0)/(\sigma/\sqrt{n})$ is standard Gaussian, so the $\alpha = 0.05$ test is: reject if $\bar{x} > \mu_0 + 1.64\sigma/\sqrt{n}$. Once again this is a U.M.P. test for all $\mu_1 > \mu_0$. The reader may observe that $\bar{x}$ is here a *sufficient statistic** for μ, so we would expect the best test to use it (*see* SUFFICIENCY).

The power function of this test is, by definition,

$$\Pr(\bar{x} > \mu_0 + 1.64\sigma/\sqrt{n} \mid \mu)$$
$$= \Pr\left(z = \frac{\bar{x} - \mu}{\sigma/\sqrt{n}} > \frac{\mu_0 - \mu}{\sigma/\sqrt{n}} + 1.64\right), \tag{5}$$

where $z = (\bar{x} - \mu)\sqrt{n}/\sigma$ now has a standard Gaussian distribution. As a function of μ, the power goes from zero at $\mu = -\infty$, to 0.05 at $\mu = \mu_0$, and to unity as μ increases to infinity.

A statistician is often asked: *How big a sample should be taken?* In the example above, if one can guess σ from previous experience, has settled on a 5% test, and is willing to pay what it costs to get a power of, say 0.90 to detect some particular μ, then (5) and the normal tables will provide the value of n that is needed. This value will increase as μ gets closer to μ_0 or as desired power is increased.

The power function (5) for fixed n is a monotonically increasing function of μ. The power functions of other plausible tests (e.g., one that uses the sample median instead of the mean) will also increase. But at no value of μ can any of these curves ever rise above curve (5)! It is the envelope of all possible power curves for this problem. This illustrates graphically how our test is optimal. If there were no U.M.P. test, no one power function would always be best (i.e., greatest for every μ).

Example 3. Let $X_1, \dots, X_n$ be i.i.d. with the Cauchy density $\pi^{-1}(1 + (x - \theta)^2)^{-1}$; let the null hypothesis be θ_0 and the alternative be $\theta_1 > \theta_0$. The likelihood ratio (L.R.) no longer simplifies—there is no single sufficient statistic for θ. The value of the L.R. for any sample depends on the values of both θ_0 and θ_1. Finding the value of k to get a test of size α would probably involve applying the central limit theorem* to the logarithm of the L.R. (*see* LIMIT THEOREMS, CENTRAL). There is no U.M.P. test—θ_1 always occurs in the test statistic.

A strategy often invoked in such cases is to seek the limit of the foregoing test as $\theta_1 \downarrow \theta_0$—the *locally most powerful* (L.M.P.) one-sided test. Using the general notation, the logarithm of the likelihood ratio is

$$\log p(\mathbf{x}, \boldsymbol{\theta}_1) - \log p(\mathbf{x}, \boldsymbol{\theta}_0).$$

When θ is real, and θ_1 is near θ_0, this is approximately proportional to

$$\partial \log p(\mathbf{x}, \theta_0)/\partial\theta_0 \tag{6}$$

and large values of this will be the L.M.P. test of θ_0 versus values of θ just greater than θ_0. Alternatively, if one thinks of all the possible power functions for tests of θ_0 vs. $\theta > \theta_0$ and chooses the test whose slope is greatest at the origin, one again derives the test statistic (6). For the power of a test based on R is

$$\gamma(\theta) = \int_R p(\mathbf{x}, \theta)\, d\mu(\mathbf{x}),$$

so the slope at θ is given by

$$\gamma'(\theta_0) = \int_R \frac{\partial p}{\partial\theta_0}(\mathbf{x}, \theta_0)\, d\mu(\mathbf{x}).$$

A similar argument to that leading to the Neyman–Pearson lemma tells us here that $\gamma'(\theta_0)$ will be a maximum, subject to $\gamma(\theta_0) = \alpha$ if R is based on large values of

$$\frac{1}{p(\mathbf{x}, \theta_0)} \frac{\partial p(\mathbf{x}, \theta_0)}{\partial\theta_0} = \frac{\partial}{\partial\theta_0} \log p(\mathbf{x}, \theta_0),$$

as we have just seen.

In the Cauchy case above, the L.M.P. test is based on large values of

$$t(\mathbf{x}) = \sum_{i=1}^{n} \frac{x_i - \theta_0}{1 + (x_i - \theta_0)^2}.$$

The asymptotic distribution of $t(\mathbf{x})$ is easily obtained from the central limit theorem.

Thinking again in terms of the graph of the power function, suppose that we have a two-sided alternative $\theta \neq \theta_0$ and attach equal importance to small deviations above and below θ_0. It is then natural to seek a critical region R such that

$$\gamma(\theta_0) = \alpha, \quad \gamma'(\theta_0) = 0, \quad \gamma''(\theta_0) = \text{maximum}.$$

A Neyman–Pearson type of argument then tells us that the optimal critical region is the set of points x such that

$$\frac{\partial^2 p(\mathbf{x},\theta_0)}{\partial\theta_0^2} \geqslant k_1 \frac{\partial p(\mathbf{x},\theta_0)}{\partial\theta_0} + k_2 p(\mathbf{x},\theta_0).$$

Such a test is said to be a locally most powerful *unbiased* test. If applied to Example 2, it leads to the 5% level test; reject if

$$\frac{|\sqrt{n}\,(\bar{x}-\theta_0)|}{\gamma} > 1.96,$$

the equal-tailed version of the U.M.P. (and hence L.M.P.) test for one-sided alterntives. If applied to Example 3, it does not lead to the two-sided version of the L.M.P. test given above.

When only one parameter is involved, much more theory is available, but most real problems involve several parameters.

SEVERAL-PARAMETER PROBLEMS

Difficulties come with several parameters. The null hypothesis is usually composite. To have a fixed significance level, we need a rejection region which is *similar* to the whole sample space in that its probability content is constant on H_0. Parameters left unspecified by H_0 are *nuisance parameters**. Even if the null hypothesis is simple, the power function of a test will vary with all the parameters. One wants it to increase rapidly in all directions away from its value at θ_0. But sacrificing slope in one direction may increase it in another! This dilemma can be resolved only by weighing the importance of each direction, according to its consequences. These difficulties were recognized very early. As a practical matter, one may use an extension of the L.R., or asymptotically equivalent procedures involving maximum likelihood estimators. There is, however, a great deal of elegant theory of optimal tests when they exist.

Example 4. Let $X_1, \ldots, X_n$ be independent and Gaussian (μ, σ^2) and let the null hypothesis be $\mu = \mu_0$ and the alternative $\mu \neq 0$. Here σ^2 is unspecified and so is a nuisance parameter. Ignoring constants,

$$\text{log likelihood} = -n\log\sigma - \frac{1}{2\sigma^2}\sum(x_i-\mu)^2. \quad (7)$$

Maximizing over all μ we find $\hat{\mu} = \bar{x}$, and over all $\sigma^2 > 0$ it yields

$$\max_{H_1}(\text{log likelihood}) = -\frac{n}{2}\log\frac{\sum(x_i-\bar{x})^2}{n} - \frac{n}{2}. \quad (8)$$

If we set $\mu = \mu_0$ in (7) and maximize, we find that

$$\max_{H_0}(\text{log likelihood}) = -\frac{n}{2}\log\frac{\sum(x_i-\mu_0)^2}{n} - \frac{n}{2} \quad (9)$$

Forming the log likelihood ratio of Neyman and Pearson but after these maximizations [i.e., (8) ÷ (9)], we get

$$-\frac{n}{2}\log\left(\sum(x_i-\bar{x})^2/\sum(x_i-\mu_0)^2\right).$$

Hence

$$+2\log \text{L.R.} = -n\log\left\{\frac{1}{1+n(\bar{x}-\mu_0)^2/\sum(x_i-\bar{x})^2}\right\} \quad (10)$$

will be large if

$$t^2 = \frac{n(\bar{x}-\mu_0)^2}{\sum(x_i-\bar{x})^2/(n-1)}, \quad (11)$$

the square of Student's t, is large. Note that for large n, (10) and (11) are essentially equal. The null distribution of t does not contain σ^2, so the t-test has similar rejection regions.

Had we done the same in the case of σ^2 known, we would have obtained

$$2\log \text{L.R.} = n(\bar{x}-\mu_0)^2/\sigma^2,$$

which on the null hypothesis has a chi-square distribution* with 1 degree of freedom. The same is asymptotically true of t^2 and so of (10). This illustrates *Wilks theorem**, which says roughly that as $n \to \infty$, $2\log$ L.R. has a chi-square distribution whose degrees of freedom are equal to the number of parameters specified by the null hypothesis.

The Pearson chi-square test may be shown to be an approximation to an L.R. test. It also produces an answer when the null hypothesis specifies several parameters, unlike Examples 3 and 4.

Example 5. Reconsider Example 4. There are two unknown parameters μ and σ^2 and $\bar{x}$ and $s^2 = \sum(x_i - \bar{x})^2/(n-1)$ are sufficient statistics for them. It follows from the theory of sufficiency* that any good test will only use these two summaries of the data. Further, any test should be unchanged if we measure the x's from a new origin and on a new scale (e.g., if they are temperatures, switch from Fahrenheit to Celsius). $\bar{x} - \mu$ and s^2 are independent of origin and $(\bar{x} - \mu)/s$ is also independent of scale. Thus, by demanding that our test should depend upon sufficient statistics and be *unchanged* under the natural group of transformations (here scale and location) which leave the problem unchanged, we arrive at the t-test. Among such invariant tests, we may now show that it has U.M.P. properties, as in Example 2 (*see* INVARIANCE PRINCIPLES IN STATISTICS). Essentially, by invoking these general principles, we are able to reduce this problem so that the Neyman–Pearson lemma can be applied.

For a long time nonparametric tests remained outside the Neyman–Pearson theory —note that all the examples above are parametric. However, by requiring invariance under the group of symmetries of the problem, they were united with optimal parametric tests.

Example 6. Test that the cumulative distributions F and G are identical [i.e., $F(x) = G(x)$], given samples $(X_1, \dots, X_m)$ and $(Y_1, \dots, Y_n)$ from F and G. When nothing is known of the common distribution, our test should be invariant under all monotonic transformations since they preserve the identity of F and G. Hence permissible tests can only depend on the ranks of the X and Y observations in the list of $m + n$ observations. Within this class of tests we can select members by Neyman–Pearson methods if we define specific alternatives to $F = G$; for example, that $G(x) = F(y - \Delta)$, $\Delta \geqslant 0$. If F is the logistic distribution*, this line of argument may be used to prove that the Wilcoxon test* is L.M.P.

One of the most surprising discoveries in the whole theory is that one need *not* pay a great price in power in order to buy a reliable significance level. To prove such assertions, one needs to introduce a definition of test efficiencies, e.g., asymptotic relative effiency*.

Example 7. Suppose that X_i are independently normal with means μ_i and variances $\sigma^2 (i = 1, \dots, n)$ and that we have an independent estimate s^2 of σ^2, distributed as $\sigma^2\chi_f^2/f$. To test a prespecified hypothesis $\sum_1^n C_i\mu_i = 0$ with $\sum_1^n C_i = 0$, we would refer $\sum C_i X_i/(s\sqrt{\sum C_i^2}\,)$ to the t-distribution* with f degrees of freedom. If, however, we began inventing such hypotheses *after* we had looked at the data, this method would be improper (e.g., we might seek the largest difference $X_i - X_j$, which certainly is not normal with mean zero and variance 2).

We may allow for all possible such comparisions of the μ_i's by using the Cauchy inequality

$$\left\{\sum_1^n C_i X_i = \sum C_i(X_i - \bar{X})\right\}^2 \leqslant \sum_1^n C_i^2 \sum_1^n (X_i - \bar{X})^2,$$

so that

$$\frac{(\sum C_i X_i)^2}{\sum C_i^2 (s^2)} \leqslant \frac{\sum_1^n (X_i - \bar{X})^2}{s^2}.$$

The left-hand side (l.h.s.) is the square of our original statistic, and the r.h.s. is $(n-1)$ times a variable distributed as F with $n-1$ and f degrees of freedom, which may thus be used to make a proper test. If, at the outset, we agree to look only at differences of pairs, a different and more stringent procedure is available.

In the inspection of data one often wants to make a rough test to see if some aspect of it is exceptional, and similar care is always required, although it is rarely possible to formulate the problem as precisely as above. Tukey has called this the *multiplicity* problem.

MISCELLANEOUS REMARKS

The account above gives only the simplest cases of the basic ideas and is biased toward application. Through the references and other entries, especially DECISION THEORY, EXPONENTIAL FAMILY, GOODNESS-OF-FIT TESTS, INVARIANCE PRINCIPLES IN STATISTICS, LIKELIHOOD RATIO TESTS, MONOTONE LIKELIHOOD RATIO, MULTIPLE COMPARISONS, NEYMAN–PEARSON LEMMA, SEQUENTIAL ANALYSIS, UNBIASEDNESS, and specific significance tests, the reader may find descriptions of those further aspects he or she seeks. The reader is fortunate that almost all the key papers have been collected, and most still make fascinating reading. Any such study will take the reader deeply into estimation* theory and the mathematical structure of probability distributions.

On logical issues, not even hinted at here, Cox and Hinkley [5] is recommended. The role of conditionality is particularly puzzling. Neyman–Pearson theory is centered on maximizing power and uses all conceivable samples. But there are often aspects of the sample that do not depend on the parameter being tested, and a good case may be made for considering only all samples with the same such aspects as our given sample. But these two principles may be conflicting. One is often tempted to use conditional tests to eliminate nuisance parameter difficulties. One practical topic closely related to hypothesis testing but not even mentioned above is interval estimation (*see* CONFIDENCE INTERVALS AND REGIONS). This arises when tests are examined to find the set of parameter values which cannot be rejected as null hypotheses. Space prevents any discussion of robust tests.

Some of these logical conundrums are eliminated by adopting a Bayesian approach to inference, which, however, brings its own difficulties. Cox and Hinkley [5] give a brief discussion (*see also* BAYESIAN INFERENCE). Box and Tiao [4] illustrate the Bayesian handling of many common statistical models. Box [3] attempts a partial synthesis of Bayesian and classical inference in which significance tests are used for model checking (model to data) but Bayesian methods are used for estimation (data to model).

References

[1] Barnett, V. (1973). *Comparative Statistical Inference*. Wiley, New York. (A nonpartisan and largely nontechnical account of all methods of statistical inference putting hypothesis testing in context.)

[2] Bickel, P. J. and Doksum, K. A. (1977). *Mathematical Statistics*. Holden-Day, San Francisco. (Introductory graduate text on classical statistic inference.)

[3] Box, G. E. P. (1980). *J. R. Statist. Soc. A*, **143**, 383–430.

[4] Box, G. E. P. and Tiao, G. C. (1973). *Bayesian Inference in Statistical Analysis*. Addison-Wesley, Reading, Mass. (A very readable text leading to practical procedures. For the philosophical and deeper issues the reader must go to their references.)

[5] Cox, D. R. and Hinkley, D. V. (1974). *Theoretical Statistics*. Wiley, New York. (The most extensive discussion of examples and counterexamples of inference procedures.)

[6] Ferguson, T. (1967). *Mathematical Statistics*. Academic Press, New York. (A decision theoretic approach.)

[7] Fisher, R. A. (1925). *Statistical Methods for Research Workers*. Oliver & Boyd, Edinburgh. (New editions of this classic still appear.)

[8] Fisher, R. A. (1971). *Collected Papers of R. A. Fisher*, Vols. 1–4, J. H. Bennett, ed. University of Adelaide Press, Adelaide, Australia.

[9] Lehmann, E. (1959). *Testing Statistical Hypotheses*. Wiley, New York. (This classic text has no modern sequel and is still the basis of all graduate courses on testing.)

[10] Neyman, J. (1967). *A Selection of the Early Statistical Papers of J. Neyman*. University of California Press, Berkeley, Calif.

[11] Neyman, J. and Pearson, E. S. (1966). *Joint Statistical Papers of J. Neyman and E. S. Pearson.* University of California Press, Berkeley, Calif. (All 10 papers deal with the development of the basic notions of testing.)

[12] Pearson, E. S. (1966). In *Research Papers in Statistics*, F. N. David, ed. Wiley, New York.

[13] Pearson, E. S. (1966). *The Selected Papers of E. S. Pearson*. Cambridge University Press, Cambridge. (A selection of papers illustrating Pearson's approach to statistical problems, including tests.)

[14] Pearson, K. (1892). *The Grammar of Science*. Adam and Charles Black, London.

[15] Pearson, K. (1900). *Philos. Mag.*, **50**, 157–175. (The paper that introduced the chi-square goodness-of-fit statistic.)

[16] "Student" (1958). *Student's Collected Papers*, E. S. Pearson and J. Wishart, eds. Cambridge University Press, Cambridge. (Papers 2 and 3 are references 1908a, 1908b; many of his papers have a modern ring.)

[17] Wald, A. (1955). *Selected Papers in Probability and Statistics*. McGraw-Hill, New York. (Wald introduced decision theory and sequential analysis and some of the modern mathematical style of statistics, so one can see here the important first steps.)

In addition to the following related entries there are numerous entries discussing specific test procedures.

(ADMISSIBILITY
BAYESIAN INFERENCE
DECISION THEORY
DEPENDENCE, TESTS OF
DISTRIBUTION-FREE METHODS
F-TESTS
GOODNESS OF FIT
INVARIANCE CONCEPTS IN STATISTICS
LIKELIHOOD RATIO TESTS
MULTIPLE COMPARISONS
NEYMAN-PEARSON LEMMA
SEQUENTIAL ANALYSIS
STATISTICAL EVIDENCE
t-TESTS
UNBIASEDNESS)

G. S. WATSON